Precalculus: Making Connections

Preliminary Edition

Lawrence Runyan
Holly Runyan
Shoreline Community College

PRENTICE HALL, UPPER SADDLE RIVER, NEW JERSEY 07458

Executive Editor: **SALLY SIMPSON**
Editorial Assistant: **SARA BETH NEWELL**
Marketing Manager: **PATRICE LUMUMBA JONES**
Marketing Assistant: **AMY LYSIK**
Editorial Director: **TIM BOZIK**
Editor-in-Chief: **JEROME GRANT**
Assistant Vice-President of Production
and Manufacturing: **DAVID W. RICCARDI**
Editorial/Production Supervision: **RICHARD DeLORENZO**
Managing Editor: **LINDA MIHATOV BEHRENS**
Executive Managing Editor: **KATHLEEN SCHIAPARELLI**
Manufacturing Buyer: **ALAN FISCHER**
Manufacturing Manager: **TRUDY PISCIOTTI**
Creative Director: **PAULA MAYLAHN**
Art Director: **JAYNE CONTE**
Cover Designer: **BRUCE KENSELAAR**

Simon & Schuster / A Viacom Company
Upper Saddle River, NJ 07458

Printed in the United States of America

10 9 8 7 6 5 4 3 2 1

ISBN 0-13-095674-0

Prentice-Hall International (UK) Limited, London
Prentice-Hall of Australia Pty. Limited, Sydney
Prentice-Hall Canada Inc., Toronto
Prentice-Hall Hispanoamericana, S.A., Mexico
Prentice-Hall of India Private Limited, New Delhi
Prentice-Hall of Japan, Inc., Tokyo
Simon & Schuster Asia Pte. Ltd., Singapore
Editora Prentice-Hall do Brasil, Ltda., Rio de Janeir
Prentice Hall, USA; Upper Saddle River, New Jersey

Contents

Prologue

The Cartesian Coordinate System and Graphing Utilities

We base our study of precalculus on graphs of equations involving two variables. In this prologue we review the Cartesian coordinate system and introduce **graphing utilities** (graphing calculators or computers equipped with graphing software). Since graphing utilities differ and the technology is constantly advancing, the presentations and applications throughout this book are independent of any particular utility. Whenever you need more specific directions, refer to the instruction manual for your graphing utility or the *Precalculus Weblet* (http://oscar.ctc.edu/precalc).

Graphing utility displays can help us recognize relationships among mathematical problems that might not otherwise be obvious. The old adage that a picture is worth a thousand words (and calculations) is graphically illustrated. Our goal is to use a graphing utility thoughtfully: (1) decide when to use a graphing utility, (2) anticipate (estimate) results, and (3) question the reasonableness of a display.

The study of precalculus includes learning some fundamental concepts of graphing to develop your geometric intuition in order to be able to anticipate the shape and position of many graphs. We begin graphing equations by plotting points in a Cartesian coordinate system to see the general shape of a graph. Graphing utilities use such point-plotting based on electronic computations to plot a graph point by point by point.

Reviewing the Cartesian Coordinate System

Recall that the Cartesian coordinate system is comprised of two real number lines. On the horizontal line, called the x axis, the numbers increase from left to right; on the vertical line, the y axis, the numbers increase from bottom to top. The first number in each ordered pair (x, y) is called the *x coordinate* or the *abscissa*. The second is the *y coordinate* or the *ordinate*. The two axes intersect at a point called the *origin*, (0, 0) (Figure 1a).

Figure 1b shows the Cartesian coordinate system on the viewing screen of a graphing utility with a window set to display values of $-10 \le x \le 10$ and $-10 \le y \le 10$. In this book we show the minimum and maximum values of x and y on the outside of the screen. Many utilities use a **standard window** that extends from -10 to 10 on both the x and y axes, as in Figure 1b.

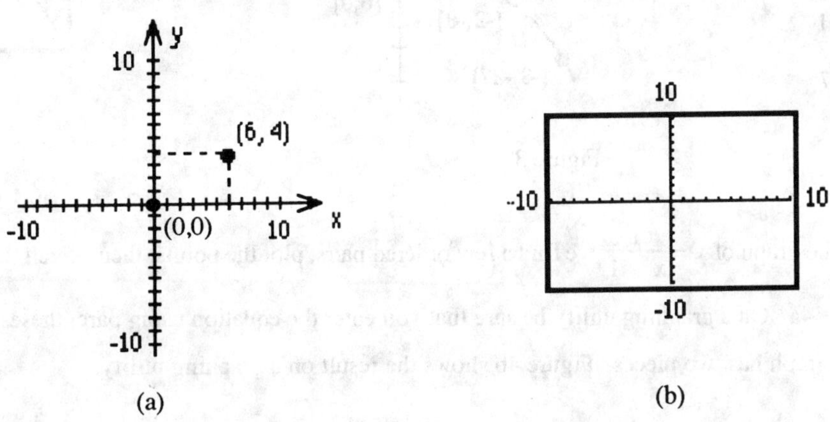

Figure 1 (a) (b)

Using Graphing Utilities

The graphing utility can be set to display any portion of the Cartesian coordinate system. Figure 2 shows a window set to display values $-2 \le x \le 10$ and $-10 \le y \le 10$. The x_{scl} and y_{scl} settings determine the distance between tick marks on each axis.

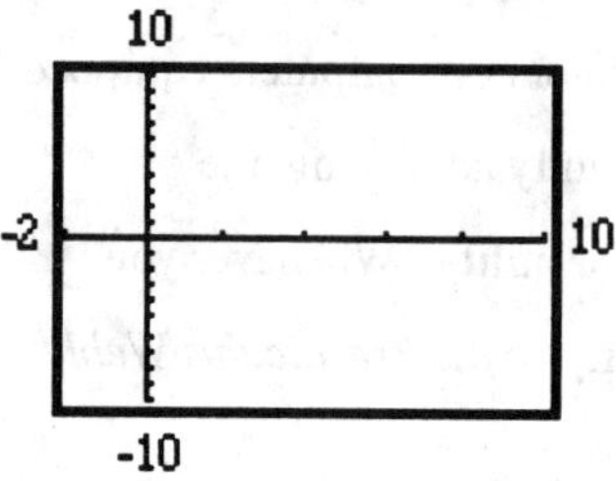

Figure 2

Display	Interpretation
$x_{min} = -2$	minimum value of x
$x_{max} = 10$	maximum value of x
$x_{scl} = 2$	distance between tick marks on the x axis is 2 units
$y_{min} = -10$	minimum value of y
$y_{max} = 10$	maximum value of y
$y_{scl} = 1$	distance between tick marks on the y axis is 1 unit

For help accessing a feature on individual graphing utilities or step-by-step procedures to do the examples in this prologue, go to the *Precalculus Weblet - Reference Center - Technology* (http://oscar.ctc.edu/precalc).

Example 1 *Sketching the graph of an equation*

Generate a table of ordered pairs that satisfy each equation, plot the points, and sketch the graph. Then recreate each graph using a graphing utility.

(a) $y = x^3$

(b) $y = \dfrac{1}{x - 1}$

Solution

(a) To sketch the graph of $y = x^3$, we find a few ordered pairs by cubing arbitrarily selected values of x, plotting the resulting ordered pairs, and then sketching the graph shown in Figure 3a. Figure 3b shows the graphing utility display of the same equation. Notice the viewing window, $-3 \le x \le 3$ and $-27 \le y \le 27$.

x	y
-3	-27
-2	-8
-1	-1
0	0
1	1
2	8
3	27

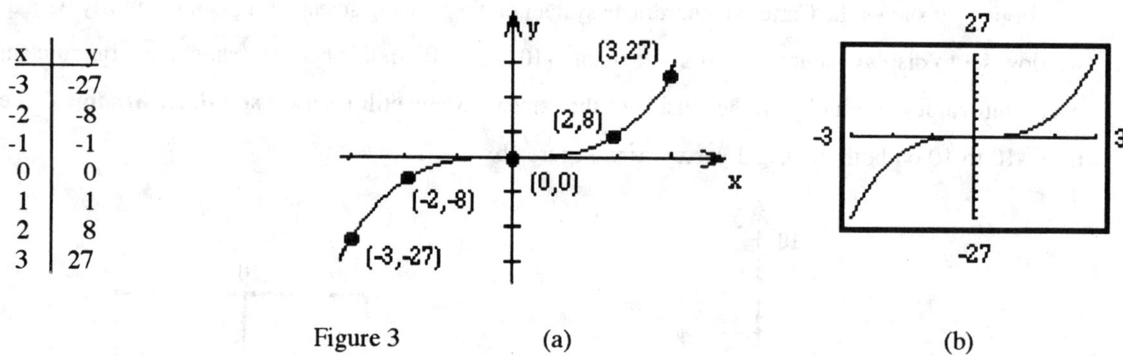

Figure 3 (a) (b)

(b) To sketch the graph of $y = \dfrac{1}{x - 1}$ we find a few ordered pairs, plot the points, then sketch the graph as shown in Figure 4a. On a graphing utility be sure that you enter the equation using parentheses, $y = 1/(x - 1)$. Notice that the graph has two pieces. Figure 4b shows the result on a graphing utility.

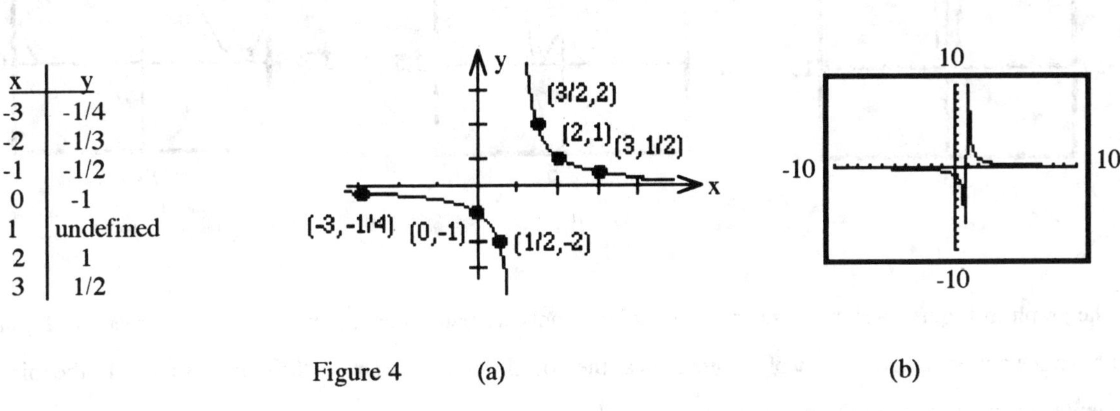

x	y
-3	-1/4
-2	-1/3
-1	-1/2
0	-1
1	undefined
2	1
3	1/2

Figure 4 (a) (b)

Did you notice how the graphs in Figures 4a and 4b differ?

Some graphing utilities may create a misleading graph for $y = \frac{1}{x - 1}$ (Figure 4b). This graphing utility created a line that is not part of the graph. Others may show the graph intersecting the x axis to the left and right, which it does not. Such results are not acceptable. We can reset the window to change the values of x and y displayed until we have an acceptable representation of the graph. We could use a dot feature to display only the ordered pairs actually calculated by the machine; misleading lines will not be drawn (Figure 5).

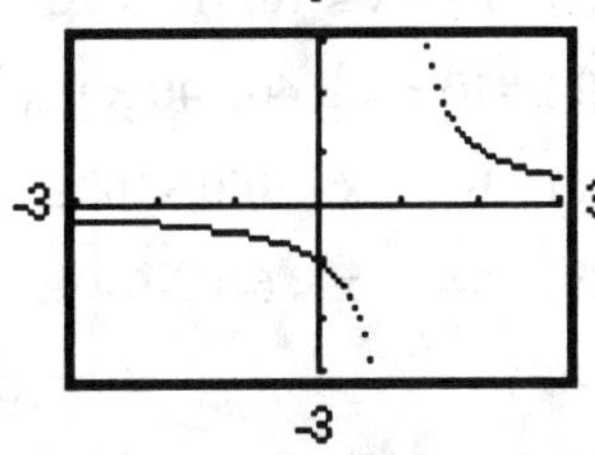

Figure 5

Viewing a Graph Locally or Globally

The zoom feature of a graphing utility acts like a zoom lens of a camera by allowing us to enlarge an area of a graph to see more detail or to back away to view more of a graph Moving in on a smaller portion of the coordinate system, we investigate the **local behavior** of a graph. Moving away to see more of a graph allows us to examine **global behavior**.

Example 2 *Examining the local behavior of a graph*

Graph $y = (4x^2 - 1)^2$ on a graphing utility for the following windows and describe your results. Either zoom in or reset the graphing utility as indicated.

(a) $-10 \leq x \leq 10$, $-10 \leq y \leq 10$ (the standard window)

(b) $-2.5 \leq x \leq 2.5$, $-2.5 \leq y \leq 2.5$ (zoom in once with a zoom factor of 4)

(c) $-0.625 \leq x \leq 0.625$, $-0.625 \leq y \leq 0.625$ (zoom in a second time)

Solution

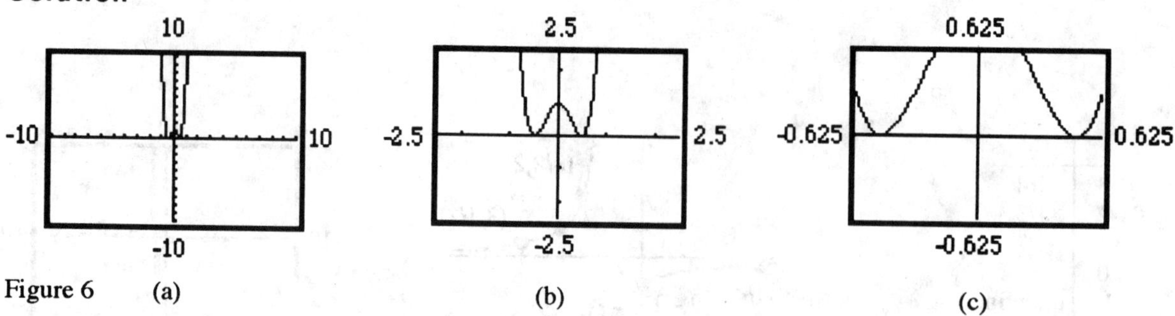

Figure 6 (a) (b) (c)

The graph in Figure 6a shows a steep curve that uses only a small horizontal portion of the screen. In Figure 6b we see a more detailed view of the graph near the origin. This view shows the correct shape. In the third view (Figure 6c) we lost the point where the graph crosses the y axis.

Example 3 *Examining the global behavior of a graph*

Graph $y = 0.05x^3 - 10x$ on a graphing utility for the following windows and describe your results. Either zoom out or reset the window as indicated.

(a) $-10 \le x \le 10$, $-10 \le y \le 10$ (the standard window)

(b) $-40 \le x \le 40$, $-40 \le y \le 40$ (zoom out once)

(c) $-160 \le x \le 160$, $-160 \le y \le 160$ (zoom out a second time)

(d) $-640 \le x \le 640$, $-640 \le y \le 640$ (zoom out a third time)

Solution

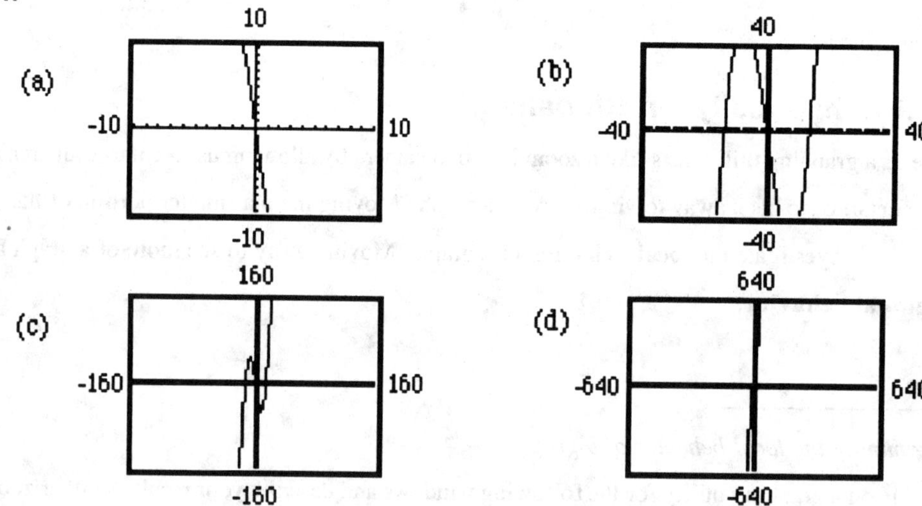

Figure 7

The graph in Figure 7a shows a steep curve. In Figure 7b we see the original curve in Figure 7a and two additional curves, suggesting that there is more to this graph. Figure 7c shows a continuous curve that is more representative of the equation. In Figure 7d we have zoomed out too far. The graph is drawn so close to the y axis that it nearly disappears.

Establishing a Rounding Convention

Calculations that are rounded produce errors. The study of rounding and other errors is an increasingly important part of our mathematics, one that is often ignored. Consider the error analysis in Table 1, where calculated amounts of money are rounded to the nearest penny. We calculate the error in each case by subtracting the approximation from the original amount.

Table 1

Original Amount	Rounded Approximation	Error
Round down:		
$75.121	$75.12	-$0.001
$75.122	$75.12	-$0.002
$75.123	$75.12	-$0.003
$75.124	$75.12	-$0.004
Round up:		
$75.126	$75.13	+$0.004
$75.127	$75.13	+$0.003
$75.128	$75.13	+$0.002
$75.129	$75.13	+$0.001

How do we round $75.125, where the digit to be rounded is a 5?

Most calculating devices round up, creating an error of +$0.005. This is larger than any other error in Table 1. In this book we use the following rule:

<u>If a 5 is the digit to be dropped, we round up when the preceding digit is odd and round down when the preceding digit is even.</u>

This method is designed to round up half the time and round down half the time. For instance, in rounding to the hundredths place, we round

4.315 up to 4.32

since the digit preceding 5 is odd, and round

12.265 down to 12.26

since the digit preceding 5 is even.

In general, waiting to round until all calculations have been completed helps reduce round-off error. Once a value with a rounding error has been created, using it in subsequent calculations generally increases the error. Also, since division often requires rounding, errors may be reduced by dividing last, when possible.

Estimating Coordinates on a Graphing Utility

Most graphing utilities have the capability of displaying the approximate coordinates of a point on a graph. After selecting the trace feature a cursor blinks on the screen. The coordinates of this blinking cursor are displayed on the bottom of the screen. When the cursor is moved, the coordinates of the new point are displayed.

Often, we need to find where a graph crosses the x or y axis, its intercepts. Consider the equation $y = 10 - x^2$ and its graph in Figure 8.

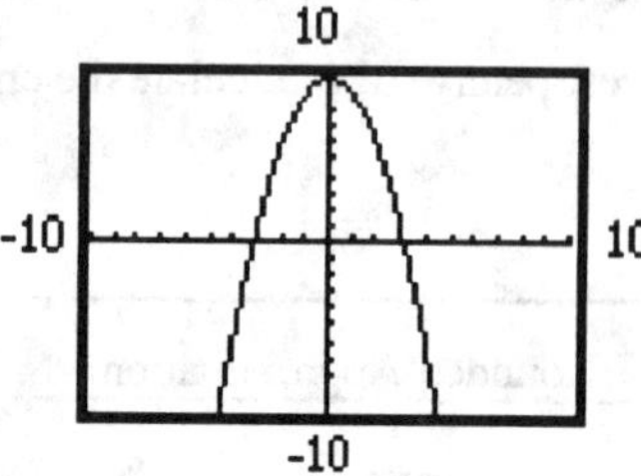

Figure 8

Since x intercepts are found by letting $y = 0$, we can use algebra to find the exact x intercepts. Substituting $y = 0$ in $y = 10 - x^2$ yields $0 = 10 - x^2$. Solving the equation $0 = 10 - x^2$ by extracting roots yields $x = \pm\sqrt{10}$.

$$0 = 10 - x^2$$
$$x^2 = 10$$
$$x = \pm\sqrt{10}$$

(See Appendix I, page A13 to review solving a quadratic equation by extracting roots.)

The y intercept is found by letting $x = 0$. If $x = 0$, then $y = 10 - x^2 = 10 - 0^2 = 10$. In the next example we use the trace feature to estimate the intercepts.

Example 4 *Estimating the intercepts of a graph*

Graph $y = 10 - x^2$ on a graphing utility with a standard window. Then use the trace feature to estimate the x and y intercepts.

Solution

From Figure 9 we approximate the x intercepts of the graph of $y = 10 - x^2$ as -3.19 and 3.19, and the y intercept as 10.

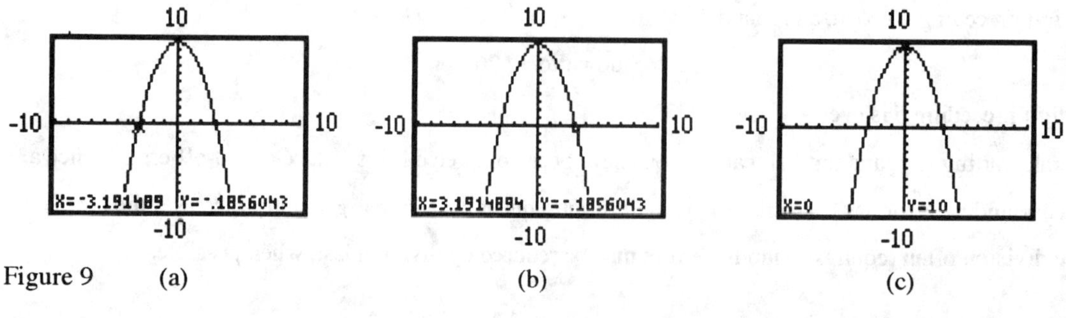

Figure 9 (a) (b) (c)

How accurate are the results in Example 4?

From the algebraic solution above we know that the x intercepts are $\pm\sqrt{10} \approx 3.16227766$. Whenever we use the trace feature to find a point on a graph, some digits in the displayed value may be approximate. We know that the values given for x intercepts in Example 4 are not exact because the displayed value of y is not exactly zero. The root or zero feature may give greater accuracy for finding the x intercepts (Figures 10a and 10b).

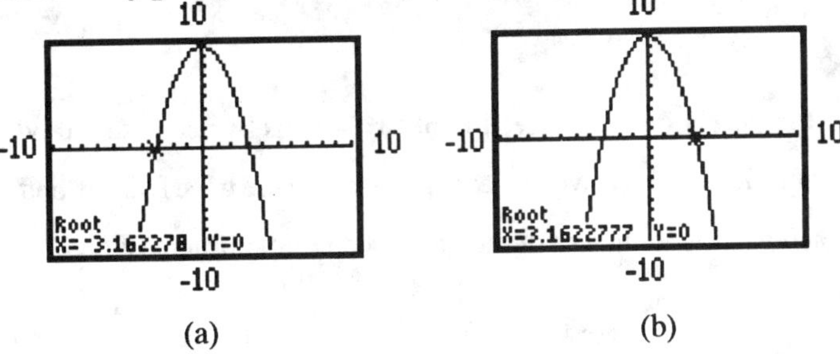

Figure 10 (a) (b)

From the algebraic solution above we know that the y intercept is 10. In Example 4, the trace feature was accurate. The value feature automatically calculates the value of a given expression for any specified value of x. It is a convenient way to find the y intercept by letting $x = 0$ (Figure 11).

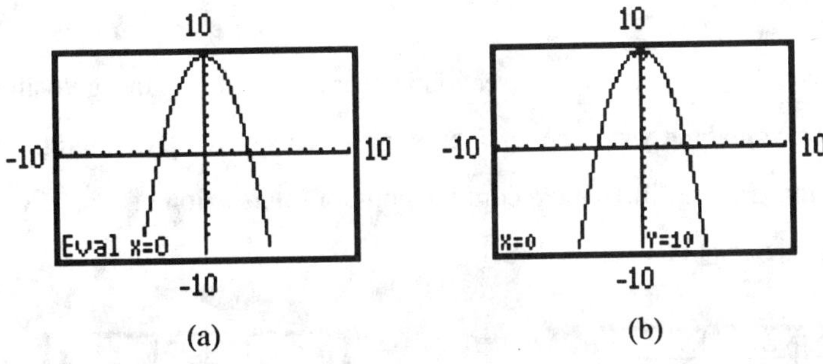

Figure 11 (a) (b)

In this book we introduce new notation: We use a **tilde** (~) to indicate that a coordinate of an ordered pair is an approximation. Writing the ordered pair $(3.\tilde{2}, 0)$ indicates that the x coordinate is rounded to one decimal place. In the ordered pair (0, 10) all digits are exact; there is no need for the tilde symbol.

NOTE: Using the ~ notation, we can label the point $(5, \sqrt{2})$ as $(5, 1.41\tilde{4})$, indicating that the y coordinate was rounded to three decimal places.

We will frequently use the maximum or minimum features to find the coordinates of high and low points on graphs (**turning points**). For example, although the point (0,36) in Figure 12a is not the highest point on the graph of $y = (x^2 - 4)(x^2 - 9)$, it is higher than nearby points. It is a **relative maximum**. The point $(-2.\tilde{5}, -6.25)$ in Figure 12b is a minimum point.

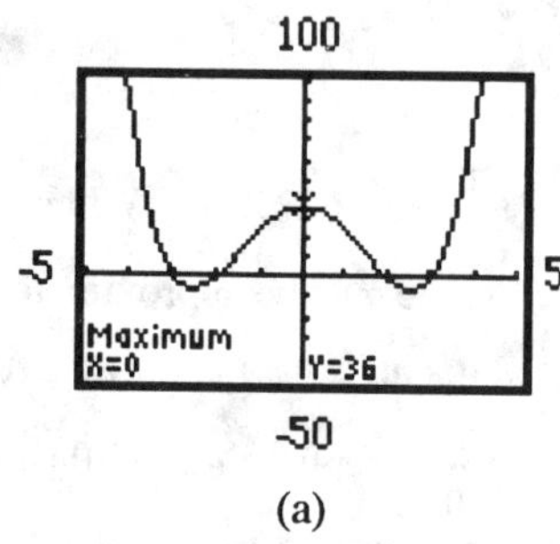

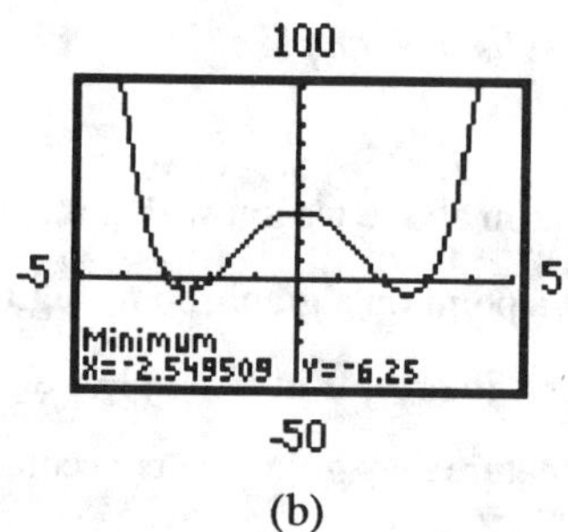

Figure 12 (a) (b)

In the next example we estimate a point of intersection of two graphs by various methods: using the table feature, which creates a table of ordered pairs (given an equation); using the trace feature; and using the intersect feature. The intersect feature gives the coordinates of the displayed point of intersection of two graphs.

Example 5 *Estimating a point of intersection*

Graph y = 2.5x - 3 and y = 0.75x + 5 on a graphing utility and estimate the point of intersection of the graphs using the table, trace, and intersect features.

Solution

Using the table feature, we see that the point of intersection is near x = 4.5, since those y values are close to each other (Figure 13a). Notice that the x values increase by increments of 0.5. Changing the increments to 0.01 refines our approximation (Figure 13b). We record the point of intersection of y = 2.5x - 3 and y = 0.75x + 5 rounded to one decimal place as $(4.\tilde{6}, 8.\tilde{4})$.

X	Y1	Y2
3	4.5	7.25
3.5	5.75	7.625
4	7	8
4.5	8.25	8.375
5	9.5	8.75
5.5	10.75	9.125
6	12	9.5

X=4.5

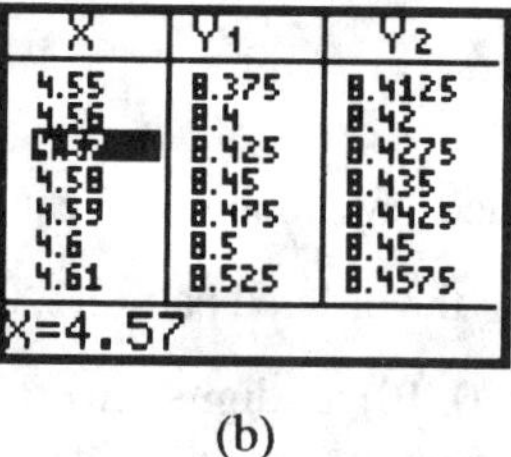

X	Y1	Y2
4.55	8.375	8.4125
4.56	8.4	8.42
4.57	8.425	8.4275
4.58	8.45	8.435
4.59	8.475	8.4425
4.6	8.5	8.45
4.61	8.525	8.4575

X=4.57

Figure 13 (a) (b)

Figure 14a shows the results using the trace feature, and Figure 14b shows a better approximation using the intersect feature.

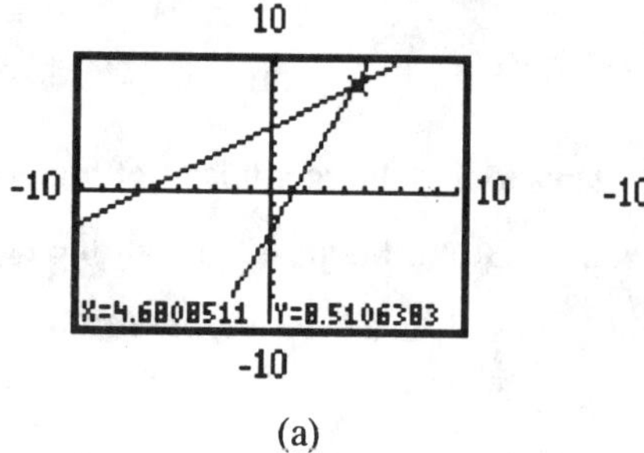

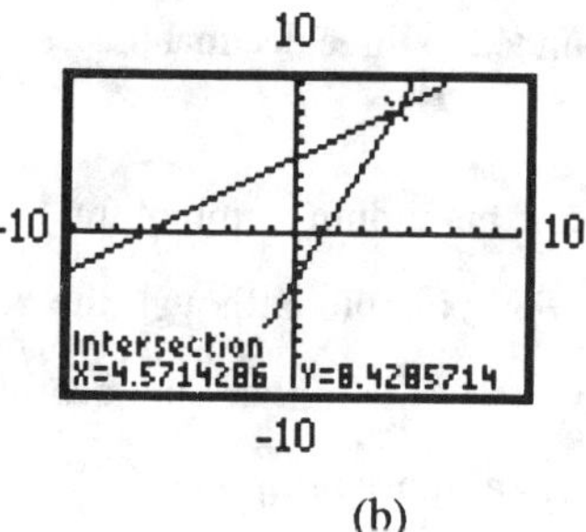

Figure 14 (a) (b)

The exact values of the point of intersection, found by algebra, are $(\frac{32}{7}, \frac{59}{7})$.

Determining Accuracy

Whenever we use a graphing utility to estimate a value of x or y, we should be aware of the possibility for inaccuracy. Any viewing screen is made up of a few hundred points called *pixels*. When we choose the dimensions of the window, each pixel is assigned one pair of coordinates. The trace feature reads the assigned coordinates of a pixel. Since the infinite number of possible coordinates in the Cartesian system cannot be represented by only a few hundred pixels, each pixel represents intervals of real numbers, horizontally and vertically.

If the solution does not fall exactly on a pixel, the nearest one will be displayed. The interval between adjacent pixels on a standard window may be over 0.2, depending on the size of the screen and the number of pixels. The interval between pixels changes as we zoom in or out. If a result is accurate to only one decimal place, it would be inappropriate to report 10 decimal places. Reconsider the displayed values of the point of intersection of the two lines in Figure 15a.

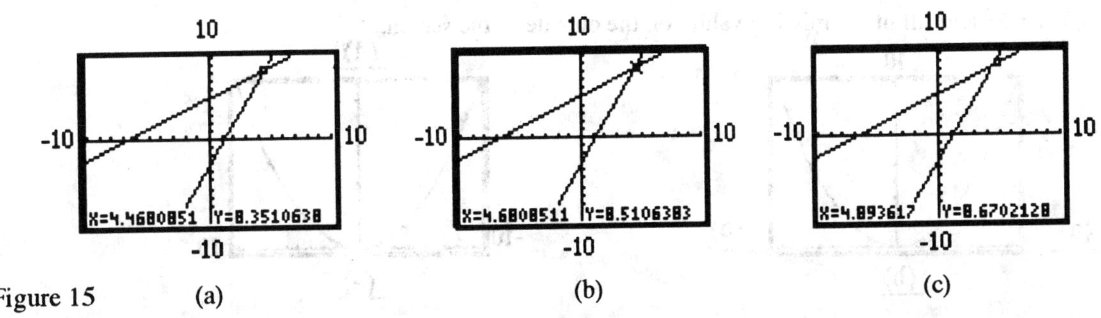

Figure 15 (a) (b) (c)

Moving the cursor one pixel to the left (Figure 15a) or one pixel to the right (Figure 15c) from the point of intersection depicted in Figure 15b changes the readout. Notice that only the units digit remained the same. This unchanged digit is accurate. We can state interval estimates of the point of intersection.

$$4.5 < x < 4.9 \quad \text{and} \quad 8.4 < y < 8.7$$

Since 4.7 and 8.55 are the midpoints of the two intervals, we write $(4.\tilde{7}, 8.\tilde{6})$ to indicate a one-decimal-place approximation of the point of intersection.

NOTE: Since the x interval is 0.4 unit in length, using the midpoint ensures an error ∈ of less than half the interval $\in < 0.2$. For y, the interval is 0.3 unit in length; the midpoint ensures that $\in < 0.15$. If either of these errors in unacceptable, we can zoom in or use another approach.

Strategy: Determining the Accuracy of Coordinates

1. Graph an equation.
2. Position the cursor at the desired point and note the displayed coordinates of that pixel.
3. Move the cursor left or right to the next pixel.
4. The digits that do not change are accurate.
5. If this accuracy is acceptable, quit. If not, zoom in and try again.

Prologue Exercises

For help with these exercises specific to individual graphing utilities, go to the *Precalculus Weblet - Reference Center - Technology* (http://oscar.ctc.edu/precalc).

1. Graph each equation on a standard window and sketch your results.

(a) $y = |x - 4|$ (b) $y = |x| - 4$

2. Graph $y = x^2$ and fill in the missing values on the outside of the screen.

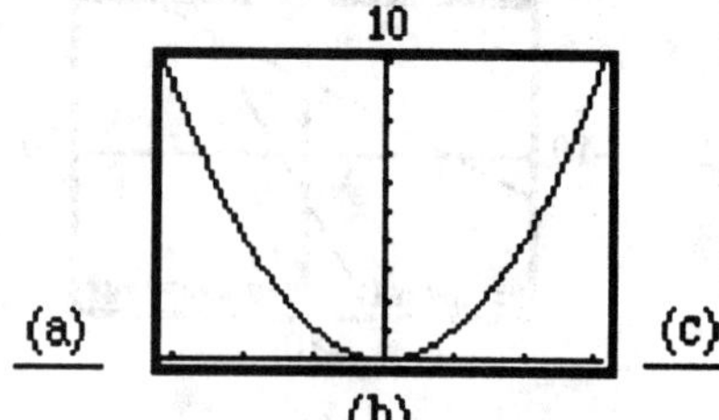

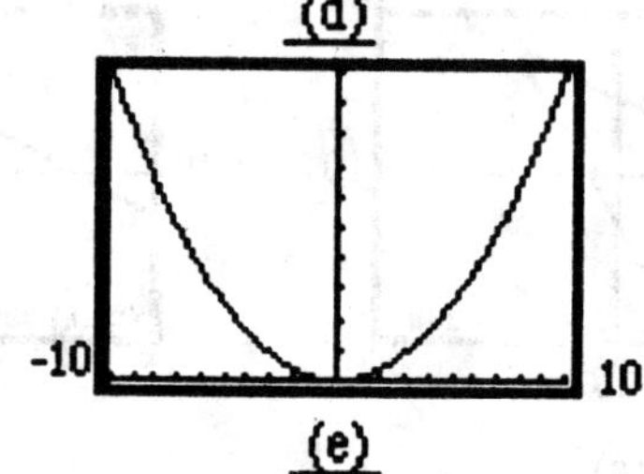

3. Graph each pair of equations on a graphing utility and approximate the point(s) of intersection.

(a) $y = x^2 + 8$
$y = 3x + 12$

(b) $y = -x^2 + 8$
$y = 2x + 7$

4. Graph $y = 27 - 3x^2$ on a graphing utility and use your results to state the coordinates of each point.

(a) x intercepts [Hint: Use the root or zero feature.]

(b) y intercept [Hint: Use the value feature with x = 0.]

(c) maximum [Hint: Use the maximum feature.]

5. Graph $y = x^3 - \frac{3}{2}x^2 - \frac{3}{2}x + 1$ on a graphing utility and use your results to state the coordinates of each point.

(a) x intercepts (b) y intercept (c) relative maximum

(d) relative minimum (e) Find the value of y when x = -5. (f) Find the value of y when x =12.

6. Graph $y = -x^3 + \frac{1}{2}x^2 + 2x - 1$ on a graphing utility and use your results to state the coordinates of each point.

(a) x intercepts (b) y intercept (c) relative maximum

(d) relative minimum (e) Find the value of y when x = 1/3. (f) Find the value of y when x = -20.

Chapter 1

Functions

Our study of functions, special relationships between two variables, began in the Prologue by plotting points in a Cartesian coordinate system and creating graphs on a graphing utility. Now we develop some of the universal concepts of functions and their graphs. In subsequent chapters we focus on individual classes of functions (i.e., linear functions, quadratic functions) and their graphs, extending the universal concepts introduced here.

The discussions in this chapter focus on functions from the Library of Reference Functions in Appendix II. These 13 reference functions represent classes of functions and form the foundation for our study. The variety of concepts and applications that can be studied and understood with this library of 13 functions includes situations in nearly every field: engineering, economics, ecology, biology, archeology, oceanography, psychology, physics, astronomy, art, and music.

One of our initial goals is to understand why the graph of a given equation has a certain size, shape, and position. This understanding is especially important in developing a questioning attitude about a display when we use graphing utilities. Otherwise, the concepts of graphing may be missed and our study risks becoming one of memorizing which keys to press rather than learning and understanding the mathematical concepts.

A Historical Note on Functions

Functions have always existed implicitly as curves, paths of moving objects, sets of ordered pairs, and relationships in words. One of the first explicit definitions of a function was given in 1667 by James Gregory. He defined a *function* as a quantity resulting from any combination of algebraic or other imaginable operations.

In 1665, Isaac Newton used the word *fluent* to describe any relationship between variables.

In 1673, Gottfried Leibniz introduced the function concept to mean any quantity varying from point to point on a curve.

In 1697, Johann Bernoulli adopted Leibniz's phrase, "function of x," which he symbolized as fx.

In 1734, Leonhard Euler defined a function and introduced the notation f(x) for a function, which has persisted to this day. Euler also introduced inverses to define logarithms, much as we will do in Chapter 4.

To learn more about the history of mathematics, go to the *Precalculus Weblet - Reference Center - History* (http://oscar.ctc.edu/precalc).

1.1 Functions and Their Properties

Defining a Function
Describing the Four Faces of Functions
Defining a One-to-One Function
Introducing the Library of Reference Functions
Describing Properties of Functions

If there is a relationship between two quantities, we can create a rule of correspondence between the quantities. Consider, for example, cooking a turkey. *When is a turkey done?* Using a meat thermometer requires monitoring its reading until the internal temperature of the bird reaches 180°F. The temperature of the turkey is related to the cooking time (Figure 1a). If the turkey has a pop-up indicator, we would have two possible values, *not done* and *done*. If we assign 0 to *not done* and 1 to *done*, we have the model in Figure 1b. Figure 1c shows another rule for cooking turkeys, "allow 15 minutes per pound," t = 15w, relating cooking time to the weight of the turkey. Each relationship describes how one quantity corresponds to another.

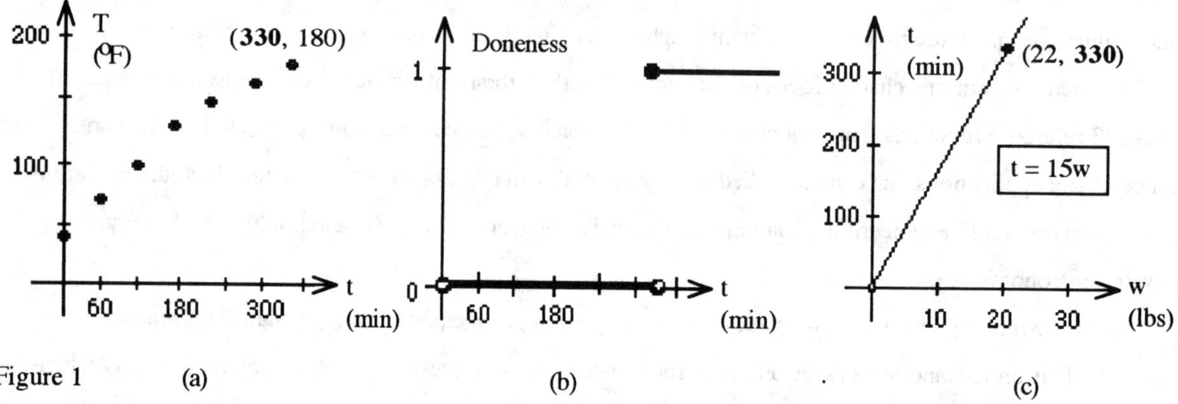

Figure 1 (a) (b) (c)
[Source: *Joy of Cooking*, I. Rombauer and M. Rombauer Becker]

Defining a Function

A correspondence between two quantities, *inputs* and *outputs*, is called a **relation**. Relations that have one and only one output for each input are called **functions**. If any input has more than one output, the relation is not a function.

The relation in Figure 2a is not a function since the input 3 is paired with both 7 and 10. However, the relation in Figure 2b is a function since each input has only one output.

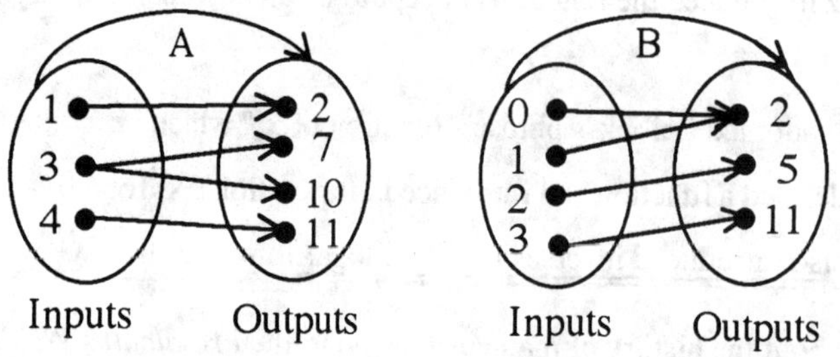

Figure 2 (a) (b)

Definition of a Function

A **function** is a rule of correspondence between two sets X and Y that assigns to each element in set X one and only one element in set Y.

The set of inputs X of a function is called the **domain** and the set of outputs Y is the **range**. The input variable is referred to as the **independent variable**. The output variable is called the **dependent variable** because its value depends on the input we choose.

The function defined by set B in Figure 2b,

$$B = \{(0,2), (1,2), (2,5), (3,11)\}$$

has domain X = {0, 1, 2, 3} and range Y = {2, 5, 11}. The following table gives common terms used to describe domain and range.

Domain	Range
values of the independent variable	values of the dependent variable
inputs	outputs
x coordinates	y coordinates
values of x	values of y (the function)

Describing the Four Faces of Functions

We have seen that functions can be described in several ways. In the preceding turkey discussions, we described cooking time in words with a rule of thumb, 15 min/lb, with data collected over time by measuring the temperature of the turkey, with the equation

$$t = 15w$$

and with several graphs. We refer to such alternative descriptions as the **four faces of functions**.

Four Faces of Functions

A function may be described:

1. in words
2. by a set of data
3. with an equation or formula
4. with a graph

If we consider the circumference C of a circle as a function of its diameter d, we can describe the function as follows.

1. In words, we write

"the circumference of a circle is equal to the product of π and its diameter"

or

"the number π is the ratio of the circumference of a circle to its diameter".

2. A set of data can be collected from actual measurements.

diameter (in.)	$1\frac{5}{8}$	$2\frac{3}{8}$	$3\frac{1}{4}$	$5\frac{3}{16}$	$10\frac{1}{8}$
circumference (in.)	$5\frac{1}{8}$	$7\frac{1}{2}$	$10\frac{1}{4}$	$16\frac{1}{4}$	$31\frac{3}{4}$

3. The equation (formula)

$$C = \pi d$$

expresses the circumference of a circle as a function of its diameter.

4. Graphs created from data or from the equation $C = \pi d$ describe this function in the Cartesian coordinate system (Figure 3).

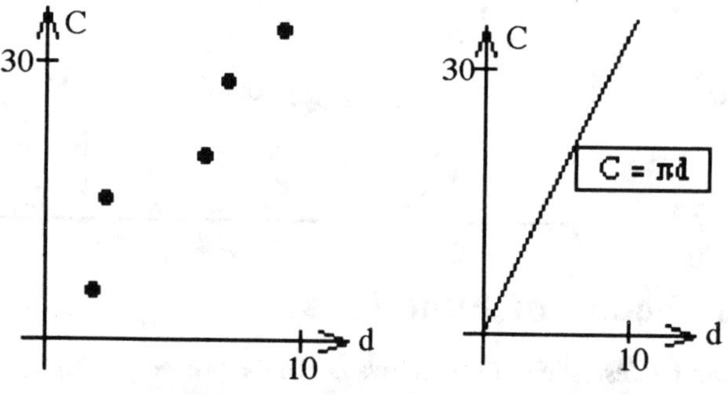

Figure 3

Why do you think the actual measurements in the table differ from values calculated from the equation $C = \pi d$?

Measurements are only as accurate as the scale on the instrument and may depend on the skill of the person making them.

To determine whether or not a small set of ordered pairs is a function, we can simply check to see if each input is paired with exactly one output. However, when a relation is given as an equation, usually we cannot check all of the resulting ordered pairs.

Can you think of a quick way to determine if an equation represents a function, by examining its graph?

<u>If a vertical line can be drawn to intersect more than one point on a graph, indicating that some value of x is associated with more than one value of y, the graph does not represent a function.</u>

For example, the graph of $x^2 + y^2 = 1$, pictured as a circle in Figure 4, fails the **vertical line test**. Notice how the vertical line associated with x = 1/2 passes through the graph at two points: $(1/2, \sqrt{3}/2)$ and $(1/2, -\sqrt{3}/2)$, two different values of y with the same value of x. This equation does not represent a function.

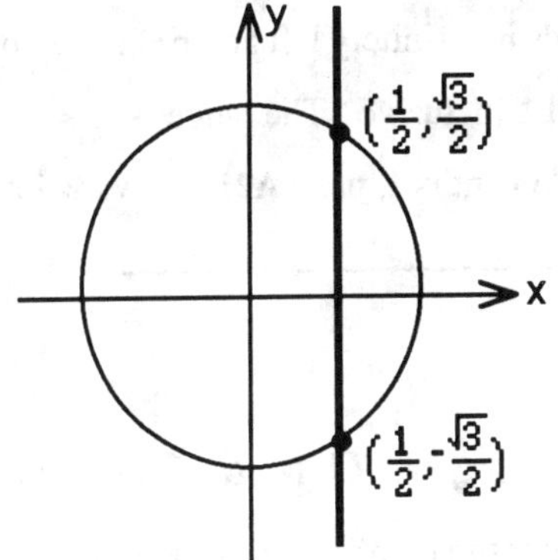

Figure 4

In the next example we see relations described in four different ways and decide which are functions. Unless otherwise stated or restricted, we assume that the domain of a function is the set of all real numbers and the range is the corresponding values of the function.

Example 1 *Identifying functions described in different ways*

Which relations describe functions? State the domain and range of the functions.

(a) The sentence "The sum of a number and 2 equals another number."

(b) The set of ordered pairs

x	1	2	2
y	2	3	5

(c) The equation $y^2 = x$

(d) The graph in Figure 5

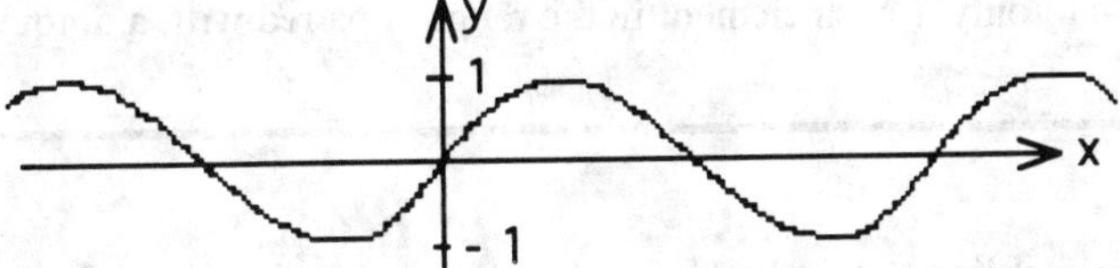

Figure 5

Solution

(a) Since adding 2 yields one output for each input, this is a function. The domain and range are all real numbers.

(b) The relation is not a function since the x coordinate 2 is paired with two different y coordinates, 3 and 5.

(c) The equation $y^2 = x$ does not express y as a function of x because each value of x does not correspond to a unique value of y. For example, $x = 4$ is paired with both $y = 2$ and $y = -2$.

(d) Since the graph in Figure 5 passes the vertical line test, it represents a function. Assuming that the graph extends indefinitely left and right, the domain is the set of all real numbers. Vertically, the graph falls between -1 and 1 inclusive. The range is $-1 \le y \le 1$, which in interval notation is [-1, 1].
(See Appendix I, page A21 to review interval notation.)

Defining a One-to-One Function

Consider the graphs of the functions

$$y = x \quad \text{and} \quad y = x^2$$

pictured in Figure 6.

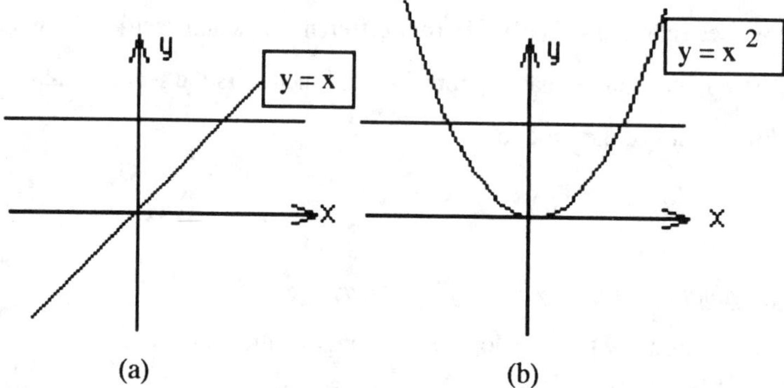

Figure 6

Both $y = x$ and $y = x^2$ are functions because each value of x is paired with one and only one value of y. For $y = x$, each value of y is paired with a unique value of x. We call the function $y = x$ a **one-to-one function.** The squaring function $y = x^2$ is not one-to-one since some values of y are paired with more than one value of x.

Definition of a One-to-One Function
A function is **one-to-one** if and only if each element in the range is paired with a unique element in the domain.

Just as we can use the vertical line test to determine if a relation is a function, we can use a **horizontal line test** to determine if a function is one-to-one (Figure 6).

<u>If any horizontal line intersects the graph more than once, the function is not one-to-one.</u>

Introducing the Library of Reference Functions

To explore the library of reference functions, go to "Explore the Concepts," Exercises E1 - E7, before continuing (page 21). For the interactive version, go to the *Precalculus Weblet - Instructional Center - Explore Concepts*.

Thirteen **reference functions** create a foundation upon which we build our precalculus study. These reference functions are presented in Appendix II, Library of Reference Functions. In Figure 7 we introduce the first seven graphs from our library. They are the linear function $y = x$, the absolute value function $y = |x|$, the square-root function $y = \sqrt{x}$, the squaring function $y = x^2$, the cubing function $y = x^3$, the reciprocal function $y = \frac{1}{x}$, and the inverse-square function $y = \frac{1}{x^2}$.

Throughout the rest of this chapter, we refer to these seven reference functions and their graphs repeatedly. (The remaining reference functions will be introduced informally in the exercises and more formally in the text as our study of functions continues.)

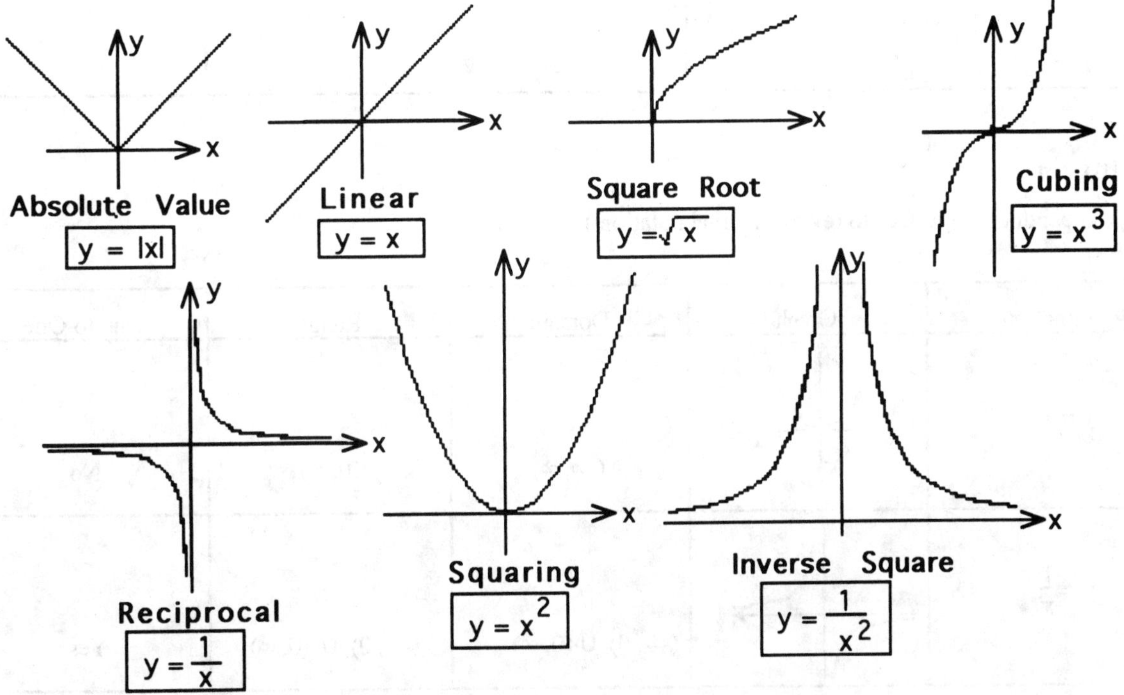

Figure 7

Each graph in Figure 7 has a shape that occurs within large families of equations. Each graph is also the simplest example of a fundamental concept. For example, the graph of $y = x$ is an example of a linear function. In Chapter 2 we will see how the graph of $y = x$ is related to the graphs of all functions of the form $y = mx + b$.

In the next example we compare reference functions using some of the terms discussed in this section.

Example 2 *Stating the domain and range of a function and deciding if it is one-to-one*

Complete the following table by

(a) drawing the reference graph,

(b) using interval notation to state the appropriate domain and range, and

(c) stating whether or not each function is one-to-one.

Function	Graph	Domain	Range	One-to-One
$y = x^2$				
$y = \frac{1}{x}$				

Solution

(See Appendix I, page A21 to review interval notation.)

Function	Graph	Domain	Range	One-to-One
$y = x^2$	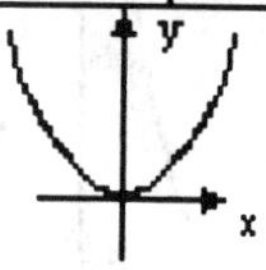 y, x	$(-\infty, \infty)$	$[0, \infty)$	No
$y = \frac{1}{x}$	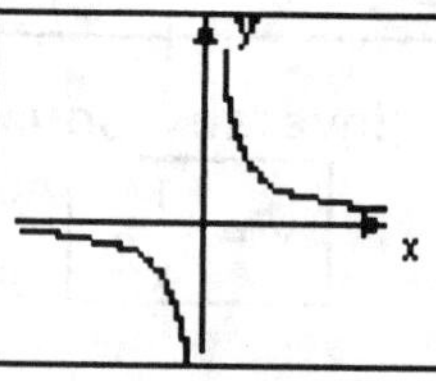 y, x	$(-\infty, 0) \cup (0, \infty)$	$(-\infty, 0) \cup (0, \infty)$	Yes

Describing Properties of Functions

We focus on intuitive descriptions of properties of functions. (These descriptions will be rigorously defined and interpreted geometrically in calculus.)

Consider the graphs in Figure 8. As we view a graph from left to right we can describe the graph as increasing or decreasing. If a graph is going up, we say that it is **increasing** (Figure 8a), and if it is going down, we say that it is **decreasing** (Figure 8b). The graph in Figure 8c is called a *constant function* since its y value remains constant, neither increasing nor decreasing. Viewing the graph in Figure 8d from left to right we notice that it is decreasing (going down) for values of $x < 0$ and increasing for values of $x > 0$. A point where a graph changes from decreasing to increasing or from increasing to decreasing is called a **turning point**. In Figure 8d the turning point is (0,0).

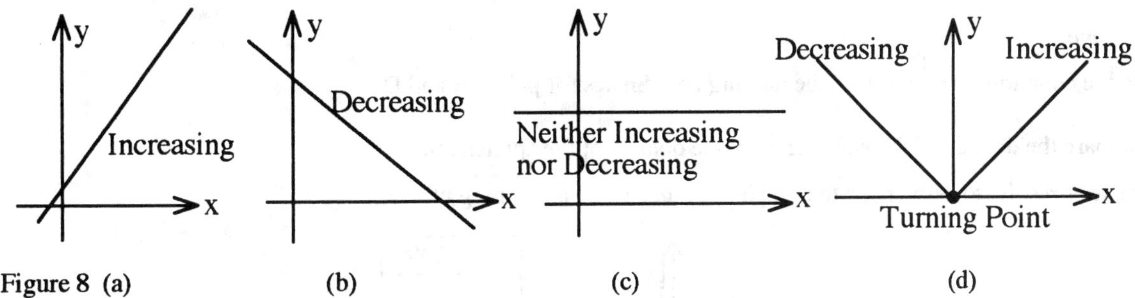

Figure 8 (a) (b) (c) (d)

Another characteristic of a function is whether or not its graph is continuous.

How would you describe a curve that is continuous?

The graph of a **continuous** function has no breaks. The graph can be drawn continuously, in one uninterrupted movement, without lifting the pencil from the paper. All the graphs in Figure 8 are continuous.

Figure 9 shows three graphs that are not continuous. None of the graphs can be drawn without lifting the pencil. They are examples of graphs that are **discontinuous**. Figure 9a represents a discontinuous function because the graph has a **hole** in it. We call the graph in Figure 9b a **step function** because it jumps from one level to the next. It is discontinuous. In Figure 9c, the graph increases without bound as x approaches 3 on the right side of the line $x = 3$, and decreases without bound as x approaches 3 on the left side of $x = 3$. The line $x = 3$ is called an **asymptote**. (We study asymptotes in detail in Chapters 3 and 4.) The graph in Figure 9c is discontinuous because it exists as two parts separated by the asymptote $x = 3$. Figure 9c also has a horizontal asymptote, the line $y = 1$.

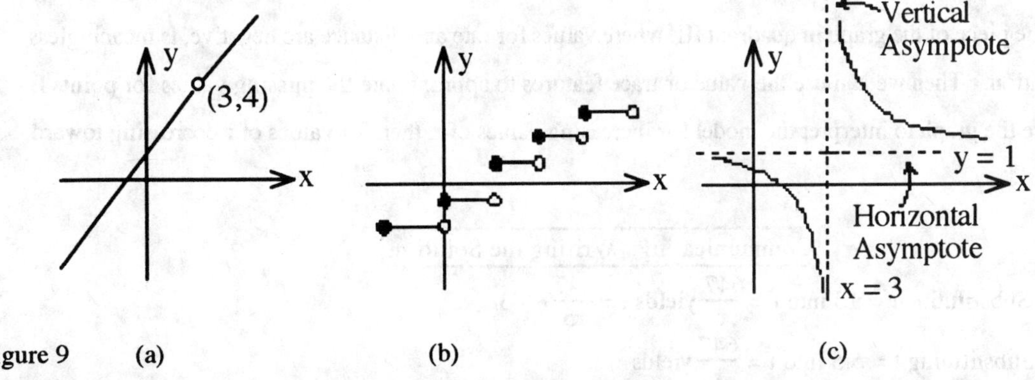

Figure 9 (a) (b) (c)

In Example 3 we use the distance formula $d = rt$, where d is distance traveled, r is rate, and t is time. We can express time as a function of rate if the distance is fixed. In the example the distance is fixed at 647 miles, yielding the function $t = \frac{647}{r}$.

Example 3 The Tourist Problem

Last August, four tourists left Seattle, Washington, for Dearborn, Montana, 647 mi away, to fish the Missouri River. The first, riding a bicycle, averaged 15 mi/hr. The second drove a car but stopped to fish every stream and averaged 30 mi/hr. The third drove straight through, averaging 60 mi/hr, and the last took only $5\frac{1}{2}$ hr in a private plane.

(a) Use the equation $t = \frac{647}{r}$ to find the missing coordinates for points A and D.

(b) Compare the domain of the equation with the domain of the application.

(c) Use your results to help create the graph in Figure 10 on a graphing utility.

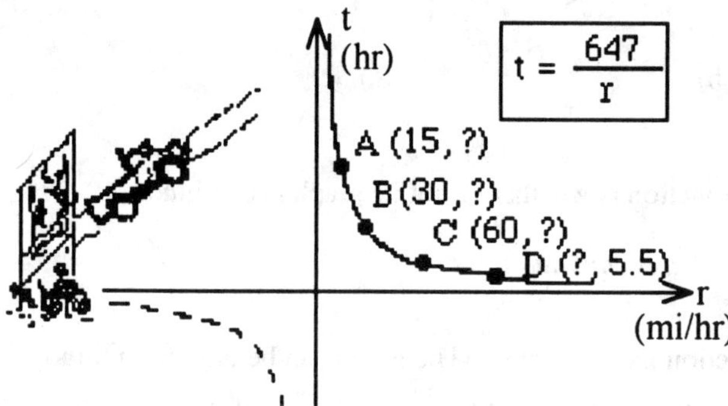

Figure 10

(d) Approximate the missing values for points B and C.

(e) What does the model indicate about the time required to complete the trip as the rate of travel increases?

(f) What does the model indicate about the time required to complete the trip as the rate of travel decreases?

Thinking: Creating a Plan

We substitute r = 15 into the equation to find the missing coordinate of point A, for the slowest traveler. We then substitute t = 5.5 to find the missing coordinate of point D, for the fastest traveler. These points help us set the screen window. Since we are interested only in positive rates and distances, we are interested in viewing quadrant I. The piece of the graph in quadrant III, where values for rate and distance are negative, is meaningless in this application. Then we can use the value or trace features to approximate the missing values for points B and C. We use the graph to interpret the model for increasing values of r, then for values of r decreasing toward zero.

Communicating: Writing the Solution

(a) Point A: Substituting r = 15 into $t = \frac{647}{r}$ yields $t = \frac{647}{15} \approx 43$

Point D: Substituting t = 5.5 into $t = \frac{647}{r}$ yields

$$5.5 = \frac{647}{r}$$
$$5.5r = 647$$
$$r \approx 118$$

(b) The domain of the equation $t = \frac{647}{r}$ is all real numbers such that $r \neq 0$. The application further restricts the domain of the function, $r > 0$.

(c) For this example, we graph $y = \frac{647}{x}$ with $0 < x < 118$ and $0 < y < 43$ (Figure 11).

(d) We approximate the missing coordinates for points B and C to be 21.6 and 10.8, respectively.

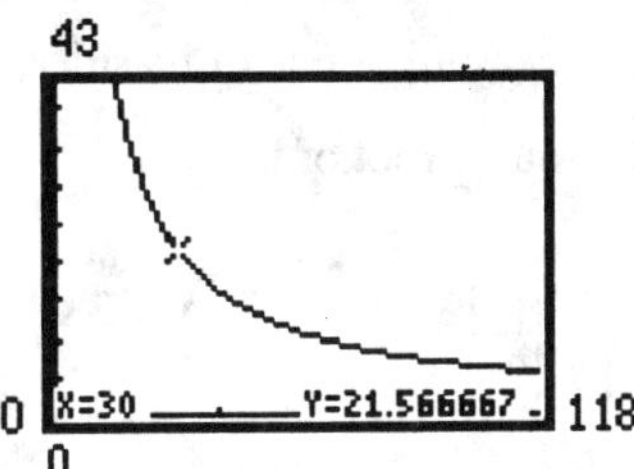

Figure 11

(e) The faster the traveler goes, the shorter the time needed to complete the trip. This graph approaches its horizontal asymptote, $t = 0$, indicating that as the rate increases, the time decreases, approaching but never reaching zero.

(f) As the graph approaches the vertical asymptote, $r = 0$, the travel time increases without bound. If the rate were $r = 0$, it would take an infinite amount of time to make the trip.

Learning: Making Connections

What happens to the travel time when the rate is doubled?

Notice from points A to B that the rate is doubled and the time is halved. From B to C the rate is again doubled and the time halved. We say that time is inversely proportional to the rate.

Explore the Concepts

Use a graphing utility to match each graph in Figure 12 with its equation.

E1. $y = \frac{1}{x}$ E2. $y = \sqrt{x}$ E3. $y = x^2$ E4. $y = |x|$ E5. $y = x$ E6. $y = x^3$ E7. $y = \frac{1}{x^2}$

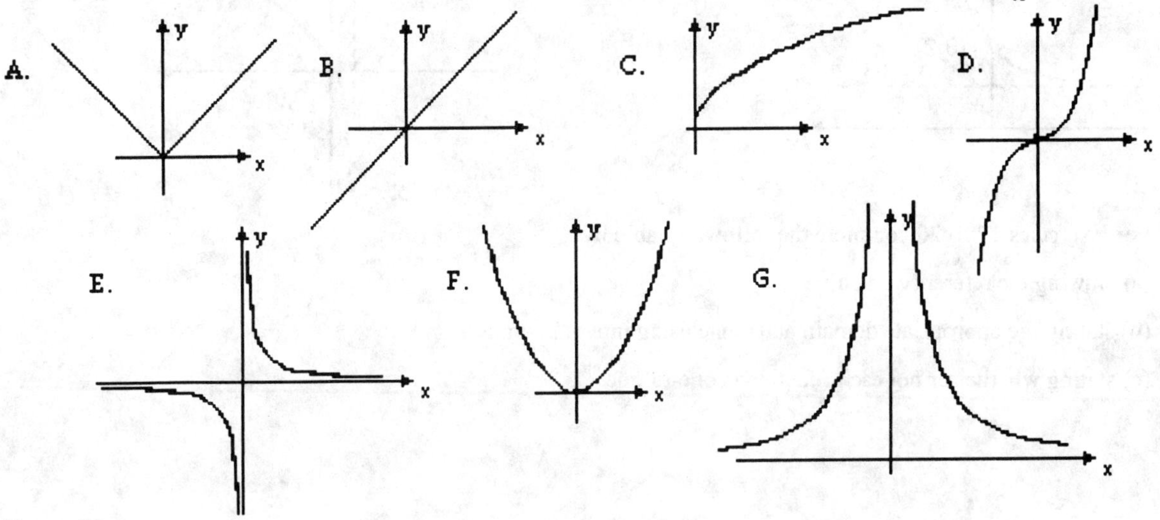

A. B. C. D. E. F. G.

Figure 12

Exercises 1.1

Develop the Concepts

For Exercises 1 to 16 determine whether each relation is a function. If it is a function, state its domain and range.

1. A = {(-27,27), (-27,-27)}

2. B = {(4,5), (-10,5)}

3. The area of a square is the square of the measure of one of its sides.

4. The measure of a side of a square is the square root of its area.

5. f :

x	3	3	-1
y	-1	0	3

6. g:

x	0	1	2
y	1	2	0

7. y = 3x

8. |y| = x

9.

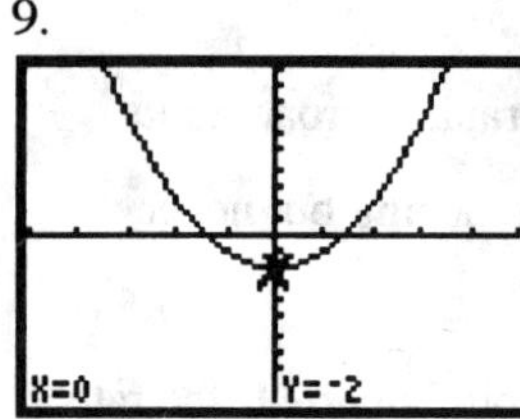

10.

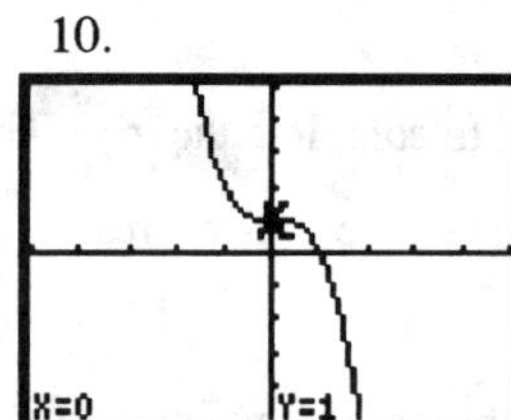

11.

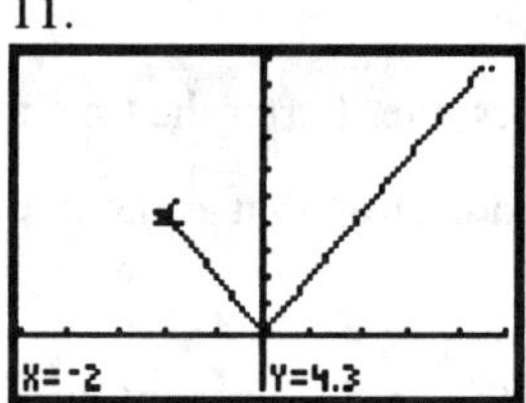

12.

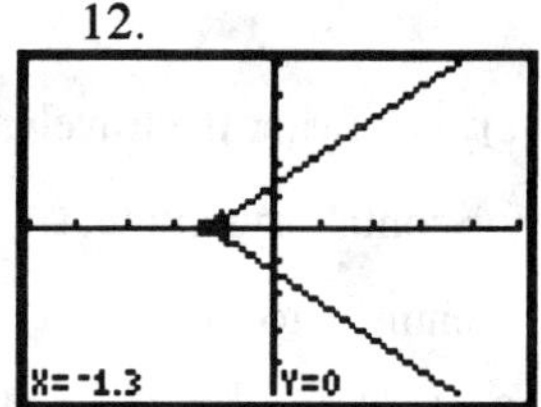

13.

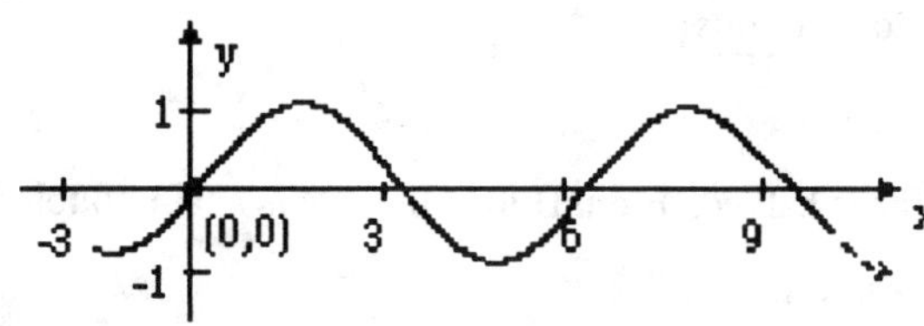

14.

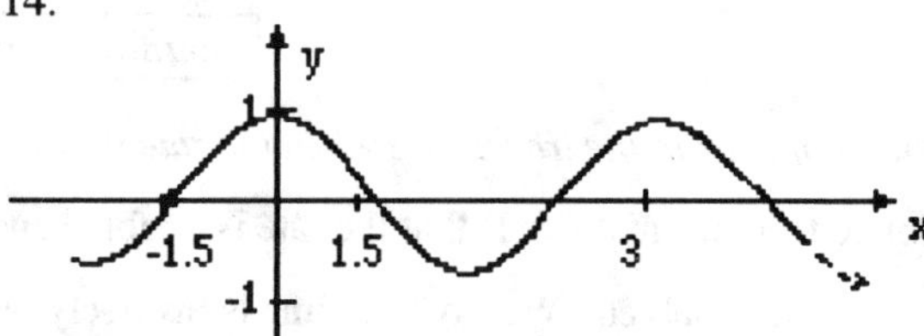

15.

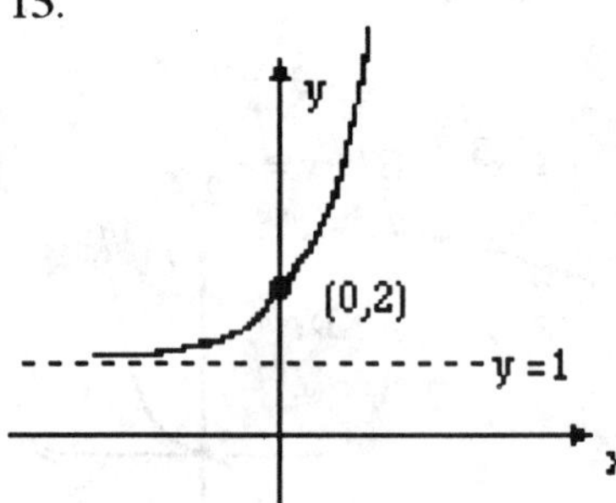

16.

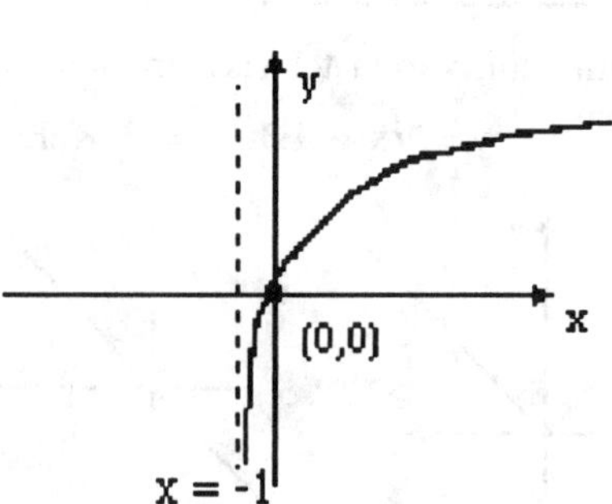

For Exercises 17 to 20 complete the following table by

(a) drawing the reference graph,

(b) stating the appropriate domain and range using interval notation, and

(c) stating whether or not each function is one-to-one.

Function	Graph	Domain	Range	One-to-One
17. $y = x$				
18. $y = \lvert x \rvert$				
19. $y = \sqrt{x}$				
20. $y = x^3$				

For Exercises 21 to 28 use the seven reference graphs (Figure 7 from Appendix II) to answer the following.

21. Which functions have all real numbers as their domain?

22. Which functions have all real numbers as their range?

23. Which of the seven functions are one-to-one functions?

24. Name three functions whose graphs are increasing over their domains.

25. Which graphs have turning points?

26. Which graphs are discontinuous?

27. Which graph is above the others for $0 < x < 1$: $y = x$, $y = x^2$, or $y = x^3$?

28. Which graph is above the others for $x > 1$: $y = x$, $y = x^2$, or $y = x^3$?

For Exercises 29 and 30 sketch a graph of a function that satisfies the following. Label the important points.

29. The domain of the function is all real numbers. The graph is continuous with turning points at (-1, 5) and (3, 3), crossing the x axis at (-2, 0). The graph is decreasing for $-1 < x < 3$. The graph is above the x axis for $x \geq -2$ and below it for $x \leq -2$.

30. The domain of the function is all real numbers and the range is $0 < y \leq 1$. The graph is increasing for $x < 0$ and decreasing for $x > 0$. There is a turning point at (0,1). The line $y = 0$ (the x axis) is a horizontal asymptote.

For Exercises 31 and 32 describe each function as increasing or decreasing for the interval stated.

*31. $y = \lvert x \rvert$ for $x < 0$

32. $y = \frac{1}{x^2}$ for $x < 0$

For detailed solutions to the exercises marked with a *, go to the *Precalculus Weblet - Instruction Center - Starred Exercises* (http://oscar.ctc.edu/precalc).

Apply the Concepts

For Exercises 33 to 38, (a) use the given equation to calculate the missing coordinates for points A and D,

(b) compare the domain of the equation with the domain of the application,

(c) use your results to help create the graph in each figure on a graphing utility,

(d) approximate the missing values for points B and C, and

(e) answer each question (e) in a complete sentence.

*33. The Hot Tub Problem The cost of owning and operating a hot tub is modeled by the equation $C = 650t + 4000$, which expresses cost C in dollars as a function of time t in years (Figure 13).

(e) Interpret point A.

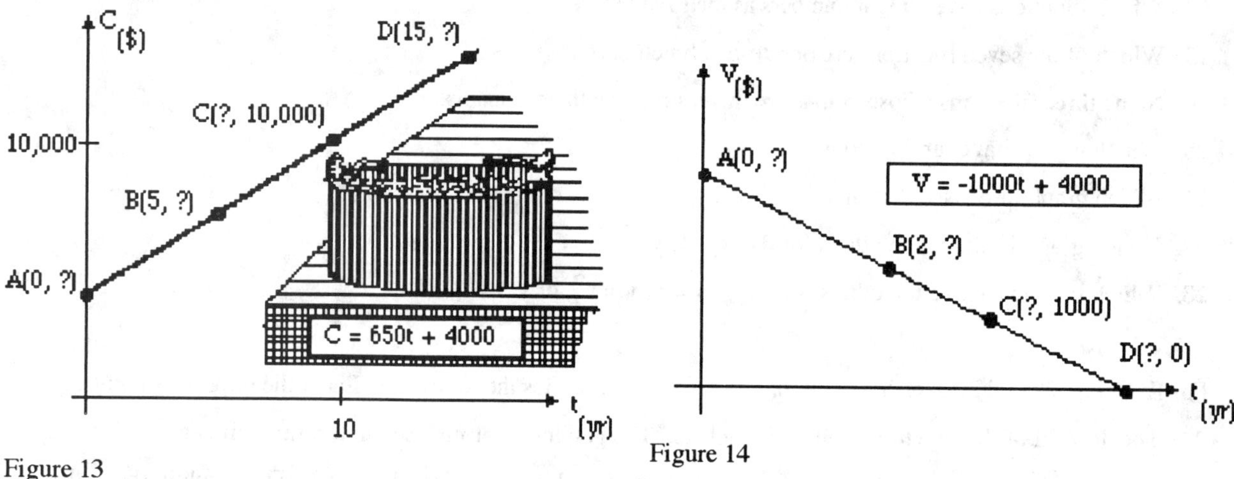

Figure 13

Figure 14

34. The Straight-Line Depreciation Problem Suppose that the Internal Revenue Service allows deducting a personal computer from taxable income based on how much the computer decreases in value each year. Assume that the equation $V = -1000t + 4000$ models the value of the computer V in dollars as a function of time t in years (Figure 14).

(e) Interpret points A and D.

For interactive help with The Rainbow Kite and Balloon Company problems, go to the *Precalculus Weblet - Assessment Center - Problem Solving Portfolio* (http://oscar.ctc.edu/precalc).

35. The Rainbow Kite and Balloon Company As owner of a kite store you realize that the demand for kites increases as the price decreases, but the profit per kite also decreases. Over a period of time you set various prices, kept track of sales, and calculated the corresponding profit. The equation

$$P = -\frac{1}{5}n^2 + 42n - 400$$

models profit P in dollars as a function of the number n of kites sold (Figure 15).

(e) Interpret points A, B, C, and D.

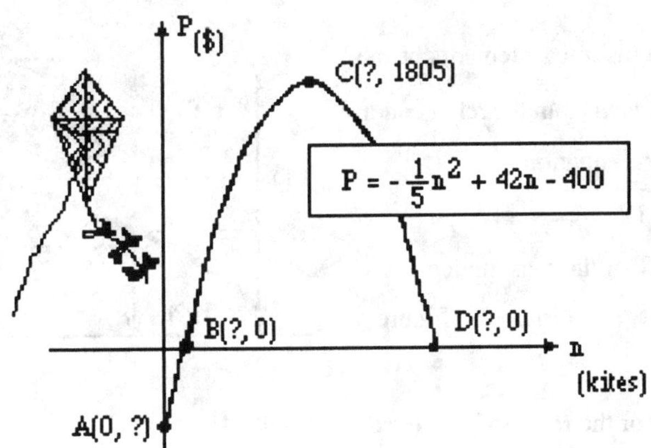

Figure 15

NOTE: Many exercise sets contain recurring scenario problems titled "The Rainbow Kite and Balloon Company" and "Wonderworld Amusement Park." The solutions to either collection of problems may be used to create a problem-solving portfolio.

36. Wonderworld Amusement Park A virtual reality version of the Indianapolis 500 race allows the paying customers to compete against the four racecar drivers represented in Figure 16. The equation t = 500/r models time t in hours as a function of average speed r in miles per hour. (Driver A won the race in 1911; driver B, Al Unser Jr., won in 1992; driver C, Al Unser Sr., won in 1978; and driver D won in 1990.) [Source: *Information Please Almanac*]

(e) Interpret points A, B, C, and D.

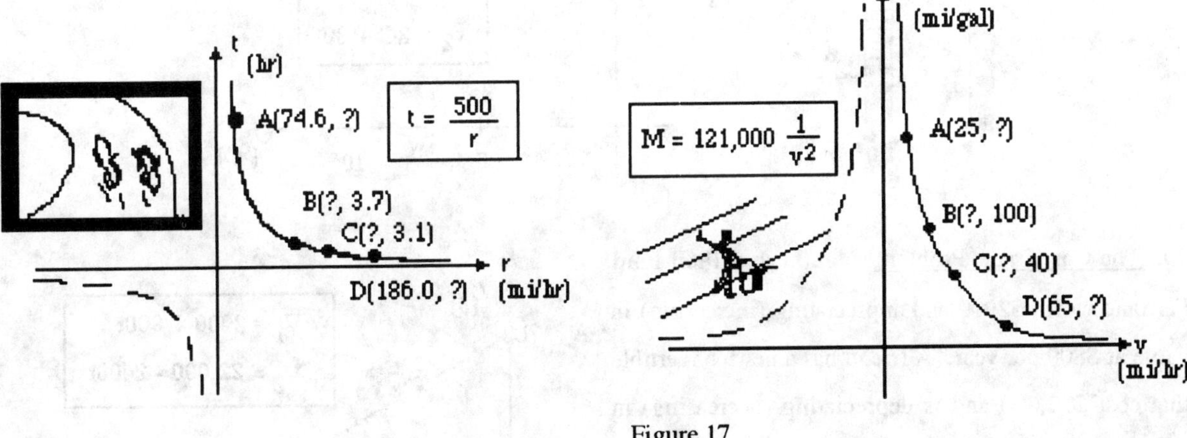

Figure 17

Figure 16

37. The Inverse-Square-Law Problem When driving, your mileage is inversely proportional to the square of your average velocity v. The equation $M = 121{,}000\frac{1}{v^2}$ models mileage M in mi/gal as a function of velocity v in mi/hr (Figure 17).

(e) Suppose that you wish to average 40 ± 10% mi/gal. How fast does this model indicate that you should drive?

38. The Roasting Problem Actual measurements of the internal temperature of a roast taken from the refrigerator and cooked in an oven lead to the equation

$$T = 120\sqrt{t + 0.12}$$

which models the temperature T of the roast in degrees Fahrenheit (°F) as a function of time t in hours (Figure 18).

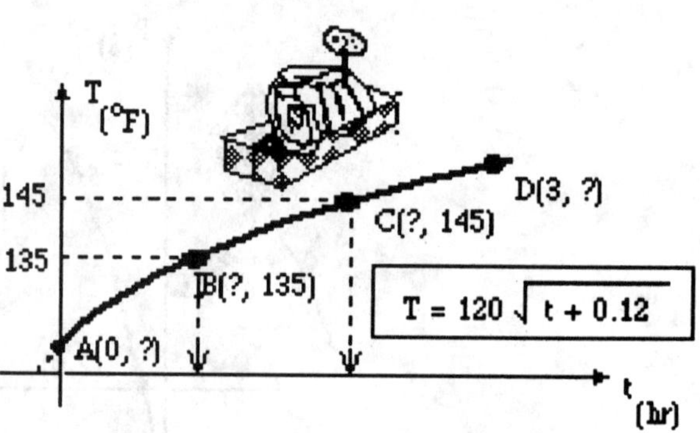

Figure 18

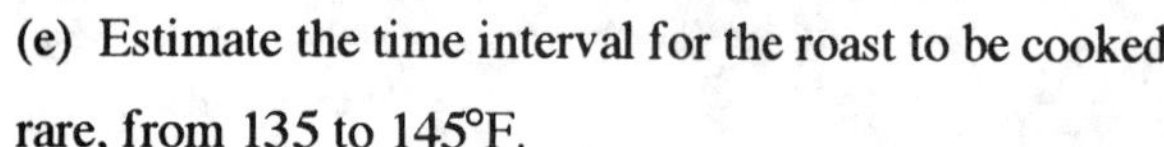

(e) Estimate the time interval for the roast to be cooked rare, from 135 to 145°F.

39. The Celsius-to-Fahrenheit Problem The formula $F_1 = \frac{9}{5}C + 32$ for changing a temperature in degrees Celsius (°C) to degrees Fahrenheit is graphed in Figure 19. The rule of thumb "double and add 30" ($F_2 = 2C + 30$) is also graphed.

(a) Recreate the graph in Figure 19 and approximate the missing values for the points.

(b) What can you conclude about the accuracy of the rule of thumb?

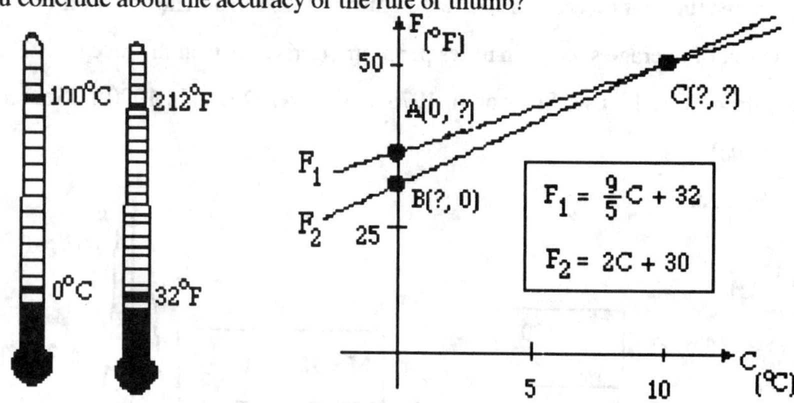

Figure 19

40. The Car Trade Problem You have a 1963 Ford Fairlane worth \$2900 and appreciating (increasing) in value at \$800 per year. A friend has a new convertible that cost \$22,000 and is depreciating (decreasing) in value at \$200 per month. Use the graphs and the equations

$$V_1(t) = 2900 + 800t$$

and

$$V_2(t) = 22{,}000 - 2400t$$

which model value V in dollars as a function of time t in years.

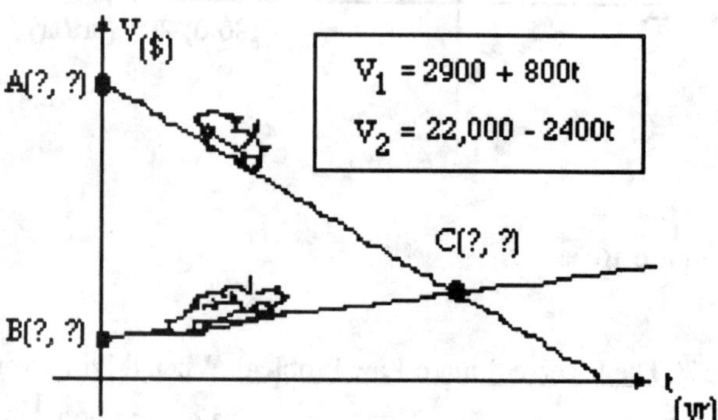

Figure 20

(a) Recreate the graph in Figure 20 and approximate the missing values for the points.

(b) Interpret the point of intersection.

Extend the Concepts

For Exercises 41 to 44 for each function defined by its graph and identified by number from Appendix II, (a) state the appropriate domain and range using interval notation, and

(b) state whether or not each function is one-to-one.

41. Graph number 5, the semicircle function.

42. Graph number 10, the exponential function.

43. Graph number 4, the logarithmic function.

44. Graph number 8, the sine function whose domain is all real numbers.

For Exercises 45 to 48 the graphing utility output for each exercise matches both equations. Decide if the output is accurate for both equations. If not, correct the graph.

45. $y = \sqrt{x^2}$ and $y = |x|$

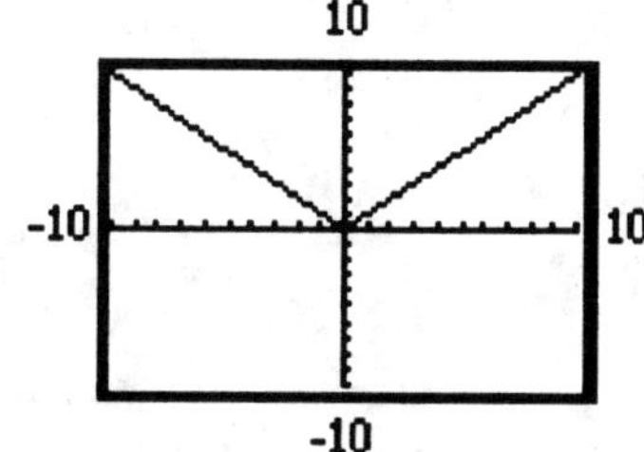

46. $y = \frac{x}{x}$ and $y = 1$

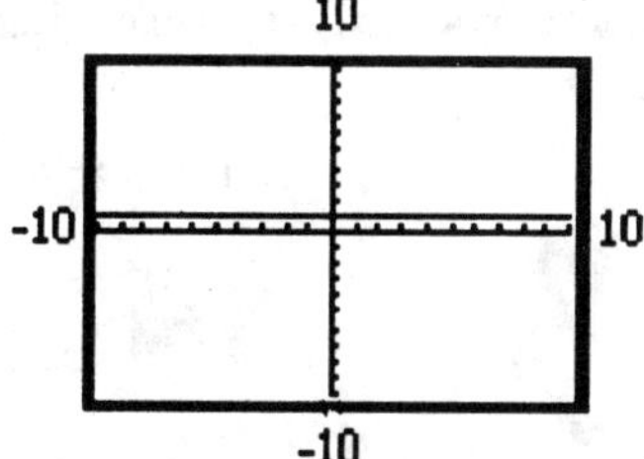

47. $y = \frac{x^2 - 1}{x + 1}$ and $y = x - 1$

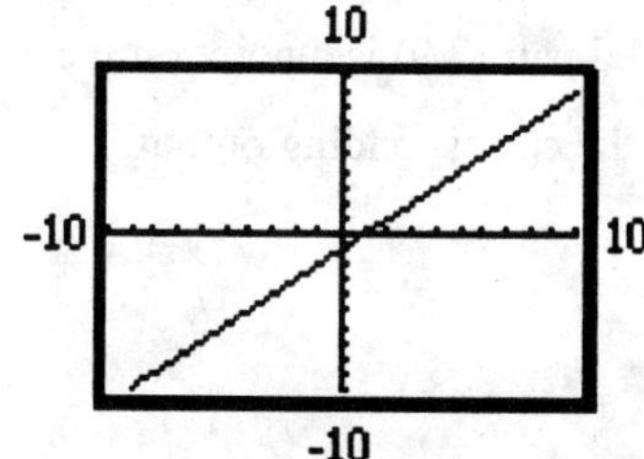

48. $y = \frac{x + 2}{(x + 2)^2}$ and $y = \frac{1}{x + 2}$

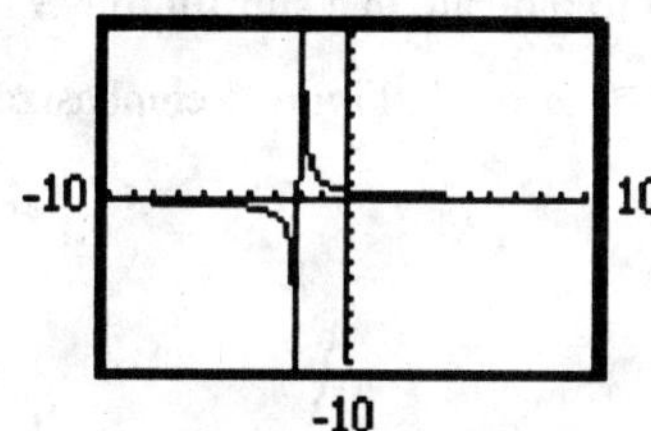

1.2 Functional Notation and Mathematical Language

Introducing Functional Notation
Using the Implication Symbol
Using the Equivalence Symbol

Throughout mathematics, functional notation is used extensively to make presentations more concise and precise. For this reason, we often see definitions and theorems given in terms of notation. To understand mathematical concepts, read and write mathematical presentations, and to continue your study of mathematics requires an understanding of mathematical language and symbolism. Modern computers, especially symbolic processors, use such language consistently.

Introducing Functional Notation

The following statements say exactly the same thing.

If $y = x^2$ and $x = 3$, then $y = 9$

and

If $f(x) = x^2$, then $f(3) = 9$

The equation $f(x) = x^2$ is written using **functional notation**, where $f(x)$ represents y. We read $f(x) = x^2$ as "f of x equals x squared" or "f at x equals x squared." The notation $y = f(x)$ emphasizes that x represents the input.

We write $f(3) = 9$ to indicate that substituting 3 for x yields $y = 9$. The ordered pair (3,9) is a point on the graph of $f(x) = x^2$. The arrow in Figure 1 emphasizes that the input $x = 3$ yields the corresponding output $f(3) = 9$.

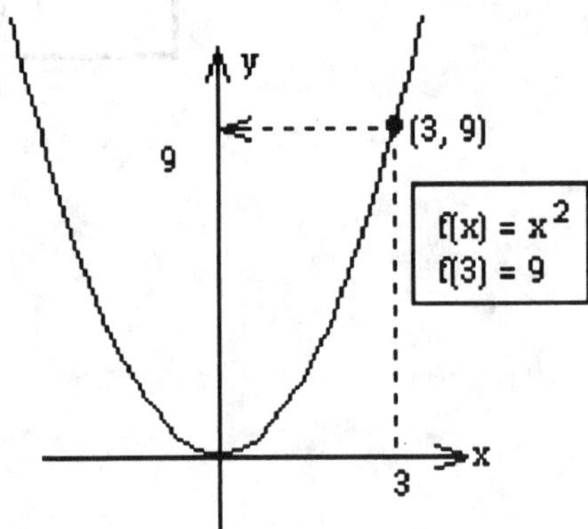

Figure 1

NOTE: The notation f(3) does not mean f times 3. However, 2f(3) means "multiply the output f(3) by 2."

In the next example we use functional notation to describe evaluating a function for different inputs.

Example 1 *Determining the value of a function*

For $f(x) = 3x + 4$, find and simplify

(a) $f(5)$ (b) $f(t)$ (c) $f(t + 1)$

Solution

(a) Finding $f(5)$ requires evaluating function f for $x = 5$.

$$f(x) = 3x + 4$$
$$f(\mathbf{5}) = 3(\mathbf{5}) + 4 = 15 + 4 = 19$$

(b) Finding $f(t)$ requires evaluating function f for $x = t$.

$$f(x) = 3x + 4$$
$$f(\mathbf{t}) = 3(\mathbf{t}) + 4 = 3t + 4$$

(c) Finding $f(t + 1)$ requires replacing x with $t + 1$.

$$f(x) = 3x + 4$$
$$f(\mathbf{t+1}) = 3(\mathbf{t+1}) + 4 = 3t + 3 + 4 = 3t + 7$$

Figure 2 shows the graph of the line $f(x) = 3x + 4$ ($y = 3x + 4$) from Example 1. The ordered pair (5, 19) represents $f(5) = 19$. The ordered pair $(t, 3t + 4)$ represents $f(t) = 3t + 4$. Once we have located t, then $t + 1$ is 1 unit larger. The ordered pair $(t + 1, 3t + 7)$ represents $f(t + 1) = 3t + 7$.

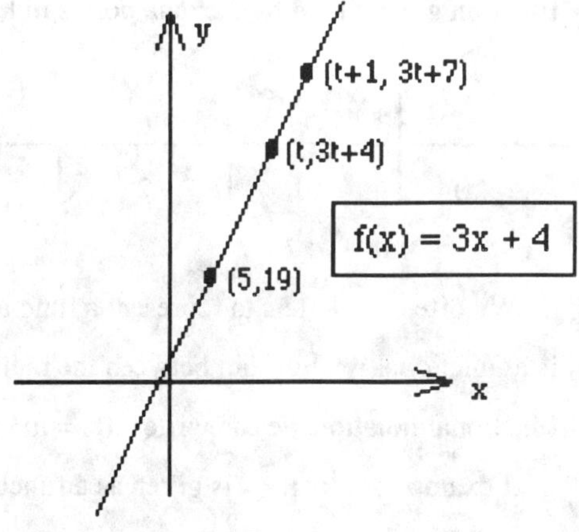

Figure 2

Any letters may be used to represent a function in functional notation, although f is most common. Regardless of what we name the dependent variable -- y, f(x), g(x), and so on -- some graphing utilities require the dependent variable to be y and the independent variable x. Therefore, we rename the variables to express y as a function of x whenever using a graphing utility. If an application involves the functions f, g, and h, we will rename them y_1, y_2, and y_3. When drawing the graphs on paper, however, we use the original names.

In the next example we name the function defined by data as "function g."

Example 2 *Reading function values*

Find the following values for function g defined by the following table.

x	-1	0	1	2	5
g(x)	5	1	3	5	10

(a) g(1) (b) 2g(1) (c) x such that g(x) = 5

Solution

Values in the "x" row are the inputs and the values in the "g(x)" row are the corresponding outputs.

(a) For g(1), the input is x = 1. Its corresponding output from the table is 3. We write

$$g(1) = 3$$

(b) Since 2g(1) means 2 times g(1) and the value of g(1) = 3, we write

$$2g(1) = 2 \cdot 3 = 6$$

(c) There are two inputs having 5 as the output, x = -1 as well as x = 2.

We often create new functions from given functions. For instance, the new function y = 2g(x), where function g is defined by the data points in Example 2, is created by multiplying all outputs by 2.

x	-1	0	1	2	5
2g(x)	10	2	6	10	20

We often use letters to represent a function that corresponds to the unknowns involved. For example, there is a functional relationship between the radius of a circle r and its area A given by the formula $A = \pi r^2$. Using functional notation we can write $A(r) = \pi r^2$ to emphasize that the area A is a function of the radius r. In the next example, velocity v is given as a function of time t.

Example 3 *Interpreting functional notation*

The function v(t) = 88 - 22t expresses the velocity in ft/sec of a braking car as a function of time in seconds. Find and interpret the following.

(a) v(4) (b) v(t) = 88

Solution

(a)

$$
\begin{aligned}
v(t) &= 88 - 22t \\
v(4) &= 88 - 22(4) \\
&= 0
\end{aligned}
$$

The braking car stopped 4 seconds after the brakes were applied.

(b) Substituting $v(t) = 88$ into $v(t) = 88 - 22t$ yields

$$
\begin{aligned}
88 &= 88 - 22t \\
22t &= 0 \\
t &= 0
\end{aligned}
$$

Initially, at the instant the brakes were applied, the velocity was 88 ft/sec = 60 mi/hr.

Using the Implication Symbol

When we write

$$\text{If } f(x) = x^2, \text{ then } f(3) = 9$$

the "If...then" statement is called an **implication**. The "if" part is the assumption (hypothesis); the "then" part is the conclusion that follows. Using the implication symbol $\Rightarrow$, we write the statement above as

$$f(x) = x^2 \Rightarrow f(3) = 9$$

Notice that the implication symbol separates the assumption $f(x) = x^2$ from the conclusion $f(3) = 9$. If the assumption is comprised of various parts, we connect them with the word *and.* In part (b) of Example 3 we could use the implication symbol to express

$$\text{substituting } v(t) = 88 \text{ into } v(t) = 88 - 22t \text{ yields } 88 = 88 - 22t$$

as

$$v(t) = 88 - 22t \text{ and } v(t) = 88 \Rightarrow 88 = 88 - 22t$$

Notice how the information preceding the $\Rightarrow$ sign emphasizes the assumptions being made.

Example 4 *Writing a conclusion that follows from a given hypothesis*

Write the conclusion that follows from each assumption.

(a) $f(x) = 2x - 1 \Rightarrow f(2) =$

(b) $f(x) = 2x - 1$, $g(x) = x + 1$, and $f(x) = g(x) \Rightarrow$

Solution

(a) $f(x) = 2x - 1 \Rightarrow f(2) = 2(2) - 1$

(b) If we assume that there is some value of x for which f(x) and g(x) are equal, we write

$$f(x) = 2x - 1,\ g(x) = x + 1 \text{ and } f(x) = g(x) \Rightarrow 2x - 1 = x + 1$$

Solving $2x - 1 = x + 1$ leads to the solution $x = 2$.

Often in this course we need to find the x and y intercepts of the graph of a function. If the points where the graph crosses the x and y axes are (a, 0) and (0, b), respectively, a is the **x intercept** and b is the **y intercept** (Figure 3).

When finding intercepts, functional notation is often used in a hypothesis to communicate the assumptions being made.

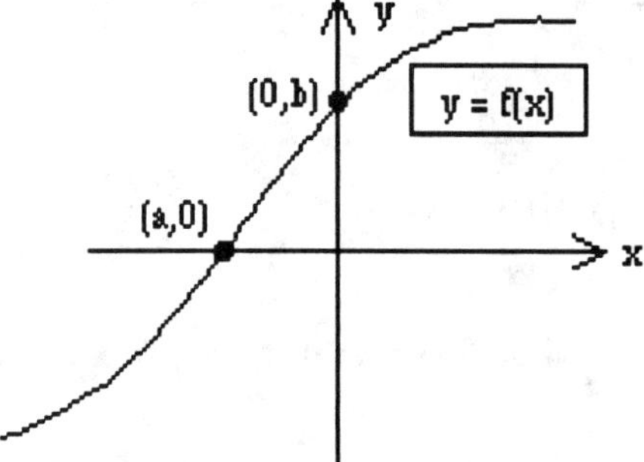

Figure 3

What point on the graph of $f(x) = 2x + 3$ is indicated by $f(x) = 2x + 3$ and $f(x) = 0 \Rightarrow x = ?$

The conclusion that follows is $2x + 3 = 0$. Solving for x yields the x intercept, $x = -3/2$ (Figure 4).

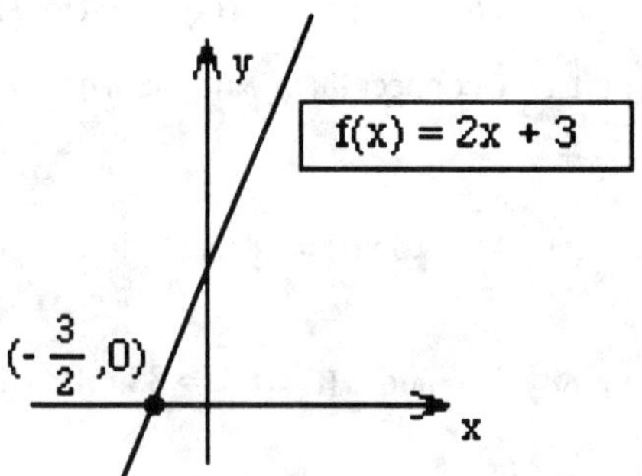

Figure 4

Example 5 *Finding x and y intercepts using function notation*

Find the x and y intercepts of $g(x) = 9 - x^2$ using appropriate notation to state the assumptions.

Solution

To find the x intercept of $g(x) = 9 - x^2$, we let $g(x) = 0$ and solve the resulting equation. Using notation, we write

$g(x) = 9 - x^2$ and $g(x) = 0 \Rightarrow$

$$\begin{aligned} 9 - x^2 &= 0 \\ 9 &= x^2 \\ x &= \pm 3 \end{aligned}$$

The x intercepts are -3 and 3.

To find the y intercept of $g(x) = 9 - x^2$, we let $x = 0$. Using notation, we write

$g(x) = 9 - x^2 \Rightarrow$

$$g(0) = 9 - 0^2 = 9$$

Notice that $g(0) = 9$ contains the assumption, $x = 0$, and the conclusion, $y = 9$. The y intercept is 9.

Figure 5 shows the graph of $g(x) = 9 - x^2$.

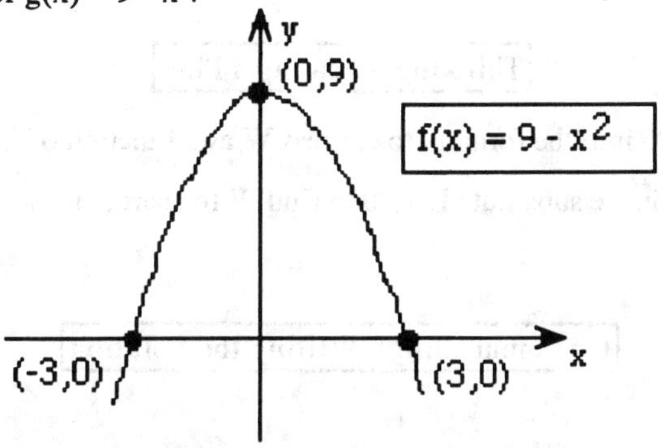

y
(0,9)
f(x) = 9 - x²
(-3,0)
(3,0)
x

Figure 5

Using the Equivalence Symbol

An equivalence is the strongest mathematical statement we can make. If two statements p and q are equivalent, each implies the other.

$$p \text{ equivalent to } q \text{ requires that } p \Rightarrow q \text{ and } q \Rightarrow p$$

An **equivalence** statement is written $p \Leftrightarrow q$.

Every definition is an equivalence statement. For example, let's define a function as follows: "A relation is a function if and only if each x in the domain is paired with a unique y in the range." The words *if and only if* indicate an equivalence statement with its two implications, $p \Rightarrow q$ and $q \Rightarrow p$, given as follows.

$p \Rightarrow q$: If a relation is a function, each x in the domain is paired with a unique y in the range.

$q \Rightarrow p$: If each x in the domain of a relation is paired with a unique y in the range, then the relation is a function.

Formulas involving any number of variables can be studied as functions by choosing specific values for all but two of the variables. Then either of the remaining variables can be considered as a function of the other, as we see in the next example.

Example 6 The Billfish Problem

The weight W in pounds of a large billfish (a fish having a large, sharp bill) is approximated by the formula

$$W = \frac{LG^2}{800}$$

where L is its length in inches and G is its girth in inches. (A fish's girth is the perimeter of its largest cross section.)

(a) Use functional notation to express weight as a function of length, W = W(L), for fish with girths of 30 in.

(b) Find the weight of the fish whose length is 20 in., W(20), for the function in part (a).

(c) Find the length of the fish whose weight is 300 lb, W(L) = 300, for the function in part (a).

Thinking: Creating a Plan

For part (a) we substitute G = 30 into the formula to express W as a function of L. Once we have weight expressed as a function of length, we substitute L = 20 to find W for part (b), and substitute W = 300 and solve for L for part (c).

Communicating: Writing the Solution

(a) $W = \frac{LG^2}{800}$ and $G = 30 \Rightarrow$

$$W(L) = \frac{L(30)^2}{800} = \frac{9}{8}L$$

(b) $W(L) = \frac{9}{8}L \Rightarrow$

$$W(20) = \frac{9}{8}(20) = 22.5$$

The weight of the fish whose length is 20 in. is 22.5 lb.

(c) $W(L) = \frac{9}{8}L$ and $W(L) = 300 \Rightarrow$

$$\frac{9}{8}L = 300$$

$$L \approx 267$$

The length of the fish is about 267 in.

Learning: Making Connections

The graph in Figure 6 summarizes our results.

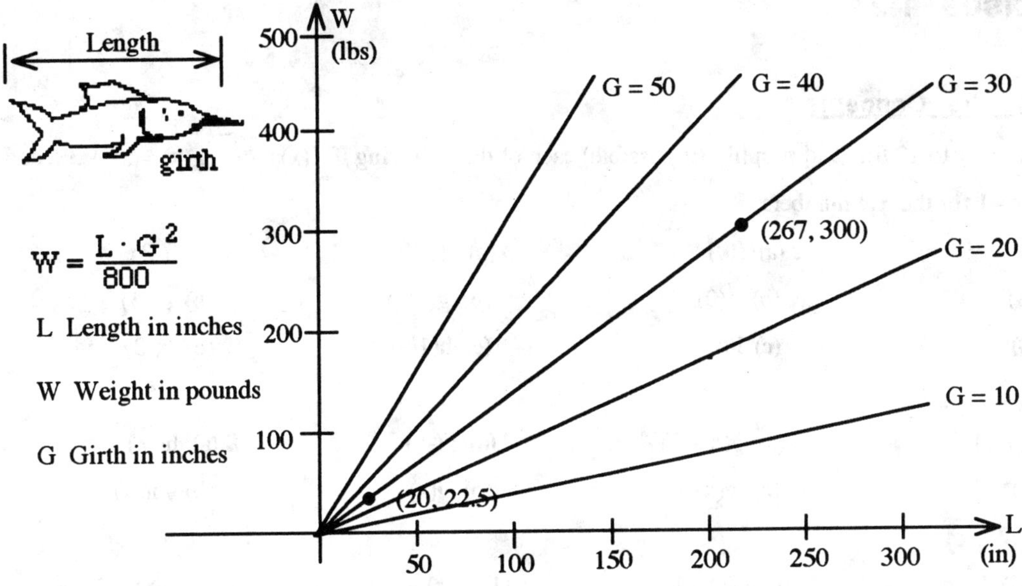

Figure 6

[Source: *Hawaii Charter Fishing Guide*]

Exercises 1.2

Develop the Concepts

For Exercises 1 to 12 find and simplify (if possible) each of the following if $f(x) = \sqrt{x}$, $g(x) = x^2 + 3x$, and $h(x) = 1/x - 1$ for the real numbers.

1.(a) f(4)
(b) g(4)
(c) h(4)

2.(a) f(0)
(b) g(0)
(c) h(0)

3.(a) f(1/9)
(b) g(1/9)
(c) h(1/9)

4.(a) f(-3)
(b) g(-3)
(c) h(-3)

5.(a) f(x + 1)
(b) f(x) + 1

6.(a) g(x + 2)
(b) g(x) + 2

7.(a) g(-x)
(b) -g(x)

8.(a) h(-x)
(b) -h(-x)

9. f(x + 4) - 1

10. -g(x) - 1

11. h(x) + 2

12. -f(-x) + 1

For Exercises 13 and 14 find the indicated values defined by the following table.

x	-10	0	10	20	30
f(x)	100	0	100	5000	5000

13. (a) f(0)
(b) -10f(10)
(c) x such that f(x) = 5000

14. (a) f(10) + f(20)
(b) f(10 + 20)
(c) x such that 2f(x) = 200

For Exercises 15 and 16 find the indicated values defined by the following table.

x	-12	-2	4	8	12
g(x)	15	10	5	-5	-10

*15.(a) g(12)
(b) g(4) + 8
(c) g(4 + 8)

16. (a) g(4 + 4)
(b) g(4) + g(4)
(c) g(4) · g(4)

For Exercises 17 and 18 approximate the indicated values from the corresponding graphs.

17. (a) h(0)
(b) h(3.5)
(c) x such that h(x) = 0

18. (a) k(-4)
(b) k(0)
(c) x such that k(x) = 0

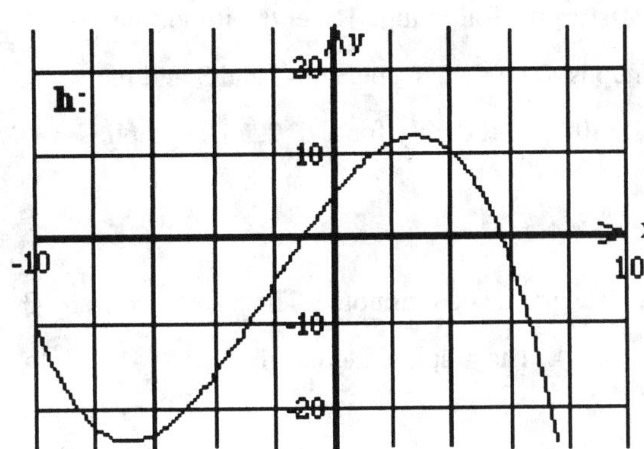

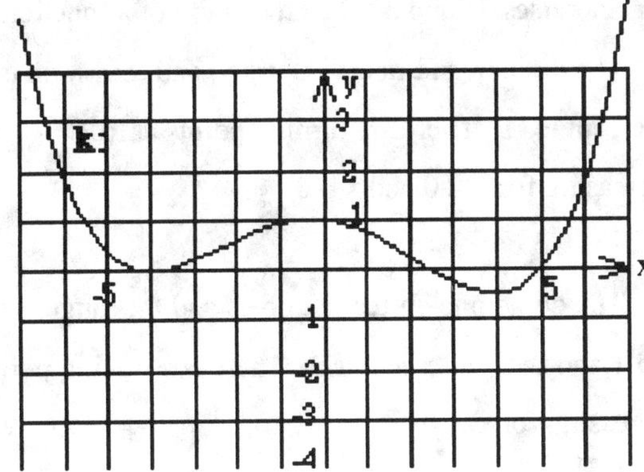

For Exercises 19 and 20 approximate the indicated values (if possible) from the corresponding graphs.

19. (a) $f_1(10)$

(b) $f_1(15)$

(c) x such that $f_1(x) = 300$

(d) x such that $f_1(x) = -150$

20. (a) $f_2(-4)$

(b) $2f_2(-4)$

(c) x such that $f_2(x) = 12.5$

(d) x such that $f_2(x) = 12$

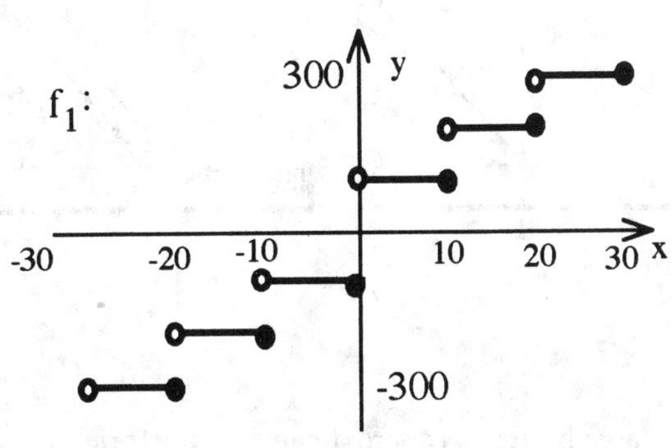

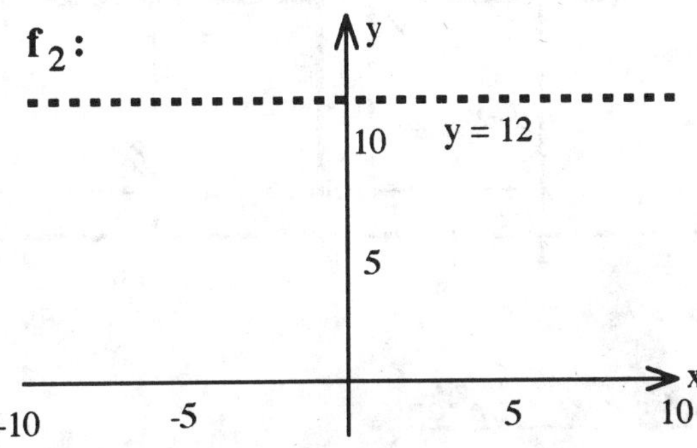

For Exercises 21 to 26 complete each implication by writing the conclusion that follows from each hypothesis.

21. $A = \pi r^2$ and $r = 10 \Rightarrow A =$ ________

22. $f(x) = x^2 + 4x - 2 \Rightarrow f(-1) =$ ________

23. $x = \dfrac{-b \pm \sqrt{b^2 - 4ac}}{2a}$ and $a = 1, b = -2, c = -3 \Rightarrow x =$ ______

24. $x = -\dfrac{b}{2a}$ and $a = 3, b = -2 \Rightarrow x =$ ________

25. $f(x) = 3x + 4$, $g(x) = 2(x + 2)$, and $f(x) = g(x) \Rightarrow$ ________

26. $f(x) = x^2 + 4$, $g(x) = x + 6$, and $f(x) = g(x) \Rightarrow$ ________

For Exercises 27 to 30 find the x and y intercepts of each function using appropriate notation to state the assumptions.

27. $f(x) = 5x - 10$

28. $g(x) = 4x^2 - 25$

29. $f(x) = x^3 - 8$

30. $g(x) = \sqrt{x + 14}$

For Exercises 31 and 32 sketch a graph of a function that satisfies the following. Label the important points.

31. The domain and range is $(-\infty, \infty)$ and the graph is continuous. The x intercepts are -2 and 3 and the y intercept is -1. There are turning points at $f(0) = -1$ and $f(3) = 0$. $f(x) > 0$ only for $x < -2$. The graph is decreasing for $x < 0$ and $x > 3$.

32. The domain is all real numbers and the range is $(-\infty,4]$. The graph is continuous. The x intercepts are -3 and 1 and the y intercept is 3. There is a turning point at $f(-1) = 4$. The graph is increasing for $x < -1$ and decreasing for $x > -1$.

For Exercises 33 to 36 find the point(s) of intersection of each pair of functions by letting $f(x) = g(x)$ and solving for x. Use your results to help match each equation with one of the graphs in Figure 7.

33. $f(x) = \sqrt{2x}$, $g(x) = x$

34. $f(x) = 2\sqrt{x}$, $g(x) = x$

35. $f(x) = \sqrt{x + 2}$, $g(x) = x$

36. $f(x) = \sqrt{x - 2}$, $g(x) = x$

A.

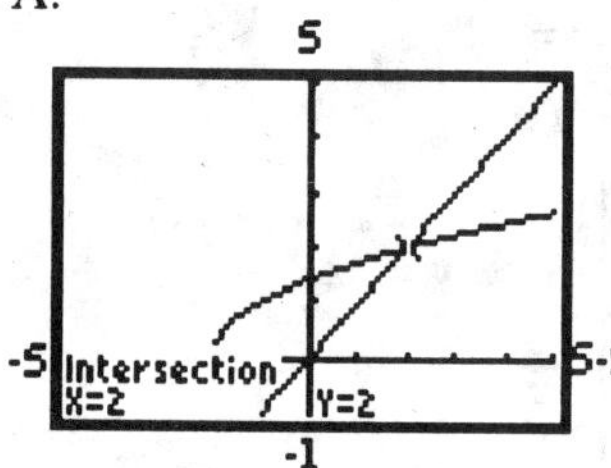

B.

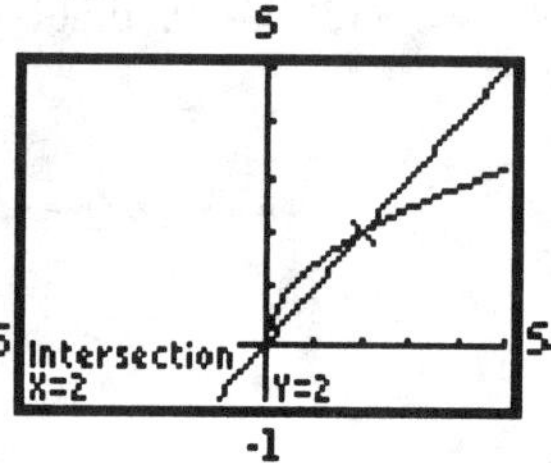

C.

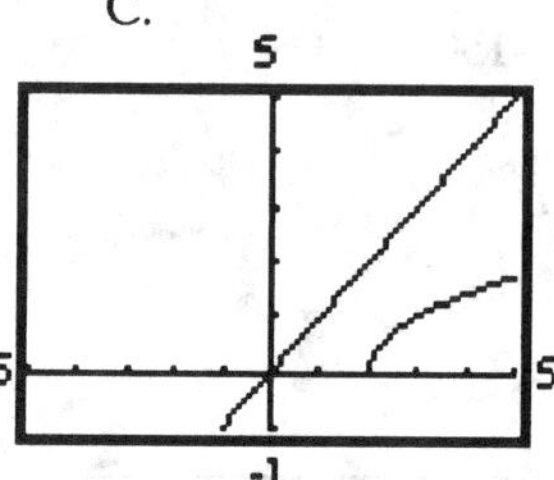

D.

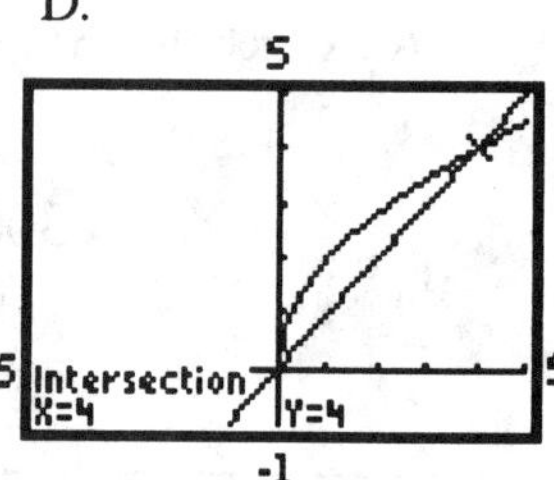

Figure 7

Apply the Concepts

37. Another Billfish Problem The weight W in pounds of a large billfish is approximated by the formula $W = \frac{LG^2}{800}$, where L is its length in inches and G is its girth in inches.

(a) Use functional notation to express the weight W of a billfish as a function of its girth G for all 20-in.-long billfish.

(b) Find the weight of a fish whose girth is 30 in., W(30), for the function in part (a).

(c) Find G such that W(G) = 45 for the function in part (a).

38. The 300-Pound Billfish Problem The weight W in pounds of a large billfish is approximated by the formula $W = \frac{LG^2}{800}$, where L is its length in inches and G is its girth in inches.

(a) Use functional notation to express the length L of a billfish as a function of its girth G for all 300-lb billfish.

(b) Find L(30) and L(60) for the function in part (a).

(c) Use functional notation to express the girth G of a billfish as a function of its length L for all 300-lb billfish.

(d) Find G(267) and G(67) for the function in part (c).

39. The Big Bucks Problem The amount of money A in a savings account after 1 year is given by the formula $A = P(1 + r)$, where r is the annual interest rate and P is the initial amount of money deposited (the principal).

(a) Use functional notation to express the amount of money as a function of the interest rate for a principal of \$500.

(b) Use the function in part (a) to find the amount of money in an account paying 7.25% interest.

(c) Express the amount of money as a function of the principal for an interest rate of 6.5%.

(d) Use the function in part (c) to find the principal in an account that has \$12,000 after 1 year.

40. Another Big Bucks Problem The amount of money A in a savings account after 2 years is given by the formula $A = P(1 + r)^2$, where r is the annual interest rate and P is the initial amount of money deposited (the principal).

(a) Use functional notation to express the amount of money as a function of the principal for an interest rate of 5.25%.

(b) Use the function in part (a) to find the principal in an account that has \$2000 after 2 years.

(c) Express the amount of money as a function of the interest rate for a principal of \$2500.

(d) Use the function in part (c) to find the interest rate in an account that has \$5000 after 2 years.

41. The Drapery Problem The length of fabric F required to make draperies is approximated by the formula $F = (l + 20)(\frac{2.5w}{f})$, where l is the length of the window, w its width in inches, and f = 54 in. is one available width of fabric.

(a) Express length of fabric as a function of window length for a window that is 36 in. wide.

(b) Find the length of fabric needed to drape a 6-ft-long window using the function in part (a).

(c) Express length of fabric as a function of window width for a window that is 72 in. long.

(d) Find the width of the window you can drape with a 300-in.-long piece of fabric using the function in part (c).

42. The Water Bill Problem The volume V of water in a cylindrical hot tub is given by the formula $V = \pi r^2 h$, where r is the radius of the tub in feet and h is the height of the water in feet.

(a) Use functional notation to express volume as a function of water height for a 6-ft-diameter hot tub.

(b) Use the function in part (a) to express your water cost C as a function of the depth of the water if water costs 5¢/100 gal. [Hint: 7.5 gal = 1 ft^3.]

43. The Rainbow Kite and Balloon Company The revenue $R(p)$ from selling kites is given in the table as a function of the price p of each kite. Use the function defined by the table to answer the following.

price, p ($/kite)	0	10	75	150
revenue, R(p) ($)	0	70	600	0

(a) $R(0)$

(b) $R(10)$

(c) $2R(10) + 10$

(d) The values of p for which $R(p) = 0$

(e) Describe how the revenue changes as the price is increased as described by the data in the table.

44. Wonderworld Amusement Park You borrow a radar gun to measure the speed of a small roller coaster. You collect the following data, which gives the speed S as a function of the time t since the ride started. Use the function defined by the table to answer the following.

time, t (min)	0	1	2	3	4	5
speed, S(t) (ft/sec)	0	22	22	44	33	0

(a) $S(1)$

(b) $S(2)$

(c) $0.85\,S(3)$

(d) The values of t such that $S(t) = 0$

(e) Describe the original ride by examining its changing speed.

Extend the Concepts

For Exercises 45 to 52 use notation to communicate finding the point(s) of intersection where $f(x) = g(x)$. Then use notation to communicate finding the corresponding y coordinates.

45. $f(x) = x^2$, $g(x) = x + 6$

46. $f(x) = -x^2$, $g(x) = -20 - x$

*47. $f(x) = \sqrt{x}$, $g(x) = -2x$

48. $f(x) = \sqrt{x}$, $g(x) = 2x$

49. $f(x) = \dfrac{4}{x + 1}$, $g(x) = \dfrac{1}{x} + 1$

50. $f(x) = \dfrac{1}{(x - 1)^2}$, $g(x) = \dfrac{-1}{x - 1}$

51. $f(x) = \dfrac{1}{\sqrt{x - 1}}$, $g(x) = \sqrt{x - 1}$

52. $f(x) = \dfrac{\sqrt{x - 1}}{x - 1}$, $g(x) = \dfrac{1}{x - 1}$

1.3 Translations and Reflections of Graphs of Functions

Graphing Translations
Graphing Reflections
Graphing Functions Involving Both Reflections and Translations
Studying Even and Odd Functions

In this section we study how certain changes in the equation of a function affect its graph in the Cartesian coordinate system. We will first look at translations, where the graphs are shifted up, down, right, or left. Then we will look at reflections, in which we create new functions by flipping a graph about one or both of the axes. Our study of translations and reflections will strengthen our geometric intuition, thus enabling us to anticipate the shape and position of a graph without depending on tedious point plotting or blindly accepting outputs from graphing utilities.

We base our study of translations and reflections on the reference functions introduced in Section 1.1. We will begin to see how each reference function is the simplest representative of an entire family of functions.

To explore how adding a constant to the x or y coordinate results in a translation of the original graph, go to "Explore the Concepts," Exercises E1 - E4 before continuing (page 53). For the interactive version, go to the *Precalculus Weblet - Instructional Center - Explore Concepts.*

Graphing Translations

All of the curves in Figure 1 have the same size and shape. Since they differ only in their location in the coordinate system, we say that such graphs are **translations** (shifts up, down, right, or left).

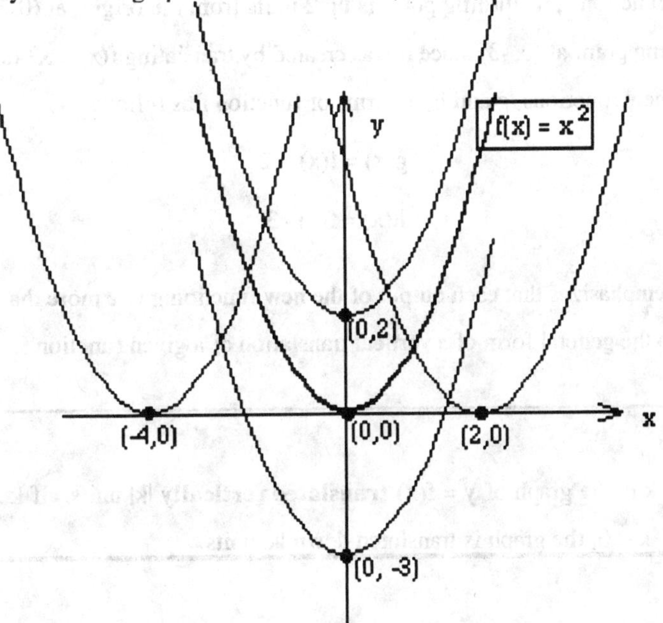

Figure 1

Let's first look at vertical translations by comparing the table of values and graphs in Figure 2. For each value of x, points on the graph of function g are 2 units above those on the graph of function f. We say that the graph of $g(x) = x^2 + 2$ is a vertical translation up 2 units of the graph of function f. Similarly, the graph of function h is a vertical translation down 3 units of the graph of function f.

x	$f(x) = x^2$	$g(x) = x^2 + 2$	$h(x) = x^2 - 3$
-2	4	6	1
-1	1	3	-2
0	0	2	-3
1	1	3	-2
2	4	6	1

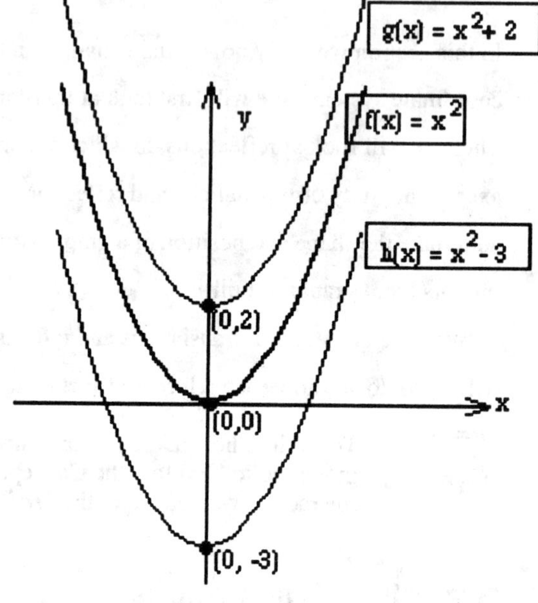

Figure 2

Notice that the turning point of $f(x) = x^2$ is (0, 0). Since the graph of function g is a vertical translation up 2 units of the graph of function f, its turning point is up 2 units from the origin, at (0,2). Similarly, the graph of function h has a turning point at (0, -3) since it was created by translating $f(x) = x^2$ down 3 units.

We can express the new functions, g and h, in terms of function f as follows:

$$g(x) = f(x) + 2$$

and

$$h(x) = f(x) - 3$$

Writing $g(x) = f(x) + 2$ emphasizes that each output of the new function g is 2 more than each output of function f. This leads to the general form of a vertical translation of a given function f.

Vertical Translation

The graph of $y = f(x) + k$ is the graph of $y = f(x)$ **translated vertically** $|k|$ units. If $k > 0$, the graph is translated up k units. If $k < 0$, the graph is translated down $|k|$ units.

We now examine horizontal translations by comparing the table of values in Figure 3. Points on the graph of function p are 4 units to the left of the corresponding ones on the graph of function f. The graph of function p is a horizontal translation of the graph of function f to the left 4 units. Similarly, the graph of $q(x) = (x - 2)^2$ is a horizontal translation of function f, right 2 units.

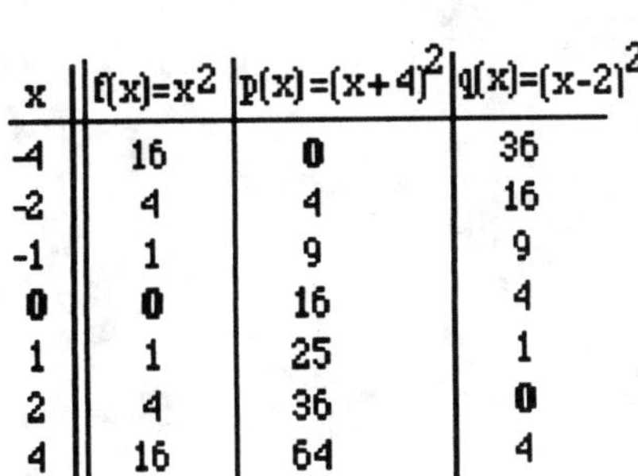

x	$f(x)=x^2$	$p(x)=(x+4)^2$	$q(x)=(x-2)^2$
-4	16	0	36
-2	4	4	16
-1	1	9	9
0	0	16	4
1	1	25	1
2	4	36	0
4	16	64	4

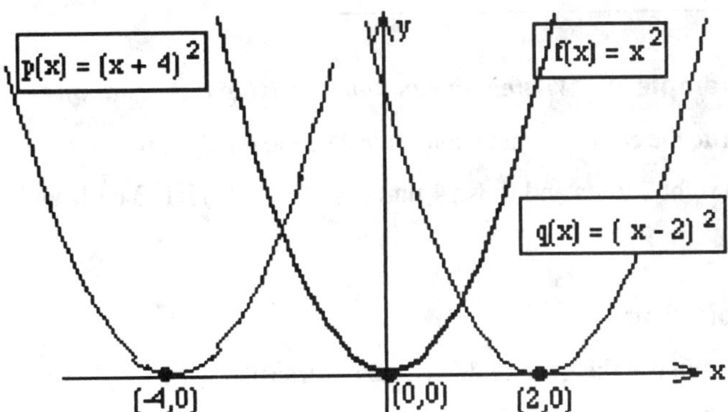

Figure 3

We can write functions p and q in terms of function f as

$$p(x) = f(x + 4)$$

and

$$q(x) = f(x - 2)$$

This leads to the general form of a horizontal translation.

Horizontal Translation

The graph of $y = f(x - h)$ is the graph of $y = f(x)$ **translated horizontally** $|h|$ units. If $h > 0$, the graph is translated right h units. If $h < 0$, the graph is translated left $|h|$ units.

Since squaring any value of x yields a nonnegative result, the smallest possible value of x^2 is zero. Notice that the lowest point on the graph of $f(x) = x^2$ in Figure 3 is (0,0). The lowest point on the graph of function p occurs when $x = -4$ since $(-4 + 4)^2 = 0^2 = 0$.

What is the smallest possible value of $(x - 2)^2$?

Since $x = 2$ makes $(x - 2)^2 = 0$, the lowest point on the graph of function q is (2,0).

Example 1 *Describing translations in words*

Describe the translation of function f in terms of right, left, up, or down.

(a) $y = f(x + 13)$

(b) $y = f(x - 100) + 0.5$

Solution

(a) For any function f, $y = f(x + 13)$ is a translation of the graph of f left 13 units.

(b) For any function f, $y = f(x - 100) + 0.5$ is a translation of the graph of f right 100 units and up 0.5 unit.

Example 2 *Writing an equation involving a translation*

Write the equation that would translate the graph of $f(x) = x^3$

(a) right 2 units and down 4 units (b) left 3 units and up $\frac{1}{2}$ unit

Solution

(a) We use the general description of translations $y = f(x - h) + k$.

$y = f(x - h) + k$ where $f(x) = x^3$ and $h = 2$, $k = -4 \Rightarrow$

$$y = (x - 2)^3 - 4$$

(b) $y = f(x - h) + k$ where $f(x) = x^3$ and $h = -3$, $k = \frac{1}{2} \Rightarrow$

$$y = [x - (-3)]^3 + \frac{1}{2}$$
$$= (x + 3)^3 + \frac{1}{2}$$

When sketching a graph on paper from a graph that was generated using a graphing utility, it is helpful to label the axes, any important points on the graph, and the values associated with the tick marks on both axes. For

$$y = \sqrt{x - 20} + 50$$

the graph of $f(x) = \sqrt{x}$ is translated 20 units to the right and up 50 units (Figure 4). We label one point to emphasize the translations involved (Figure 4b).

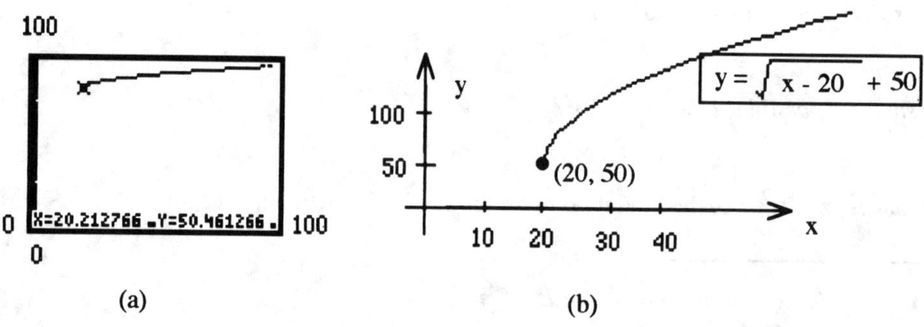

Figure 4 (a) (b)

What are the domain and range of $y = \sqrt{x - 20} + 50$*?*

Since $x - 20$ must be nonnegative,

$$x - 20 \geq 0$$
$$x \geq 20$$

The domain is $[20, \infty)$. Substituting $x = 20$ yields the smallest value of the range, $y = \sqrt{20 - 20} + 50 = 50$. The range is $[50, \infty)$. Comparing the domain and range of this function with that of the reference function $f(x) = \sqrt{x}$ reveals the translations.

Example 3 The Temperature Data Problem

The data below were collected from measurements taken at 6-hr intervals on July 4.

time t	12:00 midnight	6:00 a.m.	12:00 noon	6:00 p.m.
temperature T	54°F	63°F	97°F	85°F

(a) One month later on August 4 each of these temperatures was reached 2 hr later than on July 4. Make a table depicting this information.

(b) Each of the temperatures recorded on October 31 was 25°F lower than those on August 4. Make a table depicting this information.

(c) Use functional notation to describe the functions in parts (a) and (b) in terms of the function described for July 4.

Thinking: Creating a Plan

(a) We create a new table from the given one by adding 2 to each time entry.

(b) We create another table from the one in part (a) by subtracting 25 from each temperature entry.

(c) We call the original function $T = f(t)$, indicating that temperature T is a function of time t, and describe the new functions in parts (a) and (b).

Communicating: Writing the Solution

(a)

time, t	2:00 a.m.	8:00 a.m.	2:00 p.m.	8:00 p.m.
temperature, T	54°F	63°F	97°F	85°F

(b)

time, t	2:00 a.m.	8:00 a.m.	2:00 p.m.	8:00 p.m.
temperature, T	29°F	38°F	72°F	60°F

(c) August 4: $T = f(t - 2)$ October 31: $T = f(t - 2) - 25$

Learning: Making Connections

Of course, we can illustrate these translations in a graph (Figure 5). Reaching the temperatures 2 hr later appears as a translation to the right 2 hr. Having temperatures drop 25°F appears as a translation down 25 degrees.

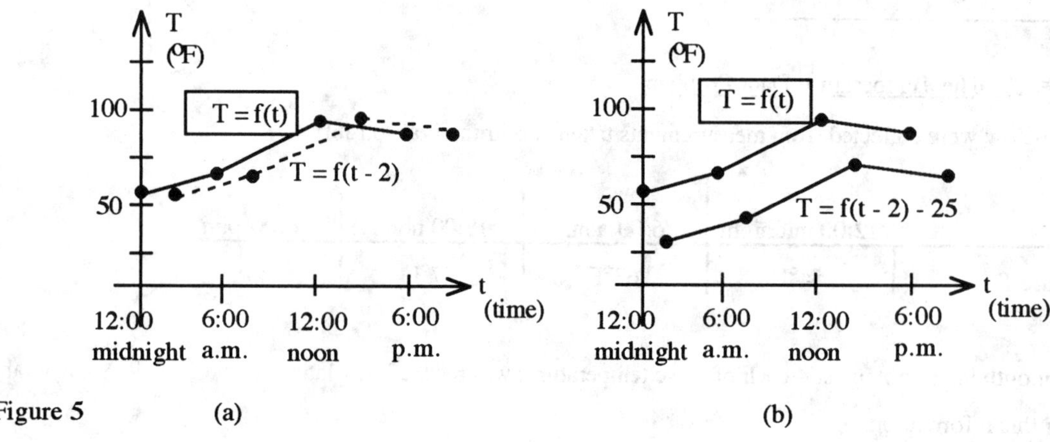

Figure 5 (a) (b)

Graphing Reflections

Figure 6 shows the graph of the reference function $f(x) = \sqrt{x}$ and three variations that occur when we make sign changes in the equation. These new graphs, called **reflections**, are mirror images of the original graph. All four graphs have exactly the same size and shape. Notice how the signs of the coordinates of the labeled points differ from the corresponding point (4, 2) on the graph of $f(x) = \sqrt{x}$.

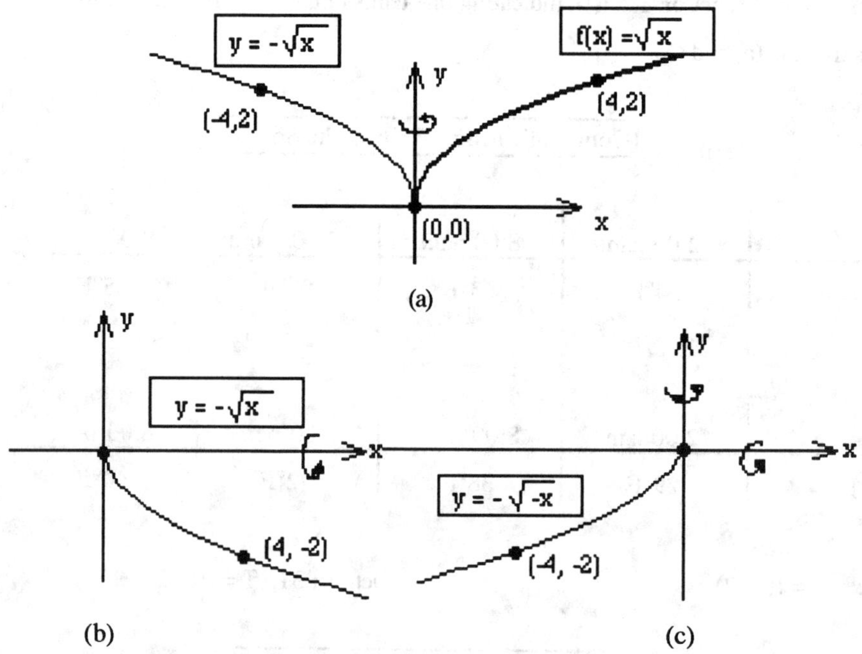

Figure 6

To explore how changing the sign of the x or y coordinate results in a reflection of the original graph, go to "Explore the Concepts," Exercises E5 - E8, before continuing (page 54). For the interactive version, go to the *Precalculus Weblet - Instructional Center - Explore Concepts.*

The negative sign in each equation causes a reflection. In $y = \sqrt{-x}$ (Figure 6a), the x values are negated, causing each point on the graph of $f(x) = \sqrt{x}$ to be reflected about the y axis. The point (4, 2) on the graph of the reference function is changed to (-4, 2) on the reflected graph. In $y = -\sqrt{x}$ (Figure 6b), we negate the y values of the reference function $f(x) = \sqrt{x}$, reflecting its graph about the x axis. The point (4, 2) is changed to (4, -2) in the reflected graph. In $y = -\sqrt{-x}$ (Figure 6c), both x and y are negated, reflecting the graph of $f(x) = \sqrt{x}$ about both axes. We describe the change in the graph as a reflection about the origin. Notice that the point (4, 2) is changed to a point directly across the origin, (-4, -2). Functional notation describes reflections.

Reflections

If $y = f(x)$, then

$y = f(-x)$ is a **reflection** of the graph of function f about the y axis.

$y = -f(x)$ is a **reflection** of the graph of function f about the x axis.

$y = -f(-x)$ is a **reflection** of the graph of function f about both the x and y axes, that is about the origin.

To graph $y = f(-x)$, we flip the template corresponding to Appendix II about its vertical axis, changing the sign of the x values, just as the notation suggests. After flipping the template, the cut corner will be at the lower right.

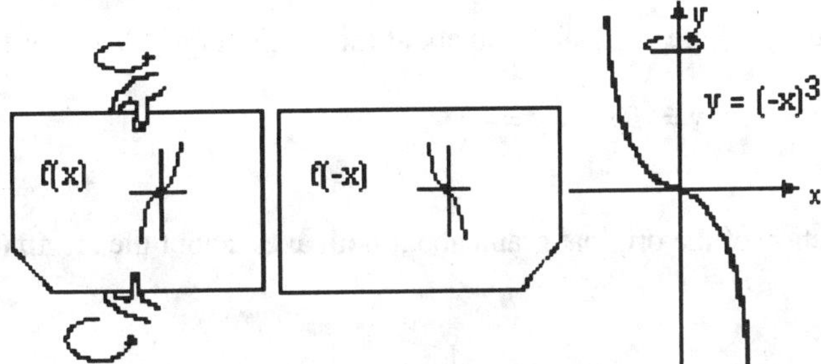

To graph $y = -f(x)$, we flip the template about its horizontal axis, changing the sign of the y values, just as the notation suggests. The cut corner will be at the upper left.

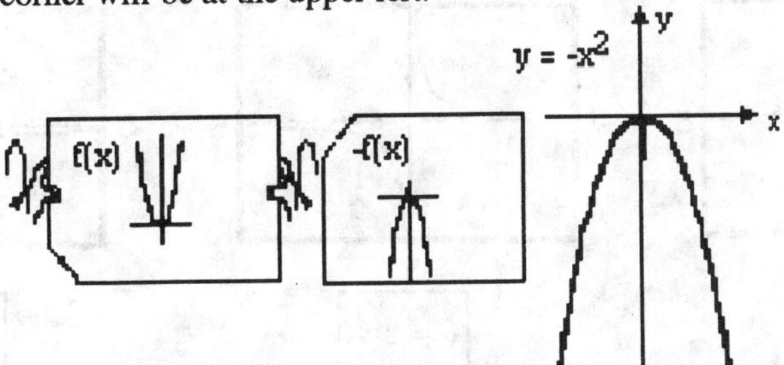

To graph $y = -f(-x)$, we flip the template about both its horizontal and vertical axes, changing the signs of both the x and y values, just as the notation suggests. The cut corner will be at the upper right.

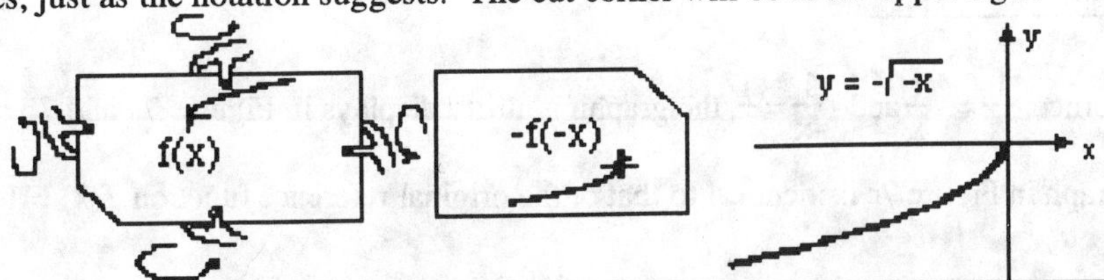

Why is the domain of $y = \sqrt{-x}$ *nonpositive real numbers?*

For -x to be greater than or equal to zero, the inputs of x cannot be positive.

$$-x \geq 0 \qquad \textit{Multiply both sides by -1.}$$
$$x \leq 0$$

Example 4 *Sketching the graphs of reflected functions*

For $f(x) = \frac{1}{x}$, state each new equation and sketch the corresponding graph.

(a) $y = f(-x)$ (b) $y = -f(x)$ (c) $y = -f(-x)$

Solution

(a) This equation is a reflection of the original graph of the reciprocal function about the y axis (Figure 7a). For $f(x) = \frac{1}{x}$,

$$y = f(-x) \Rightarrow y = \frac{1}{-x}$$

(b) This equation is a reflection of the original graph about the x axis (Figure 7b). For $f(x) = \frac{1}{x}$,

$$y = -f(x) \Rightarrow y = -\frac{1}{x}$$

(c) This equation is a reflection of the original graph about both axes, about the origin (Figure 7c). For $f(x) = \frac{1}{x}$,

$$y = -f(-x) \Rightarrow y = -\frac{1}{-x}$$

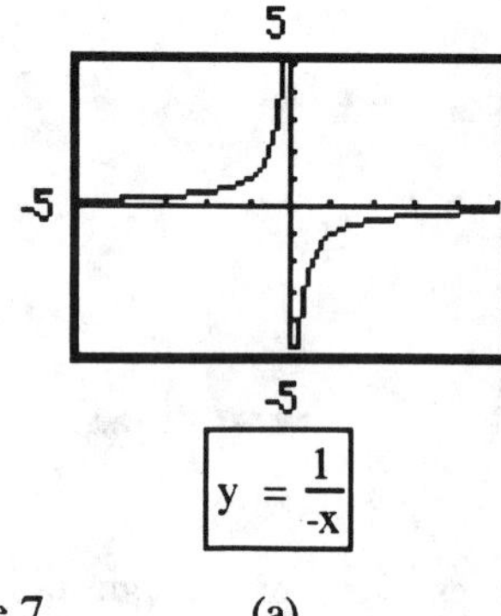

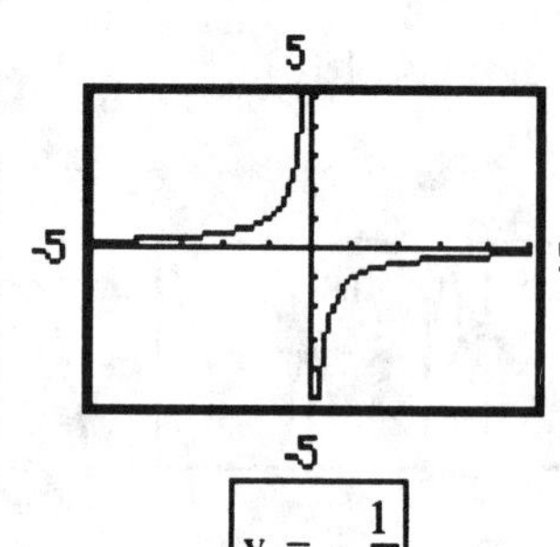

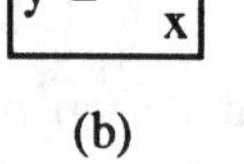

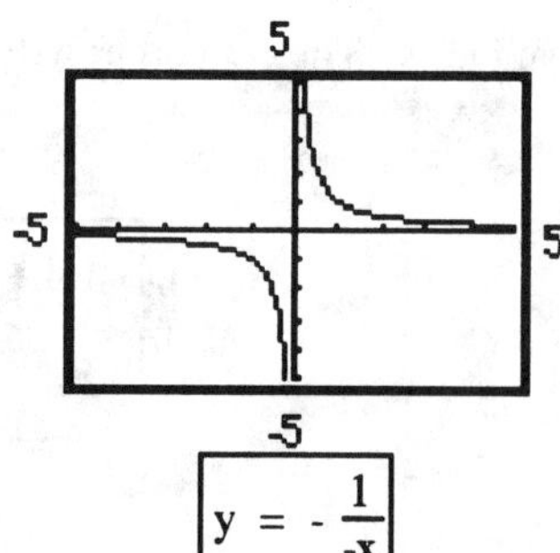

Figure 7 (a) (b) (c)

NOTE: Since $\frac{1}{-x} = -\frac{1}{x}$ and $-\frac{1}{-x} = \frac{1}{x}$, the graphing utility displays in Figures 7a and 7b are identical, and the graph in Figure 7c is identical to that of the original reference function, $f(x) = 1/x$.

Graphing Functions Involving Both Reflections and Translations

Now let's combine reflections and translations. Graphing is easier if we determine reflections first, then translations. If we know the general shape of a graph before any reflections and translations, we may not need to use a graphing utility. If we do decide to use a graphing utility, we will know if its display is reasonable.

Strategy: Graphing a Reflection and a Translation

1. Consider the general shape of the function. (Appendix II may help.)
2. Determine the reflection indicated by the equation.
3. Determine the translation indicated by the equation.
4. Determine how the reflections and translations affect the domain and range. If a graphing utility is used, set the viewing window, taking into account any reflections and translations.
5. Sketch the graph or enter the equation on a graphing utility.

Example 5 *Sketching graphs of reflected and translated reference functions*

Sketch the graph of each function.

(a) $g(x) = -|x - 12| + 11$ (b) $y = \sqrt{30 - x}$

Solution

(a) The graph of $g(x) = -|x - 12| + 11$ is a reflection and translation of the graph of the absolute value function. The graph is reflected about the x axis and translated to the right 12 units and up 11 units. This positions the turning point of the new graph at (12,11) (Figure 8a).

(b) The graph of $y = \sqrt{30 - x}$ is a reflection of the square root graph about the y axis and a translation 30 units to the right (Figure 8b).

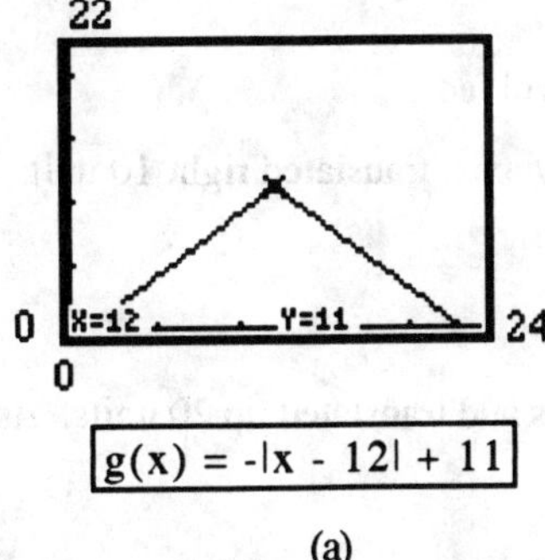

$g(x) = -|x - 12| + 11$

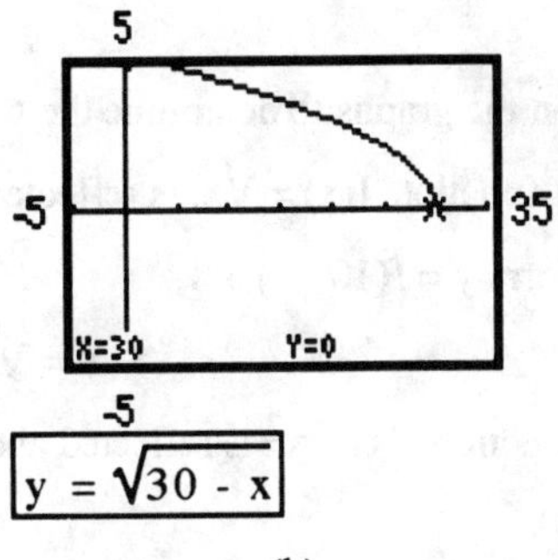

$y = \sqrt{30 - x}$

Figure 8 (a) (b)

In part (b) of Example 5, we can rewrite

$$y = \sqrt{30 - x} \quad \text{as} \quad y = \sqrt{-(x - 30)}$$

to emphasize the translation to the right 30 units.

We can determine the horizontal translation by examining the domain of the function. For $y = \sqrt{30 - x}$,

$$30 - x \geq 0$$
$$-x \geq -30$$
$$x \leq 30$$

This means that x = 30 is the largest possible value of x that will yield a real number for y.

The next example requires us to work backward from a graph to an equation. Our goal is to find a simplified equation to match a given graph.

Example 6 *Writing equations of reflected or translated graphs*

Give the simplest equation that will yield each of the graphs in Figure 9 which are reflections and/or translations of one of the reference functions.

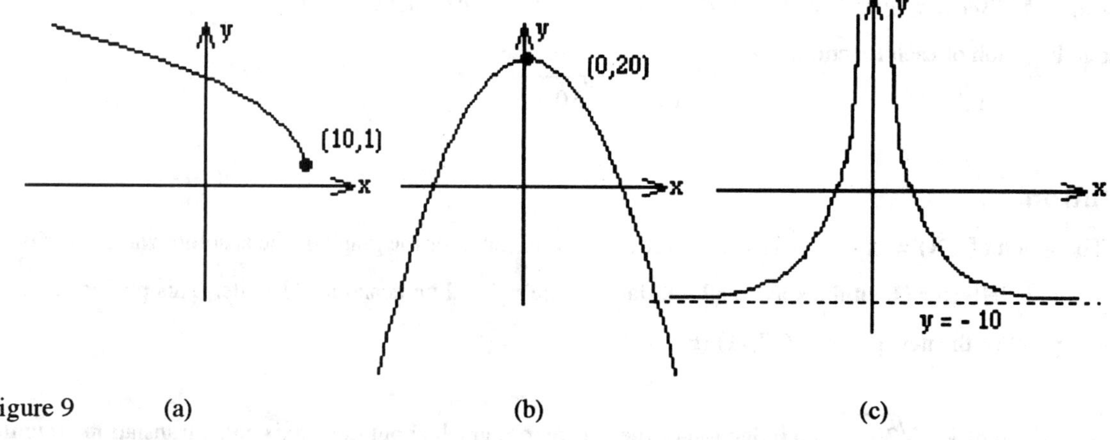

Figure 9 (a) (b) (c)

Solution

We use the labels on the graphs to determine the translation involved.

(a) The square root function, $f(x) = \sqrt{x}$, is reflected about the y axis, translated right 10 units and up 1 unit. Its equation is of the form $y = f(10 - x) + 1$,

$$y = \sqrt{10 - x} + 1$$

(b) The squaring function $f(x) = x^2$ is reflected about the x axis and translated up 20 units. Its equation is of the form $y = -f(x) + 20$,

$$y = -x^2 + 20$$

(c) This is the inverse-square function $f(x) = 1/x^2$, translated down 10 units. It is not reflected. Its equation is of the form $y = f(x) - 10$,

$$y = \frac{1}{x^2} - 10$$

Studying Even and Odd Functions

Compare the graphs of $f(x) = x^2$ and $y = f(-x) = (-x)^2$ in Figure 10.

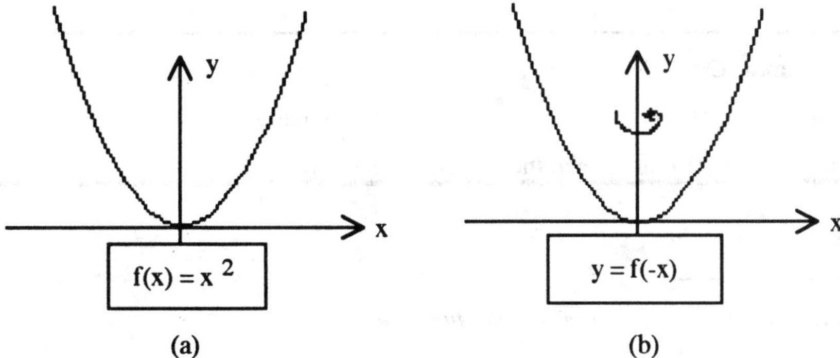

Figure 10 (a) (b)

Notice that the squaring graph did not change when reflected about the y axis because the curve on one side of the y axis is a mirror image of that on the other side. Functions whose graphs are symmetric about the y axis are called **even functions.**

How do you use functional notation to describe an even function?

A graph for which the reflection about the y axis is the same as the original requires that

$$f(-x) = f(x)$$

Algebra confirms this result for even functions. If $f(x) = x^2$, then

$$f(-x) = (-x)^2 = x^2 = f(x)$$

Now, compare the graphs of $f(x) = x^3$ and $y = -f(-x) = -(-x)^3$ in Figure 11.

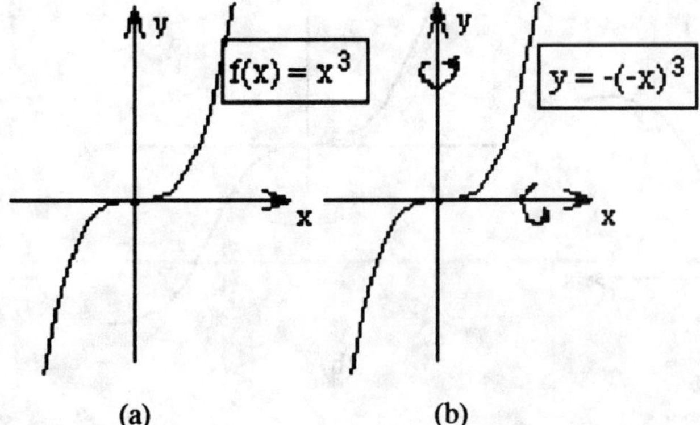

Figure 11 (a) (b)

The cubing graph did not change when reflected about both axes. A graph for which the reflection about both axes (about the origin) is the same as the original graph requires that

$$-f(-x) = f(x)$$

Functions whose graphs are symmetric about the origin are called **odd functions**. Algebra confirms this result for odd functions. If $f(x) = x^3$, then

$$-f(-x) = -(-x)^3 = (-1)(-x)(-x)(-x) = x^3 = f(x)$$

Definition of Even and Odd Functions

For an **even function**, $f(-x) = f(x)$ for all x in the domain of the function.

For an **odd function**, $-f(-x) = f(x)$ for all x in the domain of the function.

Example 7 *Sketching the graph of an even or odd function*

Complete the graph in Figure 12 so that the graph represents (a) an even function, (b) an odd function, and (c) a function that is neither even nor odd.

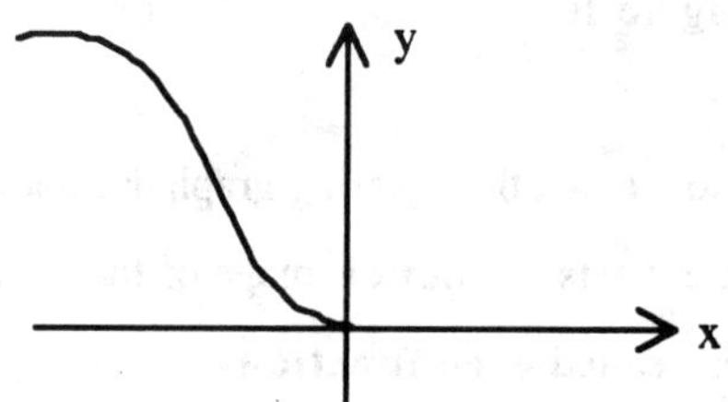

Figure 12

Solution

(a) Since the graph of an even function must be symmetric with respect to the y axis, we sketch the mirror image of the curve in quadrant I (Figure 13a).

(b) Since the graph of an odd function must be symmetric with respect to the origin, we sketch the mirror image of the curve in quadrant IV (Figure 13b).

(c) There are many ways to create a graph that does not represent either an even or an odd function. One way is to use the given graph (Figure 13c).

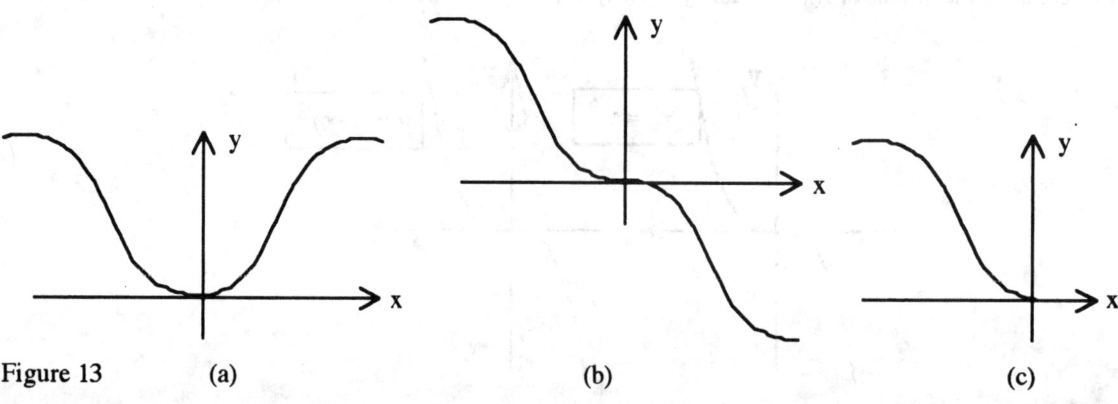

Figure 13 (a) (b) (c)

Explore the Concepts

These exercises are designed to lead you to explore how the graph of a function is *translated vertically* up or down or *horizontally* left or right. The following functions are of the form $y = f(x) + k$ or $y = f(x - h)$ where function f is one of the reference functions. For Exercises E1 and E2 write the equation of the new function. Then graph each new function on your graphing utility and state how the new function differs from the reference function.

E1.

Reference Function	Translation	New Function	Result
$f(x) = x$	$y = f(x) + 2$	$y = x + 2$	Up 2 units
$f(x) = x^3$	$y = f(x) + 1$		
$f(x) = \sqrt{x}$	$y = f(x) + 4$		
$f(x) = x^2$	$y = f(x) - 2$		
$f(x) = \lvert x \rvert$	$y = f(x) - 4$		
$f(x) = 1/x$	$y = f(x) - 10$		
$f(x) = 1/x^2$	$y = f(x) - 5$		

E2.

Reference Function	Translation	New Function	Result
$f(x) = x$	$y = f(x + 2)$	$y = (x + 2)$	Left 2 units
$f(x) = x^3$	$y = f(x + 1)$		
$f(x) = \sqrt{x}$	$y = f(x + 4)$		
$f(x) = x^2$	$y = f(x - 2)$		
$f(x) = \lvert x \rvert$	$y = f(x - 4)$		
$f(x) = 1/x$	$y = f(x - 10)$		
$f(x) = 1/x^2$	$y = f(x - 5)$		

E3. Use the results from Exercise E1 to complete the following theorem.

The graph of the function $y = f(x) + k$ is a translation of the graph of $y = f(x)$.

For $k > 0$, the translation is (up or down?) ______ units.

For $k < 0$, the translation is (up or down?) ______ units.

E4. Use the results from Exercise E2 to complete the following theorem.

The graph of the function $y = f(x - h)$ is a translation of the graph of $y = f(x)$.

For $h > 0$, the translation is (right or left?) ______ units.

For $h < 0$, the translation is (right or left?) ______ units.

These exercises are designed to explore how the graph of a function is *reflected* about the x axis, the y axis, or both axes. The following functions are of the form y = f(-x), y = - f(x), or y = -f(-x), where f is one of the reference functions. For Exercises E5 to E7 write the equation of the new function. Then graph each new function on your graphing utility and state how the new function differs from the reference function. If a graph is symmetric about a line or a point, the graph on one side is a mirror image of the graph on the other side.

E5.

Reference Function	Reflection	New Function	Result
$f(x) = \sqrt{x}$	y = f(-x)	$y = \sqrt{-x}$	about the y axis
f(x) = x	y = f(-x)		
$f(x) = x^2$	y = f(-x)		
$f(x) = x^3$	y = f(-x)		
f(x) = |x|	y = f(-x)		
f(x) = 1/x	y = f(-x)		
$f(x) = 1/x^2$	y = f(-x)		

E6.

Reference Function	Reflection	New Function	Result
$f(x) = \sqrt{x}$	y = -f(x)	$y = -\sqrt{x}$	about the x axis
f(x) = x	y = -f(x)		
$f(x) = x^2$	y = -f(x)		
$f(x) = x^3$	y = -f(x)		
f(x) = |x|	y = -f(x)		
f(x) = 1/x	y = -f(x)		
$f(x) = 1/x^2$	y = -f(x)		

E7.

Reference Function	Reflection	New Function	Result
$f(x) = \sqrt{x}$	y = -f(-x)	$y = -\sqrt{-x}$	about the origin
f(x) = x	y = -f(-x)		
$f(x) = x^2$	y = -f(-x)		
$f(x) = x^3$	y = -f(-x)		
f(x) = |x|	y = -f(-x)		
f(x) = 1/x	y = -f(-x)		
$f(x) = 1/x^2$	y = -f(-x)		

E8. Use the results from Exercises E5 to E7 to complete the following theorems.

The graph of the function y = _______ is a reflection of the graph of y = f(x) about the y axis. The graph of the function y = _______ is a reflection of the graph of y = f(x) about the x axis. The graph of the function y = _______ is a reflection of the graph of y = f(x) about the origin.

Exercises 1.3

Develop the Concepts

For Exercises 1 to 4 describe the translations of each reference function in terms of *right*, *left*, *up*, or *down*.

1. (a) $y = \sqrt{x + 2}$
   (b) $y = \sqrt{x} + 5$
2. (a) $y = (x - 3)^2$
   (b) $y = (x + 3)^2$
3. (a) $y = (x + 3)^3 - 1$
   (b) $y = 1 + (x - 3)^3$
4. (a) $y = |x - 1| - 2$
   (b) $y = |x + 4| + 3$

For Exercises 5 to 8 describe the reflection of each reference function in terms of the axis.

5. $y = \sqrt{-x}$
6. $y = -\frac{1}{x}$
7. $y = (-x)^3$
8. $y = -|-x|$

For Exercises 9 to 12 sketch the graph of (a) y = f(-x), (b) y = -f(x), and (c) y = -f(-x) for each function.

9. 10. 11. 12.

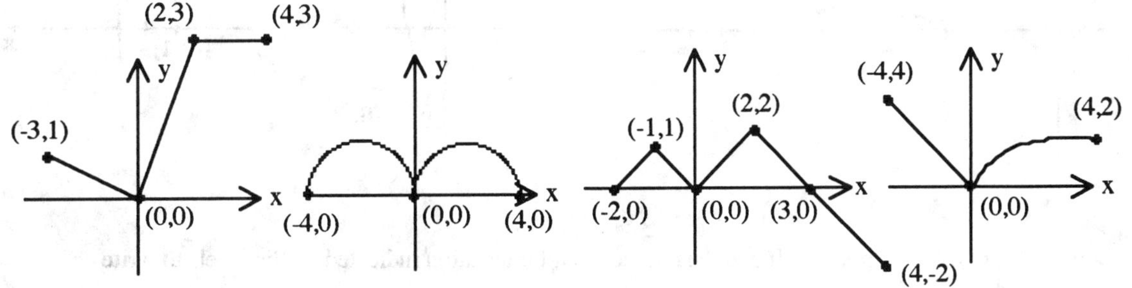

For Exercises 13 to 18 write the equation that would change the graph of $f(x) = \sqrt{x}$ as indicated.

13. (a) Translate left 1 unit.
    (b) Translate up 3 units.
14. (a) Translate right 2 units.
    (b) Translate down 1 unit.
15. Translate right 2 units and down 1 unit.
16. Translate up 3 units and left 2 units.
17. (a) Reflect about the x axis and translate up 1 unit.
    (b) Reflect about the y axis and translate right 1 unit and down 3 units.
18. (a) Reflect about the y axis and translate left 2 units.
    (b) Reflect about the origin and translate right 1 unit and down 2 units.

For Exercises 19 to 26 sketch the graph of each reflection or translation of a reference function.

19. $y = -|x - 1|$
20. $y = (1 - x)^2$
21. $y = -x^2 - 6$
22. $y = -\sqrt{x} + 6$
23. $y = -\frac{1}{(x + 22)}$
24. $y = -\frac{1}{(x + 22)^2}$
25. $y = -(-x - 3)^3 + 1$
26. $y = -|3 - x| - 1$

For Exercises 27 to 30 give the equation of each new function y = f(x - 30) + 10 for each of the functions given and sketch the graph.

*27. $f(x) = \sqrt{x}$

28. $f(x) = x^3$

29. $f(x) = |x|$

30. $f(x) = 1/x^2$

For Exercises 31 to 34 write the simplest equation that describes each graph as a translation of one of the reference functions.

31. 32. 33. 34.

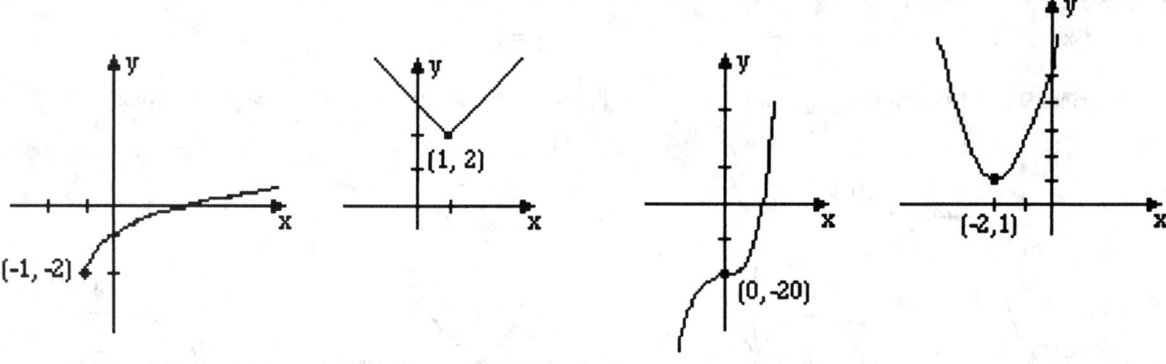

For Exercises 35 to 42 use Appendix II and the reflection and translation indicated by the labels to write an equation that matches each graph.

35. 36. 37. 38.

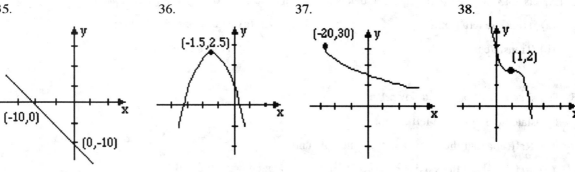

y
x
(-10,0)
(0,-10)

(-1.5,2.5)
y
x

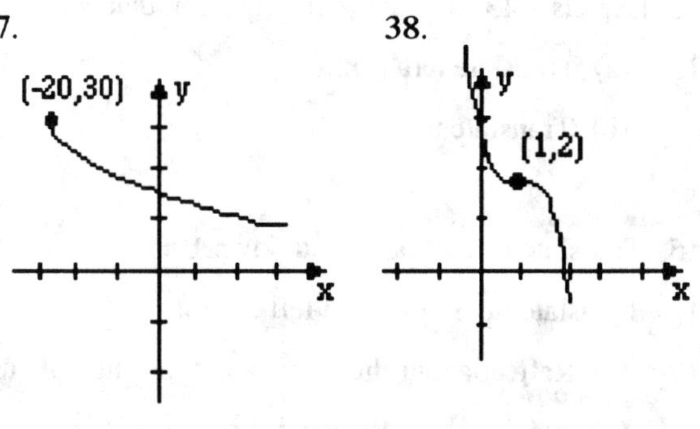

39. 40. 41. 42.

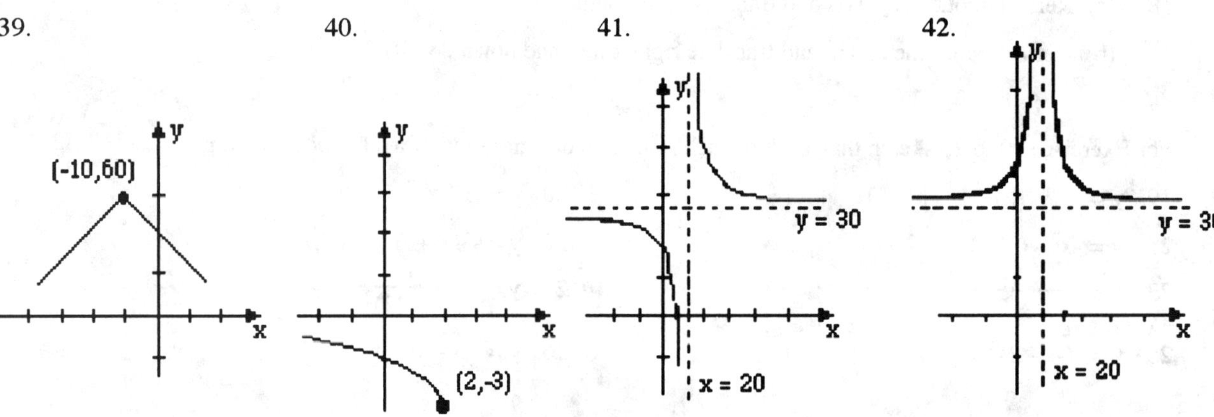

Functions f and g are defined by the following tables. For Exercises 43 to 46 give the new functions in table form created by translating or reflecting the functions.

x	-10	0	10	20
f(x)	-5	0	15	25

x	-100	-50	0	50
g(x)	500	100	0	-200

43. (a) $y = f(x) + 5$
    (b) $y = -f(x)$

44. (a) $y = g(x) - 50$
    (b) $y = g(x - 50)$

*45.(a) $y = f(-x)$
    (b) $y = f(-x) + 1$

46. (a) $y = g(x - 1)$
    (b) $y = g(1 - x)$

For Exercises 47 and 48 complete each of the following partially drawn graphs in three ways to show: (a) an even function, (b) an odd function, and (c) a function that is neither even nor odd.

47.

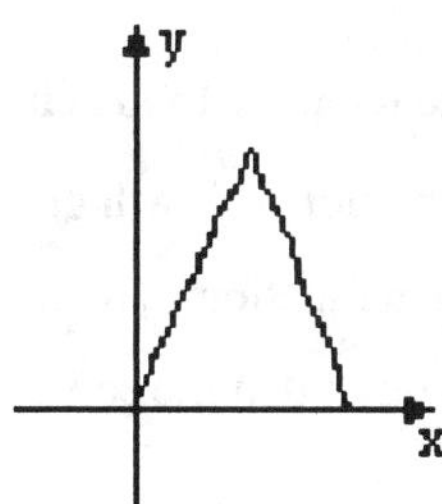

48.

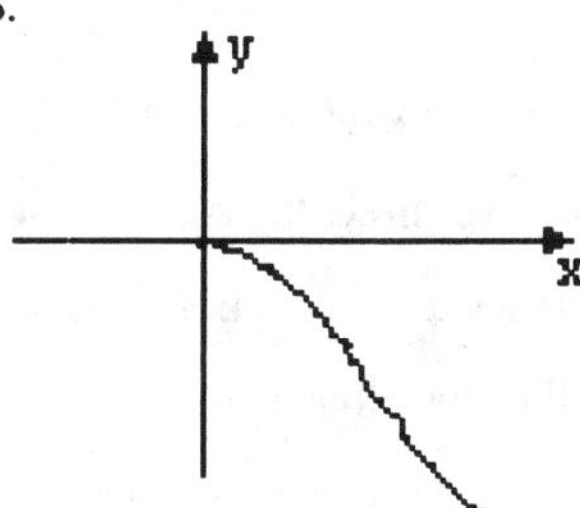

For Exercises 49 to 52 investigate graphically and confirm algebraically whether each function is an even function, odd function, or neither even nor odd.

*49. $f(x) = x^4 + 2x^2$

50. $f(x) = x^3 + x$

51. $f(x) = x - \frac{1}{x^3}$

52. $f(x) = \frac{1}{(x - 2)^2}$

Apply the Concepts

53. A Bell-Shaped Curve Problem When teachers grade students work, they often grade "on the curve". They use a graph called a bell-shaped curve, where the middle score is the class average (the mean). Just as many high grades are given as low grades and most grades fall near the middle. Figure 14 shows function f, a bell-shaped curve for a large set of test scores where the mean is 80.

(a) Sketch the graph of a test whose scores form a bell-shaped curve with a mean of 65.

(b) For a test whose mean is 65, express its equation in terms of function f described by the graph in Figure 14 using functional notation.

(c) If we let f represent the bell-shaped curve in Figure 14, the curve $y = f(x - 20)$ could represent the distribution of scores on an IQ test. What is the mean of this new distribution?

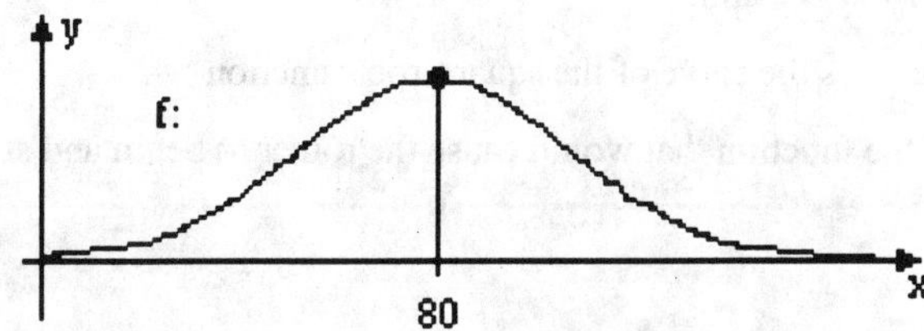

Figure 14

54. The Test Scores Problem One student decided to drop a class and repeat the class during the next term. The first five grades that the student earned originally are shown in the table below. Upon repeating the class, the student's first five grades were each 20 points higher.

(a) Complete the table by filling in the scores earned when the course was repeated.

(b) Sketch each graph showing the five individual points.

(c) Express the equation in functional notation for the scores earned during the second term if function f represents the original scores earned on each test.

test	1	2	3	4	5
original scores	80	78	70	62	58
second scores					

*55. The Template-Design Problem Suppose that you want to design a plastic template to match the Library of Reference Functions. To direct the computer-driven router to place and cut the slots for each graph, the programming engineers need the equation as well as the domain and range of each function. The computer program requires positive coordinates for cutting each graph, so the engineers request that the coordinates of the upper left corner be labeled (100, 100) (Figure 15).

(a) Write the equation that matches the curve of the absolute-value function (graph 1 of Appendix II). By actual measurement, its origin is 3.5 cm to the right and 3.5 cm down from the upper left corner. (The scale is 1 cm = 1 unit.)

(b) Determine the domain of the function that would cause the router to begin and stop cutting the absolute value function as described. [For strength, the graph (slot) must start and stop 1 cm from the edge of the template.]

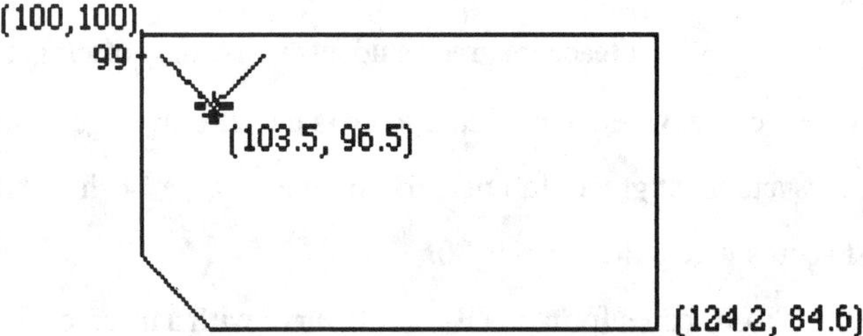

Figure 15

56. The Template-Design Problem Revisited The square-root function (graph 3 of Appendix II) has its origin located 12 cm to the right and 3 cm down from the upper left corner of the template (see Exercise 55). Assume that the scale of the graph is 1 cm = 1 unit.

(a) Write the equation that matches the curve of the square-root function.

(b) Determine the domain of the function that would cause the router to begin and stop cutting the square-root function as described.

57. The Rainbow Kite and Balloon Company The number of items that people are willing to buy is related to the price by a function called the demand function. Last year your market research determined the demand function to be $p = 43.30 - 0.21x$, where x is the quantity of kites and p is the price in dollars. The number of items that you are willing to offer for sale is related to the selling price by a function called the supply function. In the short term, you want to sell 100 kites, making the supply relation $x = 100$.

(a) Graph the demand function and supply relation on one coordinate system.

(b) A successful advertising campaign allows you to sell 120 more kites at each price. State the equation of this horizontally translated demand function.

(c) Graph the new demand function described in part (b).

(d) Compare the selling prices before and after the advertising campaign to sell 100 kites.

[Source: *Understanding Microeconomics*, R. Heilbroner and L. Thurow]

58. The Rainbow Kite and Balloon Company Last year, market research determined the demand function $p = 43.30 - 0.21x$ (Exercise 57). In the long term, you want the price per kite to average $35.50, creating a constant supply function $p = 35.50$.

(a) Graph the demand and supply functions on one coordinate system.

(b) A local tax of $2.50 per kite causes you to translate your supply function up 2.50 units. Write the equation of the new supply function.

(c) Graph the new supply function described in part (b).

(d) Compare the number of kites your customers are willing to buy before the tax when the price was $35.50 each and after the tax when the price was $38 each.

(e) How much of the tax were you able to pass on to the customer?

[Source: *Understanding Microeconomics*, R. Heilbroner and L. Thurow]

59. Wonderworld Amusement Park A market study determines the demand function for hot dogs at the park to be $p = 3.00 - 0.0025x$, where p is the price in dollars and x is the quantity of hot dogs sold.

(a) Graph this demand function and label the intercepts.

(b) What does each intercept indicate about customers' willingness to buy hot dogs?

(c) After hiring a buddy to dress up as "Frank Furter", the hot dog, you find you can sell 40 more hot dogs at each price. State the equation of this horizontally translated function.

(d) Graph the new demand function described in part (c).

60. Wonderworld Amusement Park When your advertising campaign involving Frank Furter created the new demand function $p = 3.00 - 0.0025(x - 40)$ (Exercise 59), a tax of $0.50 per hot dog translated the supply curve up 0.50 units, yielding the new supply function $p = (0.0025x + 1) + 0.50$.

(a) Graph the demand and supply functions after the campaign and new tax on one coordinate system.

(b) Find and interpret the point of intersection of the graphs in part (a).

Extend the Concepts

For Exercises 61 and 62 state the equation of a specific function to illustrate each case.

61. Vertical translations of even functions are even.

62. Some functions are neither even nor odd.

For Exercises 63 and 64 sketch a graph of a function to illustrate each case.

63. A function whose graph is not symmetric about either the y axis nor the origin -- a function that is neither even nor odd.

64. A function whose graph is symmetric about both the y axis and the origin -- a function that is both even and odd.

1.4 Changes of Scale of Graphs of Functions

Graphing Functions of the Form y = af(x)
Graphing Functions of the Form y = f(bx)
Setting the Window on a Graphing Utility
Graphing Functions with Changes of Scale, Reflections, and Translations
Applying Changes of Scale

In Section 1.1 we noted that the graph of the squaring function is a **parabola**. The three graphs in Figure 1 should all be parabolic, yet the graphing utility displays in Figures 1b and 1c are not. This is due to inappropriate windows for the equations entered.

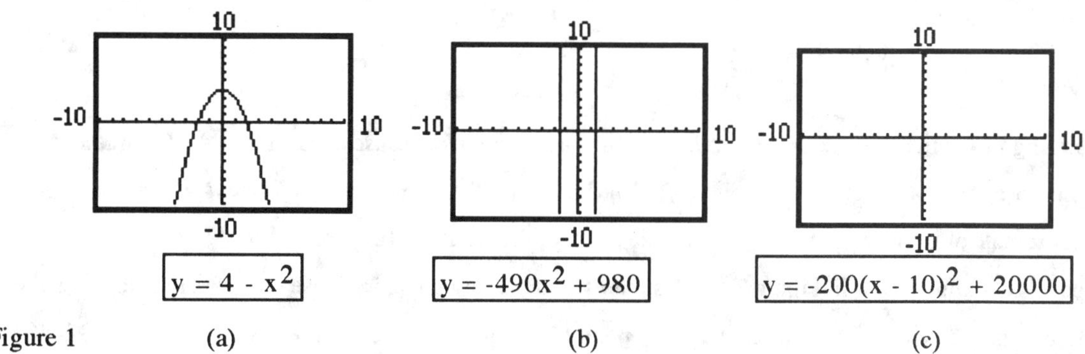

Figure 1 (a) (b) (c)

Figure 1a is the graph of $y = 4 - x^2$. It is the graph of the reference function $f(x) = x^2$ reflected about the x axis and translated up 4 units. Figure 1b, a model from physics of a free-falling object, appears as two parallel lines -- it should be another parabola. Figure 1c, a revenue function from economics, appears as one line drawn on top of the negative y axis -- it should also be a parabola.

In this section we learn how to generate a graph with a desired shape by setting the window on a graphing utility. Often, we set the window to emphasize or deemphasize part of a graph. Sometimes the window choice intentionally distorts a situation, as we will see in Example 6, where we create a misleading graph that exaggerates changing averages on SAT scores. Advertisers choose the viewing window carefully.

We consider how two new forms,

$$y = \mathbf{a}f(x) \quad \text{and} \quad y = f(\mathbf{b}x)$$

are related to a given function. The real numbers a and b are **scaling factors** that change the graph of function f in a predictable way.

To explore changes in shape of an original graph, go to "Explore the Concepts," Exercises E1 - E10, before continuing (page 73). For the interactive version, go to the *Precalculus Weblet - Instructional Center - Explore Concepts*.

Graphing Functions of the Form y = af(x)

By multiplying all of a function's outputs by a positive scaling factor a, as indicated by the notation

$$y = af(x)$$

we change the shape of the graph of $y = f(x)$ vertically, in the y direction. Figure 2a compares the graph of the square-root function $f(x) = \sqrt{x}$ with $y = a\sqrt{x}$, where the scaling factor is $a = 4$. Figure 2b compares the graph of the squaring function $g(x) = x^2$ with the graph of $y = ax^2$, where the scaling factor is $a = \frac{1}{2}$.

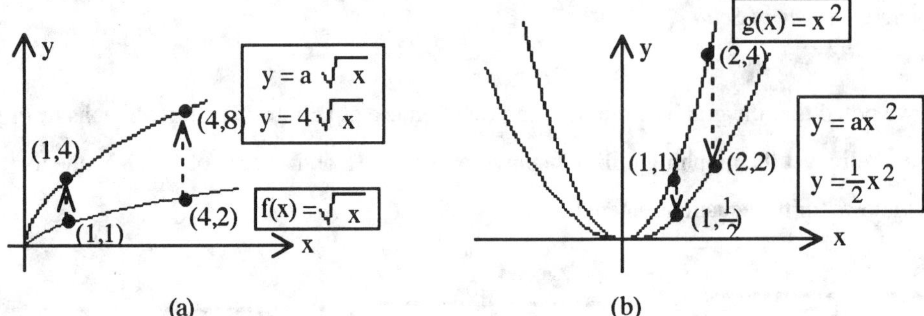

Figure 2 (a) (b)

The scaling factor in Figure 2a is greater than 1. Such scaling factors cause expansions away from the x axis, in this case, by a factor of 4. The point (4, 2) changes to (4, 8); the new y coordinate is 4 times as large as the y coordinate of the corresponding point on the graph of the reference function.

The positive scaling factor in Figure 2b is less than 1. Such scaling factors cause compressions towards the x axis, in this case, by a factor of $a = 1/2$. Here the point (2, 4) changes to (2, 2); the new y coordinate is half as large as the corresponding point on the reference graph. In both cases the changes are vertical, in the direction of the y axis.

NOTE: A scaling factor *stretches* or *shrinks* a known graph. In Figure 2a, the graph of $y = 4\sqrt{x}$ is stretched away from the x axis by a factor of 4. In Figure 2b, $y = \frac{1}{2}x^2$ is shrunk toward the x axis.

Can you explain how the graph of each of the following equations differs from the graph of its corresponding reference function?

$$y = 12|x|, \quad y = 490x^2, \quad \text{and} \quad y = \frac{0.01}{x}$$

The graph of $y = 12|x|$ is an expansion of the absolute value function away from the x axis by a factor of 12 (Figure 3a), the graph of $y = 490x^2$ is an expansion of the squaring function away from the x axis by a factor of 490 (Figure 3b), and the graph of $y = 0.01\ (1/x)$ is a compression of the reciprocal function toward the x axis by a factor of 0.01 (Figure 3c).

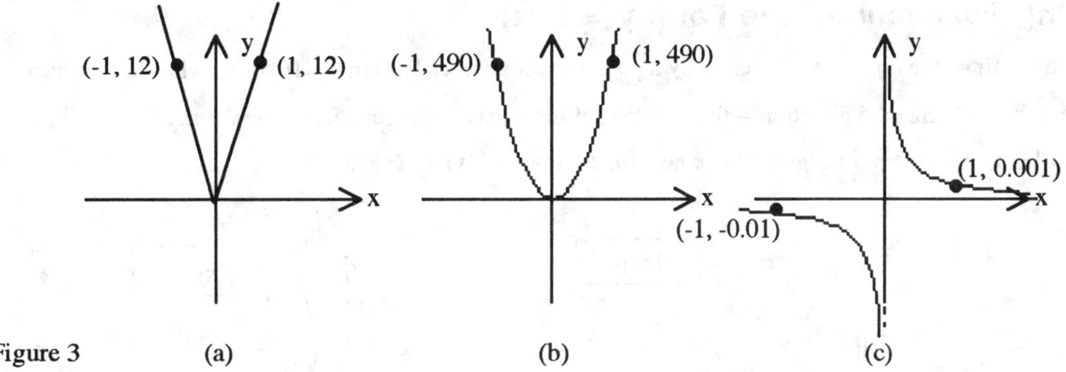

Figure 3 (a) (b) (c)

NOTE: We consider only positive scaling factors. Negative scaling factors also create a reflection.

For positive scaling factors, $a < 1$, the graph is compressed toward the x axis. For $a > 1$ the graph is expanded away from the x axis.

In the next example we sketch the graphs of the three functions in Figure 3 by rescaling the y axis. This involves renaming the tick marks on the y axis to correspond with the scaling factors. The resulting graphs match the graphs of their corresponding reference functions.

Example 1 *Rescaling the y axis*

Rescale the y axis and use the reference graphs to sketch the graph of

(a) $y = 12|x|$ (b) $y = 490x^2$ (c) $y = 0.01\,\dfrac{1}{x}$

Solution

We rescale the y axis by factors of 12, 490, and 0.01 respectively. Then we sketch the reference graphs $y = |x|$, $y = x^2$, and $y = 1/x$ on the corresponding coordinate systems (Figure 4).

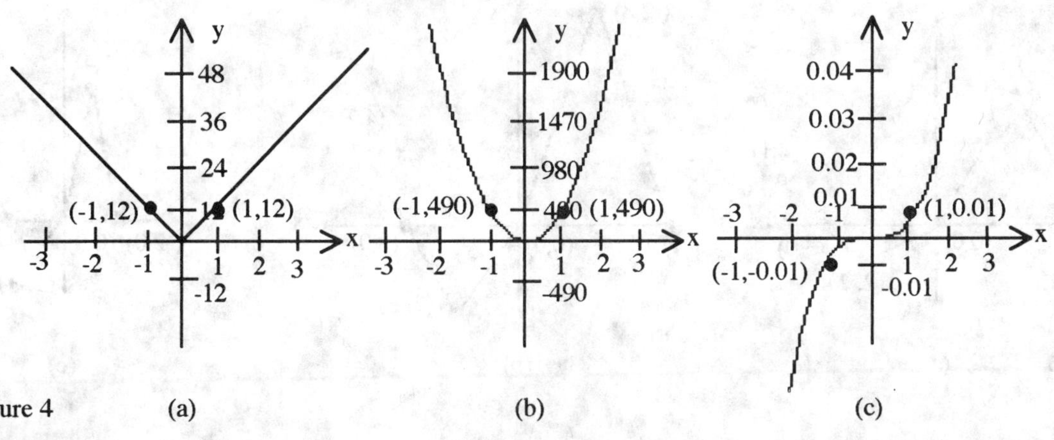

Figure 4 (a) (b) (c)

Graphing Functions of the Form y = f(bx)

If we multiply all of the inputs of a function by a positive real number b as in $y = f(bx)$, we change the shape of the graph in the horizontal direction -- the direction of the x axis. In Figure 5a, the new function $y = f(\frac{1}{2}x) = |\frac{1}{2}x|$ has $b = \frac{1}{2}$. In Figure 5b the new function $y = g(2x) = (2x)^3$ has $b = 2$.

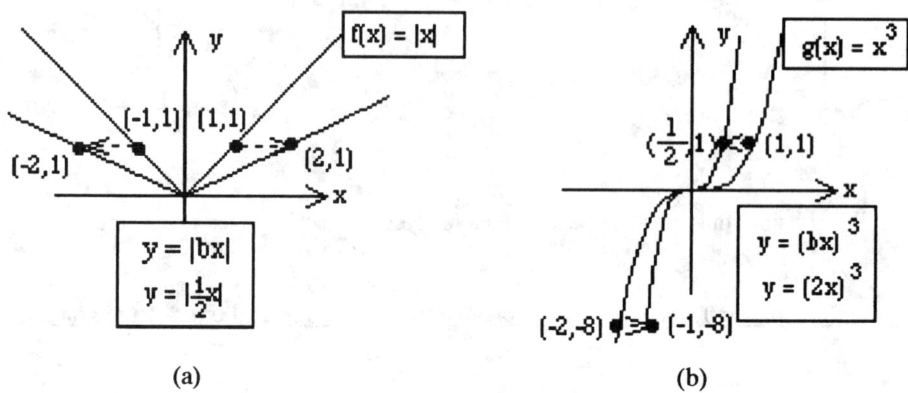

Figure 5 (a) (b)

For positive scaling factors, $b < 1$, the graph is expanded away from the y axis. For $b > 1$ the graph is compressed toward the y axis.

We can quickly sketch the graphs of the two functions in Figure 5 by rescaling the x axis. This involves renaming the tick marks on the x axis to correspond with the reciprocal of its scaling factor. For $y = |\frac{1}{2}x|$, if the label of each tick mark on the x axis is doubled, the graph will have the same size and shape as the reference function. Similarly, for $y = (2x)^3$ each tick mark on the x axis is halved.

Example 2 *Using notation to describe changes in scale*

Use functional notation to describe the changes in scale in Figures 6b and 6c given the graph of function f in Figure 6a.

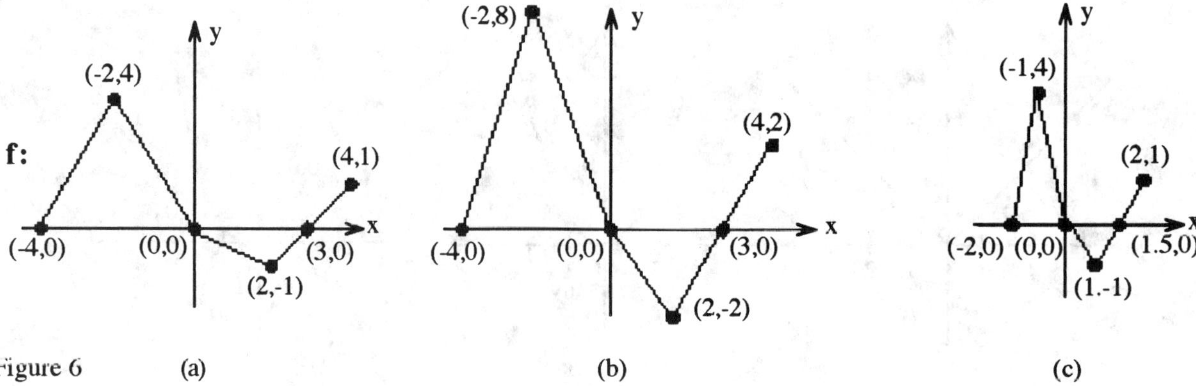

Figure 6 (a) (b) (c)

Solution

In Figure 6b the graph of function f has been expanded away from the x axis by a factor of 2; y = 2f(x). In Figure 6c the graph of function f has been compressed toward the y axis by a factor of $\frac{1}{2}$, y = f(2x).

Setting the Window on a Graphing Utility

We can reset the window on a graphing utility using the scaling factors present in an equation so that every line appears linear, every parabola looks like a parabola, absolute value are V's, and so on. For a function of the form y = af(bx), setting

$$\frac{-10}{b} \le x \le \frac{10}{b} \quad \text{and} \quad -10a \le y \le 10a$$

will yield a graph that is the same size and shape as y = f(x) in a standard window. This approach is common in applied mathematics, where scaling factors may be very large or very small numbers. Rescaling axes is also commonly used to emphasize small changes in a graph or to deemphasize large ones. We clearly label the scales on both axes to prevent creating misleading graphs. Advertisers often rescale one or both axes, and intentionally omit labeling the axes.

Changes of Scale

For a > 0, b > 0,

y = af(x) rescales the y axis by a factor of a. Reset the y_{scl} by a factor of a.

y = f(bx) rescales the x axis by a factor of 1/b. Reset the x_{scl} by a factor of 1/b.

Example 3 *Rescaling both axes for functions of the form y = af(bx)*

Enter the following functions on a graphing utility with a standard screen and describe the result. Then change the dimensions of the window so that each graph resembles the graph of the reference function $f(x) = \sqrt{x}$ as it would appear graphed using a standard window.

(a) $y = 50\sqrt{40x}$ (b) $y = 0.01\sqrt{\frac{x}{20}}$

Solution

(a) The graph of the function does not even appear using a standard window. The graph is so steep that it is drawn on top of the y axis. Since $y = 50\sqrt{40x}$ is of the form y = af(bx) with a = 50 and b = 40 (1/b = 1/40), we change the dimensions of the window to

$$-10(\mathbf{\tfrac{1}{40}}) \le x \le 10(\mathbf{\tfrac{1}{40}})$$

$$-0.25 \le x \le 0.25$$

and

$$-10(\mathbf{50}) \le y \le 10(\mathbf{50})$$

$$-500 \le y \le 500 \quad \text{(Figure 7a)}$$

(b) Again, this graph will not appear using a standard window. This graph is so flat that it is drawn on top of the x axis. Since $y = 0.01\sqrt{\frac{x}{20}}$ is of the form y = af(bx), where a = 0.01 and $b = \frac{1}{20}$ ($\frac{1}{b} = 20$), we change the window to

$$-10(\mathbf{20}) \le x \le 10(\mathbf{20})$$
$$-200 \le x \le 200$$

and

$$-10(\mathbf{0.01}) \le y \le 10(\mathbf{0.01})$$
$$-0.1 \le y \le 0.1 \quad \text{(Figure 7b)}$$

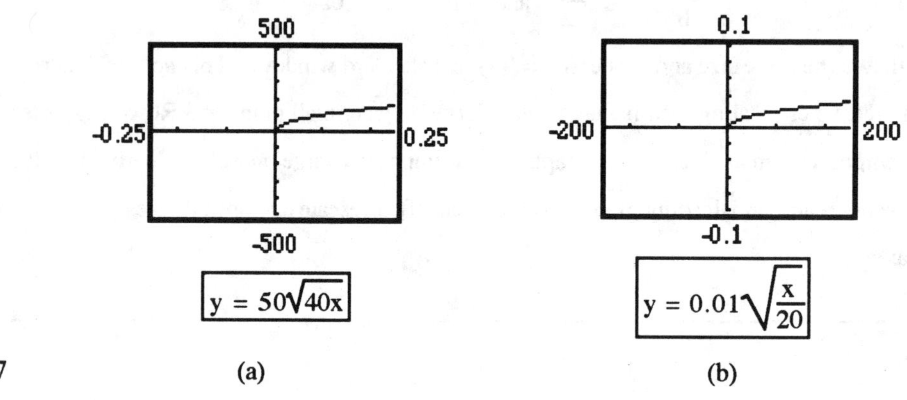

Figure 7 (a) (b)

Figures 7a and 7b both resemble the graph of the reference function $f(x) = \sqrt{x}$, except for the scales on the axes. Figure 8 shows these graphs as they would look if hand drawn on paper with each tick mark on the x axis a multiple of 1/b and each tick mark on the y axis a multiple of a.

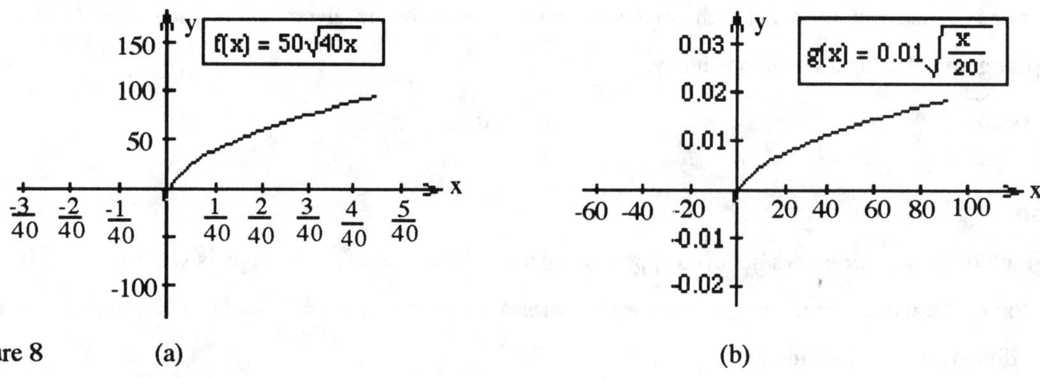

Figure 8 (a) (b)

Can you describe the changes in scale for Figures 9b and 9c using functional notation compared to Figure 9a where $f(x) = \frac{1}{x^2 + 1}$*?*

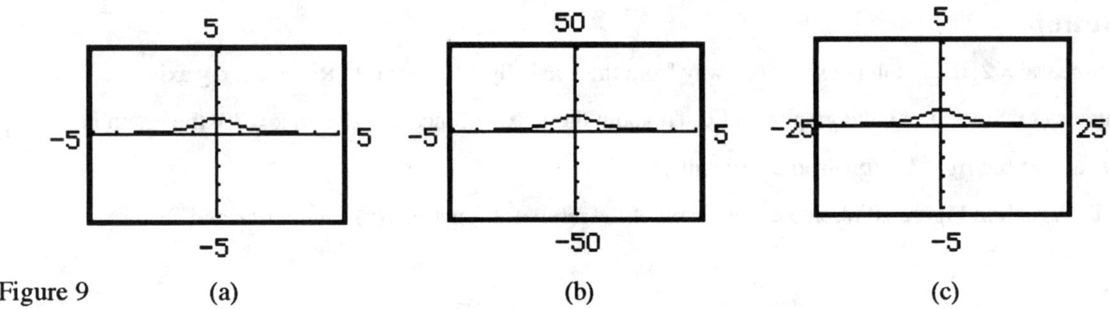

Figure 9 (a) (b) (c)

In Figure 9b, the range of the window is 10 times that of the original graph in Figure 9a, y = 10f(x). In Figure 9c, the domain of the window is 5 times the domain shown in Figure 9a, $y = f(\frac{x}{5})$. Their equations are $y = \frac{10}{x^2 + 1}$ and $y = \frac{1}{(x/5)^2 + 1}$, respectively.

Graphing Functions with Changes of Scale, Reflections, and Translations

The following summarizes all possible reflections, translations, and changes of scale of the graph of y = f(x). The values of h, k, a, and b may all affect the setting of the viewing window when using a graphing utility.

Summary: Graphing y = a·f[b(x - h)] + k

The function y = a·f[b(x - h)] + k is graphed from the graph of y = f(x) by

1. rescaling the y axis by a factor of |a|.
2. rescaling the x axis by a factor of $\frac{1}{|b|}$.
3. reflecting it about the x axis if a < 0.
4. reflecting it about the y axis if b < 0.
5. translating |h| units to the right if h > 0 or left if h < 0.
6. translating |k| units up if k > 0 or down if k < 0.

The next example does it all. There is a scaling factor present and the graph of $f(x) = x^2$ is reflected and translated.

Example 4 *Graphing a function of the form y = a·f[b(x - h)] + k*

(a) Explain how the graph of $y = -2(x - 12)^2 + 30$ relates to the graph of the squaring function, $f(x) = x^2$.

(b) Sketch the graph.

Solution

(a) Since $|a| = 2$, the graph is expanded away from the x axis by a factor of 2. Rescale the y axis to compensate. Since $a < 0$, the graph is reflected about the x axis. Since $h = 12$ and $k = 30$, the graph is translated to the right 12 units and up 30 units.

(b) The graph in Figure 10 resembles the graph of the reference function $f(x) = x^2$ in size and shape.

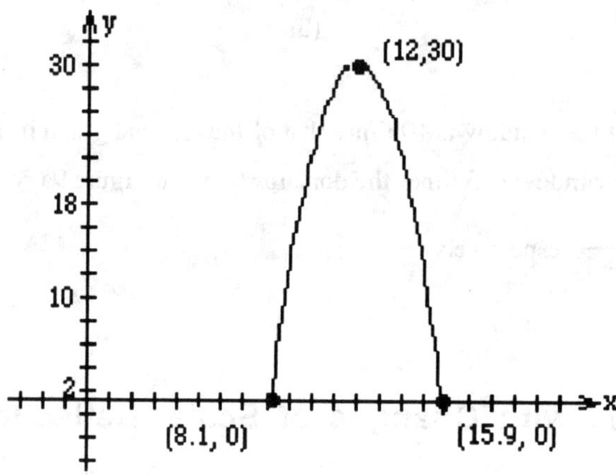

Figure 10

At the beginning of this section in Figures 1b and 1c, we viewed the graphs of the equations

$$y = -490x^2 + 980 \text{ and } y = -200(x - 10)^2 + 20{,}000$$

on a standard window. Neither appeared to be a parabola.

Could you now determine a viewing window for these functions that displays the parabolic shape?

For $y = -490x^2 + 980$, the graph of the reference function $f(x) = x^2$ is reflected about the x axis and translated up 980 units. We change the range by a factor of 490. Figure 11a shows the window in the y direction as

$$-10(490) + 980 \le y \le 10(490) + 980$$

$$-3920 \le y \le 5880$$

For $y = -200(x - 10)^2 + 20{,}000$, the graph of $f(x) = x^2$ is reflected about the x axis and translated to the right 10 and up 20,000 units. The window in the x direction in Figure 11b was set by starting with the standard window ($-10 \le x \le 10$) and adding 10.

$$-10 + \mathbf{10} \le x \le 10 + \mathbf{10}$$

$$0 \le x \le 20$$

The window in the y direction was calculated by starting with the nonnegative values, $0 \le y \le 20$.

$$0(200) \le y \le 20(200) + 20{,}000$$
$$0 \le y \le 24{,}000$$

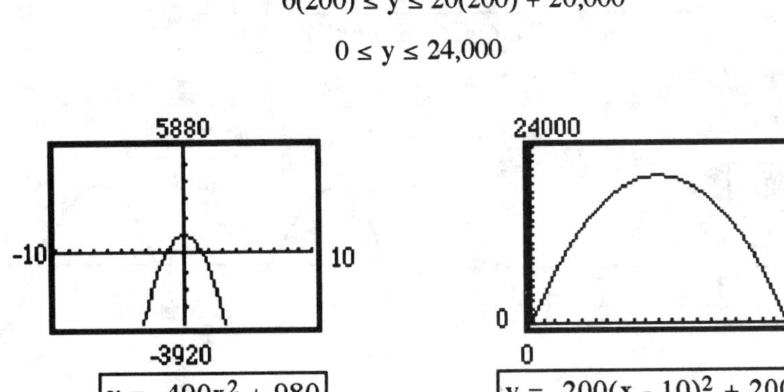

Figure 11 (a) (b)

Applying Changes of Scale

In the next example we see how changes of scale describe loudness and pitch of a musical note. Here we are studying a function defined by its graph.

Example 5 The Musical Note Problem

The musical note A created by a tuning fork vibrating at 440 cycles per second creates a graph similar to the one in Figure 12 when displayed on an electronic device called an *oscilloscope.* The height of the curve, called its *amplitude,* varies with the loudness of the sound: The louder the sound, the higher the amplitude. The number of cycles per second, called *frequency,* changes with the pitch of the sound. The pitch becomes higher as the number of cycles per second increases. Graph (a) $y = 10f(x)$ and (b) $y = f(2x)$; then describe the difference in the sound indicated by the two graphs.

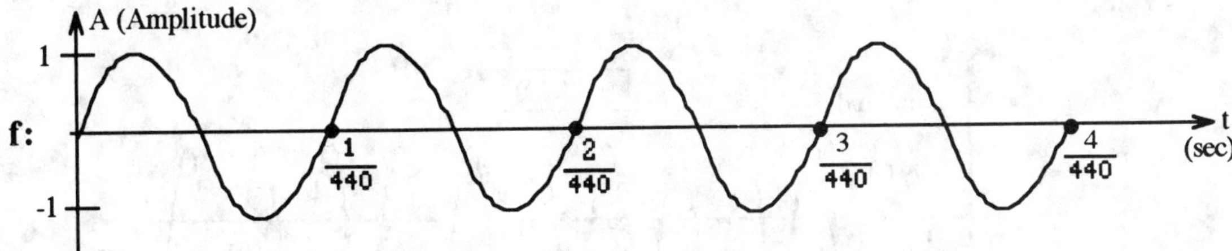

Figure 12

Thinking: Creating a Plan

In part (a), $a = 10$. Thus, to graph $y = 10f(x)$ we rescale the vertical axis by a factor of 10 (Figure 13a). Expanding the curve vertically models increasing the loudness of the sound. In part (b), $b = 2$ and $1/b = 1/2$.

Therefore, to graph $y = f(2x)$ we rescale the horizontal axis by a factor of 1/2 (Figure 13b). Compressing the curve horizontally models increasing the pitch of the sound.

Communicating: Writing the Solution

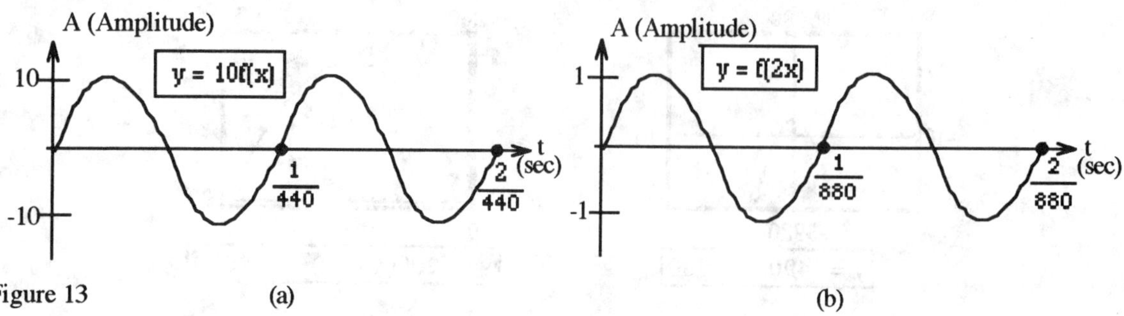

Figure 13 (a) (b)

In part (a), the sound is 10 times as loud as the original. In part (b) the pitch is twice as high as the original.

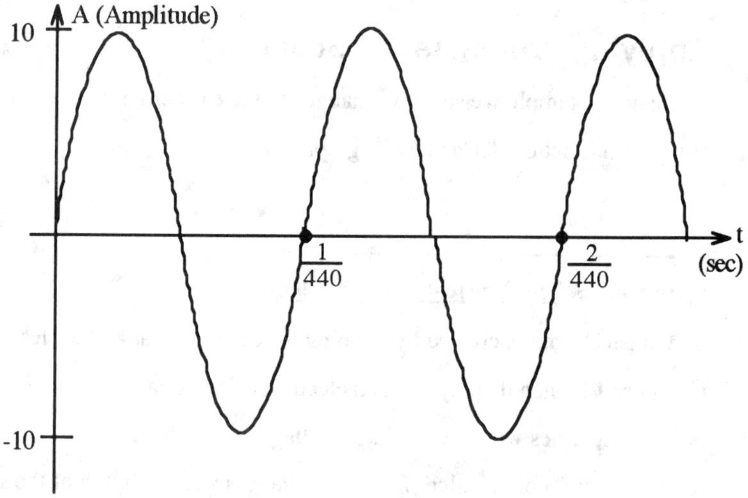

Learning: Making Connections

Letting 1 tick mark = 1 unit on the vertical axis shows the increased amplitude for part (a) (Figure 14).

Figure 14

Changing the scale in part (b) shows the change in pitch (Figure 15). Notice the labels on the x axis: $\frac{1}{880}$, $\frac{2}{880}$, and so on.

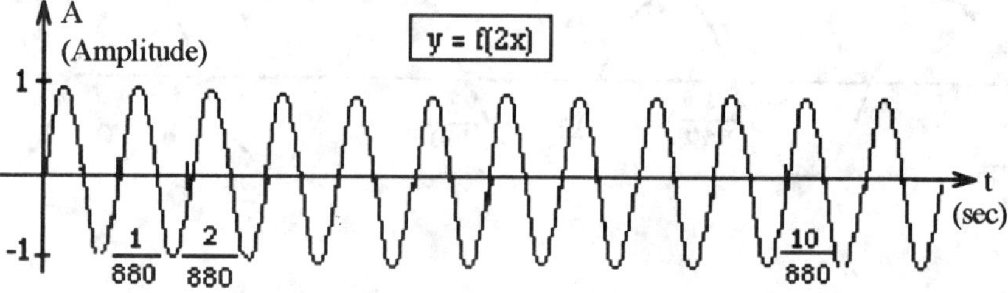

Figure 15

These graphs are related to the reference function $f(x) = \sin x$ (the sine function), which we study in Chapter 5.

Graphs are commonly used to represent data and emphasize or deemphasize attributes such as trends or changes. Sometimes graphs are drawn in ways that intentionally mislead readers in order to support a specific point of view. The next example shows a case of a misleading graph. (See the book How to Lie with Statistics by Darrell Huff for some common ways that graphs are purposely designed to misrepresent a situation.)

Example 6 The Misleading Graph Problem

The table below lists average test scores earned by students taking the Scholastic Aptitude Test in each of the years from 1986 to 1992. While the average scores are fairly uniform from year to year, it is possible to select certain scores, creating a misleading summary graph. Draw such a misleading graph and explain how the graph misrepresents the data.

	1986	1987	1988	1989	1990	1991	1992
math scores (200-800)	475	476	476	476	476	474	476
verbal scores (200-800)	431	430	428	427	424	422	423
combined scores (400-1600)	906	906	904	903	900	896	899

Thinking: Creating a Plan

If we choose the first and last years, 1986 and 1992, we should be able to draw a misleading graph that shows math scores increasing slightly and verbal scores decreasing dramatically. Selecting a range of 420 to 480, instead of 0 to 800, exaggerates the change in these averages.

Communicating: Writing the Solution

We draw a graph in which the x axis represents years and the y axis represents test scores. We plot the points representing the years 1986 and 1992, then draw lines connecting the points, as shown in Figure 16. In drawing the graph this way we are ignoring the scores for the years 1987 through 1991.

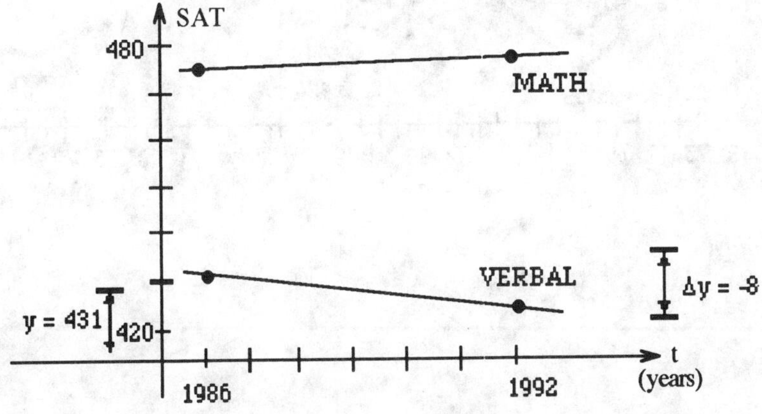

Figure 16

Learning: Making Connections

The graph can be made even more misleading if we omit the labels on the vertical axis. Then there is no way to tell that the change in test averages pictured is very slight.

Showing a portion of the range, instead of $0 \le y \le 800$, visually exaggerates the **percent change,** $\frac{\Delta y}{y}$, where y is the first y value and Δy (*delta y*) is the change in y from the first to the next. For the two verbal score points (1986,431) and (1992,423) the percent change is only

$$\frac{\Delta y}{y} = \frac{423 - 431}{431} = \frac{-8}{431} = -1.9\%$$

In the graph, this change is exaggerated because the $\Delta y = -8$ and $y = 431$ are pictured with Δy about three times the size of y. From actual measurements, the percent change shown in the drawing is

$$\frac{\Delta y}{y} = \frac{-3 \text{ cm}}{1 \text{ cm}} = -300\%$$

which is quite an exaggeration of -1.9%. This approach to scaling the axes to overemphasize small changes is very common. Selecting data that support one goal and ignoring the rest of the data is always inappropriate.

The drawings in Figure 17 zoom in on the scores that are important, to magnify changes at the expense of exaggerating the percent change.

Does the pencil on the graph in Figure 17a have any visual effect?

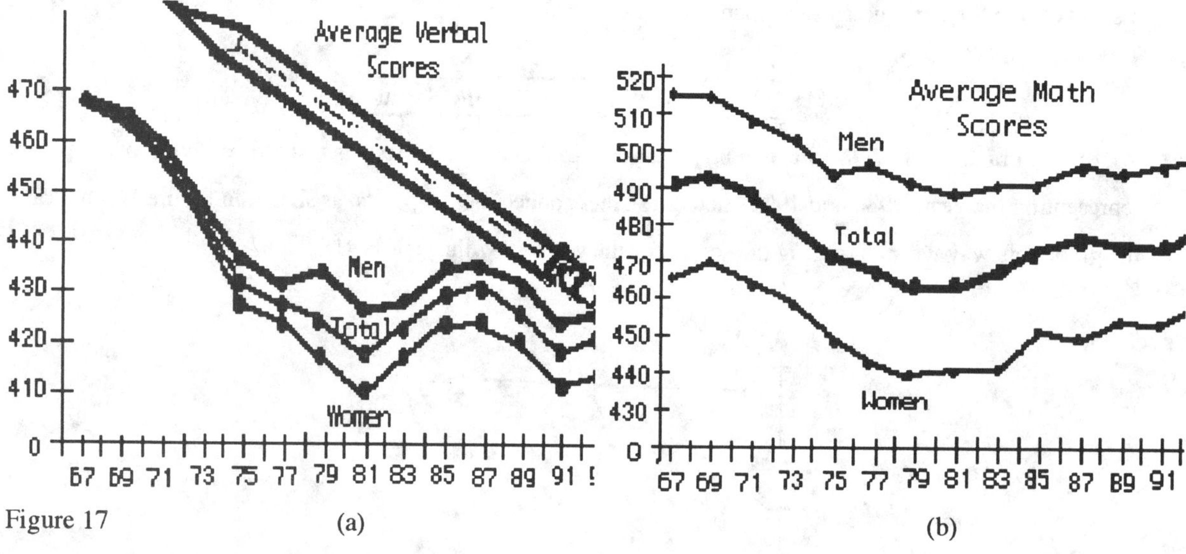

Figure 17 (a) (b)

[Source: *USA Today*]

Explore the Concepts

These exercises are designed to lead you to explore the effects of multiplying the x coordinates or the y coordinates of a given set of ordered pairs of a function by scaling factors a and b. The equations are of the form $y = af(x)$ or $y = f(bx)$. Each display has a different window, yet the graphs are identical. Notice there are 20 tick marks on both axes. Experiment with your graphing utility to help match each equation with its graphical representation.

E1. $y = 4x^2$ E2. $y = (4x)^2$ E3. $y = \frac{1}{4}x^2$ E4. $y = (\frac{1}{4}x)^2$

A.

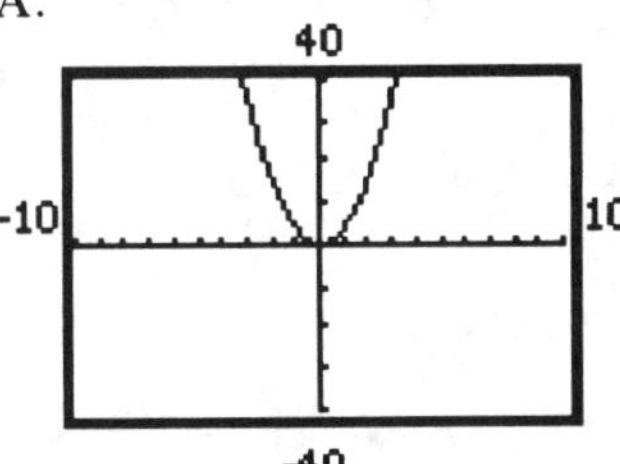

B.

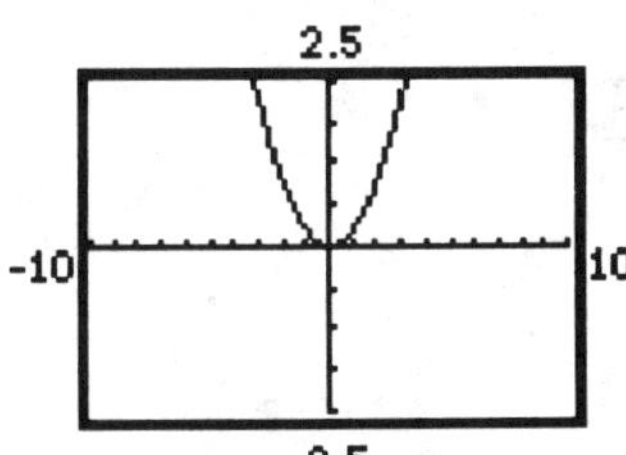

C.

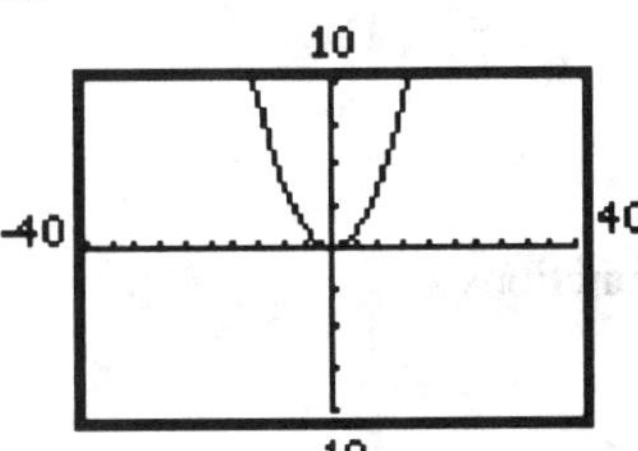

D.

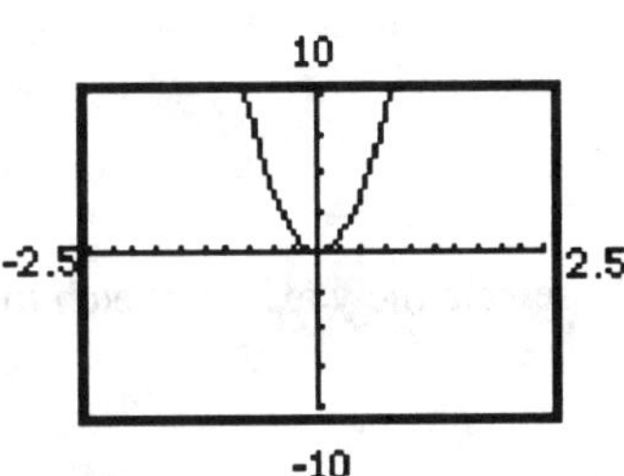

E5. $y = \frac{1}{5}|x|$ E6. $y = |\frac{1}{5}x|$ E7. $y = 5|x|$ E8. $y = |5x|$

A.

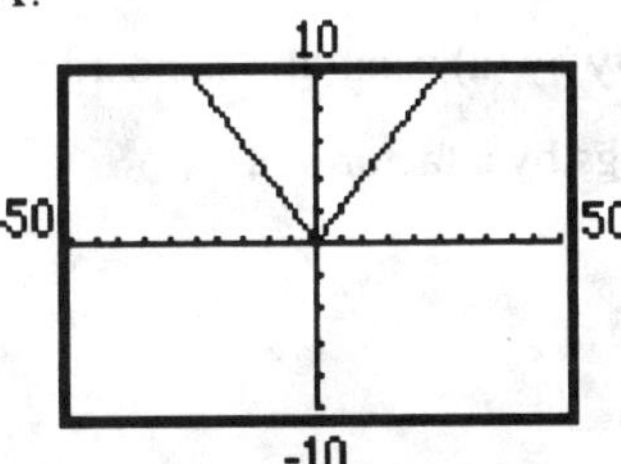

B.

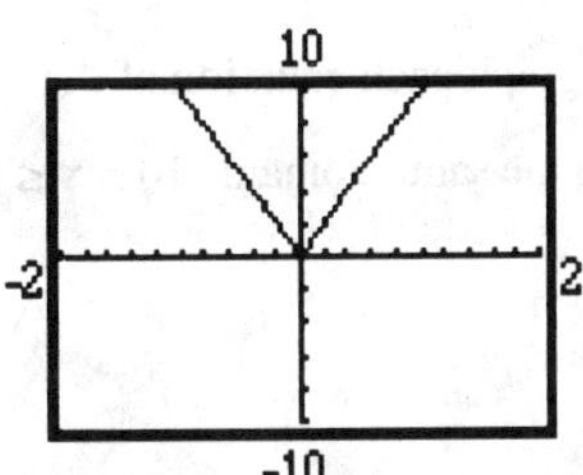

C.

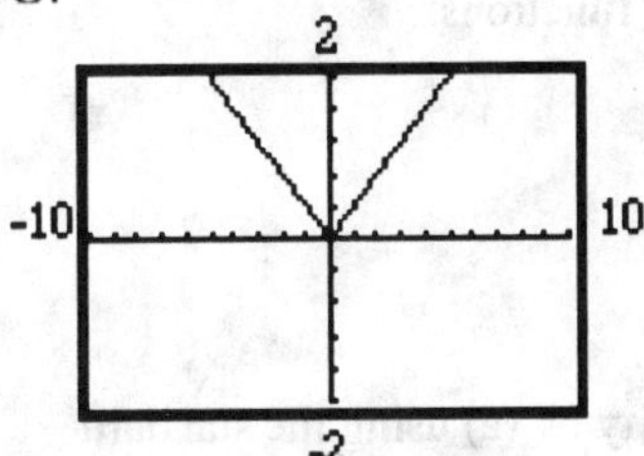

D.

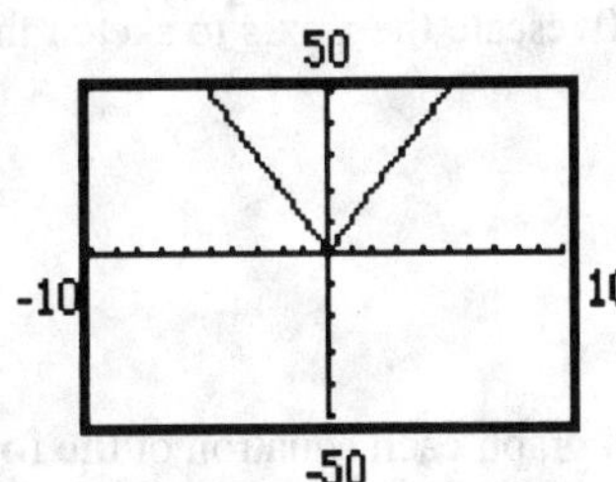

Use the results from Exercises E1 to E8 to complete the following theorems.

E9. $y = af(x)$ changes the scale on the y axis by a factor of ________.

E10. $y = f(bx)$ changes the scale on the x axis by a factor of ________.

Exercises 1.4

Develop the Concepts

For Exercises 1 to 8 graph each pair of functions on the **same** coordinate system. Plot one point on each graph that shows the effect of the scaling factor present.

1.(a) $f(x) = x^2$
(b) $g(x) = 2x^2$

2.(a) $f(x) = x^3$
(b) $g(x) = \frac{x^3}{3}$

3.(a) $f(x) = (x+1)^2$
(b) $g(x) = \frac{(x+1)^2}{3}$

4.(a) $f(x) = -x^3$
(b) $g(x) = -5x^3$

5.(a) $f(x) = \sqrt{x}$
(b) $g(x) = \sqrt{\frac{x}{2}}$

6.(a) $f(x) = |x|$
(b) $g(x) = |3x|$

7.(a) $f(x) = -|x|$
(b) $g(x) = -|3x|$

8. (a) $f(x) = \sqrt{x+5}$
(b) $g(x) = \sqrt{\frac{x+5}{2}}$

For Exercises 9 to 12 rescale the y axis to sketch the graph of one of the reference functions.

9. $y = 2\sqrt{x}$

10. $y = \frac{\sqrt{x}}{2}$

11. $y = 5|x|$

12. $y = \frac{|x|}{5}$

For Exercises 13 to 16 graph each equation of the form $y = af(x)$ on a graphing utility by (a) using the standard screen, and (b) keeping the same domain $-10 \le x \le 10$ and changing the range settings by a factor of a.

13. $y = 30x^2$

14. $y = 14\sqrt{x}$

15. $y = \frac{10}{x^2}$

16. $y = \frac{x^3}{25}$

For Exercises 17 to 20 rescale the x axis to sketch the graph of one of the reference functions.

17. $y = \sqrt{2x}$

18. $y = \sqrt{\frac{x}{2}}$

19. $y = |5x|$

20. $y = |\frac{x}{5}|$

For Exercises 21 to 24 graph each equation of the form $y = f(bx)$ on a graphing utility by (a) using the standard screen, and (b) keeping the same range $-10 \le y \le 10$ but changing the domain settings by a factor of 1/b.

21. $y = (50x)^2$

22. $y = \sqrt{24x}$

23. $y = \frac{1}{10x}$

24. $y = (25x)^3$

For Exercises 25 to 32 state the window on a graphing utility so that the graph will match the graph of the corresponding reference function (Appendix II) when graphed on a standard window.

*25. $y = 0.1\sqrt{0.1x}$

26. $y = 100\sqrt{100x}$

27. $y = 12\sqrt{x/12}$

28. $y = .0001\sqrt{.0001x}$

29. $f(x) = 20\sqrt{40x}$

30. $g(x) = 0.5\sqrt{0.01x}$

31. $y = 15\left|\frac{x}{2}\right|$

32. $y = \frac{1}{4}(12x)^3$

For Exercises 33 to 44 sketch the graph of each function by rescaling the axes if necessary.

33. $y = 3(x + 2)^2 - 1$

34. $y = \sqrt{2 - \frac{x}{2}}$

35. $y = -(2x - 1)^3$

36. $y = -2(x - 1)^2$

37. $y = \frac{1}{2}|x| - 1$

38. $y = -2(1 - x)^3 + 3$

39. $y = -\frac{20}{x^2}$

40. $y = -\frac{1}{20x^2}$

*41. $y = \frac{10}{1 - 2x}$

42. $y = \frac{-30}{x - 1}$

43. $y = 20(x - 2) + 30$

44. $y = -30(x + 2) - 10$

For Exercises 45 to 48 use the rescaled axes to help match each equation with its graph.

45. $y = 2\sqrt{x + 1}$

46. $y = \sqrt{\frac{x + 1}{2}}$

47. $y = \sqrt{2(x + 1)}$

48. $y = \frac{1}{2}\sqrt{\frac{x + 1}{2}}$

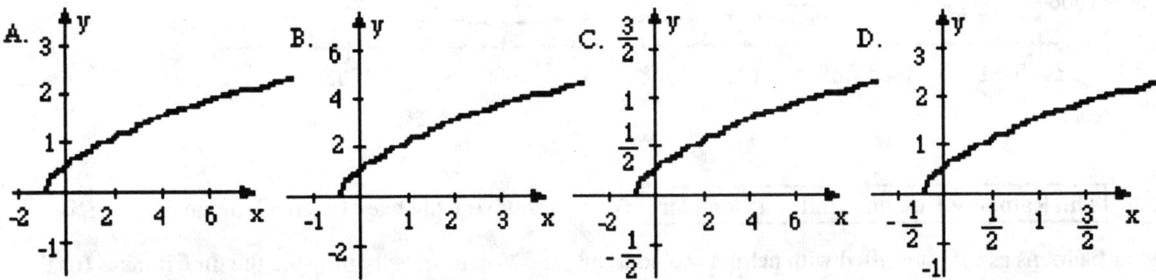

Apply the Concepts

49. The Gasoline and Oil Problem The following are average annual consumer expenditures by household on gasoline and motor oil.

(a) Sketch the graph that uses 1 unit = $100 on the vertical axis. Use t = 0 for 1988, t = 1 for 1989, and t = 2 for 1990.

(b) Design a graph that overemphasizes the rising expenditures.

year	1988	1989	1990
cost	$932	$985	$1047

[Source: *Business Almanac*, S. Goodin]

50. The Test Score Problem (a) Draw an accurate graph of the following student's test scores. (Assume that each test had a possible score of 100.)

test	1	2	3	4
score	75	80	82	85

(b) Draw a graph that overemphasizes the improvement, again assuming possible scores of 100.

*51. The Misleading Car Advertisement Problem Sketch a graph using the following data, which express car value as a function of year. Rescale the y axis so that it appears at first glance that owning this car is a really good investment. A car was purchased new in 1988 for $14,220, sold in 1989 for $14,350, and then sold again in 1990 for $14,400.

52. The Fatalities Problem Explain how the graphical presentation in Figure 18 could be misinterpreted. [Source: World Eagle Graphics]

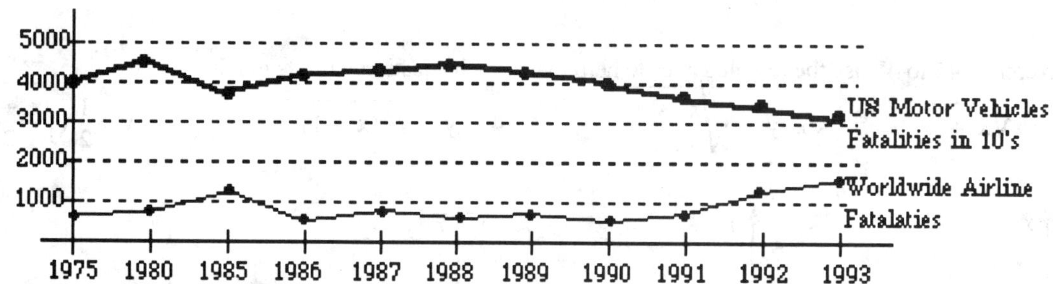

Figure 18

53. The Rainbow Kite and Balloon Company You are studying the rate of increase in the diameter of your balloons as they are filled with helium at a constant rate. You repeatedly measure the time it takes for the diameter to increase by 1 in. You are assuming that the rate of increase in the diameter R is a function of the diameter d. [R(2) = 0.32 means that the diameter increased by 0.32 in./sec when the diameter was 2 in.] Use the function described by the data in the table to answer the following.

diameter, d (in.)	2	4	6	8
rate of increase in diameter, R(d) (in./sec)	0.32	0.08	0.04	0.02

(a) Create a table for $R_2 = 2R(d)$.

(b) Describe the new function in part (a) compared to the original data.

(c) Create a table for $R_3 = R(2d)$.

(d) Describe the new function in part (c) compared to the original data.

54. Wonderworld Amusement Park The revenue from your Ferris wheel is a function of the duration of the ride. If the ride is too long, the wait is too long and people leave. If the ride is too short, people do not come back. You vary the length of the ride and count the number of riders per hour. Use the function described by the data in the table to answer the following.

length of ride, t (min)	0	2	4	6	8
number of riders, R(t) (people/hr)	0	40	150	80	20

(a) Create a table for another day when $R_2 = 0.50R(t)$.

(b) Describe the change in length of ride and number of riders in part (a) compared to the original data.

(c) Create a table for yet another day when $R_3 = R(0.50t)$.

(d) Describe the change in length of ride and number of riders in part (c) compared to the original data.

Extend the Concepts

For Exercises 55 to 58 give the equation of a reference function whose equation would call for the rescaling of the axes as pictured.

55.

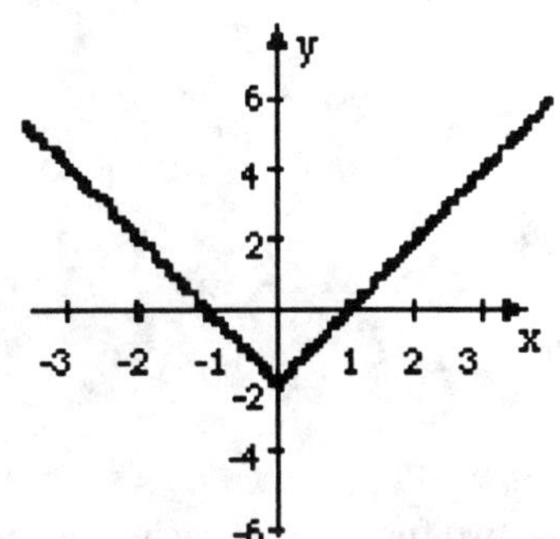

56.

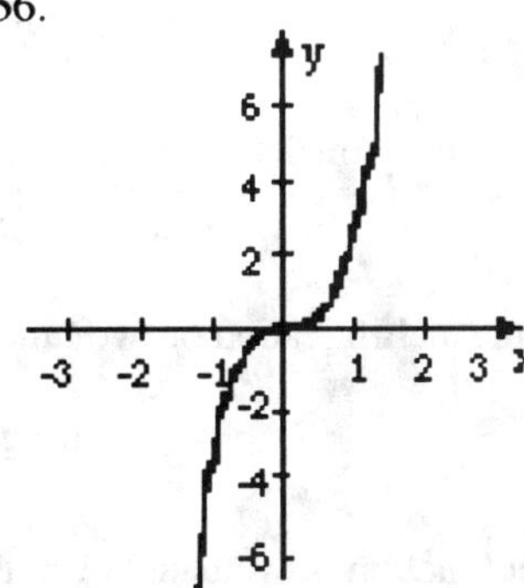

57.

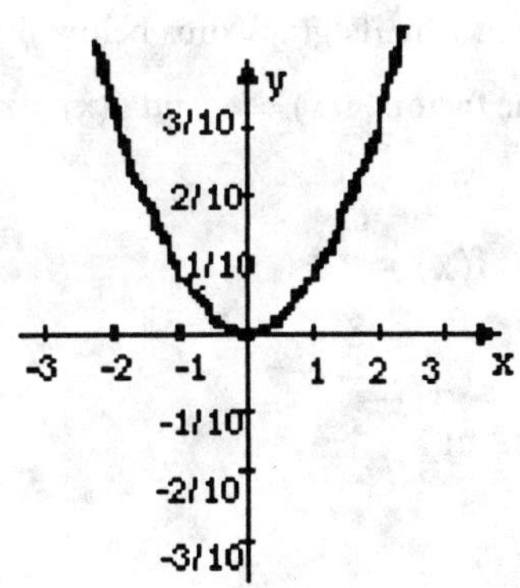

58.

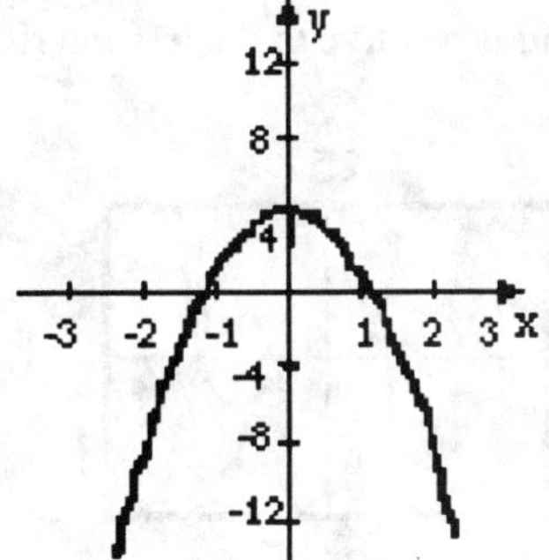

1.5 Operations, Compositions, and Inverse Functions

Adding and Multiplying Functions
Composing Functions
Finding an Inverse
Using One Graph for Both a Function and Its Inverse

We can create new functions and graphs from two known functions by adding, subtracting, multiplying, dividing, or applying them one after another. We also create new functions by reversing the input and output of a function. This allows us to apply functions to a wider variety of applications and extend our knowledge of fundamental functional concepts.

Adding and Multiplying Functions

The sum and difference of two functions f and g are defined as

$$(f + g)(x) = f(x) + g(x) \quad \text{and} \quad (f - g)(x) = f(x) - g(x)$$

These definitions indicate that sums and differences are evaluated by adding or subtracting the values of f and g at each value of x. For $f(x) = |x|$ and $g(x) = 10$, the new function $y = (f + g)(x)$ is

$$y = f(x) + g(x) = |x| + 10$$

Similarly, $y = (f - g)(x)$ is defined as

$$y = f(x) - g(x) = |x| - 10$$

We define the multiplication of two functions as

$$(f \cdot g)(x) = f(x) \cdot g(x)$$

The product function is evaluated by multiplying the values of the two functions at each value of x. Let's consider the product function $y = x^3(x - 3)$ in Figure 1a. To understand why its graph dips below the x axis for $0 < x < 3$ and is positive to the left and right requires analyzing the factors, $f(x) = x^3$ and $g(x) = x - 3$.

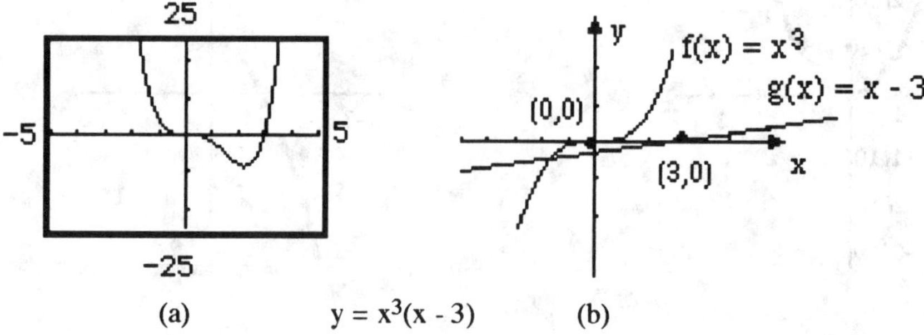

Figure 1 (a) $y = x^3(x - 3)$ (b)

The x intercepts of the factors yield the x intercepts of their product. In Figure 1b we see that the x intercepts are 0 for function f and 3 for function g. Their product function has x intercepts at these two points.

Next we look at intervals where the functions are both negative or both positive. Over these intervals, their product is positive; its graph lies above the x axis. Functions f and g are both negative for $x < 0$ and both positive for $x > 3$. Finally, we locate intervals where the two graphs are on opposite sides of the x axis. Between 0 and 3 the graph of the product function lies below the x axis because one factor is positive and the other negative. See Table 1.

Table 1

	$x < 0$	$x = 0$	$0 < x < 3$	$x = 3$	$x > 3$
$f(x) = x^3$	negative	zero	positive	positive	positive
$g(x) = x - 3$	negative	negative	negative	zero	positive
$f \cdot g(x) = x^3(x - 3)$	positive	zero	negative	zero	positive

The quotient of two functions f and g is defined as

$$\left(\frac{f}{g}\right)(x) = \frac{f(x)}{g(x)}$$

The graphs of quotients may involve values of x for which the denominator, g(x), equals zero, causing division by zero. Such functions will be considered in detail in Chapter 3.

Definitions of Operations of Functions

For two given functions f and g with x in the domain of both, we define their

sum $f + g$ $\quad (f + g)(x) = f(x) + g(x)$

difference $f - g$ $\quad (f - g)(x) = f(x) - g(x)$

product $f \cdot g$ $\quad (f \cdot g)(x) = f(x) \cdot g(x)$

quotient $\frac{f}{g}$ $\quad \left(\frac{f}{g}\right)(x) = \frac{f(x)}{g(x)},\ g(x) \neq 0$

Composing Functions

If we apply two functions one after the other, we create a new function called their **composite function.** Figure 2 shows function g acting first, creating the output g(x). This value becomes the input of function f. The final output is f(g(x)). The notation f(g(x)) is read "f of g of x." In f(g(x)) we can think of function g as the input of function f. For f(g(x)), we say that g is acted on *by* f. This requires that the function inside the parentheses, function g, be applied first. Then function f is applied to that result.

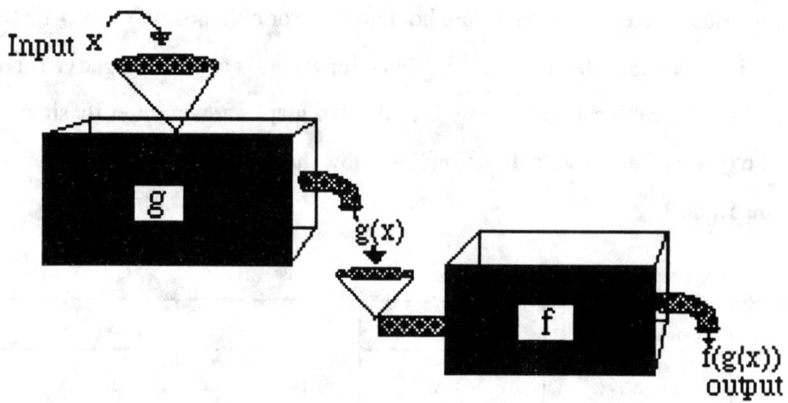

Figure 2

The function

$$y = \sqrt{x + 1}$$

can be considered a composite of two functions. For $y = \sqrt{x + 1}$ the order of operations requires that we first add 1 to x, then take the square root. We can write this composite function as

$$f(g(x)) = \sqrt{x + 1}$$

Do you think that $f(g(x)) = g(f(x))$? That is, do you believe that composition of functions is commutative?

If we reverse the order in which we apply two functions, as in g(f(x)), we may create a function that is entirely different from f(g(x)). Here, g(f(x)) takes the square root of x first, then adds 1.

$$g(\mathbf{f}(x)) = g(\sqrt{x}) = \sqrt{x} + 1$$

Function f is acted on by g. Thus, we see that for some pairs of functions

$$f(g(x)) \neq g(f(x))$$

NOTE: The compositions f(g(x)) and g(f(x)) can be written **f ∘ g** and **g ∘ f**, respectively. That is,

$$f(g(x)) = (f \circ g)(x) \quad \text{and} \quad g(f(x)) = (g \circ f)(x)$$

The results $f(g(x)) \neq g(f(x))$ can be expressed informally as $f \circ g \neq g \circ f$.

In the next example, we see that $f(g(x)) = g(f(x))$ for some functions.

Example 1 *Finding the composition of two functions*

Find f(g(x)) and g(f(x)) for the following functions.

(a) $f(x) = 2x + 3,\ g(x) = \frac{x - 3}{2}$ (b) $f(x) = x^2,\ g(x) = \sqrt{x}$

Solution

(a) $f(x) = 2x + 3$ and $g(x) = \frac{x - 3}{2} \Rightarrow$

$$f(g(x)) = f\left(\frac{x - 3}{2}\right) = 2\left(\frac{x - 3}{2}\right) + 3 = x$$

and

$$g(f(x)) = g(2x + 3) = \frac{(2x + 3) - 3}{2} = x$$

(b) $f(x) = x^2$ and $g(x) = \sqrt{x} \Rightarrow$

$$f(g(x)) = f(\sqrt{x}) = (\sqrt{x})^2 = x, \quad x \geq 0$$

and

$$g(f(x)) = g(x^2) = \sqrt{x^2} = |x| \text{ for all real numbers } x$$

Did you notice that $f(g(x)) = g(f(x))$ in part (a) but not in part (b) ?

In part (a) function f multiplies by 2, then adds 3. Function g subtracts 3, then divides by 2. Applying these two functions one after another to any input yields that value. One function undoes the operations of the other -- in the opposite order. In part (b) the restricted domain of the square root function created a restriction in the domain of the composite function.

Let's examine why the domain of the composite function

$$y = f(g(x)) = \sqrt{x^2 - 1}$$

is $x \leq -1$ or $x \geq 1$ (Figure 3).

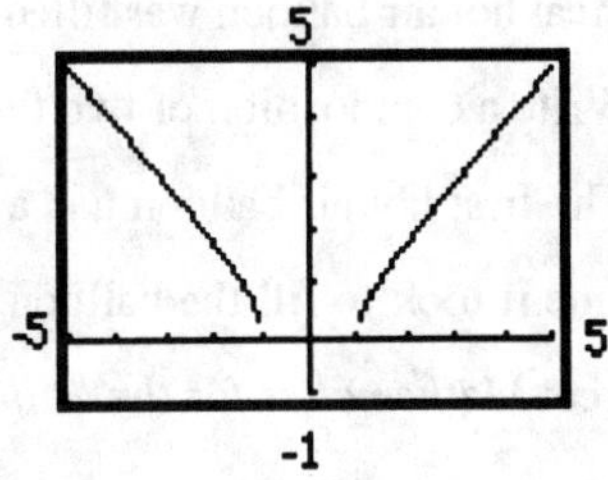

Figure 3

Consider the graph of the input function $g(x) = x^2 - 1$ (Figure 4). Since $g(x) = x^2 - 1$ is the input for the square root function, g(x) cannot be negative. Since function g is negative for $-1 < x < 1$ (Figure 4), there is no graph of the composite function f(g(x)) for $-1 < x < 1$ (Figure 3).

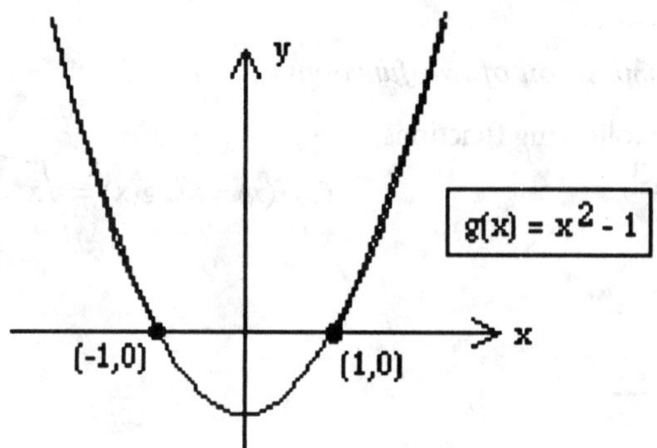

Figure 4

NOTE: Rewriting a function as a composition of two functions is called **decomposition**. We can also describe a decomposition using the variable *u* to "link" the two functions as follows.
For $f(g(x)) = \sqrt{x^2 - 1}$,

$$y = \sqrt{x^2 - 1} \Leftrightarrow y = \sqrt{u} \text{ where } u = x^2 - 1$$

This approach is used frequently in calculus. Decomposition of functions is a powerful tool in understanding the graphs of complicated functions by studying the simpler functions involved. The decomposition is not unique, although usually there is a best choice depending on the context.

Example 2 The First Hot Air Balloon Problem

A hot air balloon, launched in 1782, carried the first passengers in aviation history: a sheep, a duck, and a hen. Due to the carelessness of the sheep, the hen did not survive. Otherwise, the flight was a success. Suppose the spherical hot air balloon was filled so that the diameter was increased at 3 ft/hr.

(a) Write a composition of two functions to find volume as a function of time.

(b) The first hot air balloon had a volume of 23,430 ft^3. Use this composition function in part (a) to estimate the time it took to fill the balloon.

[Source: *Mathematics for the Millions*, L. Hogben]

Thinking: Creating a Plan

(a) The formula for the volume V of a sphere is $V = \frac{4}{3}\pi r^3$. Since V is given in terms of the radius r, we consider the diameter increasing at 3 ft/hr as the radius is increasing at 1.5 ft/hr, so that r = 1.5t, where t is the time since they began filling the balloon.

We can consider volume a function of radius, V = f(r), and radius a function of time, r = g(t). We substitute r = 1.5t into the volume formula to create the desired composite function, volume as a function of time.

$V = f(r)$ and $\mathbf{r = g(t)} \Rightarrow$

$V = f(\mathbf{g(t)})$

(b) We will substitute V = 23,430 into the formula from part (a) and solve for t.

Communicating: Writing the Solution

(a) $V = \frac{4}{3}\pi r^3$ and $r = 1.5t \Rightarrow$

$$\begin{aligned} V &= \frac{4}{3}\pi(1.5t)^3 \\ &= 4.5\pi t^3 \end{aligned}$$

(b) $V = 4.5\pi t^3$ and $V = 23{,}430 \Rightarrow$

$$\begin{aligned} 4.5\pi t^3 &= 23{,}430 \\ t^3 &= \frac{23430}{4.5\pi} \\ t &= \sqrt[3]{\frac{23430}{4.5\pi}} \\ &\approx 11.83 \end{aligned}$$

At that rate it would have taken about 12 hr to fill the balloon.

Learning: Making Connections

Follow the value of r in the next two lines, when we input a specific time, t = 12 hr.

$$r = 1.5t \text{ and } t = 12 \Rightarrow \mathbf{r = 18}$$
$$V = \frac{4}{3}\pi r^3 \text{ and } \mathbf{r = 18} \Rightarrow V = 24{,}429$$

Notice the chainlike feature of this composite function. The two lines are linked since the output r of the first line becomes the input r of the second line.

We can also study composite functions from tables of data. Notice how the values of r link the first and last tables.

input t	0	5	10	12
output r = 1.5t	**0**	**7.5**	**15**	**18**

and

input r	**0**	**7.5**	**15**	**18**
output $V = \frac{4}{3}\pi r^3$	0	1767	14,137	24,429

These tables yield the composite data that show V as a function of t.

input t	0	5	10	12
output V	0	1767	14,137	24,429

Do you see the link between the outputs of function g and the inputs of function f that create this composition function in Figure 5?

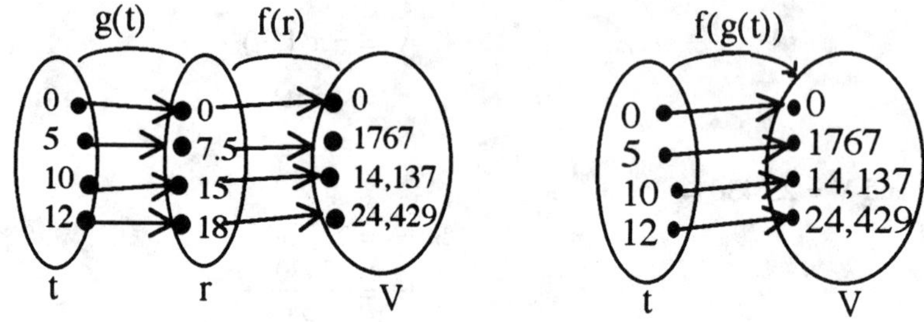

Figure 5

Finding an Inverse

The equation $F = \frac{9}{5}C + 32$ expresses the temperature in degrees Fahrenheit as a function of the temperature in degrees Celsius (°C). If we are given a temperature in degrees Celsius, we use this function to convert it to degrees Fahrenheit. If we are given temperatures in degrees Fahrenheit, however, we find the corresponding Celsius temperature using an equivalent formula, $C = \frac{5}{9}(F - 32)$. We will see why the two functions are called *inverses* of each other.

To explore inverse relations, go to "Explore the Concepts," Exercises E1 - E4 before continuing (page 91). For the interactive version, go to the *Precalculus Weblet - Instructional Center - Explore Concepts*.

For the equation $F = \frac{9}{5}C + 32$, if we input the freezing point of water, 0°C, C = 0, we find the corresponding value in degrees Fahrenheit as follows:

$$F = \frac{9}{5}C + 32 = \frac{9}{5}(0) + 32 = 32$$

Suppose that we reverse the given and the goal; given a temperature in degrees Fahrenheit, find the equal Celsius temperature. We would solve this equation for C so that F becomes the independent (input) variable.

$$F = \frac{9}{5}C + 32$$

$$F - 32 = \frac{9}{5}C$$

$$\frac{9}{5}C = F - 32$$

$$C = \frac{5}{9}(F - 32)$$

Substituting the boiling point of water, 212°F, F = 212 yields

$$C = \frac{5}{9}(F - 32) = \frac{5}{9}(212 - 32) = \frac{5}{9}(180) = 100$$

The corresponding temperature is 100°C.

The equation that expresses C as a function of F is called the **inverse** of the original function. <u>We found the equation of the inverse function simply by solving for the other variable</u>. The two equivalent functions

$$F = \frac{9}{5}C + 32 \quad \text{and} \quad C = \frac{5}{9}(F - 32)$$

are **inverses** of each other. They reverse the roles of the variables in the domain and range. The input variable of one function is the output variable of its inverse (Figure 6). The domain of the function is the range of its inverse and the range of the function is the domain of its inverse. When drawing a function or its inverse, we label the horizontal axis with the input variable and the vertical axis with the output variable.

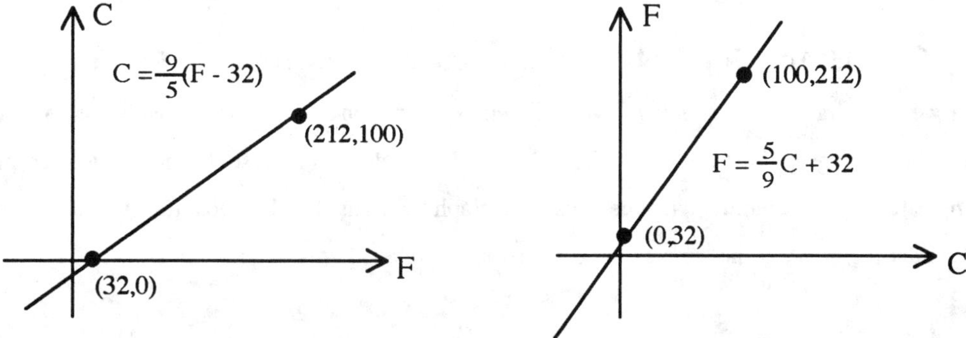

Figure 6

Figure 7 shows the graphs of this function and its inverse on a graphing utility. However, to use a graphing utility we first rename the variables to express y as two functions of x, $y_1 = \frac{9}{5}x + 32$ and $y_2 = \frac{5}{9}(x - 32)$.

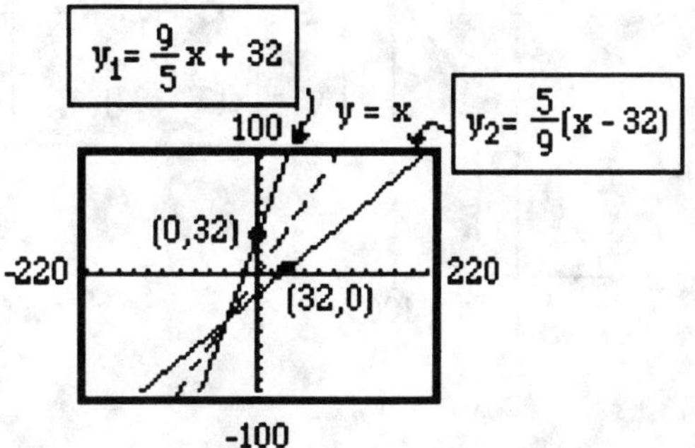

Figure 7

Notice that the two graphs are symmetric about the line y = x. By interchanging the domain and range, a point (a, b) becomes (b, a), reflecting each about the line y = x. We can find the graph of the inverse relation of any function by reflecting the graph about the line y = x.

We can find the graph of the inverse relation of all 13 reference functions by flipping the template corresponding to Appendix II on its lower left to upper right diagonal (Figure 8), flipping about the line y = x.

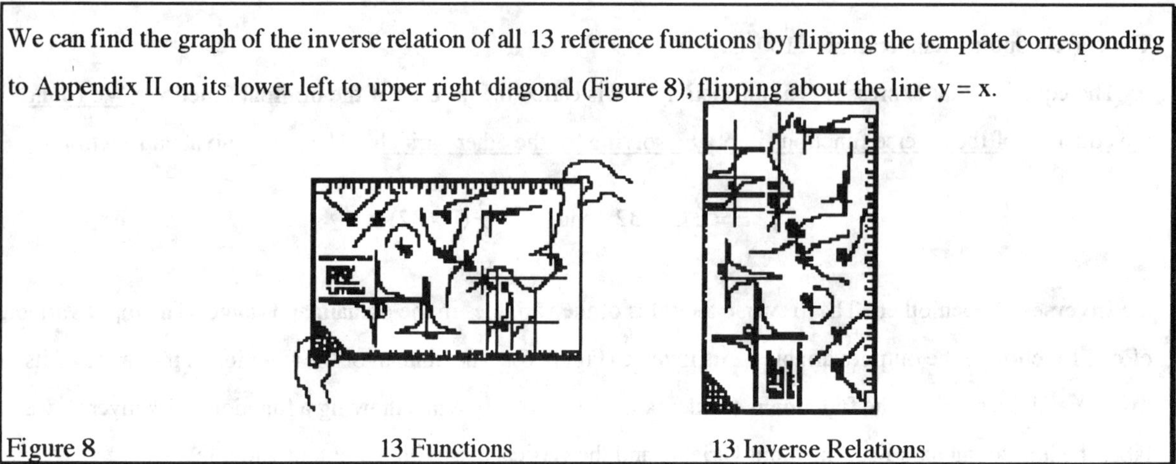

Figure 8 13 Functions 13 Inverse Relations

Using One Graph for Both a Function and Its Inverse

Thus far we have drawn two separate graphs to illustrate a function and its inverse. However, only one graph is needed. The same graph that represents the function also models its inverse. In more advanced math classes, including calculus, it is common practice to use one graph for virtually all applications involving inverses.

Reconsider the graph of the temperature conversion formula

$$F = \frac{9}{5}C + 32$$

shown in Figure 9. If we start with an input value from the horizontal axis, we can use the function $F = \frac{9}{5}C + 32$ to find its corresponding output value, its **image**. Notice in Figure 9 how the arrow

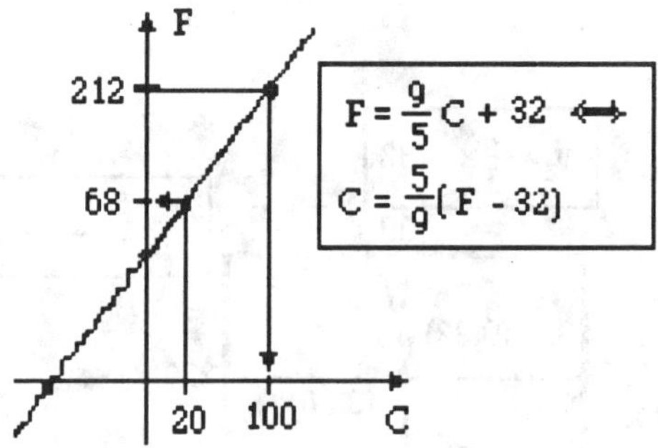

Figure 9

takes us from the input C = 20 to the graph of the function, then to the output F = 68. For this function, F = 68 is the image of C = 20.

If we input a value from the vertical axis, we can use the algebraically equivalent equation of the inverse, $C = \frac{5}{9}(F - 32)$, to find its corresponding value, its **inverse image**. Notice in Figure 9 how the arrow takes us from the input F = 212 to the graph, then to the output C = 100. For the inverse function, C = 100 is the inverse image of F = 212.

By using one graph for both a function and its inverse, we can readily see that the range of the function is the domain of its inverse. If an original function gives an output number b for input a, the **inverse** gives output a for input b (Figure 10).

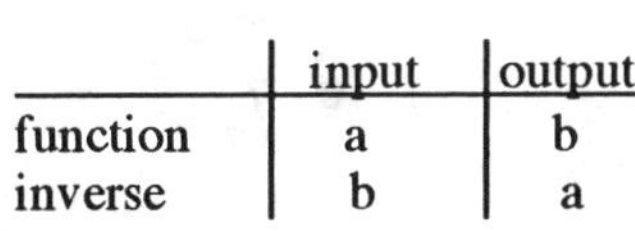

	input	output
function	a	b
inverse	b	a

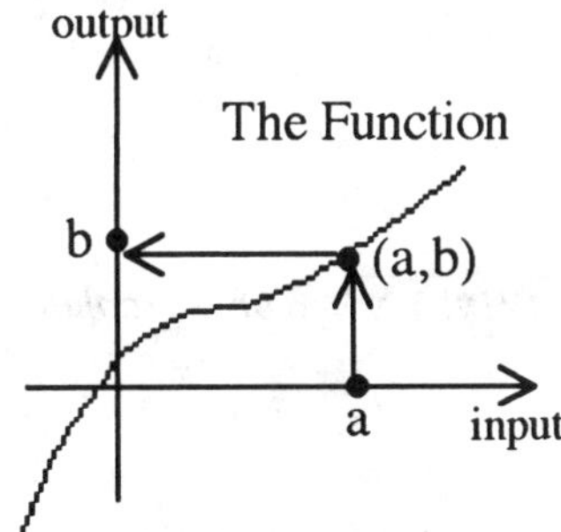

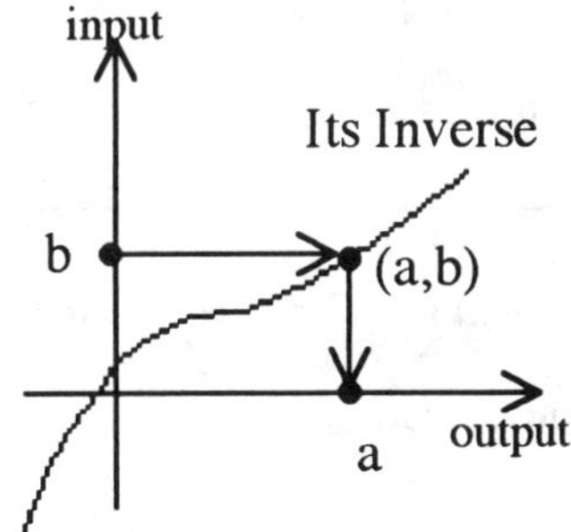

Figure 10

The notation used to denote the inverse function of a function f is

$$\mathbf{f}^{-1}$$

read "f inverse." This negative 1 is not an exponent; it is simply the way the inverse is written.

Example 3 *Finding the equation of an inverse*

(a) Find the equation of the inverse of $f(x) = \sqrt[3]{x + 8}$ by solving for x.

(b) Graph $f(x) = \sqrt[3]{x + 8}$ and use arrows on the graph to show the image of x = 19, f(19), onto the y axis and the inverse image of y = -2, $f^{-1}(-2)$, onto the x axis.

Solution

(a) Let f(x) = y and solve for x.

$$f(x) = \sqrt[3]{x + 8}$$
$$y = \sqrt[3]{x + 8}$$
$$y^3 = x + 8$$
$$x = y^3 - 8$$
$$f^{-1}(y) = y^3 - 8$$

The inverse of $f(x) = \sqrt[3]{x + 8}$ is $f^{-1}(y) = y^3 - 8$.

(b) $f(x) = \sqrt[3]{x + 8}$ $\qquad$ $f^{-1}(y) = y^3 - 8$

$f(19) = \sqrt[3]{19 + 8} = \sqrt[3]{27} = 3$ $\qquad$ $f^{-1}(-2) = (-2)^3 - 8 = -16$

See Figure 11.

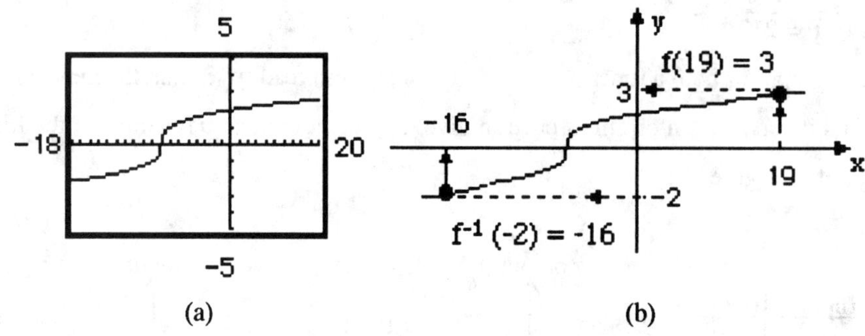

Figure 11 (a) (b)

How would we graph the inverse function $f^{-1}(y) = y^3 - 8$ on a graphing utility?

We enter $y = x^3 - 8$.

Not all inverses are functions. For example, the inverse relation of $y = x^2$ is found by taking the square roots of both sides of the equation. The result,

$$x = \pm\sqrt{y}$$

is *not* a function because positive input values of y are paired with two output values of x.

NOTE: The relation $x = \pm\sqrt{y}$ is not a function, but it can be considered in two pieces, $x = +\sqrt{y}$ and $x = -\sqrt{y}$; each piece is a function. Such relations are called *implicit functions.* In calculus, theorems for functions also hold for implicit functions.

When is an inverse relation a function?

In general, an **inverse relation** is a function if and only if the original function is one-to-one. Recall from Section 1.1 that a function to be one-to-one must pass the horizontal line test. Since $y = x^2$ is not a one-to-one function (Figure 12a), its inverse, $x = \pm\sqrt{y}$ is not a function (Figure 12b).

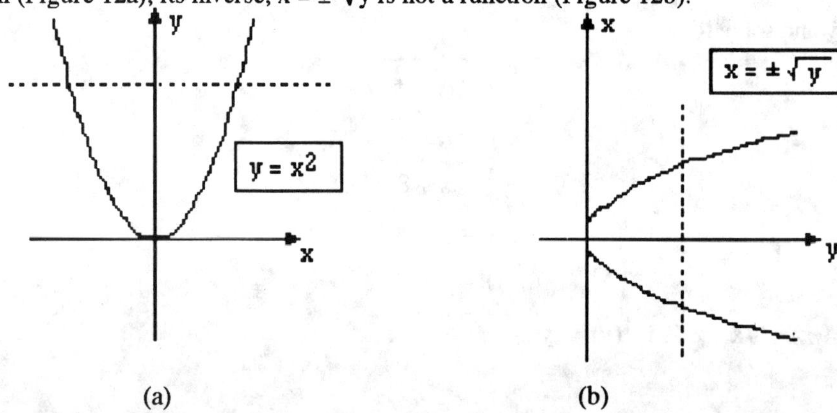

Figure 12 (a) (b)

We define two inverses: first, inverse relations for all functions, including those that are not one-to-one, then **inverse functions** for one-to-one functions.

Definition of an Inverse Relation

If f is a function, g is its **inverse relation** if and only if

$$(a, b) \in f \Leftrightarrow (b, a) \in g$$

Definition of an Inverse Function

If f is a one-to-one function, f^{-1} is its **inverse function** if and only if

$$(a, b) \in f \Leftrightarrow (b, a) \in f^{-1}$$

The following theorem involving the composition of a function and its inverse follows from the definition of an inverse function.

Theorem: Inverses

For a one-to-one function f, if $y = f(x)$ and $x = f^{-1}(y)$ are **inverses** of each other,

$$f^{-1}(f(a)) = a \quad \text{and} \quad f(f^{-1}(b)) = b$$

for all a in the domain of $f^{-1} \circ f$ and all b in $f \circ f^{-1}$

Example 4 *Using the theorem on inverses to verify inverses*

For $f(x) = \sqrt[3]{x + 8}$ and its inverse $f^{-1}(y) = y^3 - 8$ from Example 3, show that $f^{-1}(f(a)) = a$ and $f(f^{-1}(b)) = b$.

Solution

$$f^{-1}(f(a)) = f^{-1}(\sqrt[3]{a+8}\,) = (\sqrt[3]{a+8}\,)^3 - 8 = (a + 8) - 8 = a$$

and

$$f(f^{-1}(b)) = f(b^3 - 8) = \sqrt[3]{b^3 - 8 + 8} = \sqrt[3]{b^3} = b$$

In the next example we use the formula from physics for free-falling objects, which we study in detail in Section 2.3.

Example 5 The Poor Punt Problem

Suppose that a punter kicked the football so badly that the ball went straight up in the air. A tall player on the opposing team jumped up and caught the ball when it was 10 ft off the ground.

(a) Graph the function $s = -16(t - 2)^2 + 68$ that describes its height above the ground s in feet as a function of time t in seconds since the ball was kicked.

(b) Find the equation of the inverse relation and use it to find the time when the opponent caught the ball.

Thinking: Creating a Plan

(a) The graph of $s = -16(t - 2)^2 + 68$ is reflected about the t axis (the horizontal axis) and translated right 2 units and up 68 units so that the turning point is (2, 68). Let's choose $0 \le t \le 5$ and $0 \le s \le 70$ for the viewing window. (b) We find the equation of the inverse by solving the given equation for t. Then we find the inverse image(s) of $s = 10$.

Communicating: Writing the Solution

(a) See Figure 13.

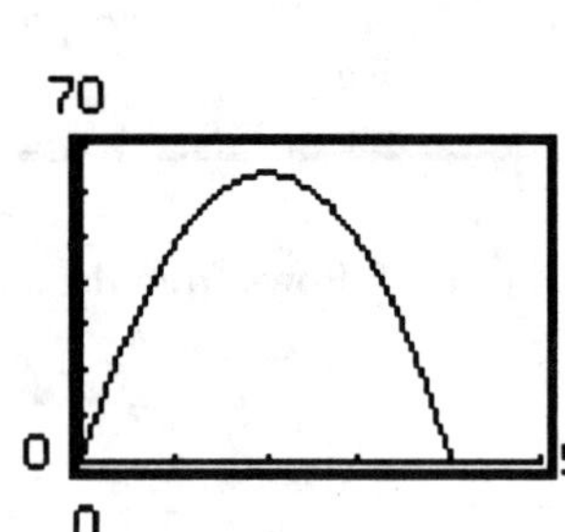

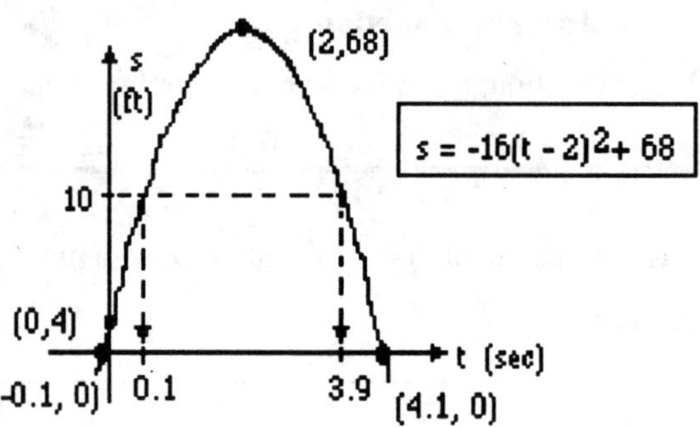

Figure 13

(b)

$$
\begin{aligned}
s &= -16(t - 2)^2 + 68 \\
16(t - 2)^2 &= 68 - s \\
(t - 2)^2 &= \frac{68 - s}{16} \qquad \textit{Take the square root of both sides.} \\
t - 2 &= \pm\frac{\sqrt{68 - s}}{4} \\
t &= 2 \pm \frac{\sqrt{68 - s}}{4}
\end{aligned}
$$

The equation of the inverse relation is $t = 2 \pm \frac{\sqrt{68 - s}}{4}$.

$t = 2 \pm \frac{\sqrt{68 - s}}{4}$ and $s = 10 \Rightarrow$

$$t = 2 \pm \frac{\sqrt{68 - 10}}{4}$$

$$t = 2 - \frac{\sqrt{68 - 10}}{4} \quad \text{or} \quad t = 2 + \frac{\sqrt{68 - 10}}{4}$$

$$t \approx 0.1 \quad \text{or} \quad t \approx 3.9$$

There are two times when the ball was 10 ft off the ground, on the way up at $t \approx 0.1$ sec and on the way down at $t \approx 3.9$ sec. The receiver probably caught the ball 3.9 sec after it was kicked, on its way down.

Learning: Making Connections

There are two inverse images for s = 10; both are found from the equation of the inverse because the inverse relation is not a function. The fact that there is more than one inverse image is important to this situation.

Explore the Concepts

These exercises explore relations and their inverses.

E1. Graph y = 2x, the function that multiplies by 2.

(a) Find the value for y when x = 5. Find x when y = 10.

(b) Find the value for y when x = 20. Find x when y = 40.

(c) What arithmetic operation finds x given y? The function that performs this operation is called the inverse function for y = 2x.

E2. Graph y = 2x + 3, the function that multiplies by 2 and adds 3.

(a) Find the value for y when x = 2. Find x when y = 7.

(b) Find the value for y when x = 20. Find x when y = 43.

(c) What arithmetic operation finds x given y? The function that performs this operation is called the inverse function for y = 2x + 3.

E3. Graph $y = \sqrt{x}$, the square-root function.

(a) Find the value for y when x = 9. Find x when y = 3.

(b) Find the value for y when x = 100. Find x when y = 10.

(c) What algebraic operation finds x given y? The function that performs this operation is called the inverse function for $y = \sqrt{x}$, $y \geq 0$.

E4. Use your results from Exercises E1 to E3 to complete the following statement.
Given a function f and its inverse f^{-1}, if the ordered pair (a, b) belongs to f, then (___,___) belongs to f^{-1}.

Exercises 1.5

Develop the Concepts

For Exercises 1 to 4 sketch the graph of functions f and g on one coordinate system. Graph the new functions f + g and f - g.

1. $f(x) = x - 3,\ g(x) = x + 4$
2. $f(x) = |x - 2|,\ g(x) = |x|$
3. $f(x) = |x + 3|,\ g(x) = |x - 3|$
4. $f(x) = \frac{1}{x},\ g(x) = -\frac{1}{x}$

For Exercises 5 to 8 graph each product function f · g. Explain the product function by analyzing the signs of each factor function, f and g.

5. $f(x) = x - 1,\ g(x) = (x + 1)^2$
6. $f(x) = x + 2,\ g(x) = (x - 2)^3$
7. $f(x) = (x - 1)^2,\ g(x) = (x + 1)^3$
8. $f(x) = x,\ g(x) = \sqrt{x + 1}$

For Exercises 9 to 16 find f(g(x)) and g(f(x)).

9. $f(x) = x^2 - 2,\ g(x) = x + 2$
10. $f(x) = \frac{2x - 1}{3},\ g(x)\frac{3x + 1}{2}$
11. $f(x) = \sqrt{x},\ g(x) = -x$
12. $f(x) = \sqrt{1 - x},\ g(x) = x + 1$
13. $f(x) = |x|,\ g(x) = 1 - x^2$
14. $f(x) = \sqrt[3]{x},\ g(x) = 2x - 2$
15. $f(x) = \sqrt{1 - x^2},\ g(x) = x + 1$
16. $f(x) = x^2 + 1,\ g(x) = \sqrt{x - 1}$

For Exercises 17 to 20 let $f(x) = \sqrt{x}$. Graph y = f(g(x)) and state the domain of the composition function.

17. $g(x) = x - 1$
18. $g(x) = 4 - x^2$

*19. $g(x) = 1 - \sqrt{x}$

20. $g(x) = 1 - |x|$

For Exercises 21 to 24 write each function as the composition of two functions f and g. (Answers may vary.)

21. $y = \sqrt{1 - 2x}$
22. $y = \sqrt[3]{x^2 - 4}$
23. $y = \sqrt{x} - 2$
24. $y = |x^2 - 4| - 5$

For Exercises 25 to 28 find the equation of the inverse by solving each equation for the other variable.

25. $C = \frac{5}{9}(F - 32)$, which changes temperatures F in degrees Fahrenheit to degrees Celsius C.
26. $c = 0.29m + 59$, which expresses the cost c of renting a car as a function of miles driven m.
27. $c = 20{,}000 + 200t$, which expresses the cost c of owning a car as a function of time t in months.
28. $v = -32(t - 2)$, which expresses the velocity v in ft/sec of a water balloon as a function of time t since it was tossed.

For Exercises 29 to 32 graph the function. Then find the image of each x and the inverse image of each y. Show the image and the inverse image on your graph. [Hint: Some are not one-to-one functions.]

29. For $f(x) = 2x - 10$ find $f(7)$ and $f^{-1}(4)$.

30. For $f(x) = \dfrac{x + 10}{2}$ find $f(2)$ and $f^{-1}(6)$.

31. For $f(x) = \sqrt[3]{x} + 1$ find $f(8)$ and $f^{-1}(2)$.

32. For $f(x) = \sqrt{1 - x^2}$ find $f(\frac{3}{5})$ and $f^{-1}(\frac{3}{5})$.

For Exercises 33 and 34 use the functions defined in the tables to find each value.

f:

x	0	3	5	7
f(x)	2	4	6	8

g:

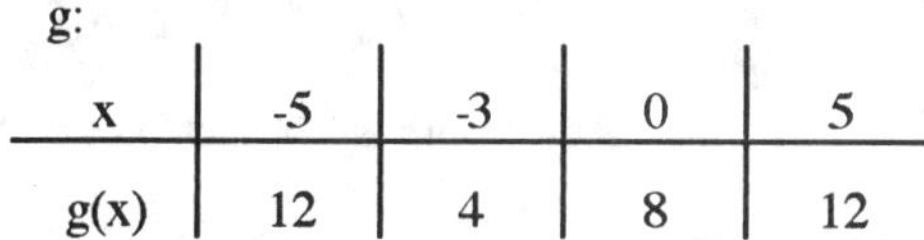

x	-5	-3	0	5
g(x)	12	4	8	12

33. (a) $f(3)$
    (b) $f^{-1}(6)$

34. (a) $g(-3)$
    (b) $g^{-1}(12)$

For Exercises 35 and 36 estimate the indicated values from the functions graphed in Figure 14.

35. (a) $f(-10)$
    (b) $f^{-1}(20)$

36. (a) $g(150)$
    (b) $g^{-1}(0)$

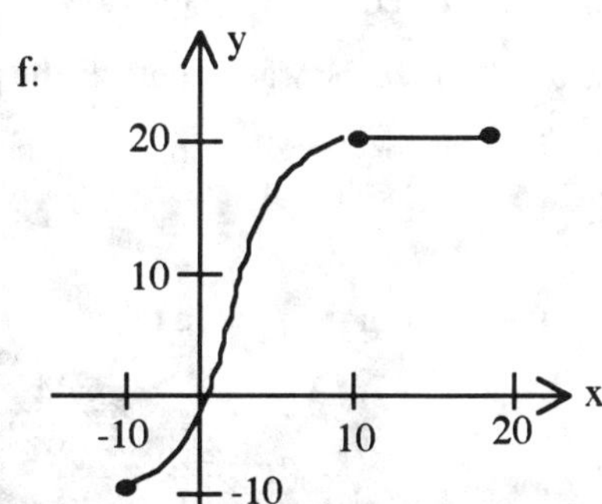

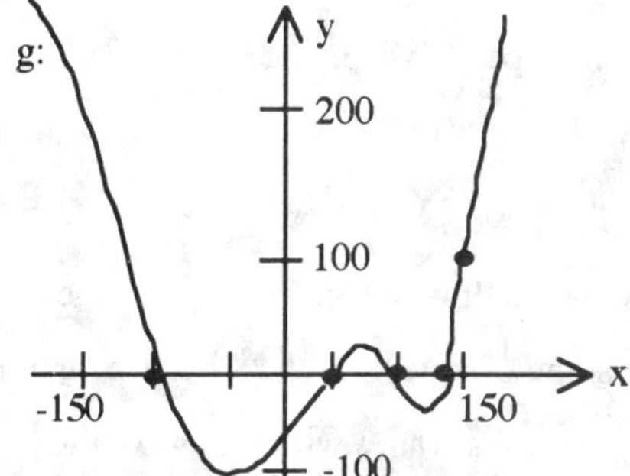

Figure 14

For Exercises 37 to 40 (a) find the equation of the inverse f^{-1} for each function; (b) prove $f^{-1}(f(a)) = a$; and (c) prove $f(f^{-1}(b)) = b$.

37. $f(x) = \dfrac{2x - 3}{4}$

38. $f(x) = 3x + 1$

39. $f(x) = 1 - x^3$

40. $f(x) = \dfrac{x^3 + 5}{4}$

Apply the Concepts

41. The Fuel Consumption Problem In boating, skippers need to know their fuel consumption in gal/hr. A skipper records the boat's fuel consumption in gallons and the distance traveled in nautical miles at various time intervals.

g: gallons	10	6	4
nautical miles	120	**80**	50

f: nautical miles	120	**80**	50
hours	6	4	2.5

Find the table of values of the composite function that links gallons to time, yielding fuel consumption in gal/hr.

42. The Clown's Kinetic Energy Problem The energy of motion, kinetic energy (KE) measured in joules, of an object is $KE = \frac{1}{2}mv^2$, where m is mass in kilograms and v is velocity in m/sec. At the circus a clown is fired from a cannon at a velocity of 30 m/sec. The clown's vertical velocity at any time t in seconds is given by $v = -9.8t + 30$. If the mass of the clown is 50 kg, find the kinetic energy at 1 sec, 3.06 sec, and 5 sec after firing. [Hint: $KE = f(v)$ and $v = g(t) \Rightarrow KE = f(g(t))$.]

Exercises 43 and 44 examine composite functions in a typical engineering problem: water flowing through a pipe. As water fills a pipe it comes in contact with the wall of the pipe, thereby slowing the flow of water through the pipe. Civil engineers must measure the wetted perimeter p (where the water contacts the pipe) to determine the flow of the water (Figure 15). The Chézy-Manning equation for fluid flowing in a pipe is

$$Q = \frac{1.486}{n} AR^{2/3}\sqrt{s}$$

where Q is the flow rate in ft^3/sec, A is the cross-sectional area of the water in the pipe in ft^2, s is the slope of pipe, R is the ratio A/p, and n is a friction factor that varies with the age of the pipe. The older the pipe, the more friction from residue. Assume that the friction factor is found using the formula

$$n = 0.015 + 0.001t$$

where t is the age of the pipe in years.

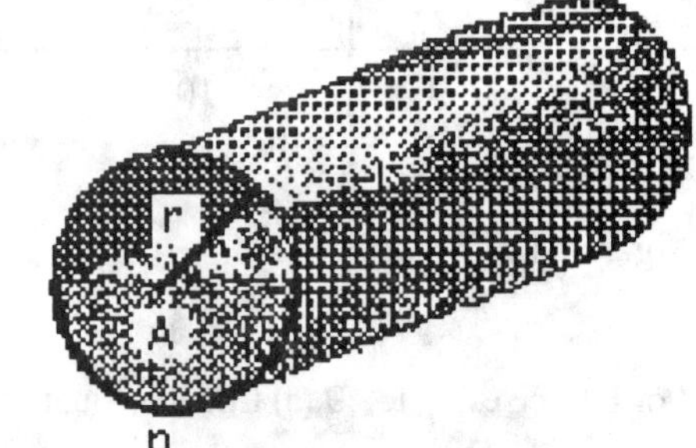

Figure 15

43. The Storm Sewer Pipe Problem Consider a half-full sewer pipe with a 1% slope, $s = 1\% = 0.01$, with a $A = 9$ ft^2 cross-sectional area of water in the pipe, and $p = \sqrt{18\pi}$ ft wetted perimeter.

(a) Find the flow rate of this pipe when it was new.

(b) Find the flow rate of this pipe when it is 10 years old.

44. The Wetted Perimeter Problem Assume that the flow rate Q in the pipe described in Exercise 43 is a constant 60 ft^3/sec.

(a) Find the equation of the function that expresses wetted perimeter p as a function of the friction factor n from the Chézy-Manning equation. [Hint: Solve the Chézy-Manning equation for p, where $R = A/p$, $s = 0.01$, and $A = 9\ ft^2$.]

(b) Calculate the wetted perimeter p for the pipe when it is new and when it is 10 years old using your equation from part (a), where $n = 0.015 + 0.001t$.

*45. The Mountain-Climbing Problem As you climbed Mt. Rainier (4250 m high), you noticed that the temperature dropped at an average rate of 5°C per 1000 m climbed. If the temperature is 15°C in nearby Seattle (sea level), the function $T = -5h + 15$ gives the Celsius temperature T as a function of height h in units of km (1000 m).

(a) Graph the function.

(b) Find the equation of the inverse function.

(c) What height is freezing level? (T = 0°C.)

(d) Show the inverse image of 10°C when you put on your long underwear.

46. The Small Craft Advisory Problem Ocean waves are usually generated by the wind. Empirical (actual) measurements show that the height H in feet of ocean waves is related to the velocity of the wind v in knots by $H = 0.025\ v^2$, $v \geq 0$.

(a) Graph the function.

(b) Find the equation of its inverse.

(c) The following categories are important to sailors. Label the categories on both axes.

Small craft advisory: From 15 to 33 knots

Gale warnings: From 33 to 48 knots

Storm warnings: From 48 to 64 knots

Hurricane warnings: Above 64 knots

47. The Rainbow Kite and Balloon Company Filling a balloon with helium at a constant rate is modeled by the function $R(r) = \frac{1}{\pi r^2}$, $r > 0$, where R is the rate and r is the radius of the balloon.

(a) Find the equation of the inverse.

(b) Is the inverse a function?

(c) Find the equation of the inverse of the new function $R_2 = 0.5R(r)$.

48. Wonderworld Amusement Park

The speed of a merry-go-round is measured in revolutions per minute (rpm). The speed in mi/hr of a rider depends on the distance from the center of the merry-go-round as described by the following data for 6 rpm.

distance from center, r (ft)	0	5	10	15	20
speed, S(r) (mi/hr)	0	2.1	4.2	6.4	8.6

(a) Create a table for the inverse.

(b) Describe the inputs and outputs for the inverse

(c) Create a table for the new function $r = S^{-1}(0.5S)$.

(d) Describe the change modeled by the equation in part (c).

Extend the Concepts

49. For $f(x) = \frac{3x - 4}{5}$ find

(a) $(cf(x))^{-1}$

(b) $(f(cx))^{-1}$

50. For $f(x) = 3x$ and $g(x) = \frac{-5x}{4}$ show

(a) $(f + g)^{-1}(y) \neq f^{-1}(y) + g^{-1}(y)$

(b) $(f \circ g)^{-1} = g^{-1} \circ f^{-1}$

For Exercises 51 and 52 (a) find the equation of $y = f(g(x))$ and (b) give the domain and range of f, g, and f(g).

51. $y = f(u) = \sqrt{u}$, where $u = g(x) = x - 1$

52. $y = f(u) = u - 1$, where $u = g(x) = \sqrt{x}$

A Guided Review of Chapter 1

1.1 Functions and Their Properties

For Exercises 1 to 6 determine whether or not the following relations are functions. For those that are functions decide whether or not they are one-to-one functions.

1. $f = \{(-1,2), (-1,3), (2,2)\}$
2. $g = \{(0,1), (1,1), (2,1)\}$
3. $y = 2x^2 - 4$
4. $y^2 = x$

5.(a) A turkey weighing 16 lb should take between $3\frac{1}{2}$ and $4\frac{1}{2}$ hr to cook.

(b) A stuffed turkey takes about 20 min/lb to cook.

6.

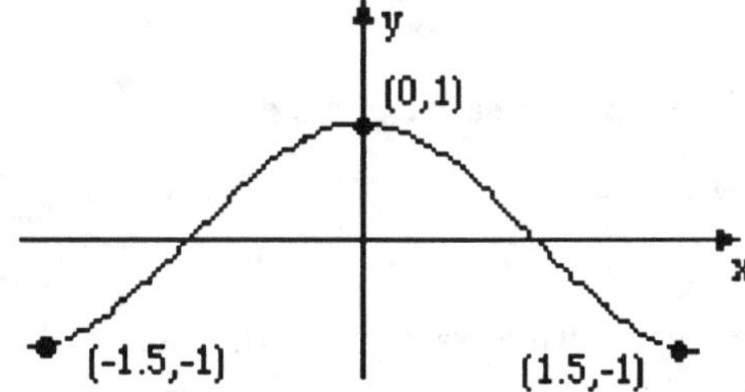

For Exercises 7 and 8 investigate the domain of each function geometrically and verify algebraically.

7. $y = \sqrt{1 - x}$
8. $y = \dfrac{1}{x - 1}$

9. The Record-Breaking Mile Run Problem The equation $T = -0.4d + 246$ models the record time T in seconds for running the mile as a function of the year d, where d = 0 corresponds to the year 1913.

(a) Find and interpret the intercepts.

(b) Use your results from part (a) to help create the graph on a graphing utility that displays the domain and range of this application and approximate the missing coordinates for points B and C.

(c) If the record has been decreasing in the past, do you think it will continue to decrease in the future?

(d) Replace the dotted portion of the graph in Figure 1 with your guess as to what the graph will probably resemble in the future.

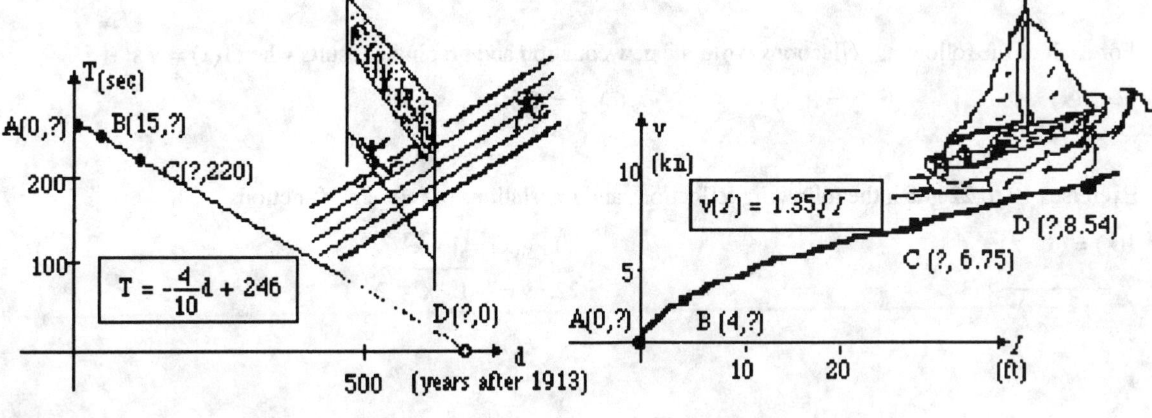

Figure 1

Figure 2

10. The Hull Speed Problem The hull speed v of a sailboat in knots is related to the waterline measurement of the length l of the boat by the function $v(l) = 1.35\sqrt{l}$ (Figure 2).

(a) Find the missing coordinates for points A and D.

(b) Use your results from part (a) to help create the graph on a graphing utility that displays the domain and range of this application if the longest sailboat is a 613-ft Club Med I.

(c) Approximate the missing values for points B and C.

(d) From points A to D, this graph is increasing at a decreasing rate. What does this say about creating faster boats by building them longer and longer?

1.2 Functional Notation and Mathematical Language

11. For $f(x) = -x^2 + 1$ use functional notation to find the simplified form of each expression.

(a) $f(3)$ (b) $f(-x)$ (c) $-f(x) + 2$ (d) $f(x + 3)$

For Exercises 12 and 13 state the conclusions that follow from the following assumptions (hypotheses).

12. $f(x) = 2x + 3$ and $f(x) = 0 \Rightarrow$

13. $x = \dfrac{-b \pm \sqrt{b^2 - 4ac}}{2a}$ and $a = 2, b = 0, c = -1 \Rightarrow$

For Exercises 14 to 16 use notation to communicate finding the point(s) of intersection.

14. $f(x) = x^2 - 4$ and $g(x) = x + 2$

15. $f(x) = \sqrt{x - 2}$ and $g(x) = x - 8$

16. $f(x) = \dfrac{1}{(x - 2)^2}$ and $g(x) = \dfrac{1}{x}$

1.3 Translations and Reflections of Graphs of Functions

17. For each of the following translations write the new equation and graph the results when $f(x) = x^2$.

(a) $y = f(x) + 2$ (b) $y = f(x + 2)$ (c) $y = f(x - 2)$ (d) $y = f(x) - 2$

18. For each of the following reflections write the new equation and graph the results when $f(x) = \sqrt{x} + 5$.

(a) $y = -f(x)$ (b) $y = f(-x)$ (c) $y = -f(-x)$

For Exercises 19 to 22 graph the following reflections and translations of reference functions.

19. $f(x) = (1 - x)^2$

20. $g(x) = 1 - x^3$

21. $y = \dfrac{-1}{(2 - x)^2} + 3$

22. $y = \sqrt{1 - x} - 2$

Functions f and g are defined by the following tables. For Exercises 23 and 24 write a table of values for each new function.

x	-5	0	5
f(x)	10	0	10

x	-100	-50	50	100
g(x)	-200	0	100	200

23.(a) $y = f(x) + 5$

(b) $y = f(x - 5)$

24.(a) $y = -g(x)$

(b) $y = -g(-x)$

For Exercises 25 to 28 find the simplest equation that matches the shape and position of each graph.

25. 26.

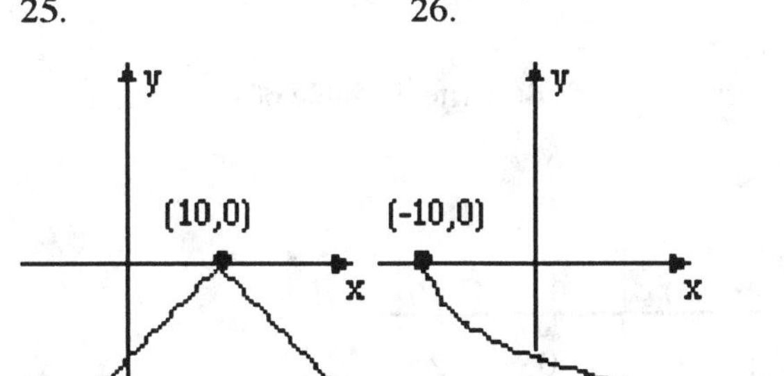

27. 28.

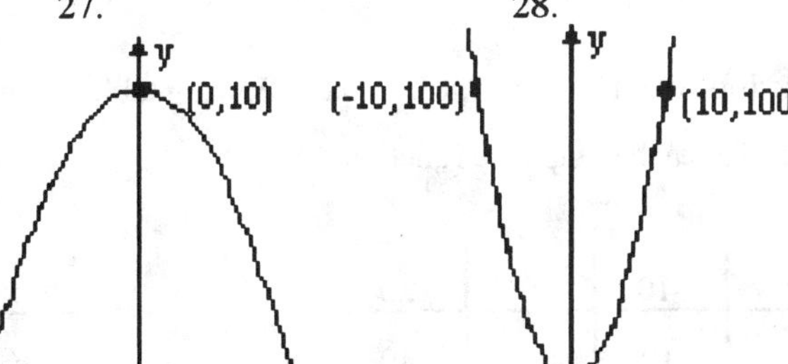

For Exercises 29 to 31 investigate graphically and confirm algebraically whether each function is even, odd, or neither even nor odd.

29. $f(x) = 4 - x^2$

30. $f(x) = |x - 4|$

31. $f(x) = x^3 - x$

1.4 Changes of Scale of Graphs of Functions

For Exercises 32 to 35 graph each function by relabeling the x or y axis as indicated by the scaling factor present in each equation.

32. $f(x) = \sqrt{15x}$

33. $g(x) = 100x^3$

34. $y = \dfrac{0.2}{x}$

35. $y = \dfrac{1}{(5x)^2}$

Function h is defined by the following table. For Exercises 36 to 38 write a table of values for each new function.

x	-2	0	2	4	8
h(x)	9	3	1	1/3	1/9

36. $y = 9h(x)$

37. $y = \dfrac{1}{3}h(x)$

38. $y = h(2x)$

39. Consider the following table that expresses mileage as a function of the age of a car.

age (yr)	1	2	3	5	10
mi/gal	26	25	24.5	23	21

(a) Create a graph that depicts this information about a car.

(b) Create a graph that exaggerates the mileage.

1.5 Operations Compositions and Inverse Functions

For Exercises 40 to 42 let $f(x) = |x - 1|$ and $g(x) = |x + 2|$.

40. Graph the sum function f + g.

41. Graph the difference function f - g.

42. Graph the product $f \cdot g$, and explain the results in terms of the individual factors, functions f and g.

For Exercises 43 and 44 find each composition for $f(x) = 14 + 2x$ and $g(x) = \frac{x}{2} - 7$.

43. $y = f(g(x))$

44. $y = g(f(x))$

Functions f and g are defined by the following tables. For Exercises 45 and 46 write the table of values that exists for each composite function.

x	-10	0	20
f(x)	10	0	40

x	-10	0	10
g(x)	0	10	20

45. $y = f(g(x))$

46. $y = g(f(x))$

For Exercises 47 and 48 let $f(x) = x^2$ and $g(x) = \sqrt{x}$.

47. Show that $f(g(b)) = b$.

48. Show that $g(f(a)) \neq a$.

49. (a) Find the inverse of $y = 9 - x^2$.

    (b) Is the inverse a function? Why or why not?

50. Draw the graph of $f(x) = 14 - 2x$. Show the image of $x = 5$ and the inverse image of $f^{-1}(10)$ on the graph.

Chapter 1 Test

For Exercises 1 to 4 sketch the graph of each function. Label one point on each graph.

1. $f(x) = |x - 20| + 5$ 2. $y = 5 - \sqrt{-x}$ 3. $g(x) = (10 - x)^3$ 4. $y = -3(x - 2)^2 - 1$

For Exercises 5 to 12 functions f and g are defined by the graphs in Figure 1.

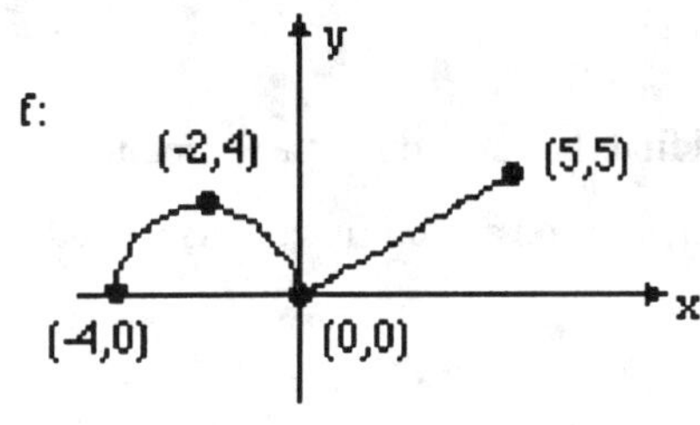

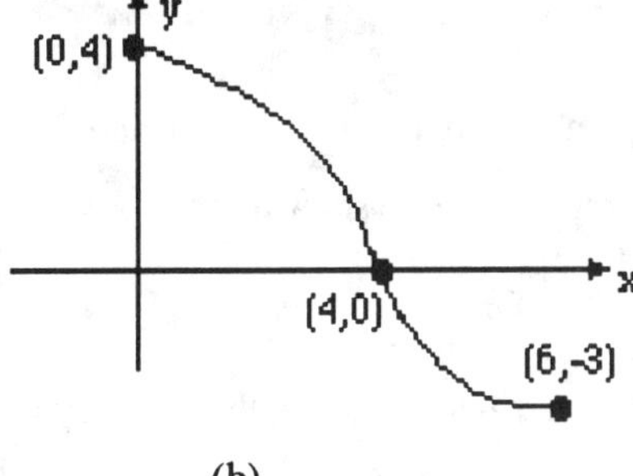

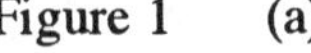

Figure 1 (a) (b)

5. Graph $y = f(-x) + 1$. 6. Graph $y = f(1 - x)$. 7. Graph $y = 2g(x)$. 8. Graph $y = g(x/2) - 1$.

9. Which function, f or g, is one-to-one?

10. State the interval(s) over which function f in increasing.

11. State the location of the turning points on the graph of function f.

12. Which function is an even function?

13. Find the x and y intercepts of $f(x) = \frac{1}{x - 4} + 2$ using appropriate notation to state the assumptions.

14. Use notation to communicate finding the point(s) of intersection of functions $f(x) = \frac{1}{x - 1} + 2$ and $g(x) = \frac{1}{x - 1}$.

15. The Fetching Problem Fetch is the uninterrupted distance the wind blows across the water, as shown in Figure 2. The maximum height H in feet of ocean waves generated by wind is a function of the maximum fetch F in nautical miles as defined by $H = 1.5\sqrt{F}$.

(a) How high a wave could be produced by wind blowing over a fetch of 25 nautical miles?

(b) The largest waves are around 60 to 70 ft high. What fetch is required to produce such waves?

[Source: *Van Nostrand's Scientific Encyclopedia*]

(c) Use the results from part (b) to graph this function for the domain of the application.

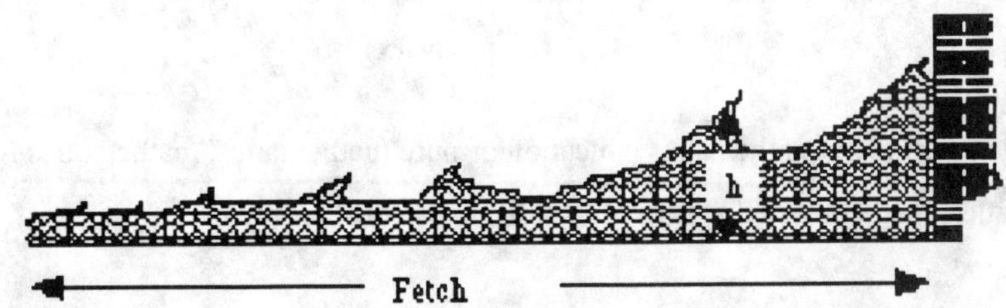

Figure 2

For Exercises 16 and 17 evaluate f(-5), g(2), f(g(2)), and g(f(-5)).

16. $f(x) = \frac{1 - x}{3}$

$g(x) = 1 - 3x$

17. $f(x) = \frac{x^2 - 5}{10}$

$g(x) = -\sqrt{10x + 5}$

18. Investigate graphically and confirm algebraically whether $f(x) = \frac{1}{x^2} - 15$ is an even function, odd function, or neither even nor odd.

19.(a) State the equations of two functions, one even and the other odd, and find the equation of their product.

(b) Graph their product from part (a) to show that the product of an even function and an odd function is an odd function.

20. Sketch the graph of a one-to-one function that is everywhere increasing and passes through the origin. Its domain is $(-\infty, \infty)$ and its range is $(-2, \infty)$. There is a horizontal asymptote at $y = -2$.

21. The Rule-of-Thumb Problem A temperature given in degrees Celsius can be converted to degrees Fahrenheit by doubling and adding 30.

(a) Write an equation for this rule of thumb.

(b) Write an equation for the inverse function.

(c) State the rule of thumb for the inverse function in words.

22. The Salmon Cookbook Problem In The Salmon Cookbook, author J. McNair recommends that salmon be cooked approximately 10 min for each inch of thickness.

(a) Write the equation that expresses cooking time T in minutes as a function of the size of the fish F in inches.

(b) Find the equation of the inverse function.

(c) Describe two situations, one in which you would use the function from part (a), and another in which you would use its inverse from part (b).

23. Functions f and g are defined by the following tables. Write the table of values that exists for each composite function.

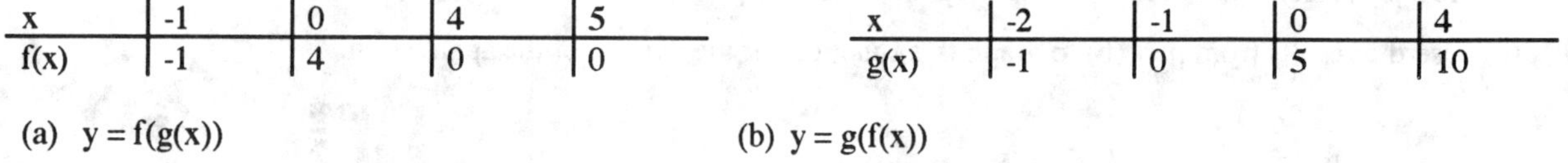

x	-1	0	4	5
f(x)	-1	4	0	0

x	-2	-1	0	4
g(x)	-1	0	5	10

(a) $y = f(g(x))$

(b) $y = g(f(x))$

24. Compare the roles in modern mathematics education of pure mathematics, mathematics for its own sake, and applied mathematics, mathematics that is used to solve real-world problems.

Chapter 2

Linear and Quadratic Functions

Chapter 2 focuses on specific attributes of linear and quadratic functions and their applications. Problem-solving examples and exercises become more open-ended and more realistic. We study models from microeconomics that include how to price a product to maximize profit or minimize cost, models from physics for free-falling objects where we find maximum height above the ground, and other such optimization problems. We find equilibrium points in situations from ecology, marketing, biology, and more. Graphing utilities allow us to find the *best-fitting* line for a given set of data as well as the *best-fitting* parabola for other sets of data.

A Historical Note on Karl Friedrich Gauss (1777-1855)

Even during his own lifetime, Gauss was considered the Prince of Mathematicians. At the age of 3, Gauss revealed his mathematical precocity when he found an error in his father's calculations of a payroll by noticing a break in the mathematical patterns.

In 1801, when the asteroid Ceres was discovered it was visible for 40 days, but astronomers then lost it. They could not compute its orbit. It took Gauss only 3 days to invent the method of finding the best-fitting line for a set of data, and calculate the orbit of Ceres. This is called the method of least squares - best fit.

Gauss also invented the telegraph, contributed to astronomy and the mathematics of magnetic fields, formalized a method of solving systems of linear equations now called *Gaussian reduction* that is still used today, and established the complex numbers by representing them as ordered pairs in the plane. He also extended the idea of prime numbers to the complex numbers, where real prime numbers such as $5 = (2 + i)(2 - i)$ are not prime complex numbers. He was the first to prove the *Fundamental Theorem of Algebra* and its corollary, which states that a polynomial equation of degree n has exactly n solutions. He pursued noneuclidean geometry, which helped Einstein formulate his theories on relativity. He also introduced the bell-shaped curve, the beginning of the study of statistics. The list of his contributions goes on and on.

Much of Gauss's work was in the field of number theory, although he worked with ease in all areas of pure and applied mathematics of his time. He said, "Whether I apply mathematics to a few clods of soil that we call planets or to purely mathematical problems, is in itself unimportant, but the latter application gives me great pleasure."

2.1 Linear Functions

Finding the Slope of a Linear Equation
Using Slope-Intercept Form of a Linear Equation
Using Point-Slope Form to Find the Equation of a Line
Graphing Piecewise-Defined Functions

In this section on linear functions we include functions defined in pieces whose equations change over portions of their domains. We use linear functions for more mathematical models than all other functions combined, even when the given information is not quite linear. In this section our applications of linear functions include consumer demand curves from economics (Figure 1a), consumer advocate models (Figure 1b), and ecosystems from ecology (Figure 1c). Economists are interested in consumer demand functions, where the number of items sold decreases as the price increases, and supply curves, where the number of items supplied increases as the price increases (Example 2). In Figure 1a, consumer advocates model various options for the costs of services, such as the two billing options from one cellular phone company (Example 5). One option has a higher monthly fee but includes 30 min of airtime. The other bills consistently at 39¢/min.

In Exercises 47 and 48 we examine the ecosystems on an island where all arachnids (spiders) have been terminated. The subsequent rate immigration from nearby populations of arachnids and the rate of extinction on the island depend on the number of species of arachnids present. Although the three models described seem unrelated, the underlying mathematics is the same.

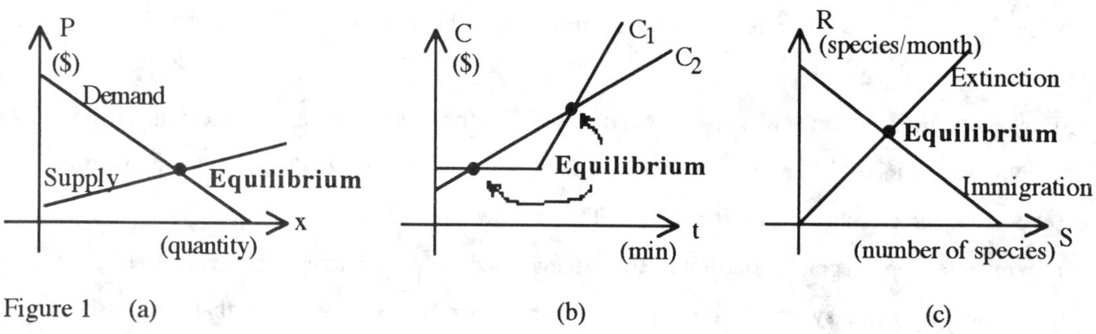

Figure 1 (a) (b) (c)

Finding the Slope of a Linear Equation

Functions that can be written in the form $Ax + By = C$, where A, B, and C are real numbers, $B \neq 0$, are called **linear functions** because their graphs are straight lines. This **standard form** is an **implicit function** because it does not express either the x or y variable explicitly as a function of the other. We often solve such an equation for y, creating an explicit function. For example, the equation $2x + 3y = 6$ can express y as a function of x explicitly as $y = -\frac{2}{3}x + 2$. We may also express x as the inverse function of y. It is our choice.

Slope measures relative rate of change between two quantities. Positive slopes describe increasing linear functions (y increases as x increases), while negative slopes describe decreasing linear functions (Figure 2). The formula for slope computes the ratio of rise to run.

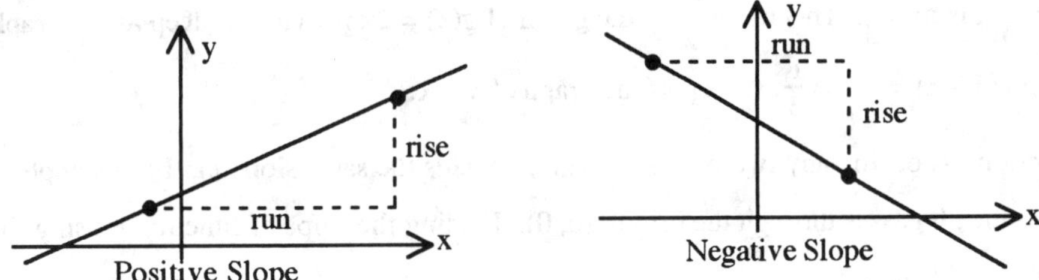

Figure 2

The *rise* is the change in y, Δy read "delta y," and the *run* is a change in x, Δx read "delta x." We use the letter *m* for slope, probably from the French word *monter*, meaning to climb.

$$m = \frac{\text{rise}}{\text{run}} = \frac{\Delta y}{\Delta x}$$

In the following definition of slope, x_1 is a specific value of x and $f(x_1)$ is the corresponding value of y. The restriction $x_1 \neq x_2$ avoids division by zero.

Definition of Slope

The **slope** m of a line containing points $(x_1, f(x_1))$ and $(x_2, f(x_2))$ is

$$m = \frac{\Delta y}{\Delta x} = \frac{y_2 - y_1}{x_2 - x_1} = \frac{f(x_2) - f(x_1)}{x_2 - x_1}, \quad x_1 \neq x_2$$

Consider the three graphs in Figure 3.

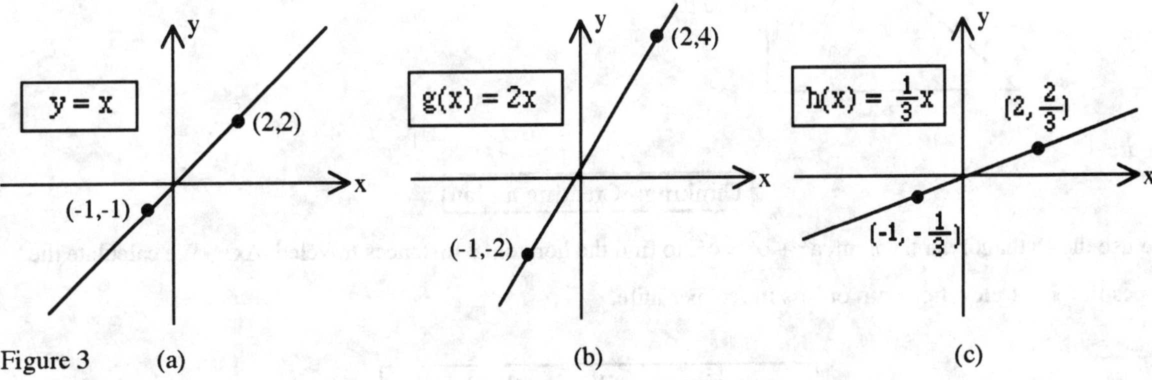

Figure 3 (a) (b) (c)

We can calculate the slope of each line in Figure 3 from the labeled points.

(a) For (-1, -1) and (2, 2), $m = \frac{\Delta y}{\Delta x} = \frac{y_2 - y_1}{x_2 - x_1} = \frac{2 - (-1)}{2 - (-1)} = \frac{3}{3} = 1$

(b) For (-1, -2) and (2,4), $m = \frac{\Delta y}{\Delta x} = \frac{g(x_2) - g(x_1)}{x_2 - x_1} = \frac{g(2) - g(-1)}{2 - (-1)} = \frac{4 - (-2)}{2 - (-1)} = \frac{6}{3} = 2$

(c) For $(-1, -\frac{1}{3})$ and $(2, \frac{2}{3})$, $m = \frac{\Delta y}{\Delta x} = \frac{h(x_2) - h(x_1)}{x_2 - x_1} = \frac{h(2) - h(-1)}{2 - (-1)} = \frac{2/3 - (-1/3)}{2 - (-1)} = \frac{1}{3}$

What is the relationship between the slope of each line in Figure 3 and its scaling factor m in y = mx?

The slope and the scaling factor are the same. The slope of y = x is m = 1, the slope of g(x) = 2x is m = 2, and the slope of $h(x) = \frac{1}{3}x$ is $m = \frac{1}{3}$. This means that the graph of g(x) = 2x is twice as steep as the graph of y = x, and the graph of $h(x) = \frac{1}{3}x$ is $\frac{1}{3}$ as steep as the graph of y = x.

For linear functions, choosing any two points on a graph yields the same slope m. For example, the graph of function g in Figure 3b passes through the origin, (0, 0). Finding the slope of function g using the points (0, 0) and (2, 4) also yields m = 2,

$$m = \frac{\Delta y}{\Delta x} = \frac{4 - 0}{2 - 0} = \frac{4}{2} = 2$$

Linear functions increase or decrease at a constant rate called *slope.* Applications of slope often include the word "per," such as traveling a constant rate of 55 mph or spending $300 per month for rent.

Example 1 The Two Trails Problem

In the Cascade Mountains, Clear Lake Trail is 2.5 mi long with a vertical rise of 800 ft (0.15 mi), while the trail to Flat Iron Lake is 0.75 mi with a vertical drop of 200 ft (0.038 mi) (Figure 4). Which trail has the steeper slope?

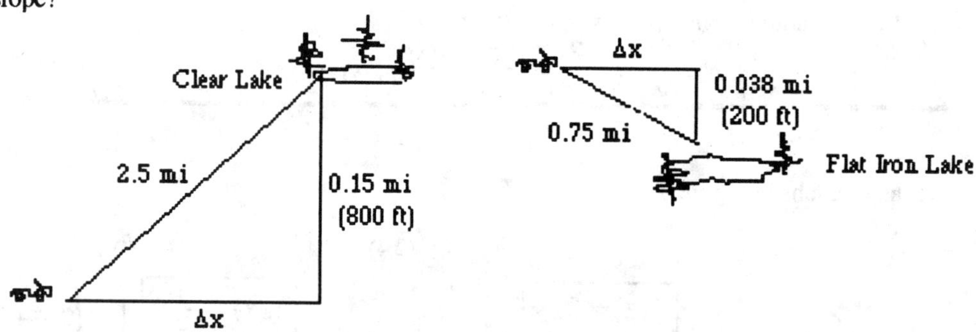

Figure 4

Thinking: Creating a Plan

We use the Pythagorean theorem, $a^2 + b^2 = c^2$, to find the horizontal distances traveled, Δx. We calculate the slopes in feet of elevation gain or loss in feet per mile.

Communicating: Writing the Solution

Clear Lake Trail

$a^2 + b^2 = c^2$ and $a = \Delta x$, $b = 0.15$, $c = 2.5 \Rightarrow$

$$(\Delta x)^2 + (0.15)^2 = 2.5^2$$

$$\Delta x \approx 2.495$$

Flat Iron Lake Trail

$a^2 + b^2 = c^2$ and $a = \Delta x$, $b = -0.038$, $c = 0.75 \Rightarrow$

$$(\Delta x)^2 + (-0.038)^2 = (0.75)^2$$

$$\Delta x \approx 0.749$$

The slope of Clear Lake Trail is

$$m = \frac{\Delta y}{\Delta x} = \frac{800 \text{ ft}}{2.495 \text{ mi}} \approx 320.6$$

This is an elevation gain of 320 ft/mi.

The slope of Flat Iron Lake Trail is

$$m = \frac{\Delta y}{\Delta x} = \frac{-200 \text{ ft}}{0.749 \text{ mi}} \approx -267$$

This is a loss in elevation of 267 ft/mi.

The Clear Lake Trail is steeper than the Flat Iron Lake Trail.

Learning: Making Connections

It is common to state slopes of hills in percentages, requiring both Δy and Δx to have the same units of measure. In this case, if we convert both Δy and Δx to miles, we find that the trail to Clear Lake has a 6% elevation gain, $m = \frac{0.15}{2.495} \approx 0.06$, and the trail to Flat Iron Lake has a 5% elevation drop, $m = \frac{-0.038}{0.749} \approx -0.05$. The choice of dimensions is an important part of the modeling process. If these trails were roads, we would refer to their slopes as *grades*.

If a trail in Example 1 were level, what would its slope be? What slope would a vertical cliff have?

The slope of a horizontal line is zero (Figure 5a); its graph neither increases nor decreases. A vertical line is as steep as a line can be; its slope is undefined (Figure 5b).

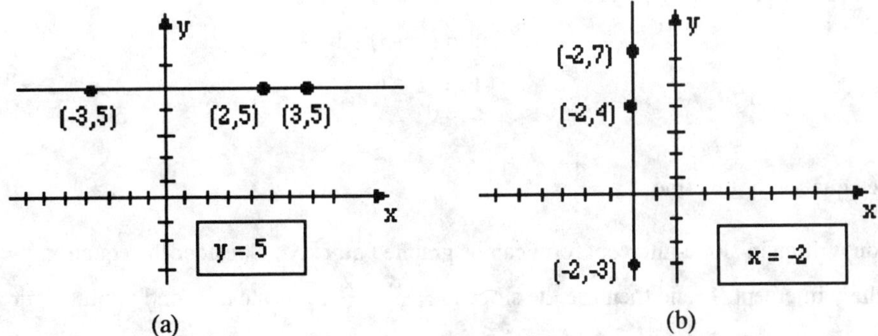

Figure 5 (a) (b)

Calculating the slopes of the lines in Figure 5 yields expected results. Using the points (2, 5) and (3, 5) on the horizontal line in Figure 5a, we have

$$m = \frac{\Delta y}{\Delta x} = \frac{5 - 5}{3 - 2} = \frac{0}{1} = 0$$

Using the points (-2, 4) and (-2, 7) on the horizontal line in Figure 5b yields

$$m = \frac{\Delta y}{\Delta x} = \frac{7 - 4}{-2 - (-2)} = \frac{3}{0} = \infty$$

Horizontal and Vertical Lines

For a fixed real number c, the equation of a horizontal line is of the form y = c and has slope m = 0. The equation of a vertical line is of the form x = c and its slope is undefined.

Using Slope-Intercept Form of a Linear Equation

When a nonvertical linear equation is solved for y, it is in slope-intercept form.

Slope-Intercept Form

A linear function written as

$$y = mx + b$$
$$f(x) = mx + b$$

is in **slope-intercept form**. The slope is m and the y intercept is b.

We can use the slope-intercept form to find the equation of a line passing through two points. We find the values of m and b. For the line passing through (-3, -1) and (6, 2), we first find the slope

$$m = \frac{\Delta y}{\Delta x} = \frac{2 - (-1)}{6 - (-3)} = \frac{3}{9} = \frac{1}{3}$$

We can find the y intercept, b, using either one of the given points. Using the point (-3, -1), we have y = mx + b and m = 1/3, x = -3, y = -1 ⇒

$$-1 = (\tfrac{1}{3})(-3) + b$$
$$-1 = -1 + b$$
$$b = 0$$

Therefore, the equation of this line is $y = \frac{1}{3}x$.

An equation written in slope-intercept form can be graphed quickly. Consider the equation y = 2x + 1. First, locate the y intercept, 1, and then use the slope $m = \frac{\Delta y}{\Delta x} = \frac{2}{1}$ to locate a second point (Figure 6).

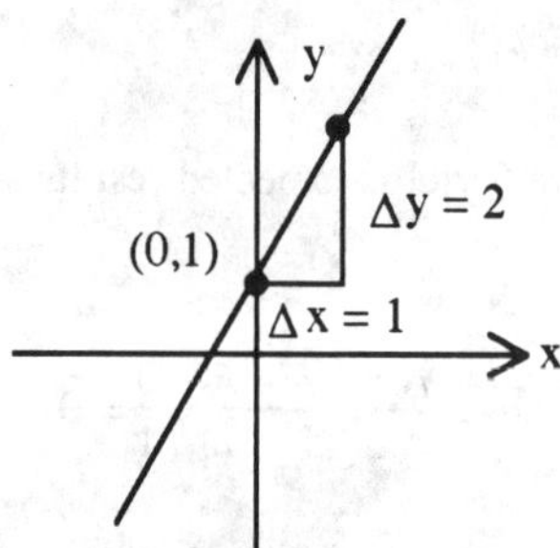

Figure 6

Notice that the equation y = 2x + 1 is a vertical translation of the graph of y = 2x up 1 unit (Figure 7a). The graphs y = 2x and y = 2x + 1 are parallel since each has a slope of 2. If the graphs of two linear equations increase (or decrease) at the same rate, they will never intersect. (See Exercise 59.)

What is the relationship of the lines defined by the equations $y = 2x$ *and* $y = -\frac{1}{2}x$*?*

The graph of the line $y = -\frac{1}{2}x$ is drawn perpendicular to the line y = 2x (Figure 7b). Its slope, m = -1/2, is the negative reciprocal of 2.

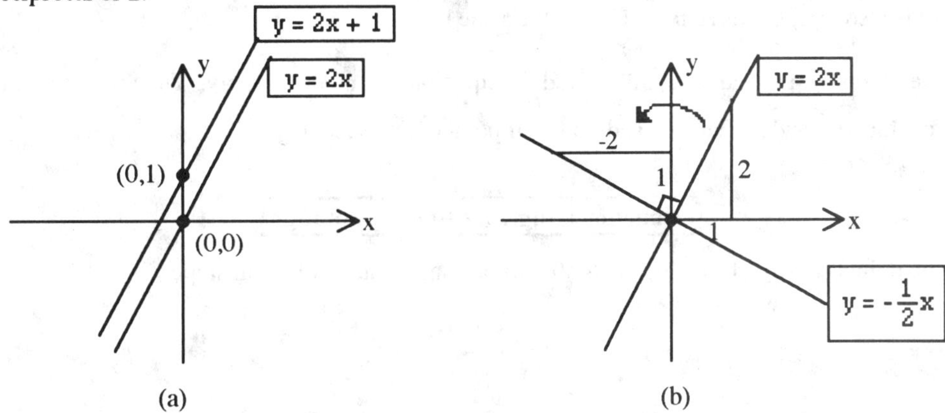

Figure 7

If we rotate the line y = 2x counterclockwise to fall on the line $y = -\frac{1}{2}x$, notice that the $\Delta y = 2$ becomes $\Delta x = -2$, and $\Delta x = 1$ becomes $\Delta y = 1$ (Figure 7b). (See Exercise 60.)

Slopes of Parallel Lines and Perpendicular Lines

Two nonvertical lines are parallel if and only if their slopes are equal.

Two nonvertical, nonhorizontal lines are perpendicular if and only if their slopes are negative reciprocals.

The next example revisits one of the three situations in the introduction. We use slope-intercept form to write the equations of two lines. We then find their point of intersection, the equilibrium point.

Example 2 The Tie-Dyed T-Shirts Problem

This summer you plan to make and sell tie-dyed T-shirts at street fairs. You have collected the following data about consumer demand, the number of shirts people are willing to buy at various prices. (Notice that if the price were $40.00, they wouldn't buy any shirts.)

quantity, x	0	150	450
price, p	40	30	10

Studying your costs leads to the number of shirts you are willing to supply at various prices. You decide that if the selling price is $10 or less, you will not make any shirts, but if the selling price is $30, you will make 600 shirts. Find market equilibrium, the price at which you and consumers are willing to supply and buy the same number of shirts.

Thinking: Creating a Plan

First we find the linear equation that models demand. Since the slope of the line passing through the points from (0, 40) and (150, 30) is $m = \frac{\Delta p}{\Delta x} = \frac{-10}{150} = -\frac{1}{15}$ and the slope of the line passing through the points (150, 30) and (450, 10) is also $-\frac{1}{15}$, the three points lie on the same line.

We use the points (0, 10) and (600, 30) to find the equation that models supply. Then we graph the equations modeling demand and supply, and find their point of intersection.

Communicating: Writing the Solution

Demand: The points (0, 40), (150, 30), and (450, 10) fall on the same line with slope

$$m = \frac{\Delta p}{\Delta x} = -\frac{1}{15}$$

Since (0, 40) is given, we know that the p intercept is 40. Substituting into the slope-intercept form yields the equation of the demand function:

$$p_1(x) = -\frac{1}{15}x + 40$$

Supply: The slope of the line passing through the points (0, 10) and (600, 30) is

$$m = \frac{\Delta p}{\Delta x} = \frac{20}{600} = \frac{1}{30}$$

and the y intercept is 10. Substituting into the slope-intercept form yields

$$p_2(x) = \frac{1}{30}x + 10$$

Figure 8 shows the point of intersection, (300, 20). Equilibrium is attained when you make 300 shirts and price them at $20.00 each.

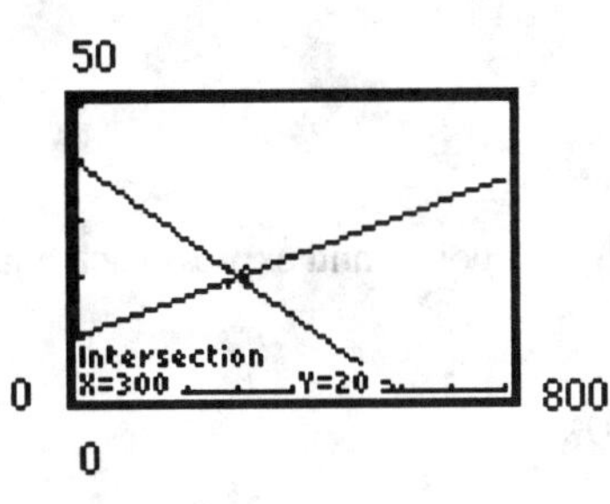

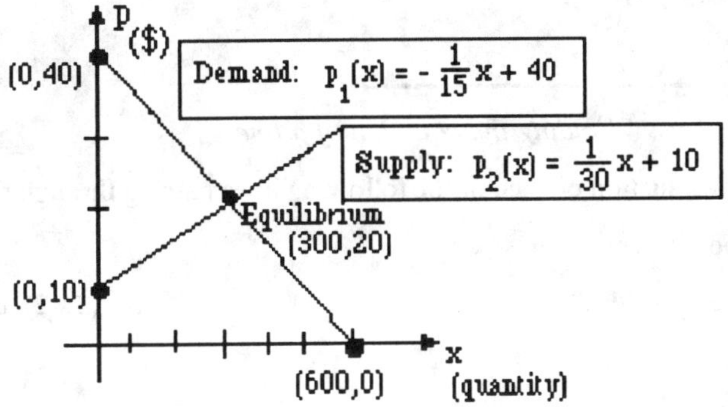

Figure 8

Learning: Making Connections

We can verify that (300, 20) is the point of equilibrium, by substitution.

$$p_1(x) = -\frac{1}{15}x + 40$$

$$p_1(300) = -\frac{1}{15}(300) + 40 = -20 + 40 = 20$$

and

$$p_2(x) = \frac{1}{30}x + 10$$

$$p_2(300) = \frac{1}{30}(300) + 10 = 10 + 10 = 20$$

Notice that if the price were above \$20, the supply curve would be to the right of demand, indicating that supply would be greater than demand -- a surplus. If the price were below \$20, the supply would be less than demand -- a shortage.

Using Point-Slope Form to Find the Equation of a Line

We can write the equation of a line given its slope m and any point on the line (x_0, y_0) in point-slope form,

$$y = m(x - x_0) + y_0$$

This says that the line y = mx is translated horizontally $|x_0|$ and vertically $|y_0|$, guaranteeing that it passes through the given point (x_0, y_0). For example, the line y = 2(x - 2) + 3 is a translation of the graph of y = 2x, to the right 2 units and up 3 units, passing through the point (2, 3).

Point-Slope Form

A linear equation written as

$$y = m(x - x_0) + y_0$$
$$f(x) = m(x - x_0) + y_0$$

is in **point-slope form**, where m is the slope and (x_0, y_0) is a point on the line.

Example 3 *Finding the equation of a line*

Find the equation of each of the following lines passing through the two given points and express each equation in slope-intercept form.

(a) (2, 3) and (4, 7)   (b) (3, -2) and (3, 4)

Solution

(a)

$$m = \frac{\Delta y}{\Delta x} = \frac{y_2 - y_1}{x_2 - x_1} = \frac{7 - 3}{4 - 2} = 2$$

We can choose either given point to find the equation. Using the point (2, 3), we have

$y = m(x - x_1) + y_1$ and $m = 2, x_1 = 2, y_1 = 3 \Rightarrow$

$$\begin{aligned} y &= 2(x - 2) + 3 \\ &= 2x - 1 \end{aligned}$$

NOTE: Using the point (4, 7) results in the same equation.

$y = m(x - x_1) + y_1$ and $m = 2, x_1 = 4, y_1 = 7 \Rightarrow$

$$\begin{aligned} y &= 2(x - 4) + 7 \\ &= 2x - 1 \end{aligned}$$

(b) Since $x_1 = x_2$, $\Delta x = 0$, and the slope is undefined. The line is vertical. Its equation, $x = 3$, cannot be written in slope-intercept form.

Graphing Piecewise-Defined Functions

In many applications the relationship between the variables changes. For example, a credit union might offer one interest rate for deposits of up to \$2000, a slightly higher interest rate for deposits of \$2000 up to \$10,000, and a still higher interest rate for deposits of \$10,000 or more. The relationship between interest earned and the balance depends on how much money has been deposited. For these cases, a different function is used for each part of the domain. The overall function can be thought of as being made up of pieces of functions; thus it is referred to as a **piecewise-defined function**. Since the interest i described above depends on the amount of money invested p, we write

$$i = \begin{cases} 0.04p, & 0 \le p < \$2000 \\ 0.05p, & \$2000 \le p < \$10{,}000 \\ 0.06p, & p \ge \$10{,}000 \end{cases}$$

to show interest rates of 4%, 5%, and 6%.

Example 4 *Graphing a piecewise-defined function*

Graph the equation

$$y = \begin{cases} 10, & x \leq 5 \\ 2x, & x > 5 \end{cases}$$

Solution

We graph the horizontal line y = 10 for x ≤ 5 and the line y = 2x for x > 5 (Figure 9).

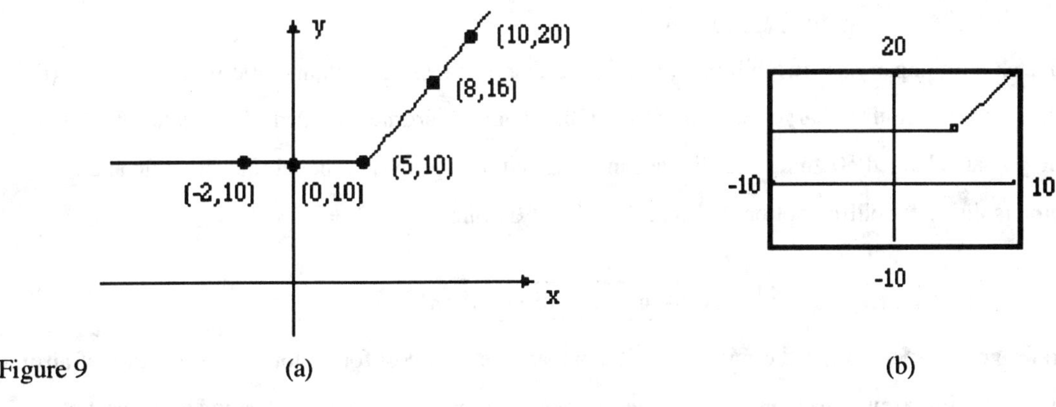

Figure 9 (a) (b)

Graphing such an equation on a graphing utility requires **Boolean expressions**, logical statements that are either true or false. (On computers the inequality symbols are entered directly from the keyboard. On graphing calculators the inequality symbols may be found using TEST.)

For the graph in Figure 9b, we enter

$$y = 10(x \leq 5) + 2x(x > 5)$$

where the expressions inside the parentheses take on a numerical value of 1 when true or a value of 0 when false. Each term is actually multiplied by 1 or 0. Table 1 shows some values of the equation. Notice that the first three entries, x = -2, x = 0, and x = 5, make x ≤ 5 true, creating a value of 1 for the first Boolean expression and a 0 for the second. The last two entries, x = 8 and x = 10, create a value of 0 for the first expression and a value of 1 for the second.

NOTE: We say that a term is "turned on" for values of the variable for which its associated Boolean factor is true and "tuned off" for values for which it is false.

Table 1

x	10(x ≤ 5) + 2x(x > 5)	y
-2	10 · **1** + 2(-2) ·	10
0	10 · **1** + 2(0) ·	10
5	10 · **1** + 2(5) · **0**	10
8	10 · **0** + 2(8) · **1**	16
10	10 · **0** + 2(10) · **1**	20

Example 5 The Car Phone Billing Problem

One cellular phone company has two billing options. One charges $19.95/month and 39¢/min for airtime (time spent on the phone). Another charges $29.95/month with 30 min of airtime included. If you talk over 30 min in 1 month, you are charged 50¢/min. Use linear functions and their graphs to model this situation and to decide which is the better billing option. [Source: U.S. West Cellular.]

Thinking: Creating a Plan

We have rates given in $/month and ¢/min. We will compare the total cost for 1 month as a function of airtime in minutes. The C intercept of each equation is the cost for using no airtime, and each slope is the cost of airtime in ¢/min. The graphs will intersect at the points where the two costs are equal.

Communicating: Writing the Solution

Option I: Let $C_1(t)$ represent the total cost as a function of airtime t, in minutes. A rate of 39¢/min represents the slope m = 0.39, and the monthly charge of $19.95 represents the C intercept. Therefore,

$$C_1(t) = 0.39t + 19.95$$

Option II: For 30 min of airtime, In $0 \le t \le 30$, the rate is fixed at $29.95. This yields the graph of a horizontal line whose equation is $C_2(t) = 29.95$. For air time over 30 min, $t > 30$, we add 50¢/min. The second part of this piecewise-defined function is $C_2(t) = 29.95 + 0.50(t - 30)$. We write

$$C_2(t) = \begin{cases} 29.95, & 0 \le t \le 30 \\ 29.95 + 0.50(t - 30), & t > 30 \end{cases}$$

See Figure 10.

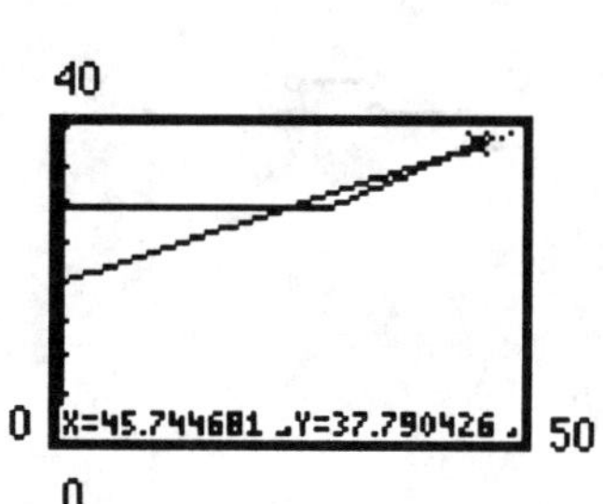

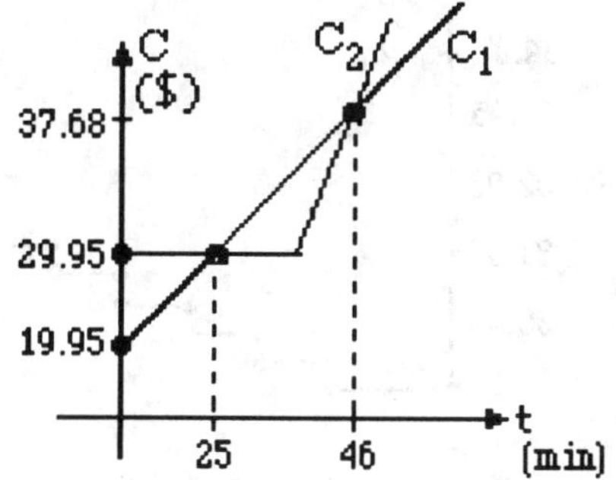

Figure 10

The points of intersection are $t \approx 25$ and $t \approx 46$. In Figure 10, the graph of the first plan lies below the second plan's graph for $t < 25$ and $t > 46$, clearly showing lower costs. If we are planning on talking less than 25 min/month or more than 46 min/month, the first option is better.

NOTE: To graph the two functions we enter

$$y_1 = 19.95 + 0.39x$$

$$y_2 = 29.95 + 0.5(x - 30)(x > 30)$$

Learning: Making Connections

To find the exact airtimes at the equilibrium points, we use algebra to find the points of intersection. (See Appendix I, page A16 to review solving a pair of linear equations by substitution.)

For $0 \le t \le 30$, $C_1(x) = C_2(x) \Rightarrow$

$$19.95 + 0.39t = 29.95$$

$$0.39t = 10$$

$$t \approx 25.64$$

For $t > 30$, $C_1(x) = C_2(x) \Rightarrow$

$$19.95 + 0.39t = 29.95 + 0.50(t - 30)$$

$$5 = 0.11t$$

$$t \approx 45.45$$

Do you think that the equation above accurately describes how phone companies charges customers?

If a phone company charged by the whole minute, a time of 30.1 min would be rounded up to 31 min. The graph in Figure 11 illustrates this refinement in the model. The charges *step up* as time passes each minute mark. The broken portions on both axes by the origin indicate those intervals are not the same size as those between other consecutive tick marks.

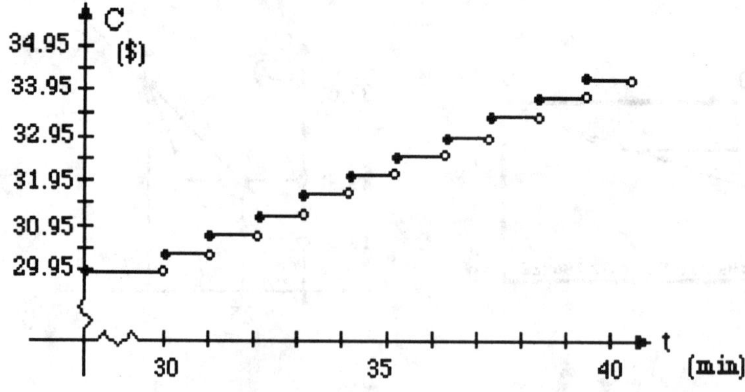

Figure 11

Recall the three equilibrium applications at the beginning of this section (Figure 12).

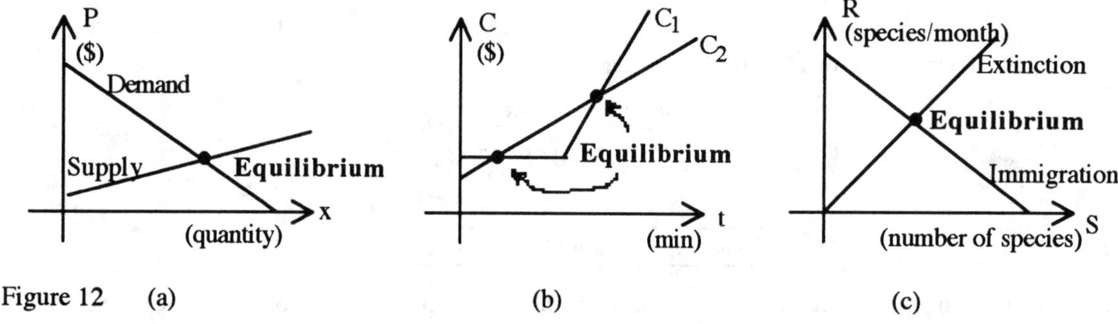

Figure 12 (a) (b) (c)

Can you explain why economists, consumer advocates, and ecologists are interested in the ***equilibrium points*** *where the two linear functions intersect?*

In Figure 12a, the point of equilibrium is the price where and the quantity bought (demanded) and sold (supplied) are equal so that there will be no shortages or surpluses, as seen in Example 2. In Figure 12b, the points of equilibrium are the times when the two billing options charge the same amount; the amounts are equal (Example 5). Finding these points allows us to make an informed decision about which option to choose for our particular calling plans. The ecologist studies immigration and extinction rates of species to aid nature at establishing stable ecosystems where the rate of increase and the rate of decrease in the number of species are equal (Figure 12c). (See Exercises 47 and 48.)

Exercises 2.1

Develop the Concepts

For Exercises 1 to 4 write each implicit function in slope-intercept form. State the slope and y intercept.

1. $2x + 3y = 12$
2. $5x - 3y = 30$
3. $0.2x - 0.5y = 1.5$
4. $\frac{x}{4} - \frac{y}{5} = 1$

For Exercises 5 to 8 use the given slope and y intercept to graph each line, then write its equation.

5. $m = 3/4$, $b = 1$
6. $m = -1/8$, $b = -3$
7. $m = 0$, $b = 2$
8. m undefined, $b = 0$

For Exercises 9 to 20 find each equation and express in slope-intercept form, if possible.

9. passing through the point $(-1, 3)$ with a slope of $m = 3$
10. passing through the point $(\frac{1}{2}, -\frac{3}{4})$ with a slope of $m = -\frac{1}{2}$
11. passing through the point $(2, 3)$ with a slope of $m = 0$
12. passing through the point $(-1, 5)$ with an undefined slope
13. satisfying $f(-1) = 5$ and $f(2) = -4$
14. satisfying $f(-5) = -30$ and $f(5) = 30$
15. having x intercept 4 and y intercept -2
16. having x intercept -1/2 and y intercept 5/4
17. passing through $(-1, -3)$ parallel to $y = \frac{2}{3}x - \frac{5}{3}$
18. passing through $(2, -5)$ parallel to $-3x + y = -4$
19. passing through $(6, -3)$ perpendicular to the x axis
20. passing through $(-\frac{1}{2}, 1)$ perpendicular to $y = \frac{3}{4}x + 1$

For Exercises 21 to 28 find the equation of each new function for $f(x) = 2x - 1$. Graph both function f and the new function on one coordinate system.

21. $y = f(x) + 1$
22. $y = f(x - 1)$
23. $y = -f(x)$
24. $y = f(-x) + 1$

*25. $y = \frac{1}{2}f(-x)$

26. $y = -2f(-x)$
27. $y = 12f(x - 3) + 5$
28. $y = 10 - f(2x - 1)$

For Exercises 29 to 32 find the point of intersection of each pair of functions graphically, and verify your results algebraically.

29. $f(x) = -\frac{1}{2}x + 3$

    $g(x) = \frac{3}{4}x - 18$

30. $f(x) = -\frac{1}{2}x + 3$

    $g(x) = 2x - 7$

31. $f(x) = 2x - 11$

$g(x) = 2x + 12$

32. $\frac{x}{3} - \frac{y}{2} = 1$

$4x - 6y = 12$

For Exercises 33 and 34 find the point of intersection of the two lines containing the given points.

33.

l_1	(-11, -21)	(11, 23)
l_2	(-11, 21)	(11, -23)

34.

l_1	(5, 100)	(25, 20)
l_2	(5, 5)	(45, 35)

For Exercises 35 to 38 graph each piecewise-defined function.

35. $y = \begin{cases} x, & x \geq 0 \\ -x, & x < 0 \end{cases}$

36. $y = \begin{cases} 2x - 1, & x > 0 \\ -2x - 1, & x \leq 0 \end{cases}$

37. $y = \begin{cases} -1, & x < 1 \\ 0, & x = 1 \\ 1, & x > 1 \end{cases}$

38. $y = \begin{cases} 2[x + 10], & -15 \leq x \leq -5 \\ 2x, & -5 < x < 5 \\ 2[x - 10], & 5 \leq x \leq 15 \end{cases}$

Apply the Concepts

39. The Car Rental Problem Arriving in Kona, Hawaii, you decide to rent a car for 1 day. One company charges a flat fee of $29.95/day with the first 100 mi included. Miles driven over 100 cost 50¢/mi. Another company charges $22.50/day and 19¢/mi. Which company should you choose?

40. The Coast Guard Problem A smuggler left a marina at midnight traveling 18 knots. Five hours later, the Coast Guard gave chase in their 40-knot cutter. If the smuggler's contact was 164 nautical miles from the marina, did the Coast Guard catch him?

41. The Triathlon Problem The distance from the start of a triathlon race (swimming, bicycling, and running) in kilometers as a function of the winning athlete's time in hours is described by the function

$$d = \begin{cases} 3t, & 0 \leq t \leq \frac{1}{3} \\ \frac{80t}{3} - \frac{71}{9}, & \frac{1}{3} < t \leq \frac{3}{2} \\ 20t + \frac{19}{9}, & \frac{3}{2} < t \leq \frac{5}{2} \end{cases}$$

What distance was covered on each leg of the race?

42. The Weight Loss Problem A dieter's personal weight loss in pounds as a function of time in weeks is described by the graph of the function in Figure 13.

(a) Describe in words what the function represents.

(b) What was the dieter's total weight loss?

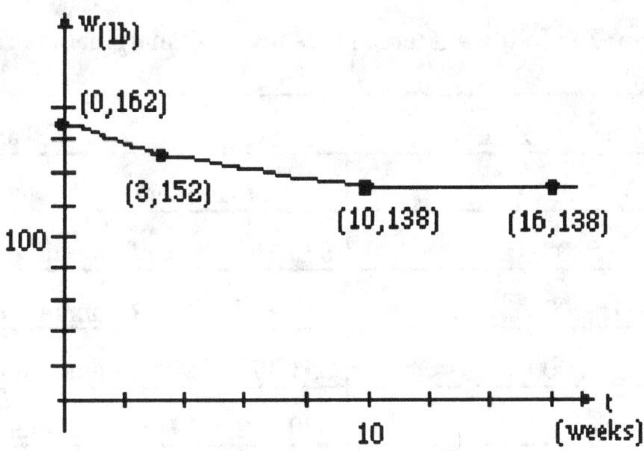

Figure 13

*43. The Weight Problem On the average, during the first 2 years of life, an infant's birth weight quadruples. Then from ages 2 to 6, the child gains about 4 or 5 lb each year. During childhood, growth is slow, taking another 8 years for the weight to double. During adolescence, growth again speeds up with a child reaching maturity weight at about age 18.

(a) If a child weighs 45 lb at age 6 and 150 lb at maturity, what was the child's birth weight?

(b) Draw a graph of this piecewise-defined function describing weight as a function of age and label the endpoints for each interval described.

[Source: *National Center of Health Statistics*]

44. The Consumer Price Index Problem The consumer price index (CPI) is the most commonly used measure of inflation. It measures the changes in price on a selection of commonly purchased goods and services relative to a base year. The following data is the consumer price index for various years based on the year 1984. A CPI of 130.7 in 1990 means it took $1.307 to have the same purchasing power as $1.039 in 1984.

year	1980	1982	1984	1986	1988	1990	1992
CPI	82.4	96.5	103.9	109.6	118.3	130.7	140.3

(a) Graph the data.

(b) Suppose that your union contract contains an inflation clause based on the CPI. If you made $15,000 in 1980, how much would you have made in 1990?

(c) Use the data for 1980 and 1992 to find a linear equation modeling the CPI.

(d) Use your equation from part (c) to predict and interpret the CPI for the year 2000.

(e) Using your equation from part (c), when will the CPI be 200?

[Source: *The Universal Almanac*]

45. The 1993 Tax Law Problem The following table gives the 1993 marginal tax rates for single taxpayers.

Bracket	Taxable Income	Tax
1	\$0 - \$22,100	15% of amount over \$0
2	\$22,100 - \$53,500	\$3,315 + 28% of amount over \$22,100
3	\$53,500 - \$115,000	\$12,107 + 31% of amount over \$53,500
4	\$115,000 - \$250,000	\$31,172 + 36% of amount over \$115,00
5	\$250,000 or over	\$79,772 + 39.6% of amount over \$250,000

(a) Find the piecewise-defined equation that expresses tax as a function of taxable income.

(b) If your taxable income was \$53,500, which tax bracket did you use to compute your taxes?

46. The 1993 Tax Law Problem The 1993 tax rates for married taxpayers filing jointly is shown in the table.

Bracket	Taxable Income	Tax
1	\$0 - \$36,900	15% of amount over \$0
2	\$36,900 - \$89,150	\$5,535.00 + 28% of amount over \$36,900
3	\$89,150 - \$140,000	\$20,165.00 + 31% of amount over \$89,150
4	\$140,000 - \$250,000	\$35,928.50 + 36% of amount over \$140,000
5	\$250,000 or over	\$75,528.50 + 39.6% of amount over \$250,000

(a) Find the piecewise-defined equation that expresses tax as a function of taxable income.

(b) Compare the taxes owed for two single people whose taxable income was \$25,000 (Exercise 45) with one married couple whose joint taxable income was \$50,000.

47. An Arachnid Problem A small island was cleared of arachnids (spiders) by ill-advised spraying. Ecologists studied the rates of immigration of various species of arachnids to the island from a nearby source where there were 420 species. The immigration rate R_i in species/month is given by the function $R_i = 0.0015(420 - S)$, where S is the number of species. Since not all species of arachnids can survive together, they also studied the rate of extinction of the immigrating species. They found that 0.3% of the species were becoming extinct each month, yielding $R_e = 0.003S$, where R_e is the extinction rate in species/month. Find and interpret the equilibrium point. [Source: *A Primer of Population Biology* , E. Wilson and W. Bossert]

48. Another Lost Species Problem In 1883, all plant and animal life on the island of Krakatoa was destroyed by a volcanic eruption. Life gradually returned to the island. The rate of immigration R_i of birds in species/month returning to the island was $R_i = \frac{27}{36}(57 - S)$, where S is the number of species. The rate of extinction R_e of birds in species/month was $R_e = r_e S$. Find r_e, the percentage of bird species lost per month at equilibrium when $R_i = R_e$ if there were 27 species of birds on the island of Krakatoa .

[Source: *A Primer of Population Biology* , E. Wilson and W. Bossert]

49. <u>The World Population Problem</u> When the world population was 5300 million, the birthrate was 138 million people/yr and the death rate was 48 million people/yr. Suppose that the birth rate remains nearly constant but the death rate is predicted to increase to 100.75 million people/yr when the population reaches 7750 million people (Figure 14).

(a) Find two linear equations: one for the birthrate and one for the death rate expressed as functions of the number of people in millions.

(b) Find and interpret the equilibrium point. [Source: *Ecology*, C. Krebs]

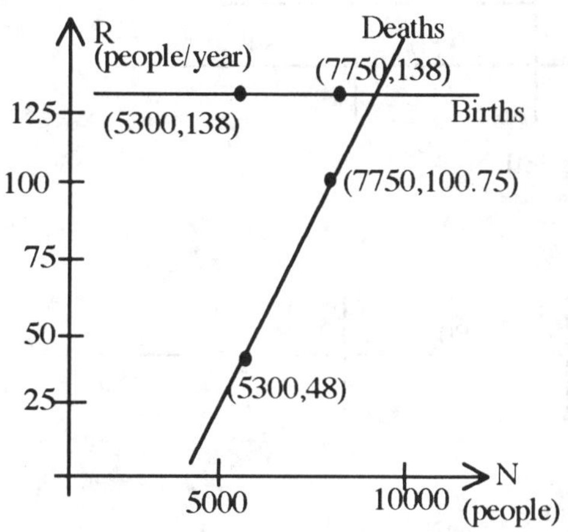

Figure 14

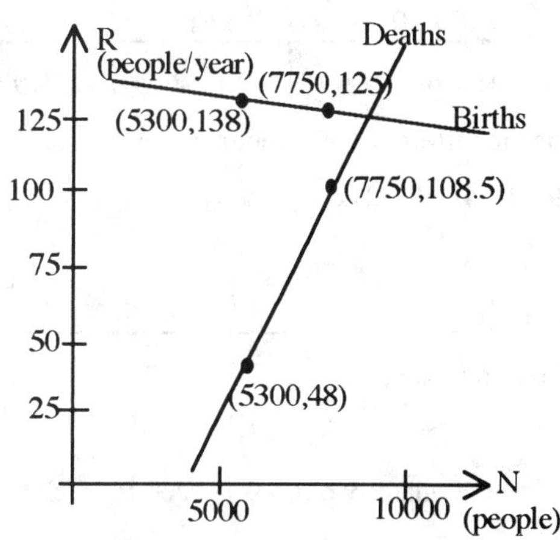

Figure 15

50. <u>Another World Population Problem</u> Suppose that the experts in Exercise 49 predict that the death rate will increase from 48 million/yr to 108.5 million/yr when the population reaches 7750 million people, and that the birthrate of 138 million/yr will drop to 125 million/yr when the population reaches 7750 (Figure 15). Compare the resulting equilibrium point with the one in Exercise 49.
[Source: *The Universal Almanac*, 1992]

51. The Rainbow Kite and Balloon Company In economics, marginal cost is the cost for producing one more item and marginal revenue is the additional revenue taken in by selling one more item. Assume that your marginal cost MC and marginal revenue MR of producing decorator windsocks are defined by the following functions, where x is the quantity of windsocks you produced and sold.

$$MC = 20, \quad MR = 100 - x$$

(a) What is the cost of each windsock?

(b) If you produce 70 windsocks, your marginal revenue will be greater than your marginal cost. Should you increase production?

(c) If you produce 90 windsocks, your marginal costs will be greater than your marginal revenue. Should you increase production?

(d) Find and interpret the equilibrium point.

52. Wonderworld Amusement Park If one book of ride tickets represents \$20.00 in costs, the marginal cost is MC = 20 (see Exercise 51). Marginal revenue MR is the additional income created by selling one more book of tickets, making marginal revenue a function of the number of books sold, x.

(a) Complete the tables to create ordered pairs (**x**, **MR**) and find the linear equation that expresses MR as a function of x. [Hint: The point (78, 21.50) from the second table is a point on the graph of the marginal revenue function.]

quantity, x	78	79	80	81	82
price, $p = 100 - 0.5x$	61	60.50	60	59.50	59
revenue, $R = px$	4758	4779.50			

MR is the difference in revenue of the 79th book of tickets and the 78th book,
4779.50 - 4758 = 21.50.

quantity, **x**	78	79	80	81
marginal revenue, **MR**	21.50			

(b) Find the quantity of books to sell to reach the equilibrium point.

[Source: *Principles of Economics*, J. Ragan Jr. and L. Thomas Jr.]

Extend the Concepts

For Exercises 53 and 54 use the given formula to find the slope of each line.

53. $m = \dfrac{f(b) - f(a)}{b - a}$ for $f(x) = 13x - 23$

54. $m = \dfrac{f(x + h) - f(x)}{h}$ for $f(x) = 13 - 27x$

*55. Use the definition of slope, $m = \dfrac{f(x_2) - f(x_1)}{x_2 - x_1}$, $x_1 \neq x_2$ to prove that the slope of $f(x) = mx + b$ is m.

56.(a) Find the x coordinate of the point of intersection of the lines whose equations are $y = m_1x + b_1$ and $y = m_2x + b_2$.

(b) When is your result defined?

57. Complete the following table for $f(x) = \frac{3}{4}x - 1$.

New Function	New Equation	Slope	Description of Resulting Slope
$y = f(x) + 2$	$y = \frac{3}{4}x + 1$	$m = \frac{3}{4}$	A vertical translation does not change the slope.
$y = 10f(x)$			
$y = f(5x)$			
$y = f(-x)$			
$x = f^{-1}(y)$			

58. Complete the following table for $f(x) = \frac{3x}{4}$ and $g(x) = 5x + 1$.

New Function	New Equation	Slope	Description of Resulting Slope
$y = f(x) + g(x)$	$y = \frac{23}{4}x + 1$	$m = \frac{23}{4}$	The slope of the sum of f and g is the sum of their slopes.
$y = f(x) - g(x)$			
$y = f(g(x))$			
$y = g(f(x))$			
$x = f^{-1}(g^{-1}(y))$			

59. In Figure 16, the x and y axes cut the two parallel lines.

(a) If l_1 is parallel to l_2, ΔAOB and ΔDOC are similar triangles, triangles having the same shape. Prove that if l_1 is parallel to l_2 ($l_1 \parallel l_2$), the slope of l_1 equals the slope of l_2.

(b) Prove that if the slopes of l_1 and l_2 are equal, $l_1 \parallel l_2$.)

[Hint: $l_1 \parallel l_2 \Leftrightarrow \angle A = \angle D$ and $\angle C = \angle B$.]

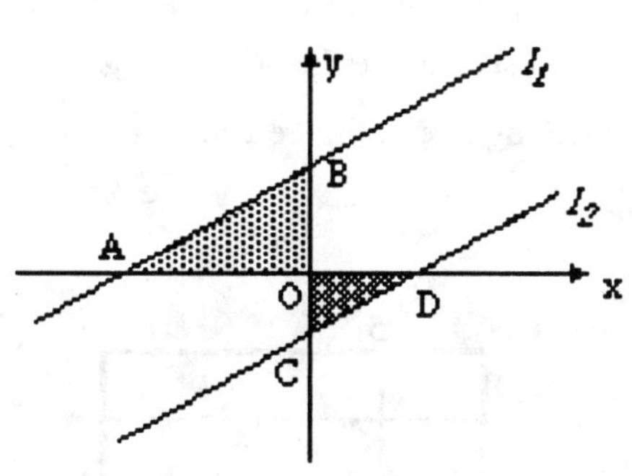

Figure 16

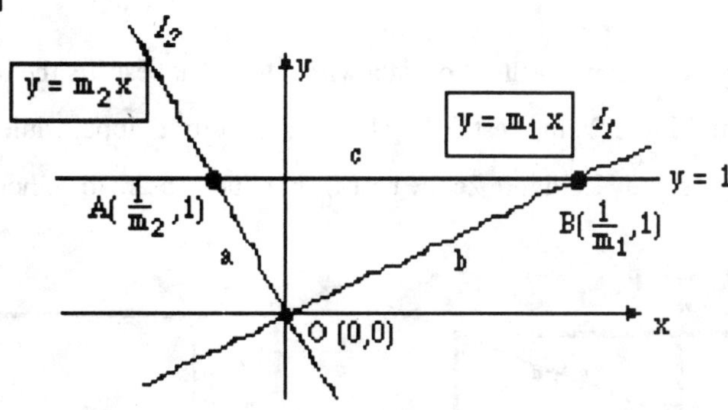

Figure 17

60.(a) Use the Pythagorean theorem and Figure 17 to prove if $m_2 = \frac{-1}{m_1}$, $\angle AOB$ is a right angle and the lines are perpendicular.

(b) If the lines are perpendicular, then $m_1 = \frac{-1}{m_2}$.

2.2 Linear Regression

Using a Graphing Utility to Find the Equation of the Best-Fitting Line
Interpreting the Correlation Coefficient and the Coefficient of Determination
Interpolating and Extrapolating from Data

Data from real-life situations seldom match any simple function exactly. Often, mathematicians can model data with a linear function. Consider the three sets of data in Figure 1.

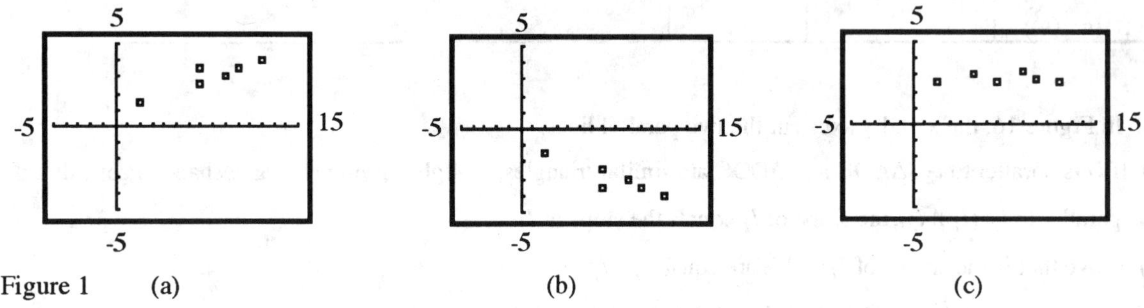

Figure 1 (a) (b) (c)

Figure 2 shows each set of data with a line that best fits the data. We call such a line the **best-fitting line**. In Figure 2a the best-fitting line has a positive slope, while the best-fitting line in Figure 2b has a negative slope. In Figure 2c the best-fitting line appears to be horizontal.

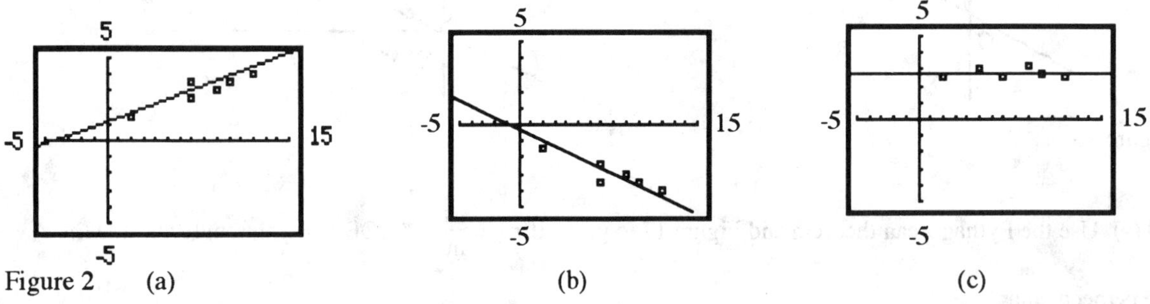

Figure 2 (a) (b) (c)

We use a graphing utility with its a linear regression program to find the equation of the best-fitting line for a given set of data. The linear regression formulas used by a graphing utility and further explanation of theory is given in Appendix III. The process of deriving those formulas is called the *method of least squares*, and the measure of how well the resulting line actually fits the data is called **regression analysis**.

To explore finding the best-fitting linear equation, go to "Explore the Concepts," Exercises E1 - E7, before continuing (page 132). For the interactive version, go to the *Precalculus Weblet - Instructional Center - Explore Concepts*.

Using a Graphing Utility to Find the Equation of the Best-Fitting Line

A researcher surveyed some students in regard to their grade-point averages (GPA) and approximate weekly study time.

weekly study time (hr)	2	7	7	9	10	12
GPA	1.5	2.5	3.5	3.0	3.5	4.0

If we had an equation that described the data points, we could use the equation to predict results not given in the original survey. For example, we could estimate the GPA of someone who averaged 11 hours of study time a week, or using the inverse, we could estimate the study time for someone who earned a 3.3 GPA. Figure 3 shows the plotted data points and the best-fitting line. We call such a line the **best-fitting line**. Notice that this set of data is not a function, and certainly not linear, but the line $y = 0.24x + 1.14$ fits the data quite well.

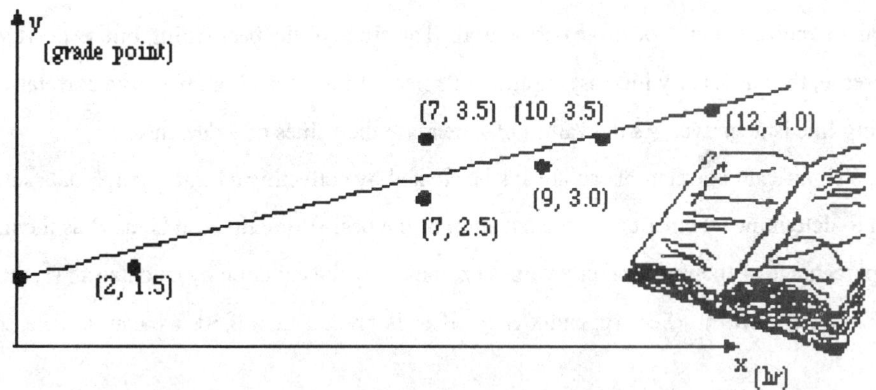

Figure 3

NOTE: The set of data in Figure 3 is not a function since two ordered pairs have the same input value and different outputs, (7, 3.5) and (7, 2.5). We can still enter the data and find the best-fitting equation.

Most graphing calculators and computer mathematics packages are programmed to find the slope and y intercept of the line that best fits a set of ordered pairs using **linear regression** (LinReg) under the statistics mode. We enter the six data points, $x_1 = 2, y_1 = 1.5, x_2 = 7, y_2 = 2.5, x_3 = 7, y_3 = 3.5, x_4 = 9, y_4 = 3.0,$ $x_5 = 10, y_5 = 3.5,$ and $x_6 = 12, y_6 = 4.0$, where the x coordinate of each point is hours studied and the y coordinate is GPA. The results after running the linear regression program may appear as shown in Figure 4.

L1	L2	L3
2	1.5	------
7	2.5	
7	3.5	
9	3	
10	3.5	
12	4	
------	------	

L1(1)=2

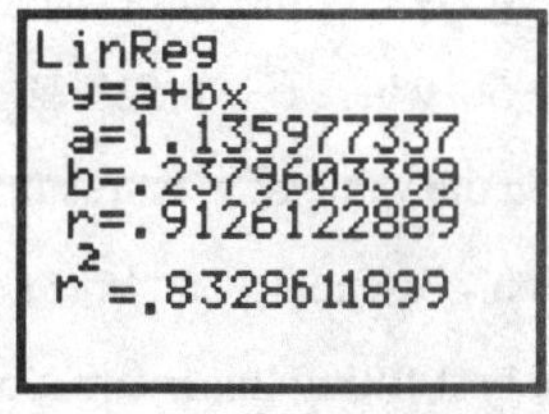

Figure 4 (a) (b)

The equation $y = a + bx$ in Figure 4b indicates that a is the y intercept of the best-fitting line and b is the slope. Therefore, the equation for the best-fitting line from the data in Figure 3 is

$$y = 1.14 + 0.24x$$

NOTE: Graphing utilities vary the form for the best-fitting linear equation; some use $y = ax + b$, others use $y = a + bx$, and still others use $y = mx + b$.

Interpreting the Correlation Coefficient and the Coefficient of Determination

In statistics the **correlation coefficient** r is a number that measures whether or not there is a significant correlation between the two variables. Values or r lie between -1 and 1, inclusive:

$$-1 \leq r \leq 1$$

Values of r near +1 indicate a high positive correlation. The slope of the best-fitting line is positive. As the values of x increase, the values of y increase. Values of r near -1 indicate a high negative correlation. The slope of the best-fitting line is negative. As the values of x increase the values of y decrease.

In statistics, a significant correlation can always be created by collecting a large enough data set. In curve fitting we need to determine whether or not the equation of the best-fitting line can be used as the model. We determine the percentage of the variation in y values explained by the equation by calculating r^2, the **coefficient of determination**. (See Appendix III.) If r^2 is greater than 0.50, we can use the equation as our model.

Figure 5 shows three sets of data with various values r^2. Notice the closer the value of r^2 is to 1, the more nearly linear the plot appears.

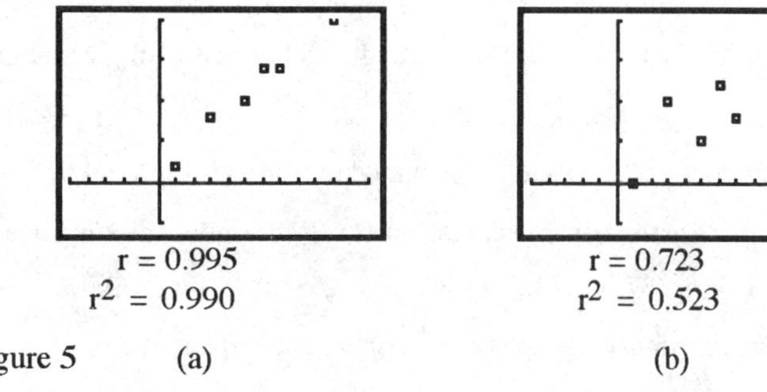

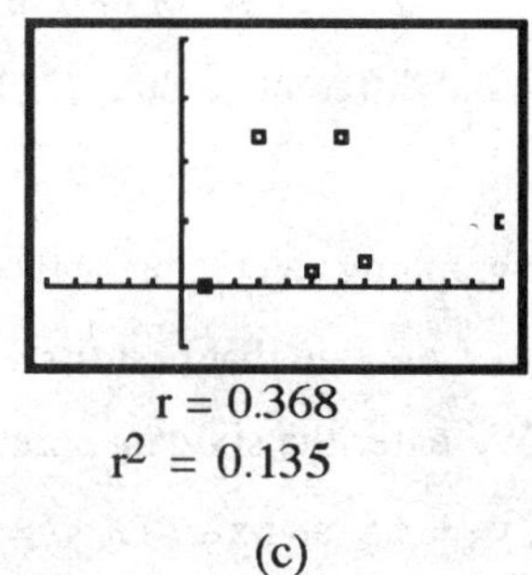

$r = 0.995$, $r^2 = 0.990$ | $r = 0.723$, $r^2 = 0.523$ | $r = 0.368$, $r^2 = 0.135$

Figure 5 (a) (b) (c)

In Figure 5a, where $r^2 = 0.990$, 99% of the variation in y is explained by the best-fitting linear equation. This is an excellent fit. In Figure 5b, where $r^2 = 0.523$, slightly over 50% of the variation is explained. This fit is barely adequate. In Figure 5c the value of r^2 is less than 50%. Only 13.5% of the variation in y is explained by the best-fitting linear equation. Essentially, no linear relationship exists between the variables. If r^2 is less than 50%, we cannot use the best-fitting linear equation to estimate other values of y. We use the average of the y values as the estimate of any other values.

Consider the following table, where the number of lawyers in the United States is increasing.

year	1960	1970	1980	1985	1988	1991
lawyers	285,933	355,242	542,205	655,191	723,189	805,872

[Source: *The American Almanac, Statistical Abstract of the United States*]

Not all of the points in Figure 6c lie on the best-fitting line. In Figure 6b we see that the correlation coefficient is r = 0.98, which results in the coefficient of determination, $r^2 = 0.96$. This means that 96% of the variation in the number of lawyers is described by our best-fitting line (Figure 6).

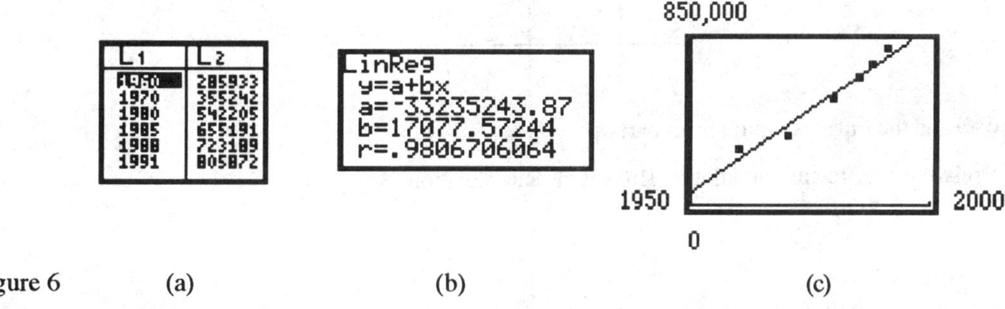

Figure 6 (a) (b) (c)

Notice the equation of the best-fitting line is y = -33,235,244 + 17,078x. The slope 17,078 indicates that the number of lawyers is increasing at a rate of 17,078 lawyers/yr.

What does a y intercept of -33,235,244 indicate?

The y intercept represents the year 0, which is outside the domain of this application. No, there were never -33 million lawyers.

Negative r values go with best-fitting lines having negative slopes. For example, the number of high school dropouts is decreasing.

year	1980	1985	1989	1990	1991	1992
dropouts	5212	4456	4109	3854	3964	3468

[Source: U.S. Bureau of the Census]

Figure 7b shows r = -0.97, which yields $r^2 = 0.95 \approx 1$. The equation is an excellent fit for the data (Figure 7).

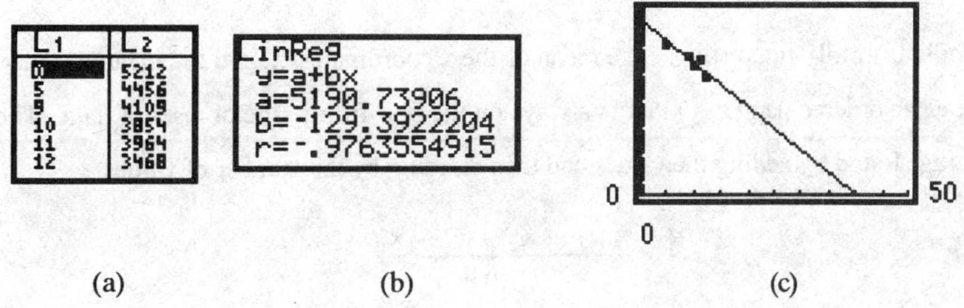

Figure 7 (a) (b) (c)

For these data, we let 0 represent the year 1980. The best-fitting equation is y = 5191 - 129x.

Why is the y intercept 5191 instead of 5212 as the data state?

Remember, the best-fitting line may not pass through all of the points given.

Example 1 *Finding the equation of the best-fitting line*

(a) Find the equation of the best-fitting line for the following data.

x	-1	2	5
y	-1	3	2

(b) Graph the data and the linear equation from part (a).

(c) State the correlation coefficient r and the coefficient of determination r^2.

Solution

(a) We enter $x_1 = -1$, $y_1 = -1$, $x_2 = 2$, $y_2 = 3$, $x_3 = 5$, and $y_3 = 2$ and choose the linear regression program (Figure 8a). The equation of the best-fitting line is

$$y = a + bx$$

$$y = 0.33 + 0.5x$$

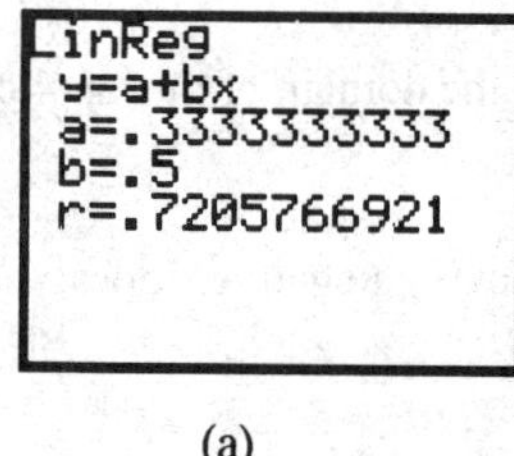

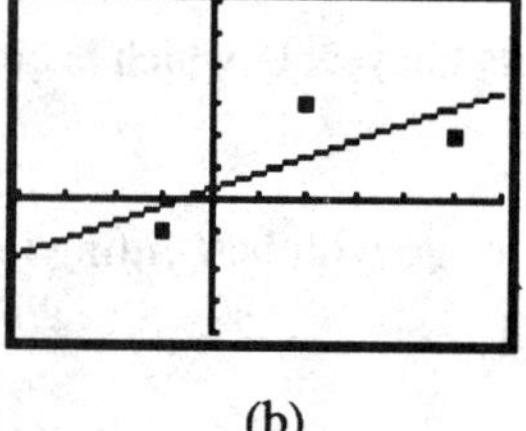

Figure 8 (a) (b)

(b) Figure 8b shows the data and the best-fitting line.

(c) The correlation coefficient is r = 0.72, and the coefficient of determination is $r^2 = 0.52$.

The next example is mildly interesting. The mean of the x coordinates, $\bar{x}$, and the mean of the y coordinates, $\bar{y}$, create an ordered pair $(\bar{x}, \bar{y})$ that is always on the best-fitting line of a set of data. The mean is the arithmetic average found by adding the values and then dividing by the number of values.

$$\bar{x} = \frac{x_1 + x_2 + \cdots + x_n}{n}$$

We will find the mean values of the x and y coordinates of the set of data from Example 1. Then we illustrate that the graph of the best-fitting line passes through the point $(\bar{x}, \bar{y})$.

Example 2 *Finding the mean values*

(a) Find the coordinates of the point $(\bar{x}, \bar{y})$ for

x	-1	2	5
y	-1	3	2

(b) Show that the point $(\bar{x}, \bar{y})$ lies on the best-fitting line, $y = \frac{1}{3} + \frac{1}{2}x$.

Solution

(a) $\bar{x} = \frac{x_1 + x_2 + x_3}{3} = \frac{-1 + 2 + 5}{3} = \frac{6}{3} = 2$ and $\bar{y} = \frac{y_1 + y_2 + y_3}{3} = \frac{-1 + 3 + 2}{3} = \frac{4}{3}$

The coordinates of the point $(\bar{x}, \bar{y})$ are $(2, \frac{4}{3})$.

(b) $y = \frac{1}{3} + \frac{1}{2}x$ and $x = 2 \Rightarrow$

$$y = \frac{1}{3} + \frac{1}{2}(2) = \frac{4}{3}$$

Therefore, the point $(\bar{x},\bar{y})$ lies on the graph of the best-fitting line $y = \frac{1}{3} + \frac{1}{2}x$.

In Section 2.1 we used the point-slope form to find the equation of a line through two given points. If there are only two data points, then linear regression can be used to find the equation.

Example 3 *Finding the equation of a line using a graphing utility*

(a) Find the equation of line that passes through the points $(\frac{1}{3}, -2)$ and $(\frac{1}{2}, 1)$ using linear regression.

(b) State the correlation coefficient r and the coefficient of determination r^2.

Solution

(a) Entering $x_1 = \frac{1}{3}$, $y_1 = -2$, $x_2 = \frac{1}{2}$, $y_2 = 1$, and using linear regression yields the equation of the best-fitting line, $y = -8 + 18x$ (Figure 9).

```
LinReg
y=a+bx
a=-8
b=18
r=1
```

Figure 9

Notice that this best-fitting line is the exact equation of the line passing through the two given points.

(b) Both the correlation coefficient and the coefficient of determination are 1 -- a perfect fit.

Interpolating and Extrapolating from Data

Using a line or function to find a new point between two given points is called *linear interpolation.* To **interpolate** means to insert between. Using a line to find a new point beyond given data is called *linear extrapolation.* To **extrapolate** means to infer that which is unknown beyond that which is known. Linear interpolation is often quite reliable, especially if the two given points are close together or if the situation is approximately linear. Extrapolating very far beyond given data can be misleading, even disastrous. Although using linear regression to make statistical extrapolations is risky, it is frequently done.

Example 4 The Blood Alcohol Problem

The following table gives the blood alcohol concentration (BAC) for a 160-lb person depending on the number of drinks.

(a) Graph the ordered pairs, treating BAC as a function of the number of drinks.

(b) Find the equation of the best-fitting line.

(c) Use the equation of the inverse to estimate the number of drinks that would result in a BAC of 0.1% for a 160-lb person.

(d) If BAC = 0.35% can cause death, how many drinks could be fatal?

[Source: Washington State Liquor Control Board]

drinks	4	5	?	9	10	?
BAC (%)	0.073	0.090	0.1	0.163	0.18	0.35

Thinking: Creating a Plan

Let x represent the number of drinks and y the BAC, then use a graphing utility to find the equation. Use the inverse equation to find the number of drinks for BACs of 0.1% and 0.35%.

Communicating: Writing the Solution

(a) See Figure 10.

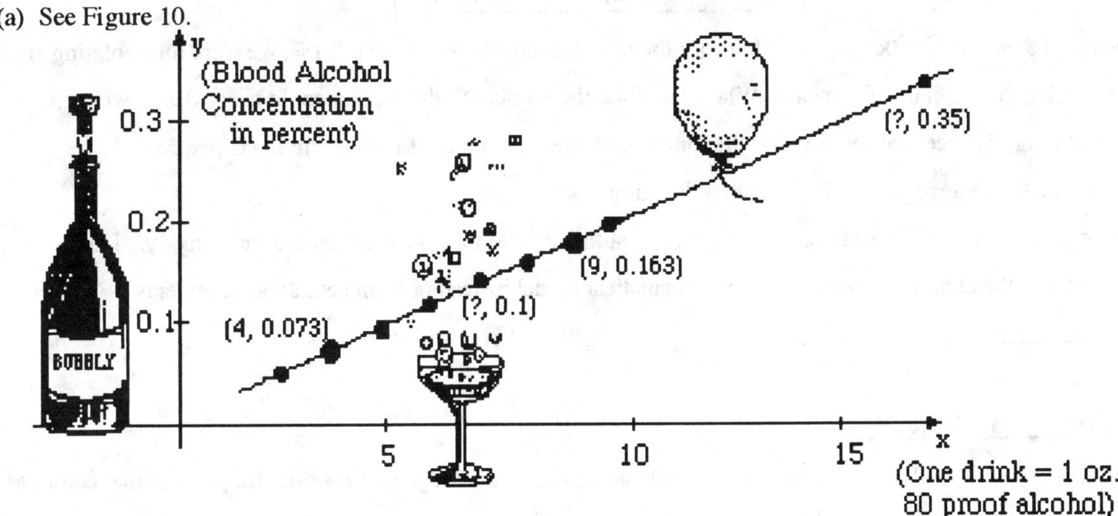

Figure 10

(b) Linear regression yields the value of a in scientific notation,

$$a = 7.6923077 \text{ E } -4 = 7.6923077 \times 10^{-4} \approx 0.000769 \approx 0.001$$

Therefore, the best-fitting line is y = 0.001 + 0.018x, where x is the number of drinks and y is the blood alcohol concentration (Figure 11).

Figure 11

(c) Finding the inverse by solving for x yields

$$\begin{aligned} y &= 0.001 + 0.018x \\ y - 0.001 &= 0.018x \\ x &= \frac{y - 0.001}{0.018} \end{aligned}$$

$x = \dfrac{y - 0.001}{0.018}$ and **y = 0.1** ⇒

$$x = \frac{0.1 - 0.001}{0.018} = 5.5$$

It would take 5.5 drinks to result in a BAC level of 0.1%.

(d) $x = \dfrac{y - 0.001}{0.018}$ and **y = 0.35** ⇒

$$x = \frac{0.35 - 0.001}{0.018} \approx 19.4$$

It would take about 19.4 drinks to result in a BAC level of 0.35%.

Learning: Making Connections

When we found the number of drinks to cause the blood alcohol level to reach 0.1%, we were interpolating since the point lies between the given data. When we found the number of drinks that could cause death, we were extrapolating. Notice that the slope of the function, 0.018, is the rate of increase in BAC per drink. The y intercept, 0.001, suggests the BAC without drinking.

Notice the value of r^2 is almost 1, indicating that the equation nearly matches the line exactly. This suggests that the data were created from a mathematical model instead of from actual measurements.

Explore the Concepts

Carefully graph the following sets of ordered pairs and use a straightedge to draw one straight line that comes as close to as many points as possible. Read the y intercept from your graph; call it a. Draw a triangle showing a specific Δy and Δx for your line. Measure Δy and Δx to calculate the slope of the line; call it b. (Do not use the data points to calculate the slope.) Write the equation of the line in the form $y = a + bx$.

E1.

x	-10	-5	2	5	8	10
y	-18	-10	-1	4	9	12

E2.

x	-10	-5	2	5	8	10
y	30	15	2	0	-10	-18

Now use the instruction book for your graphing utility to find the actual equation of the line you drew. Many utilities require the procedure outlined below.

Strategy: Best-Fitting Line on a Graphing Utility

1. Select the statistics package.
2. Enter the ordered pairs as data.
3. Select *linear regression* (LinReg) from the statistics menu.
4. Read the values for a and b from the display to create the equation $y = a + bx$.

Exercises 2.2

Develop the Concepts

For Exercises 1 to 8 decide whether there is generally a positive, negative, or no correlation between each of the following.

1. (a) age and knowledge
   (b) age and intelligence

2. (a) age and height for children
   (b) age and weight for children

3. (a) value and age of a computer
   (b) value and age of a vintage car

4. (a) education and salary
   (b) education and knowledge

5. miles per gallon of gasoline consumed and speed driven
6. hours of television watched per week and GPA
7. yield in pounds of apples per tree and density (trees/acre) of planting
8. distance required to stop a vehicle and speed traveled

For Exercises 9 to 12 use linear regression to find the equation of the line that passes through the given two points.

9. (5, 0) and (13, -2)
10. (-1, -1) and (4, 1)
11. (1, 1/5) and (2, 2/3)
12. (0.2, 0) and (0.4, 0.35)

For Exercises 13 and 14, (a) find the equation of the best-fitting line, (b) find the point $(\bar{x}, \bar{y})$, and (c) show that the point $(\bar{x}, \bar{y})$ falls on the best-fitting line.

13. {(-2, -4), (-6, -8), (-10, -14), (-18, -18)}
14. {(2, 4), (6, 8), (10, 14), (18, 18)}

For Exercises 15 to 18 find the equation of the best-fitting line described by the data. Use the equation to find the missing values.

15. Distances away from home as you return from vacation:

time (hr)	0	1	1.5	2	4.5	?
distance (mi)	3200	2500	2250	?	500	0

16. Vertical velocity of a golf ball as it leaves the ground:

time (sec)	0	1	1.5	3	3.5	?
velocity (ft/sec)	124	100	75	?	10	0

17. Price of a U.S. postage stamp:

year	1971	1975	1985	1988	1995	2000
cost (¢)	8	13	22	25	32	?

18. Consumer price index for clothing:

year	1970	1975	?	1985	1990	1995
cost ($)	59.20	72.50	100.00	105.00	124.10	?

Apply the Concepts

*19. The Calorie Problem There is a correlation between the number of calories that a person eats and weight gain. Find and interpret the slope and y intercept of the best-fitting line for the following data of a person's weight gain.

average calories per day	2000	2200	1600
average change in weight per day	0	+0.25	-0.25

20. The Calorie Maintenance Problem There is a correlation between a person's weight and the number of calories it takes to just maintain that weight. Find and interpret the slope of the best-fitting line for the following data of a person's caloric intake.

weight (lb)	105	130	180	200
calories per day	1300	1450	2000	2450

21. The Growth Study Problem During their first two years of life, a child doubles in height, from about 18 in. to 36 in. From age 2 to age 6, the average North American child grows about 3 in. From age 6 to adolescence, about 14 years of age, a child grow about 20 in. The adolescence growth spurt adds another 6 in. by maturity. Assume that no growth occurs after maturity, 18 years of age.

(a) Find the equation of the piecewise-defined function described above in words for a baby with a birth length of 18 in. who stops growing at age 18.

(b) Find the best-fitting line for the following data.

age (yr)	0	2	6	14	18
height (in.)	18	36	39	59	65

(c) Interpret the slopes of the equations from parts (a) and (b).

[Source: *Children in a Changing World*, E. Zigler and M. Stevenson]

22. The Marathon Problem In order to run a marathon in a certain time, it is necessary to run a mile is a specific amount of time. The following data show the recommended mile time and corresponding marathon time in minutes. Find and interpret the slope and y intercept of the best-fitting line.

mile	4.33	4.67	5	5.25	5.58	5.83	6	6.33	6.75	7.25
marathon	140	150	160	170	180	190	200	210	225	240

[Source: *The Self-Coached Runner*, A. Lawrence and M. Schied]

For Exercises 23 to 26 assume that each set of data can be modeled by a linear function.

23. The Airfare Problem The following table lists distances between cities in miles and corresponding airfares. Based on these data, find the price of a ticket from Anchorage, Alaska to Miami, Florida, a distance of 4004 mi.

distance, mi	175	1122	1303	1666	2894	3334
price, $	68	98	228	294	388	448

[Source: *Official Airline Guide*, Electronic Edition]

24. An AIDS Problem The following data give number of people reported as dying from AIDS in the United States from 1982 to 1992.

year	1982	1983	1985	1986	1987	1988	1989	1990	1991	1992
deaths	843	1651	6681	11,535	15,451	19,656	26,151	28,053	30,579	22,660

(a) Based on these data, compare the interpolated number of deaths in 1990 with the reported number, 28,053.

(b) Use your linear regression model to estimate the number of deaths from AIDS in the year 2000.

[Source: *AIDS*, M. Fromer; and *The World Almanac and Book of Facts*]

25. The Life Expectancy Problem The following data lists life expectancy in years for various years

year	1920	1930	1940	1950	1965	1975	1980	1987	1988	1990
life exp.	54.1	59.7	62.9	68.2	70.2	72.6	73.7	75.0	74.9	75.4

Based on this data, compare the life expectancy in the year 2010 with the projected life expectancy of 77.9 yr.

[Source: National Center for Health Statistics]

26. The Complaint Problem The following set of data gives the total number of consumer complaints against U.S. airlines as well as the number of complaints regarding baggage.

(a) Hypothesize possible relationships between general complaints as a function of the year and baggage complaints as a function of the year.

(b) Test your hypotheses from part (a).

year	1986	1987	1988	1989	1990	1991	1992	1993
general complaints	10,802	40,985	21,493	10,553	7703	6106	5639	4438
baggage complaints	2149	7438	3938	1702	1329	883	752	627

27. The Rainbow Kite and Balloon Company Suppose that your kite manufacturing business is in an area that has no competition. You survey the people in the potential market and find enough data to create a linear demand function. You study the cost involved and determine your willingness to supply kites as a function of the selling price. Find market equilibrium, the price to charge so that you will supply the same number of kites as your customers will buy.

demand:

quantity, x	0	10	20	30	40	50	60
price, p	$100	$90	$81	$75	$61	$50	$41

supply:

quantity, x	10	20	30	40	50	60	70
price, p	$0	$12	$25.50	$40	$50	$62.50	$75

28. The Tax Problem Suppose that a government tax of $2.50 is applied to each kite you sell in Exercise 27, raising the supply price, up $2.50 at every point -- translating the supply curve up 2.50 units.

(a) Find the new best-fitting linear supply function.

(b) Use the demand function from Exercise 27 and the new supply function to find the new market equilibrium.

(c) How much of the $2.50 tax was passed on the consumer?

29. Wonderworld Amusement Park Suppose that you want to start selling pies at a bakery at your amusement park. You survey the people in the potential market and find enough data to create a demand function. You study the cost involved and determine your willingness to supply pies as a function of the selling price. Find the price to charge to reach market equilibrium.

demand:

quantity , x	175	142	125	95	75	70
price, p	$1	$3	$4	$5	$6	$7

supply:

quantity, x	0	22	43	65	75	100
price, p	$4	$5	$6	$7	$8	$9

30. The Advertising Problem Suppose that your pie business in Exercise 29 benefits from an advertising campaign so that consumers are willing to buy 12 more pies at each price level, translating the demand function to the right 12 units.

(a) Find the new best-fitting linear demand function.

(b) Use the supply function from Problem 29 and the new demand function to find the new market equilibrium.

(c) How did this affect revenue? (Revenue is price times quantity.)

31. The Rainbow Kite and Balloon Company You need to calculate how much string to package with your two-stick kites. You go fly a kite with a thin string attached (Figure 12) and record the following data.

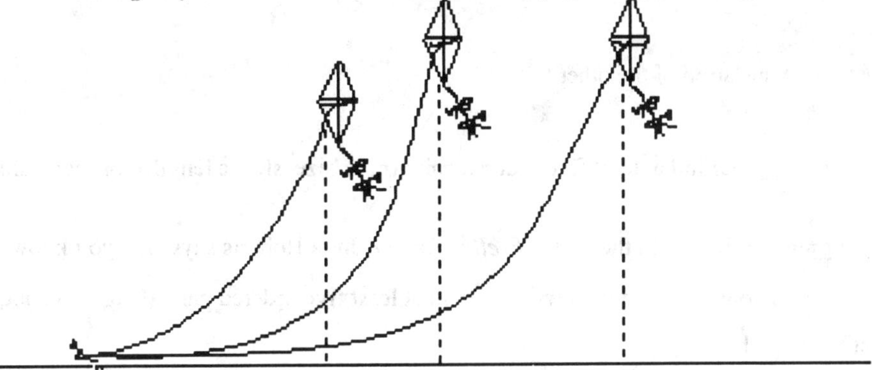

Figure 12

string length (ft)	0	50	100	250	300	500
height above the ground (ft)	0	40	85	120	140	300

(a) How much string should you package to allow the kite to fly 100 ft, 200 ft, and 300 ft off the beach?

(b) How many miles of string would be required to tie the world record of 12,471 ft above the ground?

32. Wonderworld Amusement Park In one of the bakeries at Wonderworld Amusement Park your marginal cost for scones is 0.18 since they cost 18¢ per scone to make. Your marginal revenue varies according to the quantity sold as indicated in the table. Find the quantity to sell to maximize profit by finding where marginal revenue equals marginal cost.

quantity sold	500	700	750	1000
marginal revenue ($)	1721	1261	601	0

For Exercises 33 to 36 collect your own data and find the best-fitting linear equation.

33. The Ring Size Problem Use the chart in Figure 13 to create at least five ordered pairs (circumference of ring finger, ring size) (a) for men, and (b) for women.

(c) Is one equation a translation of the other?

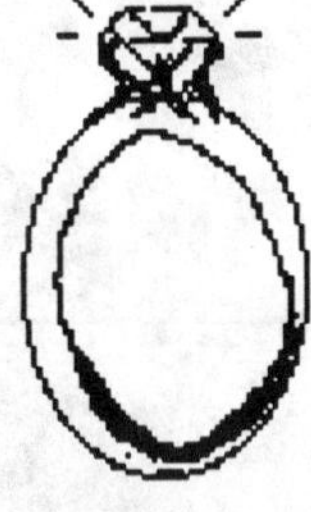

To find your ring size,
wrap a piece of string around your finger
and compare it to the chart.

Place string at this end 5 6 7 8 9 10 11 12

Figure 13

34. The Shoe Size Problem Find at least five ordered pairs (length of foot, shoe size)

(a) for men.

(b) for women.

(c) Is one equation a translation of the other?

35. The Shirt Size Problem Find at least five ordered pairs (neck size, sleeve length) for men's shirt sizes.

36. The Pretty Woman Problem In the movie *Pretty Woman*, Julia Roberts says "Do you know your foot's as big as your arm -- from your elbow to your wrist?" Find at least five ordered pairs (length of foot, length of forearm) to test the hypothesis.

Extend the Concepts

For Exercises 37 and 38 use the following table and the equation of the best-fitting line, $y = -178.9 + 4.9x$, for the average heights and ages of men between 20 and 24 years of age.

height	5'5"	5'9"	5'10"	6'	6'2"	6'4"
weight (lb)	143	163	167	176	187	198

37. Suppose that each average weight in the table dropped $2\frac{1}{2}$ lb.

(a) Make a table of data for the new function.

(b) Find the equation of the best-fitting line.

(c) Compare the resulting equation with the original best-fitting line, $y = -178.9 + 4.9x$.

38. Suppose that each average height in the given table goes up $\frac{1}{2}$ in.

(a) Make a table of data for the new function.

(b) Find the equation of the best-fitting line.

(c) Compare the resulting equation with the original best-fitting line, $y = -178.9 + 4.9x$.

For Exercises 39 to 42 use the following table, the equation of the best-fitting line, $f(x) = 4.63 + 0.30x$, and the coefficient of determination is $r^2 = 0.9657$.

x	-10	0	5	10
y	2	4	6	8

*39.(a) Make a table of data by interchanging the x and y values in the given table.

(b) Find the equation of the best-fitting line.

(c) Compare the resulting equation with the original line, $f(x) = 4.63 + 0.30x$.

(d) Compare the values of the coefficients of determination.

40.(a) Make a table of data by negating each x value.

(b) Find the equation of the best-fitting line.

(c) Compare the new equations with the original line, $f(x) = 4.63 + 0.30x$.

(d) Compare the values of the coefficients of determination.

41.(a) Make a table of data by negating each y value.

(b) Find the equation of the best-fitting line.

(c) Compare the new equation with the original line, $f(x) = 4.63 + 0.30x$.

(d) Compare the values of correlation coefficients.

42.(a) Make a table of data by multiplying each y value by 2.

(b) Make a second table from the given table by multiplying each x value by $\frac{1}{2}$.

(c) Find the equations of the best-fitting lines for parts (a) and (b).

(d) Compare the new equations with the original line, $f(x) = 4.63 + 0.30x$.

2.3 Graphs of Quadratic Functions and Optimization

Finding the Vertex of the Graph of a Quadratic Function
Finding the X and Y Intercepts of the Graph of a Quadratic Function
Applying Quadratic Functions for Free-Falling Objects
Solving Optimization Problems Involving Quadratic Functions

Quadratic functions can be written in the form

$$f(x) = ax^2 + bx + c, \; a \neq 0$$

Just as for the reference function $f(x) = x^2$, graphs of quadratic functions are parabolas. Applications of quadratic functions include problems whose goals are to find maximum or minimum values. The vertices of the parabolas represent these best values, called **optimum values**. Figure 1a shows the quantity that yields maximum profit, and Figure 1b shows the quantity that yields a minimum cost.

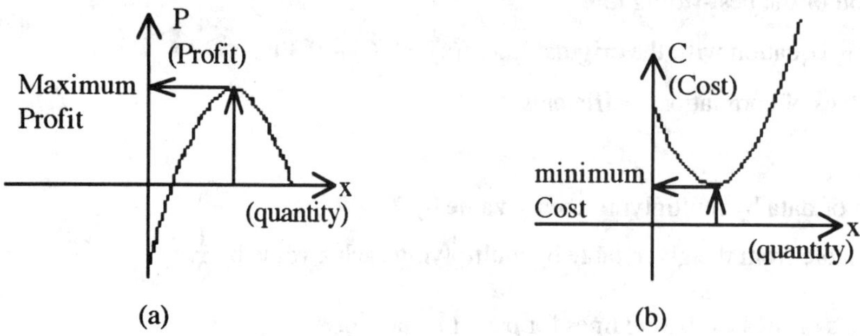

Figure 1 (a) (b)

In Example 5 we will investigate the price to charge for an item in order to maximize profit.

Arthur Laffer, from the University of Southern California, has been one of the country's most influential economists. His parabolic model (Figure 2) of how government's income tax revenue depends on the tax rate is typical of applications of quadratic functions. (Here, government revenue is the amount of money collected through income tax.)

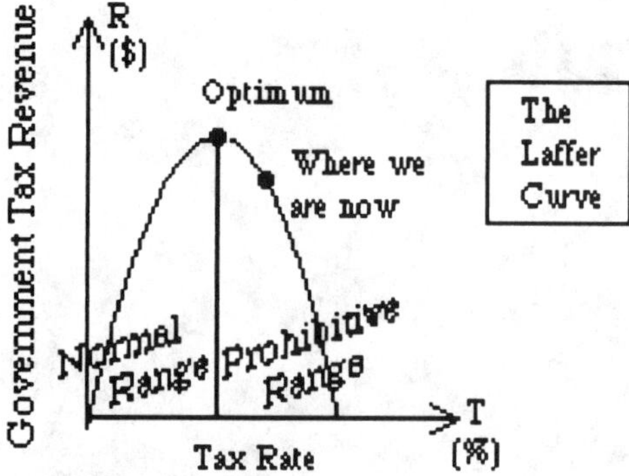

Figure 2

Notice that the Laffer curve is increasing between (0, 0) and the vertex. Increasing the income tax rate here increases tax revenue. The graph is decreasing from the vertex to (100, 0), indicating that an increase in tax rate will decrease revenue. Higher tax rates decrease people's incentive to earn more money. Of course, the government is interested in finding the tax rate that will maximize tax revenue, the vertex of Laffer's curve.

Finding the Vertex of the Graph of a Quadratic Function

To explore finding the location of the vertex of a parabola, go to "Explore the Concepts," Exercises E1 - E3, before continuing (page 154). For the interactive version, go to the *Precalculus Weblet - Instructional Center - Explore Concepts.*

An important property that describes the shape of a curve is concavity. If we think of curves forming valleys and hills; the "valley" shape in Figure 3a is **concave up** while the "hill" shape in Figure 3b is **concave down**. Notice how the graph in Figure 3c changes from concave down to concave up at (0, 0). The line pictured in Figure 3d is neither concave up nor concave down.

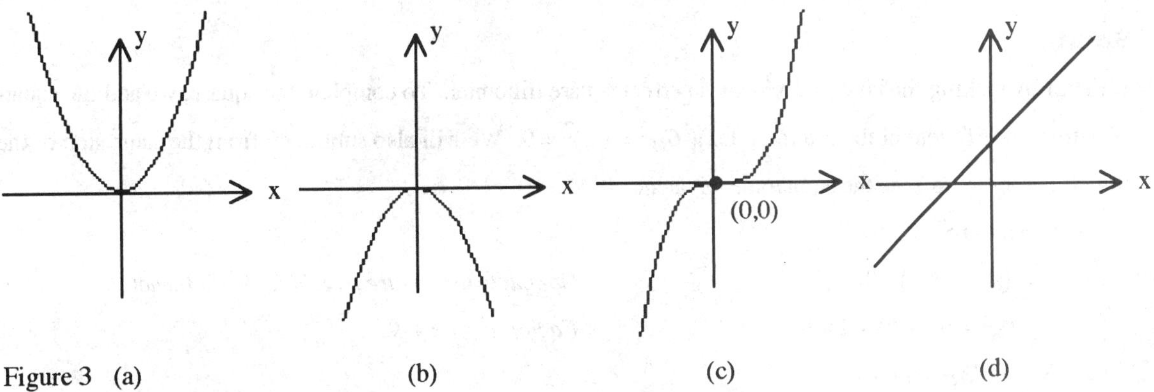

Figure 3 (a) (b) (c) (d)

Can you predict the concavity of a parabola from its equation?

Figure 4 shows the graphs of some reflections and translations of $f(x) = x^2$. In general, the parabola associated with $f(x) = ax^2 + bx + c$ is concave up for $a > 0$ (Figures 4a and 4c) and concave down for $a < 0$ (Figure 4b).

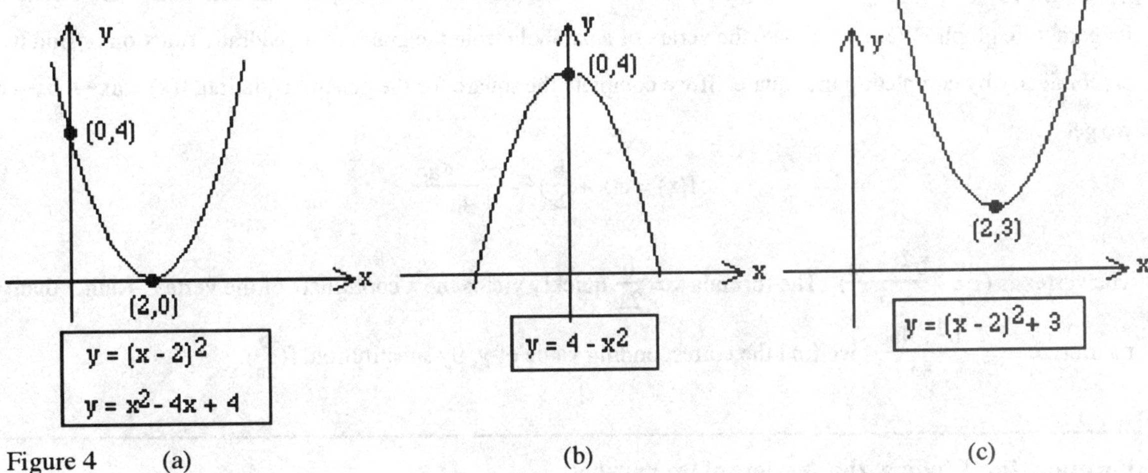

Figure 4 (a) (b) (c)

We can identify the coordinates of the vertex of a parabola if its equation is written in the form

$$f(x) = a(x - h)^2 + k$$

Since this general equation represents a translation of $y = ax^2$, $|h|$ units in the x direction and $|k|$ units in the y direction, the vertex is located at (h, k). Recall from Section 1.3 that the translation is to the right if $h > 0$ and up if $k > 0$.

In the next example, we complete the square to rewrite an equations of the form $f(x) = ax^2 + bx + c$ in the form $f(x) = a(x - h)^2 + k$. Then we identify the coordinates of the vertex of each parabola from the equation. (See Appendix I, page A14 to review completing the square.)

Example 1 *Rewriting a quadratic function in the form $y = a(x - h)^2 + k$*

Rewrite $f(x) = x^2 - 6x - 2$ in the form $y = a(x - h)^2 + k$ in order to identify the vertex.

Solution

We start by making the binomial $x^2 - 6x$ a perfect square trinomial. To complete the square, we add the square of half the coefficient of the x term, $[(1/2)(-6)]^2 = (-3)^2 = 9$. We will also subtract 9 from the same side of the equation in order to keep the equation equivalent.

$f(x) = x^2 - 6x - 2$

$= (x^2 - 6x \quad) - 2$ *Complete the square by adding 9. Subtract 9.*

$= (x^2 - 6x + \mathbf{9}) - 2 - \mathbf{9}$ *Factor $x^2 - 6x + 9$.*

$= (x - 3)^2 - 11$

The vertex is (3, -11).

Many applications involve finding the maximum or minimum value of a quadratic function -- the vertex of its parabolic graph. We can estimate the vertex of a parabola from the graph of a quadratic function or find it algebraically by completing the square. If we complete the square for the general equation, $f(x) = ax^2 + bx + c$ we get

$$f(x) = a(x + \frac{b}{2a})^2 - \frac{b^2 - 4ac}{4a}$$

The vertex is $(\frac{-b}{2a}, -\frac{b^2 - 4ac}{4a})$. The formula $x = \frac{-b}{2a}$ quickly yields the x coordinate of the vertex. Rather than memorize $y = -\frac{b^2 - 4ac}{4a}$, we find the corresponding value of y, by substitution, $f(\frac{-b}{2a})$.

Formula for Finding the Vertex of a Parabola

The vertex of the parabola $f(x) = ax^2 + bx + c$, $a \neq 0$, is

$$(\frac{-b}{2a}, f(\frac{-b}{2a}))$$

In the next example we estimate the vertex of the function from part (b) of Example 1 graphically, then verify that result with the vertex formula $x = \frac{-b}{2a}$.

Example 2 *Finding the vertex of a parabola*

(a) Graph $f(x) = -2x^2 - 6x + 1$ and estimate the coordinates of the vertex.

(b) Use the vertex formula to find the coordinates of the vertex.

Solution

(a) The vertex is located at (-1.5, 5.5) (Figure 5).

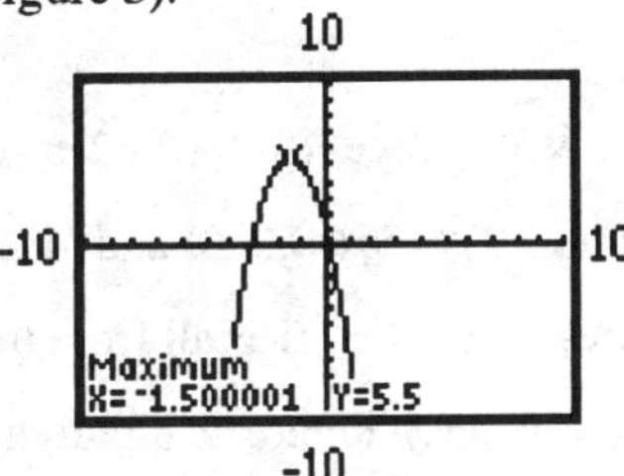

Figure 5

(b) $x = -\frac{b}{2a}$ and $a = -2, b = -6 \Rightarrow x = \frac{-(-6)}{2(-2)} = -\frac{3}{2}$

$y = f(\frac{-b}{2a}) = f(-\frac{3}{2}) = -2(-\frac{3}{2})^2 - 6(-\frac{3}{2}) + 1 = -\frac{9}{2} + 9 + 1 = \frac{11}{2}$

Finding the X and Y Intercepts of the Graph of a Quadratic Function

Many applications require finding the intercepts of a graph. Algebraically, we find the x intercepts by letting $y = 0$ and solving for x. Such solutions are called **zeros** or **roots** of the quadratic equation.

The y intercept is found by letting $x = 0$ and solving for y. For $f(x) = ax^2 + bx + c$ the y intercept is $f(0) = c$. The following summarizes how to algebraically verify the important points on the graph of a quadratic function.

Summary: Graphing a Quadratic Function of the form $f(x) = ax^2 + bx + c$

Vertex: The vertex is located at the point $(\frac{-b}{2a}, f(\frac{-b}{2a}))$.

x intercepts: Set $f(x) = 0$ and solve for x.

y intercept: Find f(0). The y intercept is c.

Example 3 *Finding the intercepts of the graph of a quadratic function*

Algebraically, find the x and y intercepts. Then verify your results graphically.

(a) $f(x) = (x - 2)^2 - 5$ (b) $g(x) = (x - 2)^2$ (c) $h(x) = (x - 2)^2 + 5$

Solution

(a) (See Appendix I, page A13 to review solving a quadratic equation by extracting roots.)

$f(x) = (x - 2)^2 - 5$ and $f(x) = 0 \Rightarrow$

$(x - 2)^2 - 5 = 0$	*Isolate the square root.*
$(x - 2)^2 = 5$	*Take the square root of both sides.*
$x - 2 = \pm\sqrt{5}$	
$x = 2 - \sqrt{5}$ or $x = 2 + \sqrt{5}$	

The exact x intercepts are $2 - \sqrt{5}$ and $2 + \sqrt{5}$. Since $f(0) = (0 - 2)^2 - 5 = -1$, the y intercept is -1. The location of the x intercepts for $f(x) = (x - 2)^2 - 5$, $(-0.\tilde{2}, 0)$ and $(4.\tilde{2}, 0)$, correspond to the decimal approximations of the intercepts in Figures 6a and 6b. (Recall from the Prologue that the symbol ~ in an ordered pair represents an approximate coordinate.) Figure 6c highlights the y intercept at (0,-1).

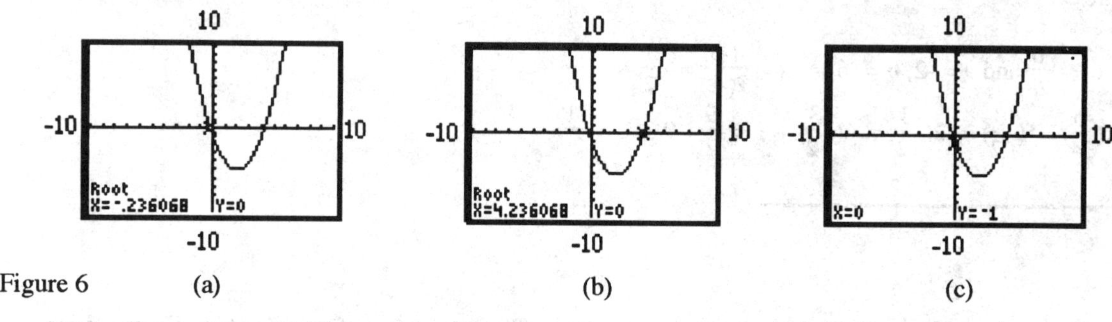

Figure 6 (a) (b) (c)

(b) $g(x) = (x - 2)^2$ and $g(x) = 0 \Rightarrow$

$(x - 2)^2 = 0$	*Take the square root of both sides.*
$x - 2 = 0$	*Solve for x.*
$x = 2$	

There is only one x intercept, 2. Since $g(0) = (0 - 2)^2 = 4$, the y intercept is 4. Figures 7a and 7b show the x and y intercepts for $g(x) = (x - 2)^2$.

Do you think the x intercept is 1.9999999 or 2?

Algebra confirmed that the exact x intercept is 2.

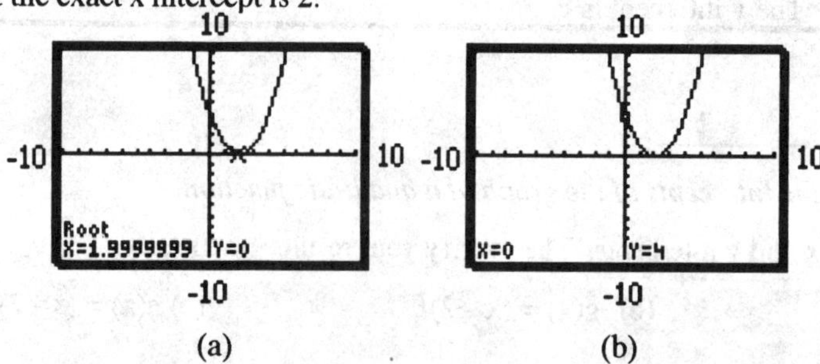

Figure 7 (a) (b)

(c) $h(x) = (x - 2)^2 + 5$ and $h(x) = 0 \Rightarrow$

$$
\begin{aligned}
(x - 2)^2 + 5 &= 0 \\
(x - 2)^2 &= -5 && \textit{Take the square root of both sides.} \\
x - 2 &= \pm\sqrt{-5} \\
x &= 2 \pm \sqrt{5}\, i
\end{aligned}
$$

There are no x intercepts since the solutions are not real numbers. Since $h(0) = (0 - 2)^2 + 5 = 9$, the y intercept is 9.

Since the vertex of the graph of $h(x) = (x - 2)^2 + 5$ is above the x axis, there are no x intercepts, as confirmed by the algebra (Figure 8).

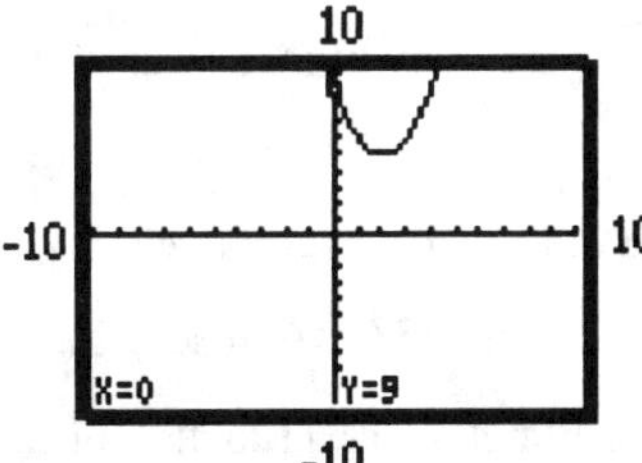

Figure 8

The expression $b^2 - 4ac$, found under the radical in the quadratic formula, is called the **discriminant**. Table 1 summarizes the results from Example 3 by indicating the value of the discriminant, the nature of the solutions, and the number of x intercepts for each quadratic function.

Table 1

Function	Function in Standard Form	Value of Discriminant $b^2 - 4ac$	Nature of Solutions	Number of x Intercepts
$f(x) = (x - 2)^2 - 5$	$f(x) = x^2 - 4x - 1$ $a = 1, b = -4, c = -1$	20	2 distinct real	two
$g(x) = (x - 2)^2$	$g(x) = x^2 - 4x + 4$ $a = 1, b = -4, c = 4$	0	1 distinct real	one
$h(x) = (x - 2)^2 + 5$	$h(x) = x^2 - 4x + 9$ $a = 1, b = -4, c = 9$	-20	no real (2 complex)	none

What is the relationship between the nature of the roots of a quadratic equation and the number of x intercepts on the graph of the corresponding quadratic function?

The Discriminant and X Intercepts

For $f(x) = ax^2 + bx + c$, $a \neq 0$

If $b^2 - 4ac > 0$, the graph has two x intercepts.

If $b^2 - 4ac = 0$, the graph has one x intercept.

If $b^2 - 4ac < 0$, the graph has no x intercepts.

Applying Quadratic Functions for Free-Falling Objects

The quadratic function

$$s(t) = -\frac{1}{2}gt^2 + v_0t + s_0$$

describes the height above the ground s as a function of time t in seconds for a free-falling object. The initial vertical velocity is v_0 and initial height above the ground is s_0. On earth, the acceleration due to gravity is

$$g = \frac{32 \text{ ft/sec}}{\text{sec}} \text{ when height is measured in feet}$$

and

$$g = \frac{9.8 \text{ m/sec}}{\text{sec}} \text{ when height is measured in meters}$$

This means that every second, a dropped object increases its velocity by 32 ft/sec (9.8 m/sec). After 1 sec, it is traveling 32 ft/sec (9.8 m/sec), after 2 sec, 64 ft/sec (19.6 m/sec), and so on. The equations that express the height above the ground measured in feet or meters as a function of time are

$$s(t) = -16t^2 + v_0t + s_0 \text{ (feet)} \quad \text{and} \quad s(t) = -4.9t^2 + v_0t + s_0 \text{ (meters)}$$

The free-falling object may be launched up from Earth, thrown down, or dropped from some height. For example, if a water balloon were thrown up at an initial vertical velocity of 20 ft/sec from a height of 10 ft, $v_0 = 20$ and $s_0 = 10$, the equation modeling its height (in feet) as a function of time is

$$s(t) = -16t^2 + 20t + 10$$

If the water balloon were thrown down at an initial velocity of 30 m/sec from a height of 100 m, its equation (in meters) is

$$s(t) = -4.9t^2 - 30t + 100$$

Notice that the initial velocity is negative for objects traveling toward Earth.

What equation would model the height of a water balloon if it were dropped from a height from 25 ft?

Since it is dropped, $v_0 = 0$. The equation is $s(t) = -16t^2 + 25$.

The equation $s(t) = -16t^2 + v_0t + s_0$ can model the height in feet of a golf ball driven off the ground. We cannot measure the horizontal length of the shot from this equation, but we can tell the time when it hits the ground and how high the ball is at any given time. Consider hitting a golf shot with an initial vertical velocity of 120 ft/sec.

$s(t) = -16t^2 + v_0t + s_0$ and $v_0 = 120$, $s_0 = 0 \Rightarrow$

$$s(t) = -16t^2 + 120t$$

From Figure 9a, the ball hit the ground (s = 0) after a flight of t = 7.5 sec. (In Figure 9a, the readout y = 1E-11 represents $1 \times 10^{-11} \approx 0$.) The golf ball reached a maximum height of 225 ft in 3.75 sec.

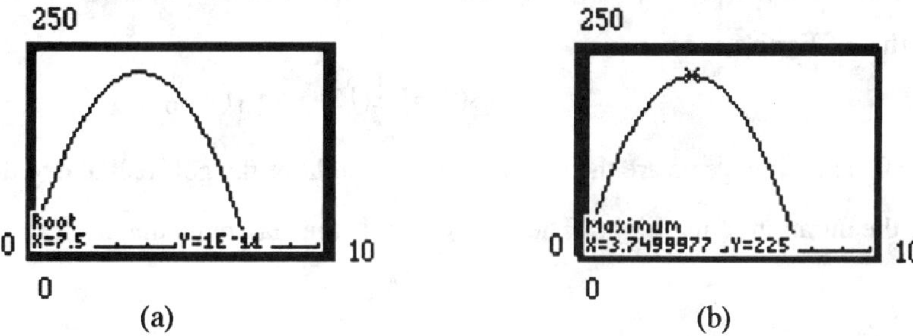

Figure 9 (a) (b)

The path of the golf ball in Figure 10b is another parabola. After 5 sec, the quadratic model, $s(t) = -16t^2 + 120t$, yields

$$s(5) = -16(5)^2 + 120(5) = 200$$

The ball was 200 ft above the ground after 5 sec.

Figure 10 (a) (b)

NOTE: Regardless of the path of the free-falling object, the free-fall function

$$s(t) = -\frac{1}{2}gt^2 + v_0t + s_0$$

measures vertical height as a function of time.

The only golf shot on the moon was hit by astronaut Alan Shepard in 1971. The golf shot on the moon in the next example uses an initial vertical velocity that would create a 300-yd drive on Earth. According to the *Guinness Book of Records*, a 300-yd golf shot on Earth would be a 1-mile drive on the moon, where the value of g is $\frac{1}{7}$ that on Earth.

Example 4 <u>The Golf Shot Problem</u>

Find the duration of flight and the maximum height for a golf ball hit on the moon with an initial vertical velocity of 120 ft/sec.

Thinking: Creating a Plan

We will create and graph the quadratic function for the moon shot. Since the acceleration due to gravity is one-seventh that of Earth's, we use

$$s(t) = -\frac{1}{7}(16)t^2 + v_0 t + s_0$$

with $s_0 = 0$. The t intercepts are the times when the height of the golf ball above the moon's surface is zero, yielding the duration of the flight. The vertex models the maximum height.

Communicating: Writing the Solution

$s(t) = -\frac{1}{7}(16)t^2 + v_0 t + s_0$ and $v_0 = 120$, $s_0 = 0 \Rightarrow$

$$s(t) = \frac{-16}{7}t^2 + 120t$$

From the graph in Figure 11a we see that the ball lands on the moon in 52.5 sec, since the horizontal intercept is 52.5. The maximum height of 1575 ft is reached 26.25 sec after the ball was hit.

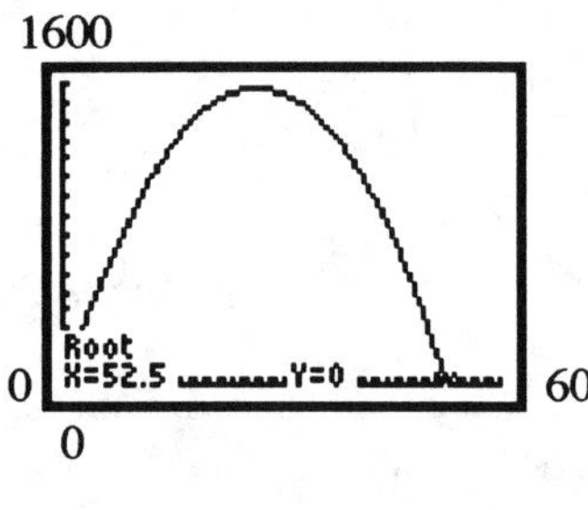

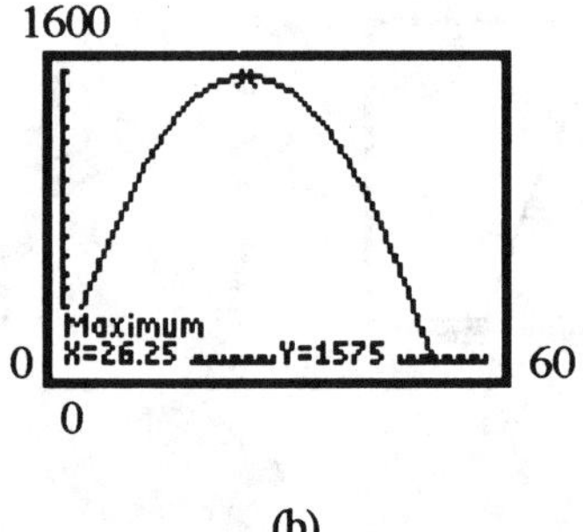

Figure 11 (a) (b)

Learning: Making Connections

Once we know the vertical component of the velocity and the initial height above the ground, we can find the maximum height and the duration of the flight. If we can find the horizontal component of the velocity, we can then find the horizontal distance. When the angle is 45°, the horizontal and vertical initial velocities are equal. We study other angles in Chapter 6. Objects thrown straight up or straight down have horizontal velocities of zero.

For this moon shot, if the angle were 45°, the horizontal velocity would be a constant 120 ft/sec and the ball would fly 6300 ft ≈ 1.2 mi, since

$$\begin{aligned} d &= rt \\ &= (120)(52.5) = 6300 \end{aligned}$$

Solving Optimization Problems Involving Quadratic Functions

The goal in business is to maximize profit. Profit is the difference between money collected from sales and the money spent. The cost of operating a business depends on both fixed cost, such as rent, and variable cost, such as the cost of purchasing materials to make a product. The variable cost depends on the quantity of items produced:

$$\text{profit} = \text{revenue} - \text{cost}$$

and

$$\text{cost} = \text{fixed cost} + \text{variable cost}$$

In a competitive market, the consumers establish the relationship between price and quantity of items sold because as price increases, consumers buy less. In the next example, market research establishes the demand function as $p(x) = 33.5 - 0.1x$, where x is the number of items sold. We will need to establish the revenue function. Since revenue is the amount of money collected from sales,

$$\text{revenue} = \text{price} \cdot \text{quantity}$$

$$R(x) = p(x) \cdot x$$

Example 5 The Kangaroo Socks Problem

Suppose that you decided to make and sell kangaroo socks, socks with a secret pocket. You surveyed the market and approximated the demand function as $p(x) = 33.5 - 0.1x$, where price p is a function of the number of packages of socks sold, x. If each package of socks costs $3.50 to make and your fixed costs for your shop are $290, what price should you charge to maximize profit?

Thinking: Creating a Plan

We create a profit function by subtracting the cost function from the revenue function. We graph the resulting profit function and estimate the coordinates of its vertex.

Communicating: Writing the Solution

Since revenue equals price times quantity and the price function is given as $p(x) = 33.5 - 0.1x$, the revenue function is

$$\begin{aligned} R(x) &= p(x) \cdot x \\ &= (33.5 - 0.1x)x \\ &= 33.5x - 0.1x^2 \end{aligned}$$

Since total cost is the fixed cost of $290 plus variable cost of 3.50x, the cost function is

$$C(x) = 290 + 3.50x$$

The profit function is

$$\begin{aligned} P(x) &= R(x) - C(x) \\ &= (33.5x - 0.1x^2) - (290 + 3.50x) \\ &= 33.5x - 0.1x^2 - 290 - 3.50x \\ &= -0.1x^2 + 30x - 290 \end{aligned}$$

The maximum profit of $1960 is achieved by selling 150 packages of socks (Figure 12). To sell 150 packages of socks you should price them according to the consumer demand function.

$$p(150) = 33.5 - 0.1x = 33.5 - 0.1(150) = 18.5$$

You should charge $18.50 for each package of socks.

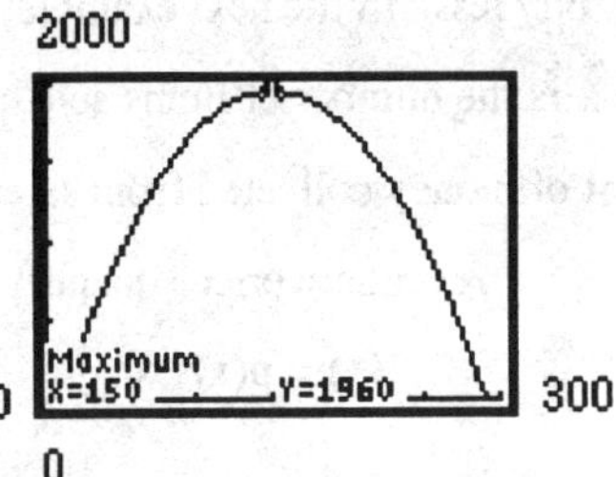

Figure 12

Learning: Making Connections

If you sell 150 packages of socks at $18.50, the revenue is $2775 since

$$R(x) = p(x) \cdot x$$

$$R(150) = [33.5 - 0.1(150)](150) = 2775$$

The total cost to produce 150 packages of socks is $815 since

$$C(150) = 290 + 3.50(150) = 815$$

Profit is revenue - cost,

$$P(150) = R(150) - C(150) = 2775 - 815 = 1960$$

In the next example we find the maximum area of a rectangular playpen. We will use the formulas for perimeter and area of a rectangle,

$$P = 2l + 2w \quad \text{and} \quad A = lw$$

where l is the length of the rectangle and w is the width. (See Appendix IV for a list of geometric formulas.)

Example 6 The Playpen Problem

A playpen is to be set up in a family's recreation room using 24-ft-long flexible fence. Find the dimensions that will yield the maximum area if a rectangular playpen is built (a) in the middle of the room, and (b) using an existing wall as one side.

Thinking: Creating a Plan

Since these are two separate problems, we will create two area functions. We use the formula for the area of a rectangle and the given length of the fence to create two equations. Substitution leads to describing area as a function of one variable. The quadratic functions have parabolic graphs whose maximums are found at the vertex. Of course, we require the domains and ranges to be nonnegative real numbers. We begin our solution with drawings of the two playpens.

Communicating: Writing the Solution

The fence for the two playpens in Figure 13 is fixed at 24 ft. In part (a) it is the perimeter of the playpen, while in part (b) the 24-ft fence consists of two widths and one length since an existing wall completes the rectangle.

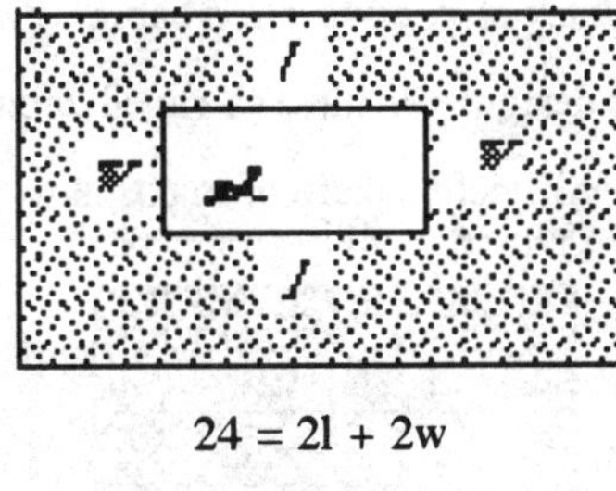

$24 = 2l + 2w$

(a)

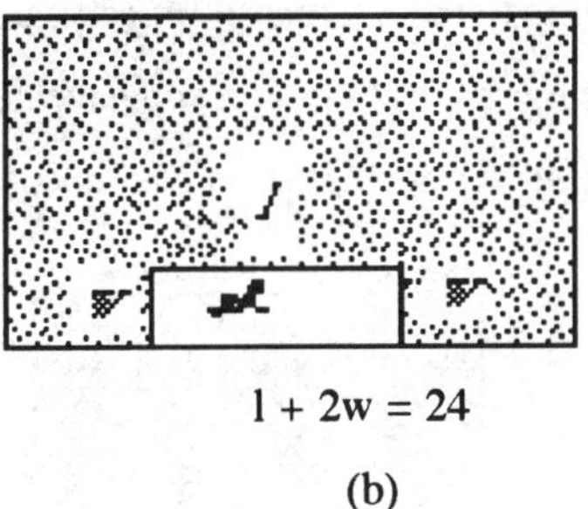

$l + 2w = 24$

(b)

Figure 13

(a) $A = lw$ where $2l + 2w = 24 \Rightarrow$

$A = lw$ where $w = 12 - l \Rightarrow$

$$A = l(12 - l)$$

The l coordinate of the vertex gives the length of the rectangle that yields the maximum area (Figure 14).

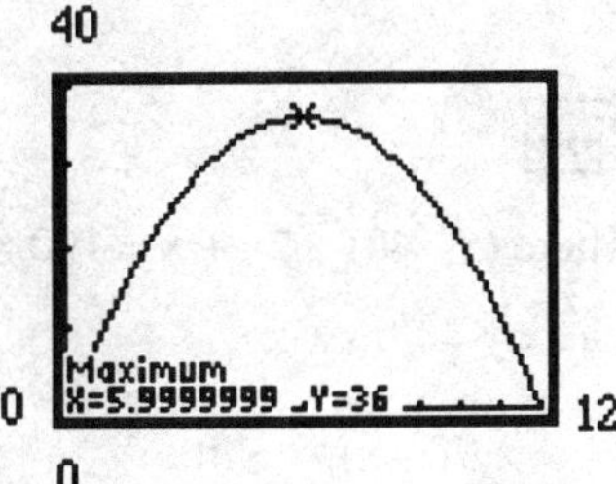

Figure 14

$l = 6 \Rightarrow$

$$w = 12 - l = 12 - 6 = 6$$

The playpen in the center of the room should be a 6-ft square.

(b) $A = lw$ where $l + 2w = 24 \Rightarrow$

$A = lw$ where $l = 24 - 2w \Rightarrow$

$$A = (24 - 2w)w$$

The w coordinate of the vertex gives the width of the rectangle that yields the maximum area (Figure 15).

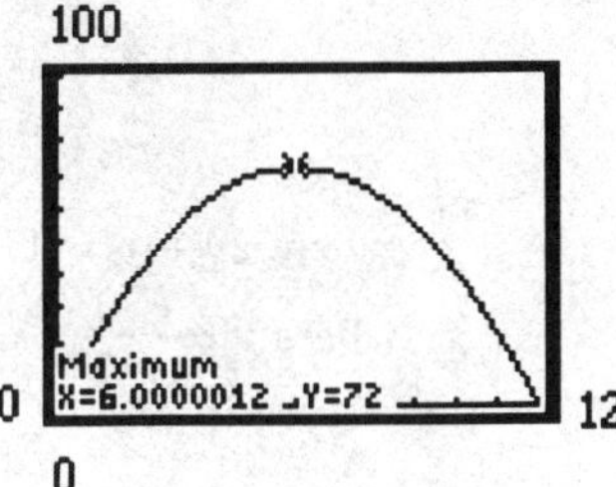

Figure 15

$w = 6 \Rightarrow$

$$l = 24 - 2w = 24 - 2(6) = 12$$

The playpen using an existing wall as one side should be 6 by 12 ft.

Learning: Making Connections

Notice that the maximum areas of the two playpens are dramatically different. The playpen in the middle of the room has a maximum area of 36 ft^2 (Figure 14), while the one using an existing wall as one side has a maximum area of 72 ft^2 (Figure 15).

The next example illustrates an important but subtle consideration that we must make whenever we solve an optimization problem.

Example 7 The Patio Problem

You are planning to build a fenced-in rectangular patio off the back of your vacation cabin. Two possible plans are shown in Figure 16. Although you have not decided on the design, you have salvaged enough pickets for a 100-ft picket fence to enclose it. What dimensions and design will yield a patio of maximum area?

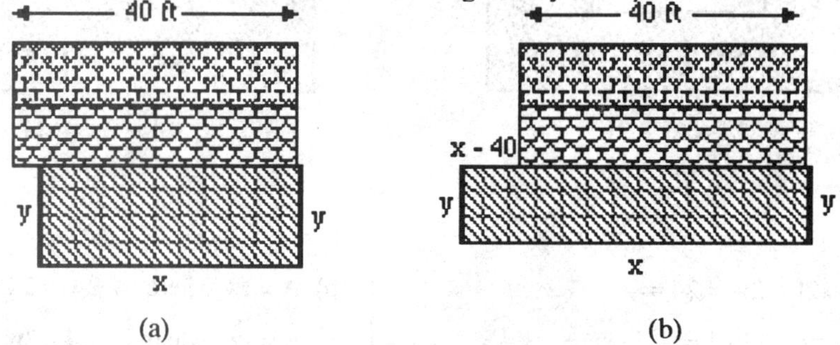

Figure 16 (a) (b)

Thinking: Creating a Plan

We find an equation for the area of each patio as a function of x, using the constraint that there is only 100 ft of fence available. Notice that x is further restricted by the size of the house. In Figure 16a, $x \leq 40$, while in Figure 16b, $x \geq 40$. We include $x = 40$ in both designs, and graph each function to find the value of x that yields the maximum area for each case.

Communicating: Writing the Solution

(a) $A = xy$ where $2y + x = 100 \Rightarrow$

$A = xy$ where $y = \dfrac{100 - x}{2} \Rightarrow$

$A = x\left(\dfrac{100 - x}{2}\right),\ x \leq 40$

(b) $A = xy$ where $(x - 40) + 2y + x = 100 \Rightarrow$

$A = xy$ where $y = 70 - x \Rightarrow$

$A = x(70 - x),\ x > 40$

Each graph in Figure 17 has the point (40, 1200) at an endpoint.

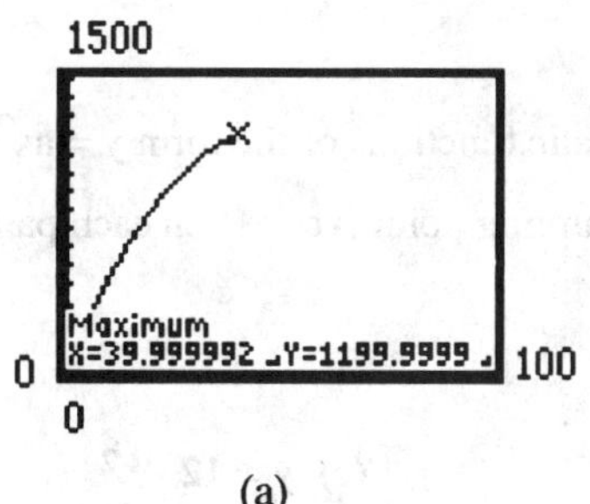

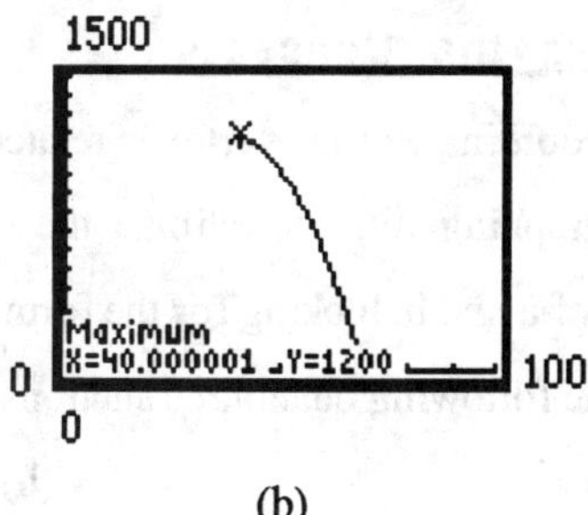

Figure 17 (a) (b)

Combining the two graphs from Figure 17 yields the maximum area of the patio at 1200 ft^2 when it is 40 ft long (Figure 18).

You should build the patio the same length as the cabin, 40 ft, and 30 ft wide.

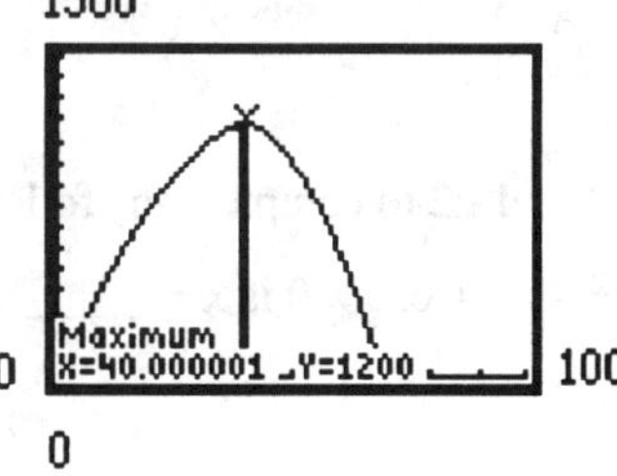

Figure 18

Learning: Making Connections

The graph in Figure 19a of the first design shows a maximum 1250 ft^2 for a 50-ft-long patio; however, the length was restricted to 40 ft or less. The graph in Figure 19b of the second design shows a maximum 1225 ft^2 for a 35-ft-long patio, but again it had the restriction $x \geq 40$.

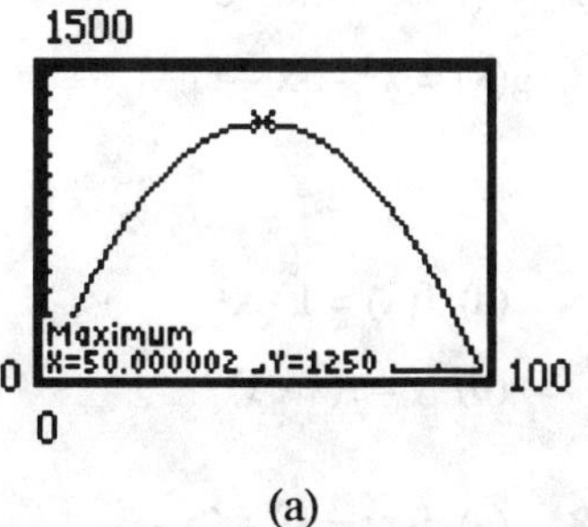

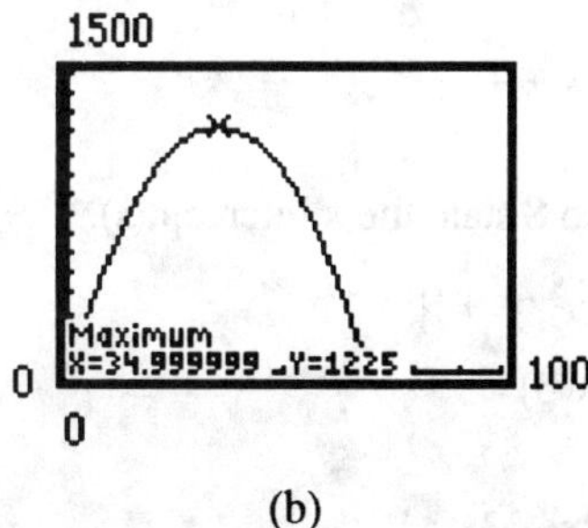

Figure 19 (a) (b)

In general, if the vertex of a parabola is not in the domain of the application, the maximum may occur at the endpoint of the graph.

Explore the Concepts

The x coordinate of the vertex is related to a and b in quadratic functions of the form $y = ax^2 + bx + c$, $a \neq 0$. Use a graphing utility to estimate the x coordinate of each turning point (vertex) on each parabola. Note the values of a and b, looking for the formula.

E1. The following quadratic functions, which have $b = 0$.

(a) $y = x^2$ (b) $y = x^2 - 10$ (c) $y = 12 - x^2$

E2. The following quadratic functions, which are all perfect square trinomials.

(a) $y = x^2 - 4x + 4 \Leftrightarrow y = (x - 2)^2$

(b) $y = 4x^2 + 4x + 1 \Leftrightarrow y = 4(x + \frac{1}{2})^2$

(c) $y = x^2 + 10x + 25 \Leftrightarrow y = (x + 5)^2$

(d) $y = 2x^2 - 20x + 50 \Leftrightarrow y = 2(x - 5)^2$

(e) $y = -2x^2 - 16x - 32 \Leftrightarrow y = -2(x + 4)^2$

(f) $y = -3x^2 + 24x - 48 \Leftrightarrow y = -3(x - 4)^2$

E3. Use your results from Exercises E1 and E2 to complete the following. The x coordinate of the vertex of the parabola whose equation is $y = ax^2 + bx + c$, $a \neq 0$ is x = _____.

Exercises 2.3

Develop the Concepts

For Exercises 1 to 4 investigate the vertex of each parabola graphically and verify algebraically by completing the square or using the vertex formula.

1. $y = x^2 - 4x - 5$
2. $y = -2x^2 - 4x + 8$
3. $g(x) = -3x^2 - x + 4$
4. $g(x) = x^2 - 3x - 4$

For Exercises 5 to 8 state the x intercept(s).

5. (a) $f(x) = x^2 - 6x + 9$
   (b) $y = f(x + 3)$
6. (a) $f(x) = 1 - x^2$
   (b) $y = f(x - 1)$
7. (a) $f(x) = -2x^2 - 3x - 1$
   (b) $y = -f(x)$
8. (a) $f(x) = x^2 + 2x - 3$
   (b) $y = -f(-x)$

For Exercises 9 to 12 state the y intercept.

9. (a) $f(x) = x^2 - 2x - 3$
   (b) $y = f(x) + 4$
   (c) $y = f(x) - 1$
10. (a) $f(x) = x^2 - 4$
    (b) $y = f(x) + 4$
    (c) $y = f(x) - 3$
11. (a) $f(x) = -x^2 - x + 2$
    (b) $y = -f(x)$
    (c) $y = f(-x)$
12. (a) $f(x) = 9 - x^2$
    (b) $y = 2f(x)$
    (c) $y = f(2x)$

For Exercises 13 to 16 graph each parabola and label the vertex and intercepts.

13. $y = -9x^2 + 30x - 25$

14. $y = x^2 - 2$

15. $f(x) = x^2 - 2x - 4$

16. $g(x) = 2 - 4x + 4x^2$

For Exercises 17 and 18 use the discriminant to find the values of k that will yield (a) two x intercepts ($b^2 - 4ac > 0$), (b) no x intercepts ($b^2 - 4ac < 0$), or (c) one x intercept ($b^2 - 4ac = 0$).

*17. $f(x) = 3x^2 - 2x + k$

18. $g(x) = x - 3kx^2 - 2$

Apply the Concepts

For Exercises 19 to 22 use the formula $s(t) = -16t^2 + v_0t + s_0$, modeling vertical distance in feet as a function of time in seconds.

19. The Skydiver Problem A skydiver dropped out of a plane some 1600 ft above the ground. The skydiver pulled the ripcord to open the parachute 10 sec later. How far above the ground was the skydiver after free-falling 10 sec?

20. The Superman Problem To save a busload of people, Superman dives straight down from an open window in the Daily Planet Building located 240 ft above the street. If his initial velocity is 200 ft/sec, how long will it take him to reach street level?

21. The Long-Shot Problem John Dailey, winner of the British Open in 1995, is one of professional golf's biggest hitters. On one of his big drives, the ball left with an initial vertical velocity of 130 ft/sec. Find the duration of flight and the maximum height for a golf ball.

22. The High-Impact Problem The longest golf shot was 437 yd hit by Jack Mann. Find the initial velocity required if the ball left the ground at a 45° angle, making its initial horizontal and vertical velocities equal.

Exercises 23 to 28 are optimization problems where their solution is found at the vertex of the parabola.

23. The Perennial Border Problem Your neighbor buys 200 ft of brick border and decides to divide his flowerbed into sections using an existing wall for one side of the garden (Figure 20). What dimensions should the neighbor use to maximize the garden area?

Figure 20

24. <u>The Aluminum Gutter Problem</u> A 10-m-long aluminum gutter is to be formed from a rectangular sheet 10 by 1000 cm by bending equal amounts up on each side as shown in Figure 21. Find the amount to bend to maximize the gutter's carrying capacity.

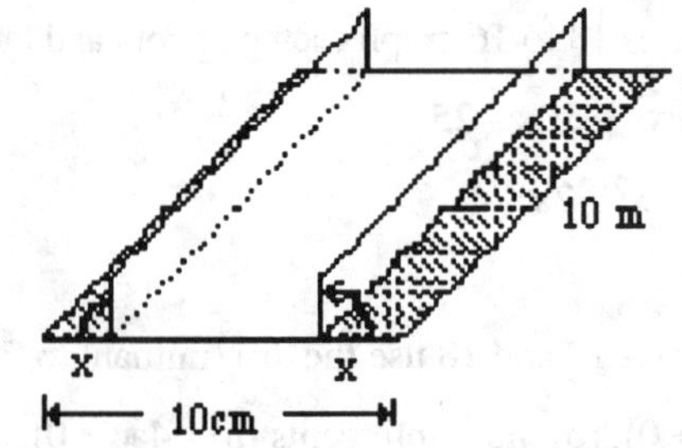

Figure 21

25. The Rainbow Kite and Balloon Company A two-stick kite is the familiar diamond kite in Figure 22. If the total length of the expensive graphite sticks is budgeted at 144 in., what dimensions provide maximum lift?

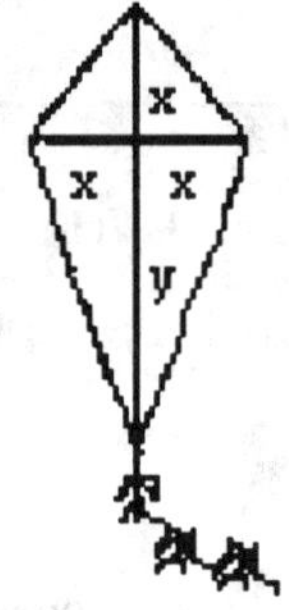

Figure 22

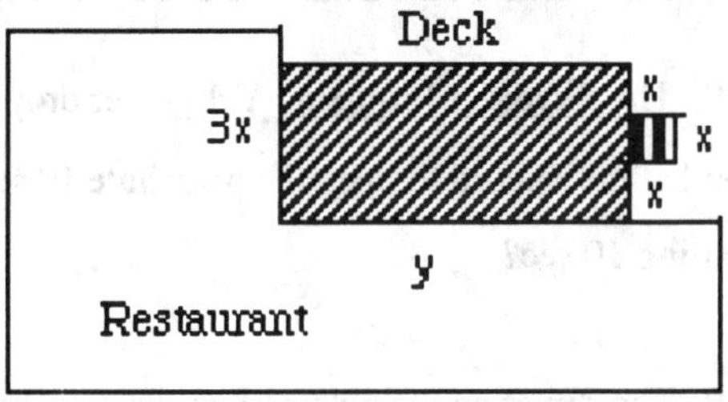

Figure 23

26. Wonderworld Amusement Park The New Orleans Cajun restaurant at Wonderworld has 120 ft of ornate wrought iron to build a dining deck as shown in Figure 23. Stairs divide the end into equal thirds. What dimensions will create the largest deck if the stairs require 6 ft of railing down each side?

27. <u>The Adjustable Playpen Problem</u> The total length of an adjustable fence that can be set up in any shape to form an enclosed playpen is 12 ft. Find the maximum area if (a) a rectangular pen is built in the corner using two existing walls for sides or (b) a pen is built in the shape of a quarter of a circle in the corner of the room.

28. <u>A Patio Problem</u> You want to build a patio surrounded by 50 ft of hedge as shown in Figure 24. An existing 12-ft section of wall is already built on the side of the planned patio as shown. Find the values of x and y that will yield the largest patio.

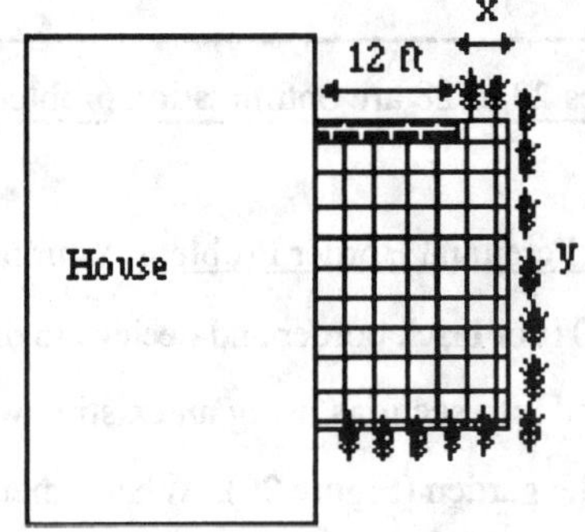

Figure 24

Exercises 29 to 32 are optimization problems where the solution is found at the endpoint of the domain of the application, not necessarily at the vertex of the parabolic model.

29. The Towering Rock Problem A rock was launched from a slingshot so that its height above the ground s in feet is given by $s = -16t^2 - 64t + 192$.

(a) What was its maximum height above the ground?

(b) Why isn't it at the vertex of the parabola?

30. Another Patio Problem Find the dimensions to build the patio in Figure 24 if there are enough plants to build a 36-ft border.

31. The Rainbow Kite and Balloon Company A dragon kite calls for two sticks as shown in Figure 25, with the total length of the expensive graphite sticks budgeted at 100 in. What dimensions provide maximum surface area?

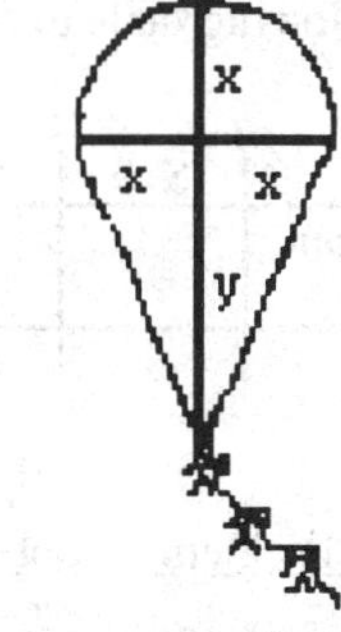

Figure 25

32. Wonderworld Amusement Park You want to design a patio with three brick planter boxes as shown in Figure 26. You have enough picket fence for a total of 16 ft of fence to be built between the planters. Find the values of x and y that will yield the largest patio.

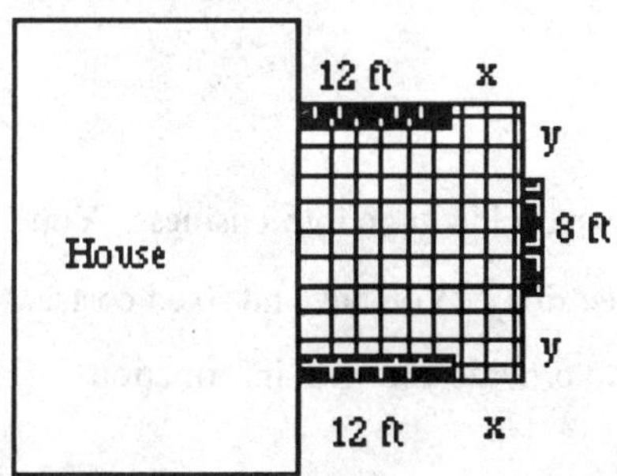

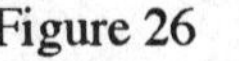

Figure 26

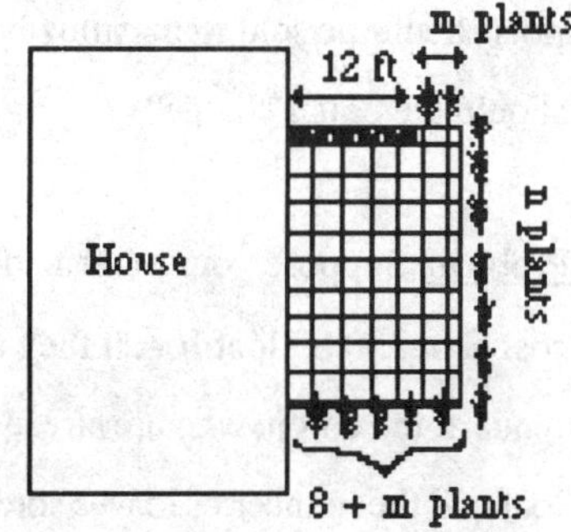

Figure 27

33. The Discrete Hedge Problem The hedge in Figure 27 consists of 36 individual plants planted in holes 18 in. on center. (No fractional parts of plants are allowed.) Complete the following table to find the number of plants on each side that will yield the maximum area.

number of plants, m	0	1	2	3	4
number of plants, m + 8					
number of plants, n					
length, l = 1.5n					
width, w = 12 + 1.5m					
area, A = lw					

34. The Tax Revenue Problem Assume that you own a monopoly. You have studied your costs and determined relationships between the selling price p and the number of items x you are willing to supply. You have also studied your customers and found the relationship between price and the number of items they are willing to buy, demand.

supply: $p_1 = 2x + 50$ demand: $p_2 = 100 - 0.1x$

(a) Find market equilibrium, the point where the supply and demand curves intersect.

(b) If a tax of t dollars per item is levied on your business, you translate your supply curve up t units to pass the additional charge on to the consumer. Find the new supply function.

(c) Complete the following table to find the tax t that will maximize tax revenue.

tax, t	\$10	\$20	\$25	\$30	\$40
new market equilibrium, x = ?					
tax revenue, TR = t·x					

*35. The Flowerpot Problem A pot thrower can make flowerpots for \$0.30 each. Fixed costs of \$27/month make the cost function $C(x) = 0.30x + 27$. A market study indicates the following relationship between the price and the number sold each day.

quantity, x	25	30	40	45
price, p	\$4	\$3	\$2.50	\$2

(a) How many pots should be sold to maximize profit?

(b) What price should the potter charge?

36. The Bread Problem Suppose your homemade bread is so good that you decide to go into business. You estimate that it cost about \$0.85/loaf for all the natural ingredients and electricity. You have no fixed costs. You ask your friends at the co-op, who are already in the bread business, to provide you with information regarding price/loaf and the number of loaves sold.

quantity, x	10	13	19	20	23
price, p	\$2.50	\$2.25	\$2.00	\$1.75	\$1.50

(a) How many loaves should you make to maximize the profit?

(b) What price should you charge?

Extend the Concepts

37. The Taxing Problem Use the supply and demand functions from Exercise 34, $p_1 = 2x + 50$ and $p_2 = 100 - 0.1x$, to find the following.

(a) Find the equation of the new supply function where $p(x) = p_1(x) + t$.

(b) Find the relationship between tax t and quantity x at market equilibrium where supply equals demand.

(c) Use the result from part (b) to express tax revenue TR as a function of quantity x, where $TR = t \cdot x$.

(d) Find the tax t that will yield maximum tax revenue.

38. The Advertising Dollar Problem Use the supply and demand functions from Exercise 34, $p_1 = 2x + 50$ and $p_2 = 100 - 0.1x$, to find the following.

(a) Find the equation of the new supply function if paying z dollars for advertising translates the supply curve p_1 up z dollars.

(b) Find the equation of the new demand function if spending z dollars/item on advertising translates the demand function p_2 right 0.50z.

(c) Find the amount to charge for the advertising that will maximize the advertising company's revenue, $R = z \cdot x$, where x is the number of items sold at the new equilibrium point.

Exercises 39 to 42 involve functions that are quadratic-in-form. Each optimum value occurs at a turning point on the graph.

*39. The Corner Playpen Problem A child's 20-ft adjustable fence is placed across a corner of her room to make a triangular play area. Show that the maximum area is achieved when the triangle is isosceles. (An isosceles triangle has two equal sides.)

40. The Cable Problem A cable TV company is running a cable along the line that lies 20 ft east of a house to a point 80 ft north of the house (Figure 28). The owners of the house can connect to the cable line cheaper if they dig the ditch connecting their house to the line of the cable themselves. The homeowners decide to find the point on the cable line that is the minimum distance from their house. Find where they should dig. [Hint: Let the coordinates of the house be (0,0) and use the distance formula $d = \sqrt{(x_2 - x_1)^2 + (y_2 - y_1)^2}$. Do not use the fact that the shortest distance from a point to a line is the perpendicular distance, because that is what your are proving.]

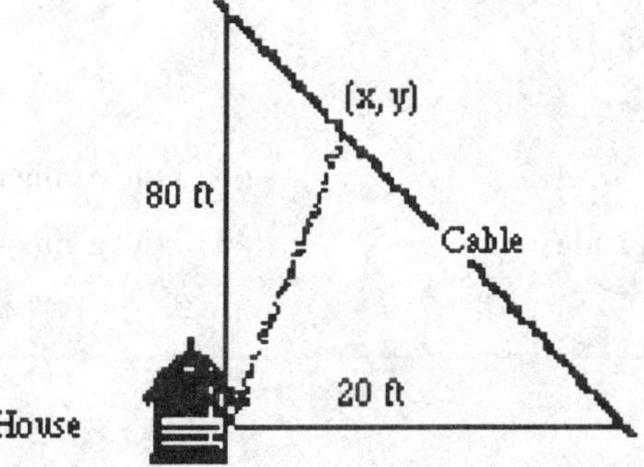

Figure 28

41. The Sailboat Problem The sailboat Lark, sailing east at 6 knots (1 knot = 1 nautical mile/hr), is initially 20 nautical miles due north of the power boat Crow's Nest, cruising north at 42 knots (Figure 29). Find the time when the two boats will be closest together.

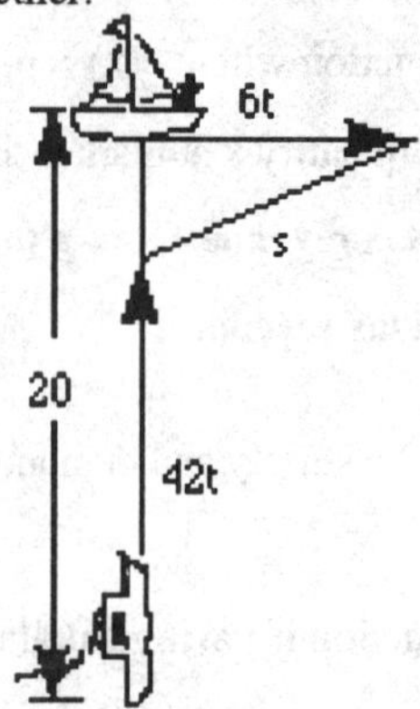

Figure 29

42. Another Sailboat Problem The sailboat Lark, sailing east, is initially 20 nautical miles due north of the power boat Crow's Nest, when the power boat suddenly loses power in one of its two engines, reducing its speed as it limps north (Figure 30a). Find the time when the two boats will be closest together. The graphs in Figures 30a and 30b indicate that each is traveling at a constant rate -- rates represented by the slopes of the two lines.

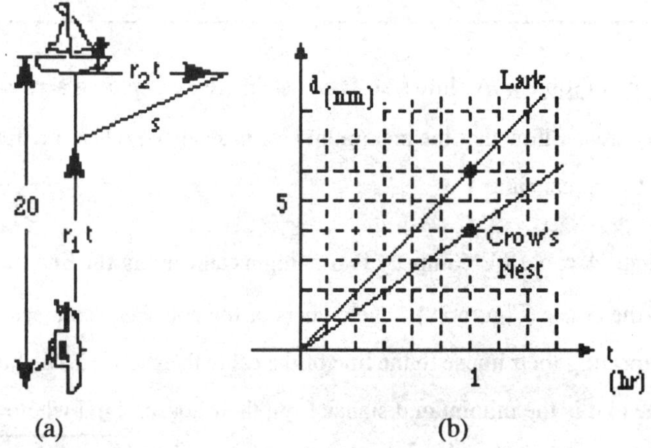

Figure 30 (a) (b)

For Exercises 43 to 46 let $r_1 = \dfrac{-b + \sqrt{b^2 - 4ac}}{2a}$ and $r_2 = \dfrac{-b - \sqrt{b^2 - 4ac}}{2a}$ be the roots of the general quadratic equation, $ax^2 + bx + c = 0$, $a \neq 0$.

43. Prove that $r_1 + r_2 = -b/a$.

44. Find a simple expression for $r_1 \cdot r_2$.

45. Find the average value of r_1 and r_2.

46. Find a simple expression for $r_1{}^2 + r_2{}^2$.

2.4 Finding Equations of Quadratic Functions

Finding the Equation of a Parabola Given the Vertex
Finding the Equation of a Parabola Given the X-Intercepts
Finding the Equation of a Parabola Given Three Points
Finding the Equation of a Parabola Using Quadratic Regression
Finding the Equation of the Best-Fitting Parabola

This section covers various techniques for finding equations of quadratic functions from given data, a process called **curve fitting**. Some of the curve-fitting techniques studied in this section generalize to other functions. Using a graphing utility, we can handle realistic data. For example, in Example 3 we find the equation of the parabolic cable on the Golden Gate Bridge in order to study its deflection (amount of bending) on Sunday, May 24, 1987, the bridge's fiftieth birthday, when 300,000 people walked onto the bridge at the same time (Figure 1).

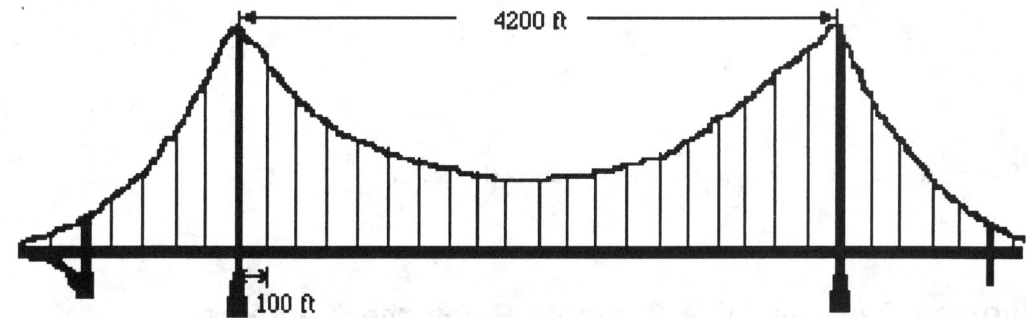

Figure 1

Finding the Equation of a Parabola Given the Vertex

Recall from Section 2.3 that a quadratic function written in the form $f(x) = a(x - h)^2 + k$ has its vertex at (h, k).

Can you find the equation of a parabola knowing only that its vertex is (1,-3)?

Knowing only the vertex of a quadratic function is not enough to allow us to find its equation since there are an infinite number of parabolas with vertex (1,-3). Figure 2 shows three such parabolas having different values for the scaling factor a.

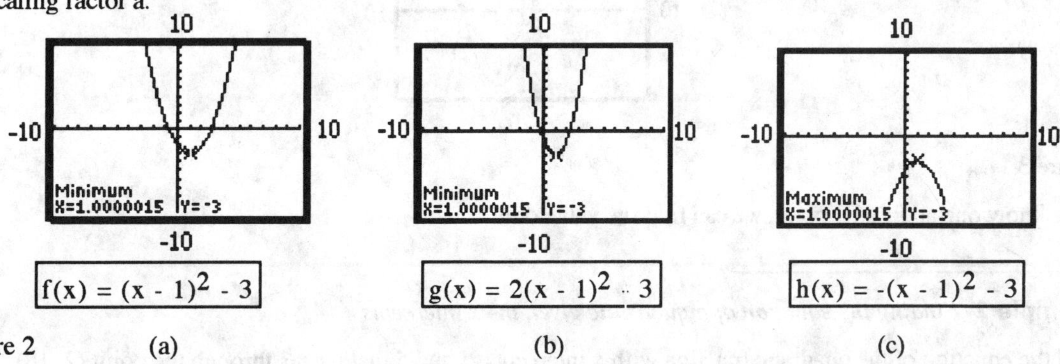

Figure 2 (a) (b) (c)

If we know the vertex of a parabola and one additional point we can find the equation of that unique parabola. The additional point allows us to find the value of the scaling factor. Notice that if $a < 0$, the parabola is concave down (Figure 2c).

Example 1 *Finding the equation of a quadratic function given the vertex*

Find the equation of the quadratic function whose graph has its vertex at (1,-3) and passes through the point (2,-2).

Solution

$f(x) = a(x - h)^2 + k$ and $h = 1, k = -3 \Rightarrow$

$$f(x) = a(x - 1)^2 - 3$$

$f(x) = a(x - 1)^2 - 3$ and $f(2) = -2 \Rightarrow$

$$-2 = a(2 - 1)^2 - 3$$

$$-2 = a - 3$$

$$a = 1$$

Hence, $f(x) = 1(x - 1)^2 - 3$ is the equation of this parabola (Figure 2a).

Finding the Equation of a Parabola Given the X Intercepts

When a quadratic function is written in factored form

$$f(x) = a(x - x_1)(x - x_2)$$

the x intercepts are x_1 and x_2. Knowing the x intercepts of a quadratic function is not enough to allow us to find its equation since there are an infinite number of parabolas having the same x intercepts but different values of a (Figure 3).

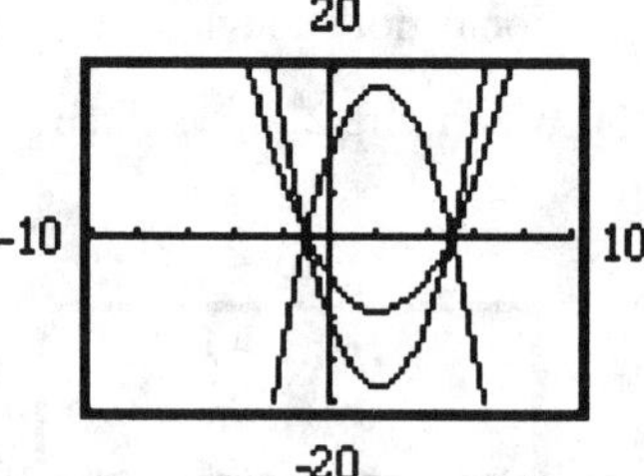

Figure 3

If we know one additional point, we can find the value of a.

Example 2 *Finding the equation of a quadratic given the x intercepts*

Find the equation of the quadratic function with x intercepts -1 and 5 that passes through the point (2, 18).

Solution

$f(x) = a(x - x_1)(x - x_2)$ and $x_1 = -1, x_2 = 5 \Rightarrow$

$$f(x) = a(x + 1)(x - 5)$$

$f(x) = a(x + 1)(x - 5)$ and $f(2) = 18 \Rightarrow$

$$18 = a(2 + 1)(2 - 5)$$

$$18 = -9a$$

$$a = -2$$

Hence, the equation of the parabola is $f(x) = -2(x + 1)(x - 5)$.

Finding the Equation of a Parabola Given Three Points

In Example 2 we were given three points, but since two of them were x intercepts the equation was easy to find. Given any three points on a parabola we can write three equations to find the three unknown coefficients a, b, and c of the parabola's equation, $f(x) = ax^2 + bx + c$. To find the equation of the parabola that passes through (1,3), (2,5), and (3,11), we substitute the coordinates of the three points into the equation $f(x) = ax^2 + bx + c$ creating three equations with three variables, a, b, and c.

$f(1) = 3 \Rightarrow$

$$3 = a(1)^2 + b(1) + c$$

$$\mathbf{3 = a + b + c}$$

$f(2) = 5 \Rightarrow$

$$5 = a(2)^2 + b(2) + c$$

$$\mathbf{5 = 4a + 2b + c}$$

$f(3) = 11 \Rightarrow$

$$11 = a(3)^2 + b(3) + c$$

$$\mathbf{11 = 9a + 3b + c}$$

The three equations with three variables are

$$a + b + c = 3$$

$$4a + 2b + c = 5$$

$$9a + 3b + c = 11$$

Multiplying both sides of the first equation by -1 and adding that result to both the second and third equations eliminates c.

$$\begin{array}{r} -a - b - c = -3 \\ \underline{4a + 2b + c = 5} \\ \mathbf{3a + b = 2} \end{array} \qquad \begin{array}{r} -a - b - c = -3 \\ \underline{9a + 3b + c = 11} \\ \mathbf{8a + 2b = 8} \end{array}$$

Solving the resulting pair of equations yields a = 2 and b = -4. Any one of the original equations yields the value of c.

$a + b + c = 3$ and $a = 2, b = -4 \Rightarrow$

$$\begin{aligned} 2 - 4 + c &= 3 \\ c &= 5 \end{aligned}$$

Therefore, the equation of the parabola passing through the points (1, 3), (2, 5), and (3, 11) is $f(x) = 2x^2 - 4x + 5$.

Table 1 summarizes the given information for each of the three possible forms for finding a quadratic equation.

Table 1

Given Information	Equation Form
vertex and a point	$y = a(x - h)^2 + k$
x intercepts and a point	$y = a(x - x_1)(x - x_2)$
any three points	$y = ax^2 + bx + c$

Finding the Equation of a Parabola Using Quadratic Regression

Most graphing utilities provide an alternative approach for finding the quadratic equation whose graph passes through three distinct, noncollinear points without repeated values of x. To find the parabola that passes through the points (1, 3), (2, 5), and (3, 11) from above, we enter the data $x_1 = 1$, $y_1 = 3$, $x_2 = 2$, $y_2 = 5$, $x_3 = 3$, and $y_3 = 11$, then use the **quadratic regression** (QuadReg) program (Figure 4).

L1	L2
1	3
2	5
3	11

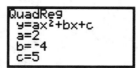

```
QuadReg
 y=ax²+bx+c
 a=2
 b=-4
 c=5
```

Figure 4

The equation is $y = 2x^2 - 4x + 5$. Its graph and the **scatter plot** of the original data are shown in Figure 5.

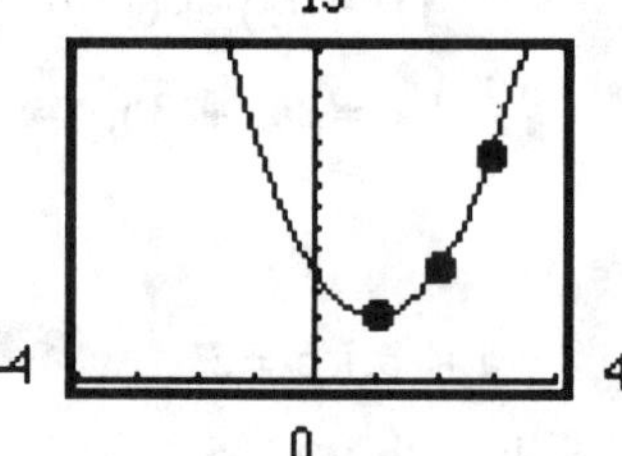

Figure 5

Example 3 The Golden Gate Bridge Birthday Party Problem

Sunday, May 24, 1987, was the fiftieth birthday of the Golden Gate Bridge in San Francisco. During the celebration, 300,000 people walked onto the bridge at the same time, uniformly loading the bridge deck and causing the roadway in Figure 6 to deflect in a parabolic shape. According to the design criterion, the bridge has a maximum downward deflection of 11 ft. A measurement made during the party showed a 1-ft deflection at a point 100 ft from the tower on the San Francisco side. Was the crowd in danger?

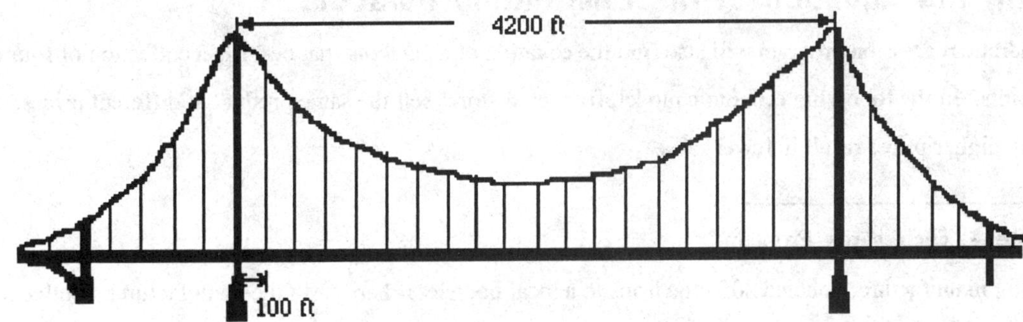

Figure 6

Thinking: Creating a Plan

Assuming that the bridge is symmetric allows us to find the equation for the parabola-shaped bridge deck during the party. For simplicity, we will place the x axis along the roadbed and the y axis in the center of the bridge. We will find the equation of the parabolic roadbed during the party using the three points in Figure 7b and quadratic regression.

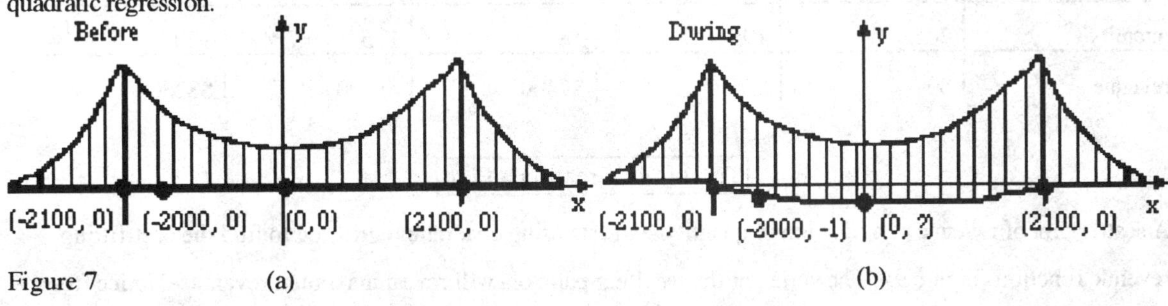

Figure 7 (a) (b)

Communicating: Writing the Solution

To find the parabola that matches the roadbed during the party, enter (-2100, 0), (-2000, -1), and (2100, 0) and use quadratic regression (QuadReg) (Figure 8).

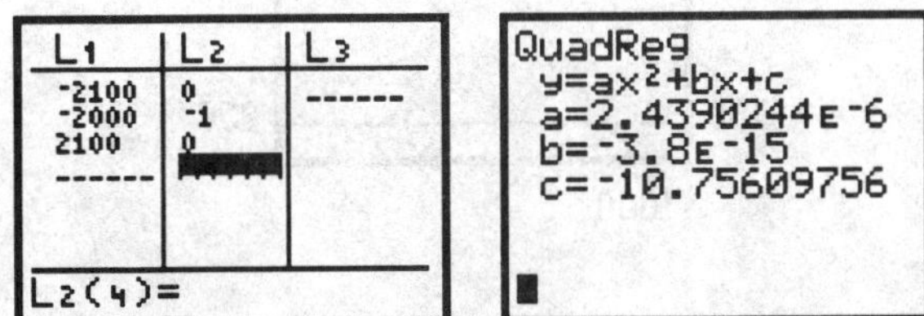

Figure 8

The equation for the roadbed was $y = 0.000002439x^2 - 0.0000000000000038x - 10.7561$. The y intercept yields the deflection, -10.756. The deflection of the bridge deck at the center was about 10 ft 9 in. The official word was that there was no danger, but according to the design specifications, 300,000 people should have been concerned.

Learning: Making Connections

In many applications, choosing the coordinate system can simplify the solution. We could have placed the x axis across the top of the towers and found two equations for the parabolic cable -- one for before and the other for during the party. Subtracting their y intercepts would yield the same result.

Finding the Equation of the Best-Fitting Parabola

The quadratic regression program will also find the equation of a parabola that best fits a collection of four or more points. In the following economic model, five retail stores sell the same product at different prices. As expected, higher prices result in fewer sales.

Example 4 The Charity Problem

A clothing manufacturer donated 205 blue jeans to a local homeless charity. In the manufacturer's outlet store, 195 pair of jeans sold for $28 each. A discount store sells 185 jeans at $40 each. The store in the mall sells 175 at $48. An exclusive store downtown sells 140 pairs at $59. The table below summarizes the revenue found by multiplying the price per pair and the quantity. Find the quantity that creates the maximum revenue.

	Charity	Outlet	Discount	Mall	Downtown
price	$0	$28	$40	$48	$59
quantity	205	195	185	175	140
revenue	$0	$5460	$7400	$8400	$8260

Thinking: Creating a Plan

A scatter plot of revenue as a function of quantity suggests using quadratic regression to find the best-fitting revenue function (Figure 9). The vertex of the resulting parabola will reveal maximum revenue. Notice from the data that raising the price of jeans increases the revenue -- to a point. After that point, sales decrease so much that revenue decreases.

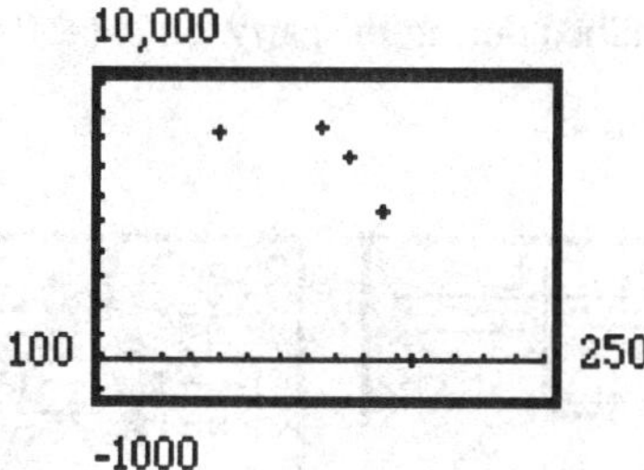

Figure 9

Communicating: Writing the Solution

We enter the data (quantity, revenue) from the table and use quadratic regression (QuadReg) (Figure 10).

L1	L2
205	0
195	5460
185	7400
175	8400
140	8260

QuadReg
y=ax²+bx+c
a=-4.757029178
b=1525.503979
c=-112180.4562

Figure 10

The best-fitting quadratic function (Figure 11) for revenue as a function of sales is approximated by

$$R = -4.76x^2 + 1525.50x - 112{,}180.46$$

The revenue is maximized by selling 160 pairs of jeans. The resulting revenue is $10,043.82.

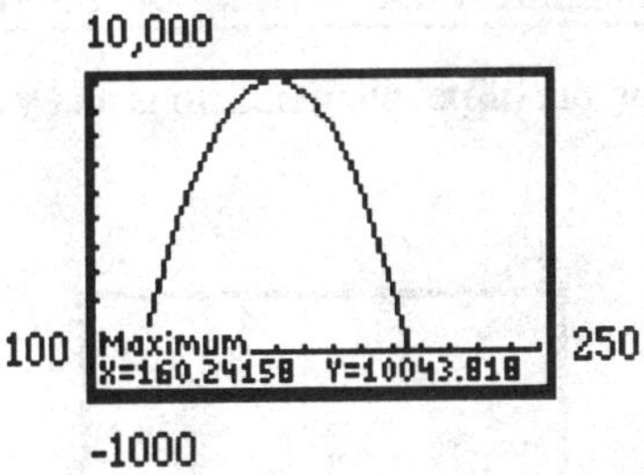

Figure 11

Learning: Making Connections

The graph of $R(x) = -4.76x^2 + 1525.50x - 112{,}180.46$ with a scatter plot of the data is shown in Figure 12.

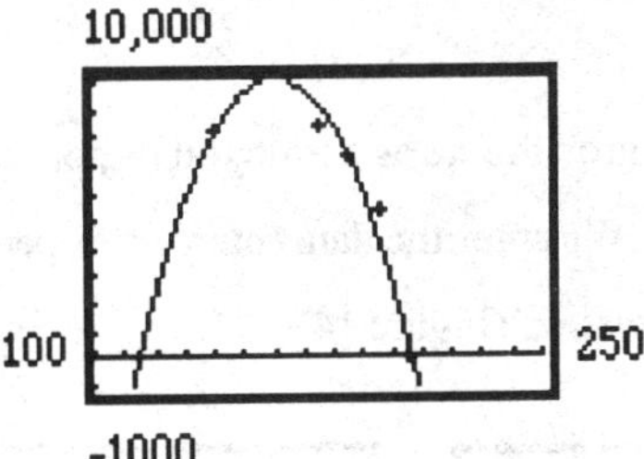

Figure 12

We use the price and quantity data to find the linear model for consumer demand. The function $p(x) = -0.792x + 177.56$ yields the price to charge to sell 160 pair of jeans, $p(160) = 50.84$. Price the jeans at $50.84 to sell 160 pairs and maximize profit.

Example 5 The ZPG Problem

Zero population growth (ZPG) occurs when the growth rate is zero. The growth rate is the difference in the birthrate and the death rate. Find the population that will yield ZPG for the U.S. population based on the following data. [Source: *The American Almanac: Statistical Abstract of the United States*]

year	1980	1985	1990	1995	2000
people (millions)	226.55	236.94	248.14	262.07	275
birthrate (millions)	2.74	3.76	4.15	4.02	3.93
death rate (millions)	1.46	2.086	2.15	2.21	2.357

Thinking: Creating a Plan

Using the table above we find the growth rate by subtracting the death rate from the birthrate.

people (millions)	226.55	236.94	248.14	262.07	275
growth rate (millions)	2.74 - 1.46 = 1.28	3.76 - 2.086 = 1.674	4.15 - 2.15 = 2.00	4.02 - 2.21 = 1.81	3.93 - 2.357 = 1.573

Then, we find the equation that expresses growth rate as a function of the population.

Communicating: Writing the Solution

A scatter plot where x represents the number of people in millions and y represents the growth rate in millions appears parabolic (Figure 13).

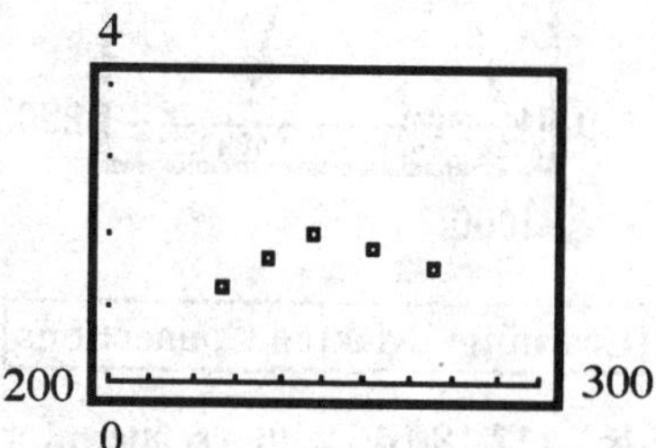

Figure 13

We use quadratic regression to find the growth rate as a function of population. The predicted ZPG is the intercept where the growth rate is zero. We enter the data (number of people in millions, growth rate in millions per year) and use quadratic regression (Figure 14).

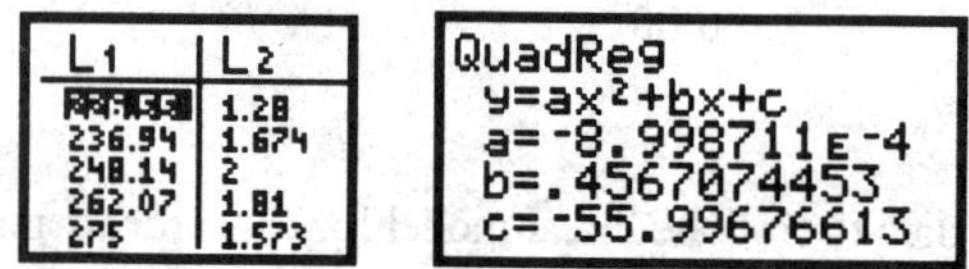

Figure 14

Therefore, the best-fitting quadratic function is

$$G(x) = -0.00089987x^2 + 0.4567x - 55.99677$$

We use the root or zero feature to find the x intercept (Figure 15).

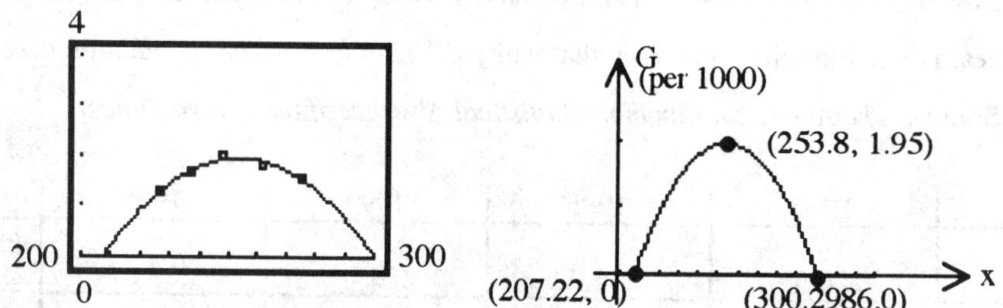

Figure 15

Based on these data and this quadratic model, the United States will reach zero population growth when its population reaches 300,298,600 people.

Learning: Making Connections

As population increases, a decreasing birth rate and an increasing death rate are not unusual. We will study this combination again in Chapter 4, where it leads to the *logistic curve* with time as the independent variable.

Exercises 2.4

Develop the Concepts

For Exercises 1 to 8 find the equation of each parabola.

1. The vertex is (-1, 0) and the y intercept is 4.
2. The vertex is (2, 0) and the y intercept is -3.
3. The vertex is (1, 2) and the graph passes through (2, 3).
4. The vertex is (-1, 3) and the graph passes through (-3, -5).
5. The x intercepts are -2 and 2. The y intercept is 8.
6. The x intercepts are -4 and 2. The y intercept is 16.

*7. The x intercepts are $1 \pm \sqrt{3}$ and the graph passes through (2, -2).

8. The zeros are $1 \pm i$ and the graph passes through (1, -2).

For Exercises 9 to 16 use quadratic regression to find the equation of each parabola.

9. (-1, 4), (0, 9), and (3, 0).

10. (-10, 63), (0, 1), and (10, 39).

11. (3, 6), (5, 12), and (9, 25).

12. (1,- 2), (2, -5), and (7, 15).

13.

x	-2	-1	0	1
y	8	2	0	2

14.

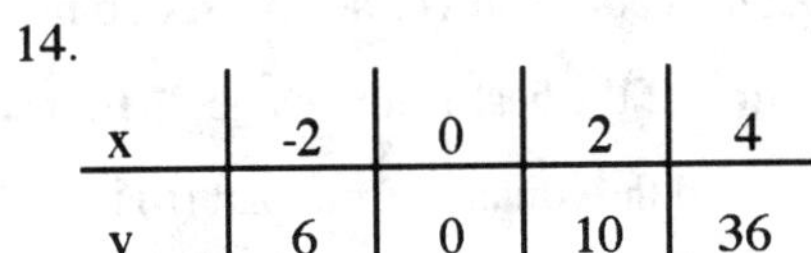

x	-2	0	2	4
y	6	0	10	36

15.

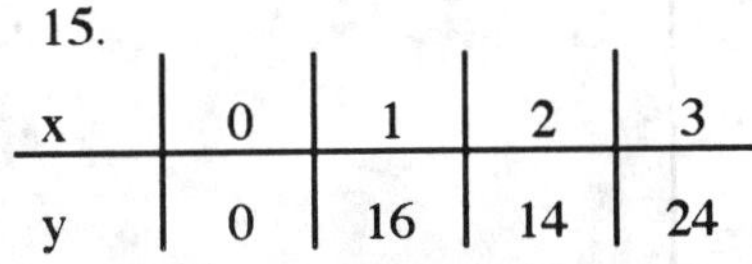

x	0	1	2	3
y	0	16	14	24

16.

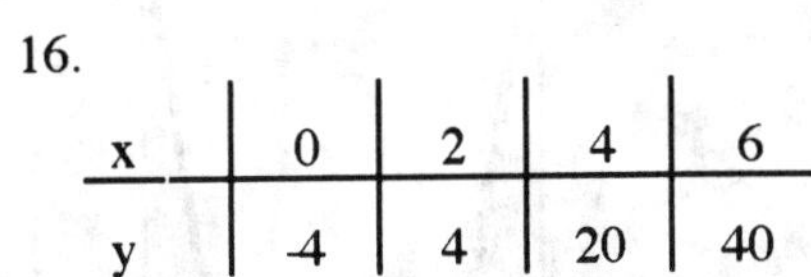

x	0	2	4	6
y	-4	4	20	40

Apply the Concepts

In Exercises 17 to 20 assume that the algebraic model is a quadratic function.

(a) Use quadratic regression to find the equation of the best-fitting parabola.

(b) Graph the best-fitting parabola and find the optimum value.

17. The Rotten Egg Problem The height above the ground of a rotten egg thrown straight up is measured as follows:

time (sec)	0	1	2	3
height (m)	0	14.7	19.6	14.7

18. Another Rotten Egg Problem The height above the ground of another rotten egg thrown straight up is measured as follows:

time (sec)	0	0.5	1	2	3	4
height (m)	0	8.5	25	30	15.5	0

19. A Profit Problem Profit from selling sunglasses was found to be:

price	\$2.95	\$4.95	\$10.95	\$12.95
profit	\$3000	\$3124	\$3128	\$2995

20. The Revenue Problem Revenue from selling loose candy was found to be:

quantity	0	45	105	150	300
revenue	\$0	\$3.65	\$5	\$3.75	-\$15

21. The Suspended Cable Problem Tape two pieces of string to a standard $8\frac{1}{2}$- by 11-in. sheet of paper as shown in Figure 16. Create a coordinate system on the paper.

(a) Find and label the coordinates of at least six points on the suspended string.

(b) Find the equation of the best-fitting parabola for each.

(c) Graph the data and the equation from part (b) to see which is nearly a parabola. (The other is called a catenary.)

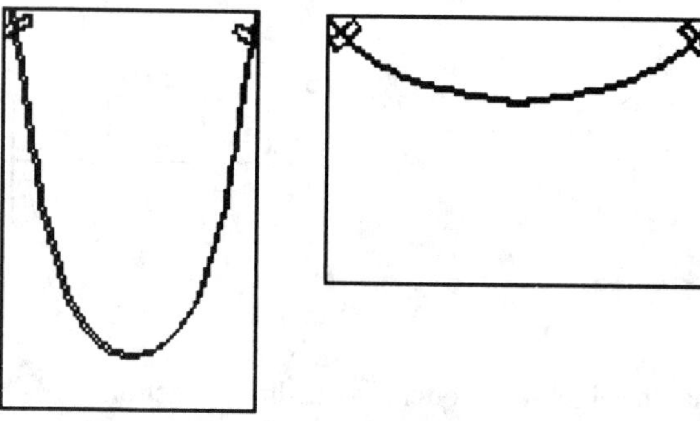

Figure 16

22. The Hand-Drawn Parabola Problem Make a freehand drawing of a parabola on a standard $8\frac{1}{2}$- by 11-in. sheet of paper. Create a coordinate system on the paper.

(a) Find and label the coordinates of at least six points on your drawing.

(b) Find the equation of the best-fitting parabola.

(c) Graph your data and the equation from part (b) to see how well your hand can draw a parabola.

23. The Circle Problem Draw four circles having different radii on a piece of graph paper. Measure the radius of each circle and count the squares on the graph paper to estimate the area of each circle. Use your measurements (radius, area) to find the equation of the best-fitting quadratic function that expresses area as a function of radius.

24. The Isosceles Triangle Problem Draw five different size isosceles right triangles on a piece of graph paper. (An isosceles right triangle is a triangle having two equal sides that form a right angle.)

(a) Measure the length of one of the equal sides and count the squares on the graph paper to estimate the area of each triangle. Use your measurements (length of side, area) to find the equation of the best-fitting quadratic function that expresses area as a function of length of side.

(b) Measure each hypotenuse. Use measurements (hypotenuse, area) to find the equation of the best-fitting quadratic function that expresses area as a function of length of the hypotenuse.

*25. The ZPG and Immigration Problem Change the population data in Example 5 by adding the immigration numbers given below to the birthrate. Find the new equilibrium point using the same death rate. [Hint: Express total growth rate as a function of N and death rate as a function of N.]

year	1980	1985	1990	1995	2000
people N (millions)	226.55	236.94	248.14	262.07	275
birthrate (millions)	2.74	3.76	4.15	4.02	3.93
immigration rate	724,000	649,000	594,000	880,000	880,000
death rate (millions)	1.46	2.086	2.15	2.21	2.357

26. The Torricelli's Law Problem A quadratic function will tell you the velocity of a speeding boat by measuring the height of the water forced up an L-shaped tube held over the side of the boat. Suppose that you took the following measurements as your friend clocked the speed.

(a) Find the quadratic equation that expresses boat speed as a function of height of water in the tube.

(b) The speed limit in narrow passages is often 6 knots. How high should the water rise in the tube?

(c) At full throttle the water rose $3\frac{1}{4}$ in. How fast can this boat go?

height (in.)	0	1	2	3
velocity (knots)	0	5.1	20.4	45.9

27. The Intercom Problem Your dream home is to have an intercom in several rooms. Each intercom is to be connected with every other intercom.

(a) Make a table of values for n = 1, 2, 3, 4, 5 intercoms and the number of required connecting lines.

(b) Find the quadratic equation expressing the number of lines as a function of the number of intercoms.

(c) Use the equation from part (b) to find how many connecting lines are required for six intercoms and 10 intercoms.

28. <u>The Triangle Numbers Problem</u> The first triangle number is 1, the second 3, the third 6, and the fourth 10 (Figure 17).

(a) Make a table of values for n = 1, 2, 3, 4, 5 dots in the bottom row and the corresponding triangle numbers.

(b) Find the quadratic equation expressing triangle numbers as a function of the dots in the bottom row of the triangle.

(c) Use the equation from part (b) to find the sixth and tenth triangle numbers.

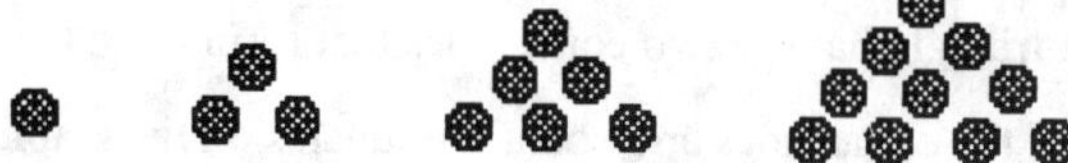

Figure 17

29. The Rainbow Kite and Balloon Company You want to make a series of windsocks with open ends and cross sections that are regular polygons or circular as shown in Figure 18. The edges at each end have to be hand-stitched, as well as one lateral seam to join the sides. The total stitching is budgeted at 72 in. Complete the following table to find the values of x and y that will yield the maximum surface area for each.

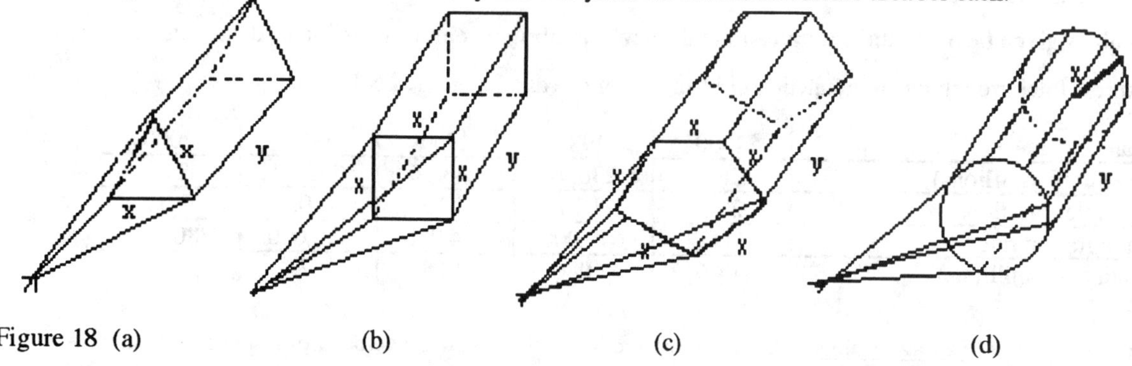

Figure 18 (a) (b) (c) (d)

Cross Section	Surface Area Function, SA	Constraint	Surface Area Subject to Constraint	Graph	Maximum Surface Area, x, y, SA

30. Wonderworld Amusement Park The revenue $R(x)$ from selling passes for unlimited rides at the park is a function of the number of passes sold x.

(a) Find the equation of the best-fitting quadratic function for the revenue data in the first table.

(b) The total cost $C(x)$ is also a function of the number of passes sold. Find the equation of the best-fitting linear function for the cost data in the second table.

(c) Use your revenue and cost functions from parts (a) and (b) to find the number of passes to sell to maximize the profit. Since profit = revenue - cost, $P(x) = R(x) - C(x)$.

passes, x	0	10	75	150
revenue, R(x)	0	70	600	0

passes, x	0	10	75
cost, C(x)	100	120	640

Extend the Concepts

31. The Book Budget Problem Suppose that you are in charge of the budget for a new textbook. There are two costs you must consider in terms of the degree of refinement (the absence of errors). To increase refinement you must increase spending on development of the manuscript. If you decrease spending, hence decrease refinement, you increase the eventual cost of having errors resulting from lost adoptions. While neither of these cost functions is necessarily quadratic, their sum, which expresses total cost as a function of the degree of refinement, is nearly quadratic. Find the optimum degree of refinement for a project with the following costs.
[Source: *An Introduction to Engineering and Engineering Design*, E. Krick]

degree of refinement, x	10%	50%	75%	90%
cost of development, y_1	$1500	$25,000	$50,000	$100,000
cost of errors, y_2	$90,000	$30,000	$10,000	$500

32. The Speed Limit Problem Suppose that you are in charge of setting the speed limit on a long bridge. As the speed limit increases, the spacing between vehicles increases. Your research team collected the following data.

speed limit, L (mi/hr)	25	30	35	40	55
spacing between vehicles, S (ft/vehicle)	13	15	18	22	53

By studying the dimensions of the variables involved, you realize that the quotient of L and S yields vehicles per hour.

$$\left[\frac{L}{S}\right] = \frac{\text{mi/hr}}{\text{mi/vehicle}} = \frac{\text{vehicles}}{\text{hr}}$$

(a) Find the table for vehicles per hour L/S as a function of the speed limit L.

[Hint: Change spacing between vehicles from feet to miles.]

(b) Find the optimum speed limit.

[Source: *An Introduction to Engineering and Engineering Design*, E. Krick]

A Guided Review of Chapter 2

2.1 Linear Functions

For Exercises 1 and 2 graph each line and write each equation in slope-intercept form.

1. The slope is -1/2 and the y intercept is -2.
2. The line is parallel to the x axis and passes through the point (4, -3).

For Exercises 3 to 6 find the equation of each line in slope-intercept form, if possible.

3. The line passes through the points (-1, -10) and (0, 10).
4. The line passes through the point (-2, 4) and is parallel to the line 2x - 5y = 10.
5. The line is perpendicular to the x axis and passes through the point (-1, 1).
6. The line whose equation satisfies f(-4) = 2 and f(5) = -5.

For Exercises 7 and 8 graph the piecewise-defined function.

7.

$$f(x) = \begin{cases} 10 - 2x, & x < 5 \\ 2x - 10, & x \geq 5 \end{cases}$$

8. $y = 1(x \leq 0) + 2(0 < x \leq 1) + 3(1 < x \leq 2) + 4(x > 2)$

9. <u>The Pyramid Problem</u> The largest pyramid monument is the Quetzalcoatl near Mexico City. It is 177 ft tall and its square base covers 45 acres. Compare the slopes of stairs built on the faces of the pyramid with those built on the edges. (One acre is 43,560 ft^2.)

10. <u>The Luxor Hotel Problem</u> The Luxor Hotel in Las Vegas, Nevada is built in the shape of a pyramid with a square base. The base is 350 ft on each side and the height of the pyramid is also 350 ft. Find the slope of the inclinators (elevators located at each corner of the pyramid that go up the edge).

11.(a) Find the equation of the inverse of f(x) = x/2 - 4.

(b) Show that $f(f^{-1}(y)) = y$. (c) Show $f^{-1}(f(x)) = x$.

12. <u>The Weather Balloon Problem</u> As a weather balloon rises, the temperature drops by 3°F per 1000 ft until the balloon passes 7 mi. Then the temperature remains nearly constant at -67°F. (Above 7 mi is the stratosphere.) The following equation gives Fahrenheit temperature T as a function of height h in feet.

$$T(h) = \begin{cases} \frac{-3}{1000} h + 60, & 0 \leq h < 36,960 \\ -67, & h \geq 36,960 \end{cases}$$

(a) Graph the function.

(b) Find the temperature on the ground.

(c) Find the temperature for h = 7 mi.

(d) Find the equation of the inverse for $0 \le h < 36{,}960$.

(e) Find the freezing level, the altitude when T = 32°F.

13. The Test Scoring Problem On a chemistry test, the class scored so poorly that the teacher decided to raise the lowest score of 45 to 65 and the highest score of 77 to 95.

(a) Find the equation that could be used to convert all other scores.

(b) To what score would a 70 convert?

(c) Interpret the y intercept.

14. The Clamming Problem In 1990 there were 10,000 clams in one clam bed and increasing by 400 clams/yr. In 1995 there were 12,500 clams in the same bed, increasing by 250 clams/yr. The harvest rate was set at 200 clams/yr.

(a) Find the equation that expresses growth rate as function of the number of clams, and the equation for the constant rate of harvest.

(b) Find and interpret the equilibrium point, the maximum sustainable yield.

2.2 Linear Regression

For Exercises 15 and 16 decide whether there is generally a positive, negative, or no correlation between each of the following.

15.(a) outside temperature and number of people at the beach

(b) outside temperature and absenteeism at work

16.(a) the number of turkeys sold and their price/lb

(b) price of ground beef/lb and the price of chicken/lb

17. Use linear regression to find the equation of the line that passes through the points (1/2, 1/10) and (-3/4, 3/5).

18. For the following set of points, show that the point $(\bar{x}, \bar{y})$ falls on the best-fitting line: {(-2, 1), (0, 4), (3, 5) (7, 2)}.

19. Find the equation of the best-fitting line. Use the equation to find the missing values.

year	1985	1986	1988	1989	1990	?
consumer expenditure by household	23,490	?	25,892	27,810	28,369	30,000

20. The Salt Concentration Problem In a science class you found the boiling temperature of water with increased concentrations of salt. You recorded the following data. Find the best-fit equation that expresses boiling point as a function of salt concentration.

salt (tsp)	0	1	2	3	4	5	6	7	8
boiling temperature (°F)	204°	206°	208°	210°	214°	214°	217°	218°	221°

21. The Skydiver Problem Two skydivers jumped from a plane at an altitude of 1000 ft, 5 min apart. Their heights above the ground were read from their wrist altimeters as recorded in the following tables. (Because of air resistance, the functions are nearly linear after a very few seconds.) Find the best-fitting linear equations and use the equations to describe the jump.

time (min)	0	2	4	5	10
height (ft)	1000	700	650	625	370

time (min)	5	7	9	10	11
height (ft)	1000	650	470	285	140

2.3 Graphs of Quadratic Functions and Optimization

For Exercises 22 and 23 graph each parabola and label the vertex and intercepts.

22. $f(x) = 2x^2 - 4x - 6$

23. $g(x) = -x^2 - 3x + 5$

For Exercises 24 and 25 use the discriminant to find the values of k that will yield (a) two real solutions ($b^2 - 4ac > 0$), (b) no real solutions ($b^2 - 4ac < 0$), and (c) one real solution ($b^2 - 4ac = 0$).

24. $2x^2 - x + k = 1$

25. $2kx^2 - 3x - 1 = 0$

26. Use the graphs of $f(x) = x^2 - 2x - 8$ and $g(x) = -x^2 + 6x - 8$ to find the points of intersection of the two parabolas. Use algebra to verify your results.

27. Show that the x intercepts of the parabola $y = a(x - h)^2 + k$ are $h \pm \sqrt{\frac{-k}{a}}$.

For Exercises 28 and 29 use the function $s = -4.9t^2 + v_0t + s_0$, which expresses height above the ground s in meters as a function of time in seconds, where v_0 is initial velocity and s_0 is initial height.

28. The Firecracker Problem The maximum height of a firecracker, 40 m, is reached 2.5 sec after it is thrown. When do its remnants hit the ground?

29. The Drum Major Problem If the drum major in the university's marching band throws his baton up at an initial velocity of 20 m/sec from a height of 2 m, how long does he have to wait for it to return to his hand 3 m off the ground?

In Exercises 30 and 31 we consider packages that are restricted by the U.S. Postal Service to having girths plus length of 108 in. Find the dimensions that will maximize the surface area.

30. The package has a square base, x in. on a side, and a length of y in. (Figure 1).

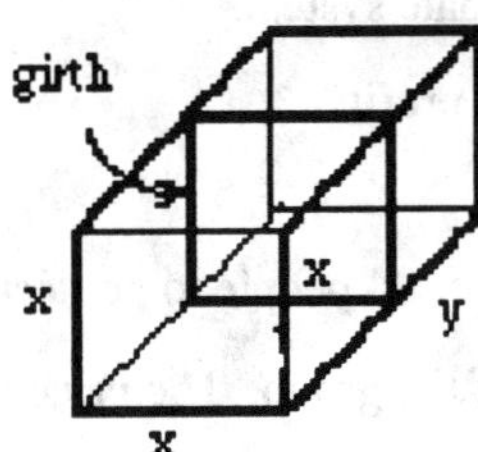

Figure 1

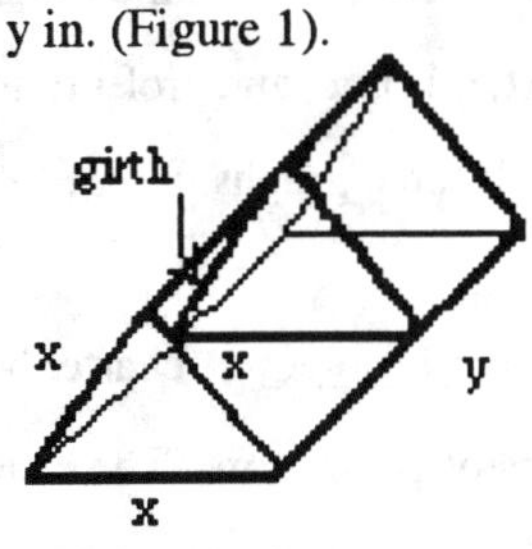

Figure 2

31. The package has a cross section that is an equilateral triangle, x in. on a side, and an area of $A = \frac{\sqrt{3}}{4}x^2$ (Figure 2).

32. <u>The Ribbon Problem</u> Find the dimensions of the box in Figure 3 that has maximum surface area if it is to be tied with a 10-ft ribbon as shown.

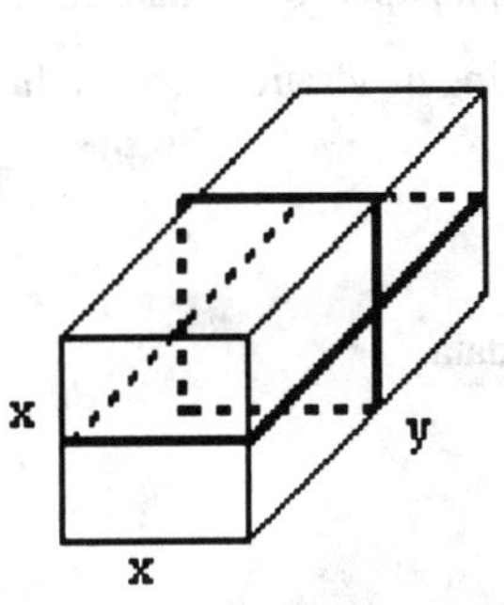

Figure 3

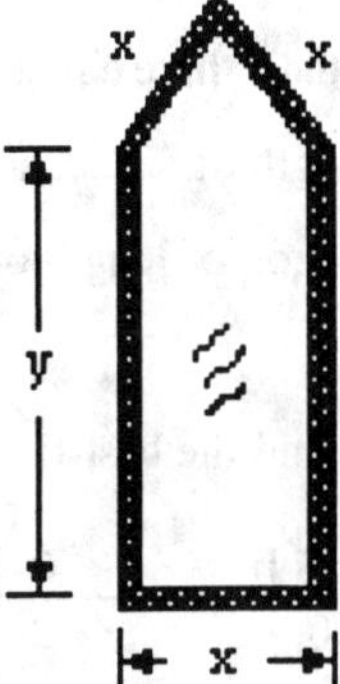

Figure 4

33. <u>The Window Problem</u> A window is to be built in the shape of a rectangle with an equilateral triangle on top (Figure 4). The surrounding frame is to be built with 6 m of gold-leaf molding. What dimensions will allow the maximum light to pass through the window?

34. <u>The Compact Disc Revenue Problem</u> The maximum revenue of $6360 is reached by selling 800 compact discs for $7.95 each.

(a) Find an equation that expresses total revenue as a function of quantity. [Hint: What revenue is obtained from selling no compact discs?]

(b) Find the quantity to sell that will maximize profit in the CD market if costs are $C(x) = 4.9x + 170$.

35. An Economic Model Problem The cost, revenue, and profit functions of an item are defined as $C = 18x + 200$, $R = -\frac{3}{4}x^2 + 75x$, and $P = -\frac{3}{4}x^2 + 57x - 200$.

(a) Use the profit function to find maximum profit.

(b) Graph the cost, revenue, and profit functions on the same coordinate system.

(c) Explain how all three graphs in part (b) fit together at maximum profit.

36. The Charter Boat Problem A charter boat can carry a maximum of 34 people to go deep-sea fishing and can be hired for groups of 10 or more. The charge depends on the size of the group. The price is $50/person for a group of 10, but for each additional person over 10 the price is reduced by $2/person. How many people would the captain take to receive the maximum revenue?

37. Sketch the graph of a quadratic function that is concave down with vertex (1, -3) and y intercept -2.

2.4 Finding Equations of Quadratic Functions

38. The Equilateral Triangle Problem Draw five different equilateral triangles on a piece of graph paper. Measure the length of their three equal sides and count the squares on the graph paper to estimate each area. Use your measurements (length of a side, area) to find the equation of the best-fitting quadratic function that expresses area as a function of length of side.

For Exercises 39 to 40 find the best-fitting quadratic equation for each set of data.

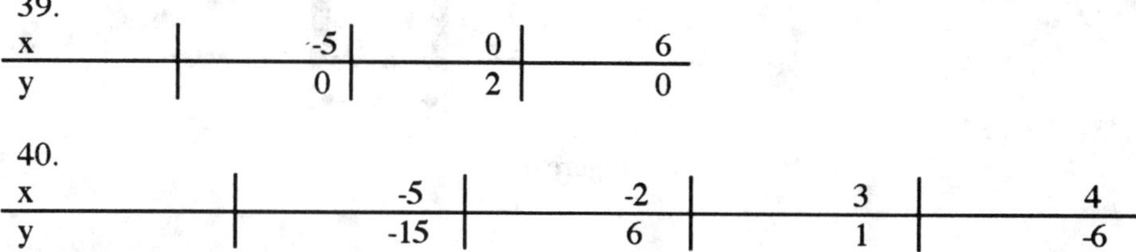

39.

x	-5	0	6
y	0	2	0

40.

x	-5	-2	3	4
y	-15	6	1	-6

41. The Tunnel Problem The following data represents the number of cars passing through a tunnel as the speed limit was varied.

(a) Use quadratic regression to find the optimum speed limit.

(b) Why does the number of cars passing through a tunnel decrease if the speed limit is too high?

speed limit	0	10	20	30	40
number of vehicles/hr	0	2000	3000	4000	4500

[Source: The Port of New York Authority]

42. <u>The Refined Project Problem</u> The following data approximate the cost of a project as a function of the amount of refinement.

(a) Find the optimum amount of refinement for this project.

(b) Why do costs increase if there is too little refinement?

degree of refinement	10%	50%	75%	90%
total cost	$92,000	$50,000	$60,000	$100,000

[Source: *An Introduction to Engineering and Engineering Design*, E. Krick]

Chapter 2 Test

For Exercises 1 to 3 sketch the graph of each function.

1. $y = 2x - 40$ 2. $x = 0$ 3. $y = -x - 15$

For Exercises 4 to 6 find the equation of each linear function and express in slope-intercept form.

4. slope $m = 1/2$ and y intercept $(0, -2)$
5. parallel to $y = 3x - 1$ passing through the point $(4, -4)$
6. perpendicular to $3x - 2y = 1$ passing through the point $(7, 3)$

7.(a) Find the equation of the line that best fits the following data.

(b) Find the missing entries.

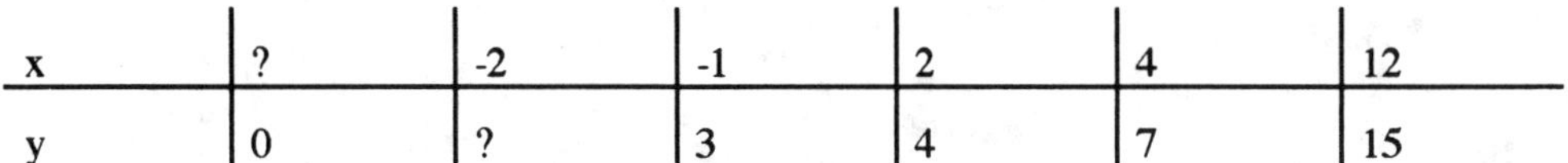

x	?	-2	-1	2	4	12
y	0	?	3	4	7	15

8. Graph the piecewise-defined function

$$f(x) = \begin{cases} 10 - x, & -10 \le x < -5 \\ 15, & -5 \le x < 5 \\ x + 10, & 5 \le x \le 10 \end{cases}$$

9.(a) Find $x = f^{-1}(y)$ for $f(x) = \dfrac{3x + 1}{4}$.

(b) Show that $f(f^{-1}(y)) = y$ and $f^{-1}(f(x)) = x$ for part (a).

(c) Sketch the graph of f and indicate $f(5)$, the image of $x = 5$, and $f^{-1}(1)$, the inverse image of $y = 1$.

For Exercises 10 and 11 graph each function. Label the x and y intercepts and the turning point.

10. $f(x) = 20x - x^2$ 11. $g(x) = -2(x - 12)^2 + 288$

12. Find the x intercepts for $f(x) = a(x + \frac{b}{2a})^2 - \frac{b^2 - 4ac}{4a}$.

13. The Equality Problem The average height of the American male is 5 ft 9 in. and has been increasing by about 0.03 in./yr since 1972. For the American female the corresponding figures are an average height of 5 ft 4 in. and annual increase of about 0.05 in. Find the equations that model this situation, graph them, and interpret.

14. The Wishing Well Problem A coin is wistfully dropped into a wishing well from a height of 3 m above the ground.

(a) Where is the coin after 2 sec?

(b) How many meters does the coin fall in 2 sec?

(c) If the splash is heard 3.5 sec after the coin is dropped, how deep is the well? (That is, how far is it to the water's surface?)

15. The Deck Problem A builder is attaching a deck to the back of a house and plans to put two trees in the corners (Figure 1). For design reasons, the builder decides to make the square-corner cutouts around the trees one-third the width of the deck. If the builder has 40 ft of railing, what dimensions should the deck be built to maximize its area?

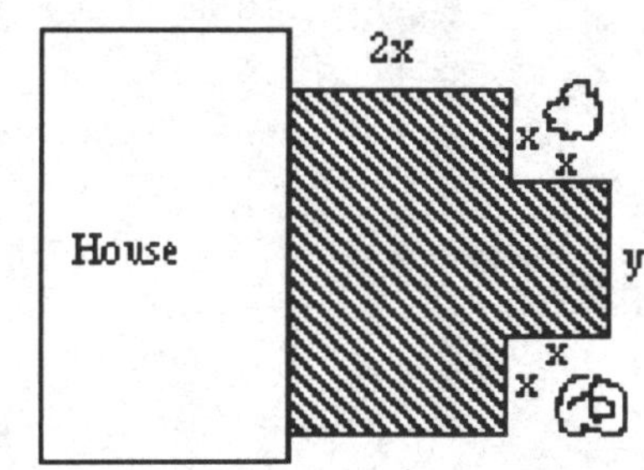

Figure 1

16. The Windsock Problem A windsock is to be made from a rectangular piece of fabric with four streamers (Figure 2a). Each exposed edge must be stitched to prevent unraveling. The design calls for triple stitching all three seams in the x direction. Of course, the 8-in. header must also be stitched in order to make the windsock circular. If the total amount of stitching is restricted by budget to 240 in., what dimensions of the rectangle will yield the maximum surface area?

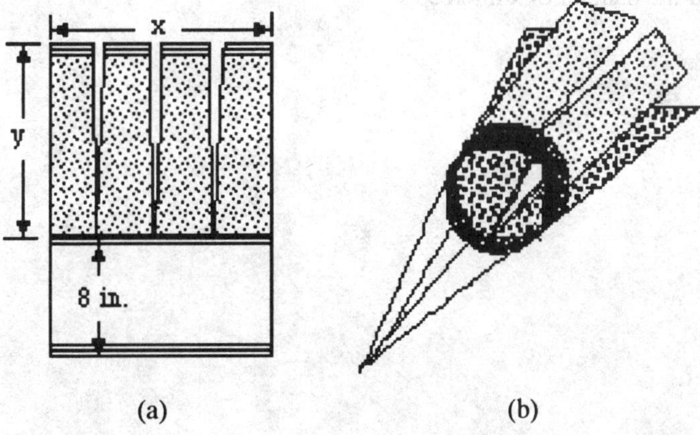

Figure 2 (a) (b)

17. The Queuing Problem The speed of a ski lift at a ski resort was varied and data were collected as to the average time that skiers spent waiting in line.

(a) Find the optimum speed of the lift. [Hint: Use quadratic regression.]

(b) Why does speeding up the lift too much result in longer waits in line?

speed of lift (ft/min)	10	20	30	40	50
average wait in line (min)	25	20	15	10	10

18. The U.S. Population Problem In 1995, the population of the United States was 262 million and growing at about 1%, including immigration. The graph in Figure 3 shows population growth rate as a function of the population.

(a) Find the quadratic equation for this model.

(b) Interpret the intercepts and vertex in terms of the population growth rate.

[Source: *The American Almanac Statistical Abstract of the United States*]

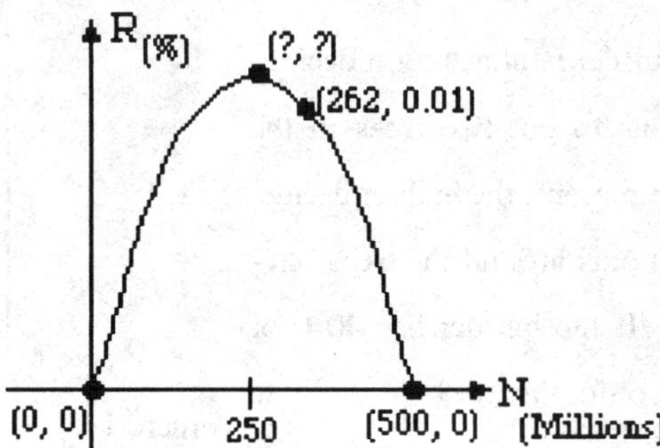

Figure 3

19. Use an application to compare the slope of a linear function and the slope of its inverse.

20. Quadratic equations have solutions (also called *roots* or *zeros*). The graphs of quadratic functions have corresponding x intercepts. Use three related applications to illustrate the possible cases concerning the number of solutions and the number of x intercepts.

Chapter 3

Polynomial and Rational Functions

In Chapter 3 we extend our knowledge of linear and quadratic functions to the theory, applications, and graphs of polynomial functions, (polynomials). We then extend this study to the quotient (ratio) of two polynomials, called rational functions. Internally, calculators and computers use polynomial and rational functions. All other functions such as exponential, logarithmic, and trigonometric are approximated with polynomial or rational functions. Polynomial and rational functions are very important.

A Historical Note on Polynomials

Two young mathematicians, Niels Abel (1802-1829) and Evariste Galois (1811-1832), contributed greatly to the theory of polynomial equations. At the age of 22, the Norwegian mathematician Abel presented a rather long, involved proof that a formula similar to the quadratic formula does not exist for polynomials of a degree higher than four. Abel, working as a substitute teacher, developed work on what are now called abelian equations and abelian groups. He was just beginning to be recognized when he died from tuberculosis on a sled trip to visit his fiancée.

The life of Galois is one of the most moving and tragic episodes in the history of mathematics. The young French revolutionist devoted only a few years to mathematics before dying at the age of 20 in a duel. At 16 he began a study of the necessary conditions for an equation to be satisfied by radicals as the quadratic equation is solved by the quadratic formula.

Galois attempted to reach the mathematical community, submitting his paper three times. The first one was lost, the second was taken home by Fourier, who died before examining it, and the third was returned as incomprehensible. Nevertheless, Galois knew fully the importance of the general principles he had developed. On the eve of the duel, anticipating death, Galois wrote a seven-page mathematical testament in a letter to his brother. This summary of his discoveries was to keep mathematicians busy for generations.

For both Abel and Galois, the golden ideas of their youth were to be left for others to develop.

3.1 Graphs of Polynomial Functions

Identifying Polynomial Functions
Examining the Global Behavior of Graphs of Polynomial Functions
Examining the Local Behavior of Graphs of Polynomial Functions
Considering the Multiplicity of Factors
Applying Polynomial Functions
Locating Turning Points and Points of Inflection

In Chapter 2 we found the optimum points on graphs of quadratic functions. Here, applications include finding relative maximum or minimums, high or low points on a graph relative to nearby points. In Example 4 we will find the dimensions of a box that maximizes its volume where the volume function is a cubic polynomial. We first graph polynomials in factored form.

Identifying Polynomial Functions

The reference functions $f(x) = x$, $f(x) = x^2$, and $f(x) = x^3$ are examples of **polynomial functions**. Functions with fractional or negative exponents such as $y = \sqrt{x} = x^{1/2}$ and $y = \frac{1}{x} = x^{-1}$ are <u>not</u> polynomials.

Definition of a Polynomial Function

$$f(x) = a_n x^n + a_{n-1}x^{n-1} + \cdots + a_1 x^1 + a_0 x^0,\ a_n \neq 0$$

is a **polynomial function** where the coefficients a_i are real numbers and the exponents are nonnegative integers.

In our study of polynomial functions we will refer to the degree of the polynomial as well as the sign of the leading coefficient. The *degree of a polynomial* is the largest exponent of the variable. We call the coefficient of the highest-degree term the *leading coefficient.* For example, the degree of the polynomial

$$f(x) = -3x^4 + 2x^3 - 7x^2 - 5x + 8$$

is 4 and its leading coefficient is -3. The 8 is called the *constant term* because it does not change for different values of x.

Examining the Global Behavior of Graphs of Polynomial Functions

We can learn about graphs of polynomials by examining the graphs of familiar odd-degree polynomials and contrasting their attributes with those for even-degree polynomials. Although the graphs extend indefinitely in both directions, for simplicity we refer to the shape of a graph to the far right and far left as the *ends* of the graph. That is, we view each graph as x approaches positive infinity and negative infinity. Studying the ends of a graph is called a *global analysis.*

To explore the behavior of the ends of a polynomial graph, go to "Explore the Concepts," Exercises E1 - E8, before continuing (page 196). For the interactive version, go to the *Precalculus Weblet - Instructional Center - Explore Concepts.*

Figure 1a shows the graphs of the odd-degree polynomials $y = x$ and $y = x^3$ with their positive leading coefficients. In a global sense, all odd-degree polynomials having a positive leading coefficient go down to the left and up to the right. Figure 1b shows the graphs of $y = -x$ and $y = -x^3$. Since their leading coefficients are negative, these graphs go up to the left and down to the right. The graphs in Figure 1b are reflections about the x axis of the graphs in Figure 1a.

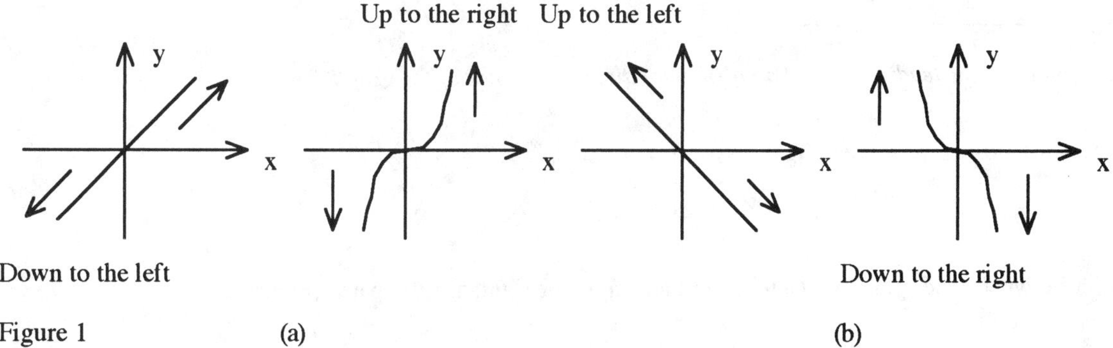

Figure 1 (a) (b)

These observations about the ends of the graphs of $y = x$, $y = x^3$, $y = -x$, and $y = -x^3$ hold for all odd-degree polynomials. When the leading coefficient is positive, the left end goes down and the right end goes up. The graph of $y = x^5$ goes down on the left and up on the right. When the leading coefficient is negative, the left end goes up and the right end goes down. All odd-degree polynomials must have at least one x intercept since the ends of their graphs are on opposite sides of the x axis.

Now we consider the behavior of even-degree polynomials by examining the graphs of $y = x^2$ and $y = x^4$ (Figure 2a). Both ends go up. Changing the sign of the leading coefficient yields $y = -x^2$ and $y = -x^4$ (Figure 2b). If the leading coefficient of an even-degree polynomial is negative, the ends go down. The graph of $y = -x^6$ goes down on both ends.

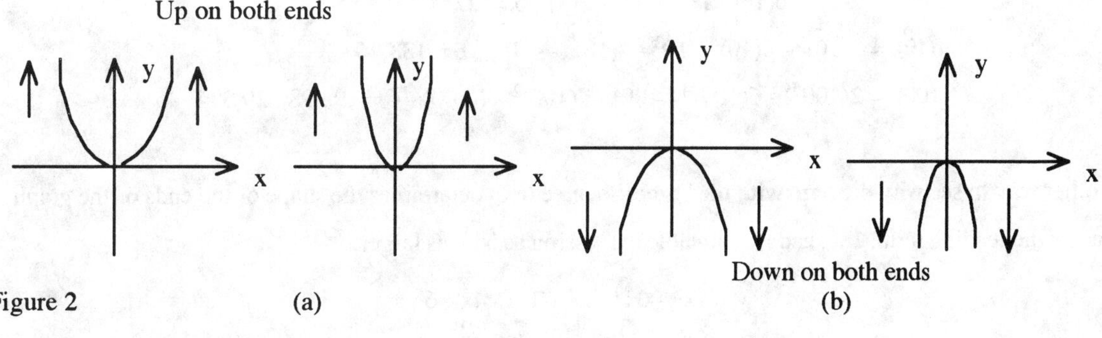

Figure 2 (a) (b)

Example 1 *Examining the global behavior of graphs*

Discuss the behavior of the ends of each graph.

(a) $f(x) = 2x^5 - 6x^4 - x^3 - 7x^2 - 4x - 6$ (b) $g(x) = -\frac{1}{2}x^4 + 2x^2 - 3x$

Solution

(a) Function f is an odd-degree polynomial (fifth degree) with a positive leading coefficient. The ends go down to the left and up to the right.

(b) Function g is an even-degree polynomial (fourth degree) with a negative leading coefficient. Both ends go down.

Why do you think the leading term of a polynomial dictates the ends of the graph?

Consider the fifth-degree polynomial from part (a) of Example 1,

$$f(x) = 2x^5 - 6x^4 - x^3 - 7x^2 - 4x - 6$$

Its graph shows the ends going down to the left and up to the right, as anticipated (Figure 3).

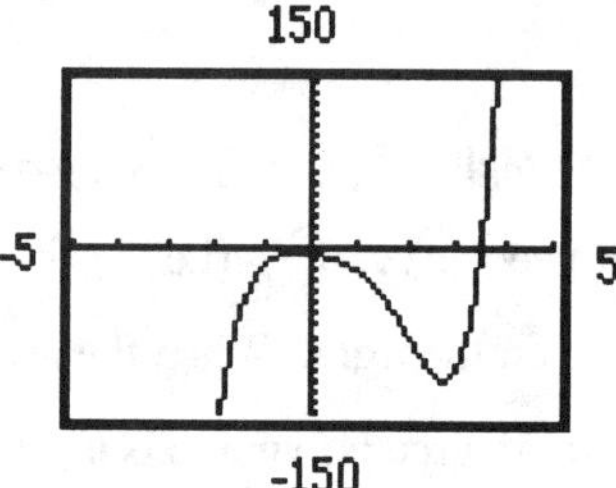

Figure 3

The leading coefficient is positive, but all others are negative. If we substitute larger and larger positive values for x, we notice the eventual dominance of the highest-powered term, $2x^5$.

$$\begin{aligned} f(x) &= 2x^5 - 6x^4 - x^3 - 7x^2 - 4x - 6 \\ f(1) &= 2(1)^5 - 6(1^4) - 1^3 - 7(1^2) - 4(1) - 6 = -22 \\ f(10) &= 2(10)^5 - 6(10)^4 - 10^3 - 7(10)^2 - 4(10) - 6 = 138{,}254 \\ f(100) &= 2(100^5) - 6(100)^4 - 100^3 - 7(100)^2 - 4(100) - 6 = 19{,}398{,}929{,}594 \end{aligned}$$

Another way to see why the term with the highest-degree term determines the shape of the ends of the graph is to factor the leading term, $2x^5$, and examine the expression as $|x|$ gets larger.

$$\begin{aligned} y &= 2x^5 - 6x^4 - x^3 - 7x^2 - 4x - 6 \\ &= 2x^5\left(1 - \frac{3}{x} - \frac{1}{2x^2} - \frac{7}{2x^3} - \frac{2}{x^4} - \frac{3}{x^5}\right) \end{aligned}$$

As values of $|x|$ get larger, the fractions approach zero, approaching

$$2x^5\,(1 - 0 - 0 - 0 - 0 - 0) = 2x^5$$

Globally, the graph of function f behaves like the graph of $y = 2x^5$.

Examining the Local Behavior of Graphs of Polynomial Functions

We now study some *local* properties of polynomial functions and their graphs by zooming in on their x and y intercepts. We can find the y intercept of any function by substituting $x = 0$. For example, if

$$f(x) = x^4 + 2x^3 - 9x^2 - 2x + 8$$

its y intercept is 8, since $f(0) = 8$. The y intercept is the constant term.

The factored form of $f(x) = x^4 + 2x^3 - 9x^2 - 2x + 8$ is $f(x) = (x + 4)(x + 1)(x - 1)(x - 2)$. (To learn how to factor a polynomial, see Appendix V.) The x intercepts of any function are found by setting $f(x) = 0$ and solving for x.

$f(x) = (x + 4)(x + 1)(x - 1)(x - 2)$ and $f(x) = 0 \Rightarrow$

$$(x + 4)(x + 1)(x - 1)(x - 2) = 0$$

$$x + 4 = 0 \quad \text{or} \quad x + 1 = 0 \quad \text{or} \quad x - 1 = 0 \quad \text{or} \quad x - 2 = 0$$

$$x = -4 \quad \text{or} \quad x = -1 \quad \text{or} \quad x = 1 \quad \text{or} \quad x = 2$$

Therefore, the x intercepts are -4, -1, 1, and 2. Recall values of x that yield $f(x) = 0$ are called roots or zeros.

Figure 4a shows the general shape of the graph obtained by connecting the intercepts with a smooth, continuous curve. Since the function is fourth degree, having a positive leading coefficient, we know that both ends go up. Notice that the turning points are rounded, not pointed. We use a graphing utility to approximate them (Figure 4b).

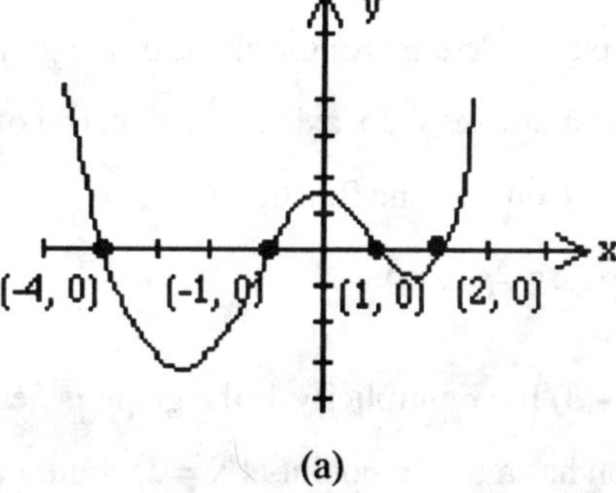

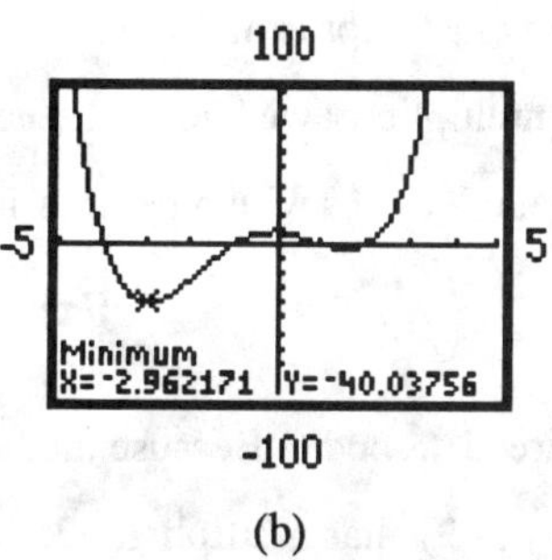

Figure 4 (a) (b)

To explore the number of x intercepts of a polynomial function, go to "Explore the Concepts," Exercises E9 - E14, before continuing (page 196). For the interactive version, go to the *Precalculus Weblet - Instructional Center - Explore Concepts.*

Recall from Chapter 2 that a linear function (a first-degree polynomial) may have one x intercept or no x intercepts if the line is horizontal. A quadratic function (a second-degree polynomial) may have two x intercepts, one x intercept, or no x intercepts.

What is the relationship between the degree of the polynomial and the number of possible x intercepts?

The degree of a polynomial determines the maximum number of x intercepts on its graph.

<u>The graph of a polynomial function of degree *n* has at most *n* x intercepts.</u>

Considering the Multiplicity of Factors

We know that the graphs of $y = (x + 3)$, $y = (x - 2)^2$, and $y = -(x - 6)^3$ have quite different shapes, especially near their x intercepts. The graph of $f(x) = (x + 3)^1$ in Figure 5 shows a linear function near its intercept, at $x = -3$. The graph of $g(x) = (x - 2)^2$ has a turning point at (2,0). The graph of $h(x) = -(x - 6)^3$ has a point of inflection at $x = 6$. A **point of inflection** is a point where a graph changes concavity -- from a valley shape to a hill shape, or vice versa.

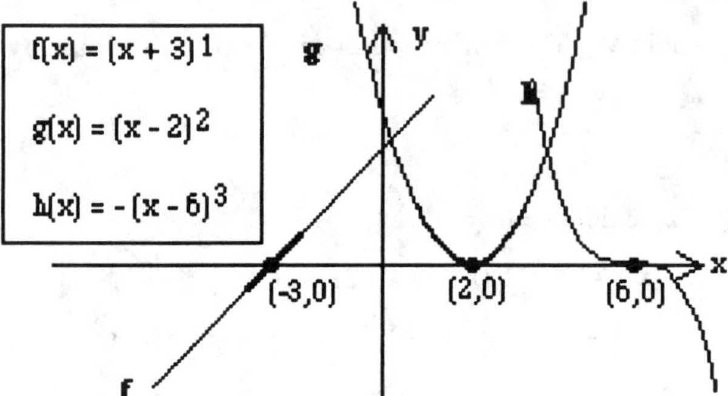

Figure 5

To explore the number of x intercepts of a polynomial function, go to "Explore the Concepts," Exercises E15 - E21, before continuing (page 194). For the interactive version, go to the *Precalculus Weblet - Instructional Center - Explore Concepts.*

The exponent on each factor defines the multiplicity of each zero of a polynomial function. **Multiplicity** is the number of times a factor is multiplied. The multiplicity determines the shape of the graph near each zero. By examining the multiplicity we know whether the graph crosses the x axis like a line, or bounces off the x axis like a parabola, or crosses with a point of inflection. Consider the function

$$f(x) = -(x + 3)(x - 2)^2(x - 6)^3$$

The x intercepts are -3, 2, and 6. Because the factor $(x + 3)$ has multiplicity 1, the graph is nearly linear near $x = -3$. The factor $(x - 2)^2$ has multiplicity 2. The graph has a turning point at $x = 2$, similar to the parabolic graph of $y = (x - 2)^2$. Near (6,0) the graph has a point of inflection as on the graph of $y = -(x - 6)^3$.

What is the degree of $f(x) = -(x + 3)(x - 2)^2(x - 6)^3$?

We do not have to multiply all the factors to find the leading term, just the coefficient and first term of each factor, $(-1)(x)(x^2)(x^3) = -x^6$. Therefore, function f is an even-degree polynomial with a negative leading coefficient. Both ends of the graph go down. The y intercept is

$$f(0) = -(0 + 3)(0 - 2)^2(0 - 6)^3 = 2592$$

We start drawing from the left, using the direction of the ends of the graph and watching the multiplicity of each factor as we approach the corresponding x intercept (Figure 6).

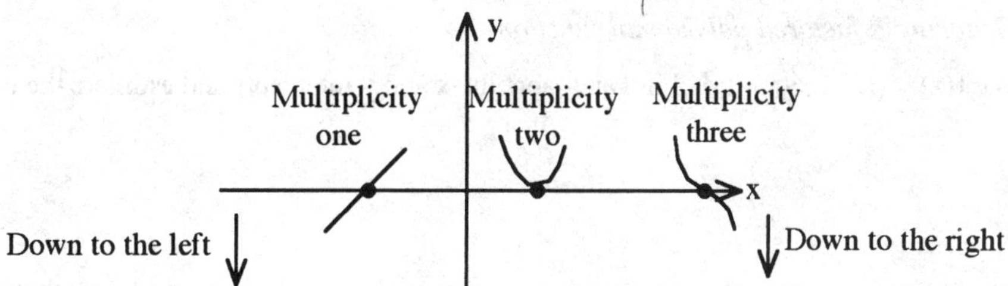

Figure 6

Notice that odd multiplicity factors change sign as the graph crosses the x axis, while even multiplicity factors do not change sign; the graph *bounces* off the x axis. The graph of the function is shown in Figure 7.

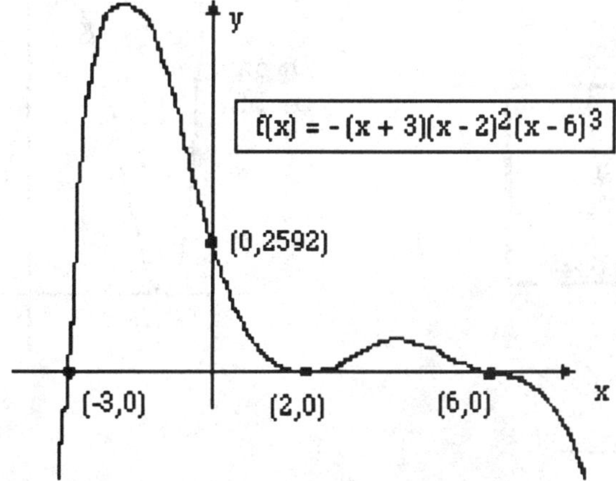

Figure 7

Why is the cubic shape near x = 6 reflected about the x axis?

Near x = 6 the other factors determine the sign. As $(x - 6)^3$ approaches zero the product approaches zero. Near x = 6 we have

$$f(x) = -(x + 3)(x - 2)^2(x - 6)^3$$
$$f(x) \approx -(6 + 3)(6 - 2)^2(x - 6)^3 = -36(x - 6)^3$$

A similar analysis near x = -3 and x = 2 will show that each of these shapes is not reflected.

It is common for graphing utility displays to omit one or both ends of a graph or to hide several x intercepts as a single point. To make sure that a display is appropriate, we may need the following algebraic information.

Summary: Graphing Polynomial Functions	
Ends of the graph:	Examine the sign of the leading coefficient and the degree of the polynomial.
y intercept:	Find f(0).
x intercepts:	Set f(x) = 0 and solve for x. Examine the multiplicity of each factor.

Example 2 *Graphing a factored polynomial function*

Sketch the graph of $f(x) = (x + 2)^2(x - 1)^2$. Find and label the x and y intercepts and estimate the turning points.

Solution

Since the function is an even-degree polynomial with a positive leading coefficient, we know that globally the graph goes up on both ends. Locally, the y intercept is 4 since f(0) = 4. The x intercepts are -2 and 1. Since both factors are of multiplicity 2, the graph will have turning points on the x axis. We use a graphing utility to estimate the coordinates of the third turning point (Figure 8).

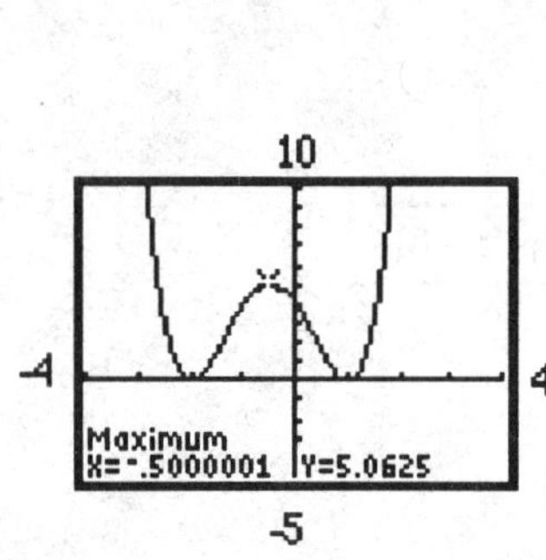

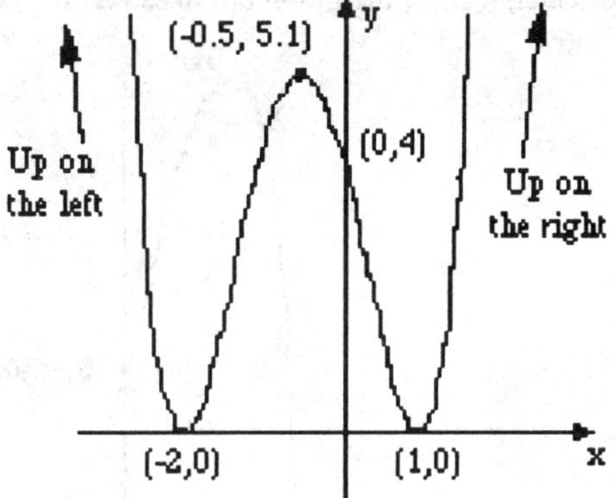

Figure 8

Example 3 *Finding an equation of a polynomial function*

Find an equation for the graph of the polynomial function in Figure 9.

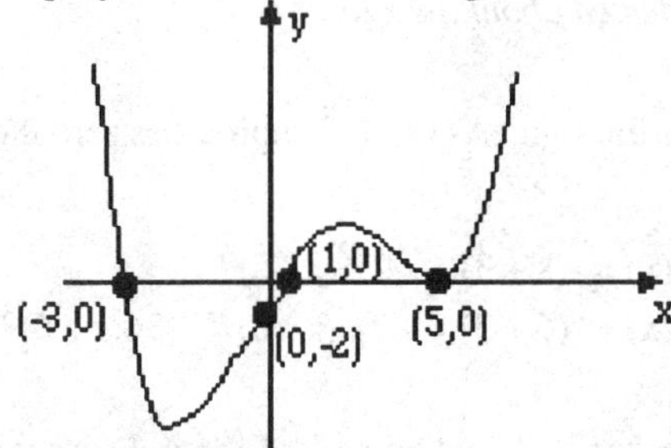

Figure 9

Solution

There are two zeros of multiplicity one, x = -3 and x = 1. They create the factors (x + 3) and (x - 1). The zero x = 5, of multiplicity two, creates the factor $(x - 5)^2$. From this information we have

$$y = a(x - 1)(x + 3)(x - 5)^2$$

where a is the scaling factor. Since the y intercept is -2, substitute x = 0 and y = -2 to find the value of a.

$y = a(x - 1)(x + 3)(x - 5)^2$ and $x = 0, y = -2 \Rightarrow$

$$\begin{aligned} a(0 - 1)(0 + 3)(0 - 5)^2 &= -2 \\ -75a &= -2 \\ a &= \frac{2}{75} \approx 0.03 \end{aligned}$$

An equation describing the graph is $y = 0.03(x - 1)(x + 3)(x - 5)^2$.

Applying Polynomial Functions

Example 4 The Big Box Problem

A square is cut from each corner of a 10-cm square of cardboard. The sides are folded up to make a box. Estimate how much should be cut to give the box maximum volume.

Thinking: Creating a Plan

We want to maximize volume with respect to the height of the box, x (Figure 10).

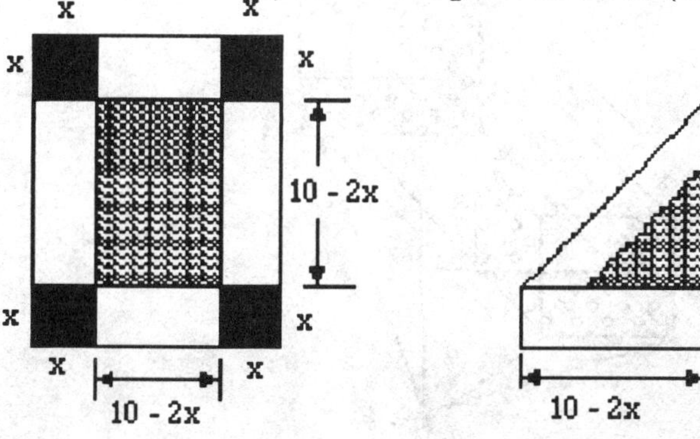

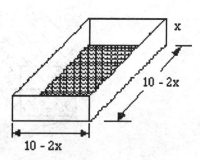

Figure 10

Since the volume, V = lwh, depends on how much we cut from each corner, we will express volume as a function of height x. We use a graphing utility to estimate the value of x that will yield the maximum volume. The domain of the application helps us set the window and decide which turning point is appropriate. The minimum value of x is x = 0. The maximum value is half the width of the cardboard, x = 5. We want to consider $0 \le x \le 5$.

Communicating: Writing the Solution

$V = lwh$ and $l = 10 - 2x$, $w = 10 - 2x$, $h = x \Rightarrow$

$$V(x) = (10 - 2x)(10 - 2x)(x), \quad 0 \le x \le 5$$

We graph $y = (10 - 2x)(10 - 2x)(x)$ and estimate the coordinates of the appropriate turning point (Figure 11). The value of x that yields the maximum volume is about 1.7 cm.

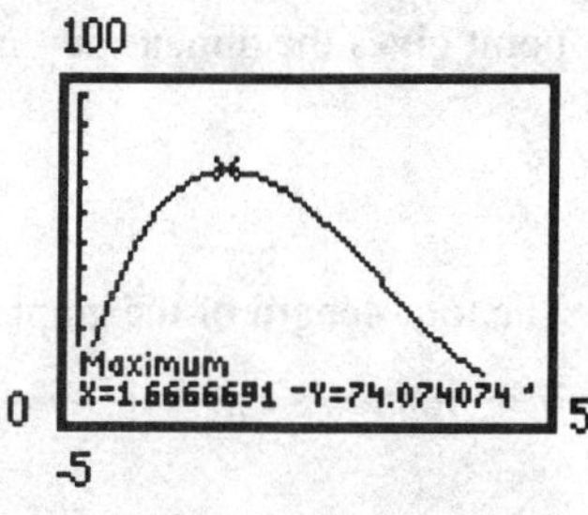

Figure 11

Learning: Making Connections

In calculus we use graphing utilities to check the plausibility of algebraic (analytic) solutions to problems such as this and to approximate other results that are irrational or otherwise beyond our grasp.

In the next example, a box kite has square cross sections. We compute its volume by multiplying its cross-sectional area by its length.

Example 5 The Box Kite Problem

A box kite is to be built with square cross sections and nylon fabric as shown in Figure 12. The frame is made of expensive graphite rods whose total length has been restricted to 216 in. Find the dimensions of the box kite that will provide maximum lift. (Assume that the larger the volume, the greater the lift.)

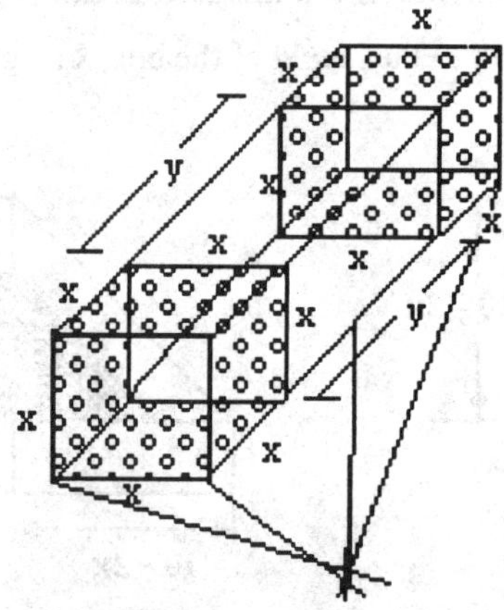

Figure 12

Thinking: Creating a Plan

We will find the values of x and y in Figure 12 that will maximize the volume subject to the restriction that the total length of the frame is 216 in. There are 24 sticks that are x inches long and 4 sticks that are y inches long. Therefore, $24x + 4y = 216$. The volume of the box kite is the area of a cross section (square with side x) times the length of the kite, $V = Al$, where $A = x^2$ and the length of the kite is $l = 2x + y$. Once we have the volume function in terms of one variable, we can graph the function and find the turning point. The turning point gives the dimension that yields the maximum volume.

Communicating: Writing the Solution

The total length of the graphite sticks is $24x + 4y = 216 \Rightarrow y = 54 - 6x$.

$V = Al$ and $A = x^2 \Rightarrow$

$$V = x^2 l \qquad \textit{Substitute } l = 2x + y.$$

$$V = x^2(2x + y)$$

$V = x^2(2x + y)$ and $\mathbf{y = 54 - 6x} \Rightarrow$

$$V = x^2(2x + \mathbf{54 - 6x})$$
$$V = x^2(54 - 4x)$$

From the graph in Figure 13, the turning point yields a maximum volume of 1458 in.^{3} when $x \approx 9$.

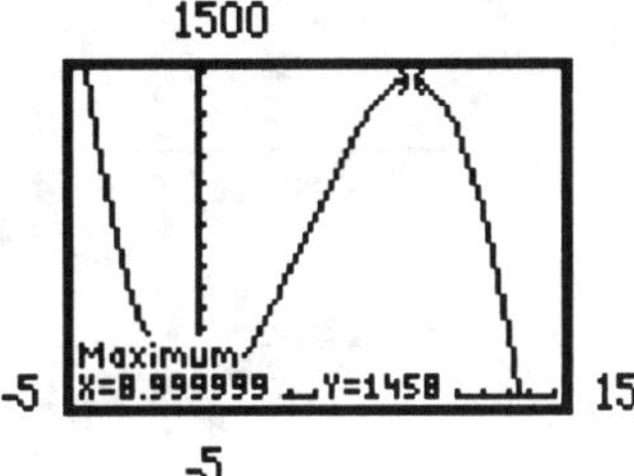

Figure 13

If x = 9, then y = 54 - 6x = 54 - 6(9) = 0. The box kite with the maximum lift (maximum volume) will have no uncovered middle section since y = 0 (Figure 14). The kite's dimensions are 9 by 9 by 18 in.

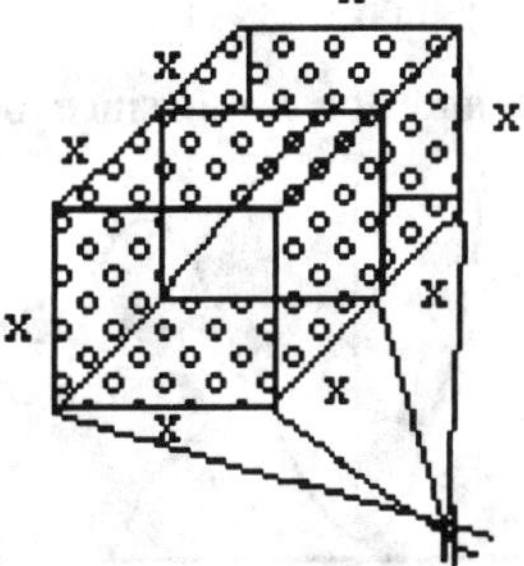

Figure 14

Learning: Making Connections

We can express volume as a function of y instead of x. Since $x = \frac{54 - y}{6}$,

$$V = x^2(2x + y) = \left(\frac{54 - y}{6}\right)^2\left[2\left(\frac{54 - y}{6}\right) + y\right]$$

We rename the variables and graph $y = \left(\frac{54 - x}{6}\right)^2\left[2\left(\frac{54 - x}{6}\right) + x\right]$ (Figure 15). Now, the maximum volume, 1458 ft^3, is on the vertical axis -- the left endpoint of the domain of the application, $x = 1.9 \times 10^{-6} \approx 0$.

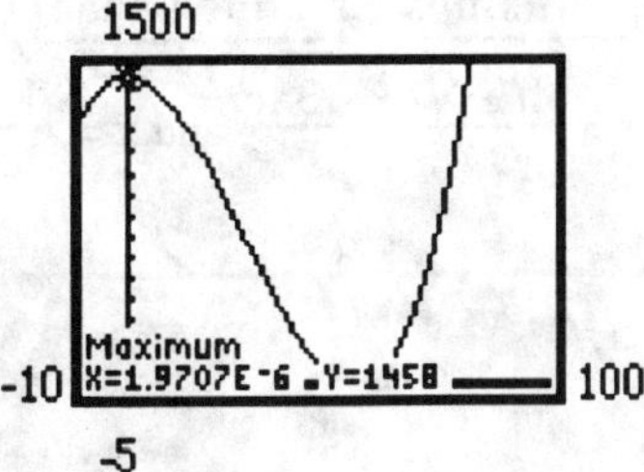

Figure 15

Locating Turning Points and Points of Inflection

Cubic polynomials often have two turning points, and always have a point of inflection. For

$$f(x) = ax^3 + bx^2 + cx + d$$

the x coordinates of the turning points and point of inflection can be estimated from the graphs or found using the following formulas:

$$\text{turning points:} \quad x = \frac{-b \pm \sqrt{b^2 - 3ac}}{3a}$$

and

$$\text{point of inflection:} \quad x = -\frac{b}{3a}$$

NOTE: These formulas may look somewhat familiar. In calculus, they are derived from the vertex and the zeros of the associated quadratic function called the derivative, $f'(x) = 3ax^2 + 2bx + c$.

At a point of inflection, a graph is increasing (or decreasing) at its maximum rate. In Figure 16, as you approach the point of inflection from the left, y increases at a faster and faster rate, since the graph is concave up. Beyond this point y increases at a slower and slower rate since the graph is concave down.

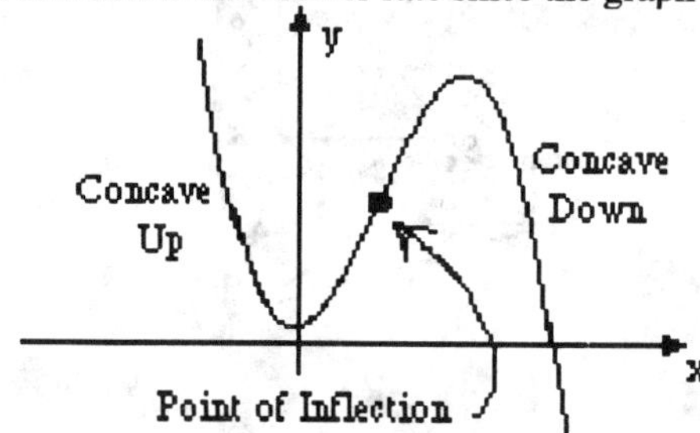

Figure 16

Example 6 The Point of Diminishing Returns Problem

The demand for your product is increased by your advertising campaign, creating the revenue function $R(z) = -z^3 + 39z^2 + 36z + 160$, where z is the weekly amount spent on advertising and R is the yearly revenue.

(a) Find the turning points and the point of inflection on the graph of R.

(b) Identify the points of diminishing returns and maximum revenue.

Thinking: Creating a Plan

For turning points we use the formulas $x = \frac{-b \pm \sqrt{b^2 - 3ac}}{3a}$ and $x = -\frac{b}{3a}$ to find the point of inflection.

Communicating: Writing the Solution

(a) $x = \dfrac{-b \pm \sqrt{b^2 - 3ac}}{3a}$ and a = -1, b = 39, c = 36 ⇒

$$x = \frac{-39 \pm \sqrt{1629}}{-3} \Rightarrow x \approx -0.45 \text{ or } x \approx 26.45$$

$$f(-0.45) = 151.78$$

and

$$f(26.45) = 9892.21$$

The turning points are $(-0.4\tilde{5}, 15\tilde{2})$ and $(26.4\tilde{5}, 989\tilde{2})$.

$x = -\dfrac{b}{3a}$ and a = -1, b = 39 ⇒

$$x = -\frac{39}{-3} = 13 \text{ and } f(13) = 5022$$

The point of inflection is (13, 5022).

(b) The maximum revenue of \$9891.34/yr results from spending \$26.60/week on advertising. The point of diminishing returns is \$13.00/week. If you spend less than \$13.00/week, your revenue will increase at a faster and faster rate. If you spend more than \$13.00/week, your revenue will increase at a slower and slower rate. At \$13.00/week, your revenue is increasing at its maximum rate.

Learning: Making Connections

Figure 17a shows one of the turning points, and Figure 17b shows the point of inflection. Although we can find the turning points using the maximum or minimum feature on a graphing utility, we can only guess where the point of inflection is from viewing the graph and use the trace feature to approximate it.

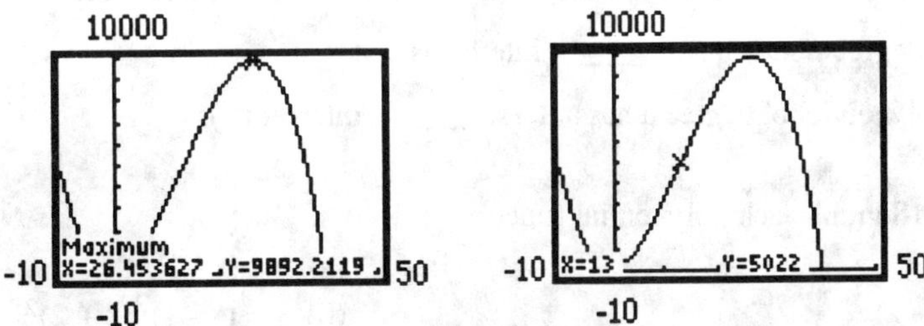

Figure 17

This is a preview of calculus where finding the turning points and points of inflection is ubiquitous.

Explore the Concepts

For Exercises E1 to E4 graph each polynomial and examine the ends of the graph.

E1. (a) $y = x$
(b) $y = 3x^3 - 10x^2$
(c) $y = 4x^5 - 10x^4 - 10x$

E2. (a) $y = -x$
(b) $y = -3x^3 + x^2 + 10$
(c) $y = -4x^5 + 10x^4$

E3. (a) $y = -x^2$
(b) $y = -3x^4 - 10x^3 - 10$
(c) $y = -4x^6 - x^5 + 10x^2 + 10$

E4. (a) $y = x^2$
(b) $y = 3x^4 - 10x^3 - 10$
(c) $y = 4x^6 + x^5 + 10x^2 + 10$

For Exercises E5 to E8 use your results from Exercises E1 to E4 to choose the correct response (positive/negative).

E5. For an odd-degree polynomial the ends of the graph go down to the left and up to the right whenever the leading coefficient is _____.

E6. For an even-degree polynomial the ends of the graph go down whenever the leading coefficient is _____.

E7. For an odd-degree polynomial the ends of the graph go up to the left and down to the right whenever the leading coefficient is _____.

E8. For an even-degree polynomial the ends of the graph go up whenever the leading coefficient is _____.

For Exercises E9 to E12 graph each polynomial and count the number of x intercepts for each graph.

E9. (a) $y = x$
(b) $y = 5x$
(c) $y = -x + 2$

E10. (a) $y = x^2$
(b) $y = x^2 - 9$
(c) $y = x^2 + 2$

E11. (a) $y = x^2(x - 1)$
(b) $y = (x + 1)^2(x - 1)$
(c) $y = x(x - 1)^2$

E12. (a) $y = x(x + 2)(x - 2)(x - 3)$
(b) $y = (x + 1)(x^2 + 2)(x - 2)$
(c) $y = x^2(x + 2)^2$

For Exercises E13 and E14 use your results from Exercises E9 to E12 to complete the following theorems.

E13. A polynomial of degree n may have _____ x intercepts.

E14. An odd-degree polynomial of degree n has at least _____ x intercept(s).

For Exercises E15 to E18 graph each polynomial function.

E15. (a) $y = x - 1$
(b) $y = (x - 1)(x + 2)$
(c) $y = x(x - 1)(x + 2)$

E16. (a) $y = (x - 1)^2$
(b) $y = (x - 1)^2(x + 2)^2$
(c) $y = x^2(x - 1)^2(x + 2)^2$

E17. (a) $y = (x - 1)^3$
(b) $y = (x - 1)^3(x + 2)^3$
(c) $y = x^3(x - 1)^3(x + 2)^3$

E18. (a) $y = x(x - 1)^2(x + 2)^3$
(b) $y = x^2(x - 1)(x + 2)^3$
(c) $y = x^3(x - 1)(x + 2)^2$

For Exercises E19 to E21 use your results from Exercises E15 to E18 to choose the correct response (linear/parabolic with a turning point/cubic with a point of inflection).

E19. If a factor has multiplicity one, then near the corresponding x intercept the graph is nearly _______.

E20. If a factor has multiplicity two, then near the corresponding x intercept the graph is nearly _______.

E21. If a factor has multiplicity three, then near the corresponding x intercept the graph is nearly _______.

Exercises 3.1

Develop the Concepts

For Exercises 1 to 4 describe the ends of the graph of each function.

1. (a) $y = x$
   (b) $y = 2x^3 - x^2$
2. (a) $y = x^2$
   (b) $y = 4x^4 - 10x$
3. (a) $y = -x^2$
   (b) $y = 5x - 4x^4$
4. (a) $y = -x$
   (b) $y = -x^5 + x^4 + 1$

For Exercises 5 to 8 find (a) the y intercept and (b) the x intercepts. (c) Discuss the effect of the multiplicity of each factor.

5. $y = (x - 1)(x + 2)^2$
6. $y = (x - 1)^2(x + 2)^3$
7. $y = (x - 2)^2(2 - x)^2$
8. $y = (x - 1)^3(1 - x)^2(2x - 2)$

For Exercises 9 to 12 match each equation with its graph in Figure 18.

9. $y = x(x + 2)(x - 2)(x - 4)$
10. $y = -x(x + 2)(x - 2)(x - 4)$
11. $y = -x(x + 2)^2(x - 2)^2(x - 4)$
12. $y = x^2(x + 2)^2(x - 2)(x - 4)$

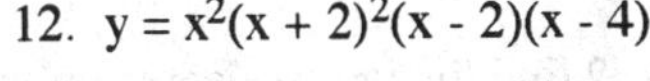

A.

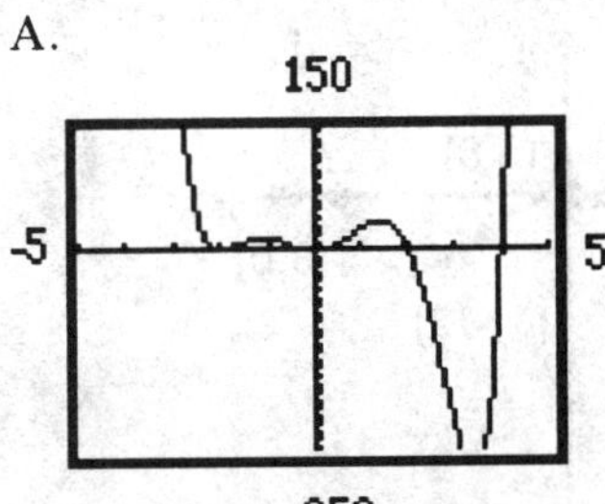

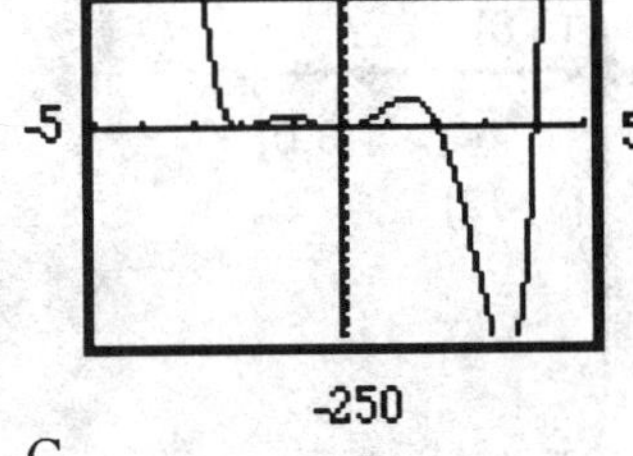

B.

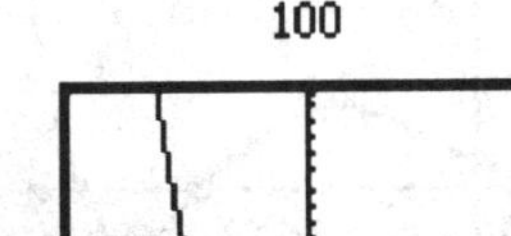

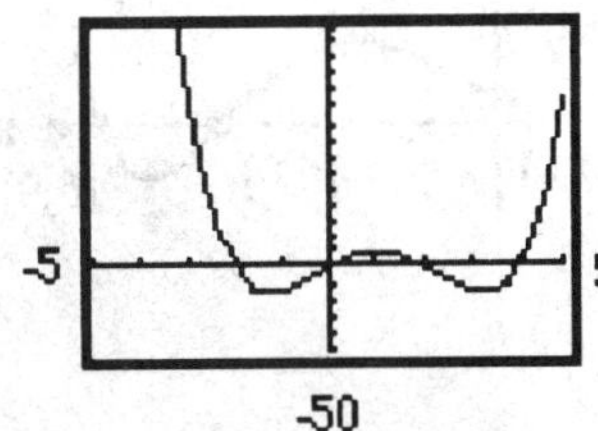

C.

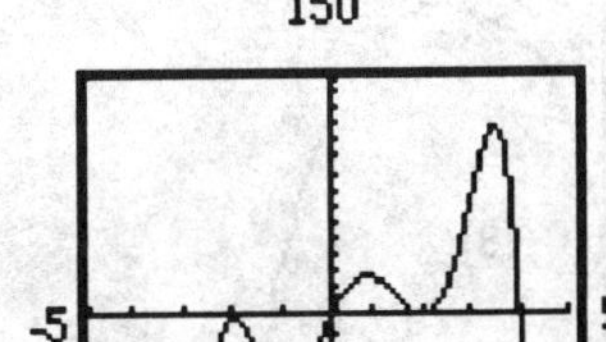

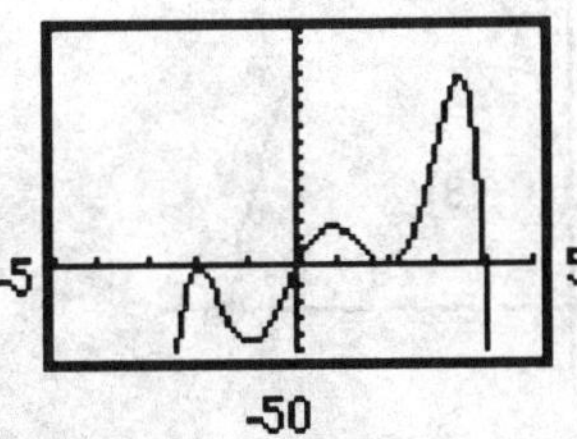

D.

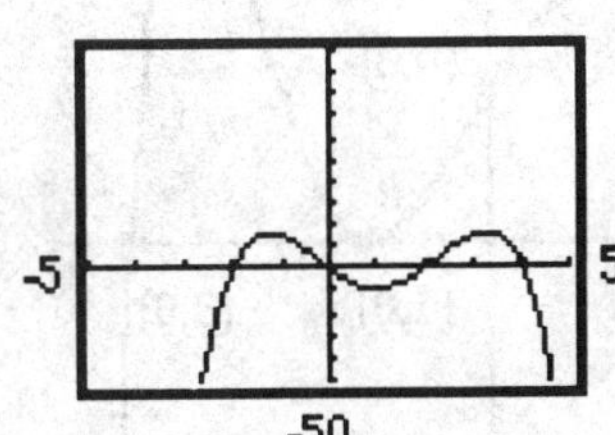

Figure 18

For Exercises 13 to 24 sketch each graph and label the x and y intercepts.

13. $y = (x + 2)(x - 2)(x - 3)$
14. $y = (x + 3)(x - 1)(x - 2)(x - 3)$
15. $y = -2x(x + 4)(x - 2)(x - 4)$
16. $y = -x(x - 1)(x + 2)$
17. $y = x^2(x + 1)$
18. $y = x(x + 1)^2$
19. $y = (x - 1)^2(x + 2)(x + 3)^2$
20. $y = -x(x + 1)^3(x + 2)(x - 3)$
21. $y = -(x + 2)^3$
22. $y = (x + 2)^4$
23. $y = (x - 2)^2(x + 1)(x + 3)^3$
24. $y = (x - 2)^3(x + 1)^2(x + 3)$

For Exercises 25 to 28 use the functions graphed in Figure 19 to sketch the graph of each reflection or translation.

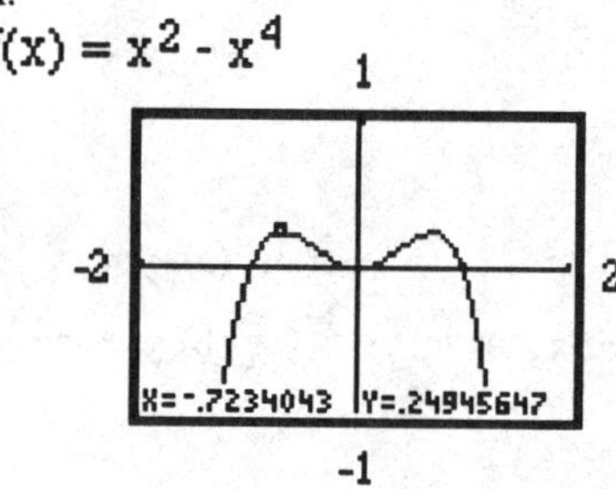

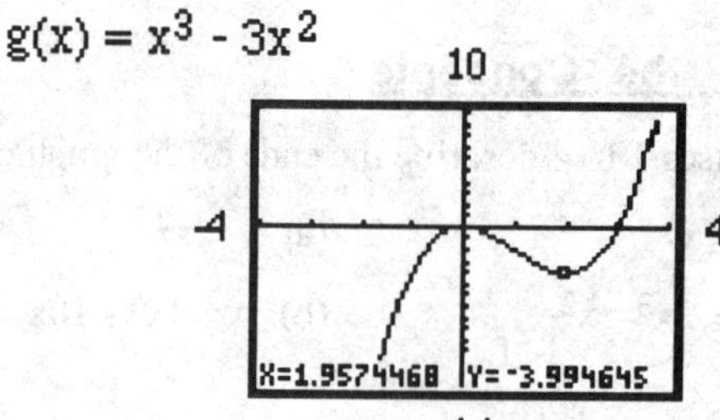

Figure 19

25. (a) $y = f(-x)$
    (b) $y = -f(x)$
    (c) $y = -f(-x)$

26. (a) $y = g(-x)$
    (b) $y = -g(x)$
    (c) $y = -g(-x)$

27. (a) $y = f(x - 1)$
    (b) $y = f(x) + 1$
    (c) $y = f(x + 1) - 1$

28. (a) $y = g(x + 10)$
    (b) $y = g(x) - 10$
    (c) $y = g(x - 10) + 10$

For Exercises 29 to 32 find the lowest-degree polynomial that describes each graph.

29.

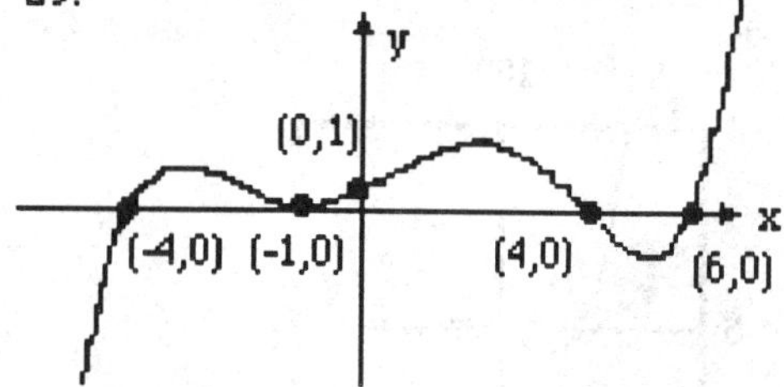

30.

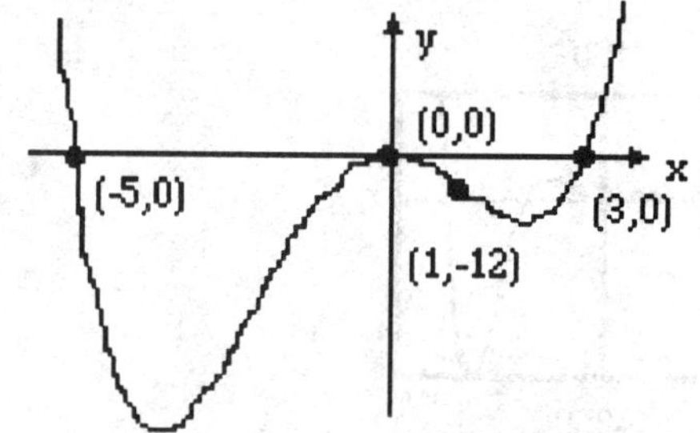

31.

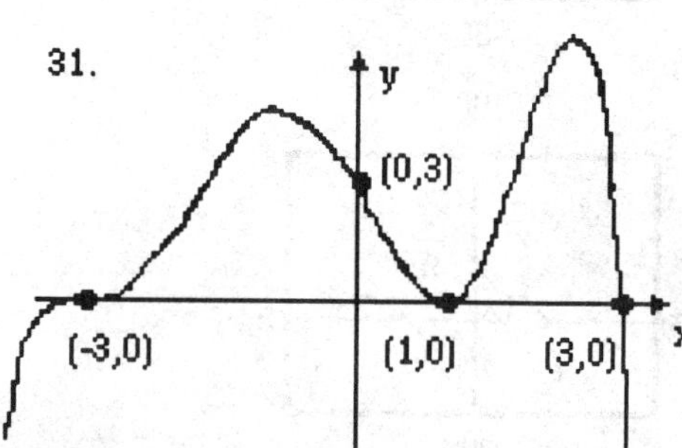

32.

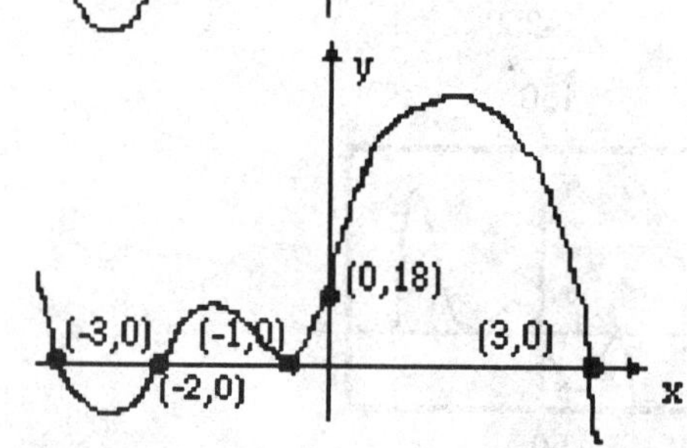

For Exercises 33 to 36 graph each pair of functions and estimate the points of intersection. Use algebra to find the exact value of the points of intersection.

*33. $f(x) = x^2(x - 1)$
     $g(x) = -(x - 1)(x + 2)(x - 3)$

34. $f(x) = -x(x + 10)^2$
    $g(x) = (x + 10)(x - 10)^2$

35. $f(x) = -(x + 10)^2(x - 10)^2$
    $g(x) = (x + 10)(x - 10)(x + 5)(x - 5)$

36. $f(x) = 3 + 3x^2 - x^4$
    $g(x) = x^4 - 10x^2 + 9$

For Exercises 37 to 40 find the turning points and the point of inflection for each cubic polynomial.

37. $y = x^3 - 18x^2 + 135$

38. $y = -2x^3 + 600x - 3744$

39. $y = -x^3 + 3x^2 + 105x$

40. $y = x^3 \cdot 3x^2 - 24x - 24$

Apply the Concepts

41. The Pizza Box Problem A box with a lid is to be constructed from a 20- by 30-in. piece of cardboard by cutting out six squares so that it can be folded as shown in Figure 20. Estimate how much should be cut to give the box maximum volume. [Hint: Try folding a piece of paper in this shape to see the location of the cutouts.]

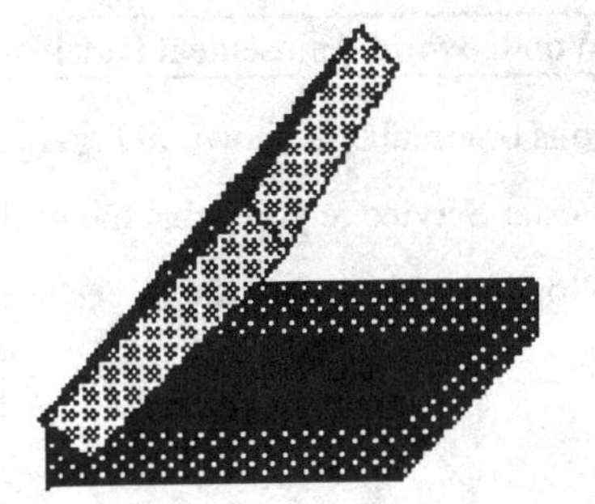

Figure 20

42. The Box-and-Lid Problem Estimate the maximum volume of the box that can be formed from a 10- by 10-in. piece of cardboard if it is cut in two pieces to make a box and separate lid (Figure 21).

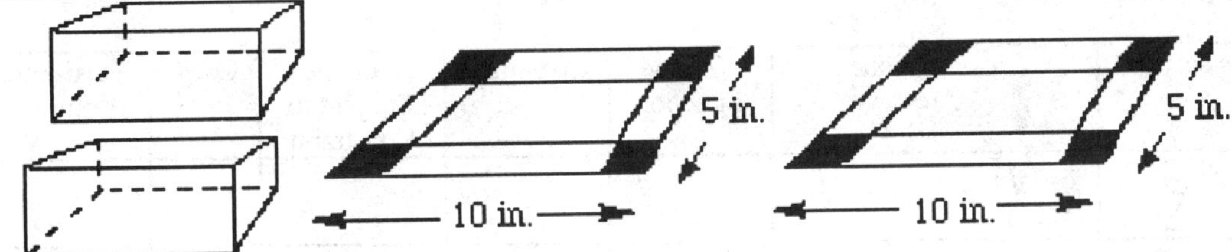

Figure 21

43. The Rainbow Kite and Balloon Company You are hired to design three windsocks, each from a single rectangular piece of nylon fabric. The edges must be hand-stitched around the perimeter of both ends as well as along the seam connecting the sides. The amount of stitching is budgeted for a total of 96 in. for each windsock. Fill in the following table and use your results to find the value of x that will yield the maximum volume in each case if the ends of the windsocks are open (Figure 22). Make and state your assumption about where the windsocks are to be stitched.

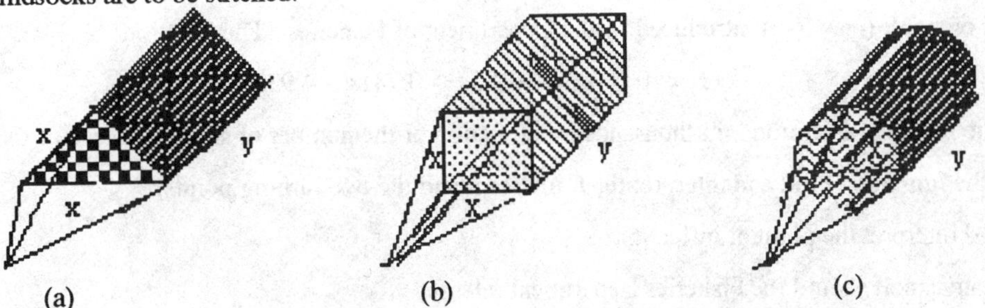

Figure 22 (a) (b) (c)

Cross Section	Cross-sectional Area	Volume Function	Constraint	Volume Subject to Constraint	Graph	Maximum Volume, x, y, V
x x x	$A = \frac{\sqrt{3}}{4}x^2$					
x x x x	$A = x^2$					
x	$A = \pi x^2$					

44. Wonderworld Amusement Park You want to make a series of boxes with cross sections that are regular polygons or circular as shown in Figure 23, to send souvenirs from the gift shop at the amusement park. The U.S. Postal Service requires that the girth plus the length of such packages be no more than 108 in. Complete the following table to find the values of x and y that yield maximum volume for each shape.

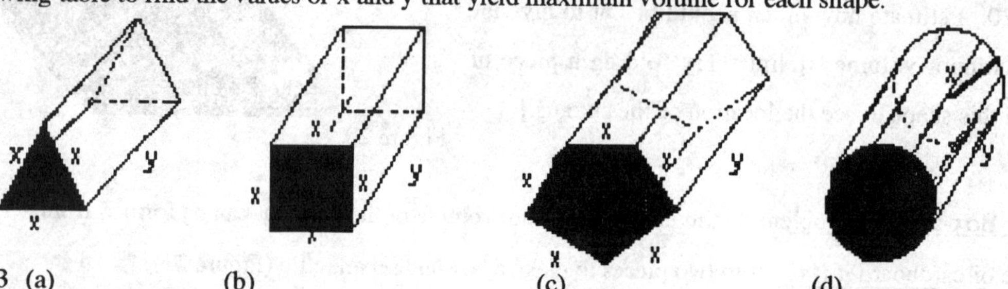

Figure 23 (a) (b) (c) (d)

Cross section	Cross-sectional Area	Volume Function	Constraint	Volume Subject to Constraint	Graph	Maximum Volume, x, y, V
	$A = \frac{\sqrt{3}}{4}x^2$					
	$A = x^2$					
	$A = \frac{x^2}{4}\sqrt{25 + 10\sqrt{5}}$					
	$A = \pi x^2$					

*45. The Fisheries Problem The number of sockeye salmon returning to the fish ladder in a dam is related to the number of smolt (new fish) introduced by the Department of Fisheries. The function

$$F = -0.006s^3 + 0.185s^2 - 0.741s + 4.933$$

describes the number of salmon F in thousands as a function of the number of smolt s in tens of thousands.

(a) Graph the function. Find and interpret the F intercept and the two turning points.

(b) Find and interpret the point of inflection.

(c) How many smolt should the Fisheries Department release?

46. The Seals and the Salmon Problem The seals are eating the returning salmon in front of a fish ladder where the salmon congregate as they search for the path upstream. The function

$$F = -3.7t^3 + 16.25t^2 - 2.5t + 10$$

gives the total number of salmon F in thousands predicted to return if the seals do not decrease their numbers as a function of time t in years from now.

(a) Find the point of inflection if the seals do not eat any salmon.

(b) Why will the salmon run not be sustained if the seals eat more than the number indicated in part (a)?

47. The Rainbow Kite and Balloon Company The function $p(t) = 0.02x^3 - 0.8x^2 + 9.0x + 12.8$ represents the price per share of a stock in the company traded "over the counter."

(a) Use the formulas for turning points and the point of inflection of a cubic polynomial to estimate the turning points and the point of inflection on the graph of function p.

(b) If you were to buy and sell this stock, what information does the points in part (a) suggest?

To find out more about stocks, go to the *Precalculus Weblet - Instructional Center - Current Events - Stock Link.*

48. Wonderworld Amusement Park The profits from the amusement park are seasonal and can be approximated with the following function, where t = 0 represents January, t = 1 is February, and so on.

$$P(t) = -240t^3 + 3250t^2 - 6640t + 1000$$

(a) Use the formulas for turning points and the point of inflection of a cubic polynomial to estimate the turning points and point of inflection.

(b) How can you use this information in your personnel office where workers are hired and fired?

Extend the Concepts

For Exercises 49 to 52 graph the following. Explain the effect of the absolute value symbols.

49. $y = | x^4 - 4x^2 |$

50. $y = | (x + 2)(2x + 1)(2x - 1) |$

51. $| y | = x^4 - 4x^2$

52. $| y | = (x + 2)(2x + 1)(2x - 1)$

For Exercises 53 to 56 graph each even function. Find and label the turning points.

53. $y = x^4 - 16$

54. $y = x^4 - 5x^2 + 4$

55. $y = -x^4 + 4x^2 + 5$

56. $y = x^4 + 4x^2 + 5$

57. Choose a third-degree polynomial in factored form. State its roots (zeros) r_1, r_2, and r_3. Rewrite the polynomial function in the form $y = x^3 + ax^2 + bx + c$. Find the relationship between each of the following and one of the coefficients, a, b, or c.

(a) $r_1 + r_2 + r_3$

(b) $r_1r_2r_3$

(c) $r_1r_2 + r_1r_3 + r_2r_3$

58. Generalize parts (a) and (b) of Exercise 57 to the four zeros of a fourth-degree polynomial of the form $y = x^4 + ax^3 + bx^2 + cx + d$.

3.2 Finite Differences

Using Finite Differences to Find a Linear Equation
Using Finite Differences to Find a Quadratic Equation
Applying Finite Differences to Higher-Degree Polynomials

In the nineteenth century, Augusta Ada Byron (Lady Lovelace) helped Charles Babbage with his invention of the first computer. His "analytic engine" used the method of *finite differences* to simplify evaluating polynomials. The computer language Ada is named in honor of Lady Lovelace, who pawned jewels to raise money to help Babbage finance the building of his analytic engine. Today, the method of finite differences is still an important one for analyzing data, evaluating polynomials, and finding the equations of certain polynomials.

We have already discussed finding best-fitting linear or quadratic equation using a graphing utility. In this section we examine a method for finding an equation for a polynomial that fits any finite collection of data where the inputs consist of consecutive integers or equally spaced integers.

Using Finite Differences to Find a Linear Equation

Consider the following data: (1, 6), (3, 30), (5, 54), and (7, 54). Notice that the inputs are equally spaced integers; x = 1, x = 3, x = 5, and x = 7 are each 2 units apart. We begin by creating a table and subtracting consecutive values of y. In Figure 1 the first differences Δy are all 24.

x	1	3	5	7
y	6	30	54	78

Δy: 30 - 6 = 24, 54 - 30 = 24, 78 - 54 = 24

Figure 1

The first differences measure the rate of change in y. Since Δx is constant, whenever the first differences are equal, the slope $\frac{\Delta y}{\Delta x}$ is constant and the relationship between x and y is linear. Next, we build another table by evaluating y = ax + b for x = 1, x = 3, x = 5, and x = 7. This form of a linear equation where a is the slope and b is the y intercept generalizes to higher-degree polynomials.

x	1	3	5	7
y = ax + b	a + b	3a + b	5a + b	7a + b

Δy: 2a, 2a, 2a

Figure 2

In Figure 2 the first differences are all 2a. To find the linear equation, we set the first differences from Figures 1 and 2 equal to each other and solve for a.

$$2a = 24$$
$$a = 12$$

The bold entries in Figure 3 create the equation 6 = a + b.

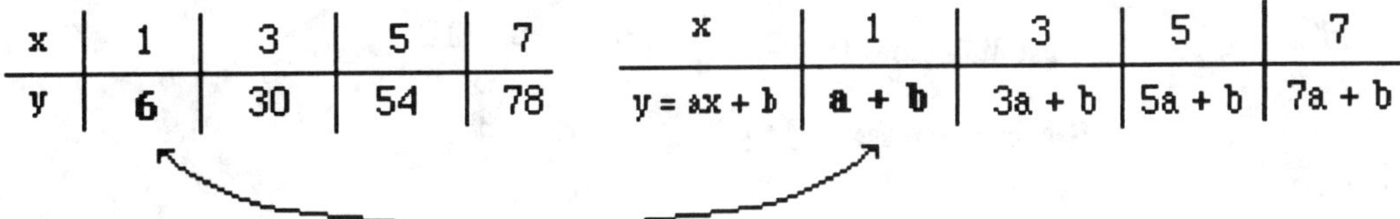

x	1	3	5	7
y	**6**	30	54	78

x	1	3	5	7
y = ax + b	**a + b**	3a + b	5a + b	7a + b

Figure 3

6 = a + b and a = 12 ⇒

$$6 = 12 + b$$

$$b = -6$$

Therefore, the linear equation that satisfies the data is $y = 12x - 6$.

Once we have found the linear equation, we can find the value of the function for any x value by substitution or by continuing the pattern in the table in Figure 1. For example, for x = 9, we add the first difference $\Delta y = 24$ to the y value associated with x = 7,

$$f(9) = 78 + 24 = 102$$

Using Finite Differences to Find a Quadratic Equation

In the next example, the first differences are not equal, but the second differences are equal. The second differences represent the rates at which the first differences are changing. When the second differences are equal, the polynomial is quadratic. We will find the quadratic equation $y = ax^2 + bx + c$.

Example 1 *Finding a quadratic equation using finite differences*

(a) Use the method of finite differences to find the quadratic function that passes through the points (1, 3), (2, 5), (3, 11), and (4, 21).

(b) Use finite differences to find f(5).

Solution

(a) We subtract the y values to find the first differences Δy, then subtract the first differences to find the second differences $\Delta^2 y$ (Figure 4).

x	1	2	3	4
y	3	5	11	21

First differences Δy: 2 6 10

Second differences $\Delta^2 y$: 4 4

Figure 4

Next, we build another table by evaluating $y = ax^2 + bx + c$ for $x = 1, x = 2, x = 3$, and $x = 4$ (Figure 5).

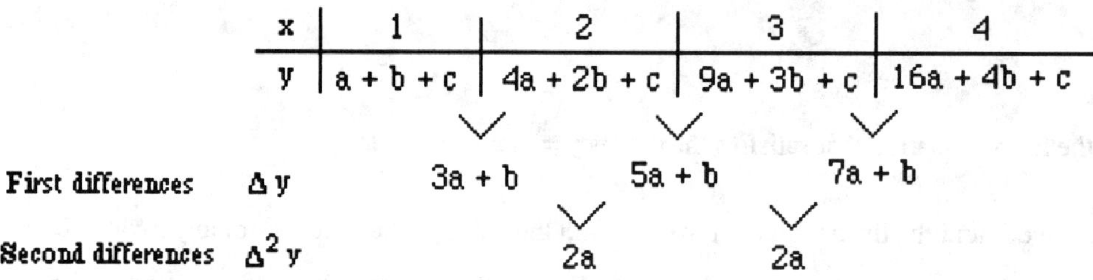

x	1	2	3	4
y	a + b + c	4a + 2b + c	9a + 3b + c	16a + 4b + c

First differences Δy: 3a + b 5a + b 7a + b

Second differences $\Delta^2 y$: 2a 2a

Figure 5

Again, the second differences are equal, $\Delta^2 y = 2a$. Matching the corresponding differences in the two tables leads to three equations (Figure 6).

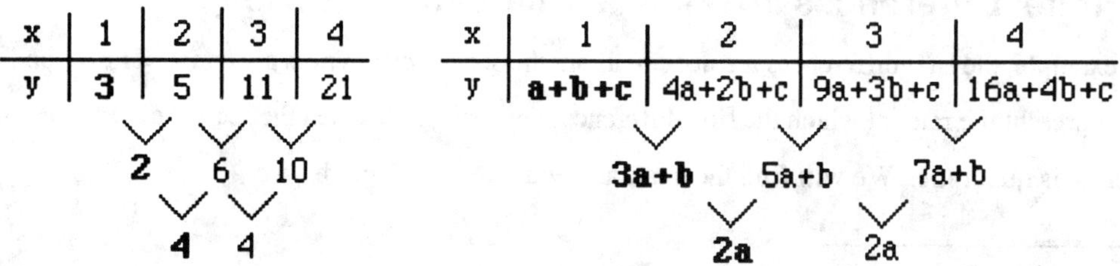

x	1	2	3	4
y	**3**	5	11	21

2 6 10

4 4

x	1	2	3	4
y	**a+b+c**	4a+2b+c	9a+3b+c	16a+4b+c

3a+b 5a+b 7a+b

2a 2a

Figure 6

The equations are $2a = 4$, $3a + b = 2$, and $a + b + c = 3$. Solving $2a = 4$ yields $a = 2$.

$3a + b = 2$ and $a = 2 \Rightarrow$

$$3(2) + b = 2$$
$$b = -4$$

$a + b + c = 3$ and $a = 2, b = -4 \Rightarrow$

$$2 + (-4) + c = 3$$
$$c = 5$$

Therefore, the equation of the quadratic function that fits these data is

$$f(x) = 2x^2 - 4x + 5$$

(b) Using finite differences to find f(5), we start at the bottom of the table and work up the table to fill in the missing column (Figure 7).

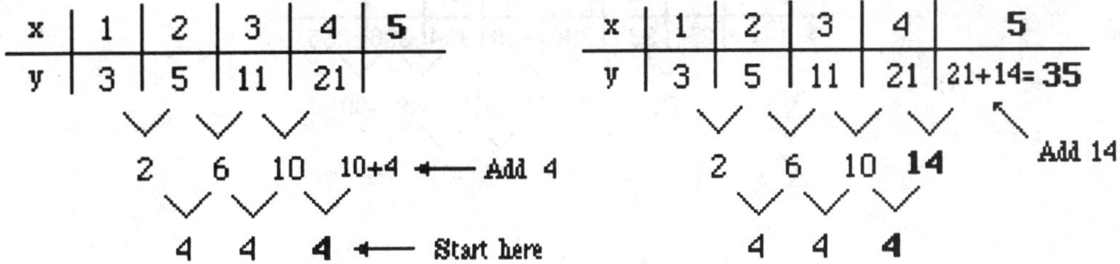

Figure 7

Therefore, f(5) = 35.

In the next example we see the advantage of finite differences over other techniques for finding equations. For moving objects, the first differences divided by Δt give the average velocity over each interval, and the second differences divided by Δt^2 gives the average acceleration. This is a preview of an important application of the derivative in calculus.

Example 2 The Free-Falling Object Problem

Find and interpret the first and second differences for the following data created by measuring the height above the ground of a free-falling object. Then find the equation of the height above the ground as a function of time.

t (sec)	1	2	3	4	5	6	7
s (ft)	256	336	384	400	384	336	256

Thinking: Creating a Plan

We construct the tables and find the first and second differences. From our experience with free-falling models, we anticipate that a quadratic function will model these data and that the second differences will be equal. We will equate the first and second differences to find the equation of motion using the general equation

$s = at^2 + bt + c$.

Communicating: Writing the Solution

Figure 8 shows the table of first and second differences for the data.

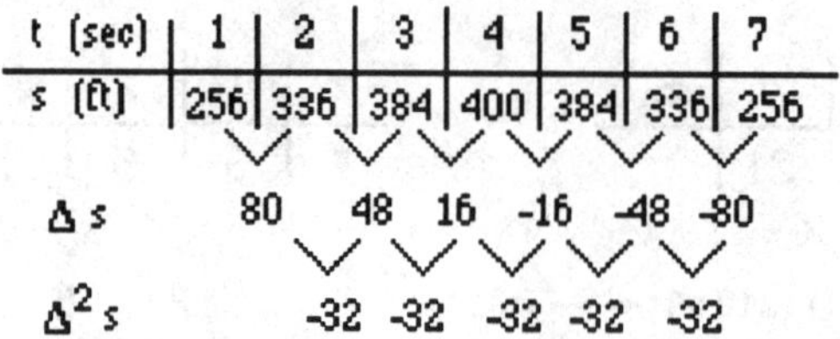

t (sec)	1	2	3	4	5	6	7
s (ft)	256	336	384	400	384	336	256

Δ s: 80 48 16 -16 -48 -80

Δ^2 s: -32 -32 -32 -32 -32

Figure 8

From t = 1 to t = 2 seconds, Δs = 80. The projectile had an average velocity of $\frac{\Delta s}{\Delta t}$ = 80 ft/sec. Over the time interval from t = 2 to t = 3 seconds, the projectile had an average velocity of 48 ft/sec. Over the time interval from t = 3 to t = 4 seconds, the projectile had an average velocity of 16 ft/sec. The object reached its highest point because the average velocity for the next time interval, from t = 4 to t = 5 seconds, is negative, -16 ft/sec. Then the object sped up as it fell, averaging -48 ft/sec, then -80 ft/sec.

The second differences were all -32, representing a constant acceleration due to gravity of

$$a = \frac{\Delta^2 s}{\Delta t^2} = -32 \text{ ft/sec}^2$$

Substituting t = 1, 2, 3, ..., 7 into the general quadratic function, $s = at^2 + bt + c$, creates the following table shown in Figure 9.

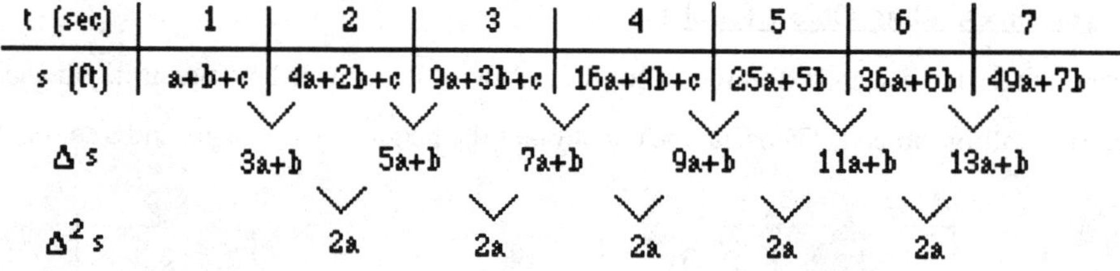

t (sec)	1	2	3	4	5	6	7
s (ft)	a+b+c	4a+2b+c	9a+3b+c	16a+4b+c	25a+5b	36a+6b	49a+7b

Δ s: 3a+b 5a+b 7a+b 9a+b 11a+b 13a+b

Δ^2 s: 2a 2a 2a 2a 2a

Figure 9

Comparing the two tables in Figure 10 creates the equations 2a = -32, 3a + b = 80, and a + b + c = 256.

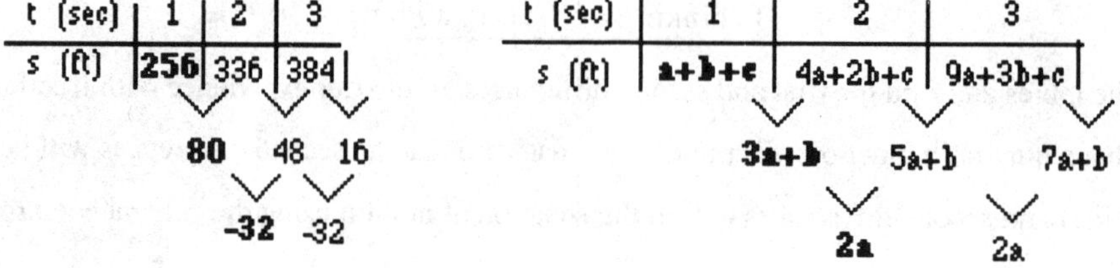

t (sec)	1	2	3
s (ft)	**256**	336	384

Δ s: **80** 48 16

Δ^2 s: **-32** -32

t (sec)	1	2	3
s (ft)	**a+b+c**	4a+2b+c	9a+3b+c

Δ s: **3a+b** 5a+b 7a+b

Δ^2 s: **2a** 2a

Figure 10

Solving 2a = -32 yields a = -16.

$3a + b = 80$ and $a = -16 \Rightarrow$

$$3(-16) + b = 80$$
$$b = 128$$

$a + b + c = 256$ and $a = -16, b = 128 \Rightarrow$

$$-16 + 128 + c = 256$$
$$c = 144$$

Therefore, the equation of the quadratic function that fits these data is

$$s(t) = -16t^2 + 128t + 144$$

Learning: Making Connections

Notice that the maximum height above the ground was attained at

$$t = \frac{-b}{2a} = \frac{-128}{2(-16)} = 4$$

The velocity at the highest point is zero, as supported by our first differences.

Applying Finite Differences to Higher-Degree Polynomials

We have seen that the first differences are equal for linear equations and the second differences are equal for quadratic equations. In the next example we extend this pattern for a cubic equation.

Example 3 *Finding the third differences for a cubic polynomial*

(a) Create a table of values for the third-degree polynomial $p(x) = 4x^3 + 3x^2 + 2x - 1$ for $x = 2, 4, 6, 8, 10$, and 12.

(b) Find and interpret the third differences for function p.

(c) Use finite differences to find p(14).

Solution

(a) Figure 11 shows the table of values found by substituting $x = 2, 4, 6, 8, 10$, and 12 into the cubic equation.

x	2	4	6	8	10	12
p(x)	49	313	985	2257	4321	7369

Figure 11

(b) Subtract as indicated in Figure 12 to find the differences. Notice that the third differences $\Delta^3 y$ are equal. The third differences are changing at a constant rate.

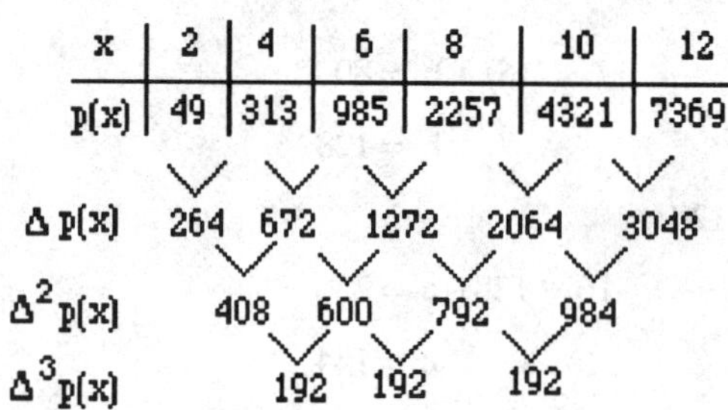

Figure 12

c) The bold entries in Figure 13 are created by starting at the bottom row and adding the appropriate values to generate f(14) = 11,593.

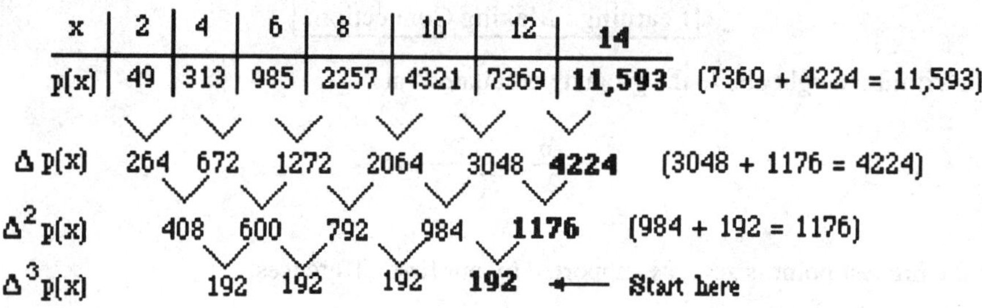

Figure 13

In the next chapter we will see that if the first differences increase at a constant percentage, the situation is modeled by exponential growth, whereas if they decrease by a constant percentage, it is modeled by exponential decay. If the second differences have a constant percent growth, the model is the logistic curve.

Exercises 3.2

Develop the Concepts

1.(a) Use the method of finite differences to find the linear equation of the form f(x) = ax + b for the following data.

x	0	1	2	3
y	3	7	11	15

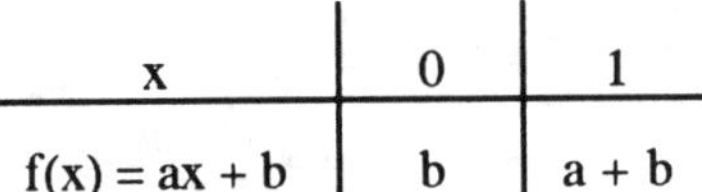

x	0	1
f(x) = ax + b	b	a + b

(b) Use finite differences to find f(4).

2.(a) Use the method of finite differences to find the linear equation of the form f(x) = ax + b for the following data.

x	10	15	20	25
y	-10	0	10	20

x	10	15
f(x) = ax + b	10a + b	15a + b

(b) Use finite differences to find f(30).

3.(a) Complete the table and use the method of finite differences to find the linear equation of the form $f(x) = ax^2 + bx + c$ for the following data.

x	0	1	2	3
y	3	5	11	21

x	0	1	2
$f(x) = ax^2 + bx + c$	c	?	?

(b) Use finite differences to find f(4).

4.(a) Complete the table and use the method of finite differences to find the linear equation of the form $f(x) = ax^2 + bx + c$ for the following data.

x	2	4	6	8
y	-2	5	11	16

x	2	4	6
$f(x) = ax^2 + bx + c$	4a + 2b + c	?	?

(b) Use finite differences to find f(10).

*5.(a) Complete the table and use the method of finite differences to find the cubic equation of the form $f(x) = ax^3 + bx^2 + cx + d$ for the following data.

x	0	1	2	3	4
y	0	-24	-42	-48	-36

x	0	1	2	3
$f(x) = ax^3 + bx^2 + cx + d$	d	?	?	?

(b) Use finite differences to find f(5).

6.(a) Complete the table and use the method of finite differences to find the cubic equation of the form $f(x) = ax^3 + bx^2 + cx + d$ for the following data.

x	-3	-1	1	3	5
y	45	3	1	-9	-75

x	-3	-1	1	3
$f(x) = ax^3 + bx^2 + cx + d$	?	?	?	?

(b) Use finite differences to find f(0).

Apply the Concepts

7. A Free-Falling Object Problem Find and interpret the first and second differences for the following data created by measuring the height above the ground of a free-falling object. Then find the equation of the height above the ground as a function of time.

t (sec)	1	3	5	7
s (ft)	256	384	384	256

8. The Baseball Toss Problem You are studying how many milk bottles it takes to build various sized pyramids for a baseball toss at a carnival (Figure 14).

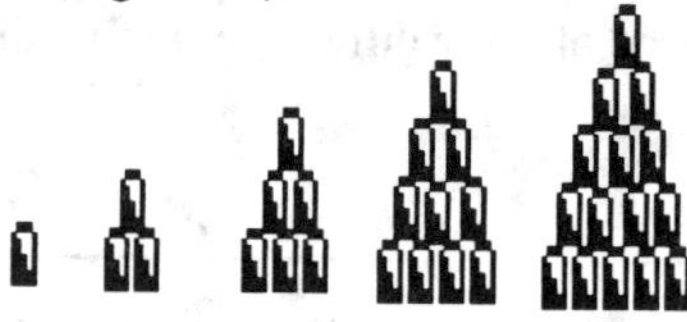

Figure 14

(a) Construct a table with x representing the number of bottles in the bottom row and y the number of bottles in the pyramid.

(b) Use finite differences to find the equation that expresses the total number of bottles as a function of the number of bottles in the bottom row.

9. The Rainbow Kite and Balloon Company We are adding new buildings at the kite company's site. The boss's son wants to connect each building to every other building with a phone line.

(a) Complete the table that expresses the number of lines needed as a function of the number of buildings.

(b) Use finite differences to find the equation of the function.

(c) Use the function from part (b) to find the number of lines needed when there are 12 buildings and when there are 20 buildings.

(d) Describe a better way to create this local area network.

buildings, n	0	1	2	3	4
lines, l	0	?	?	?	?

10. Wonderworld Amusement Park The new speedway ride at the park has cars passing through a tunnel. The following data represent the trial runs of the ride with various speed limits. The number of cars per hour passing through a tunnel as a function of various speed limits was recorded. Use finite differences to find the quadratic function for these data and determine the speed limit that allows the maximum number of cars through the tunnel.

L (mi/hr)	0	5	10	15	20
C (cars/hr)	0	450	800	1050	1200

*11. A Golf Ball Problem A golf ball's path is measured form time-lapse photos as follows.

(a) Use finite differences to find the equations for height (y) as a function of time t.

(b) Use finite differences to find the equations for horizontal distance (x) as a function of time t.

(c) Discuss the average velocity and acceleration in the y direction and in the x direction.

t (sec)	0	2	4	6	8
x (ft)	0	256	512	768	1024
y (ft)	0	192	256	192	0

12. The Golf Driver Problem A driver is a golf club used for long shots down the fairway. One with an 11° sloped face was hit an initial velocity of approximately 277 ft/sec, and the following data were recorded.

time, t (sec)	0	1	2	3
vertical distance, y (ft)	0	37	42	15
horizontal distance, x (ft)	0	272	544	816

(a) Use finite differences to find y as a function of t.

(b) Use finite differences to find x as a function of t.

(c) Find the average velocity and acceleration in each direction for each time interval.

13. The Big-Screen TV Problem You found a deal to buy a big-screen TV set on time without paying interest. Your account balance for several months is recorded below.

t (months)	4	6	8	10
b ($)	1713	1470	1227	984

(a) Use finite differences to determine when the set will be paid off.

(b) How much did the set cost?

(c) What do the first and second differences represent?

14. The Foot Size Problem You measured the foot length in inches of a few friends and recorded their shoe size.

men:

shoe size	10	12	14	16
foot length	5.25	11.25	17.25	23.25

women:

shoe size	8	9	10	11
foot length	5.25	11.25	17.25	23.25

(a) Use finite differences to find the linear equations that models foot length as a function of shoe size for men and for women.

(b) What is the relationship between the two functions in part (a)?

Extend the Concepts

15. Translate the data in Exercise 6 to the right 2 units by adding 2 to each value of x. Use finite differences to find the new function. How is this function related to the function in Exercise 6? What strategy does this suggest for finding an equation using the method of finite differences?

16. Multiply the y coordinates in Exercise 5 by 1/6 and find the equation for the third-degree polynomial that fits the new data. How does the new function compare with the function for Exercise 5? What strategy does this suggest for finding an equation using the method of finite differences?

17. The following data have equally spaced y values. Use finite differences to find the third-degree polynomial that fits these data. [Hint: Since the x values are not equally spaced, use the inverse function.]

x	-8	-1	0	1	8
y	0	1	2	3	4

18. Create your own table of data with constant third differences. Investigate reflections of the data about (a) the x axis, (b) the y axis, and (c) the origin. Find the equations using finite differences. Summarize your results.

19. Sequences such as 1, 1, 2, 3, 5, 8, ... occur on many standardized tests. Suppose that you are asked to find the next two terms, the seventh and eighth terms of the sequence.

(a) Use finite differences to find the degree of the polynomial that best fits the data.

(b) Explain why finite differences can always be used on such sequences.

(c) Find another rule for generating the *next* term in this sequence.

20. Explain how we could use finite differences to find the equation of a nearly linear set of data. Construct your own example with five data points. Compare the equation found using finite differences and the equation found using linear regression.

3.3 Interpolating and Best-Fitting Polynomial Functions

Finding the Exact-Fitting Polynomial Equation
Finding the Best-Fitting Polynomial Equation

In this section we continue our investigation of curve fitting. We find the equation of a polynomial that exactly fits a set of data or a lower-degree polynomial that approximates the data. We compare different best-fitting equations: a best-fitting **cubic polynomial** (degree 3) and a best-fitting **quartic polynomial** (degree 4) for a set of data. When observed data are collected to model a situation, it may contain inaccuracies ("noise"). The larger the sample, the more nearly the errors will be averaged out in the process. Using a least-squares polynomial to approximate the data has a smoothing effect, often creating a better model than that derived directly from the data.

Finding the Exact-Fitting Polynomial Equation

Let's start by finding the equation of the polynomial that exactly fits the data. That is, we want to find the equation that is satisfied by every given ordered pair and whose graph passes through every given point. For example, in Section 2.2 we saw that given two distinct points we can find the equation of a line that passes through the two points: a first-degree polynomial. We have also seen in Section 2.4 that given any three noncollinear points, we can find the equation of the parabola that passes through the three points: a second-degree polynomial. For n points we can find the equation of a polynomial of degree n - 1 that fits the data exactly (Table 1). We call the polynomial equation satisfied by all of the given points the **interpolating polynomial**. Interpolating polynomials fit the data exactly. In this context we use the word *interpolating* to indicate that the polynomial goes between the given points: from point to point.

Table 1

Number of Points	Degree of Interpolating Polynomial
2	first (linear)
3	second (quadratic)
4	third (cubic)
5	fourth (quartic)
n	n - 1

In Section 3.1 we found the equation of a polynomial given the x and y intercepts. In the first example we are given the zeros of three polynomials as well as an arbitrary point on each graph. We will find the interpolating polynomials algebraically using the factored form.

NOTE: Zeros that are real numbers correspond to x intercepts on a graph. Nonreal zeros do not.

Example 1 *Finding exact equations of polynomials given the zeros*

Find the equation of each curve that passes through the point (1, 48) and has the given four zeros.

(a) -3, -1, 2, and 4 (b) ± 2 and $1 \pm \sqrt{2}$ (c) $\pm i$ and $1 \pm 2i$

Solution

The lowest-degree polynomial that passes through five points is a fourth-degree polynomial. Since we are given four zeros, we begin by writing the polynomial in factored form.

(a)

$$y = a(x + 3)(x + 1)(x - 2)(x - 4) \qquad \textit{Substitute } x = 1 \textit{ and } y = 48.$$

$$48 = a(1 + 3)(1 + 1)(1 - 2)(1 - 4) \qquad \textit{Solve for } a.$$

$$48 = 24a$$

$$a = 2$$

The equation of this interpolating polynomial is

$$y = 2(x + 3)(x + 1)(x - 2)(x - 4) \Rightarrow y = 2x^4 - 4x^3 - 26x^2 + 28x + 48$$

(b)

$$y = a(x - 2)(x + 2)(x - (1 - \sqrt{2}))(x - (1 + \sqrt{2}))$$

$$y = a(x - 2)(x + 2)(x - 1 + \sqrt{2})(x - 1 - \sqrt{2}) \qquad \textit{Substitute } x = 1 \textit{ and } y = 48.$$

$$48 = a(-1)(3)(\sqrt{2})(-\sqrt{2})$$

$$48 = 6a$$

$$a = 8$$

The equation is $y = 8(x - 2)(x + 2)(x - 1 + \sqrt{2})(x - 1 - \sqrt{2}) \Rightarrow y = 8x^4 - 16x^3 - 40x^2 + 64x + 32$.

(c)

$$y = a(x - i)(x + i)(x - (1 - 2i))(x - (1 + 2i))$$

$$y = a(x - i)(x + i)(x - 1 + 2i)(x - 1 - 2i) \qquad \textit{Substitute } x = 1 \textit{ and } y = 48.$$

$$48 = a(1 - i)(1 + i)(2i)(-2i)$$

$$48 = a(2)(4)$$

$$a = 6$$

Therefore, $y = 6(x - i)(x + i)(x - 1 + 2i)(x - 1 - 2i) \Rightarrow y = 6x^4 - 12x^3 + 36x^2 - 12x + 30$.

For parts (a) and (b) of Example 1 the data can be entered in a graphing utility and the fourth-degree equation can be calculated automatically using quartic regression (QuartReg) (Figure 1).

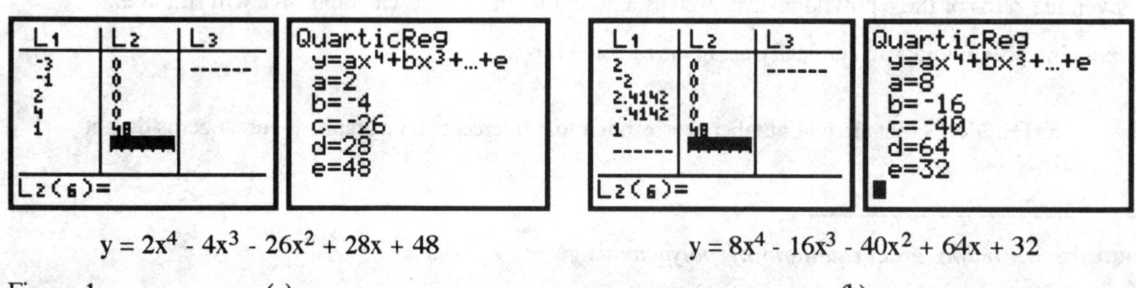

$y = 2x^4 - 4x^3 - 26x^2 + 28x + 48$ $\qquad$ $y = 8x^4 - 16x^3 - 40x^2 + 64x + 32$

Figure 1 (a) (b)

A graphing utility cannot find the equation for the points in part (c) of Example 1 because the zeros are not real numbers.

NOTE: Consult your instruction manual for your graphing utility or go to the *Precalculus Weblet - Reference Center - Technology.*

In the next example we use quadratic and cubic regression to find the equations of the interpolating polynomials that fit the data exactly.

Example 2 *Finding the exact equation of a polynomial given n points*

Find the interpolating polynomial whose graph passes through the given points. Then evaluate the missing coordinate.

(a) (1, 3), (2, -1), (2.5, ?), and (4, 3)

(b) (1, 3), (2, -1), (4, 3), (4.75, ?), and (5, 4)

Solution

(a) There are three points; the interpolating polynomial is quadratic. Entering the data and running **quadratic regression** yields the exact quadratic equation (Figure 2).

L1	L2
1	3
2	-1
4	3

QuadReg
y=ax²+bx+c
a=2
b=-10
c=11

Figure 2

Entering the equation $y = 2x^2 - 10x + 11$ into a graphing utility and evaluating the function at $x = 2.5$ yields the missing coordinate, $y = -1.5$ (Figure 3). [The point (2.5, -1.5) is the vertex of the parabola.]

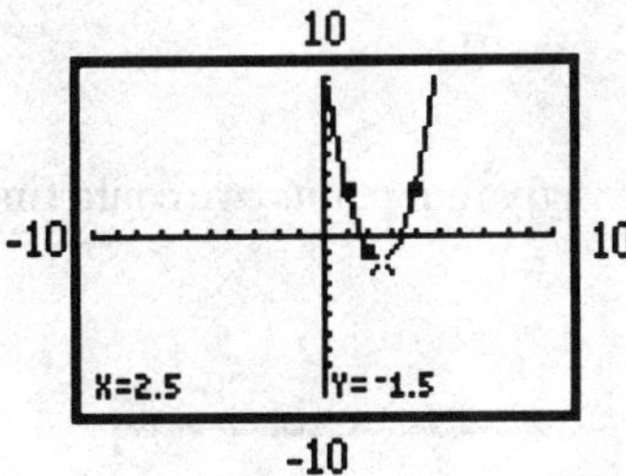

Figure 3

(b) There are four points. The interpolating polynomial is cubic. Entering the data and running **cubic regression** yields the exact cubic equation (except for the round-off error) (Figure 4).

L1	L2
1	3
2	-1
4	3
5	4

```
CubicReg
y=ax³+bx²+cx+d
a=-.5833333333
b=6.083333333
c=-18.16666667
d=15.66666667
```

$y = -0.583x^3 + 6.083x^2 - 18.167x + 15.667$

Figure 4

Entering $y = -0.583x^3 + 6.083x^2 - 18.167x + 15.667$ in a graphing utility and evaluating for $x = 4.75$ yields $y \approx 4.11$ (Figure 5).

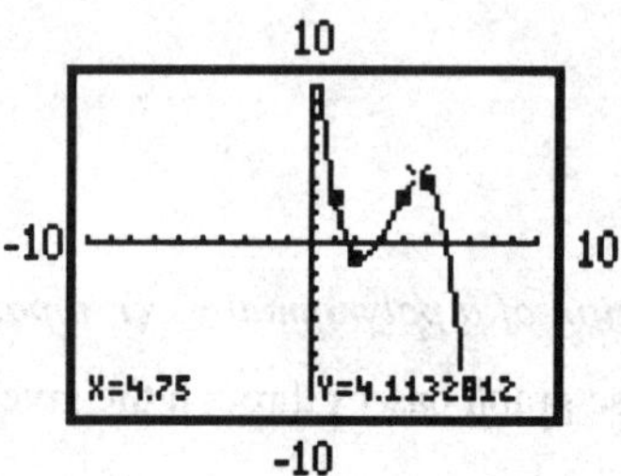

Figure 5

Finding the Best-Fitting Polynomial Equation

Consider the scatter plots in Figure 6.

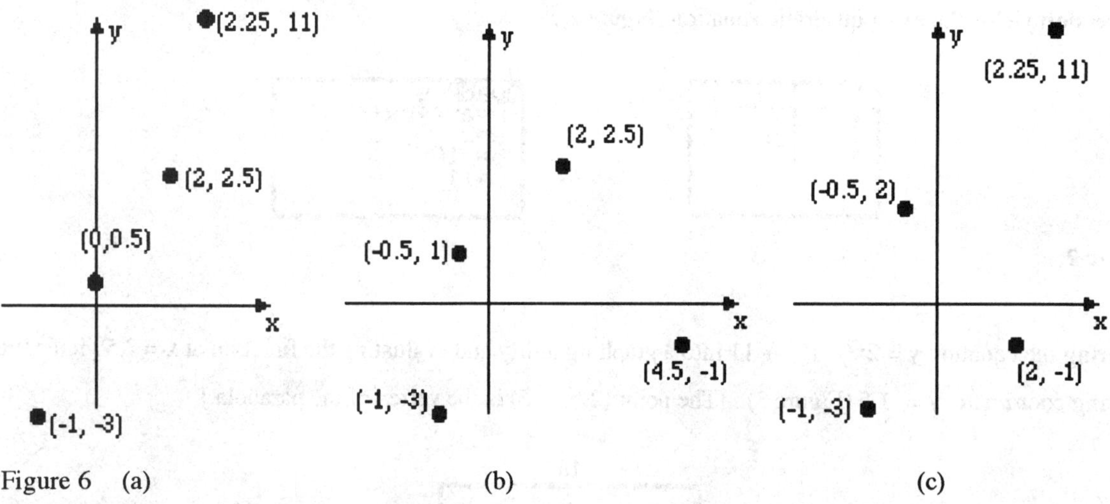

Figure 6 (a) (b) (c)

Since each of the scatter plots contains exactly four points, we could find the interpolating cubic polynomial for each set of data (Figure 7).

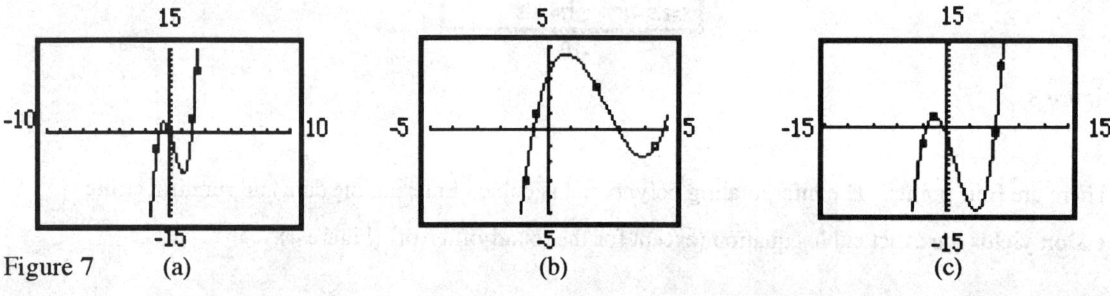

Figure 7 (a) (b) (c)

A best-fitting equation may better represent a set of data than an exact-fitting equation of having a different degree. For example, the data in Figure 6a are fairly linear, whereas the data in Figure 6b appear somewhat parabolic, with the graph going down at both ends. Perhaps the best-fitting linear equation and the best-fitting quadratic equation would be better models for the respective data than the exact-fitting cubic equation (Figure 8).

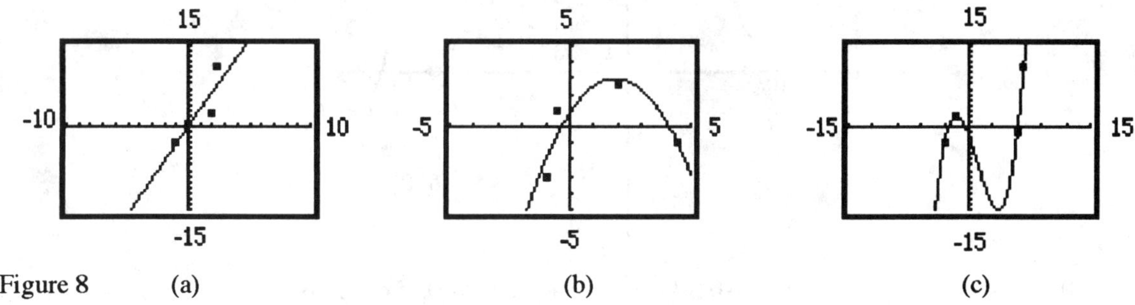

Figure 8 (a) (b) (c)

How do we decide whether to use the interpolating polynomial or the best-fitting polynomial equation?

Examining the ends of the graph via a scatter plot helps us choose the degree of the polynomial. The coefficient of determination can be calculated to determine whether a best-fitting polynomial fits well enough. (See Appendix III.)

Example 3 *Finding the best-fitting polynomial*

Use a graphing utility to find the equation of the polynomial that best fits the following data.

x	-4	-3	-2	-1	0	1	2	3	4	5	6
y	-4	0	2.5	3	2.5	1.2	1	2.5	4	2.5	-2

Solution

The graph of the data in Figure 9 indicates that a fourth-degree polynomial should fit these data well enough.

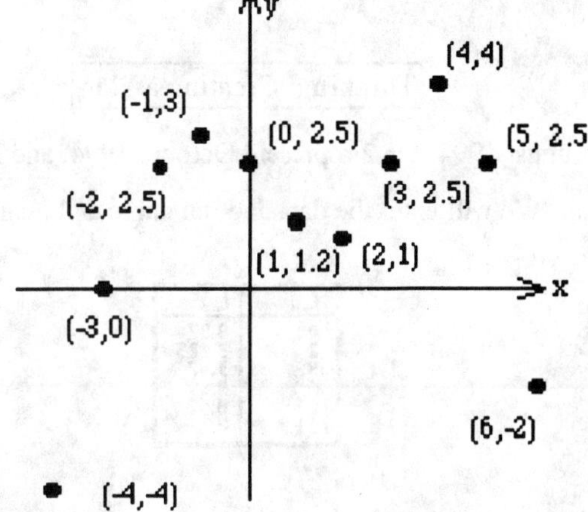

Figure 9

Entering the data in a graphing utility and running quartic regression (degree 4) yields a best-fitting polynomial whose graph is shown in Figure 10.

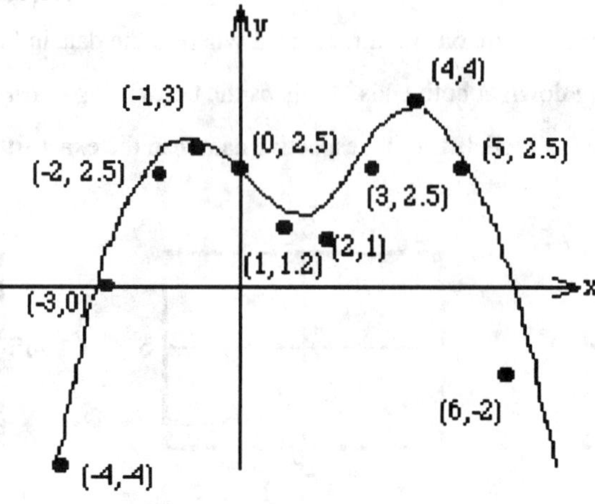

Figure 10 $y = -0.02x^4 + 0.10x^3 + 0.20x^2 - 0.55x + 1.76$

In the next example we compare a best-fitting cubic (degree 3) and a best-fitting quartic (degree 4) polynomial using cubic and quartic regression.

Example 4 The Rising and Falling Interest Problem

The following data contain the average national interest rates for 30-yr fixed-rate mortgages (home loans).

date	Aug. 1994	Oct. 1994	Dec. 1994	Feb. 1995	Apr. 1995	July 1995
interest rate	8.57%	9.125%	9.47%	9.1%	8.52%	7.5%

(a) Find the best-fitting cubic and quartic polynomials.

(b) Graph the data and each polynomial.

(c) Compare the models' estimates for interest rates in January 1995 and August 1995 with the actual rates: 9.3% in January 1995 and 8.04% in August 1995. [Sources: *Seattle Times* and HSH Association]

Thinking: Creating a Plan

We will let x = 0 represent August 1994, x = 2 represent October 1994, and so on, and find the best-fitting cubic and quartic polynomials. We will enter the data and run cubic and quartic regression (Figure 11).

L1	L2
0	8.57
2	9.125
4	9.47
6	9.1
8	8.52
11	7.5

Figure 11

To find the interest rates for January 1995 and August 1995, we evaluate the functions for x = 5 and x = 12.

Communicating: Writing the Solution

(a) <u>best-fitting cubic polynomial</u>: $y = 0.003x^3 - 0.084x^2 + 0.500x + 8.539$

<u>best-fitting quartic polynomial</u>: $y = 0.0006x^4 - 0.011x^3 + 0.004x^2 + 0.327x + 8.563$

(b) The graph of the best-fitting cubic polynomial indicates that interest rates are bottoming out near their July low of 7.5% (Figure 12a), while the best-fitting quartic polynomial indicates that rates will go much lower (Figure 12b).

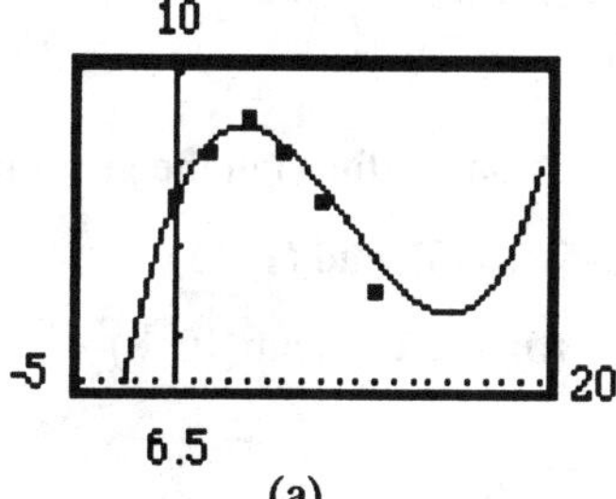

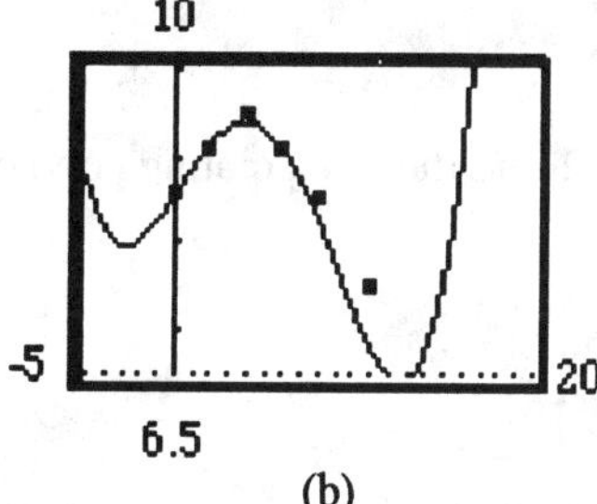

Figure 12 (a) (b)

(c) For January 1995, x = 5, both polynomials closely approximate the actual interest rate of 9.3% (Figure 13).

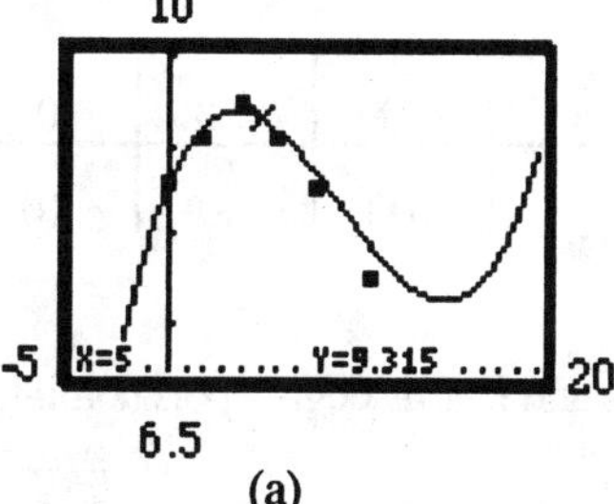

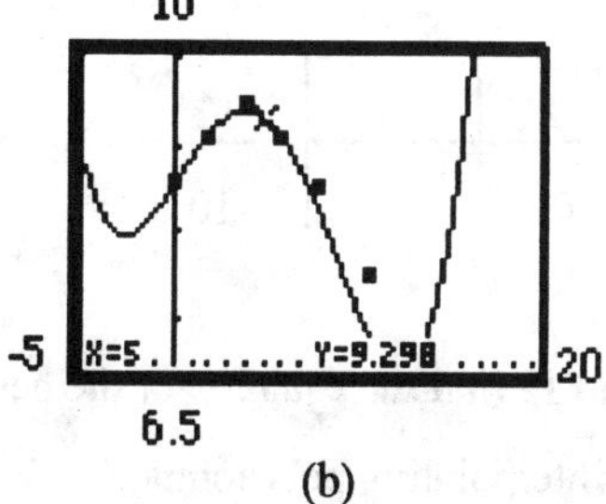

Figure 13 (a) (b)

For August 1995, x = 12, the cubic polynomial estimates an interest rate of 7.6%, while the quartic polynomial estimates an interest rate of 6.5% (Figure 14). Both models are markedly off the actual rate of 8.04%.

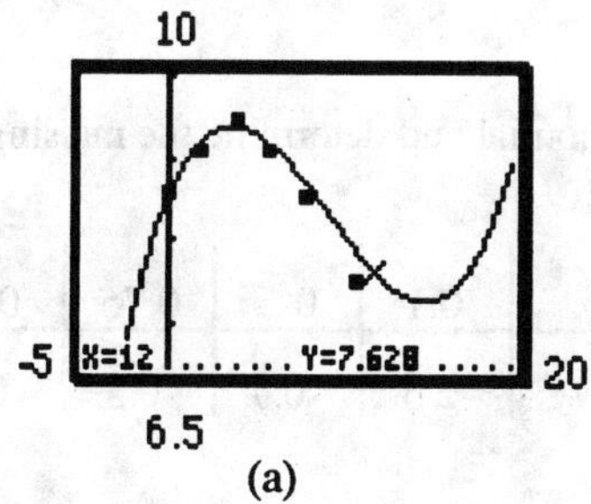

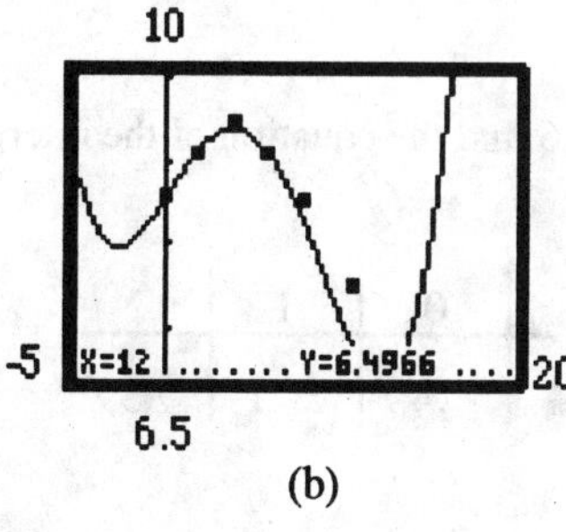

Figure 14 (a) (b)

Learning: Making Connections

From the graphs above, the cubic polynomial is a better model for this application even though the quartic polynomial appears to fit the data better. The best-fitting cubic polynomial has more of a smoothing effect than the best-fitting quartic polynomial, which more nearly approximates the data. A polynomial of degree 5, which would fit the six data points exactly but have no smoothing effect, may not be the best model for this application. For both polynomials interpolation (x = 5) was excellent, but for extrapolation (x = 12) both were markedly off target.

Exercises 3.3

Develop the Concepts

For Exercises 1 to 4 find the equation of the curve that passes through the point (0,24) and has the given zeros by first writing the polynomial in factored form.

1. -1, 2, and 4
2. 3 and $2 \pm \sqrt{2}$
3. $-1 \pm \sqrt{3}$ and $\pm 2i$
4. $-2 \pm \sqrt{3}i$ and $1 \pm \sqrt{2}i$

For Exercises 5 to 12 find the interpolating polynomial whose graph passes through the given points.

5. (-1, 2) and (3, 5)
6. (-1, -5), (0, 3), and (1, 11)
7. (-2, -5), (0, 1), and (2, -6)
8. (-10, 10), (0, 22), and (10, 8)

For Exercises 9 and 10 find the equation of the best-fitting second- and third-degree polynomials. Indicate whether each is an interpolating polynomial.

9.

x	-12	-2	0	12
y	-10	0	1	-10

10.

x	-100	-50	0	50
y	100	50	10	100

For Exercises 11 and 12 find the equation of the best-fitting third- and fourth-degree polynomials. Indicate whether each is an interpolating polynomial.

11.

x	0	10	20	30	40
y	-10	10	-10	10	20

12.

x	-120	-20	22	50	100
y	200	100	100	150	200

For Exercise 13 to 16 find the equation of the interpolating polynomial and determine the missing coordinates.

13.

x	-2	-1	0	1	2
y	0.14	0.37	?	2.71	7.39

14.

x	0.1	0.5	0.75	0.9	1
y	-2.3	-0.7	-0.3	-0.1	?

15.

x	0	1	2	3	4	6
y	0	0.84	0.91	0.14	-0.8	?

16.

x	0	1.57	3.14	4.71	6.28	7.85
y	1	0	-1	?	1	0

Apply the Concepts

17. The Interest Problem You have $5000 in an account that earns interest as indicated in the following table.

(a) Find the third-degree interpolating polynomial that fits the data.

(b) Estimate the amount of money in the account after $3\frac{1}{2}$, 4, and 5 yr.

year	0	1	2	3
amount	$5000	$5335.80	$5694.14	$6076.55

18. Another Interest Problem (a) Find the equation of the best-fitting cubic polynomial for the following data.

(b) Compare the amount of money in the account after $3\frac{1}{2}$, 4, and 5 yr with your results from Exercise 17.

year	0	1	2	3	4
amount	$5000	$5335.80	$5694.14	$6076.55	$6464.65

19. And Another Interest Problem (a) Find the equation of the fourth-degree interpolating polynomial for the data in Exercise 18.

(b) Compare the polynomial's estimate for the amount of money in the account after $3\frac{1}{2}$, 4, and 5 yr with that from Exercises 17 and 18.

20. The Richer or Poorer Problem Suppose that your bank balance was as indicated in the following table.

month	Dec. 1994	Feb. 1995	Apr. 1995	June 1995	Aug. 1995	Oct. 1995
balance	$1000	$2400	$3000	$2500	?	$2600

(a) Let December 1994 be month 0, February 1995 be month 2, and so on, in order to find the best-fitting cubic and quartic polynomials.

(b) Graph the best-fitting polynomials from part (a) and the data.

(c) Estimate the missing month's balance from each of the best-fit polynomials.

(d) What does each polynomial predict for the balance in December 1995?

(e) Discuss the smoothing effect.

*21. A Radioactive Carbon Dating Problem The following data represent the theoretical amount of carbon-14 remaining in an artifact after t years if 1000 g was present initially.

(a) Find the equation of the best-fitting third-degree polynomial.

(b) Find the equation of the fourth-degree interpolating polynomial.

(c) Calculate the amount present after 5730, 11,460, and 28,650 yr, using the polynomials from parts (a) and (b). [Hint: Enter the years in thousands.]

year	0	5000	10,000	15,000	20,000
amount (g)	1000	546	298	163	89

22. Population Growth Is a Problem The U.S. population in millions is listed in the following table.

(a) Find the equation of the best-fitting third-degree polynomial.

(b) Find the equation of the fourth-degree interpolating polynomial.

(c) Estimate the missing data using the polynomials from parts (a) and (b).

year	1970	1985	?	1990	1991	1993	1995
population (millions)	203.3	237.9	240	248.7	252.1	257.9	?

*23. The Fasting and Feasting Problem Suppose that your eating habits can be depicted as shown in Figure 15, with your hunger given on a scale from 0 to 10 and your food intake also on a scale from 0 to 10.

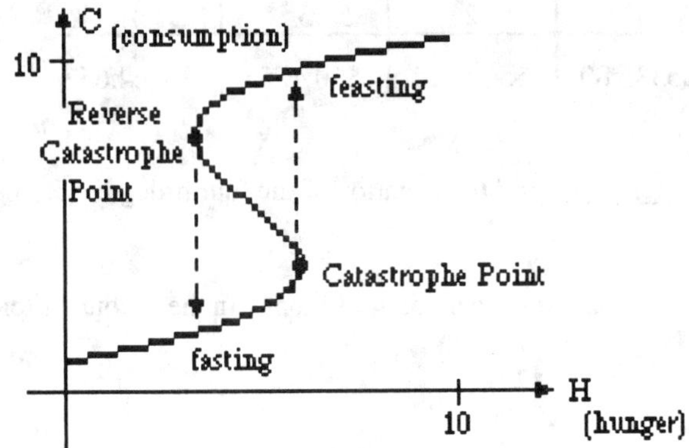

Figure 15

(a) Estimate the coordinates of five points on the graph (h, c).

(b) Reverse each ordered pair from part (a), and find the best-fitting cubic polynomial that describes hunger as a function of consumption (c, h).

(c) Estimate the catastrophe point when you are so hungry that you overeat.

(d) Estimate the reverse catastrophe point where you jump back to fasting.

[Source: *Scientific American*]

24. The Prowler Problem A prowler enters a junkyard at point A (Figure 16). The junkyard dog retreats, up to a point. At point B, a catastrophe point, the dog attacks. If this reduces the dog's fear at point C, the dog may stop attacking. A reverse catastrophe point could be reached at D when the dog reverts to retreating, only to have the cycle repeat.

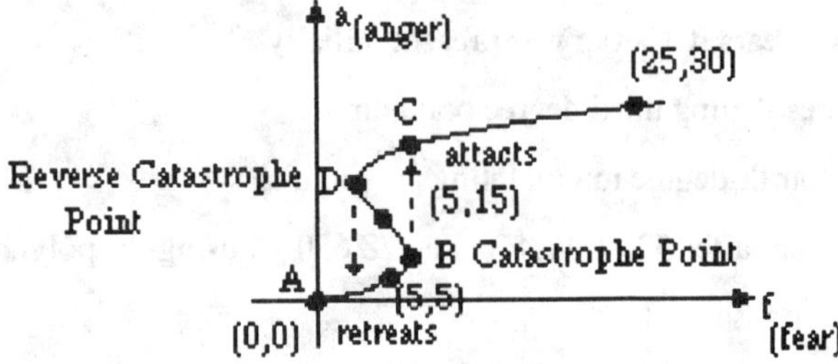

Figure 16

(a) Find the equation of the cubic polynomial whose inverse is drawn in Figure 16.

(b) Estimate the catastrophe point at B and the reverse catastrophe point at D.

Exercises 25 and 26 involve cubic splines that are made up of several third-degree polynomials that approximate a set of data by partitioning the data into overlapping subsets each containing four points.

*25. <u>The Smoothing Curve Problem</u> Use the following set of data to compare the linear and cubic splines.

x	-4	-2	0	2	4
y	0	4	8	2	4

(a) Linear Splines: Draw a straight line between each consecutive pair of points.

(b) Cubic Splines: Combine the graphs of two cubic polynomials: one passing through the first four points and the other passing through the last four points. Use the first cubic for $x < 0$ and the second for $x > 0$ to create one smooth curve passing through the given data.

26. <u>The Boatbuilder Problem</u> Boatbuilders were the first to use cubic splines to create the cross members of a ship's hull (Figure 17). Help design the illustrated cross member using four cubic splines and the measurements given.

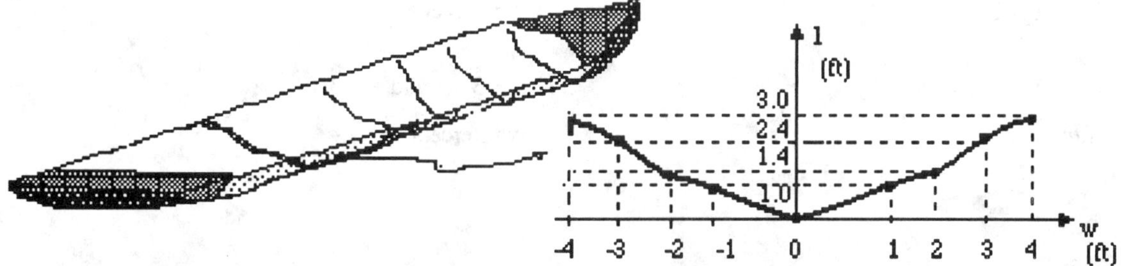

Figure 17

27. The Rainbow Kite and Balloon Company The data are the price per share of the company's stock traded "over the counter."

month, t	0	6	12	19	24
stock price, p ($)	15	35	55	30	55

(a) Find the equation of the best-fitting cubic polynomial.

(b) Use the formulas for turning points and the point of inflection of a cubic polynomial,

$$x = \frac{-b \pm \sqrt{b^2 - 3ac}}{3a} \quad \text{and} \quad x = -\frac{b}{3a}$$

to estimate the turning points and the point of inflection on the graph of the equation from part (a).

(c) If you were to buy and sell this stock, what information does the points in part (b) suggest?

28. Wonderworld Amusement Park Attendance at the amusement park varies over the course of the year as indicated by the following data.

month	Jan.	Feb.	Mar.	Apr.	May	June	July	Aug.	Sept.	Oct.	Nov.	Dec.
attend.	10,000	6370	7792	12,181	20,000	27,890	35,039	40,000	41,325	37,565	27,273	9000

(a) Find the equation of the best-fitting cubic polynomial.

(b) Use the formulas given in Exercise 27 to estimate the turning points and the point of inflection on the graph of the equation from part (a).

(c) What information does the points in part (b) suggest about hiring and firing staff members?

Extend the Concepts

For Exercises 29 to 32 find the interpolating quartic with rational coefficients having the given zeros and y intercept.

29. $x = 1 + \sqrt{2}$ and $x = \sqrt{2}$

y intercept 2

30. $x = 1 - \sqrt{3}$ and $x = 1 + \sqrt{2}$

y intercept -4

31. $x = 2 + i$ and $x = 2i$

y intercept 20

32. $x = 3 + i$ and $x = 3i$

y intercept 9

3.4 Graphs of Rational Functions

Examining Reciprocal Functions
Examining the Behavior of Graphs of Rational Functions
Examining Graphs of Rational Functions near Vertical Asymptotes
Applying Rational Functions

In this section we study graphs of functions of the form

$$f(x) = \frac{p(x)}{q(x)}, \quad q(x) \neq 0$$

If p(x) and q(x) are polynomials, their quotient is called a **rational function**. We are familiar with two rational functions from the Library of Reference Functions (Appendix II),

$$y = \frac{1}{x} \text{ and } y = \frac{1}{x^2}$$

We will use these two simple examples of rational functions to learn more about graphing rational functions and the nature of asymptotic behavior. We apply rational functions to the real world by continuing our work with optimization. In Example 4 we apply our knowledge about the graph of a rational function in order to find the dimensions of a right circular cylinder with a fixed volume that will minimize the surface area.

Examining Reciprocal Functions

Consider the function f(x) = x and its reciprocal,

$$y = \frac{1}{f(x)} = \frac{1}{x}$$

At first glance it may not appear that these graphs are related, but they are. The symmetry of f(x) = x about the origin implies that the graph of its reciprocal is also symmetric about the origin. Both are odd functions (Figure 1).

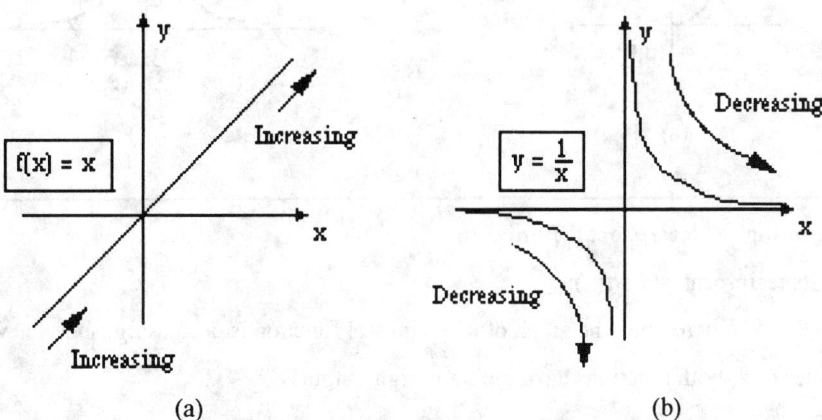

Figure 1 (a) (b)

Notice in Figure 1 that the graph of f(x) = x is everywhere increasing and each piece of the graph of y = 1/x is decreasing. For y = 1/x, as x increases, the right end of the graph decreases, approaching the x axis (Figure 1b). We say that the graph is asymptotic to the x axis (the x axis is a horizontal asymptote). Because of the symmetry of this odd function, the graph must also be asymptotic to the x axis to the far left. Substituting a few values for x shows that one graph is increasing while the other is decreasing.

x	f(x) = x	$y = \frac{1}{x}$
-10	-10	-1/10
-1	-1	-1
-1/10	-1/10	-10
0	0	undefined
1/10	1/10	10
1	1	1
10	10	1/10

Notice that taking the reciprocal of a number does not change the sign. Values for f(x) = x and y = 1/x have the same sign. At x = 0, the reciprocal function y = 1/x has no point on its graph because 1/0 is undefined.

Can you determine the relationship between the graphs of $g(x) = x^2$ *and* $y = \frac{1}{g(x)} = \frac{1}{x^2}$*?*

The graph of $y = x^2$ is symmetric about the y axis (Figure 2a). It is an even function. Therefore, its reciprocal is also an even function. As x increases to the right, $y = 1/x^2$ decreases, approaching zero from above the x axis (Figure 2b). Its graph is asymptotic to the positive x axis on the right. By the symmetry of this even function, the graph is also asymptotic to the negative x axis on the left.

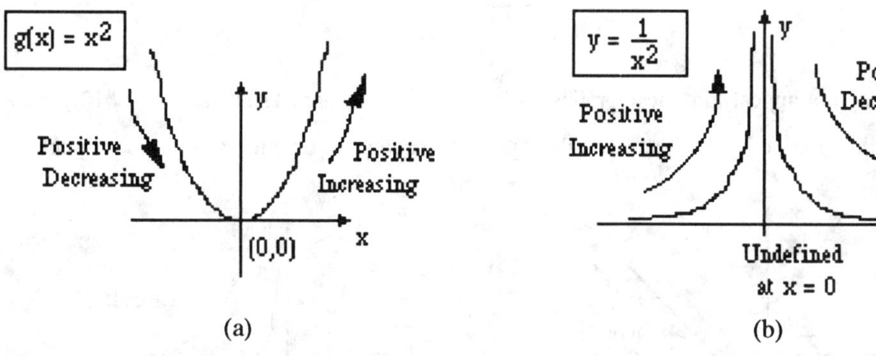

Figure 2 (a) (b)

Summary: Graphing a Reciprocal Function

For function f and its reciprocal, y = 1/f(x):

If the graph of function f is increasing, the graph of its reciprocal function is decreasing, and vice versa.

For all defined values of x, both functions have the same sign outputs.

The x intercepts of function f are vertical asymptotes on the reciprocal graph, and vice versa.

The y intercept of function f is the reciprocal of the y intercept for y = 1/f(x).

Example 1 *Graphing a reciprocal function*

Sketch the graph of $y = \frac{1}{f(x)}$ from the graph of $f(x) = -x^2 - 2$.

Solution

The graph of $f(x) = -x^2 - 2$ is a reflection about the x axis and a translation down 2 units. Its y intercept is -2. It is everywhere negative, increasing for $x < 0$ and decreasing for $x > 0$ (Figure 3).

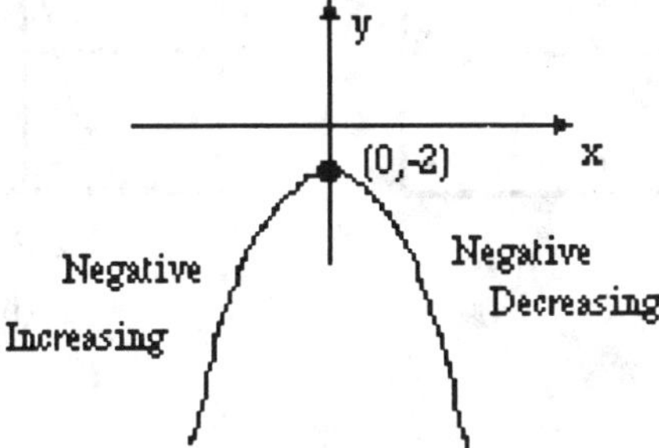

Figure 3

The graph of its reciprocal, $y = \frac{1}{-x^2 - 2} = \frac{-1}{x^2 + 2}$, will also be everywhere negative, but decreasing for $x < 0$ and increasing for $x > 0$. Its y intercept is the reciprocal of -2, -1/2 (Figure 4).

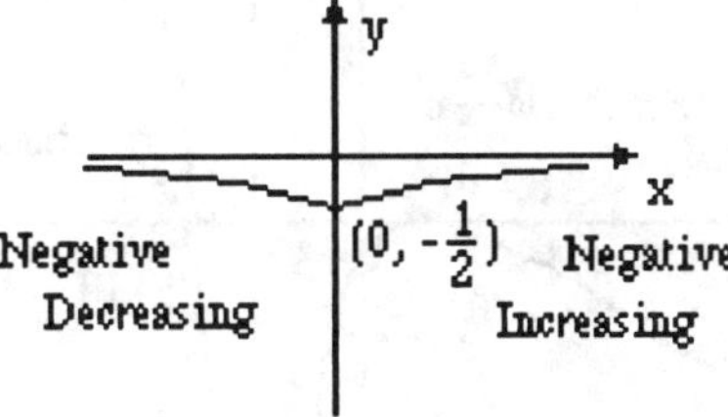

Figure 4

Examining the Behavior of Graphs of Rational Functions

In Section 3.1 we saw how the highest-degree term in a polynomial function dictates the behavior of the *ends* of a graph. Behavior of the ends of the graph of a rational function is dictated by the ratio of the leading terms in the numerator and denominator. Finding this ratio is a quick way to approximate the shape of the ends of the graph of any rational function . This ratio may not yield the exact numerical results for the values of the ordered pairs toward the ends of the graph, but it will yield the correct shape and the sign, which are generally all we need to sketch relatively good graphs quickly. For example, the ratio of the leading terms of the numerator and denominator of

$$y = \frac{2x^3 + 1}{x^2}$$

is

$$y = \frac{2x^3}{x^2} = 2x$$

This means that the ends of the graph of $y = \frac{2x^3 + 1}{x^2}$ resemble the ends of the graph of y = 2x. The ends of the graph of $y = \frac{2x^3 + 1}{x^2}$ are actually asymptotic to y = 2x, approximating its shape and position at the ends.

Figure 5a shows the graph of y = 2x with its ends going down to the left and up to the right. Figure 5b shows the graph of $y = \frac{2x^3 + 1}{x^2}$ with its ends also going down to the left and up to the right.

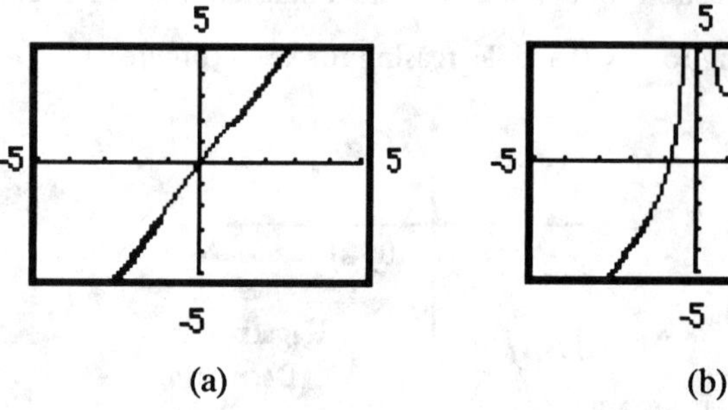

Figure 5 (a) (b)

Consider the sketch of $y = \frac{(x + 1)^2(x - 3)}{(x - 2)^2}$ in Figure 6. The ends of its graph will behave like y = x since the ratio of its leading terms is $\frac{x^3}{x^2} = x$.

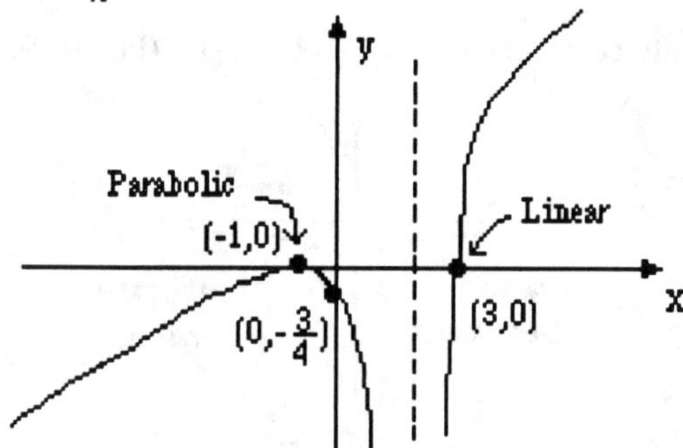

Figure 6

Just as for all other functions, knowing the x and y intercepts will help us sketch graphs of rational functions.

How do we find x and y intercepts of rational functions?

To find the y intercept we substitute x = 0. For the rational function $y = \frac{(x + 1)^2(x - 3)}{(x - 2)^2}$ the y intercept is $-\frac{3}{4}$ since

$$f(0) = \frac{(0 + 1)^2(0 - 3)}{(0 - 2)^2} = -\frac{3}{4}$$

We find the x intercepts by setting the function equal to zero and solving for x.

$y = \frac{(x + 1)^2(x - 3)}{(x - 2)^2}$ and $f(x) = 0 \Rightarrow$

$$\frac{(x + 1)^2(x - 3)}{(x - 2)^2} = 0$$ *Multiply both sides by $(x - 2)^2$.*

$$(x + 1)^2(x - 3) = 0$$

$$(x + 1)^2 = 0 \quad \text{or} \quad x - 3 = 0$$

$$x = -1 \quad \text{or} \quad x = 3$$

The x intercepts are -1 and 3.

Notice that the x intercepts of a rational function are among the zeros of its numerator. The multiplicities of the factors in the numerator determine the shape of the curve near that intercept. Just as with polynomials, a first-degree factor signals roughly linear behavior near the corresponding intercept, and a second-degree factor will cause a turning point and a roughly parabolic shape near that intercept. That is why the graph of $y = \frac{(x + 1)^2(x - 3)}{(x - 2)^2}$ in Figure 6 is parabolic near x = -1 and linear near x = 3.

Why does the graph in Figure 6 have a vertical asymptote at x = 2?

Vertical asymptotes are created by values of x that cause division by zero. If we set the expression in the denominator, $(x - 2)^2$, equal to zero and solve for x, we will have the equation of the vertical asymptote.

$$(x - 2)^2 = 0$$

$$x - 2 = 0$$

$$x = 2$$

Examining Graphs of Rational Functions near Vertical Asymptotes

Both reference functions $y = \frac{1}{x}$ and $y = \frac{1}{x^2}$ are undefined when x = 0. They have a vertical asymptotes at x = 0, the y axis. The graph of y = 1/x increases without bound as x approaches zero from the positive side and decreases without bound as x approaches zero from the negative side (Figure 7a). A vertical asymptote where the function changes sign across the asymptote is a **single pole**.

As x approaches zero from the left, the graph of $y = 1/x^2$ increases becoming asymptotic to the positive y axis. The symmetry of $y = 1/x^2$ shows that the graph be asymptotic to the positive y axis, approaching positive infinity on both sides of the y axis. Such a vertical asymptote is called a **double pole**.

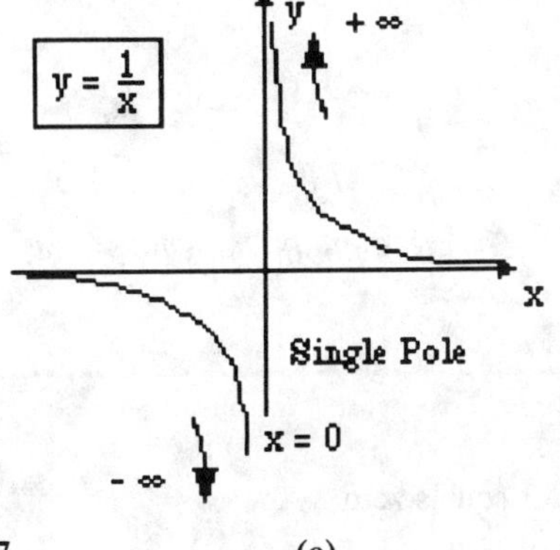

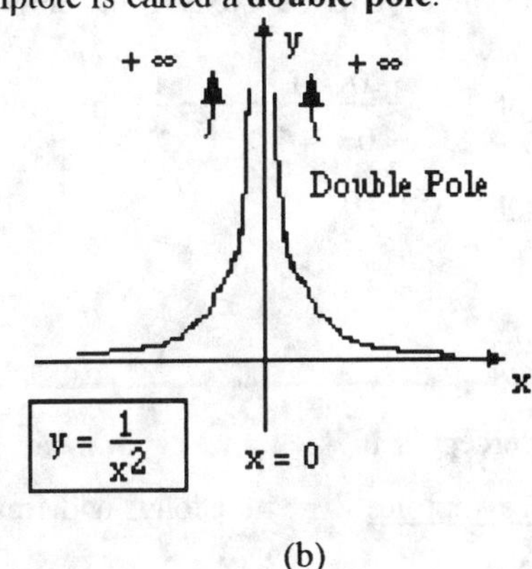

Figure 7 (a) (b)

What determines whether a vertical asymptote is a single or a double pole?

The multiplicity of the factor in the denominator dictates whether the vertical asymptote is a single or a double pole. Odd multiplicity creates a single-pole asymptote with the graph approaching positive ∞ on one side of the asymptote and -∞ on the other. Even multiplicity creates a double-pole asymptote with the graph approaching +∞ on both sides of the asymptote or -∞ on both sides.

A graph cannot cross or intersect a vertical asymptote because that value of x causes division by zero, which is undefined. As the graph approaches a vertical asymptote, it heads up toward +∞ or down toward -∞. For all other values of x, the function is defined and there must be a graph everywhere else.

Summary: Graphing Rational Functions	
Ends of the graph:	Find the ratio of the leading terms.
y intercept:	Find f(0).
x intercepts:	Set f(x) = 0 and solve for x. (Find zeros of the numerator.)
	Check the multiplicity of the factors for behavior of the graph near the x intercepts.
Vertical asymptotes:	Find values of x that create division by 0. (Find zeros of the denominator.) Check the multiplicity of the factors of the denominator to determine whether an asymptote is a single double pole.

When the x intercepts and vertical asymptotes yield the same x values, we have $f(x) = \frac{0}{0}$. This exception to the summary above is studied in Section 3.5.

Example 2 *Graphing a rational function*

Graph and label all intercepts and asymptotes for

$$f(x) = \frac{x - 1}{x^2 - 4}$$

Solution

Ends: Since the ratio of the leading terms is $\frac{x}{x^2} = \frac{1}{x}$, the ends of the graph will be similar to the graph of $y = 1/x$, asymptotic to the x axis from below the x axis on the far left and from above the x axis to the far right.

y intercept: $f(0) = \frac{0 - 1}{0^2 - 4} = \frac{-1}{-4} = \frac{1}{4}$ The y intercept is $\frac{1}{4}$.

x intercepts: $f(x) = 0 \Rightarrow$

$$\frac{x - 1}{x^2 - 4} = 0 \qquad \textit{Multiply both sides by } x^2 - 4.$$

$$x - 1 = 0$$

$$x = 1$$

The x intercept is 1.

Vertical asymptotes: The function is undefined when the denominator equals zero.

$$x^2 - 4 = 0$$
$$(x + 2)(x - 2) = 0$$
$$x + 2 = 0 \quad \text{or} \quad x - 2 = 0$$
$$x = -2 \quad \text{or} \quad x = 2$$

The vertical asymptotes are x = -2 and x = 2. They are single poles since (x + 2) and (x - 2) have multiplicity 1. The graph will approach ∞ in opposite directions on either side of these vertical asymptotes. Figure 8 shows the information.

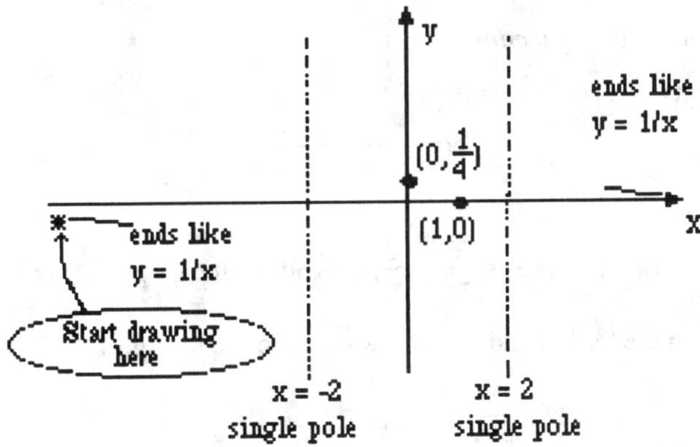

Figure 8

The final graph is shown in Figure 9.

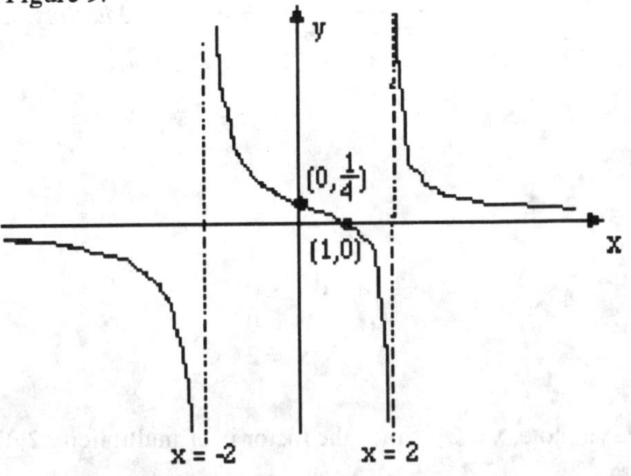

Figure 9

Graphing utilities often display misleading graphs of functions with vertical asymptotes, as shown in Figure 10. Some utilities have a *dot mode* that displays only values that are actually calculated. Lines connecting the pieces of the graph across its asymptotes are not displayed.

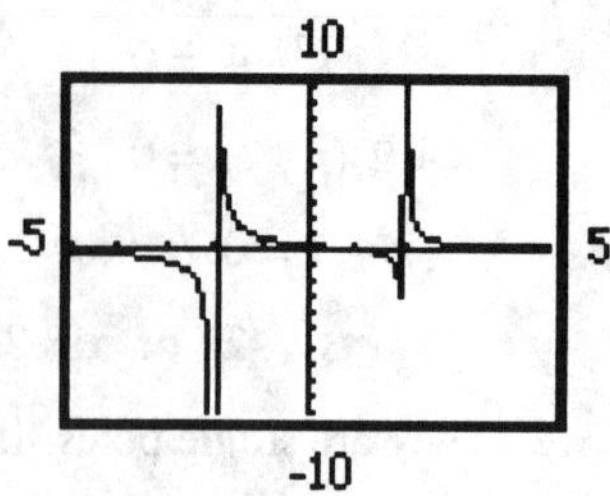

Figure 10

Example 3 *Graphing a rational function*

Graph and label all intercepts and asymptotes for

$$g(x) = \frac{1 - 2x}{x^2 - 4x + 4}$$

Solution

Ends: Since the ratio of the leading terms is $\frac{-2x}{x^2} = \frac{-2}{x}$, the ends of the graph will be similar to the graph of $y = -1/x$, above the x axis on the far left and below the x axis to the far right.

y intercept: $g(0) = \frac{1 - 2(0)}{0^2 - 4(0) + 4} = \frac{1}{4}$. The y intercept is $\frac{1}{4}$.

x intercepts: $g(x) = 0 \Rightarrow$

$$\frac{1 - 2x}{x^2 - 4x + 4} = 0 \qquad \textit{Multiply both sides by } x^2 - 4x + 4.$$

$$1 - 2x = 0$$

$$x = \frac{1}{2}$$

The x intercept is 1/2.

Vertical asymptotes:

$$x^2 - 4x + 4 = 0$$

$$(x - 2)^2 = 0$$

$$x = 2$$

There is only one vertical asymptote, $x = 2$. Since the factor is of multiplicity 2, the asymptote is a double pole. The graph does not change sign across its asymptote (Figure 11).

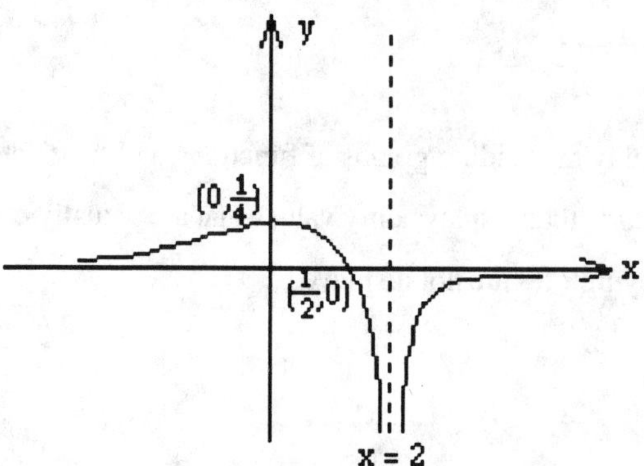

Figure 11

Applying Rational Functions

A problem similar to the next example is found in nearly every calculus text. Here, we extend our study of optimization to cubic polynomials.

Example 4 The Soda Can Problem

A bottling and can company has hired you to redesign the can for their soda pop. They want your design to minimize the surface area, yet have a fixed volume that will hold 12 fluid ounces (about 20 in^3). How should you design it?

Thinking: Creating a Plan

The volume V of a right-circular cylinder is $V = \pi r^2 h$, where r is the radius of the circular sections and h is the height of the cylinder. We must find the diameter and height of the can that will use the *least* amount of material. If we can express the surface area S as a function of one variable, say radius r, and graph the resulting function, we can estimate the optimum value from the graph.

Communicating: Writing the Solution

Volume: $V = 20$ in^3 and $V = \pi r^2 h \Rightarrow 20 = \pi r^2 h \Rightarrow \mathbf{h = \frac{20}{\pi r^2}}$

Surface area: Surface area = area of sidewall + area of two circles (Figure 12).

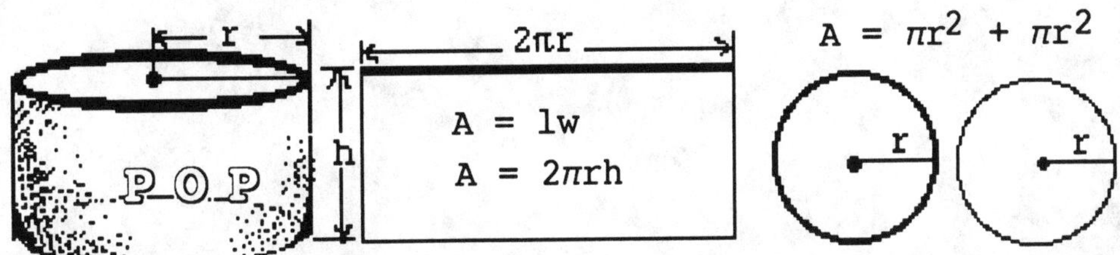

Figure 12

$$\begin{aligned} S &= (2\pi r)\mathbf{h} + 2\pi r^2 \\ &= 2\pi r\left(\frac{20}{\pi r^2}\right) + 2\pi r^2 \\ &= \frac{40}{r} + 2\pi r^2 \\ &= \frac{2\pi r^3 + 40}{r} \end{aligned}$$

Substitute $h = 20/\pi r^2$.

From the graph of $S = \dfrac{2\pi r^3 + 40}{r}$ in Figure 13, we estimate that $r = 1.5$ in. will minimize the surface area. The values of the function for $r < 0$ are not part of the domain of this application since the radius of the can is not negative.

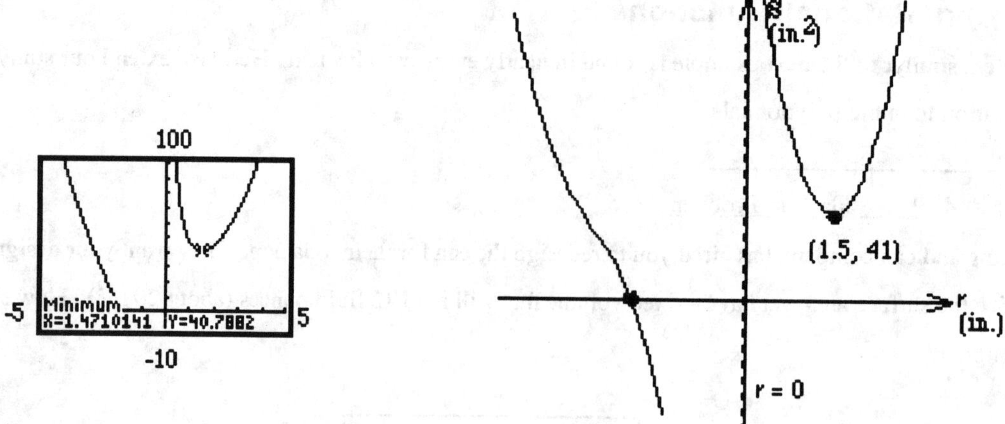

Figure 13

The diameter is d = 2r ≈ 3. The corresponding height of the can is $h = 20/\pi r^2 = 20/\pi(1.5^2) \approx 3$. The can should have a diameter of 3 in. and a height of 3 in.

Learning: Making Connections

Although your design will use the least amount of material, it may not be as aesthetically pleasing on the shelf or when held.

Exercises 3.4

Develop the Concepts

For Exercises 1 to 4 graph each pair of functions on one coordinate system. State the x intercept(s) of function f and the asymptote(s) of the reciprocal function.

1. $f(x) = x - 1,\ y = \dfrac{1}{f(x)} = \dfrac{1}{x - 1}$

2. $f(x) = 1 - x,\ y = \dfrac{1}{f(x)} = \dfrac{1}{1 - x}$

3. $f(x) = (x - 2)^2,\ y = \dfrac{1}{f(x)} = \dfrac{1}{(x - 2)^2}$

4. $f(x) = x^2 - 1,\ y = \dfrac{1}{f(x)} = \dfrac{1}{x^2 - 1}$

For Exercises 5 and 6 graph each function and its reciprocal. Label the turning points.

5. $f(x) = x^2 + 4,\ y = \dfrac{1}{f(x)} = \dfrac{1}{x^2 + 4}$

6. $f(x) = -x^2 - 1,\ y = \dfrac{1}{f(x)} = \dfrac{-1}{x^2 + 1}$

For Exercises 7 to 10 sketch the graph of the reciprocal of each function defined by its graph. Label the intercept(s) and asymptotes.

7.

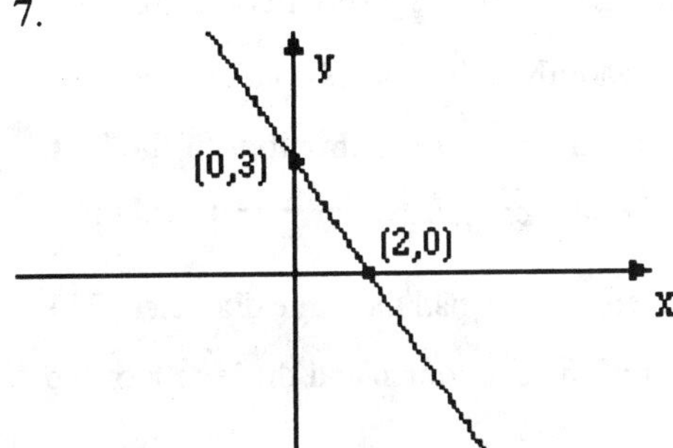

8.

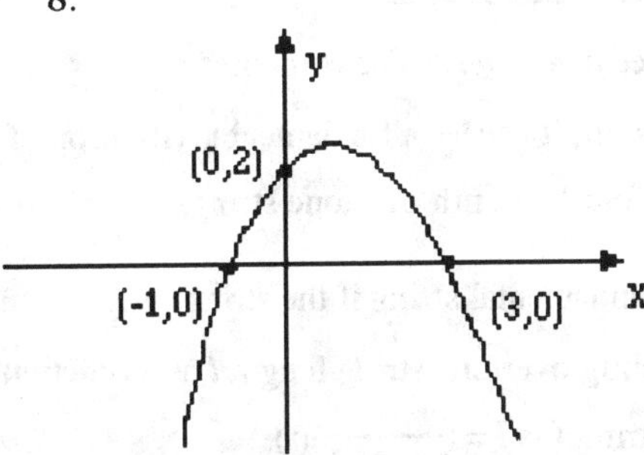

9.

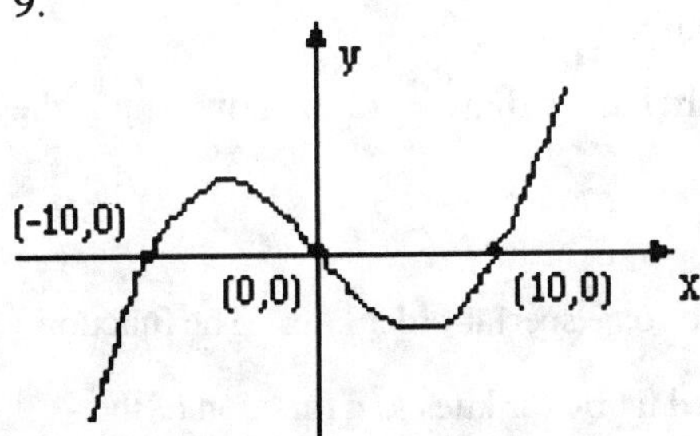

10.

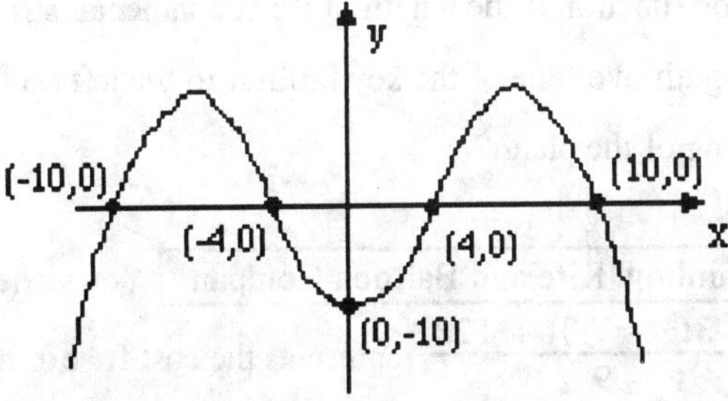

For Exercises 11 to 24 graph each rational function. Label the intercepts and vertical asymptotes.

11.(a) $y = \dfrac{x - 1}{(x + 2)^2}$

(b) $y = \dfrac{(x - 1)^2}{x + 2}$

(c) $y = \dfrac{(x - 1)^2}{(x + 2)^2}$

12.(a) $y = \dfrac{x + 3}{x - 3}$

(b) $y = \dfrac{x + 3}{(x - 3)^2}$

(c) $y = \dfrac{(x + 3)^2}{(x - 3)^2}$

13.(a) $y = \dfrac{(x - 1)(x + 2)}{x - 3}$

(b) $y = \dfrac{x - 3}{(x - 1)(x + 2)}$

14.(a) $y = \dfrac{(x + 1)^2(x - 1)}{(x - 2)^2}$

(b) $y = \dfrac{(x - 2)^2}{(x + 1)^2(x - 1)}$

15. $y = \dfrac{2x}{x^2 - 1}$

16. $y = \dfrac{-4}{x^2 - 2x - 3}$

*17. $y = \frac{9x}{x^3 - 27}$

18. $g(x) = \frac{10}{x^3 - 13x + 12}$

19. $h(x) = \frac{2x - 2}{x^2 + 4x + 3}$

20. $y = \frac{2x + 1}{x}$

21. $y = \frac{4 - 3x}{x - 1}$

22. $y = \frac{2x^2 - 5x + 3}{3x^2 + 7x + 4}$

23. $f(x) = \frac{x^2 + x - 6}{2x^2 - 2x - 12}$

24. $f(x) = \frac{x^3 - 8}{x^2 - 1}$

Apply the Concepts

*25. The Quick Weight-Loss Problem The function $W(r) = k/r^2$ gives the weight in pounds of an object r ft from the center of the Earth.

(a) Find k for a person weighing 150 lb on Earth, some 4000 mi from its center.

(b) Find the weight if the person is standing on top of a 14,000-ft mountain.

(c) Find the weight if the person is orbiting some 300 mi above Earth.

26. The Grand Piano Problem When a piano key is struck, the resulting vibrations set certain other strings vibrating (called *overtones*). The combined sound gives the quality or warmth of the note played. If the leftmost key is hit, it causes a fundamental vibration of its string, which in turn causes vibrations of the first, second, third, fourth, ..., nth overtone strings, which would respectively be $\frac{1}{2}, \frac{1}{3}, \frac{1}{4}, \frac{1}{5}, \ldots, \frac{1}{n+1}$ times the length of the fundamental string if the strings were all under the same tension and had the same diameter. The function modeling overtone string length l as a function of the number of the overtone n and the length of the fundamental string f is $l = (\frac{1}{n + 1})(f)$.

(a) Graph the function if the length of the fundamental string f is 24 in.

(b) If the eighth overtone of the key farthest to the left on the keyboard has a string $l = 13$ in., how long is the leftmost string of the piano?

27. The Rainbow Kite and Balloon Company You've decided to do some special-order kites. The function $c = \frac{0.3t^3 - 3t^2 + 7.3t + 123}{t^2 - 9}$ represents the cost/ft^2 c of hand-painted fabric for kites as a function of the delivery time t in weeks. The head of manufacturing would like the material delivered in 6 weeks. How would the price of the fabric delivered in 6 weeks compare with the lowest possible price?

28. Wonderworld Amusement Park Wonderworld is putting in a huge swimming complex and is accepting bids from outside contractors. The statisticians at your company have computed the cost c in millions of dollars to Wonderworld for building the complex in t months as described by the function $c = \frac{t^2 + 25t + 160}{2t - 2}$. If a bid must include cost and time line, what bid would you submit to Wonderworld?

29. <u>The Redesigned Soda Can Problem</u> Redesign the soda pop can from Example 4 so that its cross section is square. Find the dimensions that minimize the amount of aluminum (surface area) if the container holds 12 fluid ounces (about 20 in^3).

30. <u>The Revolutionary Soda Can Problem</u> Redesign the soda pop can from Example 4 so that its cross section is an equilateral triangle. Find the dimensions that minimize the amount of aluminum (surface area) if the container holds 12 fluid ounces (about 20 in^3).

31. <u>The Nuclear Fusion Problem</u> The function $v = \frac{100ct^2}{100t^2 + 1}$, where $c = 3 \times 10^5$ km/sec, represents the velocity v in km/hr of a rocket t hours after launching. The power is supplied by nuclear fusion and is inexhaustible.

(a) Graph the function and state the domain and range of the application.

(b) Interpret c and the asymptote.

32. <u>The Modified Nuclear-Fusion Problem</u> The following function where $c = 3 \times 10^5$ km/sec is a modification of Exercise 31. The new horizontal asymptote depicts terminal velocity. Graph the function and explain the rocket's trip.

$$v = \begin{cases} \dfrac{100ct^2}{100t^2 + 1}, & 0 \le t \le 1 \\[2ex] \dfrac{-1.01ct + 201.01c}{101t + 101}, & t \ge 1 \end{cases}$$

33. <u>The Distribution of Income Problem</u> Vilfredo Pareto (1848-1923), the Italian sociologist and economist, formulated a "natural" law of distribution of income, which states that the distribution of income will be the same regardless of taxation or the political and social structure. Recent studies have refuted the universality of this law to some extent. However, if the natural law of distribution of income is restricted to a level above subsistence, data indicate that the mathematical model for distribution of income is $N = \frac{p}{x^b}$, where b varies with the population and N equals the number of people whose income is greater than or equal to x dollars in a population of size p. Pareto's law for a population of p = 2.5 million, using b = 1/2 is $N = \frac{2.5 \times 10^6}{\sqrt{x}}$.

[Hint: Graph this function for $N \le 2.5$ million.]

(a) Find the number of millionaires, $x = \$10^6$.

(b) Find the lowest income of the top 100 incomes, N = 100.

(c) Find the number of incomes less than $8100.

34. The Distribution of Income Problem Revisited Consider a small Caribbean island, population $p = 312{,}500$. Use $b = 0.4$ for Pareto's natural law of distribution of income formula $N = \frac{p}{x^b}$.

(See Exercise 33.)

(a) Find the number of people earning over \$100,000, $x = \$10^5$.

(b) Find how may incomes are between \$100,000 and \$200,000.

(c) Find the lowest income of the top 100 incomes.

Extend the Concepts

For Exercises 35 to 38 graph the pair of functions on one coordinate system. Use your graphing utility to estimate the point(s) of intersection and algebra to find the exact values of these points.

35. $f(x) = \frac{1}{(x - 10)^2}, g(x) = \frac{1}{x - 10}$

36. $f(x) = \frac{1}{(x + 10)^2}, g(x) = \frac{1}{x + 10}$

37. $f(x) = \frac{1}{(10 - x)^2}, g(x) = \frac{1}{10 - x}$

38. $f(x) = \frac{2}{x^2 - 10}, g(x) = \frac{1}{10 - x^2}$

3.5 More Rational Functions: Limit Notation, Holes, and Oblique Asymptotes

Using Limit Notation
Locating Holes in a Graph
Examining Oblique Asymptotes

Limits and limit notation are fundamental to the study of calculus. Our study of rational functions gives us the opportunity to intuitively investigate the idea of limits and limit notation.

Using Limit Notation

Consider the graph of f(x) = 1/x (Figure 1).

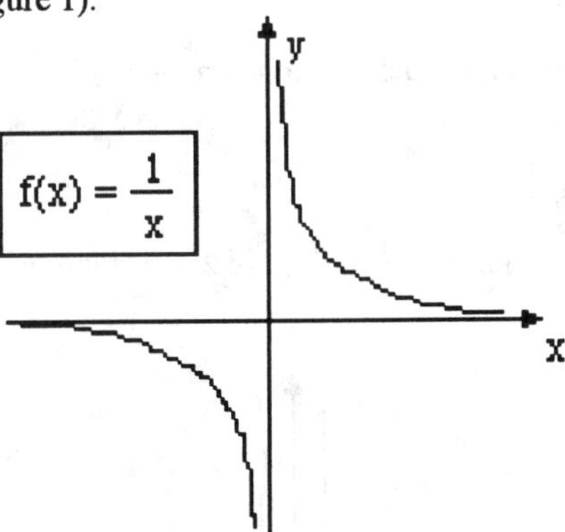

Figure 1

As we look at the far right end of the graph in Figure 1, the graph of y = 1/x approaches zero. We write y approaches 0 as $x \to +\infty$. To summarize the fact that the value of 1/x gets smaller, approaching zero as x grows larger, we write

$$\lim_{x \to +\infty} \frac{1}{x} = 0$$

This is read "the limit, as x approaches positive infinity, of $\frac{1}{x}$ equals zero." The larger the value of x, the closer $\frac{1}{x}$ is to zero.

How do you think we use limit notation to describe the left end of the graph?

We write

$$\lim_{x \to -\infty} \frac{1}{x} = 0$$

This is read "the limit, as x approaches negative infinity, of $\frac{1}{x}$ equals zero."

The y axis is another asymptote to the graph of $f(x) = \frac{1}{x}$. As x approaches zero from the right side of the y axis, values of $\frac{1}{x}$ grow larger without bound and the graph goes up. We summarize this as

$$\lim_{x \to 0^+} \frac{1}{x} = +\infty$$

read "the limit, as x approaches zero from the right side (the positive side), of 1/x is positive infinity." The positive sign in $x \to 0^+$ means to approach zero from the right. Similarly, as x approaches zero from the left, we write

$$\lim_{x \to 0^-} \frac{1}{x} = -\infty$$

The graph decreases without bound as x approaches zero from the left.

Example 1 *Writing using limit notation*

Use limit notation to describe the graph of $y = \frac{1}{x^2}$ in Figure 2 as x approaches

(a) ∞ (b) $-\infty$ (c) 0^+ (d) 0^-

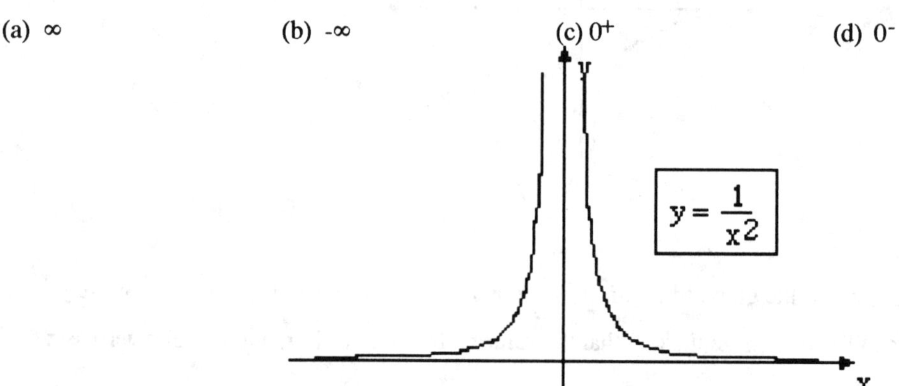

Figure 2

Solution

The graph of $y = 1/x^2$ approaches the x axis at both ends as x approaches $+\infty$ or $-\infty$. As x approaches zero from either the right or left side of the y axis, the graph of $y = 1/x^2$ approaches ∞.

(a) $\lim_{x \to +\infty} \frac{1}{x^2} = 0$ (b) $\lim_{x \to -\infty} \frac{1}{x^2} = 0$ (c) $\lim_{x \to 0^+} \frac{1}{x^2} = \infty$ (d) $\lim_{x \to 0^-} \frac{1}{x^2} = \infty$

Locating Holes in a Graph

Not all rational function have vertical asymptotes for undefined values of x. The function $f(x) = \frac{x^2 - 1}{x - 1}$ is undefined for $x = 1$.

$$f(1) = \frac{(1)^2 - 1}{1 - 1} = \frac{1 - 1}{1 - 1} = \frac{0}{0}$$

The expression $\frac{0}{0}$ is called *indeterminate.* We can examine values of the function near x = 1 to determine the nature of the graph near x = 1.

x	$f(x) = \frac{x^2 - 1}{x - 1}$
0.999	1.999
0.9	1.9
1	$\frac{0}{0}$
1.00001	2.00001
1.01	2.01
1.1	2.1

Approaching x = 1   Approaching y = 2

The closer an input value is to x = 1, the closer the output value is to y = 2. However, since x cannot equal 1, f(x) cannot equal 2. This is expressed using limit notation as

$$\lim_{x \to 1} f(x) = \lim_{x \to 1} \frac{x^2 - 1}{x - 1} = \lim_{x \to 1} \frac{(x + 1)(x - 1)}{x - 1} = \lim_{x \to 1} (x + 1) = 2$$

where $x \to 1$ means that x *approaches but does not equal* 1. Geometrically, this is shown by drawing an open circle at (1,2) to represent the one undefined value, the **hole** in the graph (Figure 3).

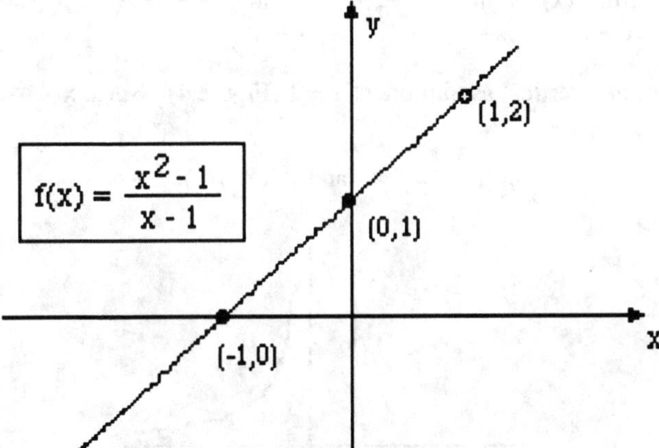

Figure 3

Since the function is undefined at x = 1, there is a hole in the graph. We write function f as

$$f(x) = \frac{(x + 1)(x - 1)}{x - 1}$$
$$= x + 1, \quad x \neq 1$$

There is not always a hole in a graph when its equation has the form $\frac{0}{0}$ for some value of x. Some have an vertical asymptote. To decide whether the indeterminate form represents a hole or a vertical asymptote, we must examine the equation using limits as in the next example. If the limit approaches a specific value, there is a hole. If it approaches infinity, there is a vertical asymptote.

Example 2 *Graphing a rational function*

Sketch the graph of $f(x) = \frac{x - 1}{(x - 1)^2}$ and use algebra and limits to label important points and vertical asymptotes.

Solution

ends: Since the ratio of the leading terms is $\frac{x}{x^2} = \frac{1}{x}$, the ends of the graph will be similar to the graph of $y = 1/x$, below the x axis on the far left and above the x axis to the far right.

y intercept: $f(0) = \frac{0 - 1}{(0 - 1)^2} = \frac{-1}{1} = -1$. The y intercept is -1.

x intercepts: $f(x) = 0 \Rightarrow$

$$\frac{x - 1}{(x - 1)^2} = 0 \qquad \textit{Multiply by } (x - 1)^2.$$
$$x - 1 = 0$$
$$x = 1$$

There is no x intercept since the function is undefined for x = 1 because x = 1 causes division by zero.

vertical asymptotes: Since $f(1) = \frac{0}{0}$, we need to consider the limit as $x \to 1$ to determine whether there is a vertical asymptote or a hole at x = 1.

$$\lim_{x \to 1} f(x) = \lim_{x \to 1} \frac{x - 1}{(x - 1)^2} = \lim_{x \to 1} \frac{1}{x - 1} = \frac{1}{0}$$

which is undefined. There is a vertical asymptote at x = 1 (Figure 4). Since $x - 1 < 0$ for $x < 1$ and $x - 1 > 0$ for $x > 1$,

$$\lim_{x \to 1^-} f(x) = -\infty \quad \text{and} \quad \lim_{x \to 1^+} f(x) = +\infty$$

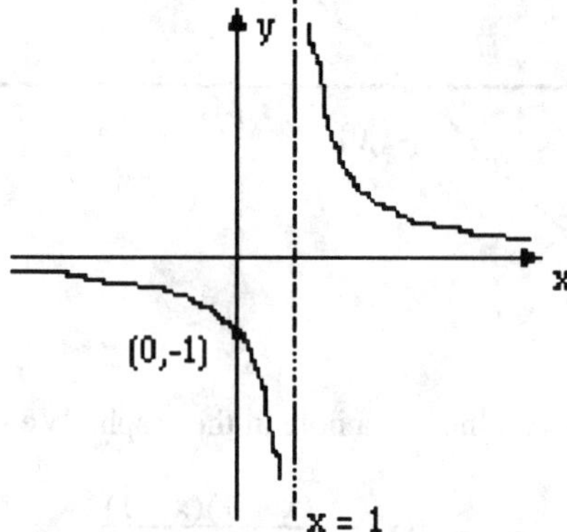

Figure 4

Examining Oblique Asymptotes

A function whose numerator is exactly one degree higher than its denominator has an oblique asymptote. Here, *oblique* simply means nonhorizontal and nonvertical. Initially, we approximated the ends of the graph of a function such as

$$f(x) = \frac{2x^2 - 1}{x + 1}$$

by finding the ratio of the leading terms ($y = \frac{2x^2}{x} = 2x$). This can be refined using long division. (See Appendix I, page A4 to review long division.)

$$\begin{array}{r l} & 2x - 2 \\ x + 1 \,\big)\!\! & \overline{2x^2 \qquad\quad -1} \\ & \underline{2x^2 + 2x} \\ & \quad -2x - 1 \\ & \quad \underline{-2x - 2} \\ & \qquad\quad 1 \end{array}$$

Therefore, $f(x) = \frac{2x^2 - 1}{x + 1}$ can be written as

$$f(x) = 2x - 2 + \frac{1}{x + 1}$$

Notice that as $x \to \pm\infty$, the remainder $\frac{1}{x + 1}$ approaches zero and f approaches $2x - 2$, making $y = 2x - 2$ the oblique asymptote on both ends (Figure 5).

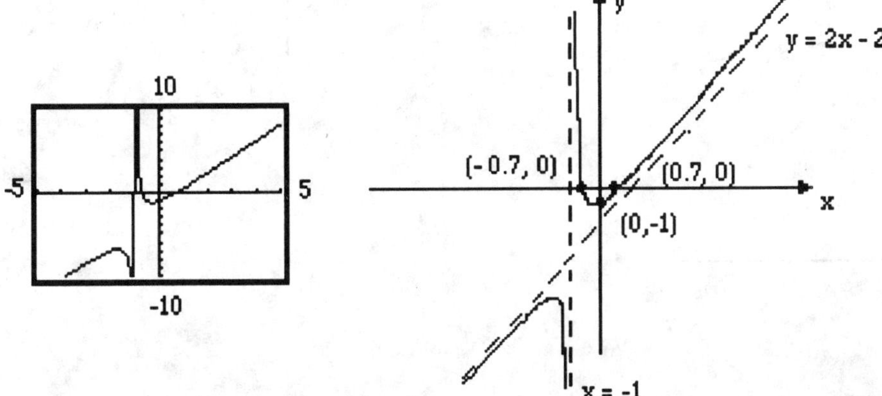

Figure 5

Often, finding the exact equation of an oblique asymptote is not necessary. Generally, a quick sketch of a graph is all that is needed to understand a question or its solution, even in calculus. However, if exact results are wanted, the equation of an oblique asymptote can be found using long division.

Example 3 *Finding the oblique asymptote*

Compare the ends of the graph of

$$f(x) = \frac{1 - x^2}{2x - 5}$$

(a) by using the ratio of the highest-degree terms to approximate the ends of the graph.

(b) by finding the exact equation of the oblique asymptote.

Solution

(a) The ends of the graph have the shape of $y = -\frac{x}{2}$ since $y = -\frac{x^2}{2x} = -\frac{x}{2}$.

(b) The exact equation of the oblique asymptote is $y = -\frac{x}{2} - \frac{5}{4}$ since

$$\begin{array}{r|llll}
 & -\frac{x}{2} & - 5/4 & & \\ \hline
2x - 5 & -x^2 & & + 1 & \\
 & -x^2 & + 5/2\,x & & \\ \cline{2-3}
 & & -5/2\,x & + 1 & \\
 & & -5/2\,x & + 25/4 & \\ \cline{3-4}
 & & & -21/4 &
\end{array}$$

The ends of the graph of $f(x) = \frac{1 - x^2}{2x - 5}$ are asymptotic to $y = -\frac{x}{2} - \frac{5}{4}$ (Figure 6).

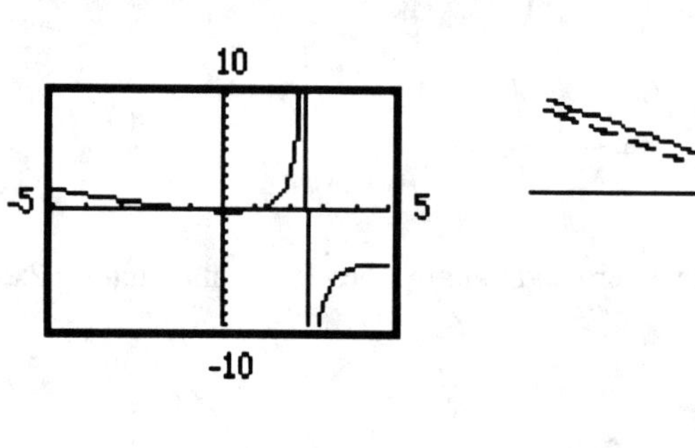

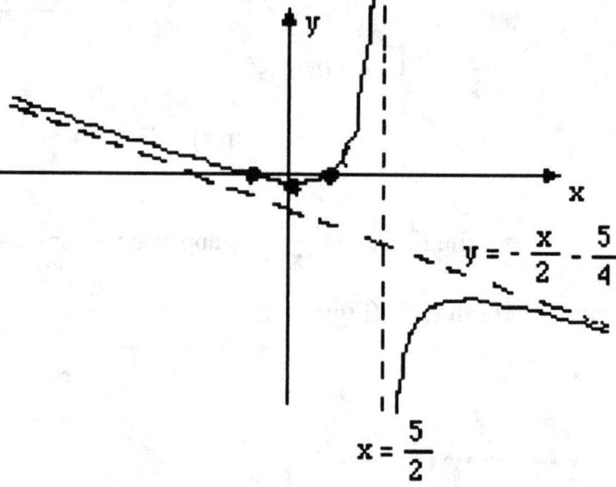

Figure 6

Exercises 3.5

Develop the Concepts

For Exercises 1 to 4 find the limits from the graphs of $f(x) = \frac{-1}{x}$ and $g(x) = \frac{-1}{x^2}$.

1. $\lim_{x \to +\infty} f(x)$

2. $\lim_{x \to +\infty} g(x)$

3. $\lim_{x \to 0^+} f(x)$

4. $\lim_{x \to 0^+} g(x)$

For Exercises 5 to 8 use limit notation to describe all the asymptotic behavior.

5. $y = \frac{-1}{x + 1}$

6. $y = \frac{-1}{(x + 1)^2}$

7. $y = \frac{-1}{x^2} + 1$

8. $y = \frac{-1}{x^2} - 1$

For Exercises 9 to 12 graph each function and label any holes in the graph. Use limits to find the coordinates of any hole.

9. $f(x) = \frac{1 - x^2}{2x - 2}$

10. $g(x) = \frac{x^2 - x - 12}{x^2 - 16}$

11.(a) $f(x) = \frac{x^2 - 2x + 1}{x - 1}$

(b) $g(x) = \frac{1}{f(x)} = \frac{x - 1}{x^2 - 2x + 1}$

12.(a) $f(x) = \frac{(x - 2)^2(x + 3)}{(x - 2)(x - 1)}$

(b) $g(x) = \frac{1}{f(x)} = \frac{(x - 2)(x - 1)}{(x - 2)^2(x + 3)}$

For Exercises 13 to 20 graph each function and label all of the important parts.

13. $y = \frac{-4}{x^2 - 2x - 3}$

14. $y = \frac{2x + 1}{x}$

15. $y = \frac{2(x - 2)}{x^2 + 4x - 3}$

16. $y = \frac{3x + 1}{3x^2 + 13x + 4}$

*17. $y = \frac{x^3 - x}{4x}$

18. $y = \frac{x^2 + 9}{x^2 - 9}$

19. $y = \frac{x^2 - 9}{x^2 + 9}$

20. $y = \frac{x^2 + x - 2}{x - 1}$

For Exercises 21 to 24 compare the shape of the ends of the graph found from the ratio of the leading terms with the exact equation of the asymptote found by long division.

21. $y = \frac{x^2}{x - 1}$

22. $y = \frac{x^3 + 8}{x^2 - x - 2}$

23. $f(x) = \frac{x^4 - 5x^2 + 4}{1 - x^3}$

24. $f(x) = \frac{x^4 - 16}{2x + 3}$

Apply the Concepts

Exercises 25 to 28 revisit previous problems. Find and interpret the exact equation of each asymptote for the ends of the graph.

*25. The Soda-Can Problem The function $S = \frac{2\pi r^3 + 40}{r}$ yields the surface area S of a can as a function of its radius r.

26. The Surface Area of a Box The function $S = \frac{2x^3 + 2592}{x}$ describes the surface area S of a box with a square base x incorporating the restriction that the volume is fixed at 648 in^3.

27. The Rainbow Kite and Balloon Company The function $c = \frac{0.3t^3 - 3t^2 + 7.3t - 123}{t^2 - 9}$ represents the cost/ft^2 c of hand-painted fabric delivered in t weeks.

28. Wonderworld Amusement Park The function $c = \frac{t^2 + 25t + 160}{2t - 2}$ yields the cost c in millions of dollars to build a swimming complex in t months.

Extend the Concepts

For Exercises 29 and 30 solve each pair of equations, graph each pair of functions on one coordinate system, and label the points of intersection.

29. $f(x) = \frac{3x^2}{x - 2}$

$g(x) = \frac{x^2 + 2}{x - 2}$

30. $f(x) = \frac{2x^3}{(3x - 6)^2}$

$g(x) = \frac{1 - 3x^2}{9x - 18}$

A Guided Review of Chapter 3

3.1 Graphs of Polynomial Functions

For Exercises 1 and 2 graph each polynomial function and label the intercepts.

1. $y = 2(x - 1)^2(x + 2)^2$

2. $y = -x^2(x + 3)$

For Exercises 3 to 6 find an equation for each graph. The x intercepts are -2, -1, 2, and 5 in all four figures. Estimate the y intercepts.

3.

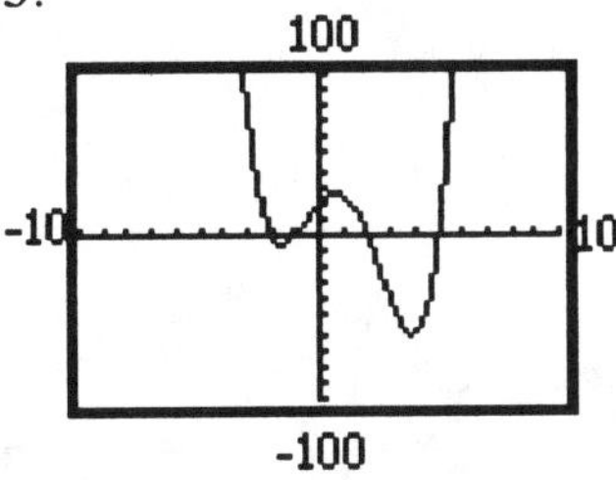

4.

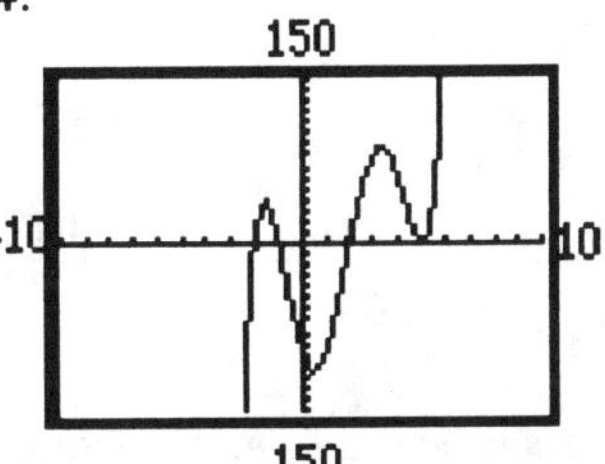

5.

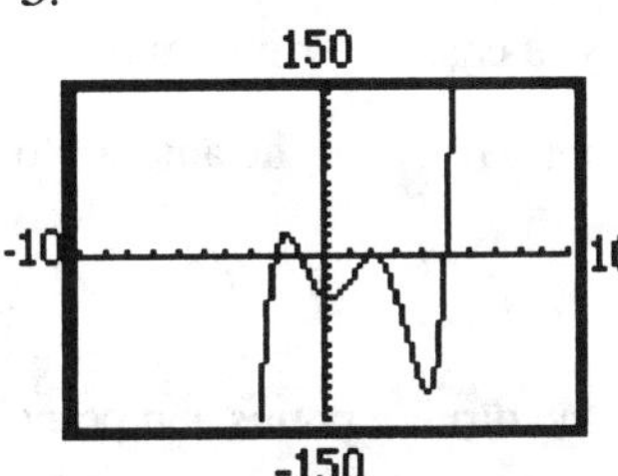

6.

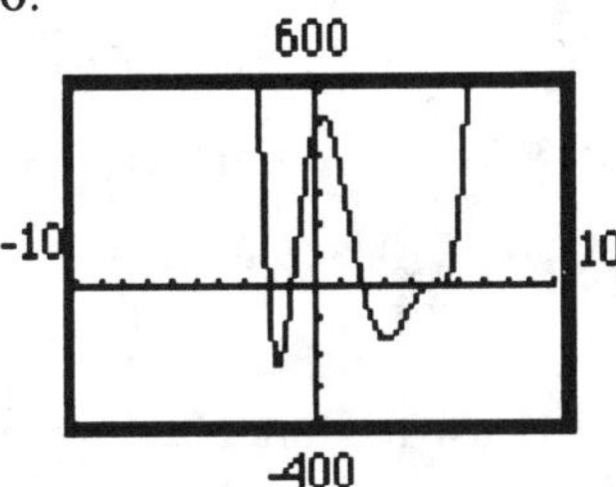

Exercises 7 to 10 are analogous because each requires finding a maximum value that is located at a turning point of the graph of a polynomial function.

7. The Largest Triangle Problem Suppose that you want to build a strawberry patch in the shape of a right triangle with exactly 12 m of rabbit fencing along its hypotenuse. What dimensions will yield the maximum area.? [Hint: Express the square of the area as a function of the length of one of the legs.]

8. A Big Box Problem Create a rectangular box by cutting out corners of a square sheet of paper a by a units as indicated (Figure 1). Complete the following table to find the relationship between a and x. For each value of a, how much should be cut, x, to create a box of maximum volume?

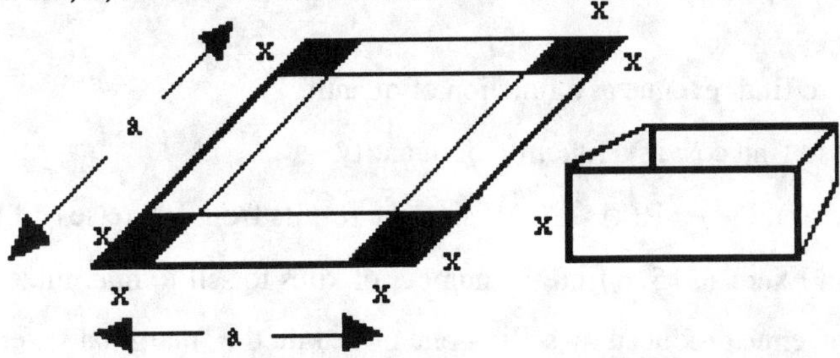

Figure 1

a	6	12	18	24
x	?	?	?	?

9. Another Big Box Problem The box in Figure 2 is made from a sheet of paper in the shape of an equilateral triangle having sides a units long. Complete the following table to find the relationship between a and x. For each value of a, how much should be cut, x, to create a box of maximum volume.

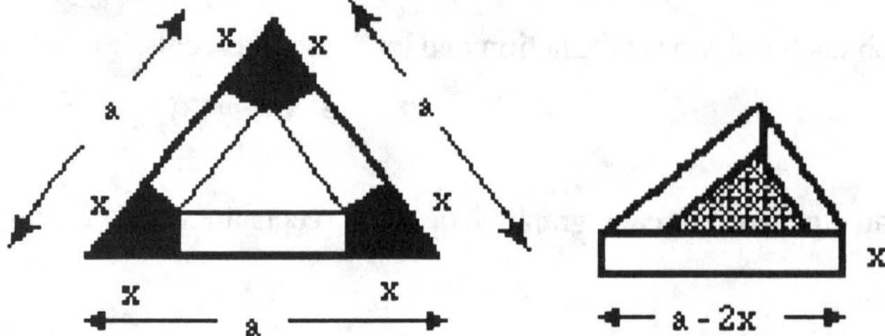

Figure 2

a	6	12	18	24
x	?	?	?	?

[Hint: The area of an equilateral triangle of side *s* units is $A = \frac{\sqrt{3}}{4}s^2$.]

10. The Big Revenue Problem The demand for your services is related to the price you charge per hour p by the function $p = 0.01x^3 - 0.1x^2 - x + 10$, where x is the number of hours you work per day. Find the amount to charge to maximize your revenue R, where R = px.

For Exercises 11 and 12 use the formulas $x = \frac{-b \pm \sqrt{b^2 - 3ac}}{3a}$ and $x = -\frac{b}{3a}$ to find the turning points and point of inflection, respectively.

11. $y = x^2 - x^3$

12. $f(x) = 2x^3 - 9x^2 - 24x - 13$

3.2 Finite Differences

For Exercises 13 to 19 use the following data showing the revenue and cost of selling a new kite.

quantity of kites, x	10	20	30	40	50
revenue, R ($)	4500	7500	9000	9000	7500
cost, C ($)	2550	3400	4250	5100	5950

13. Use finite differences to find revenue as a function of quantity.
14. Use finite differences to find cost as a function of quantity.
15. Create the profit function, P(x) = R(x) - C(x), using the results from Exercises 13 and 14.
16. Use the equation from Exercise 15 to find the number of kites to sell to maximize profit.
17. Economists call the revenue produced by selling one more kite the "marginal revenue." Use $\frac{\Delta R}{\Delta x}$ to estimate the marginal revenue for each interval.
18. Marginal cost is the cost for producing one more kite. Use $\frac{\Delta C}{\Delta x}$ to estimate the average marginal cost for each interval.

19. How are the results from Exercises 17 and 18 related to the result from Exercise 16?

3.3 Interpolating and Best-Fitting Polynomial Functions

For Exercises 20 to 23, (a) give the equation of the best-fitting cubic and quartic polynomials for each set of data and (b) plot the data and graph the polynomials.

20. (c) How can you show that the quartic exactly fits the data?

x	-4	-3	0	3	4
y	1344	-756	-144	-756	1344

21. (c) Why is the leading coefficient of the best-fitting quartic polynomial zero?

x	-5	0	5	10	15
y	-412	-12	188	1688	5988

22. (c) Which appears to be the better fit?

x	0	2	4	6	8
y	-10	0	-5	10	-10

23. (c) Does the cubic show a smoothing effect on these data?

x	-5	-2	0	2	4
y	-85	-1	4	20	100

24. The Big Revenue Problem The demand for your services is related to the price you charge per hour, where x is the number of hours you work per day.

x	0	2	3	5	10
p	10	7.65	6.35	3.75	0

(a) Find the best-fitting cubic equation that models price p in dollars as a function of x.

(b) Find the revenue function R, where R = px.

(c) Find the amount to charge to maximize your revenue.

3.4 Graphs of Rational Functions

For Exercises 25 and 26 graph each pair of functions on one coordinate system. State the x intercept(s) of function f and the asymptote(s) of the reciprocal functions.

25. $f(x) = 1 - x^2, y = \frac{1}{f(x)} = \frac{1}{1 - x^2}$

26. $f(x) = x^2 + 1, y = \frac{1}{f(x)} = \frac{1}{x^2 + 1}$

For Exercises 27 to 38 graph each rational function. Label the vertical asymptotes.

27. $y = \frac{-1}{x - 10}$

28. $y = \frac{-1}{(x - 10)^2}$

29. $y = \frac{-1}{100 - x^2}$

30. $y = \frac{-1}{x^2 - 12x}$

31. $y = \frac{10}{x^2 - x - 20}$

32. $y = \frac{-12}{x^2 - 2x - 48}$

33. $y = \frac{1 - x^2}{x^3 - 64}$

34. $y = \frac{x - 1}{x^3 - 16x}$

35. $y = \frac{x^3 + x - 2}{x^2}$

36. $f(x) = \frac{x^4 - 16}{x^2 - 9}$

37. $g(x) = \frac{9x - x^3}{x^2 - x - 110}$

38. $f(x) = \frac{2x^4 - 10x^2 + 8}{x^4 - 2x^3 - 15x^2}$

39. The Vegetable Garden Problem The fertilizer you want to use on your new vegetable garden covers 2000 ft^2 per bag. You've decided to build a rectangular garden that requires exactly 3 bags of fertilizer. Naturally you'll need to fence the garden. Find the dimensions to build the garden to minimize the cost of fencing if the fence cost \$1.69/ft.

40. A U.S. Post Office Box Problem Suppose that the U.S. Postal Service decides to increase the size of the largest rectangular box it will accept so that the volume can be 54 ft^3. Find the dimensions of the box that will have the minimum surface area if the cross section is a square.

41. Another U.S. Post Office Box Problem Suppose that the U.S. Postal Service decides to increase the size of the largest rectangular box it will accept so that the volume can be 54 ft^3. Find the dimensions of the box that will have the minimum surface area if the width is to be twice the height.

42. The Least-Cost Expediting Problem The total cost C for a large project is a function of the time t in days the project must be completed according to the function $C(t) = \frac{0.04t^3}{(t - 100)^2}$.

(a) Find the completion time that will yield the minimum cost.

(b) Find and interpret the vertical asymptote.

(c) Find and interpret the oblique asymptote.

3.5 More Rational Functions: Limit Notation Holes and Oblique Asymptotes

43. Use the graph of $f(x) = \frac{-1}{x - 1}$ to find the following limits.

(a) $\lim_{x \to 1^+} f(x)$

(b) $\lim_{x \to +\infty} f(x)$

44. Use the graph of $f(x) = \frac{x - 1}{x^2 - 1}$ to find the following limits.

(a) $\lim_{x \to 1} f(x)$

(b) $\lim_{x \to +\infty} f(x)$

45. Give the equation of the oblique asymptote for $f(x) = \frac{2x^4 - x^3}{x^3 - x}$.

46. Give the coordinates of the hole in the graph of $f(x) = \frac{2x^2 - x - 3}{x^3 - x}$.

Chapter 3 Test

1. Graph $f(x) = -(x + 2)^2(x + 1)^3(x - 1)$ and label the intercepts.

For Exercises 2 and 3 find an equation of the lowest-degree polynomial for each graph.

2.

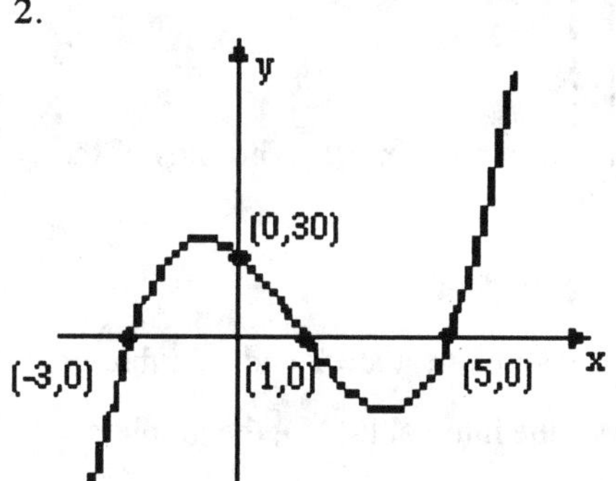

3.

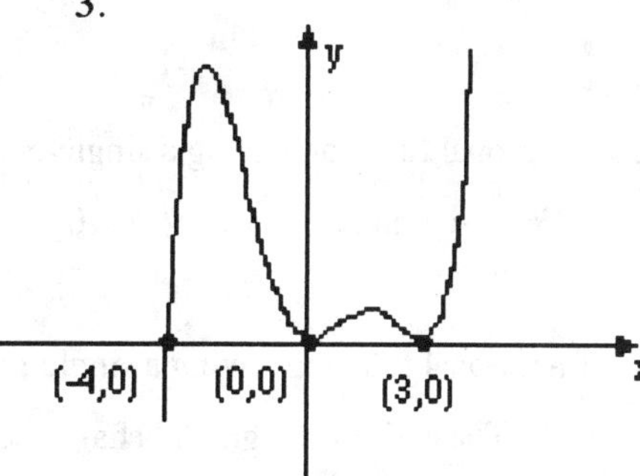

For Exercises 4 to 6 consider the polynomial $p(x) = -2x^4 + 7x^3 + 23x^2 - 43x + 15$.

4. Describe the ends of the graph.

5. State the y intercept.

6. Describe the number of possible x intercepts.

7. Complete the table and use the method of finite differences to find the linear equation of the form $f(x) = ax + b$ for the following data.

x	2	4	6	8
y	-1	0	1	2

x	2	4
$f(x) = ax + b$	?	?

8. Complete the table and use the method of finite differences to find the linear equation of the form $f(x) = ax^2 + bx + c$ for the following data.

x	-1	0	1	2
y	6	3	4	9

x	-1	0	1
$f(x) = ax^2 + bx + c$	?	?	?

9. Approximate the irrational zeros for $f(x) = x^4 - 3x^2 + 2x - 5$.

10. Factor the $f(x) = x^4 - 6x^3 + 8x^2 + 4x - 4$ given that one zero is $x = 1 + \sqrt{3}$ and the other is $x = 1 - \sqrt{3}$.

For Exercises 11 and 12 graph each rational function. Label the intercepts and asymptotes

11. $f(x) = \dfrac{-10}{10 - x}$

12. $g(x) = \dfrac{x^2}{16 - x^2}$

For Exercises 13 and 14 give an equation of a rational function that matches each graph.

13.

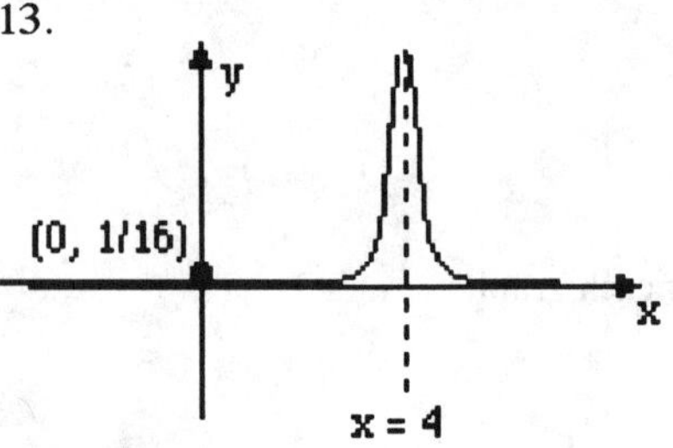

14.

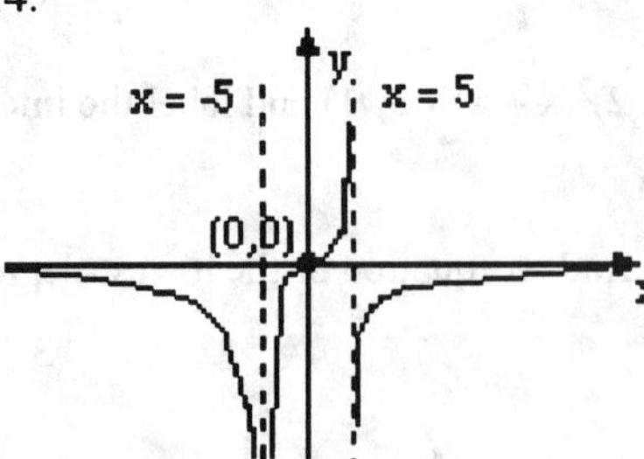

15. Sketch the graph of a rational function having a single-pole vertical asymptote at x = 1. The ends of the graph behave like y = 1. For x < 1 and for x > 3, f(x) > 0.

16. Sketch the graph of a rational function having a single-pole vertical asymptote at x = 2 and a double-pole vertical asymptote at x = -1. The ends of the graph behave like y = 0. For the interval (-1, 2), the graph is concave up, having a turning point at (1, 3). For x > 2, f(x) < 0.

17. The Tent Problem A tent is to be made as shown in Figure 1 with two ends. Since the reinforced seams are costly to manufacture, they are fixed at 400 in.

(a) Use the graph of the volume function to find the dimensions that will optimize volume.

(b) Use the graph of the surface area function to find the dimensions that will optimize the surface area.

[Hint: The area of an equilateral triangle is $A = \frac{\sqrt{3}}{4}x^2$.]

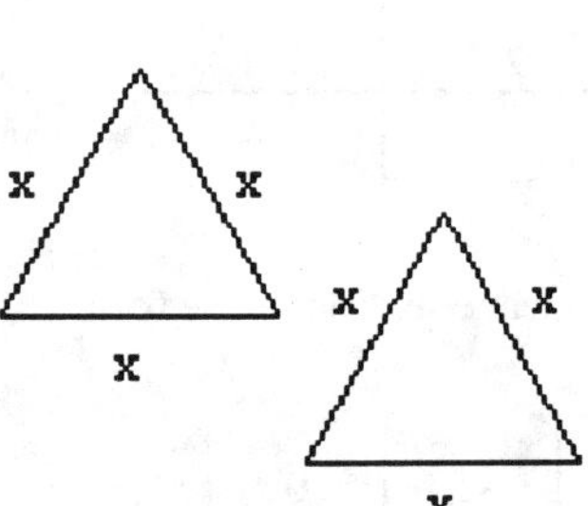

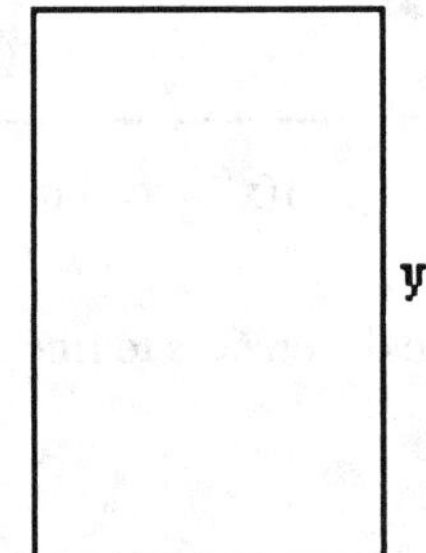

3x

y

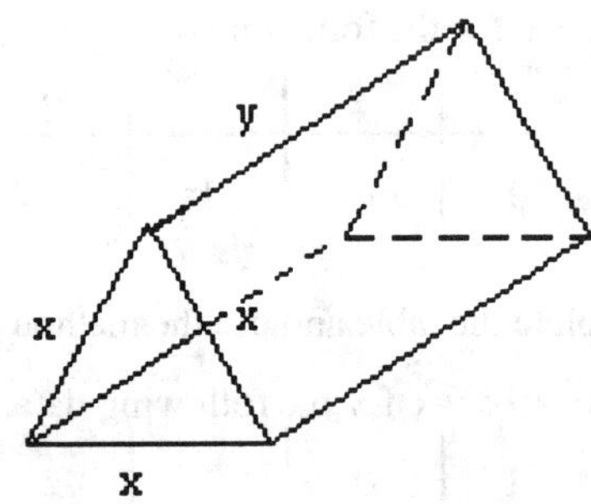

Figure 1

18. Another Tent Problem Suppose that the tent in Figure 1 is to have a volume of 66 ft^3. Find the dimensions that will optimize the following.

(a) the amount of material required to make the tent (its surface area)

(b) the amount of stitching required on the seams

Chapter 4

Exponential and Logarithmic Functions

Chapter 4 deals with exponential functions and their inverses, logarithmic functions. Exponential and logarithmic functions are found in the mathematics of finance and navigation, both of which originally motivated the invention of logarithms by two different people. As we will see, applications of these functions include population growth studies in sociology, radioactive decay in physics, and carbon dating in archaeology. We will also investigate logarithmic scales, including the Richter scale for earthquakes and the decibel scale for sound, as well as in curve fitting with exponential, logarithmic, and power functions in log-log and semi-log coordinate systems that apply logarithmic scales.

A Historical Note on Logarithms

John Napier invented logarithms in 1594 as an aid to arithmetic. His logarithms used $1/e \approx (1 - 10^{-7})^{10^7}$ as their base. Napier's logarithms were not readily or universally accepted.

In 1600, Joost Burgi independently compiled a table of logarithms, which he published 20 years later, along with the first table of antilogs. He used 1.0001 as his base. Henry Briggs computed the first table of base 10 logarithms in 1615 by tedious repeated multiplications and extractions of square roots. For example, Briggs calculated base 10 logarithms by first taking successive roots of 10, $\sqrt{10}$, $\sqrt{\sqrt{10}}$, $\sqrt{\sqrt{\sqrt{10}}}$, ...
until he found a number *slightly* larger than 1. It took 54 successive square roots. Calling this number x he had $10^{(1/2)^{54}} = x$ and $\log x = (1/2)^{54}$ as a starting point for the computations. Both Napier and Burgi founded their inventions on a comparison of arithmetic and geometric sequences, since the terms of an arithmetic sequence are the logarithms of the corresponding terms of the geometric sequence:

$$\ldots, -2, -1, 0, 1, 2, 3, \ldots \quad \text{and} \quad \ldots, a^{-2}, a^{-1}, a^0, a^1, a^2, a^3, \ldots$$

The intriguing idea was that multiplication and division in the geometric sequence correspond to simple addition and subtraction in the arithmetic sequence. Napier chose the name **logarithm**, meaning *ratio number*, to refer to a common ratio in a sequence of numbers used in calculating his tables. The modern definition of a logarithmic function as the inverse of an exponential function is attributed to Euler, over 150 years later. He also introduced the natural base, e, and extended logarithms to negative and complex numbers in 1747. The French mathematician Augustin Cauchy (1789-1857) finally completed the basic theory of logarithms in 1821.

4.1 Exponential Functions

Solving Exponential Equations
Defining the Natural Base e
Exploring Compound Interest
Graphing Exponential Functions
Applying Exponential Growth
Introducing Exponential Decay

Functions of the form

$$y = b^x$$

where $b > 0$ and $b \neq 1$ such as

$$y = 10^{x+1} \text{ and } y = 3^{2x}$$

are called **exponential functions**. Graphs of exponential functions increase (or decrease) faster than other functions. In this section we define the number e, a number present in problems involving continuous growth or decay. In Examples 2 and 3 we see applications of exponential equations for calculating the growing amount of money in a saving account and in Example 5 we see an exponential growth model for the population of Mexico. In Example 6 we use a model for exponential decay to calculate the depreciating value of a truck.

Solving Exponential Equations

Exponential functions of the form $y = b^x$ are one-to-one functions. If two expressions with the same base are equal, they must have equal exponents. We can use the following to solve exponential equations in the same base.

$$b^s = b^t \Leftrightarrow s = t$$

Example 1 *Solving exponential equations with like bases*

Solve each equation by writing both sides of the equation as an exponential expression having the same base.

(a) $2^{x+1} = 8$ (b) $3^{1-x} = \frac{1}{9}$ (c) $10^{2x} = 1$

Solution

(a)
$$2^{x+1} = 8$$
$$2^{x+1} = 2^3$$
$$x + 1 = 3$$
$$x = 2$$

(b)
$$3^{1-x} = \frac{1}{9}$$
$$3^{1-x} = 3^{-2}$$
$$1 - x = -2$$
$$x = 3$$

(c)
$$10^{2x} = 1$$
$$10^{2x} = 10^0$$
$$2x = 0$$
$$x = 0$$

Defining the Natural Base e

Whenever we do problems involving continuous growth or decay, the irrational number e occurs, naturally. The **natural base** e is an irrational number nearly as prominent as π.

To explore the value of the irrational number e, go to "Explore the Concepts," Exercises E1 - E3, before continuing (page 263). For the interactive version, go to the *Precalculus Weblet - Instructional Center - Explore Concepts.*

The number e is approximately 2.718. Leonhard Euler (pronounced "Oiler") named the irrational number e for *exponent.* The value of **e** is approximated by the expression $(1 + 1/x)^x$ for large positive values of x.

$$e \approx (1 + 1/x)^x$$

Using limit notation gives

$$e = \lim_{x \to \infty} (1 + 1/x)^x$$

Since e is an irrational number, continued calculations with larger and larger values of x yield closer approximations of e but will never yield a terminating nor repeating decimal value. Figure 1 shows the graph of $f(x) = (1 + 1/x)^x$ with the value for x = 4, x = 100, and x = 10,000 displayed on the bottom of the respective windows.

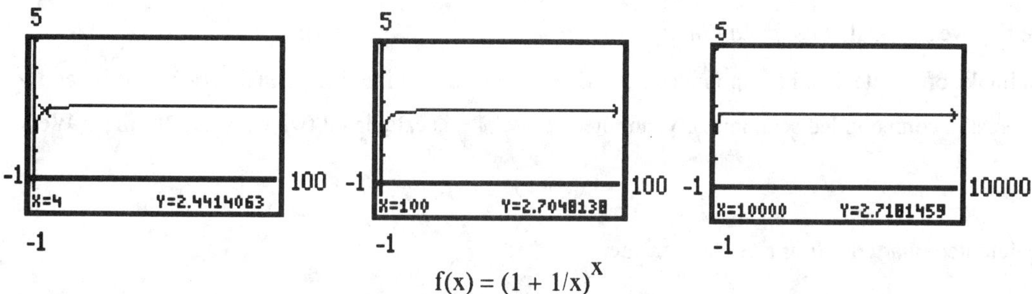

$f(x) = (1 + 1/x)^x$

Figure 1

Notice that for larger and larger values of x the corresponding values of y agree to more and more decimal places, with the 10-decimal-place estimate

$$e \approx 2.7182818285$$

Exploring Compound Interest

Recall the formula for computing simple interest, i = Prt, for an initial principal P invested at an annual interest rate r for t years. For example, the simple interest earned on \$40 invested for 1 year at 5% is \$8.

$$\begin{aligned} i &= Prt \\ &= (40)(0.05)(1) = 8 \end{aligned}$$

The amount of money in an account A consists of the principal investment and the earned interest.

$$\begin{aligned} A &= P + i \\ &= P + Prt \end{aligned}$$

After 1 year (t = 1), the amount of money in the account is

$$\begin{aligned} A_1 &= P + Prt,\ t = 1 \Rightarrow \\ &= P + Pr \\ &= P(1 + r) \end{aligned}$$

Next, consider the amount of money in the account after one <u>more</u> year ($t = 1$ again). The new principal for the second year is the ending amount from year one, $A_1 = P(1 + r)$.

$$A_2 = A_1 + A_1rt,\ t = 1 \Rightarrow$$
$$= A_1 + A_1r$$
$$= A_1(1 + r) \qquad \textit{Substitute } A_1 = P(1 + r).$$
$$= P(1 + r)(1 + r)$$
$$= P(1 + r)^2$$

Repeating this procedure over t years yields the formula for annually compounded interest.

$$A = P(1 + r)^t$$

Earning interest on the principal plus previously paid interest, as in $A_2 = A_1 + A_1r$ where A_1 includes interest from the first year, is called *compounding*.

Most banks offer interested compounded several times a year; semiannually, quarterly, monthly, and so on. If the interest is compounded semiannually, an interest rate of $\frac{r}{2}$ is calculated twice a year, 2t times. We have

$$A = P(1 + \frac{r}{2})^{2t}$$

If it is calculated quarterly, four times a year, then

$$A = P(1 + \frac{r}{4})^{4t}$$

This generalizes to the form for interest compounded k times a year,

$$A = P(1 + \frac{r}{k})^{tk}$$

Example 2 <u>The Savings Account Problem</u>

Find the amount of money in an account with a principal of \$1000 invested for 1 year at an annual interest rate of 5% compounded (a) quarterly, (b) monthly, and (c) daily using the formula $A = P(1 + \frac{r}{k})^{tk}$.

Thinking: Creating a Plan

Since the principal, rate, and time are the same for the three parts,

$$A = P(1 + \frac{r}{k})^{tk} \text{ and } P = 1000, r = 0.05, t = 1 \Rightarrow A = 1000(1 + \frac{0.05}{k})^k$$

and substitute k = 4 for part (a), k = 12 for part (b), and k = 365 for part (c).

Communicating: Writing the Solution

(a) $A = 1000(1 + \frac{0.05}{k})^k$ and $k = 4 \Rightarrow$

$$A = 1000(1 + \frac{0.05}{4})^4$$
$$\approx 1050.945337$$

The account will have about $1050.95 after 1 year.

(b) $A = 1000(1 + \frac{0.05}{k})^k$ and $k = 12 \Rightarrow$

$$A = 1000(1 + \frac{0.05}{12})^{12}$$
$$\approx 1051.161898$$

The account will have about $1051.16 after 1 year.

(c) $A = 1000(1 + \frac{0.05}{k})^k$ and $k =365 \Rightarrow$

$$A = 1000(1 + \frac{0.05}{365})^{365}$$
$$\approx 1.051.267496$$

The account will have about $1051.27 after 1 year.

Learning: Making Connections

Table 1 shows the amount of money in the account for various compounding periods. Notice how the amount of money in the account increases with the number of compounds, but not at a constant rate.

Table 1

Compounding Period	**k**	$A = P(1 + \frac{r}{k})^{tk}$	**Amount after 1 year**
Annually	1	$A = 1000(1 + \frac{0.05}{1})^1$	$1050.00
Semiannually	2	$A = 1000(1 + \frac{0.05}{2})^2$	$1050.62
Quarterly	4	$A = 1000(1 + \frac{0.05}{4})^4$	$1050.95
Monthly	12	$A = 1000(1 + \frac{0.05}{12})^{12}$	$1051.16
Weekly	52	$A = 1000(1 + \frac{0.05}{52})^{52}$	$1051.24
Daily	365	$A = 1000(1 + \frac{0.05}{365})^{365}$	$1051.27

One way to appreciate the number e is to see how naturally it occurs. Let's extend Table 1 to include the amount in the same account if we compound it every minute for 1 year. That is, k = 365(24)(60) = 525,600.

$A = 1000(1 + \frac{0.05}{k})^k$ and $k = 525,600 \Rightarrow$

$$A = 1000(1 + \frac{0.05}{525600})^{525600(1)}$$
$$\approx 1.051.271107$$

Increasing the number of compounding periods increases the amount of money in the account -- but by smaller and smaller amounts. When we round to the nearest cent, compounding every minute yields $1051.27, the same amount as daily compounding.

The definition for e as

$$e = \lim_{x \to \infty} \left(1 + \frac{1}{x}\right)^x$$

suggests that as the number of compounds approaches infinity, $k \to \infty$, the formula $A = P(1 + \frac{r}{k})^{tk}$ approaches the formula for continuously compounded interest,

$$A = Pe^{rt}$$

NOTE: The expression $(1 + \frac{r}{k})^{tk}$ becomes e^{rt} as follows. $(1 + \frac{r}{k})^{tk} = (1 + \frac{r}{k})^{\frac{rtk}{r}}$.
Let $x = \frac{k}{r}$ and $\frac{1}{x} = \frac{r}{k}$. Then $(1 + \frac{r}{k})^{\frac{rtk}{r}} = [(1 + \frac{1}{x})^{k/r}]^{rt} = [(1 + \frac{1}{x})^{x}]^{rt}$ which approaches e^{rt} as $x \to \infty$.

Compound Interest Formulas

If a principal P (present value) is invested at an annual interest rate r, the Amount A (Future Value) in t years is

$A = P(1 + r)^t$ for annually compounded interest

$A = P(1 + \frac{r}{k})^{tk}$ for interest compounded k times a year

$A = Pe^{rt}$ for continuously compounded interest

Example 3 *Calculating continuously compounded interest*

(a) Use the formula for continuously compounded interest $A = Pe^{rt}$ to find the amount of money in an account after 1 year if the principal is $1000 and the interest rate is 5%.

(b) Compare your result from part (a) with the amounts calculated in Example 1.

Solution

(a) $A = Pe^{rt}$ and $P = 1000$, $r = 0.05$, and $t = 1 \Rightarrow$

$$A = 1000e^{(0.05)(1)}$$
$$\approx 1051.27$$

The amount in the account is about $1051.27.

(b) The amounts in Table 1 approach the amount for continuously compounded interest as the compounding periods increase.

Graphing Exponential Functions

Figure 2 shows the graphs of three exponential functions: $y = 2^x$, $y = 3^x$, and the reference function $y = e^x$.

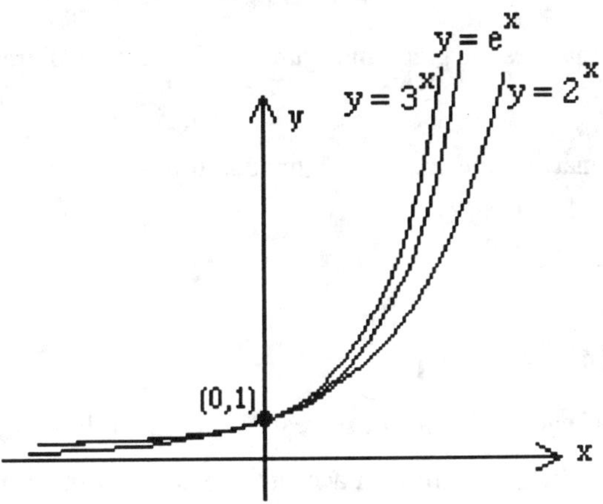

Figure 2

The graph of $y = e^x$ is between the graphs of $y = 2^x$ and $y = 3^x$ because the number $e \approx 2.7182818285$ is between 2 and 3.

Why do these graphs have the same y intercept, 1?

For $y = b^x$, $b > 0$, $b^0 = 1$.

Everything that we have learned about reflections, translations, and changes of scale apply to exponential functions. In the next example we graph reflections and translations of the reference function $y = e^x$.

Example 4 *Graphing exponential functions*

Match each graph in Figure 3 with one of the equations $f(x) = e^{-x}$, $g(x) = 1 + e^x$, and $h(x) = 1 - e^x$, and explain how each graph is related to the graph of the reference function $y = e^x$ in Figure 2.

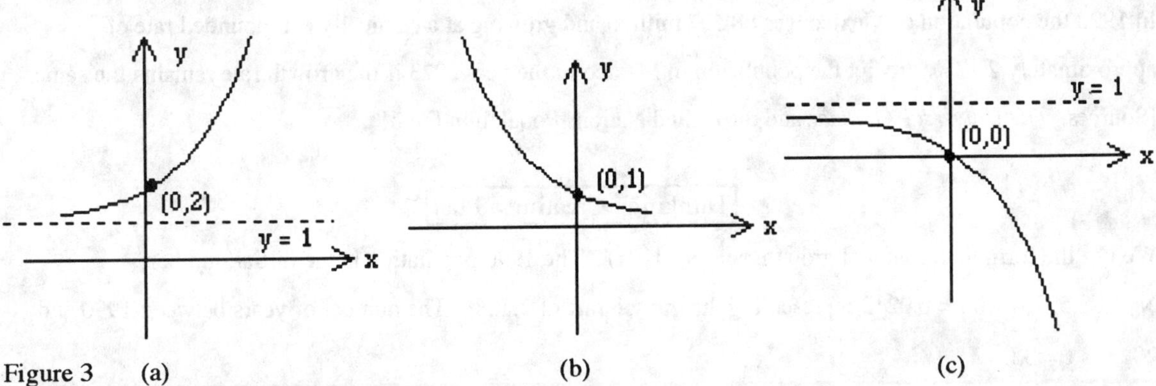

Figure 3 (a) (b) (c)

Solution

(a) The graph of $f(x) = e^{-x}$ matches the graph in Figure 3b. It is a reflection of the graph of $y = e^x$ about the y axis.

(b) The graph of $g(x) = 1 + e^x$ matches the graph in Figure 3a. It is a vertical translation of the graph of $y = e^x$ translated up 1 unit.

(c) The graph of $h(x) = 1 - e^x$ matches the graph in Figure 3c. It is a reflection of the graph of $y = e^x$ about the x axis translated up 1 unit.

Applying Exponential Growth

Exponential growth occurs whenever the rate of increase depends on the amount present. For example, if you have more money to invest, the growth in your account is greater. If there are more bacteria present, the rate at which they increase is greater.

We have seen how exponential functions such as

$$A = P(1 + r)^t$$

are used to calculate the amount of money in an account. Now, we extend our study to exponential functions of the form

$$N = N_0(1 + r)^t$$

The two formulas are analogous. Problems involving the mathematics of finance use the formula with A and P. Other applications involving exponential growth use N and N_0, where N is the number present after some period of time t in years, N_0 is the number present initially at $t = 0$, and r is the annual growth rate. The next example models the population growth of Mexico.

Example 5 The Population of Mexico Problem

In 1990 the population of Mexico was 88.597 million and growing at an annually compounded rate of approximately 2.12%. Predict the population in Mexico in the year 2023 if the growth rate remains the same. [Sources: *The Universal Almanac* and the United Nations Population Fund]

Thinking: Creating a Plan

We use the formula for annual growth, $N = N_0(1 + r)^t$. The 1990 population is the initial number, $N_0 = 88.597$, with $r = 0.0212$ representing the growth rate of 2.12%. The number of years between 1990 and 2023 is $t = 33$.

Communicating: Writing the Solution

$N = N_0(1 + r)^t$ and $N_0 = 88.597$, $r = 0.0212$, and $t = 33 \Rightarrow$

$$N = 88.597(1 + 0.0212)^{33}$$
$$\approx 177.042$$

The population will be about 177 million in the year 2023.

Learning: Making Connections

Notice that the population doubled in 33 years from about 88.5 million to 177 million. *Doubling time* is the time it takes to double the initial value.

Just as $A = P(1 + r)^t$ is analogous to $N = N_0(1 + r)^t$, the formula for continuously compounded interest, $A = Pe^{rt}$, is analogous the generic formula for continuous exponential growth

$$N = N_0e^{kt}$$

Here, k is the proportionality constant, the growth rate.

Let's reconsider the population growth of Mexico from Example 5 with a continuously compounded growth rate of 2.098% to predict the population in the year 2023.

$N = N_0e^{kt}$ and $N_0 = 88.597$, $k = 0.02098$, $t = 33 \Rightarrow$

$$N = 88.597e^{(0.02098)(33)} \approx 177.05$$

The exponential function $N = 88.579(1.0212)^t$ from Example 5 is nearly equivalent to this exponential function $N = 88.597e^{0.02098t}$. The continuously compounded growth rate of 2.098% has an effective annual yield of 2.12% (Figure 4). We will explore changing the base of an exponential expression in the next section.

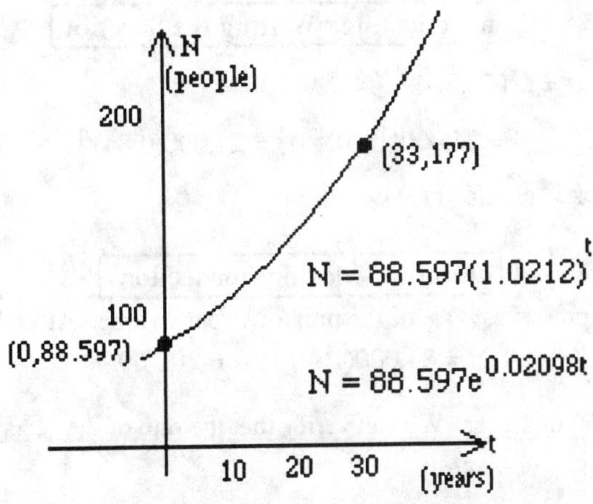

Figure 4

Introducing Exponential Decay

The graph of $y = 2^{-x}$ is the reflection of the graph of $f(x) = 2^x$ about the y axis (Figure 5). Notice that as x increases, y decreases, approaching zero. This decreasing function models **exponential decay**.

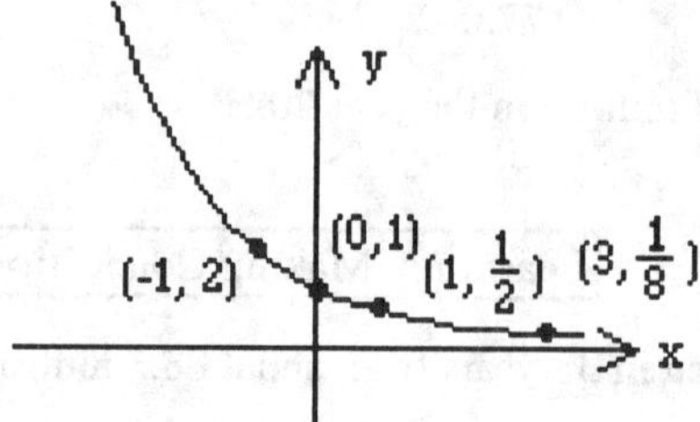

Figure 5

We can express $y = 2^{-x}$ as an equivalent function with a base of 1/2.

$$y = 2^{-x} \Leftrightarrow y = (\frac{1}{2})^x$$

Exponential equations with a base such that $0 < b < 1$, are decreasing when the exponent is positive.

For this reason, formulas of the form

$$N = N_0(1 - r)^t \text{ and } N = N_0e^{-kt}$$

both describe exponential decay. The expression $(1 - r)$ in the formula $N = N_0(1 - r)^t$ is between 0 and 1. Such exponential decay occurs whenever the rate of decrease is proportional to the amount present, as in the next example.

Example 6 The Depreciating Truck Problem

Suppose that you bought a pickup truck for $24,000. If the truck is expected to depreciate by about 16% each year, what will it be worth in 4 years?

Thinking: Creating a Plan

Since the truck is depreciating at an annual rate of 0.16, we use $N = N_0(1 - r)^t$.

Communicating: Writing the Solution

$N = N_0(1 - r)^t$ and $N_0 = 24{,}000$, $r = 0.16$, $t = 4 \Rightarrow$

$$N = 24{,}000(1 - 0.16)^4 = 24{,}000(0.84)^4 \approx 11{,}948.91$$

In 4 years, the value of the truck is about $11,949.

Learning: Making Connections

A decrease of 16% per year preserves 84% of the previous year's value. After 1 year,

$$N = 24{,}000(1 - 0.16)^1 \approx 20{,}160$$

The truck depreciated to $20,160 in 1 year. We can verify that the rate of depreciation was 16% by calculating the percent change, $\frac{\Delta v}{v_0}$.

$$\frac{\Delta v}{v_0} = \frac{20160 - 24000}{24000} = -0.16$$

Plotting these points in the Cartesian coordinate system with the horizontal axis time and the vertical axis value, we see that the value of the truck is decreasing, approaching zero (Figure 6). Notice the value of this car decreased by about half in 4 years.

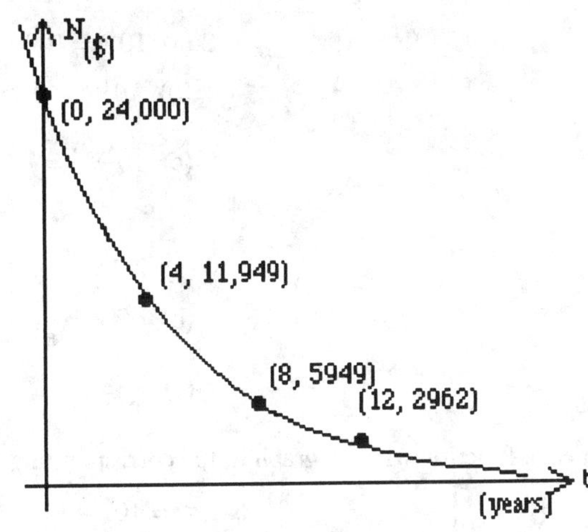

Figure 6

Explore the Concepts

E1. Graph $y = (1 + \frac{1}{x})^x$ for $0 < x \leq 100{,}000$.

E2. Complete the following table.

x	2	4	8	10	100	1000	10,000	100,000
y								

E3. Use your results from Exercises E1 and E2 to complete the following. $e = \lim_{x \to \infty} (1 + \frac{1}{x})^x \approx$ __________.

Exercises 4.1

Develop the Concepts

For Exercises 1 to 4 solve each exponential equation.

1.(a) $5^{x-1} = 25$
(b) $5^{2x} = \frac{1}{5}$
(c) $5^x = 1$

2.(a) $10^{-x} = 1000$
(b) $10^{1-x} = 0.1$
(c) $10^{2x+1} = 1$

3.(a) $e^{2x} = 1$
(b) $e^{1-x} = \frac{1}{e}$
(c) $e \cdot e^x = 1$

4.(a) $2e^x = e^x + 1$
(b) $e^{2x-1} = \frac{1}{e^2}$
(c) $\frac{e^x}{e} = e^{2x}$

For Exercises 5 to 8 match each function with its graph in the corresponding figure.

5. $f(x) = 10^{-x}$
$g(x) = -10^x$
$h(x) = -10^{-x}$

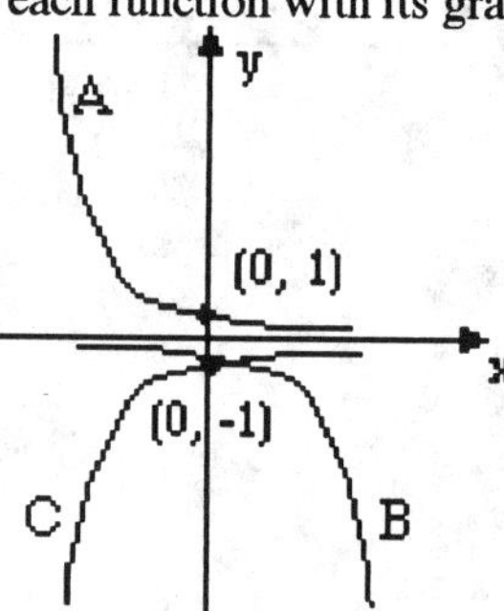

6. $f(x) = 10^x + 1$
$g(x) = 10^{x+1}$
$h(x) = 10^x - 1$

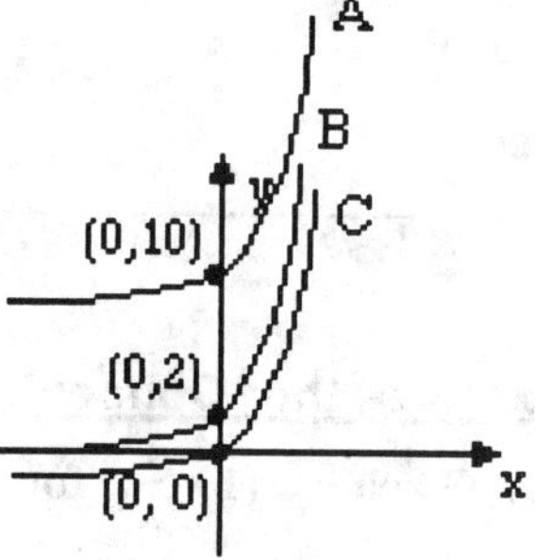

7. $f(x) = e^x - 3$
$g(x) = 3 - e^x$
$h(x) = e^{-x} - 3$

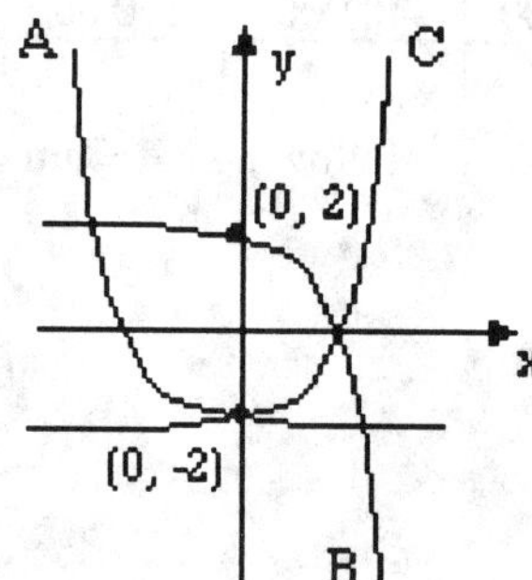

8. $f(x) = e^{x-3} - 3$
$g(x) = e^{3-x} - 3$
$h(x) = 3 - e^{x-3}$

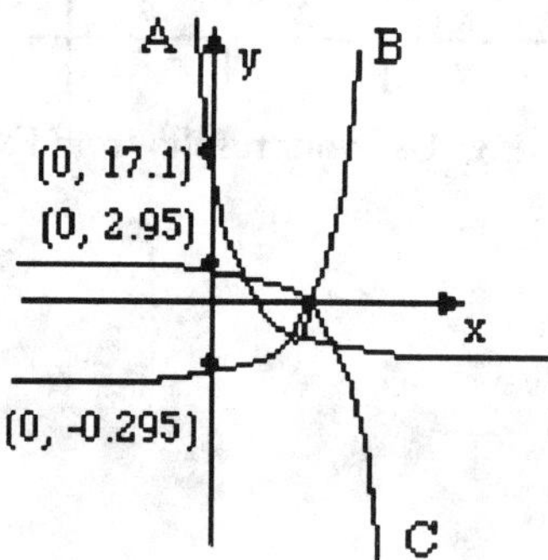

For Exercises 9 to 12 graph each function and state the equation of the asymptote.

9.(a) $y = e^x + 1$
(b) $y = e^{x+1}$

10.(a) $y = e^{-x}$
(b) $y = -e^{-x}$

11.(a) $y = -10^x$
(b) $y = 10^{-x}$

12.(a) $y = 10^{x-2}$
(b) $y = 10^x - 2$

For Exercises 13 to 16 graph each function, state the equation of the asymptote, and label the intercepts.

13. $y = 1 - e^x$

14. $y = 10^{-x} + 10$

15. $f(x) = -e^{(-x-2)} + 1$

16. $f(x) = e^{-x} + 1$

For Exercises 17 to 20 graph each pair of functions on one coordinate system. Find and label the point of intersection.

17. $f(x) = 2^{-x}$
    $g(x) = 2^{x+1}$

18. $f(x) = 3^{1-x}$
    $g(x) = 9^{x}$

19. $f(x) = e^{x} - 1$
    $g(x) = 1 - e^{x}$

20. $f(x) = 2e^{x}$
    $g(x) = e^{x} + 1$

21. Find the amount of money in an account if $1500 is invested at 4.5% interest compounded as indicated for 3 years.
(a) annually (b) quarterly (c) monthly (d) continuously

22. Use the formula $A = P(1 + r/k)^{tk}$ to find the amount of money after 1 year if $1 is invested at 100% interest compounded as indicated.
(a) monthly (b) weekly (c) daily (k = 365) (d) hourly

For Exercises 23 and 24 study $A = Pe^{rt}$ by changing the goal. Assume that you want to start a college account for a child.

23. If you deposit $10,000 today, how much will the child have for college? (Make up your own values for t and r.)

24. If you think the child will need $40,000 to go to college, how much must you deposit today? (Make up your own values for t and r.)

Apply the Concepts

25. The Interest Comparison Problem Compare interest earned at 8% annually compounded interest with 8% continuously compounded interest on a principal of $5000 for $8\frac{1}{2}$ years.

26. Another Interest Comparison Problem Compare interest earned at 10% annually compounded interest with 10% continuously compounded interest on a principal of $100,000 for 16 months.

*27. An Effective Annual Yield Problem Find the effective annual yield (EAY) for 10% continuously compounded interest by finding r in the formula $A = P(1 + r)^{t}$ that will yield exactly the same amount (future value A) as $A = Pe^{0.10t}$ for every P and any time. [Hint: Let P = $1.00 and t = 1 year, and solve $P(1 + r)^{t} = Pe^{0.10t}$ for r.]

28. Another Effective Annual Yield Problem Find the effective annual yield (EAY) for 10% interest compounded (a) quarterly, (b) daily, and (c) hourly. See Exercise 27.

29. The Rainbow Kite and Balloon Company For the last several days the kite factory has earned an average daily profit of 0.1% ($r/365 = 0.001$) return on investment. Find the effective annual yield (EAY). See Exercise 27.

30. Wonderworld Amusement Park Your cousin who owns The Rainbow Kite and Balloon Factory was boasting about profit. You felt great about Wonderworld's 1% return on investment earned last month ($r/12 = 0.01$). Find the effective annual yield for 1% monthly compounded interest. (It is not 12%.) See Exercise 27.

31. The U.S. Population Problem On July 1, 1994 the U.S. population was 260,349,838. On July 1, 1995 it was 262,755,270. If this annual growth rate continued, what would the population be in the year 2000? [Source: U.S. Census Bureau]

32. The Nevada Population Problem On July 1, 1994 the population of Nevada was 1,462,026. On July 1, 1995 it was 1,530,108. If this annual growth rate continued, what would the population be in the year 2000? [Source: U.S. Census Bureau]

*33. The Immigration Problem The U.S. population in 1995 was about 263 million people, with a birthrate of 15 per 1000 and a death rate of 9 per 1000. If 8000 immigrants enter the country each year, what will the total population be in the year 2000? [Source: ZPG]

34. The Zero Population Growth Web Site Problem On September 8, 1997 the Zero Population Growth Organization on their web site (http://www.igc.apc.org/zpg/index.html) estimated the world population to be 5924036680 and growing at about 1.5% annually. How many people will be added to this number in 24 hours?

35. The Depreciating Pickup Truck Problem A new Ford F150 Lariat pickup had a suggested retail price of $24,170. A 1-year-old comparable model had a value of $14,625.
(a) Use the depreciation formula $N = N_0(1 - r)^t$ to find r for this 1-year depreciation.
(b) Use your results from part (a) to estimate the value of the new pickup when it is 4 months old.
[Source: *Kelley Blue Book*, http://www.kbb.com]

36. The Depreciating Porche Problem A 2-year-old Porche 911 had a trade-in value of $51,250- and a comparable 3-year-old model had a trade-in value of $46,425.
(a) Use the depreciation formula $N = N_0(1 - r)^t$ to find r for this 1-year depreciation.
(b) Use your results from part (a) to estimate the value of a comparable 7-year-old Porsche 911.
[Source: *Kelley Blue Book*, http://www.kbb.com]

37. The Game of Chess Problem The game of chess was invented for a Prussian king who was so excited about the game that he told the inventor he could have anything he wanted as a reward. The wily inventor asked for a few grains of rice: 1 for the first square on the chess board, 2 for the second, 4 for the third, 8 for the fourth, and so on, until the board was covered. How many grains of rice were needed for the last square - the 64th square?

38. The Samurai Sword Problem In his book The Ascent of Man, J. Bronowski describes the process of making Samurai swords. These legendary Japanese sword smiths created flexible, yet strong swords by folding the metal in half and pounding the heated metal out flat again, repeating the process 15 times. How many layers of steel are there when the sword is done? (Incidentally, the shape of a portion of the Samurai sword is identical to the graph of an exponential function.)

39. The Origami Problem Suppose that you start with a 10- by 10-in. square piece of paper. You fold it in half, then in half again and again.

(a) If the paper is 0.006 cm thick, how thick would the folded paper be after the twenty-fourth fold?

(b) How many times can you fold a piece of paper in half?

40. The Dahlia Problem You decide to grow flowers called dahlias. You start by planting one tuber (root). In the fall you dig up the tubers and find four tubers, which you separate, store over winter, and replant the tubers in spring. If this growth continues for 10 years, how many dahlia tubers will you have 10 years after you initially planted that first tuber?

41. The Amoeba Population Problem The population growth of amoebas is modeled by $N = N_0e^{kt}$, where t is measured in hours and the original amoeba count is $N_0 = 300$. The growth rate is $k = 4.16$.

(a) How many amoeba will there be in 1 hr?

(b) When will the number reach 1 billion?

42. The Depreciated Computer Problem A computer originally cost $5000.

(a) Find the value of a computer after 10 years if it depreciated at a continuous rate of 10%.

(b) When will the value be $2500?

43. The Baby Boom Problem In 1987, when the population of the United States was 250 million, a baby boom was experienced with an average annual population growth rate of 1%.

(a) Find the values of N_0 and k for the model $N = N_0e^{kt}$ for the number of people as a function of time.

(b) What will the population be in the year 2000?

(c) How long will it take for the population to reach 500 million?

44. The Husk of Hares Problem At the beginning of a study there were 1000 hares, and they were increasing at an average rate of 7.2%.

(a) Use the formula $N_h = N_0e^{kt}$ to find how many hares were there 1 year before the study began.

(b) When there were 1000 hares ($N_h = 1000e^{kt}$), the number of foxes was increasing so that $N_f = 74e^{0.08t}$. How many years will it take for the number of foxes to reach 10% of the number of hares, $N_f = 0.10N_h$?

Extend the Concepts

For Exercises 45 to 48 solve each equation.

45. $2^{2^x} = 2$

46. $(2^x)^{-x} = \frac{1}{2}$

47. $e^x(e^x - 1) = 0$

48. $e^{2x} - 2e^{x+1} + e^2 = 0$

For Exercises 49 and 50 use $e^x \approx 1 + \frac{x}{1!} + \frac{x^2}{2!} + \frac{x^3}{3!} + \cdots + \frac{x^n}{n!}$ to approximate e for the given values of n.

49. n = 3

50. n = 4

4.2 Logarithmic Functions

Defining Logarithms as Exponents
Defining Logarithms in Base 10 and Base e
Solving Equations Using the Definition of a Logarithm
Graphing Logarithmic Functions
Applying Exponential Decay

Functions such as

$$y = \log x - 1 \quad \text{and} \quad y = \frac{1}{2}\log_3 x$$

are called **logarithmic functions**. As we will see, the inverse of an exponential function is a logarithmic function and the inverse of a logarithmic function is an exponential function. In this section we extend our study of exponential decay to include problems involving carbon dating where the unknown time is an exponent.

Defining Logarithms as Exponents

The expression

$$y = \log_b x$$

is read "y equals the log base b of x," where "log" is an abbreviation for *logarithm*. The following definition shows that a **logarithm** is an exponent. Notice the equivalent equations are inverses of each other. One inputs x and outputs y: the other inputs y and outputs x.

Definition of a Logarithm in Base b

For $x > 0$ and $b > 0$, $b \neq 1$

$$y = \log_b x \Leftrightarrow x = b^y$$

Regardless of the base, when an equation in logarithmic form is written as its equivalent equation in exponential form, the value of the logarithm is the exponent (Table 1).

Table 1

Logarithmic Form	Exponential Form
$\log_2 8 = \mathbf{3}$	$2^{\mathbf{3}} = 8$
$\log_{10} \frac{1}{10} = \mathbf{-1}$	$10^{\mathbf{-1}} = \frac{1}{10}$
$\log_3 1 = \mathbf{0}$	$3^{\mathbf{0}} = 1$

NOTE: We restrict the definition of a logarithm to positive bases, to avoid imaginary results. For example, $y = \log_{(-3)} x$ for $y = 1/2$ requires that $x = \sqrt{-3}$, which is imaginary.

In the following example we find the value of each logarithm by first applying the definition of a logarithm and then rewriting each statement in its equivalent exponential form.

Example 1 *Finding the value of logarithms*

Find the value of each logarithm.

(a) $y = \log_5 125$ (b) $y = \log_{10} \frac{1}{100}$ (c) $y = \log_b 1, b > 0, b \neq 1$

Solution

(a)

$$y = \log_5 125$$
$$5^y = 125$$
$$5^y = 5^3$$
$$y = 3$$

(b)

$$y = \log_{10} 100$$
$$10^y = 100$$
$$10^y = 10^2$$
$$y = 2$$

(c)

$$y = \log_b 1$$
$$b^y = 1$$
$$b^y = b^0$$
$$y = 0$$

Defining Logarithms in Base 10 and Base e

Your calculator is programmed to find logarithms in two very important bases.

Do you know the two bases?

Calculators have a [LOG] key that is used to find a logarithm in base 10 and a [LN] key that is used to find logarithms in base e. The equations

$$y = \log_{10} x \Leftrightarrow y = \log x$$

are both read "y equals the log base 10 of x." Logarithms in base 10 are called **common logarithms**. The equations

$$y = \log_e x \Leftrightarrow y = \ln x$$

are read "y equals the logarithm base e of x." Logarithms in base e are called **natural logarithms**. Logarithms in base e will prove useful in calculus where using base e creates simpler functions and calculations. Notice in the following definition of a logarithm that y is the *exponent* with a base of 10 for $y = \log x$ and is an *exponent* with base e for $y = \ln x$.

> **Definition of a Logarithm**
> For $x > 0$,
>
> $$y = \log x \Leftrightarrow x = 10^y$$
>
> and
>
> $$y = \ln x \Leftrightarrow x = e^y$$

Example 2 *Switching from logarithmic to exponential form*

Use the definition of logarithm to rewrite each of the following in exponential form.

(a) $\log \frac{1}{10} = -1$ (b) $\log 1{,}000{,}000 = 6$ (c) $\ln 1 = 0$ (d) $\ln 3 \approx 1.099$

Solution

The base in parts (a) and (b) is 10, while the base in parts (c) and (d) is e.

(a) $\log \frac{1}{10} = -1 \Leftrightarrow 10^{-1} = \frac{1}{10}$

(b) $\log 1{,}000{,}000 = 6 \Leftrightarrow 10^6 = 1{,}000{,}000$

(c) $\ln 1 = 0 \Leftrightarrow e^0 = 1$

(d) $\ln 3 \approx 1.0986 \Leftrightarrow e^{1.0986} \approx 3$

Example 3 *Evaluating logarithms*

Evaluate the following.

(a) log 2 (b) log (-4) (c) ln 24 (d) ln 0

Solution

(a) $\log 2 \approx 0.3010$

(b) log (-4) does not exist.

(c) $\ln 24 \approx 3.178$

(d) ln 0 does not exist.

Parts (b) and (d) of Example 3 are both undefined since we can only find the logarithm of positive values. There is no real value of x such that 10^x is negative and no value of x such that e^x is 0.

Does the result in Example 3a, log 2 ≈ 0.3010, seem reasonable?

Any logarithm can be understood by examining its corresponding exponential statement. In this case, $\log 2 \approx 0.3010 \Leftrightarrow 10^{0.3010} \approx 2$ means that

$$10^{0.3010} \approx 10^{3/10} = \sqrt[10]{10^3} = \sqrt[10]{1000} \approx 2$$

Solving Equations Using the Definition of a Logarithm

Some logarithmic equations can be solved by writing the equation in exponential form using the definition of a logarithm. For example, the logarithmic equation

$$\log x = -2$$

can be solved by writing it in its equivalent exponential form.

$$10^{-2} = x$$
$$x = 1/100$$

Can you solve the logarithmic equation ln x = -2 by applying the definition of a logarithm?

Writing ln x = -2 in its equivalent exponential form yields

$$e^{-2} = x$$
$$x = 1/e^2$$
$$x \approx 0.1353$$

Example 4 *Solving logarithmic equations using the definition of a logarithm*

Solve the following equations.

(a) $2 \log x = 18$ (b) $\ln 2x = -2$

Solution

(a)

$2 \log x = 18$		*Divide both sides by 2.*
$\log x = 9$	$\Leftrightarrow$	*Use the definition of a logarithm to rewrite in exponential form.*
$x = 10^9$		

(b)

$\ln 2x = -2$	$\Leftrightarrow$	*Use the definition of a logarithm to rewrite in exponential form.*
$e^{-2} = 2x$		*Approximate e^{-2}.*
$0.1353 \approx 2x$		
$x \approx 0.0677$		

We can also use the definition of a logarithm to solve an exponential equation in base 10 or base *e*.

Example 5 *Solving exponential equations using the definition of a logarithm*

Solve the following equations.

(a) $10^{x+1} = 2.9$ (b) $1 - e^x = 0.5$

Solution

(a)

$10^{x+1} = 2.9$	$\Leftrightarrow$	*Use the definition of a logarithm to rewrite in logarithmic form.*
$\log 2.9 = x + 1$		*Approximate log 2.9.*
$0.4624 \approx x + 1$		*Subtract 1 from both sides.*
$x \approx -0.5376$		

(b)

$1 - e^x = 0.5$		*Subtract 1 from both sides.*
$-e^x = -0.5$		
$e^x = 0.5$	$\Leftrightarrow$	*Use the definition of a logarithm to rewrite in logarithmic form.*
$\ln 0.5 = x$		*Approximate ln 1.5.*
$x \approx -0.6931$		

Graphing Logarithmic Functions

Next we examine reflections, translations, and changes of scale of the logarithmic functions. Figure 1a shows the graph of $y = \log x$, and Figure 1b shows the graph of $y = \ln x$.

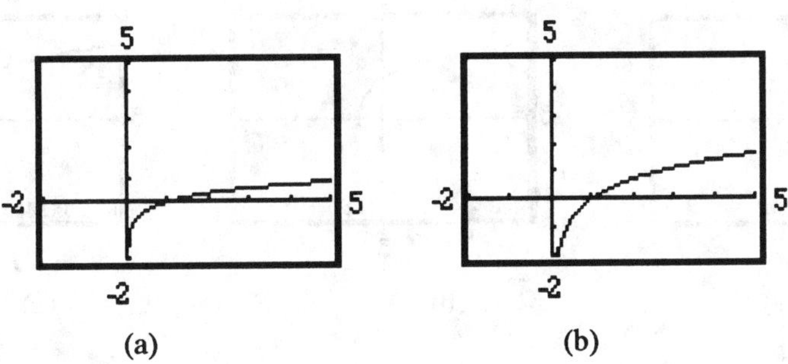

Figure 1 (a) (b)

Recall from Example 3 that logarithms are defined only for positive numbers regardless of the base. That is why the domain for both functions is x > 0. Both graphs have the y axis as a vertical asymptote. Also, both graphs pass through the point (1, 0) because log 1 = ln 1 = 0 ($10^0 = 1$ and $e^0 = 1$).

Example 6 *Graphing logarithmic functions*

Identify the reflections and translations and then graph (a) y = 1 - log x and (b) y = ln (x + 1). State the equations of the asymptotes.

Solution

(a) This is a reflection about the x axis and translation up 1 unit of the graph of y = log x (Figure 2a). The vertical asymptote is x = 0.

(b) This is a translation of the graph of y = ln x left 1 unit (Figure 2b). The vertical asymptote is x = -1.

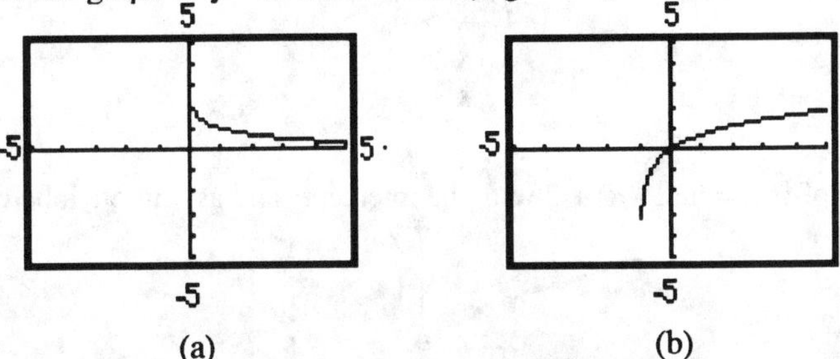

Figure 2 (a) (b)

Example 7 *Graphing a logarithmic function*

(a) Graph the function f(x) = ln (1 - x) + 2 and estimate the x and y intercepts, and the asymptote from your graph.

(b) Verify your results from part (a) algebraically.

Solution

(a) Figure 3 shows the graph of f(x) = ln (1 - x) + 2. The x intercept is x ≈ 0.9 (Figure 3a), the y intercept is y = 2 (Figure 3b), and the vertical asymptote appears to be very close to x = 1. The display shows no y coordinate for x > 1 (Figure 3c).

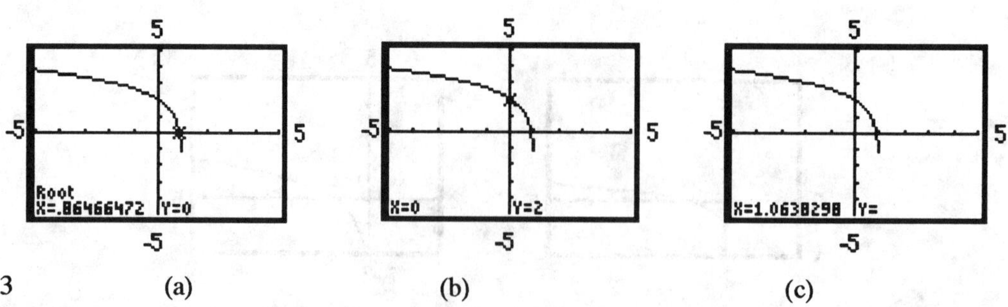

Figure 3 (a) (b) (c)

(b) <u>x intercept</u>: $f(x) = \ln(1 - x) + 2$ and $f(x) = 0 \Rightarrow$

$$0 = \ln(1 - x) + 2 \qquad \textit{Add -2 to both sides.}$$

$$\ln(1 - x) = -2 \Leftrightarrow \qquad \textit{Apply the definition of a logarithm.}$$

$$1 - x = e^{-2} \qquad \textit{Solve for x.}$$

$$x = 1 - e^{-2}$$

$$x \approx 0.8647$$

The x intercept is approximately 0.86.

<u>y intercept</u>: $f(x) = \ln(1 - x) + 2 \Rightarrow$

$$f(0) = \ln(1 - 0) + 2 = 0 + 2 = 2$$

The y intercept is 2.

<u>Asymptote</u>: The domain of the function requires the input values to be positive. Therefore,

$$1 - x > 0$$
$$-x > -1$$
$$x < 1$$

Figure 4 shows a sketch of $f(x) = \ln(1 - x) + 2$ with the intercepts and asymptote labeled.

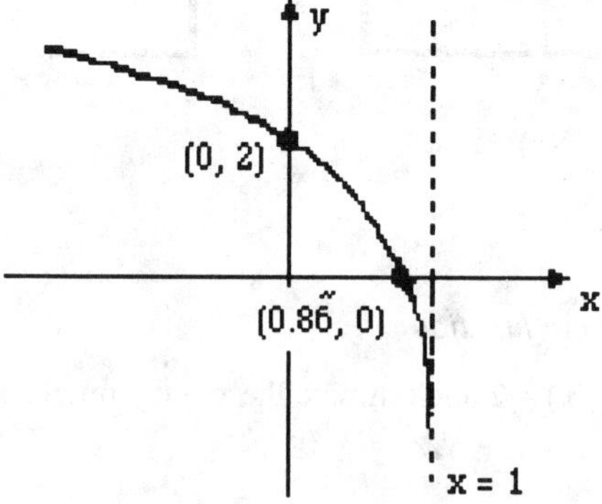

Figure 4

What is the relationship between the graphs of exponential and logarithmic functions?

To explore the relationship between the graphs of exponential and logarithmic functions, go to "Explore the Concepts," Exercises E1 - E4, before continuing (page 278). For the interactive version, go to the *Precalculus Weblet - Instructional Center - Explore Concepts.*

Exponential and logarithmic functions are inverses of each other. One graph is the reflection of the other about the line y = x of the other. We enter $y = e^x$ (Figure 5a) and y = ln x (Figure 5b).

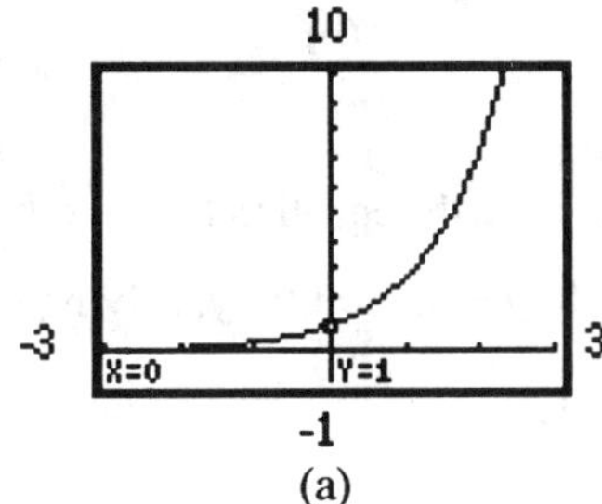

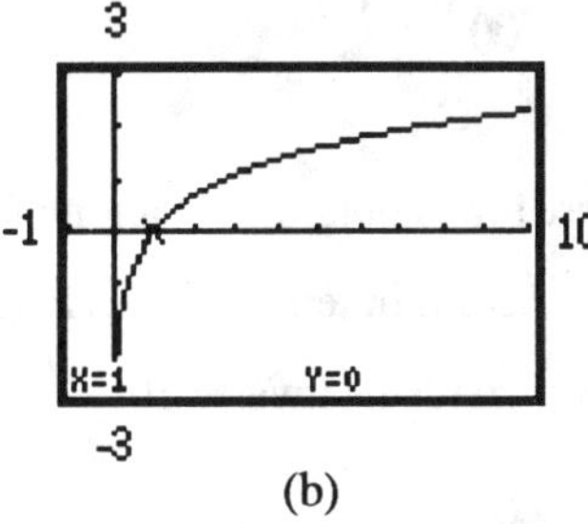

Figure 5 (a) (b)

Compare the domains and ranges of the exponential and logarithmic functions.

Function:	$y = e^x$	y = ln x
domain	$(-\infty,\infty)$	$(0,\infty)$
range	$(0,\infty)$	$(-\infty,\infty)$

Hold the template that corresponds to Appendix II by the corners as shown in Figure 6, flip it across the diagonal, and see the graphs of the inverses of $y = e^x$ and y = log x.

Figure 6

The four functions shown in Figure 7 are two pairs of inverse functions. The reference function $y = e^x$ and its inverse x = ln y (y = ln x on a graphing utility) are base e. The reference function y = log x and its inverse $x = 10^y$ ($y = 10^x$ on a graphing utility) are base 10.

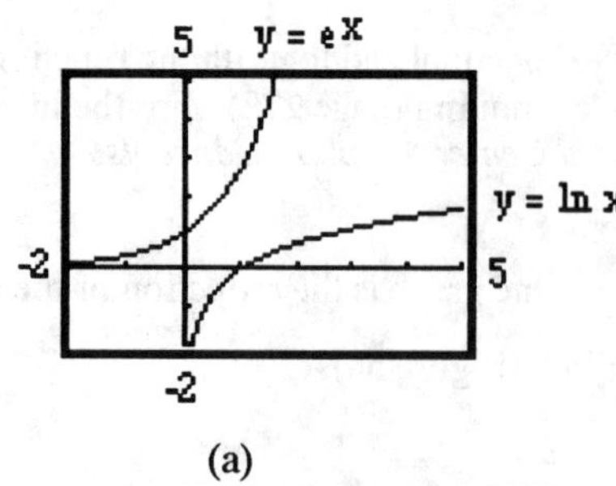

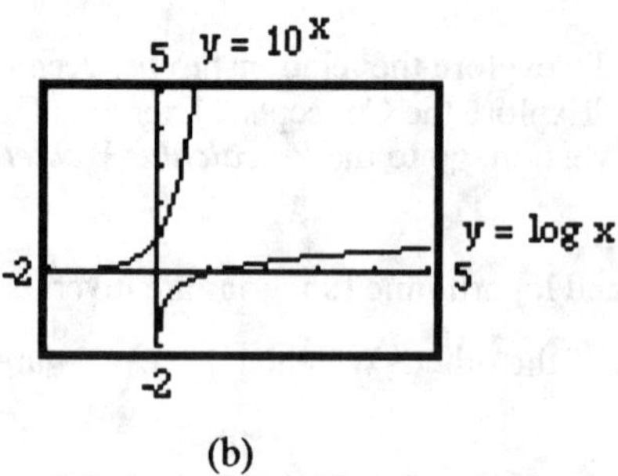

Figure 7 (a) (b)

Recall the procedure for finding the equation of an inverse; solve the equation for the other variable. For example, solving $y = e^x$ for x in terms of y yields its inverse, $\ln y = x$. Similarly, solving the logarithmic function $y = \log x$ for x yields its inverse $10^y = x$.

Example 8 *Finding the equation of an inverse*

Find the equation of each inverse.

(a) $y = e^{-x}$ (b) $y = 1 - \log x$

Solution

(a)

$y = e^{-x} \Leftrightarrow$ *Apply the definition of a logarithm.*

$\ln y = -x$ *Solve for x.*

$x = -\ln y$

(b)

$y = 1 - \log x$

$\log x = 1 - y \Leftrightarrow$ *Apply the definition of a logarithm.*

$x = 10^{1-y}$

Applying Exponential Decay

Exponential decay of radioactive elements is modeled by decay functions of the form $N(t) = N_0 e^{-kt}$, where N_0 is the initial count, k is a proportionality constant ($k > 0$), and t is time. In 1946, Willard Libby of the University of Chicago won a Nobel prize in chemistry for his discovery of carbon dating. Cosmic rays entering our atmosphere continually create carbon-14, replacing the carbon-14 as it decays. This small but constant amount of carbon-14 combines with oxygen to form carbon dioxide, which is absorbed by plants and hence all living things. Libby reasoned that since carbon-14 absorption ceases at death, the amount originally present will reduce by half every 5730 ± 40 years, the *half-life of carbon-14.*

We can find the proportionality constant k for carbon-14 as follows. After one half-life (5730 years) the amount N of carbon-14 present will be half of the original amount, $N = \frac{1}{2}N_0$.

$N = N_0e^{-kt}$ and $t = 5730$, $N = \frac{1}{2}N_0 \Rightarrow$

$$\frac{1}{2}N_0 = N_0e^{-k(5730)} \qquad \textit{Divide both sides by } N_0.$$

$$0.5 = e^{-5730k} \qquad \textit{Apply the definition of a logarithm.}$$

$$\ln 0.5 = -5730k$$

$$-0.6931 \approx -5730k$$

$$k \approx 0.000121$$

Therefore, the carbon-14 exponential decay model is defined by the function $N = N_0e^{-0.000121t}$ (Figure 8).

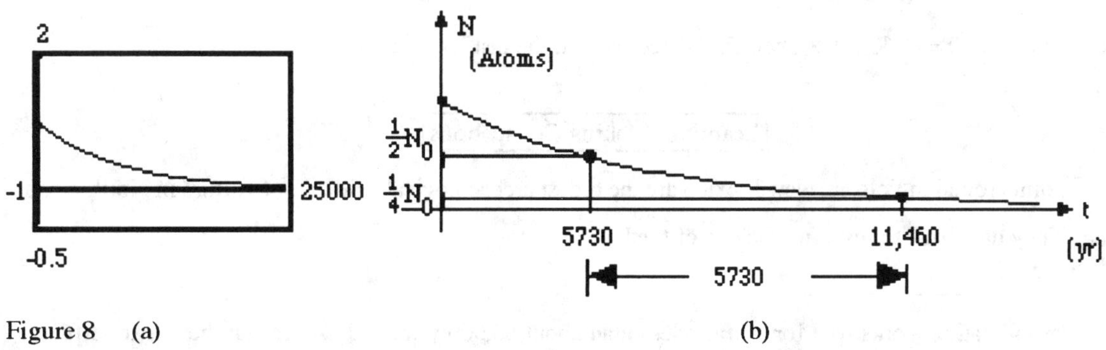

Figure 8 (a) (b)

A sample to be dated must be collected and cleaned carefully to prevent contamination. Then it is chemically treated and placed in a specially designed radiation detector along with an artificial control sample from the U.S. National Bureau of Standards which contains the amount of carbon-14 in living tissue, N_0. The laboratory reports the ratio of carbon-14 in the sample to that of the standard, $\frac{N}{N_0}$.

Since the test yields values of $\frac{N}{N_0}$, the inverse of the exponential function is used to date an artifact. That is, to find the time t since death, we solve $N = N_0e^{-kt}$ for t.

$$N = N_0e^{-kt}$$

$$\frac{N}{N_0} = e^{-kt} \Leftrightarrow \qquad \textit{Apply the definition of logarithm.}$$

$$-kt = \ln\frac{N}{N_0} \qquad \textit{Divide both sides by -k.}$$

$$t = \frac{-1}{k}\ln\frac{N}{N_0} \qquad k = 0.000121.$$

$$t = -8264.46 \ln\frac{N}{N_0}$$

Example 9 The Archaeologists' Find Problem

A team of archaeologists discovered spear points in remains of now extinct bison (ancestors of buffalo) near the site of an ancient campfire. From the other prehistoric remains and geological clues of the area, they expect the spear points to date back to just after the last Ice Age, nearly 8500 B.C. The team sent a piece of charcoal from the campfire to the lab for analysis. (Since spear points were never alive, they cannot be carbon dated.) The report said that 27.2% of the original carbon-14 remains. Did the team's find date back as far as they expected?

Thinking: Creating a Plan

We found the model for carbon-14 to be $t = -8264.46 \ln \frac{N}{N_0}$. We will let $\frac{N}{N_0} = 0.272$ to see whether they have found evidence of very early civilization in North America.

Communicating: Writing the Solution

$t = -8264.46 \ln \frac{N}{N_0}$ and $\frac{N}{N_0} = 0.272 \Rightarrow t = -8264.46 \ln 0.272 \approx 10{,}760$

This is 10,760 - 2000 = 8760 B.C. Their find dated back to the last Ice Age.

Learning: Making Connections

Projectile points found in Folsom, New Mexico are the oldest archaeological evidence of culture in North America. They date back to just after the end of the Ice Age.

Carbon dating works well for artifacts less than about 50,000 years old. After that there is too little carbon-14 left to be separated from background radiation with the usual methods. A new technique involving an AMS (accelerator mass spectrometer) allows accurate carbon-14 dating back 70,000 years, but the method is quite expensive. [Source: *The Practical Archaeologist*, J. MacIntosh] For older specimens, such as moon rocks, radioactive elements with longer half-lives are used.

Explore the Concepts

E1.(a) Graph $y = 10^x$ on your graphing utility using $-5 \le x \le 5$ and $0 \le y \le 100{,}000$.

(b) Complete the following table of values.

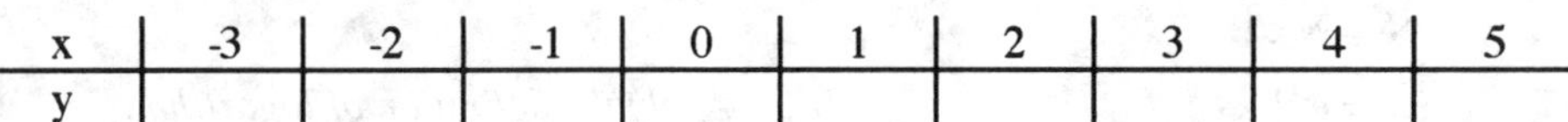

x	-3	-2	-1	0	1	2	3	4	5
y									

E2.(a) Graph $y = \log x$ on your graphing utility using $0 \le x \le 100{,}000$ and $-5 \le y \le 5$.

(b) Complete the following table of values.

x	0.001	0.01	0.1	1	10	100	1000	10,000
y								

E3. Use your results from Exercises E1 and E2 to complete the following: if (a, b) is a point on the graph of $y = 10^x$, ____is a point on the graph of $y = \log x$.

E4. Use your results from Exercises E1 to E3 to complete the following: Reflecting the graph of $y = 10^x$ about the line _____ creates the graph of $y = \log x$.

Exercises 4.2

Develop the Concepts

For Exercises 1 and 2 write each exponential equation as a logarithmic equation.

1.(a) $10^{-3} = 0.001$

(b) $e^0 = 1$

2.(a) $10^0 = 1$

(b) $e^{0.5} \approx 1.6$

For Exercises 3 and 4 write each logarithmic equation as an exponential equation.

3.(a) $\log \frac{1}{10} = -1$

(b) $\ln 64 \approx 4.2$

4.(a) $\log 1 = 0$

(b) $\ln 10 \approx 2.3$

For Exercises 5 to 8 evaluate the logarithms.

5.(a) log 2.48

(b) log 248

(c) log 2,480,000

6.(a) ln 3.9

(b) ln 0.002

(c) ln 27

7.(a) log 0

(b) log (-45)

8.(a) ln 0

(b) ln (-2)

For Exercises 9 to 12 solve each logarithmic equation by using the definition of a logarithm.

9. $\log x = 4$

10. $\ln(-3x) = 0.5$

11. $\log x^2 = 0$

12. $\ln \sqrt{x} = 1$

For Exercises 13 to 18 solve each exponential equation by using the definition of a logarithm.

13. $10^x = 9$

14. $10^x = -9$

15. $e^x = 2$

16. $10^{-2x} - 1 = 0$

* 17. $2e^{4x} = 20$

18. $2(10^{-x}) = 1$

For Exercises 19 to 22 graph each function and state the equation of the asymptote.

19.(a) $y = \log(x - 1)$

(b) $y = \log x - 1$

20.(a) $y = -\log(-x)$

(b) $y = \log(-x)$

21.(a) $y = -\ln(-x)$

(b) $y = -\ln x$

22.(a) $y = \ln x + 2$

(b) $y = \ln(x + 2)$

For Exercises 23 to 26 graph each function, state the equation of the asymptote, and label the intercepts.

23. $y = -\log(x + 3)$

24. $y = \ln(1 - x)$

25. $f(x) = \log(2 - x) + 1$

26. $f(x) = 1 - \ln(x + 2)$

For Exercises 27 to 34 find the equation of the inverse of each function by using the definition of a logarithm to solve for x.

27. $y = e^{x-2}$

28. $y = 1 - e^x$

29. $y = \ln (x - 2)$

30. $y = 2 - \ln x$

31. $f(x) = 10^x - 3$

32. $f(x) = -10^x - 1$

33. $f(x) = \log (x + 1) - 2$

34. $f(x) = -\log x + 3$

For Exercises 35 to 38 graph each pair of functions on the same coordinate system and label the intercepts and asymptotes.

*35.(a) $y = -e^x$

(b) $y = \ln (-x)$

36.(a) $y = -e^{-x}$

(b) $y = -\ln (-x)$

37.(a) $y = e^{x+1}$

(b) $y = \ln x - 1$

38.(a) $y = e^{-x/2}$

(b) $y = -2 \ln x$

For Exercises 39 and 40 study $A = Pe^{rt}$ by changing the goal. Assume that you want to start a college account for a child.

39. If you have \$10,000 to deposit and want \$40,000, how long will it take? (Make up your own value for r.)

40. If you deposit \$10,000 today and want enough money for the child to attend a private college, what interest rate would you have to earn? (Find your own values for A and t.)

Apply the Concepts

41. The Rainbow Kite and Balloon Company Sales of your newest kites are growing exponentially. Each summer you sell twice as many as two summers ago, indicating that $N = N_0 (2^{t/2})$. You sold 12 the first season.

(a) At this rate of increase, how many kites will be sold 10 years after their debut?

(b) At this rate of increase, how many kites will be sold 20 years after their debut?

(c) How long before you sell 10,000 kites?

42. Wonderworld Amusement Park Initially, our ride on your Ferris wheel lasted 15 minutes. Since then you have been cutting the ride time 10% each season.

(a) How long will the ride last six seasons after this pattern of reduction started?

(b) If the pattern continued, how long before the ride essentially has no length?

43. <u>The Fossil Problem</u> Using traditional methods, the upper limit to carbon-14 dating is approximately 50,000 years.

(a) How much of the original amount of carbon-14 will still be present in a fossil that died 50,000 years ago? Assume that $N = N_0e^{-0.000121t}$.

(b) A more modern technique (AMS method) allows accurate carbon-14 dating back 70,000 years. How much of the original carbon-14 is left in such a sample? Assume that $N = N_0e^{-0.000121t}$.

[Source: *The Practical Archaeologist*, J. McIntosh]

44. <u>The Radioactive Decay Problem</u> The equation modeling the radioactive decay of 20 g of carbon-14 is $N = 20e^{-0.000121t}$

(a) How much of the original amount of carbon-14 will be left in 5000 years?

(b) How long will it take before there are 2 g left?

45. <u>The Rural Population Problem</u> The current population in a small farming town is 350 people and is decreasing at an average annual rate of 12% .

(a) Use the formula $N = N_0e^{-kt}$ to find the equation that models this population.

(b) If people continue to move away from the town at the same rate, estimate when the mayor and the sheriff will be the only ones left.

46. <u>The Endangered Species Problem</u> If there were 200 spotted owls in 1980 in the primitive forests, and if destruction of their habitat had reduced their number to 90 in 1991, find the value of k and use the formula $N = N_0e^{-kt}$ to determine when they will be virtually extinct.

47. <u>The Wrong Half-Life Problem</u> Willard Libby estimated the half-life of carbon-14 as 5570 years. He was 160 years short of 5730 years.

(a) Find the value of k in $N = N_0e^{-kt}$ for a half-life of $t = 5570$ years.

(b) Compare the results in part (a) with the value of k associated with $t = 5730$ years ($k \approx 0.000121$) to find the correction factor to find the correct time for artifacts dated with Libby's incorrect half-life.

48. <u>The Chernobyl Accident Problem</u> At 1 A.M. on April 26, 1986 a nuclear disaster occurred in Chernobyl in the Soviet Ukraine.

(a) Radioactive iodine (^{131}I) with a half-life of 8 days was present in the fallout. Milk from cows that ate grass contaminated with iodine had to be destroyed. If the amount of ^{131}I present just after meltdown had to be reduced by 95% for the milk to be safe, how long will it take?

(b) Cesium-137, with a half-life of 30 years, was also present. It tends to concentrate in muscles and is the most dangerous component of the fallout. It was also on grass eaten by cows and became concentrated in their milk. How long will it take to reduce the amount of cesium-137 by 95%?

49. The Lifesaving Radiation Problem The half-life of cobalt-60 is 5.3 years. Cobalt is used in radiation treatment for cancer. How long will it take to reduce the level absorbed by 40%?

50. The Error in Carbon Dating Problem In the text the half-life of carbon-14 was given as 5730 ± 40 years. How does ± 40 on the half-life of carbon-14 affect the possible age of an artifact? [Hint: Find the two values for k, then date an archaeological find for which there is 10% of the original carbon-14 left, $N = 0.10N_0$, using each value of k.]

*51. The Effects-or-the-Error-in-Carbon-Dating Problem To understand the sensitivity of carbon dating, examine the error caused by mismeasuring the amount of carbon-14, $N = 0.15N_0$, as $N = 0.18N_0$.

52. The Radium Problem (a) Derive the model of the form $N = N_0e^{-kt}$ for radium, a radioactive substance discovered by Marie Curie in 1898, if radium's half-life is approximately 1690 years.

(b) Express t as a function of N by finding the inverse from part (a).

(c) Use your results from part (b) to find the age of a sample that contains 10% of the original amount of radium.

Extend the Concepts

For Exercises 53 to 56 solve each logarithmic equation that has a base other than 10 or e.

53. $\log_x 32 = 5/2$

54. $\log_x 81 = -4/3$

55. $\log_5 (x - 1) = 2$

56. $\log_7 (x^2 - 11x + 31) = 0$

For Exercises 57 to 60 solve each equation.

57. $\ln e = x$

58. $\log_x x^{4x} = 8$

* 59. $\log[\log(\log x)] = 1$

60. $x^{(\log_x 10x)} = 1/10$

For Exercises 61 and 62 use the expression $\ln(1 + x) \approx x - \frac{x^2}{2} + \frac{x^3}{3} - \frac{x^4}{4} \cdots + \frac{x^n}{n}$ to approximate $\ln(1 + x)$.

61. $n = 3$ and $x = 1$ to approximate $\ln 2$

62. $n = 5$ and $x = 1$ to approximate $\ln 2$

4.3 Theorems on Logarithms

Applying the Change of Base Formula
Introducing Theorems on Logarithms

In Section 4.3 we continue our study of solving exponential and logarithmic equations. We apply the change of base formula to solve exponential equations having bases other that 10 or e and to extend the use of our graphing utilities. Have you ever wondered why a croissant is so flaky, yet rich in taste? We will use the change of base formula to answer that question in Example 3. Then we will learn three important theorems on logarithms that we apply in equation solving and proving the change of base formula. We continue to apply functions and graphs to problems involving exponential growth and decay.

Applying the Change of Base Formula

Suppose that we want to find the value of $\log_3 10{,}000$. Our calculators do not have a logarithm key for base 3. The following formula changes any base to 10 or e.

Change of Base Formula

$$\log_b x = \frac{\ln x}{\ln b} = \frac{\log x}{\log b}, \quad b > 0,\ b \neq 1$$

Now we can calculate $\log_3 10{,}000$ using either base e or base 10.

$$\log_3 10{,}000 = \frac{\ln 10000}{\ln 3} \approx 8.3836$$

and

$$\log_3 10{,}000 = \frac{\log 10000}{\log 3} \approx 8.3836$$

In the first example we use the change of base formula to create a logarithmic graph in base 3 on a graphing utility.

Example 1 *Using the change of base formula*

(a) Apply the change of base formula to graph $y = \log_3 (x + 2)$ on a graphing utility and estimate the y intercept.

(b) Find the y intercept of the graph algebraically.

Solution

(a) Graph $y = \log_3 (x + 2)$ by entering $y = \dfrac{\ln (x + 2)}{\ln 3}$ (Figure 1).

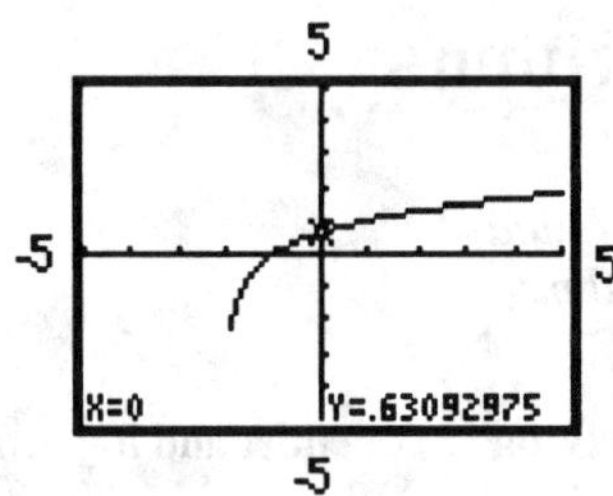

Figure 1

The y intercept is approximately 0.6.

(b) $y = \log_3 (x + 2)$ and $x = 0 \Rightarrow$

$y = \log_3 2$ *Apply the change of base formula.*

$y = \dfrac{\ln 2}{\ln 3}$

$y \approx 0.631$

We use the change of base formula for solving an exponential equation when the base in not 10 or e.

Example 2 Doubling-Time Problem

Find the doubling time of an amount of money invested at 10% compound interest compounded annually.

Thinking: Creating a Plan

Since the interest is compounded annually we use the formula $A = P(1 + r)^t$. To find doubling time, let $A = 2P$.

Communicating: Writing the Solution

$A = P(1 + r)^t$ and $A = 2P$, $r = 0.10 \Rightarrow$

$2P = P(1 + 0.10)^t$ *Divide both sides by P.*

$2 = (1.10)^t$ *Write in logarithmic form.*

$t = \log_{1.10} (2)$ *Apply the change of base formula.*

$t = \dfrac{\log 2}{\log 1.10}$

$t \approx 7.273$

The doubling time is about 7 years, 3 months, and 8 days.

Learning: Making Connections

Bankers use the rule of 72 to estimate the doubling time d, where $d = \dfrac{72}{100r}$. For continuously compounded interest they use the rule of 69, $d = \dfrac{69}{100r}$. For r = 10%, $d = \dfrac{72}{10} = 7.2$ and $d = \dfrac{69}{10} = 6.9$. It takes less time to double your money if the number of compounding periods is increased.

In the next example, we use the change of base formula to solve for the exponent in an exponential function in base 3.

Example 3 The Croissant Problem

In volume 2 of Julia Child's Mastering the Art of French Cooking, the following process for making puff pastry for croissants is repeated six times. The dough is rolled out, spread evenly with butter, folded in thirds, let rest, and then rolled out and folded again (Figure 2). Julia claims that this creates hundreds of layers of butter and pastry.

(a) How many layers are there exactly?

(b) How many folding repetitions would it take to have 10,000 layers?

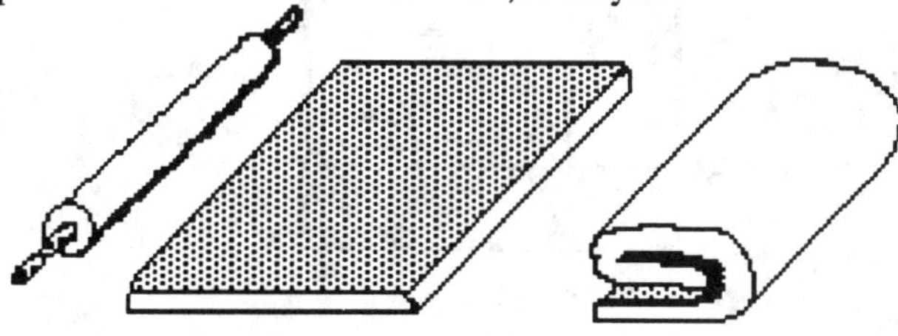

Figure 2

Thinking: Creating a Plan

(a) The pattern of layering creates 3, 9, 27, ... , 3^n layers of pastry. We evaluate the function $L = 3^n$ for $n = 6$, where L is the number of layers of pastry and n is the number of repetitions.

(b) Set $L = 10{,}000$ and use the definition of a logarithm and the change of base formula to solve for n.

Communicating: Writing the Solution

(a) $L = 3^n$ and $n = 6 \Rightarrow$

$$L = 3^6 = 729$$

Julia was right; there are hundreds of layers of pastry, 729 to be exact, making the baked croissant light and flaky.

(b) $L = 3^n$ and $L = 10{,}000 \Rightarrow$

$$10{,}000 = 3^n \Leftrightarrow \qquad \textit{Apply the definition of a logarithm.}$$

$$n = \log_3 10{,}000 \qquad \textit{Apply the change of base formula.}$$

$$n = \frac{\ln 10000}{\ln 3} \approx 8.3836$$

Nine repetitions creates over 10,000 layers of pastry.

Learning: Making Connections

Since we can't roll the pastry a fraction of a time, we round the decimal 8.3836 up to 9. We could also say that after eight repetitions there were nearly 10,000 layers. A graph of the layers as a function of the number of folds, $L = 3^n$, is a **discrete** set of points (Figure 3).

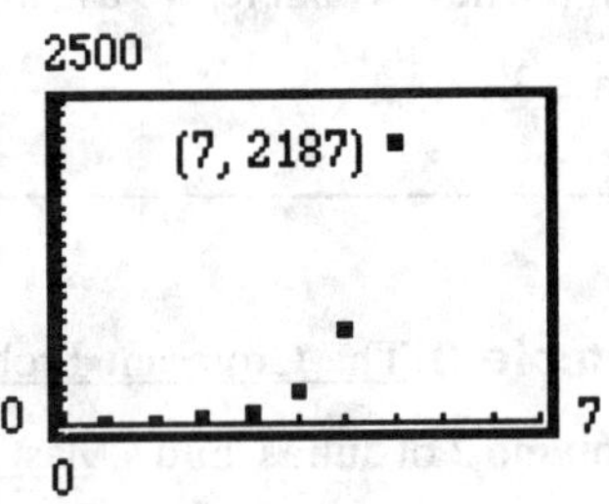

Figure 3

The change of base formula can also be used to explain the difference between the graphs of y = log x and y = ln x (Figure 4).

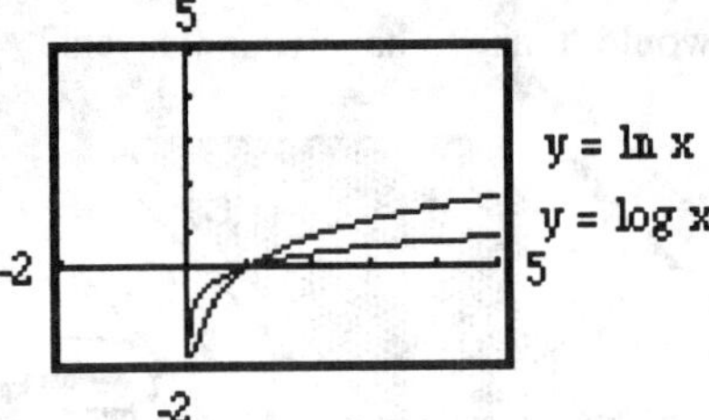

Figure 4

What is the connection between "change of scale" and the graphs of y = log x and y = ln x?

Using the change of base formula, we can show that $y = \ln x \approx 2.3 \log x$.

$$y = \ln x \Leftrightarrow y = \frac{\log x}{\log e} \Leftrightarrow y \approx \frac{\log x}{0.43} \Leftrightarrow y \approx 2.3 \log x$$

The graph of y = ln x increases about 2.3 times faster than that of y = log x.

Introducing Theorems on Logarithms

To explore the theorems of logarithms, go to "Explore the Concepts," Exercises E1 - E4, before continuing (page 290). For the interactive version, go to the *Precalculus Weblet - Instructional Center - Explore Concepts*.

When John Napier invented logarithms, his motivation was to make arithmetic calculations easier to perform. Theorems on logarithms are based on the following theorems on exponents. To multiply, divide, and raise a base to a power, the exponents are added, subtracted, and multiplied, respectively.

$$b^s \cdot b^t = b^{s+t}$$

$$\frac{b^s}{b^t} = b^{s-t}$$

$$(b^s)^p = b^{sp}$$

Since logarithms are exponents, we can restate the three theorems of exponents in terms of logarithms.

Theorems:	**Logarithms**
(1)	$\log_b MN = \log_b M + \log_b N$
(2)	$\log_b \dfrac{M}{N} = \log_b M - \log_b N$
(3)	$\log_b M^p = p \cdot \log_b M$

The proofs of these three theorems follow from the theorems on exponents. The following is the proof of theorem (2), $\log_b \frac{M}{N} = \log_b M - \log_b N$. [See Exercises 49 and 50 for the proofs of theorems (1) and (3).]

Proof of theorem (2):

Let $s = \log_b M$ and $t = \log_b N$, so that $\log_b M + \log_b N = s + t$.

$M = b^s$ and $N = b^t \Rightarrow$

$$\frac{M}{N} = \frac{b^s}{b^t} \qquad \textit{Subtract exponents.}$$

$$\frac{M}{N} = b^{s-t} \Leftrightarrow \qquad \textit{Write in log form.}$$

$$\log_b \frac{M}{N} = s - t$$

Therefore, $\log_b \frac{M}{N} = \log_b M - \log_b N$.

We can use theorem (3) to prove the change of base formula, $\log_b x = \frac{\ln x}{\ln b}$.

$$y = \log_b x \qquad \textit{Write in exponential form.}$$

$$x = b^y \qquad \textit{Take the logarithm of both sides.}$$

$$\ln x = \ln b^y \qquad \textit{Apply theorem (3) to the right side.}$$

$$\ln x = y \cdot \ln b \qquad \textit{Divide both sides by ln b.}$$

$$y = \frac{\ln x}{\ln b}$$

In the next example we solve logarithmic equations by first rewriting complicated logarithmic expressions into a simpler ones using the theorems on logarithms.

Example 4 *Solving a logarithmic equation*

Solve $\log_3 (x + 2) + \log_3 (x + 1) - \log_3 (x + 6) = 1$.

Solution

The first two logarithmic expressions are added and the third expression is subtracted. We rewrite the first two as a product using theorem (1) and the last expression as a quotient using theorem (2).

$\log_3 (x + 2) + \log_3 (x + 1) - \log_3 (x + 6) = 1$	*Apply theorems (1) and (2).*
$\log_3 \dfrac{(x + 2)(x + 1)}{x + 6} = 1 \Leftrightarrow$	*Apply the definition of a log.*
$\dfrac{(x + 2)(x + 1)}{x + 6} = 3^1$	*Multiply both sides by x + 6.*

$$x^2 + 3x + 2 = 3(x + 6)$$
$$x^2 + 3x + 2 = 3x + 18$$
$$x^2 = 16$$
$$x = \pm 4$$
$$x = 4$$

The solution x = -4 must be rejected because it is not in the domain of all the logarithmic functions involved in the original problem. For $\log_3 (x + 2)$, x must be greater than -2. For $\log_3 (x + 1)$, x must be greater than -1, and for $\log_3 (x + 6)$, x must be greater than -6. Any solution must meet all these restrictions.

Example 5 *Solving for a point of intersection*

Graph f(x) = ln (x - 1) and g(x) = ln x - 1. Estimate the point of intersection. Verify the estimate of the point of intersection algebraically.

Solution

The graphs appear to intersect at (1.58, -0.54) (Figure 5).

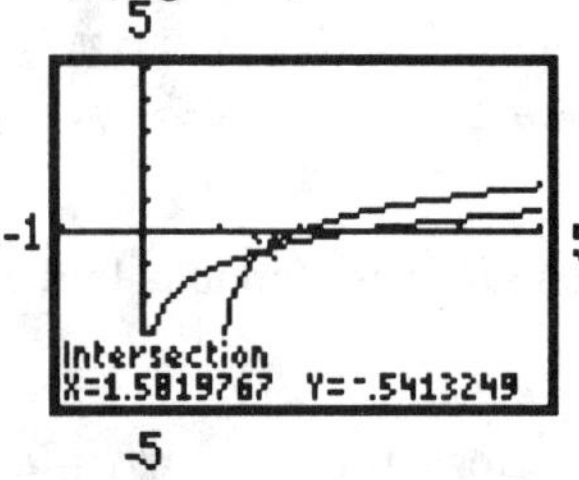

Figure 5

Solving f(x) = ln (x - 1) and g(x) = ln x - 1 simultaneously yields the point of intersection.

$f(x) = g(x) \Rightarrow$

$\ln (x - 1) = \ln x - 1$	
$\ln x - \ln (x - 1) = 1$	*Apply theorem (2).*
$\ln \dfrac{x}{x - 1} = 1 \Leftrightarrow$	*Apply the definition of logarithm.*
$\dfrac{x}{x - 1} = e^1$	*Multiply both sides by (x - 1).*
$x = ex - e$	*Collect the x's on the left side.*
$x - ex = -e$	*Factor the common factor, x.*
$(1 - e)x = -e$	*Divide both sides by (1 - e).*
$x = \dfrac{-e}{1 - e} \approx \dfrac{2.72}{1.72} \approx 1.58$	

and

$$y \approx f(1.58) = \ln (1.58 - 1) \approx -0.54$$

The intersection of the graphs of f(x) = ln(x - 1) and g(x) = ln x - 1 is the point (1.58,-0.54).

In Section 4.2 we solved exponential equations in base 10 and base e using the definition of a logarithm. In Examples 1 to 3 we used the change of base formula to solve exponential equations having a base other than 10 or e. To solve an exponential equation involving different bases such as

$$5^{x+1} = 3^x$$

we take the logarithm of both sides of the equation and apply theorem (3).

$$5^{x+1} = 3^x \qquad \textit{Take the logarithm of both sides.}$$

$$\ln 5^{x+1} = \ln 3^x \qquad \textit{Apply theorem (3).}$$

$$(x + 1) \ln 5 = x \ln 3$$

$$(x + 1)(1.6) \approx x(1.1)$$

$$1.6x + 1.6 \approx 1.1x$$

$$0.5x \approx -1.6$$

$$x \approx -3.2$$

In the following application we have an equation involving exponential expressions in two bases. We take the logarithm of both sides and apply the theorems on logarithms to solve it.

Example 6 The North American Population Problem

In 1998, Mexico's population was 98,553,000 and growing at an annual rate of 1.8%. The U.S. population was 270,312,000 in 1998, with a birthrate of 14 per 1000 and a death rate of 9 per 1000, creating a continuously compounded growth rate of 5 per 1000. If the growth rates continue, how long before Mexico's population catches up with the U.S. population? [Source: U.S. Census Bureau, http://www.census.gov/cgi-bin/ipc/www/idbsum.html]

Thinking: Creating a Plan

Since the annual growth rate is given, we use $N = N_0(1 + r)^t$ to model the population growth of Mexico. We are given that the U.S. growth is continuous, so we use $N = N_0e^{kt}$ to model its population We set the equations equal and solve for time t.

Communicating: Writing the Solution

Mexico: $N_1 = N_0(1 + r)^t$ and $N_0 = 98{,}553$, $r = 0.018 \Rightarrow N = 98{,}553(1.018)^t$

U.S.: $N_2 = N_0e^{kt}$ and $N_0 = 270{,}312$, $r = 0.005 \Rightarrow N = 270{,}312e^{0.005t}$

$N_1 = N_2 \Rightarrow$

$$98{,}553(1.018)^t = 270{,}312e^{0.005t}$$
$$1.018^t = 2.7428e^{0.005t}$$
$$\ln 1.018^t = \ln(2.7428e^{0.005t})$$
$$t \ln 1.018 = \ln 2.7428 + 0.005t \cdot \ln e$$
$$0.01784t \approx 1.0090 + 0.005t$$
$$0.01284t \approx 1.0090$$
$$t \approx 78.58$$

In the year 2077, Mexico's population will surpass that of the United States if present growth rates continue.

Learning: Making Connections

We can also make a change of base in an exponential expression. To change 1.018^t to base e, we write

$$y = 1.018^t$$
$$\ln y = t \cdot \ln 1.018$$
$$\ln y = 0.0178t$$
$$y = e^{0.0178t}$$

In general, $y = b^x \Leftrightarrow y = e^{x \ln b}$.

Explore the Concepts

For Exercises E1 to E4, graph the pairs of functions and use the results to help complete each statement.

E1. $y = \log 10x$ and $y = 1 + \log x$ for $1 \le x \le 1000$ and $-5 \le y \le 5$

Since $\log 10 = 1$, $\log 10x = \log$ ______ $+ \log$ ______.

E2. $y = \log \frac{x}{10}$ and $y = \log x - 1$ for $1 \le x \le 1000$ and $-5 \le y \le 5$

Since $\log \frac{1}{10} = \log 10^{-1} = -1$, $\log \frac{x}{10} = \log$ ______ $+ \log$ ______.

E3. $y = \ln x$, $y = \ln(ex)$, and $y = \ln(\frac{x}{e})$ for $0 < x \le 10$ and $-5 \le y \le 5$

Since $\ln e = 1$ and $\ln(\frac{1}{e}) = \ln e^{-1} = -1$,

$\ln(ex) = \ln$ ______ $+ \ln$ ______ and $\ln(\frac{x}{e}) = \ln$ ______ $- \ln$ ______.

E4. $y = \log x^2$ and $y = 2\log x$; $y = \ln x^3$ and $y = 3 \ln x$

For $x > 0$, $\log_b x^n =$ ______.

Exercises 4.3

Develop the Concepts

For Exercise 1 to 4 use the change of base formula to calculate each logarithm.

1.(a) $\log_2 10$
(b) $\log_2 3$
(c) $\log_2 30$

2.(a) $\log_3 100$
(b) $\log_3 5$
(c) $\log_3 20$

3.(a) $\log_5 3$
(b) $\log_5 9$
(c) $\log_5 27$

4.(a) $\log_{12} 1/144$
(b) $\log_7 1/49$
(c) $\log_4 1/64$

For Exercises 5 to 8, (a) algebraically find the x and y intercepts and (b) write each function in base e and use the graph to check your results from part (a).

5. $y = \log_2 (x + 3)$
6. $y = \log_4 (x - 1)$
7. $y = \log_2 x + 3$
8. $y = \log_4 x - 1$

For Exercises 9 to 24 solve each equation for x. Check the solution to make sure that it is in the domain of the functions involved.

9. (a) $\ln x = 1$
(b) $\ln x = 0$
(c) $\ln x = -1$

10.(a) $\log_2 x = 3$
(b) $\log_2 (-x) = 3$
(c) $\log_2 (x + 1) = 3$

11.(a) $2^x = 32$
(b) $e^x = 32$
(c) $3^x = 32$
(d) $10^x = 32$

12.(a) $\log x = -1$
(b) $\log (x + 2) = -1$
(c) $\log_2 (x^2 - x) = 1$
(d) $\log_2 (x^2 + 8x) = 2$

13. $\log_3 x + \log_3 (x - 2) = 1$

14. $\log_2 x + \log_2 (x + 3) = 2$

15. $\log(x + 2) = -1 - \log x$

16. $\log_2 x - \log_2 (x - 1) = 1$

17. $\log_2 2x^2 - \log_2 (2x + 3) = 1$

18. $\log_3 9 - \log_3 (x - 2) = 1$

*19. $\ln x = \ln (x - 6) - \ln (x - 4)$

20. $1 + 2 \log_2 x = \log_2 (5x - 3)$

21. $\log_2 3x - \log_2 (x + 3) = \log_2 (x - 1) + 1$

22. $\log_2 \sqrt{x + 6} - \log_2 (x - 1) = 3/2$

23. $\frac{1}{2}\ln (x + 5) = \ln (x - 1)$

24. $\log (\ln x) = 0$

25. Solve the exponential equation $10^x = 0.001$ by the first step described.

(a) Rewrite both sides of the equation as powers of 10.

(b) Use the definition of a logarithm to rewrite the equation.

(c) Take the common logarithm of both sides and apply theorem (3).

(d) Take the natural logarithm of both sides and apply theorem (3).

26. Solve the exponential equation $e^x = \frac{1}{\sqrt{x}}$ by the first step described.

(a) Rewrite both sides of the equation as powers of e.

(b) Use the definition of a logarithm to rewrite the equation.

(c) Take the common logarithm of both sides and apply theorem (3).

(d) Take the natural logarithm of both sides and apply theorem (3).

For Exercises 27 to 34 graph each pair of functions on one coordinate system and label the point(s) of intersection.

27. $f(x) = \log(x - 3)$, $g(x) = 1 - \log x$

28. $f(x) = \log_2 x - 4$, $g(x) = \log_2(2x - 31)$

29. $f(x) = \log_3(\frac{1}{x} + 1)$, $g(x) = 2 - \log_3(x + 4)$

30. $f(x) = \log_2(x - 4)$, $g(x) = 3 - \log_2(x + 3)$

*31. $f(x) = \ln^2 x$, $g(x) = \ln x^2$

32. $f(x) = \ln^2 x$, $g(x) = \ln x$

33. $f(x) = \ln^2 x$, $g(x) = 2 - \ln x$

34. $f(x) = \ln^2 x$, $g(x) = \ln^3 x$

Apply the Concepts

35. The Millionaire Problem How long would it take to become a millionaire if \$10,000 is invested at 7% compounded quarterly?

36. The Paper Cutting Problem If a paper molecule is 10^{-9} cm in diameter, how many times could you cut a 20- by 30- cm piece of paper in half without destroying the paper molecule?

37. The Rule of 72 Problem The rule of 72 gives the approximate doubling time for 18% annually compounded interest as $x = \frac{72}{100r} = \frac{72}{18} = 4$ yr. Use logarithms and the formula $A = P(1 + r)^t$ to check this result.

38. The Rule of 69 Problem The rule of 69 gives $d = \frac{69}{100r} = \frac{69}{10} \approx 7$ years as the doubling time for 10% continuously compounded interest. Check this time using $A = Pe^{rt}$.

39. The Growing Population Problem According to the U.S. National Academy of Science, 10 billion is nearly the limiting number of people that Earth can support. In 1990, the world population was 5.333 billion and increasing at 1.7%. How long does Earth has left? [Source: *The Universal Almanac*]

40. The Growing Population Problem Revisited Exponential growth is impossible to maintain for long. The figures in Exercise 39 were given in 1990. Today the growth rate is estimated at less than 1.5%, which changes the situation. With a growth rate of 1.5%, how many years from 1990 does Earth have before its population reaches 10 billion?

41. The Rainbow Kite and Balloon Company The formula

$$B(t) = -135{,}500e^{0.005625t} + 177{,}500$$

models the monthly principal balance on an installment loan for two company vans.

(a) Find and interpret the intercepts.

(b) Find the time it will take to reduce the balance by half.

(c) If the monthly payments are $998.44/month, how much interest will you pay over the duration of the loan?

42. Wonderworld Amusement Park You've allocated $15,000 to the advertising department to update their computer equipment and get "on-line." Your accountant has advised you to trade in equipment as soon as it depreciates by half. Find when each depreciation method reaches $7500.

(a) Depreciate the equipment at a constant rate of $1500/year.

(b) Depreciate the equipment at a continuous rate of 10%/year.

43. The E. Coli Problem An Escherichia coli cell divides in two every 20 minutes. Write two exponential models assuming that initially there are 2 million cells present such that (a) $N = N_0e^{kt}$ and (b) $N = N_0\,2^{t/20}$. (c) Compare each model's prediction of the time it would take for the E. coli population to reach 10 billion. [Source: *Biology*, H. Curtis]

44. The Growing School Districts Problem In the city of Yakima there are two rival high schools. One has 3000 students and is doubling in size every 6 years, while the other has 4500 students and is growing at 5.5%. Use the formula $N = N_0e^{kt}$ to determine when their enrollments will be the same.

Extend the Concepts

For Exercises 45 to 48 solve each equation and check the solution.

45. $\log_x x^{4x} = 8$

46. $x^{(\log_x 10x)} = \frac{1}{10}$

47. $\log|x + 1| = -3$

48. $|\log(x - 1)| = 1$

49. Prove: $\log_b MN = \log_b M + \log_b N$

50. Prove: $\log_b M^p = p\log_b M$

4.4 More on Exponential and Logarithmic Functions

Solving Pairs of Exponential and Logarithmic Functions by Graphing
Studying More Mathematics of Finance
Finding the Best-Fitting Exponential and Logarithmic Equations

In this section we focus on solving exponential equations geometrically using graphs of exponential functions. We continue our study of the mathematics of finance by calculating the future value of an annuity account, and finding the monthly payment for a loan. We end this section with curve fitting and finding the best-fitting exponential and logarithmic functions. We use exponential and logarithmic regression programs on a graphing utility to find equations of the form $y = ab^x$ and $y = a + b\ln x$.

Solving Pairs of Exponential and Logarithmic Functions by Graphing

In the first example we compare solving a pair of exponential function by graphing and with algebra. We will see some applications that can only be solved using graphing.

Example 1 *Solving a pair of exponential equations*

(a) Find the point of intersection of $y = 125e^{0.5x}$ and $y = 25e^{-0.25x}$.

(b) Verify your results from part (a) algebraically.

Solution

(a) The x coordinate of the point of intersection of the graphs, $x \approx -2.145917$ is the solution to the equation (Figure 1).

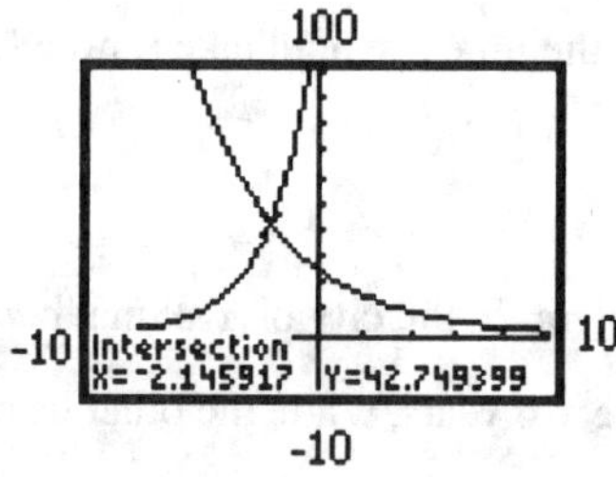

Figure 1

(b) $y = 125e^{0.5x}$ and $y = 25e^{-0.25x} \Rightarrow$

$$125e^{0.5x} = 25e^{-0.25x} \qquad \textit{Divide by 125.}$$

$$e^{0.5x} = 0.2e^{-0.25x} \qquad \textit{Divide by } e^{-0.25x}.$$

$$\frac{e^{0.5x}}{e^{-0.25x}} = 0.2 \qquad \textit{Subtract exponents.}$$

$$e^{0.75x} = 0.2 \qquad \textit{Apply the definition of a logarithm.}$$

$$0.75x = \ln 0.2$$

$$x \approx -2.1459$$

$y = 125e^{0.5x}$ and $x = -2.1459 \Rightarrow y = 125e^{0.5(-2.1459)} \approx 42.749.$

Example 2 The Two-Populations Problem

In 1990, the population of the state of Washington was 4.867 million and increasing at a rate of 8%. The state of New York's population was 18.241 million people and growing at 1.1%. If the growth rates continue, when will the two states have the same populations?

[Source: *The American Almanac Statistical Abstract of the United States*]

Thinking: Creating a Plan

For population growth we use the exponential growth model, $N(t) = N_0e^{kt}$. Since we are given N_0 and k for each state, we can write the equations modeling their population growths. We will graph these functions, $N_w(t) = 4.867e^{0.08t}$ and $N_n(t) = 18.241e^{0.011t}$, and estimate the point of intersection of their graphs where the populations are equal.

Communicating: Writing the Solution

From Figure 2, the state of Washington will catch New York state in a little over 19 years from 1990, in the year 2009, when each population will reach 22.5 million.

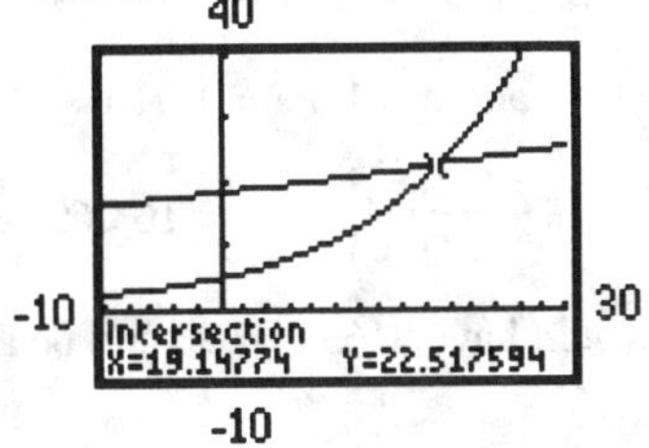

Figure 2

Learning: Making Connections

In 1995, the growth rate of the state of Washington dropped to 2%. If the population of Washington state was 7.26 million, then

$N_w(t) = N_n(t) \Rightarrow$

$$7.26e^{0.02t} = 18.241e^{0.011t}$$

$$t \approx 102.4$$

It would take over a century from 1995 for the populations to be equal.

Studying More Mathematics of Finance

Our knowledge of exponents and logarithms allows us to extend our study of the mathematics of finance for annuities and loans using two new finance formulas.

Suppose you want to set up a college account for a newborn child. Assume that the child will need the money in 18 years. You found a mutual fund in which you could earn 12% interest per year compounded monthly, and you are prepared to deposit $200 per month into the account.

Such an investment is called a *simple annuity*, an account where regular payments are made n times per year and interest is compounded n times per year. In an annuity, each deposit earns interest for a different time period. In this case, the first deposit of $200 earns interest for 216 months. The final deposit earns interest for one month only. The future value of this account is

$$S = 200(1 + \frac{0.12}{12})^{216} + 200(1 + \frac{0.12}{12})^{215} + \cdots + 200(1 + \frac{0.12}{12})^{1}$$

The formula for such a sum leads to the formula for any simple annuity,

$$S = R(\frac{a^n - 1}{i}), \quad a = 1 + i$$

where S is the future value of the money after n regular deposits of R dollars each at an interest rate i per saving period. For this college fund, the annual interest of $r = 12\%$ is compounded monthly, making $i = \frac{r}{12} = \frac{0.12}{12} = 0.01$. At this rate of saving, let's find out what college the child will be able to attend.

$S = R(\frac{a^n - 1}{i})$, $R = 200$, $n = 216$, $i = 0.01$, and $a = 1 + i = 1.01 \Rightarrow$

$$S = 200(\frac{1.01^{216} - 1}{0.01}) = 163{,}697.90$$

With the large return of 12%, the child would have $163,697.90 and be able to attend practically any college.

Example 3 *Using the formula for an annuity*

Find the number of deposits it would take to have a future value of $1,000,000 if you deposit $200 per month into the 12% per year mutual fund described above.

Solution

Since the annual interest is 12%, $i = \frac{0.12}{12} = 0.01$.

$S = R(\frac{a^n - 1}{i})$, $S = 1{,}000{,}000$, $R = 200$, $i = 0.01$, and $a = 1 + i = 1.01 \Rightarrow$

$$1{,}000{,}000 = 200(\frac{1.01^n - 1}{0.01})$$ *Divide both sides by 200.*

$$5000 = (\frac{1.01^n - 1}{0.01})$$ *Multiply both sides by 0.01.*

$$50 = 1.01^n - 1$$ *Add 1 to both sides.*

$$51 = 1.01^n$$ *Write in logarithmic form.*

$$n = \log_{1.01} 51$$ *Apply the change of base formula.*

$$n = \frac{\log 51}{\log 1.01}$$

$$n \approx 395.14$$

It would take about 395 month, about 33 years, for the annuity to grow to \$1 million. This sounds like a good retirement annuity for someone entering the workforce.

A loan is similar to an annuity, with the future value being the total amount P of the loan where the regular deposits are monthly payments M. The formula for finding the monthly payment of a loan taken for n months is

$$M = \frac{iP}{1 - a^{-n}}, \quad a = 1 + i$$

with interest per month $i = \frac{r}{12}$.

Example 4 *Finding the number of loan payments*

Find the number of payments necessary to pay off a new car loan of \$23,000 if you can pay \$400 per month and found a bank offering a 4.9% annual interest rate.

Solution

Since the annual interest rate is 4.9%, $i = \frac{0.049}{12} = 0.004$.

$M = \frac{iP}{1 - a^{-n}}$, M = 400, P = 23,000, i = 0.004, and a = 1 + i = 1.004 ⇒

$$400 = \frac{0.004(23000)}{1 - 1.004^{-n}}$$

$$400 = \frac{92}{1 - 1.004^{-n}} \qquad \textit{Multiply both sides by } 1 - 1.004^{-n}.$$

$$400(1 - 1.004^{-n}) = 92 \qquad \textit{Divide both sides by 400.}$$

$$1 - 1.004^{-n} = 0.23$$

$$0.77 = 1.004^{-n} \qquad \textit{Write in logarithmic form.}$$

$$\log_{1.004} 0.77 = -n \qquad \textit{Apply the change of base formula.}$$

$$-n = \frac{\ln 0.77}{\ln 1.004}$$

$$n \approx 65.47$$

It will take about $5\frac{1}{2}$ years to pay off the car.

Formulas: Annuities and Loans

Annuities: The future value S of a simple annuity after n regular deposits of R dollars each at an interest rate per saving period i is

$$S = R\left(\frac{a^n - 1}{i}\right), \quad a = 1 + i$$

Loan payments: The monthly payments M of a loan of P dollars taken for n months at an interest rate per borrowed period i is

$$M = \frac{iP}{1 - a^{-n}}, \quad a = 1 + i$$

Finding the Best-Fitting Exponential and Logarithmic Equations

Most graphing utilities can find the best-fitting exponential and logarithmic equations of the form

$$y = ab^x \text{ and } y = a + b \ln x$$

for a set of data using **exponential regression** (ExpReg) and **logarithmic regression** (LnReg) programs, respectively. Each of these regression programs has a restriction on the data. Since the range of the exponential function $y = b^x$ is $y > 0$ and the domain of the logarithmic function $y = \ln x$ is $x > 0$, the exponential regression program will run only when all the y values are greater than zero and the logarithmic regression program will run only when all the x values are greater than zero.

In the next example we will find the best-fitting exponential equation.

Example 5 The Franchise Problem

The following table indicates the number of fast-food restaurants in a franchise for various years.

year	1954	1957	1959	1968	1978	1980	1985	1990	1992	1997
restaurants	1	14	100	1000	5000	7259	9911	11,803	13,093	21,000

(a) Find the best-fitting exponential function that models the number of restaurants as a function of the year.

(b) Draw the scatter plot and graph of the equation from part (a).

(c) Assuming that the exponential growth continues, use your equation to estimate the number of restaurants in the year 2025.

Thinking: Creating a Plan

To avoid calculator round-off errors from figures that are too large, we will enter the year corresponding to 1954 as year 1, 1957 as year 4, and so on. After we have the best-fitting exponential equation for the data, we will extrapolate to find the number of restaurants in the year 2025 by letting $x = 72$.

year	1954	1957	1959	1968	1978	1980	1985	1990	1992	1997
x, (year)	1	4	6	15	25	27	32	37	39	44
y, (no. restaurants)	1	14	100	1000	5000	7259	9911	11,803	13,093	21,000

Communicating: Writing the Solution

(a) Enter the data and access ExpReg (Figure 3).

L1	L2	L3
1	1	------
4	14	
6	100	
15	1000	
25	5000	
27	7259	
32	9911	

L1(1)=1

Figure 3

The best-fitting exponential equation is $y = 9.99(1.22)^x$.

(b) The scatter plot and best-fitting exponential curve are shown in Figure 4.

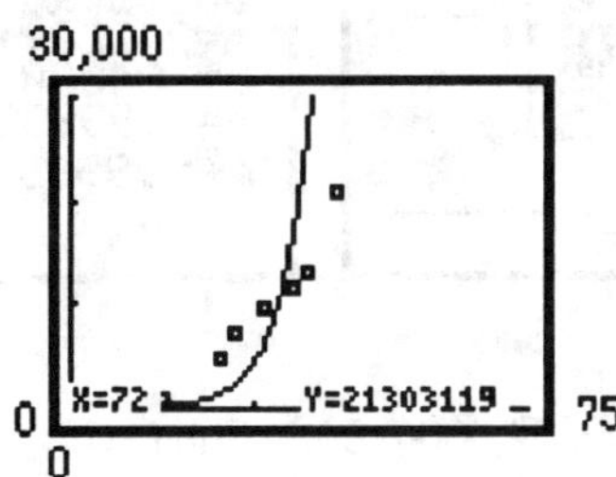

Figure 4

Learning: Making Connections

Since the value of r^2 is about 0.86, 86% of the variation in y is explained by this equation. This is a good fit. We can change any exponential equation of the form $y = ab^x$ to an equation in base e,

$$y = ab^x \Leftrightarrow y = ae^{x \cdot \ln b}$$

Therefore, $y = 9.99(1.22)^x \Leftrightarrow y = 9.00e^{1.199x}$. Changing to base e reveals that the growth rate is almost 120%.

Example 6 The Depreciating Truck Problem Revisited

The following data for a depreciating truck is in 4-year intervals.

t (yr)	0	4	8	12
N ($)	24,000	11,949	5949	2962

(a) Find the best-fitting exponential equation that models the truck's value as a function of its age of the form $y = ab^x$.

(b) Express the equation in part (a) as an exponential equation in base e.

Thinking: Creating a Plan

We use exponential regression (ExpReg) to find the best-fitting equation. Then for part (b) we use $y = ab^x \Leftrightarrow y = ae^{x \cdot \ln b}$ to express the equation in base e.

Communicating: Writing the Solution

L1	L2	L3
1E-4	24000	------
4	11949	
8	5949	
12	2962	
------	------	

L1(1)=1E-4

```
ExpReg
 y=a*b^x
 a=24000.10712
 b=.840001919
 r=-.9999999997
 r²=.9999999994
```

Figure 5

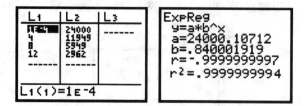

The best-fitting exponential equation is $y = 24{,}000(0.84)^x$ (Figure 6).

$y = ae^{x \cdot \ln b}$ and $a = 24{,}000$, $b = 0.84 \Rightarrow$

$$y = 24{,}000e^{(\ln 0.84)x}$$

$$y = 24{,}000\, e^{-0.174x}$$

See Figure 6.

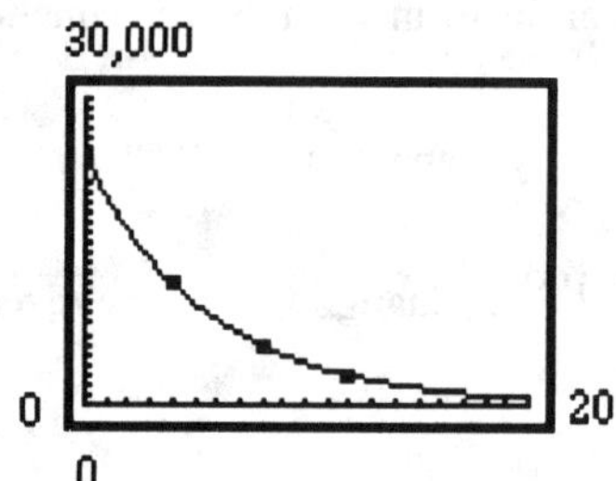

Figure 6

Learning: Making Connections

Notice that the input data is in equal increments. Let's calculate the first differences.

0	4	8	12
24,000	11,949	5949	2962

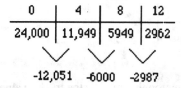

-12,051 -6000 -2987

Now let's calculate the percent change $\frac{\Delta N}{N_0}$ in the first differences over each interval.

$$\frac{-12051}{24000} \approx -0.502,\ \frac{-6000}{11949} \approx -0.502, \text{ and } \frac{-2987}{5949} \approx -0.502$$

The percent decrease is the same, -50%. Data that have a constant percent change in the first differences represent exponential growth when the differences are positive and exponential decay when they are negative. Evaluating the formula $N = N_0(1 - r)^t$ for $t = 0$ and $t = 4$ leads to the formula for exponential depreciation given the initial values and one value some time later.

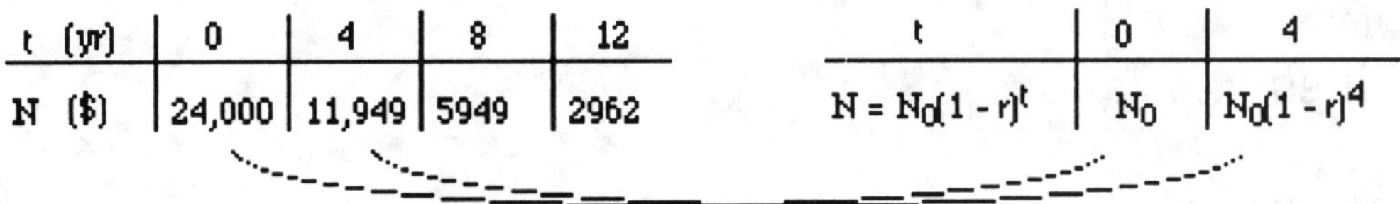

t (yr)	0	4	8	12
N ($)	24,000	11,949	5949	2962

t	0	4
$N = N_0(1 - r)^t$	N_0	$N_0(1 - r)^4$

Equating corresponding entries yields $N_0 = 24{,}000$ and $N_0(1 - r)^4 = 11{,}949$.

$N_0(1 - r)^4 = 11{,}949$ and $N_0 = 24{,}000 \Rightarrow$

$$24{,}000(1 - r)^4 = 11{,}949$$

$$(1 - r)^4 \approx 0.497875$$

$$1 - r \approx \sqrt[4]{0.497875}$$

$$1 - r \approx 0.84$$

$$r \approx 0.16$$

Therefore, the exponential equation using finite differences is $N = 24{,}000(0.84)^t$.

Exercises 4.4

Apply the Concepts

1. The Equality Problem Two people plan to get married (to each other) as soon as their savings accounts have the same amount of money. He now has $25,000 invested at 5.5% continuously compounded interest. She only has $13,000, but her investments are earning 11% interest compounded annually. Express the amount of money A (future value) in each account as a function of time in years, and graph both functions to estimate how long they will wait to be married.

2. The Tune Town Problem The population of "Toons" in Tune Town was 24,000 in 1960 and growing at an average annual rate of 1.7% The population of humans was 124,000 in 1960 and decreasing at an average annual rate of 2.2%. Use $N = N_0(1 + r)^t$ and $N = N_0(1 - r)^t$ to estimate how many years from 1960 the two populations will be the same.

3. The Matching Population Problem In 1990 the population of Iraq was approximately 18.919 million and growing at about 34 per 1000. The population of Iran was approximately 56.585 million and growing at 33 per 1000. Use the formula $N = N_0e^{kt}$ to determine when their combined populations will equal that of the United States whose population in 1990 was 249.2325 million and growing at 6 per 1000?
[Source: United States Population Fund]

4 The North America Population Problem The 1998 population of Canada was 30,675,000, with an annual growth rate of 1.1%. The 1998 population of Mexico was 98,553,000, with a birthrate of 25 per 1000 and a death rate of 5 per 1000. The 1998 population of US was 270,312,000 in 1998 with a birthrate of 14 per 1000 and a death rate of 9 per 1000. When will the combined population of Canada and Mexico equal the population of the United States?
[Source: US Census Bureau, http://www.census.gov/cgi-bin/ipc/www/idbsum.html]

For Exercises 5 to 8 use the monthly payment formula $M = \frac{iP}{1 - a^{-n}}$, where $a = 1 + i$ describes the monthly payment M of a loan of P dollars for n months with interest per payment period i.

5. The Mortgage Problem (a) Compare the monthly payments for buying a $100,000 home with annual percentage rate (APR) of $8\frac{1}{2}$% over 15 years and over 30 years.

(b) Compare the total paid for the house in each case by multiplying the monthly payments and the number of payments.

6. The Total Interest Problem Compare the total interest paid on a $100,000 home loan over 20 years if the annual percentage rate in 1998 was 8% and the annual percentage rate in 1994 was 9 1/2%.

7. The Pay off Problem If you pay off your 20-year mortgage in 10 years, do you pay half as much interest? (Choose your own annual percentage rate and loan amount.) Is this fair? Explain why or why not.

8. The Car Loan Problem Find the number of payments n required to pay back a car loan of $12,500 for the annual percentage rate 7.5% if (a) M = $350 and (b) M = $400.

For Exercises 9 to 11 use the monthly payment formula $S = R(\frac{a^n - 1}{i})$ where a = 1 + i describes the future value S of an annuity with n regular deposits of R dollars each an interest rate i per saving period.

9. The Rainbow Kite and Balloon Company Find the future value of an employee's simple annuity if $250 per month (half from the company and half from the employee) is deposited into an account earning 10% monthly compounded interest after 10 years and after 20 years.

10. Wonderworld Amusement Park Find the number of months it takes the annuity in Exercise 9 to be worth $200,000 and $400,000.

11. An Annuity Problem Find the number of payments n required to create an annuity worth $1,000,000 if you deposit $250/month and earn 12% per year interest.

12. The Monthly Balance Problem The monthly balance formula $P_k = P(\frac{a^n - a^{k-1}}{a^n - 1})$ gives the balance at the beginning of the kth month for a loan of P dollars for n months with $a = 1 + i = 1 + \frac{r}{12}$. Find the number of months it takes to reduce a $100,000 loan scheduled for 20 years to reduce by half if the APR = 8%. (This is called the half-life of the loan.)

*13. The Population Projections Problem The following table shows the projected world population in billions people for various years.

year	1990	2000	2010	2020
population projection	5.333	6.291	7.255	8.281

(a) Check the projection for the year 2000 from the 1990 population of 5.333 billion using the 1990 growth rate of 1.7%.

(b) Find the growth rate k necessary to hit the 2010 projection of 7.255 billion from the 2000 projection of 6.291 billion.

(c) Use your rate from part (b) to check the projection for the year 2020.

(d) Use exponential regression to find the best-fitting equation of the form $y = ab^x$.

(e) Change the equation in part (d) to be in base e, and find the growth rate. [Source: *The Universal Almanac*]

14. The Threatened Bald Eagle Problem The following data give the number of eagle pairs for various years. On July 12, 1995, the bald eagle was removed from the endangered species list and listed as a threatened species.

year	1963	1984	1994
pairs of eagles	333	1800	4500

(a) Find the best-fitting exponential and logarithmic equations to model the growing population.

(b) Choose the equation from part (a) with the highest coefficient of determination to estimate the number of pairs of eagles in the year 2000.

[Source: U.S. Fish and Wildlife Service, Division of Endangered Species]

*15. The Wind Chill Problem The following data give the wind speed in miles per hour and the wind chill factor when the temperature is 30°F.

(a) Find the best-fitting logarithmic equation.

(b) Fill in the missing entries.

wind speed (mi/hr)	1	5	10	15	20	25	?	30	35	40	45
wind chill (°F)	?	27	16	9	4	1	0	-2	-4	-5	-6

16. The Good Night's Sleep Problem The following data represent the percentage of nonsense syllables recalled t hours after studying them.

(a) Decide the nature of the graph.

(b) Find the best-fitting logarithmic equation that models the data.

(c) Use your graphs and their equations to test the hypothesis that a good night's rest is essential.

[Source: Exploring Psychology by D. Myers]

time asleep or awake after learning, hr	0	1	2	3	4	5	6	7	8
% recalled (asleep)	100	75	58	57	57	58	59	59.5	59.6
% recalled (awake)	100	45	25	22	19	18	16	15	8

Extend the Concepts

The graphs of the functions in Exercises 17 to 20 depict the *damping* effect of exponential functions. State the behavior of the graph as $x \to \infty$.

17. $y = xe^{-x}$

18. $y = 100xe^{-x}$

19. $y = x^2e^{-x}$

20. $y = xe^{-x}$

For Exercises 21 to 24 graph each function and describe the symmetry.

21. $y = \ln |x|$

22. $y = \ln x^2$

23. $y = e^{|x|}$

24. $y = e^{-x^2}$

Applications involving wave theory in the field of oceanography utilize *hyperbolic* functions $y = \sinh x$ and $y = \cosh x$. For Exercises 25 and 26 let $y_1 = \frac{1}{2}e^x$ and $y_2 = \frac{1}{2}e^{-x}$, and state the relationship between the graphs described by the equations in part (a) and (b).

25.(a) $f(x) = y_1 - y_2$

(b) $y = \sinh x$

26.(a) $f(x) = y_1 + y_2$

(b) $y = \cosh x$

27. Show that $(\sinh x)^2 - (\cosh x)^2 = -1$.

28. Show that $1 - (\frac{\sinh x}{\cosh x})^2 = \frac{1}{(\cosh x)^2}$.

4.5 More Applications Including Logistic Equations

Applying Newton's Law of Cooling
Depreciating Equipment with Residual Value
Defining a Logistic Function
Defining a Surge Function

This section is devoted to the study of some important forms of exponential and logarithmic functions. We first study Newton's law of cooling, which models a cooling object such as a cup of coffee. Then in Example 2 we examine the depreciating value of a new car to some residual value. Both of these examples decrease to a value other than zero. Then we introduce the logistic function, a function whose graph appears exponential, but then changes and approaches a horizontal asymptote. We use a logistic function in Example 4 to model a spreading rumor. The number of people who hear the rumor grows exponentially at first, then slows as fewer people remain who have not heard the rumor, and then finally approaches a limiting number, the number of people who could hear the rumor.

We also introduce the surge function to model medicine intake and removal from our bodies. Initially, the absorption is rapid after a drug is administered. The concentration in the bloodstream soon peaks, and then falls off as the drug is eliminated.

Applying Newton's Law of Cooling

The rate at which an object cools is proportional to the difference between the object's temperature and the temperature of its surrounding environment, its ambient temperature. For example, a hot cup of coffee in a cold room cools rapidly approaching the ambient temperature. This situation leads to Newton's law of cooling (or heating), which expresses the temperature T of an object as a function of time t such that

$$T = De^{-kt} + T_a$$

Let's investigate this formula by considering the changing temperature of a cup of coffee that is 180°F cooling in a 80°F room.

$$\text{As } t \to \infty,\ De^{-kt} \to 0 \text{ and } T \to T_a$$

This says that as time increases, the expression De^{-kt} approaches zero and the temperature of the coffee approaches room temperature. Therefore, T_a represents the temperature of the surrounding environment, the ambient temperature of 80°F.

$$T_a = 80$$

Now we can find and interpret D knowing that at time t = 0, the coffee was 180°F.

$T = De^{-kt} + 80$, $t = 0$, and $T = 180 \Rightarrow$

$$180 = De^{-k(0)} + 80$$

$$180 = D(1) + 80$$

$$D = 180 - 80$$

$$D = 100$$

We see that D represents the difference in temperature between the object and the surrounding environment.

To find the value of k, we need more information. Suppose that the coffee cooled to 130°F after 1 min. $T = 100e^{-kt} + 80$, $t = 1$, and $T = 130 \Rightarrow$

$$130 = 100e^{-k(1)} + 80 \qquad \textit{Subtract 80 from both sides.}$$
$$50 = 100e^{-k} \qquad \textit{Divide both sides by 50.}$$
$$e^{-k} = .5 \Leftrightarrow \qquad \textit{Apply the definition of a logarithm.}$$
$$\ln 0.5 = -k$$
$$k = -\ln(0.5)$$
$$k \approx 0.693$$

The proportionality constant is $k \approx 0.693$. The function modeling this cooling cup of coffee is

$$T = 100e^{-0.693t} + 80$$

The graph in Figure 1 shows a horizontal asymptote at $T = 80$, room temperature.

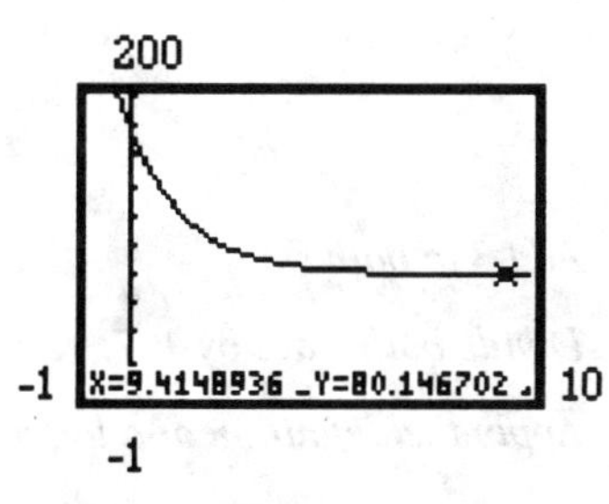

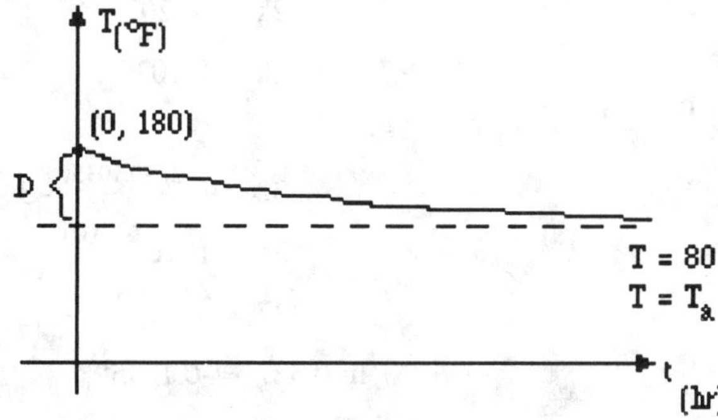

Figure 1

Formula: Newton's Law of Cooling

The formula

$$T = De^{-kt} + T_a$$

gives the temperature of an object T as a function of time t, where T_a is the ambient temperature, D is the difference of the object's original temperature and the ambient temperature, $T_0 - T_a$, and k is the proportionality constant.

Example 1 The Ice Cold Hot Tub Problem

An outdoor hot tub filled with water heated to 104°F was accidentally turned off at midnight when the outside temperature was -3°F. If the temperature of the hot tub has dropped to 84°F by 2:00 a.m., when will it freeze?

Thinking: Creating a Plan

We will use Newton's law of cooling, $T = De^{-kt} + T_a$, and let midnight represent $t = 0$. Since the ambient temperature is -3°F, $T_a = -3$ and $D = 104 - (-3) = 107$. Then we find k by substituting $t = 2$ and $T = 84$. Finally, we will find the time when $T = 32$ (freezing temperature is 32°F).

Communicating: Writing the Solution

$T = De^{-kt} + T_a$, $T_a = -3$, and $D = 107 \Rightarrow$

$$T = 107e^{-kt} - 3$$ *Let t = 2 and T(2) = 84.*

$$84 = 107e^{-k(2)} - 3$$ *Add 3 to both sides.*

$$87 = 107e^{-2k}$$ *Divide both sides by 107.*

$$e^{-2k} \approx 0.813 \Leftrightarrow$$ *Apply the definition of a logarithm.*

$$\ln 0.813 \approx -2k$$

$$-2k \approx -0.207$$

$$k \approx 0.1035$$

$T = 107e^{-0.1035t} - 3$ and $T = 32 \Rightarrow$

$$32 = 107e^{-0.1035t} - 3$$ *Add 3 to both sides.*

$$35 = 107e^{-0.1035t}$$ *Divide both sides by 107.*

$$0.327 \approx e^{-0.1035t} \Leftrightarrow$$ *Apply the definition of a logarithm.*

$$\ln 0.327 \approx -0.1035t$$

$$-1.12 \approx -0.1035t$$

$$t \approx 10.8$$

Since the water will freeze in about 10 hr and 48 min, the water in the hot tub will be frozen at 10:48 a.m.

Learning: Making Comments

The saving point for the hot tub owner is that the ambient temperature may not stay constant until 10:48 a.m. By then, if the temperature rises, this model would no longer apply.

Why would D be negative if a cool object were in a warm environment?

The graph would be increasing approaching the horizontal asymptote at $T = T_a$, and $D = T_0 - T_a$ would be negative (Figure 2).

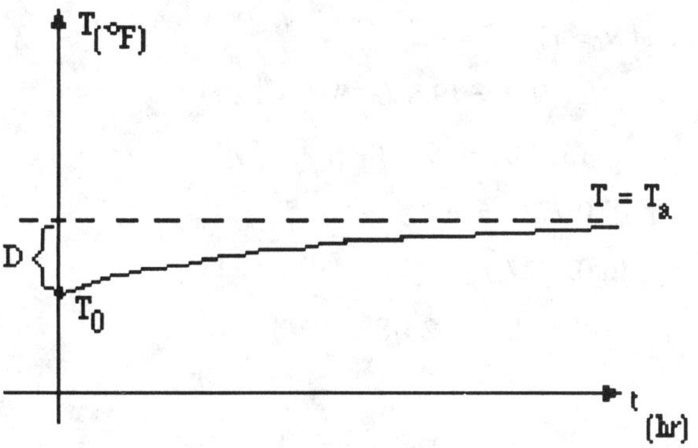

Figure 2

Depreciating Equipment with Residual Value

When we buy a new car it begins to depreciate, but not to a value of zero. All cars have some base value below which its value will not drop, called its *residual value*. The rate at which the car depreciates is proportional to the difference between its initial value and its residual. An expensive car with a low residual depreciates rapidly.

Formula: Depreciated Equipment

The formula

$$V = De^{-kt} + V_r$$

gives the value of an item V as a function of time t, where V_r is the ambient temperature, D is the difference of the item's original value and its residual value, $V_0 - V_r$, and k is the proportionality constant.

Example 2 The New Car Problem

Consider a new car costing $22,000 that is worth $17,500 one year later. When will it be worth half its original value if its residual value is only $2200?

Thinking: Creating a Plan

To find the function modeling the depreciating value of the car, we will use the depreciating formula $V = De^{-kt} + V_r$ with $V_r = 2200$ and $D = V_0 - V_r = 22{,}000 - 2200 = 19{,}800$. We use the value after one year to find k. Once we have the function, we can find when the car is worth half its original value, $11,000.

Communicating: Writing the Solution

$V = De^{-kt} + V_r$, $V_r = 2200$, and $D = 19{,}800 \Rightarrow$

$$V = 19{,}800e^{-kt} + 2200$$

$V = 19{,}800e^{-kt} + 2200$, $t = 1$ and $V = 17{,}500 \Rightarrow$

$$17{,}500 = 19{,}800e^{-k(1)} + 2200$$

$$15{,}300 = 19{,}800e^{-k}$$

$$0.7727 = e^{-k}$$

$$\ln 0.7727 = -k$$

$$k \approx 0.2578$$

Therefore, the function is

$$V = 19{,}800e^{-0.2578t} + 2200$$

Half the original value of \$22,000 is \$11,000.

$V = 19{,}800e^{-0.2578t} + 2200$and $V= 11{,}000 \Rightarrow$

$$11{,}000 = 19{,}800e^{-0.2578t} + 2200$$

$$8800 = 19{,}800e^{-0.2578t}$$

$$0.4400 = e^{-0.2578t}$$

$$\ln 0.4400 \approx -0.2578t$$

$$0.2578t \approx 0.8109$$

$$t \approx 3.14$$

In about 3 years the car will depreciate by half.

Learning: Making Connections

The graph of $V = 19{,}800e^{-0.2578t} + 2200$ in Figure 3 shows the initial value and the value approaching the residual value asymptotically.

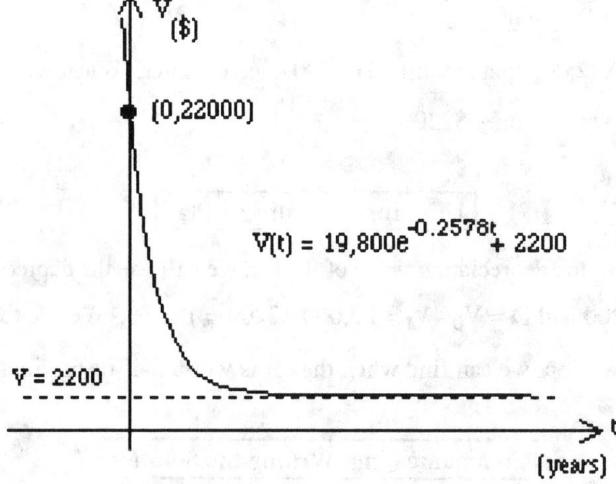

Figure 3

Defining a Logistic Function

A **logistic function**, a function of the form

$$N(t) = \frac{c}{1 + ae^{-bt}}$$

expresses the number present N as a function of time t to model a wide variety of growth problems that "logically" approach some limiting value. The graph of a logistic function increases or decreases like an exponential function initially, but then approaches a horizontal asymptote (Figure 4).

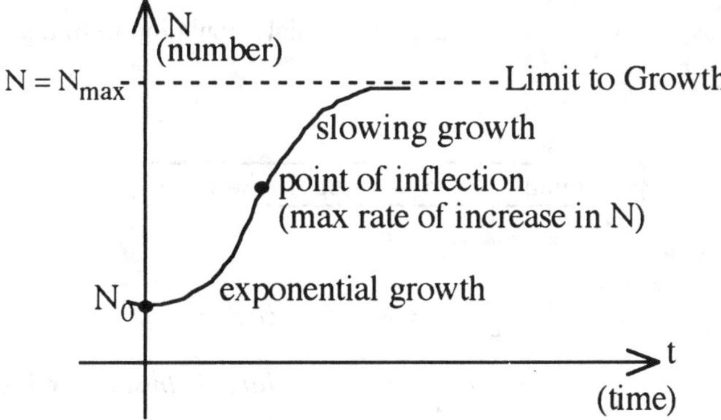

Figure 4

The logistic curve was named by P. F. Verhulst, a Belgian demographer who applied it in 1838 to study population growth in the United States. The derivation of the logistic function is based on the assumption that if the birthrate remains constant (increasing the population every year), the death rate will increase as N increases. If the death rate increases as the number of people increase, the population will approach a limiting value where the death rate and the birthrate are equal. At this equilibrium point, we have zero population Growth (Figure 5).

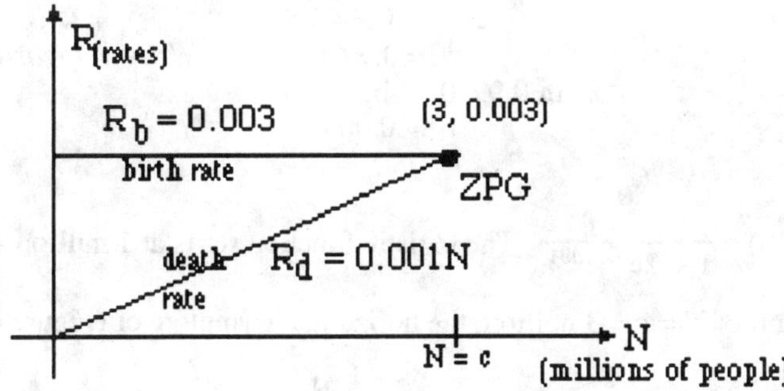

Figure 5

Why does the domain of this application end at the ZPG point in Figure 5?

Once the birthrate and death rate are equal, there will be no further increase in population. That is why this point represents the limit to growth.

In the next example we discover that N = c is the horizontal asymptote, the limit to growth.

Example 3 The ZPG Problem

Use $N(t) = \frac{c}{1 + ae^{-bt}}$ to find the logistic function that models a population of 1 million people initially with a birthrate of 2 per 1000 and a limit to growth of 3 million people.

Thinking: Creating a Plan

For $N(t) = \frac{c}{1 + ae^{-bt}}$, as t approaches ∞, $e^{-bt} = \frac{1}{e^{bt}}$ approaches zero, and N(t) approaches $\frac{c}{1 + a(0)} = c$. Since the limit to growth is given as 3 million, c = 3. We use the initial population to find a. Finally, we use N(1) = 1 + 0.002(1) = 1.002 million to find b.

Communicating: Writing the Solution

$N(t) = \frac{c}{1 + ae^{-bt}}$ and N(0) = 1 and c = 3 ⇒

$$\frac{3}{1 + ae^0} = 1 \qquad \textit{Since } e^0 = 1,\ ae^0 = a.$$

$$\frac{3}{1 + a} = 1 \qquad \textit{Multiply both sides by } 1 + a.$$

$$1 + a = 3$$

$$a = 2$$

$N(t) = \frac{3}{1 + ae^{-bt}}$ and a = 2 ⇒ $N(t) = \frac{3}{1 + 2e^{-bt}}$

N(1) = 1.002 and t = 1 ⇒

$$1.002 = \frac{3}{1 + 2e^{-b}} \qquad \textit{Multiply both sides by } 1 + 2e^{b}.$$

$$1.002(1 + 2e^{-b}) = 3 \qquad \textit{Divide both sides by } 1.002.$$

$$1 + 2e^{-b} = \frac{3}{1.002}$$

$$1 + 2e^{-b} = 2.9940$$

$$2e^{-b} = 1.9940$$

$$e^{-b} = 0.9970 \qquad \textit{Write in logarithmic form.}$$

$$\ln 0.9970 = -b$$

$$b \approx 0.003$$

The logistic function is $N(t) = \frac{3}{1 + 2e^{-0.003t}}$. The logistic function starts at 1 million and increases toward its asymptotic population limit of N = c = 3 million, the horizontal asymptote of (Figure 6).

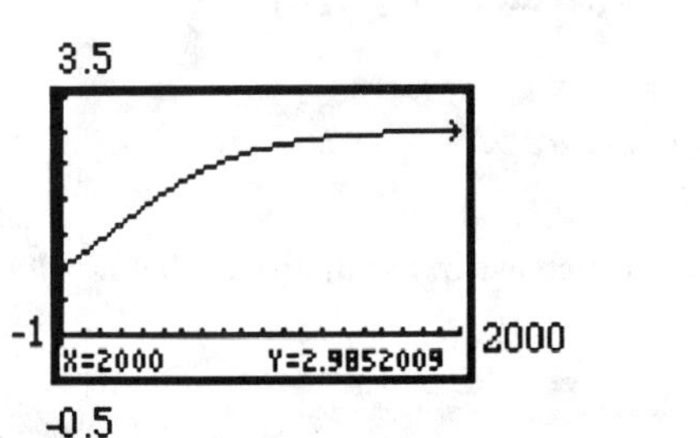

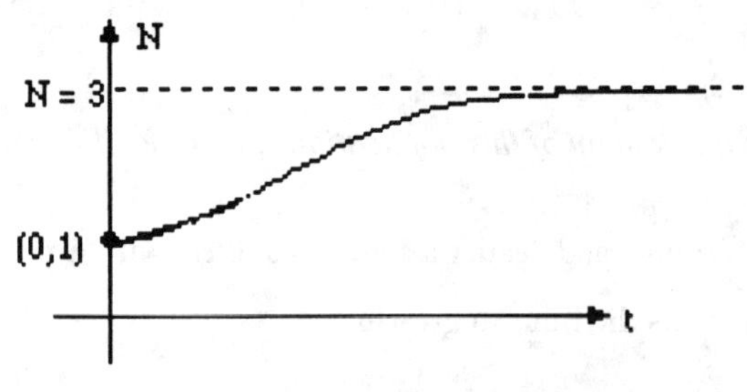

Figure 6

Learning: Making Connections

Compare Figures 5 and 6. In Figure 5, growth rates are functions of the number of people. In Figure 6, the number of people is a function of time.

Can you create a new function that expresses the death rate as a function of time using composition of the death rate function $R_d = 0.001N$ *from Figure 5 and the logistic function* $N(t) = \dfrac{3}{1 + 2e^{-0.003t}}$ *from Figure 6?*

For death rate $R_d = 0.001\mathbf{N}$ and $\mathbf{N} = \dfrac{3}{1 + 2e^{-0.003t}} \Rightarrow R_d = \dfrac{0.003}{1 + 2e^{-0.003t}}$. Considering $t \to \infty$ for this new function confirms the limit of growth. The death rate $R_d \to 0.003$ as $t \to \infty$, matching the birthrate from Figure 5 -- approaching ZPG.

Example 4 The Spreading Rumor Problem

The function $N = \dfrac{c}{1 + ae^{-bt}}$ models the number of people who have heard a rumor in t hours if c is the total number of people who could hear the rumor. Suppose that one student starts a rumor that a certain professor is an easy grader and there are a total of 7000 students and faculty members.

(a) Show that a represents the number of people who initially did not know the rumor.

(b) Find b if the rumor spreads to 15 people in 1 hr.

(c) If we assume that no one will tell the professor about the rumor, how long will it take for everyone on campus except the professor to hear it?

Thinking: Creating a Plan

For part (a) we find a by letting $N = 1$ at time $t = 0$. We expect this number to be $a = 6999$ since it represents the number of people who had not heard the rumor at the start. Then we substitute $t = 1$ and $N = 15$ in the equation to find the value of b. For part (c) we find t when $N = 6999$.

Communicating: Writing the Solution

(a) $N = \dfrac{c}{1 + ae^{-bt}}$ and $c = 7000, N = 1, t = 0 \Rightarrow$

$$1 = \frac{7000}{1 + ae^{-b(0)}}$$

$$1 = \frac{7000}{1 + a} \qquad \textit{Multiply both sides by } (1 + a).$$

$$1 + a = 7000$$

$$a = 6999$$

Initially, 6999 of the 7000 people had not heard the rumor.

(b) $N = \dfrac{c}{1 + 6999e^{-bt}}$ and c = 7000, N = 15, t = 1, a = 6999 ⇒

$15 = \dfrac{7000}{1 + 6999e^{-b(1)}}$	*Multiply both sides by $(1 + 6999e^{-b}$.*
$15(1 + 6999e^{-b}) = 7000$	*Divide both sides by 15.*
$1 + 6999e^{-b} \approx 466.67$	*Subtract 1 from both sides.*
$6999e^{-b} \approx 465.67$	*Divide both sides by 6999.*
$e^{-b} \approx 0.0665 \Leftrightarrow$	*Apply the definition of logarithm.*
$\ln 0.0665 \approx -b$	
$b \approx 2.71$	

The constant of proportionality is approximately 2.71.

(c) $N = \dfrac{7000}{1 + 6999e^{-2.71t}}$ and N = 6999 ⇒

$6999 = \dfrac{7000}{1 + 6999e^{-2.71t}}$	
$6999(1 + 6999e^{-2.71t}) = 7000$	*Divide both sides by 6999.*
$1 + 6999e^{-2.71t} \approx 1.00014$	*Subtract 1 from both sides.*
$6999e^{-2.71t} \approx 0.00014$	*Divide both sides by 6999.*
$e^{-2.71t} \approx 0.00000002 \Leftrightarrow$	*Apply the definition of logarithm.*
$\ln 0.00000002 \approx -2.71t$	
$-2.71t \approx -17.728$	
$t \approx 6.5$	

In approximately $6\frac{1}{2}$ hours, everyone but the professor has heard the rumor.

Learning: Making Connections

If we sketch the graph, we see that this model is another logistic function with the asymptote, the limit to growth, representing the maximum number of people who could hear the rumor (Figure 7).

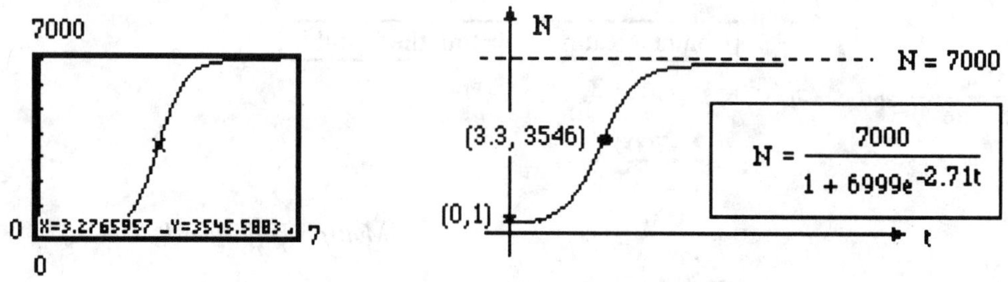

Figure 7

What do you think the point of inflection as (3.3, 3546) represents?

This is when the rumor is spreading at its maximum rate. After this point, where the graph becomes concave down, so many people have heard the rumor that it becomes harder to find someone to tell.

The Logistic Function
The function

$$N(t) = \frac{c}{1 + ae^{-bt}}$$

represents the varying number N as a function of time t where c is the limit to growth, $a = \frac{c}{N_0} - 1$, and b is the proportionality constant.

Defining a Surge Function

Functions of the form

$$f(t) = ate^{-bt}$$

are called **surge functions**, because for $t \geq 0$ they begin with a rapid, yet decreasing absorption rate, peak, and then decrease toward zero (Figure 8).

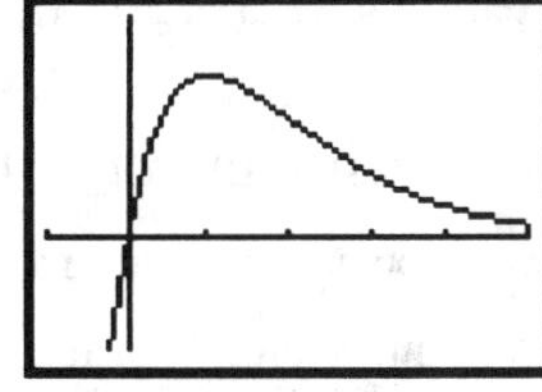

Figure 8

Example 5 *Graphing surge functions*

Graph each surge function of the form $f(t) = ate^{-bt}$ and estimate the value of t that yields the maximum for each graph.

(a) $f(t) = te^{-t}$ (b) $g(t) = te^{-2t}$ (c) $h(t) = 5te^{-t}$

Solution

Enter $y = xe^{-x}$, $y = xe^{-2x}$, and $y = 5xe^{-x}$ (Figure 9).

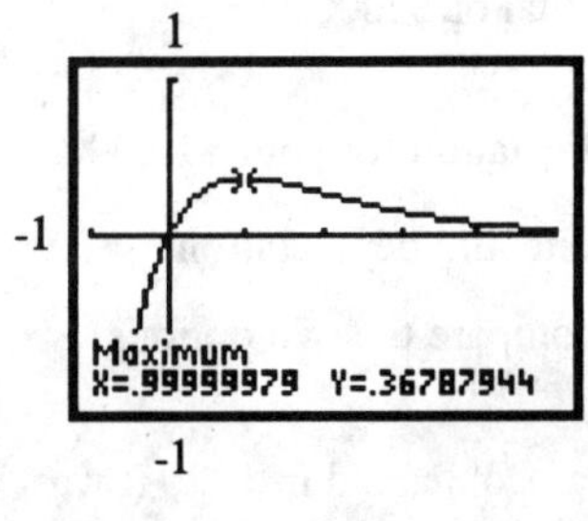

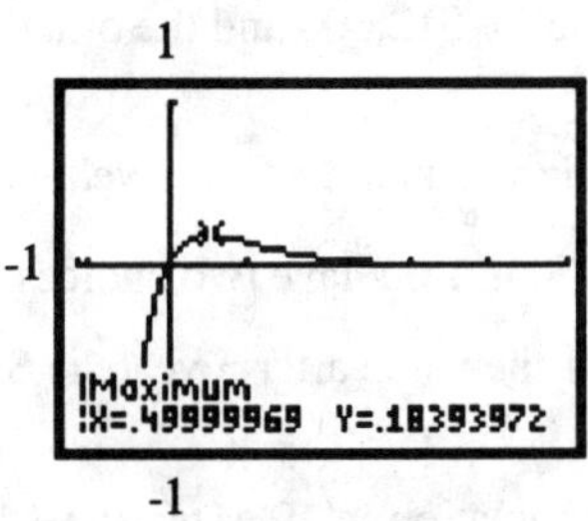

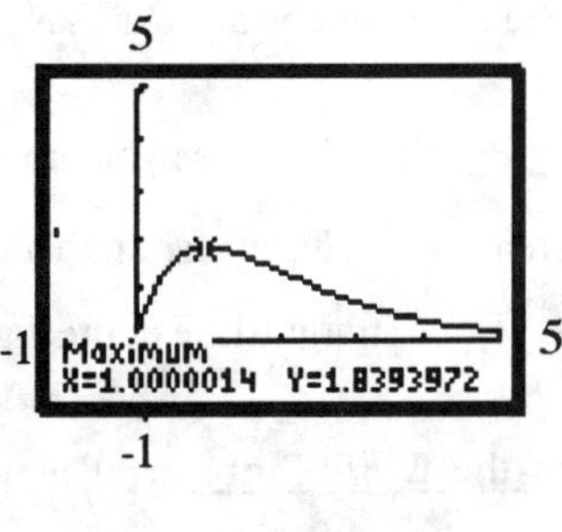

Figure 9 (a) (b) (c)

The maximums are 1, 0.5, and 1. In each case $t = \frac{1}{b}$. The corresponding value of y depends on a and b.

$f(t) = ate^{-bt}$ and $t = 1/b \Rightarrow$

$$f(1/b) = \frac{a}{b}e^{-b(1/b)} = \frac{a}{b}e^{-1}$$

As we will see in the exercises, surge functions are excellent models for medicine -- intake and removal. Pharmaceutical chemists use the published value of the half-life for a given drug to determine the frequency and dosage.

Exercises 4.5

Apply the Concepts

1. The Cooling Coffee Problem If a 190°F cup of coffee cools to 185°F in 10 min in a 68°F room, find the temperature of the coffee in 1/2 hr. Use Newton's law of cooling formula $T = De^{-kt} + T_a$.

2. The Turkey Cooking Problem If the oven temperature is 325°F and the initial temperature of a turkey is 40°F, find the time it will take for the internal temperature of the turkey to be between 175° and 185°F, knowing that the temperature after 30 min was 60°F. Use Newton's law of "heating" formula $T = T_a - De^{-kt}$.

*3. The Hot and Iced Coffee Problem You have an iced latté (40°F) and your friend has a hot latté (180°F) in different containers. You are engaged in a discussion about Newton's law of cooling and completely forget your coffee drinks. Ten minutes later, your iced latté reached 56°F and your friend's hot latté cooled to 158°F. Yours is outside, where the temperature is 80°F, and your friend's is inside, where room temperature is 68°F. Use a graphing utility to estimate when the coffees will reach the same temperature.

4. The Foul Play Problem Sherlock Holmes arrived at the scene of the crime and found the corpse of the victim. He immediately took the temperature of the corpse (88°F) and of the room (68°F). When the coroner arrived, 1 hr later, she congratulated him on his actions and took another body temperature (80°F). When did the victim die? [Hint: Normal body temperature is 98.6°F.]

5. The Two-Cars Problem Compare the values of two cars in 5 and 10 years if each costs $18,600 initially and is worth $14,600 after one year. One has a residual of $12,600 and the other a residual of $2000.

6. The Taxing Problem Suppose you are allowed to deduct the first 3 years of depreciation for your $10,000 computer system, which has a residual value of $500. You have two choices: exponential depreciation $V(t) = De^{-kt} + V_r$, or straight-line depreciation. If the value after 5 years is $505, compare the two options.

7. The End-of-the-World Problem The world population in 1990 was 5.01 billion people, and in 1996 it was 5.8 billion.

(a) If the limit to growth is 10 billion people, find the equation of the logistic function.

(b) How many people does the model predict for the years 2000 and 3000?

8. The Threshold to Avoid Extinction Problem In ecological studies of endangered species, knowing whether or not the population N_0 is above or below the *threshold* is critical. Consider the logistic equation

$$N(t) = \frac{N_0 T}{N_0 + (T - N_0)e^{rt}}$$

where T is the threshold. Graph the populations described below and interpret your results.

(a) $N_0 = 3, T = 2, r = 0.01$ $(T < N_0)$

(b) $N_0 = 1, T = 2, r = 0.01$ $(T > N_0)$

*9. The Epidemic Problem The spread of diseases has a natural limit, the number of people susceptible. The rate at which a disease spreads is high initially, then slow as fewer people are left unexposed. This is naturally an application of the logistic curve. Find the length of time before virtually everyone in the United States (250 million people) has been exposed to a virus if initially 1 person brought it into the country, and two days later 20 people had the disease. [Hint: Use the formula $N = \frac{c}{1 + ae^{-bt}}$.]

10. The Advertising Problem Advertising executives use the logistic function to model the effect of advertising campaigns. Suppose that a company wants to promote a new fat substitute. Initially only 10 people knew of the product, but 10 months after the campaign blitz began, a survey indicated that 10% of the population of 2.5 million citizens had heard of it. How long would it take to reach 50% of the people? [Hint: Use the formula $N = \frac{c}{1 + ae^{-bt}}$.]

11. The Rainbow Kite and Balloon Company The data represent the number of 6,070,810 people in a state who have heard your new advertising campaign.

time (months)	0	2	4	6
people who have heard (thousands)	958.5	1294.4	1651.7	2130

(a) Find the equation of the logistic function of the form $N(t) = \frac{c}{1 + ae^{-bt}}$ for these data.

(b) According to the logistic function from part (a), how long will it take before two-thirds of the people in the state will have heard the campaign?

12. Wonderworld Amusement Park Attendance at the theme park has a limit of 22,000 people. This season attendance averaged 18,200, 12% more than last season. When will the attendance virtually reach its limit?

13. The Barometric Pressure Problem Barometric pressure, measured in millimeters of mercury (Hg) is given for any altitude h in kilometers by $P = P_0e^{-kh}$.

(a) Find P_0 if the barometric pressure at sea level today is P = 760 mmHg.

(b) Find k if P = 750 mmHg at Clearview, whose elevation is 160 m (h = 0.16 km) above sea level.

(c) Graph the resulting function and label the following four points: barometric pressure at sea level; on top of Mt. Rainier (4250 m high); in Denver, the "mile-high city"; and at a height where the barometric pressure is 1000 mmHg.

14. The Multiplication Tables Problem Suppose that you are learning to multiply in base 2, where the only digits are 0 and 1. You recorded how many multiplication facts f you knew as a function of time t in hours as follows:

hours, t	0	1	2
facts, f	0	20	36

Assume that your learning curve is modeled by the equation $f(t) = a(1 - e^{-kt})$.

(a) What does the function estimate the number of multiplication facts you will know in 1 hr?

(b) How long will it take to learn 50 facts?

[Hint: Solve two equations with two variables, letting $u = e^{-k}$ and $u^2 = e^{-2k}$.]

15. The Heron Problem In our fish pond the birthrate of fish is decreasing at a rate proportional to the number of fish, and the death rate, a result of being eaten by the neighborhood heron, remains constant. The logistic function for the fish population is $N(t) = \frac{a}{b + e^{ct}}$. Let a = 103, b = 0.03, and c = 0.003.

(a) What was the initial number of fish?

(b) How many fish are left in 10 days and in 100 days?

(c) What does the model predict as t approaches infinity?

16. The Ballooning Bullfrog Population Problem If we model the population of African bullfrogs with the function $N(T) = \frac{a}{b - e^{ct}}$, where a = 20, b = 11, and c = 0.003, what does the model predict for value(s) of t in days for which the denominator equals 0?

Exercises 17 to 20 model the amount of medicine in the bloodstream (plasma) as a function of time since taken. Surge functions of the form $f(t) = ate^{-bt}$ model the absorption elimination time.

17. The Peak Time Problem Let a = 10 in the formula $f(t) = ate^{-bt}$.

(a) Create four surge functions by substituting b = 1, 2, 4, and 10 into the equation $f(t) = 10te^{-bt}$.

(b) Find the time to maximize concentration, and complete the following table.

(c) Find a simple formula for t as a function of b.

b	1	2	4	10
t at maximum concentration	?	?	?	?

(d) Explain why setting a = 10 does not change the formula.

18. The Aspirin Acid Problem Aspirin (acetylsalicylic acid) was first marketed by Bayer around 1899. U.S. consumption is about 10,000 tons/yr. Absorption and elimination times vary from person to person, but generally the concentration is $C(t) = 30{,}000te^{-6.7t}$, where the maximum concentration for two 325-mg tablets in a person whose total quantity of blood is 4 liters is commonly reported in milligrams/deciliter as follows:

$$C_{max} = \frac{650 \text{ mg}}{4 \text{ L}} = 1625 \text{ mg/DL}$$

(a) Find and interpret the maximum point on the graph of the function $C(t) = 30{,}000te^{-6.7t}$.

(b) Find the half-life of aspirin. [Hint: let $C = (1/2)(1625)$ mg/DL $= 812.5$ mg/DL and solve for t.]

19. The Nicotine in the Bloodstream Problem The surge function $C(t) = 14.6te^{-5.36t}$ models the concentration of nicotine in the bloodstream t hours after smoking a cigarette. Discuss the situation.
[Source: *Clinical Pharmacology Online*, http://www.cponline.gsm.com]

20. The Caffeine Problem Use the surge function for caffeine concentration as a function of time in minutes $C(t) = 910\ te^{-0.67t}$ to answer the following.

(a) Find the time to maximum concentration.

(b) Find the half-life of caffeine.

[Source: *Clinical Pharmacology Online*, http://www.cponline.gsm.com]

4.6 Logarithmic Scales and Curve Fitting

Creating a Logarithmic Scale
Applying Log Scales to Earthquakes and Sound
Applying Linear Regression to Semi-Log and Log-Log Coordinate Systems
Massaging and Adjusting Data

In this section we examine logarithmic scales. Their applications include the decibel scale for measuring the intensity of sound and the Richter scale for measuring earthquakes. We use logarithms to examine what different Richter readings really mean in terms of the size of an earthquake.

We also look at data plotted in a coordinate system with one or both axes scaled using logarithms. In some cases, graphs of certain nonlinear data appear nearly linear on these new coordinate systems. This will help us understand the connection between linear regression and curve fitting with exponential functions of the form $y = ab^x$, power functions of the form $y = ax^b$, and logarithmic functions of the form $y = a + b \ln x$. By studying logarithmic scales we will understand why some data create errors when we are trying to find certain best-fitting equations. This section is the heart of our curve-fitting study, where we will use our intuition and technology to create a reasonable mathematical model from data.

Creating a Logarithmic Scale

Figure 1 shows the graph of $y = \log x$. The vertical line on the right is a logarithmic scale (log scale). Notice the tic marks on the log scale are not equally spaced; the integers get closer together. Follow a point from the x axis to the graph to the log scale. The log scale is labeled with the corresponding x coordinate.

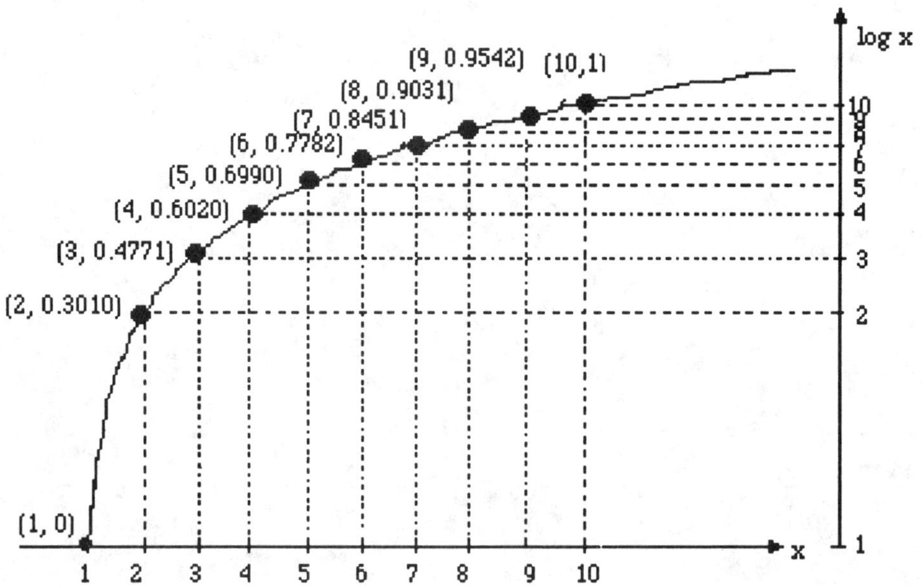

Figure 1

Figure 2 shows an extended logarithmic scale in base 10 below the line. The numbers on top are exponents (logarithms). Notice that 100 is located twice as far from the number 1 as it is from the number 10. Since

$$\log 10 = 1 \text{ and } \log 100 = 2$$

the tick mark 10 is located 1 unit from the origin, while the tick mark 100 is 2 units from the origin.

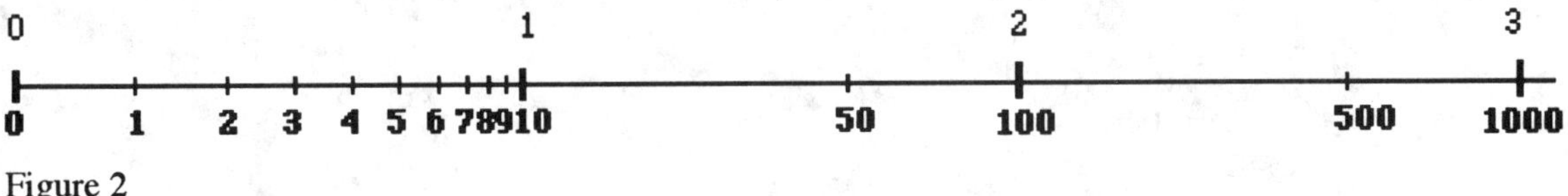

Figure 2

Applying Log Scales to Earthquakes and Sound

A familiar use of a log scale is the Richter scale used for measuring earthquakes. Charles F. Richter established a "zero" earthquake to which all others could be compared. This zero earthquake was arbitrarily defined to be the smallest earthquake that had been recorded. The curves that represent the amplitude of the ground motion at varying distances from the epicenter of the earthquake as recorded on seismographs are generally parallel and nearly logarithmic. This means that the vertical distance between the graph of seismic readings and the graph of the zero earthquake is constant. This constant is called the *magnitude* M of earthquake A (Figure 3).

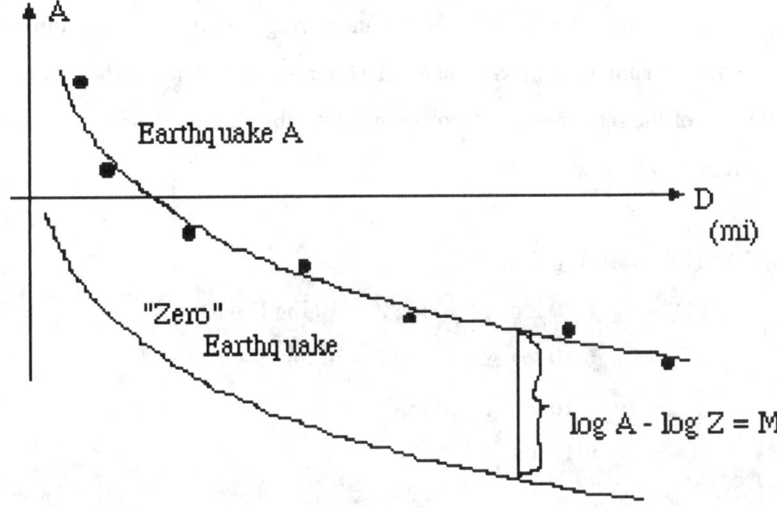

Figure 3

The difference in the magnitudes for two earthquakes, M_A and M_B, is the logarithm of the quotient of their amplitudes, A and B.

$M_A = \log A - \log Z$ and $M_B = \log B - \log Z \Rightarrow$

$$\begin{aligned} M_A - M_B &= (\log A - \log Z) - (\log B - \log Z) \\ &= \log A - \log B \\ &= \log \frac{A}{B} \end{aligned}$$

How much stronger is an earthquake that measures 6 than one that measures 4?

For two earthquakes whose magnitudes are $M_A = 6$ and $M_B = 4$, their difference in Richter readings is only 2, yet the earthquake with magnitude 6 is 100 times as strong as the one with magnitude 4.

$$M_A - M_B = \log\frac{A}{B} \qquad \textit{Let } M_A = 6 \textit{ and } M_B = 4.$$
$$2 = \log\frac{A}{B} \Leftrightarrow \qquad \textit{Apply the definition of logarithm.}$$
$$10^2 = \frac{A}{B}$$
$$A = 100B$$

The log scale in Figure 4 shows that an earthquake with a magnitude of 6 is 1,000,000 times stronger than the zero earthquake and an earthquake with a magnitude of 4 is 10,000 times stronger than the zero earthquake.

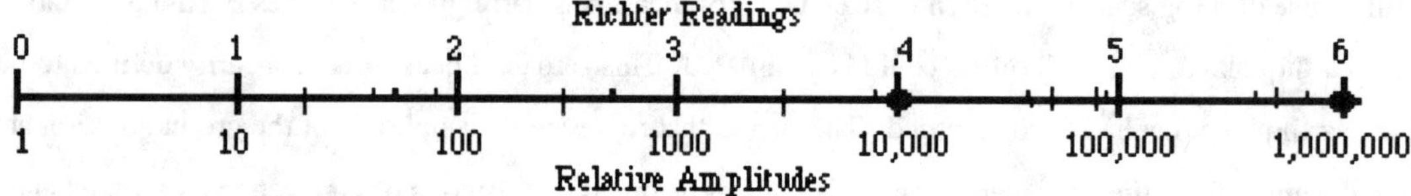

Figure 4

Analogous to the Richter scale is the decibel scale for measuring the intensity of a sound relative to the least audible sound. The decibel reading for a given sound is 10 times the difference of the logarithms of that sound's intensity and the intensity of the zero sound. "Zero" sound is the threshold of hearing for youths. (The "deci" in *decibel* means multiplied by 10.)

$D_A = 10(\log A - \log Z)$ and $D_B = 10(\log B - \log Z) \Rightarrow$

$$D_A - D_B = 10 \log A - 10 \log Z - (10 \log B - 10 \log Z)$$
$$D_A - D_B = 10 \log A - 10 \log Z - 10 \log B + 10 \log Z$$
$$= 10 \log A - 10 \log B$$
$$= 10 \log\frac{A}{B}$$

Example 1 The Exercise Class Problem

Compare the intensity of two sounds, an aerobic exercise class measuring 124 dB and an ordinary conversation measuring 70 dB.

Thinking: Creating a Plan

We will let $D_A = 124$ and $D_B = 70$ and use the formula $D_A - D_B = 10 \log\frac{B}{A}$.

Communicating: Writing the Solution

$D_A - D_B = 10 \log \frac{A}{B}$ and $D_A = 124$ and $D_B = 70 \Rightarrow$

$$124 - 70 = 10 \log \frac{A}{B}$$

$$54 = 10 \log \frac{A}{B}$$

$$5.4 = \log \frac{A}{B} \Leftrightarrow$$

Apply the definition of logarithm.

$$\frac{A}{B} = 10^{5.4}$$

$$A \approx 251{,}189\, B$$

Therefore, the sounds of the exercise class, although only 54 dB higher than the sound of a normal conversation, is 250,000 times as intense.

Learning: Making Connections

A 70-dB increase, such as comparing an ordinary conversation (70 dB) and some headphones (140 dB), is an increase in intensity by a factor of $10^7 = 10{,}000{,}000$. Some common decibel readings are shown on the following logarithmic scale. [Source: *Hope Health Letter*]

	192 Possible fatal internal injury
	150 Possible permanent deafness, depending on duration of exposure
Painful	140 Jet plane
	130 Threshold of pain
Deafening	120 Rock music, thunder
	110 Riveter, full symphony fortissimo
	100
Very loud	90 Noisy kitchen, freeway
	80 Garbage disposal, TV
Loud	70 Ordinary conversation
	60 Distant telephone
	50 City street, quiet music
Moderate	40 Soft music
faint	30 A large room
	20 A ticking watch, whisper at 5 ft
Very faint	10 Average threshold of hearing
	0 Least audible sound

Applying Linear Regression to Semi-Log and Log-Log Coordinate Systems

Recall that curve fitting is defined as finding the equation of the function that models a situation or collection of data. Now we will apply our knowledge of logarithms to see the connection between linear regression and curve fitting with exponential, power, and logarithmic functions. Table 1 summarizes four models.

Table 1

Model	Mode	Equation
Linear	LinReg	$y = a + bx$
Logarithmic	LnReg	$y = a + b \cdot \ln x$
Exponential	ExpReg	$y = ab^x$
Power	PwrReg	$y = ax^b$

Two questions need to be addressed when curve fitting:

1. Which family of functions will best describe the situation or data?
2. How well does the resulting function actually fit the data?

Recall from Section 2.2 that the coefficient of determination, r^2, describes the variation in y explained by the best-fitting equation. We can use r^2 to determine how well an equation fits the data by choosing the model whose r^2 is closest to 1. If $r^2 < 0.50$, which means that less than 50% of the variation in y is explained by the best model, the model is not acceptable. In such cases we use the average y value, $\bar{y}$, to estimate missing data.

The data in the next example where created by drawing circles on graph paper and recording the radius and area of each circle. The area was determined by *counting* the squares inside five such circles (Figure 5).

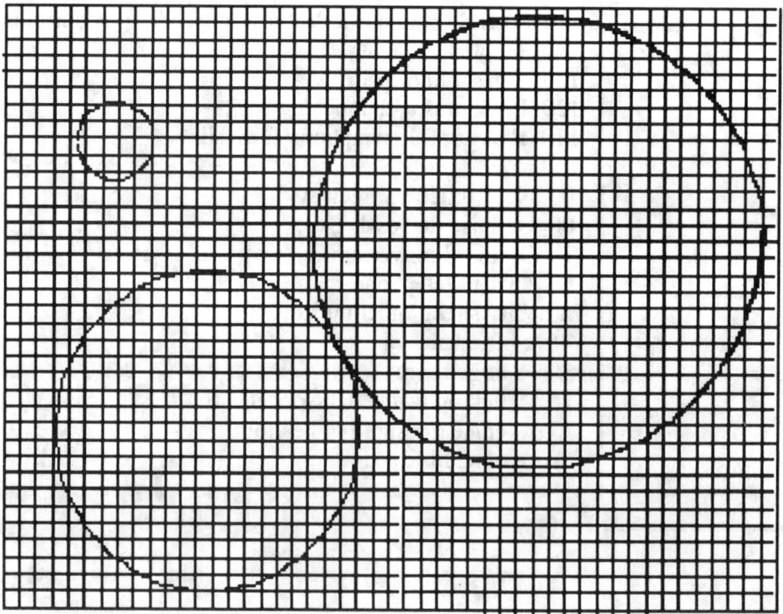

Figure 5

Example 2 The Area of a Circle Problem

Use a graphing utility to determine the function that best fits the following measurements of radii and area of five circles by running linear, logarithmic, exponential, and power regression.

radius of each circle, x	2	4	5	9	11
area of each circle, y	12	53	79	236	376

Thinking: Creating a Plan

We enter the data in a graphing utility and use the coefficient of determination r^2 to decide whether the best-fitting equation is linear, logarithmic, exponential, or power function. We choose the model with r^2 closest to 1.

Communicating: Writing the Solution

Table 2 summarizes the outputs from the graphing utility.

Table 2

Linear $y = a + bx$	**Logarithmic** $y = a + b\ln x$	**Exponential** $y = ab^x$	**Power** $y = ax^b$
a = -98.1350	a = -186.8803	a = 9.5335	a = 3.1489
b = 40.2153	b = 204.05623	b = 1.4234	b = 1.9897
r = 0.98171	r = 0.91409	r = 0.96626	r = 0.99935
$r^2 = 0.96376$	$r^2 = 0.83556$	$r^2 = 0.93367$	$r^2 = 0.99870$

Since the highest value of r^2 is for a power function ($r^2 = 0.99870$), the data best fits the equation of the form $y = ax^b$. The equation is $y = 3.1489x^{1.9897}$

Learning: Making Connections

Since the formula of a circle is $A = \pi r^2$, a perfect fit would be $y = \pi x^2$. (A perfect fit is impossible to achieve by actual measurement.) Notice that $a = 3.1489 \approx \pi$ and $b = 1.9897 \approx 2$.

We know that the graph of $y = mx + b$ is a straight line in the Cartesian coordinate system where the scales on both axis have uniform constant distances between tic marks. We call such scales *linear scales*. A coordinate system with one axis a logarithmic scale and the other a linear scale is called a **semi-log coordinate system**. A coordinate system with both axes logarithmic scales is called a **log-log coordinate system**. Figures 6a and 6b show two different semi-log coordinate systems, and Figure 6c shows a log-log coordinate system where $u = \ln x$ and $v = \ln y$.

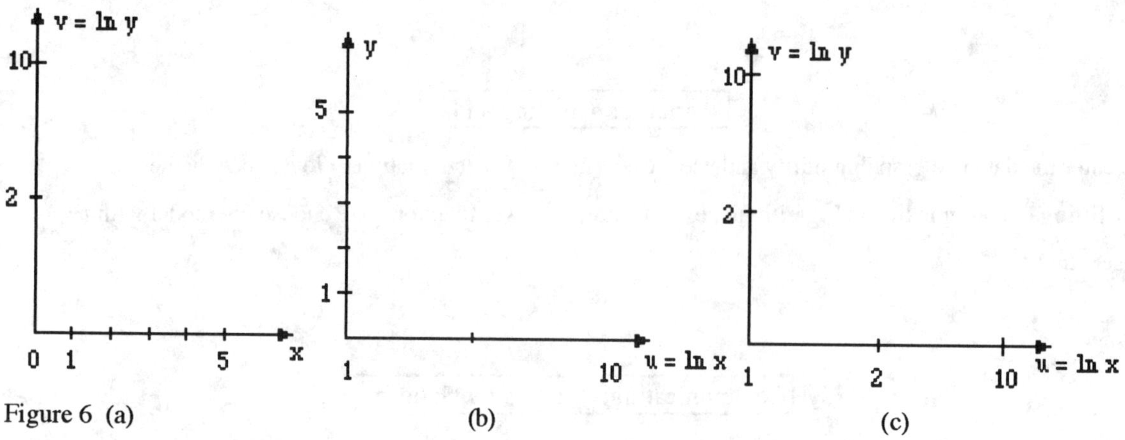

Figure 6 (a) (b) (c)

Let's graph the data from Example 2 on various coordinate systems.

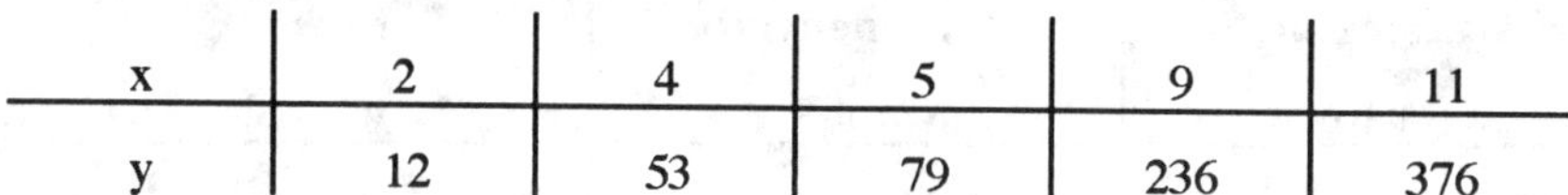

x	2	4	5	9	11
y	12	53	79	236	376

As expected, the graph of the data on the Cartesian coordinate system appears parabolic for $x > 0$ since our best-fitting equation, $y = 3.1489x^{1.9897}$, is nearly parabolic (Figure 7a). The data are also plotted on the two semi-log coordinate systems (Figures 7b and 7c) and the log-log coordinate system (Figure7d).

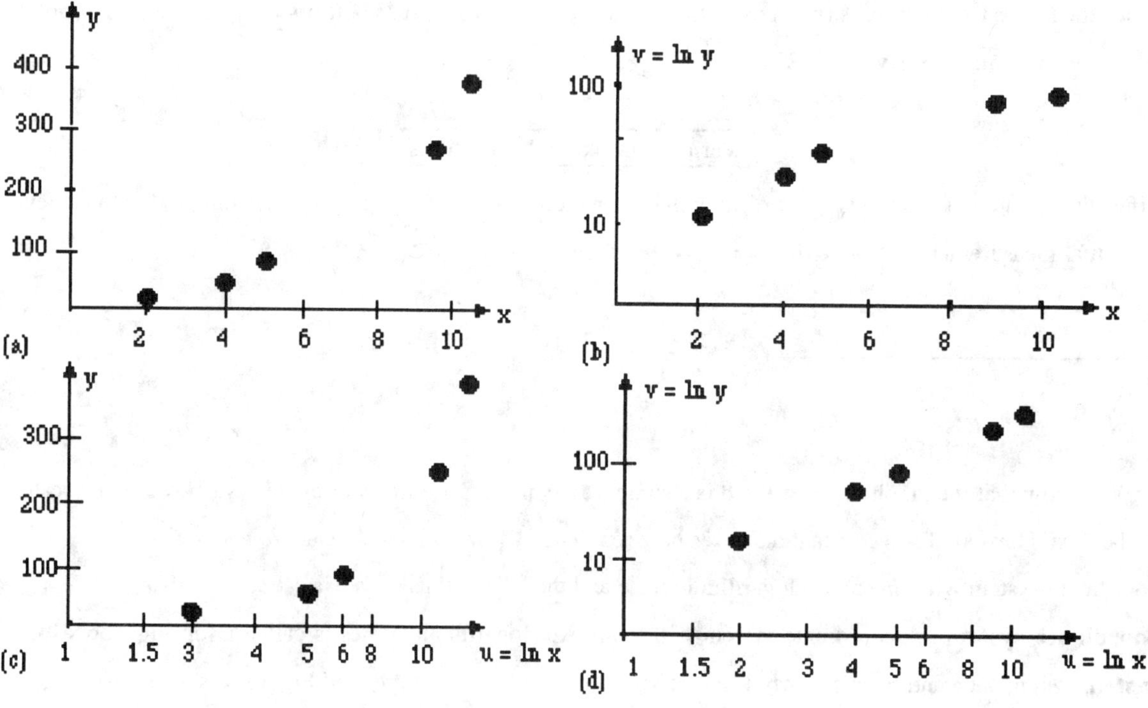

Figure 7

Notice that the graph on the log-log coordinate system in Figure 7d appears most linear. We can show that the graphing utility is applying linear regression to the logarithms of the data. Let's enter the data as logarithms and apply linear regression to find the equation of the line in Figure 7d.

ln x	0.6931	1.3863	1.6094	2.1972	2.3979
ln y	2.4849	3.9703	4.3694	5.4638	5.9296

The result is y = 1.1471 + 1.9897x (Figure 8). Notice that $r^2 = (0.99935)^2 = 0.9987$ is the same for the power function using the original data.

```
LinReg
 y=a+bx
 a=1.147149684
 b=1.989672929
 r=.9993519919
```

Figure 8

Since the axes were labeled ln x and ln y instead of x and y, we write

$$y = 1.1471 + 1.9897x$$
$$\ln y = 1.1471 + 1.9897 \ln x$$

Solving this equation for y yields our previous result.

$$\ln y = 1.1471 + 1.9897 \ln x$$

$$\ln y - 1.9897 \ln x = 1.1471 \qquad \textit{Apply theorem (3) on logarithms.}$$

$$\ln y - \ln x^{1.9897} = 1.1471 \qquad \textit{Apply theorem (2) on logarithms.}$$

$$\ln \frac{y}{x^{1.9897}} = 1.1471 \Leftrightarrow \qquad \textit{Apply the definition of a logarithm.}$$

$$e^{1.1471} = \frac{y}{x^{1.9897}} \qquad \textit{Multiply both sides by } x^{1.9897}.$$

$$y = e^{1.147} x^{1.9897} \qquad \textit{Evaluate } e^{1.147}.$$

$$y = 3.1487\, x^{1.9897}$$

Next, we consider a logarithmic equation of the form y = a + b ln x, which appears linear when plotted in a semi-log coordinate system with a logarithmic scale on the horizontal axis. For example, Figure 9a shows the graph of y = 1 + 2 ln x in the Cartesian coordinate system. If we let u = ln x, we have the equation

$$y = 1 + 2u$$

whose graph is linear in Figure 9b.

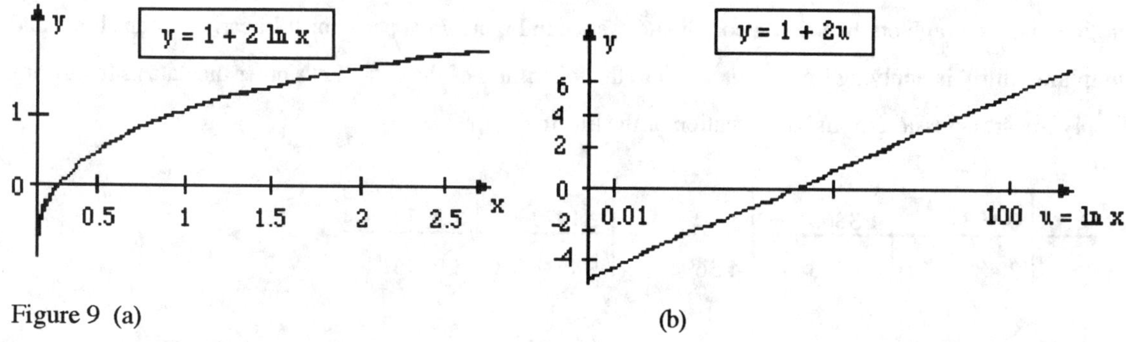

Figure 9 (a) (b)

This means that the least-squares method for finding the equation of the best-fitting linear equation can be used to find the equation of the best-fitting logarithmic equation, since the data are linear when graphed in this semi-log system.

How do you think we change an exponential function into a linear function?

Consider an exponential equation of the form $y = ab^x$, such as

$$y = 2(20^x)$$

whose graph is shown in Figure 10a. Figure 10b shows the graph on a semi-log coordinate system where $v = \ln y$. This means that linear regression can be applied to find the best-fitting exponential equation. Taking the logarithm of both sides of $y = 2(20^x)$ creates the linear equation in Figure 10b.

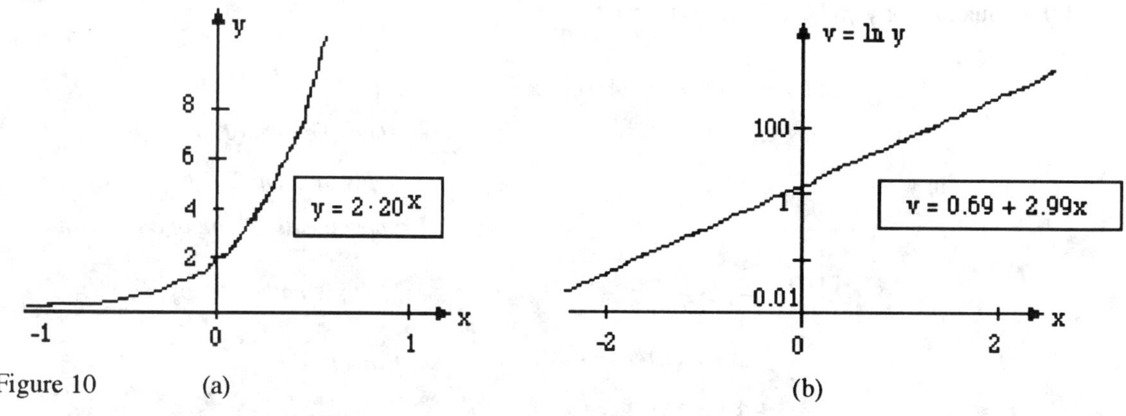

Figure 10 (a) (b)

Table 3 summarizes the coordinate system that create linear functions of particular equations.

Table 3

Model	Equation	Coordinate System	x Axis	y Axis	Linear Equation
Linear	$y = a + bx$	Cartesian	x	y	$y = a + bx$
Logarithmic	$y = a + b \ln x$	semi-log	$u = \ln x$	y	$y = a + bu$
Exponential	$y = ab^x$	semi-log	x	$v = \ln y$	$v = a' + b'x$
Power	$y = ax^b$	log-log	$u = \ln x$	$v = \ln y$	$v = a' + bu$

(Note that ln a and ln b are constants.)

Massaging and Adjusting Data

In Example 2 we did not include a circle with radius zero that has an area of zero square units. Adding the point (0,0) to the data creates the following table.

x	0	2	4	5	9	11
y	0	12	53	79	236	376

An error message on the graphing utility for logarithmic, exponential, and power functions prevents us from considering any of them as the best-fitting curve.

Linear $y = a + bx$	**Logarithmic** $y = a + b\ln x$	**Exponential** $y = ab^x$	**Power** $y = ax^b$
$a = -51.610$ $b = 204.05623$ $r = 0.9614$ $r^2 = 0.9243$	ERROR	ERROR	ERROR

Why do we get an error message?

The error message is because they use semi-log and log-log systems, and log 0 is undefined.

From the preceding discussion, we know that logarithmic and exponential data appear linear when graphed in a semi-log system. We use a log scale for the x axis for logarithmic regression and a log scale for the y axis for exponential regression. Since logarithms are defined for positive values only, using logarithmic regression requires $y > 0$, while using exponential regression requires $x > 0$. Data that are best fit by a power function appear linear in a log-log system requiring both $x > 0$ and $y > 0$.

Regression	Restrictions
Logarithmic	$x > 0$
Exponential	$y > 0$
Power	$x > 0,\ y > 0$

What can we do to avoid getting an error message?

In such cases we **massage** the data by slightly changing a data point so that it will not create an error. For example, if we want to include the data point (0,0), we could enter a point that is very close to (0, 0), say (0.001, 0.001). Since log 0.001 is defined, we no longer have an error.

We can also consider translations of the logarithmic, power, and exponential functions

$$y = a + b\ln x,\quad y = ab^x,\ \text{and}\ \ y = ax^b$$

Consider the following data graphed in Figure 11a, which appears logarithmic. Notice that each x value is negative or zero. Since x must be positive to find the best-fitting logarithmic equation, we **adjust** the data to translate the curve 10 units to the right (Figure 11b) by adding 10 to each x value.

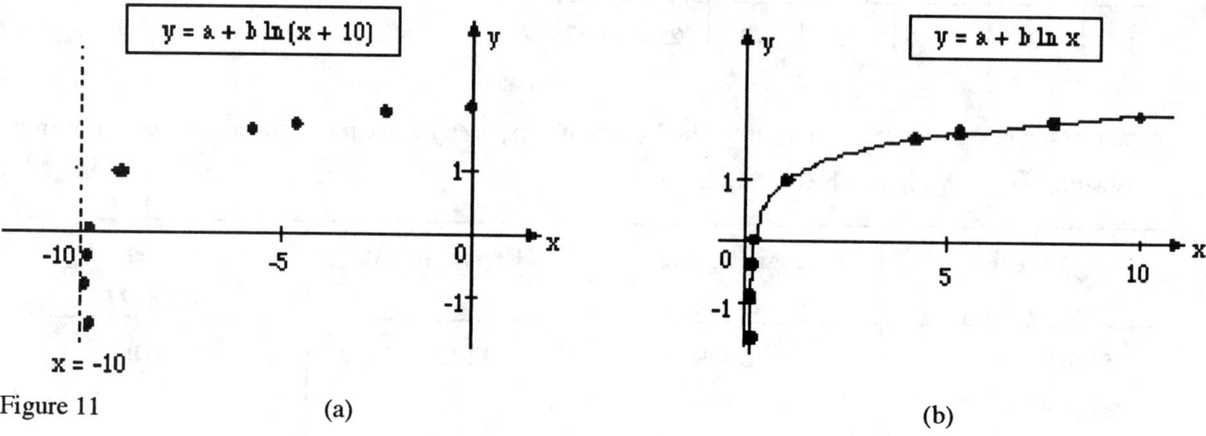

Figure 11 (a) (b)

We enter the data in a graphing utility as (x+10, y) and find the best-fitting logarithmic equation. To accommodate the data adjustment, the resulting equation from the graphing utility is translated left 10 units.

$$y = a + b \ln (x + 10)$$

Similarly, for the data in Figure 12a that appear to be exponential, we add 10 to each y value to accommodate the vertical translation (Figure 12b).

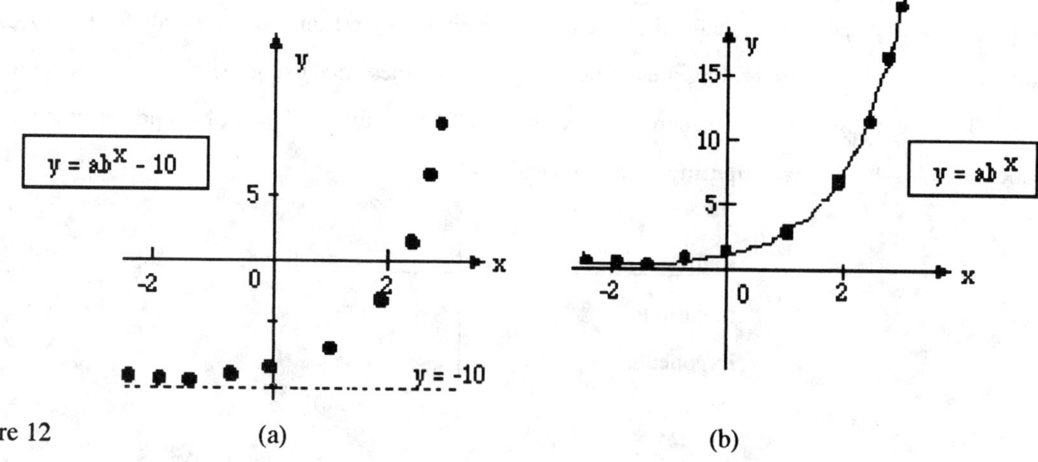

Figure 12 (a) (b)

We enter the data in a graphing utility as (x, y+10) and find the best-fitting exponential equation. To accommodate the data adjustment, the resulting equation from the graphing utility is translated back down 10 units, $y = ab^x - 10$.

Example 3 The Cooling Cup of Tea Problem

A cup of tea was heated to 120°F and placed in a room whose temperature was constant. The tea's temperature was measured at various time intervals, yielding the following results.

time (min)	0	3	11	34	57	64	95	109	135
temp (°F)	122	120	110	93	84	82	76	74	73.5

(a) Plot the data and massage or adjust the data if needed.

(b) Find the equation that best fits the data.

(c) Use your equation to predict the temperature of the water in 3 hr.

Thinking: Creating a Plan

We plot the data in a Cartesian coordinate system and anticipate the nature of the graph. We decide if a translation is involved and adjust and massage the data if necessary. Then, for part (c), since 3 hr = 180 min, we will use our equation to find the temperature that corresponds with 180 min.

Communicating: Writing the Solution

(a) The scatter plot in Figure 13 and the original data suggest that the graph is an exponential function translated up 73 units. The best-fitting equation should be of the form $y = ab^x + 73$.

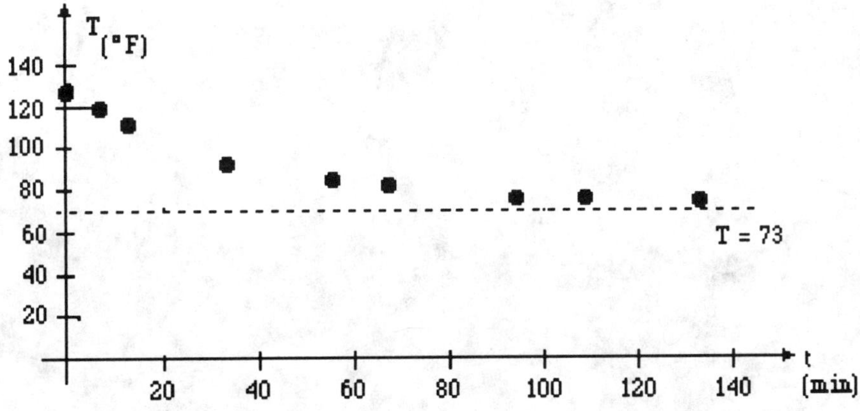

Figure 13

Since the first x entry is not defined for ln x, we massage the data by letting x = 0.001 instead of 0. We adjust the data by subtracting 73 from each y value. We enter the following data in a graphing utility.

x	0.001	3	11	34	57	64	95	109	135
y - 73	49	47	37	20	11	9	6	1	0.5

(b) We enter these data on a graphing utility and find the best-fitting exponential function. Since a = 58.7 and b = 0.968, the equation is $y = 58.7(0.968)^x + 73$. Notice the translation of the graph up 73 units.

(c) $y = 58.7(0.968)^x + 73$ and $x = 180 \Rightarrow$

$$y = 58.7(0.968)^{180} + 73$$

$$y \approx 73.168$$

Therefore, the tea will cool to about 73.17° after 3 hr.

Learning: Making Connections

The horizontal asymptote, y = 73, is the ambient temperature -- the temperature of the room. The temperature of the tea will eventually remain constant at room temperature.

Since e is the preferred base, we can express the equation $y = 58.72(0.968)^x + 73$ in base e as

$$y = 58.72(0.968)^x + 73$$

$$T = 58.72(e^{\ln 0.968})^t + 73$$

$$T = 58.7\, e^{-0.0325t} + 73$$

Exercises 4.6

Apply the Concepts

1. The Earthquake Problem How much stronger is an earthquake that measures 8 on the Richter scale than one that measures 4?

2. The 1906 San Francisco Earthquake Problem Show that San Francisco's 1906 earthquake, which measured 8.4 on the Richter scale, was 250 times stronger than one measuring 6.

3. The 1989 San Francisco Earthquake Problem The first reports of the San Francisco earthquake of 1989 indicated that its Richter reading was 6.9. Later, scientists reported that the reading was actually 7.1. How much stronger is a 7.1 earthquake than a 6.9 earthquake?

4. The Earthquake of the 1990s Problem After the San Francisco earthquake of 1989, scientists predicted that a 9.0 earthquake could occur in the 1990s somewhere along the Pacific coast. How much stronger is a 9.0 earthquake than San Francisco's 7.1 earthquake?

5. The Rustling Leaves Problem Leaves rustling in a breeze have a sound intensity of 20 dB. How many times louder is that than a soft whisper of 10 dB?

6. The Niagara Falls Problem The sound intensity of Niagara Falls is 1 billion times greater than the least audible sound. How many decibels is this?

*7. Show the change of variables that change the power function $y = 100x^{-4}$ into a linear function in a log-log coordinate system. Graph the function on the rectangular coordinate system (Figure 14a) and on the log-log coordinate system (Figure 14b).

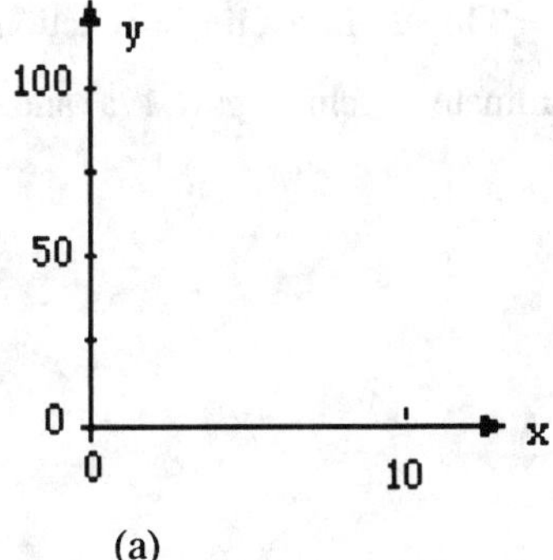

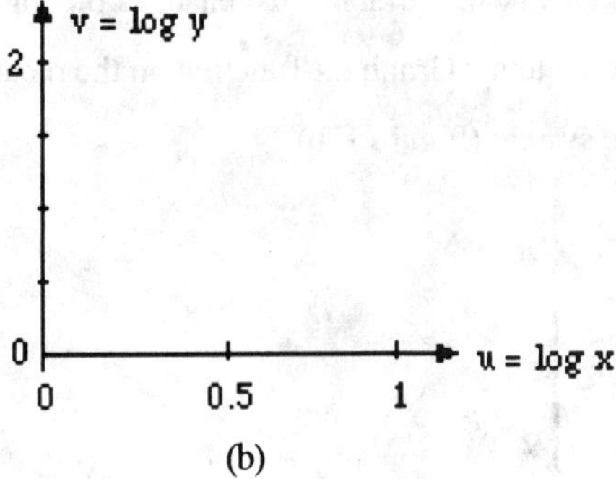

Figure 14 (a) (b)

8. Show the change of variables that changes the power function $y = 2\sqrt{x}$ into a linear function in a log-log coordinate system. Graph the function on the rectangular coordinate system (Figure 15a) and on the log-log coordinate system (Figure 15b).

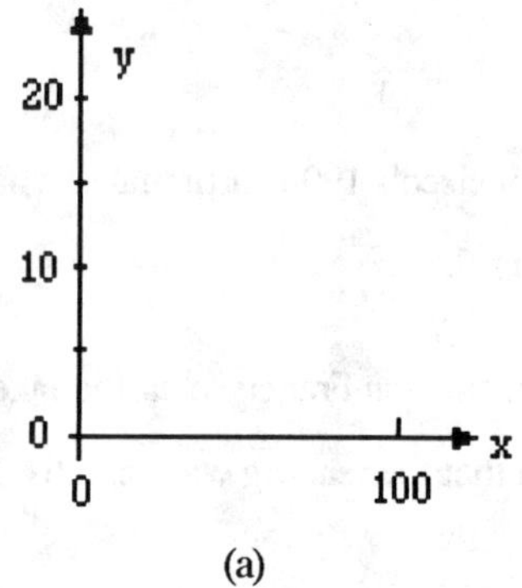

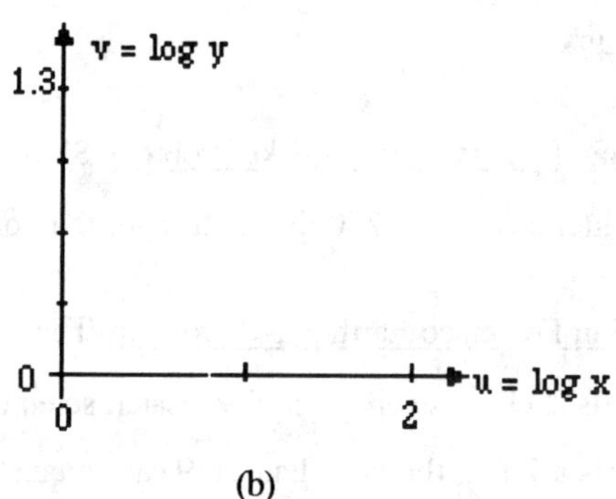

Figure 15 (a) (b)

9. Show the change of variables that changes the exponential function $y = 2e^{0.5x}$ into a linear function in a semi-log coordinate system. Graph the function on the rectangular coordinate system (Figure 16a) and on the semi-log coordinate system (Figure 16b).

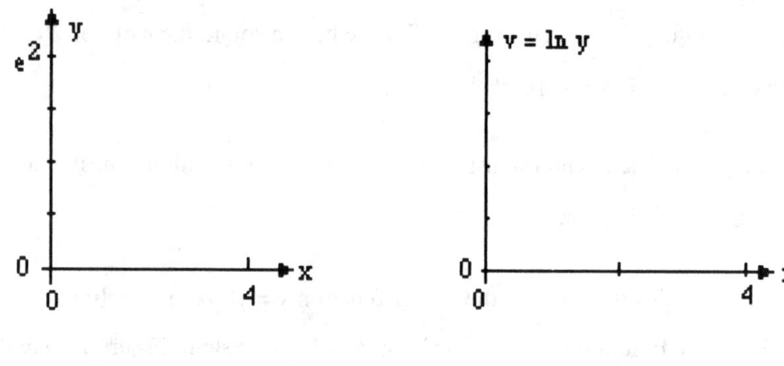

Figure 16 (a) (b)

10. Show the change of variables that changes the log function $y = 3 \log 2x$ into a linear function in a semi-log coordinate system. Graph the function on the rectangular coordinate system (Figure 17a) and on the semi-log coordinate system (Figure 17b).

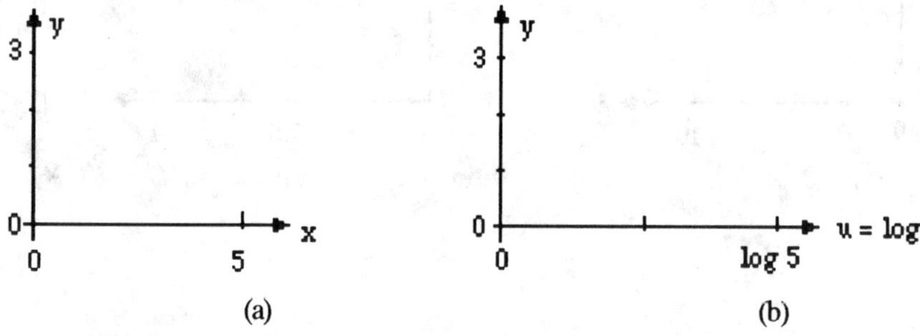

Figure 17 (a) (b)

11. The Beanie Baby Problem Many children and adults collect Beanie Babies, small stuffed animals created by the TY Company. Some of the beanies have been retired, resulting in higher values. The butterfly beanie named Flutter originally cost $5 in 1995, was retired in 1996, and had a value of $300 in 1997. According to *The Beanie Baby Handbook*, it is anticipated that Flutter may be worth $2000 by the year 2007.

(a) Find the best-fitting equation that this source may have used to model Flutter's value as a function of time.

(b) Use your equation from part (a) to approximate Flutter's value in the year 2027.

12. The Suggested Weight Problem The following data give the suggested weight in pounds for adults from 19 to 34 years of age of various heights.

(a) Find the best-fitting equation for women's weight as a function of height.

(b) Find the best-fitting equation for men's weight as a function of height.

(c) Compare the suggested weight for women and men who are 5 ft 7 in. tall.

(d) Compare the suggested heights of women and men who weigh 135 lb.

height	5'	5'2"	5'4"	5'6"	5'8"	5'10"	6'	6'2"	6'4"
women's weight (lb)	97	104	111	118	125	132	140	148	156
men's weight (lb)	128	137	146	155	164	174	184	195	205

13. The Taxes By Year Problem The following data give the tax owed in various years for a single person earning $20,000.

(a) Find the best-fitting equation.

(b) Estimate the taxes that a single person earning $20,000 will pay in 1995 if this trend continues.

year	1975	1991	1992	1993
taxes	$4153	$2168	$2115	$2093

[Source: *Information Please Almanac Atlas and Yearbook*]

14. The Taxable Income Problem The following data give the tax owed from taxable incomes in the year 1993.

(a) Find the best-fitting equation for someone filing a single return.

(b) Find the best-fitting equation for someone filing a joint return.

(c) Compare the tax paid by two people earning $25,000 if they are unmarried filling separately with a married couple earning $50,000 filing jointly.

taxable income	$10,000	$20,000	$30,000	$40,000	$50,000
single return	$593	$2093	$3833	$6633	$9433
joint return	-$1511	$236	$2160	$3660	$5160

15. The Snowboarding Problem Suppose you own a ski resort and have noticed that the number of skiers is decreasing and the number of snowboarders is increasing. You are thinking of designating half the runs for skiers and half the runs for snowboarders. According to the following data, in what year will the number of skiers and snowboarders be the same?

Year	Skiers	Snowboarders
1990	11,354,000	1,455,000
1991	10,427,000	1,577,000
1992	10,782,000	1,227,000
1993	10,495,000	1,841,000
1994	10,620,000	2,061,000
1995	10,140,000	2,254,000
1996	10,466,000	3,711,000

[Source: *1997 SIA Snow Sports Facts*]

16. The Kepler's Law Problem The following data give the length of the orbit in earth years and average distance from the sun in astronomical units, AU. (1 AU = 93 million miles, the distance from Earth to the sun.)

(a) Find the best-fitting equation that expresses length of the orbit in terms of average distance from the sun.

(b) The planet Uranus was discovered in 1781 by the amateur astronomer Sir William Hershel using his homemade telescope. Compare the length of its year as calculated by your equation in part (a) with that given by Kepler's law, $y^2 = x^3$, for x = 18.97 AU.

planet	Mercury	Venus	Earth	Mars	Jupiter	Saturn	Uranus	Neptune	Pluto
distance from sun (AU)	0.39	0.72	1.0	1.52	5.20	9.54	18.97	30.06	39.53
length of orbit (yr)	0.24	0.62	1.0	1.88	11.87	29.48	84.07	164.91	248.6

*17. The Marriage and Divorce Problem The following data give the number of marriages and divorces in the United States for various years.

(a) Find the best-fitting equation for marriages

(b) Find the best-fitting equation for divorces.

(c) When do the equations predict that the number of marriages equal the number of divorces?

year	1950	1960	1970	1980	1990
marriages	1,667,231	1,523,000	2,158,802	2,406,708	2,448,000
divorces	385,144	393,000	708,000	1,182,000	1,175,000

18. The U.S. Population Problem The following data give the U.S. population rounded to millions of people for various years.

(a) Find the best-fitting power and best-fitting exponential equations.

(b) Use your equations from part (a) to estimate the missing entries.

year	1900	1910	1920	1930	1940	1950	1960	1970	1980	1990	1995
population	76	92	106	123	132	151	?	203	227	249	?

[Source: Department of Commerce, Bureau of the Census]

19. The Olympic Winners Problem The following data give the men's and women's winning times in seconds for the 100-m dash for various years. Find the best-fitting equation for (a) men and (b) women.

year	1900	1920	1932	1952	1960	1972	1980	1992
men	11.0	10.8	10.3	10.4	10.2	10.14	10.25	9.96

year	1928	1932	1952	1960	1972	1980	1992
women	12.2	11.9	11.5	11.0	11.07	11.06	10.82

[Source: *Sports Illustrated Sports Almanac*]

20. The Education/Income Problem The following data give the annual earnings in 1995 for men and women categorized by the amount of education completed.

(a) Find the best-fitting equation for men.

(b) Find the best-fitting equation for women.

(c) Estimate the amount of education required for a woman's income to match a man's.

Education	Men	Women
no high school diploma	$16,748	$9,790
high school graduate	$26,333	$15,970
some college or associate's degree	$29,851	$17,962
bachelor's degree	$46,111	$26,841
advanced degree	$69,588	$37,813

[Source: *World Almanac, 1998*]

21. The Rainbow Kite and Balloon Company You are considering a new advertising campaign. The following data give the percent of U.S. households with television sets in various years.

year	1950	1955	1960	1965	1970	1975	1978	1979	1980	1981	1982
% having TV	9	65	87	93	95	97	98	98	98	98	98

[Source: *United States Statistical Abstracts*]

(a) Decide the nature of the graph. Massage and/or adjust the data as needed.

(b) Find the best-fitting equation that models the data.

(c) Use your equation to find the percent of U.S. households that were able to watch the landing of *Apollo XII* on the moon in 1969.

(d) Use your equation to find the percent of U.S. households that will watch your advertising commercial when Dick Clark bring in the new millennium in the year 2001. (The second millennium ends on Dec. 31, 2000.)

22. Wonderworld Amusement Park A new feature at the park is paintball battles. When a burst of compressed nitrogen is forced against a metal bolt held in place by a spring in a paintball gun, the paintball is propelled at a given speed. The speed can be adjusted by turning a velocity knob on the gun. When turned clockwise this knob tightens the spring around the bolt, giving the bolt more potential energy and increasing the force of each shot. An electrochronograph recorded the speed of the paintball in ft/sec for various numbers of revolutions of the velocity knob.

revolutions	0	1	2	3	4	5	6	7	8	9	10
speed (ft/sec)	15	20	40	80	170	250	340	390	405	409	410

(a) Decide the nature of the graph.

(b) Massage and/or adjust the data.

(c) Find the best-fitting equation that models the data.

23. <u>The Energy from Earthquakes Problem</u> The formula $\log E = 11.4 + 1.5M$ gives the energy E in ergs released by an earthquake of magnitude M on the Richter scale. Graph E as a function of M and label the following points.

(a) Seattle's 1964 quake, $M = 6.2$

(b) Alaska's 1964 quake, $M = 7.6$

(c) Managua's 1973 quake, $M = 7$

(d) A tornado that release energy $E = 4 \times 10^{18}$ ergs

(c) A 1-megaton nuclear bomb, $E = 4.2 \times 10^{15}$ ergs

24. The Actual-Intensity-of-Sound Problem The Weber-Fechner law expresses the perceived amplitude A of a sound relative to the actual intensity S and the least audible (zero) sound $Z \approx 10^{-12}$ watt per square meter (W/m^2) as $A = 10 \log \left(\frac{S}{Z}\right)$. Graph A as a function of S and label the points modeling the following sounds.

10^{-11}	Average threshold of hearing
10^{-3}	Freeway sounds
1	Rock music
10	Threshold of pain

25. The Age and Height Problem Collect data showing your age in months and corresponding height in inches.

(a) Fill in the table and plot the data in a Cartesian coordinate system.

age (months)	-9						
height (in.)	0						

(b) From the graph in part (a), decide the nature of the graph.

(c) Massage and/or adjust the data.

(d) Find the best-fitting equation that models your age and height.

(e) Use your equation to test the hypothesis that adult height is twice a 2-year-old's height.

26. The Price Problem Collect data showing the price of one commodity (such as a particular brand of pop) and the corresponding per unit price for various sizes of containers.

(a) Fill in the table and plot the data in a Cartesian coordinate system.

size	0						
per unit price	0						

(b) From the graph in part (a), decide the nature of the graph.

(c) Massage and/or adjust the data.

(d) Find the best-fitting equation the data.

(e) Use your equation to test the hypothesis that the giant economy size saves money.

A Guided Review of Chapter 4

4.1 Exponential Functions

1. The Stacked Paper Problem Consider cutting a piece of paper in half and stacking the pieces. How many cuts are required to create a stack of paper 1 m high if paper is 0.006 cm thick?

2. Find the amount of money in an account if $1000 is invested at 5.5% interest compounded as indicated for 3 years.

(a) annually (b) monthly (c) weekly (d) hourly (e) continuously

4.2 Logarithmic Functions

For Exercises 3 and 4 write each exponential equation as a logarithmic equation.

3.(a) $10^{-1} = 0.1$ 4.(a) $10^{0.30} \approx 2$

(b) $e^{-1} \approx 0.37$ (b) $e^{0.69} \approx 2$

For Exercises 5 and 6 write each logarithmic equation as an exponential equation.

5.(a) $\log \frac{1}{100} = -2$ 6.(a) $\log y = 0$

(b) $\ln 10 \approx 2.30$ (b) $\log y = -1$

For Exercises 7 to 10 solve each logarithmic equation by using the definition of a logarithm.

7. $\log x = -1$ 8. $\ln (1 - 3x) = 0$

9. $\log x^2 = 2$ 10. $\ln \sqrt{1 - x} = 0$

For Exercises 11 to 16 solve each exponential equation by using the definition of a logarithm.

11. $10^x = 0.0001$ 12. $10^{-x} = e$

13. $e^x = 2.718$ 14. $10^{-x} - 10 = 0$

15. $2e^x = 20$ 16. $10(10^{-2x}) = 1$

For Exercises 17 to 20 graph each function, and label asymptote and intercept(s).

17. $y = 1 - e^{-x}$ 18. $y = -\ln(x - 1)$

19. $f(x) = 1 - \log(x - 10)$ 20. $f(x) = e - e \ln (-x)$

For Exercises 21 and 22 find the equation of the inverse.

21. $y = e^{-x} - 3$ 22. $y = \ln(ex)$

23. The Big Bucks Problem Assume you have $7500 invested in an account that is paying 9.0% continuously compounded interest. I have $15,000 invested in an account that is paying 4.5% continuously compounded interest. Use the formula $A = Pe^{rt}$ to find out how long it will take before we have the same amount of money.

4.3 Theorems on Logarithms

24. The Populations of France and Germany Problem In 1990, France had a population growth rate of nearly 0.4% and a population of approximately 56.173 million.

(a) How long would it take for France's population to double?

(b) How long would it take France's population to match West Germany's with a 1990 population of 60.539 million, a birthrate of 10 per 1000, and a death rate of 11 per 1000?

(c) How long would it take France's population to match all of Germany's population if East Germany had a population in 1990 of 16.649 million and zero population growth?

[Source: United Nations Population Fund]

For Exercises 25 to 32 solve each equation for x. Check the solution to make sure that it is in the domain of each function.

25. (a) $\ln x = -2$
    (b) $\ln x = 0$
    (c) $\ln x = 2$

26. (a) $\log_3 x = 2$
    (b) $\log_3 (-x) = -2$
    (c) $\log_3 (1 - x) = 1/2$

27. (a) $10^x = 2$
    (b) $e^x = 2$

28. (a) $10^{1-x} = e$
    (b) $e^{x-1} = 10$

29. $\log_2 x + \log_2 (x - 3) = 2$

30. $\log x + \log (x + 3) = 1$

31. $\log(x + 2) = \log x - 2$

32. $\ln x - \ln (x - 1) = 1$

For Exercises 33 to 36 solve the exponential equation $e^x = 1/e$ using the first step described.

33. Rewrite both sides of the equation as powers of e.

34. Use the definition of a logarithm to rewrite the equation.

35. Take the common logarithm of both sides.

36. Take the natural logarithm of both sides.

For Exercises 37 to 40 use the graphs of each pair of functions to find the point(s) of intersection.

37. $f(x) = \ln (x + 3)$, $g(x) = 1 + \ln x$

38. $f(x) = \log (x + 4)$, $g(x) = \log(2x - 4)$

39. $f(x) = \log_3(\frac{1}{x})$, $g(x) = 2 - \log_3(x + 4)$

40. $f(x) = \log_5(x + 4)$, $g(x) = 1 - \log_5 x$

4.4 More on Exponential and Logarithmic Functions

41. The Loan Problem The function $M = \frac{iP}{1 - a^{-n}}$, where a = 1 + i, describes the monthly payment M of a loan of P dollars for n months with interest per payment period i. Find the number of payments n required to pay back a car loan of $22,500 if the annual percentage rate is 6.5% if M = $450. (Let i = 0.065/12.)

42. An Annuity Problem The function $S = R(\frac{a^n - 1}{i})$, where a = 1 + i describes the future value S of an annuity with n regular deposits of R dollars each and an interest rate i per saving period. Find the number of payments n required to create an annuity worth $1,000,000 if you deposit $300/month and earn 10%/year interest. (Let i = 0.10/12.)

4.5 More Applications Including Logistic Equations

43. The Cooling Coffee Problem If a 180°F latté cools to 175°F in 10 min in a 68°F room, find the temperature of the coffee in 1/2 hr. Use Newton's law of cooling formula $T = De^{-kt} + T_a$.

44. The Turkey Cooking Problem If the oven temperature is 350°F and the initial temperature of a roast is 35°F, find the time it will take for the internal temperature of the turkey to be between 165 and 170°F, knowing that the temperature after 30 min was 50°F. Use Newton's law of "heating", $T = T_a - De^{-kt}$.

45. The Epidemic Problem The spread of diseases has a natural limit, the number of people susceptible. This is an application of the logistic curve. Find the length of time before virtually everyone on a campus of 25,000 students has been exposed to a virus if initially one person brought it to school and 1 week later 12 people had been exposed.

46. The Cooled-off Corpse Problem You discover a murder victim. You note the room temperature as 70°F and the temperature of the corpse as 90°F. The body was instantly taken to the morgue, where the room temperature was 40°F and the corpse temperature is still 90°F. One hour later, the corpse temperature was 80°F. Use the function

$$T(t) = \begin{cases} Ae^{-kt} + 70, & t \leq 0 \\ Ce^{-kt} + 40, & t \geq 0 \end{cases}$$

where T(0) = 90 is the time you took the first temperatures, to find the time of death as follows:

(a) Use $T(t) = Ce^{-kt} + 40$ with T(0) = 90 to find C.

(b) Find k using your results from part (a).

(c) Use $T(t) = Ae^{-kt} + 70$ and T(0) = 90 to find the value of A.

(d) Use your results from part (c) to find the time of death.

[Hint: The value of k is the same in both equations as it is determined by the characteristics of the cooling body.]

47. The Emerging Nation Problem In 1800, the U.S. population was about 5 million people and the population of the United Kingdom was over 16 million. The US population, growing exponentially, doubled in 20 years. The United Kingdom's population, increasing nearly linearly, doubled in 65 years.

(a) Write an equation for the population of each country as a function of time.

(b) Estimate the year when the U.S. population caught that of the United Kingdom.

[Source: *The US Statistical Atlas, Census of 1900*]

48. The Radioactive Decay Problem Suppose that you are to estimate the half-life of an unknown radioactive substance to determine a safe waste treatment. You find initially that the substance is emitting 15 disintegrations per gram per minute, $N_0 = 15$. Six months later it is emitting 14 disintegrations per gram per minute. How long will the substance have to be stored before 90% has decayed?

4.6 Logarithmic Scales and Curve Fitting

49. The following gives the annual tuition and required fees for full-time students attending public 2- and 4-year colleges for the academic year ending in the given year.

year	1980	1985	1990	1991	1992	1993
2-year college	355	584	756	824	937	1018
4-year college	662	1117	1608	1707	1933	2190

[Source: U.S. National Center for Educational Statistics]

(a) Find the best-fitting equation for tuition at a 2-year college as a function of year.

(b) Find the best-fitting equation for tuition at a 4-year college as a function of year.

(c) Use the equations to predict the annual college fee for a child born in 1996.

50. The following data give the price of a first-class U.S. postage stamp for various years.

year	1971	1975	1985	1988	1995
cost (¢)	8	13	22	25	32

(a) Find the best-fitting equation.

(b) Predict the price of a stamp in the year 2000.

(c) If the U.S. Postal Service reduces the price of a stamp to 30¢ in the year 1996, predict the price of a stamp in the year 2000.

51. The following data give the manufacturers' value in millions of dollars of recording media for various years.

year	1980	1985	1990	1991
long-play phonograph albums	2290.3	1280.5	86.5	29.4
prerecorded cassettes	776.4	2411.5	3472.4	3019.6
compact discs	–	389.5	3451.6	4337.7

(a) Find the year when the value from producing long-play phonograph albums and prerecorded cassettes was about the same.

(b) Find the year when the value from producing prerecorded cassettes and compact discs was about the same. [Source: *Inside the Recording Industry: A Statistical Overview*]

52. Have four or five friends stand with their arms extended at their sides and measure their "wing span" from finger tip to finger tip. Measure their height. Then find the best-fitting equation that relates wing span to height.

Chapter 4 Test

For Exercises 1 and 2 solve each logarithmic equation by using the definition of a logarithm.

1.(a) $\log x = -\frac{1}{2}$

(b) $\log x^2 = -1$

2.(a) $\ln(ex) = 2$

(b) $\ln\sqrt{\frac{x}{e}} = -1$

For Exercises 3 and 4 solve each exponential equation by using the definition of a logarithm.

3.(a) $10^x = 200$

(b) $e^x = 200$

4.(a) $10^x = e$

(b) $e^x = 10$

For Exercises 5 and 6 graph each function and label the intercepts and asymptotes.

5. $f(x) = e^{-x} + 1$

6. $f(x) = \ln(1 - x)$

For Exercises 7 to 10 solve each equation.

7. (a) $e^{2x} - e^x = 0$

(b) $e^{2x} - 2e^x + 1 = 0$

8.(a) $\log_x 16 = 4$

(b) $\log_x \frac{1}{16} = 2$

9. $\log x - \log(x + 1) = -2$

10. $\log_3 x - \log_3(x + 1) = \log_3(x + 4) - 2$

11. Graph the following functions on the same coordinate system. Find and label the point(s) of intersection.

$$y = \ln x + 1$$
$$y = \ln(x + 1)$$

For Exercises 12 and 13 find the equation of the inverse.

12. $y = e^{x-1}$

13. $y = 1 - 2\ln x$

14. The Doubling and Halving Times Problem

(a) Find the doubling time for an investment that was worth \$10,000 one year ago and is expected to worth \$12,000 in 2 years if the investment is earning continuously compounded interest.

(b) Find the half-life for a new car that will be worth \$22,500 in one year and \$19,000 one year later if the depreciation is exponential decay.

15. The Decibel Scale Problem The decibel reading for a sound with intensity A is $D_A = 10\log\frac{A}{Z}$, where Z is the zero sound, the threshold of hearing; $Z = 10^{-12}$ W/m^2.

(a) Find the decibel readings for the threshold of hearing, $A = 10^{-12}$ W/m^2, and for the threshold pain, $A = 1$ W/m^2.

(b) What is the ratio of the intensities of a rock band playing at 120 dB on stage at the gorge, and two blocks away, where the intensity is only 85 dB?

16. The Ebola Virus Problem Find the logistics function for the data concerning the deadly Ebola virus, which kills 50 to 90% of those infected. On Wednesday, October 4, 1989 a shipment of research monkeys arrived at a Reston, Virginia quarantine center. One monkey brought the virus into the center, where there were 450 monkeys. On November 1 there were 29 infected monkeys. How long before virtually every monkey was sick?
[Source: *The Hot Zone*, R. Preston]

17. The Loan Problem The function $M = \frac{iP}{1 - a^{-n}}$, where a = 1 + i describes the monthly payment M of a loan of P dollars for n months with interest per payment period i. Find the number of payments n required to pay back a college loan of $20,000 if i = 0.005% (6%/yr) and the payments are M = $500/month.

18. The Annuity Problem The function $S = R(\frac{a^n - 1}{i})$, where a = 1 + i describes the future value S of an annuity with n regular deposits of R dollars each and an interest rate i per saving period. Find the number of payments n required to create an annuity worth $20,000 if i = 0.005% (6%/yr) and you are depositing $500/month.

19. The Folded-Paper Problem Suppose you fold a piece of paper in half, then punch a hole in it. Fold it again and punch another hole.
(a) Repeat for four folds and record the number of holes after each fold.
(b) Find the equation that best fits your data.
(c) How many folds and punches would it take to have 1024 holes?

20. The Emerging Nation Population Problem Find the equation of the best-fitting function for each country's population in millions of people. Find the year when the U.S. population caught that of the United Kingdom.

year	1800	1830	1850	1890	1900
United States	5	13	23	64	76
United Kingdom	17	27	28	38	42

[Source: *The U.S. Statistical Atlas, Census of 1900*]

Chapter 5

Trigonometry and the Circular Functions

The functions we have studied so far approach $+\infty$ or $-\infty$ on their ends, or are asymptotic. In this chapter we study functions whose graphs repeat periodically. To model repeating motion such as orbiting space vehicles and planets, swinging pendulums, vibrating strings, changing tides, and recurring economic patterns, we need the **circular functions**. They include the sine, cosine, and tangent functions and their reciprocals, cosecant, secant, and cotangent.

We begin our study of these functions as they evolved historically with the study of right triangle trigonometry. Then we create the trigonometric functions and apply them in mathematical models involving triangles. We extend the properties of the trigonometric functions to the circular functions. With our modern approach and the use of technology, we are able to deemphasize much traditional vocabulary in favor of graphs and applications -- the heart of the modeling process.

A Historical Note on the Origin of the Word "sine"

The following quote is from The Other Side of the Equation, by Howard W. Eves (Boston: Prindle, Weber & Schmidt, Inc., 1969) p. 19.

"The origin of the word *sine* is curious. Āryabhata called it *ardhā-jyā* ("half chord") and also *jyā-ardhā* ("chord half"), and then abbreviated the term by simply using *jyā* ("chord"). From *jyā* the Arabs phonetically derived *jîba*, which, following the Arabian practice of omitting vowel symbols, was written as *jb*. Now *jîba*, aside from its technical significance, is a meaningless word in Arabic. Later writers, coming across *jb* as an abbreviation for the meaningless *jîba* decided to substitute *jaib* instead, which contains the same letters and is a good Arabian word meaning "cove" or "bay." Still later, Gherardo of Cremona (ca. 1150), when he made his translations from the Arabic, replaced the Arabic *jaib* by its Latin equivalent, *sinus*, whence came our present word *sine*."

5.1 Right Triangle Trigonometry

Describing Angles and Triangles
Defining Sine, Cosine, and Tangent
Introducing Two Special Triangles
Defining Angles of Elevation and Depression

Geometry is the study of plane figures, and **trigonometry** is specifically the study of triangles. In this section we begin by reviewing some terms from geometry that are essential to trigonometry. Then we define sine, cosine, and tangent (abbreviated sin, cos, and tan) as ratios of the sides of a right triangle. The study of right triangle trigonometry dates back to A.D. 500, when the sine function was introduced.

In Example 5 of this section we apply trigonometry to find how much of the top of Mt. St. Helens in Washington State blew off during its volcanic eruption in 1980.

Describing Angles and Triangles

In geometry angles are measured in degrees, where a complete circle measures 360 degrees. Traditionally, each degree is divided into 60 minutes and each minute into 60 seconds. In notation, an angle whose measure is 35 degrees, 30 minutes is written $35°30'$. Our calculators use decimal degrees. Since 30 minutes is one-half a degree, $35°30' = 35.5°$.

We classify angles according to their angle measurement. Figure 1a shows an acute angle, $\angle A$, and Figure 1b shows a right angle, $\angle B$. The measure of an acute angle is between $0°$ and $90°$, while the measure of a right angle is $90°$.

Figure 1 (a) (b)

The small square at the vertex of $\angle B$ in Figure 1b indicates that $\angle B$ is right angle.

Figure 2a shows an equilateral triangle, a triangle having three equal sides and three equal angles. Figure 2b shows a right triangle, a triangle containing a $90°$ angle and Figure 2c shows an isosceles triangle, a triangle having two equal sides. The angles opposite the equal sides are also equal. The hash marks indicate equal sides or equal angles.

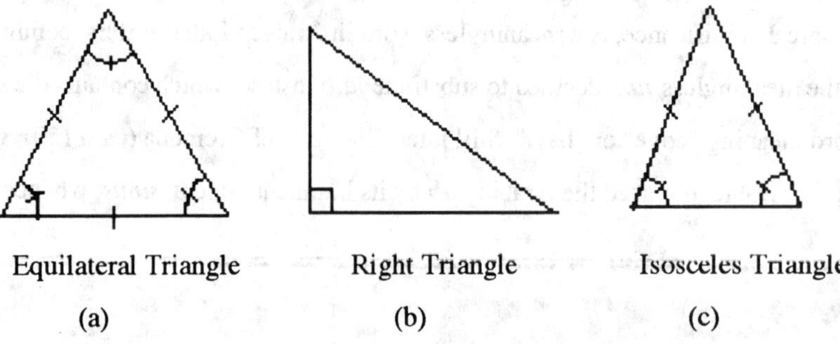

Equilateral Triangle Right Triangle Isosceles Triangle

Figure 2 (a) (b) (c)

Figure 3 shows ΔABC, a right triangle where c represents the length of the **hypotenuse**, the side opposite the right angle. The other two sides, a and b, are referred to as the legs of the right triangle. The vertices of a triangle are usually labeled with capital letters, while the corresponding side opposite each angle is typically labeled using the same lowercase letter.

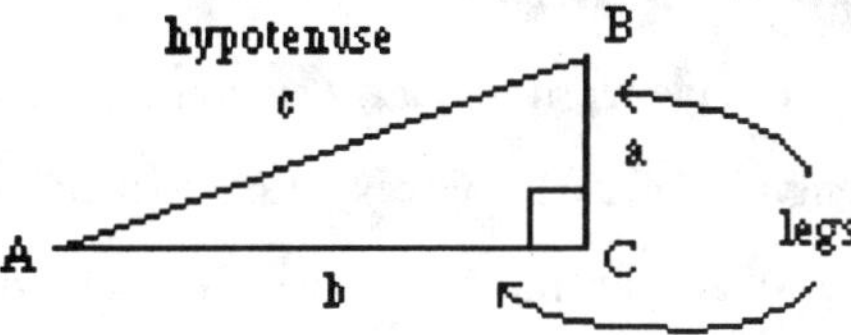

Figure 3

The relationship between the lengths of the sides of such a right triangle are stated in the Pythagorean theorem,

$$a^2 + b^2 = c^2$$

Given two sides of a right triangle, the Pythagorean theorem can be used to find the length of the third side. To find the measure of the third angle in a triangle, we will use the fact that the sum of the three angles of a triangle equals 180°,

$$\angle A + \angle B + \angle C = 180°$$

In this context, $\angle A + \angle B + \angle C$ refers to the sum of the measures of the three angles.

Example 1 *Solving a triangle*

Find the length of side c and the measure of $\angle A$ in ABC (Figure 4).

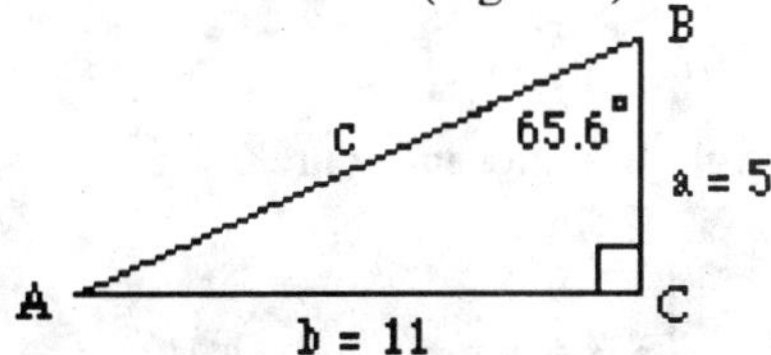

Figure 4

Solution

$$a^2 + b^2 = c^2 \qquad \textit{Substitute } a = 5 \textit{ and } b = 11.$$

$$5^2 + 11^2 = c^2$$

$$146 = c^2 \qquad \textit{Take the square root of both sides.}$$

$$c = \sqrt{146} \approx 12.08 \qquad \textit{Choose the positive value.}$$

$\angle A + \angle B + \angle C = 180°$ and $\angle B = 65.6°$, $\angle C = 90° \Rightarrow$

$$\angle A + 65.6° + 90° = 180°$$

$$\angle A = 24.4°$$

To explore the relationship between the sides of a right triangle, go to "Explore the Concepts," Exercises E1 - E4, before continuing (page 355). For the interactive version, go to the *Precalculus Weblet - Instructional Center - Explore Concepts.*

Defining Sine, Cosine, and Tangent

In the study of right triangle trigonometry, the words *adjacent side* and *opposite side* refer to sides of a right triangle relative to one of the acute angles. Often, the Greek letters θ (theta), α (alpha), or β (beta) are used to represent angles. In Figure 5 the sides of a right triangle are named in relation to the location of angle θ. The hypotenuse is always the side across from the right angle. The adjacent side is next to angle θ, and the opposite side is across from θ.

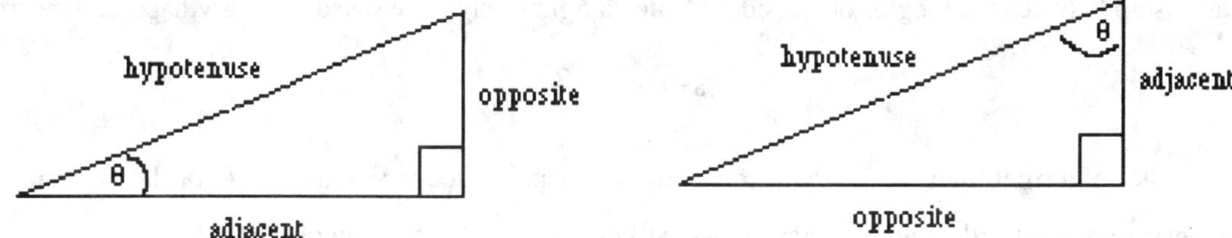

Figure 5

The following definitions of **sine, cosine** and **tangent** describe ratios of the sides of a right triangle.

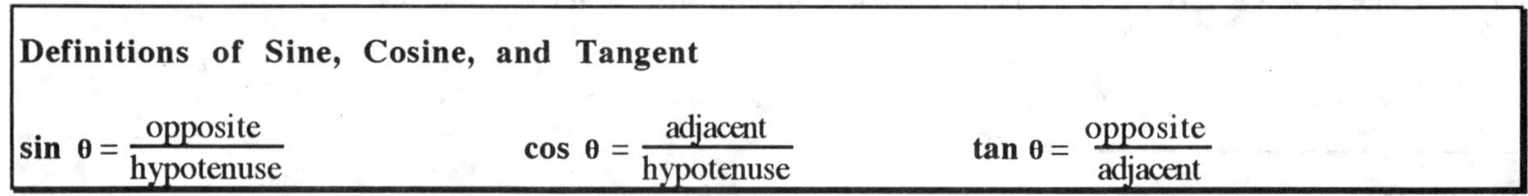

Definitions of Sine, Cosine, and Tangent

$\sin\theta = \dfrac{\text{opposite}}{\text{hypotenuse}}$ $\cos\theta = \dfrac{\text{adjacent}}{\text{hypotenuse}}$ $\tan\theta = \dfrac{\text{opposite}}{\text{adjacent}}$

Example 2 *Using the definition of sine*

Find the length of sides b and c in the right triangles in Figure 6.

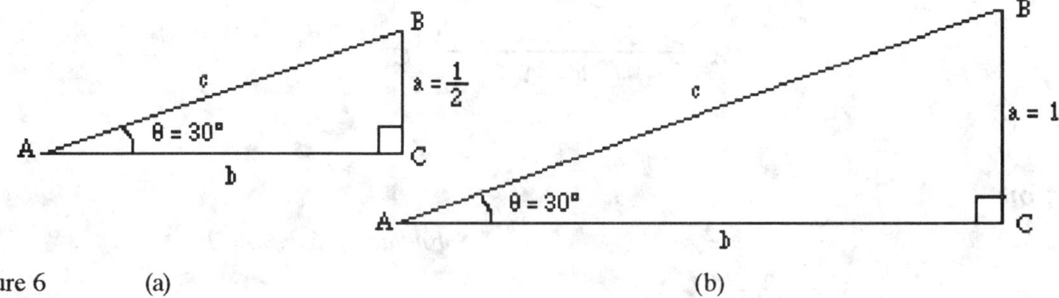

Figure 6 (a) (b)

Solution

(a)

$$\sin\theta = \frac{\text{opposite}}{\text{hypotenuse}}$$
$$\sin 30° = \frac{1/2}{c}$$
$$0.5 = \frac{1/2}{c}$$
$$0.5c = 1/2$$
$$c = 1$$

(b)

$$\sin\theta = \frac{\text{opposite}}{\text{hypotenuse}}$$
$$\sin 30° = \frac{1}{c}$$
$$0.5 = \frac{1}{c}$$
$$0.5c = 1$$
$$c = 2$$

Using the Pythagorean theorem yields the length of side b.

$a^2 + b^2 = c^2$	$a^2 + b^2 = c^2$
$(\frac{1}{2})^2 + b^2 = 1^2$	$1^2 + b^2 = 2^2$
$b^2 = 1 - \frac{1}{4}$	$b^2 = 4 - 1$
$b = \sqrt{\frac{3}{4}} = \frac{\sqrt{3}}{2}$	$b = \sqrt{3}$

The triangles in Example 2 have the same shape. They are called *similar* triangles.

What do you think the relationship is between the size of corresponding angles and lengths of corresponding sides of similar triangles?

Corresponding angles of similar triangles are equal and the ratios of corresponding sides are equal. That means the trigonometric ratios of equal angles in similar triangles always have the same value. Notice that $\sin 30° = 0.5$ in both triangles of Example 2.

In the next example we find the angle whose cosine is 0.6. The input is the value of the cosine ratio, the output is the measure of the corresponding angle. The labels on the keys vary from calculator to calculator, but the process is called finding the **inverse cosine**.

NOTE: The $\boxed{\text{COS}^{-1}}$ key is the inverse key. It does not mean $\frac{1}{\cos\theta}$.

Example 3 *Using the definitions of cosine and sine*

(a) Find the measure of angle θ using the definition of cosine (Figure 7).

(b) Find the length of side y two ways: using the definition of sine and using the Pythagorean theorem.

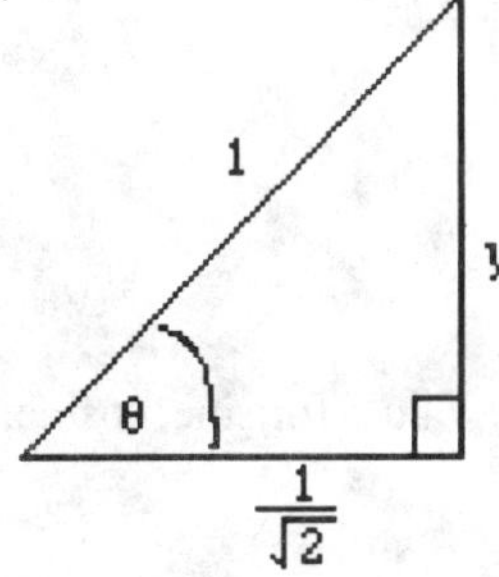

Figure 7

Solution

(a)

$$\cos\theta = \frac{\text{adjacent}}{\text{hypotenuse}}$$

$$\cos\theta = \frac{1/\sqrt{2}}{1}$$

$$\theta = 45°$$

(b)

$$\sin\theta = \frac{\text{opposite}}{\text{hypotenuse}}$$
$$\sin 45^\circ = \frac{y}{1}$$
$$y \approx 0.7071$$

$$a^2 + b^2 = c^2$$
$$\left(\frac{1}{\sqrt{2}}\right)^2 + y^2 = 1^2$$
$$\frac{1}{2} + y^2 = 1$$
$$y^2 = \frac{1}{2}$$
$$y = \frac{1}{\sqrt{2}} \approx 0.7071$$

Both methods for finding the value of y in part (b) yield the same result, $1/\sqrt{2} \approx 0.7071$. Since the legs of the triangle in Figure 7 both equal $1/\sqrt{2}$, the triangle is isosceles. The measure of the other acute angle is also 45°.

Introducing Two Special Triangles

In mathematical modeling involving triangles, we can often set up the models so that the angles involved are familiar ones whose exact values of sine, cosine, and tangent are known. Such right triangles, whose acute angles measure 30° and 60° (30-60-90 triangles), and those whose acute angles both measure 45° (45-45-90 triangles) have exact values for sine, cosine, and tangent. In Example 2 we found the length of the sides of two 30-60-90 triangles. They were in the ratio $1 : \frac{1}{2} : \frac{\sqrt{3}}{2}$. In Example 3 we found the length of the sides of a 45-45-90 triangle. They were in the ratio $1 : \frac{1}{\sqrt{2}} : \frac{1}{\sqrt{2}}$. These ratios are summarized in Figure 8.

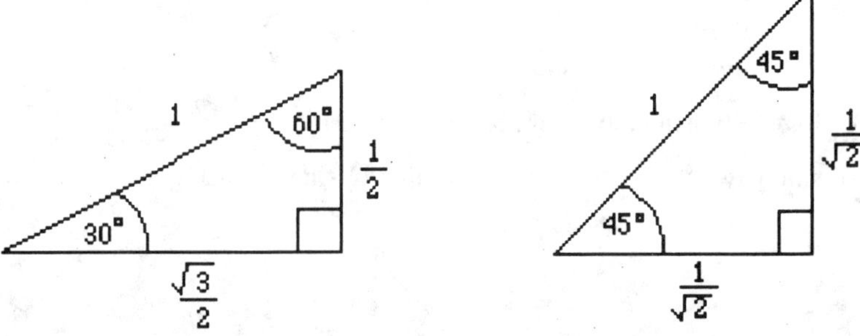

Figure 8

From Figure 8 we can determine the exact values for sine, cosine, and tangent of these angles without using a calculator (Table 1).

Table 1

Angle	$\sin\theta = \frac{\text{opposite}}{\text{hypotenuse}}$	$\cos\theta = \frac{\text{adjacent}}{\text{hypotenuse}}$	$\tan\theta = \frac{\text{opposite}}{\text{adjacent}}$
30°	$\frac{1}{2}$	$\frac{\sqrt{3}}{2}$	$\frac{1}{\sqrt{3}}$
45°	$\frac{1}{\sqrt{2}}$	$\frac{1}{\sqrt{2}}$	1
60°	$\frac{\sqrt{3}}{2}$	$\frac{1}{2}$	$\sqrt{3}$

Textbook problems here and in calculus are often set up to use these familiar angles in order to simplify arithmetic.

Defining Angles of Elevation and Depression

Many applications involving angle measure describe an angle of elevation or an angle of depression. An **angle of elevation** is an angle measured from a horizontal line upward, while an **angle of depression** is measured from a horizontal line downward (Figure 9).

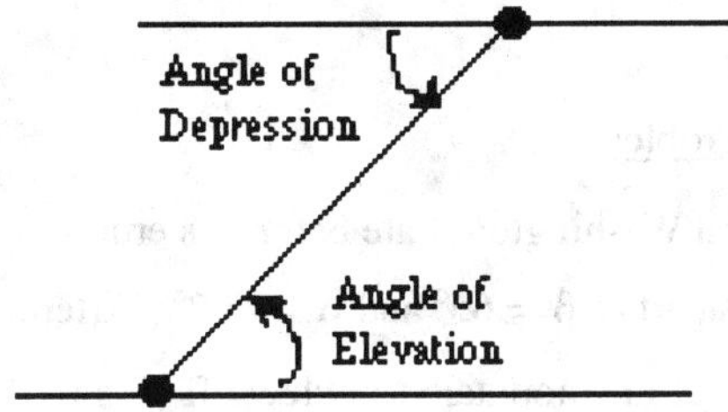

Figure 9

Example 4 The Kite Problem

Suppose you let out 500 ft of kite string and planted the spool in the sand on the beach, noticing that the angle of elevation of the kite is approximately 45°. How high is the kite flying above the ground?

Thinking: Creating a Plan

A sketch shows the kite string is the hypotenuse of the right triangle (Figure 10).

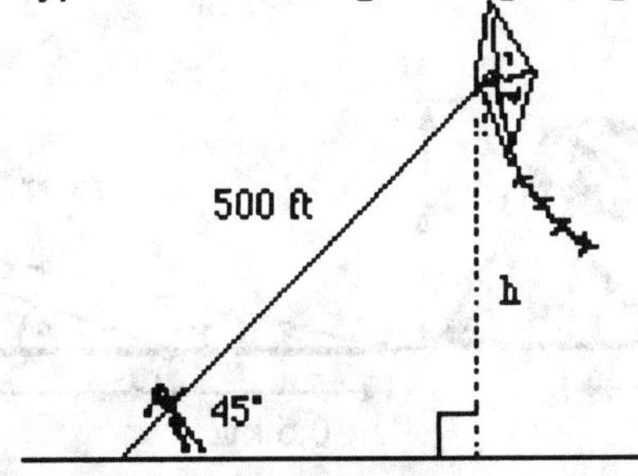

Figure 10

Communicating: Writing the Solution

$$\sin\theta = \frac{\text{opposite}}{\text{hypotenuse}}$$

$$\sin 45° = \frac{y}{500}$$

$$y = 500 \cdot \sin 45°$$

$$y \approx 353.6$$

The kite is flying about 354 ft above the ground.

Learning: Making Connections

Since the triangle in Figure 10 is a 45-45-90 triangle, the ratios of the side to the hypotenuse is $\frac{1}{\sqrt{2}}$: 1.

For this special triangle we could have found y using ratios.

$$\frac{\frac{1}{\sqrt{2}}}{1} = \frac{y}{500}$$

$$y = \frac{500}{\sqrt{2}} \approx 353.6$$

Example 5 The Mt. St. Helens Problem

To find the height of Mt. St. Helens in Washington State before its eruption in 1980, you measured the angles of elevation from two points 0.5 km apart as $\beta = 65°$ and $\alpha = 73.7°$. After the eruption you measured $\beta = 61.6°$ and $\alpha = 71.2°$. How much of the mountain top blew off (Figure 11)?

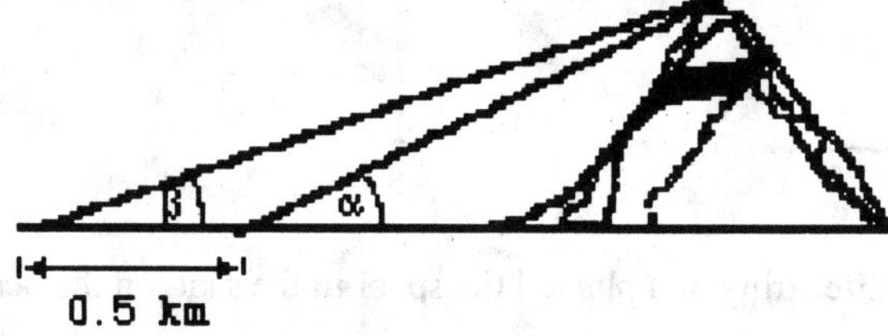

Figure 11

Thinking: Creating a Plan

We use tangent to create two equations with two variables, x and h_1 before the eruption, and x and h_2 after the eruption (Figure 12).

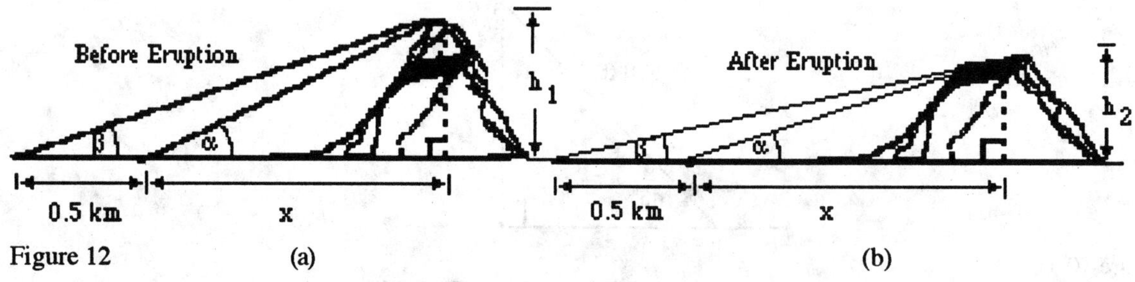

Figure 12 (a) (b)

Communicating: Writing the Solution

Before the eruption:

$$\tan\beta = \frac{h_1}{x + 0.5} \quad \text{and} \quad \tan\alpha = \frac{h_1}{x}$$

$$\tan 65° = \frac{h_1}{x + 0.5} \quad \text{and} \quad \tan 73.7° = \frac{h_1}{x}$$

$$2.14 = \frac{h_1}{x + 0.5} \quad \text{and} \quad 3.42 = \frac{h_1}{x}$$

$$2.14x + 1.07 = h_1 \qquad 3.42x = h_1$$

Therefore,

$$\begin{aligned} 2.14x + 1.07 &= 3.42x \\ 1.07 &= 1.28x \\ x &\approx 0.84 \end{aligned}$$

$h_1 = 3.42x$ and $x = 0.84 \Rightarrow$

$$h_1 = 3.42(0.84) \approx 2.9$$

The mountain was 2.9 km high.

After the eruption:

$\tan \alpha = \frac{h_2}{x}$ and $\alpha = 71.2°$, $x = 0.84 \Rightarrow$

$$\begin{aligned} \tan 71.2° &= \frac{h_2}{0.84} \\ 2.94 &\approx \frac{h_2}{0.84} \\ h_2 &\approx 2.5 \end{aligned}$$

$h_1 - h_2 = 2.9 - 2.5 = 0.4$

About 0.4 km = 400 m of the mountain blew off in the eruption of 1980.

Learning: Making Connections

Observations made before and after an event are often compared to study the event itself. Equations of the form $y = (\tan \theta)x$ are linear functions with slope $m = \tan \theta = \frac{\Delta y}{\Delta x}$. In this example we solved two systems of linear equations.

Explore the Concepts

For Exercises E1 to E4 use the scale provided to measure the lengths of the sides of each triangle. Calculate the ratios

$$\frac{y}{r}, \frac{x}{r}, \text{ and } \frac{y}{x}$$

which are called the sine, cosine, and tangent, respectively.

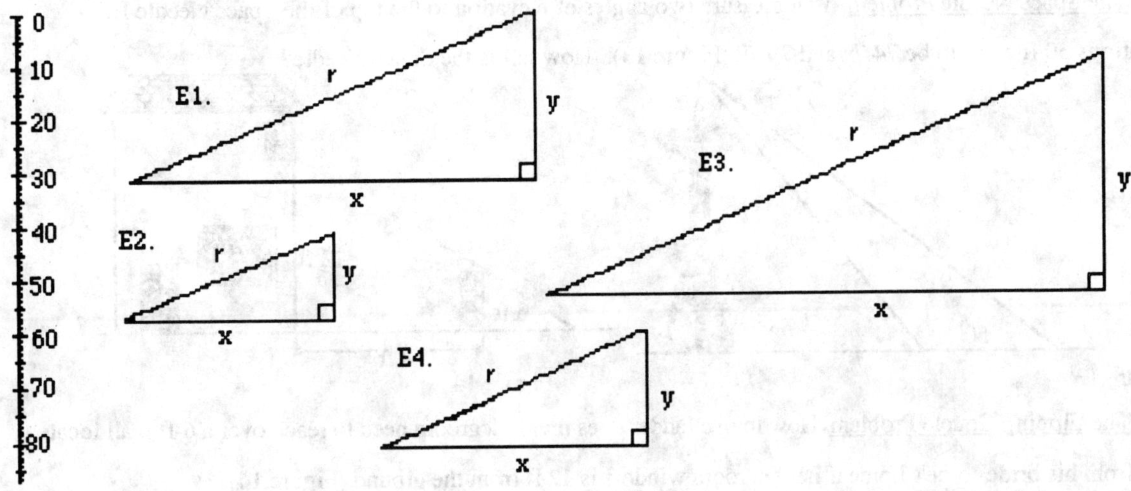

Exercises 5.1

Develop the Concepts

For Exercises 1 to 4 find the value of each expression.

1. (a) sin 45°
   (b) cos 45°
   (c) tan 45°

2. (a) sin 90°
   (b) cos 90°
   (c) tan 90°

3. (a) sin 50°
   (b) sin 130°
   (c) sin 230°
   (d) sin 310°

4. (a) cos 10°
   (b) cos 170°
   (c) cos 190°
   (d) cos 350°

For Exercises 5 to 8 find θ if $0° \leq \theta \leq 90°$.

5. (a) $\sin \theta = 0.5$
   (b) $\cos \theta = 0.5$

6. (a) $\sin \theta = 1$
   (b) $\cos \theta = 1$

7. (a) $\tan \theta = 1$
   (b) $\tan \theta = 0$

8. (a) $\sin \theta = \frac{\sqrt{2}}{2}$
   (b) $\cos \theta = \frac{\sqrt{2}}{2}$

For Exercises 9 to 12 use the Pythagorean theorem and/or the definitions of sine, cosine, and tangent to solve the triangles by finding the length of the missing sides and the measure of the missing angles.

9.

11.

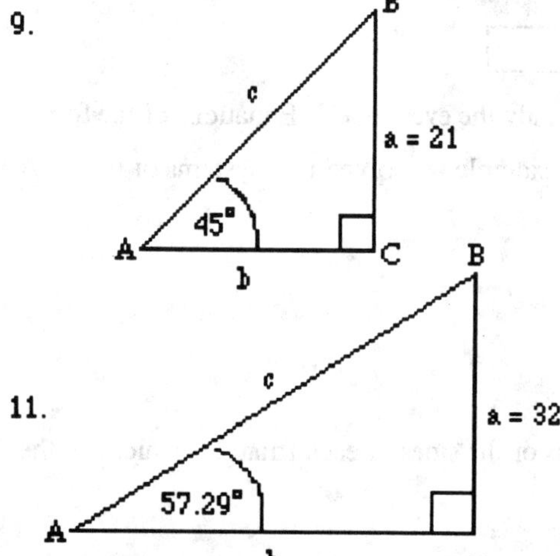

10.

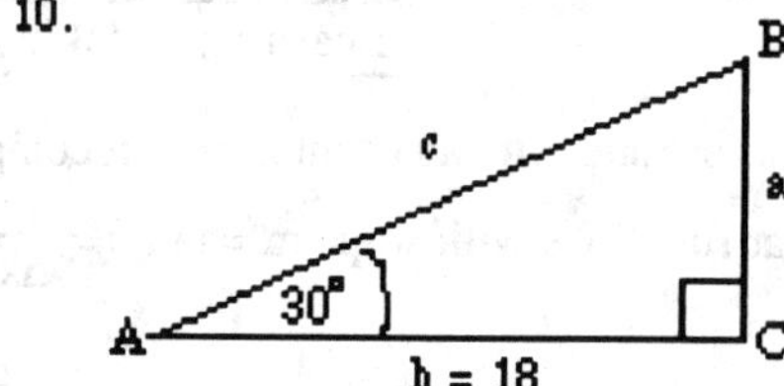

12.

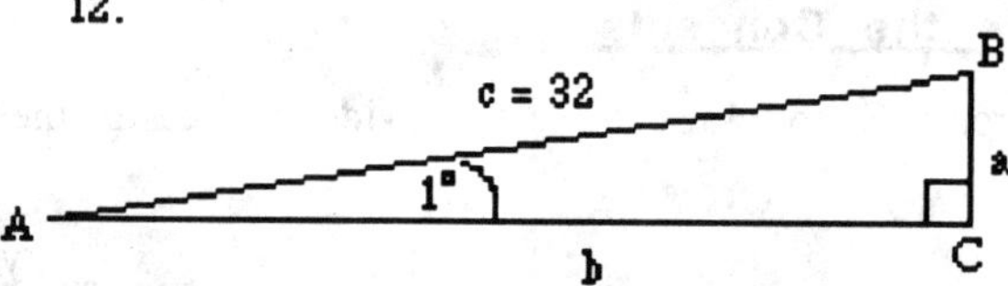

Apply the Concepts

13. The Space Needle Problem You measure two angles of elevation to the top of the Space Needle from locations 50 ft apart to be 74.7° and 79.7° (Figure 13). How tall is the Space Needle?

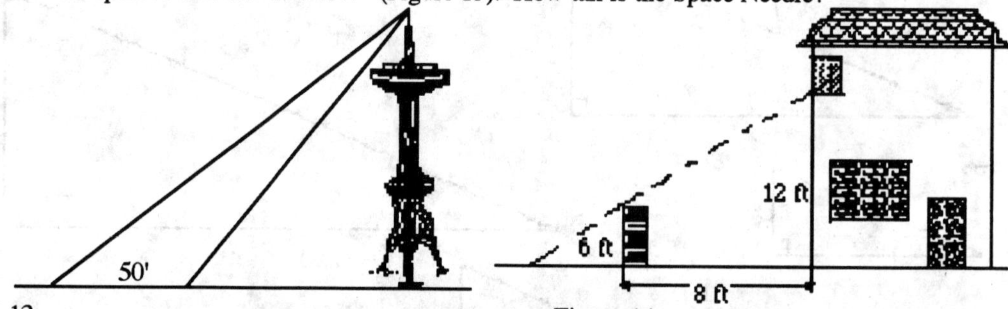

Figure 13

Figure 14

14. The Eloping Couple Problem How long a ladder does the bridegroom need to reach over a 6-ft wall located 8 ft from his bride-to-be's home if her bedroom window is 12 ft from the ground (Figure 14)?

15. The Yellowstone Fire Problem At Yellowstone National Park, Yogi Bear spotted a fire from the top of a 30-ft tree. If the angle of depression was 15.5°, how far from the base of the tree was the fire?

16. The Trapezoid Problem Find the area of trapezoid ABCD in Figure 15 using the formula $A = \left(\frac{b_1 + b_2}{2}\right)h$, where b_1 and b_2 represent the parallel bases and h represents the altitude, if $\angle A = 25°$, AB = 12, BC = 28, and AD = 40.

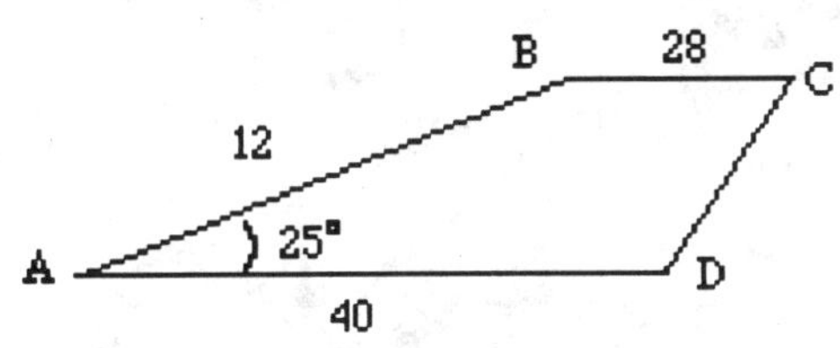

Figure 15

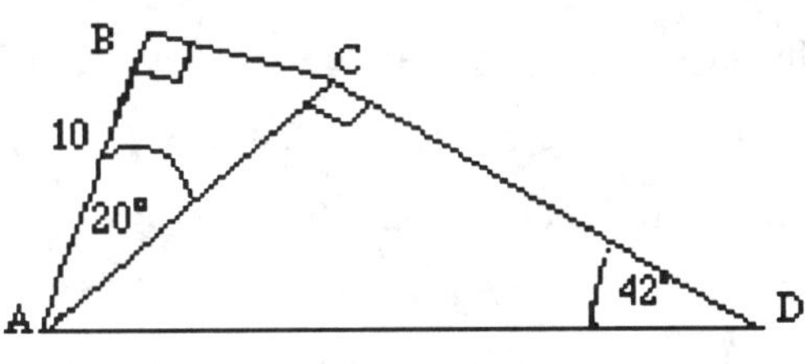

Figure 16

17. The Hypotenuse Problem Find the length of hypotenuse AD in Figure 16 if AB = 10, $\angle BAC = 20°$, and $\angle D = 42°$.

18. The Flagpole Problem A 100-ft flagpole to be secured on the rooftop of a building in the windy city of Chicago is to be supported by three cables attached 10 ft from the top of the pole and anchored 20 ft from its base on top of the building.

(a) Use trigonometry to calculate how much cable is necessary to secure the pole.

(b) Verify your result from part (a) using the Pythagorean theorem.

*19. The Ninth-Hole Problem If the ninth hole at Camelback Golf Course has a 90° dogleg to the right, the suggested line to drive the ball is 230 yd straight ahead toward the palm trees and then 180 yd to the green, as shown in Figure 17. If the best shot you have ever driven is 285 yd, should you try a direct hit across the lake from the tee to the hole? If not, at what angle to the fairway should you hit your best shot?

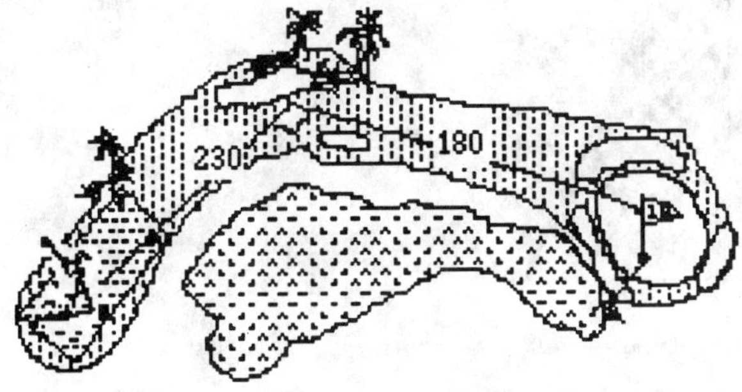

Figure 17

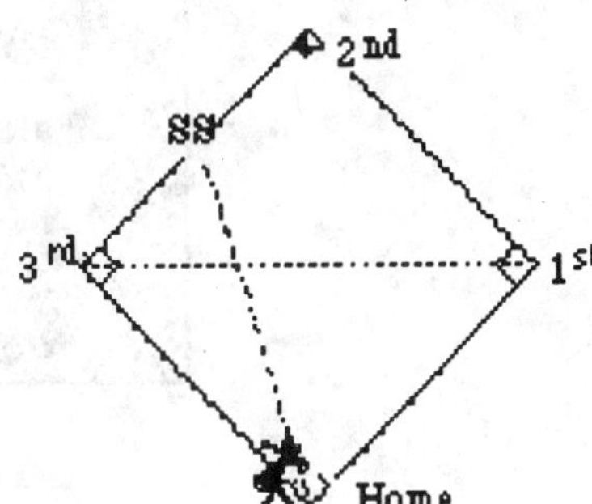

Figure 18

20. The Baseball Diamond Problem If the sides of a baseball diamond measure 90 ft, how far is it from third base to first base? How far is it from shortstop to home if shortstop is located midway between second and third base (Figure 18)?

21. The Airport Problem Find the angle of descent (depression) that a jet approaching an airport needs to fly if the plane is 5 mi (ground distance) from touchdown and flying at an altitude of 5000 ft.

22. The Derby Island Problem An airplane charting islands in Peaceful Sound is flying 300 ft above the sound approaching Derby Island (Figure 19). The angle of depression is 25° to the closest side of the island and 15° to the farthest side. How long is the island?

Figure 19

23. The Lighthouse Problem The angles of depression from the top of a lighthouse 150 ft above the sea to two boats are 10.2° and 20.3° (Figure 20). How far apart are the two boats? How far are the boats from the lighthouse?

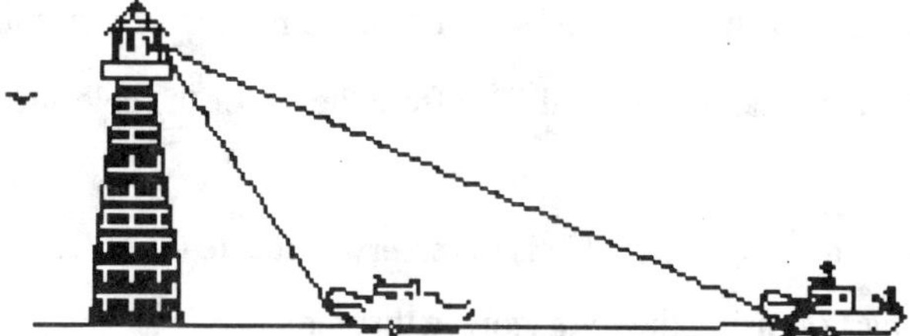

Figure 20

24. The Folded-Paper Problem An $8\frac{1}{2}$- by 11-in. sheet of paper is folded over as shown in Figure 21 so that the upper right-hand corner touches the edge of the other side. Find the angle θ formed at the fold.

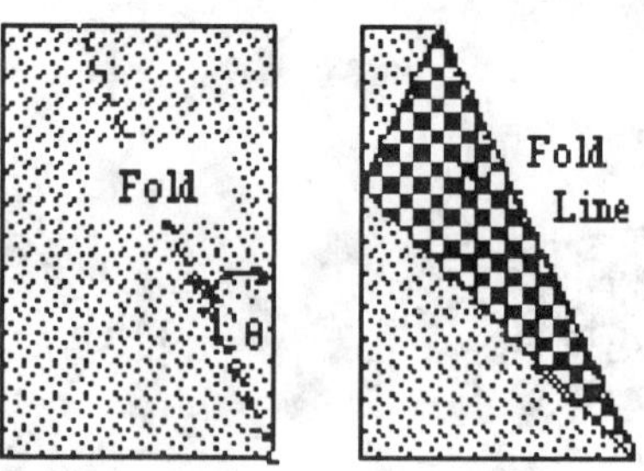

Figure 21

25. The Rainbow Kite and Balloon Company What is the angle of elevation of a kite that is two-fifths as far down the beach as it is above the beach?

26. Wonderworld Amusement Park You have built a new 100-ft-tall ride at the park. The sun is directly overhead at noon and sets at 7:30 p.m. If the ride casts a 16-ft shadow, what time in the afternoon is it?

Extend the Concepts

*27. The Solar-Heated Room Problem A solar-heated room with a solid floor is designed to be heated by the winter sun as pictured in the Figure 22. On December 21, $\alpha = 23°$ and on June 21, $\beta = 72°$. The height of the ceiling in this room is to be 7 ft 6 in.

(a) Find x, the overhang required to prevent the summer sun from hitting the floor.

(b) Use your answer from part (a) to find y, the greatest depth the sun will reach into the room on December 21.

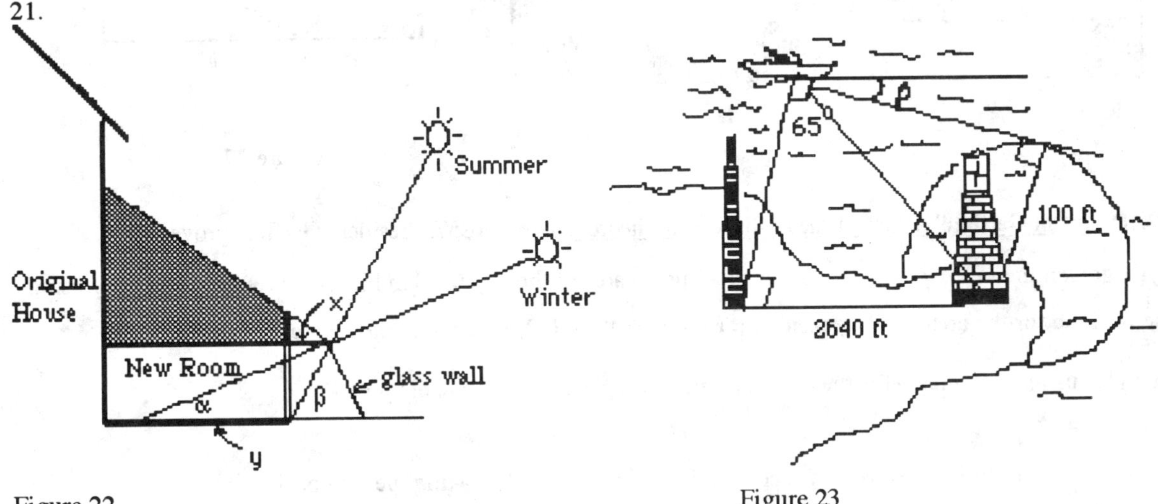

Figure 22

Figure 23

28. The Circle of Danger Problem When directly offshore of a smokestack (Figure 23) you can see the lighthouse at Point-No-Point. From your chart you see that the stack and the lighthouse are 2640 ft apart. You calculate the angle from your boat to the lighthouse as 65°. You would like to pass the lighthouse as close to shore as possible but know from your charts that you must pass outside the "circle of danger." Find β, your new course, in order to pass the lighthouse 100 ft offshore.

29. The Rectangular Solid Problem Given the rectangular solid in Figure 24, find the lengths of DG and DF if DC = 8, BC = 6, and CG = 12.

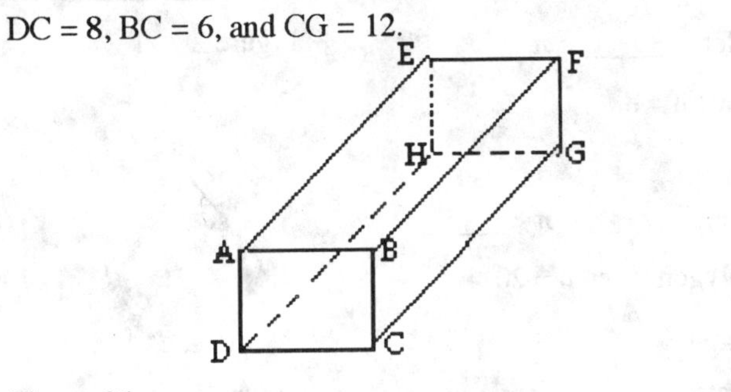

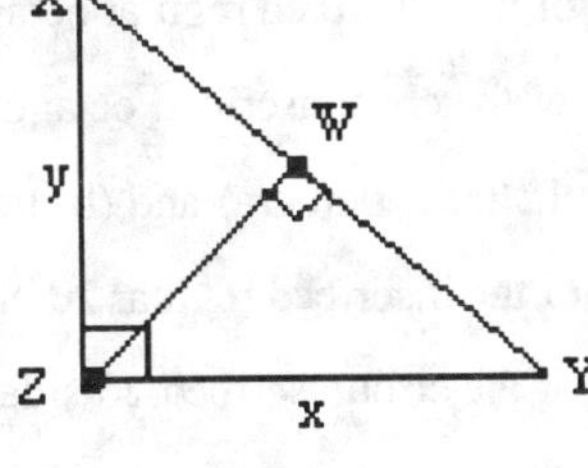

Figure 24

Figure 25

30. The Altitude Problem Find the length of WZ (Figure 25) in terms of x and y. [Hint: WZ is the altitude to the hypotenuse and the formula for finding the area of a triangle is $A = \frac{1}{2}bh$.]

31. Pythagoras' Proof of the Pythagorean Theorem It is believed that Pythagoras proved the Pythagorean theorem as shown in Figure 26. Study the figure and explain how he proved that $a^2 + b^2 = c^2$.

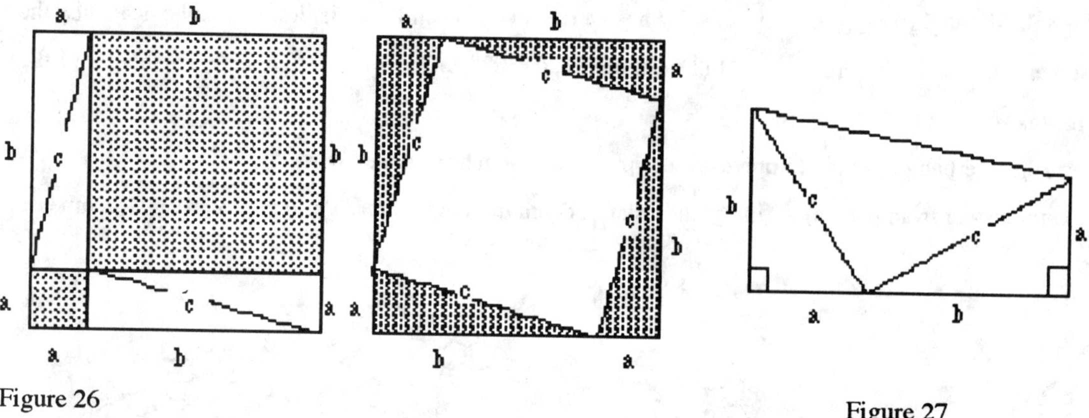

Figure 26

Figure 27

32. President Garfield's Proof of the Pythagorean Theorem About 1857, President Garfield proved the Pythagorean theorem $a^2 + b^2 = c^2$ by calculating the area of the trapezoid in Figure 27 two ways: (1) by using the formula for the area of a trapezoid $A = (\frac{b_1 + b_2}{2})h$ and (2) by computing the areas of the three right triangles using the formula for the area of a triangle $A = \frac{1}{2}bh$. Recreate his proof.

For Exercises 33 to 36, the circle in each exercise has radius $r = 1$, making the area of the circle $A = \pi r^2 = \pi(1)^2 = \pi$. In 240 B.C., Archimedes used a "method of exhaustion" to define π by finding the areas of regular polygons inscribed and circumscribed around a circle of radius 1. The more sides n the polygon has, the closer its areas come to the area of the circle, $A = \pi$.

33.(a) Find the area of the inscribed equilateral triangle when $n = 3$ (Figure 28).

(b) Find the area of the circumscribed triangle.

(c) Use your results from parts (a) and (b) to complete: ____ $< \pi <$ ____

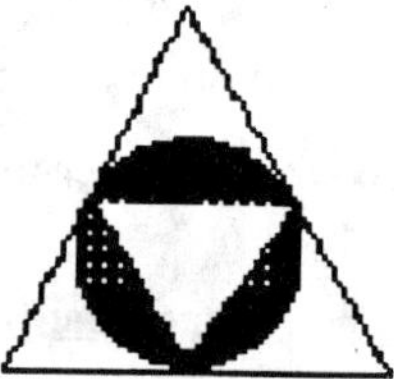

Figure 28

34.(a) Find the area of the inscribed regular pentagon when $n = 5$ (Figure 29).

(b) Find the area of the circumscribed pentagon.

(c) Use your results from parts (a) and (b) to complete: ____ $< \pi <$ ____

35.(a) Find the area of the inscribed regular octagon when $n = 8$.

(b) Find the area of the circumscribed octagon.

(c) Use your results from parts (a) and (b) to complete: ____ $< \pi <$ ____

36.(a) Find the area of the inscribed regular 20-sided polygon when $n = 20$.

(b) Find the area of the circumscribed 20-sided polygon.

(c) Use your results from parts (a) and (b) to complete: ____ $< \pi <$ ____

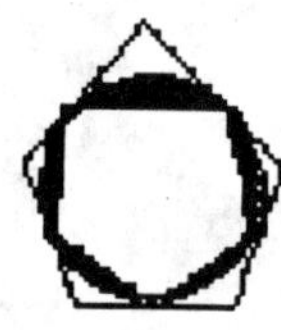

Figure 29

5.2 The Circular Functions

Measuring Angles in Radians
Locating Reference Points
Defining the Circular Functions
Finding Convenient Reference Points

Regardless of the size of the angle, your calculator can display defined values of trigonometric functions. For example, sin 745° ≈ 0.42 and cos (-180°) = -1. Yet, we cannot draw a right triangle having a 745° angle nor one having a negative angle. Because our study of trigonometry in Section 5.1 was restricted to right triangles, the angles where restricted to $0° < \theta < 90°$. Now, we extend the definition of the trigonometric functions to the **circular functions**, functions that are described in terms of points on a circle of radius r.

Measuring Angles in Radians

We can measure angles using real numbers called *radians*. We define a radian in terms of a **central angle**, an angle with its vertex at the center of a circle.

> **Definition of a Radian**
> One **radian** is the measure of a central angle whose intercepted arc length equals the radius of the circle (Figure 1a).

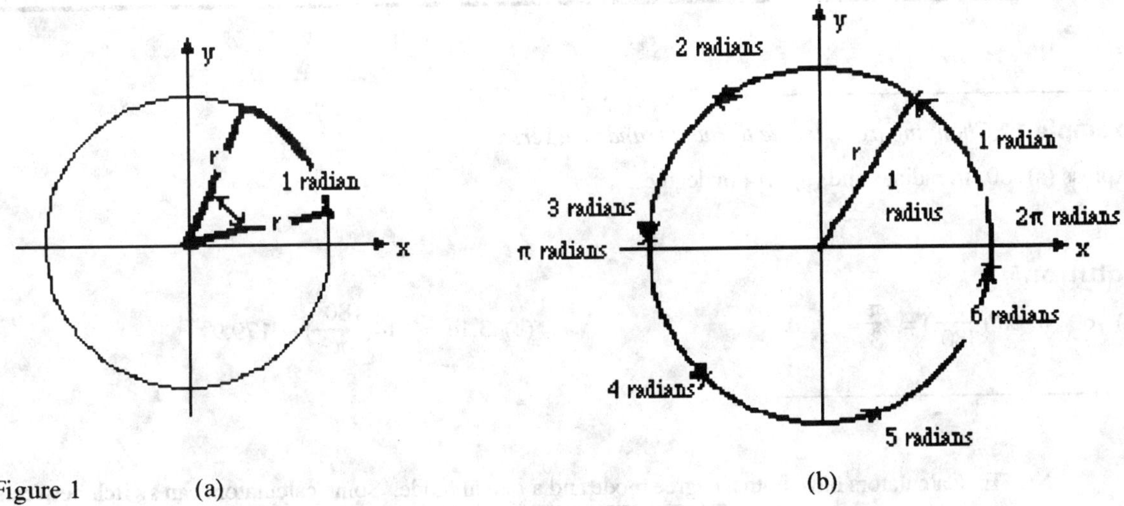

Figure 1 (a) (b)

Notice in Figure 1a that one radian is associated with an arc length equal to one radius of the circle.

The degree measure and radian measure of any angle are related. The formula for the circumference of a circle provides the connection. One complete revolution of a circle has an arc length equal to its circumference, $C = (2\pi)r$. That means there are exactly 2π radians (about 6 radians) in one revolution of a circle (Figure 1b).

Can you change an angle measured in degrees to radian measure, and vice versa?

Any angle measured in degrees can be changed to its equivalent radian measure, or vice versa, by multiplying by a form of the multiplicative identity 1. Since one revolution of a circle is 2π radians (360°), we have

$$2\pi = 360°$$

which reduces to $\pi = 180°$. We will use the conversion factor $\frac{\pi}{180°} = 1$ to change an angle measured in degrees to radian measure.

$$45° = 45°\left(\frac{\pi}{180°}\right) = \frac{\pi}{4} \approx 0.78$$

To change angles measured in radians into degrees, we multiply by the conversion factor $\frac{180°}{\pi} = 1$.

$$\frac{1}{6}\pi = \frac{\pi}{6}\left(\frac{180°}{\pi}\right) = 30° \quad \text{or} \quad 1.75\left(\frac{180°}{\pi}\right) \approx 100.3°$$

The degree symbol is present for an angle measured in degrees, but no symbol is used with radian measure since radians are simply real numbers. In calculus we use radian measure, but sometimes we switch from radians to degrees since the concept of degrees is more familiar.

Angle Conversions

To change an angle measured in degrees to radians, multiply by $\frac{\pi}{180°} = 1$.

To change an angle measured in radians to degrees, multiply by $\frac{180°}{\pi} = 1$.

Example 1 *Changing from degree to radian, and vice versa*

Express (a) -60° in radians and (b) 3.14 in degrees.

Solution

(a) $-60° = -60°\left(\frac{\pi}{180°}\right) = -\frac{\pi}{3} \approx -1.05$

(b) $3.14 = 3.14\left(\frac{180°}{\pi}\right) \approx 179.9°$

NOTE: Calculators have both a degree mode and a radian mode. Some calculators can switch between degrees and radians by pressing [2nd] [DRG].

Some formulas are simpler when an angle is measured in radians. For example, consider finding an **arc length**, a portion of the circumference of a circle. Figure 2 shows an arc of a circle of radius r with the central angle measuring 30°.

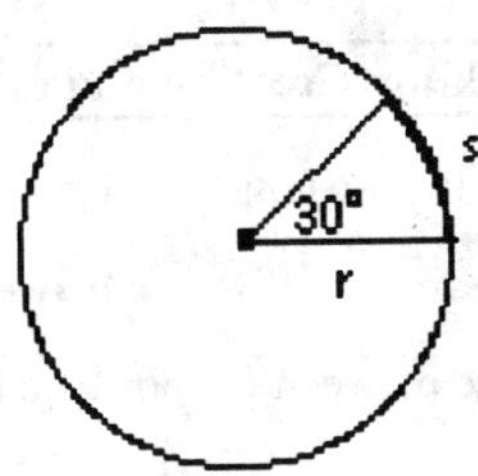

Figure 2

The length of the arc for the entire circle is its circumference, $C = 2\pi r$. We can find the arc length s by finding the portion of the circumference cut by the 30° angle.

$$\begin{aligned} s &= \frac{30^\circ}{360^\circ}(2\pi r) \\ &= \frac{1}{12}(2\pi r) \\ &= \frac{\pi}{6}r \end{aligned}$$

In radian measure, $30^\circ = \pi/6$. Therefore, the arc length is simply $s = r\theta$ if θ is measured in radians.

Arc Length Formula

For any circle, the arc length s associated with its central angle θ (Figure 3) measured in radians is

$$s = r\theta$$

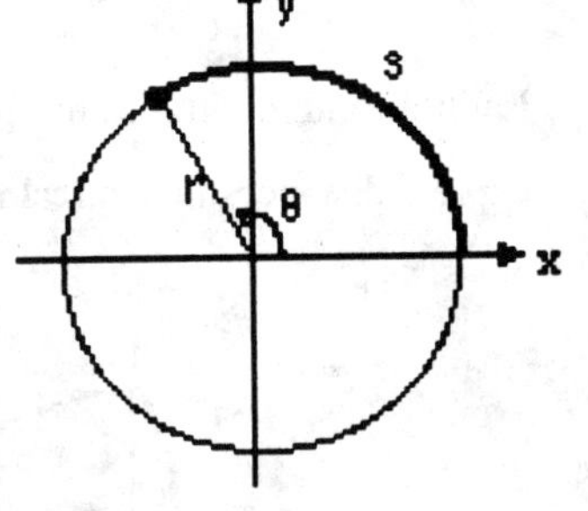

Figure 3

Example 2 The Latitude Problem

The latitude of point P on the surface of earth (radius about 4000 mi) is the measure of angle α in Figure 4.

(a) Seattle is 47°N latitude since $\alpha = 47^\circ$. How far north of the equator is Seattle?

(b) Part of the U.S.-Canadian border is 49°N latitude. How far south of the border is Seattle?

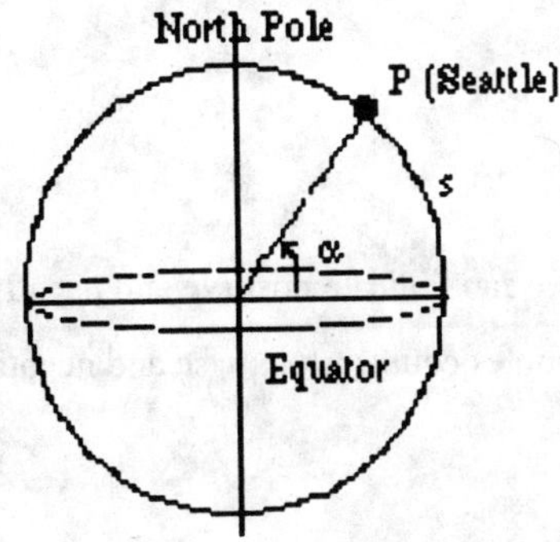

Figure 4

Thinking: Creating a Plan

First we change 47° to radian measure and find the arc length using $s = r\theta$. This tells us how the distance between the equator and Seattle. For part (b) we find the distance between the equator and the Canadian border. The difference of these two results is the distance between the border and Seattle.

Communicating: Writing the Solution

(a)

$$47° = 47° \left(\frac{\pi}{180°}\right) = \frac{47\pi}{180} \approx 0.8203$$

$s = r\theta$ and $r = 4000$, $\theta = 0.8203 \Rightarrow s = 4000(0.8203) \approx 3281$

Seattle is about 3281 mi north of the equator.

(b)

$$49° = 49° \left(\frac{\pi}{180°}\right) = \frac{49\pi}{180} \approx 0.8552$$

$s = r\theta$ and $r = 4000$, $\theta = 0.8552 \Rightarrow s = 4000(0.8552) \approx 3421$

The U.S.-Canadian border is about 3421 mi north of the equator. Since 3421 - 3281 = 140, Seattle is 140 mi south of the border.

Learning: Making Connections

Latitude and longitude are used to describe a position on the surface of earth (Figure 5). It is very convenient to express distances in nautical miles since 1 degree of arc equals 1 nautical mile.

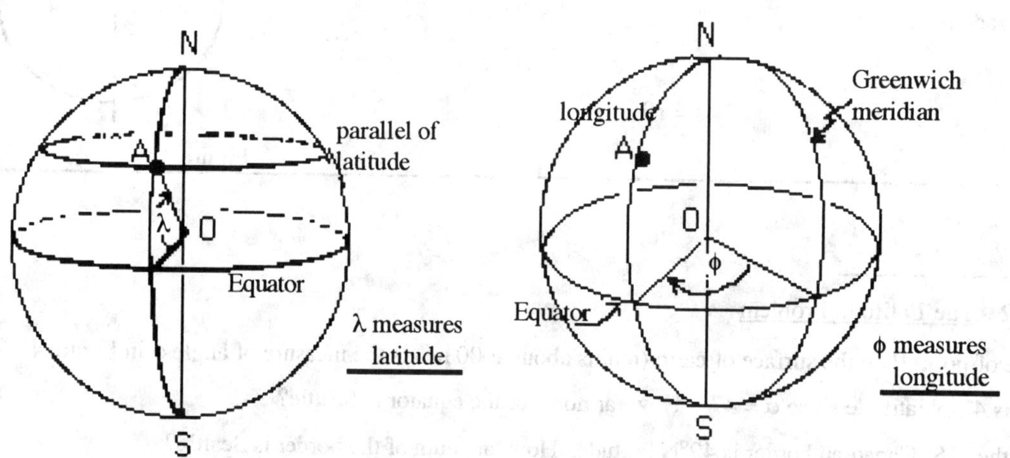

Figure 5

Locating Reference Points

Whether using degree or radian measure, we can indicate positive and negative angles of any size in the Cartesian coordinate system. We measure positive angles counterclockwise and negative angles clockwise from the positive x axis (Figure 6).

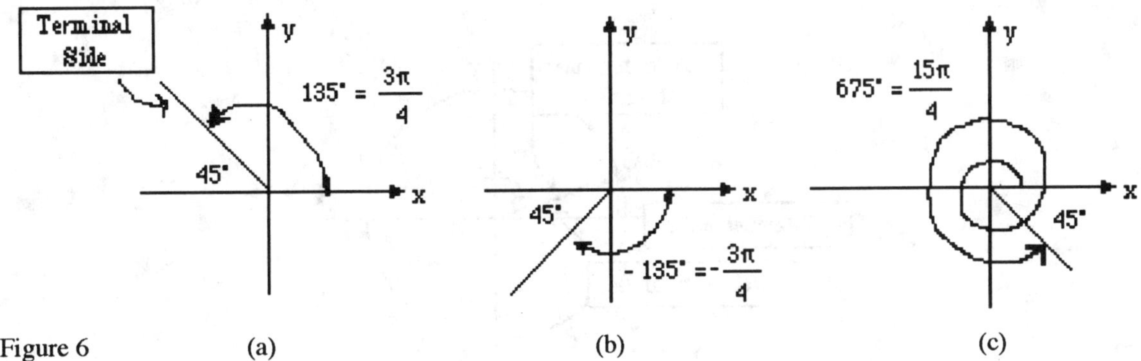

Figure 6 (a) (b) (c)

Notice that the terminal side of 135° lies in quadrant II (Figure 6a), while the terminal side of -135° lies in quadrant III (Figure 6b) and the terminal side of 675° lies in quadrant IV (Figure 6c). The **reference angle** of any angle positioned in a quadrant of the Cartesian coordinate system is the positive acute angle between the terminal side of the angle and the x axis. The reference angle for each angle in Figure 6 is 45°(π/4).

Example 3 *Identifying reference angles*

Sketch each of the following angles in the Cartesian coordinate system and identify the corresponding reference angle.

(a) 400° (b) $-\frac{3\pi}{2}$ (c) 6π (d) 1

Solution

See Figure 7.

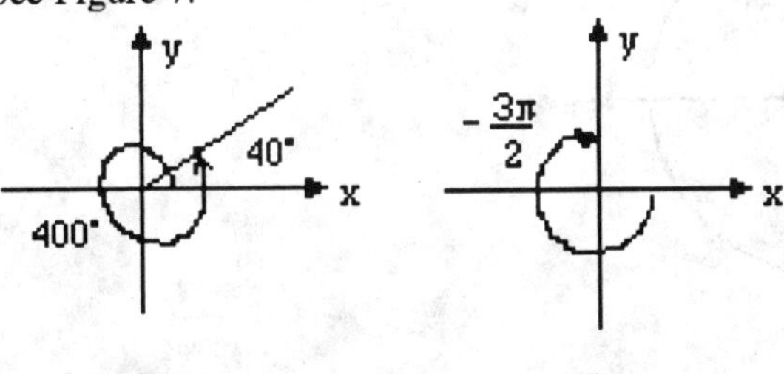

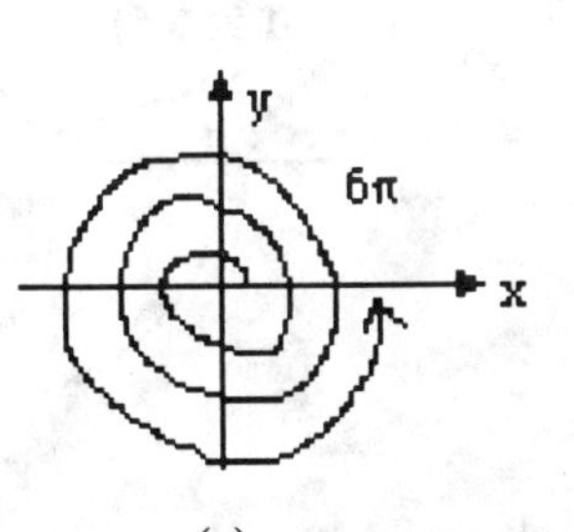

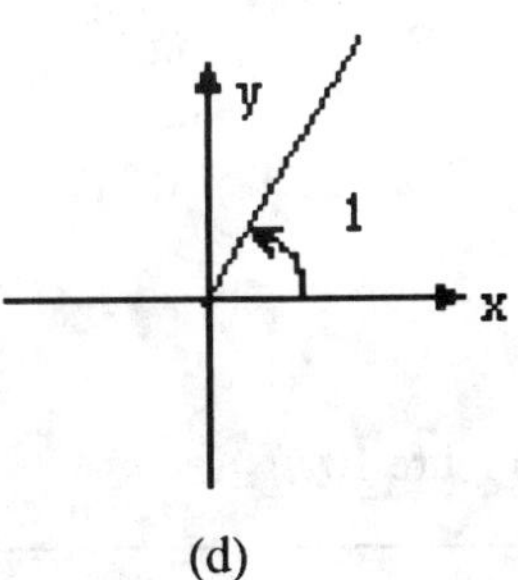

Figure 7 (a) (b) (c) (d)

(a) The reference angle for 400° is 40°.

(b) The reference angle for $-\frac{3\pi}{2}$ is $\frac{\pi}{2}$.

(c) The reference angle for 6π is 0.

(d) The reference angle for 1 is 1 (1 radian).

Defining the Circular Functions

In order to extend our knowledge of trigonometry to include angles of all sizes, we locate reference points and reference angles in a circle in the Cartesian coordinate system. The **reference point** of an angle is the point where the terminal side of the angle intersects a circle (Figure 8).

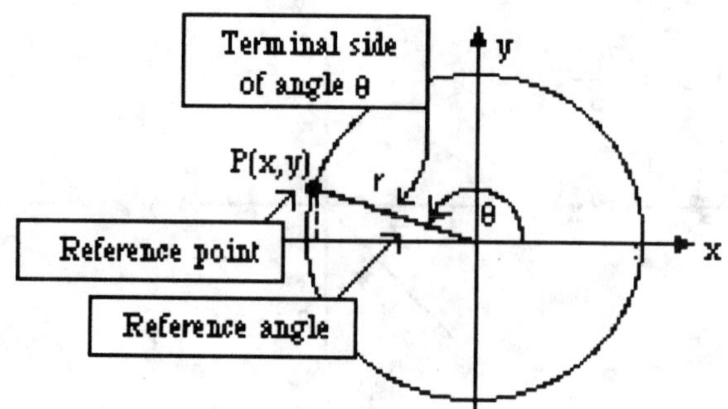

Figure 8

Example 4 *Finding a reference point*

Measure x and y in Figure 9 in centimeters, to find the coordinates of the reference point, P.

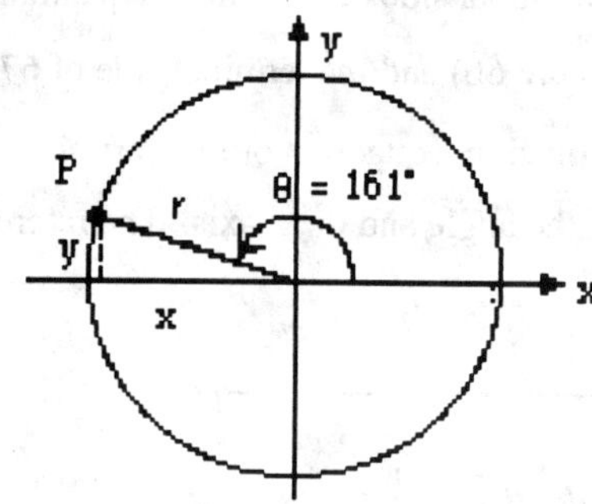

Figure 9

Solution

By actual measurement, $x \approx 1.9$ cm and $y \approx 0.6$ cm. Since point P is located in quadrant II, its coordinates are (-1.9, 0.6) (Figure 10).

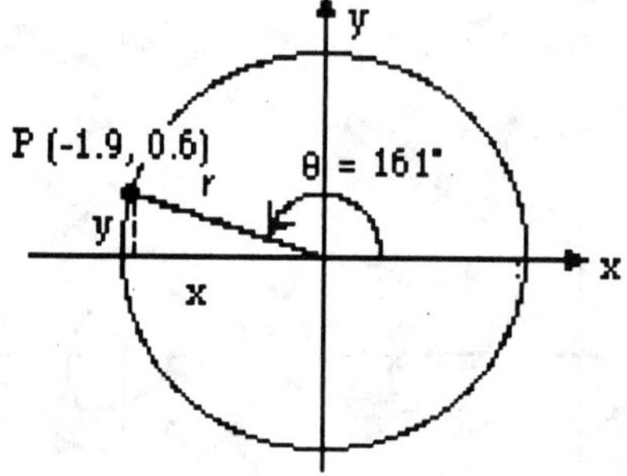

Figure 10

Example 5 *Comparing measured ratios with calculator values*

(a) Calculate the ratios $\frac{y}{r}$, $\frac{x}{r}$, and $\frac{y}{x}$ for $x = -1.9$, $y = 0.6$, and $r = 2$ (Figure 10).

(b) Find calculator values for sin 161°, cos 161°, and tan 161°.

Solution

(a) $\frac{y}{r} = \frac{0.6}{2} = 0.3$ $\qquad \frac{-x}{r} = \frac{-1.9}{2} = -0.95$ $\qquad \frac{y}{x} = \frac{0.6}{-1.9} \approx -0.32$

(b) $\sin 161° \approx 0.32$ $\qquad \cos 161° \approx -0.94$ $\qquad \tan 161° \approx -0.34$

How are the ratios in Example 5 related to the values of sine, cosine, and tangent?

As the actual measurements suggest, there is a relationship between the coordinates of a reference point and the values of sine, cosine, and tangent. It appears that $\sin\theta = \frac{y}{r}$, $\cos\theta = \frac{x}{r}$, and $\tan\theta = \frac{y}{x}$.

Compare the definitions of the trigonometric functions in terms of a right triangle from Section 5.1 with that of the circular functions defined in terms of the coordinates of the reference point on a circle and the radius of the circle in Table 1.

Table 1

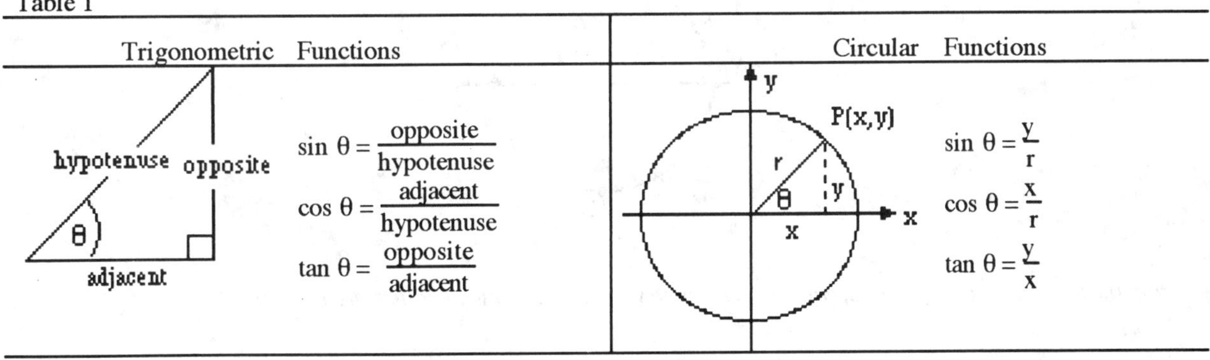

The definition of the circular functions extends the definition of the trigonometric functions to include all angles, including negative angles and angles greater than or equal to 90°. The definitions of the circular functions are further simplified if we consider a unit circle, a circle having radius r = 1.

What are the definitions of sine, cosine, and tangent for a unit circle?

If r = 1, then $\sin\theta = \frac{y}{r} = \frac{y}{1} = y$, $\cos\theta = \frac{x}{r} = \frac{x}{1} = x$, and $\tan\theta = \frac{y}{x}$. Thinking of sine as the y coordinate of the reference point and cosine as the x coordinate of the reference point helps us determine the sign of their values. The unit circle in Figure 11 shows the signs of x and y in each quadrant.

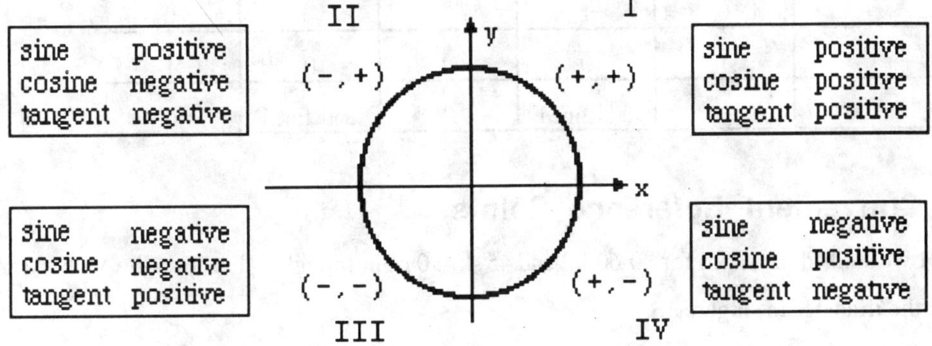

Figure 11

Note that since y is positive in quadrants I and II and negative in III and IV, values of sine are positive for any angle whose reference point falls on the top half of the circle and negative for those on the bottom half. Likewise, since x is positive in quadrants I and IV and negative in II and III, values of cosine are positive for any angle whose reference point falls on the right half of the circle and negative for those on the left half. Tangent, which is the ratio of y and x is, positive when x and y have the same sign (quadrants I and III) and negative when the signs are different (quadrants II and IV). Note that r is always considered positive since it is the radius of a circle.

Some reference points lie on a coordinate axes. The unit circle in Figure 12 shows the reference points for 0°, 90°, 180°, 270°, and 360° ($0, \frac{\pi}{2}, \pi, \frac{3\pi}{2}$, and 2π) are (1, 0), (0, 1), (-1, 0), (0, -1), and (1, 0), respectively.

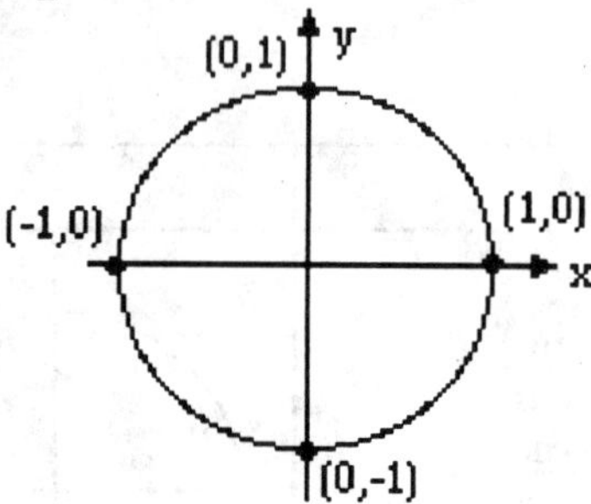

Figure 12

From the unit circle in Figure 12, can you determine why certain tangent values are undefined?

Since $\tan\theta = \frac{y}{x}$ is undefined when x = 0, any angle whose reference point lies on the y axis causes division by zero. Therefore, tan 90°, tan 270°, tan 450°, tan (-90°), and so on are all undefined. Table 2 shows the values of sine, cosine, and tangent for multiples of 90° for $0° \le \theta \le 360°$.

Table 2

Angle (degrees):	0°	90°	180°	270°	360°
(radians):	0	$\pi/2$	π	$3\pi/2$	2π
reference point on a unit circle	(1,0)	(0,1)	(-1,0)	(0,-1)	(1,0)
$\sin\theta = y/r$	0	1	0	-1	0
$\cos\theta = x/r$	1	0	-1	0	1
$\tan\theta = y/x$	0	undefined	0	undefined	0

Finding Convenient Reference Points

In Section 5.1 we studied the ratios for 30-60-90 and 45-45-90 right triangles. For triangles with hypotenuse 1 unit we have the triangles in Figure 13.

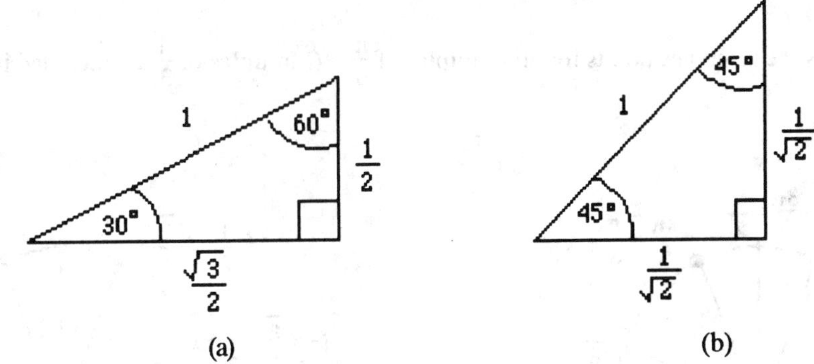

Figure 13 (a) (b)

Using the ratios for 30-60-90 and 45-45-90 triangles allows us to find reference points for angles that are multiples of 30°, 45°, and 60° without using a calculator. Once we know these reference points, we know their exact sine, cosine, and tangent values. These angles are used extensively in this text and in calculus texts to simplify calculations and to eliminate rounding errors. Since work in calculus and beyond is done with angles measured in radians and because such arithmetic is generally simpler using radian measure, we need to think of these familiar angles as

$$30° = \frac{\pi}{6}, \; 45° = \frac{\pi}{4}, \text{ and } 60° = \frac{\pi}{3}$$

Figure 14 shows the reference points for $\frac{\pi}{6}$ (Figure 14a), $\frac{\pi}{4}$ (Figure 14b), and $\frac{\pi}{3}$ (Figure 14c).

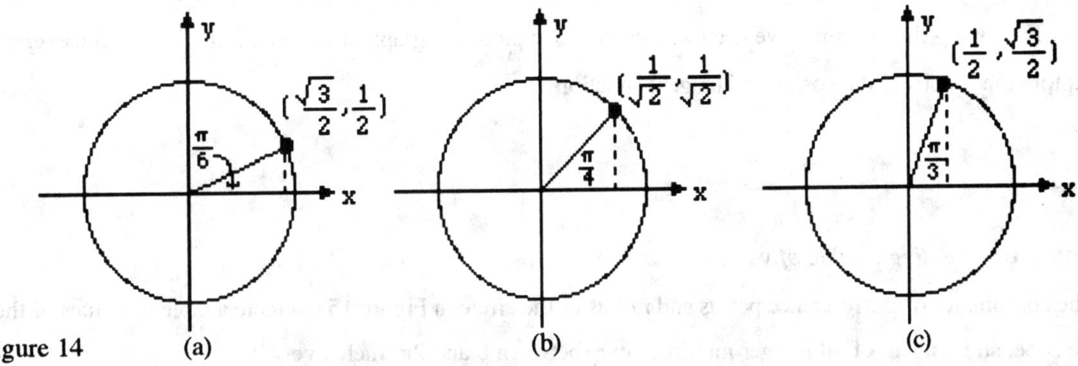

Figure 14 (a) (b) (c)

Now we can label reference points for all angles that are multiples of these special angles, where the sign of each the coordinate is determined by the quadrant in which each angle falls. Figure 15 shows the reference points on a unit circle for the multiples of $\theta = \frac{\pi}{4}$.

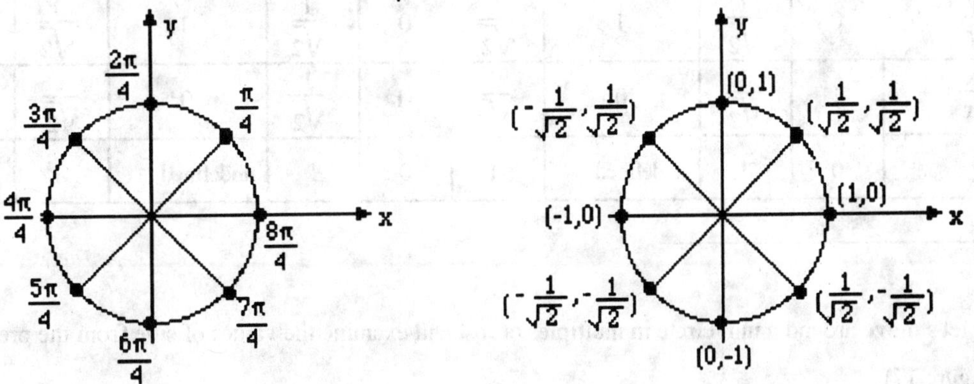

Figure 15

Figure 16 shows the reference points for all multiples of $\frac{\pi}{6}$. (Multiples of $\frac{\pi}{3}$ are included in the multiples of $\frac{\pi}{6}$.)

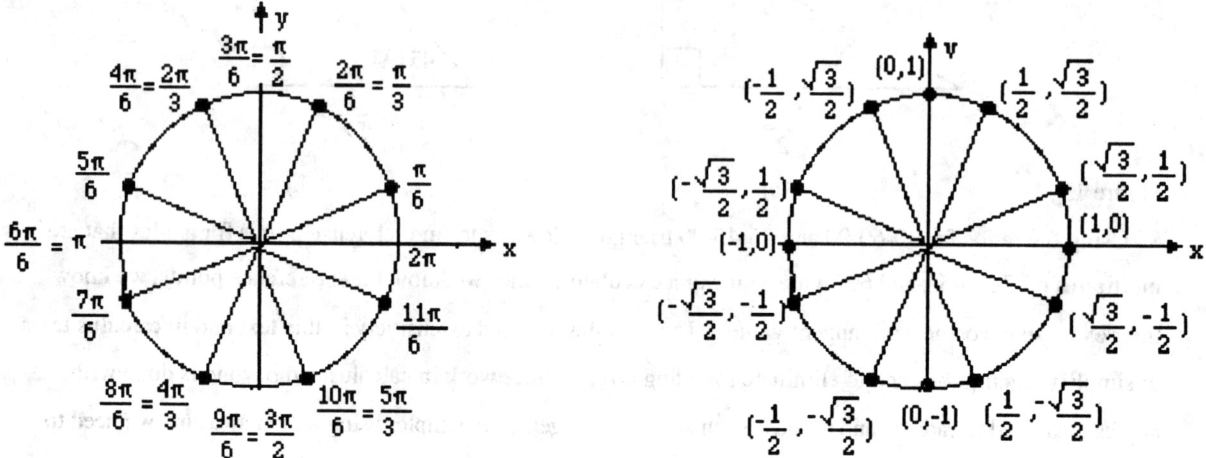

Figure 16

In the next example we state the coordinates of reference points for all integer multiples of $\pi/4$ between 0 and 2π. Following this example, we use these values to explore the graph of the sine function. In the exercises we explore the graphs of the cosine and tangent function.

Example 6 *Creating a table of values*

Use the coordinates of the reference points and radius of the circle in Figure 15 to create a table of values of the sine, cosine, and tangent of all integer multiples of $\frac{\pi}{4}$ between 0 and 2π, inclusive.

Solution

θ	0	$\frac{\pi}{4}$	$\frac{\pi}{2}$	$\frac{3\pi}{4}$	π	$\frac{5\pi}{4}$	$\frac{3\pi}{2}$	$\frac{7\pi}{4}$	2π
$\sin\theta = \frac{y}{r}$	0	$\frac{1}{\sqrt{2}}$	1	$\frac{1}{\sqrt{2}}$	0	$-\frac{1}{\sqrt{2}}$	-1	$-\frac{1}{\sqrt{2}}$	0
$\cos\theta = \frac{x}{r}$	1	$\frac{1}{\sqrt{2}}$	0	$-\frac{1}{\sqrt{2}}$	-1	$-\frac{1}{\sqrt{2}}$	0	$\frac{1}{\sqrt{2}}$	1
$\tan\theta = \frac{y}{x}$	0	1	undefined	-1	0	1	undefined	-1	0

Now let's move around a unit circle in multiples of $\pi/4$ and examine the values of sine from the preceding table (Figure 17).

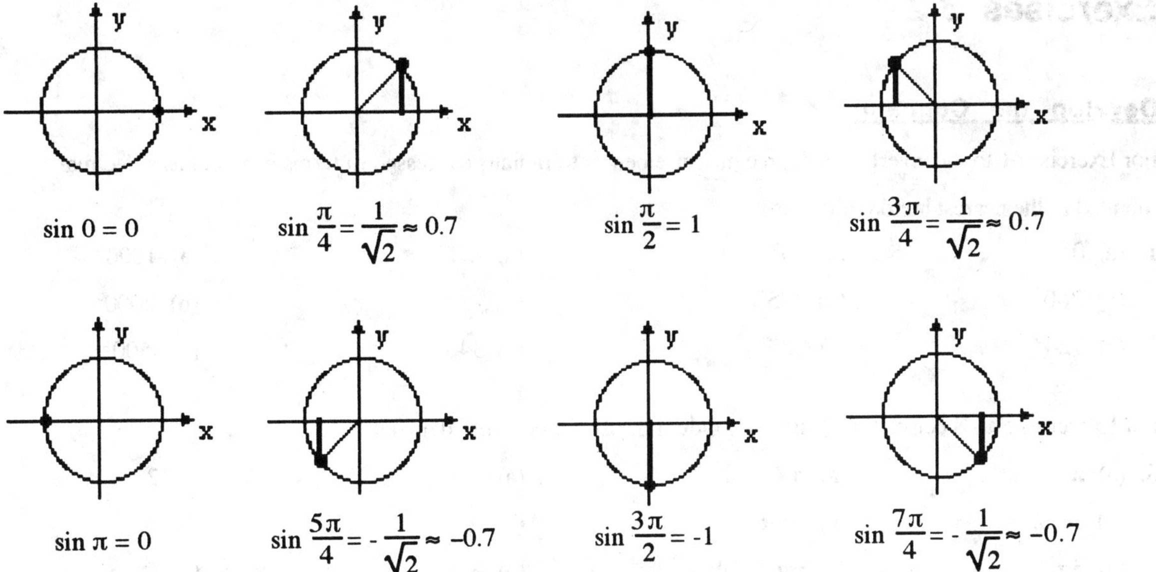

Figure 17

The graph of the sine function reveals its shape as we stack the values from Figure 17 along the x axis from x = 0 to x = 2π (Figure 18).

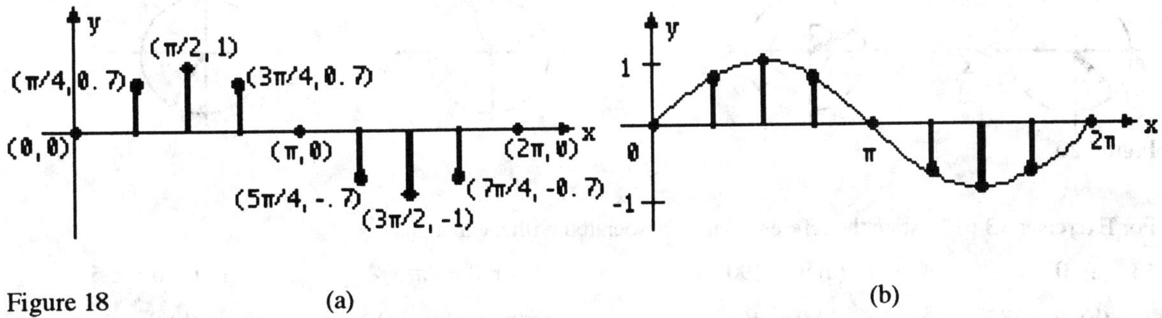

Figure 18 (a) (b)

A graphing utility display for -2π ≤ x ≤ 4π shows a repeating pattern (Figure 19). For sine, the pattern repeats every 2π (360°). We call 2π the period. We discuss such details further in the next section.

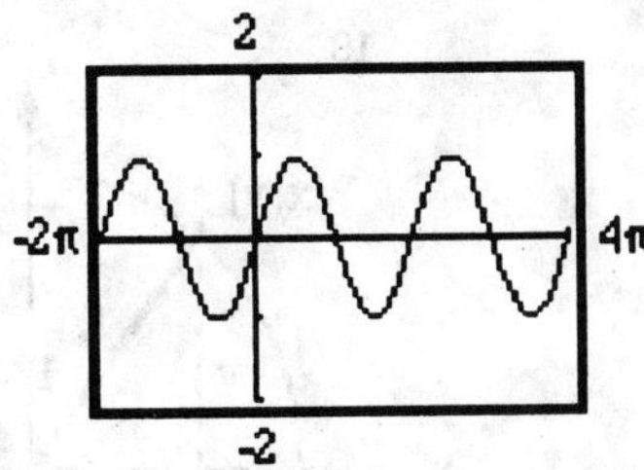

Figure 19

Exercises 5.2

Develop the Concepts

For Exercises 1 to 4 convert each degree measurement into radians expressed in terms of π and as a decimal rounded to the nearest hundredth.

1. (a) 0°	2. (a) 45°	3. (a) -20°	4. (a) -1000°
(b) 360°	(b) 135°	(b) -70°	(b) 2000°
(c) 720°	(c) 225°	(c) 340°	(c) 500°

For Exercises 5 to 8 convert the following radian measurements into degrees.

5. (a) $\pi/2$	6. (a) $\pi/6$	7. (a) 1.5	8. (a) 2
(b) $3\pi/2$	(b) $5\pi/6$	(b) 2.5	(b) 4
(c) $5\pi/2$	(c) $-\pi/6$	(c) 3.5	(c) 6

For Exercises 9 to 12 match the angle with its approximate measure in radians as shown in Figure 20.

9. -5 10. -1 11. 4 12. 9

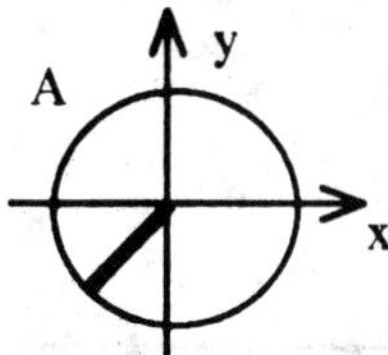

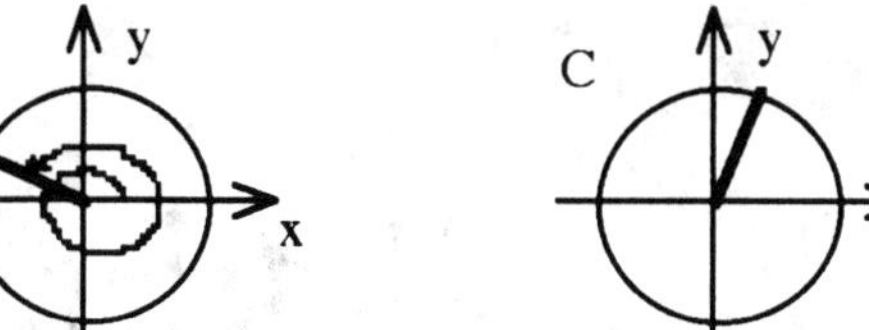

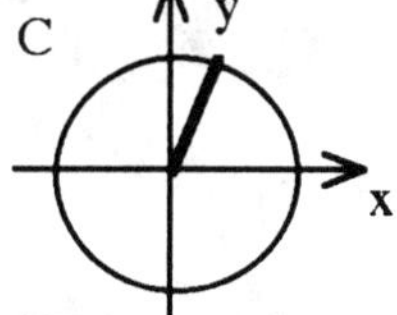

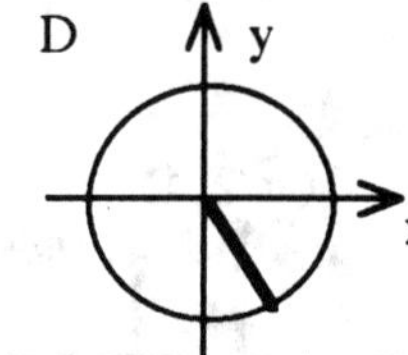

Figure 20

For Exercises 13 to 16 state the reference angle associated with each angle.

13. (a) $\theta = 105°$	14.(a) $\theta = -190°$	15.(a) $\theta = 7\pi/6$	16.(a) $\theta = 8.5$
(b) $\theta = 285°$	(b) $\theta = -100°$	(b) $\theta = -11\pi/12$	(b) $\theta = -17$

For Exercises 17 to 20 measure x, y, and r in centimeters and (a) state the reference point; (b) calculate the ratios, y/r, x/r, and y/x; and (c) find the calculator values for sine, cosine, and tangent of the indicated angle.

17.

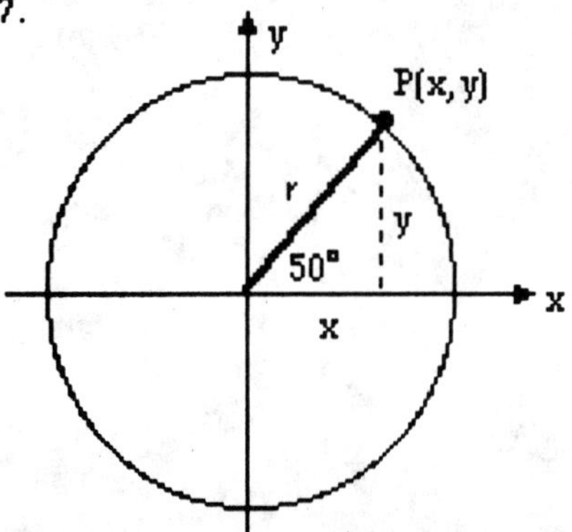

18.

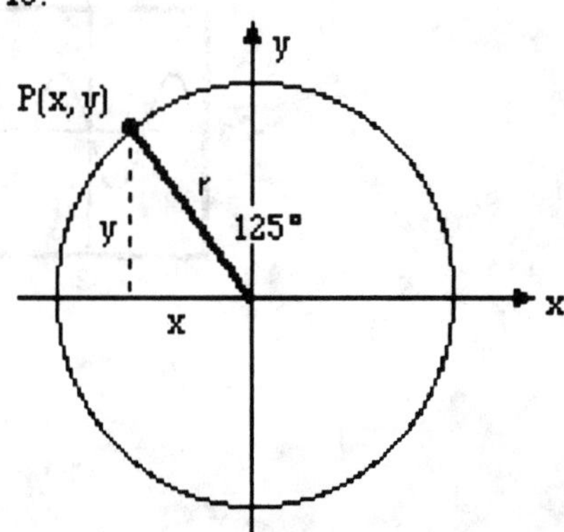

19.

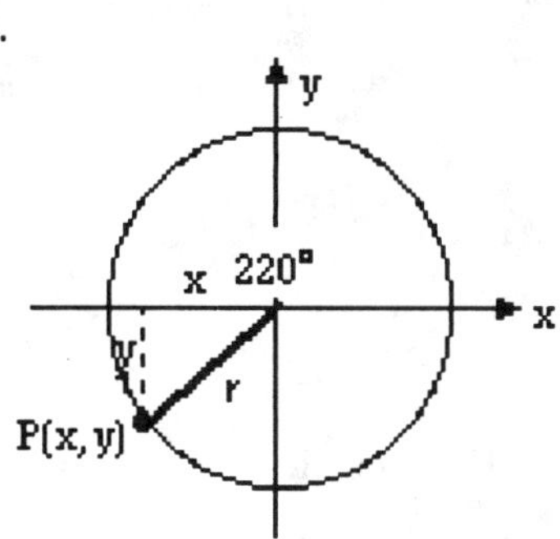

20.

y
x
x
-25°
y
r
P(x, y)

For Exercises 21 to 24 (a) draw a circle with radius r = 1 unit and label the reference point for each angles, and (b) use the reference point to state the exact values for sin θ, cos θ, and tan θ for each angle.

21. $\theta = -\pi$ 22. $\theta = 7\pi/2$ 23. $\theta = -2\pi/3$ 24. $\theta = -\pi/6$

For Exercises 25 and 26 find the value of each circular function if $\cos\theta = 5/13$ for (a) θ in quadrant I and (b) θ in quadrant IV.

25. sin θ 26. tan θ

For Exercises 27 and 28 find the value of each circular function if $\tan\theta = -6/8$ for (a) θ in quadrant II and (b) θ in quadrant IV.

27. sin θ 28. cos θ

Apply the Concepts

In Exercises 29 to 32 approximate the radius of the earth as 4000 mi.

29. The Rainbow Kite and Balloon Company A hot air balloon trip passes over three cities located on longitude 90°. Find the distance between Dubuque, Iowa (latitude 42°30'), St. Louis, Missouri (latitude 38°35'), and Memphis, Tennessee (latitude 35°9').

30. Wonderworld Amusement Park You want to create an island adventure at the park that simulates the view of the Galápagos Islands from a distant satellite. The Galápagos Islands lie on the equator. Find the rotational speed of an iguana sitting on a rock in the islands by dividing the distance traveled in miles by the time it takes to make one revolution in hours.

*31. The Spinning Earth Problem

(a) Find a formula for the rotational speed of a person standing on Earth at latitude λ.

(b) State the domain and range of your formula in part (a).

32. The Flying with the Sun Problem

(a) Find the distance from Reno, Nevada (longitude 120°W) and Athens, Greece (longitude 24°E) if both cities have approximately the same latitude, 39°.

(b) Find the rotational speed of a person standing on Earth at latitude 39°N.

(c) If a plane left Athens at sunrise and landed in Reno at sunrise, how fast was it flying?

Extend the Concepts

33. Use the coordinates of the reference points and radius of the circle in Figure 13 to complete the table of values of the sine, cosine, and tangent of all integer multiples of $\pi/6$ between 0 and 2π, inclusive.

θ	0	$\frac{\pi}{6}$	$\frac{\pi}{3}$	$\frac{\pi}{2}$	$\frac{2\pi}{3}$	$\frac{5\pi}{6}$	π	$\frac{7\pi}{6}$	$\frac{4\pi}{3}$	$\frac{3\pi}{2}$	$\frac{5\pi}{3}$	$\frac{11\pi}{6}$	2π
$\sin\theta$	0	$\frac{1}{2}$			$\frac{\sqrt{3}}{2}$			$-\frac{1}{2}$					
$\cos\theta$			$\frac{1}{2}$			$-\frac{\sqrt{3}}{2}$			$-\frac{1}{2}$				
$\tan\theta$				und.				$\frac{1}{\sqrt{3}}$			$-\sqrt{3}$		0

34. Move around a circle in multiples of $\pi/6$ using the values of $\sin\theta = y/r$ to sketch a graph of the sine function (see Exercise 33).

35. Move around a circle in multiples of $\pi/4$ and $\pi/6$ using the values of $\cos\theta = x/r$ to sketch a graph of the cosine function (see Example 6 and Exercise 33).

36. Move around a circle in multiples of $\pi/4$ and $\pi/6$ using the values of $\tan\theta = y/x$ to sketch a graph of the tangent function (see Example 6 and Exercise 33).

*37. A degree can be divided into 60 minutes (60'), and each minute divided into 60 seconds (60").

(a) Express 1 radian in degrees, minutes, and seconds.

(b) Convert 1 degree, 1 second, and 1 minute each to radian measure.

38. A *grad* is another unit of measure for angles. There are 400 grads in one revolution. Convert the following angles to grads.

(a) 90° (b) 30° (c) 1000° (d) 20π

For Exercises 39 to 42 use a graphing utility to sketch the graph of each function. Use the table of values from Example 6 to help label the indicated points with their exact values.

39. $y = 2\sin\theta$ for $0 \le \theta \le 2\pi$, $\theta = \pi/2$ and $\theta = 3\pi/2$

40. $y = 20\cos 2\theta$ for $-\pi \le \theta \le \pi$, $\theta = -\pi/2$ and $\theta = \pi/2$

41. $y = -2 \tan \theta$ for $-\pi \le \theta \le \pi$, $\theta = -\pi/4$ and $\theta = \pi/4$

42. $y = \frac{1}{20} \sin \frac{\theta}{2}$ for $-4\pi \le \theta \le 4\pi$, $\theta = -\pi$ and $\theta = \pi$

43. Explain why the coordinates of the reference point for $\frac{\pi}{6}$ are $(\frac{\sqrt{3}}{2}, \frac{1}{2})$ using a unit circle (Figure 21).

[Hint: Establish that ΔPOP' is an equilateral triangle.]

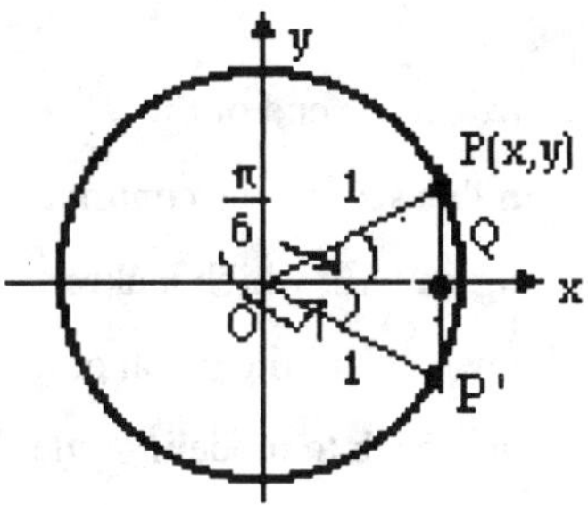

Figure 21

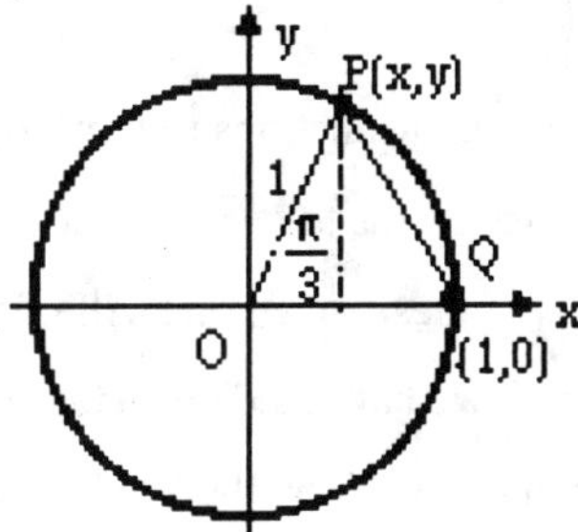

Figure 22

44. Explain why the coordinates of the reference point for $\pi/3$ are $(\frac{1}{2}, \frac{\sqrt{3}}{2})$ using a unit circle (Figure 22).

[Hint: The perpendicular line PQ is a perpendicular bisector.]

5.3 Graphs of the Circular Functions

Defining Periodicity
Reflecting and Translating Graphs of Circular Functions
Changing Amplitude and Period
Graphing One Period of a Circular Function
Finding the Equation of a Periodic Graph

In Section 5.2 we defined the circular functions in terms of reference points on a circle. At the end of the section we used various reference points to generate the graph of the sine function. In this section we continue our investigation of the graphs of these functions. All of these graphs have a repeating pattern, which makes them different from any function we have studied so far. In Example 7 we use our knowledge of the graph of a sine function to model the changing tides on the Vietnam-China coast. The analytic approach to modeling tides in the ocean is based on the assumption that complex tides at any port are composed of simple tides with varying period. Many modern applications of the circular functions use time as the independent variable. Such applications require using radian measurement.

Defining Periodicity

To explore the repeating pattern of the circular functions, go to "Explore the Concepts," Exercises E1 - E6, before continuing (page 385). For the interactive version, go to the *Precalculus Weblet - Instructional Center - Explore Concepts.*

Figure 1 shows graphs of portions of the graphs of the reference functions y = sin x, y = cos x, and y = tan x, respectively.

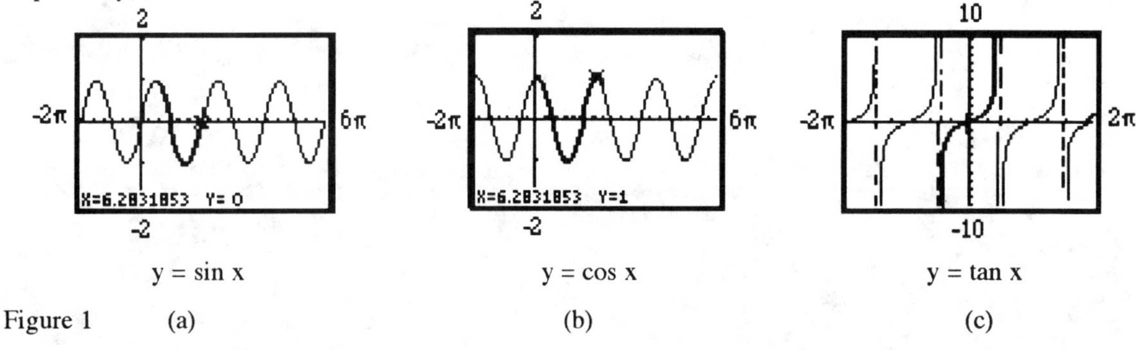

y = sin x — y = cos x — y = tan x

Figure 1 (a) (b) (c)

Notice that the viewing windows in Figure 1 are in radians (real numbers). For this reason we call the input variable x. For example, we write y = sin x rather than y = sin θ.

The domain of y = sin x and y = cos x is all real numbers, while the domain of y = tan x is all real numbers that are not multiples of $\frac{\pi}{2}$. The range of both the sine and cosine function is [-1, 1], while the range of the tangent function is all real numbers.

Each of these graphs has a repeating pattern. We call this repeating property *periodicity*. For sine and cosine, the pattern repeats every 2π. We say that sine and cosine have a **period** of 2π. The graphs are repetitions of the highlighted portions $0 \leq x < 2\pi$ (Figures 1a and 1b). The graph of the tangent function repeats every π units. Its period is π. One period is highlighted for $-\pi/2 < x < \pi/2$ (Figure 1c).

Reflecting and Translating Graphs of Circular Functions

Reflections of the circular functions behave exactly like those for all other functions. In the first example we examine reflections of the sine function.

Example 1 *Graphing reflections of the circular functions*

Sketch the graphs of

(a) y = -sin x (b) y = sin (-x)

Solution

(a) The graph of y = -sin x is a reflection of the graph of y = sin x about the x axis (Figure 2).

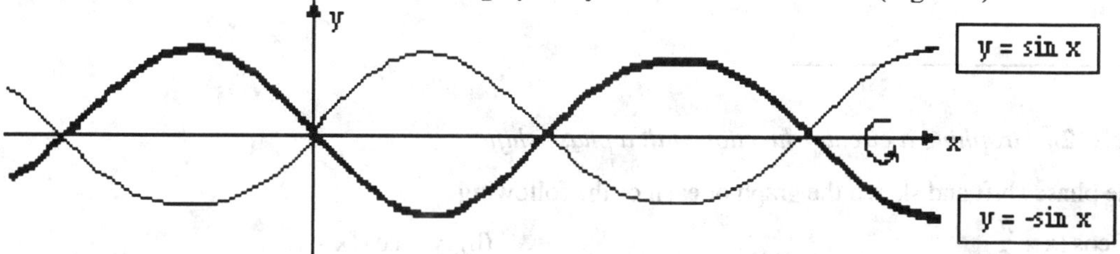

Figure 2

(b) The graph of y = sin(-x) is a reflection of the graph of y = sin x about the y axis (Figure 3).

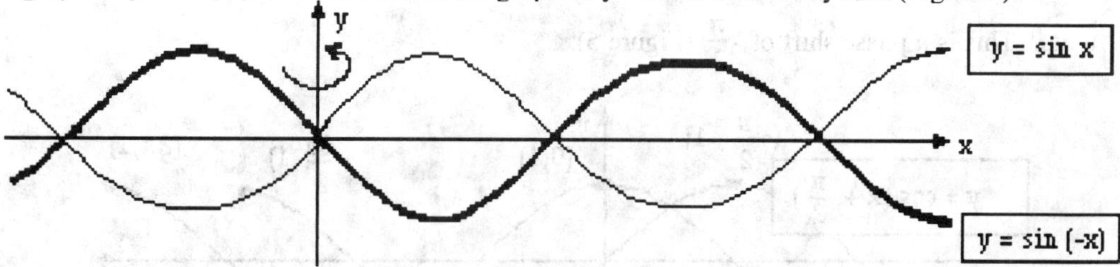

Figure 3

The results in Figures 2 and 3 are identical. This illustrates the fact that sine is an odd function,

$$\sin(-x) = -\sin x$$

Translations of the circular functions also behave as those for all functions. Figure 4 shows the graphs of y = tan x + 1 and y = tan x - 2 as vertical translations of the graph of y = tan x, up 1 and down 2 units, respectively.

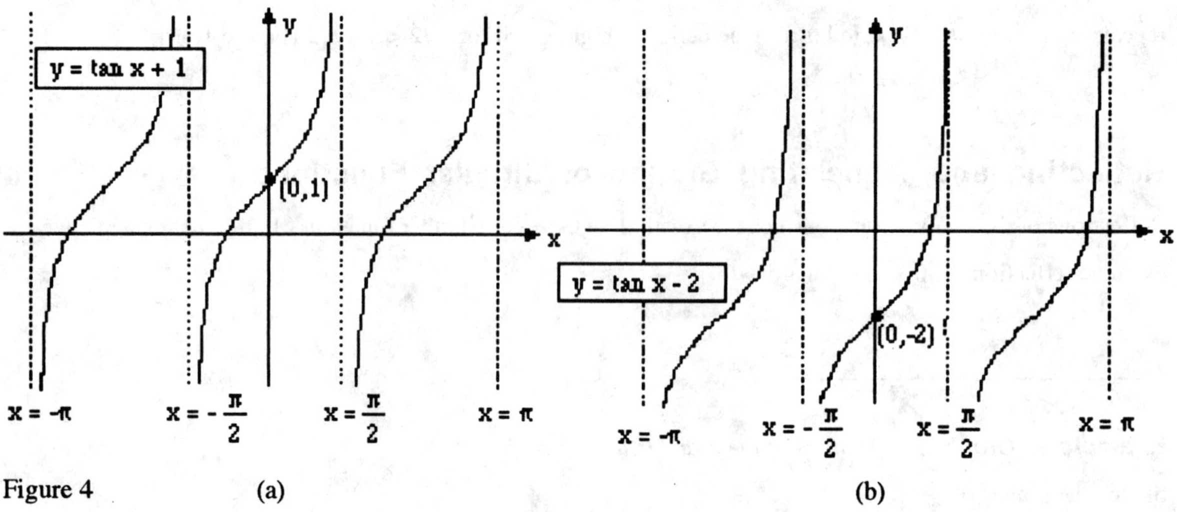

Figure 4 (a) (b)

When dealing with the circular functions, we refer to a horizontal translation as a **phase shift**. A positive phase shift is a translation to the right; a negative phase shift is a translation to the left.

Example 2 *Graphing a circular function with a phase shift*

State the phase shift and sketch the graph of each of the following.

(a) $y = \cos(x + \frac{\pi}{2})$ (b) $y = \cos(x - 4)$

Solution

(a) The graph of $y = \cos(x + \frac{\pi}{2})$ is a horizontal translation of the graph of $y = \cos x$ to the left $\frac{\pi}{2} \approx 1.57$ units. This is a phase shift of $-\frac{\pi}{2}$ (Figure 5).

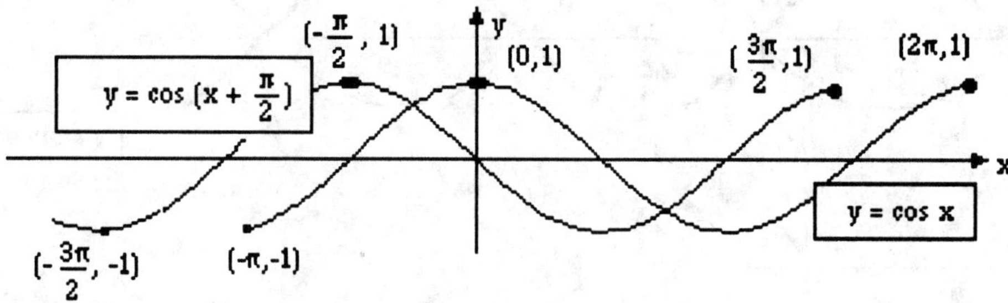

Figure 5

(b) We translate the graph of $y = \cos(x - 4)$ to the right 4 units. Its phase shift is 4 (Figure 6).

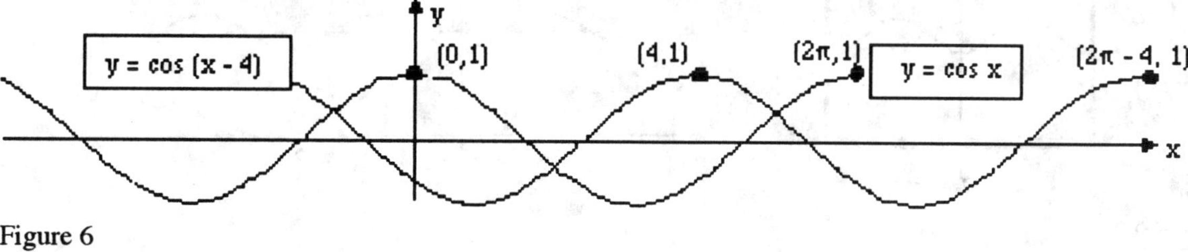

Figure 6

Changing Amplitude and Period

To explore the effects of a scaling factors on the circular functions, go to "Explore the Concepts," Exercises E7 - E10, before continuing (page 385). For the interactive version, go to the *Precalculus Weblet - Instructional Center - Explore Concepts*.

For $y = A \sin x$ and $y = A \cos x$, the scaling factor $|A|$ is called the **amplitude**. Notice in Figure 7 how the peaks and valleys of $y = 2 \sin x$ are twice as high and twice as low as those on the graph of $y = \sin x$. We say that the amplitude of $y = \sin x$ is 1, while the amplitude of $y = 2 \sin x$ is 2.

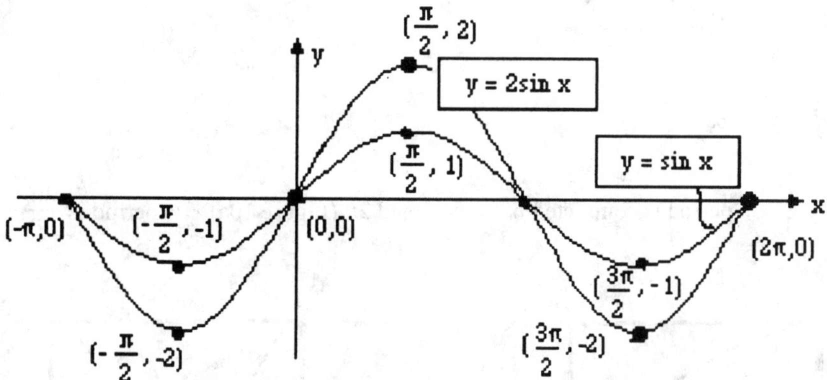

Figure 7

In the next example we graph a cosine function. The absolute value of its scaling factor is its amplitude.

Example 3 *Graphing a circular function with a change of amplitude*

Describe the graph of $y = -\frac{1}{2}\cos x + 3$ in terms of the graph of $y = \cos x$, and sketch its graph.

Solution

The graph of $y = \cos x$ has been reflected about the x axis and translated up 3 units. It has an amplitude of 1/2 (Figure 8).

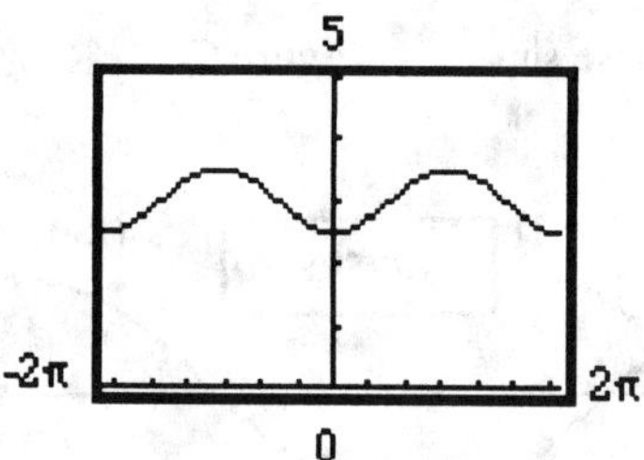

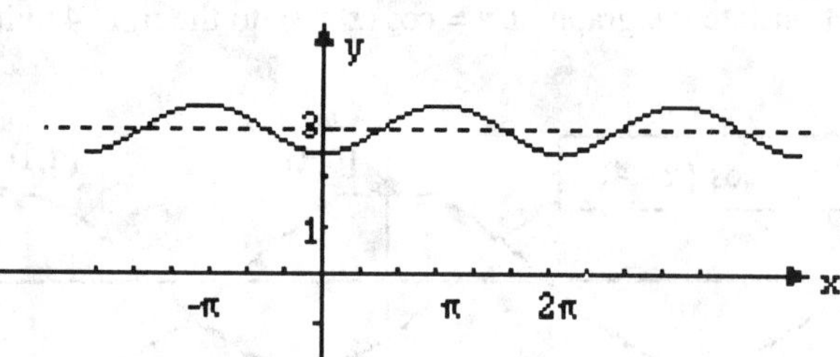

Figure 8

Consider the graphs of y = cos 4x (Figure 9). The scaling factor 4 changes the period of the graph. There are four complete periods of the cosine function between 0 and 2π. Its period is $\frac{2\pi}{4} = \frac{\pi}{2}$.

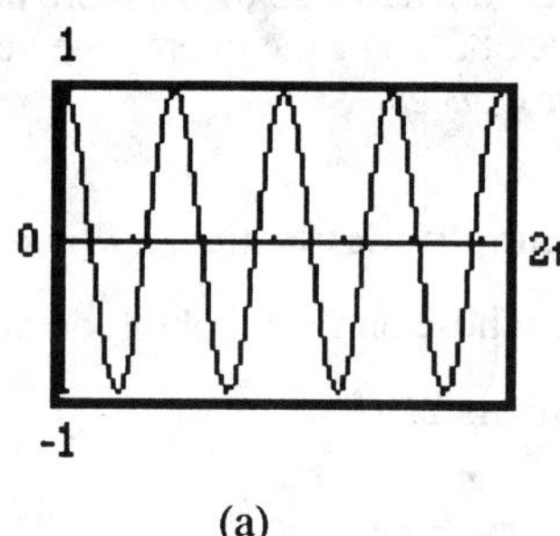

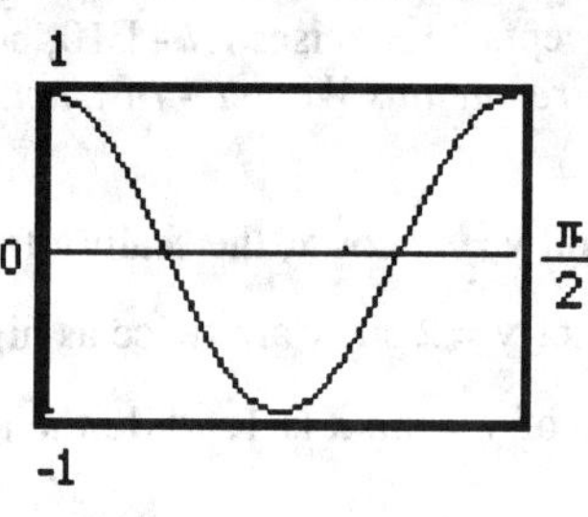

Figure 9 (a) (b)

What is the period of $y = \cos\frac{x}{3}$?

For $y = \cos\frac{x}{3}$, only $\frac{1}{3}$ of a period is completed between 0 and 2π (Figure 10a). Its period is $\frac{2\pi}{1/3} = 6\pi$ (Figure 10b).

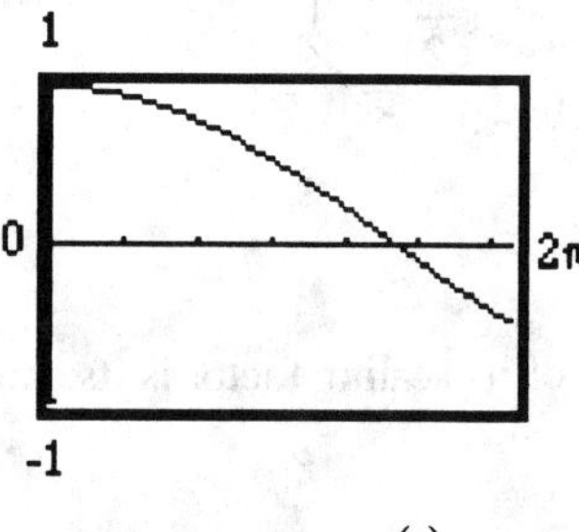

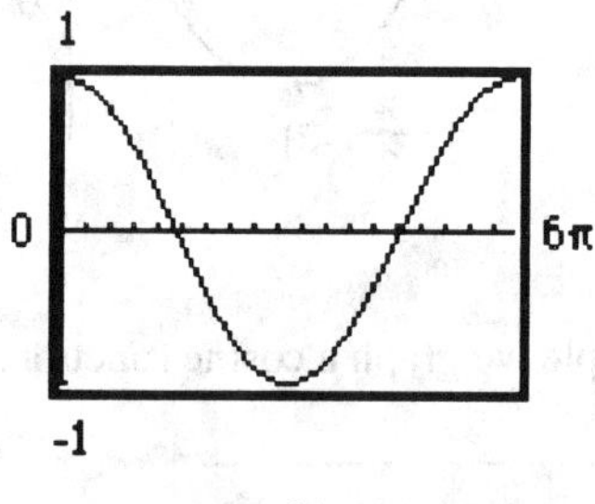

Figure 10 (a) (b)

Summary: Graphing Sine and Cosine

For all real values x, the graphs of y = A sin (Bx + C) + D and y = A cos (Bx + C) + D have

amplitude:	$\lvert A\rvert$
period:	$\frac{2\pi}{\lvert B\rvert}$
phase shift:	$x = -\frac{C}{B}$
vertical translation:	$\lvert D\rvert$

If D is positive, the translation is up, and if D is negative, the translation is down $\lvert D\rvert$ units.

Example 4 *Identifying amplitude, period, and phase shift*

Find the amplitude, period, and phase shift of $y = 4 \cos (3x - \frac{3\pi}{2})$ and sketch its graph.

Solution

Comparing $y = 4 \cos (3x - \frac{3\pi}{2})$ with $y = A \cos (Bx + C) + D$ shows that $A = 4$, $B = 3$, $C = -\frac{3\pi}{2}$, and $D = 0$. Therefore, the amplitude is $|A| = |4| = 4$, the period is $\frac{2\pi}{|B|} = \frac{2\pi}{3}$, and the phase shift is $\frac{-C}{B} = -\frac{-3\pi/2}{3} = \frac{\pi}{2}$. Figure 11 shows one period of the function highlighted from $(\frac{\pi}{2}, 4)$ to $(\frac{7\pi}{6}, 4)$.

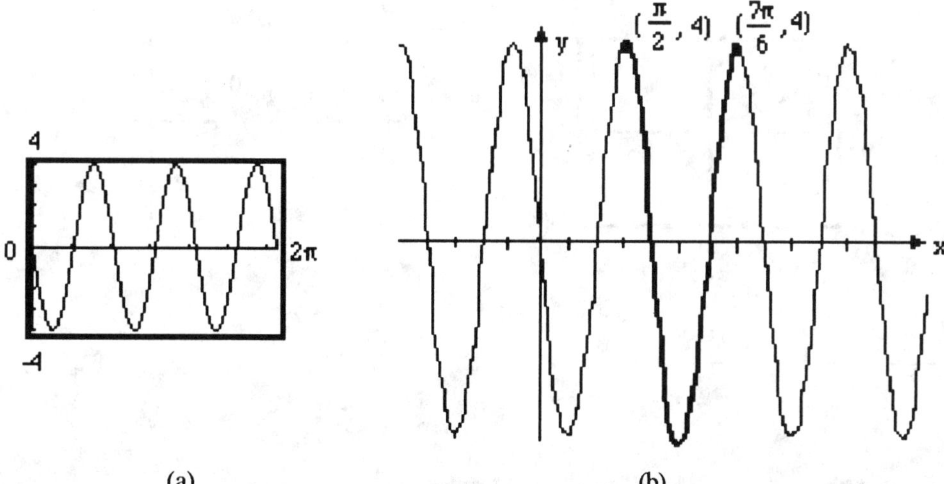

Figure 11 (a) (b)

The values of A, B, C, and D for $y = A \tan (Bx + C) + D$ outlined below are analogous for the graphs of the sine and cosine functions with two exceptions. The tangent function does not have *amplitude*. The value of $|A|$ is a scaling factor that stretches the graph if $|A| > 1$ and flattens it for $0 < |A| < 1$. Since tangent has a period of π rather than 2π, its period is determined by $\frac{\pi}{|B|}$.

Summary: Graphing Tangent

For all real values x, the graph of $y = A \tan (Bx + C) + D$ has

scaling factor:	$\lvert A \rvert$
period:	$\frac{\pi}{\lvert B \rvert}$
phase shift:	$x = -\frac{C}{B}$
vertical translation:	$\lvert D \rvert$

If D is positive, the translation is up, and if D is negative, the translation is down $|D|$ units.

Example 5 *Sketching a tangent graph*

Sketch the graph of $y = -2\tan\frac{x}{2} + 1$.

Solution

Comparing $y = -2\tan\frac{x}{2} + 1$ with $y = -A\tan(Bx + C) + D$ shows that $A = -2$, $B = \frac{1}{2}$, $C = 0$, and $D = 1$. The graph is reflected about the x axis. The scaling factor is $|A| = |-2| = 2$. The period is $\frac{\pi}{|B|} = \frac{\pi}{1/2} = 2\pi$ and there is no phase shift but the graph is translated up 1 unit (Figure 12).

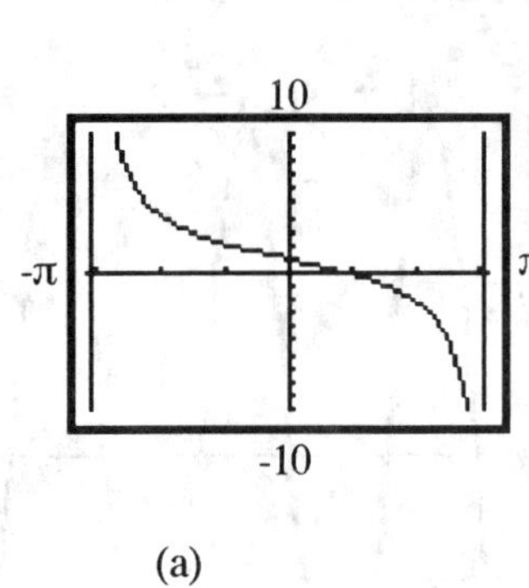

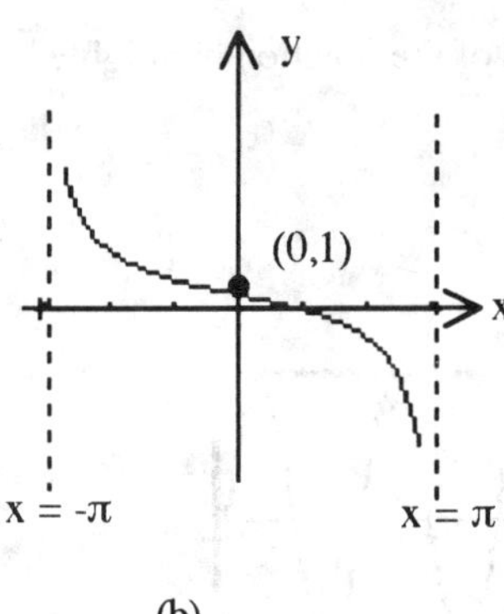

Figure 12 (a) (b)

Graphing One Period of a Circular Function

Often we graph only one period of a circular function since the graphs of these functions simply repeat. It is essential to remember that these graphs repeat periodically even though they may not be drawn that way. Reflections, vertical translations, and amplitudes are easily determined from the equation, but changes of period and phase shift are more difficult to identify. A quick way to sketch one period or to find the setting for a graphing utility window that will display one period is to examine the argument of the function between the beginning and ending points of one period. For example, one complete period of the graph of $y = \cos x$ occurs between 0 and 2π, inclusive. For $y = 4\cos(3x - \frac{3\pi}{2})$, the argument

$$3x - \frac{3\pi}{2}$$

must take on all values between 0 and 2π. Solving a compound inequality for x reveals the phase shift and the period.

$$0 \le 3x - \frac{3\pi}{2} \le 2\pi \qquad \textit{Add } \frac{3\pi}{2} \textit{ to each part of the inequality.}$$

$$\frac{3\pi}{2} \le 3x \le \frac{7\pi}{2} \qquad \textit{Divide each part by 3.}$$

$$\frac{\pi}{2} \le x \le \frac{7\pi}{6}$$

The graph of $y = 4\cos(3x - \frac{3\pi}{2})$ will complete one period for $\frac{\pi}{2} \le x \le \frac{7\pi}{6}$ (Figure 13).

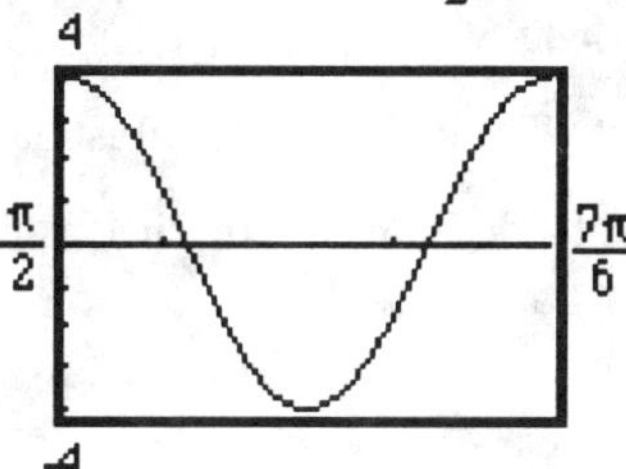

Figure 13

In the next example we graph one period of the tangent function. We will examine the argument of the function between two asymptotes. For example, one complete period of the graph of $y = \tan x$ occurs between $-\frac{\pi}{2}$ and $\frac{\pi}{2}$.

Example 6 *Graphing one period of a circular function*

Find the setting for a graphing utility window that will display one complete period of the graph of $y = 3\tan(2x - \frac{\pi}{2})$.

Solution

The tangent graph does not have amplitude. The scaling factor 3 makes the graph steeper. One period of the reference tangent function occurs between $-\frac{\pi}{2}$ and $\frac{\pi}{2}$. Hence, the argument of the function, $2x - \frac{\pi}{2}$, must take on all values between $-\frac{\pi}{2}$ and $\frac{\pi}{2}$.

$-\frac{\pi}{2} < 2x - \frac{\pi}{2} < \frac{\pi}{2}$ *Add $\frac{\pi}{2}$ to each part of the inequality.*

$0 < 2x < \pi$ *Divide each part by 2.*

$0 < x < \frac{\pi}{2}$

The graph of this function will have one complete period between $x = 0$ and $x = \pi/2$ (Figure 14).

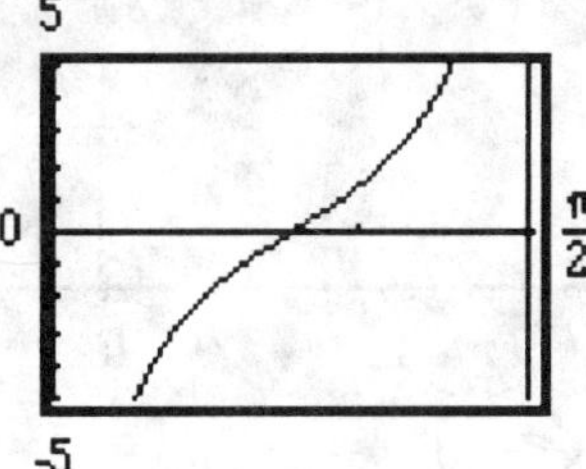

Figure 14

Finding the Equation of a Periodic Graph

The general equations $y = A \sin(Bx + C) + D$, $y = A \cos(Bx + C) + D$, and $y = A \tan(Bx + C) + D$ can be used to find equations of periodic functions. The next example illustrates the importance of period and amplitude in modeling situations that repeat a definite pattern, such as tides.

Example 7 The Vietnam Tide Problem

The Gulf of Tonkin off the Vietnam-China coast has a diurnal tide with one high tide and one low tide each day. On a given day, if Vietnam's high tide is +14 ft above average and the low tide is +6 ft, find an equation that models this daily event and sketch one period of its graph.

Thinking: Creating a Plan

The simplest function that periodically repeats a high and low is of the form

$$h(t) = A \sin Bt + D$$

whose amplitude is $|A|$, period is $\frac{2\pi}{|B|}$, and vertical translation is D. We will let t represent time in hours and h the height of the water in feet. The period of this diurnal tide is 24 hr.

Communicating: Writing the Solution

The tide range from 6 to 14 ft requires a model with an amplitude of 4 units since the difference between the high and low is 8 ft. To create a minimum of 6, we translate the graph of $y = 4 \sin t$ up 10 units. The period of 24 hr requires $\frac{2\pi}{B} = 24 \Leftrightarrow B = \frac{2\pi}{24} = \frac{\pi}{12}$. The equation modeling the Vietnam tide is $h(t) = 4 \sin \frac{\pi}{12} t + 10$ (Figure 15).

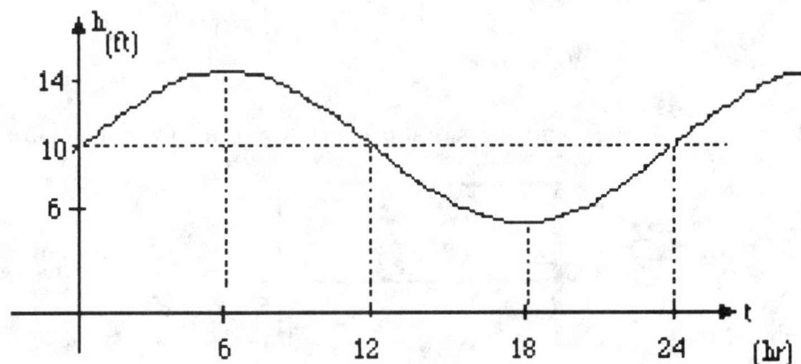

Figure 15

Learning: Making Connections

The cosine function could also have been used as a model for this situation, but it would have required a translation to start increasing at (0, 10).

The moon is the major cause of changing tides. Figure 16 shows the moon's strong gravitational pull on its side of earth. On the opposite side, the influence is much less because it is farther away. Gravitational attraction of the moon causes the ocean to bulge.

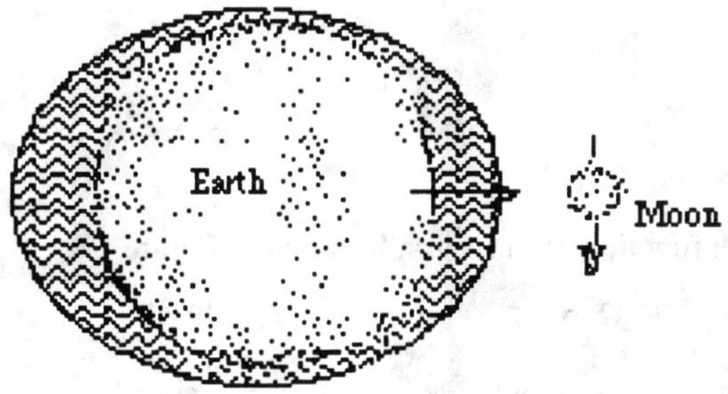

Earth

Moon

Figure 16

The actual period of a diurnal tide is 24 hr and 50 min, because the moon passes over any point on earth every 24 hr and 50 min. Tides are also affected by the gravitational attraction of the sun. Since it is so far away, the sun has only $\frac{5}{11}$ the influence of the moon.

Explore the Concepts

E1. Graph y = sin x over each of the following domains.

(a) $0 \le x \le 2\pi$ (b) $0 \le x \le 4\pi$ (c) $-2\pi \le x \le 2\pi$ (d) $-6\pi \le x \le 6\pi$

E2. The graphs of y = sin x in Exercise E1 show that the sine curve repeats itself at regular intervals. What is this interval?

E3. Graph y = cos x over each of the following domains.

(a) $0 \le x \le 2\pi$ (b) $0 \le x \le 4\pi$ (c) $-2\pi \le x \le 2\pi$ (d) $-6\pi \le x \le 6\pi$

E4. The graphs of y = cos x in Exercise E3 show that the cosine curve repeats itself at regular intervals. What is this interval?

E5. Graph y = tan x over each of the following domains.

(a) $0 \le x \le \pi$ (b) $0 \le x \le 2\pi$ (c) $-\pi \le x \le \pi$ (d) $-6\pi \le x \le 6\pi$

E6. The graphs of y = tan x in Exercise E5 show that the tangent curve repeats itself at regular intervals. What is this interval?

E7. Graph the following equations of the form y = A sin x for $0 \le x \le 2\pi$.

(a) y = sin x (b) y = 2 sin x (c) y = 3 sin x (d) $y = \frac{1}{3}\sin x$

E8. Use your results from Exercise E7 to explain the geometric interpretation of A in y = A sin x.

E9. Graph the following equations of the form y = sin Bx for $0 \le x \le 2\pi$.

(a) y = sin 2x (b) y = sin 3x (c) y = sin 4x (d) y = sin 6x

(e) $y = \sin \frac{1}{2}x$ (f) $y = \sin \frac{1}{4}x$

E10. Use your results from Exercise E9 to explain the geometric interpretation of B in y = sin Bx.

Exercises 5.3

Develop the Concepts

For Exercises 1 to 8 match each function with its graph shown in Figure 17.

1. $y = \cos(x - \frac{\pi}{2})$
2. $y = \sin(x + \frac{\pi}{2})$
3. $y = \sin 2x$
4. $y = \sin 3x$
5. $y = \cos \frac{x}{2}$
6. $y = \cos(x + \frac{\pi}{2})$
7. $y = \frac{1}{2} \sin x$
8. $y = 2 \cos x$

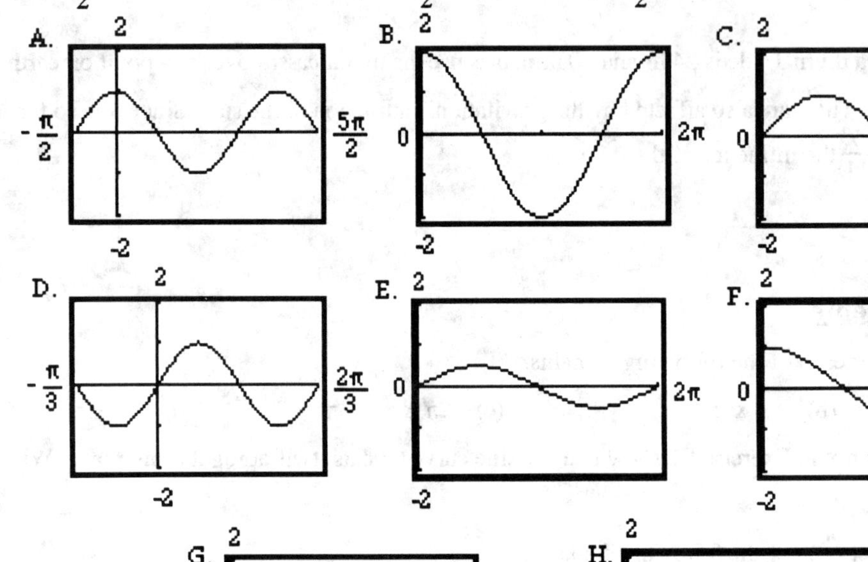

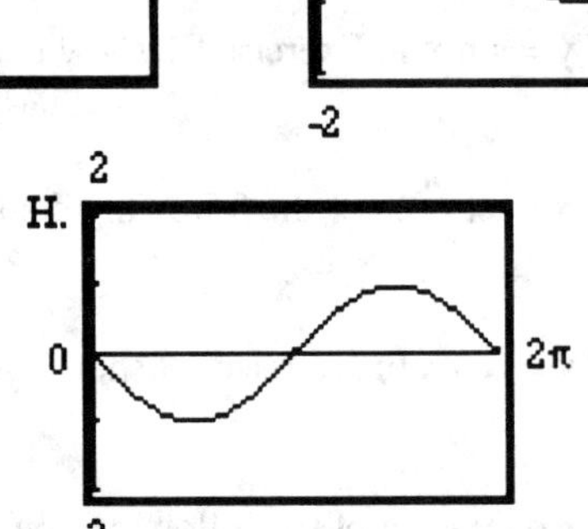

Figure 17

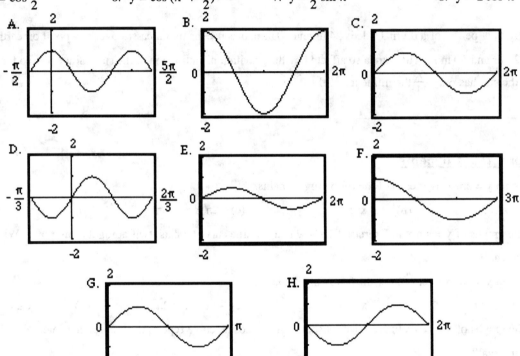

For Exercises 9 to 18 graph one period for each circular function.

9. (a) $y = \sin 2x$
   (b) $y = \sin \frac{x}{2} + 1$

10. (a) $y = \cos 4x$
    (b) $y = \cos \frac{x}{4} - 2$

11. (a) $y = \cos(x + \frac{\pi}{4})$
    (b) $y = \cos(x - \frac{\pi}{2})$

12. (a) $y = -\sin(x + \pi)$
    (b) $y = -\sin(x - \pi)$

13. $y = 2 \sin(3x - \frac{\pi}{2})$

14. $y = 3 \cos 3(x - \frac{\pi}{6})$

15. $y = -\tan \frac{x}{10}$

16. $y = 2 \tan 3x$

17. $y = 3 \tan 3(x - \frac{\pi}{4})$

18. $y = \tan(2x + \frac{\pi}{2}) - 2$

For Exercises 19 to 22 state the period of each function.

19. $y = \sin \frac{x}{2}$

20. $y = \sin \frac{2x}{3}$

21. $y = -3 \tan (2x + \pi)$

22. $y = 2 \tan 3x$

For Exercises 23 to 26 state the y intercept for each function.

23. $y = 2 \cos x$

24. $y = -3 \sin x$

25. $y = -\tan x + 2$

26. $y = \tan (x + \frac{\pi}{2})$

For Exercises 27 and 28 state the equations of three asymptotes on each graph.

27. $y = \tan 2x$

28. $y = \tan (2x - \frac{\pi}{4})$

For Exercises 29 to 36 state an equation that best describes the graph.

29.

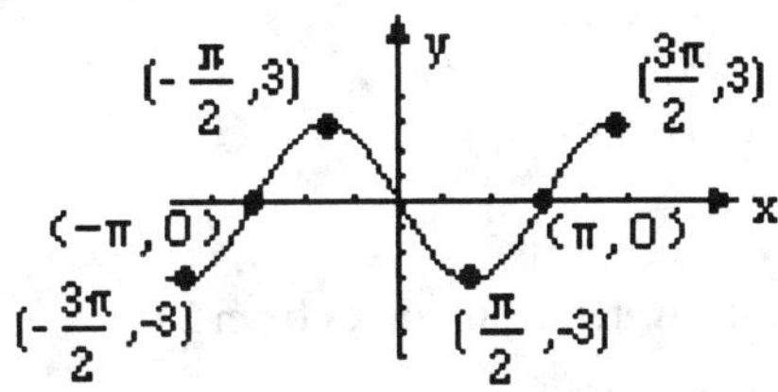

30.

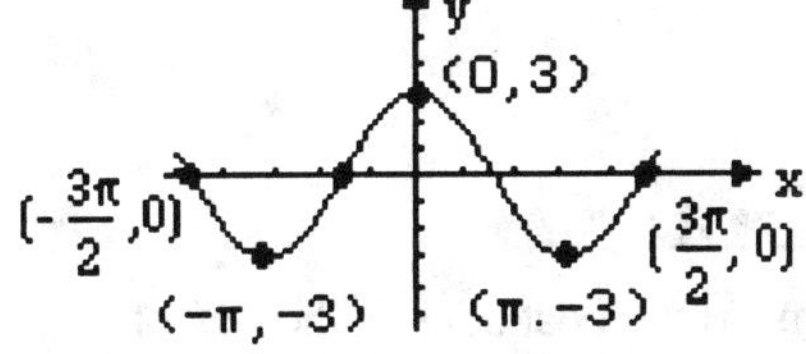

31.

32.

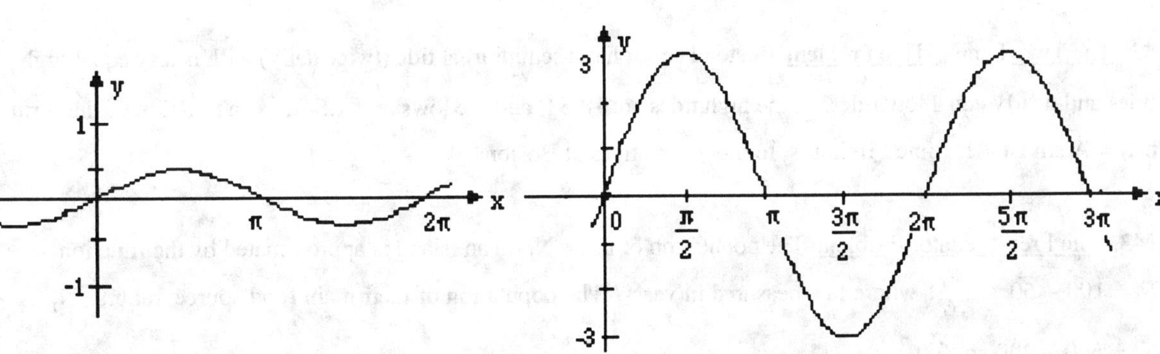

33.

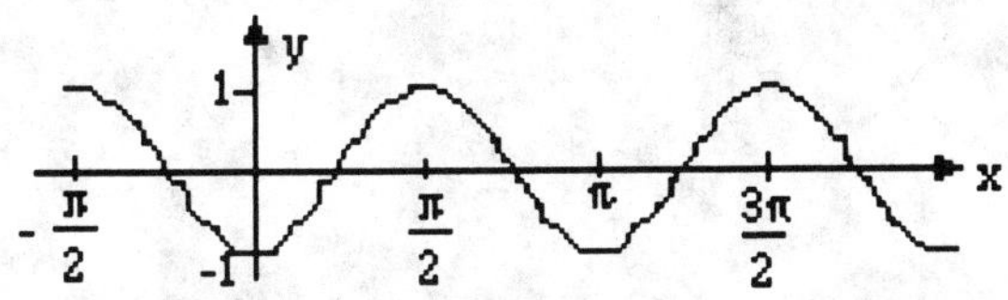

34.

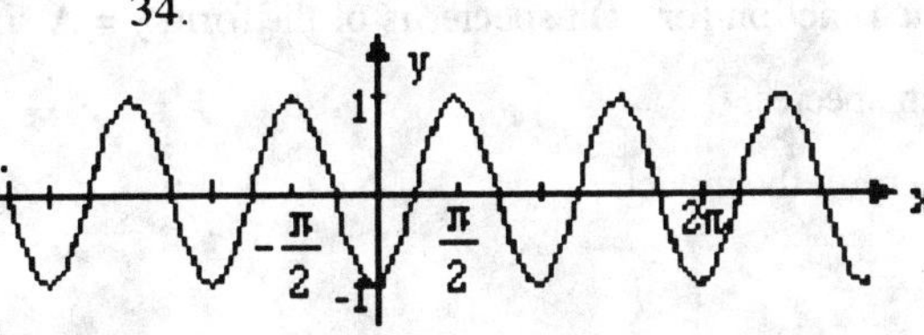

35. 36.

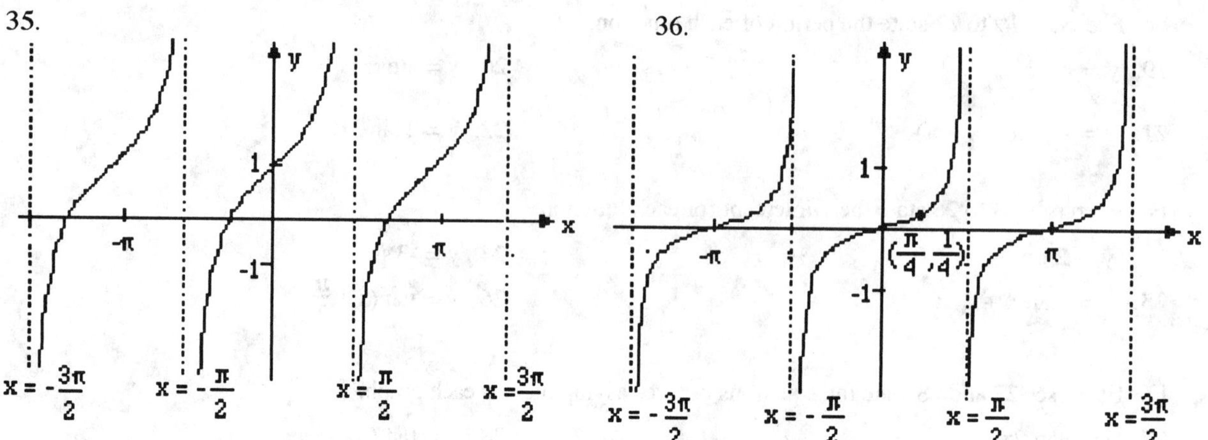

For Exercises 37 to 40 use the graph of each function to determine whether it is even, odd, or neither.

37. $f(x) = 2 - \sin x$

38. $f(x) = 2 + \cos x$

39. $f(x) = \tan(-x)$

40. $f(x) = -\tan x$

Apply the Concepts

41. A Tide Problem At the boat dock in the Bay of Fundy in Nova Scotia the water depth varies from 2 ft deep to 58 ft deep. The average interval between successive high tides is 12 hr and 25.5 min. Write a function of the form $h(t) = A \sin Bt + D$ that models these two conditions.

42. The Two-Timing-Tide Problem Boston Harbor has a semidiurnal tide (twice daily) with nearly equal high tides and nearly equal low tides. If the high tides are +8.5 ft and the lows are -1.5 ft, use a function of the form $h(t) = A \sin Bt + D$, time t in hours, to model the tides at Boston.

*43. The Prey-Predator Problem The population of foxes N_f on an island is approximated by the function $N_f = 100 + 50 \sin \frac{\pi}{6}t$, where t is measured in years. The population of their main food source, rabbits, N_r, is $N_r = 500 + 200 \sin (\frac{\pi}{6}t - \pi)$.

(a) Graph both functions.

(b) The function for each species is of the form $y = A \sin (Bx + C) + D$. Interpret the value of A, B, C, and D for each species.

(c) Interpret the model.

44. <u>The Home Fires Burning Problem</u> Consider how the varying temperature outside T_o affects the temperature inside your home T_i, where $T_o = 14 + 14 \sin(\frac{\pi}{12}t)$ and $T_i = 28 + 2 \sin(\frac{\pi}{12}t - 1)$, with temperature measured in degrees Celsius and t in hours.

(a) Graph the functions over a 48-hr period.

(b) The function for outside temperature and the function for inside temperature is of the form $y = A \sin(Bx + C) + D$. Interpret the value of A, B, C, and D for outside and inside temperatures.

(c) Interpret the model.

45. The Rainbow Kite and Balloon Company Figure 18 shows the new wedge kite you've designed in the shape of an isosceles triangle.

(a) Express the area of the kite as a function of angle θ.

(b) Use the graph of the area functions from part (a) to find the measure of θ that will create a kite with the maximum surface area.

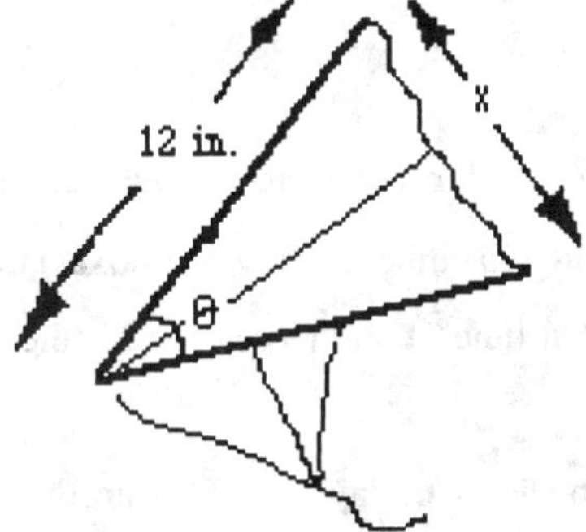

Figure 18

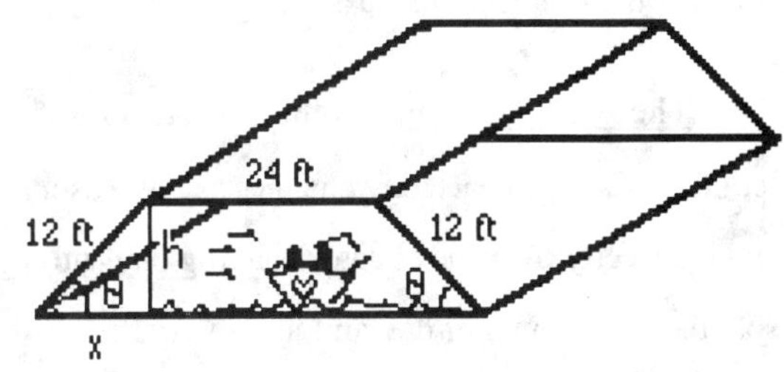

Figure 19

46. Wonderworld Amusement Park You are to design a "tunnel of love" ride at the park, with the love boats entering the tunnel as depicted in Figure 19.

(a) Express the area of the trapezoidal entrance as a function of θ.

(b) Use the graph of the area functions from part (a) to find the value of θ that will maximize the entrance area.

47. <u>The Biorhythms Problem</u> Three cycles may influence our daily lives-- cycles called biorhythms. The physical cycle has a period of 23 days. The emotional cycle has a period of 28 days, and the intellectual cycle has a period of 33 days.

(a) Write a function of the form $y = \sin Bx$ for each cycle.

(b) Graph the three functions for $0 \leq x \leq 60$.

(c) According to the graphs, when should you accept any challenge and when should you just stay in bed?

(d) How long does it take for the physical and emotional cycles to meet again at $y = 0$ when both cycles are neutral -- neither high nor low?

(e) How long does it take for all three cycles to return to $y = 0$, a neutral state?

48. The Circadian Rhythms Problem Certain human activities such as sleeping are affected by the sun's rising and setting periodically. The graph in Figure 20 represents one revolution of the earth having 12 hr of daylight and 12 hr of darkness.

(a) Indicate on the graph the interval that represents your sleep time if you like to go to bed early and wake at daybreak.

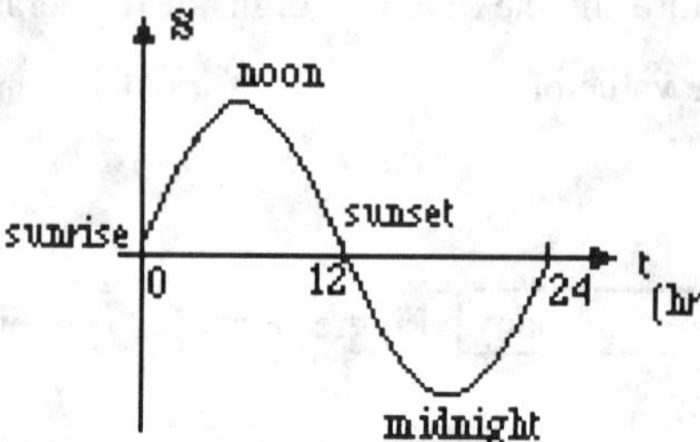

Figure 20

(b) Translate the graph to represent another town whose time zone is 3 hr ahead.

(c) Use the graphs to discuss jetlag.

*49. The Earth Problem As the Earth takes 365 days, 5 hr, 48 min, and 46 sec to revolve around the sun, the length of daylight each day changes as the seasons change. On March 21 and September 22 (the equinoxes) the sun is directly over the equator, making day and night have equal lengths. On June 21 and December 21 (the solstices) the hours of day and night vary the most.

(a) Write a sine function of the form $H = A \sin Bt + D$ that approximates the hours of daylight H for northern Finland, where the longest day is 24 hr.

(b) Indicate the equinoxes and solstices on your graph. [Hint: Let $t = 0$ correspond to one of the equinoxes.]
[Source: *Weather: an Exploration of the Forces that Drive the World's Weather*, P. Lafferty]

50. The Eternal Spring Problem In the land of eternal spring, Quito, Ecuador, the low temperature rarely falls below 8°C or exceeds 22°C.

(a) Graph the function and interpret its turning points.

(b) Make the assumptions necessary to express the temperature in paradise as a function of time of the form $T = A \sin Bt + D$ for a typical year.
[Source: *Weather: an Exploration of the Forces that Drive the World's Weather*, P. Lafferty]

Extend the Concepts

For Exercises 51 to 54 use the graph of each of the following functions to graph its reciprocal $y = \frac{1}{f(x)}$.

51. $f(x) = \sin x$

52. $f(x) = -\sin x$

53. $f(x) = 2 \sin x$

54. $f(x) = \sin 2x$

5.4 Reciprocals, Sums, and Products of the Circular Functions

Defining the Reciprocal Circular Functions
Graphing the Reciprocal Circular Functions
Graphing Sum and Product Functions Involving Circular Functions

Many applications of circular functions involve reciprocals, sums, and products of circular functions. Some examples of periodic motion include the sun setting, a pendulum swinging, a piston moving, a weight bouncing on a spring, and the vibrating string of a musical instrument. These are periodically recurring events (motion), each of which has a period (the length of time in one cycle) and a frequency (the number of cycles per unit time interval). For example, 365 sun settings per year means that the frequency of the setting sun is 365 times per year and its period is 1/365 yr ≈ 1 day since it takes 1 day for the rising and setting sun to complete one cycle.

Defining the Reciprocal Circular Functions

The reciprocals of sine, cosine, and tangent form three other circular functions, **cosecant**, **secant**, and **cotangent** (abbreviated csc, sec, and cot).

Definitions of the Reciprocal Circular Functions

$$\csc x = \frac{1}{\sin x},\ \sin x \neq 0$$

$$\sec x = \frac{1}{\cos x},\ \cos x \neq 0$$

$$\cot x = \frac{1}{\tan x},\ \tan x \neq 0$$

Calculators only have keys for sine, cosine, and tangent. Suppose we want to find the value for csc $\frac{\pi}{4}$. We can enter $1 \div \sin\frac{\pi}{4}$ or use the reciprocal key $\boxed{x^{-1}}$, $(\sin\frac{\pi}{4})^{-1}$. Care must be taken to apply the reciprocal key since

$$\frac{1}{\sin(\pi/4)} \neq \sin\frac{1}{\pi/4}$$

Example 1 *Finding values of reciprocal functions*

Find the exact value of each of the following and the calculator approximation.

(a) $\sec\frac{5\pi}{4}$ (b) $\cot\frac{7\pi}{4}$

Solution

(a) Since $\cos\frac{5\pi}{4} = -\frac{\sqrt{2}}{2}$, $\sec\frac{5\pi}{4} = -\frac{2}{\sqrt{2}} \approx -1.414$. [Enter $(\cos\frac{5\pi}{4})^{-1}$.]

(b) Since $\tan\frac{7\pi}{4} = -1$, $\cot\frac{7\pi}{4} = -1$. [Enter $(\tan\frac{7\pi}{4})^{-1}$.]

Figure 1a shows the reference point for $\frac{5\pi}{4}$ in quadrant III and Figure 1b shows the reference point for $\frac{7\pi}{4}$ in quadrant IV. The signs for cosecant, secant, and cotangent can be determined by the quadrant in which their reference point falls just as the signs of their respective reciprocal functions.

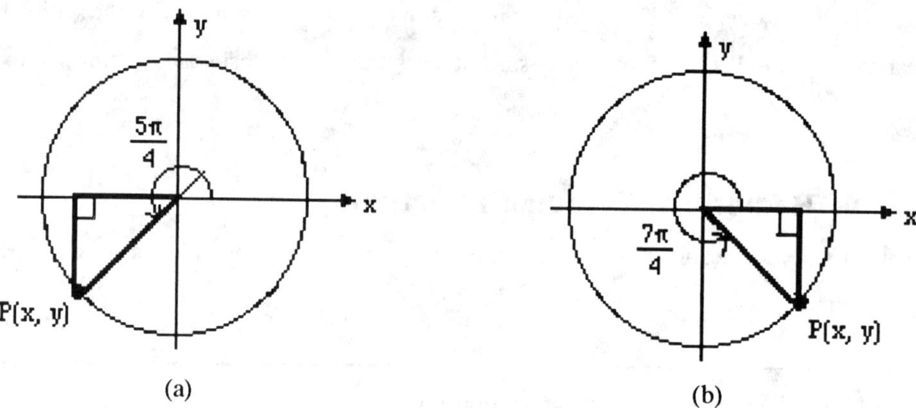

Figure 1 (a) (b)

Graphing the Reciprocal Circular Functions

In Section 3.4 we graphed reciprocal functions. Recall that a graph of a reciprocal has a vertical asymptote wherever the original function has an x intercept, and vice versa. A graph of a function and its reciprocal are both positive or negative over the same interval, but where one graph is increasing, the graph of the reciprocal function is decreasing. Figure 2 shows graphs of portions of the sine, cosine, and tangent graphs, respectively, along with the graphs of their reciprocals -- cosecant, secant, and cotangent.

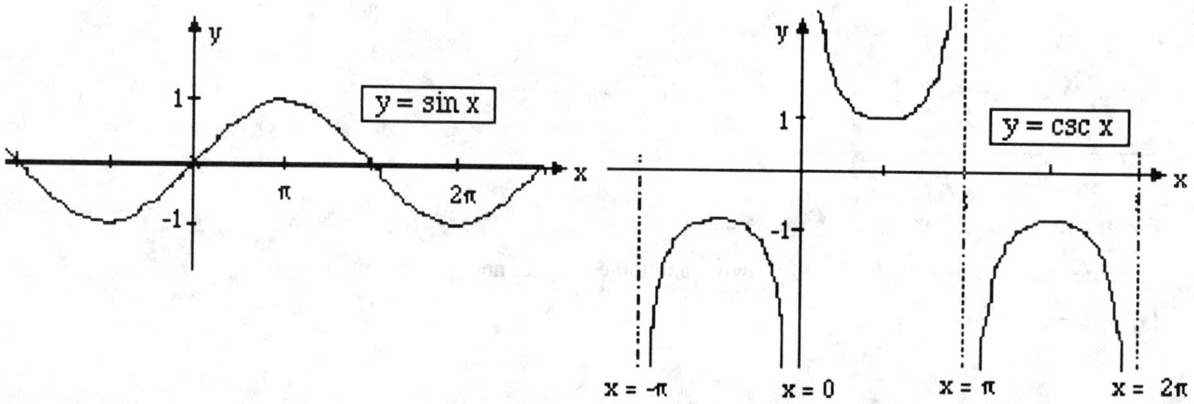

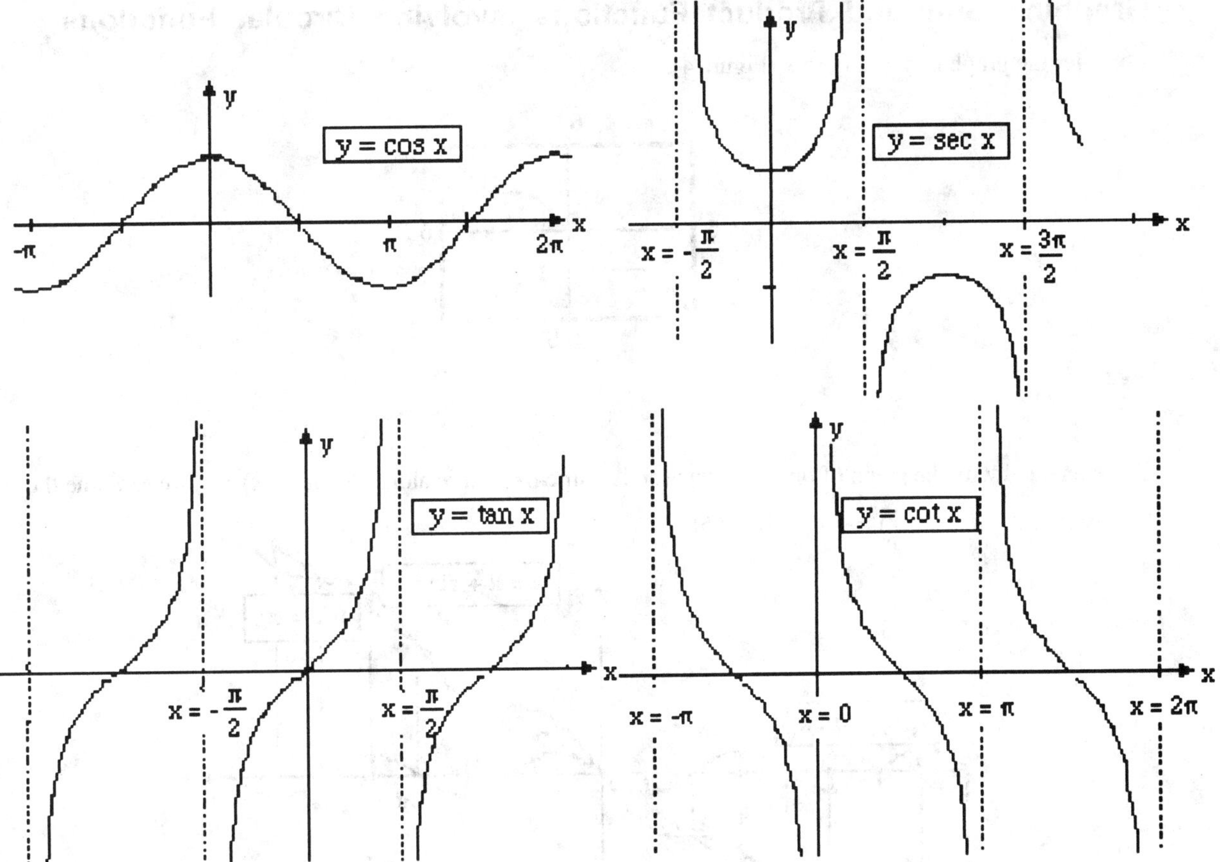

Figure 2

Example 2 *Graphing a reciprocal function*

Describe the graph of y = 20 sec x - 10 in terms of the graph of y = sec x. Graph one period of the function.

Solution

The graph of y = sec x has been translated down 10 units. It has turning points 10 units above and 30 units below the x axis (Figure 3).

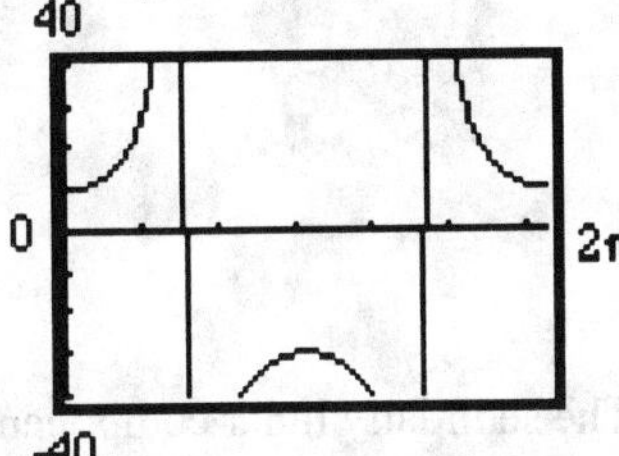

Figure 3

Graphing Sum and Product Functions Involving Circular Functions

Consider the graph of $y = x + \sin x$ (Figure 4).

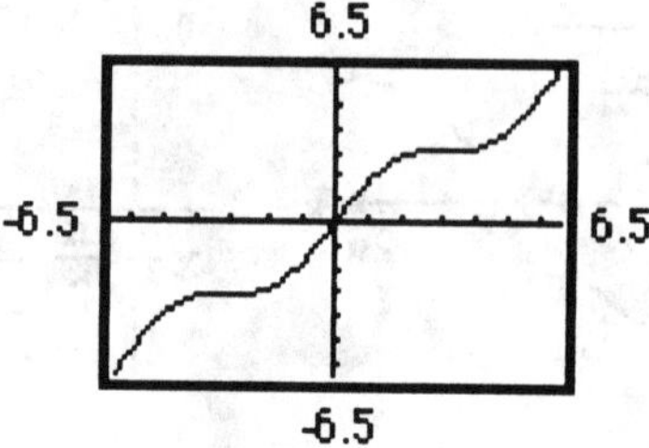

Figure 4

To understand why the graph of this sum function is a sine-like curve along the line $f(x) = x$, we examine the graphs of $f(x) = x$ and $g(x) = \sin x$ (Figure 5).

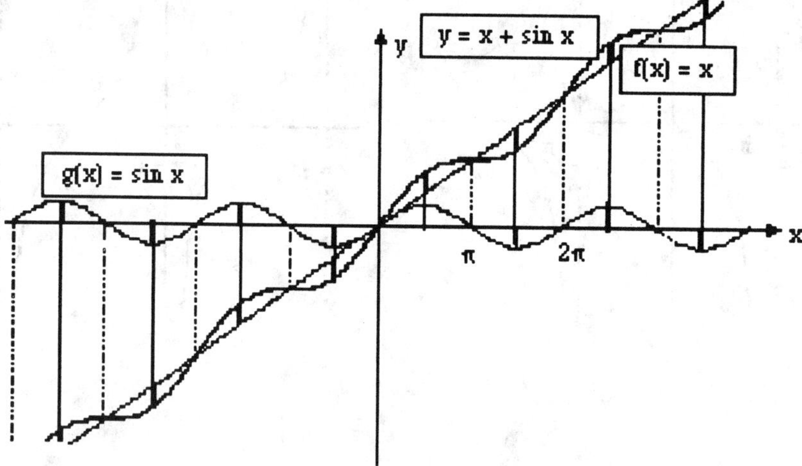

Figure 5

As x increases the y values of function $g(x) = \sin x$ are added, then subtracted, from the y values of $f(x) = x$ as the sine curve changes periodically from positive to negative. Every time the value of the sine function is zero, the graph of the sum crosses the line $f(x) = x$.

Periodic sine functions can be added to model the total change as a sum of its parts. The next example adds the effects of the sun and the moon to find a periodic model of changing tides.

Example 3 The Port O'Call Problem

There are two main tides at Port O'Call. The semidaily lunar component has a relative amplitude of 1.0 ft and a period of 12 hr 25 min (twice daily). The daily solar component has a relative amplitude of 0.584 ft and a period of 23 hr 50 min (once daily).

(a) Find circular functions that model each component and graph their sum.

(b) If the best time to go crabbing is during an incoming tide, when should we set the crab pots?

Thinking: Creating a Plan

Assume that $t = 0$ corresponds to midnight. Each of the two equations requires finding the value of B in $y = A \sin Bx$. Recall $|A|$ is the amplitude and that $\frac{2\pi}{|B|}$ equals the period of the function. Then we interpret the graph of the sum function, the times when the tide is low and when the tide is incoming, to find the time intervals from low tide to high tide.

Communicating: Writing the Solution

(a) Semidaily lunar component:

The amplitude of the semidaily component is 1 and the period is 12 hr and 25 min ≈ 12.42 hr. Therefore,

$$\frac{2\pi}{|B|} = 12.42$$

$$2\pi = 12.42\,|B|$$

$$B \approx 0.506$$

$h_1 = A \sin Bt$ and $A = 1, B = 0.506 \Rightarrow h_1(t) = 1.0 \sin 0.506t$

Daily solar component:

The amplitude of the daily component is 0.584 and the period is 23 hr 50 min ≈ 23.83 hr.

$$\frac{2\pi}{|B|} = 23.83$$

$$2\pi = 23.83\,|B|$$

$$B \approx 0.264$$

$h_2 = A \sin Bt$ and $A = 0.584, B = 0.264 \Rightarrow h_2(t) = 0.584 \sin 0.264t$

The sum function $y = h_1 + h_2$ models height of the water in feet as a function of time in hours, $y = 1.0 \sin 0.506t + 0.584 \sin 0.264t$ (Figure 6).

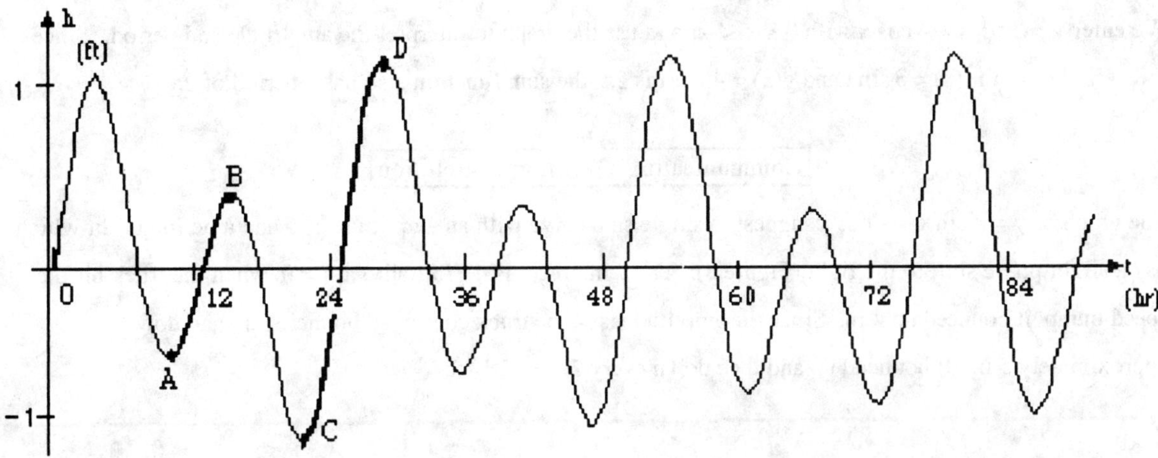

Figure 6

(b) The best time for crabbing is from point A to B or better yet, from C to D. From the graphs in Figure 7 we can set the pots at point A, 9.8 hr after midnight, at about 9:48 a.m. or at point C, 21.25 hr after midnight, at 9:15 p.m.

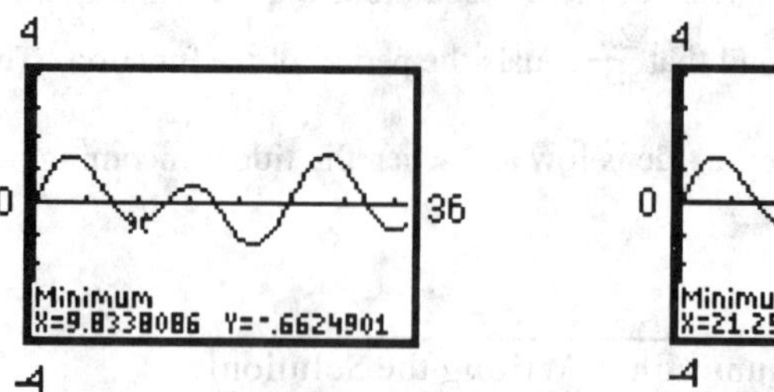

Figure 7

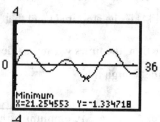

Learning: Making Connections

The sun and moon can both pull the ocean the same way, causing very high tides. This reinforcement occurs during full and new phases of the moon. The sun and the moon can also nearly cancel the effects of each other, causing smaller high tides. This occurs during quarter moons. There may be as many as 62 tidal constituents used to predict tides.

In the next example we see that the sum of two circular functions can create a single sine curve.

Example 4 The Truck Springs Problem

An equation for modeling the shock absorbers of a truck after it hits a speed bump is $y = 3 \sin t + 4 \cos t$, $t \geq 0$, where t is time in seconds and y is vertical movement in inches. Graph the function and interpret the amplitude and period of the sum function.

Thinking: Creating a Plan

We enter $y = 3 \sin x + 4 \cos x$ with $0 \leq x \leq 4\pi$ and use the graph to interpret the amplitude and period. Since the period of both $f(x) = 3 \sin t$ and $g(x) = 4 \cos t$ is 2π, the sum function also has a period of 2π.

Communicating: Writing the Solution

The graph of $y = 3 \sin x + 4 \cos x$ suggests a single sine curve with an amplitude of 5 and a period of 2π which has a slight phase shift to the right (Figure 8). The y intercept is 4. Initially, at $t = 0$, when the truck hit the speed bump, it bounced up 4 in. Since the amplitude is 5, the truck continued bouncing up and down approximately 5 in. It bounced up and then down every $2\pi \approx 6.28$ sec.

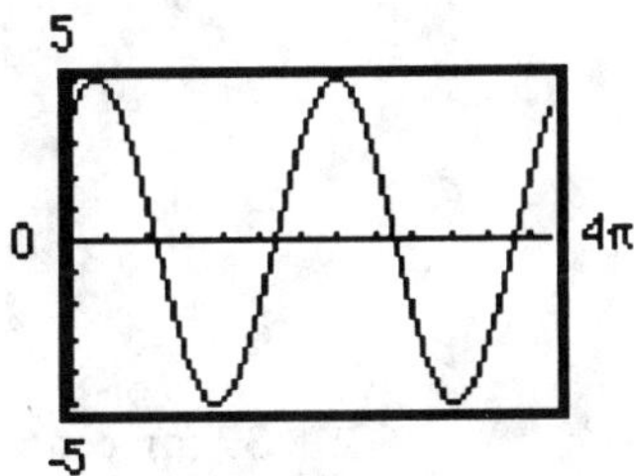

Figure 8

Learning: Making Connections

This is an example of undamped periodic motion. In reality, even the worst shock absorbers would soon stop vibrating.

According to the model describing the shock absorbers of a truck in Example 4, the truck continues to bounce forever. The product function $y = 2e^{-x/2} \sin 6x$ in the next example exemplifies how a good shock absorber works.

Example 5 *Graphing the product of two function*

Explain the damping effect of a shock absorber in Figure 9 described by the product graph of $y = 2e^{-x/2} \sin 6x$ in terms of $f(x) = 2e^{-x/2}$ and $g(x) = \sin 6x$.

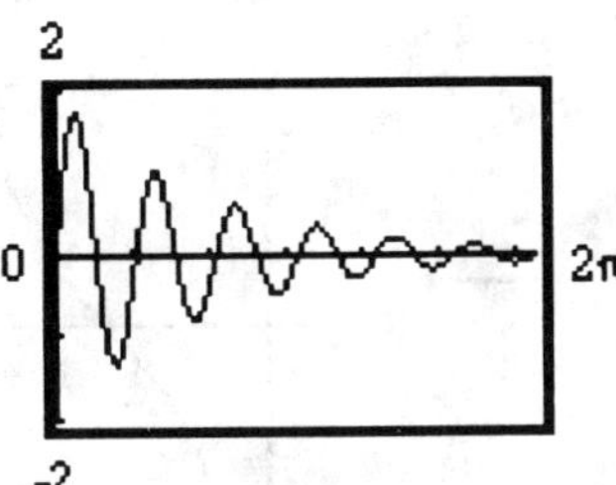

Figure 9

Solution

The decreasing exponential function $f(x) = 2e^{-x/2}$ (Figure 10a) has a damping effect as the amplitude of the product function approaches zero (Figure 10b).

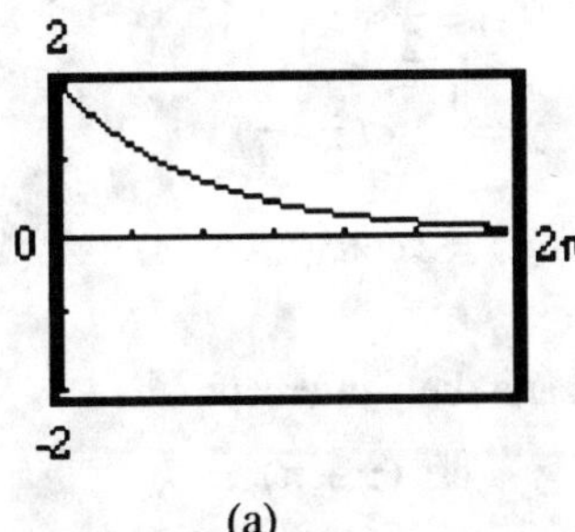

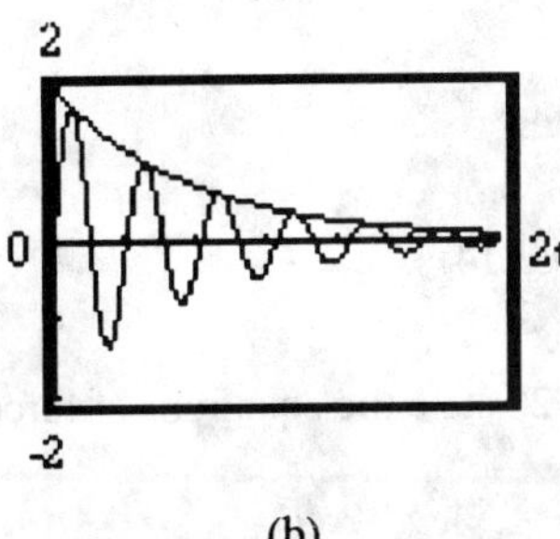

Figure 10 (a) (b)

Exercises 5.4

Develop the Concepts

For Exercises 1 to 4 find each value.

1. (a) $\sin 25°$ (b) $\csc 25°$
2. (a) $\cos 100°$ (b) $\sec 100°$
3. (a) $\cot 5.2$ (b) $\cot(-\frac{\pi}{2})$
4. (a) $\cot 3.14$ (b) $\cot \pi$

For Exercises 5 to 8 state the equation of the reciprocal function that best describes it.

5. 6.

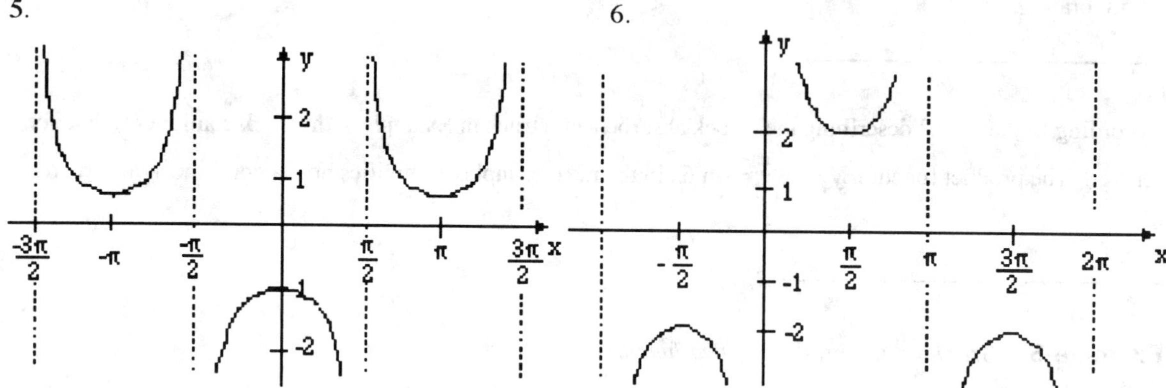

7.

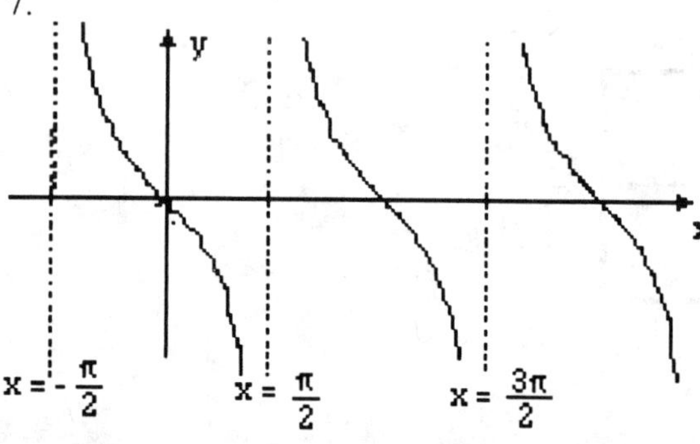

8.

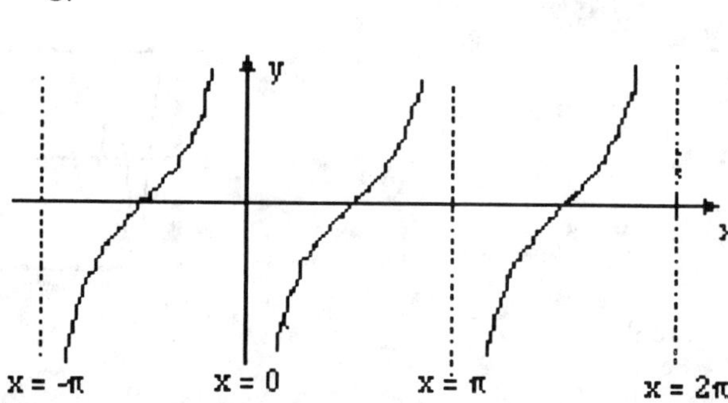

For Exercises 9 to 16 graph one period for each circular function.

9. $y = \sec 2x$
10. $y = -\sec \frac{x}{5}$
11. $f(x) = -2 \sec x - 5$
12. $f(x) = \frac{1}{2} \sec x + 4$
13. $y = 2 \csc 2(x - \frac{\pi}{3})$
14. $y = \frac{1}{3} \csc(\frac{x}{2} - \frac{\pi}{4})$
15. $f(x) = 12 \cot(x - 3.14)$
16. $f(x) = 2 \cot 3(x - 6.28)$

For Exercises 17 to 20 state the equations of three asymptotes and the y intercept.

17. $y = \sec(x - \frac{\pi}{2})$
18. $y = \csc(\frac{x}{2} + \pi)$
19. $y = \csc(x + \frac{\pi}{2})$
20. $y = \cot(\frac{x - 1.5}{2})$

For Exercises 21 to 24 use the graph of each function to explain whether it is even, odd, or neither.

21. $f(x) = \sec x$

22. $f(x) = \cot x$

23. $f(x) = \sec x \tan x$

24. $f(x) = \csc x \cot x$

For Exercises 25 to 32 describe the general shape of the graph.

25. $y = 1 + \cos x$

26. $y = x^2 + 0.1 \cos x$

27. $y = \frac{1}{x} + \sin 2x$

28. $y = -x + \sin 4x$

29. $y = \sin x + \frac{\sin 3x}{3} + \frac{\sin 5x}{5} + \frac{\sin 7x}{7}$

30. $y = \frac{\sin 2x}{2} + \frac{\sin 4x}{4} + \frac{\sin 6x}{6} + \frac{\sin 8x}{8}$

31. $y = \sin x - \frac{\sin 2x}{2} + \frac{\sin 3x}{3} - \frac{\sin 4x}{4}$

32. $y = \cos x + \frac{\cos 3x}{9} + \frac{\cos 5x}{25} + \frac{\cos 7x}{49}$

For Exercises 33 to 38, discuss the amplitude of each product function.

33. $y = x \sin x$

34. $y = x^2 \sin x$

35. $y = x^3 \sin x$

36. $y = \frac{1}{x} \sin x$

37. $y = f(x)g(x)$, where $f(x) = e^{-x}$ and $g(x) = \sin x$

38. $y = f(x)g(x)$, where $f(x) = e^{-x}$ and $g(x) = \cos 2x$

For Exercises 39 to 42 describe their apparent relationship between the graph of each pair of functions.

39. (a) $y = \sin x \cos x$
(b) $y = \sin 2x$

40. (a) $y = \cos^2 x - \sin^2 x$
(b) $y = \cos 2x$

41. (a) $y = 1 + \tan^2 x$
(b) $y = \sec^2 x$

42. (a) $y = 1 - \csc^2 x$
(b) $y = \cot^2 x$

Apply the Concepts

*43. The Pothole Problem When a vehicle hits a pothole, it bounces up and down. The displacements in inches of a cab and a bus t seconds after hitting a New York pothole are given by $d_1 = (10 - t) \sin 2\pi t$ and $d_2 = \frac{10}{t} \sin 2\pi t$, respectively. Compare the two rides.

44. The Swing Problem Consider two children swinging side by side on two playground swings so that each displacement from vertical (in feet) is given by $d_1 = 6 \sin 3\pi t$ and $d_2 = 12 \sin \pi t$, where t is time in minutes. Compare the two rides from the graph of $d = d_1 - d_2$.

45. Another Tide Problem Suppose that the tides in Example 3 have periods of 12 hr and 24 hr, with respective amplitudes 1 and 0.50.

(a) Find circular functions that model each component.

(b) Use the graph of the sum of the functions from part (a) to estimate the period and amplitude of the tides.

46. The Sedan Problem Consider a luxury sedan that hits a pothole in such a way that its height above its normal position (in feet) is given by the equation $y = 0.7 \sin t + 0.7 \cos t$. Discuss the period and amplitude of the bouncing car.

47. The Ducks and Geese Problem In 1964, the duck population in the United States was 30 million. Eight years later it reached its peak of 40 million. In 1988, it reached its lowest population of 20 million. The goose population in 1964 was about 10 million. By 1970 it peaked at 15 million. The goose population cycle repeats every 20 years.

(a) Find an equation of the form $y = A \sin Bx + D$ that models the periodically changing duck population.

(b) Find an equation of the form $y = A \sin Bx + D$ that models the periodically changing goose population.

(c) Use the graph of the sum of these two functions to model the total water fowl population.

[Source: *Sports Afield*]

48. The Destructive-Interference Problem A new muffler has been invented that uses destructive interference to quiet the car. If the equation of the sound produced by the car is $y = 10 \sin 659t$, what function f of the form $f(x) = A \cos (Bt + C)$ would represent the sound produced by the muffler so that the combined sound has zero amplitude? (The two sounds are out of phase.)

49. The AM Radio Problem Listening to someone play middle C (261.6 Hz) over an AM (amplitude modulation) radio station allocated to 1000 cycles/sec is modeled by $y = \sin 523.2\pi t (\sin 2000\pi t)$.

(a) Graph the product.

(b) Show that the varying amplitude is described by $y = \sin 512\pi t$.

50. The FM Radio Problem FM radios use frequency modulation to carry the message. The voltage equation for a frequency-modulated signal is $v = v_c \sin(\omega_c t + m_f \sin \omega_m t)$. Graph this function for $v_c = 1$, $\omega_c = 2$, $\omega_m = 3$, and $m_f = 1000$ showing the frequency modulation.

Extend the Concepts

For Exercises 51 to 54 estimate the period, amplitude, and phase shift of each function.

*51. $y = 2 \sin x + 3 \cos x$

52. $y = 2 \sin x - 3 \cos x$

53. $y = \sin 2x + \cos 2x$

54. $y = \cos \frac{x}{4} - \sin \frac{x}{4}$

5.5 Inverses of the Circular Functions

Finding the Inverse Relation of a Circular Function
Finding a Calculator Value of an Inverse Circular Function

The equations y = sin x, y = cos x, and y = tan x express y as a function of x -- given a value of x we know how to find the unique corresponding values of y. Given a value of y and finding the corresponding values of x is a different problem. Since the circular functions are not one-to-one functions, the x values corresponding to a given y value are *not* unique. There are two ways to handle this difficulty. (1) Notice the pattern of repetition of the particular function and give all possible solutions, the inverse images of x. (2) Restrict the domain of each function to create a one-to-one function. Then each value of y will have a unique inverse image. This is the way a calculator finds an inverse image. After addressing these two methods for evaluating inverses of the circular function, we use inverses to solve equations involving circular functions.

Finding the Inverse Relation of a Circular Function

Recall that if (a, b) is a point on the graph of a function, then (b, a) is a point on the graph of its inverse. An inverse simply reverses the roles of the domain and range. The graphs of the inverse sine, inverse cosine, and inverse tangent relations are shown in Figure 1. Notice that these graphs are not functions since each input has more than one output. (Use *inverse* command on a graphing utility to graph these inverse relations.)

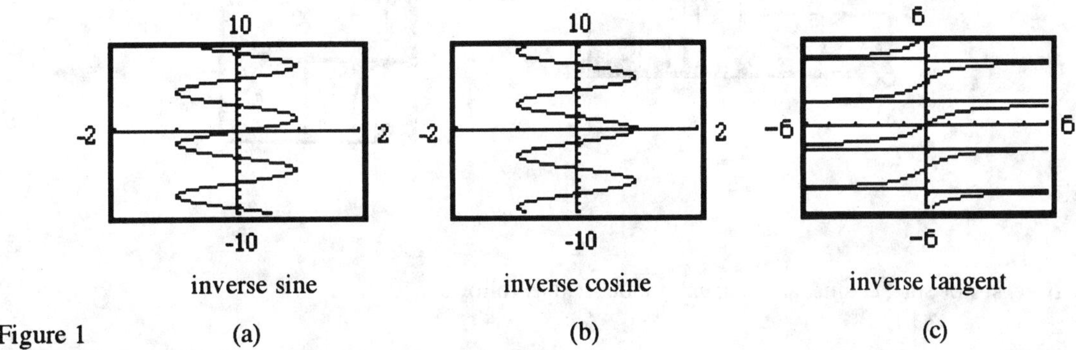

inverse sine, inverse cosine, inverse tangent

Figure 1 (a) (b) (c)

Figure 2 shows a hand-drawn sketch of these graphs with familiar tick marks on the axes.

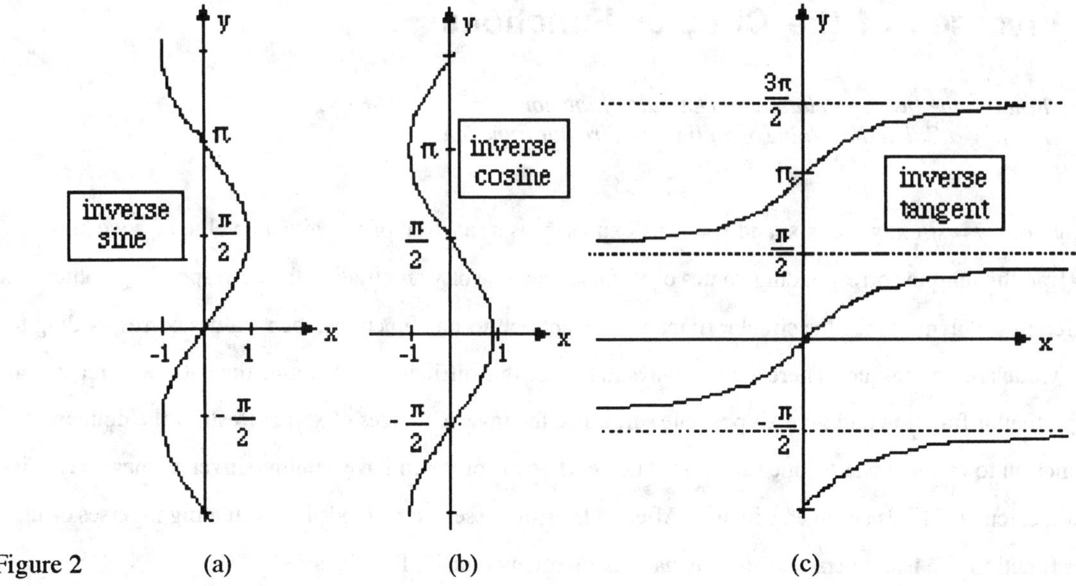

Figure 2 (a) (b) (c)

NOTE: When Appendix II is held in its vertical position, we see the graphs of the inverse sine, inverse cosine, and inverse tangent relations (Figure 3).

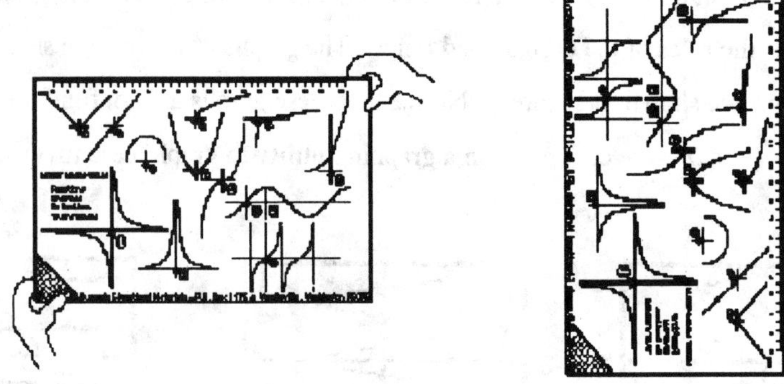

Figure 3

The **inverses** of sine, cosine, and tangent are defined as follows.

Definition of Inverses of the Circular Functions

The inverse of $y = \sin x$ is $x = \sin^{-1} y$.

The inverse of $y = \cos x$ is $x = \cos^{-1} y$.

The inverse of $y = \tan x$ is $x = \tan^{-1} y$.

The equation

$$x = \sin^{-1} y$$

is read "x equals the inverse sine of y." The superscript -1 is <u>not</u> an exponent but simply part of the name of this new relation.

Consider $x = \sin^{-1}(1/2)$. Finding the value of the inverse sine of 1/2 leads to an infinite number of outputs, as indicated on the horizontal line at $y = 1/2$ (Figure 4).

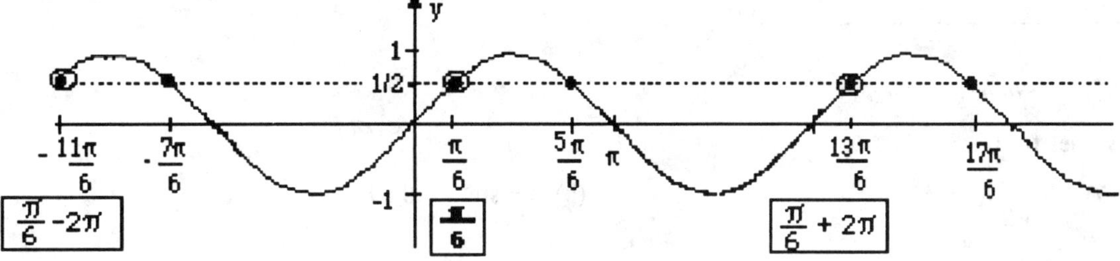

Figure 4

Figure 4 shows the repeating pairs of the outputs every 2π,

$$x = \ldots, \frac{\pi}{6} - 2\pi, \frac{\pi}{6}, \frac{\pi}{6} + 2\pi, \frac{\pi}{6} + 4\pi, \ldots$$

or

$$x = \ldots, \frac{5\pi}{6} - 2\pi, \frac{5\pi}{6}, \frac{5\pi}{6} + 2\pi, \frac{5\pi}{6} + 4\pi, \ldots$$

How can we state concisely all the multiple values for x that satisfy $x = \sin^{-1}(1/2)$?

We write

$$x = \frac{\pi}{6} + 2k\pi \quad \text{or} \quad x = \frac{5\pi}{6} + 2k\pi, \text{ k an integer}$$

Although an infinite number of solutions exist for $x = \sin^{-1}(1/2)$, a calculator displays only one output. Entering $\sin^{-1}(1/2)$ with your calculator in radian mode yields $0.5235987756 \approx \pi/6$. From the graph in Figure 4 we see that $\pi - \frac{\pi}{6} \approx 2.6180$ is another solution. Using these approximations, we express the complete solution as

$$x \approx 0.5236 + 2k\pi \quad \text{or} \quad x \approx 2.6180 + 2k\pi, \text{ k an integer}$$

How do we decide whether to give exact values of approximate ones?

When possible we use exact answers rather than calculator approximations. Recall the exact values for the special angles derived from 30-60-90 and 45-45-90 right triangles (Table 1).

Table 1

Angle (rad):	0	$\pi/6$	$\pi/4$	$\pi/3$	$\pi/2$
sine	0	$\frac{1}{2}$	$\frac{1}{\sqrt{2}}$	$\frac{\sqrt{3}}{2}$	1
cosine	1	$\frac{\sqrt{3}}{2}$	$\frac{1}{\sqrt{2}}$	$\frac{1}{2}$	0
tangent	0	undefined	1	undefined	undefined

In the next example we find all solutions $0 \le x < 2\pi$. Since sine and cosine have a period of 2π, often this is often sufficient.

Example 1 *Finding inverses*

Find all values for $0 \le x < 2\pi$.

(a) $x = \cos^{-1}(\frac{1}{\sqrt{2}})$ (b) $x = \sin^{-1}(-0.69)$

Solution

(a) Since $\cos^{-1}(\frac{1}{\sqrt{2}})$ has an exact value, we do not need to use a decimal approximation.

$$x = \cos^{-1}(\frac{1}{\sqrt{2}}) = \frac{\pi}{4}$$

Examining the cosine graph in Figure 5 for $0 \le x < 2\pi$ indicates a second solution.

$$x = 2\pi - \frac{\pi}{4} = \frac{7\pi}{4}$$

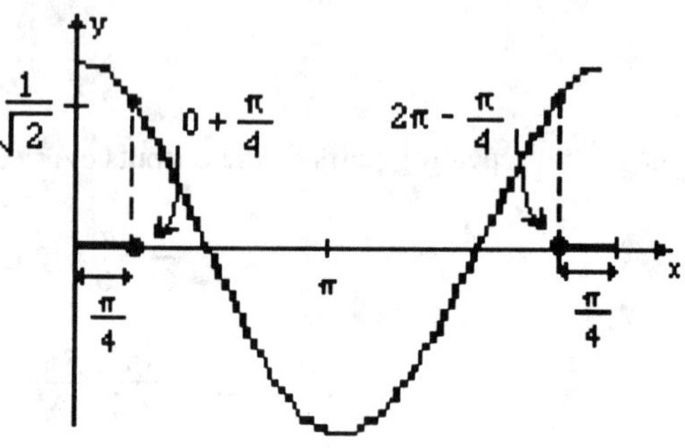

Figure 5

Using a calculator to approximate $\frac{\pi}{4}$ and $\frac{7\pi}{4}$ yields $x \approx 0.78$ and $x \approx 2\pi - 0.78 \approx 5.5$.

(b) $x = \sin^{-1}(-0.69) \approx -0.76$

Since the sine graph is negative between π and 2π,

$$x = \pi + 0.76 \approx 3.90$$

and

$$x = 2\pi - 0.76 \approx 5.52$$

are the two values for $0 \le x < 2\pi$ (Figure 6).

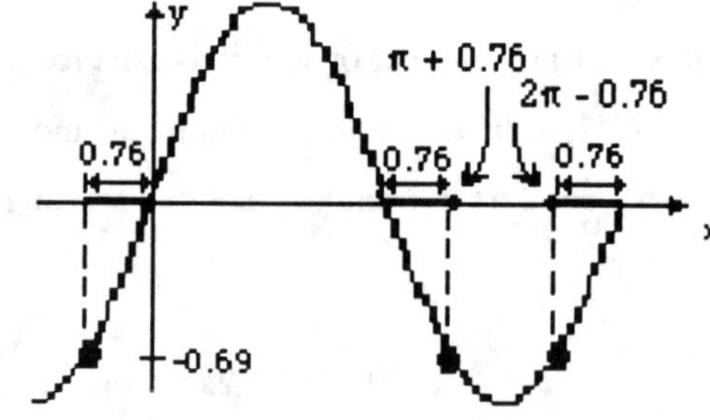

Figure 6

Finding a Calculator Value of an Inverse Circular Function

We have seen that calculators yield a single output for each input. For inverse sine, calculators restrict the domain of the sine function to

$$\frac{-\pi}{2} \le x \le \frac{\pi}{2}$$

creating a "new" sine function that is one-to-one. We designate this function with a capital "S":

$$y = \text{Sin } x$$

The inverse function of $y = \text{Sin } x$ is written

$$x = \text{Sin}^{-1} y$$

The inverse cosine and inverse tangent functions are created by restricting the domains of the cosine and tangent functions as follows

$$y = \text{Cos } x, \quad 0 \le x \le \pi$$

and

$$y = \text{Tan } x, \quad -\frac{\pi}{2} < x < \frac{\pi}{2}$$

Figure 7 shows the graphs of the functions $y = \text{Sin}^{-1} x$, $y = \text{Cos}^{-1} x$, and $y = \text{Tan}^{-1} x$.

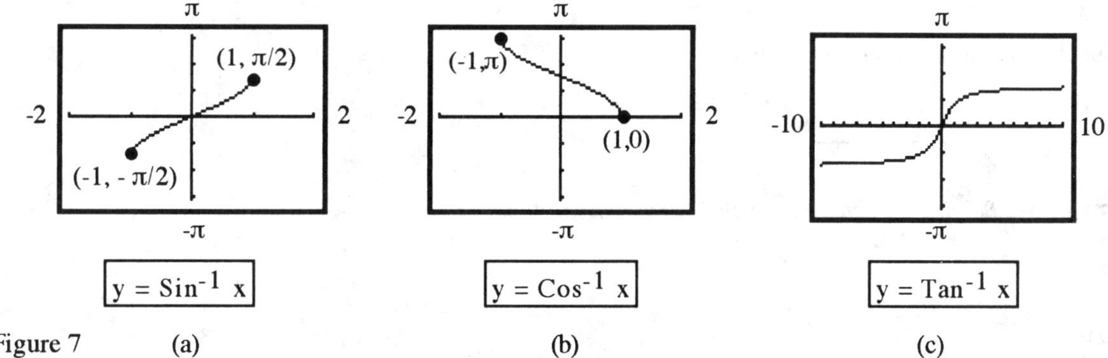

Figure 7 (a) (b) (c)

Recall that the domain of a function is the range of its inverse and the range of a function is the domain of its inverse.

Inverse Circular Functions

For $y = \text{Sin}^{-1} x$, the domain is $-1 \le x \le 1$ and the range is $\frac{-\pi}{2} \le y \le \frac{\pi}{2}$.

For $y = \text{Cos}^{-1} x$, the domain is $-1 \le x \le 1$ and the range is $0 \le y \le \pi$.

For, $y = \text{Tan}^{-1} x$, the domain is all real numbers and the range is $\frac{-\pi}{2} < y < \frac{\pi}{2}$.

NOTE: Some graphing utilities use the word "Arc" to represent the inverse functions as $y =$ **Arcsin** x, $y =$ **Arccos** x, and $y =$ **Arctan** x.

Example 2 *Finding values for inverse sine*

Find the following values of x.

(a) $x = \text{Sin}^{-1} 0.8415$ (b) $x = \sin^{-1} 0.8415$ for $0 \le x < 2\pi$ (c) $x = \sin^{-1} 0.8415$

Solution

(a) There is only one solution, $x \approx 1$.

(b) There are two results for $0 \le x < 2\pi$ (Figure 8), $x \approx 1$ or $x \approx \pi - 1 \approx 2.14$.

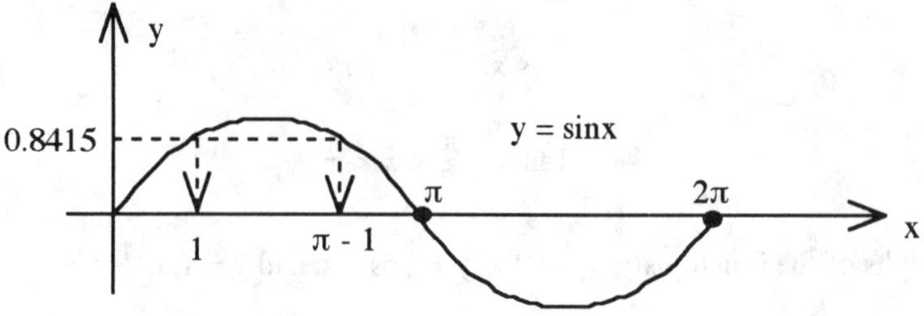

Figure 8

(c) There are an infinite number of solutions.

$$x \approx 1 + 2k\pi \quad \text{or} \quad x \approx 2.14 + 2k\pi, \ \text{k an integer}$$

Example 3 *Solving an equation for the calculator solution*

Solve $2 \text{ Sin } 2x = 1$.

Solution

$$2 \text{ Sin } 2x = 1 \qquad \textit{Isolate the sine function.}$$

$$\text{Sin } 2x = \frac{1}{2}$$

$$2x = \text{Sin}^{-1}\frac{1}{2} \qquad \text{Sin}^{-1}\frac{1}{2} = \frac{\pi}{6} \approx 0.5236.$$

$$2x = \frac{\pi}{6} \qquad \textit{Divide both sides by 2.}$$

$$x = \frac{\pi}{12} \approx 0.2618$$

The next example features a right triangle and the inverse tangent function. Because of the context, the angle θ is measured in degrees.

Example 4 The Frightening Foggy Flight Problem

Airports must restrict takeoffs and landings during fog. They measure the *ceiling* by shining a spotlight straight up at a known distance from the tower. A device called a ceiliometer is used to measure the angle of elevation θ to the lighted spot on the clouds (Figure 9). Graph θ as a function of the ceiling *c* for a spotlight 100 ft from a ceiliometer positioned from the vertex of θ. Find θ for each minimum ceiling.

type of plane	A	B	C	D
minimum landing speed (mi/hr)	0-90	91-120	121-140	141-165
minimum ceiling (ft)	1000	1540	1640	1740

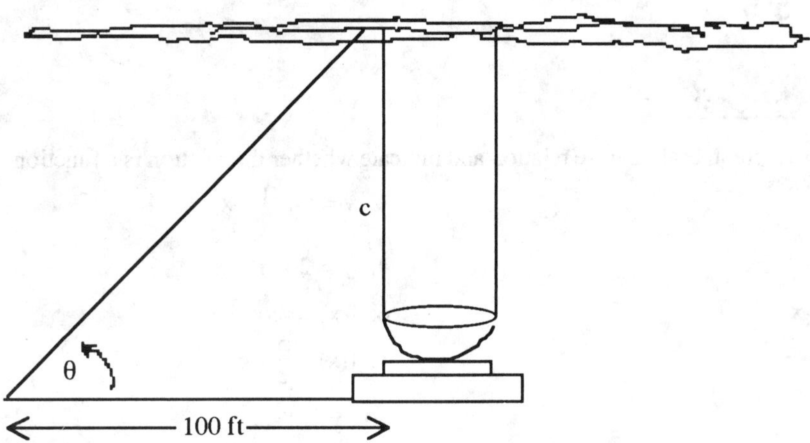

Figure 9

Thinking: Creating a Plan

From Figure 9 we see $\tan\theta = \frac{c}{100}$. The inverse tangent function expresses θ as a function of c, $\theta = \mathrm{Tan}^{-1}\left(\frac{c}{100}\right)$.

Communicating: Writing the Solution

We will find θ for each point in Figure 10.

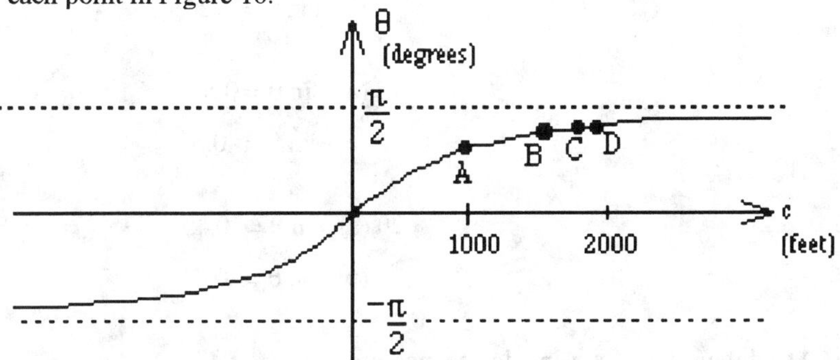

Figure 10

Point A: $\theta = \mathrm{Tan}^{-1}\left(\frac{c}{100}\right)$ and $c = 1000 \Rightarrow \theta = \mathrm{Tan}^{-1}\left(\frac{1000}{10}\right) \approx 84.3°$

Point B: $\theta = \mathrm{Tan}^{-1}\left(\frac{c}{100}\right)$ and $c = 1540 \Rightarrow \theta = \mathrm{Tan}^{-1}\left(\frac{1540}{100}\right) \approx 86.3°$

Point C: $\theta = \mathrm{Tan}^{-1}\left(\frac{c}{100}\right)$ and $c = 1640 \Rightarrow \theta = \mathrm{Tan}^{-1}\left(\frac{1640}{100}\right) \approx 86.5°$

Point D: $\theta = \mathrm{Tan}^{-1}\left(\frac{c}{100}\right)$ and $c = 1740 \Rightarrow \theta = \mathrm{Tan}^{-1}\left(\frac{1740}{100}\right) \approx 86.7°$

Learning: Making Connections

Since these angles of elevation are all nearly 90°, the change in θ is small. Flights are restricted by visibility, which measures horizontal sight distance, as well as by the ceiling which measures vertical sight distance.

Exercises 5.5

Develop the Concepts

For Exercises 1 to 12 graph each inverse relation and indicate whether the relation is a function.

1. $y = \sin^{-1}\frac{x}{2}$
2. $y = \mathrm{Sin}^{-1}(-x)$
3. $y = -\mathrm{Cos}^{-1} x$
4. $y = 2\tan^{-1} x$
5. $f(x) = \sin^{-1}(-x)$
6. $f(x) = -\cos^{-1} x$
7. $f(x) = \sin^{-1} x - \frac{\pi}{2}$
8. $f(x) = \frac{\pi}{2} - \cos^{-1} x$
9. $y = \mathrm{Tan}^{-1} 2x$
10. $y = \mathrm{Tan}^{-1}\frac{x}{2}$
11. $y = \mathrm{Tan}^{-1}(x - 1)$
12. $y = \mathrm{Tan}^{-1}(1 - x)$

For Exercises 13 to 20 find two angles such that $0 \le \theta < 2\pi$ that have the given values.

13.(a) $\sin\theta = 0.7$
(b) $\sin\theta = -0.7$

14.(a) $\cos\theta = 0.5$
(b) $\cos\theta = -0.5$

15.(a) $\tan\theta = 5.1$
(b) $\tan\theta = -5.1$

16.(a) $\tan\theta = 12$
(b) $\tan\theta = -12$

17.(a) $\cos\theta = 0.7$
(b) $\cos\theta = -0.7$

18.(a) $\sin\theta = 0.5$
(b) $\sin\theta = -0.5$

19.(a) $\sin\theta = 0.66$
(b) $\cos\theta = 0.66$

20.(a) $\sin\theta = -0.45$
(b) $\tan\theta = -0.45$

For Exercises 21 to 24 find the value of x using degree measure.

21.(a) $x = \mathrm{Cos}^{-1}\frac{1}{2}$
(b) $x = \cos^{-1}\frac{1}{2}$
(c) $x = \cos^{-1}\frac{1}{2}$ for $0° \le x < 360°$

22.(a) $x = \mathrm{Sin}^{-1}(-1)$
(b) $x = \sin^{-1}(-1)$
(c) $x = \sin^{-1}(-1)$ for $0° \le x < 360°$

23.(a) $x = \mathrm{Tan}^{-1}(-2)$
(b) $x = \mathrm{Cos}\text{-}1\ (-2)$

24.(a) $x = \mathrm{Cos}^{-1}(-0.88)$
(b) $x = \mathrm{Tan}^{-1}(-0.88)$

For Exercises 25 to 28 find the value of x using radian measure.

25.(a) $x = \mathrm{Tan}^{-1} 100$
(b) $x = \tan^{-1} 100$
(c) $x = \tan^{-1} 100$ for $0 \le x < 2\pi$

26.(a) $x = \cos^{-1} 0$ for $0 \le x < 2\pi$
(b) $x = \sin^{-1} 0$ for $0 \le x < 2\pi$
(c) $x = \tan^{-1} 0$ for $0 \le x < 2\pi$

27.(a) $x = \sin^{-1} 0.75$ for $0 \le x < 2\pi$

(b) $x = \sin^{-1} (-0.75)$ for $0 \le x < 2\pi$

28.(a) $x = \text{Tan}^{-1} (-1.4)$

(b) $x = \tan^{-1} (-1.4)$

For Exercises 29 to 32 use inverses to solve each equation for x in radians.

29.(a) $\text{Sin } x = 1/2$

(b) $\text{Sin } 2x = 1/2$

(c) $\text{Sin } \frac{x}{2} = \frac{1}{2}$

30.(a) $\text{Sin } x = -0.25$

(b) $\text{Sin } (2x + 1) = -0.25$

(c) $\text{Sin} \frac{x + 1}{2} = -0.25$

31.(a) $\text{Cos } x = 0.6$

(b) $\text{Cos } 2x = 0.6$

(c) $2 \text{ Cos } \frac{x}{2} = 0.6$

32.(a) $\text{Tan } x = -1$

(b) $2 \text{ Tan } x = -1$

(c) $2 \text{ Tan } 2x = -1$

For Exercises 33 to 38 find each value.

33. If $x = \sin^{-1} \frac{4}{5}$, find cos x.

34. If $x = \sin^{-1} (-\frac{4}{5})$, find cos x.

35. If $x = \text{Sin}^{-1} (\frac{4}{5})$, find tan x.

36. If $x = \text{Sin}^{-1} (-\frac{4}{5})$, find tan x.

37. If $x = \cos^{-1} (-\frac{2}{5})$, find sin x.

38. If $x = \tan^{-1} \frac{5}{12}$, find cos x.

For Exercises 39 to 46 solve each equation for x in radians.

39. $\frac{1}{2} \text{Sin}^{-1} \frac{\sqrt{3}}{2} = 2x$

40. $\frac{1}{4} \text{Cos}^{-1} (-\frac{\sqrt{3}}{2}) = x + 1$

41. $\frac{1}{3} \text{Tan}^{-1} (-\sqrt{3}) = x - 1$

42. $\frac{1}{4} \text{Sin}^{-1} (-\frac{\sqrt{3}}{2}) = 2x$

*43. $\text{Sin}^{-1} x = \text{Tan}^{-1} x$

44. $\text{Cos}^{-1} x = \text{Sin}^{-1} 2x$

45. $\text{Cos}^{-1} x^2 = \frac{\pi}{2} - \text{Sin}^{-1} x$

46. $\frac{\pi}{2} - \text{Sin}^{-1} x = \text{Tan}^{-1} 2x$

Apply the Concepts

47. The Ferris Wheel Ride Problem A Ferris wheel has a 50-ft diameter, completes one revolution every 15 sec, and makes 10 rev/ride.

(a) Find an equation of the form $y = A \sin (Bt + C) + D$ that models the height above the ground as a function of time.

(b) Find the equation of the inverse.

(c) Approximate the times when a car is 25 ft off the ground.

48. The Merry-Go-Round Problem A merry-go-round 10 m in diameter completes 1 revolution every 20 sec.

(a) Find the equation of the form $y = A \sin (Bt + C) + D$ that models the situation.

(b) Find the equation of the inverse.

(c) If a ride lasts 2 min, approximate the times when a rider passes his starting point.

49. The Rainbow Kite and Balloon Company You are to design a theater with a 24-ft-high screen (Figure 11).

(a) Express the viewing angle θ as a function of the perpendicular distance x from the screen's wall.

(b) Find the value of x that maximizes the viewing angle.

(c) Discuss the impact this information could have on the design of the theater.

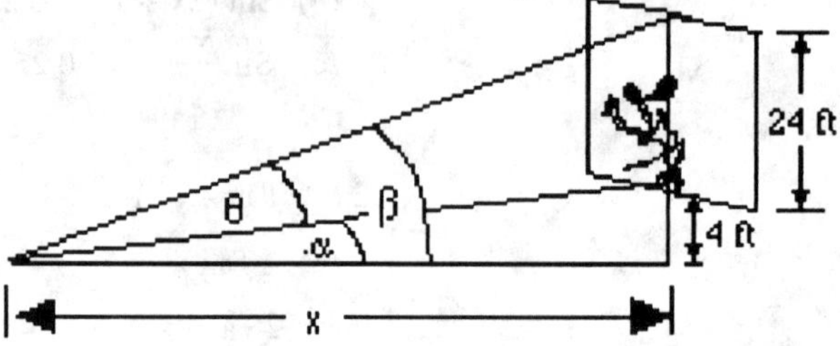

Figure 11

50. Wonderworld Amusement Park You are in charge of designing a billboard to advertise the factory. People start slowing down for a traffic signal at point A (Figure 12). You want to build the billboard at the position B that maximizes the viewing angle α.

(a) Write α as a function of x.

(b) Find the value of x that maximizes the viewing angle α.

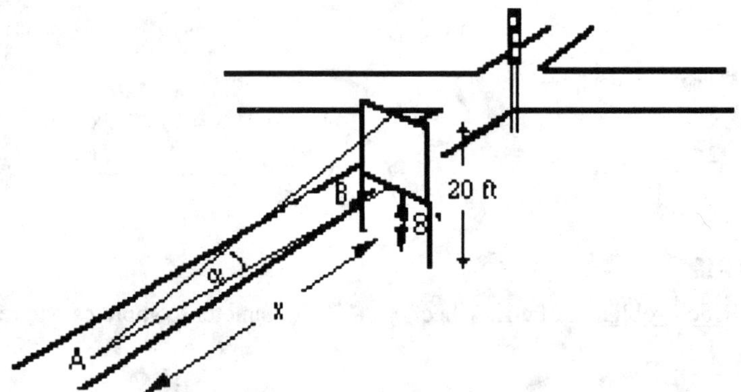

Figure 12

*51. The Ocean Swells Problem You are fishing for salmon and the skipper of the boat detects a school of fish at 60 ft on his electronic fish-finder. You let out 65 ft of line as shown in Figure 13.

(a) If the ocean waves swell to 10 ft every 80 sec, find the equation in the form h = A sin Bt that models the wave action of the boat at any time t.

(b) Find the equation of the form h = A sin Bt + D that models the location of the hook at any time t.

(c) Use the graph of the equation from part (b) to estimate how long your bait is within 5 ft (vertically) of the school of fish.

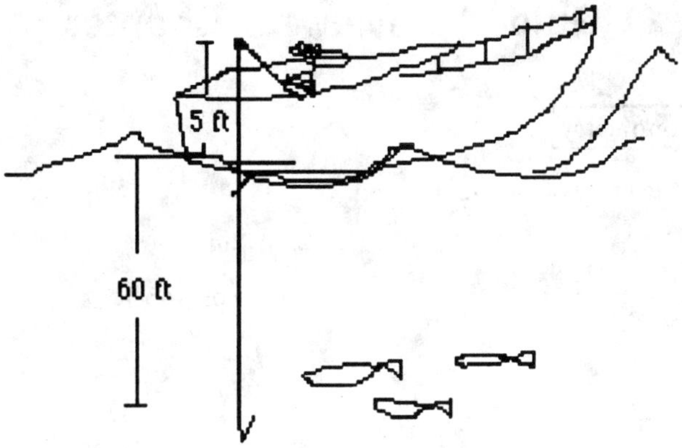

Figure 13

52. A Ferris Wheel Problem While riding a Ferris wheel at the Evergreen Fair you are surprised to notice that you can see your car over the top of the horse barn when you approach the top of the ride, as indicated in Figure 14. The Ferris wheel completes one revolution every $2\frac{1}{2}$ min.

(a) Express your height above the ground as a function of time.

(b) Find the equation of the inverse.

(c) How long can you see your car on each revolution of the wheel?

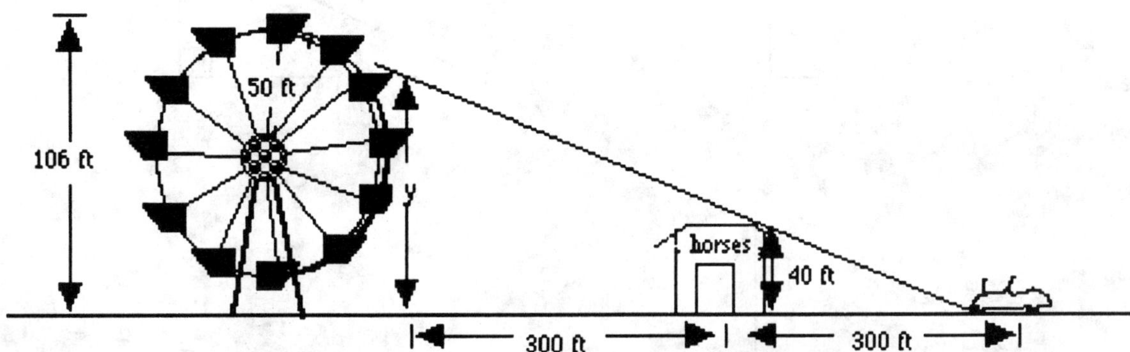

Figure 14

Extend the Concepts

For Exercises 53 to 56 graph each pair of functions on one coordinate system and find the points of intersection.

53. $f(x) = \text{Sin}^{-1} x$, $g(x) = \text{Cos}^{-1} x$

54. $f(x) = \text{Sin}^{-1} x$, $g(x) = \text{Tan}^{-1} x$

55. $f(x) = 2\,\text{Sin}^{-1} x$, $g(x) = \text{Cos}^{-1} x$

56. $f(x) = \text{Sin}^{-1} x$, $g(x) = \frac{1}{2}\text{Cos}^{-1} x$

A Guided Review of Chapter 5

5.1 Right Triangle Trigonometry

For Exercises 1 and 2 find the value of each expression.

1.(a) sin 30°

(b) cos 30°

(c) tan 90°

2.(a) sin 45°

(b) cos (-45°)

(c) tan 135°

For Exercises 3 and 4 find θ if $0° \le \theta < 90°$.

3.(a) $\sin\theta = \sqrt{3}/2$

(b) $\cos\theta = 1/2$

4.(a) $\sin\theta = 0$

(b) $\cos\theta = 1$

For Exercises 5 to 8 use the Pythagorean theorem and/or the definitions of sine, cosine, and tangent to solve the following triangles by finding the length of the missing sides and the measure of the missing angles.

5.

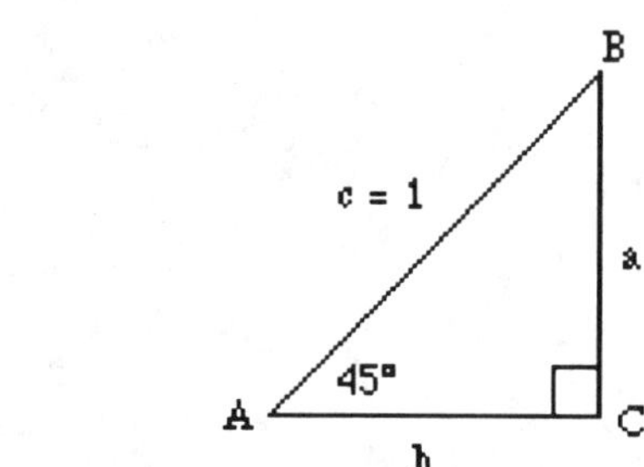

6.

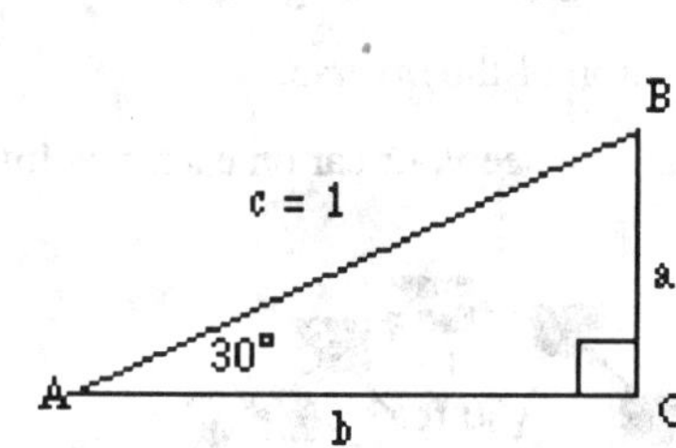

7.

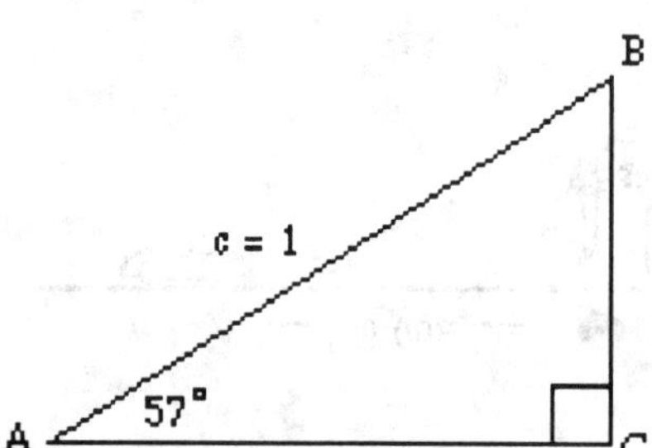

8.

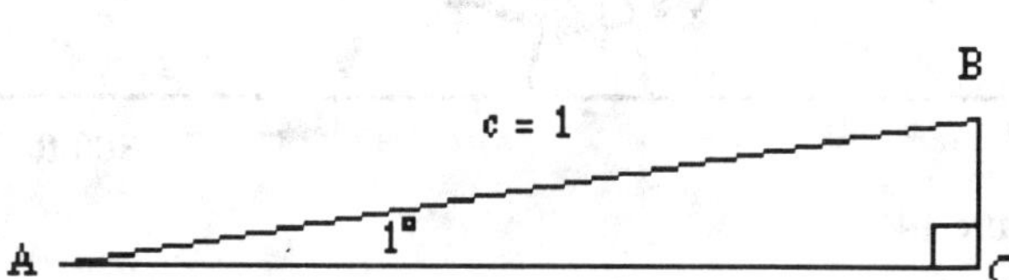

9. <u>The Short Path on a Box Problem</u> The box in Figure 1 measures 3 by 4 by 5 in.

(a) How long is the path from A to X to B on the surface of the box in Figure 2 if x = 0?

(b) How long is the diagonal from A to B?

(c) Find x that will result in the shortest path from A to X to B on the surface of the box.

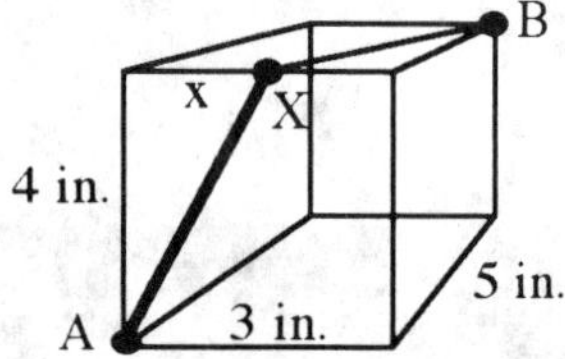

Figure 1

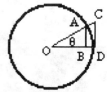

Figure 2

10. The Old-Fashioned Sine Problem Originally, the trigonometric functions were defined in terms of tangents and secants of a circle. Find AB, OB, and CD in terms of θ if the radius of the circle is 1 unit (Figure 2).

11. The Bald Eagle Problem A great bald eagle flying over an open field can spot a mouse if angle α is 1', where 1 minute = 1/60 of 1 degree (Figure 3).

(a) How high can the eagle fly and still see the mouse if it is 3 in. long?

(b) Find a function that models the height of the eagle above the mouse as a function of angle α.

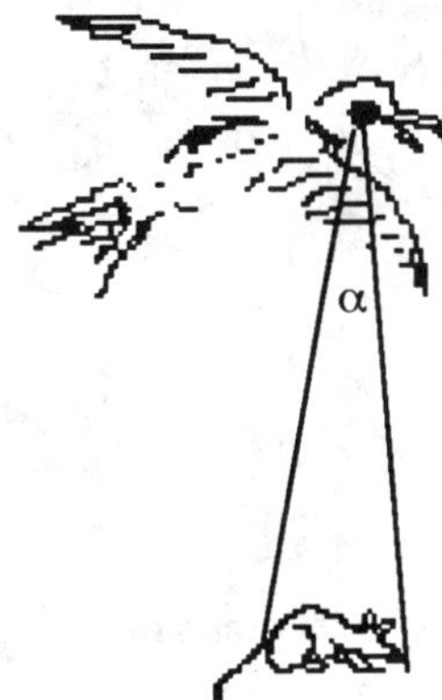

Figure 3

12. The Luxor Hotel Problem The Luxor Hotel in Las Vegas is built in the shape of a pyramid. The square base is 350 ft on each side. The inclinator is an elevator that goes up the corner of the structure at an angle of 39°. How high does the inclinator rise above ground level? (The building rises 350 ft, and the inclinator does not reach the top.)

5.2 The Circular Functions

For Exercises 13 and 14 convert each degree measurement into radians expressed in terms of π and as a decimal rounded to the nearest hundredth.

13. (a) 120° (b) 240° (c) 300°

14. (a) -30° (b) -150° (c) -210°

For Exercises 15 and 16 convert each radian measurement into degrees.

15. (a) $\frac{\pi}{3}$ (b) $\frac{4\pi}{3}$ (c) $\frac{5\pi}{3}$

16. (a) $\frac{-5\pi}{4}$ (b) $\frac{-7\pi}{4}$ (c) - $\frac{-13\pi}{4}$

17. The CD Problem A compact disc runs at a constant linear speed of 125 cm/sec as it spins faster.

(a) Find the number of rev/sec when the laser is reading the innermost tract, 4.5 cm in diameter.

(b) Find the number of rev/sec when the laser is reading the outermost tract, 11.6 cm in diameter.

18. The Speeding Satellite Problem Three satellites in geosynchronous orbits 22,300 mi above the equator can cover the Earth. (The radius of Earth is about 4000 mi.)

(a) Find the measure of each angle α in Figure 4.

(b) Use $s = r\,\theta$ to help find the velocity in mi/hr of the satellites.

[Source: *Van Nostrand's Scientific Encyclopedia*]

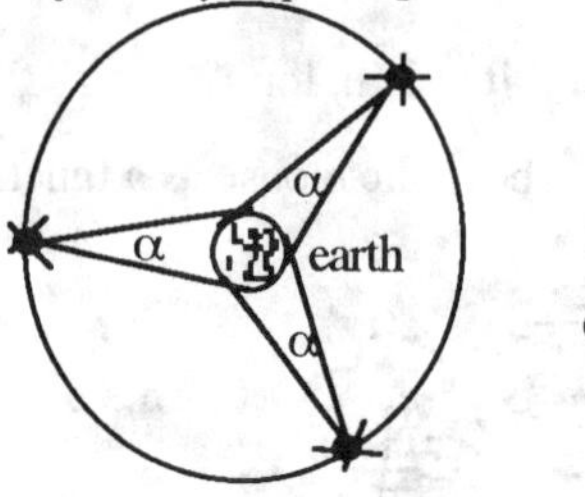

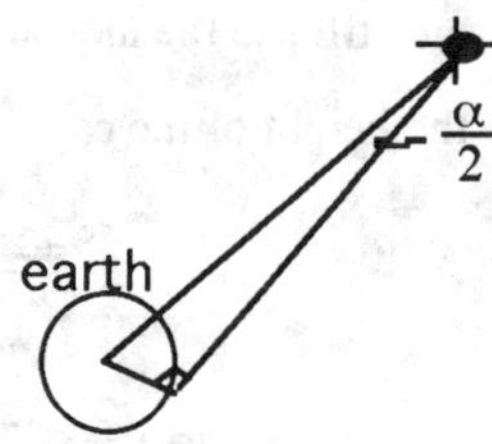

Figure 4

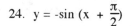

5.3 Graphs of the Circular Functions

For Exercises 19 to 24 match each function with its graph in Figure 5.

19. $y = 2\cos\left(x - \frac{\pi}{2}\right)$

20. $y = -2\sin\left(x + \frac{\pi}{2}\right)$

21. $y = \frac{1}{2}\cos 2x$

22. $y = \frac{1}{2}\sin 2x$

23. $y = -\cos\frac{x}{2}$

24. $y = -\sin\left(x + \frac{\pi}{2}\right)$

A.

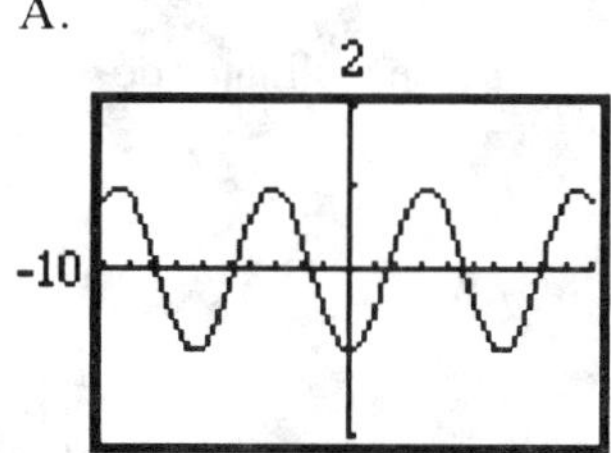

B.

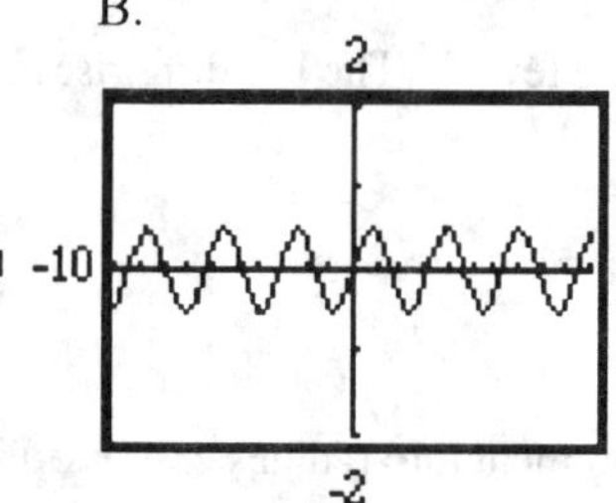

C.

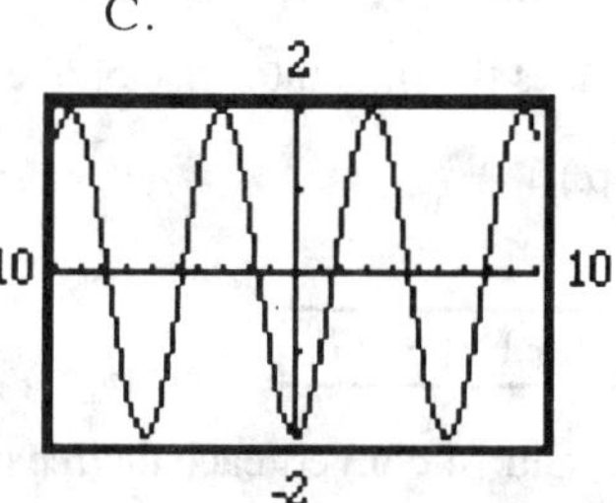

D.

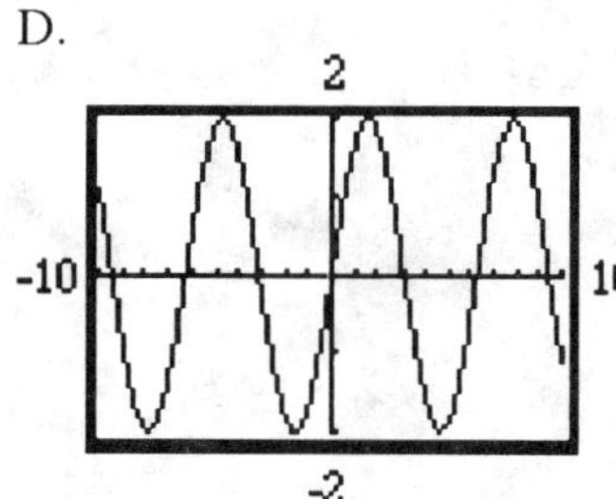

E.

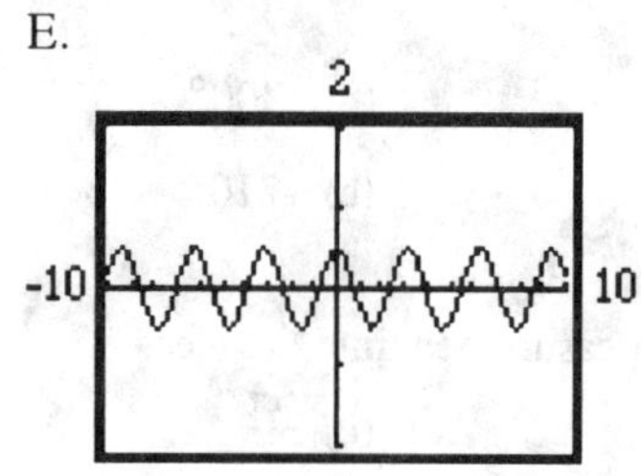

F

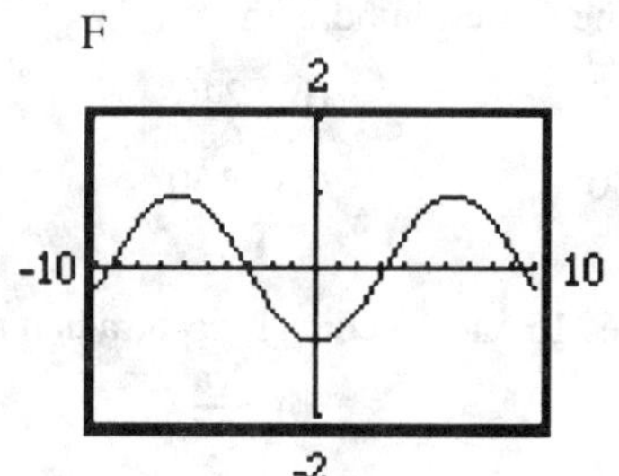

Figure 5

For Exercises 27 to 30 graph one period for each circular function.

25.(a) $y = \sin\left(\frac{x}{2}\right)$

(b) $y = \sin\left(\frac{x}{2} + \frac{\pi}{2}\right)$

26.(a) $y = \cos(-2x)$

(b) $y = \cos(2 - 2x)$

27.(a) $y = 10 \cos (x + \frac{\pi}{4})$

(b) $y = 0.1 \cos (x - \frac{\pi}{2})$

28.(a) $y = -10 \sin (x + \frac{\pi}{4})$

(b) $y = -10 \sin (x - \frac{\pi}{4})$

For Exercises 29 and 30 state the period of each function.

29.(a) $y = \sin \frac{3x}{2}$

(b) $y = -3 \tan (\frac{2x}{3} + \pi)$

30.(a) $y = -2 \cos \frac{-4x}{3}$

(b) $y = -2 \tan 10x$

For Exercises 31 to 36 state an equation that best describes each graph.

31.

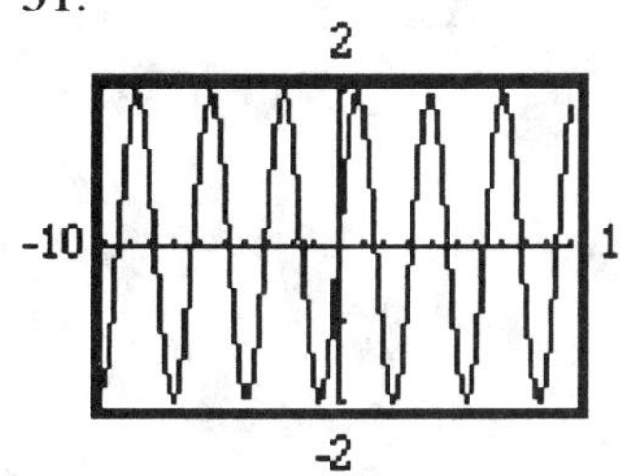

32.

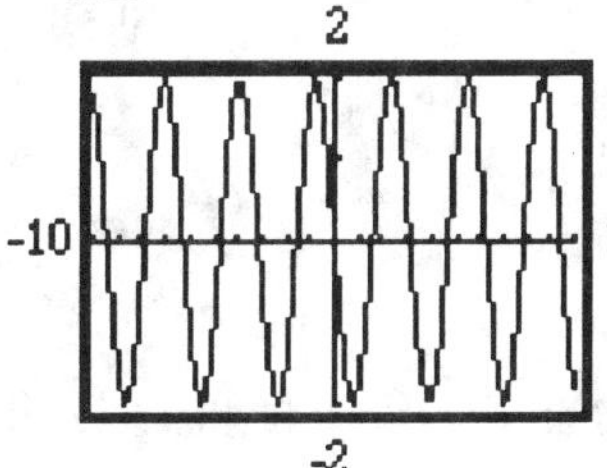

33.

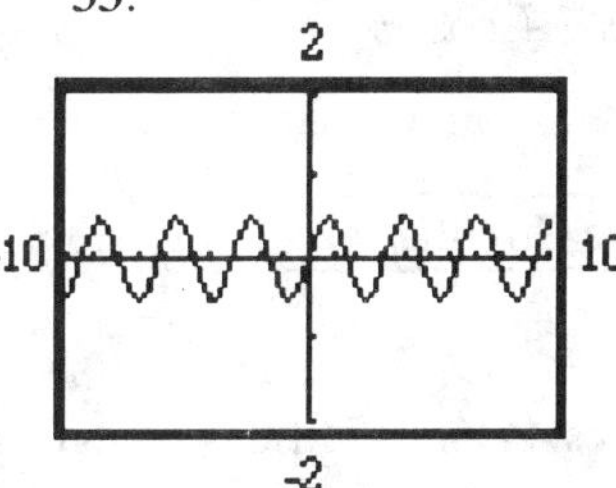

34.

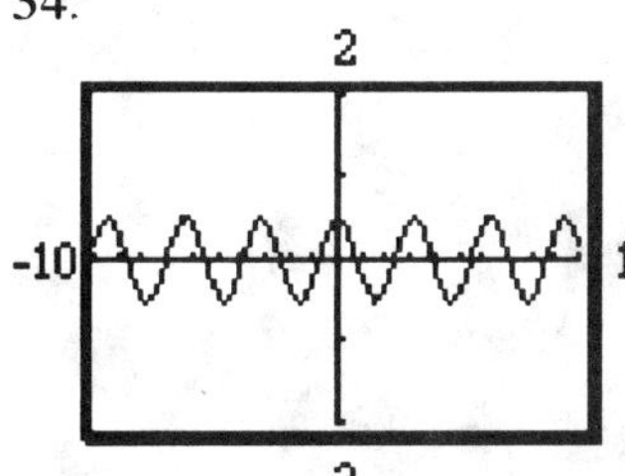

35.

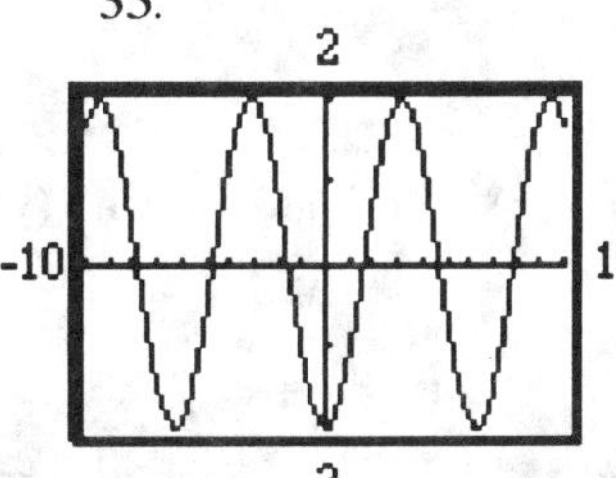

36.

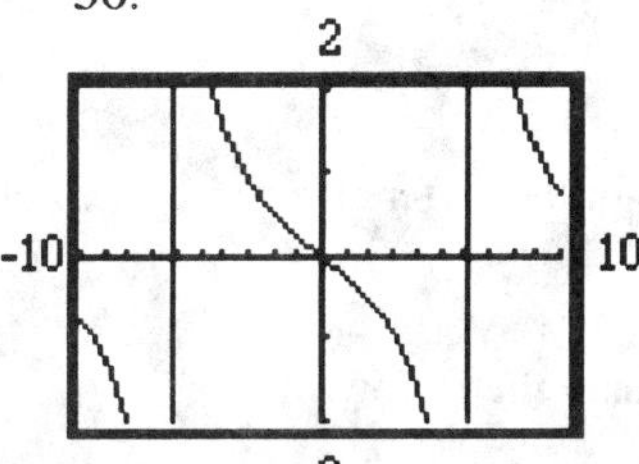

37. The High-Temperature Problem Following is the normal daily maximum temperatures for Seattle, Washington. The temperature pattern is nearly periodic. Use the largest and smallest entries and the period to find an equation of the form $T = A \sin Bx + D$ that approximates these temperatures. Use your equation to fill in the missing entries. [Hint: Call November month 0, December month 1, etc.]

Nov.	Dec.	Jan.	Feb.	Mar.	Apr.	May	June	July	Aug.	Sept.	Oct.
50.5	45.1	45.0	49.5	?	57.2	63.9	69.9	75.2	75.8	?	59.7

38. The Wet Feet Problem Following is the normal monthly rainfall for Seattle. The pattern is nearly periodic. Use the largest and smallest entries and the period to find an equation of the form $y = A \sin Bx + D$ that approximates these rainfalls. Use your equation to fill in the missing entries. [Hint: Call October month 0, etc.]

Oct.	Nov.	Dec.	Jan.	Feb.	Mar.	Apr.	May	June	July	Aug.	Sept.
3.23	?	5.91	5.38	3.99	3.54	2.33	1.70	1.50	?	1.14	1.88

5.4 Reciprocals Sums and Products of the Circular Functions

39. Find the value of each reciprocal function.

(a) $\sec \pi$ (b) $\cot 125°$ (c) $\csc 5.4$

For Exercises 40 and 41 graph one period of each reciprocal function.

40. $y = -2 \csc x + 1$ 41. $y = \cot 4x$

For Exercises 42 and 43 describe the general shape of each sum or product.

42. $y = \cos \frac{x}{2} + \cos \frac{x}{3}$ 43. $y = (\sin x)(\sin x)$

5.5 Inverse Circular Functions

For Exercises 44 and 45 find two angles such that $0° \le \theta < 360°$ that have the given values.

44.(a) $\sin \theta = 0.866$ 45.(a) $\cos \theta = 0.866$

(b) $\sin \theta = -0.866$ (b) $\cos \theta = -0.866$

46. Explain the difference in inverse notation: $x = \sin^{-1}(y)$ and $x = \text{Sin}^{-1}(y)$.

47. Evaluate the following.

(a) $x = \cos^{-1}(0.7)$ (b) $x = \text{Cos}^{-1}(0.7)$ (c) $x = \cos^{-1}(0.7)$ for $0 \le x < 2\pi$

For Exercises 48 and 49 solve each equation for x in radians.

48. $\text{Cos}^{-1}(-0.75) = 2x$ 49. $3\,\text{Sin}^{-1}\,0.6 = x - 1$

50. <u>The Classroom Board Problem</u> The bottom of the 48-in.-tall board in the front of the classroom is 30 in. above eye level (Figure 6). Find the distance from the front of the classroom to sit in order to maximize the viewing angle, θ.

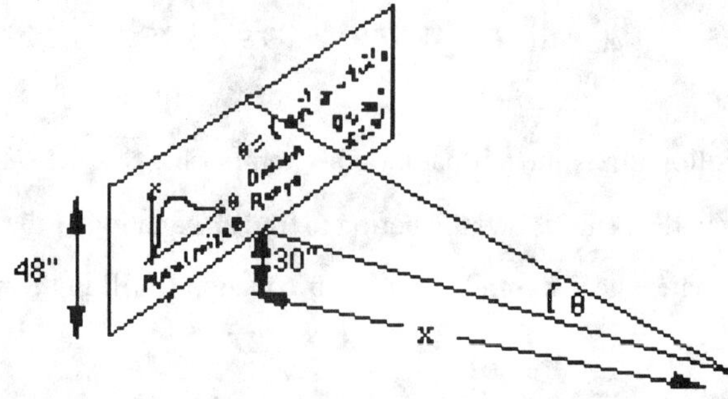

Figure 6

Chapter 5 Test

1. Fill in each blank with one of the symbols <, >, or =.

(a) $\sin 30°$ ___ $\cos 30°$ (b) $\sin 30°$ ___ $\cos 60°$

(c) $\sin 120°$ ___ $\cos 120°$ (d) $\sin 240°$ ___ $\cos 300°$

2.(a) Convert 315° to radian measure. (b) Convert $-\frac{5\pi}{4}$ to degree measure.

3. State the coordinates of the reference point for $-\pi$.

For Exercises 4 and 5 use the Pythagorean theorem and/or the definitions of sine, cosine, and tangent to solve each triangle by finding the length of the missing sides and the measure of the missing angles.

4.

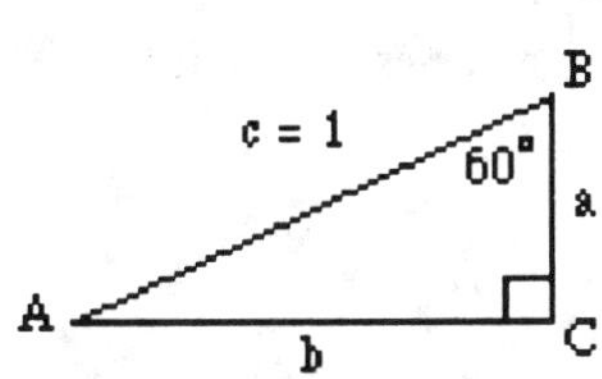

5.

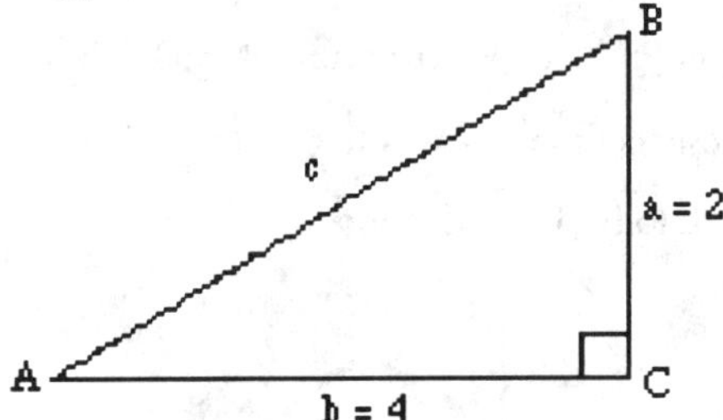

For Exercises 6 to 9 match each function with its graph in Figure 1

6. $y = \sin\left(2x - \frac{\pi}{2}\right)$

7. $y = -2\sin\left(2x + \frac{\pi}{2}\right)$

8. $y = \frac{1}{2}\cos\left(2x - \frac{\pi}{2}\right)$

9. $y = \frac{1}{2}\cos 2x$

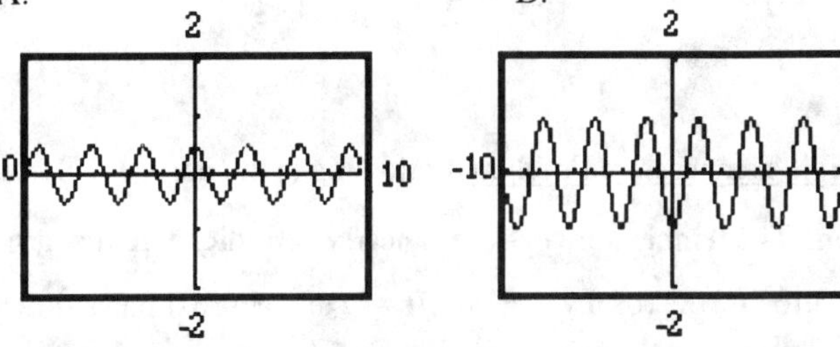

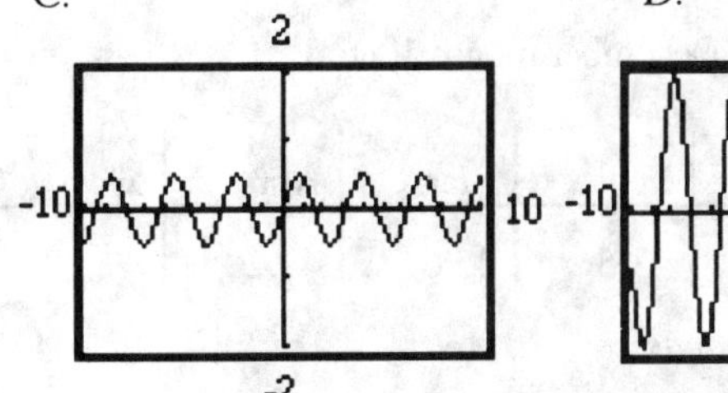

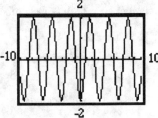

Figure 1

10. <u>The Golfers' Dilemma Problem</u> Golfers teeing off on the twelfth tee at Snohomish Golf Course cannot see the group in front of them because of a hill. A mirror is mounted 20 ft off the ground on a tree, at an angle of 34° to the trunk. If you move to a point 10 ft from the trunk, you see the group ahead of you. Can you hit your tee shot? The angles of incidence and reflection are equal (Figure 2).

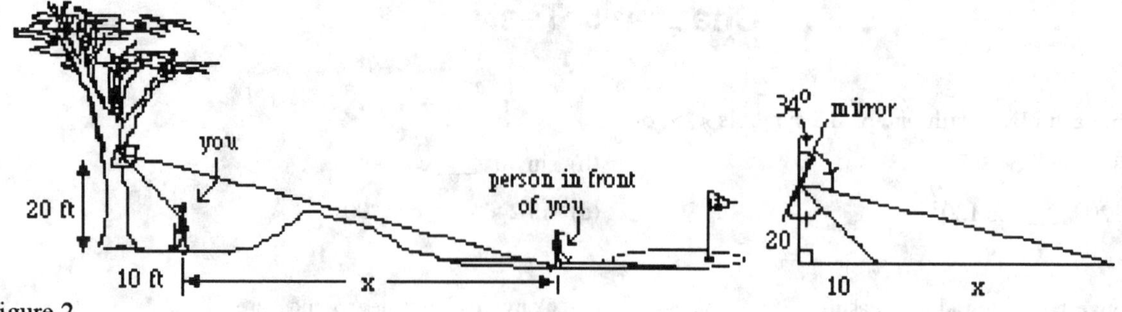

Figure 2

11. The Solar Orbit Problem The Earth's orbit around the sun is nearly circular, with an average radius of 93 million miles.

(a) Find Earth's angular velocity in revolutions per year.

(b) Find Earth's linear velocity as it travels around the sun in miles per hour.

12. The Household Current Problem Electric current in U.S. households is modeled by the function graphed in the Figure 3, where t represents time and V voltage. It is called alternating current because it reverses direction periodically. Use the graph to find the equation in the form V = A sin Bt that models the current.

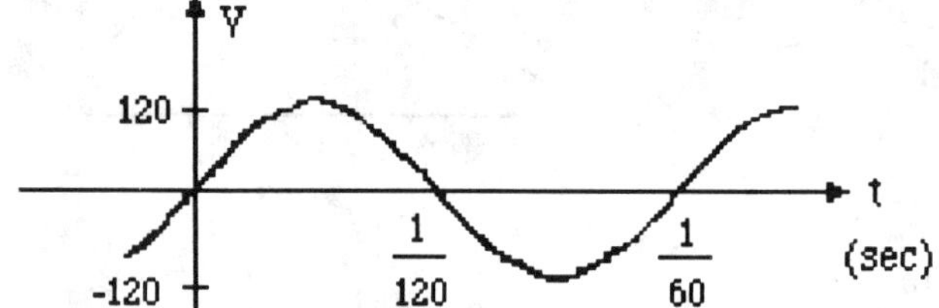

Figure 3

13. The High-Temperature Problem Following is the normal daily average temperatures for Seattle, Washington. The temperature pattern is nearly periodic. Use the largest and smallest entries and the period to find an equation of the form y = A sin Bt + D that approximates these temperatures. Use your equation to fill in the missing entry.

[Let November be month 0, December month 1, etc.]

Nov.	Dec.	Jan.	Feb.	Mar.	Apr.	May	June	July	Aug.	Sept.	Oct.
45.3	40.5	40.1	43.5	45.6	49.2	?	60.9	65.2	65.5	60.6	52.8

[Source: *The American Almanac*]

14. The Diameter of the Moon Problem The moon's distance from Earth varies from 221,460 to 252,700 mi causing the moon's apparent diameter to vary. Find the measure of the angle subtended by the diameter of the moon, angles α and β in Figure 4, if the diameter of the moon is 2158 mi.

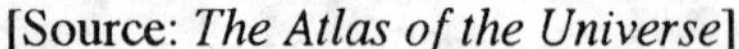
[Source: *The Atlas of the Universe*]

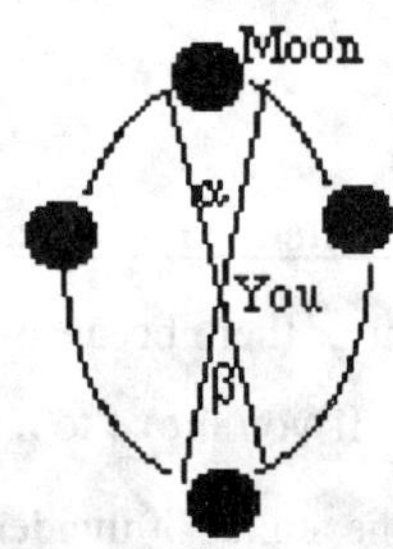

Figure 4

15. The Golf Ball Problem Revisited The range R of a golf shot in feet is given by $R = \frac{v_0^2}{32}\sin 2\theta$, where v_0 is the initial velocity.

(a) Graph R as a function of θ for a ball hit with an initial velocity of 120 ft/sec.

(b) Find the angle θ of the club face that will yield the longest drive.

(c) What is the longest drive?

16. The Art Exhibit Problem You are attending an art exhibit that is showing three paintings having heights h of 24 and 48 in. (Figure 5). The 24-in. painting is hung so that the bottom of the frame is 12 in. above eye level. The 48 in. painting is hung so that the bottom of the frame is at eye level. Express the viewing angle θ as a function of the distance x in front of each painting. Use the graph of your equation to explore the size of the viewing angle as the distance x approaches zero for the following paintings.

(a) 24-in. painting

(b) 48-in. painting

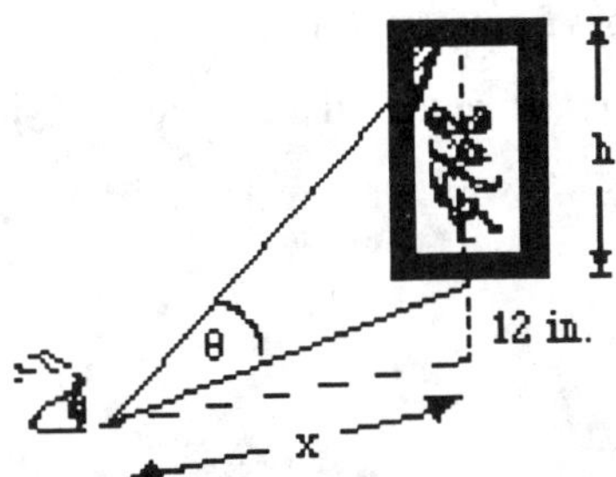

Figure 5

17. The Inexpensive Gutters Problem You are installing continuous gutters on your new house. You have two choices for the shape of the gutter (Figures 6 and 7). Find an equation for each gutter that expresses the carrying capacity of the gutter as a function of angle θ. Use the graph of your equation to find the value of θ that will maximize the carrying capacity for each of the gutters.

(a) gutter in Figure 6

(b) gutter in Figure 7

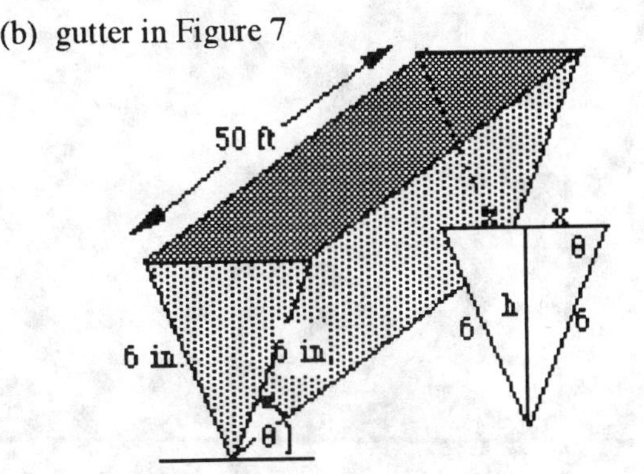

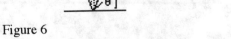

Figure 6

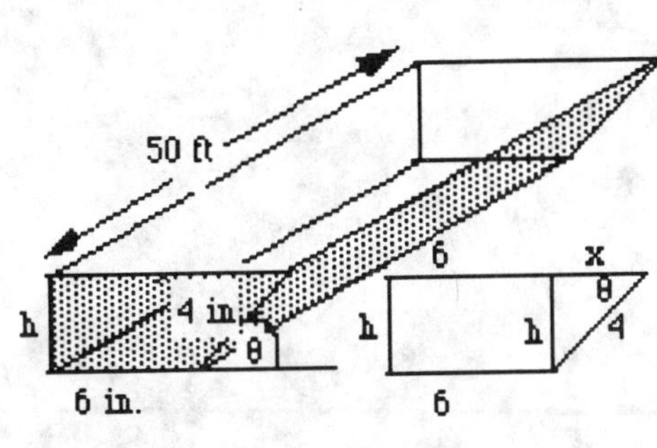

Figure 7

18. <u>The Season Problem</u> Since your kite factory is located on the beach along with several competing companies, the demand for kites depends on the season as well as the price (Figure 8).

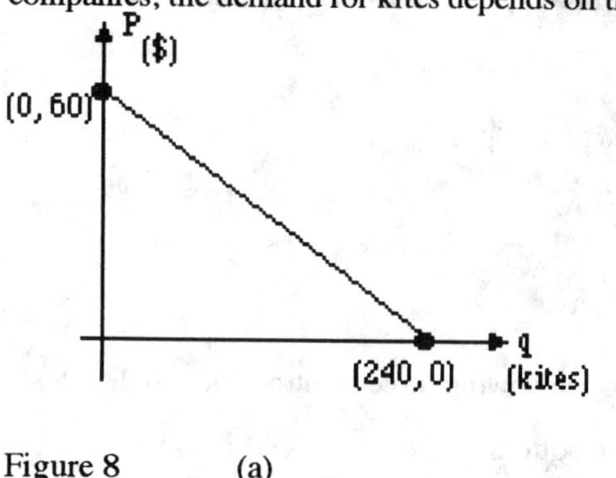

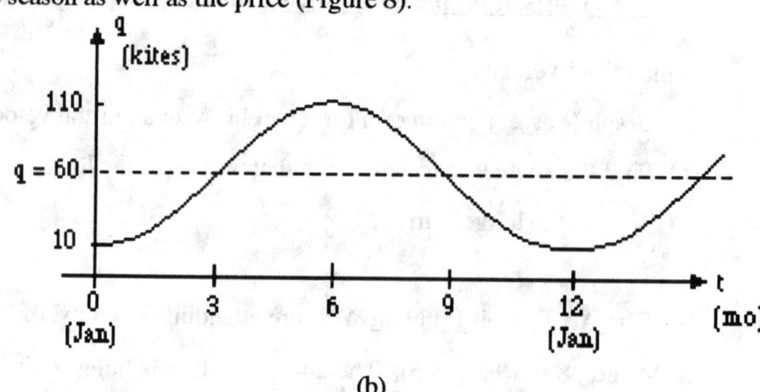

Figure 8 (a) (b)

(a) Find the equation for the graph in Figure 8a that expresses price as a function of quantity.

(b) Find the equation for the sine graph in Figure 8b that expresses quantity as a function of time.

(c) Use the equations from parts (a) and (b) to find the function that expresses price as a function of time. (It is the composition function of the functions from parts (a) and (b).)

(d) Find and graph the revenue function in order to find the month that yields maximum revenue.

Chapter 6

Trigonometric Equations and Applications

In this chapter we study equation solving as well as applications involving triangles. Some of the more important applications involve vectors, quantities that have both magnitude and direction, such as wind speed, current velocity, high- and low-pressure areas on weather maps, and forces involved in moving objects. We introduce vectors in Section 6.5 and study them in more detail in Chapter 10.

A Historical Note on the Trigonometric Identities

The trigonometric identities did not emerge from the mathematical community as one block of knowledge but were discovered by different mathematicians over about 900 years. The following summary of their development is based on A Short History of Mathematics by Vera Sanford, Houghton Mifflin, Boston, 1958, p. 300.

	Mathematician	date
Relation between trigonometric functions:		
$\sin^2 x + \cos^2 x = 1$	Hipparchus	c.140 B.C.
$\tan x = \dfrac{\sin x}{\cos x}$		
$\sec x = \sqrt{1 + \tan^2 x}$	Abû'l-Wefâ	c.980
$\csc x = \sqrt{1 + \cot^2 x}$		
$\sec x = \dfrac{1}{\cos x}$	Rhaeticus	1551
Functions of the sum or difference of two angles:		
$\sin (x - y)$	(?) Hipparchus	
$\cos (x - y)$	Ptolemy	c.150
$\sin 2x$	Abû'l-Wefâ	c.980
$\sin 3x$	Vieta	1591
$\sin nx$ and $\cos nx$ (series)	Newton	1676
$\tan nx$	de Lagny	c.1710
$\sec nx$		
$(\cos \phi + i \sin \phi)^n$	DeMoivre	1722
$\tan 2x$	Euler	1748
$\cot 2x$		
$\sin (x/2)$	Ptolemy	c.150
Formulas used in the solution of triangles:		
Sine law (law of sines)	Ptolemy	c.150
Cosine law (law of cosines)	Euclid	c.300 B.C.
Tangent law:		
$\dfrac{\sin A + \sin B}{\sin A - \sin B} = \dfrac{\tan (1/2)(A + B)}{\tan (1/2)(A - B)}$	Regiomontanus	1464
$\dfrac{a + b}{a - b} = \dfrac{\tan (1/2)(A + B)}{\tan (1/2)(A - B)}$	Fincke	1583
	Vieta	1580
Area of a triangle = $(1/2)ab \sin C$	Regiomontanus	1464

6.1 Trigonometric Identities

Introducing the Fundamental and Pythagorean Identities
Proving Identities
Disproving Identities

An **identity** is an equation that holds for all values of the variable for which the expressions are defined. Identities are used to change complicated expressions into simpler, equivalent ones. Some of the identities proved in this section are used to solve equations involving trigonometric functions. The decision as to the letters used as variables is often an individual choice. Initially, we use θ for the argument of a function when we interpret the situation on a unit circle and x when we are interpreting the situation on a graph.

Introducing the Fundamental and Pythagorean Identities

The *fundamental identities* follow directly from the definitions of the six circular functions. The fundamental identities allow us to change expressions involving any of the circular functions into expressions involving sines and cosines.

$$\csc\theta = \frac{1}{\sin\theta},\quad \sec\theta = \frac{1}{\cos\theta},\quad \tan\theta = \frac{\sin\theta}{\cos\theta},\quad \cot\theta = \frac{1}{\tan\theta} = \frac{\cos\theta}{\sin\theta}$$

Another collection of identities are called the *Pythagorean identities*. They follow directly from the Pythagorean theorem. Figure 1 shows a unit circle with the sides of a reference triangle labeled x, y, and r = 1.

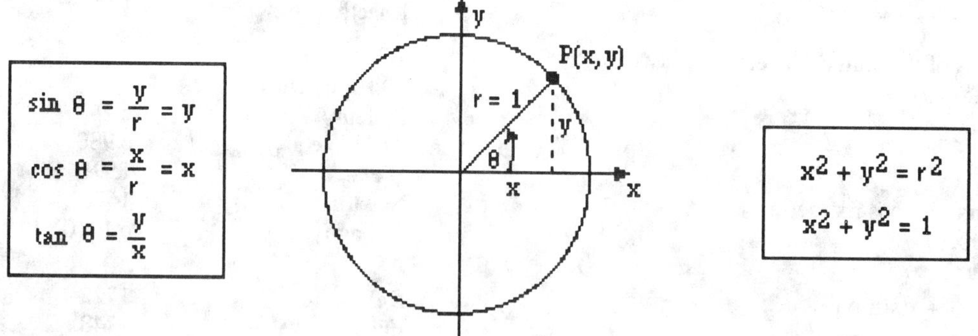

Figure 1

From the Pythagorean theorem we have $x^2 + y^2 = 1$. Replacing x with $\cos\theta$ and y with $\sin\theta$ yields

$$(\cos\theta)^2 + (\sin\theta)^2 = 1$$

We write

$$\cos^2\theta + \sin^2\theta = 1$$

NOTE: Squaring a circular function does not square the angle, $(\sin\theta)^2 = \sin^2\theta \neq \sin\theta^2$.

The Pythagorean identities allow us to write $\sin^2\theta$ in terms of $\cos^2\theta$ and $\cos^2\theta$ in terms of $\sin^2\theta$. Solving $\cos^2\theta + \sin^2\theta = 1$ for $\sin^2\theta$ and $\cos^2\theta$ yields

$$\sin^2\theta = 1 - \cos^2\theta$$

and

$$\cos^2\theta = 1 - \sin^2\theta$$

Other Pythagorean identities can be derived from $\cos^2\theta + \sin^2\theta = 1$ as follows.

$$\begin{aligned} \cos^2\theta + \sin^2\theta &= 1 \\ \frac{\cos^2\theta}{\cos^2\theta} + \frac{\sin^2\theta}{\cos^2\theta} &= \frac{1}{\cos^2\theta} \\ 1 + \tan^2\theta &= \sec^2\theta \end{aligned}$$

Divide both sides by $\cos^2\theta$.

$$\begin{aligned} \cos^2\theta + \sin^2\theta &= 1 \\ \frac{\cos^2\theta}{\sin^2\theta} + \frac{\sin^2\theta}{\sin^2\theta} &= \frac{1}{\sin^2\theta} \\ \cot^2\theta + 1 &= \csc^2\theta \end{aligned}$$

Divide both sides by $\sin^2\theta$.

Figure 2 shows three right triangles that model the Pythagorean identities. Using the three right triangles we have $\sin^2\theta + \cos^2\theta = 1$ (Figure 2a), $1 + \tan^2\theta = \sec^2\theta$ (Figure 2b), and $\cot^2\theta + 1 = \csc^2\theta$ (Figure 2c).

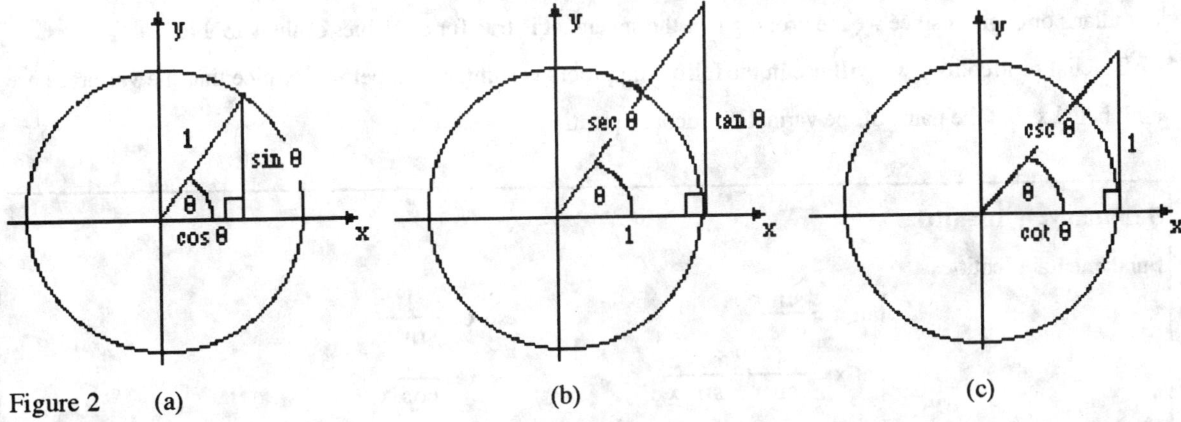

Figure 2 (a) (b) (c)

In the next example we use a graphing utility to investigate a Pythagorean identity.

Example 1 *Graphing an identity*

Use the graphs of $y = \tan^2 x$ and $y = \sec^2 x$ (Figure 3) to explain the identity $1 + \tan^2 x = \sec^2 x$.

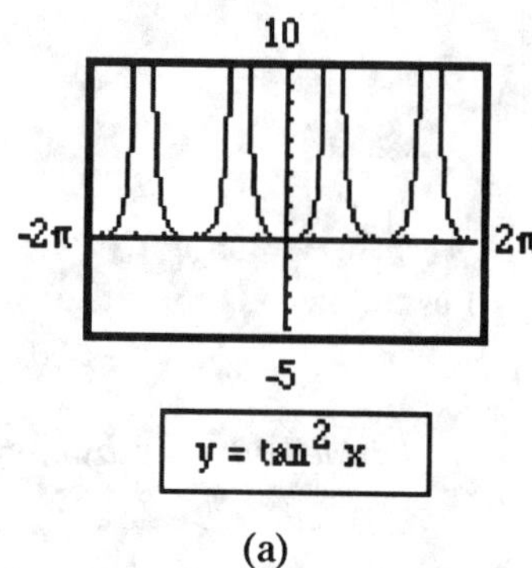

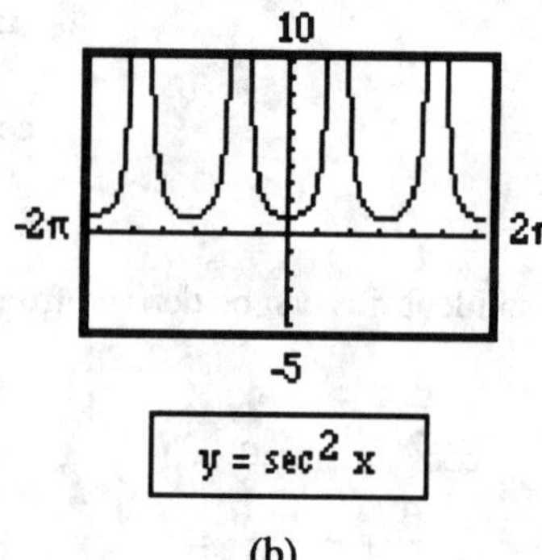

Figure 3 (a) (b)

Solution

Translating the graph of $y = \tan^2 x$ (Figure 3a) up 1 unit results in the graph of $y = \sec^2 x$ (Figure 3b), since

$$1 + \tan^2 x = \sec^2 x$$

Proving Identities

Proving identities involves showing algebraically that the two sides of a given equation are equal for all values of the variable for which the expressions involved are defined. The variable used in the proof may be measured in radians or degrees since we are proving that the statement is true for all values of the variable.

The list of identities we will use in the following proofs is summarized below. Notice that the variables are all labeled x, but the name of the variable is inconsequential.

Summary: Identities

Fundamental Identities

$$\tan x = \frac{\sin x}{\cos x} \qquad \csc x = \frac{1}{\sin x}$$

$$\cot x = \frac{1}{\tan x} = \frac{\cos x}{\sin x} \qquad \sec x = \frac{1}{\cos x}$$

Pythagorean Identities

$$\sin^2 x + \cos^2 x = 1 \Leftrightarrow \sin^2 x = 1 - \cos^2 x \Leftrightarrow \cos^2 x = 1 - \sin^2 x$$

$$1 + \tan^2 x = \sec^2 x \Leftrightarrow \tan^2 x = \sec^2 x - 1 \Leftrightarrow \sec^2 x - \tan^2 x = 1$$

$$\cot^2 x + 1 = \csc^2 x \Leftrightarrow \cot^2 x = \csc^2 x - 1 \Leftrightarrow \csc^2 x - \cot^2 x = 1$$

The rules of logic determine valid presentations of all proofs, including proving trigonometric identities.

We discuss two valid organizations of proofs of identities. The first approach is to start with one side of the equation, usually the more complex side, and use established identities and algebra to change it into the other side. Often we begin by using the fundamental identities to rewrite expressions in terms of sines and cosines. For example, we can rewrite $\sin x \cot x$ as $\cos x$ as follows:

$$\sin x \cot x = \sin x \left(\frac{\cos x}{\sin x}\right) = \cos x$$

Example 2 *Proving an identity*

Prove the following identities.

(a) $\csc x - \cot x \cos x = \sin x$

(b) $\sin x \cdot \tan x + \cos x = \sec x$

Solution

We start with the left side and create a chain of equalities that end with the right side. The valid conclusion is that the expression on the left side is equal to the expression on the right side, establishing our original equation as an identity.

(a)

$$\begin{aligned} &\csc x - \cot x \cos x = \\ &\frac{1}{\sin x} - \left(\frac{\cos x}{\sin x}\right)\cos x = \\ &\frac{1-\cos^2 x}{\sin x} = \\ &\frac{\sin^2 x}{\sin x} = \sin x \end{aligned}$$

Therefore, $\csc x - \cot x \cos x = \sin x$

(b)

$$\begin{aligned} &\sin x \tan x + \cos x = \\ &\sin x\left(\frac{\sin x}{\cos x}\right) + \cos x = \\ &\sin x\left(\frac{\sin x}{\cos x}\right) + \cos x = \\ &\frac{\sin^2 x}{\cos x} + \frac{\cos^2 x}{\cos x} = \\ &\frac{1}{\cos x} = \sec x \end{aligned}$$

Therefore, $\sin x \tan x + \cos x = \sec x$

The second method of organizing a proof is to work with each side of the equation separately, changing each into a common statement. For example, to prove $\sin x \cot x = \cos^2 x \sec x$, we write

$$\sin x \cot x = \sin x \left(\frac{\cos x}{\sin x}\right) = \cos x$$

and

$$\cos^2 x \sec x = \cos^2 x \left(\frac{1}{\cos x}\right) = \cos x$$

Hence,

$$\sin x \cot x = \cos^2 x \sec x$$

NOTE: Notice that our conclusion is not $\cos x = \cos x$. We emphasize our conclusion with words such as *hence* or *therefore.*

The first method is used when one side of the equation is relatively simple. The second method is commonly used when both sides of the equation are relatively complicated, as in the next example, where we change each side into $\sec^2 x$.

Example 3 *Proving an identity*

Prove the identity

$$\sin^2 x + \sec^2 x - \tan^2 x \cos^2 x = 1 + \tan^2 x$$

Solution

$$\sin^2 x + \sec^2 x - \tan^2 x \cos^2 x =$$
$$\sin^2 x + \sec^2 x - \left(\frac{\sin^2 x}{\cos^2 x}\right)\cos^2 x =$$
$$\sin^2 x + \sec^2 x - \sin^2 x = \mathbf{\sec^2 x}$$

and

$$1 + \tan^2 x = \mathbf{\sec^2 x}$$

Therefore,

$$\sin^2 x + \sec^2 x - \tan^2 x \cos^2 x = 1 + \tan^2 x$$

Disproving Identities

To prove that the equation

$$\cos x = \sqrt{1 - \sin^2 x}$$

is not an identity, we find one **counterexample** -- one value of the variable that does not satisfy the equation. For example, when $x = \pi$,

$$\cos \pi = \sqrt{1 - \sin^2 \pi}$$
$$-1 = \sqrt{1 - 0}$$
$$-1 = 1$$

Since this last line is false, the assumption that led to it must be invalid. Thus,

$$\cos x \neq \sqrt{1 - \sin^2 x}$$

Exercises 6.1

Develop the Concepts

For Exercises 1 to 4 use the graphs of functions f and g in (a) to explain each identity in (b).

1. (a) $f(x) = \sin^2 x$, $g(x) = \cos^2 x$
   (b) $\sin^2 x = 1 - \cos^2 x$

2. (a) $f(x) = \sec^2 x$, $g(x) = \tan^2 x$
   (b) $\tan^2 x = \sec^2 x - 1$

3. (a) $f(x) = \cot^2 x$, $g(x) = \csc^2 x$
   (b) $\csc^2 x = 1 + \cot^2 x$

4. (a) $f(x) = \cot^2 x$, $g(x) = \csc^2 x$
   (b) $-\cot^2 x = 1 - \csc^2 x$

For Exercises 5 to 30 prove or disprove that each equation is an identity.

5. $\sin x \cot x = \cos x$

6. $\dfrac{\tan x \cot x}{\sin x} = \csc x$

7. $\tan x \sin x \cos x + \sin x \cos x \cot x = 1$

8. $\tan x \sin x = \sec x - \cos x$

*9. $\sin 2x = 2 \sin x$

10. $\dfrac{2 + 3 \cos x}{\sin x} = 2 \csc x + 3 \cot x$

11. $\dfrac{1}{\sec^2 x} + \dfrac{1}{\csc^2 x} = 1$

12. $\sin^2 x = \sin x^2$

13. $\dfrac{\sin x}{1 + \cos x} = \dfrac{1 - \cos x}{\sin x}$

14. $(2 + \cos x)(2 - \cos x) = 3 + \sin^2 x$

15. $\sin x \tan x + \cos x = \sec x$

16. $\sin(\alpha + \beta) = \sin \alpha + \sin \beta$

17. $\dfrac{\sec x \cot x \sin x}{\cos x \csc x} = \tan x$

18. $\cos^2 x \tan^2 x = 1 - \cos^2 x$

19. $\sec^2 x = \tan x + 1$

20. $1 - \cos^2 x = (\tan x \cos x)^2$

21. $\dfrac{\cos^2 x}{\sin x (1 + \csc x)} = 1 - \sin x$

22. $\dfrac{1 - \cos x}{\sin x} = \dfrac{\sin x}{1 + \cos x}$

23. $\dfrac{\tan x}{\sec x - 1} = \dfrac{\sec x + 1}{\tan x}$

24. $\cot x \sin x = \cos^2 x$

25. $\dfrac{1 - \cos x}{\sin x} + \dfrac{\sin x}{1 - \cos x} = \dfrac{2}{\sin x}$

26. $\dfrac{1}{1 + \sin x} + \dfrac{1}{1 - \sin x} = 2 \sec^2 x$

27. $(\sin x + \cos x)^2 + (\sin x - \cos x)^2 = 2$

28. $(2 \sec x - 2 \tan x)(2 \sec x + 2 \tan x) = 4$

29. $\sin^2 x + 2 \sin x \cos x + \cos^2 x = 1$

30. $\sin^4 x + 2 \sin^2 x \cos^2 x + \cos^4 x = 1$

Extend the Concepts

For Exercises 31 to 34 graph each pair of functions to guess whether or not they are identical.

*31. $f(x) = 2\cos\left(x - \frac{\pi}{3}\right)$

$g(x) = \cos x + \sqrt{3} \sin x$

32. $f(x) = 2 \sin\left(x + \frac{\pi}{6}\right)$

$g(x) = \sqrt{3} \sin x + \cos x$

33. $f(x) = 1 + \cos x$

$g(x) = 2 \cos^2 \frac{x}{2}$

34. $f(x) = 1 - \cos 2x$

$g(x) = 2 \sin^2 x$

For Exercises 35 to 38 graph the functions to discover whether the functions are even, odd, or neither.

35. $f(x) = \sin x \cos x$

36. $f(x) = \tan (x/2)$

37. $f(x) = x^2 \cos x$

38. $f(x) = x^2 \sin x$

39.(a) Complete the table by stating the equation of each reflection.

	y = f(-x)	y = -f(x)	y = -f(-x)
f(x) = sin x			

(b) From your table identify which equations are equal.

40.(a) Complete the table by stating the equation of each reflection.

	y = f(-x)	y = -f(x)	y = -f(-x)
f(x) = cos x			

(b) From your table identify which equations are equal.

41.(a) Complete the table by stating the equation of each reflection.

	y = f(-x)	y = -f(x)	y = -f(-x)
f(x) = tan x			

(b) From your table identify which equations are equal.

42.(a) Complete the table by stating the equation of each reflection.

	y = f(-x)	y = -f(x)	y = -f(-x)
f(x) = sec x			

(b) From your table identify which equations are equal.

For Exercises 43 to 46 graph the equations and identify which ones are equal.

43.(a) $y = \sin x$

(b) $y = \sin(\pi + x)$

(c) $y = \sin(2\pi - x)$

(d) $y = \sin(\pi - x)$

44.(a) $y = \tan x$

(b) $y = \tan(\pi + x)$

(c) $y = \tan(2\pi - x)$

(d) $y = \tan(\pi - x)$

45.(a) $y = \cos x$

(b) $y = \cos(\pi + x)$

(c) $y = \cos(2\pi - x)$

(d) $y = \cos(\pi - x)$

46.(a) $y = \sec x$

(b) $y = \sec(\pi + x)$

(c) $y = \sec(2\pi - x)$

(d) $y = \sec(\pi - x)$

For Exercises 47 to 50 graph the equations and identify whether or not they are identical.

47.(a) $y = \sin x$

(b) $y = \cos(\frac{\pi}{2} - x)$

48.(a) $y = \cos x$

(b) $y = \sin(\frac{\pi}{2} - x)$

49.(a) $y = \cot(\pi - x)$

(b) $y = \tan(x - \frac{\pi}{2})$

50.(a) $y = \sec(x + \frac{\pi}{2})$

(b) $y = \csc(-x)$

6.2 More Trigonometric Identities

Introducing the Sum and Difference Formulas
Introducing the Double- and Half-Angle Formulas
Proving More Identities

In this section we study the sum and difference, double-angle, and half-angle formulas -- identities that are essential in calculus. It is appropriate for us to examine their derivations and become familiar with, if not memorize, them. The table on the inside back cover summarizes the more important trigonometric identities and formulas.

Introducing the Sum and Difference Formulas

Consider the expression $\cos(\alpha - \beta)$. It is very easy to guess that the distributive property yields $\cos\alpha - \cos\beta$. However, a counterexample quickly proves that this is not the case.

$\cos(\alpha - \beta) = \cos\alpha - \cos\beta$ and $\alpha = \frac{\pi}{3}$ and $\beta = \frac{\pi}{6}$ $\Rightarrow$

$$\cos(\alpha - \beta) = \cos\alpha - \cos\beta$$
$$\cos\left(\frac{\pi}{3} - \frac{\pi}{6}\right) = \cos\frac{\pi}{3} - \cos\frac{\pi}{6}$$
$$\cos\left(\frac{2\pi}{6} - \frac{\pi}{6}\right) = \frac{1}{2} - \frac{\sqrt{3}}{2}$$
$$\cos\frac{\pi}{6} = \frac{1 - \sqrt{3}}{2}$$
$$\frac{\sqrt{3}}{2} = \frac{1 - \sqrt{3}}{2}$$

Hence,

$$\cos(\alpha - \beta) \neq \cos\alpha - \cos\beta$$

The cosine of the difference of two angles is

$$\cos(\alpha - \beta) = \cos\alpha \ \cos\beta + \sin\alpha \ \sin\beta$$

Figure 1 shows reference points for α, β, and $\alpha - \beta$ as P_1, P_2, and P_3, respectively.

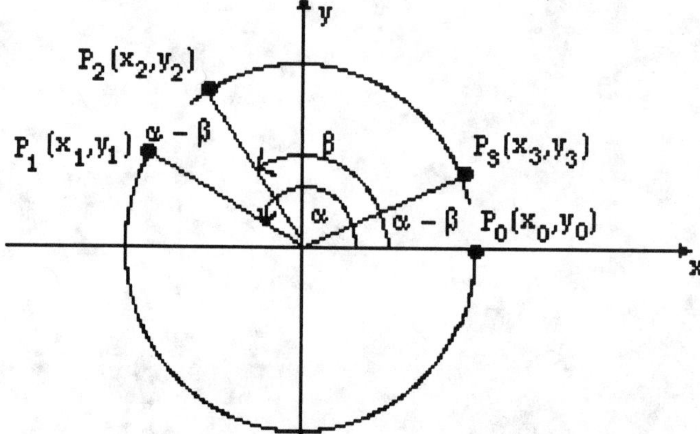

Figure 1

Notice that the arcs P_1P_2 and P_0P_3 are equal, since they cut off equal arcs of the circle. We can prove that $\cos(\alpha - \beta) = \cos\alpha \cos\beta + \sin\alpha \sin\beta$ by setting the distance between points P_1 and P_2 and the distance between P_0 and P_3 equal.

$$\sqrt{(x_2 - x_1)^2 + (y_2 - y_1)^2} = \sqrt{(x_3 - 1)^2 + (y_3 - 0)^2}$$

Squaring both sides and collecting like terms yields

$$(x_2 - x_1)^2 + (y_2 - y_1)^2 = (x_3 - 1)^2 + (y_3 - 0)^2$$

$$x_2^2 - 2x_1x_2 + x_1^2 + y_2^2 - 2y_1y_2 + y_1^2 = x_3^2 - 2x_3 + 1 + y_3^2$$

$$(x_1^2 + y_1^2) + (x_2^2 + y_2^2) - 2x_1x_2 - 2y_1y_2 - 1 - (x_3^2 + y_3^2) = -2x_3$$

Since $x^2 + y^2 = 1$ for all points on a unit circle (radius $r = 1$), this equation simplifies to

$$1 + 1 - 2x_1x_2 - 2y_1y_2 - 1 - 1 = -2x_3$$

$$2x_3 = 2x_1x_2 + 2y_1y_2$$

$$x_3 = x_1x_2 + y_1y_2$$

Substituting $x_1 = \cos\alpha$, $y_1 = \sin\alpha$, $x_2 = \cos\beta$, $y_2 = \sin\beta$, and $x_3 = \cos(\alpha - \beta)$ yields

$$\cos(\alpha - \beta) = \cos\alpha \cos\beta + \sin\alpha \sin\beta$$

Similarly, the cosine of the sum of two angles is

$$\cos(\alpha + \beta) = \cos\alpha \cos\beta - \sin\alpha \sin\beta$$

Consider the graph of $f(x) = 3\cos x + 4\sin x$ in Figure 2. This function appears to be a translated sine or cosine function.

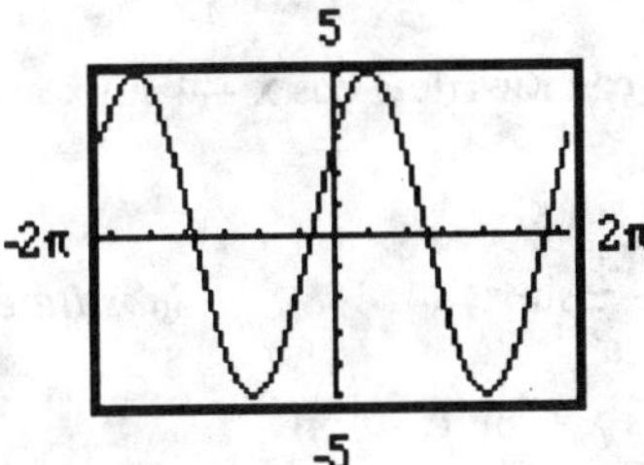

Figure 2

Using the sum formula for cosine, we can express this equation as a function of the form $f(x) = A\cos(Bx + C)$. In the next example we use the sum formula for cosine to determine the period, amplitude, and phase shift of $f(x) = 3\cos x + 4\sin x$. In applications in courses such as differential equations, physics, and engineering, we rewrite such expressions.

Example 1 *Applying the sum formula for cosine*

Express f(x) = 3 cos x + 4 sin x in the form f(x) = A cos (Bx + C). Then state the amplitude, period, and phase shift of the graph of this function (Figure 2).

Solution

Comparing

$$\cos\alpha\cos\beta + \sin\alpha\sin\beta$$

and

$$3\cos x + 4\sin x$$

we notice that the patterns almost match. But there is no angle for which cosine is 3 and sine is 4. Figure 3 shows a circle of radius r with the reference point (3, 4).

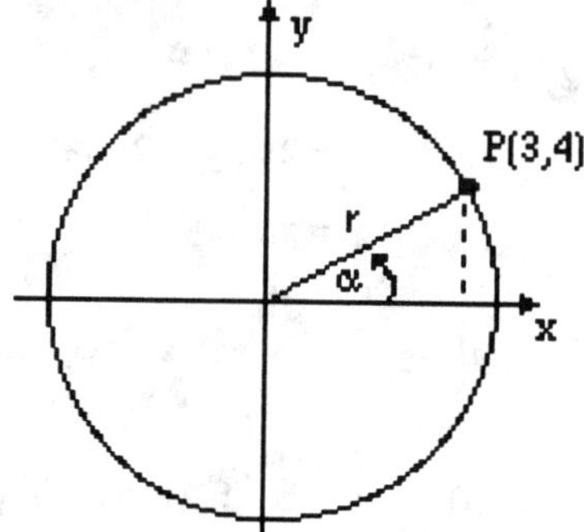

Figure 3

We need to find an angle α such that $\cos\alpha = \frac{3}{r}$ and $\sin\alpha = \frac{4}{r}$. Using the Pythagorean theorem yields r = 5.

$$3^2 + 4^2 = r^2$$
$$r^2 = 25$$
$$r = 5$$

Therefore, $\cos\alpha = \frac{3}{5}$ and $\sin\alpha = \frac{4}{5}$. We can rewrite 3 cos x + 4 sin x as $5\left(\frac{3}{5}\cos x + \frac{4}{5}\sin x\right)$.

$$f(x) = 3\cos x + 4\sin x$$

$$f(x) = 5\left(\frac{3}{5}\cos x + \frac{4}{5}\sin x\right)$$ *Substitute cos α for 3/5 and sin α for 4/5.*

$$f(x) = 5(\cos\alpha\cos x + \sin\alpha\sin x)$$

$$f(x) = 5\cos(\alpha - x) = 5\cos(x - \alpha)$$

Since $\alpha = \tan^{-1}\left(\frac{4}{3}\right) \Rightarrow \alpha \approx 0.9$.

$$f(x) = 5\cos(x - 0.9)$$

The amplitude is 5, the period is 2π, and the phase shift is 0.9.

The addition formula for the sine function can be derived from the addition formula for cosine. Recall that translating the graph of the cosine function right $\pi/2$ units yields the graph of the sine function. Therefore, the sine of angle θ is equal to the cosine of its complement, $\pi/2 - \theta$.

$\sin\theta = \cos\left(\frac{\pi}{2} - \theta\right)$	*Let $\theta = \alpha + \beta$.*
$\sin(\alpha + \beta) = \cos\left[\frac{\pi}{2} - (\alpha + \beta)\right]$	
$= \cos\left(\frac{\pi}{2} - \alpha - \beta\right)$	
$= \cos\left[\left(\frac{\pi}{2} - \alpha\right) - \beta\right]$	*Apply cosine of the difference of two angles.*
$= \cos\left(\frac{\pi}{2} - \alpha\right)\cos\beta + \sin\left(\frac{\pi}{2} - \alpha\right)\sin\beta$	*Substitute $\sin\theta$ for $\cos(\pi/2 - \alpha)$.*
$= \sin\alpha\cos\beta + \cos\alpha\sin\beta$	

Therefore, $\sin(\alpha + \beta) = \sin\alpha\cos\beta + \cos\alpha\sin\beta$.

The sine of the difference of two angles is $\sin(\alpha - \beta) = \sin\alpha\cos\beta - \cos\alpha\sin\beta$. These results are summarized below.

Sum and Difference Formulas

$$\cos(\alpha + \beta) = \cos\alpha\cos\beta - \sin\alpha\sin\beta$$
$$\cos(\alpha - \beta) = \cos\alpha\cos\beta + \sin\alpha\sin\beta$$
$$\sin(\alpha + \beta) = \sin\alpha\cos\beta + \cos\alpha\sin\beta$$
$$\sin(\alpha - \beta) = \sin\alpha\cos\beta - \cos\alpha\sin\beta$$

The formulas for the tangent of the sum and difference of two angles are summarized below, then applied in the following example.

Sum and Difference Formulas for Tangent

$$\tan(\alpha + \beta) = \frac{\tan\alpha + \tan\beta}{1 - \tan\alpha\tan\beta}$$
$$\tan(\alpha - \beta) = \frac{\tan\alpha - \tan\beta}{1 + \tan\alpha\tan\beta}$$

Introducing the Double- and Half-Angle Formulas

The **double-angle formulas** involve α and 2α. We can derive the formula for $\sin 2\alpha$ using the addition formula $\sin(\alpha + \beta) = \sin\alpha\cos\beta + \cos\alpha\sin\beta$ as follows by letting $\beta = \alpha$.

$\sin(\alpha + \beta) = \sin\alpha\cos\beta + \cos\alpha\sin\beta$ and $\beta = \alpha \Rightarrow$

$$\sin(\alpha + \alpha) = \sin\alpha\cos\alpha + \cos\alpha\sin\alpha$$
$$\sin 2\alpha = 2\sin\alpha\cos\alpha$$

Similarly, we derive the formula for $\cos 2\alpha$ using the addition formula $\cos(\alpha + \beta) = \cos\alpha\cos\beta - \sin\alpha\sin\beta$ as follows.

$\cos(\alpha + \beta) = \cos\alpha\cos\beta - \sin\alpha$ and $\beta = \alpha \Rightarrow$

$$\cos(\alpha + \alpha) = \cos\alpha\cos\alpha - \sin\alpha\sin\alpha$$

$$\cos 2\alpha = \cos^2\alpha - \sin^2\alpha$$

Substituting $1 - \sin^2\alpha$ for $\cos^2\alpha$ yields another form for $\cos 2\alpha$, $\cos 2\alpha = 1 - 2\sin^2\alpha$. Substituting $1 - \cos^2\alpha$ for $\sin^2\alpha$ yields a third form for $\cos 2\alpha$, $\cos 2\alpha = 2\cos^2\alpha - 1$.

Double-Angle Formulas for Sine and Cosine

$$\sin 2\alpha = 2\sin\alpha\cos\alpha$$

$$\cos 2\alpha = \cos^2\alpha - \sin^2\alpha$$

$$= 1 - 2\sin^2\alpha$$

$$= 2\cos^2\alpha - 1$$

The **half-angle formulas** involve α and $\frac{\alpha}{2}$. We derive the half-angle formulas from the double-angle formulas above by making a change of variables, α for $\frac{\alpha}{2}$, and solving for $\sin\frac{\alpha}{2}$, then for $\cos\frac{\alpha}{2}$.

Half-Angle Formulas for Sine and Cosine

$$\sin\frac{\alpha}{2} = \pm\sqrt{\frac{1 - \cos\alpha}{2}} \qquad \cos\frac{\alpha}{2} = \pm\sqrt{\frac{1 + \cos\alpha}{2}}$$

Proving More Identities

For a list of identities and formulas, see the inside back cover.

Example 2 *Proving an identity*

Prove the identity

$$\frac{\sin 2x}{1 - \cos 2x} = \cot x$$

Solution

$$\frac{\sin 2x}{1 - \cos 2x} =$$
$$\frac{2 \sin x \cdot \cos x}{1 - (\cos^2 x - \sin^2 x)} =$$
$$\frac{2 \sin x \cdot \cos x}{(1 - \cos^2 x) + \sin^2 x} =$$
$$\frac{2 \sin x \cos x}{\sin^2 x + \sin^2 x} =$$
$$\frac{2 \sin x \cos x}{2 \sin^2 x} = \frac{\cos x}{\sin x} = \cot x$$

Therefore, $\dfrac{\sin 2x}{1 - \cos 2x} = \cot x$.

The identity in the next example probably gave Napier the idea for logarithms. Early logarithmic tables were called tables of sines. Notice that this identity expresses a product as an identical sum.

Example 3 *Proving an identity*

Prove the identity $\sin \alpha \cos \beta = \frac{1}{2} \sin(\alpha + \beta) + \frac{1}{2} \sin (\alpha - \beta)$.

Solution

We apply the sum and difference formulas for sine to turn the right side into the left side.

$$\frac{1}{2} \sin (\alpha + \beta) + \frac{1}{2} \sin (\alpha - \beta) =$$
$$\frac{1}{2} [\sin \alpha \cos \beta + \cos \alpha \sin \beta] + \frac{1}{2} [\sin \alpha \cos \beta - \cos \alpha \sin \beta] =$$
$$\frac{1}{2} \sin \alpha \cos \beta + \frac{1}{2} \sin \alpha \cos \beta = \sin \alpha \cos \beta$$

Therefore, $\sin \alpha \cos \beta = \frac{1}{2} \sin (\alpha + \beta) + \frac{1}{2} \sin (\alpha - \beta)$.

Exercises 6.2

Develop the Concepts

For Exercises 1 to 20 prove or disprove that each equation is an identity.

1. $\dfrac{\cos(\alpha - \beta)}{\cos(\alpha + \beta)} = \dfrac{1 + \tan\alpha\tan\beta}{1 - \tan\alpha\tan\beta}$

2. $\dfrac{\sin(\alpha + \beta)}{\sin(\alpha - \beta)} = \dfrac{\tan\alpha + \tan\beta}{\tan\alpha - \tan\beta}$

3. $\cot(x - y) = \dfrac{\cot x \cot y + 1}{\cot y - \cot x}$

4. $\dfrac{1 + \tan x}{1 - \tan x} = \dfrac{\cos 2x}{1 - \sin 2x}$

5. $\dfrac{\sin 2x}{\sin x} - \dfrac{\cos 2x}{\cos x} = \sec x$

6. $\cot 2x = \dfrac{\cot^2 x - 1}{2\cot x}$

7. $\csc 2x + \cot 2x = \cot x$

8. $\sin 3x = 2\cos x - \cos^3 x$

[Hint: Test $x = 0$.]

9. $\dfrac{\sin 3x - \sin x}{\cos 3x + \cos x} = \tan x$

10. $2\sin 2x\cos x - \sin 3x = \sin x$

11. $4\sin x\cos^2 x - \sin 3x = \sin x$

12. $\cos 3x = 4\cos^3 x - 3\cos x$

13. $\dfrac{\sin 4x}{2\sin 2x} = \cos 2x$

14. $\tan 2x = \dfrac{2\tan x}{1 - \tan^2 x}$

15. (a) $\tan\dfrac{x}{2} = \dfrac{\sin x}{1 + \cos x}$

(b) $\dfrac{\sin x}{1 + \cos x} = \dfrac{\tan x}{\sec x + 1}$

16. (a) $\tan x = \dfrac{1 - \cos 2x}{\sin 2x}$

(b) $\dfrac{1 - \cos 2x}{\sin 2x} = \dfrac{\sec 2x - 1}{\tan 2x}$

17. (a) $\cot\dfrac{x}{2} = \dfrac{\sin x}{1 - \cos x}$

(b) $\dfrac{\sin x}{1 - \cos x} = \dfrac{1}{\csc x - \cot x}$

18. (a) $\cot x = \dfrac{1 + \cos 2x}{\sin 2x}$

(b) $\dfrac{1 + \cos 2x}{\sin 2x} = \dfrac{\sec 2x + 1}{\tan 2x}$

19. $\dfrac{2\sin^2\frac{x}{2}}{\sin^2 x} = \dfrac{1}{1 + \cos x}$

20. $\dfrac{2\tan\frac{x}{2}}{1 + \tan^2\frac{x}{2}} = \sin x$

21. (a) $\tan x = \dfrac{\sin 2x}{1 + \cos 2x}$

(b) $\cot x = \dfrac{\sin 2x}{1 - \cos 2x}$

22. (a) $\dfrac{\cos\frac{x}{2} + \sin\frac{x}{2}}{\cos\frac{x}{2} - \sin\frac{x}{2}} = \dfrac{1 + \sin x}{\cos x}$

(b) $\dfrac{\cos\frac{x}{2} - \sin\frac{x}{2}}{\cos\frac{x}{2} + \sin\frac{x}{2}} = \dfrac{1}{\sec x + \tan x}$

For Exercises 23 and 24 use the identity $\sin(x + y) = \sin x\cos y + \cos x\sin y$ to write each expression in the form $f(t) = \sin(x + \alpha)$, where α is the phase shift (the horizontal translation).

23. $f(t) = \dfrac{1}{\sqrt{2}}\sin t + \dfrac{1}{\sqrt{2}}\cos t$

24. $f(t) = \dfrac{\sqrt{3}}{2}\sin t + \dfrac{1}{2}\cos t$

For Exercises 25 and 26 prove each identity. These identities are useful in calculus to develop the derivatives of the sine and cosine functions.

25. $\sin x + \sin y = 2 \sin\left(\frac{x + y}{2}\right) \cos\left(\frac{x - y}{2}\right)$

26. $\cos x + \cos y = 2 \cos\left(\frac{x + y}{2}\right) \cos\left(\frac{x - y}{2}\right)$

Extend the Concepts

27. Consider

$$\theta = \tan^{-1}\frac{6.5}{x} - \tan^{-1}\frac{4}{x}$$

Use the identity for tan(x - y) to show that $\theta = \tan^{-1}\frac{2.5x}{x^2 + 26}$.

28. Use the identity $\sin x + \sin y = 2 \sin\frac{x + y}{2} \cos\frac{x - y}{2}$ to represent the sum functions $f(t) = \sin 440\pi t + \sin 450\pi t$ as a product.

For Exercises 29 to 32 express each function as a single function using the double-angle formulas. Then graph the function.

*29. $f(x) = \sin x \cos x$

30. $g(x) = \sin 2x \cos 2x$

31. $f(x) = \cos^2 x - \sin^2 x$

32. $g(x) = \cos^2 2x - \sin^2 2x$

For Exercises 33 to 36 express each function without a squared term using the half-angle formulas. Then graph the function.

33. $f(x) = \sin^2 \frac{x}{2}$

34. $f(x) = \cos^2 x$

35. $f(x) = \sin^2 2x$

36. $f(x) = \cos^2 2x$

For Exercises 37 to 40 rewrite each function as a single identical function using the sum or difference formulas. Then graph the function.

*37. $f(x) = \frac{1}{\sqrt{2}}(\sin x + \cos x)$

38. $f(x) = \frac{1}{\sqrt{2}}(\sin x - \cos x)$

39. $f(x) = \frac{\sqrt{3}}{2}\sin x - \frac{1}{2}\cos x$

40. $f(x) = 6 \sin x + 8 \cos x$

For Exercises 41 to 44 solve each equation for x.

41. $2 \sin^{-1} x = \cos^{-1} 1$

42. $2 \tan^{-1} x = \sin^{-1}\frac{4}{5}$

43. $2 \sin^{-1} x = \sin^{-1} 2x$

44. $2 \cos^{-1} x = \cos^{-1} x$

45. Use the graphs of $f(x) = \tan\frac{x}{2}$ and $g(x) = \frac{\sin\frac{x}{2}}{\cos\frac{x}{2}} = \pm\sqrt{\frac{1 - \cos x}{1 + \cos x}} = \pm\frac{\sqrt{1 - \cos^2 x}}{1 + \cos x}$ to explain why $\tan\frac{x}{2} = \frac{\sin x}{1 + \cos x}$ without the ± sign.

6.3 Equations Involving the Circular Functions

Solving Equations
Applying Equation Solving

In this section we solve equations that contain one or more circular functions. All the rules of arithmetic and algebra apply in solving equations involving circular functions. Because of the periods of the circular functions, these equations may possess an infinite number of solutions. Of course, graphing utilities can be used to estimate the solutions.

Solving Equations

We refer to the solutions given by the calculator as the **calculator solutions** and all possible solutions as the **general solutions**. The next three examples illustrate this distinction.

Example 1 *Solving an equation for calculator solutions*

Find the particular solution of $3 \cos x = 1$.

Solution

$$3 \cos x = 1 \qquad \textit{Divide both sides by 3.}$$

$$\cos x = \frac{1}{3} \qquad \textit{Apply the definition of inverse sine.}$$

$$x = \cos^{-1}\left(\frac{1}{3}\right)$$

$$x \approx 1.23$$

Since the periods of sine and cosine are 2π, often finding solutions between o and 2π is sufficient.

Example 2 *Solving an equation for exact solutions*

Find the exact solutions for $2 \sin^2 x + \sin x - 1 = 0$ for $0 \le x < 2\pi$.

Solution

$$2 \sin^2 x + \sin x - 1 = 0$$

$$(2 \sin x - 1)(\sin x + 1) = 0$$

$$2 \sin x - 1 = 0 \text{ or } \sin x + 1 = 0$$

$$\sin x = \frac{1}{2} \text{ or } \sin x = -1$$

$$x = \sin^{-1}\frac{1}{2} \text{ or } x = \sin^{-1}(-1)$$

$$x = \frac{\pi}{6} \text{ or } x = \pi - \frac{\pi}{6} = \frac{5\pi}{6} \text{ or } x = \frac{3\pi}{2}$$

How could we state the decimal solutions for Example 2?

The decimal solutions for $0 \leq x < 2\pi$ are

$$x \approx 0.52 \quad \text{or} \quad x = \pi - 0.52 \approx 2.62 \quad \text{or} \quad x \approx -1.57 = 2\pi - 1.57 \approx 4.71$$

Example 3 *Solving an equation for general solutions*

Find the general solutions for $2 \sin^2 x + \cos x = 2$.

Solution

We will use the identity $\sin^2 x = 1 - \cos^2 x$ to eliminate $\sin^2 x$ so that all terms involve only one function, cos x.

$$2 \sin^2 x + \cos x = 2 \qquad \textit{Substitute the identity } \sin^2 x = 1 - \cos^2 x.$$

$$2(1 - \cos^2 x) + \cos x = 2 \qquad \textit{Use the distributive property.}$$

$$2 - 2\cos^2 x + \cos x = 2 \qquad \textit{Subtract 2 from both sides.}$$

$$2\cos^2 x - \cos x = 0$$

$$\cos x(2\cos x - 1) = 0$$

$$\cos x = 0 \quad \text{or} \quad 2\cos x - 1 = 0$$

$$\cos x = 0 \quad \text{or} \quad \cos x = 1/2$$

$$x = \frac{\pi}{2} \quad \text{or} \quad x = \frac{3\pi}{2} \quad \text{or} \quad x = \frac{\pi}{3} \quad \text{or} \quad x = \frac{5\pi}{3}$$

The general solutions are $x = \frac{\pi}{2} + k\pi$ or $x = \frac{\pi}{3} + 2k\pi$ or $x = \frac{5\pi}{3} + 2k\pi$, where k is an integer. The decimal solutions are $x \approx 1.57 + k\pi$ or $x \approx 1.05 + 2k\pi$ or $x \approx 5.24 + 2k\pi$ where k is an integer.

Summary: Solving Equations Involving Circular Functions

1. Use identities to transform the equation into an equivalent equation involving only one circular function, with the same argument.
2. Use the basic techniques of algebra to solve for the single circular function.
3. Use inverse circular functions to find the calculator solutions. If necessary, specify the general solutions.

Example 4 *Finding points of intersection*

Graph $f(x) = 2 \sin x$ and $g(x) = -\cos x$ on one coordinate system. Find the point(s) of intersection algebraically.

Solution

Figure 1 shows the graphs of functions f and g.

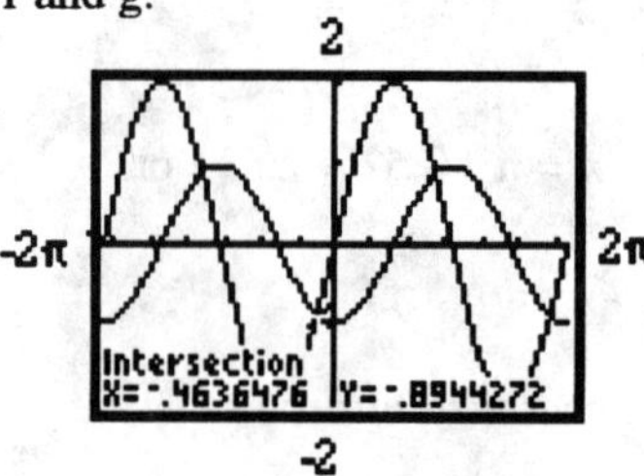

Figure 1

At the points of intersection, the y values are equal.

$f(x) = g(x) \Rightarrow$

$$2 \sin x = -\cos x \qquad \textit{Divide both sides by } 2 \cos x.$$

$$\frac{\sin x}{\cos x} = -\frac{1}{2} \qquad \textit{Substitute } \frac{\sin x}{\cos x} = \tan x.$$

$$\tan x = -\frac{1}{2}$$

$$x = \tan^{-1}(-\frac{1}{2}) \approx -0.46$$

$$x \approx -0.46 + k\pi,\ \ k \text{ an integer}$$

$$y = f(-0.46) \approx -0.89$$

Figure 2 shows a sketch with the points of intersection labeled.

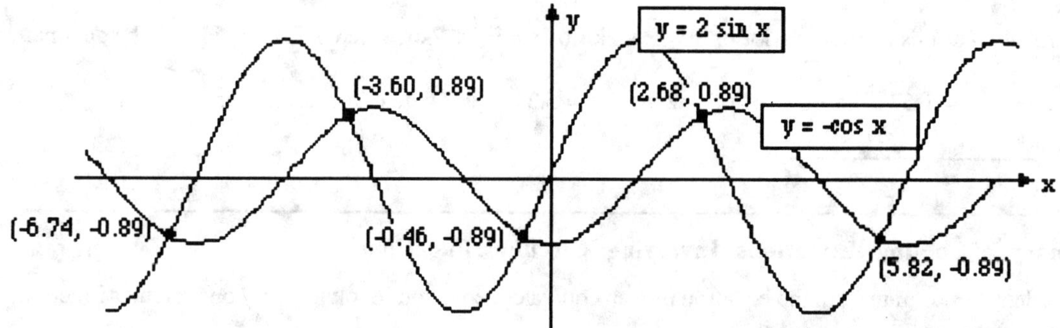

Figure 2

The next example illustrates how to solve equations where the functions do not have the same argument. In this case the double-angle formula is used to make all arguments x.

Example 5 *Solving an equation for general solutions*

Solve $\sin 2x = 2 \sin x$ for the general solution.

Solution

$\sin 2x = 2 \sin x$	*Substitute* $\sin 2x = 2 \sin x \cos x$.
$2 \sin x \cos x = 2 \sin x$	*Subtract* $2 \sin x$ *from both sides.*
$2 \sin x \cos x - 2 \sin x = 0$	*Divide both sides by 2.*
$\sin x \cos x - \sin x = 0$	*Apply the distributive property.*
$\sin x(\cos x - 1) = 0$	

$$\sin x = 0 \quad \text{or} \quad \cos x - 1 = 0$$
$$x = k\pi \quad \text{or} \quad \cos x = 1$$
$$x = k\pi \quad \text{or} \quad x = 2k\pi$$
$$x = k\pi, \ k \text{ an integer}$$

The solutions to Example 5 are the x intercepts of the graphs of $y = \sin 2x$ and $y = 2 \sin x$ (Figure 3).

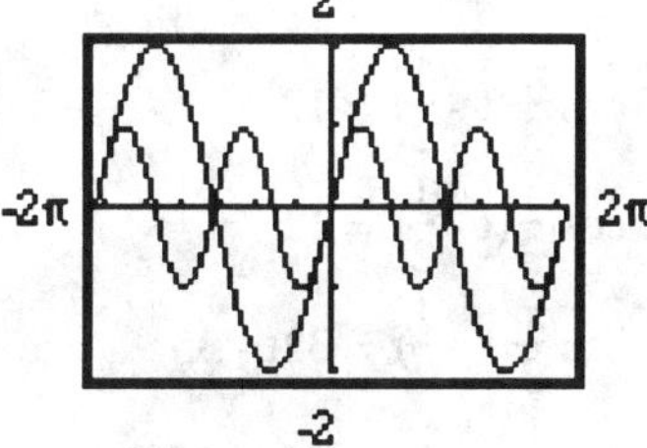

Figure 3

Notice that $\sin 2x \neq 2 \sin x$ for all values of x. They are equal only for multiples of π. The y coordinate of these points of intersection are all 0 since for $x = k\pi$,

$$\sin 2x = \sin 2k\pi = 0$$

and

$$2 \sin x = 2 \sin k\pi = 0$$

Applying Equation Solving

In the next example the equation cannot be solved algebraically. We use a graphing utility to estimate the solution after simplifying the expressions.

Example 6 The Container Ship Problem

A container ship is being loaded so that it sinks farther into the ocean at a constant rate of 16 in. every half hour. Initially, its hull is 27 ft above the seafloor. Suppose that the tide is coming in (but will soon be going out) according to the function

$$d = 20 + 5\cos\frac{\pi}{3}(t - \frac{1}{2})$$

where t is time in hours from now and d is the depth of the water under the hull in feet. If the ship must leave while she has 12 ft of water under her hull, how long do they have to finish loading her?

Thinking: Creating a Plan

We will find the equation of the linear function that represents the depth of the water under the hull as the ship is being loaded. Graphing the linear function and the given cosine function yields an approximation of when the sum is 12 ft.

Communicating: Writing the Solution

$d_1 = 27 - 16(2t)$ and $d_2 = 20 + 5\cos\frac{\pi}{3}(t - \frac{1}{2})$

$d_1 + d_2 = 12 \Rightarrow$

$$27 - 16(2t) + 20 + 5\cos\frac{\pi}{3}(t - \frac{1}{2}) = 12$$

$$47 - 32t + 5\cos\frac{\pi}{3}(t - \frac{1}{2}) = 12$$

$$5\cos\frac{\pi}{3}(t - \frac{1}{2}) = 32t - 35$$

Figure 4 shows the graphs of $y = 5\cos\frac{\pi}{3}(t - \frac{1}{2})$ and $y = 32t - 35$.

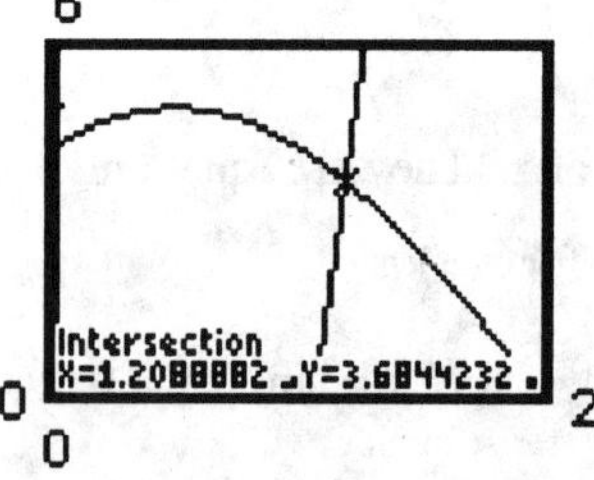

Figure 4

The point of intersection indicates that t = 1.2. They have about 1.2 hr (1 hr 12 min) to finish loading the container.

Learning: Making Connections

The equation $d_1 + d_2 = 12$ cannot be solved algebraically. We can only estimate the solution.

Exercises 6.3

Develop the Concepts

For Exercises 1 to 18 solve each equation for the calculator solution of x in radians.

1. $\tan x = 25$
2. $12 \cos x = -5$
3. $\sin^2 x - \sin x = 0$
4. $\cos x - \cos^2 x = 0$
5. $\sec^2 x - \tan x = 1$
6. $\csc^2 x - \cot x = 1$
7. $\cot x \sin x = \cos^2 x$
8. $\sin^2 x = \cos x + 1$
9. $\tan^3 x - \tan^2 x = 3 \tan x - 3$
10. $\cot^3 x - \cot^2 x = 3 \cot x - 3$

*11. $2 \cos^4 x = 2 - 3 \sin^2 x$

12. $\dfrac{2 \sin^2 x + 3 \cos x}{\cos^2 x} = 0$
13. $\sqrt{\tan x + 3} = 2 \tan x$
14. $\sqrt{\sin x - 1} = 2 \cos x$
15. $\tan x \sin x = \dfrac{3}{2}$
16. $3 \sec x \tan x = 2$
17. $2 \sin x \cos x - \cos x = 0$
18. $2 \sin x \tan x - 2 \sin x + \tan x - 1 = 0$

For Exercises 19 to 26 solve each equation for $0 \le x < 2\pi$.

19. $\sin 2x = 2 \sin x$
20. $\cos 2x = \cos x$
21. $\cos 2x = \cos^2 x$
22. $\sin 2x = \sin^2 x$
23. $\sin 2x - \cos x = 0$
24. $\sin x \sin 2x = \cos x$
25. $\cos 2x = \sin x$
26. $\cos x \cos 2x = \sin 2x \sin x$

For Exercises 27 to 32 solve each equation for the general solutions of x in radians.

27. $\sec x \csc x = \tan x + \cot x$
28. $\sin x \cos x = \cos 2x$
29. $\sin x - \csc x = 1$
30. $\dfrac{\tan x}{\cot x} = 1$
31. $10 \sin x + 2 \csc x = 9$
32. $\cos 2x = 2 \cos x$

For Exercises 33 to 36 graph each pair of functions on one coordinate system and estimate the point(s) of intersection for $0 \le x < 2\pi$. Algebraically, find the exact coordinates of the points.

33. $f(x) = 2 \cos^2 x$
    $g(x) = 1 - \sin x$
34. $f(x) = \cos 2x$
    $g(x) = -\sin^2 x$

*35. $f(x) = \sin 2x$
    $g(x) = 2 \sin x$

36. $f(x) = \sin 3x$
    $g(x) = -\sin x$

Extend the Concepts

For Exercises 37 to 40 solve each equation for the calculator solution.

37. $\cos 4x \cos 1 + \sin 4x \sin 1 = 1$
38. $\sin x^2 \cos x - \cos x^2 \sin x = 0$
39. $\sin (x + y) + \sin (x - y) = \cos y$
40. $\cos (x + y) + \cos (x - y) = \sqrt{3} \cos y$

6.4 Applications of the Circular Functions to Triangles

Applying the Law of Cosines
Applying the Law of Sines
Examining the Ambiguous Case of the Law of Sines
Finding Areas of Triangles

In Section 5.1 we solved problems involving right triangles. In this section we introduce the law of sines and the law of cosines, formulas used to solve any triangle, even ones without a right angle. In this section we use the law of sines and law of cosines in applications involving azimuths, headings, and bearings. Azimuths and headings are two names for the same system of direction that measures angles clockwise from north as shown in Figure 1.

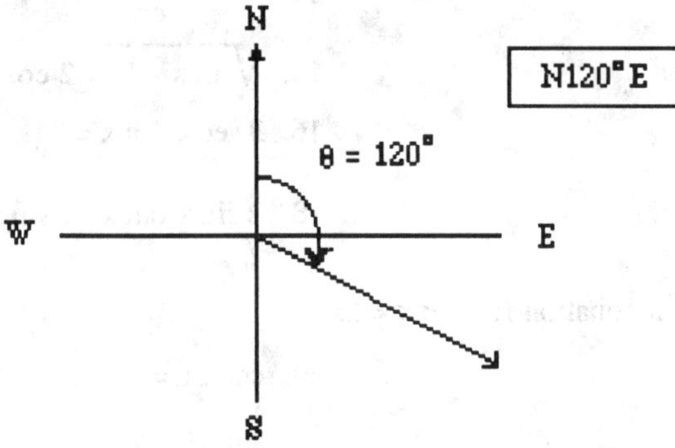

Figure 1

Azimuths and bearings are used in surveying, and headings and bearings are used in navigation. Figure 1 shows an azimuth of 120°, a heading of 120°, and a bearing of N120°E, read "north 120 degrees east." If you were on a ship or in a plane, you could be moving at a heading of 120°. Your destination might be at a bearing of N120°E from your present position. In both navigation and surveying, angles are measured clockwise from north.

Applying the Law of Cosines

We usually use the *law of cosines* to find the measure of any angle of a triangle if the lengths of the three sides are known or to find a side when two sides and an included angle are known: that is, when side-side-side (SSS) or side-angle-side (SAS) are known. One form of the law of cosines is

$$a^2 = b^2 + c^2 - 2bc \cos A$$

The proof of the law of cosines follows from the Pythagorean theorem. Figure 2 shows ΔABC with altitude y perpendicular to side AC.

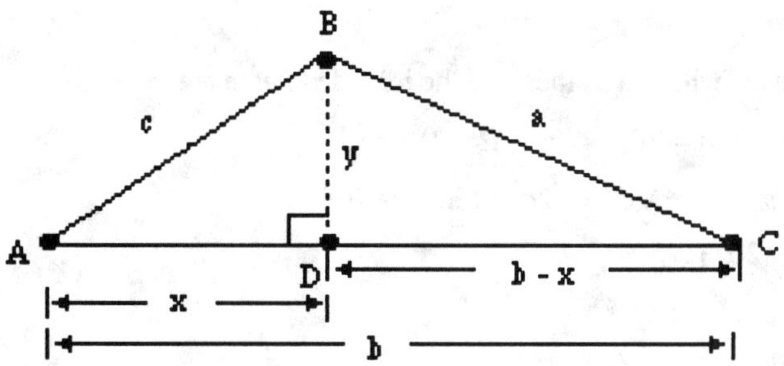

Figure 2

Altitude y divides ΔABC into two right triangles. We begin by finding expressions for x and y.

For ΔABD,

$$\sin A = \frac{y}{c} \Rightarrow y = c \sin A$$

and

$$\cos A = \frac{x}{c} \Rightarrow x = c \cos A$$

Using the Pythagorean theorem with the sides of ΔBDC yields

$$y^2 + (b - x)^2 = a^2 \qquad \textit{Substitute } c \sin A \textit{ for } y \textit{ and } c \cos A \textit{ for } x.$$

$$(c \sin A)^2 + (b - c \cos A)^2 = a^2$$

$$c^2 \sin^2 A + b^2 - 2bc \cos A + c^2 \cos^2 A = a^2$$

$$c^2 \sin^2 A + c^2 \cos^2 A - 2bc \cos A + b^2 = a^2$$

$$c^2(\sin^2 A + \cos^2 A) + b^2 - 2bc \cos A = a^2 \qquad \textit{Substitute } \sin^2 A + \cos^2 A = 1.$$

$$a^2 = b^2 + c^2 - 2bc \cos A$$

NOTE: If $A = 90°$ then $\cos A = 0$ and the law of cosines becomes the Pythagorean theorem, $a^2 = b^2 + c^2 - 2bc \cos 90° \Rightarrow a^2 = b^2 + c^2$.

Law of Cosines

$$a^2 = b^2 + c^2 - 2bc \cos A$$

$$b^2 = a^2 + c^2 - 2ac \cos B$$

$$c^2 = a^2 + b^2 - 2ab \cos C$$

Example 1 *Using the law of cosines*

Find the length of side a if $A = 40°$, $b = 20$, and $c = 50$ (Figure 3).

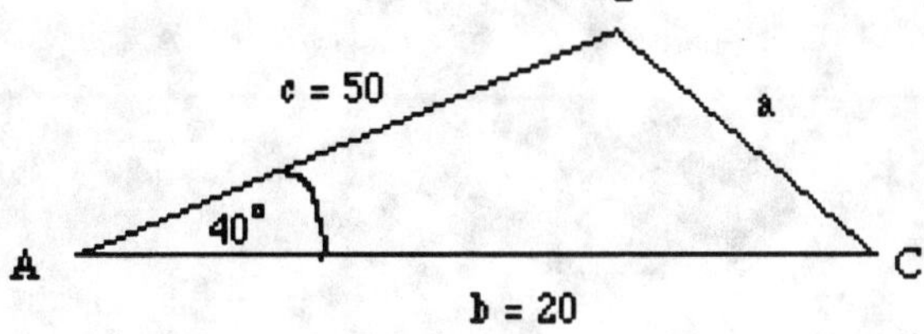

Figure 3

Solution

The law of cosines can be used when two sides and the included angle are given.

$a^2 = b^2 + c^2 - 2bc \cos A$ and $b = 20, c = 50, A = 40° \Rightarrow$

$$a^2 = 20^2 + 50^2 - 2(20)(50) \cos 40°$$

$$a^2 = 1367.9$$

$$a \approx 36.99$$

Example 2 A Whale of a Problem

A large tourist yacht is loaded with whale watchers. A small inflatable boat with an observer is 100 m from the yacht at a heading of 30°. The observer spots a pod of killer whales 500 m ahead at a heading of 110°.

(a) At that moment, how far are the whales from the yacht?

(b) What heading should the yacht take to approach the whales?

Thinking: Creating a Plan

First, we draw a picture and label it with the information given (Figure 4).

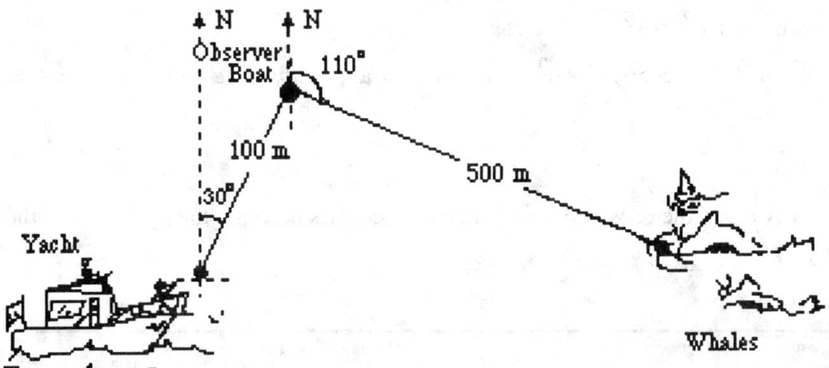

Figure 4

Notice that the two dashed north-south lines are parallel. This means that alternate interior angles are equal, establishing the 30° angle by the inflatable boat (Figure 5). The sum of the two angles on the right of the north-south line by the inflatable boat is 180°, establishing the 70° angle.

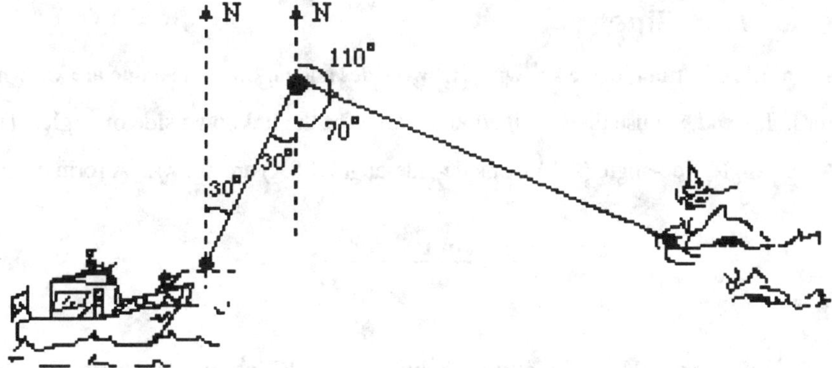

Figure 5

We form a triangle and use the law of cosines to find the heading and distance from the yacht to the whales.

Communicating: Writing the Solution

(a) Figure 6 shows ΔABC with the sides and angles labeled.

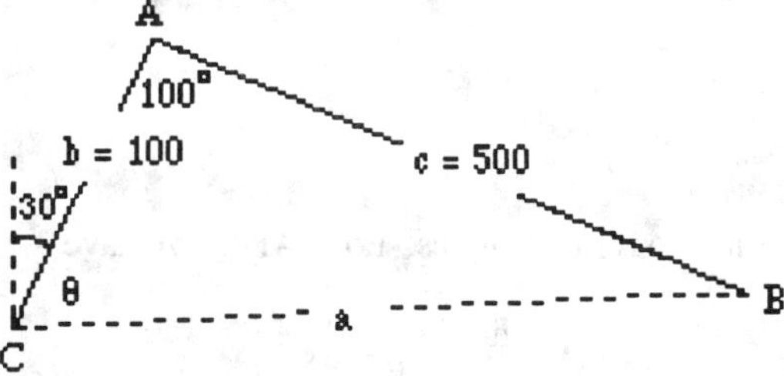

Figure 6

Side a represents the distance from the yacht to the whales.

$a^2 = b^2 + c^2 - 2bc \cos A$ and $b = 100, c = 500, A = 100° \Rightarrow$

$$a^2 = 500^2 + 100^2 - 2(100)(500)(\cos 100°)$$

$$a \approx 526.6$$

The distance to the whales is approximately 526.6 m.

(b) To find the heading, we find first find θ.

$c^2 = a^2 + b^2 - 2ab \cos C$ and $a = 526.6, b = 100, c = 500, C = \theta \Rightarrow$

$$500^2 = 526.6^2 + 100^2 - 2(526.6)(100) \cos \theta$$

$$\cos \theta \approx 0.3542$$

$$\theta = \cos^{-1} 0.3542$$

$$\theta \approx 69.2°$$

Hence, the heading from the yacht to the whales is $30° + \theta = 30° + 69.2° = 99.2°$.

Learning: Making Connections

The bearing to the whales is N99.2°E. If the whales are moving, this calculation may need to be repeated so that the yacht can change its heading.

Applying the Law of Sines

If any two angles and a side of a triangle are known or if two sides and an opposite angle are known, we cannot use the law of cosines. Instead, we use the *law of sines* to solve for the unknown side or angle. That is, when angle-angle-side (AAS), angle-side-angle (ASA), or side-side-angle (SSA) are known. A form of the law of sines is

$$\frac{\sin A}{a} = \frac{\sin B}{b} = \frac{\sin C}{c}$$

To prove $\frac{\sin A}{a} = \frac{\sin C}{c}$, we draw ΔABC with altitude h dividing ΔABC into two right triangles (Figure 7).

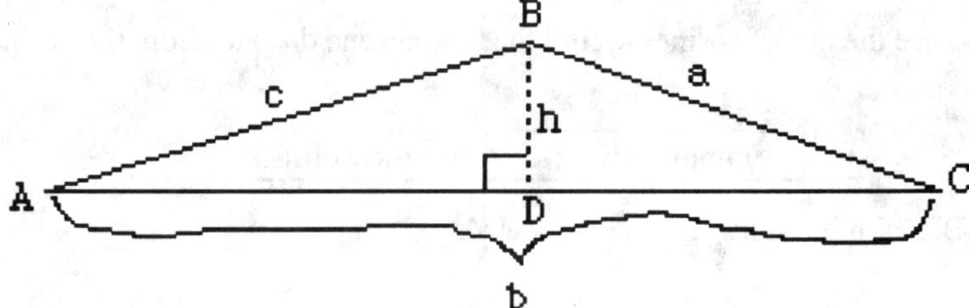

Figure 7

We find an expression for h using the two right triangles. For ΔABD, we have

$$\sin A = \frac{h}{c} \Rightarrow h = c\sin A$$

For Δ BCD we have

$$\sin C = \frac{h}{a} \Rightarrow h = a\sin C$$

$h = c\sin A$ and $h = a\sin C \Rightarrow$

$$c\sin A = a\sin C$$

$$\frac{\sin A}{a} = \frac{\sin C}{c}$$

A similar argument works to show $\frac{\sin A}{a} = \frac{\sin B}{b}$.

Law of Sines

$$\frac{\sin A}{a} = \frac{\sin B}{b} = \frac{\sin C}{c}$$

Any two of the three ratios together form an equation that can be used to find a missing side or angle.

Sometimes the given information lends itself to using either the law of cosines or the law of sines. In the next example we find θ from Example 2 using the law of sines instead of the law of cosines.

Example 3 *Using the law of sines*

Use the law of sines to find θ in Figure 8.

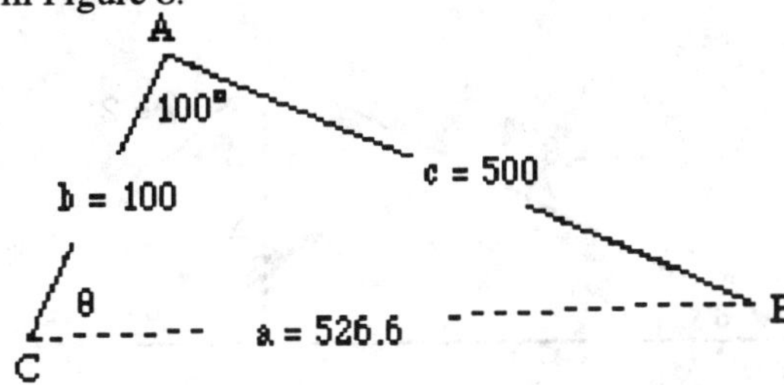

Figure 8

Solution

$\frac{\sin A}{a} = \frac{\sin \theta}{c}$ and $a = 526.6$, $A = 100°$, $c = 500$ ⇒

$$\frac{\sin 100°}{526.6} = \frac{\sin \theta}{500}$$

$$\sin \theta \approx 0.9351$$

$$\sin^{-1} 0.9351 \approx \theta$$

$$\theta \approx 69.2°$$

Examining the Ambiguous Case of the Law of Sines

In a problem where two sides and an angle of a triangle are given, using the law of sines to find a second angle of the triangle may have surprising results. Given two sides and an angle that is not includes may lead to two possible triangles. This is called the **ambiguous case** of the law of sines.

In the next example we use the law of sines and find that no triangle exists for the given information.

Example 4 *Using the law of sines*

Use the law of sines to find angle C if $A = 30°$, $a = 2$, and $c = 8$.

Solution

$\frac{\sin A}{a} = \frac{\sin C}{c}$ and $A = 30°$, $a = 2$, $c = 8$ ⇒

$$\frac{\sin 30°}{2} = \frac{\sin C}{8}$$

$$\sin C = 2$$

$$C = \sin^{-1} 2$$

Angle C does not exist. Figure 9 shows a sketch of the triangle. Side a is drawn across from A. Its length can be compared with the h height of the triangle. Side a is too short since $a < h$.

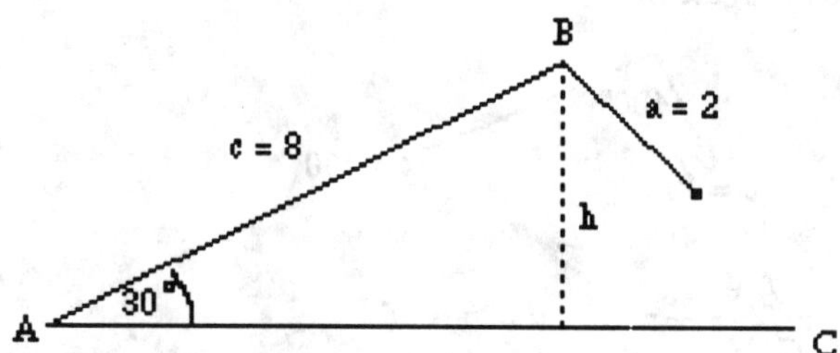

Figure 9

How can we tell in advance whether or not a triangle is possible?

We can determined if a triangle is possible by comparing the length of side a with the height of the triangle h. For the triangle in Figure 9, we have

$\sin A = \frac{h}{c}$ and $A = 30^\circ, c = 8 \Rightarrow$

$$\sin 30^\circ = \frac{h}{8}$$
$$h = 4$$

This result means that side a must be at least 4 units long. Since a = 2 in Example 4, the triangle is not possible.

What if the length of side a was more than 4 units?

If $a > 4$, there would have been two solutions. The next example shows how two different triangles solve a problem. Recall from geometry that SSA is <u>not</u> sufficient information to prove that two triangles are congruent. Congruent triangles have the same size and shape. Figure 10 shows two triangles that have two equal sides and an equal angle but are not the same size or shape. They are not congruent.

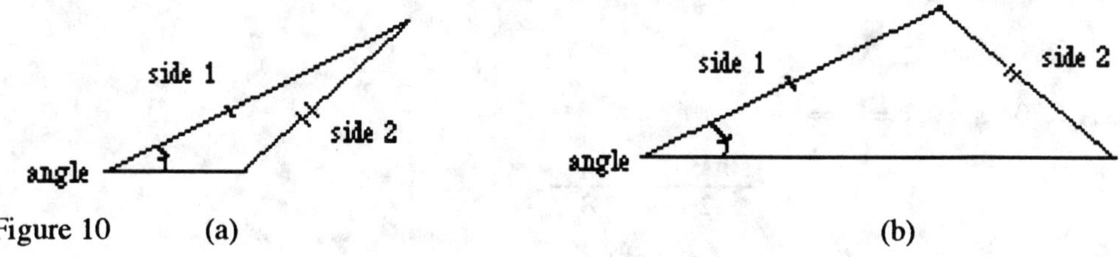

Figure 10 (a) (b)

It is possible to have two different triangles that have two equal sides and one equal angle.

Example 5 *Using the law of sines*

Use the law of sines to find the missing parts of a triangle if A = 30°, a = 6, and c = 8 (Figure 11).

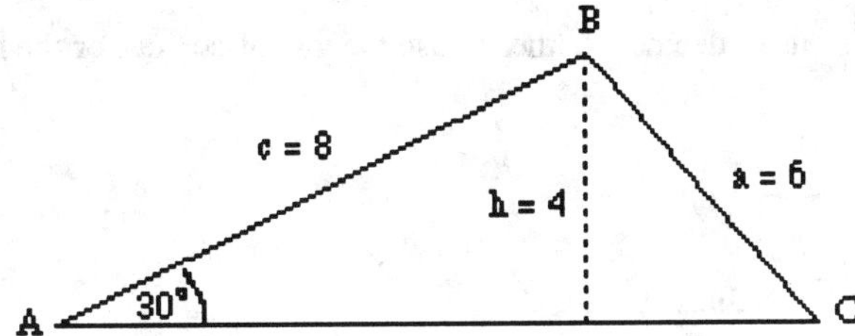

Figure 11

Solution

$\frac{\sin A}{a} = \frac{\sin C}{c}$ and A = 30°, a = 6, c = 8 ⇒

$$\frac{\sin 30°}{6} = \frac{\sin C}{8}$$
$$\frac{0.5}{6} = \frac{\sin C}{8}$$
$$6 \sin C = 4$$
$$\sin^{-1}\frac{2}{3} = C$$

There are two angles whose sine is 2/3.

$$C \approx 41.8° \text{ or } C \approx 180° - 41.8° = 138.2°$$

B = 180° - (A + C) and C ≈ 41.8° ⇒ B ≈ 108.2°

$\frac{\sin A}{a} = \frac{\sin B}{b}$ and A = 30°, a = 6, B = 108.2° ⇒

$$\frac{\sin 30°}{6} = \frac{\sin 108.2°}{b}$$

b ≈ 11.40

B = 180° - (A + C) and C ≈ 138.2° ⇒ B ≈ 11.8°

$\frac{\sin A}{a} = \frac{\sin B}{b}$ and A = 30°, a = 6, B = 11.8° ⇒

$$\frac{\sin 30°}{6} = \frac{\sin 11.8°}{b}$$

b ≈ 2.45

Figure 12 shows the two possible triangles.

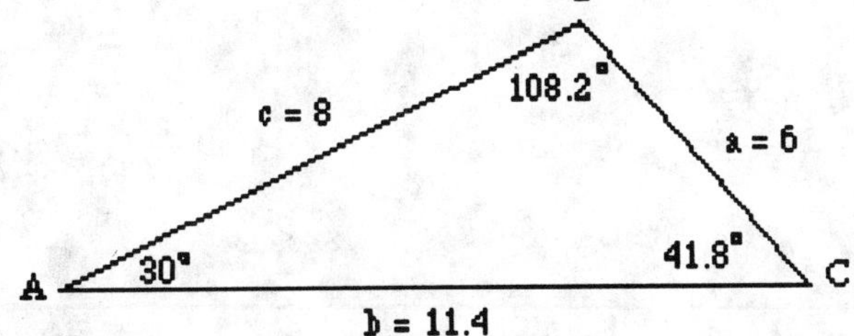

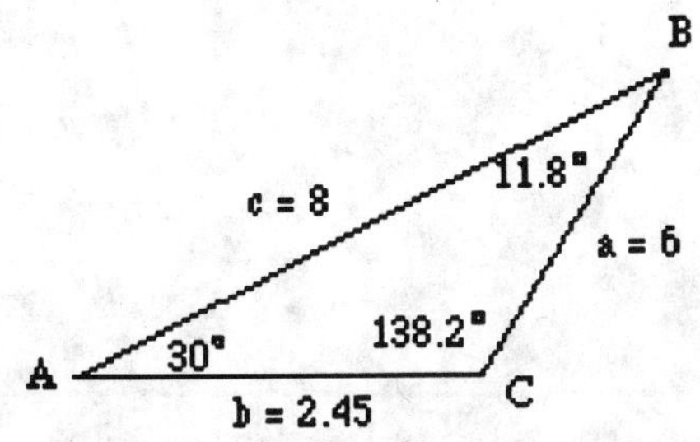

Figure 12

When the given information is ambiguous, such as in Example 5, the complete solution consists of all the measurements for both triangles. In an application, often we have enough other information to select the appropriate triangle.

The next example requires us to decide whether to use the law of cosines or the law of sines.

Example 6 The Railroad Crossing Problem

A proposed highway will have a bearing of S15°W from town. The highway will cross an existing railroad track. The bearing of the track is N70°E (Figure 13). From an arbitrary point on the tracks 2 mi from town, the town is at a bearing of N30°E. The surveying crew must plant a stake at the intersection of the railroad tracks and the proposed highway. Where should they put the stake?

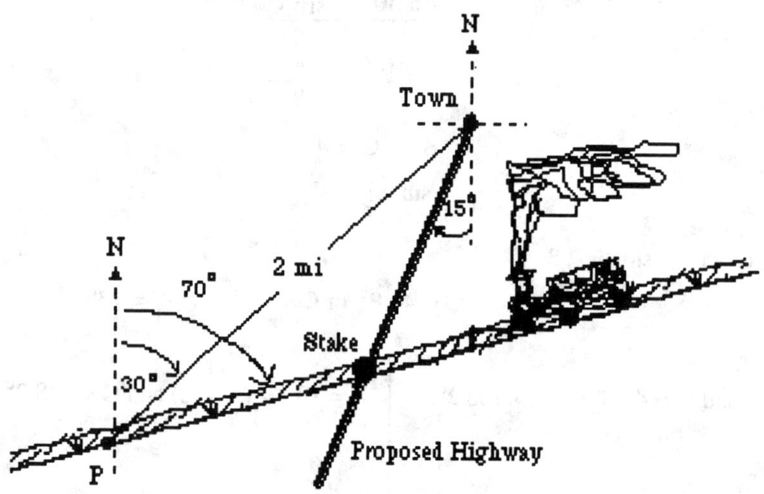

Figure 13

Thinking: Creating a Plan

The two dashed north-south lines are parallel making the alternate interior angles equal establishing the 30° angle by town (Figure 14). The difference between 70° and 30° creates the 40° angle by point P on the track.

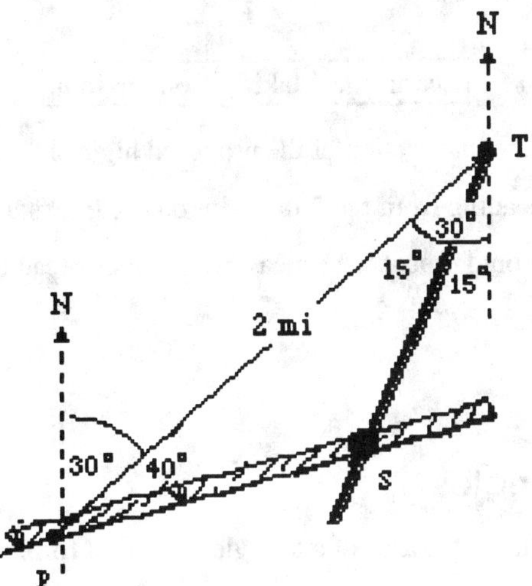

Figure 14

Since the sum of the angles of a triangle is 180°, ∠S = 180° - (40° + 15°) = 125°. We want to find the distance from point P to point S (Figure 15).

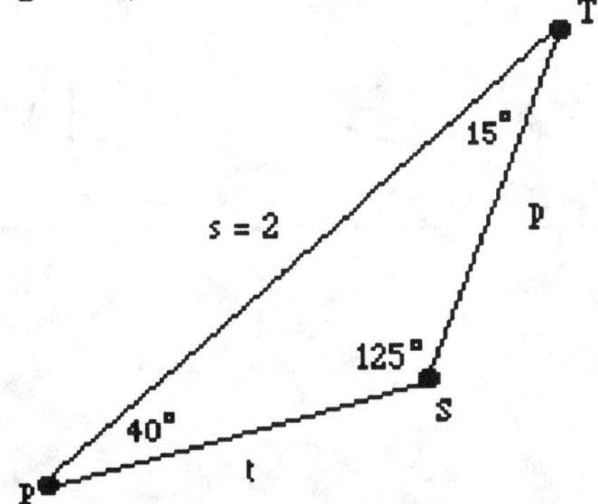

Figure 15

Communicating: Writing the Solution

$\frac{\sin P}{p} = \frac{\sin S}{s}$ and P = 40°, S = 125°, s = 2 ⇒

$$\frac{\sin 40°}{p} = \frac{\sin 125°}{2}$$

$$p \approx 1.57$$

$\frac{\sin T}{t} = \frac{\sin S}{s}$ and T = 15°, S = 125°, s = 2 ⇒

$$\frac{\sin 15°}{t} = \frac{\sin 125°}{2}$$

$$t \approx 0.63$$

Measuring from town, the stake should be set 1.57 mi at a bearing of S15°W. From the 2-mi point on the tracks it should be set 0.63 mi at a bearing of N70°E.

Learning: Making Connections

The surveying crew can now find the intersection of the proposed highway and the tracks by measuring from town. They can use the relative heading from the 2-mi point on the tracks to check their position. For increased accuracy, the distances would probably be measured in feet instead of miles.

Finding Areas of Triangles

Next we introduce formulas for finding the area of a triangle. The first formula

$$A = \frac{1}{2}ab \sin C$$

is used when the information given includes the lengths of two sides and the included angle. In ΔABC, the height h is drawn perpendicular to side AC shown in Figure 16. Then the area formula $A = \frac{1}{2}bh$ is applied.

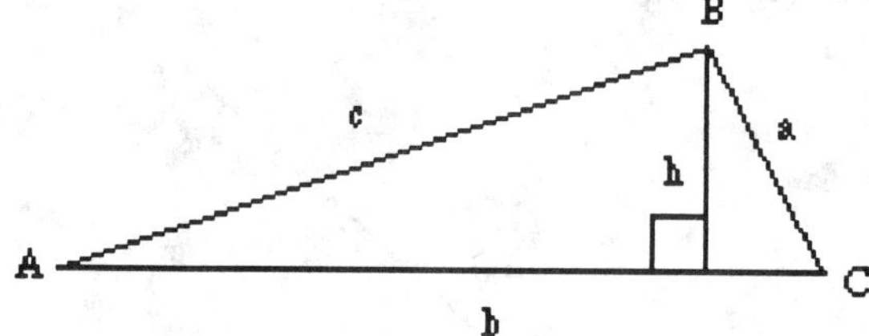

Figure 16

From Figure 16 we have

$$\sin C = \frac{h}{a} \Rightarrow h = a \sin C$$

Substituting h = a sin C into the formula for area of a triangle, $A = \frac{1}{2}bh$ yields our new formula.

$$A = \frac{1}{2}b\,(a \sin C) \Rightarrow A = \frac{1}{2}ab \sin C$$

The second formula is known as *Heron's formula*..

$$A = \sqrt{s(s - a)(s - b)(s - c)} \text{ where } s = \frac{1}{2}(a + b + c)$$

It is used when the given information includes the length of the three sides, a, b, and c. The value of s is half the perimeter of ΔABC. Its proof uses the law of cosines.

Formulas: Area of a Triangle

$$A = \frac{1}{2}ab \sin C$$

$$A = \sqrt{s(s - a)(s - b)(s - c)} \text{ where } s = \frac{1}{2}(a + b + c)$$

Exercises 6.4

Develop the Concepts

For Exercises 1 to 4 use the law of cosines to find the measure of the missing side.

1. $A = 40°$, $C = 24°$, $c = 31.6$, $a = 50$
2. $A = 65.5°$, $B = 25°$, $a = 28$, $c = 20$
3. $A = 25°$, $C = 115°$, $a = 13$, $b = 20$
4. $B = C = 20°$, $b = c = 40$

For Exercises 5 to 8 use the law of sines to find the measure of the missing side.

5. $A = 10°$, $a = 50$, $b = 287$
6. $C = 140°$, $a = 197$, $c = 288$
7. $A = 15°$, $b = 200$, $a = 151$
8. $A = 20°$, $a = 200$, $b = 335$

For Exercises 9 to 16 use the laws of sine and/or cosines to find all the unknown parts of the triangle. If no triangle is possible, write "no solution." If two triangles are possible, give both solutions.

9. $A = 40°$, $b = 28$, $c = 13$
10. $B = 120°$, $a = c = 100$
11. $A = 10°$, $a = 100$, $b = 288$
12. $A = 15°$, $b = 200$, $a = 51$
13. $A = 30°$, $a = 40$, $c = 100$
14. $A = 30°$, $a = 60$, $c = 100$
15. $B = 50°$, $b = 110$, $c = 140$
16. $C = 50°$, $a = 140$, $c = 100$

For Exercises 17 to 20 use the appropriate formula to find the area of each triangle.

17. ΔABC, where $B = \pi/6$, $a = 4$, and $c = 10$.
18. ΔXYZ, where $x = 1$, $y = 2$, $X = \pi/3$, and $Y = \pi/5$.
19. ΔABC, where $a = 4$, $b = 6$, and $c = 8$.
20. ΔABC, where $a = 12$, $b = 6\sqrt{3}$, and $c = 6$.

Apply the Concepts

21. The Trisected Angle Problem Equilateral triangle ACF has sides 30 in. long. If angle A is trisected (divided into three equal angles), find the lengths of CD, DE, and EF in Figure 17.

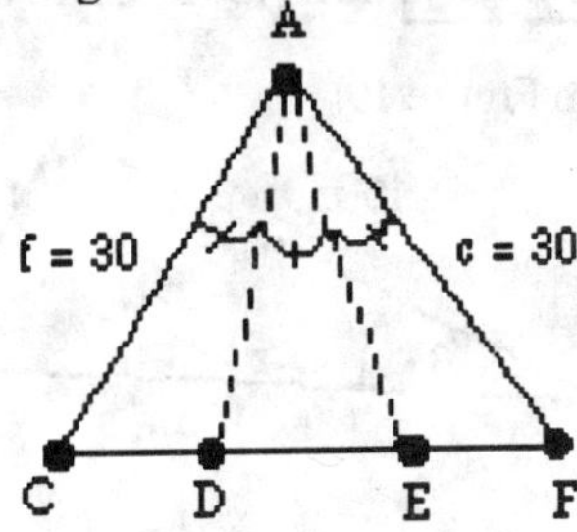

Figure 17

22. The Bisected-Angle Problem Figure 18 shows ΔABC with AD bisecting angle A (angle A is divided into 2 equal angles). Show that AD divides side CB into segments proportional to the two adjacent sides of ΔABC. That is, show that $\frac{f}{g} = \frac{b}{c}$.

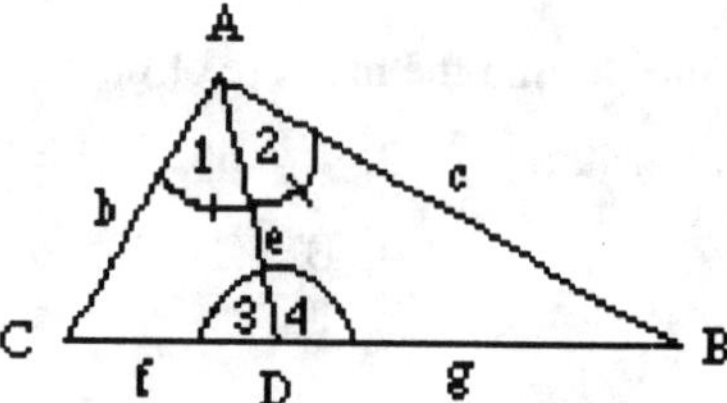

Figure 18

*23. The Inscribed Octagon Problem Find the area of a regular octagon (eight-sided polygon) inscribed in a circle of radius 10 cm.

24. The Circumscribed Octagon Problem Find the area of a regular octagon circumscribed around a circle of radius 10 cm.

*25. The Racing-Sloop Problem From atop two lighthouses situated 100 nautical miles apart, race officials spot a sloop at noon. From lighthouse A its bearing is N50°W, and from lighthouse B its bearing is N30°E. At 2:00 p.m. the new bearing from A is N70°W and from B is N10°W. Find the speed of the racing sloop if lighthouse A is due east of lighthouse B.

26. The Air Controller Problem An air controller notices two planes flying at the same altitude on a collision course with each other. From the control tower, one plane is located 18 mi away at a bearing of N30°W, flying at 300 mi/hr. The other plane is 24 mi from the tower at a bearing of N45°W, traveling at 450 mi/hr.

(a) How far apart are the two planes?

(b) How much time do the pilots have to change course to avoid collision?

27. The Rainbow Kite and Balloon Company Find the measure of angle A and side a that will yield the maximum surface area of the wedge kite in Figure 19.

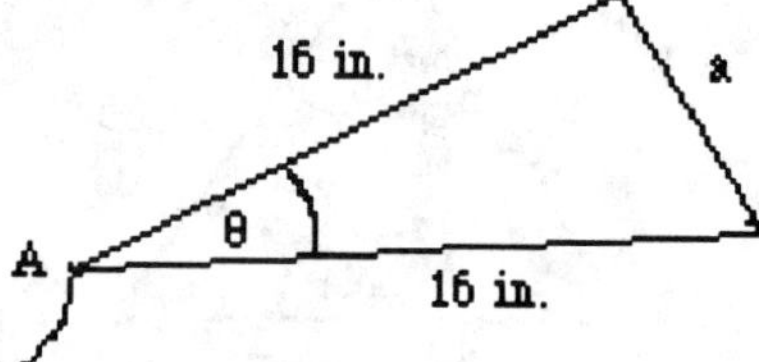

Figure 19

28. Wonderworld Amusement Park A bridge is to be constructed at the amusement park connecting Pearl Isle to Black Diamond Cafe. Surveyors have taken the measurements as shown in Figure 20. How long is the intended bridge?

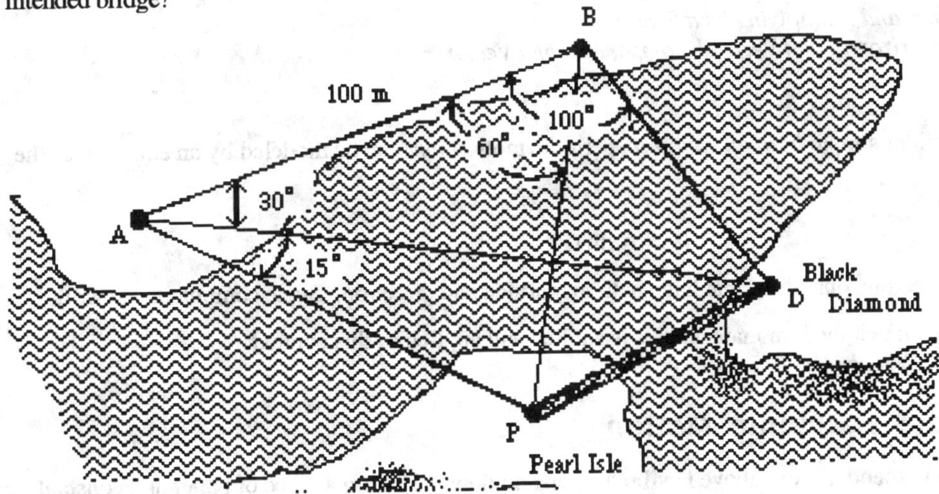

Figure 20

6.5 More Applications: Simple Harmonic Motion and Vectors

Applying Simple Harmonic Motion
Defining a Vector
Adding Vectors and Multiplying by a Scalar
Finding the Horizontal and Vertical Components of a Vector

In this section we consider simple *harmonic motion*, periodic motion that can be modeled by an equation of the form

$$y = A \sin (Bt + C)$$

We also introduce vectors, quantities that have both magnitude and direction. Since computers can work with ease using vectors, much modern mathematics is studied and applied in this context.

Applying Simple Harmonic Motion

Suppose that a weight suspended from above is vibrating up and down. Pulling a piece of paper at a constant rate past a marker on the weight records the path of the weight (Figure 1). This is simple harmonic motion.

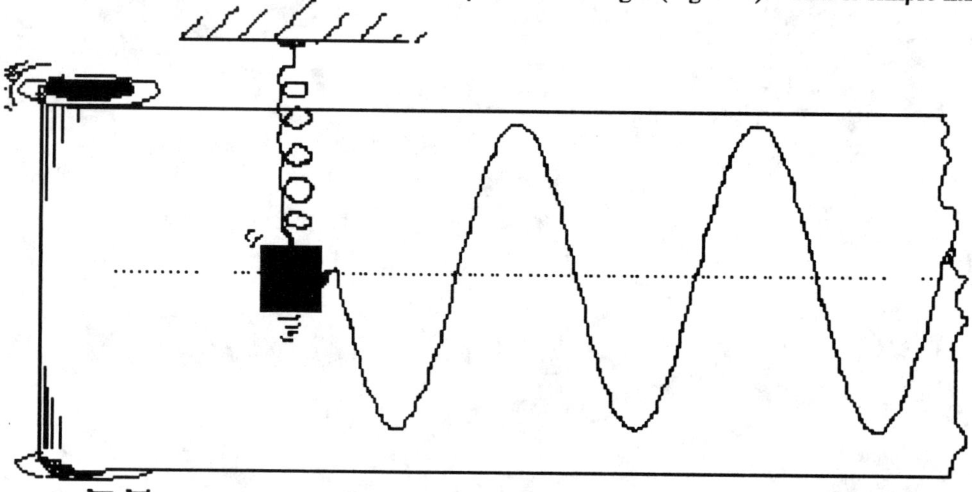

Figure 1

If there were no friction, the weight would repeat one cycle periodically (down, up, and back to its starting position).

Frequency is the number of cycles/unit time. In this case, frequency is the number of vibrations/sec (vps) measured in hertz, where 1 Hz = 1 vps. Assume that for some weight and a particular spring, the equation for its vertical motion is $y = 4 \sin 2t$, where y is measured in centimeters and t in seconds. Since the amplitude of the function is 4, the weight travels 8 cm vertically. Since B = 2, the period t is

$$t = \frac{2\pi}{B}$$

$$t = \frac{2\pi}{2} = \pi \approx 3$$

It takes about 3 sec for the spring to move down and back up. The frequency f of vibration is the reciprocal of the period,

$$f = \frac{1}{t}$$

$$f = \frac{1}{\pi} \approx \frac{1}{3}$$

This means that the spring completes about one-third of a complete cycle each second.

Next, let's consider harmonic motion modeled by the sum of two sine functions whose periods are close to each other, such as

$$f(x) = \sin 2x$$

with period π and

$$g(x) = \sin 2.5x$$

with period 0.8π (Figure 2).

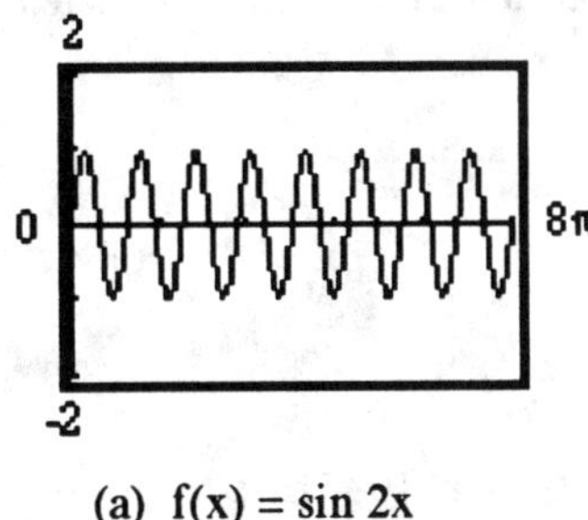

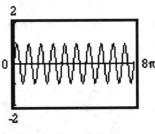

Figure 2 (a) $f(x) = \sin 2x$ (b) $g(x) = \sin 2.5x$

Do you think the sum of $y = f(x) + g(x)$ is also the graph of a sine function?

The sum is not a sine function (Figure 3).

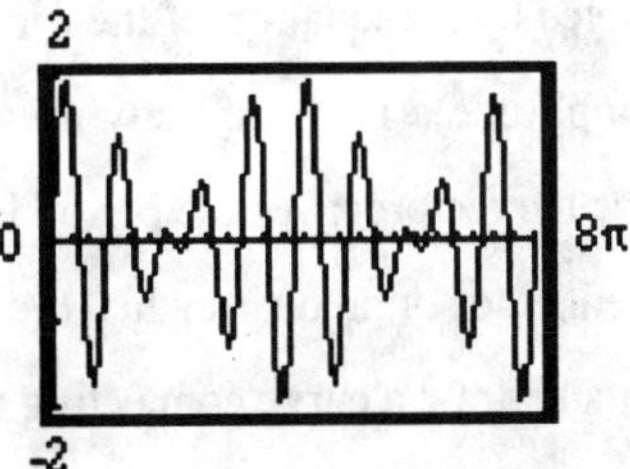

Figure 3

However, there are periodic swells of amplitude. Connecting the peaks and valleys of the sum function results is two sine curves (Figure 4).

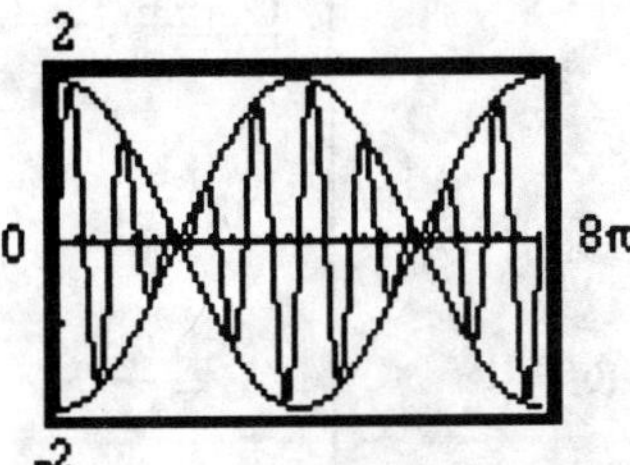

Figure 4

In music, the sum results in the beat *phenomenon*. The changing amplitude results in swells of loudness that repeat periodically. Suppose that this sum function represents the result of a guitar sounding middle A slightly out of tune with an in-tune piano sounding middle A. Periodically, their sounds are *out of phase*, canceling each other; then *in phase*, reinforcing each other (Figure 5).

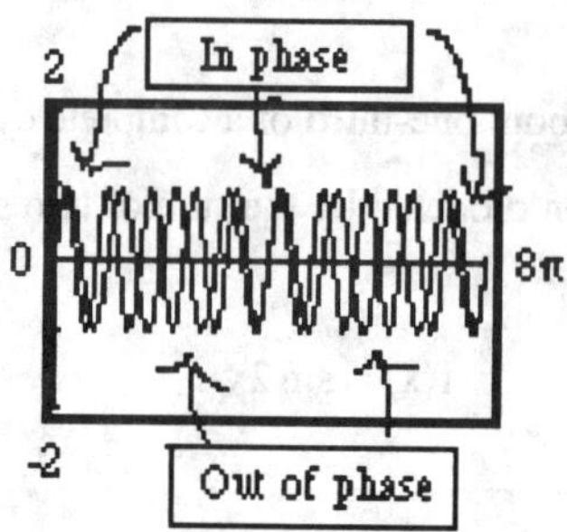

Figure 5

The guitarist tunes the guitar by adjusting the strings' tensions until the beat is eliminated. The next example models how musicians tune their instruments.

Example 1 The Beating Guitar Problem

Suppose that a piano has its E string below middle C tuned to vibrate at 164 Hz. The guitarist plays the same note at 160 Hz. Estimate the period and frequency of the beat.

Thinking: Creating a Plan

We will assume that each sound is modeled by an equation of the form $y = A \sin Bt$, and that both instruments are being played at the same volume (amplitude). For simplicity we will let $A = 1$. We find the period of each function from the given frequencies since the reciprocal of the period is the frequency.

The sum of the two functions represents the sound heard when the two instruments are played simultaneously. We will graph the sum and draw a curve connecting the peaks and valleys. We can approximate the period and frequency of the beat from the graph of the sum function.

Communicating: Writing the Solution

Piano: $f = 1/t$ and $f = 164 \Rightarrow$

$$164 = 1/t$$
$$t = \frac{1}{164}$$

$$B = \frac{2\pi}{t} \text{ and } t = \frac{1}{164} \Rightarrow B = \frac{2\pi}{1/164} = 328\pi$$

$y = A \sin Bt$ and $A = 1, B = 328\pi \Rightarrow$

$$y = \sin 328\pi t$$

Guitar: $f = 1/t$ and $f = 160 \Rightarrow$

$$160 = 1/t$$
$$t = \frac{1}{160}$$

$$B = \frac{2\pi}{t} \text{ and } t = \frac{1}{160} \Rightarrow B = \frac{2\pi}{1/160} = 320\pi$$

$y = A \sin Bt$ and $A = 1, B = 320\pi \Rightarrow$

$$y = \sin 320\pi t$$

The graph of their sum, $y = \sin 328\pi t + \sin 320\pi t$, is shown in Figure 6 with a curve that represents the beat created by connecting the peaks and valleys.

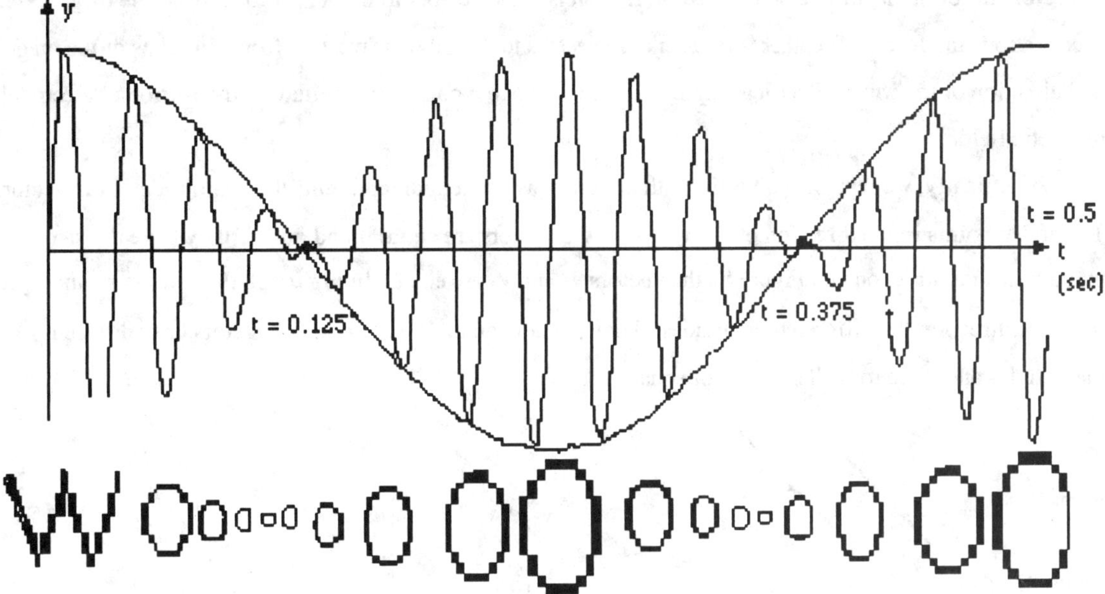

Figure 6

The period of the beat function appears to be 0.5 sec, making the frequency $f = 1/t = 2$; the beat repeats 2 times/sec.

Learning: Making Connections

Figure 7 shows the individual sine waves, $y = \sin 328\pi t$ and $y = \sin 320\pi t$, that produced the beat.

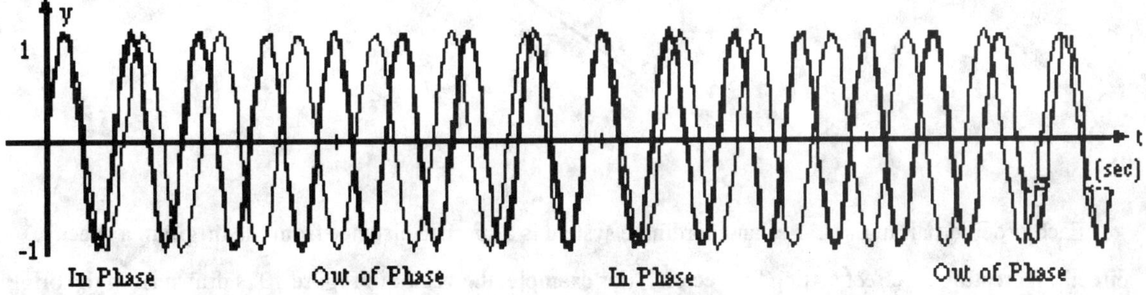

Figure 7

In calculus and physics we rewrite the sum as a product,

$$y = \sin 328\pi t + \sin 320\pi t = 2\cos 4\pi t\,(\sin 324\pi t)$$

This can be thought of as a sine wave having a varying amplitude. The amplitude is the function $2\cos 4\pi t$.

Defining a Vector

A **vector** has both magnitude and direction. Physicists use vectors to model position from the origin, velocity, acceleration, and forces. Weather forecasters use vectors to describe the weather fronts in terms of the magnitude and direction of the wind. Electrical engineers use vectors to model the magnitude and direction of electrical and magnetic fields.

Geometrically, we represent a vector with an arrow whose length represents the magnitude of the vector. The arrow points in the direction of the vector. Two vectors are **equal** if and only if they have the same magnitude and direction. In Figure 8a the vectors $\mathbf{v}_1$ and $\mathbf{v}_2$ are equal. In Figure 8b the vectors $\mathbf{v}_1$ and $\mathbf{v}_2$ have the same direction but different magnitudes. They are not equal. In Figure 8c the vectors have different magnitudes and directions. They are not equal.

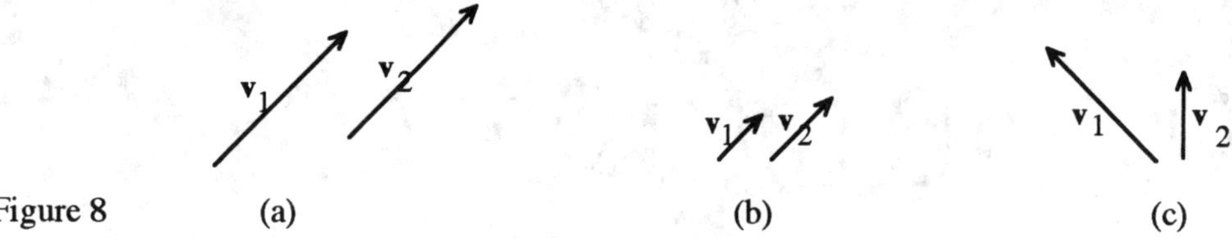

Figure 8 (a) (b) (c)

We represent a vector in print using a boldface letter, **v**. By hand we draw an arrow over the letter used to name the vector, $\vec{v}$. The magnitude of a vector is denoted $\|\mathbf{v}\|$ in print and $\|\vec{v}\|$ by hand.

We often represent vectors in terms of two points, its initial point (the point at its tail) and its terminal point (the point at its tip). Figure 9 shows the vector from point P to point Q,

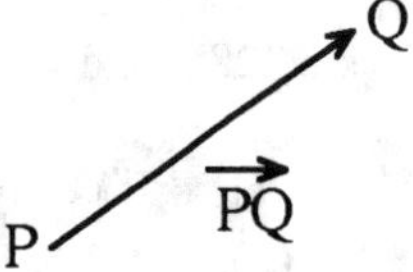

Figure 9

Each ordered pair in the Cartesian coordinate system is a specific distance from the origin in a specific direction, creating a perfect setting for vectors. For example, the vector in Figure 10 is drawn from the origin to the ordered pair (3, 4). We name this vector with the coordinates of the point at its tip, $\mathbf{v} = (3, 4)$. The magnitude of this vector, $\|\mathbf{v}\|$, is the length of the arrow.

Figure 10 (a) (b)

We find the magnitude of **v** = (3, 4) using the Pythagorean theorem since the length of the vector is the hypotenuse of a right triangle whose legs measure 3 and 4.

$$\|\mathbf{v}\| = \sqrt{3^2 + 4^2} = \sqrt{25} = 5$$

The direction of v is found using $\tan\theta = \frac{b}{a}$ with a = 3 and b = 4.

$$\tan\theta = \frac{4}{3}$$

$$\tan^{-1}\frac{4}{3} = \theta$$

$$\theta \approx 53.13°$$

Definitions of Vector, Magnitude, and Direction

A vector **v** has both magnitude and direction. If **v** = (a, b), its magnitude is

$$\|\mathbf{v}\| = \sqrt{a^2 + b^2}$$

and its direction is

$$\theta = \tan^{-1}\left(\frac{b}{a}\right),\ a \neq 0$$

y
(a,b)
b
v = (a,b)
θ
a
x

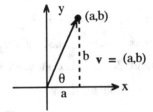

Since the inverse tangent is not a function, there are many choices for θ. We choose one value $0 \le \theta < 2\pi$ $(0° \le \theta < 360°)$ consistent with the quadrant in which the tip of the vector lies.

Example 2 *Finding the length and direction of a vector*

Find the magnitude and direction of (a) $\mathbf{v}_1 = (6, 8)$ and (b) $\mathbf{v}_2 = (-6, 8)$.

Solution

(a) $\|\mathbf{v}\| = \sqrt{a^2 + b^2}$ and a = 6, b = 8 ⇒

$$\|\mathbf{v}_1\| = \sqrt{6^2 + 8^2} = 10$$

$\theta = \tan^{-1}\frac{b}{a}$ and a = 6, b = 8 ⇒

$$\theta_1 = \tan^{-1}\frac{8}{6}$$

$$\theta_1 \approx 0.93$$

$$\theta_1 \approx 53.1°$$

(b) $\|\mathbf{v}\| = \sqrt{a^2 + b^2}$ and a = -6, b = 8 ⇒

$$\|\mathbf{v}_2\| = \sqrt{6^2 + (-8)^2} = 10$$

$\theta = \tan^{-1}\frac{b}{a}$ and a = -6, b = 8 ⇒

$$\theta_2 = \tan^{-1}\left(-\frac{8}{6}\right)$$

$$\theta_2 \approx -0.927 + \pi$$

From Figure 11, we see the tip of $\mathbf{v}_2$ in quadrant II; therefore,

$$\theta_2 \approx 2.21$$

$$\theta_2 \approx 126.9°$$

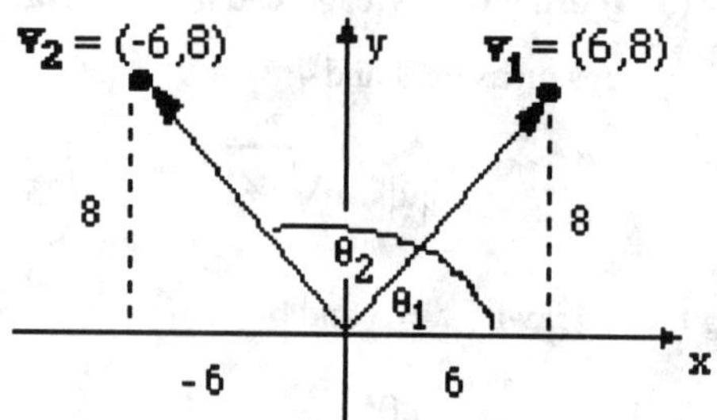

Figure 11

Adding Vectors and Multiplying by a Scalar

The **zero vector O** = $\vec{O}$ is a vector having zero magnitude and therefore no specific direction. All other vectors have magnitude and direction. If we multiply a vector by some positive real number, we change its magnitude. In this context we refer to such real numbers as **scalars**. For example, multiplying $\mathbf{v}_1$ by the scalar 2 doubles its magnitude (Figure 12a). Multiplying $\mathbf{v}_2$ by -2 not only doubles the magnitude but also reverses its direction (Figure 12b).

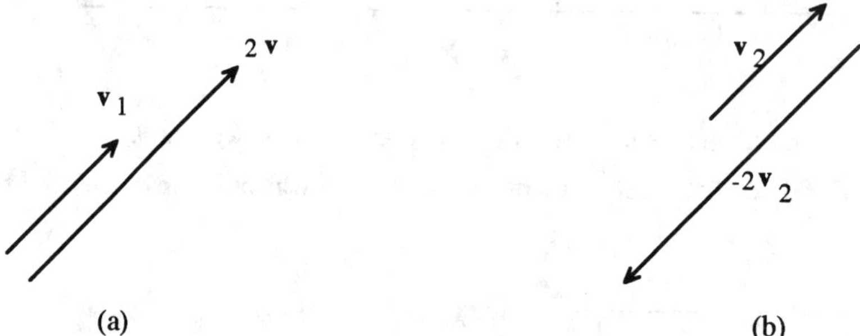

Figure 12 (a) (b)

The **sum** of two vectors is another vector, called the **resultant**. We can find the resultant vector of $\mathbf{v}_1 + \mathbf{v}_2$ using the **parallelogram law** The sum of two vectors is the diagonal vector in the parallelogram formed by those two vectors and two equal vectors drawn at their tips (Figure 13a).

We can also represent the resultant vector using the **tail-to-tip method**. We draw $\mathbf{v}_2$ from the tip of $\mathbf{v}_1$. Then **r** is drawn from the tail of $\mathbf{v}_1$ to the tip of $\mathbf{v}_2$ (Figure 13b). The tail-to-tip method works well when more than two vectors are to be added.

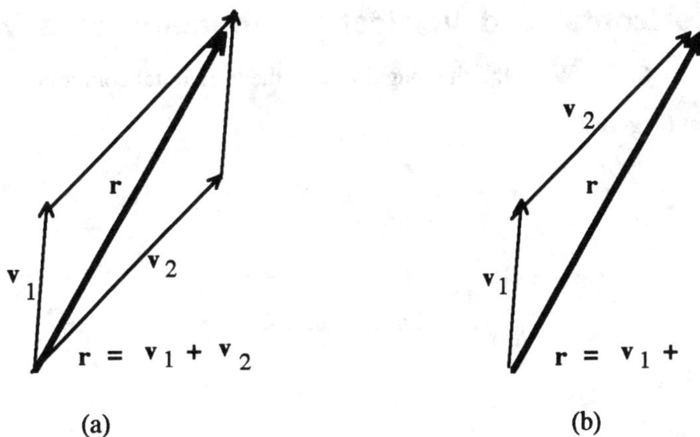

Figure 13 (a) (b)

The sum of two vectors is found by adding their corresponding components, as stated in the next definition.

Definition of the Sum of Two Vectors

If $\mathbf{v}_1$ and $\mathbf{v}_2$ are vectors such that $\mathbf{v}_1 = (a_1, b_1)$ and $\mathbf{v}_2 = (a_2, b_2)$, their sum is

$$\mathbf{v}_1 + \mathbf{v}_2 = (a_1, b_1) + (a_2, b_2) = (a_1 + a_2, b_1 + b_2)$$

Definition of Multiplication by a Scalar

If $\mathbf{v}_1$ is a vector such that $\mathbf{v}_1 = (a_1, b_1)$ and c is a real number, then

$$c\mathbf{v}_1 = (ca_1, cb_1)$$

Example 3 *Finding the sum of two vectors*

(a) Find **r**, the sum of $\mathbf{v}_1 = (5, 1)$ and $\mathbf{v}_2 = (2, 6)$.

(b) Sketch $\mathbf{v}_1$, $\mathbf{v}_2$, and $\mathbf{r} = \mathbf{v}_1 + \mathbf{v}_2$ using the parallelogram law.

Solution

(a)
$$\mathbf{r} = \mathbf{v}_1 + \mathbf{v}_2 = (5, 1) + (2, 6) = (7, 7)$$

(b) Notice that $\mathbf{r} = (7, 7)$ is the diagonal of the parallelogram (Figure 14).

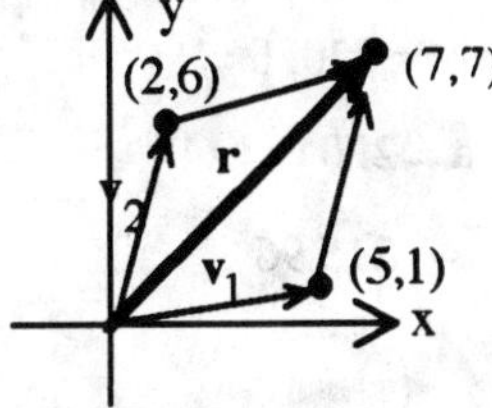

Figure 14

Finding the Horizontal and Vertical Components of a Vector

Figure 15 shows that $\mathbf{v} = (a, b)$. We call a the magnitude of the horizontal component and b the magnitude of the vertical component (Figure 15).

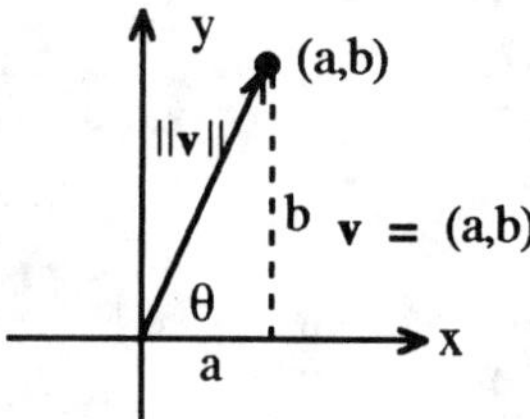

Figure 15

If we know the magnitude and direction of vector **v**, *how do we find its horizontal and vertical components?*

From Figure 15 we have

$$\sin\theta = \frac{b}{\|\mathbf{v}\|} \Rightarrow b = \|\mathbf{v}\|\sin\theta$$

$$\cos\theta = \frac{a}{\|\mathbf{v}\|} \Rightarrow a = \|\mathbf{v}\|\cos\theta$$

In the next example we find the horizontal and vertical components of a velocity vector representing a golf ball hit at an angle of 11° with an initial velocity of 220 ft/sec. In Section 7.3 we use these components to model the flight.

Example 4 *Finding the horizontal and components of a vector*

Find the horizontal and vertical components of a vector whose magnitude is 220 at an angle of 11° with the horizontal (Figure 16).

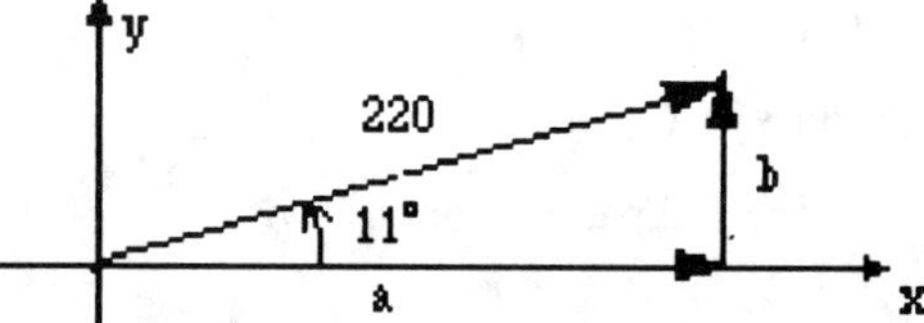

Figure 16

Solution

Horizontal component: $a = \|\mathbf{v}\|\cos\theta$ and $\|\mathbf{v}\| = 220,\ \theta = 11° \Rightarrow$

$$a = 220\cos 11°$$

$$a \approx 215.96$$

Vertical component:

$b = \|\mathbf{v}\|\sin\theta$ and $\|\mathbf{v}\| = 220,\ \theta = 11° \Rightarrow$

$$b = 220\sin 11°$$

$$b \approx 41.98$$

In the next example we have two dogs pulling on a rug in different directions with different forces. The resultant force is the force that would cause the same reaction as if a single bigger dog were pulling on the rug in the direction indicated.

Example 5 The Dog Fight Problem

Two dogs, little Fouffie and big Samos, are pulling on a rug with respective forces of 10 and 20 lb in the directions 53° and -60° (Figure 17). Find the direction and magnitude of the resultant force.

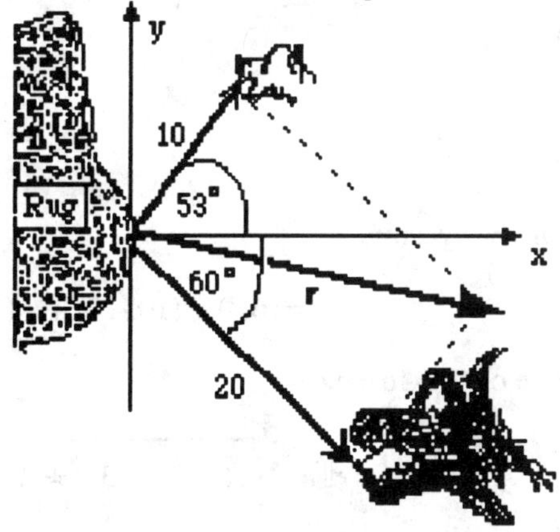

Figure 17

Thinking: Creating a Plan

The vector representing the force of the small dog has a magnitude of 10 and the other has a magnitude of 20. Using the magnitude and direction of the two vectors, we can find the horizontal and vertical components. Then we can find the sum of the two vectors, the resultant vector.

Communicating: Writing the Solution

Figure 18 shows the vector representing the force of the small dog.

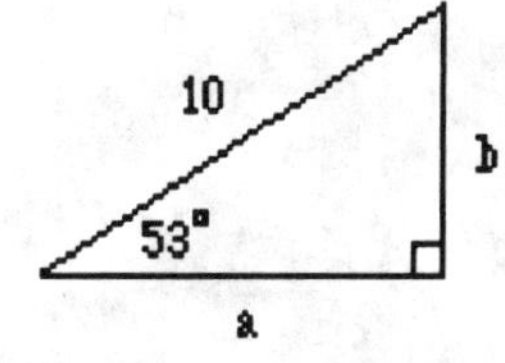

Figure 18

Horizontal component:

$a = \|\mathbf{v}\| \cos \theta$ and $\|\mathbf{v}\| = 10, \theta = 53° \Rightarrow$

$$a = 10 \cos 53°$$

$$a \approx 6.0$$

Vertical component:

$b = \|\mathbf{v}\| \sin \theta$ and $\|\mathbf{v}\| = 10, \theta = 53° \Rightarrow$

$$b = 10 \sin 53°$$

$$b \approx 8.0$$

Therefore, $\mathbf{v}_1 = (6, 8)$.

Figure 19 shows the vector representing the force of the big dog.

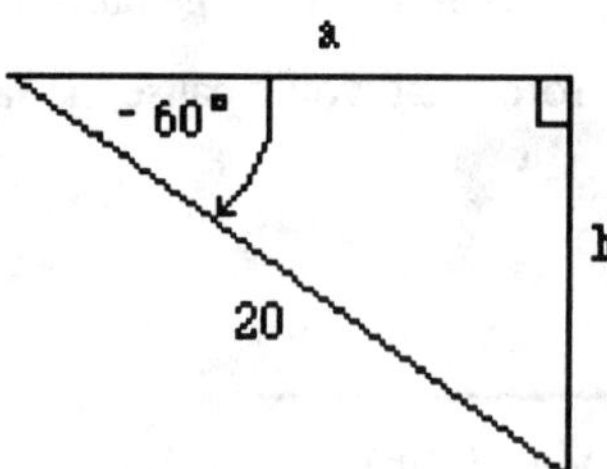

Figure 19

Horizontal component:

$a = \|\mathbf{v}\| \cos\theta$ and $\|\mathbf{v}\| = 20,\ \theta = -60° \Rightarrow$

$a = 20 \cos(-60°)$

$a = 10$

Vertical component:

$b = \|\mathbf{v}\| \sin\theta$ and $\|\mathbf{v}\| = 20,\ \theta = -60° \Rightarrow$

$b = 10 \sin(-60°)$

$b \approx -17.3$

Therefore, $\mathbf{v_2} = (10, -17.3)$.

The resultant vector is the sum of the two vectors

$$\mathbf{r} = (6,8) + (10,-17.3) = (16, -9.3)$$

The resultant force has the magnitude of 18.5 lb, since

$$\|\mathbf{r}\| = \sqrt{16^2 + (-9.3)^2} \approx 18.5$$

The direction of the resultant is $\theta = \tan^{-1}\left(\frac{-9.3}{16}\right) \approx -30°$.

Making Connections

When Galileo discovered the parallelogram law long before the Cartesian coordinate system was invented, the only way he could find the magnitude and direction of the resultant vector was to draw the two given vectors to scale, then draw and measure the resulting vector and the angle involved. It was actually the invention of complex numbers that led to the use of ordered pairs to model vectors.

Exercises 6.5

Develop the Concepts

For Exercises 1 and 2 perform the vector operations for $\mathbf{V}_1 = (3, 4)$ and $\mathbf{V}_2 = (5, 12)$.

1. (a) $\mathbf{V}_1 + \mathbf{V}_2$
   (b) $\mathbf{V}_2 + \mathbf{V}_1$

2. (a) $\mathbf{V}_1 + (\mathbf{V}_2 + \mathbf{V}_1)$
   (b) $(\mathbf{V}_1 + \mathbf{V}_2) + \mathbf{V}_1$

For Exercises 3 and 4 find the magnitudes for $\mathbf{V}_1 = (3, 4)$ and $\mathbf{V}_2 = (5, 12)$.

3. (a) $\|\mathbf{v}_1\|$ and $\|\mathbf{v}_2\|$
   (b) $\|\mathbf{v}_1 + \mathbf{v}_2\|$ and $\|\mathbf{v}_1\| + \|\mathbf{v}_2\|$

4. (a) $\|2\mathbf{v}_1\|$ and $2\|\mathbf{v}_2\|$
   (b) $\|2\mathbf{v}_1 + 3\mathbf{v}_2\|$ and $2\|\mathbf{v}_1\| + 3\|\mathbf{v}_2\|$

For Exercises 5 to 8 find the magnitude and direction of each vector.

5. (a) $\mathbf{v}_1 = (5, 12)$
   (b) $\mathbf{v}_2 = (-12, -5)$

6. (a) $\mathbf{v}_1 = (-5, 12)$
   (b) $\mathbf{v}_1 = (-50, 120)$

7. (a) $\mathbf{v}_1 = (\frac{1}{\sqrt{2}}, \frac{1}{\sqrt{2}})$
   (b) $\mathbf{v}_2 = (\frac{-1}{\sqrt{2}}, \frac{-1}{\sqrt{2}})$

8. (a) $\mathbf{v}_1 = (\frac{\sqrt{3}}{2}, \frac{1}{2})$
   (b) $\mathbf{v}_2 = (\frac{-1}{2}, \frac{\sqrt{3}}{2})$

For Exercises 9 to 12 find the horizontal and vertical components, a and b, of a vector whose magnitude $\|\mathbf{v}\|$ and angle with the horizontal θ are given (Figure 20).

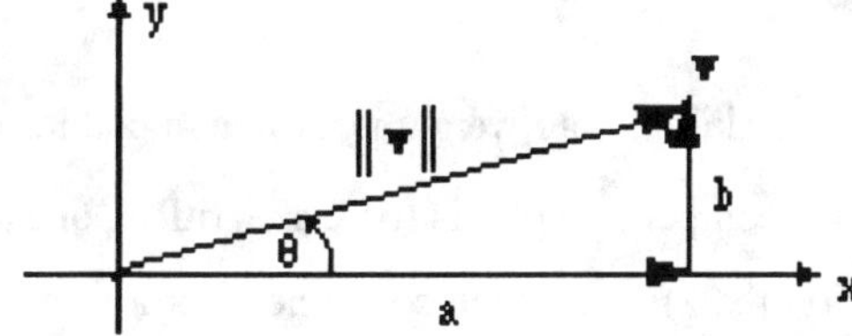

Figure 20

9.(a) $\|\mathbf{v}\| = 1, \theta = 0$
   (b) $\|\mathbf{v}\| = 1, \theta = \frac{\pi}{2}$

10.(a) $\|\mathbf{v}\| = 1, \theta = \frac{3\pi}{4}$
   (b) $\|\mathbf{v}\| = 1, \theta = \frac{-\pi}{4}$

11.(a) $\|\mathbf{v}\| = 240, \theta = 1$
   (b) $\|\mathbf{v}\| = 240, \theta = 2$

12.(a) $\|\mathbf{v}\| = 5, \theta = 210°$
   (b) $\|\mathbf{v}\| = 10, \theta = -120°$

Apply the Concepts

13. The Musical Fifth Problem Find the sine function that models striking each of the keys on a piano. (These two notes represent a *fifth* on our musical scale.)
(a) Middle C if its string vibrates at 261.6 Hz. (Use amplitude a = 1.)
(b) G above middle C, whose string vibrates at 392 Hz.
(c) Graph each function and their harmonious sum.

14. The Octave Problem Find the sine function that models striking middle C (261.6 Hz) and the C below middle C on a piano. (The frequency of the lower C is half that of middle C.) Graph their sum. Use the graph of the sum to determine whether or not a *beat* occurs when the two keys are played simultaneously.

15. The Beating Guitar Problem A piano has its G above middle C tuned to vibrate at 164 Hz. A guitarist plays the same note at 160 Hz. Determine the frequency of the beat created by playing the two instruments together.

16. The Out-of-Tune Duet Problem Revisited The musical note E above middle C on a piano is tuned 329.6 Hz while a slightly out-of-tune guitar has its E string tuned to 320 Hz. Find the two sine curves and express their sum as a product. Determine the frequency of the beat.

*17. The Wagon Pulling Problem When pulling a wagon we can imagine the force we are exerting as acting in the direction of the wagon's tongue (Figure 21). Find the horizontal and vertical components of a force of 100 lb for $\theta = \pi/6$ and $\theta = \pi/3$. Explain the roles of each component in the movement of the wagon.

Figure 21

18. The Steep Slope Problem When skiing, your weight is a force in pounds acting in a vertical direction, toward the center of the Earth. Two components of this force are critical, one acting down the hill, the other acting perpendicular to the hill (Figure 22). Find each component for a person weighing 150 lb ($\|\mathbf{F}\| = 150$) skiing down hills with slopes of (a) $\theta = 15°$ and (b) $\theta = 45°$. Explain the roles of each component in the movement of the skier.

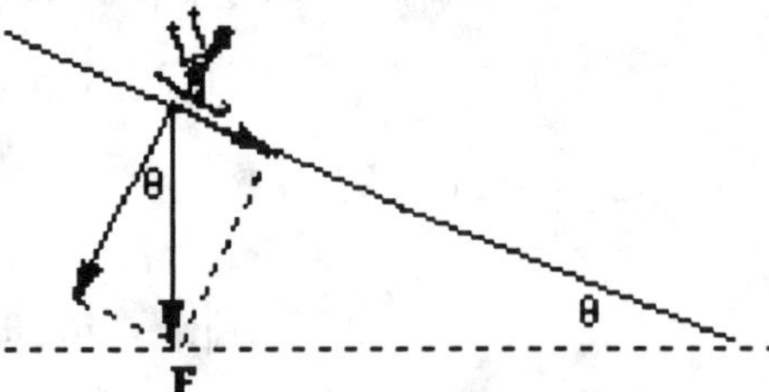

Figure 22

19. The Rainbow Kite and Balloon Company Two movers are helping move a large billboard advertising the company. If they apply forces of 75 and 95 lb at an angle of 35° to each other, find the resultant force.

20. Wonderworld Amusement Park You wish to cross the river at the park on a raft where the current is 6 mi/hr. If you can paddle the raft at 4 mi/hr, how far will you actually travel if the trip takes 90 min?

*21. <u>The Santa Claus Problem</u> The wind is gusting from the north pole at 40 mi/hr and Santa wishes to go directly from Chicago, Illinois to Portland, Oregon in 1 hr. Portland is some 2400 mi from Chicago at a bearing of 275°. (Recall that bearing is measured clockwise from due north.) At what heading should Rudolph lead the eight tiny reindeer?

22. <u>The Steamship Problem</u> A ship steams 60 mi north, turns to a heading of 45° and steams 20 mi, then turns south and steams 75 mi. How far is the ship from its starting place, and what is its bearing from its starting place? [Hint: Use the tail-to-tip method of adding vectors to draw the sketch.]

A Guided Review of Chapter 6

6.1 Trigonometric Identities

For Exercises 1 and 2 use the graphs of functions f and g in part (a) to explain each identity in (b).

1. (a) $f(x) = \cos 2x$, $g(x) = \cos^2 x$
   (b) $2\cos^2 x - 1 = \cos 2x$

2. (a) $f(x) = \cos 2x$, $h(x) = \sin^2 x$
   (b) $1 - 2\sin^2 x = \cos 2x$

For Exercises 3 to 6 prove or disprove that each of the following is an identity.

3. $\tan x \cot x = \sin x \csc x$

4. $\tan x \sin x - \sec x + \cos x = 0$

5. $(\sin x - \cos x)^2 + \sin 2x = 1$

6. $\tan x + \cot x = \sec x \csc x$

For Exercises 7 and 8 graph the following equations and identify whether or not they are identical.

7. (a) $y = \sin(2\pi + x)$
   (b) $y = \sin(\pi - x)$

8. (a) $y = \sin x$
   (b) $y = \cos(\frac{\pi}{2} - x)$

6.2 More Trigonometric Identities

For Exercises 9 to 14 prove or disprove that each equation is an identity.

9. $\sin(\alpha + \beta) + \sin(\alpha - \beta) = 2\sin\alpha\sin\beta$

10. $\dfrac{\cos(\alpha - \beta)}{\sin(\alpha - \beta)} = \dfrac{\cot\alpha\cot\beta - 1}{\cot\beta + \cot\alpha}$

11. $\dfrac{1 - \tan x}{1 + \tan x} = \dfrac{1 - \sin 2x}{\cos 2x}$

12. $\cot x - \tan x = \dfrac{2}{\tan 2x}$

13. $\cos(\alpha + \beta)\cos(\alpha - \beta) = \sin^2\alpha - \sin^2\beta$

14. $\cos u - \cos v = -2\sin(\frac{u + v}{2})\sin(\frac{u - v}{2})$

6.3 Equations Involving the Circular Functions

For Exercises 15 and 16 solve each equation for the calculator solution of x.

15. $\tan x - \sin x = 0$

16. $\sin x \tan x = \sec x - \cos x$

For Exercises 17 to 20 solve each equation for $0 \le x < 2\pi$.

17. $\cos 2x = 2\cos x$

18. $\cos 2x = \sin 2x$

19. $4\tan x \sin^2 x = 3\tan x$

20. $\sin 2x = \cos^2 2x$

For Exercises 21 and 22 solve each equation for the general solutions of x.

21. $\cos^2 x - \sin^2 x = 1$

22. $\sin\frac{x}{2}\cos\frac{x}{2} = \cos x$

For Exercises 23 and 24 graph each pair of functions on one coordinate system and estimate the point(s) of intersection for $0 \le x < 2\pi$. Algebraically, find the exact coordinates of the points.

23. $f(x) = \cos^4 x$
    $g(x) = \cos^2 x$

24. $f(x) = -\cos 2x$
    $g(x) = \cos^2 x$

6.4 Applications of the Circular Functions to Triangles

For Exercises 25 to 28 find the measure of the missing angles and sides for ΔABC. If no triangle is possible write "no solution." If two triangles are possible, give both solutions.

25. B = 120°, a = c = 100

26. A = B = 60°, c = 17.6

27. B = 40°, a = 31.6, c = 50

28. A = 32°, B = 58°, c = 34

29. The Titanic II Problem On the Titanic II's maiden voyage to New York, to avoid a crash the captain turned starboard (right) at an angle of 32° to his original course and traveled 15 nautical miles. He then turned the ship to port (left) and returned to his original course, intersecting it at an angle of 40°. How far off course did the ship travel (Figure 1)?

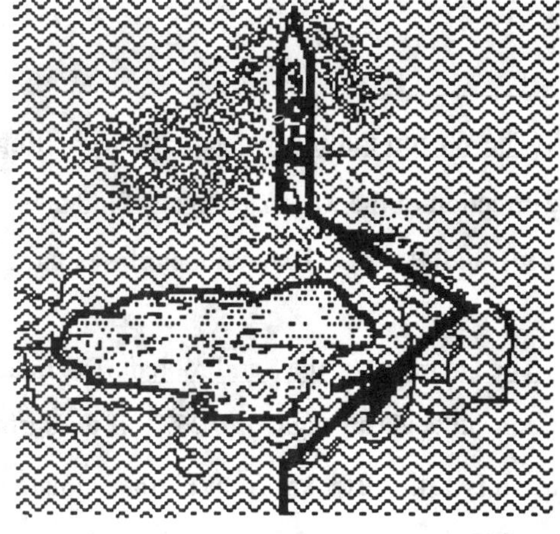

Figure 1

6.5 More Applications: Simple Harmonic Motion and Vectors

For Exercises 30 and 31 perform the indicated vector operations for $\mathbf{V}_1 = (-7, 24)$ and $\mathbf{V}_2 = (14, -24)$.

30.(a) $\mathbf{V}_1 + \mathbf{V}_2$

(b) $\mathbf{V}_2 + \mathbf{V}_1$

31.(a) $\mathbf{V}_1 + (\mathbf{V}_2 + \mathbf{V}_1)$

(b) $(\mathbf{V}_1 + \mathbf{V}_2) + \mathbf{V}_1$

For Exercises 32 and 33 find the magnitudes for $\mathbf{V}_1 = (-7, 24)$ and $\mathbf{V}_2 = (14, -24)$.

32.(a) $\|\mathbf{V}_1\|$

(b) $\|\mathbf{V}_2\|$

33.(a) $\|\mathbf{V}_1 + \mathbf{V}_2\|$

(b) $\|\mathbf{V}_1\| + \|\mathbf{V}_2\|$

For Exercises 34 and 35 find the horizontal and vertical components, a and b, of a vector whose magnitude $\|\mathbf{v}\|$ and angle with the horizontal θ are given.

34. $\|\mathbf{v}\| = 24$, $\theta = \pi/6$

35. $\|\mathbf{v}\| = 24$, $\theta = -\pi/6$

36. The Out-of-Tune Duet Problem Revisited The musical note middle C on a piano is tuned 440 Hz, while a slightly out-of-tune guitar has its C string tuned to 450 Hz. Find the two sine curves and express their sum as a product. Determine the frequency of the beat.

37. The English Channel Problem Suppose that you wanted to swim the English Channel from Dover to Calais (Figure 2). What direction should you swim if you can swim 8 km/hr in still water but there is a westward current of 5 km/hr?

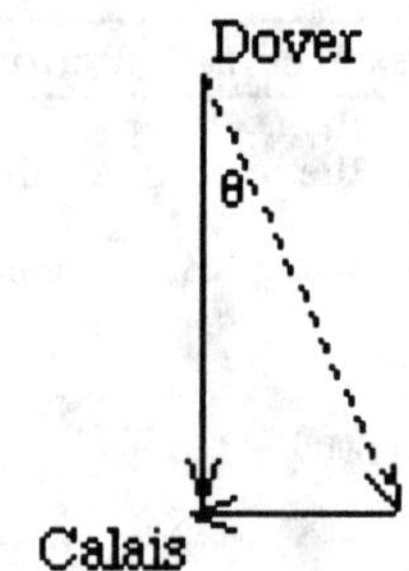

Figure 2

38. The Wagon Pulling Problem When pulling a wagon we can imagine the force we are exerting as acting in the direction of the wagon's tongue. Find the horizontal and vertical components of a force of 100 lb for $\theta = \pi/6$ and $\theta = \pi/3$. Explain the roles of each component in the movement of the wagon.

39. The Diameter of the Sun Problem The radius of the sun subtends an angle of 0.267°. If the surface of the sun is 1.49×10^8 km from Earth, estimate the diameter of the sun. (See Figure 3.)

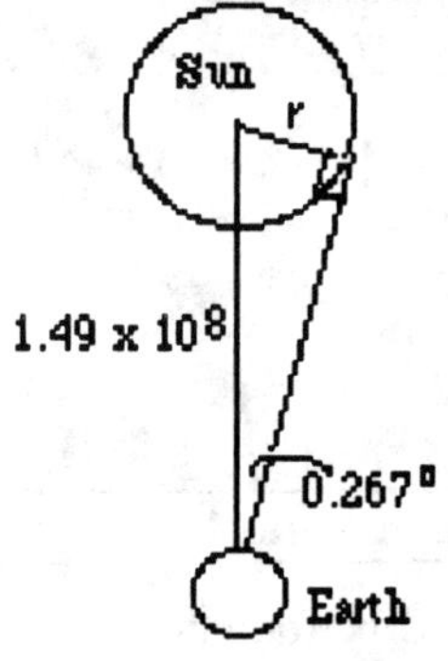

Figure 3

40. The Dead Radio Problem A pilot is heading due north at 100 mi/hr in his Cessna jet. After flying for 1 hr in powerful crosswinds he calls back to the control tower and hears before the radio goes dead: "You are traveling at a bearing of 30° and the winds are blowing at a steady 60 mi/hr from the .... ." What is the location of the jet?

Chapter 6 Test

1. Use the graphs of functions f and g in part (a) to explain the identity in (b).

(a) $f(x) = \cos^2 2x$, $g(x) = \sin^2 2x$

(b) $1 - \sin^2 2x = \cos^2 2x$

2. Prove or disprove that each of the following is an identity.

(a) $\tan 2x \cos 2x = \sin 2x$

(b) $\dfrac{\tan^2 x}{1 - \sec^2 x} + \cos^2 x = -\sin^2 x$

(c) $\sin 2x = 2 \sin x$

(d) $2 \sin^2 x = \dfrac{\sin^2 2x}{1 + \cos 2x}$

3. Solve each equation for $0 \le x < 2\pi$.

(a) $1 - \sin x = \cos x$

(b) $\cos^2 x - \sin x = 1$

4. Solve $2 \sin^2 x + 3 \sin x - 1 = 0$ for both calculator and general solutions.

5. Perform the indicated vector operations and calculate the magnitudes as indicated for $\mathbf{V}_1 = (-21, 72)$ and $\mathbf{V}_2 = (72, -21)$.

(a) $\|\mathbf{V}_1 + \mathbf{V}_2\|$

(b) $\|\mathbf{V}_1\| + \|\mathbf{V}_2\|$

6. Find the horizontal and vertical components, a and b, of a vector whose magnitude $\|\mathbf{v}\|$ and angle with the horizontal θ are $\|\mathbf{v}\| = 2$, and $\theta = \pi/3$.

7. Find the measure of the missing angles and sides for ΔABC. If no triangle is possible, write "no solution." If two triangles are possible, give both solutions.

(a) $C = 30°$, $a = 10$, $b = 6$

(b) $A = 43°$, $B = 58°$, $c = 27.5$

8. <u>The D.B. Cooper Caper Problem</u>

The airline highjacker D.B. Cooper parachuted from a plane and got away with $250,000. If the officials had planted an electronic location-finder device in the parachute, the chute could have been found. Figure 1 shows the headings from a nearby airport to a police helicopter and D.B.'s parachute. Help the police find D.B.'s chute.

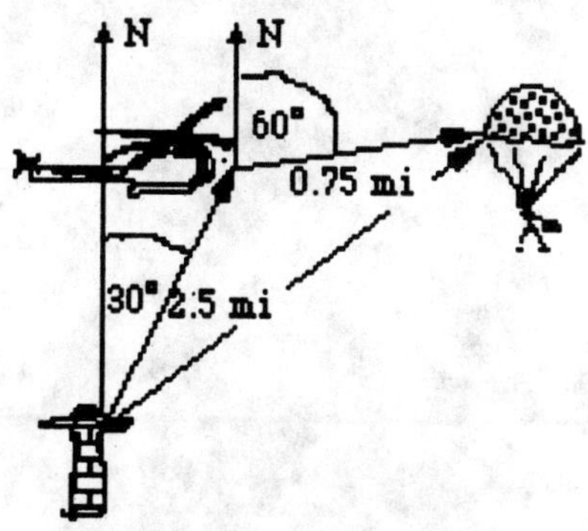

Figure 1

9. The Out-of-Tune Duet Problem The musical note E above middle C on a piano is tuned to 329.6 Hz (vibrations/sec). A guitar slightly out of tune has an E string vibrating at 320 Hz.

(a) Find an equation of the form $y = \sin Bt$ for each instrument.

(b) Estimate the period of the beat from the graph of their sum.

10. The Super Cannon Shot A cannon designed to launch satellites was positioned on a mountain 1 mi (5280 ft) above sea level. Satellites were launched horizontally. In order to go into orbit, the height above the surface of the earth had to remain at 1 mi after it had fallen a vertical distance of 1 mi (Figure 2). Find the initial velocity required.

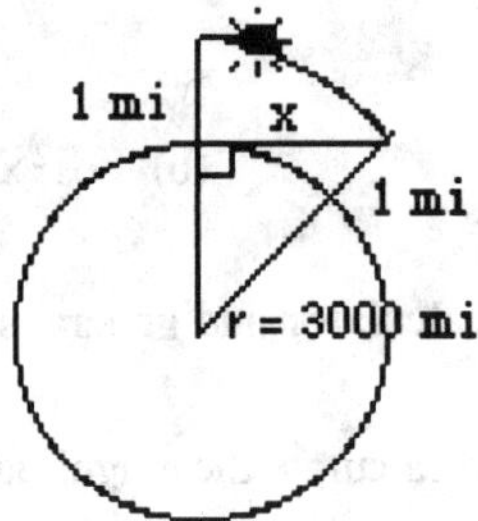

Figure 2

Chapter 7

Polar Coordinates and Parametric Equations

Many applications motivate the study of polar coordinates and parametric equations. Generalizing some of the concepts of rectangular coordinates to polar coordinates will help us understand both systems. Changing to polar coordinates greatly simplifies some models and the associated calculations. For example, if there is symmetry about a point, as in a circle, equations written in polar form are often simpler than in rectangular form.

Polar coordinates are intriguing in their own right. As we will see, some of their graphs are beautiful. Many curves in space, motion in the real world, and rotations in computer graphics can be simplified using polar coordinates or parametric equations.

A Historical Note on the Bernoulli Family

The Bernoulli family consisted of three generations of mathematicians, starting in the first generation with three brothers.

Jacob I 1654 -1705   Nicolaus I 1622-1716   Johannes I 1667-1748

Nicolaus II 1687-1726   Nicolaus III 1695-1726   Daniel 1700-1782   Johannes II 1710-1790

Johannes III 1746-1807   Jacob II 1759-1789

Jacob I and Johannes I worked with Leibniz to develop many of the ideas of the calculus and its applications. Jacob also made significant contributions to other areas of mathematics, including polar coordinates and probability theory. Johannes, originally a physician, seems to have been more quarrelsome than Jacob, especially over mathematics. Their letters to each other were often quite hot, and Johannes attempted to steal some of Jacob's ideas. Johannes and one of his sons entered a contest sponsored by the French Academy of Science. The son won the prize.

Of the other Bernoullis, Daniel, sometimes called the founder of mathematical physics, was perhaps the most influential. Like his father, he became a physician before turning to mathematics. His work was in calculus, differential equations, probability, vibrating strings, and hydrodynamics. He developed the concept of conservation of energy and worked with the motion of fluids.

7.1 Introduction to Polar Coordinates

Plotting Points in a Polar Coordinate System
Graphing Lines and Circles
Graphing Roses
Graphing Limaçons
Graphing Systems of Polar Graphs

In this section we study and apply a second method of describing points in the plane. Instead of using the rectangular coordinate system (Cartesian coordinate system), we use the polar coordinate system. A **polar coordinate system** is a graphing system that resembles a spider web with radial lines emerging from the center and concentric circles connecting the lines. Figure 1 shows a polar coordinate system drawn on top of a rectangular coordinate system, with its x and y axes, to demonstrate the relationship between the two coordinate systems.

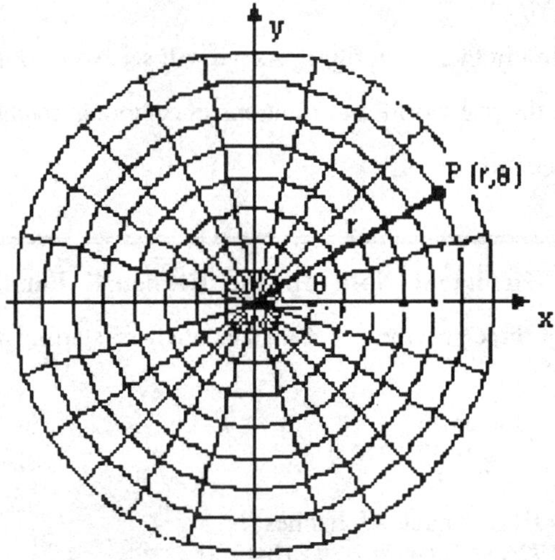

Figure 1

Polar coordinates are used when directions are related to a single point, called the pole. For example, if you tell someone where you live by pointing toward your home and telling them a distance, you are using polar coordinates. If, on the other hand, you tell them to go 10 blocks east and 4 blocks north, you are using rectangular (Cartesian) coordinates. Plots of land in the United States are laid out on a rectangular coordinate system. In Europe, the location of a plot of land is described given an angle and distance from a specific point, a polar coordinate system.

Plotting Points in a Polar Coordinate System

To plot points in the polar coordinate system, we use the **polar coordinates (r, θ)**, where r is the distance from the **pole** (origin) and θ is the angle measured from the positive horizontal axis.

An angle θ is measured from the positive horizontal axis (our old x axis) in a counterclockwise direction for positive θ and clockwise for negative θ. The distance r is measured from the pole. If r is positive, the distance is out along the terminal side of the angle. If r is negative, we measure along the line formed by extending the terminal side of angle θ back through the pole. For example, to plot the point $(5, \pi/4)$ on a polar coordinate system, we draw the terminal side of $\theta = \pi/4$ and put a point along that ray 5 units from the pole. To plot $(-5, \pi/4)$, we measure 5 units backward through the pole (Figure 2).

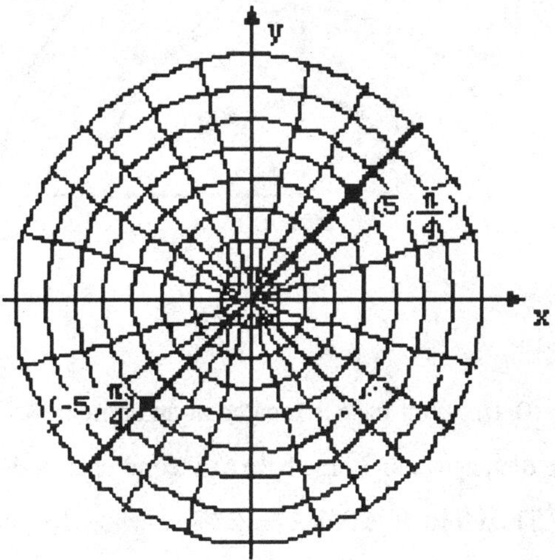

Figure 2

Can you think of other names for the point $(5, \frac{\pi}{4})$*?*

There are an infinite number of labels for any point in a polar coordinate system. Some other names for the point $(5, \frac{\pi}{4})$ are $(5, 45°)$, $(-5, \frac{5\pi}{4})$, $(-5, -\frac{3\pi}{4})$, $(5, 0.78)$, and $(5, \frac{9\pi}{4})$.

Consider a baseball diamond, which is a square measuring 90 ft on each side (Figure 3). Because directions in the game of baseball are given relative to home plate, a polar coordinate system is implied.

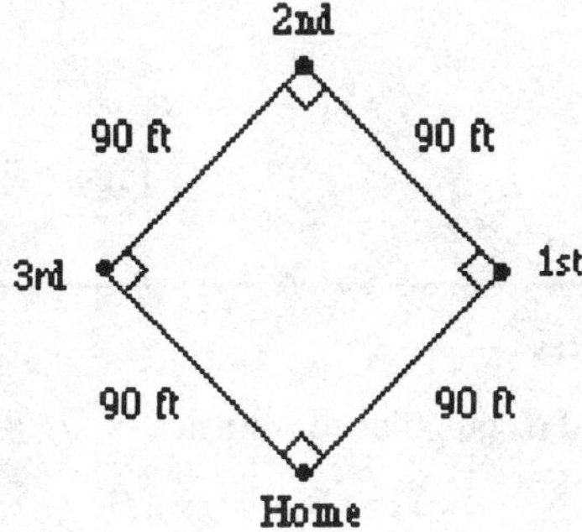

Figure 3

Example 1

Find the coordinates for all four bases in the baseball diamond in Figure 4.

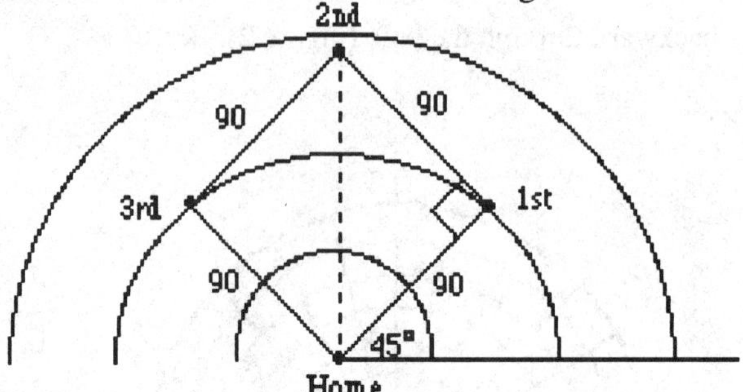

Figure 4

Solution

Home plate is located at the pole (0, 0). First base is located at an angle of 45°, 90 ft away from home plate. Its coordinates are (90, 45°). The distance from home base to second base is the hypotenuse of the 45-45-90 right triangle having sides 90 ft. Therefore, it is $90\sqrt{2} \approx 127.3$ ft away. Its polar coordinates are (127.3, 90°). Third base is located at an angle of 135°, 90 ft away; (90, 135°).

The Pythagorean theorem and the definitions of sine, cosine, and tangent enable us to change points and equations from one form to the other. Using the right triangle in Figure 5, we have

$$\sin\theta = \frac{y}{r} \Rightarrow y = r\sin\theta$$

$$\cos\theta = \frac{x}{r} \Rightarrow x = r\cos\theta$$

and

$$\tan\theta = \frac{y}{x} \Rightarrow \theta = \tan^{-1}\left(\frac{y}{x}\right)$$

which lead to the following.

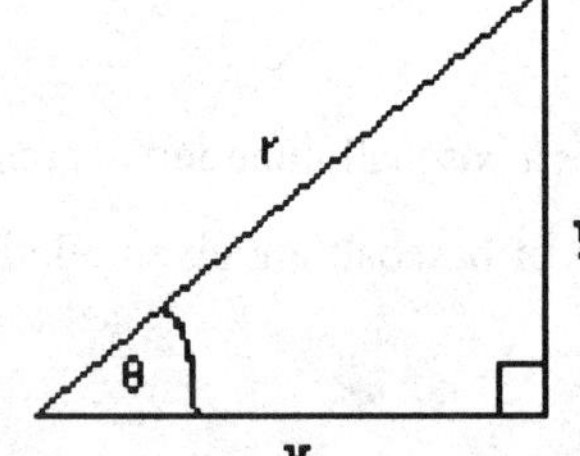

Figure 5

Polar and Rectangular Coordinates

If point P has polar coordinates (r, θ) and rectangular coordinates (x, y), then

$$x = r\cos\theta,\ y = r\sin\theta$$

$$x^2 + y^2 = r^2$$

$$\theta = \tan^{-1}\left(\frac{y}{x}\right),\quad x \neq 0$$

In the next example we change a point from rectangular into polar form and back again.

Example 2 *Switching coordinate systems*

(a) Change the rectangular coordinates (-3, 4) into polar coordinates.

(b) Check your results from part (a) by changing them back to rectangular coordinates.

Solution

(a) $x^2 + y^2 = r^2$ and $x = -3, y = 4 \Rightarrow$

$$9 + 16 = r^2$$
$$r = \pm 5$$

$\theta = \tan^{-1}(\frac{y}{x})$ and $x = -3, y = 4 \Rightarrow$

$$\theta = \tan^{-1}(\frac{4}{-3}) \approx -0.93$$

Since the point (-3, 4) is located in quadrant II, one pair of polar coordinates is (-5, -0.93). In degree measure this is (-5, -53.1°).

(b) To change (-5, -0.93) back into rectangular coordinates, we have

$x = r\cos\theta, y = r\sin\theta$ and $r = -5, \theta = -0.93 \Rightarrow$

$x = -5\cos(-0.93)$	$y = -5\sin(-0.93)$
$x = -5(0.600)$	$y = -5(-0.7999)$
$x = -3$	$y = 4$

Thus, the point has rectangular coordinates (-3, 4).

Graphing Lines and Circles

Some curves have simple equations in polar form, but this is not true for all curves. We will use polar coordinates only when they are helpful. If polar coordinates make the equation more complicated than the rectangular equations, we will them. One of our goals is to learn when it is appropriate to use polar coordinates. In general, since polar coordinates radiate out from the pole, graphs that have point symmetry have simple polar equations. Graphs of polar equations can be created or verified by plotting points. For example, consider the graph of the polar equation $\theta = \pi/3$ in Figure 6.

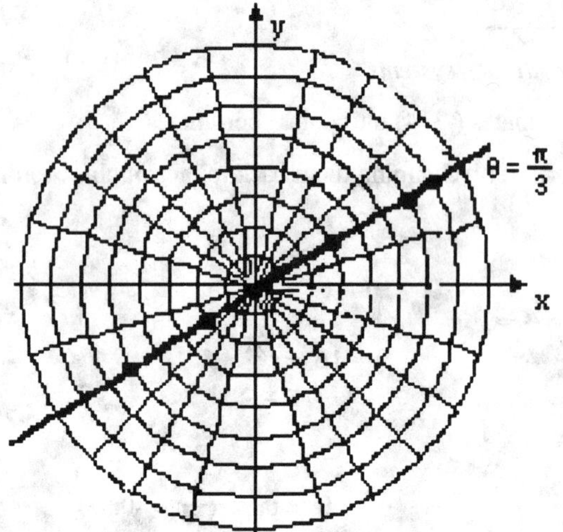

Figure 6

The points that satisfy $\theta = \pi/3$ all fall on a line.

θ	$\pi/3$	$\pi/3$	$\pi/3$	$\pi/3$	$\pi/3$	$\pi/3$
r	-5	-2	0	3	6	7

Can you think of some other polar equations whose graph results in the same line?

Other equations whose graphs are the line in Figure 4 are $\theta = 4\pi/3$ and $\theta = -150°$.

Lines

A polar equation of the form $\theta = c$

where c is a constant is a line through the pole, a radial line.

Example 3 *Switching from polar to rectangular form*

Change the polar equation $\theta = \pi/3$ into a rectangular equation in x and y.

Solution

$\theta = \pi/3$ and $\tan\theta = y/x \Rightarrow$

$$\tan(\pi/3) = \frac{y}{x}$$

$$\sqrt{3} = \frac{y}{x}$$ *Multiply both sides by x.*

$$y = \sqrt{3}\,x$$

Notice that $y = \sqrt{3}\,x$ is equivalent to a rectangular equation of the form $y = mx$, where $m = \tan\theta$.

The next example has r fixed at a constant value for all values of θ.

Example 4 *Graphing a polar equation*

Graph r = 2 by making a table of values and plotting points in a polar coordinate system.

Solution

All values of θ are paired with r = 2.

θ	0	$\frac{\pi}{6}$	$\frac{\pi}{3}$	$\frac{\pi}{2}$	$\frac{2\pi}{3}$	$\frac{5\pi}{6}$	π	$\frac{7\pi}{6}$	$\frac{4\pi}{3}$	$\frac{3\pi}{2}$	$\frac{5\pi}{3}$	$\frac{11\pi}{6}$	2π
r	2	2	2	2	2	2	2	2	2	2	2	2	2

Graphing these points leads to the graph of a circle (Figure 7).

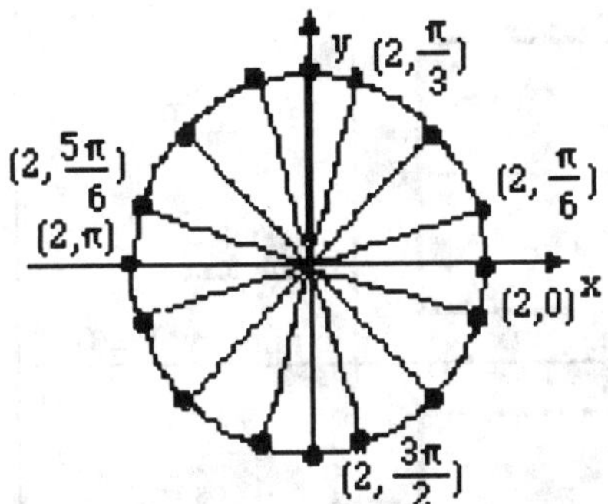

Figure 7

Polar equations can be graphed by generating a table of values, plotting the points, and then connecting them in order as in Example 4. Most graphing utilities have a polar coordinate option. The window must be set for θ as well as for x and y. Figure 8a shows the window that will graph the equation r = 2 from θ = 0 to θ = 2π in increments (steps) of θ = π/6, just as the table in Example 4. The resulting graph of r = 2 is shown in Figure 8b. For values of θ beyond 2π, the graph traces over the same circle, periodically.

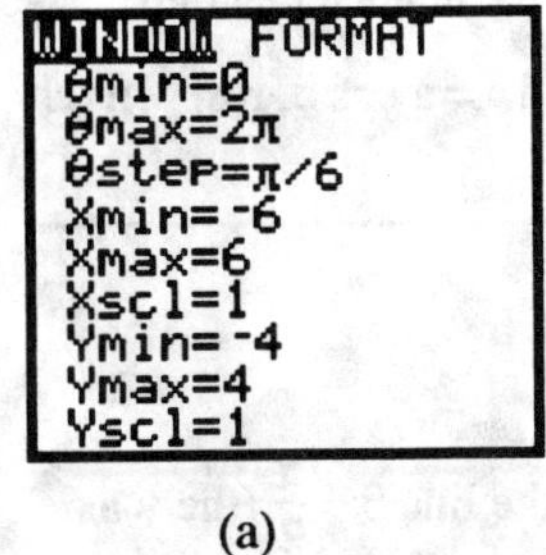

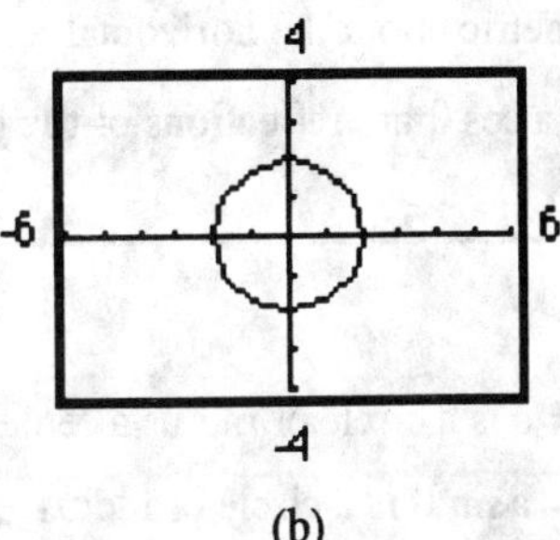

Figure 8 (a) (b)

Equations of the form r = a (Example 4), as well as equations of the form r = a sin θ and r = a cos θ are all circles. Table 1 shows some equations of circles of the form r = a sin θ and r = a cos θ and their graphs.

Table 1

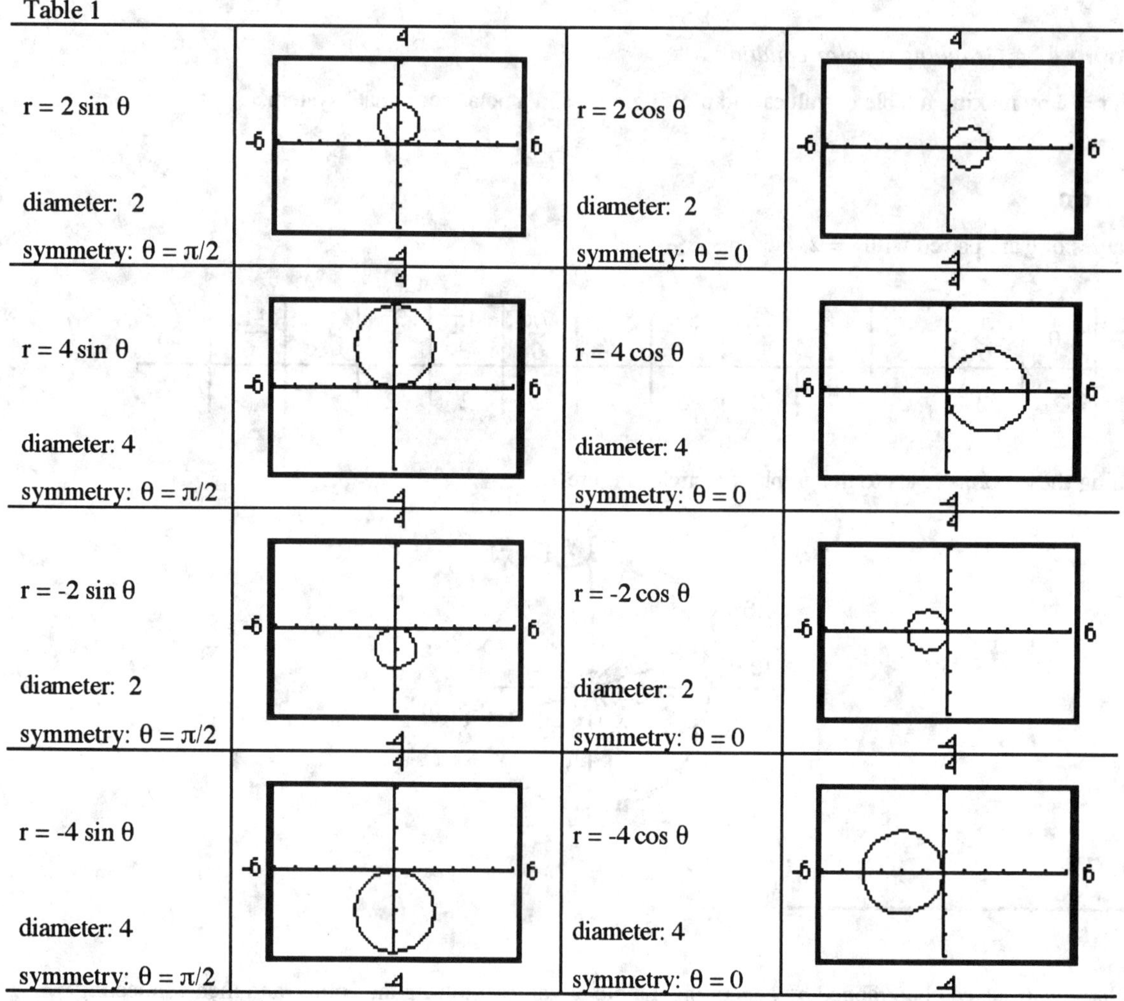

Equation	Graph	Equation	Graph
r = 2 sin θ diameter: 2 symmetry: θ = π/2		r = 2 cos θ diameter: 2 symmetry: θ = 0	
r = 4 sin θ diameter: 4 symmetry: θ = π/2		r = 4 cos θ diameter: 4 symmetry: θ = 0	
r = -2 sin θ diameter: 2 symmetry: θ = π/2		r = -2 cos θ diameter: 2 symmetry: θ = 0	
r = -4 sin θ diameter: 4 symmetry: θ = π/2		r = -4 cos θ diameter: 4 symmetry: θ = 0	

Which graphs in Table 1 are symmetric about the vertical axis (θ = π/2) ? Which are symmetric about the horizontal axis (θ = 0) ? Where are the diameters found in each polar equation?

All the graphs of the form r = a sin θ are symmetric about the vertical axis, while all the graphs of the form r = a cos θ are symmetric about the horizontal axis. All the circles have diameter |a| units. Notice also that r = -a sin θ and r = -a cos θ are reflections of the graphs r = a sin θ and r = a cos θ, respectively.

Circles
If a > 0 then

r = a is a circle of radius a centered at the pole.
r = a sin θ is a circle of radius $\frac{|a|}{2}$ symmetric about the line $\theta = \frac{\pi}{2}$ (the y axis).
r = a cos θ is a circle of radius $\frac{|a|}{2}$ symmetric about the line θ = 0 (the x axis).

Graphing Roses

Another group of polar equations is called **roses**. Their equations are of the form

$$r = a \sin n\theta \quad \text{or} \quad r = a \cos n\theta$$

The period of these polar functions creates repeated sections of the graphs called petals. Table 2 shows particular equations with their graphing utility displays.

Table 2

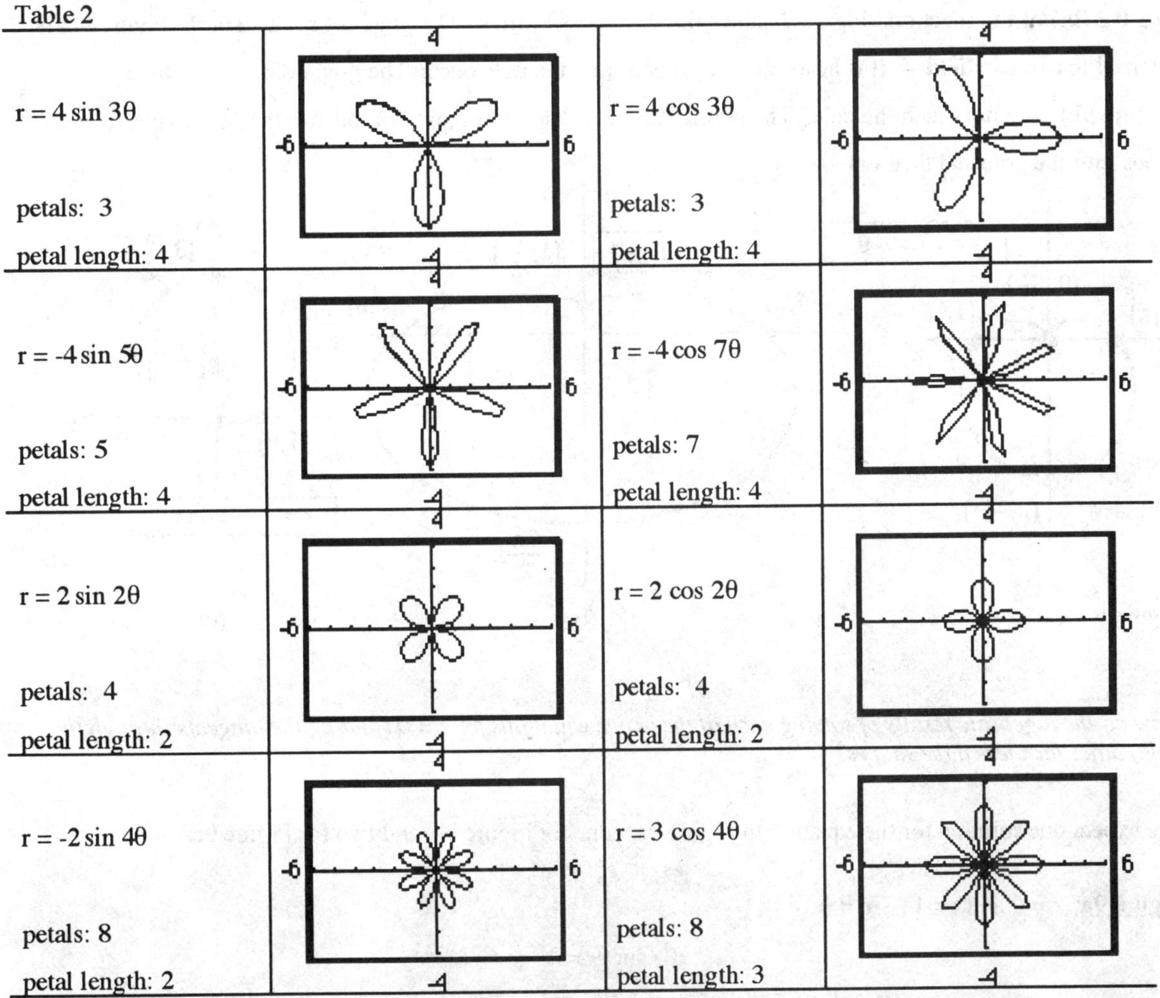

$r = 4 \sin 3\theta$ petals: 3 petal length: 4	$r = 4 \cos 3\theta$ petals: 3 petal length: 4
$r = -4 \sin 5\theta$ petals: 5 petal length: 4	$r = -4 \cos 7\theta$ petals: 7 petal length: 4
$r = 2 \sin 2\theta$ petals: 4 petal length: 2	$r = 2 \cos 2\theta$ petals: 4 petal length: 2
$r = -2 \sin 4\theta$ petals: 8 petal length: 2	$r = 3 \cos 4\theta$ petals: 8 petal length: 3

How do we determine the number of petals in a rose from its equation?

Studying Table 2, we notice there are n petals if n is odd and 2n petals if n is even. Graphs of the form $r = a \sin n\theta$ are symmetric about $\theta = \pi/2$, while all the graphs of the form $r = a \cos n\theta$ are symmetric about the $\theta = 0$.

Roses

A rose has a polar equation of the form

$$r = a \sin n\theta \quad \text{or} \quad r = a \cos n\theta$$

where there are n petals if n is odd and 2n petals if n is even. The length of the petals is $|a|$ units.

Graphing Limaçons

Limaçons have equations of the form

$$r = a \pm b \sin \theta \quad \text{or} \quad r = a \pm b \cos \theta$$

Just as for circles and roses, limaçons are symmetric about $\theta = \pi/2$ for sine and about $\theta = 0$ for cosine. The relative sizes of $|a|$ and $|b|$ determine how many times the limaçon goes through the pole in one period ($0 \le \theta \le 2\pi$) of the function. Figure 7 shows the three possibilities. The graph of $r = 1 - \sin \theta$ (Figure 9a) is referred to as a **cardioid** --- it is heart-shaped. It goes into the pole once. The graph of $r = 4 - 3 \sin \theta$ (Figure 9b) does not touch the pole. The graph of $r = 1 + 2 \sin \theta$ (Figure 9c) touches the pole twice, as it loops into the pole and then out.

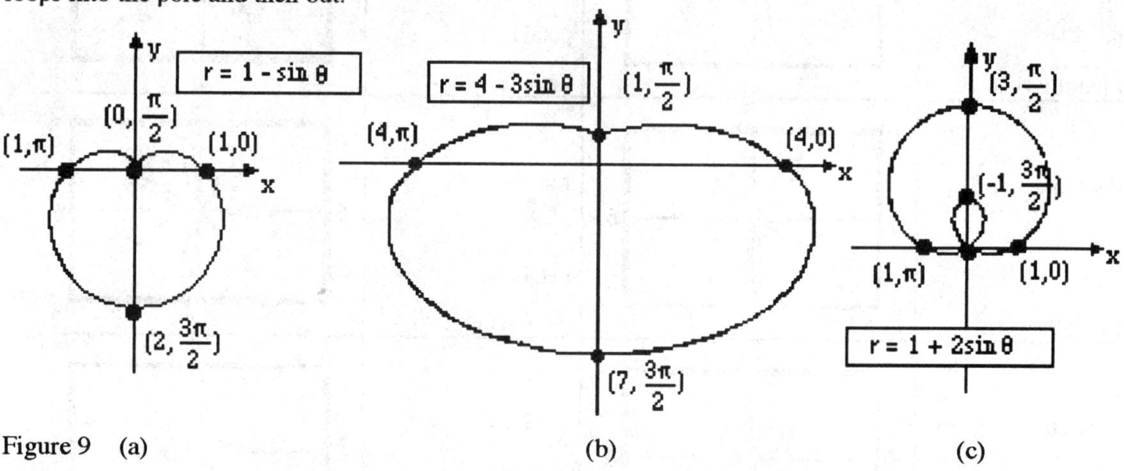

Figure 9 (a) (b) (c)

How do the algebraic results of solving each of the above equations for $r = 0$ (finding the tangents through the pole) differ for these three shapes?

We expect one solution for the equation in Figure 9a, none for Figure 9b, and two for Figure 9c.

Figure 9a: $r = 0$ and $r = 1 - \sin \theta \Rightarrow$

$$\begin{aligned} 1 - \sin \theta &= 0 \\ \sin \theta &= 1 \\ \theta &= \frac{\pi}{2} \end{aligned}$$

if $|a| = |b|$ and there is exactly one solution between 0 and 2π.

Figure 9b: $r = 0$ and $r = 4 - 3 \sin \theta \Rightarrow$

$$\begin{aligned} 4 - 3 \sin \theta &= 0 \\ \sin \theta &= \frac{4}{3} \end{aligned}$$

no solution

if $|a| > |b|$ and there is no solution.

Figure 9c: $r = 0$ and $r = 1 + 2\sin\theta \Rightarrow$

$$1 + 2\sin\theta = 0$$

$$\sin\theta = -1/2$$

$$\theta = \frac{7\pi}{6} \text{ or } \theta = \frac{11\pi}{6}$$

if $|a| < |b|$ and there are two solutions between 0 and 2π.

For the equation $r = 1 + 2\sin\theta$ in Figure 9c, we find the turning points $(3, \pi/2)$ and $(-1, 3\pi/2)$ by examining the maximum and minimum values of the sine function. (Recall that the sine function has a maximum value of 1 at $\theta = \pi/2$ and a minimum of -1 at $\theta = 3\pi/2$.)

$$\theta = \frac{\pi}{2} \Rightarrow r = 1 + 2\sin\frac{\pi}{2} = 3$$

and

$$\theta = \frac{3\pi}{2} \Rightarrow r = 1 + 2\sin\frac{3\pi}{2} = -1$$

If we superimpose a Cartesian coordinate system, the x intercepts $(1, 0)$ and $(1, \pi)$ are found by letting $\theta = 0$; then $\theta = \pi$.

$$\theta = 0 \Rightarrow r = 1 + 2\sin 0 = 1 + 0 = 1$$

and

$$\theta = \pi \Rightarrow r = 1 + 2\sin\pi = 1 + 0 = 1$$

See Figure 10.

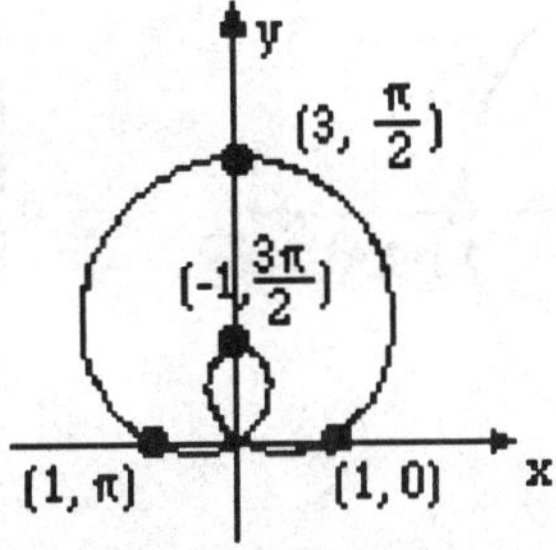

Figure 10

Limaçons

A limaçon has an equation of the form

$$r = a \pm b\sin\theta \quad \text{or} \quad r = a \pm b\cos\theta$$

If $|a| = |b|$, it is a cardioid.

If $|a| > |b|$, it does not pass through the pole.

If $|a| < |b|$, it loops through the pole.

NOTE: Actually, there are four cases for shapes of limaçons. The middle case, where $|a| > |b|$, can be refined. If $|a| < 2|b|$, the graph has a dimple as shown in Figure 8b. If $|a| \geq 2|b|$, there is no dimple and the graph is called a *convex limaçon*.

In the next example we graph a limaçon of the form $r = a \pm b\cos\theta$ which is symmetric about the line $\theta = 0$. In this case we examine minimum and maximum values for cosine, $\theta = 0$ and $\theta = \pi$. The intercepts are found by setting $\theta = \pi/2$ and $\theta = 3\pi/2$.

Example 5 *Graphing a limaçon*

Sketch the graph of $r = -2 + 2\cos\theta$.

Solution

Since $|-2| = |2|$, the limaçon is a cardioid. The cosine graph is symmetric about the line $\theta = 0$ (Figure 11).

<u>Maximum cosine value</u>: $\theta = 0 \Rightarrow r = -2 + 2\cos 0 = -2 + 2(1) = 0$

<u>Minimum cosine value</u>: $\theta = \pi \Rightarrow r = -2 + 2\cos\pi = -2 + 2(-1) = -4$

<u>Other intercepts</u>:

$\theta = \frac{\pi}{2} \Rightarrow r = -2 + 2\cos\frac{\pi}{2} = -2 + 0 = -2$

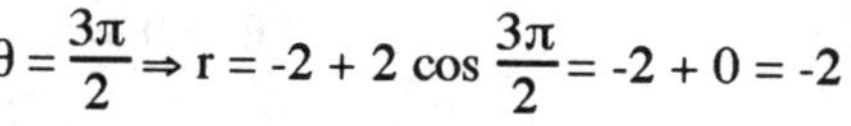

$\theta = \frac{3\pi}{2} \Rightarrow r = -2 + 2\cos\frac{3\pi}{2} = -2 + 0 = -2$

y
$(-2, \frac{3\pi}{2})$
$(0,0)$
$(-4, \pi)$
x
$(-2, \frac{\pi}{2})$

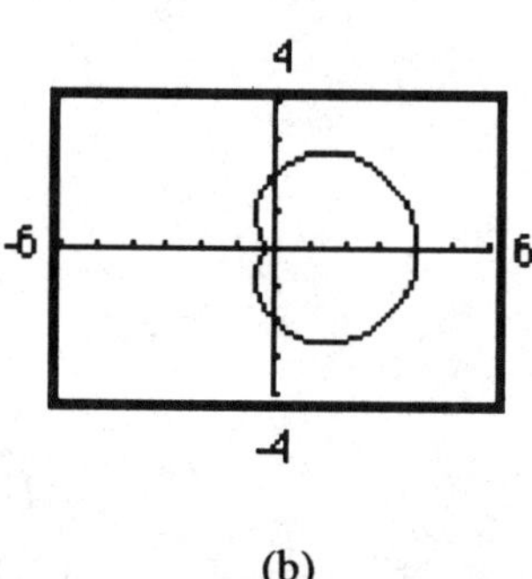

Figure 11 (a) (b)

Graphing Systems of Polar Graphs

The next example shows the simultaneous solutions of two polar equations and their graphs. We will find and label their points of intersection. This problem occurs in calculus, where the points of intersection must be found in order to find the area of any region bounded by such graphs. Since a point lying on both curves may use different values of r and θ for each curve (remember that each point has many names), such algebra must be accompanied by a geometric representation of the graphs involved.

Example 6 *Graphing a system of polar equations*

Find the points of intersection of the polar equations $r = \cos 2\theta$ and $r = \cos\theta$. Compare the geometric and algebraic results.

Solution

A four-petal rose and a circle intersect at the origin and at three other points (Figure 12).

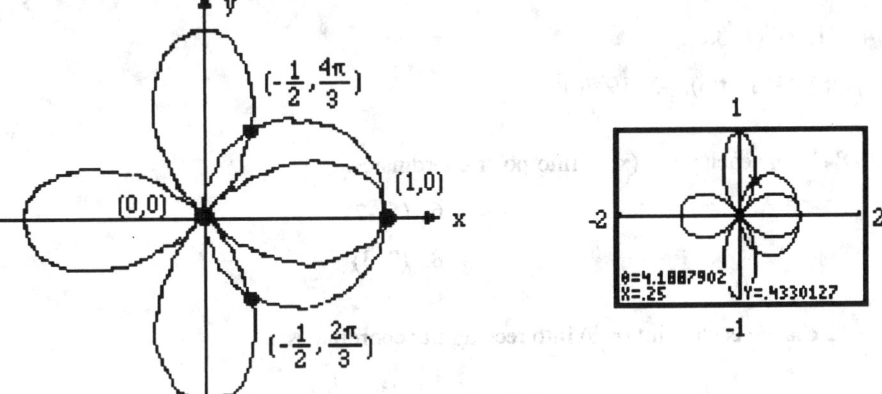

Figure 12

$$\cos 2\theta = \cos\theta \qquad \textit{Substitute } \cos 2\theta = 2\cos^2\theta - 1.$$

$$2\cos^2\theta - 1 = \cos\theta$$

$$2\cos^2\theta - \cos\theta - 1 = 0$$

$$(2\cos\theta + 1)(\cos\theta - 1) = 0$$

$$\cos\theta = -1/2 \text{ or } \cos\theta = 1$$

$$\theta = \frac{2\pi}{3} \text{ or } \theta = \frac{4\pi}{3} \text{ or } \theta = 0$$

$$\theta = \frac{2\pi}{3} \text{ and } r = \cos\theta \Rightarrow r = \cos\frac{2\pi}{3} \Rightarrow r = -\frac{1}{2}$$

$$\theta = \frac{4\pi}{3} \text{ and } r = \cos\theta \Rightarrow r = \cos\frac{4\pi}{3} \Rightarrow r = -\frac{1}{2}$$

$$\theta = 0 \text{ and } r = \cos\theta \Rightarrow r = \cos 0 \Rightarrow r = 1$$

The points of intersection are $(-\frac{1}{2}, \frac{2\pi}{3})$, $(-\frac{1}{2}, \frac{4\pi}{3})$, (1, 0), and (0, 0). Notice that the point (0, 0) did not appear in the algebraic solution because the rose goes through the pole at $(0, \frac{\pi}{4})$ and the circle at $(0, \frac{\pi}{2})$.

Exercises 7.1

Develop the Concepts

For Exercises 1 to 4 plot the points (r, θ) on a polar graph.

1. $(1, \pi/3), (-1, -\pi/3), (-1, \pi/3), (1, -\pi/3)$
2. $(3, \pi/6), (-2, \pi/6), (0, \pi/6), (7, \pi/6)$
3. $(1, 0), (1, \pi/4), (1, \pi), (1, 3\pi/2)$
4. $(3, \pi/6), (-3, 7\pi/6), (3, 13\pi/6), (-3, 19\pi/6)$

For Exercises 5 to 8 change each point (x, y) into polar coordinates.

5. $(-5, 2)$
6. $(5, -2)$
7. $(-4, 0)$
8. $(0, 3)$

For Exercises 9 to 12 change each point (r, θ) into rectangular coordinates.

9. $(1, \pi/4)$
10. $(-2, \pi/3)$
11. $(4, \pi/2)$
12. $(1, -3\pi/4)$

For Exercises 13 to 16 change each rectangular equation to an equation in polar form and simplify.

13. $x^2 + y^2 = 25$
14. $y = \pm\sqrt{16 - x^2}$

*15. $(x - 5)^2 + y^2 = 25$

16 $x^2 + (y - 6)^2 = 36$

For Exercises 17 to 22 change each polar equation into an equation in rectangular form and simplify.

17. $r = -3$
18. $r = 5$
19. $\theta = \pi/3$
20. $\theta = -\pi/2$
21. $r = 2 \sin \theta$
22. $r = -10 \cos \theta$

For Exercises 23 to 40 identify each curve and sketch its graph.

23. $\theta = \pi/12$
24. $\theta = -\pi/4$
25. $r = -1$
26. $r = 4$
27. $r = 3 \sin \theta$
28. $r = -3 \cos \theta$
29. $r = 4 \cos \theta$
30. $r = 4 \sin \theta$
31. $r = 2 \cos 4\theta$
32. $r = \cos 2\theta$
33. $r = 4 \sin 3\theta$
34. $r = -2 \sin 2\theta$
35. $r = 2 + 2 \sin \theta$
36. $r = 1 + \cos \theta$
37. $r = 2 + 3 \sin \theta$
38. $r = 1 + 2 \cos \theta$
39. $r = 3 + 2 \cos \theta$
40. $r = 3 - 2 \cos \theta$

For Exercises 41 to 50 solve each system of equations.

41. $r = 4 \cos \theta$

$r = 4 \sin \theta$

42. $r = 1 + 2 \cos \theta$

$r = \cos \theta$

*43. $r = \sin 2\theta$

$r = \sin \theta$

44. $r = \sin 2\theta$

$r = \cos 2\theta$

45. $r = 6$

$r = 6 \cos \theta$

46. $r = 2 \sin \theta$

$r = 2\sqrt{3} \cos \theta$

47. $r = \sin \theta$

$r = 1 - \sin \theta$

48. $r = 2$

$r = 4 \cos \theta$

49. $r = \cos 2\theta$

$r = -\cos \theta$

50. $r = 1 - 2\cos \theta$

$r = 4 + \cos \theta$

Apply the Concepts

51. The Rainbow Kite and Balloon Company You created a new water balloon baseball game for the company picnic. Find the polar coordinates for the pitcher's mound in the baseball diamond in Example 1 if it is located 60 ft 6 in. from home plate along the line connecting home plate and second base.

52. Wonderworld Amusement Park Consider the baseball diamond in Example 1.

(a) If the center fielder caught a fly ball at polar coordinates (270, 1.6), where was he? [Hint: The 1.6 is measured in radians.]

(b) What are the polar coordinates for a ball hit 270 ft down the right-field line?

Extend the Concepts

For Exercises 53 to 58 sketch each curve.

53. $r = \theta$

54. $r = e^{\theta}$

55. $r^2 = 2 \cos 2\theta$

56. $r^2 = 8 \sin 2\theta$

*57. $r = \dfrac{6}{2 - 2 \sin \theta}$

58. $r = \dfrac{4}{1 - 2 \cos \theta}$

7.2 Polar Form of Complex Numbers and DeMoivre's Theorem

Expressing a Complex Number in Polar Form
Multiplying Complex Numbers and Raising to a Power
Finding Roots of a Complex Number

Recall that a complex number of the form a + bi, where $i = \sqrt{-1}$. Some operations of arithmetic of complex numbers such as raising to powers and extracting roots are easier to perform in polar form. In this section we learn how to express a complex number in polar form. We will also find all the complex roots of polynomial equations. Most graphing utilities perform these operations automatically.

Expressing a Complex Number in Polar Form

A complex number can be graphed in a complex plane where the horizontal axis represents the real part of the complex number and the vertical axis represents the imaginary part (Figure 1). We will express a complex number as $z = x + yi$.

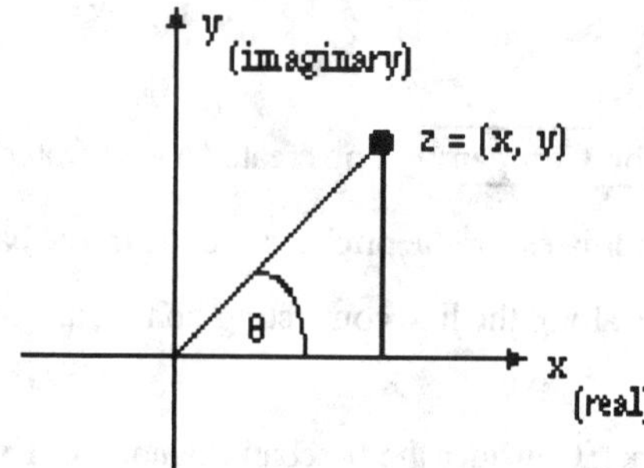

Figure 1

From Figure 1 we have

$$\cos\theta = \frac{x}{r} \quad \text{and} \quad \sin\theta = \frac{y}{r}$$

To write any complex number $z = x + yi$ in polar form we use the polar equations

$$x = r\cos\theta \quad \text{and} \quad y = r\sin\theta$$

and substitute for x and y as follows.

$$z = x + yi$$

$$z = r\cos\theta + (r\sin\theta)i$$

$$z = r(\cos\theta + i\sin\theta)$$

Using shortened notation, we express $z = r(\cos\theta + i\sin\theta)$ as $z = r \text{ cis } \theta$. If the graph of $z = x + yi$ falls in quadrant I or IV, $\theta = \tan^{-1}(\frac{y}{x})$. If the graph of z is in quadrant II or III, $\theta = \tan^{-1}(\frac{y}{x}) + \pi$.

In the following example we practice changing a complex number in rectangular coordinates into its equivalent polar form. To express $z = x + yi$ in the form $z = r \text{ cis } \theta$, we need to find values for θ and r.

Example 1 *Expressing a complex number in polar form*

Write the polar form for each complex number.

(a) $z = 2 + 2i$

(b) $z = -5\sqrt{3} + 5i$

Solution

(a) $\theta = \tan^{-1}(\frac{y}{x})$ and $x = 2, y = 2 \Rightarrow$

$$\theta = \tan^{-1} 1$$

$$\theta = \pi/4 \approx 0.78$$

Using the Pythagorean theorem,

$r^2 = x^2 + y^2$ and $x = 2, y = 2 \Rightarrow$

$$r^2 = 2^2 + 2^2$$

$$r = \sqrt{8} \approx 2.83$$

Therefore, the polar form is $z = \sqrt{8}\ \text{cis}\ (\frac{\pi}{4}) \approx 2.83\ \text{cis}\ (0.78)$.

(b) Since the graph of z is in quadrant II,

$\theta = \tan^{-1}(\frac{y}{x}) + \pi$ and $x = -5\sqrt{3}, y = 5 \Rightarrow$

$$\theta = \tan^{-1} \frac{5}{-5\sqrt{3}}$$

$$= \frac{-\pi}{6} + \pi \approx -0.52 + \pi$$

$$= \frac{5\pi}{6} \approx 2.62$$

$r^2 = x^2 + y^2$ and $x = -5\sqrt{3}, y = 5 \Rightarrow$

$$r^2 = (-5\sqrt{3})^2 + 5^2$$

$$r = 10$$

Therefore, the polar form is $z = 10\ \text{cis}\ (\frac{5\pi}{6}) \approx 10\ \text{cis}\ (2.62)$.

We can check our results by changing the polar form of the numbers back into rectangular form. For part (a) of Example 1,

$$z = 2\sqrt{2}\,(\cos\frac{\pi}{4} + i\sin\frac{\pi}{4})$$ *Evaluate* $\cos\frac{\pi}{4}$ *and* $\sin\frac{\pi}{4}$.

$$= 2\sqrt{2}\,(\frac{\sqrt{2}}{2} + \frac{\sqrt{2}}{2}i)$$ *Apply the distributive property.*

$$= 2 + 2i$$

Multiplying Complex Numbers and Raising to a Power

The following theorem shows a quick way to multiply two complex numbers. Its proof follows directly from the trigonometric identities for sin $(\alpha + \beta)$ and cos $(\alpha + \beta)$.

Multiplying Two Complex Numbers

If z_1 and z_2 are two complex numbers,

$$z_1 = r_1(\cos \alpha + i \sin \alpha) \quad \text{and} \quad z_2 = r_2(\cos \beta + i \sin \beta)$$

their product is

$$z_1 z_2 = r_1 r_2[\cos (\alpha + \beta) + i \sin (\alpha + \beta)] = r_1 r_2 \text{ cis } (\alpha + \beta)$$

Proof: We multiply the polar forms of z_1 and z_2, collect like terms, and simplify.

$$\begin{aligned} z_1 z_2 &= r_1 (\cos \alpha + i \sin \alpha)\, r_2(\cos \beta + i \sin \beta) \\ &= r_1 r_2 (\cos \alpha \cos \beta + i \cos \alpha \sin \beta + i \sin \alpha \cos \beta + i^2 \sin \alpha \sin \beta) \\ &= r_1 r_2 [(\cos \alpha \cos \beta - \sin a \sin b) + (\cos \alpha \sin \beta + \sin \alpha \cos \beta)\, i] \\ &= r_1 r_2 [\cos (\alpha + \beta) + i \sin (\alpha + \beta)] \end{aligned}$$

Example 2 *Finding the product of two complex numbers*

(a) Find the product $z_1 z_2$ for $z_1 = 2 + 2i$ and $z_2 = -5\sqrt{3} + 5i$.

(b) Then express $2 + 2i$ and $-5\sqrt{3} + 5i$ in polar form and find their product using $z_1 z_2 = r_1 r_2 \text{ cis } (\alpha + \beta)$.

Solution

(a)

$$\begin{aligned} z_1 z_2 &= (2 + 2i)(-5\sqrt{3} + 5i) \\ &= -10(1 + \sqrt{3}) + 10(1 - \sqrt{3})i \\ &\approx -27.3 - 7.3i \end{aligned}$$

(b) From Example 1 we know that $z_1 \approx 2.83 \text{ cis } 0.78$ and $z_2 \approx 10 \text{ cis } (2.62)$.

$z_1 z_2 = r_1 r_2 \text{ cis}(\alpha + \beta)$ and $r_1 = 2.83, r_2 = 10, \alpha = 0.78, \beta = 2.62 \Rightarrow$

$$\begin{aligned} z_1 z_2 &\approx (2.83)(10) \text{ cis } (0.78 + 2.62) \\ &\approx 28.3 \text{ cis } (3.4) \\ &\approx 28.3(\cos 3.4) + 28.3\, i(\sin 3.4) \\ &\approx -27.4 - 7.2i \end{aligned}$$

Using $z_1 z_2 = r_1 r_2 \text{ cis } (\alpha + \beta)$ with $\alpha = \beta = \theta$ and $r_1 = r_2 = r$ generates a formula for finding the square of complex number.

$$\begin{aligned} z^2 &= r \cdot r \text{ cis } (\theta + \theta) \\ &= r^2 \text{ cis } 2\theta \end{aligned}$$

Example 3 *Squaring a complex number*

(a) Calculate z^2 for $z = -5\sqrt{3} + 5i$ by squaring z.

(b) Express $z = -5\sqrt{3} + 5i$ in polar form and evaluate z^2 using $r^2 = \text{cis } 2\theta$.

Solution

(a) $z = -5\sqrt{3} + 5i \Rightarrow$

$$\begin{aligned} z^2 &= (-5\sqrt{3} + 5i)^2 \\ &= 25(3) - 50\sqrt{3}i + 25i^2 \\ &= 75 - 50\sqrt{3}i - 25 \\ &= 50 - 50\sqrt{3}i \\ &\approx 50 - 86.6i \end{aligned}$$

(b) From part (b) of Example 1, $z \approx 10 \text{ cis } 2.62$.

$z^2 = r^2 \text{ cis } 2\theta$ and $r = 10, \theta \approx 2.62 \Rightarrow$

$$\begin{aligned} z^2 &\approx 10^2 \text{ cis } 2(2.62) \\ &\approx 100 \text{ cis } (5.24) \\ &\approx 50.3 - 86.4i \end{aligned}$$

The real advantage in using polar form is apparent when we raise a complex number to a power greater than 2.

Since $z^2 = r^2 \text{ cis } 2\theta$, are there analogous formulas for z^3 and z^4?

DeMoivre's theorem gives the formula for raising a complex number to any power that is a natural number.

DeMoivre's Theorem

If z is a complex number whose polar form is

$$z = r(\cos\theta + i\sin\theta) = r \text{ cis } \theta$$

then for any natural number n,

$$z^n = r^n (\cos n\theta + i \sin n\theta) = r^n \text{ cis } (n\theta)$$

Example 4 *Using DeMoivre's theorem*

Find $(1 + i)^4$ using DeMoivre's theorem where $z = 1 + i = \sqrt{2}\ \text{cis}\ (\frac{\pi}{4})$.

Solution

$z^n = r^n$ cis $n\theta$ and $r = \sqrt{2}, \theta = \frac{\pi}{4} \Rightarrow$

$$\begin{aligned} z^4 &= (\sqrt{2})^4 \text{ cis } 4(\tfrac{\pi}{4}) \\ &= 4 \text{ cis } \pi \\ &= 4 \cos \pi + i \sin \pi \\ &= 4(-1 + 0i) = -4 \end{aligned}$$

Finding Roots of a Complex Number

DeMoivre's theorem also allows us to find all the roots of a polynomial equation of the form

$$z^3 = a$$

There are three complex solutions, the three cube roots of *a*. In the next example we use the period of the sine and cosine functions, 2π, to find three values of θ such that $\sin\theta = 0$. We write $\theta = 0 + 2k\pi$, for $k = 0, 1$, and 2.

Example 5 *Finding cube roots of a number*

Solve $z^3 = 8$ using DeMoivre's theorem.

Solution

For $8 + 0i$,

$$\theta = \tan^{-1}(\tfrac{0}{8}) = 0$$

$r^2 = x^2 + y^2$ and $x = 8, y = 0 \Rightarrow$

$$r = 8$$

The polar form of 8 is 8 cis (0).

By DeMoivre's theorem,

$z^3 = r^3$ cis 3θ and $z^3 = 8$ cis $(0) \Rightarrow$

$$r^3 \text{ cis } 3\theta = 8 \text{ cis } (0)$$

Therefore,

$$r^3 = 8 \text{ and } 3\theta = 0 + 2k\pi \text{ for } k = 0, 1, 2$$
$$r = 2 \text{ and } \theta = 0 + \frac{2k\pi}{3} \text{ for } k = 0, 1, 2$$

$z = r \text{ cis } \theta$ and $r = 2, \theta = \frac{2k\pi}{3} \Rightarrow$

$$z = 2 \text{ cis } (\frac{2k\pi}{3})$$
$$= 2(\cos \frac{2k\pi}{3} + i \sin \frac{2k\pi}{3})$$

$k = 0 \Rightarrow$

$$z = 2(\cos 0 + i \sin 0) = 2$$

$k = 1 \Rightarrow$

$$z = 2(\cos \frac{2\pi}{3} + i \sin \frac{2\pi}{3}) = 2(-\frac{1}{2} + \frac{\sqrt{3}}{2} i) = -1 + \sqrt{3}i$$

$k = 2 \Rightarrow$

$$z = 2[\cos(4\pi/3) + i \sin(4\pi/3)] = 2(-\frac{1}{2} - \frac{\sqrt{3}}{2} i) = -1 - \sqrt{3}i$$

The three solutions are 2, $-1 + \sqrt{3}i$, and $-1 - \sqrt{3}i$.

Exercises 7.2

Develop the Concepts

For Exercises 1 to 4 express each complex number in polar form. Then check the results by changing it back into rectangular form.

1. $z = 1 + i$

2. $z = 1 - i$

3.(a) $z = 1$

(b) $z = i$

4.(a) $z = \sqrt{3} + i$

(b) $z = 1 - \sqrt{3}\, i$

For Exercises 5 to 8 find $z_1 z_2$ and change the result into polar form.

5. $z_1 = 2\sqrt{3} + 2i$

$z_2 = 2\sqrt{3} - 2i$

6. $z_1 = 2(\cos \frac{-\pi}{6} + i \sin \frac{-\pi}{6})$

$z_2 = 2(\cos \frac{\pi}{3} + i \sin \frac{\pi}{3})$

7. $z_1 = 5\,[\cos (\tan^{-1} \frac{-4}{3} + i \sin (\tan^{-1} \frac{-4}{3})]$

$z_2 = 2\sqrt{2}(\cos \frac{\pi}{4} + i \frac{\sin \pi}{4})$

8. $z_1 = -6 + 8i$

$z_2 = -4 - 4i$

For Exercises 9 to 16 solve each of the following using DeMoivre's theorem.

9. $z^3 = 1$

10. $z^3 = 8 + 8i$

11. $z^5 = i$

12. $z^4 = 81i$

13. $z^3 = 1 + \sqrt{3}\, i$

14. $z^5 = -8 - 8i$

*15. $z^3 = -27 - 27i$

16. $z^4 = (2 + 2\sqrt{3}i)^3$

Extend the Concepts

17. Find the fourth roots of unity using DeMoivre's theorem.

18. Find $\sqrt{z}$ for $z = 3 + 4i$ by solving $a + bi = \sqrt{3 + 4i}$ for a and b.

19. Interpret the geometric result of multiplying $z = 1 + \sqrt{3}i$ by i.

20. Interpret the geometric result of multiplying $z = r(\cos \theta + i \sin \theta)$ by $z = -i = \cos \frac{3\pi}{2} + i \sin \frac{3\pi}{2}$.

21. Express $z_1 = 3i$, $z_2 = \sqrt{3} + i$ and their quotient z_1/z_2 in polar form. What do you notice?

22. Euler's formula, which takes calculus to derive, states that a complex number can be written as $x + yi = r(\cos \theta + i \sin \theta) = re^{i\theta}$. Write $z = 1 + i$ in exponential form.

*23. Extend the domain of $y = \ln x$ to the complex numbers by showing that $\ln z = \ln re^{i\theta} = \ln r + i\theta + 2k\pi i$, $k = 0, 1, 2, \ldots$.

24. Find $\ln (1 + i)$.

*25. Show that i^i is a real number. [Hint: Use logarithms and Euler's formula from Exercise 22.]

7.3 Parametric Equations

Using Parametric Equations to Model Motion
Eliminating the Parameter
Changing Polar Equations into Parametric Equations
Applying Parametric Equations

A **parametric equation** is an equation that expresses the x and y coordinates of any point in terms of another variable, called a **parameter**. For example, visualize the flight of a golf ball from the time it leaves the tee to the time it lands on the fairway. The flight involves both horizontal and vertical distances, as well as the distance along its parabolic flight path. We describe the horizontal distance with an equation of the form

$$x = x(t)$$

and the vertical distance with an equation of the form

$$y = y(t)$$

where t represents time. Such a pair of functions are called parametric equations and t is called a parameter. Any value of t generates the x and y coordinates of the position of the ball at any time t. Graphing the ordered pairs (x, y) for all t in the domain creates the geometric model of the actual flight path of the ball.

Using Parametric Equations to Model Motion

We usually use the word *velocity* to represent the vector **v** whose magnitude is the speed of the object and whose direction is the direction of the motion of the object. Now that we have the language of vectors we can be more precise. First, let's consider the vertical aspect of motion of a free-falling object. The formula

$$y = -\frac{1}{2}gt^2 + \|\mathbf{v}_y\|\,t + y_0$$

describes the height above the ground at any time t in seconds, where g is the acceleration due to gravity, $\|\mathbf{v}_y\|$ is the initial vertical velocity, and y_0 is the initial height above the ground.

Why does the magnitude $\|\mathbf{v}_y\|$ determine the duration of flight?

The vertical component of the initial velocity determines the duration of flight because the larger the initial vertical component of the velocity, the longer the object will stay in the air. Free-falling objects are only accelerated or decelerated by gravity --- gravity only acts vertically, toward the center of the earth.

Next, let's consider the horizontal aspect of motion for any free-falling object. The horizontal component of the velocity is a constant. It is unaffected by gravity since gravity only acts vertically. This horizontal component of the initial velocity is the value of r in the formula

$$d = rt$$

where d is distance, r is a constant rate, and t is time.

In Section 6.5 we established the horizontal and vertical components of **v** = (a, b) as

$$a = \|\mathbf{v}\| \cos \theta \quad \text{and} \quad b = \|\mathbf{v}\| \sin \theta$$

The magnitudes of the horizontal component $\mathbf{v}_x$ and the vertical component $\mathbf{v}_y$ of a velocity vector $\mathbf{v}_0$ are

$$\|\mathbf{v}_x\| = \|\mathbf{v}_0\| \cos \theta \quad \text{and} \quad \|\mathbf{v}_y\| = \|\mathbf{v}_0\| \sin \theta$$

We can model the flight of an object launched at an angle θ with the horizon if we know the initial velocity of the object and the acceleration due to gravity.

Suppose that a golf ball leaves the tee at an initial velocity of 220 ft/sec at an angle of elevation of θ = 13°. In the vertical direction, the height of the ball at any time is

$$y = -\frac{1}{2}gt^2 + \|\mathbf{v}_y\|\, t + y_0$$

where g, the acceleration due to gravity, is 32 ft/sec^2.

$y(t) = -\frac{1}{2}t^2 + \|\mathbf{v}_y\|\, t + y_0$ and $g = 32$, $\|\mathbf{v}_y\| = \|\mathbf{v}_0\| \sin \theta \Rightarrow$

$$y = -16t^2 + \|\mathbf{v}_0\| \sin \theta\, t + y_0$$

$y(t) = -16t^2 + \|\mathbf{v}_0\| \sin \theta\, t + y_0$ and $v_0 = 220, \theta = 13°, y_0 = 0 \Rightarrow$

$$y(t) = -16t^2 + 49.49t + 0$$

This equation models the height of the golf ball as a function of time.

The velocity in the x direction is constant during the entire flight. Hence,

$$d = rt$$

$$x(t) = \|\mathbf{v}_x\|\, t = (\|\mathbf{v}_0\| \cos \theta)\, t$$

$x(t) = (\|\mathbf{v}_0\| \cos \theta)\, t$ and $\|\mathbf{v}_0\| = 220$, $\theta = 13° \Rightarrow$

$$x(t) = 220(\cos 13°)\, t$$
$$\approx 214.36t$$

This equation models the horizontal distance traveled by the golf ball. The position of the ball at any time t is given by the vector

$$\mathbf{p} = (214.36t, -16t^2 + 49.49t)$$

Most graphing utilities have the capability to graph functions given as parametric equations. The viewing window must be set for t as well as for x and y. Figure 1 shows the window that will calculate ordered pairs (x(t), y(t)) for t = 0 to t = 4 in increments of 0.1.

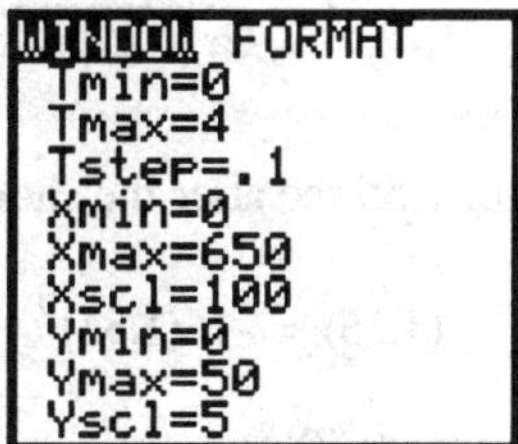

Figure 1

Figure 2 shows the resulting graph of $x(t) = 214.36t$ and $y(t) = -16t^2 + 49.49t$.

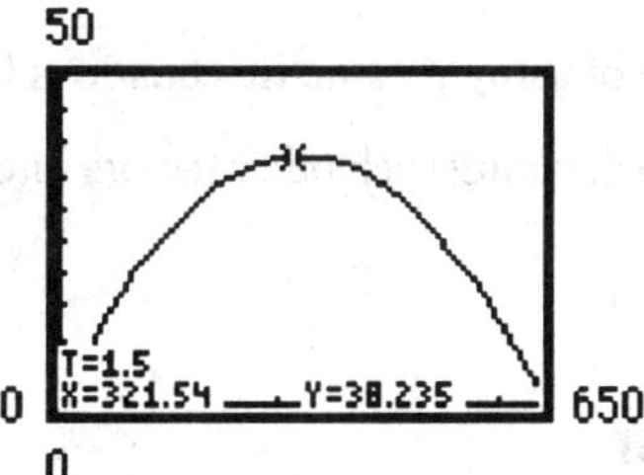

Figure 2

Example 1 <u>The Flight of the Golf-Ball Problem</u>

Discuss the flight of the golf ball hit at an angle of elevation of 13° with an initial velocity of 220 ft/sec.

Thinking: Creating a Plan

The duration of the flight is found by setting $y(t) = 0$ to find the time when the height above the ground is 0. The length of the drive is found by substituting this time into function x(t). The height of the ball is found by finding the maximum value of y, using $t = -\frac{b}{2a}$ for quadratic equations.

Communicating: Writing the Solution

<u>Time of flight</u>: $y(t) = -16t^2 + 49.49t$ and $y(t) = 0 \Rightarrow$

$$-16t^2 + 49.49t = 0$$

$$t(-16t + 49.49) = 0$$

$$t = 0 \text{ or } -16t + 49.49 = 0$$

$$t = 0 \text{ or } t = 3.09$$

The flight took about 3.09 sec.

<u>Length of shot</u>:

$$x(t) = 214.36t$$

$$x(3.09) \approx 214.36(3.09) = 662.37$$

The length of the shot was about 662.37 ft (221 yd).

Height of shot: The maximum value of $y(t) = -16t^2 + 49.49t$ occurs when

$$t = \frac{-b}{2a} = \frac{-49.49}{2(-16)} \approx 1.55$$

Hence, the maximum height is attained about 1.55 sec after the shot. Its maximum height is

$$y(1.55) = -16(1.55)^2 + 49.49(1.55) = 38.2695$$

Therefore, the maximum height was approximately 38 ft.

Learning: Making Connections

This model is still a little oversimplified because wind resistance and the shape of the golf ball (the dimples) were ignored. In calculus, the real value of using parametric equations for such models becomes even more clear when velocity and acceleration are studied, taking such other factors into account.

Eliminating the Parameter

Example 2 shows how to eliminate the parameter t by substitution. This often helps identify the path described by the parametric equations.

Example 2 *Eliminating the parameter*

Change the pair of equations for the golf ball in Example 1,

$$x = 214.36t \quad \text{and} \quad y = -16t^2 + 49.49t$$

into one equation in x and y by eliminating the parameter t.

Solution

We solve the first equation for t as a function of x and substitute the result into the second equation.

$x = 214.36t \Rightarrow$

$$t = \frac{x}{214.36}$$

Then

$$y = -16t^2 + 49.49t$$

$$y = -16\left(\frac{x}{214.36}\right)^2 + 49.49\left(\frac{x}{214.36}\right)$$

$$y = -0.00035x^2 + 0.2309x$$

What do you think the equation $y = -0.00035x^2 + 0.2309x$ represents?

This equation is the actual path of the golf ball described above. It is a concave-down parabola whose intercepts are (0,0) and (659.7, 0). Its vertex is (329.86, 38.08) (Figure 3).

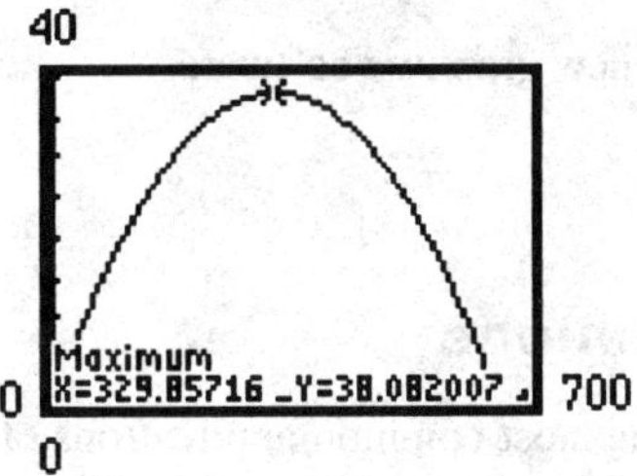

Figure 3

These results approximate those in Example 1 except that the rectangular form does not yield the time for any of the ball's positions.

Changing Polar Equations into Parametric Equations

Every polar equation $r = f(\theta)$ can be quickly changed to parametric equations using

$$x = r\cos\theta \text{ and } y = r\sin\theta$$

to create the parametric equations

$$x = f(\theta)\cos\theta \text{ and } y = f(\theta)\sin\theta$$

where θ is now the parameter. This important modern technique allows computers to graph relations that are not functions. This approach is used by graphing utilities to graph polar equations in a rectangular coordinate grid.

In the next example we use the double-angle formula $\sin 2\theta = 2\sin\theta\cos\theta$ to write a product as a single sine function.

Example 3 *Changing polar equations to parametric equations*

Change the polar equation of a circle $r = \cos\theta$ into two parametric equations.

Solution

$x = r\cos\theta$ and $r = \cos\theta \Rightarrow$

$$x = (\cos\theta)(\cos\theta)$$

$$x = \cos^2\theta$$

$y = r\sin\theta$ and $r = \cos\theta \Rightarrow$

$$y = \cos\theta\sin\theta \Rightarrow y = \frac{1}{2}\sin 2\theta$$

Changing a polar equation to parametric equations creates ordered pairs (x, y) for values of θ. In Example 3, the ordered pairs are of the form $(\cos^2 \theta, \frac{1}{2}\sin 2\theta)$, which can be plotted on a graphing utility in a rectangular coordinate system without creating any new graphing software.

Applying Parametric Equations

The next two examples are typical of the most common applications of parametric equations. Similar problems are studied in every calculus text.

Example 4 The Ferris-Wheel Problem

Write the parametric equations of a ride on a Ferris wheel having a diameter of 240 ft if 1 revolution takes 30 sec. Graph $x = x(t)$ and $y = y(t)$ and discuss the model with θ measured in radians.

Thinking: Creating a Plan

The radius of the Ferris wheel is 120 ft. From the right triangle in Figure 4, we express x and y as functions of θ, where θ is the angle through which the circular wheel has turned.

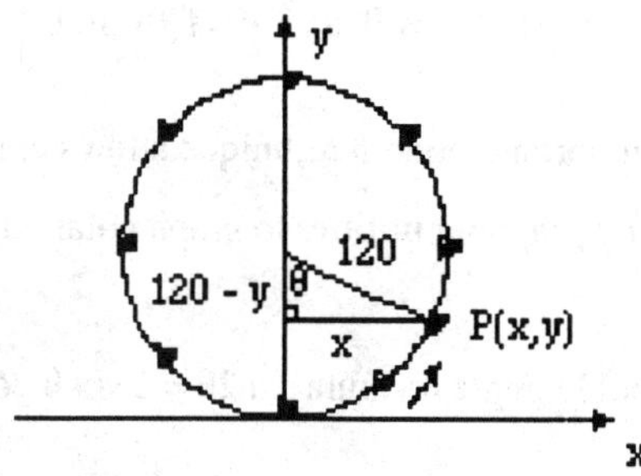

Figure 4

Then we find θ as a function of time and use substitution to complete the derivation of the parametric equations. Since the velocity of the Ferris wheel is constant, the distance traveled s is given by $d = rt$.

Communicating: Writing the Solution

From the right triangle in Figure 4,

$$\sin\theta = \frac{\text{opposite}}{\text{hypotenuse}} \qquad \cos\theta = \frac{\text{adjacent}}{\text{hypotenuse}}$$

$$\sin\theta = \frac{x}{120} \qquad \cos\theta = \frac{120 - y}{120}$$

$$x = 120\sin\theta \qquad y = 120 - 120\cos\theta$$

We have just expressed x and y as functions of θ. For example, if $\theta = \frac{\pi}{2}$, the car on the Ferris wheel is halfway up, since

$$x\left(\frac{\pi}{2}\right) = 120 \sin\left(\frac{\pi}{2}\right) = 120$$

and

$$y\left(\frac{\pi}{2}\right) = 120 - 120 \cos\left(\frac{\pi}{2}\right) = 120$$

Next we will express θ as a function of time. Assume that a constant rate of speed allows us to use the formula d = rt for rate r. The distance d is the circumference of the circle, 240π ft, since

$$d = 2\pi r = 240\pi$$

One revolution requires 30 sec. The rate of the Ferris wheel is 8π ft/sec, since

$$r = \frac{d}{t} \Rightarrow r = \frac{240\pi}{30} = 8\pi$$

The distance traveled is the arc length s,

$$d = s = r\theta = 120\theta$$

and r = 8π ft/sec. Thus we have the relationship between θ and time.

$$\begin{aligned} d &= rt \\ 120\,\theta &= (8\pi)t \\ \theta &= \frac{8\pi}{120}t = \frac{\pi}{15}t \end{aligned}$$

By substitution,

$$x = 120 \sin\theta \text{ becomes } x = 120\sin\frac{\pi}{15}t$$

and

$$y = 120 - 120\cos\theta \text{ becomes } y = 120 - 120\cos\frac{\pi}{15}t$$

The graph of the parametric equations for $0 \le t \le 30$ is shown in Figure 5.

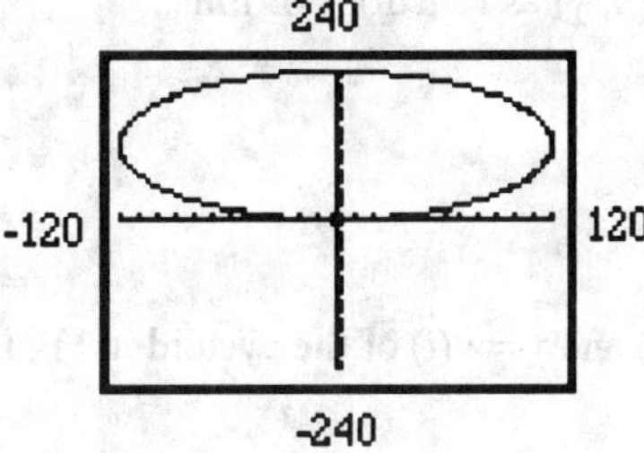

Figure 5

Learning: Making Connections

The graphs of each of these parametric equations are shown in Figure 6.

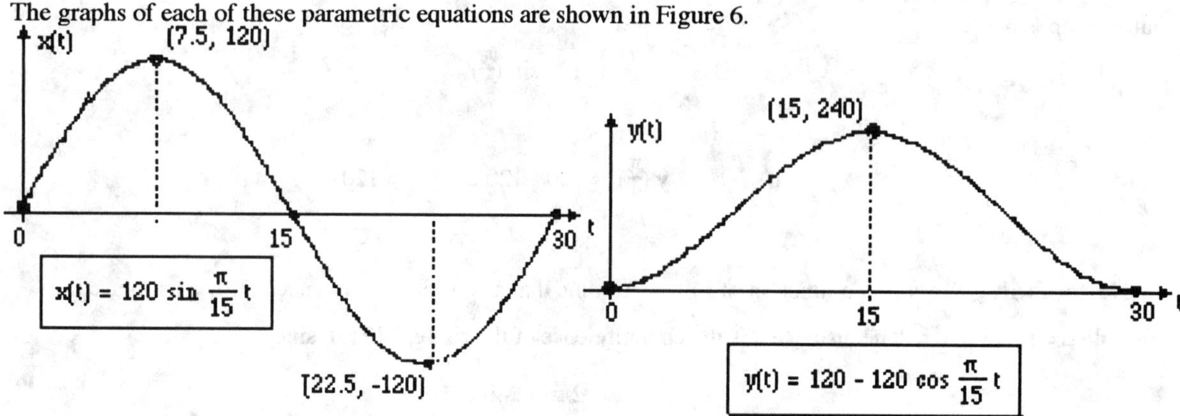

Figure 6

Notice that x(0) = y(0) = 0 since we placed the origin of our coordinate system at the starting point on the ground. As t increases, both x and y increase. The maximum x value is reached in 7.5 sec, when the car is halfway up to the top of the wheel. Then the car moves in a negative x direction (to the left). At t = 15, the maximum height is reached, before the car begins to come down. The car is back on the ground at t = 30 sec. The y value is never negative because the car is always above the ground.

The history of the parametric equations of a cycloid in the next example involves many famous mathematicians. A **cycloid** is the path traced out by one point on a circle as the circle rolls along the ground. A portion of an upside-down cycloid is the best shape for children's slides because it is the fastest path under gravity from the top of a slide to a fixed point on the ground. This property gives it the name *brachistochrone*, meaning "path of least time." Also, if one child starts from the top and another from any other point on a cycloid slide, they will reach the ground at the same time. This property gives the curve the name *tautochrone*, where *tauto* means "same." This derivation of the parametric equations of the cycloid is based on expressing the coordinates of the point on the circle P(x, y) as functions of time.

Example 5 The Cycloid Problem

Derive the parametric equations x = x(t) and y = y(t) of the cycloid in Figure 7.

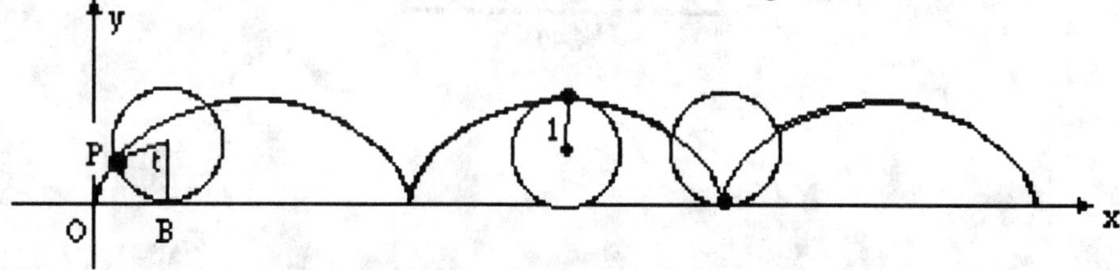

Figure 7

Thinking: Creating a Plan

Consider a circle of radius 1 rolling from point O to point B (Figure 8). The distance OB on the x axis equals the arc length s, OB = s.

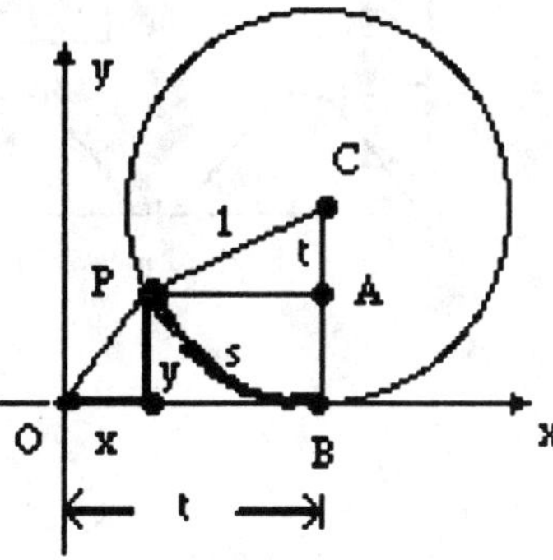

Figure 8

For the circle in Figure 8, $s = r\theta$ and $r = 1, \theta = t \Rightarrow s = t$. Hence, OB = t. The right triangle PCA will yield x and y as functions of t.

Communicating: Writing the Solution

$$\sin t = \frac{AP}{1} \Rightarrow AP = \sin t$$

$$\cos t = \frac{CA}{1} \Rightarrow CA = \cos t$$

From Figure 8 we find one of the parametric equations of the cycloid.

$OB = x + AP$ — *Substitute t = OB and AP = sin t.*

$t = x + \sin t$ — *Solve for x.*

$\mathbf{x = t - \sin t}$

Since CB is a radius of a circle, CB = 1.

$CA + y = CB$ — *Substitute 1 = CB and CA = cos t.*

$\cos t + y = 1$ — *Solve for y.*

$\mathbf{y = 1 - \cos t}$

Learning: Making Connections

The graphs in Figure 9 show x and y as separate functions of t. Notice how x increases without bound but not at a constant rate. Sometimes the point is moving backward relative to the wheel. Notice also that y travels from 0 to 2 repeatedly with period $t = 2\pi$ since the circumference of the circle is $C = 2\pi r = 2\pi(1) = 2\pi$.

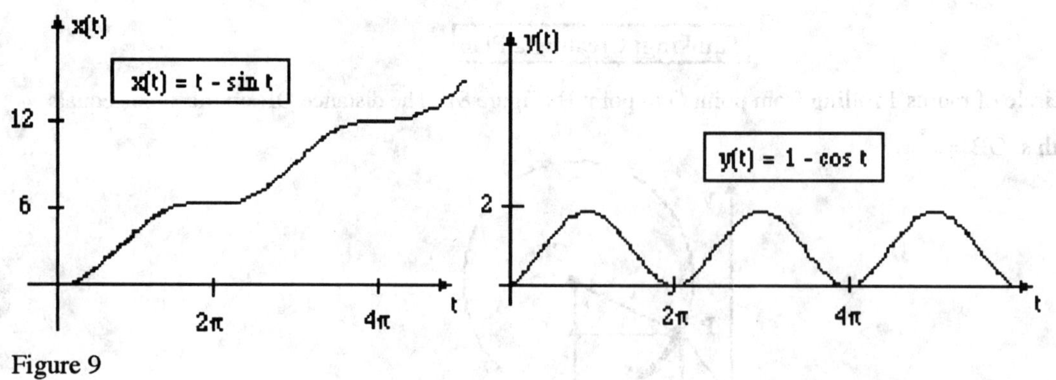

Figure 9

Exercises 7.3

Develop the Concepts

For Exercises 1 to 8 change the pair of equations into one equation in x and y by eliminating the parameter. Then sketch the graph.

1. $x = \sin t$, $y = \cos t$
2. $x = \cos t$, $y = \sin t$
3. $x = 2 \sin t$, $y = 4 \cos t$
4. $x = \sin t$, $y = 2 \cos t$
5. $x = 1 - \cos t$, $y = 1 - \sin t$
6. $x = t - 1$, $y = t$
7. $x = e^t$, $y = e^{2t}$
8. $x = \cos t$, $y = \cos 2t$

Exercises 9 to 12 are parametric equations for $y = x^2$. If t is time, determine where each graph starts at $t = 0$ and whether it represents all or part of the path $y = x^2$ for $t \geq 0$.

9. $x = t$, $y = t^2$
10. $x = \cos t$, $y = \cos^2 t$
11. $x = e^{t/2}$, $y = e^t$
12. $x = \sin t$, $y = (1 - \cos 2t)/2$

For Exercises 13 to 16 express x and y as functions of the parameter t.

13. $x^2 + y^2 = 1$
[Hint: Let $x = \cos t$.]

14. $x^2/1 + y^2/4 = 1$
[Hint: Let $y = 2 \sin t$.]

*15. $x^2/9 + y^2/4 = 1$

16. $x + y^2 = 1$

17. Show that $x = 4 - t$ and $y = 2 + t$ are parametric equations for a line.

18. Show that $x = 2 - 3t$ and $y = 1 + 2t$ are parametric equations for a line.

*19. Find parametric equations for the line through (0,4) with a slope of $m = 3/4$. (There is more than one answer.)

20. Find parametric equations for the line through (0,4) and (5,0). (There is more than one answer.)

Apply the Concepts

21. <u>The Longest Drive Golf Ball Problem</u> You hit your drive with an initial velocity of 30 m/sec at an angle of elevation of 30° and your friend hit his at 60° with an initial velocity of 30 m/sec. Who won the longest-drive contest? [Hint: In the metric system, the acceleration due to gravity is 9.8 m/sec.]

22. <u>The Golf on the Moon Problem</u> In February 1971, *Apollo 14* landed on the moon and astronaut Alan Shepard hit several golf balls. There was no air resistance since there was no air. If the moon's gravity is about one-sixth that of Earth's, discuss the possibilities.

*23. The Merry-Go-Round Problem Write the parametric equations for a playground ride on a merry-go-round that is spinning at 1 rev every 10 sec. Assume that the diameter is 10 m. Graph $x = x(t)$ and $y = y(t)$ and discuss the model.

24. The Angular Velocity Problem Show that the parametric equations for circular motion with velocity ω radians per second are $x = r \cos \omega t$ and $y = r \sin \omega t$.

Exercises 25 and 26 are generalizations of the cycloid model.

25. The Wheel and its Hubcap Problem Consider a tire positioned on a road such that its hubcap just touches the curb (Figure 10). Point A is on the top of the tire and point B is on the top of the hubcap. The wheel is 2 ft in diameter, and the hubcap is 1 ft in diameter. Use parametric equations to model the situation if the wheel is turning at 1 rev/sec. Compare x(t) for 1 revolution of the wheel. (The graph of the point on the hubcap is called a curtate cycloid.)

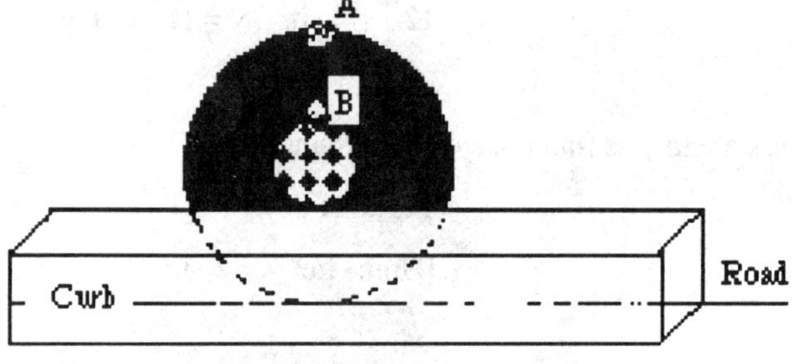

Figure 10

26. The Hypocycloid Problem Graph the hypocycloid whose parametric equations are $x(t) = 4 \cos^3 t$ and $y(t) = 4 \sin^3 t$.

Extend the Concepts

27. Prove that the parametric equations $x(t) = (1 - t)x_1 + tx_2$ and $y(t) = (1 - t)y_1 + ty_2$ represent the line through (x_1, y_1) and (x_2, y_2).

28. Find parametric equations for the parabola whose intercepts are $(-r, 0)$ and $(r, 0)$ and whose vertex is $(0, r^2)$.

A Guided Review of Chapter 7

7.1 Introduction to Polar Coordinates

For Exercises 1 to 4 identify each curve and sketch its graph.

1. $\theta = \pi/12$
2. $r = 12$
3. $r = -12 \sin \theta$
4. $r = -12 \cos \theta$

For Exercises 5 to 8 change the following rectangular equations to equations in polar form and simplify.

5. $y = -\dfrac{1}{\sqrt{2}}x$
6. $y = -\sqrt{3}\,x$
7. $x^2 + y^2 = 144$
8. $y = \pm\sqrt{4 - x^2}$

For Exercises 9 to 16 change the following polar equations into equations in rectangular form and simplify.

9. $r = 12$
10. $r = 4$
11. $\theta = 5\pi/3$
12. $\theta = -3\pi/2$
13. $r = -12 \sin \theta$
14. $r = -12 \cos \theta$
15. $r = \dfrac{1}{1 - \sin \theta}$
16. $r = \dfrac{1}{2 - 2\cos \theta}$

For Exercises 17 to 28 identify each curve and sketch its graph.

17. $\theta = 5\pi/4$
18. $r = \theta$
19. $r = -6 \sin \theta$
20. $r = -6 \cos \theta$
21. $r = 12 - 12 \cos \theta$
22. $r = 6 - 12 \cos \theta$
23. $r = 12 + 12 \sin \theta$
24. $r = 6 + 2 \cos \theta$
25. $r = 12 \cos 4\theta$
26. $r = 12 \cos 2\theta$
27. $r^2 = 4 \cos \theta$
28. $r^2 = -9 \sin 2\theta$

For Exercises 29 and 30 solve each system of equations.

29. $r = 12 + 12 \cos \theta$
    $r = 12 - 12 \sin \theta$
30. $r^2 = 24 \cos \theta$
    $r = 3 \sin \theta$

7.2 Polar Form of Complex Numbers and DeMoivre's Theorem

For Exercises 31 to 34 express each complex number in polar form. Then check the results by changing it back into rectangular form.

31. $z = 12 - 5i$
32. $z = 5 + 12i$

33.(a) $z = -i^2$
   (b) $z = (-i)^2$

34.(a) $z = \sqrt{3} - i$
   (b) $z = 1 + \sqrt{3}\,i$

For Exercises 35 and 36 find $z_1 z_2$ in polar form.

35. $z_1 = \sqrt{3} + i$

$z_2 = \sqrt{3} - i$

36. $z_1 = 2(\cos \frac{-5\pi}{6} + i \sin \frac{-5\pi}{6})$

$z_2 = 2(\cos \frac{2\pi}{3} + i \sin \frac{2\pi}{3})$

For Exercises 37 to 40 solve using DeMoivre's theorem.

37. $z^3 = 125$

38. $z^3 = 125 + 125i$

39. $z^5 = -i$

40. $z^4 = (1 + i)^3$

7.3 Parametric Equations

For Exercises 41 to 44 change the pair of equations into one equation in x and y by eliminating the parameter. Sketch the graph.

41. $x = \sin 2t$, $y = \cos 2t$

42. $x = 2 \cos t$, $y = 2 \sin t$

43. $x = 2 \sin t$, $y = 4 \cos t$

44. $x = e^{t/2}$, $y = e^t$

Exercise 45 and 46 are parametric equations for $y = \sqrt{x}$. If t is time, determine where each graph starts at $t = 0$ and whether it represents all or part of the path $y = \sqrt{x}$, for $t \geq 0$.

45. $x = t$, $y = \sqrt{t}$

46. $x = \cos^2 t$, $y = \cos t$

For Exercises 47 and 48 express x and y as functions of the parameter t.

47. $x^2 + y^2 = 4$

48. $r = 2 + 2 \sin \theta$

49. The Golf on the Moon Problem In February 1971, *Apollo 14* landed on the moon and astronaut Alan Shepard hit several golf balls. There was no air resistance since there was no air. If the moon's gravity is about one-sixth that of Earth's, so that $g = \frac{1}{6}(32)$ ft/sec, how far would a ball hit at an angle of 30° fly if its initial velocity were 100 ft/sec?

50. The Angular Velocity Problem Show that the parametric equations for circular motion with velocity $\omega = 120$ ft/sec are $x = \cos 120t$ and $y = \sin 120t$. What is the radius of this circular motion?

Chapter 7 Test

1. Identify the curve and sketch its graph.

$$\theta = 1$$

2. Change the rectangular equations $x^2 + y^2 = 10{,}000$ to an equation in polar form and graph.

3. Change the following polar equations into equations in rectangular form and simplify.

(a) $r = 100$

(b) $r = -2 \sin \theta$

4. Let $f(\theta) = 2 \cos 3\theta$. State the equation of the new function and sketch its graph.

(a) $r = f(-\theta)$

(b) $r = f(\frac{\pi}{4} - \theta)$

5. Identify each curve and sketch its graph.

(a) $r = - \sin \theta$

(b) $r = 1 - \cos \theta$

(c) $r = 4 - 2 \cos \theta$

(d) $r^2 = \cos 2\theta$

(e) $r = 2\theta$

6. Solve each system of equations.

*(a) $r = 2 \cos 2\theta$

$r = 2 + 4 \sin \theta$

(b) $r^2 = \sin 2\theta$

$r = 1$

7. Find $z_1 z_2$ in polar form.

(a) $z_1 = i$, $z_2 = 1 + i$

(b) $z_1 = 1 - i\sqrt{3}$, $z_2 = \sqrt{3} - i$

8. Change the pair of equations $x = 2 \sin 2t$ and $y = \cos 2t$ into one equation in x and y by eliminating the parameter.

9. For $r = 2 \sin \theta$ express x and y as functions of the parameter t.

10. The Golf on the Moon Problem In February 1971, *Apollo 14* landed on the moon and astronaut Alan Shepard hit several golf balls. If the moon's gravity is about one-sixth that of Earth's, so that $g = \frac{1}{6}(32)$ ft/sec, how far would a ball hit at an angle of 60°° fly if its initial velocity were 200 ft/sec?

11. The Angular Velocity Problem (a) Show that the parametric equations for circular motion with velocity 100 ft/sec and the radius of the circular path is 1 ft are $x(t) = \cos(100t)$, $y(t) = \sin(100t)$.

(b) Find the equations for circular motion if the radius is 2 ft and the velocity is still 100 ft/sec/sec.

Chapter 8

Conic Sections and Quadric Surfaces

Conic sections are the cross section of a right circular cone cut by a plane (Figure 2).

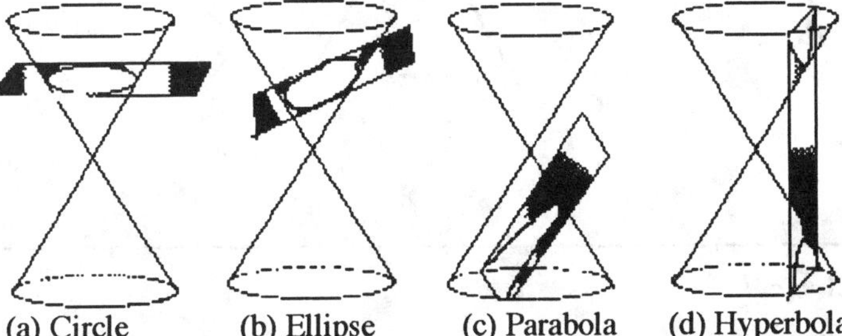

Figure 2 (a) Circle (b) Ellipse (c) Parabola (d) Hyperbola

We begin the chapter by studying the four shapes described in Figure 1 and extend the study to include rotated conic sections and conic sections in three dimensions called *quadric surfaces.*

A Historical Note on Edmund Halley

Edmund Halley was born October 29, 1656, twenty-six years before sighting the comet that bears his name. By age 19, he was recognized as an ingenious youth well versed in all parts of mathematics. From the start of his education at Queen's College, Oxford, his abilities in mathematics, the classics, and astronomy were evident. He cataloged 341 stars in the southern hemisphere.

On a visit to Paris in 1682, he first saw his comet, which then disappeared from view and returned a few days later. (It went behind the sun.) Using data from the Paris Observatory, Halley tried to calculate the comet's course, but he "made a hash of it." A later visit to Sir Isaac Newton led to Halley's eventual correct prediction of when the comet would return. Halley proposed that the comet's path was a result of an inverse-square law of gravitational attraction, stating that the force of attraction is inversely proportional to the square of the comet's distance from the sun. Halley could not prove his hypothesis, but Newton did. The resulting orbit is a skinny ellipse with the sun at one focus. Halley contributed to Newton's famous Mathematical Principles of Natural Philosophy by correcting errors, preparing drawings, and financing the printing. The book contains Newton's concept of the universal law of gravitation as well as Newton's calculus. Without the influence of Halley, we might never have learned this from Newton. As a creative map maker, Halley invented *level curves* to show contours on a map to indicate elevation.

8.1 The Distance Formula and the Circle

Deriving the Equation of a Circle
Finding the Equation of a Circle Given Three Points
Applying the Circle

In this section we study the **circle**, the cross section formed by cutting a cone with a horizontal plane as shown in Figure 1.

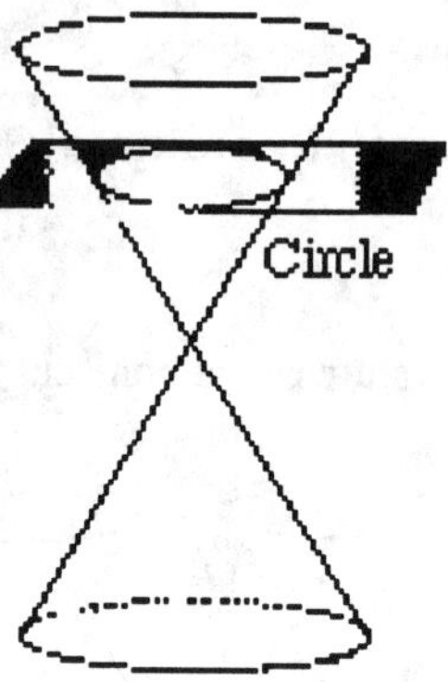

Figure 1

> **Definition of a Circle**
>
> A circle is the set of all points that are a given distance from a fixed point, its center. This constant distance is called its **radius**.

Deriving the Equation of a Circle

To derive the equation of a circle, we use the *distance formula*. Consider the points (x_1, y_1) and (x_2, y_2) in Figure 2a.

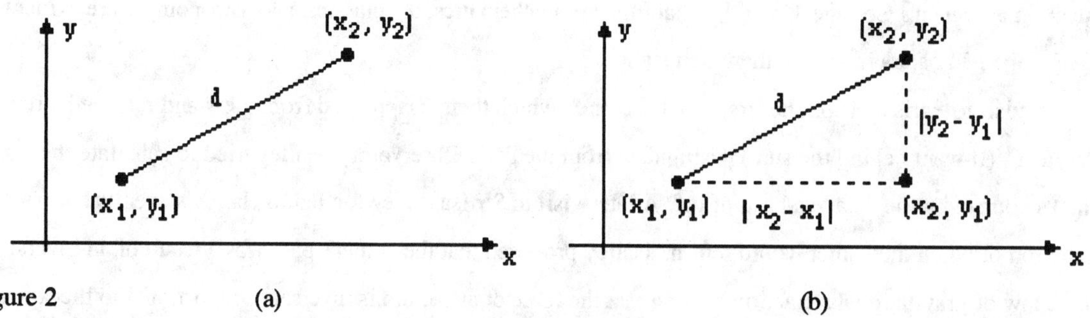

Figure 2 (a) (b)

To find the distance d we draw a right triangle whose hypotenuse is the distance between the points (Figure 2b). Notice that the horizontal distance is $|x_2 - x_1|$ and the vertical distance is $|y_2 - y_1|$. Using the Pythagorean theorem, we have

$$d^2 = (x_2 - x_1)^2 + (y_2 - y_1)^2$$
$$d = \sqrt{(x_2 - x_1)^2 + (y_2 - y_1)^2}$$

NOTE: Since any real number squared yields a nonnegative result, the absolute-value symbols are eliminated from the distance formula.

Distance Formula

The distance between two points (x_1, y_1) and (x_2, y_2) is

$$d = \sqrt{(x_2 - x_1)^2 + (y_2 - y_1)^2}$$

In the first example we find the distance between several points. To identify which distances we are finding, we identify the endpoints of the segments. For example, the notation $d(P_1, O)$ means that we are finding the distance between points P_1 and O.

Example 1 *Using the distance formula*

Find the distances between $P_1(3, 4)$, $P_2(-5, 0)$, $P_3(-3, -4)$, and $P_4(4, -3)$ and the origin O(0, 0) (Figure 3).

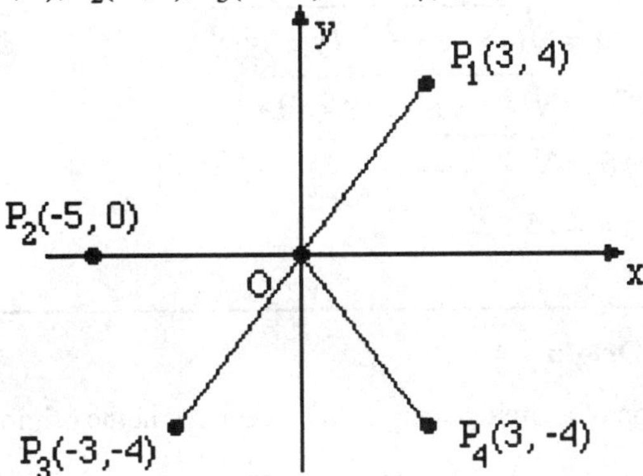

Figure 3

Solution

$$d = \sqrt{(x_2 - x_1)^2 + (y_2 - y_1)^2}$$

$$d(P_1, O) = \sqrt{(3 - 0)^2 + (4 - 0)^2} = \sqrt{9 + 16} = \sqrt{25} = 5$$

$$d(P_2, O) = \sqrt{(-5 - 0)^2 + (0 - 0)^2} = \sqrt{25 + 0} = \sqrt{25} = 5$$

$$d(P_3, O) = \sqrt{(-3 - 0)^2 + (-4 - 0)^2} = \sqrt{9 + 16} = \sqrt{25} = 5$$

$$d(P_4, O) = \sqrt{(4 - 0)^2 + (-3 - 0)^2} = \sqrt{16 + 9} = \sqrt{25} = 5$$

These points are all 5 units from the origin.

In Example 1 the four points lie on a circle of radius 5 (Figure 4).

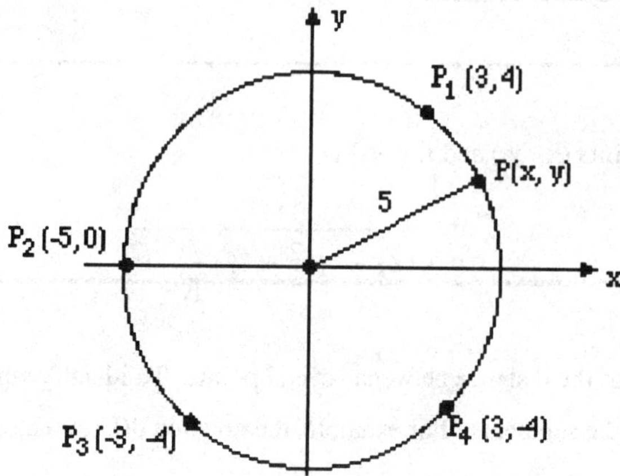

Figure 4

We can find the equation that describes all points P(x, y) that are exactly 5 units from the center of the circle using the distance formula.

$$d = \sqrt{(x_2 - x_1)^2 + (y_2 - y_1)^2}$$

$$d(P,O) = \sqrt{(x - 0)^2 + (y - 0)^2} \qquad \textit{Substitute 5 for } d(P,O).$$

$$5 = \sqrt{x^2 + y^2} \qquad \textit{Square both sides.}$$

$$x^2 + y^2 = 25$$

A Circle Centered at the Origin

The standard form of the equation of a circle having radius r centered at the origin is

$$x^2 + y^2 = r^2$$

We can translate a circle so that its center is a point other than the origin. The equation changes in a predictable way.

Example 2 *Finding the equation of a circle*

Find the equation of the circle of radius r = 5 whose center is C(-2, 1).

Solution

Figure 5 shows an arbitrary point P(x, y) located 5 units from C(-2, 1).

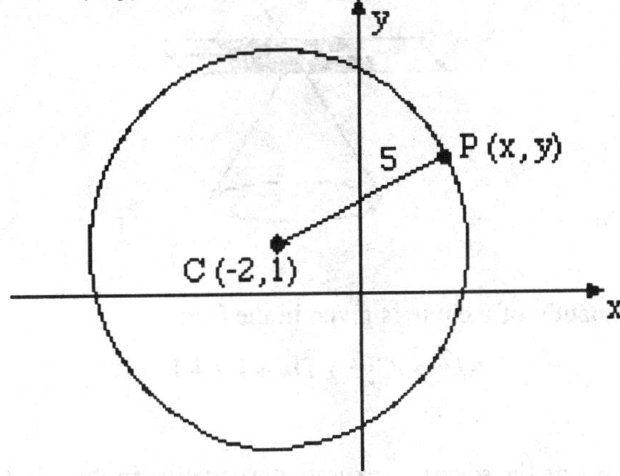

Figure 5

$d = \sqrt{(x_2 - x_1)^2 + (y_2 - y_1)^2}$, and $d = 5$, $x_1 = -2$, $x_2 = x$, $y_1 = 1$, $y_2 = y \Rightarrow$

$$5 = \sqrt{(x - (-2))^2 + (y - 1)^2}$$
$$5 = \sqrt{(x + 2)^2 + (y - 1)^2}$$
$$25 = (x + 2)^2 + (y - 1)^2$$

Therefore, the equation of the circle centered at (-2,1) with radius $r = 5$ is $(x + 2)^2 + (y - 1)^2 = 25$.

A Circle Centered at (h, k)

The standard form of the equation of a circle having radius r centered at (h, k) is

$$(x - h)^2 + (y - k)^2 = r^2$$

Can you describe the graph of $x^2 + y^2 = 0$?

The circle having radius $r = 0$ centered at the origin is a point at (0, 0). We call this *degenerate* conic section a **point circle** or a **null circle** (Figure 6).

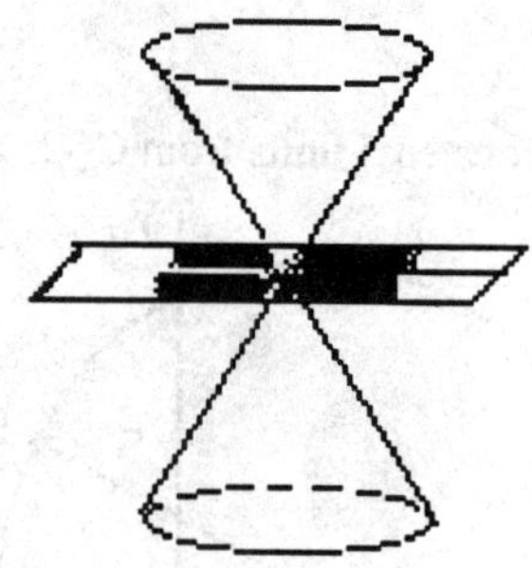

Figure 6

In the next example the equation of a circle is given in the form

$$Ax^2 + Cy^2 + Dx + Ey + F = 0$$

where $A = C \neq 0$. We will complete the square to express the equation in the standard form of a circle,

$$(x - h)^2 + (y - k)^2 = r^2$$

so that we can readily identify the center and radius of the circle.

Example 3 *Expressing the equation of a circle in standard form*

Express the equation $x^2 + y^2 + 4x - 2y - 20 = 0$ in standard form.

Solution

$$x^2 + y^2 + 4x - 2y - 20 = 0 \qquad \textit{Rearrange the terms.}$$

$$(x^2 + 4x + \ \ ) + (y^2 - 2y + \ \ ) = 20 \qquad \textit{Complete the square.}$$

$$(x^2 + 4x + \mathbf{4}) + (y^2 - 2y + \mathbf{1}) = 20 + \mathbf{4} + \mathbf{1}$$

$$(x + 2)^2 + (y - 1)^2 = 25$$

This equation describes the circle in Example 2 with center (-2, 1) and radius $r = \sqrt{25} = 5$. (This graph of this equation is shown in Figure 5.)

Circles are implicit functions because they can be considered in pieces that are functions. We graph a circle on a graphing utility by entering the equations of the top half of the circle and the top and bottom halves separately. To graph the circle described in Example 3, we would first solve the equation for y.

$$(x + 2)^2 + (y - 1)^2 = 25$$

$$(y - 1)^2 = 25 - (x + 2)^2 \qquad \textit{Take the square root of both sides.}$$

$$y - 1 = \pm\sqrt{25 - (x + 2)^2}$$

$$y = 1 \pm \sqrt{25 - (x + 2)^2}$$

Enter $y_1 = 1 + \sqrt{25 - (x + 2)^2}$ and $y_2 = 1 - \sqrt{25 - (x + 2)^2}$ separately (Figure 7).

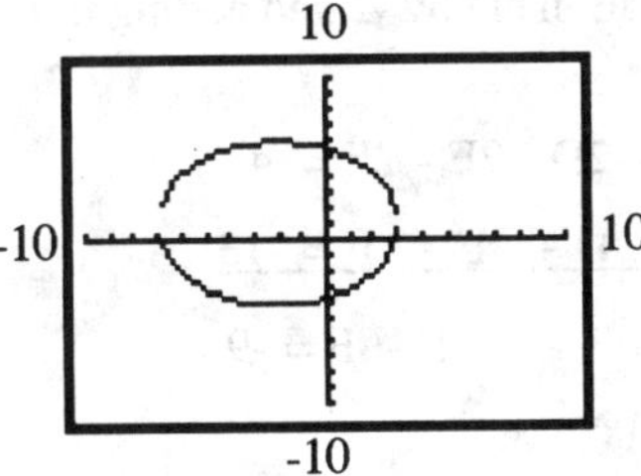

Figure 7

Finding the Equation of a Circle Given Three Points

Many applications require finding the radius of a circle through three given points. It is called the *radius of curvature.* Highway curves are labeled in the engineer's drawing with the radius of curvature to help the survey crew stake out the curve on the roadbed, as we will see in Exercises 37 and 38. The next example illustrates the process of finding the equation of the circle given any three noncollinear points. We will use the general equation

$$x^2 + y^2 + Dx + Ey + F = 0$$

and substitute the coordinates of each of the three points in order to create three linear equations in three variables.

Example 4 *Finding the equation of a circle through three given points*
Find the equation of the circle that passes through the points $P_1(1, 1)$, $P_2(2, 3)$, and $P_3(0, -2)$.

Solution

For P_1, $x^2 + y^2 + Dx + Ey + F = 0$ and $x = 1$, $y = 1 \Rightarrow$

$$1 + 1 + D + E + F = 0$$

$$D + E + F = -2 \qquad \textit{Call this equation (1).}$$

For P_2, $x^2 + y^2 + Dx + Ey + F = 0$ and $x = 2$, $y = 3 \Rightarrow$

$$4 + 9 + 2D + 3E + F = 0$$

$$2D + 3E + F = -13 \qquad \textit{Call this equation (2).}$$

For P_3, $x^2 + y^2 + Dx + Ey + F = 0$ and $x = 0$, $y = -2 \Rightarrow$

$$0 + 4 + 0D - 2E + F = 0$$

$$-2E + F = -4 \qquad \textit{Call this equation (3).}$$

Since P_3 yields an equation with two variables, we use equations (1) and (2) to find a second equation with the variables E and F. Multiplying equation (1) by -2 and adding it to equation (2) yields

$$\begin{aligned} -2D - 2E - 2F &= 4 \\ \underline{2D + 3E + F} &\underline{= -13} \\ E - F &= -9 \end{aligned}$$

Call the sum equation (4).

Then, adding equations (3) and (4) yields

$$\begin{aligned} -2E + F &= -4 \\ \underline{E - F} &\underline{= -9} \\ -E \quad &= -13 \end{aligned}$$

Therefore, E = 13. We find the value of F by substituting E = 13 into equation (4).

$E - F = -9$ and $E = 13 \Rightarrow$

$$\begin{aligned} 13 - F &= -9 \\ F &= 22 \end{aligned}$$

We find the value of D by substituting E = 13 and F = 22 into equation (1).

$D + E + F = -2$ and $E = 13, F = 22 \Rightarrow$

$$\begin{aligned} D + 13 + 22 &= -2 \\ D &= -37 \end{aligned}$$

Therefore, the equation of the circle is $x^2 + y^2 - 37x + 13y + 22 = 0$.

By completing the square, we can express the equation in Example 4 in standard form.

$$x^2 + y^2 - 37x + 13y + 22 = 0 \Rightarrow (x - \frac{37}{2})^2 + (y + \frac{13}{2})^2 = 362.5$$

The radius of curvature passing through these three points is $r = \sqrt{362.5} \approx 19.04$. The center of the circle is $(\frac{37}{2}, -\frac{13}{2})$.

Applying the Circle

Example 5 The Demasted Sailboat Problem

A circular arch bridge has 40 ft clearance in the center and spans 100 ft across the water (Figure 8). If an approaching sailboat requires 35 ft clearance, can the skipper clear the bridge under power while staying in the right lane?

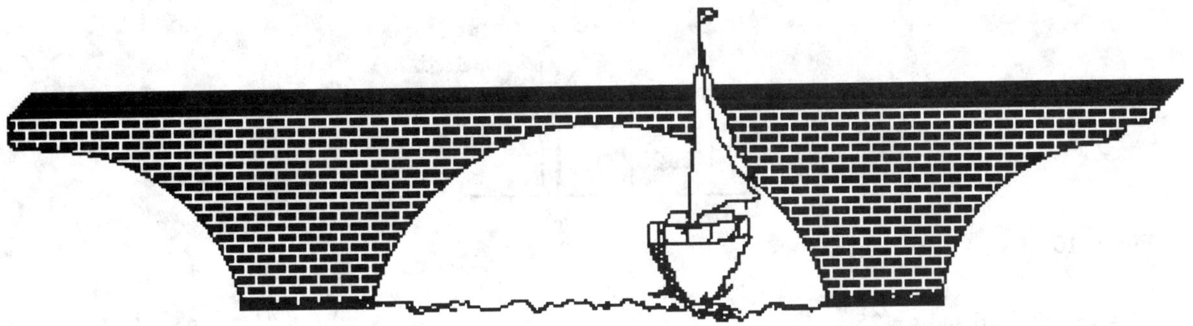

Figure 8

Thinking: Creating a Plan

A sketch of the circle in the Cartesian coordinate system with the center of the arch on the y axis is helpful to find the equation of the circle (Figure 9). Note that the center of the circle (located at the origin) lies below the surface of the water. We can use the Pythagorean theorem to find the equation of the circle.

If we assume that there are two lanes for boat traffic and the center of one lane is at $x = 25$, the height of the bridge above the water, h, can be determined. If it is more than 35 ft, the skipper can pass under the bridge while staying in his lane.

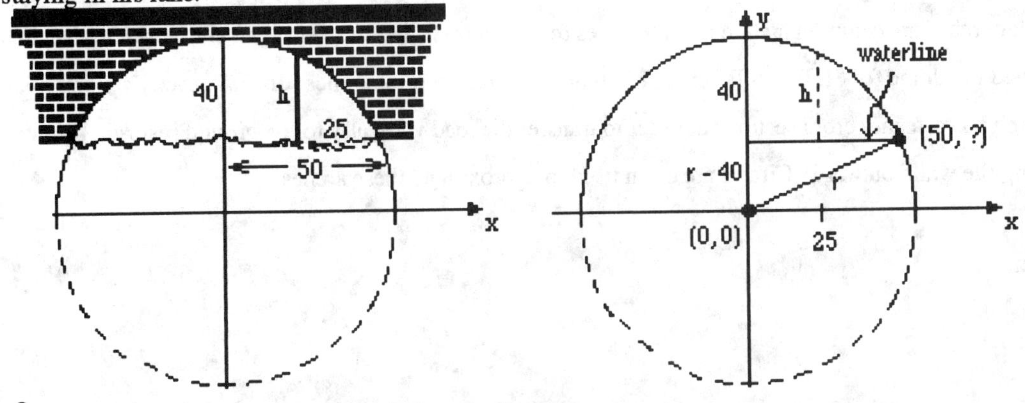

Figure 9

Communicating: Writing the Solution

$a^2 + b^2 = c^2$ and $a = r - 40$, $b = 50$, $c = r \Rightarrow$

$$(r - 40)^2 + 50^2 = r^2$$

$$r^2 - 80r + 1600 + 2500 = r^2$$

$$-80r = -4100$$

$$r = 51.25$$

The equation of the circle containing the arch of the bridge is

$$x^2 + y^2 = (51.25)^2$$

$$x^2 + y^2 = 2626.6$$

Therefore, the center of the circle is 11.25 ft below the waterline since $r - 40 = 51.25 - 40 = 11.25$ (Figure 10).

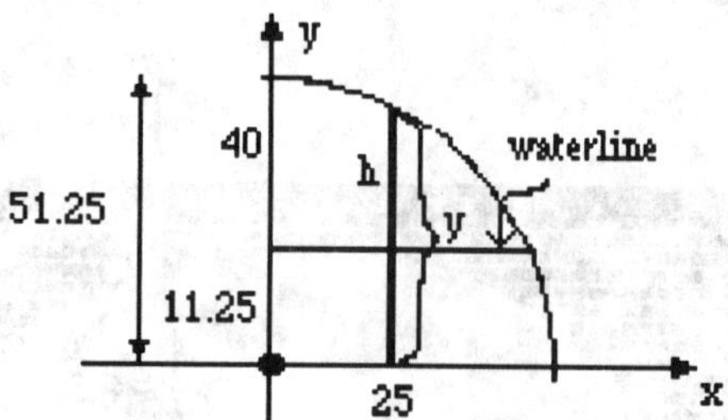

Figure 10

$x^2 + y^2 = 2626.6$ and $x = 25 \Rightarrow$

$$25^2 + y^2 = 2626.6$$
$$y^2 = 2001.6$$
$$y = \pm 44.74$$
$$y = 44.74$$

$h = y - 11.25$ and $y = 44.74 \Rightarrow h = 44.74 - 11.25 = 33.49$. The skipper cannot make it under the bridge staying in this lane since there is only 33.49 ft clearance.

Learning: Making Connections

Circular arches are common architectural features for entrances, windows, and even entire rooms. Michelangelo designed the dome for St. Peters Basilica, but like most architectural domes it is not exactly semicircular. A true arch has catenary cross sections in order to transfer the load vertically to the ground instead of horizontally, pushing the walls outward. Circles are often used to approximate these arches.

Exercises 8.1

Develop the Concepts

For Exercises 1 to 4 use the distance formula to find the distance between points A and B.

1. A(1, -1), B(4, 3)
2. A(-1, -3), B(2, 1)
3. A(3, -5), B(7, -1)
4. A(3, -5), B(5, -3)

For Exercises 5 and 6 use the distance formula to determine whether or not the points are collinear because they lie on the same straight line. [Hint: If d(A,B) + d(B,C) = d(A,C), then points A, B, and C are collinear.]

5. A(0, -7), B(1, -2), C(4, 13)
6. $A(\frac{1}{2}, \frac{1}{2})$, $B(\frac{3}{2}, 0)$, $C(-1, -\frac{5}{4})$

For Exercises 7 and 8 use the distance formula and the Pythagorean theorem to decide whether the points form a right triangle.

7. A(1, 0), B(2, 5), C(3, 2)
8. A(0, 1), $B(\frac{63}{5}, -\frac{11}{5})$, C(3, 5)

For Exercises 9 to 12 find the equation of each circle having the given center and radius.

9. C(0, 0), r = 5
10. C(1, 2), r = 3
11. C(-2, -1), $r = \sqrt{2}$
12. C(0, 2), r = 1/2

For Exercises 13 to 20 state the center and radius of each circle and sketch the graph.

13. $x^2 + y^2 = 1$
14. $x^2 + y^2 = 4$
15. $x^2 + y^2 = 0$
16. $x^2 + y^2 = 8$
17. $(x - 1)^2 + (y - 3)^2 = 1$
18. $(x + 1)^2 + (y + 3)^2 = 1$
19. $(x - 1)^2 + y^2 = 1$
20. $(x - 1)^2 + (y - 3)^2 = 0$

For Exercises 21 to 28 express each equation in the standard form for a circle and state the center and radius.

21. $x^2 + y^2 + 2x + 4y = 11$
22. $x^2 + y^2 - 2x - 4y = 11$
23. $x^2 + y^2 - 4x + 2y + 6 = 0$
24. $x^2 + y^2 + 6x - 4y - 3 = 0$
25. $x^2 - 4x + y^2 = 0$
26. $x^2 + y^2 = 2y$
27. $x^2 + y^2 - 3x + y = 0$
28. $3x^2 + 3y^2 - 12x + 6y - 2 = 0$

For Exercises 29 to 32 graph each semicircle.

29. $y = \sqrt{1 - x^2}$
30. $y = -\sqrt{16 - x^2}$
31. $y = 1 + \sqrt{1 - x^2}$
32. $y = 2 - \sqrt{9 - x^2}$

For Exercises 33 to 36 find the equation of each circle in standard form that passes through the given three points.

33. (-5, 0), (0, 5), (3, 4)
34. (5, 2), (1, 6), (1, -2)
35. (1, 2), (0, 5), (-1, 3)
36. (7, 2), (-1, 2), (3, -2)

Apply the Concepts

37. The Rainbow Kite and Balloon Company Engineers are laying out a circular curve for the driveway to the factory using a Cartesian coordinate system. The beginning of the curve has coordinates (100, 250), the end has coordinates (400, 250), and the radius of the curvature is 150 ft. Find the center of the circle that includes this curve. [Hint: Use $(x - h)^2 + (y - k)^2 = 150^2$ to find two equations with two variables, h and k.]

38. Wonderworld Amusement Park Highway engineers need to know the radius of curvature for a curve of a roller coaster designed to go through the points (400, 500), (500, 400), and (600, 100). If these three points are drawn on the Cartesian coordinate system superimposed over their blueprints, what is the radius of curvature?

*39. The Tangent-to-the-Circle Problem Find the points of tangency for the tangents drawn from (0, 5) to the circle $x^2 - 2x + y^2 + 4y = 20$. [Hint: The radius drawn to the point of tangency is perpendicular to the tangent.]

40. The Intersecting Circles Problem Find the equation of the circle that passes through the point (-33, 4) and the intersection of the circles $x^2 + y^2 = 25$ and $(x - 6)^2 + y^2 = 25$.

*41. The Inscribed Triangle Problem A triangle inscribed in a circle creates a right triangle with its hypotenuse the diameter of the circle (Figure 11).

(a) If the radius of the semicircle is 2 units, find an equation that expresses area of the triangle as a function of the x coordinate of point P.

(b) Use the graph of the function in part (a) to find the coordinates of P that maximize the area of the triangle.

Figure 11

Figure 12

42. The Inscribed Rectangle Problem A rectangle is inscribed in a circle (Figure 12).

(a) If the radius of the circle is 2 units, find the equation that expresses the area of the rectangle as a function of the x coordinate of point P.

(b) Use the graph of the function in part (a) to find the coordinates of P that maximize the area of the rectangle.

Extend the Concepts

For Exercises 43 to 46 sketch the solution set for each system of inequalities.

43. $x^2 + y^2 \leq 9$
$x^2 + y^2 \geq 4$

44. $x^2 + y^2 > 0$
$x^2 + y^2 \leq 1$

45. $(x + 1)^2 + y^2 \le 4$
$(x - 1)^2 + y^2 \le 4$

46. $x^2 + y^2 \ge 9$
$x^2 + y^2 \le 4$

47. Prove that M is the midpoint between $A(x_1,y_1)$ and $B(x_2, y_2)$ where $M = (\frac{x_1 + x_2}{2}, \frac{y_1 + y_2}{2})$.

48. Prove that $d(A,B) = d(B,A)$ and $d(A,A) = 0$ for $A(x_1, y_1)$ and $B(x_2, y_2)$.

49. Verify the triangle inequality $d(AC) + d(CB) \ge d(AB)$ for A(1,-1), B(7,7), and C(7,-1). Why is it called the triangle inequality?

50. Prove that the diagonals of the rectangle with vertices A(a, b), B(c, b), D(c, d), and D(a, d) are equal.

For Exercises 51 to 54 use the distance formula to find the equation that describes the set of all points P(x, y) such that each statement is true.

51. $d(P,A) = 4$ for A(0, 0)

52. $d(P,A) = 4$ for A(3, -4)

53. $d(P,A) + d(P,B) = 10$ for A(-4, 0) and B(4, 0)

54. $d(P,A) + d(P,B) = 10$ for A(0, -5) and B(0, 5)

8.2 The Ellipse and the Hyperbola

Defining an Ellipse
Defining a Hyperbola

In this section we study two more conic sections: the **ellipse**, which is the cross section formed when a tilted plane cuts through a cone (Figure 1a), and the **hyperbola**, which is the cross section of a cone formed when a vertical plane cuts the cone. The hyperbola has two branches (Figure 1b).

(a)

(b)

Figure 1

Although the shapes of these two conic sections look nothing alike, their definitions, derivations, and equations are analogous, making it natural to study them together. Probably the most important application of the ellipse that we study is orbits. Earth travels around the sun in an elliptical orbit (Exercise 29). In Example 5 we discuss Halley's comet, which also travels in an elliptical orbit.

An application involving hyperbolas is the long-range navigation system (Loran). It uses hyperbolas to enable ships and planes to establish their positions. Loran uses one master and two slave stations paired off to transmit pulses at two different frequencies. The *master station and slave station I* are the foci for one family of hyperbolas, and the *master station and slave station II* are the foci for a second family of hyperbolas. Ships monitor the delay time in microseconds between the signals sent by the master and those sent by the slaves to determine which pair of hyperbolas will give their location (Figure 2).

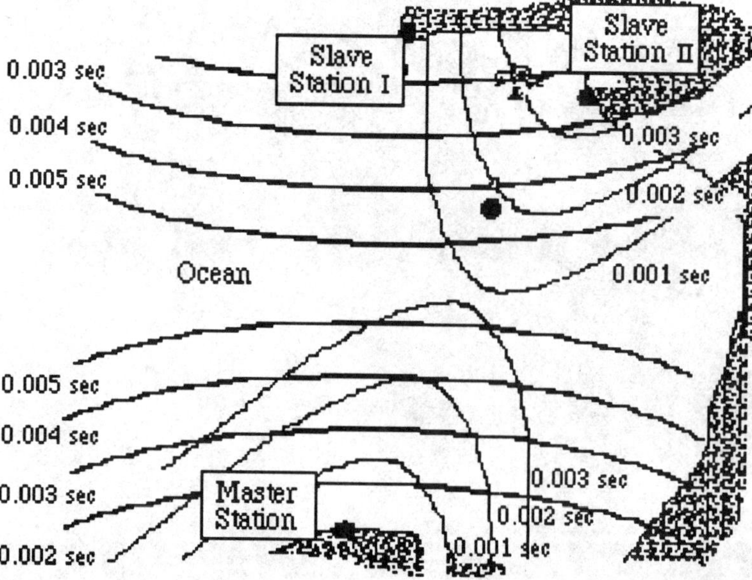

Figure 2

Defining an Ellipse

The description of an ellipse involves distances from two fixed points called **foci**. Imagine driving two nails in a board, tying the ends of a string to them, and positioning a pencil as shown in Figure 3. The shape drawn by the pencil traveling around the nails with the string kept taut is an ellipse.

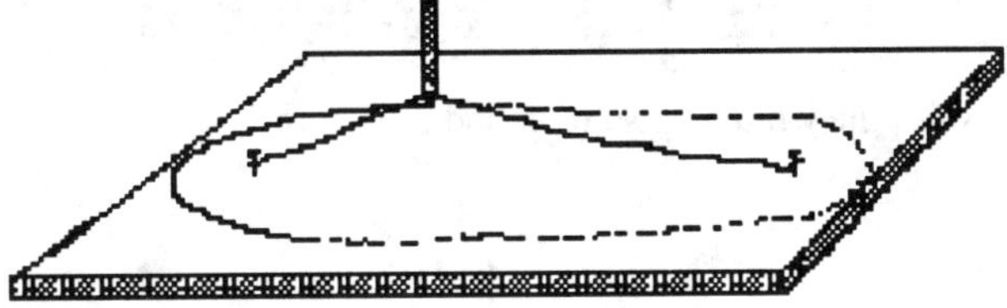

Figure 3

Definition of an Ellipse

An ellipse is the set of all points; the sum of whose distances from two fixed points F and F' equals a constant. The two fixed points are called **foci**.

Suppose that the length of the string in Figure 3 is 10 cm and the nails are 8 cm apart. The distance from any point on the ellipse (a position of the pencil) to one nail plus the distance from that same point to the other nail must always equal 10 cm. Figure 4 shows the ellipse positioned in the Cartesian coordinate system, with the origin at the center of the ellipse. The foci (nails) are located 8 units apart, at (-4, 0) and (4, 0).

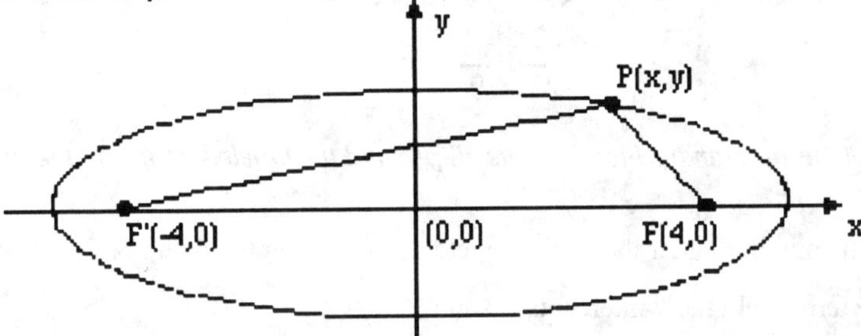

Figure 4

We can find the equation of the ellipse using the fact that the sum of the distances from P to the foci is 10. That is, the sum of the **focal radii** d(P,F') and d(P,F) is 10.

$$d(P,F') + d(P,F) = 10 \qquad \textit{Apply the distance formula.}$$

$$\sqrt{(x-(-4))^2 + (y-0)^2} + \sqrt{(x-4)^2 + (y-0)^2} = 10 \qquad \textit{Isolate one radical.}$$

$$\sqrt{(x+4)^2 + y^2} = 10 - \sqrt{(x-4)^2 + y^2} \qquad \textit{Square both sides.}$$

$$(x+4)^2 + y^2 = 100 - 20\sqrt{(x-4)^2 + y^2} + (x-4)^2 + y^2$$

$$x^2 + 8x + 16 + y^2 = 100 - 20\sqrt{(x-4)^2 + y^2} + x^2 - 8x + 16 + y^2 \qquad \textit{Isolate the radical and simplify.}$$

$$4x - 25 = -5\sqrt{(x-4)^2 + y^2} \qquad \textit{Square both sides.}$$

$$16x^2 - 200x + 625 = 25[(x-4)^2 + y^2]$$

$$9x^2 + 25y^2 = 225$$

Using the equation $9x^2 + 25y^2 = 225$, we can find the x and y intercepts of the graph in Figure 4.

$$9x^2 + 25y^2 = 225 \text{ and } y = 0 \Rightarrow x = \pm 5$$

and

$$9x^2 + 25y^2 = 225 \text{ and } x = 0 \Rightarrow y = \pm 3$$

The x intercepts are -5 and 5 and the y intercepts are -3 and 3 (Figure 5).

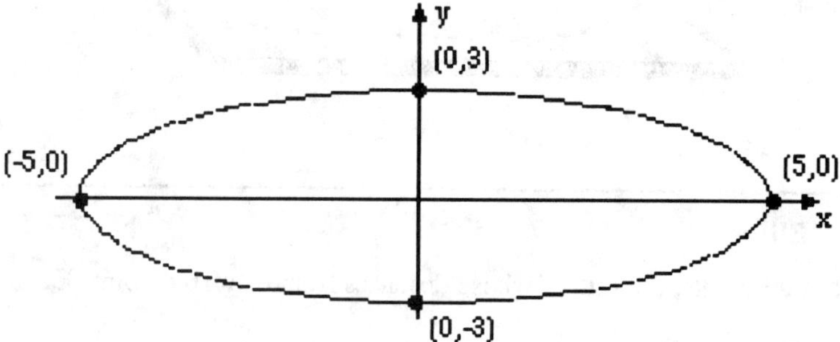

Figure 5

If we divide both sides of $9x^2 + 25y^2 = 225$ by 225, we find the standard form for the equation of this ellipse.

$$9x^2 + 25y^2 = 225$$

$$\frac{9x^2}{225} + \frac{25y^2}{225} = \frac{225}{225}$$

$$\frac{x^2}{25} + \frac{y^2}{9} = 1$$

What do you notice about the standard form of this ellipse and the labeled points on the graph in Figure 5?

The square roots of the number under the x^2 term yield the x intercepts, -5 and 5. The square roots of the number under the y^2 term yield the y intercepts, -3 and 3.

Standard Form of an Ellipse Centered at the Origin

The standard form of an ellipse centered at the origin with x intercepts -a and a and y intercepts -b and b is

$$\frac{x^2}{a^2} + \frac{y^2}{b^2} = 1$$

The location of the foci does not appear in the equation but can easily be found. Figure 6 shows the string held taut in a horizontal position. Notice that the length of the "string" is 2a.

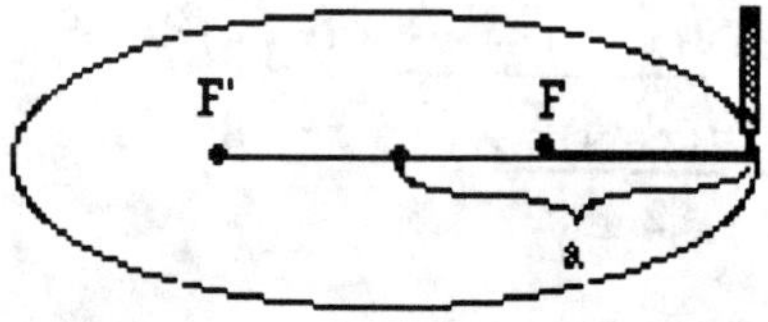

Figure 6

Figure 7 shows the string in a different position. If we let c be the distance from the center to one of the foci, we have the relationship $b^2 + c^2 = a^2 \Rightarrow c^2 = a^2 - b^2$.

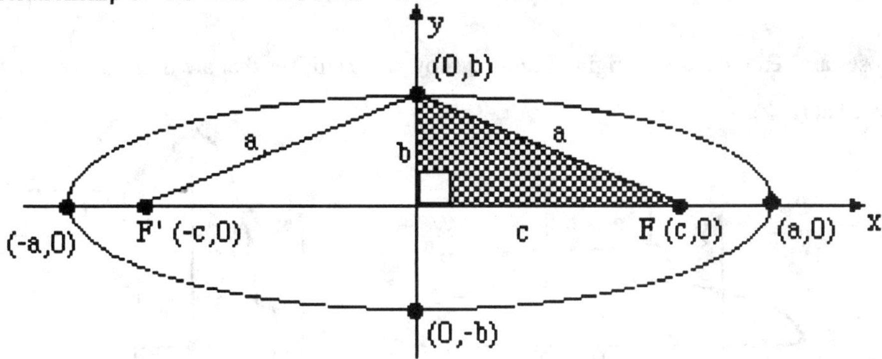

Figure 7

The ellipse has a major axis and a minor axis. The foci are always located on the major (longer) axis. In Figure 8a, the horizontal axis of the ellipse is called its **major axis** and the vertical axis is called its **minor axis**, making the major axis 2a units long and the minor axis 2b units long.

What is the relationship between a, b, and c if the major axis of an ellipse is vertical?

Figure 8b shows an ellipse in which $b > a$. Then $c^2 = b^2 - a^2$.

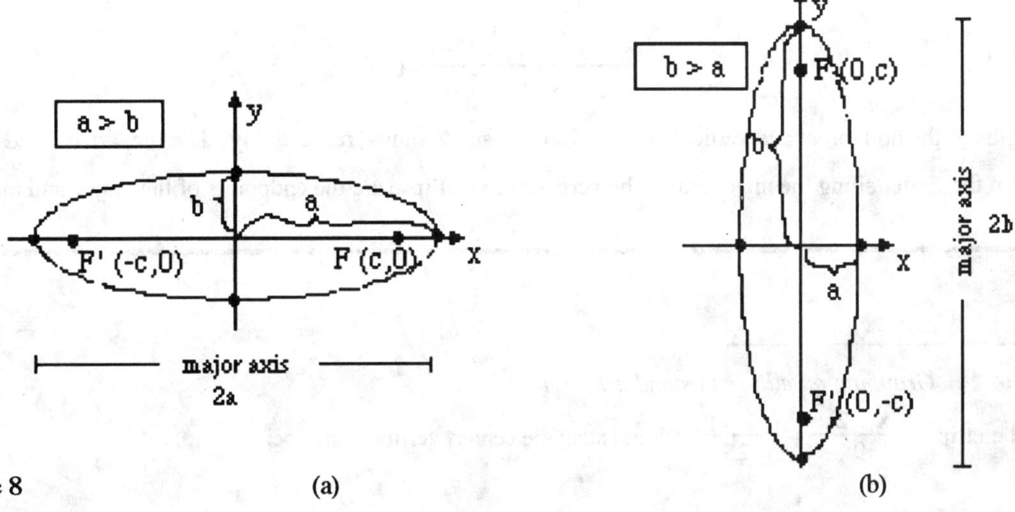

Figure 8 (a) (b)

Foci of an Ellipse

The foci of an ellipse centered at the origin having equation

$$\frac{x^2}{a^2} + \frac{y^2}{b^2} = 1$$

are located at (-c, 0) and (c, 0) if $a > b$ where $c^2 = a^2 - b^2$ and (0, -c) and (0, c) if $b > a$, where $c^2 = b^2 - a^2$.

NOTE: If $a = b$, the ellipse is a circle and $c = \sqrt{a^2 - b^2} = 0$, since a circle's focus is its center.

Not all ellipses are centered at the origin. Figure 9 shows two ellipses that are translated so that their centers are located at (h, k).

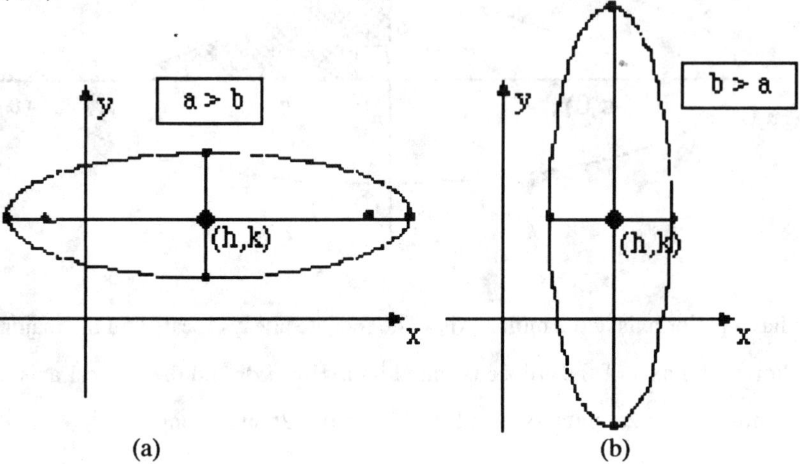

Figure 9 (a) (b)

Standard Form of an Ellipse Centered at (h, k)

The standard form of the equation of an ellipse centered at (h, k) is

$$\frac{(x - h)^2}{a^2} + \frac{(y - k)^2}{b^2} = 1$$

The lengths of the horizontal and vertical axes are 2a units and 2b units, respectively. The foci are located c units from the center along the major axis. The vertices of an ellipse are the endpoints of the major and minor axes.

Example 1 *Graphing an ellipse in standard form*

Graph the ellipse $\frac{(x + 2)^2}{9} + \frac{(y - 1)^2}{25} = 1$ and state the center, vertices, and foci.

Solution

The center of the ellipse is (-2, 1). From the equation we know that $a^2 = 9$ and $b^2 = 25$. The major axis is vertical since $b > a$. The vertices are $b = 5$ units vertically from the center at (-2, -4) and (-2, 6) and $a = 3$ units horizontally from the center at (-5, 1) and (1, 1) (Figure 10a). The foci are located 4 units from the center at (-2, -3) and (-2, 5) since $c^2 = b^2 - a^2 \Rightarrow c = 4$ (Figure 10b).

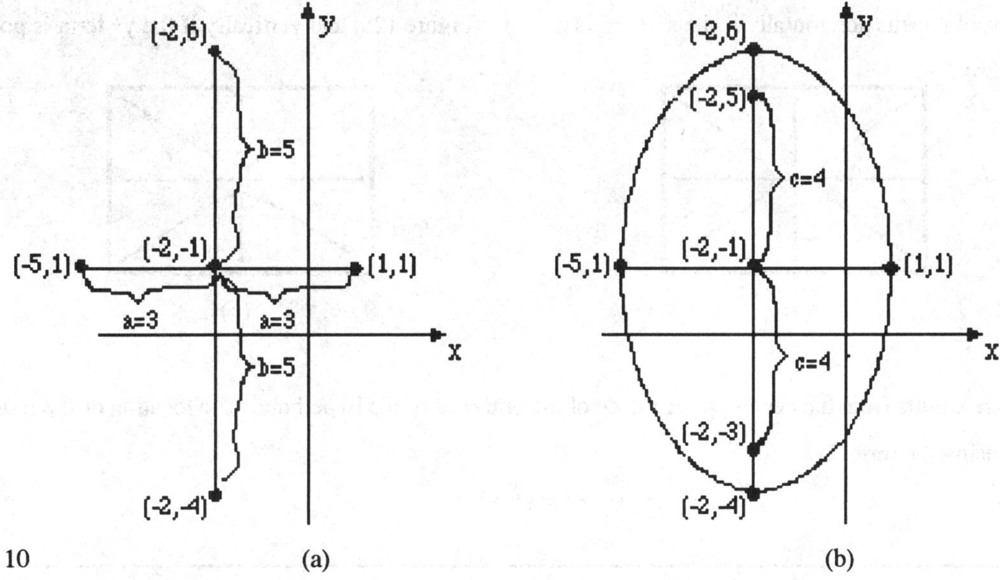

Figure 10 (a) (b)

Defining a Hyperbola

For the ellipse, the sum of the focal radii is a constant. For a hyperbola, the *difference* between the focal radii, d(P,F) - d(P,F'), is a constant (Figure 11).

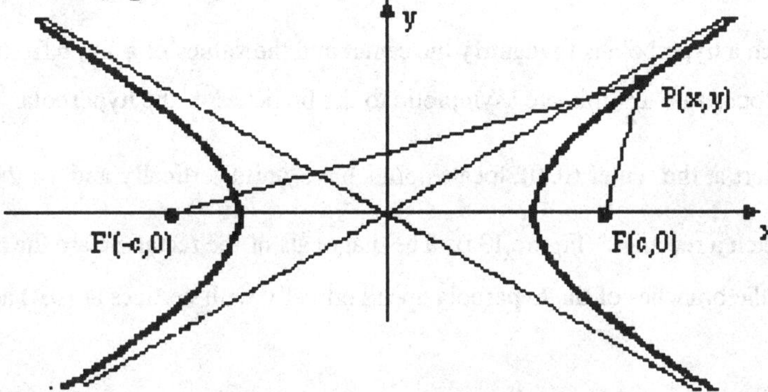

Figure 11

> **Definition of a Hyperbola**
> A hyperbola is the set of points such that the differences of their distances from any point two fixed points F and F' equals a constant. The two fixed points are called the foci.

The graph of a hyperbola has two separate pieces that look like parabolas but are not parabolas. The equation for a hyperbola is derived just as we derived the equation for an ellipse using the distance formula and some fixed value for the difference of d(P,F) and d(P,F'). There are two resulting forms for the hyperbola centered at the origin. They are

$$\frac{x^2}{a^2} - \frac{y^2}{b^2} = 1 \quad \text{and} \quad \frac{y^2}{b^2} - \frac{x^2}{a^2} = 1$$

The hyperbola opens horizontally if the x^2 term is positive (Figure 12a) and vertically if the y^2 term is positive (Figure 12b).

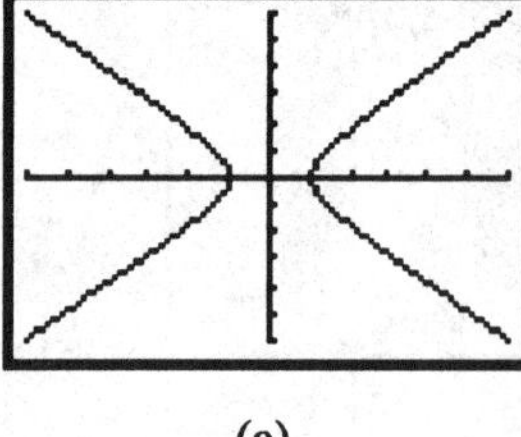

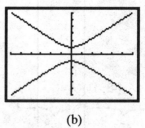

Figure 12 (a) (b)

The foci are c units from the center on the inside of the branches of the hyperbola. The location of the foci can be found using the formula

$$c^2 = a^2 + b^2$$

Standard Form of a Hyperbola Centered at the Origin

The equation of a hyperbola centered at the origin with asymptotes $y = \pm \frac{b}{a}x$, where $c^2 = a^2 + b^2$, is

$$\frac{x^2}{a^2} - \frac{y^2}{b^2} = 1 \text{ if it opens horizontally and } \frac{y^2}{b^2} - \frac{x^2}{a^2} = 1 \text{ if it opens vertically.}$$

A quick way to sketch a hyperbola is to identify the center and the values of a and b from the equation, and to create lines having slopes $m = \pm\frac{b}{a}$ that are asymptotic to the branches of the hyperbola. For example, to graph $\frac{y^2}{9} - \frac{x^2}{4} = 1$, we start at the center (0, 0), locate points b = 3 units vertically and a = 2 units horizontally from the center, and sketch a rectangle (Figure 13a). The diagonals of the rectangle are the asymptotes. Since the y^2 term is positive, the branches of the hyperbola open vertically with vertices at (0,4) and (0,-4) (Figure 13b).

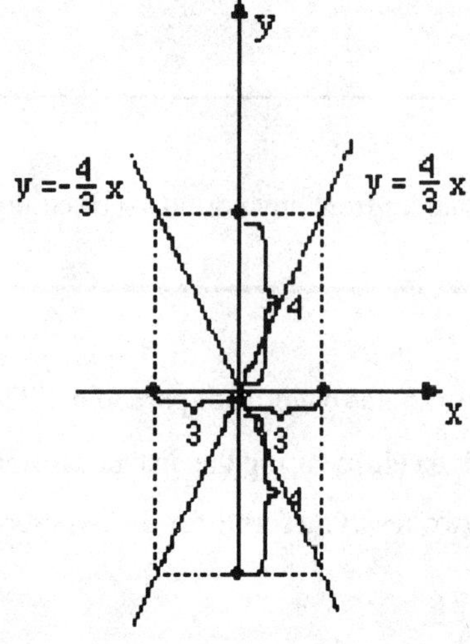

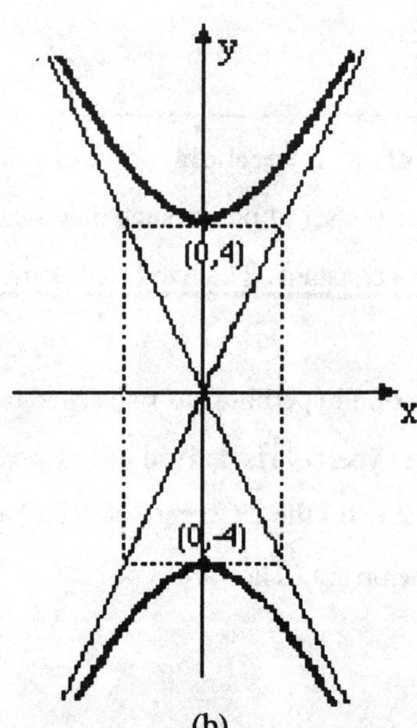

Figure 13 (a) (b)

We could locate the foci, $c^2 = b^2 + a^2$ and $b = 4$, $c = 3 \Rightarrow c^2 = 25$. The foci are 5 units from the center at (0, 5) and (0, -5).

Translations of hyperbolas are graphed in the same manner with center located at (h, k).

Standard Form of a Hyperbola Centered at (h, k)

The standard form of a hyperbola is

$$\frac{(x - h)^2}{a^2} - \frac{(y - k)^2}{b^2} = 1 \text{ if it opens horizontally}$$

and

$$\frac{(y - k)^2}{b^2} - \frac{(x - h)^2}{a^2} = 1 \text{ if it opens vertically}$$

The foci are located c units from (h, k) on the same axis as the vertices where $c^2 = a^2 + b^2$.

In the next example we graph two conic sections written in the form

$$Ax^2 + Cy^2 + Dx + Ey + F = 0$$

If neither A nor C is 0 and A and C have the same sign, it is the equation for an ellipse. The equation represents a hyperbola if A and C have opposite signs. (Recall from Section 8.1 that if A and C have the same sign and A = C, the equation represents a circle.) We complete the square to write the equation in standard form.

Example 2 *Expressing equations in standard form*

Express the following in standard form, graph the resulting equations, and label the important points.

(a) $9x^2 + 25y^2 + 36x - 50y - 164 = 0$ (b) $9x^2 - 16y^2 - 18x - 64y - 199 = 0$

Solution

(a) Since the x^2 and y^2 terms have the same sign and their coefficients are not equal, the equation is an ellipse.

$$\begin{aligned} 9x^2 + 25y^2 + 36x - 50y - 164 &= 0 \\ 9(x^2 + 4x\ ) + 25(y^2 - 2y\ \ ) &= 164 \\ 9(x^2 + 4x + \mathbf{4}) + 25(y^2 - 2y + \mathbf{1}) &= 164 + 9(\mathbf{4}) + 25(\mathbf{1}) \\ 9(x + 2)^2 + 25(y - 1)^2 &= 225 \\ \frac{(x + 2)^2}{25} + \frac{(y - 1)^2}{9} &= 1 \end{aligned}$$

The center is (-2, 1). The vertices are 5 units left and right and 3 units above and below the center at (-7, 1), (3, 1), (-2, 4), and (-2, -2). The foci are 4 units left and right of the center at (-6, 1) and (2, 1) since $c^2 = a^2 - b^2 \Rightarrow c^2 = 16$ (Figure 14).

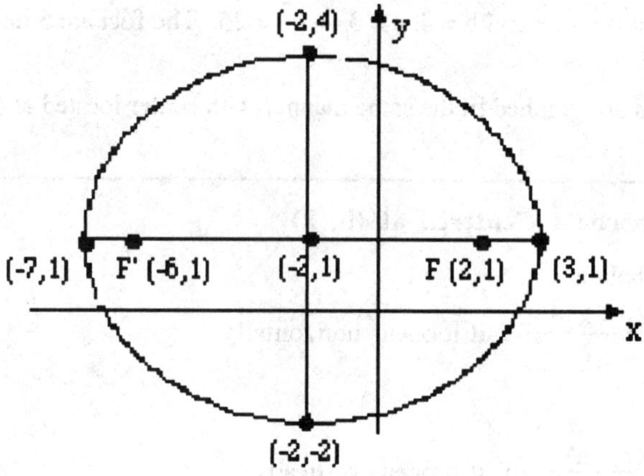

Figure 14

(b) Since the x^2 and y^2 terms have different signs, the equation is a hyperbola.

$$
\begin{aligned}
9x^2 - 16y^2 - 18x - 64y - 199 &= 0 \\
9x^2 - 18x - 16y^2 - 64y &= 199 \\
9(x^2 - 2x + \mathbf{1}) - 16(y^2 + 4y + \mathbf{4}) &= 199 + \mathbf{9} - \mathbf{64} \\
9(x - 1)^2 - 16(y + 2)^2 &= 144 \\
\frac{(x - 1)^2}{16} - \frac{(y + 2)^2}{9} &= 1
\end{aligned}
$$

The center is (1, -2). The value of a is $\sqrt{16} = 4$, the value of b is $\sqrt{9} = 3$, and the asymptotes pass through the center and have slopes of $m = \pm 3/4$. The foci are 5 units from the center (Figure 15).

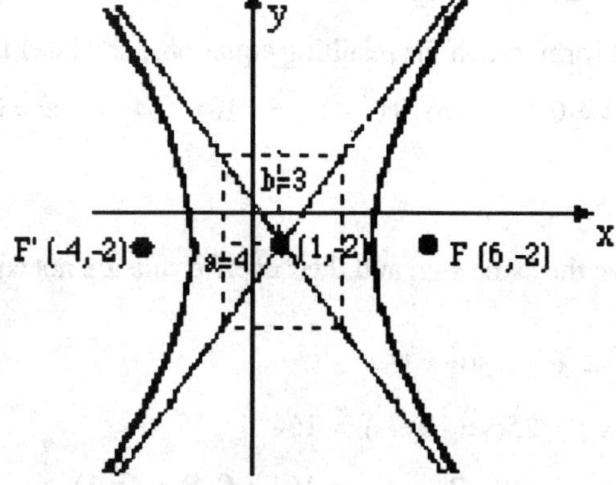

Figure 15

Recall from Section 8.1 that $x^2 + y^2 = 0$ is the equation of a point circle whose graph is a point at the origin. Similarly, the graph of $\frac{x^2}{25} + \frac{y^2}{9} = 0$ is a point at the origin since the only ordered pair that satisfies the equation is (0, 0). The graph is called a *point ellipse.* It is a degenerate conic section.

What is the graph of $\frac{y^2}{9} - \frac{x^2}{16} = 0$*?*

It is two intersecting lines since $\frac{y^2}{9} - \frac{x^2}{16} = 0 \Rightarrow y = \pm\frac{3}{4}x$.

The next example deals with the orbit of Halley's comet. The units of measure are astronomical units (AU), where 1 AU = 92,957,200 miles, the mean distance from Earth to the sun.

Example 3 The Halley's Comet Problem

Halley's comet travels in an elliptical orbit (Figure 16).

(a) Find the equation of the orbit if its distance from the sun, which is located at one focus, varies from 0.6 AU at perihelion (further distance from a focus on the major axis) to 35.3 AU at aphelion (shortest distance from a focus).

(b) Calculate $T^{2/3}$ for T = 76.08 yr. Find this value in the equation of the elliptical orbit (this is Kepler's law).

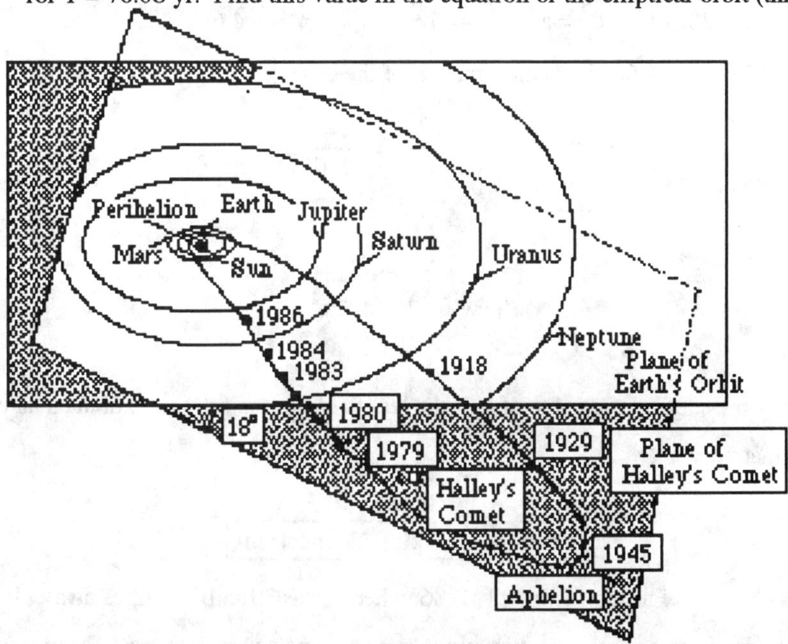

Figure 16

Thinking: Creating a Plan

If the comet's elliptical orbit is positioned in the Cartesian coordinate system so that the center is at the origin,

$$\frac{x^2}{a^2} + \frac{y^2}{b^2} = 1 \text{ and } a^2 - b^2 = c^2$$

where 2a is the length of the major axis, 2b is the length of the minor axis, and c is the distance from the center to each focus, the one on the left being the position of the sun (Figure 17).

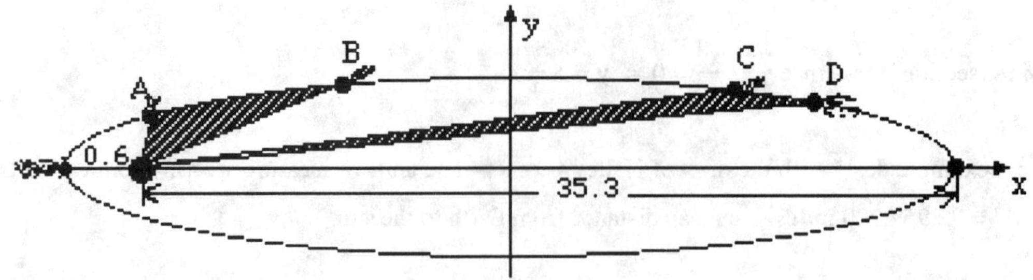

Figure 17

Finding a and b will yield the equation of the orbit.

Communicating: Writing the Solution

(a) The major axis measures $2a = 0.6 + 35.3 = 35.9 \Rightarrow a = 17.95$ AU.

The distance between foci measures $2c = 35.3 - 0.6 \Rightarrow c = 17.35$ AU.

The minor axis measures 2b, where $b^2 = a^2 - c^2 \Rightarrow b^2 = 21.18 \Rightarrow b = 4.60$ AU.

$\frac{x^2}{a^2} + \frac{y^2}{b^2} = 1$ and $a = 17.95$, $b = 4.60$ yields the equation of the orbit,

$$\frac{x^2}{322.20} + \frac{y^2}{21.18} = 1$$

(b) $T^{2/3}$ and $T = 76.08 \Rightarrow$

$$T^{2/3} \approx 17.9547902$$

This is approximately the same as the value of half the length of the major axis, a, of such an elliptical orbit. Kepler's law states that $T^2 = a^3$.

Learning: Making Connections

Halley's comet was closest to Earth on April 11, 1986, when it was 39 million miles away. It will return in the year 2062. Since the gravitational attraction of the sun on comets varies inversely as the square of their distance from the sun, a graph of the resulting elliptical orbit shows the comet sweeping out equal slices of area in equal time. Because the shaded areas in Figure 16 are nearly equal, they are swept out in the same time, but since arc AB is greater than arc CD, the velocity of the comet is faster nearer the sun.

Exercises 8.2

Develop the Concepts

In Exercises 1 to 6 express each equation in standard form for an ellipse and sketch its graph. State the center, the length of the major and minor axes, and the foci.

1. $\frac{x^2}{9} + \frac{y^2}{4} = 1$
2. $\frac{x^2}{4} + \frac{y^2}{9} = 1$
3. $\frac{(x - 1)^2}{4} + \frac{y^2}{9} = 1$
4. $\frac{(x - 1)^2}{9} + \frac{y^2}{4} = 1$
5. $4(x + 2)^2 + 9(y - 1)^2 = 36$
6. $9(x - 1)^2 + 4(y + 2)^2 = 36$

For Exercises 7 and 8 find the equation of each ellipse in standard form.

7. The foci are F(1, 0) and F'(-1, 0), and the length of the minor axis is 6.
8. The foci are F(-2, 4) and F'(-2, -4), and the endpoints of the major axis are (-2, 10) and (-2, -10).

For Exercises 9 to 16 express each equation in standard form for a hyperbola and sketch its graph. State the center, vertices, and foci.

9. $\frac{x^2}{25} - \frac{y^2}{16} = 1$
10. $\frac{y^2}{16} - \frac{x^2}{25} = 1$
11. $x^2 - y^2 = 4$
12. $(y - 2)^2 - \frac{x^2}{4} = 1$
13. $\frac{(x - 2)^2}{4} - \frac{(y + 3)^2}{16} = 1$
14. $16(y - 3)^2 - 9(x + 2)^2 = 144$
15. $3(y + 1)^2 - (x - 1)^2 = 16$
16. $5x^2 - (y + 5)^2 = -25$

For Exercises 17 and 18 find the equation for each hyperbola.

17. center at (1, -2), a focus at (6, -2), and an asymptote defined as $y = \frac{4}{3}x - \frac{10}{3}$
18. opens horizontally with asymptotes $y = \pm\frac{1}{3}x$

For Exercises 19 to 26 write each equation in standard form and sketch the graph.

19. $x^2 + 4y^2 + 24y + 20 = 0$
20. $x^2 + 4y^2 - 2x + 1 = 0$

*21. $27x^2 + 20y^2 - 216x + 80y = 28$

22. $x^2 - y^2 - 8x = 9$
23. $x^2 + 5y^2 + 5x + 5y = 0$
24. $-16x^2 + 9y^2 + 32x + 36y = -164$
25. $25x^2 - 100x - 9y^2 - 18y + 166 = 0$
26. $16x^2 + 4y^2 - 32x + 16y - 32 = 0$

Apply the Concepts

27. The Rainbow Kite and Balloon Company How long a string should be used to lay out an elliptical display room that is to be 20 ft long and 16 ft wide? Where should the ends of the string be staked?

28. **Wonderworld Amusement Park** A carpenter wants to construct a semielliptical arch in the Tunnel of Love. The opening is to be 2 ft high in the center and 5 ft wide along the hearth. The contractor first makes a stencil by drawing the ellipse using two tacks and a string. How long a string is necessary, and how far apart should the tacks be set?

*29. The Earth's-Orbit Problem Earth travels around the sun in an elliptical orbit; the minimum distance from the sun (one of the foci) is 147.5×10^6 km and the maximum is 152.5×10^6 km. Find the equation of the orbit. (These distances are called the perigee and the apogee, respectively, of the earth's orbit.)

30. The Sputnik I Problem If the equation of the orbit of the satellite *Sputnik I* in kilometers was described as $\frac{x^2}{(6968)^2} + \frac{y^2}{(6958.7)^2} = 1$ with Earth as one of the foci, find the perigee and apogee.

31. The Jupiter Problem Jupiter, the largest planet (its volume is 1312 times that of Earth), travels between 506.8×10^6 and 459.8×10^6 mi from the sun. Find the equation of its elliptical orbit.

32. The Area and Circumference of an Ellipse Problem Evaluate the area and circumference of a circle having radius r using the following formulas using $A = \pi ab$ and $C \approx \pi\sqrt{2(a^2 + b^2)}$. [Hint: A circle is a special ellipse.]

33. The Length of the Orbit Problem Find the distance traveled by Halley's comet in one revolution. See Example 3. [Hint: Use the formula from Exercise 32.]

34. The Stereo Problem If the sound source is at one focus of an elliptical listening room, the sound waves are reflected off the curve through the other focus. How far from the center of the 15- by 8-ft room should the stereo speaker and a chair be set for best stereo listening?

Extend the Concepts

35. Find the x and y intercepts of $\frac{(x-2)^2}{4} + \frac{(y+1)^2}{9} = 1$.

36. Write the equation of the set of all points P(x, y) for which the sum of the distances from the points (4, 0) and (-4, 0) is always 10.

37. Prove that the length of a chord passing through the origin and the points $P_1(x_1, y_1)$ and $P_2(x_2, y_2)$ on the ellipse, $\frac{x^2}{a^2} + \frac{y^2}{b^2} = 1$, is given by $s = 2\sqrt{x_1^2 + y_2^2}$.

38. Prove that the length of a chord joining any two points (x_1, y_1) and (x_2, y_2) on the ellipse $\frac{x^2}{a^2} + \frac{y^2}{b^2} = 1$ is

$$s = \sqrt{1 + \frac{(\Delta y)^2}{(\Delta x)^2}}\,|\Delta x|, \text{ where } \Delta x = x_2 - x_1 \text{ and } \Delta y = y_2 - y_1.$$

39. Write the equation of the asymptotes for the hyperbola $\frac{(y-2)^2}{4} - \frac{(x+1)^2}{3} = 1$.

40. Find the x intercepts of $\frac{y^2}{25} - \frac{(x-2)^2}{9} = 1$.

For Exercises 41 to 44 graph each pair of relations on one coordinate system. Find and label the points of intersection.

41. $x^2 + y^2 = 20$

$4x^2 + 2y^2 = 48$

42. $y^2 - x^2 = 5$

$2y = x$

43. $y^2 - x^2 = 0$

$y = x$

44. $x^2 - y^2 = 4$

$y = x/4$

For Exercises 45 to 52 sketch the solution set for each system of inequalities.

45. $\frac{x^2}{16} + \frac{y^2}{9} \le 1$

$\frac{x^2}{9} + \frac{y^2}{16} \le 1$

46. $x^2 + y^2 \ge 1$

$y^2 - \frac{x^2}{4} \le 1$

47. $y^2 - x^2 \le 9$

$x^2 - y^2 \le 9$

48. $\frac{x^2}{4} + \frac{(y-2)^2}{9} \ge 1$

$\frac{x^2}{4} + \frac{(y+2)^2}{9} \ge 1$

49. $y^2 - x^2 \ge 9$

$x^2 - y^2 \ge 9$

50. $\frac{(x-1)^2}{4} + \frac{y^2}{25} \le 1$

$\frac{(x-1)^2}{4} + \frac{y^2}{9} \ge 1$

51. How does the fact that hyperbolas are not functions restrict the placement of the stations in the Loran system?

52. Create a system of navigation based on ellipses analogous to the Loran system and explain how it works.

8.3 The Parabola and a Review of Conic Sections

Defining a Parabola
Identifying Conic Sections

In this section we study the **parabola**, which is the conic section formed when a plane cuts the cone parallel to an edge of the cone (Figure 1). The parabola has already been discussed in Chapter 2 as a quadratic function of the form $y = ax^2 + bx + c$. Here we define the parabola as a set of points equidistant from a given point and a line.

Figure 1

The next time you are driving over a suspended bridge, notice the vertical columns. On some bridges they are equally spaced like those shown in Figure 2a, while on others they get farther and farther apart until you reach the low point of the cable and then get closer and closer, as shown in Figure 2b.

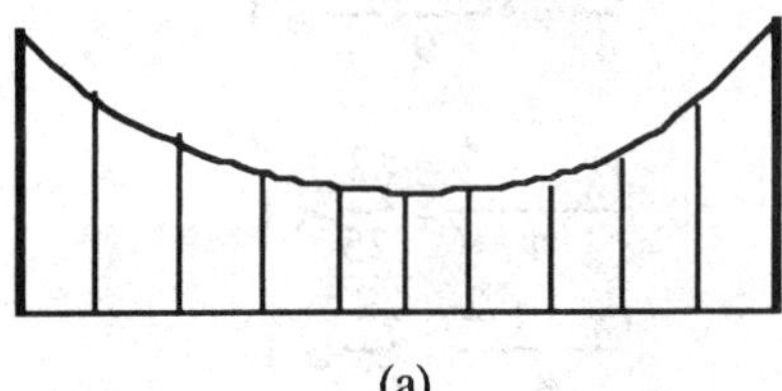

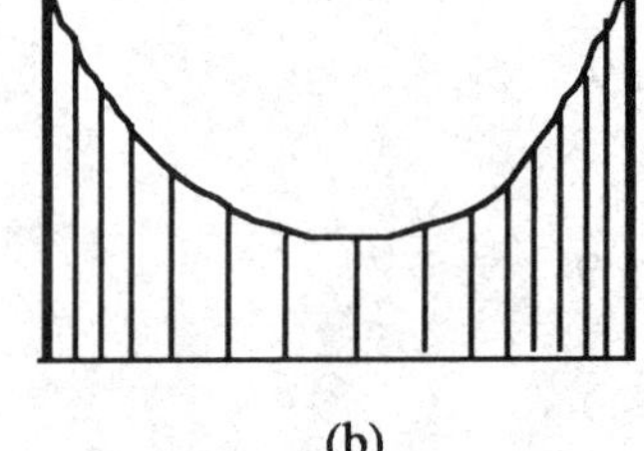

Figure 2 (a) (b)

The cable in Figure 2a is a parabola because it carries a uniform load (spacing on the roadbed between uprights is equal). The one in Figure 2b is a catenary and the load is equally distributed on the cable itself (uprights connect equal distances measured along the curve). The equation of a catenary is much more complicated than the equation of a parabola. If the catenary and parabola look relatively the same, it is often easier to estimate the mathematical model for the catenary curve using an equation of a parabola. We will investigate this in Example 3.

At the end of this section we review all four conic sections: the circle, the ellipse, the hyperbola, and the parabola.

Defining a Parabola

Consider a fixed point at (0, 1), a given line described by the equation $y = -1$, and several points that are equidistant from the given point and line (Figure 3a). Connecting all points equidistant from the point F(0, 1) and the line $y = -1$ yields the concave-up parabola with vertex V(0, 0) (Figure 3b).

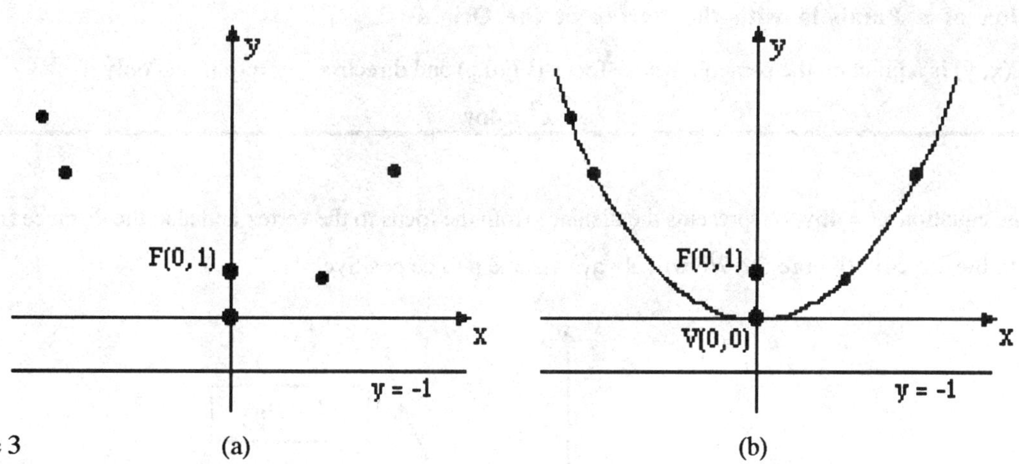

Figure 3 (a) (b)

> **Definition of a Parabola**
>
> A parabola is the set of points P(x, y) equidistant from a fixed point, called its **focus**, and a fixed line, called its **directrix**.

We can derive the equation of the parabola in Figure 4 by equating the distance between the focus F(0, 1) and any point P(x, y) and the distance between the point P(x, y) and the line y = -1.

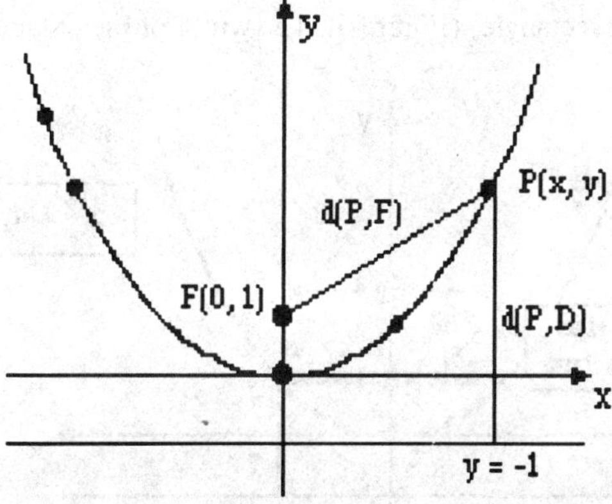

Figure 4

$$
\begin{aligned}
d(P,F) &= d(P,D) \\
\sqrt{(x-0)^2 + (y-1)^2} &= |\, y - (-1) \,| \\
\sqrt{x^2 + (y-1)^2} &= y + 1 \qquad \textit{Square both sides.} \\
x^2 + (y-1)^2 &= (y+1)^2 \\
x^2 + y^2 - 2y + 1 &= y^2 + 2y + 1 \\
x^2 &= 4y
\end{aligned}
$$

That is, the equation of the parabola having vertex (0, 0), focus (0, 1), and directrix y = -1 is $x^2 = 4y$. This leads to the standard form of a concave-up parabola with the vertex at the origin.

Equation of a Parabola with the Vertex at the Origin

Point P(x, y) is a point on the parabola whose focus is F(0,p) and directrix is y = -p if and only if

$$x^2 = 4py$$

In the equation $x^2 = 4py$, p represents the distance from the focus to the vertex and also the distance from the vertex to the directrix (Figure 5). We will always assume p to be positive.

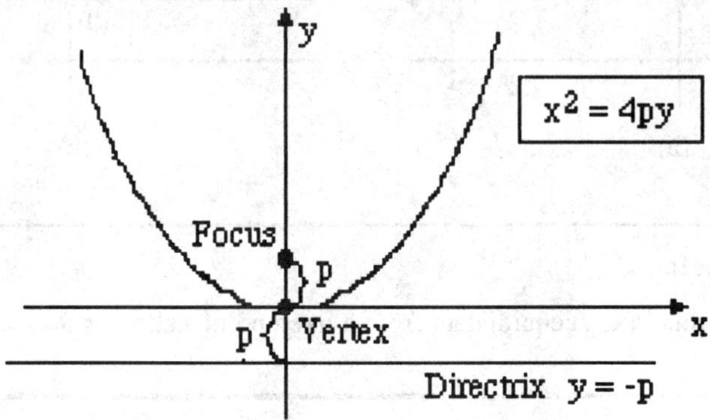

Figure 5

The chord drawn through the focus parallel to the directrix is called the **latus rectum** (*latus* means line, and *rectum* means right angle, as in a rectangle) (Figure 6). The width of the parabola along the latus rectum is 4p.

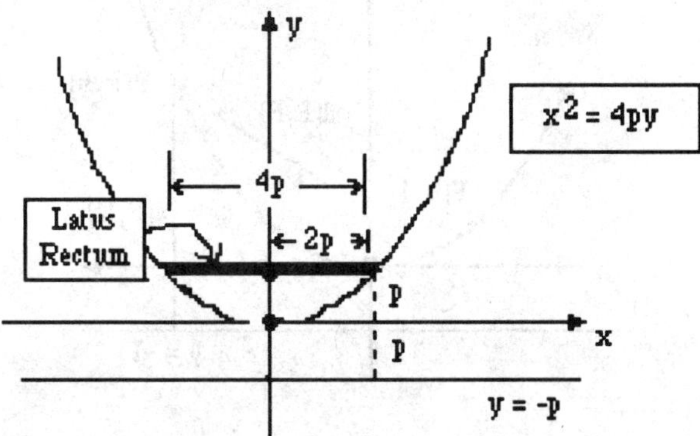

Figure 6

By the definition of a parabola, any point on the parabola must be equidistant from the focus and the directrix.

Can you see how the width of the parabola can easily be determined if we know the value of p?

The distance from the vertex to the focus is p units. Then, from the focus to the parabola along the latus rectum is a distance of 2p. In the next example we find the value of p and use it to sketch the graph of a parabola.

Example 1 *Graphing a parabola*

Graph the parabola $x^2 = 8y$.

Solution

Comparing the equation $x^2 = 8y$ to the general equation $x^2 = 4py$ reveals that

$$4p = 8 \Rightarrow p = 2$$

The vertex is located at the origin and the focus is $p = 2$ units above the vertex. A point on the parabola is $2p = 4$ units from the focus along the latus rectum (Figure 7).

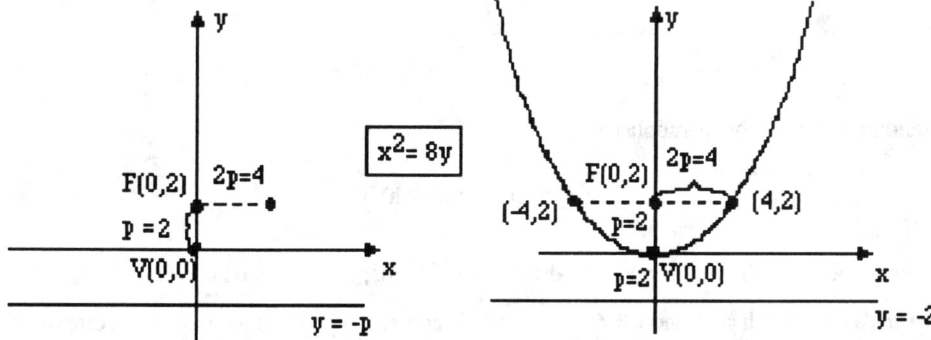

Figure 7

The standard form of a parabola for each of its orientations is summarized below. Notice that equations having an x^2 term open up or down, and those having a y^2 term open to the left or right.

Standard Form of a Parabola with Vertex (h, k)	
$(x - h)^2 = 4p(y - k)$	graph opens up
$(x - h)^2 = -4p(y - k)$	graph opens down
$(y - k)^2 = 4p(x - h)$	graph opens right
$(y - k)^2 = -4p(x - h)$	graph opens left

Example 2 *Finding the equation of a parabola*

Find the equation of a parabola having directrix $x = -4$ and focus $(2, 6)$.

Solution

A quick sketch of the given information shows that the focus is to the right of the directrix. The parabola opens to the right (Figure 8).

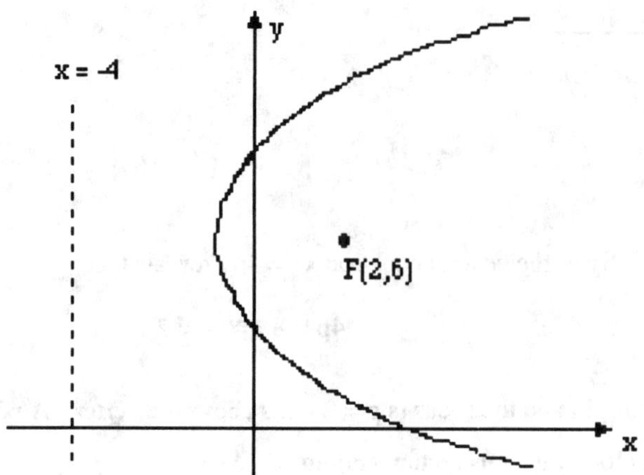

Figure 8

Hence, the standard form for this parabola is

$$(y - k)^2 = 4p(x - h)$$

Since the vertex is located halfway between the directrix and focus, on the same horizontal line as the focus, the coordinates of the vertex are h = -1 and k = 6. Also, p = 3 represents the distance from the vertex to the focus. Therefore, the equation is $(y - 6)^2 = 12(x + 1)$.

We often see architectural shapes that appear to be parabolas. Some are parabolas, but many are not. For example, cables suspended between two supports, such as telephone poles, are the shape is a *catenary*. The equation of a catenary is the form

$$y = \cosh x = \frac{e^x + e^{-x}}{2}$$

where cosh is called the *hyperbolic cosine*. The next example compares the two shapes, a catenary and a parabola.

Example 3 The Catenary Cable Problem

Assume an engineer found that the catenary f(x) = cosh x models a cable hung between two supports 20 ft apart. Find the equation of an approximating parabola and compare it to the exact catenary function.

Thinking: Creating a Plan

We will find three ordered pairs on the graph of f(x) = cosh x using the values x = 0, x = -10, and x = 10 (Figure 9).

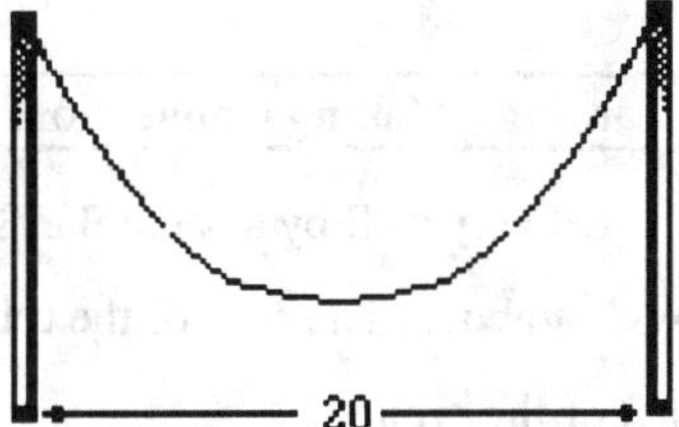

Figure 9

Since the catenary is symmetric about a line through its lowest point, a quadratic equation will fit the catenary exactly at x = 0 and x = ± 10. Then we will compare the equations at some value, say x = 2.

Communicating: Writing the Solution

For $f(x) = \cosh x = \dfrac{e^{x} + e^{-x}}{2}$,

$$f(0) = 1$$
$$f(-10) \approx 11{,}013.23$$
$$f(10) \approx 11{,}013.23$$

Therefore, the catenary passes through the points (0, 1), (-10, 11013.2$\widetilde{3}$), and (10, 11013.2$\widetilde{3}$).

Entering the three points and running the quadratic regression program (Figure 10) yields

$$y = 110.42x^2 + 2.99x + 1$$

L1	L2	L3
0	1	------
-10	11013	
10	11073	
------	------	

L1(1)=0

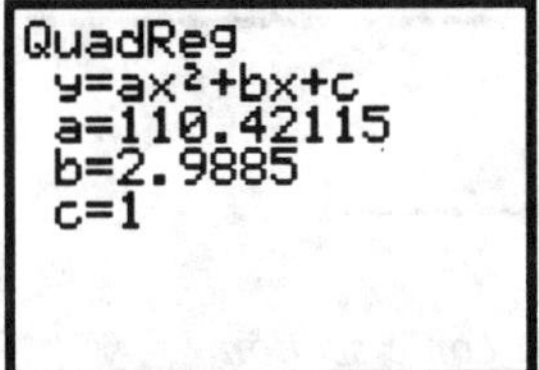

Figure 10

The parabola (Figure 11a) passes through the three known points on the catenary (Figure 11b). Although the equation for the parabola is much simpler than the equation of the catenary, the error created by approximating this catenary with a parabola is over 400 feet at x = 2.

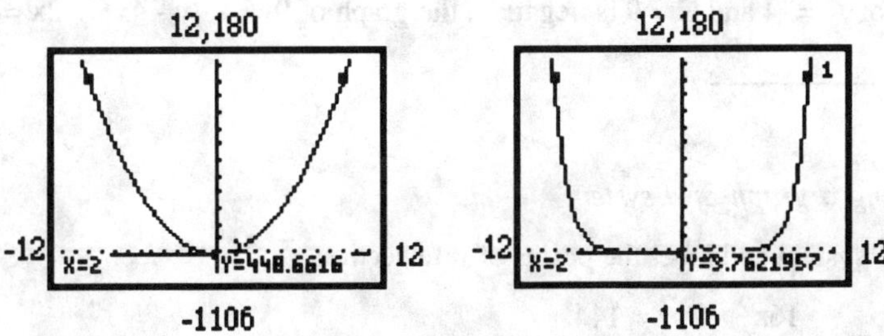

Figure 11

Learning: Making Connections

Sometimes a catenary cannot be approximated very well by a parabola. This catenary was not pulled tight enough to lend itself to approximation by a parabola. The ends of the cable are supported by 11,013-ft towers 20 ft apart and the cable sags to within 1 ft of the ground.

Identifying Conic Sections

We have seen that the values of A and C in the equation

$$Ax^2 + Cy^2 + Dx + Ey + F = 0$$

determine which conic section is described and that the values of D and E determine its translation. If there is only one second-degree term in an equation (only Ax^2 or Cy^2 is present), the graph is a parabola. The following summary is helpful in identifying conic sections by examining the values of A and C in the equation.

Summary: Conic Sections

$Ax^2 + Cy^2 + Dx + Ey + F = 0$ is

a circle if $A = C \neq 0$.

a parabola if $A = 0$ or $C = 0$ but not both.

an ellipse if $AC > 0$ and $A \neq C$.

a hyperbola if $AC < 0$.

Example 4 *Identifying a conic section*

Identify each of the following as the equation of a circle, parabola, ellipse, or hyperbola.

(a) $9x^2 - 2x + 9y^2 + 4y = 1$ (b) $9y^2 + 4y - 4x^2 + 2x = 10$

Solution

(a) Since $A = C = 9$, the graph of $9x^2 - 2x + 9y^2 + 4y = 1$ is a circle.

(b) Since the product of $A = -4$ and $C = 9$ is negative, the graph of $9y^2 + 4y - 4x^2 + 2x = 10$ is a hyperbola.

Example 5 *Sketching a graph of a system*

Sketch the graph of the system and label the points of intersection.

$$16x^2 - 9y^2 = 144$$
$$49x^2 + 24y^2 = 1176$$

Solution

Rewriting each equation in standard form yields a hyperbola and an ellipse.

$$\frac{x^2}{9} - \frac{y^2}{16} = 1 \text{ and } \frac{x^2}{24} + \frac{y^2}{49} = 1$$

The intersection of a hyperbola and an ellipse will have at most four points in common. Solving each equation for y^2 yields

$$16x^2 - 9y^2 = 144 \Rightarrow y^2 = \frac{16}{9}x^2 - 16$$

and

$$49x^2 + 24y^2 = 1176 \Rightarrow y^2 = 49 - \frac{49}{24}x^2$$

$y^2 = \frac{16}{9}x^2 - 16$ and $y^2 = 49 - \frac{49}{24}x^2 \Rightarrow$

$$\begin{aligned} \frac{16}{9}x^2 - 16 &= 49 - \frac{49}{24}x^2 \\ 384x^2 - 3456 &= 10{,}584 - 441x^2 \\ 825x^2 &= 14{,}040 \\ x^2 &\approx 17.018 \\ x &\approx \pm 4.125 \end{aligned}$$

$x^2 \approx 17.018$ and $y^2 = \frac{16}{9}x^2 - 16 \Rightarrow$

$$\begin{aligned} y^2 &= \frac{16}{9}(17.018) - 16 \\ y^2 &\approx 14.254 \\ y &\approx \pm 3.776 \end{aligned}$$

See Figure 12.

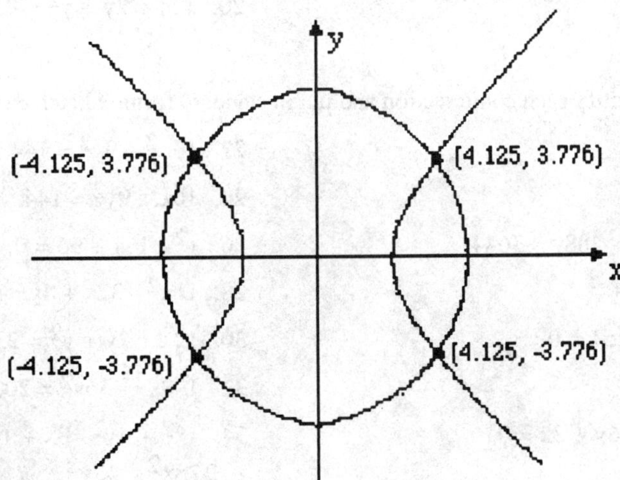

Figure 12

Exercises 8.3

Develop the Concepts

For Exercises 1 to 4 indicate the direction in which each parabola opens and state the value of p.

1. $x^2 = 20y$
2. $y^2 = -12x$
3. $y^2 = 8x$
4. $x^2 = -10y$

For Exercises 5 to 12 express each equation in the standard form for a parabola and sketch the graph. State the vertex, focus, and directrix.

5. $x^2 = 16y$
6. $x^2 = -16y$
7. $y^2 - 4x = 0$
8. $\frac{y^2}{4} = -x$
9. $\frac{(x - 1)^2}{8} = y$
10. $x^2 = 8(y - 1)$
11. $(y - 1)^2 = -8(x + 2)$
12. $y^2 + 2y = -4x + 7$

For Exercises 13 to 16 find an equation in standard form for each parabola.

13. Vertex is (0, 0), focus (0, -3)
14. Vertex is (6, -1), focus (3, -1)

*15. Focus is (0, 0), directrix $y = 3$

16. Focus (-1, 3), directrix $x = 5$

For Exercises 17 to 20 write each equation in standard form for a parabola and sketch each graph.

17. $x^2 + 2y - 12 = 0$
18. $y^2 + 3x - 6y = 0$
19. $x^2 - 6x - 10y - 1 = 0$
20. $12x + 2y + y^2 = 23$

For Exercises 21 to 36 identify each conic section and put in standard form. Sketch each graph.

21. $16x^2 - 9y^2 = 144$
22. $16x^2 + 9y^2 = 144$
23. $16x^2 - 9y = 144$
24. $16x + 9y^2 = 144$
25. $25x^2 - 100x - 144y^2 + 288y = 3644$
26. $y^2 - 16x + 96 = 0$
27. $64x^2 - 36y^2 = 2304$
28. $16x^2 - 32x + 4y^2 + 16y = 32$
29. $y^2 + 6y + 4x^2 + 8x - 51 = 0$
30. $12x + 2y + y^2 = 23$
31. $x^2 - y^2 = 0$
32. $100x^2 - 36y^2 = 200x + 144y + 269$
33. $16x^2 - 32x + 4y^2 + 16y + 32 = 0$
34. $4x^2 + 4y^2 - 8x + 16y + 11 = 0$
35. $4x^2 + 16x - 9y^2 + 18y = 9$
36. $\frac{25x^2}{4} + \frac{49y^2}{9} = 1$

Apply the Concepts

37. The Rainbow Kite and Balloon Company Spot lights are positioned at the company's entrance. Ideally, the reflector of a spot light is parabolic. The light bulb must be located at the focus so that the light will travel parallel to the axis of symmetry. Locate the bulb for the cross section in Figure 13.

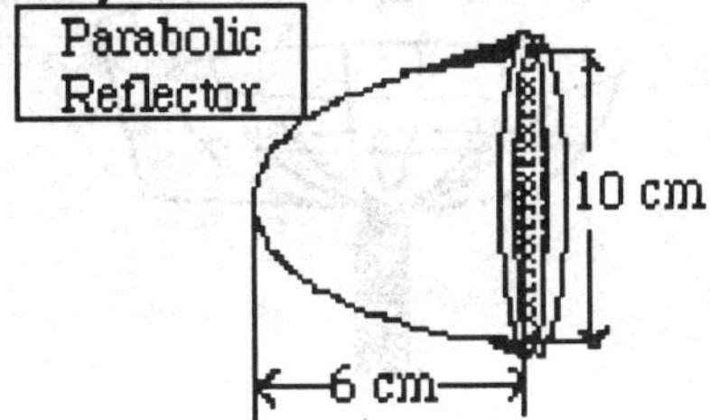

Figure 13

38. Wonderworld Amusement Park On the track of the racecar ride is a parabolic curve to ensure a smooth ride in and out of a valley. Find the location of the culvert for the valley in Figure 14.

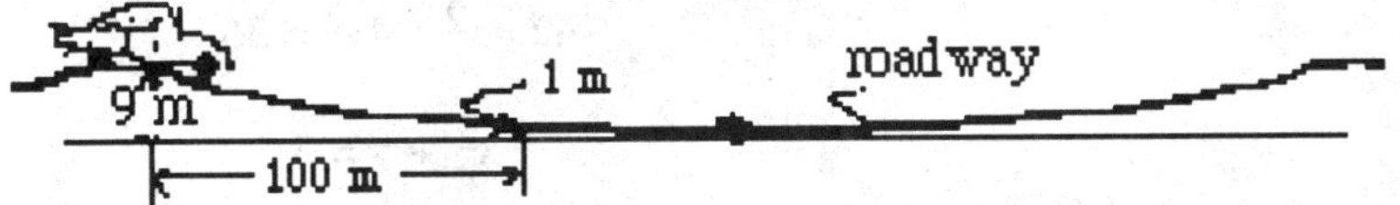

Figure 14

*39. The Newtonian Telescope Problem The Newtonian telescope shown in Figure 15 has a 200-in. diameter. At one end of the telescope there is a parabolic reflector (mirror). The parabolic mirror curves up 2.5 in. on each side. Where should a camera be mounted if it is to be located at the focus of the parabolic mirror?

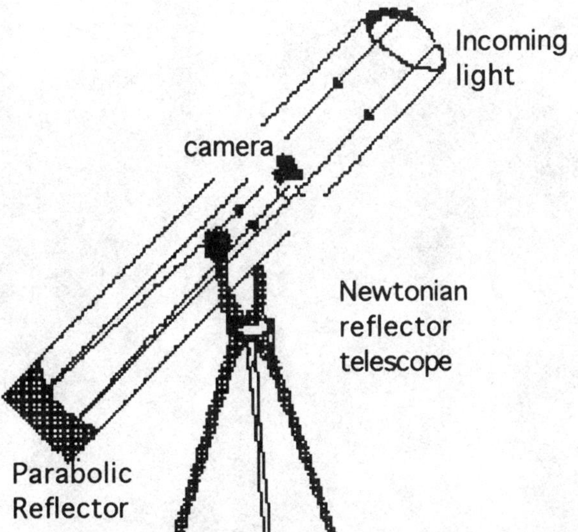

Figure 15

40. The Satellite Dish Problem A parabolic satellite dish captures incoming signals by reflecting them to one point, where a device called a "horn" captures the signal (Figure 16). Where should the horn be placed if the diameter of the dish is 8 ft and its height is 2 ft?

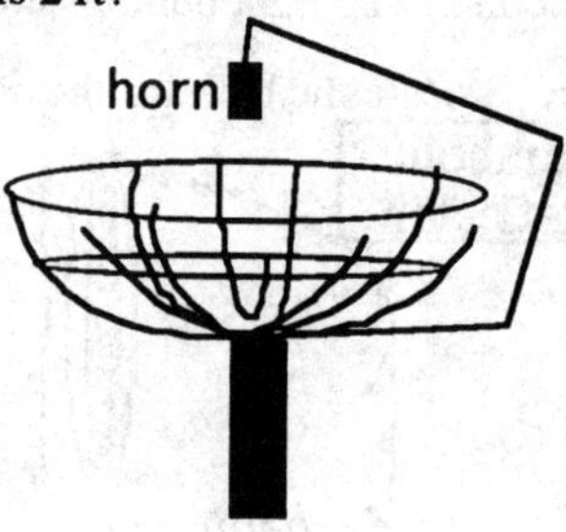

Figure 16

Extend the Concepts

For Exercises 41 to 44 graph each pair of relations on one coordinate system. Find and label the points of intersection.

41. $\frac{x^2}{9} + \frac{y^2}{4} = 1$

$x^2 + y^2 = 4$

42. $x^2 + 4y^2 = 36$

$x^2 + y = 3$

43. $x^2 + y^2 = 8$

$x - y = 1$

44. $(x - 1)^2 + (y + 2)^2 = 13$

$2x - y = -3$

8.4 Rotations and Conic Sections

Rotating the Coordinate Axes
Finding the Angle of Rotation
Predicting and Graphing Rotated Conics

In this section we complete the study of the graphs of any second-degree equation in the Cartesian coordinate system by studying the effect of the Bxy term in

$$Ax^2 + Bxy + Cy^2 + Dx + Ey + F = 0$$

If $B \neq 0$, the graph of the conic section has been rotated through some angle θ.

Rotating the Coordinate Axes

Consider the point P whose coordinates are (x, y) in the xy coordinate system and (u, v) in a new uv coordinate system created by rotating the xy coordinate system through an angle θ (Figure 1).

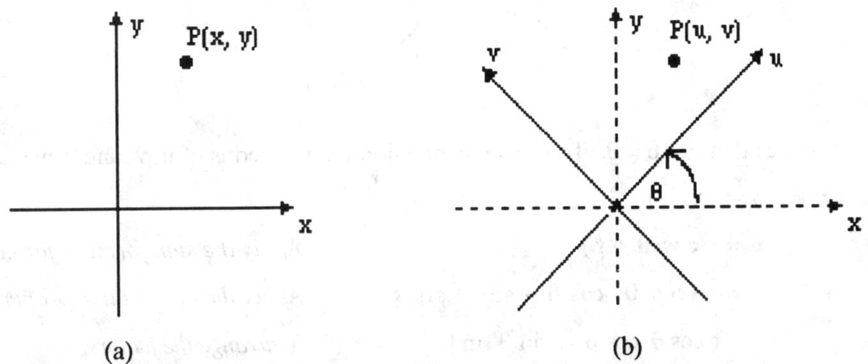

Figure 1 (a) (b)

What is the relationship between the uv coordinates and the xy coordinates?

To find the relationship requires the sum formulas for sine and cosine from Section 6.3,

$$\sin(\alpha + \beta) = \sin\alpha\cos\beta + \cos\alpha\sin\beta$$

and

$$\cos(\alpha + \beta) = \cos\alpha\cos\beta - \sin\alpha\sin\beta$$

First let's draw a right triangle in the xy coordinate system (Figure 2a) and a right triangle in the uv coordinate system (Figure 2b), with the same point P at one vertex. From Figure 2a we have

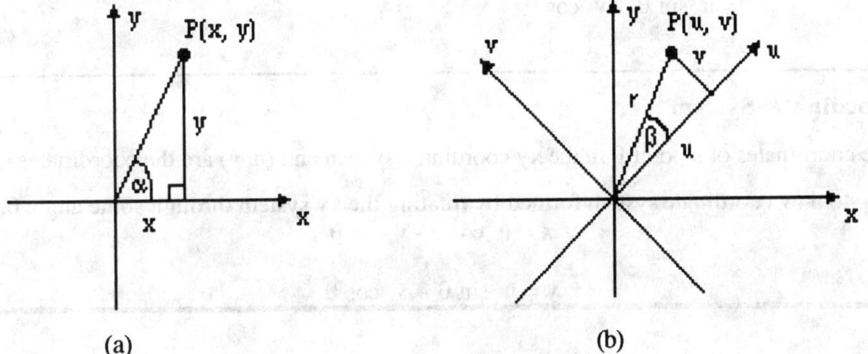

Figure 2 (a) (b)

$$\sin \alpha = \frac{y}{r} \Rightarrow y = r \sin \alpha$$

and

$$\cos \alpha = \frac{x}{r} \Rightarrow x = r \cos \alpha$$

From Figure 2b we have

$$\sin \beta = \frac{v}{r} \Rightarrow v = r \sin \beta$$

and

$$\cos \beta = \frac{u}{r} \Rightarrow u = r \cos \beta$$

Next, we need the relationship between α, β, and the angle of rotation θ (Figure 3).

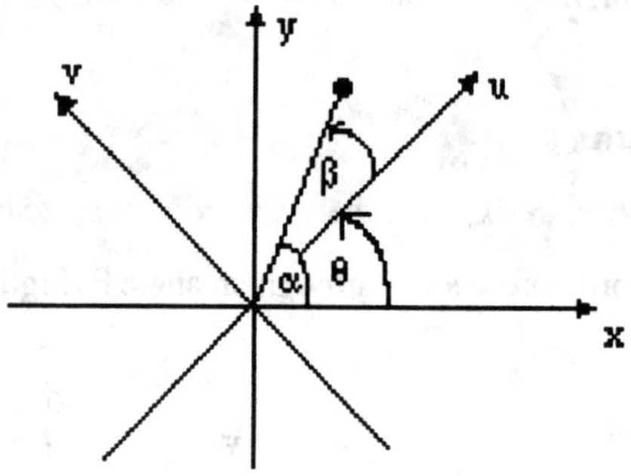

Figure 3

From Figure 3 we see that $\alpha - \beta = \theta$. To find an expression for x in terms of u, v, and θ, we have $x = r\cos\alpha$ and $\alpha = \theta + \beta \Rightarrow$

$x = r\cos(\theta + \beta)$	*Apply the sum formula for cosine.*
$= r(\cos\theta\cos\beta - \sin\theta\sin\beta)$	*Apply the distributive property.*
$= r\cos\theta\cos\beta - r\sin\theta\sin\beta$	*Rearrange the factors.*
$= \cos\theta(r\cos\beta) - \sin\theta(r\sin\beta)$	*Substitute u for r cos β and v for r sin β.*
$= u\cos\theta - v\sin\theta$	

Likewise, to find an expression for y in terms of u, v, and θ we have

$y = r\sin\alpha$ and $\alpha = \theta + \beta \Rightarrow$

$y = r\sin(\theta + \beta)$	*Apply the sum formula for sine.*
$= r(\sin\theta\cos\beta + \cos\theta\sin\beta)$	*Apply the distributive property.*
$= r\sin\theta\cos\beta + r\cos\theta\sin\beta$	*Rearrange the factors.*
$= \sin\theta(r\cos\beta) + \cos\theta(r\sin\beta)$	*Substitute u for r cos β and v for r sin β.*
$= u\sin\theta + v\cos\theta$	

Rotated Coordinate System

If (x, y) are the coordinates of a point P in the xy coordinate system and (u, v) are the coordinates of the same point P relative to a uv coordinate system formed by rotating the xy system through some angle θ, then

$$x = u\cos\theta - v\sin\theta$$

and

$$y = u\sin\theta + v\cos\theta$$

Example 1 *Finding an equation in a rotated system*

(a) Find the equation of the reference function $y = 1/x$ in the uv coordinate system with $\theta = \pi/4$.

(b) In the uv system, graph the equation from part (a), and label the vertices and foci.

Solution

(a) We will use the equations $x = u \cos \theta - v \sin \theta$ and $y = u \sin \theta + v \cos \theta$ with $\theta = \frac{\pi}{4}$ to create an equation in u and v.

$x = u \cos \theta - v \sin \theta$ and $\theta = \frac{\pi}{4} \Rightarrow$

$$x = u \cos \pi/4 - v \sin \pi/4$$

$$x = \frac{u}{\sqrt{2}} - \frac{v}{\sqrt{2}}$$

$y = u \sin \theta + v \cos \theta$ and $\theta = \frac{\pi}{4} \Rightarrow$

$$y = u \sin \pi/4 - v \cos \pi/4$$

$$y = \frac{u}{\sqrt{2}} + \frac{v}{\sqrt{2}}$$

We substitute the expressions for x and y in the equation of the reference function $y = 1/x$.

$$y = \frac{1}{x}$$

$$xy = 1$$

$$\left(\frac{u}{\sqrt{2}} - \frac{v}{\sqrt{2}}\right)\left(\frac{u}{\sqrt{2}} + \frac{v}{\sqrt{2}}\right) = 1$$

$$\frac{u^2}{2} - \frac{v^2}{2} = 1$$

Therefore, the equation $\frac{u^2}{2} - \frac{v^2}{2} = 1$ indicates that the graph is a hyperbola in the uv coordinate system.

(b) $c^2 = a^2 + b^2$, $a = \sqrt{2}$ and $b = \sqrt{2} \Rightarrow c^2 = 2 + 2 = 4$

The distance from the origin to the foci on the u axis is $c = 2$. Therefore, the uv coordinates of the foci are $(-2, 0)$ and $(2, 0)$ (Figure 4).

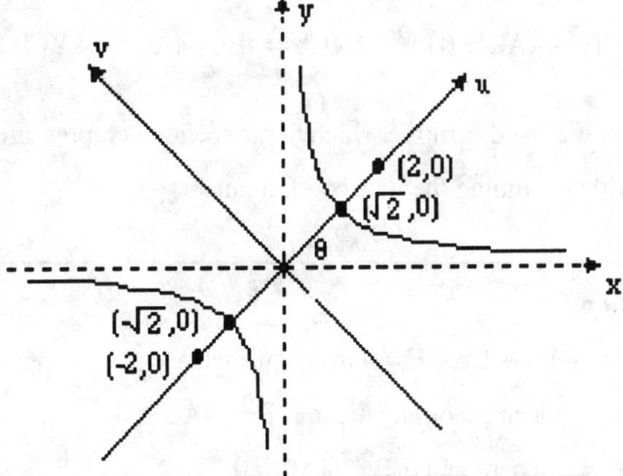

Figure 4

Finding the Angle of Rotation

Substituting the equations

$$x = u\cos\theta - v\sin\theta$$

and

$$y = u\sin\theta + v\cos\theta$$

into the general equation for a rotated conic section

$$Ax^2 + Bxy + Cy^2 + Dx + Ey + F = 0,\ B \neq 0$$

results in the formula

$$\theta = \frac{1}{2}\tan^{-1}\frac{B}{A - C} \quad \text{for } A \neq C$$

by requiring the coefficient of the uv term to equal zero. If A = C, then $\theta = \frac{\pi}{4}$.

Angle of Rotation to Eliminate the xy Term

The angle of rotation of a conic section θ described by the equation $Ax^2 + Bxy + Cy^2 + Dx + Ey + F = 0$, $B \neq 0$ is

$$\theta = \frac{1}{2}\tan^{-1}\frac{B}{A - C} \quad \text{if } A \neq C \text{ and } \theta = \pi/4 \text{ if } A = C$$

for $0 < \theta < \pi/2$.

Predicting and Graphing Rotated Conics

If we rotate a conic section in the xy coordinate system through the appropriate angle θ we create a new equation in the uv coordinate system that eliminates the uv term. The equation

$$Ax^2 + Bxy + Cy^2 + Dx + Ey + F = 0$$

becomes

$$A'u^2 + \mathbf{0}uv + C'v^2 + D'u + E'v + F' = 0$$

There are relationships between the original coefficients and the new ones. For example, the constant terms are equal F = F', the sum A + C is the same as the sum A' + C', and $B^2 - 4AC = (B')^2 - 4A'C'$. If B' = 0, then

$$B^2 - 4AC = (B')^2 - 4A'C' \Rightarrow B^2 - 4AC = -4A'C'$$

The following summarizes how we can determine which conic section is represented by a given equation using the values of A, B, and C -- without finding the new coefficients.

The Rotated Conic Section

The equation $Ax^2 + Bxy + Cy^2 + Dx + Ey + F = 0$ may represent:

a circle or an ellipse if $B^2 - 4AC < 0$

a parabola if $B^2 - 4AC = 0$

a hyperbola if $B^2 - 4AC > 0$

If an equation does not represent a circle, an ellipse, a parabola, or a hyperbola, then it represents either one or two straight lines, a point, or no ordered pairs of real numbers.

For the equation in the next example, $x^2 + 2xy + y^2 + \sqrt{2}\,x - \sqrt{2}\,y = 0$, A = 1, B = 2, and C = 1 yields $B^2 - 4AC = 0$. The graph is probably a parabola.

Example 2 *Graphing a rotated conic section*

(a) Find the angle through which the axes must be rotated to remove the xy term in the equation

$$x^2 + 2xy + y^2 + \sqrt{2}\,x - \sqrt{2}\,y = 0$$

(b) Express the equation in terms of u and v.

(c) Sketch the graph in the uv coordinate system.

Solution

(a) Since $A = C = 1$, $\theta = \frac{\pi}{4}$.

(b) Substituting $\theta = \frac{\pi}{4}$ yields

$$x = u\cos\theta - v\sin\theta = u\left(\frac{1}{\sqrt{2}}\right) - v\left(\frac{1}{\sqrt{2}}\right) = \frac{1}{\sqrt{2}}(u - v)$$

and

$$y = u\sin\theta + v\cos\theta = u\left(\frac{1}{\sqrt{2}}\right) + v\left(\frac{1}{\sqrt{2}}\right) = \frac{1}{\sqrt{2}}(u + v)$$

Substituting $\frac{1}{\sqrt{2}}(u - v)$ for x and $\frac{1}{\sqrt{2}}(u + v)$ for y transforms the original equation into the quadratic equation in terms of u and v.

$$\begin{aligned} x^2 + 2xy + y^2 + \sqrt{2}\,x - \sqrt{2}\,y &= 0 \\ \frac{1}{2}(u - v)^2 + (u - v)(u + v) + \frac{1}{2}(u + v)^2 + (u - v) - (u + v) &= 0 \\ (u^2 - 2uv + v^2) + (2u^2 - 2v^2) + (u^2 + 2uv + v^2) + (2u - 2v) - (2u + 2v) &= 0 \\ 4u^2 - 4v &= 0 \\ v &= u^2 \end{aligned}$$

(c) Since $4p = 1$, the focus is $(0, \frac{1}{4})$ in the uv coordinate system (Figure 5).

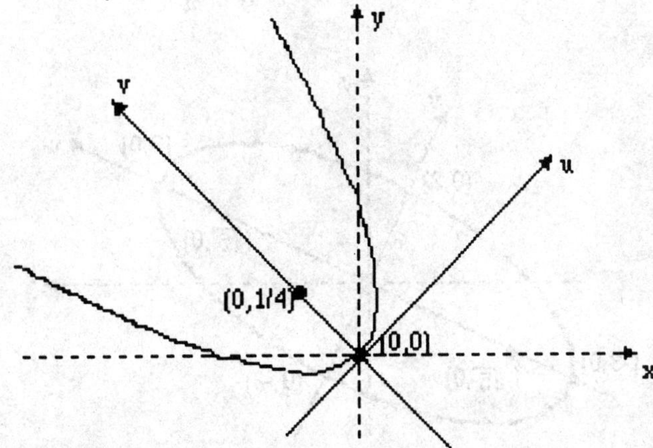

Figure 5

In the next example we use the half-angle formulas to find the exact values of sin and cos θ from tan 2θ. Finding the exact value of sine and cosine of an angle once tan 2θ is known is also used in other applications.

Example 3 *Graphing a rotated conic section*

Graph $5x^2 - 4xy + 8y^2 - 36 = 0$ in a rotated coordinate system.

Solution

Since $B^2 - 4AC = 16 - 160 = -144$ and $-144 < 0$, the conic section is an ellipse.

$$\tan 2\theta = \frac{B}{A - C} = \frac{-4}{5 - 8} = \frac{4}{3}$$

$$\sin^2\theta = \frac{1 - \cos 2\theta}{2} = \frac{1 - 3/5}{2} = \frac{1}{5} \Rightarrow \sin\theta = \frac{1}{\sqrt{5}}$$

and

$$\cos^2\theta = \frac{1 + \cos 2\theta}{2} = \frac{1 + 3/5}{2} = \frac{4}{5} \Rightarrow \cos\theta = \frac{2}{\sqrt{5}}$$

$$x = u\cos\theta - v\sin\theta = u\left(\frac{2}{\sqrt{5}}\right) - v\left(\frac{1}{\sqrt{5}}\right) = \frac{1}{\sqrt{5}}(2u - v)$$

and

$$y = u\sin\theta + v\cos\theta = u\left(\frac{1}{\sqrt{5}}\right) + v\left(\frac{2}{\sqrt{5}}\right) = \frac{1}{\sqrt{5}}(u + 2v)$$

$5x^2 - 4xy + 8y^2 - 36 = 0 \Leftrightarrow$

$$\frac{5}{5}(2u - v)^2 - \frac{4}{5}(2u - v)(u + 2v) + \frac{8}{5}(u + 2v)^2 = 36$$

$$5(4u^2 - 4uv + v^2) - 4(2u^2 + 3uv - 2v^2) + 8(u^2 + 4uv + 4v^2) = 180$$

$$20u^2 + 45v^2 = 180$$

$$\frac{u^2}{9} + \frac{v^2}{4} = 1$$

See Figure 6.

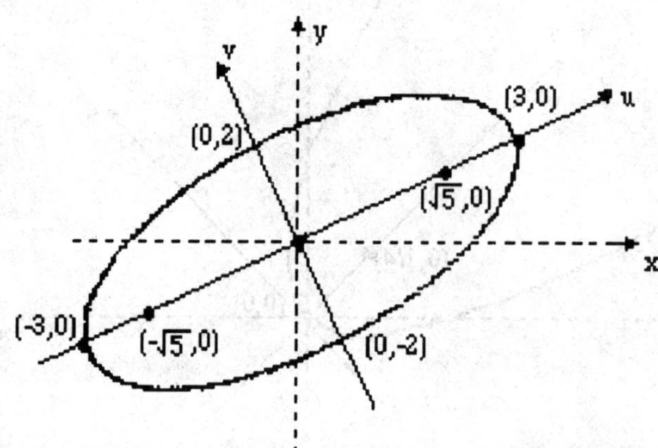

Figure 6

Equations of rotated conic sections can be treated as implicit functions by solving for either x or y. The two resulting equations will involve square roots. For the ellipse in Example 3, we can solve

$$5x^2 - 4xy + 8y^2 - 36 = 0$$

for y using the quadratic formula as follows. First, we rearrange the equation in descending powers of y.

$$8y^2 - 4xy + (5x^2 - 36) = 0$$

$$y = \frac{-b \pm \sqrt{b^2 - 4ac}}{2a}, \; a = 8, \; b = -4x, \; c = (5x^2 - 36) \Rightarrow$$

$$y = \frac{4x \pm \sqrt{(-4x)^2 - 4(8)(5x^2 - 36)}}{2(8)}$$

$$y = \frac{x \pm \sqrt{72 - 9x^2}}{4}$$

See Figure 7.

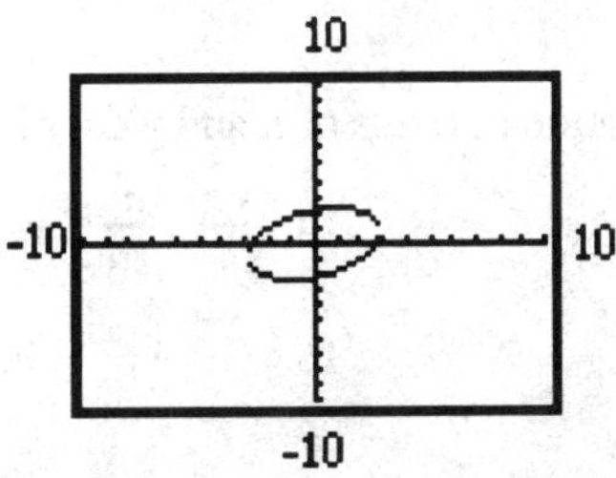

Figure 7

NOTE: Since graphing rotated conics by solving for x or y does not create an equation in standard form, we cannot determine the vertices, foci, or asymptotes as we did in previous sections.

Exercises 8.4

Develop the Concepts

For Exercises 1 to 4 use the theorem for rotated conic sections to determine which conic section is probably given by each of the following equations.

1. $x^2 - 2xy + y^2 - 1 = 0$
2. $3x^2 - 2xy - 4y^2 + 4y = 2$
3. $2x^2 + 5xy + 2y^2 - 3y = 4$
4. $x^2 - 3xy + 3y^2 + 4x = 1$

For Exercises 5 to 8 determine the angle through which each conic section has been rotated.

5. $3x^2 + 4xy - y^2 - 4x + 6y - 3 = 0$
6. $2x^2 - \sqrt{3}xy + 3y^2 + 6y - 2\sqrt{3} = 0$
7. $2\sqrt{3}x^2 + xy + 3\sqrt{3}y^2 + \sqrt{3}x - \sqrt{3} = 0$
8. $5x^2 + 6xy + 5y^2 + 6y - 5 = 0$

For Exercises 9 to 12 graph each conic section in the uv coordinate system rotated through the given angle.

*9. $\theta = 45°, \frac{u^2}{9} + \frac{v^2}{16} = 1$

10. $\theta = 60°, \frac{v^2}{4} - \frac{u^2}{16} = 1$

11. $\theta = \frac{\pi}{6}, (v - 1)^2 = 4u$

12. $\theta = \frac{\pi}{4}, \frac{(u - 1)^2}{4} + \frac{(v - 2)^2}{9} = 1$

For Exercises 13 to 16 express each equation in terms of x and y coordinates.

13. $\frac{u^2}{9} + \frac{v^2}{16} = 1, \theta = 45°$
14. $\frac{v^2}{4} - \frac{u^2}{16} = 1, \theta = 60°$
15. $\frac{(u - 1)^2}{4} + \frac{(v - 2)^2}{9} = 1, \theta = \pi/4$
16. $(v - 1)^2 = 4u, \theta = \pi/6$

For Exercises 17 to 20 state each angle of rotation, write each equation in uv form, and graph each rotated conic section.

*17. $10x^2 + 52xy + 10y^2 = 576$

18. $15x^2 + 40xy - 15y^2 + 625 = 0$

19. $101x^2 + 30xy + 29y^2 = 104$

20. $x^2 - 2\sqrt{3}\,xy + 3y^2 = 0$

Extend the Concepts

For Exercises 21 to 24 graph and label each rotated and translated conic section.

21. $4x^2 + 8xy + 4y^2 - 7\sqrt{2}\,x - 9\sqrt{2}\,y + 6 = 0$
22. $10x^2 - 12xy + 10y^2 - 36\sqrt{2}\,x - 28\sqrt{2}\,y + 4 = 0$
23. $-6x^2 + 20xy - 6y^2 - 20\sqrt{2}\,x + 44\sqrt{2}\,y - 92 = 0$
24. $x^2 + 2\sqrt{3}\,xy - y^2 + 2\sqrt{3}\,x + 2y = 0$

8.5 Graphs in a Three-Dimensional Coordinate System

Plotting Points and Graphing Planes in R^3
Graphing Cylinders

In this section we graph points and equations in a three-dimensional coordinate system called R^3. Figure 1 shows the coordinate axes as drawn in R^3, with the x axis extending front and back, the y axis right and left, and the z axis up and down.

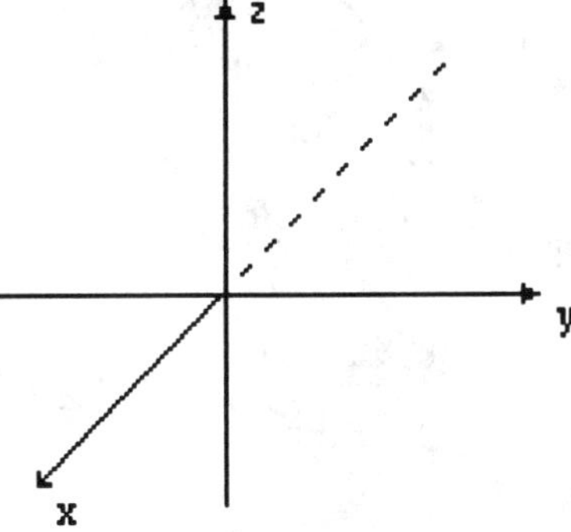

Figure 1

While the two-dimensional system has four quadrants, the three-dimensional system has eight **octants**. In the first octant x, y, and z values are positive. Often, the first octant view of a surface in R^3 is adequate for visualizing the entire object, especially when the surface is symmetric.

Plotting Points and Graphing Planes in R^3

To plot a point in the xy coordinate system (R^2) requires an ordered pair (x, y).

How is a single point located in R^3?

Since R^3 incorporates three axes, locating a single point requires an **ordered triple (x, y, z)**. Figure 2 shows some points graphed in R^3.

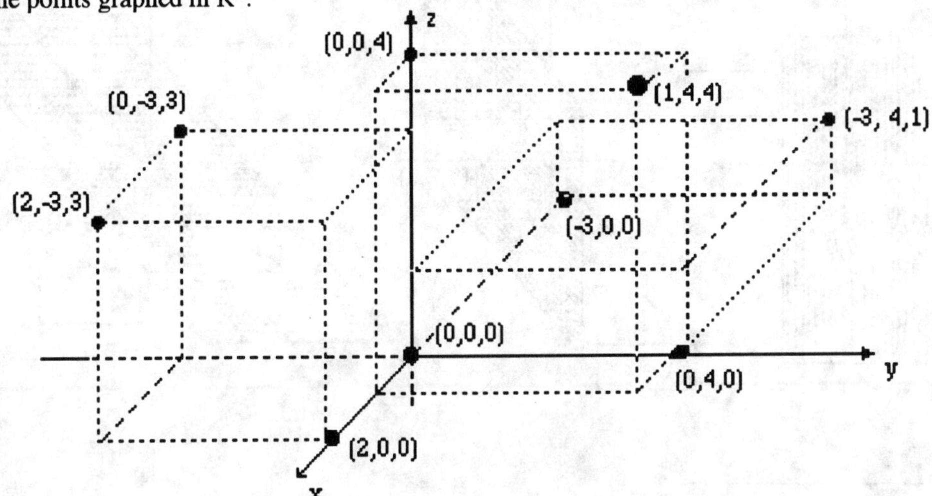

Figure 2

Linear equations are graphed as planes in R^3. There are three **coordinate planes**, x = 0, y = 0, and z = 0 (Figure 3). The ordered triples in the yz plane (Figure 3a) have x coordinates of zero, while the ones in the xz plane (Figure 3b) have y coordinates of zero, and those in the xy plane (Figure 3c) have z coordinates of zero.

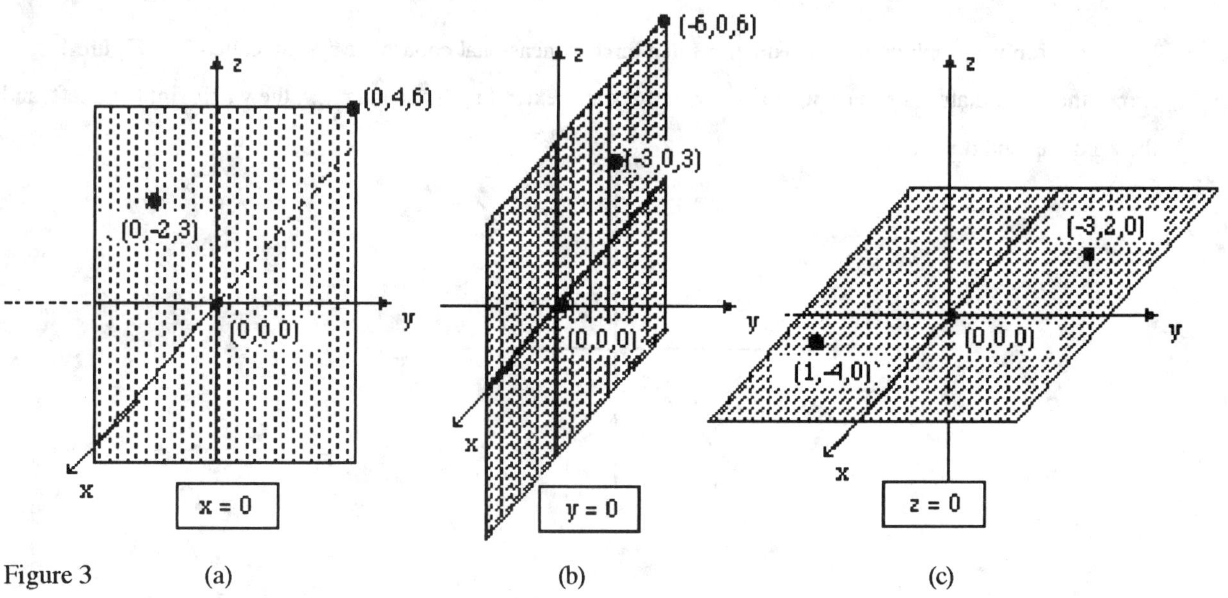

Figure 3 (a) (b) (c)

Example 1 *Graphing planes*

Graph the following planes in R^3.

(a) x = 3 (b) y = 2 (c) z = -1

Solution

The graph of x = 3 is a translation of the yz plane 3 units forward, the graph of y = 2 is a translation of the xz plane 2 units to the right, and the graph of z = -1 is a translation of the xy plane 1 unit down (Figure 4).

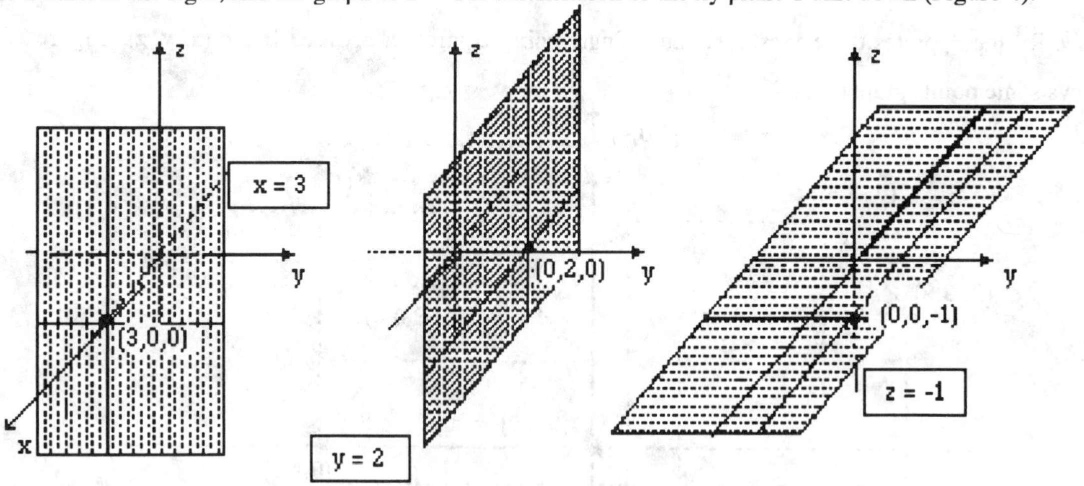

Figure 4

Often we sketch the graph of a plane by drawing the triangular region of the plane created by connecting its x, y, and z intercepts. The plane actually extends indefinitely, but this portion is sufficient to visualize the orientation and position of the plane.

Example 2 *Graphing a representation of a plane*

Graph the first-octant portion of $2x + 3y + 4z = 12$.

Solution

We find the x, y, and z intercepts. These three points are the vertices of the triangular region where the plane cuts through the first octant.

x intercept: $2x + 3y + 4z = 12$ and $y = 0, z = 0 \Rightarrow x = 6$

y intercept: $2x + 3y + 4z = 12$ and $x = 0, z = 0 \Rightarrow y = 4$

z intercept: $2x + 3y + 4z = 12$ and $x = 0, y = 0 \Rightarrow z = 3$

Connecting the intercepts located at (6, 0, 0), (0, 4, 0), and (0, 0, 3) with yields a triangular portion of the plane (Figure 5).

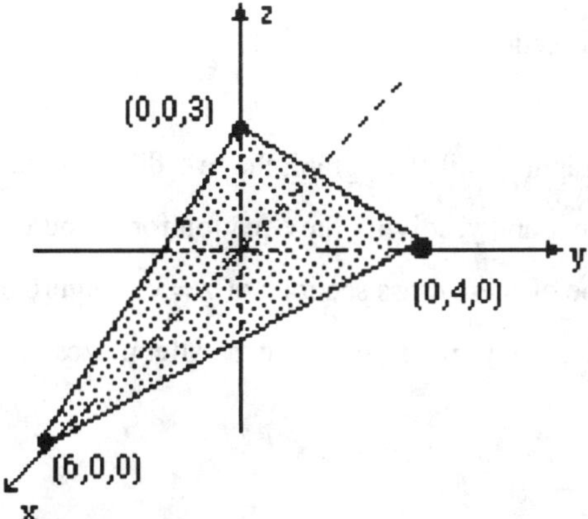

Figure 5

Graphing Cylinders

An important approach used to analyze equations in R^3 is to examine cross sections of the graph in each of the three coordinate planes. These cross sections are called **traces**. Three-dimensional figures in which parallel cross sections in one direction yield the congruent traces are called **cylinders**. Figure 6 shows the graphs of a linear cylinder with lines as cross sections (Figure 6a), a right circular cylinder with circular cross sections (Figure 6b), and a parabolic cylinder with parabolic cross sections (Figure 6c).

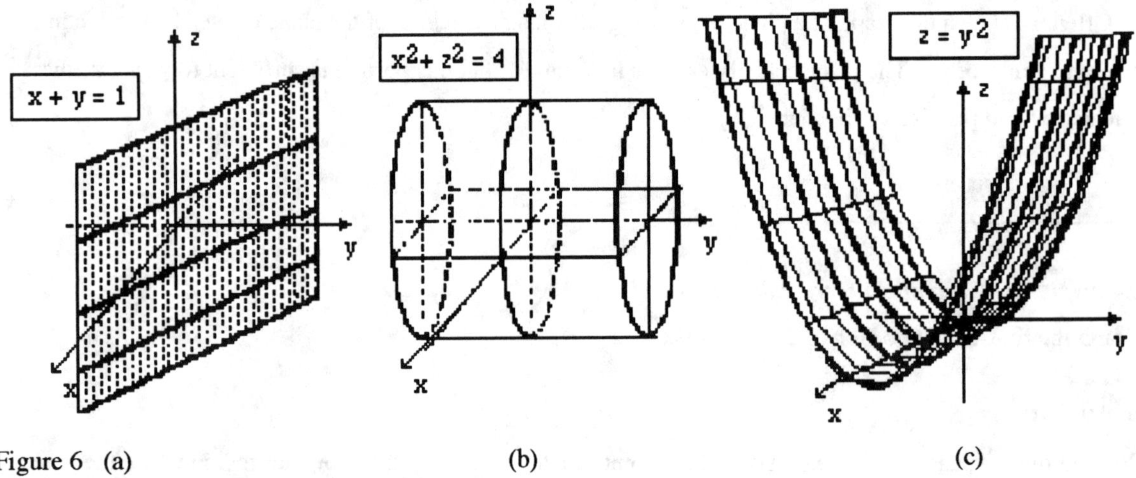

Figure 6 (a) (b) (c)

Such cylinders are easily recognized because their equations contain only two variables.

Do you notice a relationship between the missing variable in the equations in Figure 6 and the direction of the parallel congruent traces in each drawing?

The parallel congruent traces are in the direction of the missing variable. For $x + y = 1$, parallel lines are stacked in the z direction, and for $x^2 + z^2 = 4$, parallel circles are lined up in the y direction. For $z = y^2$, the parallel traces are found in the x direction.

NOTE: We can use a graphing utility to graph the two-dimensional cross sections of a cylinder by changing the variables to x and y, solving the equation for y, and entering the resulting equation. For example, to see the shape of these cross sections of $z = y^2$ (Figure 6c), we enter $y = x^2$. The resulting curve is then transferred to the yz plane and a few parallel planes.

Example 3 *Graphing a circular cylinder*

Graph $x^2 + y^2 = 25$ in R^3.

Solution

Since the z variable is missing in the equation, the graph is a cylinder. When $z = 0$, the trace in the xy plane is a circle with radius 5 centered at the origin, because

$$z = 0 \text{ and } x^2 + y^2 = 25 \Rightarrow x^2 + y^2 = 25$$

The trace in any plane parallel to the xy plane is the same circle (Figure 7). For example,

$$z = 3 \text{ and } x^2 + y^2 = 25 \Rightarrow x^2 + y^2 = 25$$

and

$$z = -4 \text{ and } x^2 + y^2 = 25 \Rightarrow x^2 + y^2 = 25$$

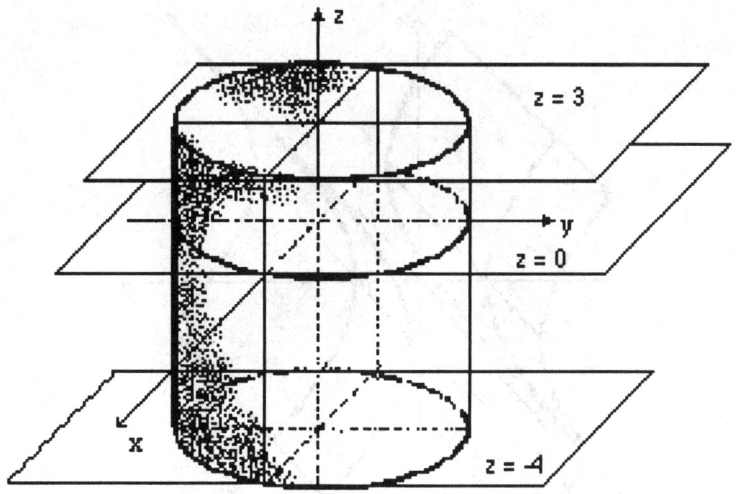

Figure 7

Example 4 *Graphing cylinders*

Graph the following cylinders in R^3.

(a) $\frac{x^2}{4}+\frac{y^2}{9}=1$ (b) $\frac{y^2}{4}-\frac{z^2}{9}=1$

Solution

(a) $\frac{x^2}{4}+\frac{y^2}{9}=1$ is an elliptical cylinder whose traces in the z direction are the same ellipse (Figure 8).

$z=1$ and $\frac{x^2}{4}+\frac{y^2}{9}=1 \Rightarrow \frac{x^2}{4}+\frac{y^2}{9}=1$

$z=0$ and $\frac{x^2}{4}+\frac{y^2}{9}=1 \Rightarrow \frac{x^2}{4}+\frac{y^2}{9}=1$

$z=-3$ and $\frac{x^2}{4}+\frac{y^2}{9}=1 \Rightarrow \frac{x^2}{4}+\frac{y^2}{9}=1$

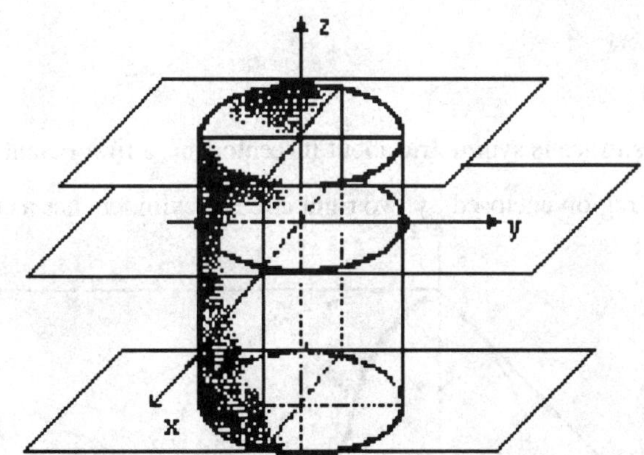

Figure 8

(b) $\frac{y^2}{4}-\frac{z^2}{9}=1$ is a hyperbolic cylinder whose traces in the x direction are the same hyperbola (Figure 9).

$x=2$ and $\frac{y^2}{4}-\frac{z^2}{9}=1 \Rightarrow \frac{y^2}{4}-\frac{z^2}{9}=1$

$x=0$ and $\frac{y^2}{4}-\frac{z^2}{9}=1 \Rightarrow \frac{y^2}{4}-\frac{z^2}{9}=1$

$x=-3$ and $\frac{y^2}{4}-\frac{z^2}{9}=1 \Rightarrow \frac{y^2}{4}-\frac{z^2}{9}=1$

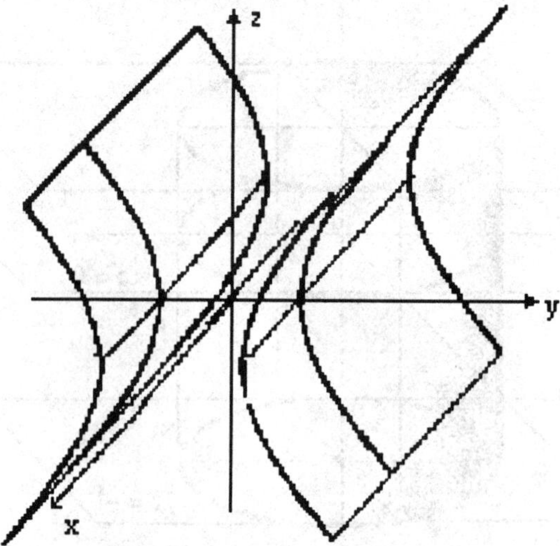

Figure 9

The next example illustrates a drawing skill that is useful in calculus and related applications where the volume is enclosed by more than one surface.

Example 5 *Finding the points of intersection of two cylinders in* R^3

Sketch the first octant graph of the cylinders $x^2 + z^2 = 1$ and $y^2 + z^2 = 1$ on one three-dimensional coordinate system. Then draw the line of intersection of the surfaces.

Solution

Since each surface is symmetric about its centerline, a first-octant sketch is sufficient to visualize the entire shape. The region enclosed by two right circular cylinders has a curved line of intersection (Figure 10).

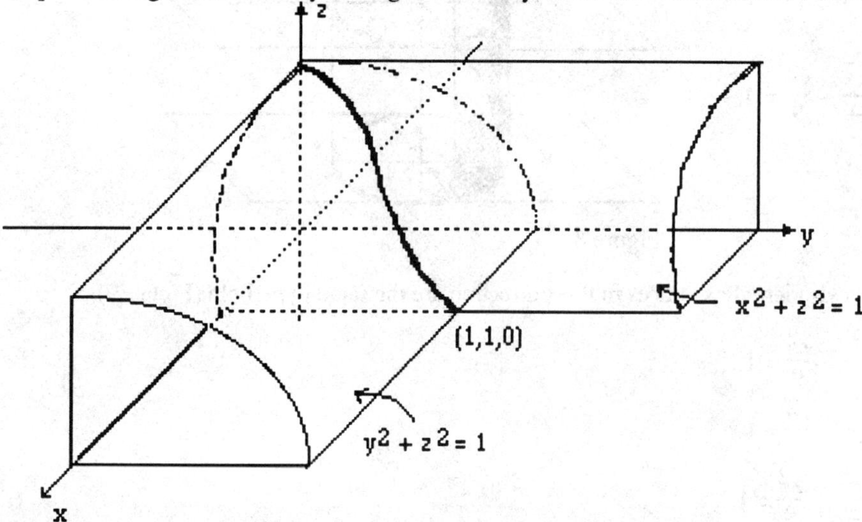

Figure 10

Exercises 8.5

Develop the Concepts

For Exercises 1 to 4 refer to Figure 11 to find the coordinates of the missing points.

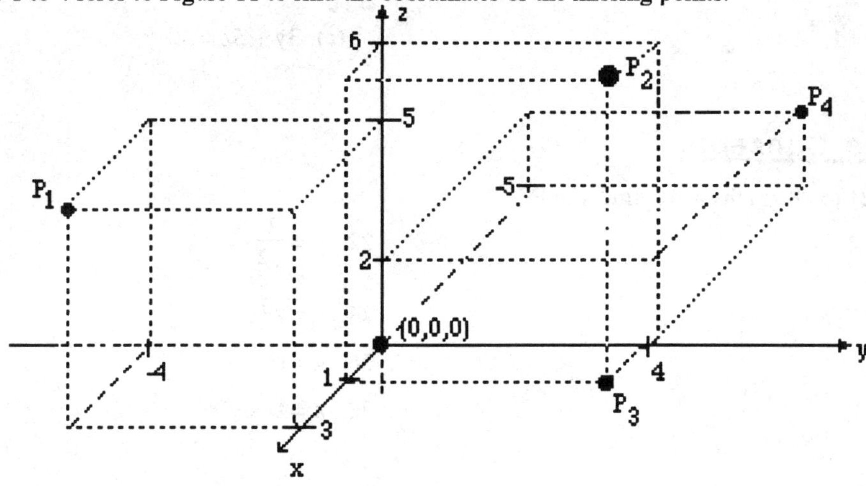

Figure 11

1. Find the coordinates for P_1
2. Find the coordinates for P_2
3. Find the coordinates for P_3
4. Find the coordinates for P_4

For Exercises 5 to 8 draw a sketch in R^3 of each plane.

5. $y = 2$
6. $x = 2$
7. $z = 2$
8. $y = -2$

For Exercises 9 to 12 find the x, y, and z intercepts of each plane. Use them to draw the triangular portion that falls in octant I of a three-dimensional coordinate system.

9. $3x + y + 4z = 12$
10. $2x + 3y + z = 6$
11. $x + 4y + z = 8$
12. $3x + 4y + z = 24$

For Exercises 13 to 20 graph the cylinders in R^3.

13.(a) $x + y = 1$
(b) $x + z = 1$
(c) $y + z = 1$

14.(a) $x^2 + y^2 = 25$
(b) $y^2 + z^2 = 25$
(c) $x^2 + z^2 = 25$

15.(a) $4x^2 + z^2 = 16$
(b) $4y^2 + z^2 = 16$
(c) $4x^2 + y^2 = 16$

16.(a) $x^2 - y^2 = 1$
(b) $y^2 - x^2 = 1$
(c) $z^2 - x^2 = 1$

17.(a) $y = x^2$
(b) $z = x^2$
(c) $y = z^2$

18.(a) $x = y^2$
(b) $x = z^2$
(c) $z = y^2$

19.(a) $\frac{x^2}{4} + \frac{y^2}{9} = 1$

(b) $\frac{y^2}{9} + \frac{z^2}{25} = 1$

(c) $\frac{x^2}{4} - \frac{y^2}{9} = 1$

20.(a) $2x + 3y = 6$

(b) $4x - 2z = 1$

(c) $-3y + 5z = 30$

Extend the Concepts

For Exercises 21 to 28 graph each cylinder in R^3.

21. $z = \frac{1}{x}$

22. $y = \frac{1}{x^2}$

23. $y = |z|$

24. $y = z^3$

*25. $z = \ln y$

26. $y = e^z$

27. $z = \sin y$

28. $x = \cos y$

For Exercises 29 to 32 draw two sketches, one in R^2 and the other in R^3.

29. $y = x^3$

30. $y = \sqrt{x}$

31. $y = \tan x$

32. $y = \text{Sin}^{-1} x$

For Exercises 33 to 25 graph each translated cylinders in R^3.

33. $(x - 2)^2 + z^2 = 9$

34. $z = (y + 1)^2$

35. $\frac{(y + 2)^2}{4} - \frac{(z - 1)^2}{4} = 1$

36. $\frac{(x - 1)^2}{4} + \frac{(y - 2)^2}{9} = 1$

For Exercises 37 to 40 sketch each inequality in R^3.

37. $y + z \leq 4$

38. $z^2 - x^2 \geq 1$

39. $x^2 + y^2 \leq 9$

40. $4y^2 + 9z^2 \geq 36$

For Exercises 41 to 44 sketch the three-dimensional curve of intersection of the systems of cylinders in octant I.

41. $x^2 + y^2 = 1$
$x^2 + z^2 = 1$

42. $x^2 + 16z^2 = 16$
$y^2 + 9z^2 = 9$

43. $y = 9 - z^2$
$x^2 + y^2 = 9$

44. $x^2 + y^2 = 1$
$4x + z = 4$

8.6 Quadric Surfaces

Graphing Surfaces of Revolution
Graphing Other Quadric Surfaces
Graphing Using Contour Maps

In this section we study **quadric surfaces**, which are three-dimensional generalizations of second-degree equations. The points that satisfy these second-degree equations are points on the surface of the three-dimensional graph, not interior points. Some of the cylinders covered in Section 8.5 are considered quadric surfaces since their equations are second degree. In this section we discover the shape of the graph by considering the x, y, and z traces.

Graphing Surfaces of Revolution

When a conic section is spun about an axis, the result is a **quadric surface** called a **surface of revolution**. Figure 1 shows a line, circle, ellipse, parabola, and hyperbola that have been spun about one of the axes. The resulting solid is called a cone, sphere, ellipsoid, paraboloid, hyperboloid of one sheet, or hyperboloid of two sheets, respectively

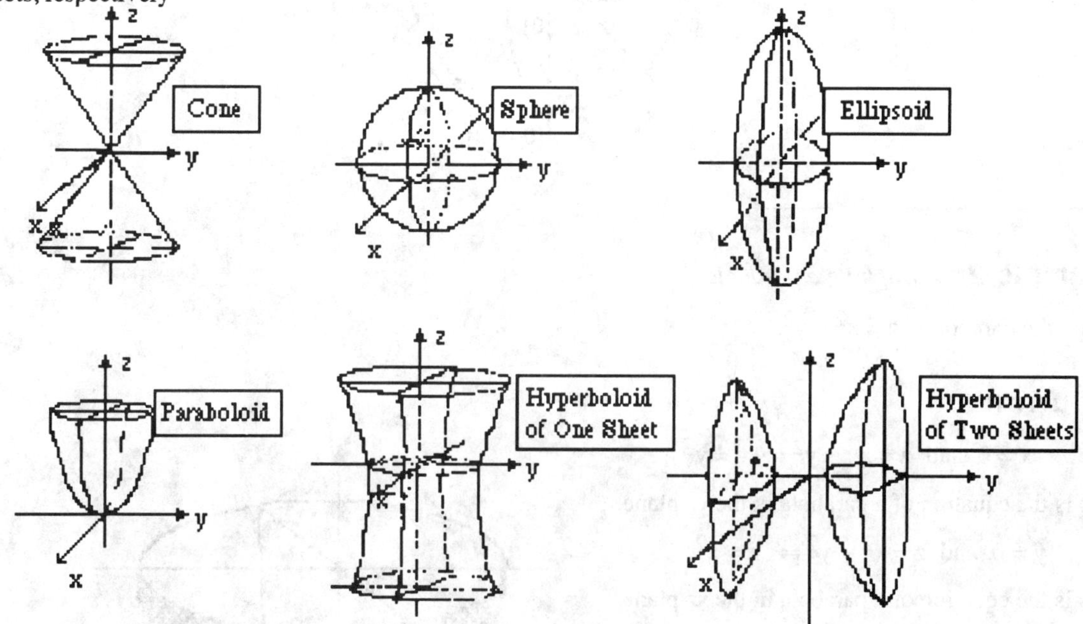

Figure 1

The shape of the solid is found by considering the x, y, and z traces.

Example 1 *Graphing a sphere*

Graph $x^2 + y^2 + z^2 = 4$ in R^3.

Solution

$$x = 0 \text{ and } x^2 + y^2 + z^2 = 4 \Rightarrow y^2 + z^2 = 4$$

This equation is a circle in the yz plane centered at (0, 0) with radius r = 2.

$$y = 0 \text{ and } x^2 + y^2 + z^2 = 4 \Rightarrow x^2 + z^2 = 4$$

This equation is a circle in the xz plane centered at (0, 0) with radius r = 2.

$$z = 0 \text{ and } x^2 + y^2 + z^2 = 4 \Rightarrow x^2 + y^2 = 4$$

This equation is a circle in the xy plane centered at (0, 0) with radius r = 2.

The resulting surface is a sphere centered at (0, 0, 0) with radius r = 2 (Figure 2).

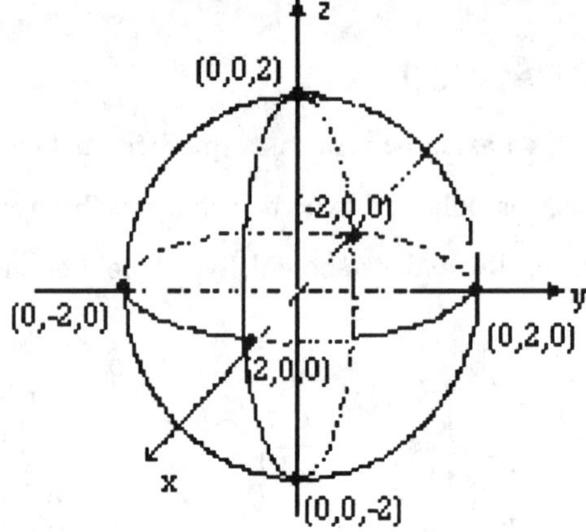

Figure 2

Example 2 *Graphing a paraboloid*

Graph the paraboloid $z = x^2 + y^2$.

Solution

$$x = 0 \text{ and } z = x^2 + y^2 \Rightarrow z = y^2$$

This is the equation of a parabola in the yz plane.

$$y = 0 \text{ and } z = x^2 + y^2 \Rightarrow z = x^2$$

This is the equation of a parabola in the xz plane.

$$z = 0 \text{ and } z = x^2 + y^2 \Rightarrow x^2 + y^2 = 0$$

This is the equation of a point circle. It yields the point (0, 0, 0).

$$z = 4 \text{ and } z = x^2 + y^2 \Rightarrow x^2 + y^2 = 16$$

This is the equation of a circle in the z = 4 plane with radius r = 4. Since two traces yield a parabola, the surface is referred to as a paraboloid (Figure 3).

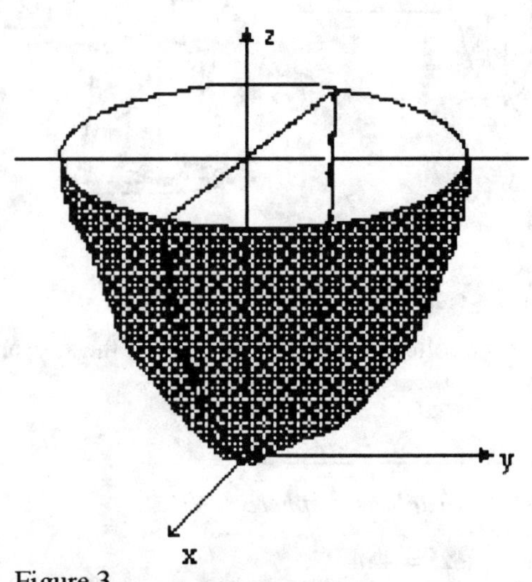

Figure 3

Example 3 *Graphing a hyperboloid*

Graph the hyperboloid $\frac{x^2}{4} - \frac{z^2}{1} - \frac{y^2}{1} = 1$.

Solution

$$x = 0 \text{ and } \frac{x^2}{4} - \frac{z^2}{1} - \frac{y^2}{1} = 1 \Rightarrow -z^2 - y^2 = 1$$

This equation yields no real points in the yz plane.

$$x = \pm 2 \text{ and } \frac{x^2}{4} - \frac{z^2}{1} - \frac{y^2}{1} = 1 \Rightarrow z^2 + y^2 = 0$$

This is a point circle.

$$y = 0 \text{ and } \frac{x^2}{4} - \frac{z^2}{1} - \frac{y^2}{1} = 1 \Rightarrow \frac{x^2}{4} - \frac{z^2}{1} = 1$$

The equation is a hyperbola in the xz plane.

$$z = 0 \text{ and } \frac{x^2}{4} - \frac{z^2}{1} - \frac{y^2}{1} = 1 \Rightarrow \frac{x^2}{4} - \frac{y^2}{1} = 1$$

The equation is a hyperbola in the xy plane. The equation $\frac{x^2}{4} - \frac{z^2}{1} - \frac{y^2}{1} = 1$ is a hyperboloid of two sheets (Figure 4).

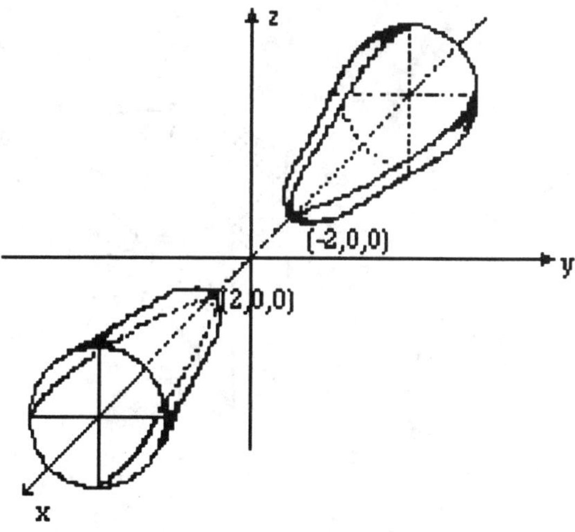

Figure 4

Notice that in each of the examples of a surface of revolution, one set of parallel traces consists of circles with varying radii.

Graphing Other Quadric Surfaces

Some quadric surfaces are not surfaces of revolution of a conic section but have conic sections as traces in their three coordinate planes. Two such surfaces are the **elliptic paraboloid** and the **hyperbolic paraboloid**. Both have second-degree equations. The elliptic paraboloid shown in Figure 5 has traces in the xz and yz planes that are parabolas, while traces above the xy plane are ellipses.

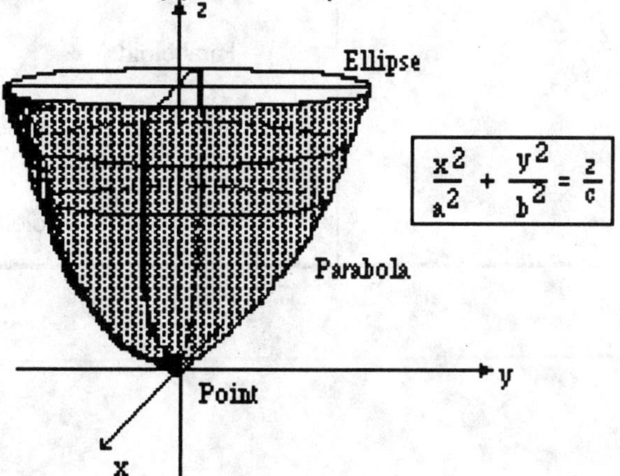

Figure 5

Figure 6 shows a hyperbolic paraboloid, often referred to as a *saddle.* It has hyperbolas as traces in planes parallel to the xy plane and in planes parallel to the xz plane. It has parabolas in the planes parallel to the yz plane.

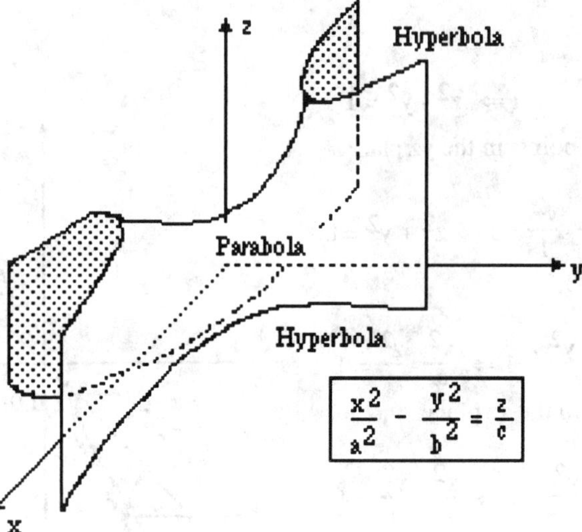

Figure 6

The following summarizes second-degree equations and the corresponding quadric surface.

Summary: Quadric Surfaces

General Equation	Quadric Surface
$x^2 + y^2 - c^2z^2 = 0$	Right circular cone
$(x - h)^2 + (y - k)^2 + (z - l)^2 = a^2$	Sphere
$\frac{x^2}{a^2} + \frac{y^2}{b^2} + \frac{z^2}{c^2} = 1$	Ellipsoid
$\frac{x^2}{a^2} + \frac{y^2}{b^2} - \frac{z^2}{c^2} = 1$	Hyperboloid of one sheet
$\frac{x^2}{a^2} - \frac{y^2}{b^2} - \frac{z^2}{c^2} = 1$	Hyperboloid of two sheets
$\frac{x^2}{a^2} + \frac{y^2}{a^2} = \frac{z}{c}$	Paraboloid
$\frac{x^2}{a^2} + \frac{y^2}{b^2} = \frac{z}{c}$	Elliptic paraboloid
$\frac{x^2}{a^2} - \frac{y^2}{b^2} = \frac{z}{c}$	Hyperbolic paraboloid

---

Example 3 The Bread and Butter Problem

Suppose that you decide to make herb butters and bread for a local gourmet food store. The demand for each is estimated from competitors to be as follows. You can sell 5 kg of your herb butter at \$5 per kilogram and 7 kg when priced at \$3 per kilogram. You sell 6 dozen loaves when priced at \$9 per dozen and 4 dozen at \$10 per dozen. Your costs are estimated to be $C(x, y) = \frac{1}{2}y^2 + 4x + 4y + 10$. Find the amount to produce daily to maximize your profit.

Thinking: Creating a Plan

The prices are established by the demand equations as functions of x or y. For butter we write the demand function (price as a function of quantity) using the ordered pairs (5, 5) and (7, 3). For bread we use the ordered pairs (6, 9) and (4, 10). We will write the revenue function where revenue is the product of price and quantity. We will write profit as a function of x and y, by finding the difference of the revenue and cost functions. We can estimate the maximum profit from the graph of the function.

Communicating: Writing the Solution

Butter: The linear demand function expressing price as a function of amount of butter is $p(x) = 10 - x$.
Bread: The linear demand function expressing price as a function of amount of bread is $q(y) = 12 - \frac{1}{2}y$.

Therefore, the revenue is

$$R(x, y) = [p(x)]x + [q(y)]y = (10 - x)x + (12 - \frac{1}{2}y)y$$

and cost is given by

$$C(x, y) = \frac{1}{2}y^2 + 4x + 4y + 10$$

Since profit = revenue - cost, we have

$$\begin{aligned} P(x, y) &= R(x, y) - C(x, y) \\ &= (10 - x)x + (12 - \frac{1}{2}y)y - (\frac{1}{2}y^2 + 4x + 4y + 10) \\ &= -x^2 + 6x - y^2 + 8y - 10 \\ &= 15 - (x - 3)^2 - (y - 4)^2 \end{aligned}$$

The graph in Figure 7 shows that the profit function is a concave-down paraboloid translated so that its vertex is (3, 4, 15). This point represents maximum profit achieved by selling 3 kg of herb butter and 4 loaves of bread each day.

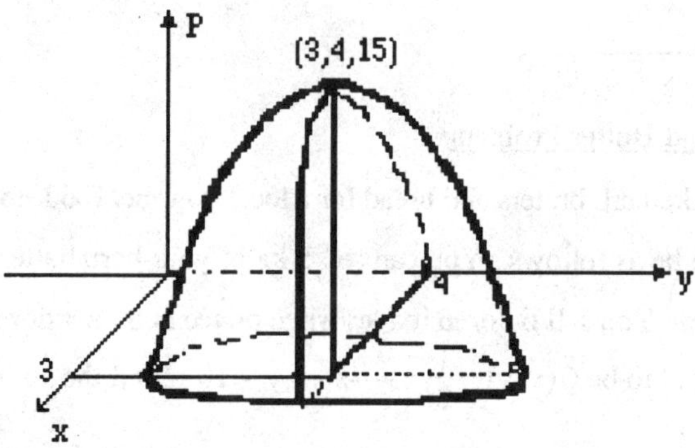

Figure 7

Communicating: Making Connections

Completing the square on this second-degree equation makes it easy to recognize the graph. This is a common approach to translated quadric surfaces. The method applied here to functions of two variables is readily extended to more products. A demand equation for cost is found from market surveys. The total cost function is provided by the company accountants. The optimum is generally found using techniques from calculus.

Graphing Using Contour Maps

A graphing utility can be used to graph a two-dimensional representation of a three-dimensional surface called a *contour map.* A contour map can represent hills (Figure 8a) or valleys (Figure 8b) as indicated by the labels on the level curves. A **level curve** consists of points (x, y) that have the same value of z. For contour maps locations on the same level curve have the same elevation.

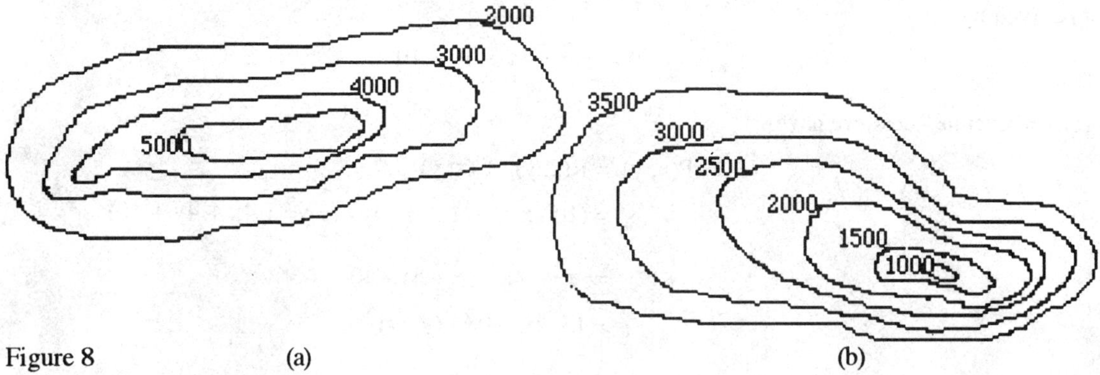

Figure 8 (a) (b)

When adjacent level curves are close together, the slope of the hill or valley is steep. When adjacent level curves are far apart, the slope is less steep. At any point on a level curve the direction to head to cause the greatest change in elevation is perpendicular to the level curve. Skiers call this direction "down the fall line." There will be no change in elevation along a level curve.

In Example 4 we had profit as a function of x and y such that

$$P = 15 - (x - 3)^2 - (y - 4)^2$$

Level curves are created from such functions by selecting values of one variable to create an equation in two variables whose graph is in a plane. For example,

$P = 0 \Rightarrow$

$$15 - (x - 3)^2 - (y - 4)^2 = 0$$
$$(x - 3)^2 + (y - 4)^2 = 15$$

$P = 5 \Rightarrow$

$$15 - (x - 3)^2 - (y - 4)^2 = 5$$
$$(x - 3)^2 + (y - 4)^2 = 10$$

$P = 10 \Rightarrow$

$$15 - (x - 3)^2 - (y - 4)^2 = 10$$
$$(x - 3)^2 + (y - 4)^2 = 5$$

These three equations represent circles centered at (3,4) with radii $r = \sqrt{15}$, $r = \sqrt{10}$, and $r = \sqrt{5}$, respectively.

$P = 15 \Rightarrow$

$$15 - (x - 3)^2 - (y - 4)^2 = 15$$
$$(x - 3)^2 + (y - 4)^2 = 0$$

This is the equation of a point circle located at (3, 4).

The resulting contour map clearly shows the maximum profit of $P = 15$ at $x = 3$, $y = 4$ (Figure 9).

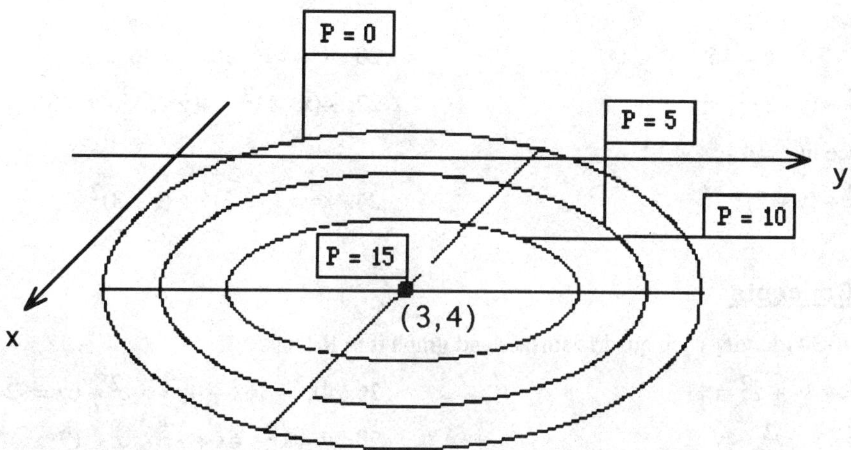

Figure 9

Any graphing utility can be used to draw such two-dimensional level curves. Each equation would be solved for y as a function of x. If the equations were implicit functions, more than one function would be entered. For example, for $P = 0$ above, we enter

$$y = 4 - \sqrt{15 - (x - 3)^2} \quad \text{and} \quad y = 4 + \sqrt{15 - (x - 3)^2}$$

on a graphing utility.

Exercises 8.6

Develop the Concepts

For Exercises 1 to 12 identify the quadric surface by examining the equations of the traces. Draw the xy, xz, and yz traces of each equation on one coordinate system in R^3.

1. $x^2 + y^2 + z^2 = 4$
2. $x^2 + y^2 + z^2 = 169$
3. $4x^2 + y^2 + z^2 = 16$
4. $x^2 + 4y^2 + 16z^2 = 16$
5. $x^2 + y^2 - z^2 = 1$
6. $x^2 - y^2 + z^2 = 4$

*7. $x^2 - 4y^2 - 4z^2 = 16$

8. $-x^2 - y^2 + z^2 = 4$
9. $y = x^2 + z^2$
10. $z = 4x^2 + 4y^2$
11. $z^2 = 4x^2 + 4y^2$
12. $y^2 = x^2 + z^2$

For Exercises 13 to 24 identify each quadric surface and sketch its graph in R^3.

13. $(x - 1)^2 + y^2 + z^2 = 9$
14. $(x - 1)^2 + (y + 2)^2 + (z - 1)^2 = 9$
15. $9(x - 1)^2 + y^2 + 4z^2 = 36$
16. $4(x - 1)^2 + 9(y - 2)^2 + (z + 1)^2 = 36$
17. $4x^2 - y + 4z^2 = 16$
18. $z = (x - 1)^2 + (y - 3)^2$ [Hint: Try the trace in $z = 4$.]
19. $(x - 1)^2 + (y - 3)^2 - z = 16$
20. $x + 4y^2 + 9z^2 = 36$
21. $z = 4(x - 1)^2 + (y - 3)^2$
22. $4(x - 1)^2 + 4(y - 5)^2 - z = 36$

[Hint: Try the trace in $z = 4$.]

23. $z^2 = (x - 1)^2 + (y + 2)^2$
24. $x^2 = (y - 2)^2 + (z + 3)^2$

Extend the Concepts

For Exercises 25 to 34 identify each quadric surface and graph it in R^3.

25. $x^2 - 2x + y^2 + 4y + z^2 = 11$
26. $4x^2 - 16x + 4y^2 + z^2 + 6z = -24$
27. $z = 11 - x^2 + 2x - y^2 - 4y$
28. $y = x^2 - 6x + z^2 - 8z + 17$
29. $x = 8 - 4y^2 + 8y - z^2 + 4z$
30. $4x^2 + 9z^2 - y^2 = 144$
31. $9x^2 - y^2 - 16z^2 = 144$
32. $y^2 - 2y - x^2 + z^2 = 0$
33. $x^2 - y^2 + z = 1$
34. $x + 4y^2 - z^2 = 4$

A Guided Review of Chapter 8

8.1 The Distance Formula and the Circle

1. Use the distance formula to determine whether or not the points A(-1, -8), B(1, - 2), C(4, 7) are collinear because they lie on the same straight line. [Hint: If d(A,B) + d(B,C) = d(A,C),points A, B, and C are collinear.]

2. Use the distance formula and the Pythagorean theorem to decide whether the points A(-10, 10), B(40, 130), C(-10, -110) form a right triangle.

For Exercises 3 and 4 find the equation of each circle having the given center and radius.

3. C(0, 0), r = 4

4. C(-1, 2), r = 4

For Exercises 5 to 8 state the center and radius.

5. $x^2 + y^2 = 144$

6. $x^2 + y^2 = 0$

7. $x^2 + (y - 12)^2 = 144$

8. $(x + 1)^2 + (y - 1)^2 = 1$

For Exercises 9 to 12 express each equation in the standard form for a circle and state the center and radius.

9. $x^2 + y^2 + 2x - 4y = 20$

10. $x^2 + y^2 - 2x + 4y = 20$

11. $x^2 - 12x + y^2 = 0$

12. $x^2 + y^2 = 12y$

For Exercises 13 and 14 graph each semicircle.

13. $y = -\sqrt{4 - x^2}$

14. $y = 12 + \sqrt{9 - x^2}$

15. Find the equation of the circle in standard form that passes through the points (0, 15), (3, 12), and (-3, 12).

16. The Curved Route Problem Engineers are laying out a circular curve on Route 66 using a Cartesian coordinate system. The beginning of the curve has coordinates (100, 100), the end has coordinates (400, 100), and the radius of the curvature is 100 ft. Find the center of the circle that includes this curve. [Hint: Use $(x - h)^2 + (y - k)^2 = 100^2$ to find two equations with two variables, h and k.]

8.2 The Ellipse and the Hyperbola

For Exercises 17 and 18 express each equation in standard form for an ellipse and sketch its graph. State the center, the length of the major and minor axes, and the foci.

17. $4(x - 12)^2 + (y - 12)^2 = 36$

18. $(x - 12)^2 + 4(y + 12)^2 = 36$

19. Find the equation of the ellipse whose foci are F(-2, 4) and F'(2, 4), and the endpoints of the major axis are (-4, 4) and (4, 4).

For Exercises 20 and 21 express each equation in standard form for a hyperbola and sketch its graph. State the center, vertices, and foci.

20. $16x^2 - y^2 = 16$

21. $16(y - 2)^2 - (x + 2)^2 = 16$

For Exercise 22 to 23 find the equation for each hyperbola.

22. Center at (1, 1), a focus at (0, 1), and an asymptote defined as $y = x$

23. Opens vertically with asymptotes $y = \pm x$

For Exercises 24 and 25 write each equation in standard form, sketch the graph, and label the important points.

24. $x^2 + 12y^2 + 24y = 13$

25. $x^2 - y^2 - 12x = 0$

26. The Garden Problem How long a string should be used to lay out an elliptical garden that is to be 64 ft long and 36 ft wide? Where should the ends of the string be staked?

27. The Earth's Orbit Problem Earth travels around the sun in an elliptical orbit; the minimum distance from the sun (one of the foci) is 147.5×10^6 km and the maximum 152.5×10^6 km. (These distances are called the perigee and the apogee, respectively, of Earth's orbit.) Find the lengths of the major and minor axes of the elliptical orbit.

8.3 The Parabola and a Review of Conic Sections

For Exercises 28 and 29 express each equation in the standard form for a parabola. State the vertex, focus, and directrix.

28. $y = x^2$

29. $x^2 = 16 - 16y$

For Exercises 30 and 31 find an equation in standard form for each parabola.

30. vertex is (0, 0), focus (4, 0)

31. focus is (1, 1), directrix $y = 3$

For Exercises 32 and 33 find an equation in standard form for a parabola.

32. $4x^2 + 12y = 12$

33. $12x + 12y + y^2 = 0$

For Exercises 34 to 41 identify each conic section, write it in standard form, and sketch the graph.

34. $x^2 - y^2 = 144$

35. $x^2 + y^2 = 144$

36. $4x^2 - y = 144$

37. $4x + 9y^2 = 144$

38. $64x^2 - 36y^2 = 396$

39. $y^2 - 16x^2 = 0$

40. $12x^2 - 144x - 144y^2 + 288y = 432$

41. $16x^2 + 64x + 4y^2 + 24y = 540$

42. The Flashlight Problem Ideally, the reflector of a flashlight is parabolic. The light bulb must be located at the focus so that the light will travel parallel to the axis of symmetry. Locate the bulb for a flashlight with an opening 2.5 in. wide and a depth of 4 in.

43. The Big Rectangle Problem Find the dimensions of the largest rectangle that can be inscribed in the ellipse $4x^2 + y^2 = 4$.

8.4 Rotations and Conic Sections

For Exercises 44 and 45 use the theorem for rotated conic sections to determine which conic section is probably given by each of the following equations.

44. $x^2 + 2xy + y^2 + 12 = 0$

45. $4x^2 - 12xy - y^2 + 4y = 0$

For Exercises 46 and 47 determine the angle through which each conic section has been rotated.

46. $x^2 + 4xy + y^2 - 4x + 6y - 3 = 0$

47. $3x^2 + \sqrt{3}xy + 2y^2 - 42 = 0$

For Exercises 48 and 49 graph each conic section in the uv coordinate system.

48. $x^2 + 4xy + y^2 - 3 = 0$

49. $3x^2 + \sqrt{3}xy + 2y^2 - 42 = 0$

8.5 Graphs in a Three-Dimensional Coordinate System

For Exercises 50 and 51 draw a sketch in R^3 of each plane.

50. $y = -12$

51. $x = 12$

Find the x, y, and z intercepts of each plane. Use them to draw the triangular portion that falls in octant I of a three-dimensional coordinate system.

52. $x + 6y - 4z = 12$

53. $12x - 3y + 4z = 36$

For Exercises 54 and 55 graph the cylinders in R^3.

54.(a) $x + y = 4$

(b) $x^2 + z^2 = 4$

(c) $y^2 + z = 1$

55.(a) $x^2 - y^2 = 4$

(b) $y^2 - z^2 = 4$

(c) $x^2 - z^2 = 25$

For Exercises 56 and 57 sketch the three-dimensional curve of intersection of the systems of cylinders in octant I.

56. $x^2 + y^2 = 4$
$x^2 + z^2 = 4$

57. $4x^2 + z^2 = 16$
$y^2 + z^2 = 9$

8.6 Quadric Surfaces

For Exercises 58 to 63 identify each quadric surface by examining the equations of the traces. Draw the xy, xz, and yz traces of each equation on one coordinate system in R^3.

58. $x^2 + y^2 + z^2 = 25$

59. $x^2 + 4y^2 + 4z^2 = 16$

60. $x^2 - y^2 - z^2 = 1$

61. $x^2 + y^2 - z^2 = 4$

62. $z^2 = x^2 - 4y^2$

63. $x = y^2 - z^2$

For Exercises 64 and 65 identify each quadric surface and sketch its graph in R^3.

64. $(x - 12)^2 - (y + 12)^2 - (z - 12)^2 = 144$

65. $9(x - 1)^2 + y + 4z^2 = 36$

Chapter 8 Test

1. Express each equation in the standard form for a circle and state the center and radius.

(a) $x^2 + y^2 - 6x = 16$ (b) $x^2 + y^2 = 2y$

2. Graph each semicircle.

(a) $y = -\sqrt{100 - x^2}$ (b) $y = 10 + \sqrt{100 - x^2}$

3. Express each equation in standard form for an ellipse and sketch its graph. State the center, the length of the major and minor axes, and the foci.

(a) $x^2 + 9y^2 = 81$ (b) $9(x - 2)^2 + y^2 = 81$

4. Find the equation of each ellipse in standard form if the foci are F(0,-4) and F'(0,4), and the length of the minor axis is 12.

5. Express each equation in standard form for a hyperbola and sketch its graph. State the center, vertices, and foci.

(a) $4x^2 - y^2 = 16$ (b) $36(y + 2)^2 - x^2 = 36$

6. Find an equation in standard form for the parabola with its vertex is (1, 0) and its focus at (5, 0).

7. Write each equation in standard form for a parabola and sketch each graph. Label the important parts.

(a) $x^2 + 12y = 24$ (b) $x + 2y + y^2 = 0$

8. Identify each conic section, and write its equation in standard form.

(a) $x^2 = y^2 + 64$ (b) $x^2 = 36 - y^2$

(c) $4x^2 = 144 - 144y$ (d) $6x^2 + 24x + y^2 = 24$

9. Determine the angle through which $2\sqrt{3}x^2 + xy + \sqrt{3}y^2 - x + y = 1$ has been rotated.

10. Find the equation of $x^2 + xy + y^2 = 0$ in the rotated uv coordinate system.

11. Draw a sketch in R^3 of each plane.

(a) $y = -12$ (b) $2x + 3y + 6z = 12$

12. Graph the following cylinders in R^3.

(a) $x + y = 10$

(b) $x^2 + y^2 = 100$

13. Identify each quadric surface by examining the equations of the traces. Draw the xy, xz, and yz traces of each of the following on one coordinate system in R^3.

(a) $x^2 + 4y^2 + z^2 = 16$

(b) $y = 16 - x^2 - 4z^2$

14. The Big Rectangle Problem Find the dimensions of the largest rectangle that can be inscribed in the ellipse $x^2 + 4y^2 = 4$.

15. The Garden Problem How long a string should be used to lay out an elliptical garden that is to be 124 ft long and 26 ft wide? Where should the ends of the string be staked?

Chapter 9

Matrices, Determinants, and Vectors

This chapter is a brief introduction to a vast area of study and applications called *linear algebra*. Applying matrices to solving systems of linear equations and inequalities is important enough alone to warrant their study; but matrices also describe functions that transform one coordinate system into another, such as the rotation of axes, making their study fundamental to both algebra and calculus. Matrices are also fundamental to discrete mathematics, computer science, and probability, making their application increasingly more common. In Section 9.4 we apply vectors, which are column matrices, to finding equations of lines and planes in three dimensions. Graphing utilities can, of course, perform the matrix operations and evaluate determinants.

A Historical Note on William Hamilton

William Hamilton, through the intuitively based representation of complex numbers as directed line segments, led the search for three-dimensional complex numbers, later called vectors. Representing three-dimensional vectors as ordered triples (x, y, z) was naturally considered, but the algebra of complex numbers that went with two-dimensional vectors could not be extended to three dimensions.

Hamilton spent years trying to define a three-dimensional analog of complex numbers because the algebra (vector algebra or linear algebra) was not to have the commutative law under multiplication. This was totally unprecedented. Finally, Hamilton defined **quaternions** as four-dimensional numbers of the form

$$3 + 4\mathbf{i} + 6\mathbf{j} + 2\mathbf{k}$$

where $\mathbf{i} = \begin{bmatrix}1 & 0 & 0\end{bmatrix}$, $\mathbf{j} = \begin{bmatrix}0 & 1 & 0\end{bmatrix}$, and $\mathbf{k} = \begin{bmatrix}0 & 0 & 1\end{bmatrix}$. This resulting noncommutative algebra led to the algebra of vectors.

One evening as Hamilton was walking across the Brougham Bridge in Dublin, the idea of this new algebra came to him in a flash. Having no pencil or paper, he took out his pocketknife and scratched his idea on a stone on the bridge. Today a plaque on the bridge commemorates his discovery. It was a revolutionary point in the history of mathematics. His invention of a noncommutative algebra set a precedent for creativity in mathematics.

9.1 Introduction to Matrices

Defining a Matrix
Performing Operations with Matrices
Transforming and Rotating Matrices
Applying Matrices

In this section we learn the basic definitions, notation, and operations of matrices. Matrices offer a convenient way to organize data and perform operations on the data such as adding and multiplying. In Example 6 we will see how matrices are used to inventory items in a store.

Defining a Matrix

A **matrix** is a rectangular array of entries, called *elements*, arranged in horizontal rows and vertical columns. The **order** of a matrix gives the number of rows and the number of columns in the matrix. For example,

$$\begin{bmatrix} 1 & 2 & 3 \end{bmatrix}$$

is a matrix of order 1 x 3 (read "1 by 3"), since it has one row and three columns. We call it a **row matrix** because it has only one row. The matrix

$$\begin{bmatrix} 1 \\ -1 \\ 4 \end{bmatrix}$$

is an example of a 3 x 1 matrix since it has three rows and one column. We call it a **column matrix** (or vector).

What are the orders of the two matrices M and N?

$$M = \begin{bmatrix} 4 & \pi \\ 0 & 2 \\ 1 & -3 \end{bmatrix} \quad \text{and} \quad N = \begin{bmatrix} 4 & \frac{-1}{2} & 2 \\ 7 & 11 & 0 \end{bmatrix}$$

M is a 3 x 2 matrix and N is a 2 x 3 matrix. In matrix M, the entry π is in the first row and second column. We write $a_{12} = \pi$. In general, a_{ij} is the entry in the ith row and jth column. A general 3 x 2 matrix looks like

$$A = \begin{bmatrix} a_{11} & a_{12} \\ a_{21} & a_{22} \\ a_{13} & a_{23} \end{bmatrix}$$

Definition of a Matrix

An $m \times n$ matrix is a rectangular array of elements having m horizontal rows and n vertical columns.

$$A = \{a_{ij}\}, \quad 1 \le i \le m \ \text{ and } \ 1 \le j \le n$$

NOTE: Most graphing utilities can handle large matrices. The elements are entered one at a time after the order of the matrix is established.

Two matrices are **equal** if and only if they have the same order and their corresponding elements are equal. The following example illustrates matrix equality.

Example 1 *Setting matrices equal*

Find the value of x and y that will make the matrices equal.

$$\begin{bmatrix} x \\ 2y \end{bmatrix} = \begin{bmatrix} 4 \\ 6 \end{bmatrix}$$

Solution

$$\begin{bmatrix} x \\ 2y \end{bmatrix} = \begin{bmatrix} 4 \\ 6 \end{bmatrix} \Rightarrow x = 4 \ \text{ and } \ y = 3$$

Performing Operations with Matrices

Addition of two matrices is quite natural. Matrix addition can be performed only on matrices of the same order. To add the matrices, we add the corresponding elements in the two matrices as shown in Example 2.

NOTE: Enter the matrices in the following examples using your graphing utility for practice. It is probably as important to automate matrix operations as it is to learn them since, in practice, matrices are too large to manipulate by pencil and paper.

Example 2 *Adding matrices*

Add the following matrices if possible.

(a) $\begin{bmatrix} 2 & 3 \\ 1 & 4 \end{bmatrix} + \begin{bmatrix} -1 & 0 \\ 3 & 4 \end{bmatrix}$ (b) $\begin{bmatrix} 1 & 2 & -1 \\ 3 & 4 & 0 \end{bmatrix} + \begin{bmatrix} 2 & -1 & 0 \\ 4 & -3 & 1 \end{bmatrix}$ (c) $\begin{bmatrix} 1 & 2 \\ 3 & 4 \\ 4 & 3 \end{bmatrix} + \begin{bmatrix} 2 & -1 & 3 \\ 4 & 0 & 1 \end{bmatrix}$

Solution

(a) $\begin{bmatrix} 2 & 3 \\ 1 & 4 \end{bmatrix} + \begin{bmatrix} -1 & 0 \\ 3 & 4 \end{bmatrix} = \begin{bmatrix} 2+(-1) & 3+0 \\ 1+3 & 4+4 \end{bmatrix} = \begin{bmatrix} 1 & 3 \\ 4 & 8 \end{bmatrix}$

(b) $\begin{bmatrix} 1 & 2 & -1 \\ 3 & 4 & 0 \end{bmatrix} + \begin{bmatrix} 2 & -1 & 0 \\ 4 & -3 & 1 \end{bmatrix} = \begin{bmatrix} 3 & 1 & -1 \\ 7 & 1 & 1 \end{bmatrix}$

(c) These matrices cannot be added since they do not have the same order. The first is a 3 x 2 matrix and the second a 2 x 3 matrix.

There are two kinds of multiplication to consider for matrices. The first is called *scalar multiplication,* shown in Example 3. Here we define a **scalar** to be any real number. Each entry of the matrix is multiplied by the given real number.

Example 3 *Multiplying by a scalar*

Find the scalar products.

(a) $4\begin{bmatrix} 2 & 3 \\ -1 & 2 \end{bmatrix}$

(b) $\frac{1}{3}\begin{bmatrix} 3 & 9 \\ 7 & 12 \end{bmatrix}$

Solution

(a) $4\begin{bmatrix} 2 & 3 \\ -1 & 2 \end{bmatrix} = \begin{bmatrix} 4\cdot 2 & 4\cdot 3 \\ 4\cdot(-1) & 4\cdot 2 \end{bmatrix} = \begin{bmatrix} 8 & 12 \\ -4 & 8 \end{bmatrix}$

(b) $\frac{1}{3}\begin{bmatrix} 3 & 9 \\ 7 & 12 \end{bmatrix} = \begin{bmatrix} 1 & 3 \\ \frac{7}{3} & 4 \end{bmatrix}$

The second type of matrix multiplication involves two matrices. Multiplication of matrices is defined for matrices if the number of columns in the first matrix is the same as the number of rows in the second. For example, we can multiply a 2 x 3 matrix by a 3 x 4 matrix since the first matrix has three columns and the second has three rows. The product is a 2 x 4 matrix (Figure 1).

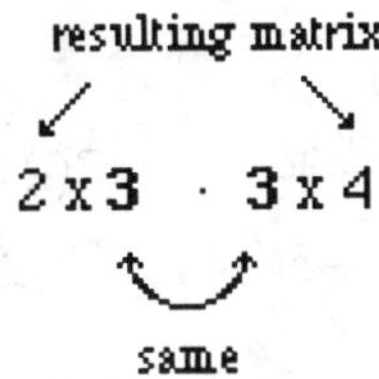

Figure 1

For example, to multiply the matrices

$$\begin{bmatrix} 1 & 3 & 4 \\ 2 & 1 & 3 \end{bmatrix} \text{ and } \begin{bmatrix} 2 & 1 \\ -3 & 4 \\ 0 & 5 \end{bmatrix}$$

we find the entries of the product matrix as follows:

$$\begin{bmatrix} \mathbf{1} & \mathbf{3} & \mathbf{4} \\ 2 & 1 & 3 \end{bmatrix} \begin{bmatrix} \mathbf{2} & 1 \\ \mathbf{-3} & 4 \\ \mathbf{0} & 5 \end{bmatrix} = \begin{bmatrix} \mathbf{1}(\mathbf{2})+\mathbf{3}(\mathbf{-3})+\mathbf{4}(\mathbf{0}) & 1(1)+3(4)+4(5)) \\ 2(2)+1(-3)+3(0) & 2(1)+1(4)+3(5) \end{bmatrix} = \begin{bmatrix} -7 & 33 \\ 1 & 21 \end{bmatrix}$$

To find the row 1 column 1 entry of the product matrix, the entries in the first row of the 2 x 3 matrix are paired with those in the first column of the 3 x 2 matrix. The three products are added to get the row 1, column 1 entry, -7. The position of the product matrix is determined by the row selected in the first matrix and the column selected in the second. For example, for the product of two 3 x 3 matrices

$$c_{11} = a_{11}b_{11} + a_{12}b_{21} + a_{13}b_{31} = \Sigma\, a_{ik}b_{kj}$$

Matrices are said to be not **conformable** to matrix multiplication if the number of columns of the first does not equal the number of rows of the second.

Matrix Multiplication

In matrix multiplication the matrices must have orders

m x **p** and **p** x n

to be conformable. Their product will then have order m x n.

Example 4 *Multiplying matrices*

Multiply the following matrices.

(a) $\begin{bmatrix} 2 & 4 & 3 \\ -2 & 1 & 5 \end{bmatrix} \begin{bmatrix} 1 & 2 \\ 3 & 0 \\ -4 & 1 \end{bmatrix}$ (b) $\begin{bmatrix} 1 & 2 \\ 3 & 0 \\ -4 & 1 \end{bmatrix} \begin{bmatrix} 2 & 4 & 3 \\ -2 & 1 & 5 \end{bmatrix}$ (c) $\begin{bmatrix} 2 & 3 & 1 \\ 1 & -3 & 2 \end{bmatrix} \begin{bmatrix} 4 & 5 \\ 8 & 7 \end{bmatrix}$

Solution

(a) $$\begin{bmatrix} 2 & 4 & 3 \\ -2 & 1 & 5 \end{bmatrix}\begin{bmatrix} 1 & 2 \\ 3 & 0 \\ -4 & 1 \end{bmatrix} = \begin{bmatrix} 2(1)+4(3)+3(-4) & 2(2)+4(0)+3(1) \\ -2(1)+1(3)+5(-4) & -2(2)+1(0)+5(1) \end{bmatrix} = \begin{bmatrix} 2 & 7 \\ -19 & 1 \end{bmatrix}$$

(b) $$\begin{bmatrix} 1 & 2 \\ 3 & 0 \\ -4 & 1 \end{bmatrix}\begin{bmatrix} 2 & 4 & 3 \\ -2 & 1 & 5 \end{bmatrix} = \begin{bmatrix} 1(2)+2(-2) & 1(4)+2(1) & 1(3)+2(5) \\ 3(2)+0(-2) & 3(4)+0(1) & 3(3)+0(5) \\ -4(2)+1(-2) & -4(4)+1(1) & -4(3)+1(5) \end{bmatrix} = \begin{bmatrix} -2 & 6 & 13 \\ 6 & 12 & 9 \\ -10 & -15 & -7 \end{bmatrix}$$

(c) A 2 x 3 matrix and a 2 x 2 matrix are not conformable.

Example 5 *Testing the commutative property for matrices*

Find the products, AB and BA, to see if the commutative property for matrices holds.

$$A = \begin{bmatrix} 2 & 1 \\ 3 & -1 \end{bmatrix} \text{ and } B = \begin{bmatrix} 1 & 2 \\ 0 & -3 \end{bmatrix}$$

Solution

$$AB = \begin{bmatrix} 2 & 1 \\ 3 & -1 \end{bmatrix}\begin{bmatrix} 1 & 2 \\ 0 & -3 \end{bmatrix} = \begin{bmatrix} 2 & 1 \\ 3 & 9 \end{bmatrix} \qquad BA = \begin{bmatrix} 1 & 2 \\ 0 & -3 \end{bmatrix}\begin{bmatrix} 2 & 1 \\ 3 & -1 \end{bmatrix} = \begin{bmatrix} 8 & -1 \\ 3 & 9 \end{bmatrix}$$

The fact that these two products are not equal proves that matrix multiplication is not commutative.

The results of Example 5 do not imply that the matrix product AB never equals BA. As we will see, some special pairs of matrices do commute.

Summary: Matrix Operations

For matrices A and B, if

$A = \{a_{ij}\}$ and $B = \{b_{ij}\}$, then $A + B = \{c_{ij}\}$ where $c_{ij} = a_{ij} + b_{ij}$

$A = \{a_{ij}\}$, then $\alpha A = \{c_{ij}\}$ where $c_{ij} = \alpha a_{ij}$

$A = \{a_{ik}\}$ and $B = \{b_{kj}\}$, then $AB = \{c_{ij}\}$ where $c_{ij} = \Sigma a_{ik} b_{kj}$

Transforming and Rotating Matrices

We have already seen several transformations of sets of ordered pairs, where new functions were created from given ones by reflecting them about either axis or the origin. Matrices can be used to change one set of ordered pairs into another. Such matrices are called **transformations**. Consider the four points (-1, 1), (1, 1), (1, 2), and (-1, 2). We can represent the four points in a matrix with each point a column.

$$\begin{bmatrix} -1 & 1 & 1 & -1 \\ 1 & 1 & 2 & 2 \end{bmatrix}$$

Multiplying by the matrix

$$\begin{bmatrix} 1 & 0 \\ 0 & -1 \end{bmatrix}$$

changes the sign of each y coordinate.

$$\begin{bmatrix} 1 & 0 \\ 0 & -1 \end{bmatrix}\begin{bmatrix} -1 & 1 & 1 & -1 \\ 1 & 1 & 2 & 2 \end{bmatrix} = \begin{bmatrix} -1 & 1 & 1 & -1 \\ -1 & -1 & -2 & -2 \end{bmatrix}$$

The geometric interpretation is a reflection of the four points across the x axis (Figure 2). An entire finite set of points can be reflected in one multiplication.

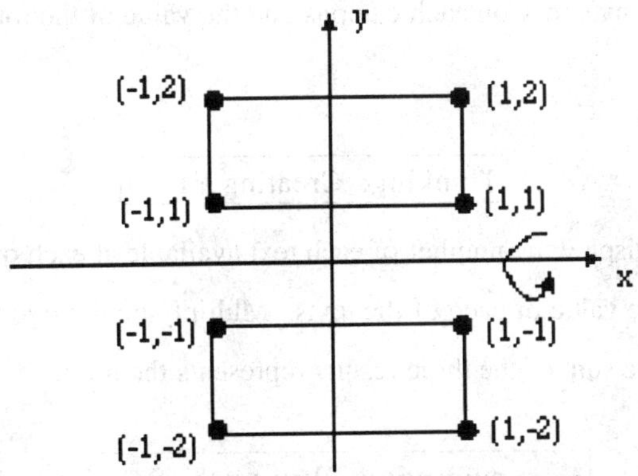

Figure 2

The other reflections and changes of scales can be performed similarly with matrix multiplication, however translations cannot.

We saw in Section 8.4 that rotating the xy coordinate system through an angle θ created a uv coordinate system, where

$$u = x \cos \theta + y \sin \theta$$
$$v = -x \sin \theta + y \cos \theta$$

Rotation of axes is naturally performed by multiplying by a transformation matrix. This is equivalent to the matrix multiplication

$$\begin{bmatrix} u \\ v \end{bmatrix} = \begin{bmatrix} \cos\theta & \sin\theta \\ -\sin\theta & \cos\theta \end{bmatrix}\begin{bmatrix} x \\ y \end{bmatrix}$$

For a given value of θ, the 2 x 2 matrix is the transformation that rotates the coordinate system. For input values of x and y, multiplying by this matrix yields the corresponding coordinates in the rotated uv system.

The inverse of this matrix switches form the uv system back to the xy system. This is how computer graphics programs rotate an image on the screen.

Applying Matrices

The following example is a typical matrix application in a situation where a table of data is involved. Matrices become even more important when hundreds of entries must be handled by a computer.

Example 6 The Inventory Problem

A textbook business sells three different textbooks on three different campuses. North Campus has 14 copies of book A, 10 of book B, and 12 of book C. Central Campus has none of book A, 7 of book B, and 5 of book C. South Campus has 15 of book A, 4 of book B, and 10 of book C. Book A sells for $10, book B for $12, and book C for $18. Find the inventory on each campus and the value of the total inventory by setting up the data in matrices.

Thinking: Creating a Plan

We will use a 3 x 3 matrix to display the number of each text available at each of the three branches of the store and a 3 x 1 matrix to assign the value of each of the texts. Multiplying the two matrices yields the value of the inventory on each campus. The sum of the three results represents the total.

Communicating: Writing the Solution

$$\begin{array}{l} \\ \textit{North} \\ \textit{Central} \\ \textit{South} \end{array} \begin{array}{c} \textit{Book} \\ \begin{array}{ccc} A & B & C \end{array} \\ \begin{bmatrix} 14 & 10 & 12 \\ 0 & 7 & 5 \\ 15 & 4 & 10 \end{bmatrix} \end{array} \begin{bmatrix} \$10 \\ \$12 \\ \$18 \end{bmatrix} = \begin{bmatrix} \$476 \\ \$174 \\ \$378 \end{bmatrix}$$

The top entry in the column matrix, $476, represents the total inventory of all three books at the North Campus bookstore:

$$(14\ \cancel{\text{books}})\frac{\$10}{\cancel{\text{book}}} + (10\ \cancel{\text{books}})\frac{\$12}{\cancel{\text{book}}} + (12\ \cancel{\text{books}})\frac{\$18}{\cancel{\text{book}}} = \$476$$

The total value of the column matrix, $476 + $174 + $378 = $1028, represents the inventory on all three campuses.

Learning: Making Connections

Arrays like these take advantage of the ability of matrices to organize complex sets of data. An important part of computer science involves storing, retrieving, manipulating, and performing calculations on such arrays.

The next example illustrates how matrices are applied to the study of networks such as a salesperson's possible routes through several cities. Generally, such networks consist of vertices (the cities) connected by edges (the roads between them). When the number of cities is large, finding the shortest route for the salesperson is an unsolved problem in mathematics because all known approaches take too long to compute, even with the fastest computers. The a_{ij} entry in the matrix representation of a network is the number of edges joining vertices i and j.

Example 7 *Creating a matrix representation of a network*

Create a matrix representation of a network with four vertices connected with five edges (Figure 3).

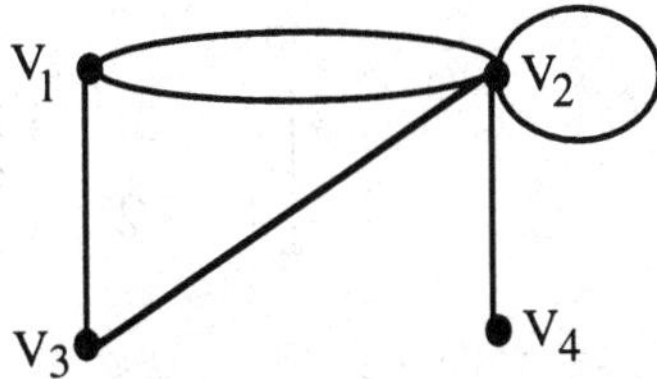

Figure 3

Solution

We will count the number of edges joining the appropriate vertices and enter the count in the appropriate spot in the matrix. For example, the number of edges joining v_1 and v_1 is 0. This entry goes in the a_{11} position. The number of edges joining v_1 and v_2 is 2. This entry goes in the a_{12} position.

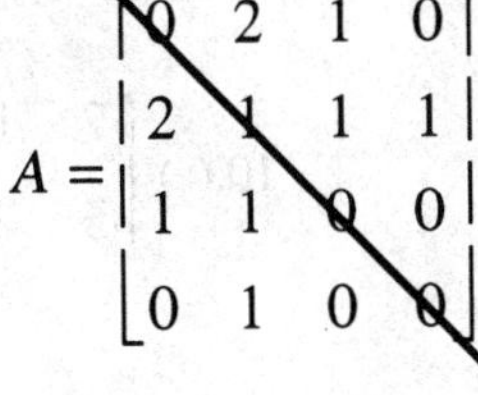

$$A = \begin{bmatrix} 0 & 2 & 1 & 0 \\ 2 & 1 & 1 & 1 \\ 1 & 1 & 0 & 0 \\ 0 & 1 & 0 & 0 \end{bmatrix}$$

This matrix is symmetric about the diagonal because the a_{ij} and a_{ji} entries must be the same (there are the same number of entries joining v_i to v_j as there are joining v_j to v_i). The network can be reconstructed from its matrix. For example, a loop is present whenever there is a 1 on the main diagonal. The 1 is in the second row, second column, a_{22}, so there was a single path called a loop at v_2. A zero in the a_{11} position indicates that there is no loop at v_1.

Exercises 9.1

Develop the Concepts

For Exercises 1 to 18 perform the operations indicated whenever possible. If not possible, write, "not conformable."

1.(a) $\begin{bmatrix} 2 & -1 \\ 3 & 4 \end{bmatrix} + \begin{bmatrix} 1 & -3 \\ 2 & 7 \end{bmatrix}$

(b) $\begin{bmatrix} 1 & -3 \\ 2 & 7 \end{bmatrix} + \begin{bmatrix} 2 & -1 \\ 3 & 4 \end{bmatrix}$

2.(a) $\left\{ \begin{bmatrix} 1 & 2 & 3 \\ -2 & 4 & 7 \end{bmatrix} + \begin{bmatrix} 1 & 3 & 7 \\ 2 & -1 & 4 \end{bmatrix} \right\} + \begin{bmatrix} 2 & 8 & -11 \\ 5 & -7 & 4 \end{bmatrix}$

(b) $\begin{bmatrix} 1 & 2 & 3 \\ -2 & 4 & 7 \end{bmatrix} + \left\{ \begin{bmatrix} 1 & 3 & 7 \\ 2 & -1 & 4 \end{bmatrix} + \begin{bmatrix} 2 & 8 & -11 \\ 5 & -7 & 4 \end{bmatrix} \right\}$

3. $5\begin{bmatrix} 1 \\ 0 \\ 0 \end{bmatrix} + 7\begin{bmatrix} 0 \\ 1 \\ 0 \end{bmatrix} - 10\begin{bmatrix} 0 \\ 0 \\ 1 \end{bmatrix}$

4. $\begin{bmatrix} 2 & 3 & 4 \\ 1 & -2 & 1 \end{bmatrix} + \begin{bmatrix} 2 & -1 \\ 3 & -2 \\ 4 & 1 \end{bmatrix}$

5. $\begin{bmatrix} 4 & 7 & 1 \\ 3 & 2 & -2 \\ 1 & 5 & 7 \end{bmatrix} + \begin{bmatrix} 0 & 0 & 0 \\ 0 & 0 & 0 \\ 0 & 0 & 0 \end{bmatrix}$

6. $\begin{bmatrix} -2 & 1 & 3 \\ 1 & -7 & 4 \\ -2 & 3 & -1 \end{bmatrix} + \begin{bmatrix} 2 & -1 & -3 \\ -1 & 7 & -4 \\ 2 & -3 & 1 \end{bmatrix}$

7. $\frac{1}{12}\begin{bmatrix} 4 & 2 \\ 6 & 6 \end{bmatrix}$

8. $\frac{-1}{2}\begin{bmatrix} 2 & 3 \\ -2 & -4 \end{bmatrix}$

9.(a) $\begin{bmatrix} 2 & 3 \\ 1 & 4 \end{bmatrix}\begin{bmatrix} -2 & 3 \\ -2 & 1 \end{bmatrix}$

(b) $\begin{bmatrix} -2 & 3 \\ -2 & 1 \end{bmatrix}\begin{bmatrix} 2 & 3 \\ 1 & 4 \end{bmatrix}$

10.(a) $\begin{bmatrix} 2 & 1 \\ 3 & 2 \end{bmatrix}\begin{bmatrix} 2 & -1 \\ -3 & 2 \end{bmatrix}$

(b) $\begin{bmatrix} 2 & -1 \\ -3 & 2 \end{bmatrix}\begin{bmatrix} 2 & 1 \\ 3 & 2 \end{bmatrix}$

11.(a) $\begin{bmatrix} 1 & 2 & 3 \end{bmatrix}\begin{bmatrix} 4 \\ 2 \\ -3 \end{bmatrix}$

(b) $\begin{bmatrix} 4 \\ 2 \\ -3 \end{bmatrix}\begin{bmatrix} 1 & 2 & 3 \end{bmatrix}$

12.(a) $2\left\{\begin{bmatrix} 3 & 1 \\ -2 & -1 \end{bmatrix}\begin{bmatrix} 2 & 1 \\ -3 & 1 \end{bmatrix}\right\}$

(b) $\left\{2\begin{bmatrix} 3 & 1 \\ -2 & -1 \end{bmatrix}\right\}\begin{bmatrix} 2 & 1 \\ -3 & 1 \end{bmatrix}$

13.(a) $\begin{bmatrix} 1 & 3 & 4 \\ -1 & 0 & 2 \\ 1 & 4 & -1 \end{bmatrix}\left\{\begin{bmatrix} 0 & 3 & -2 \\ 1 & -1 & -2 \\ 2 & 2 & 1 \end{bmatrix}\begin{bmatrix} 1 & 2 & 0 \\ 1 & 3 & -1 \\ 2 & -2 & 2 \end{bmatrix}\right\}$

(b) $\left\{\begin{bmatrix} 1 & 3 & 4 \\ -1 & 0 & 2 \\ 1 & 4 & -1 \end{bmatrix}\begin{bmatrix} 0 & 3 & -2 \\ 1 & -1 & -2 \\ 2 & 2 & 1 \end{bmatrix}\right\}\begin{bmatrix} 1 & 2 & 0 \\ 1 & 3 & -1 \\ 2 & -2 & 2 \end{bmatrix}$

14.(a) $\left\{4\begin{bmatrix} 1 & 3 & 0 \\ -1 & 2 & 1 \end{bmatrix}\right\}\left\{2\begin{bmatrix} 1 & 2 \\ -1 & -2 \\ 3 & 4 \end{bmatrix}\right\}$

(b) $(2\cdot 4)\left\{\begin{bmatrix} 1 & 3 & 0 \\ -1 & 2 & 1 \end{bmatrix}\begin{bmatrix} 1 & 2 \\ -1 & -2 \\ 3 & 4 \end{bmatrix}\right\}$

15.(a) $\begin{bmatrix} 2 & 3 \\ 3 & 7 \end{bmatrix}\begin{bmatrix} 1 & 0 \\ 0 & 1 \end{bmatrix}$

(b) $\begin{bmatrix} 3 & -1 & 2 \\ 4 & 7 & -11 \\ 10 & 9 & 8 \end{bmatrix}\begin{bmatrix} 1 & 0 & 0 \\ 0 & 1 & 0 \\ 0 & 0 & 1 \end{bmatrix}$

16.(a) $\begin{bmatrix} 1 & 3 \\ 2 & 7 \end{bmatrix}\begin{bmatrix} 7 & -3 \\ -2 & 1 \end{bmatrix}$

(b) $\begin{bmatrix} 1 & 2 & 0 \\ 4 & 4 & 1 \\ 3 & 0 & 1 \end{bmatrix}\begin{bmatrix} 2 & -1 & -1 \\ \frac{-1}{2} & \frac{1}{2} & \frac{-1}{2} \\ -6 & 3 & -2 \end{bmatrix}$

17. $\begin{bmatrix} 3 & -2 & -3 & -1 \\ -2 & 1 & 0 & -2 \\ -4 & 2 & 1 & 0 \end{bmatrix}\begin{bmatrix} 4 & -3 \\ 9 & -6 \\ -2 & 1 \end{bmatrix}$

18. $\begin{bmatrix} 4 & -3 \\ 9 & -6 \\ -2 & 1 \end{bmatrix}\begin{bmatrix} 3 & -2 & -3 & -1 \\ -2 & 1 & 0 & -2 \\ -4 & 2 & 1 & 0 \end{bmatrix}$

In Exercises 19 and 20, compare the two parts.

19.(a) $\{2+3\}\begin{bmatrix} 1 & 2 & 3 \\ 4 & 5 & 6 \end{bmatrix}$

(b) $2\begin{bmatrix} 1 & 2 & 3 \\ 4 & 5 & 6 \end{bmatrix}+3\begin{bmatrix} 1 & 2 & 3 \\ 4 & 5 & 6 \end{bmatrix}$

20.(a) $4\left\{\begin{bmatrix} 1 & -2 \\ 3 & 1 \end{bmatrix}+\begin{bmatrix} 2 & 0 \\ -3 & 1 \end{bmatrix}\right\}$

(b) $4\begin{bmatrix} 1 & -2 \\ 3 & 1 \end{bmatrix}+4\begin{bmatrix} 2 & 0 \\ -3 & 1 \end{bmatrix}$

21. Compare αA and $A\alpha$ for $\alpha = -3$ and $A = \begin{bmatrix} 1 & 2 \\ 3 & 4 \end{bmatrix}$.

22. Compare $(\alpha\beta)A$ and $\alpha(\beta A)$ for $\alpha = 3$, $\beta = 8$, and $A = \begin{bmatrix} 1 & 2 \\ 3 & 4 \end{bmatrix}$.

Apply the Concepts

In Exercises 23 to 28, multiply the matrices A and B where $B = \begin{bmatrix} -1 & 1 & 1 & -1 \\ 1 & 1 & 2 & 2 \end{bmatrix}$ represents the box in Figure 1. Describe the transformation of the box modeled by matrix A.

*23. $A = \begin{bmatrix} -1 & 0 \\ 0 & 1 \end{bmatrix}$

24. $A = \begin{bmatrix} -1 & 0 \\ 0 & -1 \end{bmatrix}$

25. $A = \begin{bmatrix} 0 & 1 \\ 1 & 0 \end{bmatrix}$

26. $A = \frac{1}{\sqrt{2}}\begin{bmatrix} 1 & 1 \\ -1 & 1 \end{bmatrix}$

27. $A = \begin{bmatrix} \frac{1}{2} & \frac{\sqrt{3}}{2} \\ -\frac{\sqrt{3}}{2} & \frac{1}{2} \end{bmatrix}$

28. $A = \begin{bmatrix} \frac{\sqrt{3}}{2} & \frac{1}{2} \\ -\frac{1}{2} & \frac{\sqrt{3}}{2} \end{bmatrix}$

29. **The Rainbow Kite and Balloon Company** You inventory three different kites: the diamond kite, a box kite, and a dragon kite at three different outlets -- the beach store, the uptown store, and the downtown store. The beach store has 120 diamond kites, 100 box kites, and 50 dragon kites. The uptown store has 220 diamond kites, 100 box kites, and 150 dragon kites. The downtown store has 20 diamond kites, 10 box kites, and 5 dragon kites. The diamond kites cost \$27.50, the box kites cost \$69.50, and the dragon kites cost \$12.50. Find the inventory in each store and the value of the total inventory by setting up the data in matrices.

30. **The Rainbow Kite and Balloon Company** Multiply the inventory matrix in Exercise 29 by the following matrix and describe the shift in inventory.

$$A = \begin{bmatrix} 0 & 1 & 0 \\ 1 & 0 & 0 \\ 0 & 0 & 1 \end{bmatrix}$$

31. **Wonderworld Amusement Park** There are three ways to purchase tickets for the three big rides at the park: the Ferris wheel, the roller coaster, and the merry-go-round. Riders can buy tickets one at a time, a book of 10 tickets, or a book of 25 tickets. The Ferris wheel took in \$500 from 5 single tickets, 20 ten-ticket-books, and 5 twenty-five-ticket books. The roller coaster took in \$200 from 10 single tickets, 5 ten-ticket-books, and 0 twenty-five-ticket-books. The merry-go-round took in \$400 from 0 single tickets, 10 ten-ticket-books, and 20 twenty-five-ticket-books. Find the price of each ticket option by setting up a matrix equation.

32. **Wonderworld Amusement Park** Multiply the ticket distribution matrix from Exercise 31 by the following matrix and describe the results.

$$A = \begin{bmatrix} 0.20 & 0 & 0 \\ 0 & 2 & 0 \\ 0 & 0 & 0.10 \end{bmatrix}$$

In Exercises 33 to 36, write the matrix that represents each circuit.

33.

34.

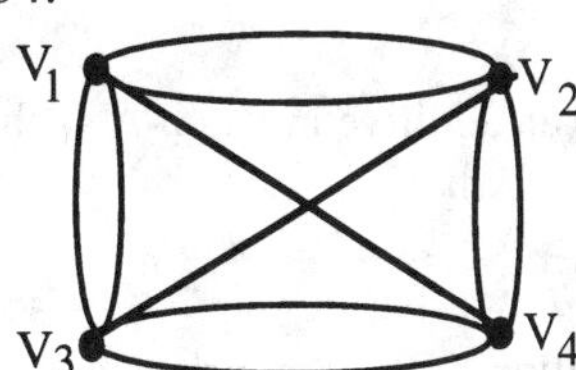

35.

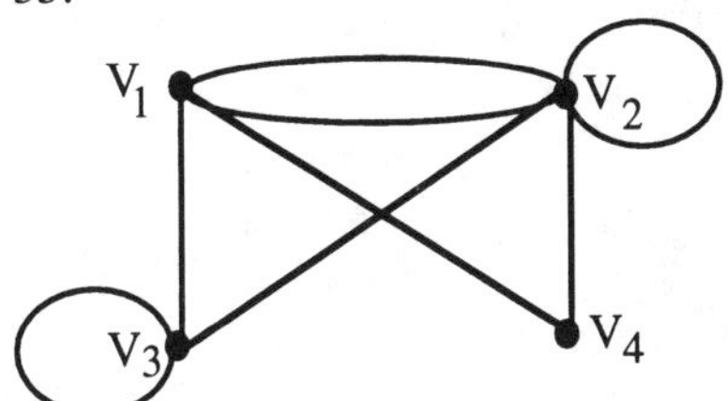

36.

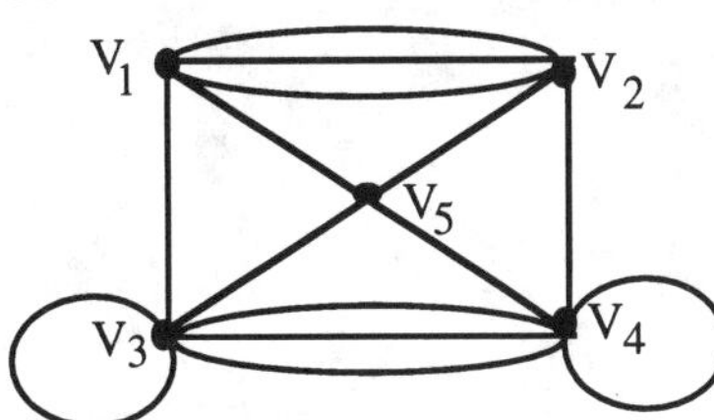

In Exercises 37 to 40, draw the circuit modeled by each matrix.

*37. $\begin{bmatrix} 1 & 2 & 2 & 2 \\ 2 & 1 & 2 & 2 \\ 2 & 2 & 1 & 2 \\ 2 & 2 & 2 & 1 \end{bmatrix}$

38. $\begin{bmatrix} 0 & 1 & 1 & 1 \\ 1 & 0 & 1 & 1 \\ 1 & 1 & 0 & 1 \\ 1 & 1 & 1 & 0 \end{bmatrix}$

39. $\begin{bmatrix} 0 & 2 & 2 & 0 \\ 2 & 0 & 0 & 2 \\ 2 & 0 & 0 & 2 \\ 0 & 2 & 2 & 0 \end{bmatrix}$

40. $\begin{bmatrix} 1 & 1 & 2 & 0 \\ 1 & 1 & 0 & 2 \\ 2 & 0 & 1 & 1 \\ 0 & 2 & 1 & 1 \end{bmatrix}$

Extend the Concepts

The transpose A^T of matrix A is found by making row 1 of A column 1 of A^T, row 2 of A column 2 of A^T, and so on. Compare A^TB^T and B^TA^T with $(AB)^T$, and $(A + B)^T$ with $A^T + B^T$.

41. $A = \begin{bmatrix} 1 & 3 \\ 2 & -4 \end{bmatrix}$ and $B = \begin{bmatrix} 2 & 4 \\ -1 & 3 \end{bmatrix}$

42. $A = \begin{bmatrix} 1 & 2 & -1 \\ 3 & -2 & 0 \\ 1 & 3 & 1 \end{bmatrix}$ and $B = \begin{bmatrix} -1 & 2 & 2 \\ 3 & 4 & 5 \\ 2 & -1 & 0 \end{bmatrix}$

For Exercises 43 to 45 consider the set of all 2 x 2 matrices under normal matrix addition.

43. Is the set closed under matrix addition? In other words, is the sum of two 2 x 2 matrices always a 2 x 2 matrix?

44. Is there an additive identity O in the set such that A + O = O + A = A for all 2 x 2 matrices A? If so, what is it?

45. Does each matrix A have an additive inverse -A such that A + (-A) = (-A) + A = O? Give an example with a 2 x 2 matrix.

46. Illustrate that matrix addition of 2 x 2 matrices is commutative.

47. Illustrate that matrix addition of 2 x 2 matrices is associative.

9.2 Introduction to Determinants

Evaluating a 2 x 2 Determinant
Evaluating a 3 x 3 Determinant
Examining Properties of Determinants

In this section we study determinants and their properties. A **determinant** is a function that assigns a real number to a square matrix. A **square matrix** is a matrix having the same number of rows and columns. The domain of this function is the set of square matrices, and the range is the set of real numbers. The application and theory of determinants are found in studies of systems of linear equations and of matrices that represent transformations.

Evaluating a 2 x 2 Determinant

If A is the square 2 x 2 matrix

$$A = \begin{bmatrix} a_{11} & a_{12} \\ a_{21} & a_{22} \end{bmatrix}$$

then the value of the determinant of A is written

$$\det A = \begin{vmatrix} a_{11} & a_{12} \\ a_{21} & a_{22} \end{vmatrix}$$

The two vertical lines indicate that the rectangular array is a determinant, in this case a second-order determinant. To find the value of a 2 x 2 determinant we find the difference of two products as shown in Figure 1.

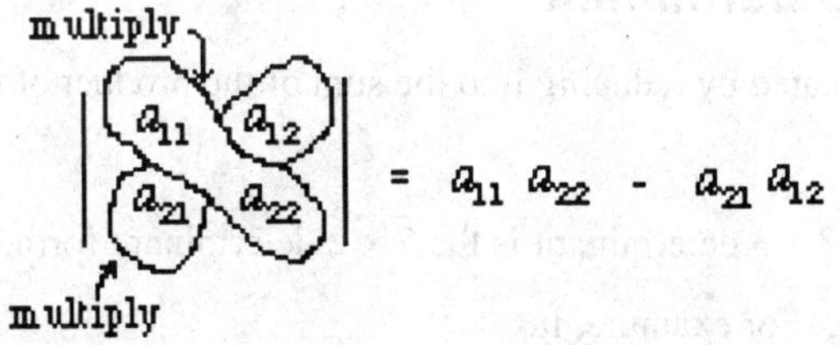

multiply

$$\begin{vmatrix} a_{11} & a_{12} \\ a_{21} & a_{22} \end{vmatrix} = a_{11}\,a_{22} - a_{21}\,a_{12}$$

multiply

Figure 1

For the matrix $A = \begin{bmatrix} 2 & 3 \\ -1 & 4 \end{bmatrix}$ the value of its corresponding determinant is

$$\det A = \begin{vmatrix} 2 & 3 \\ -1 & 4 \end{vmatrix} = 2(4) - (-1)(3) = 11$$

Definition of the Value of a 2 x 2 Determinant

The 2 x 2 determinant $\begin{vmatrix} a_{11} & a_{12} \\ a_{21} & a_{22} \end{vmatrix}$ has the value $a_{11}a_{22} - a_{21}a_{12}$.

NOTE: As with matrices, learning how to automate evaluating determinants is probably as important as learning their properties. In practice, determinants are far too large to evaluate by hand. In fact, one drawback of determinants is that they can easily get too large to evaluate with a machine.

Example 1 *Evaluating 2 x 2 determinants*

Evaluate the determinants of $B = \begin{bmatrix} 3 & 2 \\ 4 & -1 \end{bmatrix}$ and $3B = \begin{bmatrix} 9 & 6 \\ 12 & -3 \end{bmatrix}$.

Solution

$$\text{If } B = \begin{bmatrix} 3 & 2 \\ 4 & -1 \end{bmatrix}, \ \det B = \begin{vmatrix} 3 & 2 \\ 4 & -1 \end{vmatrix} = 3(-1) - 4(2) = -11$$

and

$$\det 3B = \begin{vmatrix} 9 & 6 \\ 12 & -3 \end{vmatrix} = 9(-3) - 12(6) = -99$$

Notice in Example 1 that det 3B = 9(det B). Such results will be examined further and generalized in the exercises. If we interchange the columns of matrix B we have a new matrix, $A = \begin{bmatrix} 2 & 3 \\ -1 & 4 \end{bmatrix}$. Notice that the value of det A is 11 and the value of det B = -11.

Evaluating a 3 x 3 Determinant

A 3 x 3 determinant can be evaluated by reducing it to the sum of the product of three scalars and three 2 x 2 determinants.

A **minor** of an element of a 3 x 3 determinant is the 2 x 2 determinant formed by eliminating the row and column containing that element. For example, in

$$\det A = \begin{vmatrix} 1 & 4 & 3 \\ 3 & 1 & 5 \\ -1 & 0 & 2 \end{vmatrix}$$

the minor of $a_{12} = 4$ is found by eliminating both the first row and second column containing the element 4. The minor of **4** in

$$\begin{vmatrix} 1 & 4 & 3 \\ 3 & 1 & 5 \\ -1 & 0 & 2 \end{vmatrix} \quad \text{is} \quad \begin{vmatrix} 3 & 5 \\ -1 & 2 \end{vmatrix}$$

The minor of $a_{33} = \mathbf{2}$ is found by eliminating the third row and column. The minor of **2** in

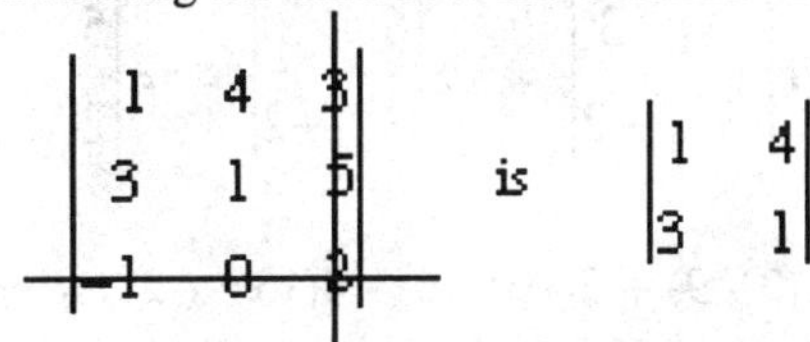

To evaluate the determinant of a 3 x 3 matrix, we reduce it to three 2 x 2 determinants, which we then evaluate. These three 2 x 2 determinants are the three minors of the elements of any row or column.

Evaluating a 3 x 3 Determinant

1. Choose a row or a column.
2. Find the minor for each element in that row or column.
3. Multiply each element by the value of its minor.
4. Add or subtract these three products according to the following checkerboard pattern of signs:

$$\begin{vmatrix} + & - & + \\ - & + & - \\ + & - & + \end{vmatrix}$$

This process for evaluating a 3 x 3 determinant generalizes quite naturally to any other square matrix.

Example 2 *Evaluating a 3 x 3 determinant*

Find the value of the determinant of the 3 x 3 matrix (a) using the second row and (b) using the third column.

$$\begin{bmatrix} 5 & -1 & 3 \\ 1 & 2 & 4 \\ 2 & 3 & -2 \end{bmatrix}$$

Solution

(a) The elements in row 2 are $a_{21} = 1$, $a_{22} = 2$, and $a_{23} = 4$. The checkerboard pattern of signs for row 2 are -, +, and -. We add the minors as shown in Figure 2.

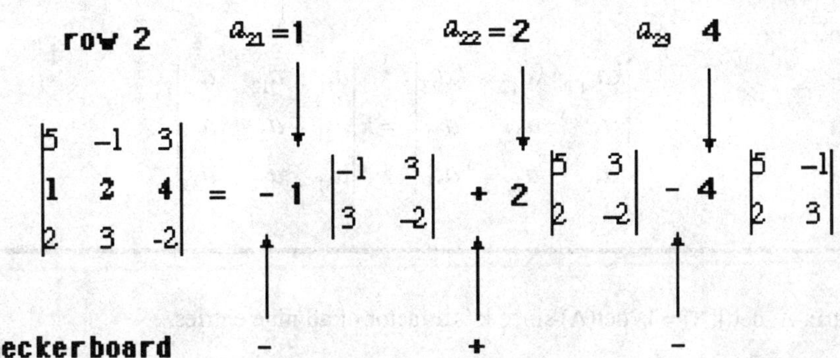

Figure 2

$$\begin{vmatrix} 5 & -1 & 3 \\ \mathbf{1} & \mathbf{2} & \mathbf{4} \\ 2 & 3 & -2 \end{vmatrix} = \mathbf{-1}\begin{vmatrix} -1 & 3 \\ 3 & -2 \end{vmatrix} + \mathbf{2}\begin{vmatrix} 5 & 3 \\ 2 & -2 \end{vmatrix} - \mathbf{4}\begin{vmatrix} 5 & -1 \\ 2 & 3 \end{vmatrix}$$

$$= \mathbf{-1}(2 - 9) + \mathbf{2}(-10 - 6) - \mathbf{4}(15 - (-2)) = -93$$

(b) The elements in column 3 are $a_{13} = 3$, $a_{23} = 4$, and $a_{33} = -2$. The checkerboard pattern of signs for column 3 are + , - , and +.

$$\begin{vmatrix} 5 & -1 & \mathbf{3} \\ 1 & 2 & \mathbf{4} \\ 2 & 3 & \mathbf{-2} \end{vmatrix} = +\mathbf{3}\begin{vmatrix} 1 & 2 \\ 2 & 3 \end{vmatrix} - \mathbf{4}\begin{vmatrix} 5 & -1 \\ 2 & 3 \end{vmatrix} + \mathbf{(-2)}\begin{vmatrix} 5 & -1 \\ 1 & 2 \end{vmatrix}$$

$$= \mathbf{3}(3 - 4) \mathbf{-4}(15 - (-2)) \mathbf{-2}(10 - (-1)) = -93$$

Notice that finding the value of the 3 x 3 determinant different ways yields the same result.

Are there other ways to evaluate the 3 x 3 determinant in Example 2?

We could have used row 1, row 3, column 1, or column 2 with the same result.

Examining Properties of Determinants

Fortunately, the process of evaluating determinants can be greatly simplified by making use of their properties. We demonstrate these properties using 3 x 3 determinants, but the theorems hold for all determinants.

The first property tells us how to factor a common value of a row or column of a determinant.

Factoring a Determinant

Multiplying a row or column by a constant (scalar) multiplies the value of the determinant by that constant. That is, for a row,

$$\begin{vmatrix} ka_{11} & ka_{12} & ka_{13} \\ a_{21} & a_{22} & a_{23} \\ a_{31} & a_{32} & a_{33} \end{vmatrix} = k\begin{vmatrix} a_{11} & a_{12} & a_{13} \\ a_{21} & a_{22} & a_{23} \\ a_{31} & a_{32} & a_{33} \end{vmatrix}$$

For a 3 x 3 matrix A, $\det(kA) = k^3\det(A)$ since k is a factor of all nine entries.

Example 3 *Comparing determinants*

Show that

$$\begin{vmatrix} \mathbf{2} & 1 & 1 \\ \mathbf{4} & 3 & 4 \\ \mathbf{6} & 4 & 1 \end{vmatrix} = \mathbf{2} \cdot \begin{vmatrix} 1 & 1 & 1 \\ 2 & 3 & 4 \\ 3 & 4 & 1 \end{vmatrix}$$

by evaluating both determinants.

Solution

$$\begin{vmatrix} \mathbf{2} & \mathbf{1} & \mathbf{1} \\ 4 & 3 & 4 \\ 6 & 4 & 1 \end{vmatrix} = \mathbf{2}\begin{vmatrix} 3 & 4 \\ 4 & 1 \end{vmatrix} - \mathbf{1}\begin{vmatrix} 4 & 4 \\ 6 & 1 \end{vmatrix} + \mathbf{1}\begin{vmatrix} 4 & 3 \\ 6 & 4 \end{vmatrix} = 2(-13) - 1(-20) + 1(-2) = -8$$

$$2\begin{vmatrix} 1 & 1 & 1 \\ 2 & 3 & 4 \\ 3 & 4 & 1 \end{vmatrix} = 2\left\{1\begin{vmatrix} 3 & 4 \\ 4 & 1 \end{vmatrix} - 1\begin{vmatrix} 2 & 4 \\ 3 & 1 \end{vmatrix} + 1\begin{vmatrix} 2 & 3 \\ 3 & 4 \end{vmatrix}\right\} = 2[1(3 - 16) - 1(2 - 12) + 1(8 - 9)] = 2(-4) = -8$$

Consider the following matrix.

$$A = \begin{bmatrix} 2 & \mathbf{4} & 2 \\ 3 & \mathbf{0} & -1 \\ 1 & \mathbf{0} & 3 \end{bmatrix}$$

Notice how quickly we can find the value of its determinant using the second column, which contains two zeros.

$$\det(A) = -4\begin{vmatrix} 3 & -1 \\ 1 & 3 \end{vmatrix} + \mathbf{0}\begin{vmatrix} 2 & -2 \\ 1 & 3 \end{vmatrix} + \mathbf{0}\begin{vmatrix} 2 & 2 \\ 3 & -1 \end{vmatrix}$$

$$= -4(10) + \mathbf{0} + \mathbf{0} = -40$$

We need to evaluate only one 2 x 2 determinant to find the value of this 3 x 3 determinant. The next theorem tells us how we can create zeros in rows and columns of determinants.

Adding Rows and Columns of a Determinant

Any constant multiple of a row (or column) can be added to any other row (or column) without changing the value of the determinant. For a column, for example,

$$\begin{vmatrix} a_{11} & a_{12} & a_{13} \\ a_{21} & a_{22} & a_{23} \\ a_{31} & a_{32} & a_{33} \end{vmatrix} = \begin{vmatrix} a_{11} & a_{12} + ka_{11} & a_{13} \\ a_{21} & a_{22} + ka_{21} & a_{23} \\ a_{31} & a_{32} + ka_{31} & a_{33} \end{vmatrix}$$

Example 4 *Comparing the value of determinants*

Of the following three determinants, the second represents the first determinant with the second row added to the third row, $R_2 + R_3$. Their sum replaces the third row. The last determinant represents the first with twice the second column added to both the first column and the third column, $2C_2 + C_1$ and $2C_2 + C_3$. Check the value of each determinant.

$$\begin{vmatrix} 2 & -1 & 2 \\ 3 & 2 & 0 \\ 1 & 5 & 3 \end{vmatrix} = \begin{vmatrix} 2 & -1 & 2 \\ 3 & 2 & 0 \\ \mathbf{4} & \mathbf{7} & \mathbf{3} \end{vmatrix} = \begin{vmatrix} \mathbf{0} & -1 & \mathbf{0} \\ \mathbf{7} & 2 & \mathbf{4} \\ \mathbf{11} & 5 & \mathbf{13} \end{vmatrix}$$

Solution

$$\begin{vmatrix} 2 & -1 & 2 \\ 3 & 2 & 0 \\ 1 & 5 & 3 \end{vmatrix} = +2\begin{vmatrix} 3 & 2 \\ 1 & 5 \end{vmatrix} - 0\begin{vmatrix} 2 & -1 \\ 1 & 5 \end{vmatrix} + 3\begin{vmatrix} 2 & -1 \\ 3 & 2 \end{vmatrix}$$

$$= +\mathbf{2}(13) - \mathbf{0}(11) + \mathbf{3}(7) = 47$$

$$\begin{vmatrix} 2 & -1 & 2 \\ 3 & 2 & 0 \\ \mathbf{4} & \mathbf{7} & \mathbf{3} \end{vmatrix} = +2\begin{vmatrix} 3 & 2 \\ 4 & 7 \end{vmatrix} - 0\begin{vmatrix} 2 & -1 \\ 4 & 7 \end{vmatrix} + 3\begin{vmatrix} 2 & -1 \\ 3 & 2 \end{vmatrix}$$

$$= +\mathbf{2}(13) - \mathbf{0}(18) + \mathbf{3}(7) = 47$$

$$\begin{vmatrix} \mathbf{0} & -1 & \mathbf{0} \\ \mathbf{7} & 2 & \mathbf{4} \\ \mathbf{11} & 5 & \mathbf{13} \end{vmatrix} = 0\begin{vmatrix} 3 & 2 \\ 1 & 5 \end{vmatrix} - (-1)\begin{vmatrix} 2 & -1 \\ 1 & 5 \end{vmatrix} + 0\begin{vmatrix} 2 & -1 \\ 3 & 2 \end{vmatrix}$$

$$= +\mathbf{0}(6) - (\mathbf{-1})(47) + \mathbf{0}(13) = 47$$

The third determinant was the easiest to compute since the first row had only one nonzero entry. This is how we apply the preceding theorem. The next theorem states another important property of determinants.

The Determinant of the Product of Two Matrices

If A and B are both n x n matrices, then

$$\det AB = (\det A)(\det B)$$

Example 5 *Verifying the product of matrices*

Verify that det AB = (det A)(det B) for

$$A = \begin{bmatrix} 2 & 1 & 1 \\ 4 & 3 & 4 \\ 6 & 4 & 1 \end{bmatrix} \text{ and } B = \begin{bmatrix} 2 & -1 & 2 \\ 3 & 2 & 0 \\ 1 & 5 & 3 \end{bmatrix}$$

Solution

From Examples 3 and 4, we know that det A = -8 and det B = 47. Hence, (det A)(det B) = -8(47) = -376.

The product AB is a third matrix.

$$AB = \begin{bmatrix} 2 & 1 & 1 \\ 4 & 3 & 4 \\ 6 & 4 & 1 \end{bmatrix} \begin{bmatrix} 2 & -1 & 2 \\ 3 & 2 & 0 \\ 1 & 5 & 3 \end{bmatrix} = \begin{bmatrix} 8 & 5 & 7 \\ 21 & 22 & 20 \\ 25 & 7 & 15 \end{bmatrix}$$

The value of det AB is

$$\det AB = \begin{vmatrix} 8 & 5 & 7 \\ 21 & 22 & 20 \\ 25 & 7 & 15 \end{vmatrix} = 8\begin{vmatrix} 22 & 20 \\ 7 & 15 \end{vmatrix} - 5\begin{vmatrix} 21 & 20 \\ 25 & 15 \end{vmatrix} + 7\begin{vmatrix} 21 & 22 \\ 25 & 7 \end{vmatrix}$$

$$= 8(190) - 5(-185) + 7(-403) = -376$$

Exercises 9.2

Develop the Concepts

For Exercises 1 to 8 find the determinants of the matrices.

1.(a) $\begin{bmatrix} 1 & 2 \\ 3 & 4 \end{bmatrix}$ (b) $\begin{bmatrix} 2 & 2 \\ 6 & 4 \end{bmatrix}$ (c) $\begin{bmatrix} -1 & 6 \\ 3 & 12 \end{bmatrix}$ (d) $\begin{bmatrix} -2 & 6 \\ 3 & 12 \end{bmatrix}$

2.(a) $\begin{bmatrix} 1 & 0 & 0 \\ 2 & 3 & 8 \\ 4 & 5 & 10 \end{bmatrix}$ (b) $\begin{bmatrix} -1 & 0 & 0 \\ -2 & 3 & 8 \\ -4 & 5 & 10 \end{bmatrix}$ (c) $\begin{bmatrix} 1 & 0 & 0 \\ -2 & -3 & -8 \\ 4 & 5 & 10 \end{bmatrix}$

3.(a) $\begin{bmatrix} 0 & 2 & 3 \\ 0 & 1 & 1 \\ 3 & 4 & 2 \end{bmatrix}$ (b) $\begin{bmatrix} 0 & 2 & 3 \\ 3 & 4 & 2 \\ 0 & 1 & 1 \end{bmatrix}$ (c) $\begin{bmatrix} 0 & 3 & 2 \\ 0 & 1 & 1 \\ 3 & 2 & 4 \end{bmatrix}$

4.(a) $\begin{bmatrix} 1 & 2 \\ -3 & -2 \end{bmatrix}$ (b) $\begin{bmatrix} 1 & -3 \\ 2 & -2 \end{bmatrix}$

5.(a) $\begin{bmatrix} 1 & 2 & -3 \\ -2 & 0 & 1 \\ 3 & 0 & 2 \end{bmatrix}$ (b) $\begin{bmatrix} 1 & -2 & 3 \\ 2 & 0 & 0 \\ -3 & 1 & 2 \end{bmatrix}$

6.(a) $\begin{bmatrix} 1 & 3 \\ -2 & 4 \end{bmatrix}$ (b) $\begin{bmatrix} 1 & 5 \\ -2 & 0 \end{bmatrix}$ (c) $\begin{bmatrix} 1 & 3 \\ 0 & 10 \end{bmatrix}$ (d) $\begin{bmatrix} 16 & 3 \\ 18 & 4 \end{bmatrix}$

7.(a) $\begin{bmatrix} 3 & 0 & 0 \\ -10 & 2 & 0 \\ 11 & 3 & 4 \end{bmatrix}$ (b) $\begin{bmatrix} 3 & 5 & 9 \\ 0 & 2 & 10 \\ 0 & 0 & 4 \end{bmatrix}$ (c) $\begin{bmatrix} 3 & 0 & 0 \\ 0 & 2 & 0 \\ 0 & 0 & 4 \end{bmatrix}$

8.(a) $\begin{bmatrix} 1 & 2 & 1 \\ 2 & 2 & 2 \\ 3 & 3 & 3 \end{bmatrix}$ (b) $\begin{bmatrix} 0 & 2 & 9 \\ 0 & 1 & 3 \\ 0 & -4 & 5 \end{bmatrix}$ (c) $\begin{bmatrix} 2 & 3 & 2 \\ 4 & -9 & 4 \\ 6 & 4 & 6 \end{bmatrix}$

For Exercises 9 to 12 evaluate each pair of determinants, using shortcuts when available.

9. (a) $\begin{bmatrix} 1 & 2 \\ -1 & 3 \end{bmatrix}$ (b) $\begin{bmatrix} 4 & 2 \\ -4 & 3 \end{bmatrix}$

10. (a) $\begin{bmatrix} 1 & -4 \\ 2 & 3 \end{bmatrix}$ (b) $\begin{bmatrix} 3 & -12 \\ 2 & 3 \end{bmatrix}$

11.(a) $\begin{bmatrix} 1 & 0 & 0 \\ -1 & 2 & 2 \\ 4 & 1 & 3 \end{bmatrix}$ (b) $\begin{bmatrix} 3 & 0 & 0 \\ -3 & 2 & 2 \\ 12 & 1 & 3 \end{bmatrix}$

12.(a) $\begin{bmatrix} 0 & 3 & 0 \\ 1 & 2 & 3 \\ -2 & 4 & 1 \end{bmatrix}$ (b) $\begin{bmatrix} 0 & 3 & 0 \\ 1 & 2 & 3 \\ 4 & -8 & -2 \end{bmatrix}$

For Exercises 13 to 16 evaluate each determinant by first creating two 0s in a row or column. Then use that row or column to evaluate.

*13. $\begin{bmatrix} 2 & 1 & 1 \\ 2 & -1 & -4 \\ -4 & 3 & 8 \end{bmatrix}$ 14. $\begin{bmatrix} 3 & 6 & 3 \\ 2 & 4 & 2 \\ 4 & 3 & -1 \end{bmatrix}$

15. $\begin{bmatrix} 1 & 3 & 2 \\ 2 & -1 & -4 \\ -4 & 3 & 8 \end{bmatrix}$ 16. $\begin{bmatrix} 1 & -1 & 2 \\ 3 & 1 & 4 \\ 2 & -3 & 1 \end{bmatrix}$

For Exercises 17 to 24 find each determinant for

$$A = \begin{bmatrix} 4 & -2 \\ 2 & -3 \end{bmatrix} \quad B = \begin{bmatrix} 4 & 0 & -2 \\ 1 & 3 & 1 \\ 2 & 1 & 0 \end{bmatrix} \quad \text{and } C = \begin{bmatrix} 3 & 1 & -2 \\ -10 & 0 & 4 \\ 2 & -1 & 2 \end{bmatrix}$$

17.(a) det A

(b) det A^T

(A^T is the transpose of A. If A = $\{a_{ij}\}$, then $A^T = \{a_{ji}\}$.)

18.(a) (det B)(det C)

(b) det BC

19.(a) det (B + C)

(b) det B + det C

20.(a) det 4A

(b) 4det A

21. det(2B)

22. det 3B

23. det[3B(C)]

24. 27 (det B)(det C)

Extend the Concepts

For Exercises 25 to 28 prove each of the following without expanding the determinants.

25. $$\begin{vmatrix} a_{11} & a_{12} \\ a_{21} & a_{22} \end{vmatrix} = \begin{vmatrix} a_{11} + a_{12} & a_{12} \\ a_{21} - a_{11} + a_{22} - a_{12} & a_{22} - a_{12} \end{vmatrix}$$

26. $$\begin{vmatrix} a_{11} & a_{12} & a_{13} \\ a_{21} & a_{22} & a_{23} \\ a_{31} & a_{32} & a_{33} \end{vmatrix} = \begin{vmatrix} a_{11} - 2a_{11} & \frac{1}{2}a_{12} & a_{12} + 2a_{13} \\ a_{21} + 2a_{22} & \frac{1}{2}a_{22} & a_{22} + 2a_{23} \\ a_{31} + 2a_{32} & \frac{1}{2}a_{32} & a_{32} + a_{33} \end{vmatrix}$$

27. $$\begin{vmatrix} a_{11} & a_{12} & a_{13} \\ a_{21} & a_{22} & a_{23} \\ a_{31} & a_{32} & a_{33} \end{vmatrix} = \begin{vmatrix} a_{22} & a_{23} & a_{21} \\ a_{32} & a_{33} & a_{31} \\ a_{12} & a_{32} & a_{11} \end{vmatrix}$$

28. $$\begin{vmatrix} 3a_{11} & 6a_{12} & 3a_{13} \\ a_{21} & a_{22} & a_{23} \\ a_{31} & a_{32} & a_{33} \end{vmatrix} = 6\begin{vmatrix} a_{11} & a_{12} & a_{13} \\ a_{21} & a_{22} & a_{23} \\ a_{31} & a_{32} & a_{33} \end{vmatrix}$$

9.3 Vectors in Two and Three Dimensions

Introducing Unit and Position Vectors
Introducing Three-Dimensional Vectors
Multiplying Vectors
Applying Vectors

Recall that a vector in two dimensions, R^2, can be represented by the ordered pair at its tip when its tail is at the origin (Figure 1). These vectors have magnitudes and direction represented by the length and direction of the arrow.

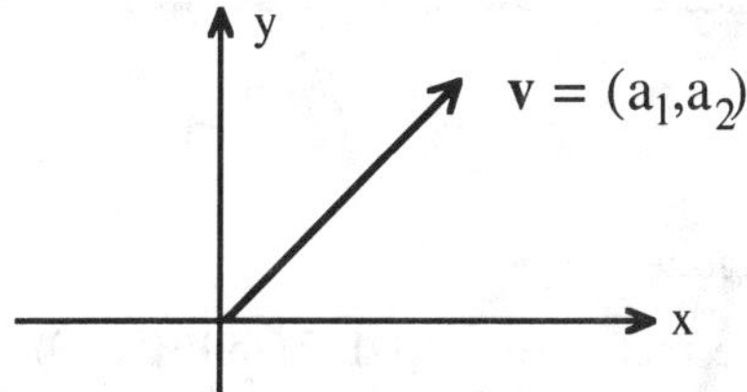

Figure 1

Introducing Unit and Position Vectors

It is often convenient to represent vectors in terms of **unit vectors**, vectors whose magnitudes are 1. Figure 2a shows two unit vectors,

$$\mathbf{i} = (1, 0) \text{ and } \mathbf{j} = (0, 1)$$

Any vector can be written as the sum of scalar multiples of these unit vectors:

$$\mathbf{v} = (a_1, a_2) = a_1\mathbf{i} + a_2\mathbf{j}$$

That is, $\mathbf{v}$ is the sum of $a_1\mathbf{i}$ and $a_2\mathbf{j}$. Since $a_1\mathbf{i}$ is a scalar multiple of $\mathbf{i}$ on the x axis, a_1 is called the *i component* of vector $\mathbf{v}$. Since $a_2\mathbf{j}$ is a scalar multiple of $\mathbf{j}$ on the y axis, a_2 is called the *j component* of vector $\mathbf{v}$ (Figure 2b).

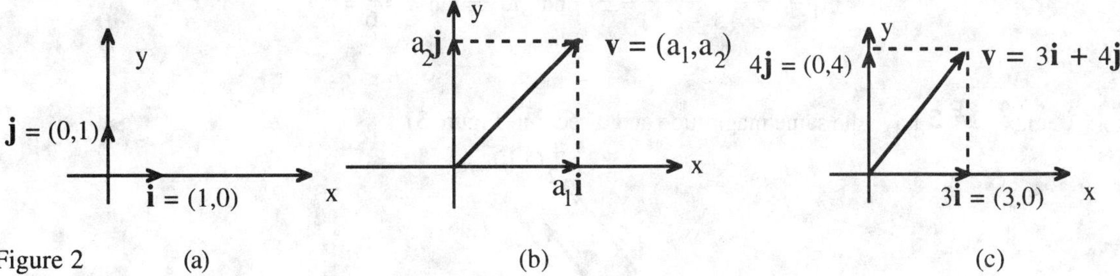

Figure 2 (a) (b) (c)

The vector $\mathbf{v} = 3\mathbf{i} + 4\mathbf{j}$ can be interpreted as the diagonal of the parallelogram formed by $3\mathbf{i}$ and $4\mathbf{j}$ (Figure 2c).

Vectors can also be written as directed line segments $\overrightarrow{P_1P_2}$, between two points, P_1 and P_2. Such vectors are equal to a vector from the origin found by subtracting the coordinates of the tail P_1 from those of the tip P_2, as indicated in the next theorem. Vectors with their tails at the origin are called **position vectors**.

Definition of a Position Vector

The position vector from $P_1(x_1, y_1)$ to $P_2(x_2, y_2)$ is

$$\mathbf{v} = \overrightarrow{P_1P_2} = (x_2 - x_1, y_2 - y_1) = (x_2 - x_1)\,\mathbf{i} + (y_2 - y_1)\,\mathbf{j}$$

Example 1 *Finding a position vector*

(a) Find the position vector equal to the vector from $P_1(-1, 2)$ to $P_2(5, 10)$.

(b) Show that the position vector from part (a) has the same magnitude and direction as the vector from P_1 to P_2.

Solution

(a) $\mathbf{v} = \overrightarrow{P_1P_2} = (5 - (-1), 10 - 2) = (6,8)$

(b) Notice that $\|\mathbf{v}\| = \sqrt{6^2 + 8^2} = \sqrt{100} = 10$ and the distance from P_1 to P_2 by the distance formula is

$$\sqrt{(x_2 - x_1)^2 + (y_2 - y_1)^2} = \sqrt{[5 - (-1)]^2 + (10 - 2)^2} = 10$$

Hence, $\mathbf{v} = (6, 8)$ and $\overrightarrow{P_1P_2}$ have the same magnitude. To show that they have the same direction, we compute the slope of the lines from $P_1(-1, 2)$ to $P_2(5, 10)$ and from (0, 0) to (6, 8).

$$m_1 = \frac{\Delta y}{\Delta x} = \frac{10 - 2}{5 - (-1)} = \frac{4}{3} \quad \text{and} \quad m_2 = \tan\theta = \frac{8}{6} = \frac{4}{3}$$

Vectors **V** and $\overrightarrow{P_1P_2}$ have the same magnitude and direction (Figure 3).

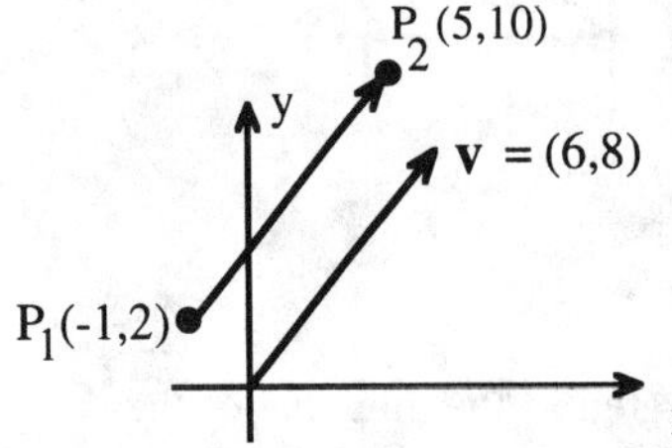

Figure 3

The vectors **i** and **j** are not the only unit vectors. We can find a unit vector in the direction of any vector by dividing that vector by its magnitude.

Unit Vector

The unit vector in the direction of **v** is **u**, where

$$\mathbf{u} = \frac{\mathbf{v}}{\|\mathbf{v}\|}, \quad \mathbf{v} \neq 0$$

Example 2 *Finding a unit vector*

Find a unit vector in the direction of **v** in Figure 9.

Solution

Since $\mathbf{v} = (6,8)$ and $\|\mathbf{v}\| = 10$, the unit vector in the direction of **v** is

$$\mathbf{u} = \frac{\mathbf{v}}{\|\mathbf{v}\|} = \frac{1}{10}(6, 8) = (\frac{6}{10}, \frac{8}{10})$$

Notice since $\|\mathbf{v}\| = 10$, $\mathbf{u} = \frac{1}{10}\mathbf{v}$ is one-tenth as long as **v**. Thus, $\mathbf{u} = \frac{1}{10}\mathbf{v}$ is 1 unit in length.

Introducing Three-Dimensional Vectors

A vector in three dimensions, R^3, can be represented by an ordered triple, the coordinates of the point at its tip when its tail is at the origin (Figure 4).

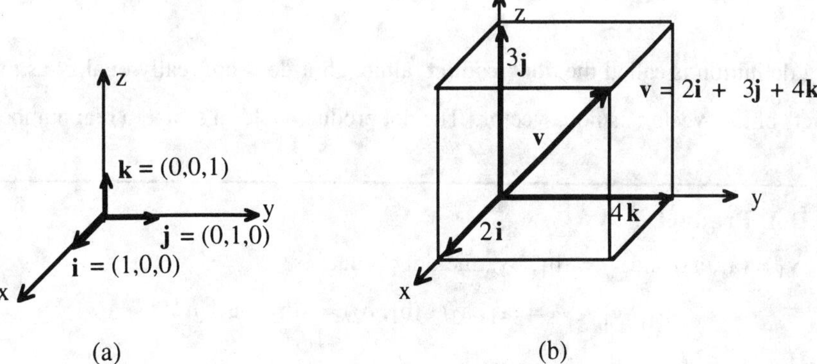

Figure 4 (a) (b)

Figure 4a shows three **unit vectors** in R^3, **i** , **j**, and **k**, in the direction of the coordinate axes.

Any vector in R^3 can be written as a liner combination of **i**, **j**, and **k** . For example, $\mathbf{v} = (2, 3, 4)$ can be expressed as

$$\mathbf{v} = 2\mathbf{i} + 3\mathbf{j} + 4\mathbf{k}$$

Vector **v** in Figure 4b is the diagonal of a box whose sides are 2**i**, 3**j**, and 4**k**.

The following definitions in R^3 are analogous to definitions in R^2.

Definition of Magnitude of a Vector in R^3

If **v** is a vector in R^3 such that $\mathbf{v} = (a_1, a_2, a_3)$ then the magnitude of **v** is

$$\|\mathbf{v}\| = \sqrt{a_1^2 + a_2^2 + a_3^2}$$

Definition of the Sum of Two Vectors in R^3

If $\mathbf{v}_1$ and $\mathbf{v}_2$ are vectors in R^3 such that $\mathbf{v}_1 = (a_1, a_2, a_3)$ and $\mathbf{v}_2 = (b_1, b_2, b_3)$, the sum of the vectors is

$$\mathbf{v}_1 + \mathbf{v}_2 = (a_1, a_2, a_3) + (b_1, b_2, b_3) = (a_1 + b_1, a_2 + b_2, a_3 + b_3)$$

Multiplying Vectors

We are now ready to put vectors to work. There are three different *multiplications* involving vectors: multiplications by a scalar, dot products, and cross products. Multiplying by a scalar α changes the magnitude of the vector by the factor $|\alpha|$.

Definition of Multiplication by a Scalar

For scalar $\alpha \in R$, and $\mathbf{v}_1 = (a_1, a_2, a_3)$ in R^3,

$$\alpha\mathbf{v}_1 = (\alpha a_1, \alpha a_2, \alpha a_3)$$

The following definition is called the **dot product**, although it does not really qualify as multiplication since this "product" of two vectors is not a vector. The dot product yields a scalar, a real number.

Definition of Dot Product

For $\mathbf{v}_1, \mathbf{v}_2 \in R^2$, $\mathbf{v}_1 = (a_1, a_2)$ and $\mathbf{v}_2 = (b_1, b_2)$, the dot product is

$$\mathbf{v}_1 \cdot \mathbf{v}_2 = (a_1, a_2) \cdot (b_1, b_2) = a_1b_1 + a_2b_2$$

For $\mathbf{v}_1, \mathbf{v}_2 \in R^3$, $\mathbf{v}_1 = (a_1, a_2, a_3)$ and $\mathbf{v}_2 = (b_1, b_2, b_3)$, the dot product is

$$\mathbf{v}_1 \cdot \mathbf{v}_2 = (a_1, a_2, a_3) \cdot (b_1, b_2, b_3) = a_1b_1 + a_2b_2 + a_3b_3$$

Example 3 *Finding scalar products*

Find the scalar products $\mathbf{v}_1 \cdot \mathbf{v}_2$ and $\mathbf{v} \cdot \mathbf{v}_1$ for

(a) $\mathbf{v}_1 = (3, 4)$ and $\mathbf{v}_2 = (-1, 2)$ (b) $\mathbf{v}_1 = (-1, 3, 4)$ and $\mathbf{v}_2 = (0, 4, -3)$

Solution

(a) $\mathbf{v}_1 \cdot \mathbf{v}_2 = (3, 4) \cdot (-1, 2) = 3(-1) + 4(2) = -3 + 8 = 5$

$\mathbf{v}_1 \cdot \mathbf{v}_1 = (3, 4) \cdot (3, 4) = 3(3) + 4(4) = 9 + 16 = 25$

(b) $\mathbf{v}_1 \cdot \mathbf{v}_2 = (-1, 3, 4) \cdot (0, 4, -3) = 0 + 12 - 12 = 0$

$\mathbf{v}_1 \cdot \mathbf{v}_1 = (-1, 3, 4) \cdot (-1, 3, 4) = 1 + 9 + 16 = 26$

In Example 3 with $\mathbf{v}_1 = (3, 4)$,

$$\mathbf{v}_1 \cdot \mathbf{v}_1 = 25$$

and the magnitude of this vector is

$$\|\mathbf{v}_1\| = \sqrt{3^2 + 4^2} = 5$$

For $\mathbf{v}_1 = (-1, 3, 4)$

$$\mathbf{v}_1 \cdot \mathbf{v}_1 = 26$$

and

$$\|\mathbf{v}_1\| = \sqrt{(-1)^2 + 3^2 + 4^2} = \sqrt{26}$$

The next theorem extends the dot product by relating it to the magnitude of a vector.

> If $\mathbf{v}$ is an element of R^2 or R^3, then
>
> $$\mathbf{v} \cdot \mathbf{v} = \|\mathbf{v}\|^2$$

<u>Proof:</u>

$\mathbf{v} = (a_1, a_{12}, a_3) \Rightarrow \|\mathbf{v}\| = \sqrt{a_1{}^2 + a_2{}^2 + a_3{}^2}$

and

$$\mathbf{v} \cdot \mathbf{v} = a_1 a_1 + a_2 a_2 + a_3 a_3 = \|\mathbf{v}\|^2$$

Notice in part (b) of Example 3, $\mathbf{v}_1 \cdot \mathbf{v}_2 = 0$. These vectors are said to be **orthogonal** because the angle between them measures 90°. The following theorem is used to prove that if $\theta = 90°$, then $\mathbf{v}_1 \cdot \mathbf{v}_2 = 0$ since $\cos 90° = 0$.

If θ is the angle between $\mathbf{v}_1$ and $\mathbf{v}_2$, then

$$\mathbf{v}_1 \cdot \mathbf{v}_2 = \|\mathbf{v}_1\| \, \|\mathbf{v}_2\| \cos\theta$$

The proof of this theorem follows from the law of cosines.

The next example illustrates using the dot product to find the angle between two vectors. This method is much shorter than the alternative, using the law of cosines.

Example 4 *Finding the measure of an angle*

Find the measure of the angle between $\mathbf{v}_1 = (1, 3, 4)$ and $\mathbf{v}_2 = (4, 3, 0)$.

Solution

By the preceding theorem,

$$\cos\theta = \frac{\mathbf{v}_1 \cdot \mathbf{v}_2}{\|\mathbf{v}_1\| \, \|\mathbf{v}_2\|} = \frac{1(4) + 3(3) + 4(0)}{\sqrt{26}\sqrt{25}} = \frac{13}{\sqrt{650}}$$

$$\theta \approx 59.3°$$

In Section 6.4, Example 2, A Whale of a Problem, we used the law of cosines to find the angle between two sides of a triangle. Now we will relabel two sides of that triangle as vectors and use the theorem

$$\mathbf{v}_1 \cdot \mathbf{v}_2 = \|\mathbf{v}_1\| \, \|\mathbf{v}_2\| \cos\theta$$

to find θ and the heading from the tourist's yacht to the whales.

Example 5 *Finding the angle between two vectors*

Find the angle between $\mathbf{v}_1 = (50, 50\sqrt{3})$ and $\mathbf{v}_2 = (519.8, -84.4)$ in Figure 5.

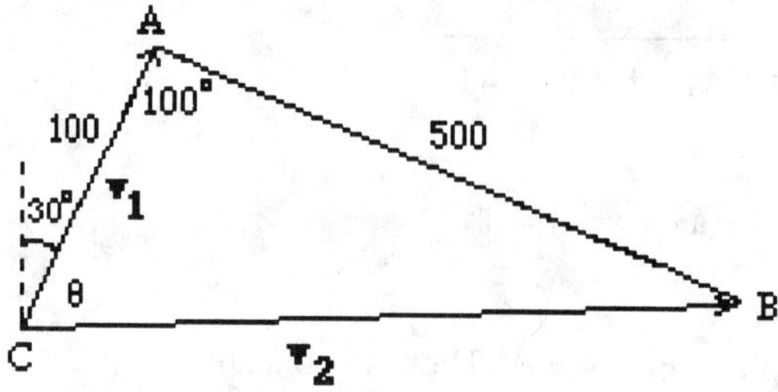

Figure 5

Solution

We use the dot product to find the angle between two vectors.

$$\mathbf{v}_1 \cdot \mathbf{v}_2 = 50(519.8) + (50\sqrt{3})(-84.4) \approx 18{,}681$$

Since

$$\|\mathbf{v}_1\| = \sqrt{50^2 + (50\sqrt{3})^2} = 100 \text{ and } \|\mathbf{v}_2\| = \sqrt{(519.8)^2 + (-84.4)^2} \approx 526.6$$

we have

$$\mathbf{v}_1 \cdot \mathbf{v}_2 = \|\mathbf{v}_1\|\,\|\mathbf{v}_2\| \cos\theta$$

$$\cos\theta = \frac{\mathbf{v}_1 \cdot \mathbf{v}_2}{\|\mathbf{v}_1\|\,\|\mathbf{v}_2\|}$$

$$\cos\theta = \frac{18681}{100(526.6)}$$

$$\theta = \cos^{-1} 0.3547$$

$$\theta \approx 69.2^\circ$$

Hence, the heading from the yacht to the whales is $\theta + 30^\circ \approx 99.2^\circ$.

The next theorem deals with orthogonal vectors for which the angle between them is $\frac{\pi}{2} = 90^\circ$.

Orthogonal Vectors

Two vectors $\mathbf{v}_1$ and $\mathbf{v}_2$ are orthogonal vectors if and only if

$$\mathbf{v}_1 \cdot \mathbf{v}_2 = 0,\ \mathbf{v}_1 \neq 0,\ \mathbf{v}_2 \neq 0$$

Proof:

$\mathbf{v}_1 \cdot \mathbf{v}_2 = \|\mathbf{v}_1\|\,\|\mathbf{v}_2\| \cos\theta$ and $\mathbf{v}_1 \cdot \mathbf{v}_2 = 0 \Rightarrow$

$$\|\mathbf{v}_1\|\,\|\mathbf{v}_2\| \cos\theta = 0$$

$$\cos\theta = 0$$

$$\theta = \pi/2,\ 3\pi/2,\ \ldots$$

The converse of the theorem states that

$$\text{if } \theta = \pi/2 \text{ then } \mathbf{v}_1 \cdot \mathbf{v}_2 = \|\mathbf{v}_1\|\,\|\mathbf{v}_2\| \cos \pi/2 = \|\mathbf{v}_1\|\,\|\mathbf{v}_2\|\,(0) = 0$$

Applying Vectors

Consider the children in Figure 6 trying to pull wagons.

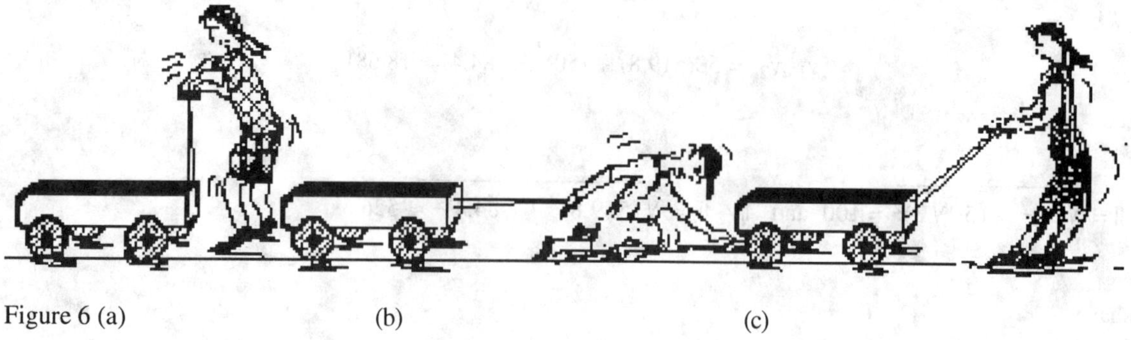

Figure 6 (a) (b) (c)

The child in Figure 6a is pulling straight up on the tongue and the wagon does not roll. All the force is applied to lift the wagon. There is no component in the desired direction of motion. The child in Figure 6b is pulling horizontally. All of his force is in the direction of motion. The child in Figure 6c is pulling at some angle θ. Some of her force moves the wagon and some of the force tries to lift it. We define the **work** done in such a situation as the product of the component of the force in the direction of motion, **p**, and the distance traveled. We need to find **p** as shown in Figure 7.

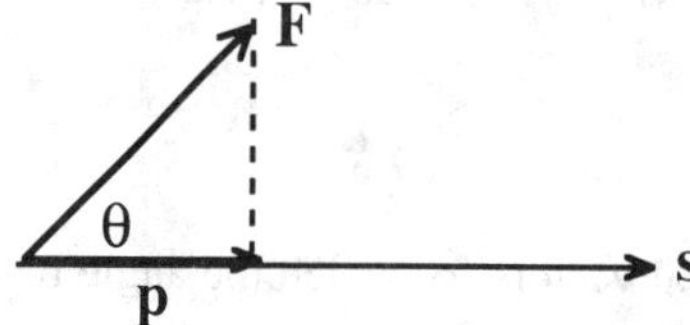

Figure 7

In Figure 7 we see that a force **F** applied at an angle θ to the direction of motion given by **s** has a component in the direction of motion. Since the component of **F** that does the work is

$$\|\mathbf{p}\| = \|\mathbf{F}\| \cos \theta$$

and

$$\text{work} = (\text{force})(\text{distance})$$
$$W = (\|\mathbf{F}\| \cos \theta)\, \|\mathbf{s}\|$$
$$W = \mathbf{F} \cdot \mathbf{s}$$

Example 6 *Finding the amount of work*

Find the work done by the girl in Figure 6c if she exerts a force of 35 lb at an angle of 60° and moves the wagon 10 ft.

Solution

We find the horizontal and vertical components of the force vector, then take the dot product.

$$x = \|\mathbf{F}\| \cos 60° = \frac{35}{2} = 17.5$$

$$y = \|\mathbf{F}\| \sin 60° = \frac{35\sqrt{3}}{2} \approx 30.3$$

Figure 8 shows a representation of the work.

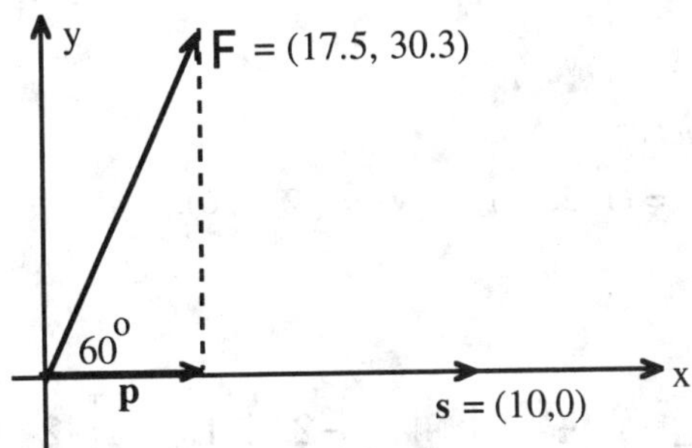

Figure 8

$$\begin{aligned} W &= \mathbf{F}\cdot\mathbf{s} \\ &= \left(\frac{35}{2}, \frac{35\sqrt{3}}{2}\right)\cdot(10,0) \\ &= \left(\frac{35}{2}\right)(10) + \left(\frac{35\sqrt{3}}{2}\right)(0) = 175 \end{aligned}$$

The amount of work done is 175 ft-lb.

Without vectors and the dot product, the law of cosines could have been used to find the component of **F** in the direction of motion.

The following definition of the **cross product** of a pair of vectors is a genuine multiplication of two vectors. This product yields a vector as the result, but as we shall see, this multiplication is unusual because it is not commutative. This lack of commutativity was a major stumbling block for the mathematicians who eventually contributed to the creation of vectors. Since commutativity is so basic to mathematics, no one thought a system could fail to have the commutative property.

Definition of Cross Product

For vectors $\mathbf{v}_1 = (a_1, a_2, a_3)$ and $\mathbf{v}_2 = (b_1, b_2, b_3)$ in R^3 the cross product $\mathbf{v}_1 \times \mathbf{v}_2$ is

$$\mathbf{v}_1 \times \mathbf{v}_2 = \begin{vmatrix} \mathbf{i} & \mathbf{j} & \mathbf{k} \\ a_1 & a_2 & a_3 \\ b_1 & b_2 & b_3 \end{vmatrix}$$

$$= (a_2b_3 - a_3b_2)\mathbf{i} - (a_1b_3 - a_3b_1)\mathbf{j} + (a_1b_2 - a_2b_1)\mathbf{k}$$

Example 7 *Finding cross products*

Find $\mathbf{v}_1 \times \mathbf{v}_2$ and $\mathbf{v}_2 \times \mathbf{v}_1$ for $\mathbf{v}_1 = (1, 3, -2)$ and $\mathbf{v}_2 = (-2, 4, 0)$.

Solution

$$\mathbf{v}_2 \times \mathbf{v}_1 = \begin{vmatrix} \mathbf{i} & \mathbf{j} & \mathbf{k} \\ -2 & 4 & 0 \\ 1 & 3 & -2 \end{vmatrix} = \mathbf{i}\begin{vmatrix} 4 & 0 \\ 3 & -2 \end{vmatrix} - \mathbf{j}\begin{vmatrix} -2 & 0 \\ 1 & -2 \end{vmatrix} + \mathbf{k}\begin{vmatrix} -2 & 4 \\ 1 & 3 \end{vmatrix}$$

$$= (-8, -4, -10)$$

$$\mathbf{v}_1 \times \mathbf{v}_2 = \begin{vmatrix} \mathbf{i} & \mathbf{j} & \mathbf{k} \\ 1 & 3 & -2 \\ -2 & 4 & 0 \end{vmatrix} = \mathbf{i}\begin{vmatrix} 3 & -2 \\ 4 & 0 \end{vmatrix} - \mathbf{j}\begin{vmatrix} 1 & -2 \\ -2 & 0 \end{vmatrix} + \mathbf{k}\begin{vmatrix} 1 & 3 \\ -2 & 4 \end{vmatrix}$$

$$= (8, 4, 10)$$

$\mathbf{v}_2 \times \mathbf{v}_1 = -(\mathbf{v}_1 \times \mathbf{v}_2)$

Notice that $\mathbf{v}_1 \times \mathbf{v}_2 \neq \mathbf{v}_2 \times \mathbf{v}_1$, which proves that this multiplication is not commutative. The cross product $\mathbf{v}_1 \times \mathbf{v}_2$ is orthogonal to both vectors $\mathbf{v}_1$ and $\mathbf{v}_2$. Consider the dot products of $\mathbf{v}_1$ and $\mathbf{v}_2$ with the cross product $\mathbf{v}_1 \times \mathbf{v}_2$.

$$\mathbf{v}_1 \cdot (\mathbf{v}_1 \times \mathbf{v}_2) = (1, 3, -2) \cdot (8, 4, 10) \quad \textit{Substitute } (v_1 \times v_2) = (8,4,10).$$
$$= 8 + 12 - 20$$
$$= 0$$

and

$$\mathbf{v}_2 \cdot (\mathbf{v}_1 \times \mathbf{v}_2) = (-2, 4, 0) \cdot (8, 4, 10)$$
$$= -16 + 16 + 0$$
$$= 0$$

The cross product is said to be the **normal** to the plane formed by $\mathbf{v}_1$ and $\mathbf{v}_2$. Notice that $\mathbf{v}_2 \times \mathbf{v}_1 = -(\mathbf{v}_1 \times \mathbf{v}_2)$, which makes $\mathbf{v}_2 \times \mathbf{v}_1$ a normal to the same plane but going in the opposite direction. If one cross product is above the plane, the other is below.

Figure 9 shows that the direction of $\mathbf{v}_1 \times \mathbf{v}_2$ is found by curling the fingers of your right hand from $\mathbf{v}_1$ toward $\mathbf{v}_2$. Then your thumb points in the direction of the normal vector $\mathbf{N} = \mathbf{v}_1 \times \mathbf{v}_2$.

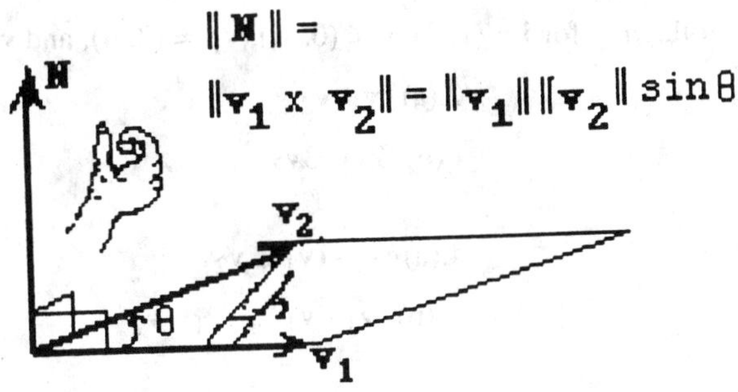

Figure 9

Notice that $\mathbf{v}_2 \times \mathbf{v}_1$ requires curling your fingers from $\mathbf{v}_2$ to $\mathbf{v}_1$ and your thumb points in the opposite direction.

The next theorem is analogous to the one for dot products.

$$\|\mathbf{v}_1 \times \mathbf{v}_2\| = \|\mathbf{v}_1\|\,\|\mathbf{v}_2\|\,|\sin\theta|$$

Consider the area of the parallelogram in Figure 9.

$$A = bh \text{ and } b = \|\mathbf{v}_1\| \Rightarrow A = \|\mathbf{v}_1\|\,h$$

and

$$\sin\theta = \frac{h}{\|\mathbf{v}_2\|} \Rightarrow h = \|\mathbf{v}_2\| \sin\theta$$

Hence, substituting $h = \|\mathbf{v}_2\| \sin\theta$ yields

$$A = \|\mathbf{v}_1\|\,\|\mathbf{v}_2\| \sin\theta$$

Although the value of the magnitude of the cross product of two vectors is the same number as the area of the parallelogram formed by the vectors, the units of measure are different. While area is measured in square inches, square centimeters, and so on, $\mathbf{v}_1 \times \mathbf{v}_2$ has no dimensions.

Exercises 9.3

Develop the Concepts

For Exercises 1 to 4 evaluate the following for $\mathbf{i} = (1, 0)$, $\mathbf{j} = (0, 1,)$, $\mathbf{v}_1 = (3, 4)$, and $\mathbf{v}_2 = (-2, 5)$.

1.(a) $\mathbf{i} \cdot \mathbf{i}$

(b) $\mathbf{i} \cdot \mathbf{j}$

2.(a) $\mathbf{v}_1 \cdot \mathbf{v}_2$

(b) $2\mathbf{v}_1 \cdot 3\mathbf{v}_2$

3.(a) $\mathbf{v}_1 \cdot \mathbf{v}_2$

(b) $\mathbf{v}_2 \cdot \mathbf{v}_1$

4.(a) $\mathbf{v}_1 \cdot (\mathbf{v}_1 + \mathbf{v}_2)$

(b) $\mathbf{v}_1 \cdot \mathbf{v}_1 + \mathbf{v}_1 \cdot \mathbf{v}_2$

For Exercises 5 to 8 evaluate the following for $\mathbf{i} = (1, 0, 0)$, $\mathbf{j} = (0, 1, 0)$, $\mathbf{k} = (0, 0, 1)$, $\mathbf{v}_1 = (1, -1, 3)$ and $\mathbf{v}_2 = (0, 3, 1)$.

5.(a) $\mathbf{i} \cdot \mathbf{j}$

(b) $\mathbf{i} \cdot \mathbf{k}$

(c) $\mathbf{j} \cdot \mathbf{k}$

6.(a) $2(\mathbf{v}_1 \cdot \mathbf{v}_2)$

(b) $(2\mathbf{v}_1) \cdot (\mathbf{v}_2)$

(c) $(2\mathbf{v}_1) \cdot (3\mathbf{v}_2)$

7.(a) $\mathbf{i} \times \mathbf{j}$

(b) $\mathbf{j} \times \mathbf{k}$

(c) $\mathbf{i} \times \mathbf{k}$

8.(a) $\mathbf{v}_1 \times \mathbf{v}_2$

(b) $\mathbf{v}_2 \times \mathbf{v}_1$

(c) $\mathbf{v}_1 \times \mathbf{v}_1$

For Exercises 9 to 16 let $\mathbf{v}_1 = (1, 6, 4)$, $\mathbf{v}_2 = (-1, 3, -2)$ and $\mathbf{v}_3 = (1, -1, 2)$. Compare the results in parts (a) and (b).

9.(a) $\mathbf{v}_1 \cdot (\mathbf{v}_2 + \mathbf{v}_3)$

(b) $\mathbf{v}_1 \cdot \mathbf{v}_2 + \mathbf{v}_1 \cdot \mathbf{v}_3$

10.(a) $\mathbf{v}_1 \times (\mathbf{v}_2 + \mathbf{v}_3)$

(b) $\mathbf{v}_1 \times \mathbf{v}_2 + \mathbf{v}_1 \times \mathbf{v}_3$

11.(a) $2(\mathbf{v}_1 \cdot \mathbf{v}_2)$

(b) $(2\mathbf{v}_1) \cdot (2\mathbf{v}_2)$

12.(a) $2(\mathbf{v}_1 \times \mathbf{v}_2)$

(b) $(2\mathbf{v}_1) \times \mathbf{v}_2$

13.(a) $\mathbf{v}_1 \cdot (\mathbf{v}_2 - \mathbf{v}_3)$

(b) $\mathbf{v}_1 \cdot \mathbf{v}_2 - \mathbf{v}_1 \cdot \mathbf{v}_3$

14.(a) $\mathbf{v}_1 \times (\mathbf{v}_2 + \mathbf{v}_3)$

(b) $(\mathbf{v}_1 \times \mathbf{v}_2) + (\mathbf{v}_1 \times \mathbf{v}_3)$

15.(a) $\mathbf{v}_1 \cdot (\mathbf{v}_2 \cdot \mathbf{v}_3)$

(b) $(\mathbf{v}_1 \cdot \mathbf{v}_2) \cdot \mathbf{v}_3$

16.(a) $\mathbf{v}_1 \cdot (\mathbf{v}_2 \times \mathbf{v}_3)$

(b) $(\mathbf{v}_1 \cdot \mathbf{v}_2) \times \mathbf{v}_3$

For Exercises 17 to 24 let $\mathbf{v}_1 = (1, -2, 3)$, $\mathbf{v}_2 = (0,1,-2)$, $\mathbf{v}_3 = (-2, -1, 4)$, $\mathbf{0} = (0, 0, 0)$, $s = 3$, and $t = 2$ to illustrate each property.

17. $\mathbf{v}_1 \times \mathbf{v}_2 = -\mathbf{v}_2 \times \mathbf{v}_1$

18. $\mathbf{v}_1 \times (\mathbf{v}_2 + \mathbf{v}_3) = (\mathbf{v}_1 \times \mathbf{v}_2) + (\mathbf{v}_1 \times \mathbf{v}_3)$

19. $(\mathbf{v}_1 + \mathbf{v}_2) \times \mathbf{v}_3 = (\mathbf{v}_1 \times \mathbf{v}_3) + (\mathbf{v}_2 \times \mathbf{v}_3)$

20. $(s\mathbf{v}_1) \times (t\mathbf{v}_2) = (st)(\mathbf{v}_1 \times \mathbf{v}_2)$

21. $\mathbf{0} \times \mathbf{v}_1 = \mathbf{v}_1 \times \mathbf{0} = \mathbf{0}$

22. $(\mathbf{v}_1 \times \mathbf{v}_1) = \mathbf{0}$

23. $\mathbf{v}_1 \cdot (\mathbf{v}_2 \times \mathbf{v}_3) = (\mathbf{v}_1 \times \mathbf{v}_2) \cdot \mathbf{v}_3$

24. $\mathbf{v}_1 \cdot (\mathbf{v}_2 + \mathbf{v}_3) = \mathbf{v}_1 \cdot \mathbf{v}_2 + \mathbf{v}_1 \cdot \mathbf{v}_3$

For Exercises 25 to 28 find the value(s) of k such that each condition is satisfied.

25. $\mathbf{v}_1 = (3, 2)$, $\mathbf{v}_2 = (k, -2)$, $\mathbf{v}_1 \cdot \mathbf{v}_2 = 0$ (They are orthogonal.)

26. $\mathbf{v}_1 = (1, -2k, 3)$, $\mathbf{v}_2 = (2, -1, 4)$, $\mathbf{v}_1 \cdot \mathbf{v}_2 = 0$

27. $\mathbf{v}_1 = (2k, k, 3)$, $\mathbf{v}_2 = (2, 1, 3)$, $\mathbf{v}_1 \times \mathbf{v}_2 = \mathbf{0}$

28. $\mathbf{v}_1 = (1, k, -2)$, $\mathbf{v}_2 = (2, -1, 1)$, $(\mathbf{v}_1 \times \mathbf{v}_2) \cdot \mathbf{v}_1 = 0$

For Exercises 29 to 36 find the component of $\mathbf{v}_1$ in the direction of $\mathbf{v}_2$.

*29. $\mathbf{v}_1 = (3, 5)$, $\mathbf{v}_2 = (1, 0)$

30. $\mathbf{v}_1 = (3, 5)$, $\mathbf{v}_2 = (0, 1)$

31. $\mathbf{v}_1 = (1, 3, 7)$, $\mathbf{v}_2 = (1, 0, 0)$

32. $\mathbf{v}_1 = (1, -2, 1)$, $\mathbf{v}_2 = (-1, 0, 3)$

33. $\mathbf{v}_1 = (1, 3)$, $\mathbf{v}_2 = (5, 12)$

34. $\mathbf{v}_1 = (1, 1)$, $\mathbf{v}_2 = (5, 1)$

35. $\mathbf{v}_1 = (1, 3, 5)$, $\mathbf{v}_2 = (0, 0, 5)$

36. $\mathbf{v}_1 = (1, 7, -2)$, $\mathbf{v}_2 = (2, -1, 3)$

Apply the Concepts

*37. The Swing Problem You are swinging on a swing attached with 18-ft chains. If you jumps off when you are 8 ft off the ground and your velocity is given by $\mathbf{v} = (12,8)$ ft/sec, how far will you fly?

38. The Rope-and-Ball Problem Suppose that you are practicing for a sports competition by throwing a weighted ball on the end of a 36-in. rope. When you let the rope go, the velocity of the ball in ft/sec is given by the vector $\mathbf{v} = (-17,16)$. How far will the ball fly if it is released 6 ft above the ground?

39. The Work Problem Your friend applies a force of 50 lb by pulling on the tongue of a wagon at an angle of $\theta = 30°$ with the horizontal direction of motion. Find the work done if he moves the wagon 10 ft. [Hint: The projection of **F** onto **S** is the component of **F** in the direction of motion.]

40. The More-Work Problem Find the work done in pulling a wagon up a 15° hill that is $\frac{1}{4}$ mi long if a force of 2000 lb is applied vertically.

Extend the Concepts

For Exercises 41 to 44 prove the following theorems for three dimensional vectors $\mathbf{v}_1$, $\mathbf{v}_2$ and $\mathbf{v}_3$, by examining the components of the results. Let $\mathbf{v}_1 = (a_1, a_2, a_3)$, $\mathbf{v}_2 = (b_1, b_2, b_3)$, and $\mathbf{v}_3 = (c_1, c_2, c_3)$.

41. $\mathbf{v}_1 \cdot (\mathbf{v}_2 + \mathbf{v}_3) = \mathbf{v}_1 \cdot \mathbf{v}_2 + \mathbf{v}_1 \cdot \mathbf{v}_3$

42. $\mathbf{v}_1 \times \mathbf{v}_1 = \mathbf{0}$

43. $\|\mathbf{v}_1\|^2 = \mathbf{v}_1 \cdot \mathbf{v}_1$

44. $(\mathbf{v}_1 \times \mathbf{v}_2) \cdot \mathbf{v}_1 = 0$

9.4 Equations of Lines and Planes in Three Dimensions

Finding Equations of Planes in Three Space
Finding Equations of Lines in Three Space

Now we are ready to study linear models in three dimensions. To build these models we first study finding the equations of lines and planes.

Finding Equations of Planes in Three Space

In the last section we saw that the cross product of two vectors is normal to the plane formed by those vectors and to every parallel plane. Using this normal

$$\mathbf{N} = \mathbf{v_1} \times \mathbf{v_2} = (A, B, C)$$

we can describe all the points $P(x, y, z)$ in the plane through $P_0(x_0, y_0, z_0)$. If

$$\mathbf{v_1} = \mathbf{P_0P} = (x - x_0, y - y_0, z - z_0)$$

is a vector in the plane, it must be orthogonal to $\mathbf{N}$ since $\mathbf{N}$ is orthogonal to every vector in the plane. Then

$$\begin{aligned} \mathbf{N} \cdot \mathbf{v_1} &= 0 \\ (A, B, C) \cdot (x - x_0, y - y_0, z - z_0) &= 0 \\ A(x - x_0) + B(y - y_0) + C(z - z_0) &= 0 \end{aligned}$$

This completes the proof of the following theorem.

> The equation of the plane through $P_0(x_0, y_0, z_0)$ with $\mathbf{N} = (A, B, C)$ is
> $$A(x-x_0) + B(y-y_0) + C(z-z_0) = 0$$

Example 1 *Finding the equation of a plane*

Find the equation of the plane containing the points $P_0(40, 0, 0)$, $P_1(10, 5, 0)$, and $P_2(0,0,8)$.

Solution

First we need a normal $\mathbf{N}$ to the plane. Let $\mathbf{v_1} = \mathbf{P_0P_1}$ and $\mathbf{v_2} = \mathbf{P_0P_2}$. Then

$$\mathbf{v_1} \times \mathbf{v_2} = \begin{vmatrix} \mathbf{i} & \mathbf{j} & \mathbf{k} \\ -30 & 5 & 0 \\ -40 & 0 & 8 \end{vmatrix}$$

$$= \mathbf{i}\begin{vmatrix} 5 & 0 \\ 0 & 8 \end{vmatrix} - \mathbf{j}\begin{vmatrix} -30 & 0 \\ -40 & 8 \end{vmatrix} + \mathbf{k}\begin{vmatrix} -30 & 5 \\ -40 & 0 \end{vmatrix}$$

$$= (40, 240, 200)$$

We will use $\mathbf{N} = (\frac{1}{40})(\mathbf{v_1} \times \mathbf{v_2}) = (1, 6, 5)$ for convenience, since any nonzero scalar multiple of a normal is still normal. Multiplying by a scalar does not change a vector's direction. The equation of the plane is

$$\begin{aligned} A(x - x_0) + B(y - y_0) + C(z - z_0) &= 0 \\ 1(x - 40) + 6(y - 0) + 5(z - 0) &= 0 \\ x + 6y + 5z &= 40 \end{aligned}$$

Finding Equations of Lines in Three Space

We know from our previous work with systems of equations that a system of two equations with three unknowns cannot have a unique solution. Two linear equations representing two planes may intersect in a line. This line contains the solutions to the system consisting of their two equations. We can find the equation of this line by solving the system of equations.

Example 2 *Finding the equation of a line of intersection*

Find the equation of the line of intersection of the planes

$$x + 3y + z = 12 \qquad (1)$$

$$3x + 4y - 2z = 26 \qquad (2)$$

Solution

$$-3x - 9y - 3z = -36 \qquad \textit{Multiply equation (1) by -3.}$$

$$\underline{3x + 4y - 2z = 26} \qquad \textit{Add the equations.}$$

$$-5y - 5z = -10$$

$$\mathbf{y + z = 2} \qquad \textit{Call this is equation (3).}$$

$$x + 3y + z = 12 \qquad \textit{This is equation (1).}$$

$$\underline{-3y - 3z = -6} \qquad \textit{Multiply equation (3) by -3.}$$

$$\mathbf{x - 2z = 6} \qquad \textit{Add the equations to create equation (4).}$$

This last step is as far as we can go. If we want one of the points on the line, we choose a value for z and solve for x and y. For example, if z = 0, then

$$\mathbf{x - 2z = 6} \Rightarrow x = 2z + 6 = 2(0) + 6 = 6$$

and

$$\mathbf{y + z = 2} \Rightarrow y = -z + 2 = 0 + 2 = 2$$

We can rewrite equations (3) and (4) by solving for z:

$$z = \frac{x - 6}{2} \quad \text{and} \quad z = \frac{y - 2}{-1}$$

This creates the **symmetric form** for the equation of a line

$$\frac{x - 6}{2} = \frac{y - 2}{-1} = \frac{z - 0}{1}$$

The symmetric form for the equation of the line in Example 2 is written in the form

$$\frac{x - x_0}{a} = \frac{y - y_0}{b} = \frac{z - z_0}{c}$$

where $\mathbf{v} = (a, b, c)$ is a vector in the direction of the line and $P_0 (x_0, y_0, z_0)$ is a point on the line. In Example 2 the vector $\mathbf{v}_1 = (a, b, c) = (2, -1, 1)$ is a vector in the direction of our line. The point $P_0 (x_0, y_0, z_0) = P_0 (6, 2, 0)$ is a point on the line. Figure 1 shows this line as the intersection of two planes.

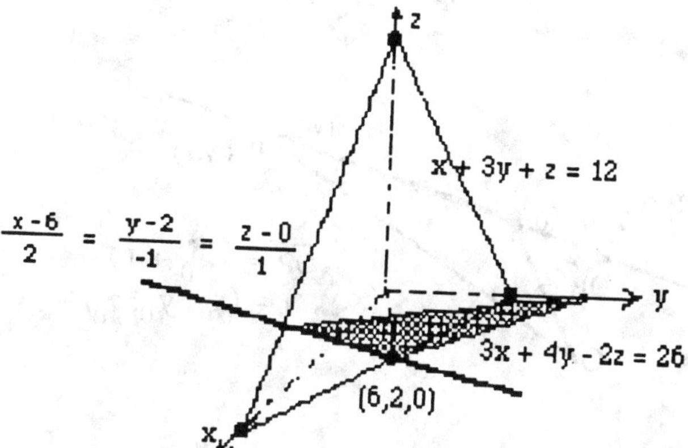

Figure 1

We could repeat this process for any value of z, but instead we will let z be a fixed but arbitrary value t. If z = t, then

$$x = 2z + 6 = 2t + 6$$

and

$$y = -z + 2 = -t + 2$$

Now we know all points on the line. They are

$$(2t + 6, -t + 2, t)$$

where t is any real number. Since t is a parameter, this resulting form for the equation of line is called the **parametric form**.

The next theorems summarize these ideas.

Symmetric Form

The symmetric form for the equation of a line through $P_0(x_0, y_0, z_0)$ in the direction of $\mathbf{v}_1 = (a, b, c)$ is

$$\frac{x - x_0}{a} = \frac{y - y_0}{b} = \frac{z - z_0}{c}$$

Parametric Form

The parametric form for the equation of a line through $P_0(x_0, y_0, z_0)$ in the direction of $\mathbf{v}_1 = (a, b, c)$ is

$$x = x_0 + at$$
$$y = y_0 + bt$$
$$z = z_0 + ct$$

Figure 2 shows a point on the line P_0 and its associated position vector. Notice the vector $\mathbf{v}$ in the direction of the line. If point P(x, y, z) is any point on the line, there exists a scalar t such that

$$\overrightarrow{OP} = \overrightarrow{OP_0} + t\,\mathbf{v}$$

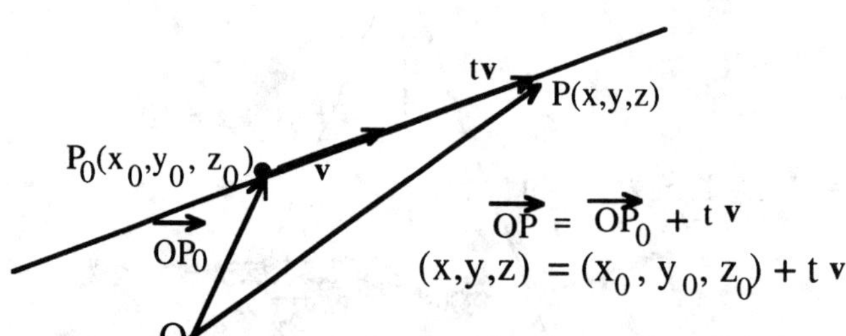

Figure 2

Exercises 9.4

Develop the Concepts

For Exercises 1 to 4 find the equation of each plane.

1. The plane contains P(1, 2, 3) with normal **N** = (2, 4, 4).
2. The plane contains P(1, 2, 3) with normal **N** = (-1, -2, -2).
3. The plane contains points $P_1(2, 1, 3)$, $P_2(-2, 1, 1)$, and $P_3(4, -2, -2)$.
4. The plane contains points $P_1(0, 0, 0)$, $P_2(3, -1, 2)$, and $P_3(-2,-4, 5)$.

For Exercises 5 to 8 find the symmetric and parametric forms of the equation of each of the following lines. Find two additional points on each line.

*5. The line passing through (1, 2, 3) in the direction of **v** = (2, 4,- 2).

6. The line passing through (1, 2, 3) in the direction of **v** = (-1, -2, 1).
7. The line containing the points $P_1(0, 0, 0)$ and $P_2(3, 1, 2)$.
8. The line containing the points $P_1(1, -2, 1)$ and $P_2(3, -1, 2)$.

For Exercises 9 to 12 find the equation of each line.

9. The line in the direction of the x axis passing through the point (2, 1, 3).

10. The line that is the intersection of the two planes 2x - 3y + 4z = 1 and 3x + 5y - 3z = 4.

11. The line through the point (1, -1, 1) and perpendicular to the plane 3x - 2y - z = 4.

12. The line through the point (1, -1, 2) in the direction of the normal to the plane containing the three points $P_1(1, -1, 2)$, $P_2(0, 1, 3)$, and $P_3(1, 1, 2)$.

For Exercises 13 to 20 find the equation of each plane.

13. The plane perpendicular to the line $\frac{x - 1}{2} = \frac{y + 3}{4} = \frac{z - 1}{-1}$ that contains the point (1, -3,1).

14. The plane perpendicular to the line x = 1 - 2t, y = 3 + t, and z = 4 that contains the point (-1, 4, 4).

*15. The plane parallel to the plane 3x - y + 4z = 5 that contains the point (1, -1, 3).

16. The plane that contains the lines $\frac{2 - x}{3} = \frac{y + 1}{4} = \frac{z - 2}{1}$ and $\frac{x + 1}{2} = \frac{y - 1}{3} = \frac{2z - 1}{4}$.

17. The plane that is perpendicular to the plane 3x - 2y + z = 3 and contains the points $P_1(1,- 1, 0)$ and $P_2(-3, 1, 2)$.

18. The plane that contains the point (-3, -1, 1) and the line x = 3 - 2t, y = 4 + t, and z = 2t - 5.

19. The plane that contains $\frac{x - 1}{3} = \frac{3 - y}{4} = \frac{z + 1}{5}$ and is parallel to $\frac{x - 1}{3} = \frac{2y - 3}{4} = \frac{z + 3}{5}$.

20. The plane perpendicular to the planes $2x - y - 3z = 1$ and $x + 2y - z = 3$ that also contains the point (-1, 0, -2).

Extend the Concepts

21. Find the measure of the smaller angle formed by the two intersecting planes $3x - y + z = 5$ and $2x + y - 3z = 1$.

22. Find the measure of the two supplementary angles ($a_1 + a_2 = 180°$) formed by the planes $3x + 2y - z = 7$ and $2x + y = 2$.

23. Find the distance from the point (1, -1, 3) to the xy plane (plane $z = 0$).

24. Find the distance from the point (-2, 1, 0) to the plane $2x - y + z = 1$.

25. Find the perpendicular distance between the two planes $3x - 2y + 5z = 3$ and $6x - 4y + 10z = -1$.

26. Find the perpendicular distance from the origin to the line $\frac{2x - 1}{4} = \frac{y - 3}{5} = \frac{1 - z}{2}$.

27. Draw the graph of the plane $3x + 2y + z = 18$ and a normal to the plane from the point (2, 4, 4).

28. Draw the graph of the line through $P_1(4, 1, 3)$ and $P_2(1, 8, 5)$. Then draw and label a vector in the direction of the line from the point $P_1(4, 1, 3)$.

29. Prove that if a, b, and c are nonzero x, y, and z intercepts, respectively, of a plane, its equation is $\frac{x}{a} + \frac{y}{b} + \frac{z}{c} = 1$.

30. Prove that the perpendicular distance from the point $P_0(x_0, y_0, z_0)$ to the plane $ax + by + cz + d = 0$ is given by

$$\frac{|ax_0 + by_0 + cz_0 + d|}{\sqrt{a^2 + b^2 + c^2}}$$

31. Prove that the perpendicular distance between the two parallel planes $ax + by + cz = d_1$ and $ax + by + cz = d_2$ is given by

$$\frac{|d_1 - d_2|}{\sqrt{a^2 + b^2 + c^2}}$$

A Guided Review of Chapter 9

9.1 Introduction to Matrices

For Exercises 1 to 4 perform the indicated operations whenever possible. If not possible, write "not conformable."

1.(a) $\begin{bmatrix} 0 & -1 \\ -1 & -5 \end{bmatrix} + \begin{bmatrix} 1 & 1 \\ 2 & 5 \end{bmatrix}$

(b) $\begin{bmatrix} 1 & 1 \\ 2 & 5 \end{bmatrix} + \begin{bmatrix} 0 & -1 \\ -1 & 5 \end{bmatrix}$

2.(a) $\left\{\begin{bmatrix} -2 & 0 & 0 \\ -2 & 0 & 2 \end{bmatrix} + \begin{bmatrix} 1 & 2 & 1 \\ 2 & -1 & -2 \end{bmatrix}\right\} + \begin{bmatrix} -2 & 4 & -1 \\ -2 & -1 & 2 \end{bmatrix}$

(b) $\begin{bmatrix} -2 & 0 & 0 \\ -2 & 0 & 2 \end{bmatrix} + \left\{\begin{bmatrix} 1 & 2 & 1 \\ 2 & -1 & -2 \end{bmatrix} + \begin{bmatrix} -2 & 4 & -1 \\ -2 & -1 & 2 \end{bmatrix}\right\}$

For Exercises 3 and 4, compare the two parts.

3. Compare αA and $A\alpha$ for $\alpha = 12$ and $A = \begin{bmatrix} 12 & 0 \\ -12 & 12 \end{bmatrix}$

4. Compare $(\alpha\beta)\,A$ and $\alpha(\beta\, A)$ for $\alpha = 2$, $\beta = 5$, and $A = \begin{bmatrix} -1 & -1 \\ 0 & -1 \end{bmatrix}$

9.1 Introduction to Matrices

For Exercises 5 and 6 find the determinants of the given matrix.

5.(a) $\begin{bmatrix} 1 & 0 \\ 0 & 1 \end{bmatrix}$

(b) $\begin{bmatrix} 3 & 2 \\ 6 & 4 \end{bmatrix}$

6.(a) $\begin{bmatrix} 5 & 0 & 0 \\ -1 & 2 & 0 \\ -2 & 15 & 12 \end{bmatrix}$

(b) $\begin{bmatrix} 5 & 0 & 0 \\ -2 & 15 & 12 \\ -1 & 2 & 0 \end{bmatrix}$

For Exercises 7 to 10 find each determinant for

$A = \begin{bmatrix} 2 & -2 \\ 2 & 2 \end{bmatrix}$ $B = \begin{bmatrix} 0 & 1 & 1 \\ 1 & 2 & -1 \\ -2 & 1 & -1 \end{bmatrix}$ and $C = \begin{bmatrix} 0 & 1 & -2 \\ 1 & 2 & 1 \\ 1 & -1 & -1 \end{bmatrix}$

(A^T is the transpose of A. If $A = \{a_{ij}\}$ then $A^T = \{a_{ji}\}$. That is, the rows of A are the columns of A^T.)

7.(a) det A

(b) det A^T

8.(a) (det B)(det C)

(b) det BC

9. (a) (det C)(det B)

(b) det(CB)

10.(a) $\det(A + A^T)$

(b) $\det A + \det A^T$

9.3 Vectors in Two and Three Dimensions

For Exercises 11 to 14 evaluate for $\mathbf{i} = (1, 0)$, $\mathbf{j} = (0, 1,)$, $\mathbf{v}_1 = (-7, 4)$, and $\mathbf{v}_2 = (-4, 2)$.

11.(a) $\mathbf{i} \cdot \mathbf{i}$

(b) $\mathbf{i} \cdot \mathbf{j}$

12.(a) $\mathbf{v}_1 \cdot \mathbf{v}_2$

(b) $2\mathbf{v}_1 \cdot 3\mathbf{v}_2$

13.(a) $\mathbf{v}_1 \cdot \mathbf{v}_2$

(b) $\mathbf{v}_2 \cdot \mathbf{v}_1$

14.(a) $\mathbf{v}_1 \cdot (\mathbf{v}_1 + \mathbf{v}_2)$

(b) $\mathbf{v}_1 \cdot \mathbf{v}_1 + \mathbf{v}_1 \cdot \mathbf{v}_2$

For Exercises 15 to 18 evaluate the following for $\mathbf{i} = (1, 0, 0)$, $\mathbf{j} = (0, 1, 0)$, $\mathbf{k} = (0,0,1)$, $\mathbf{v}_1 = (4, 0, 3)$, and $\mathbf{v}_2 = (4, 1, 0)$.

15.(a) $\mathbf{i} \cdot \mathbf{j}$

(b) $\mathbf{i} \cdot \mathbf{k}$

(c) $\mathbf{j} \cdot \mathbf{k}$

16.(a) $2(\mathbf{v}_1 \cdot \mathbf{v}_2)$

(b) $(2\mathbf{v}_1) \cdot (\mathbf{v}_2)$

(c) $(2\mathbf{v}_1) \cdot (3\mathbf{v}_2)$

17.(a) $\mathbf{i} \times \mathbf{j}$

(b) $\mathbf{j} \times \mathbf{j}$

(c) $\mathbf{k} \times \mathbf{k}$

18.(a) $\mathbf{v}_1 \times \mathbf{v}_2$

(b) $\mathbf{v}_2 \times \mathbf{v}_1$

(c) $\mathbf{v}_1 \times \mathbf{v}_1$

For Exercises 19 and 20 find the value(s) of k such that each condition is satisfied.

19. $\mathbf{v}_1 = (1, 2)$, $\mathbf{v}_2 = (-2, k)$, $\mathbf{v}_1 \cdot \mathbf{v}_2 = 0$ (They are orthogonal.)

20. $\mathbf{v}_1 = (2, -1, 3k)$, $\mathbf{v}_2 = (3, -2, 1)$, $\mathbf{v}_1 \cdot \mathbf{v}_2 = 0$

For Exercises 21 and 22 find $\dfrac{\mathbf{v}_1 \cdot \mathbf{v}_2}{\|\mathbf{v}_2\|^2}\mathbf{v}_2$.

21. $\mathbf{v}_1 = (1, 1)$, $\mathbf{v}_2 = (6, 8)$

22. $\mathbf{v}_1 = (0, 3, 0)$, $\mathbf{v}_2 = (1, 1, 1)$

23. <u>The Swing Problem</u> You are swinging on a swing attached with 20-ft chains. If you jump off when you are 5 ft off the ground and your velocity is given by $\mathbf{v} = (4, 3)$ ft/sec, how far will you fly?

24. <u>The Work Problem</u> You apply a force of 100 lb by pulling on the tongue of a wagon at an angle of $\theta = 45°$ with the horizontal direction of motion. Find the work done if you move the wagon 10 ft.

9.4 Equations of Lines and Planes in Three Dimensions

For Exercises 25 to 28 find the equation of each plane.

25. The plane contains $P(1, 2, 1)$ with normal $\mathbf{N} = (1, 1, 0)$.

26. The plane contains $P(1, 2, 1)$ with normal $\mathbf{N} = (-1, -1, 0)$.

27. The plane contains points $P_1(0, 1, 0)$, $P_2(1, 0, 0)$, and $P_3(0, 0, 1)$.

28. The plane contains points $P_1(0, 0, 0)$, $P_2(1, 0, 0)$, and $P_3(0, 0, 1)$.

For Exercises 29 to 32 find the symmetric and parametric forms of the equation of each line. Find two additional points on each line.

29. The line passing through (1, 1, 1) in the direction of $\mathbf{v} = (1, \sqrt{3}, 0)$.

30. The line passing through (1, 2, 3) in the direction of $\mathbf{v} = (-\sqrt{3}, 0, 1)$.

31. The line containing the points $P_1(0, 0, 0)$ and $P_2(1, 1, 1)$.

32. The line containing the points $P_1(1, 0, 1)$ and $P_2(0, -1, 0)$.

For Exercises 33 and 34 find the equation of each plane.

33. The plane perpendicular to the line $\frac{x+1}{-2} = \frac{y-1}{2} = z - 1$ that contains the point (1, 0, 1).

34. The plane perpendicular to the line $x = t$, $y = 1 + t$, and $z = 1$ that contains the point (-1, 1, 0).

Chapter 9 Test

For Exercises 1 and 2 perform the indicated operations whenever possible.

1. $\begin{bmatrix} 2 & -4 \\ -2 & -4 \end{bmatrix} + \begin{bmatrix} -2 & 4 \\ 2 & 4 \end{bmatrix}$ 2. $\begin{bmatrix} -2 & 2 & -2 \\ 2 & 2 & -2 \end{bmatrix} + \begin{bmatrix} 2 & -2 & 2 \\ -2 & -2 & 2 \end{bmatrix} + \begin{bmatrix} -2 & 2 & -2 \\ -2 & -2 & 2 \end{bmatrix}$

For Exercises 3 and 4, find the determinant of the matrix.

3.(a) $\begin{bmatrix} 1 & 1 \\ 1 & 1 \end{bmatrix}$ (b) $\begin{bmatrix} 6 & 9 \\ 2 & 3 \end{bmatrix}$ 4.(a) $\begin{bmatrix} 2 & 1 & 2 \\ 1 & 2 & 1 \\ -2 & 1 & -2 \end{bmatrix}$ (b) $\begin{bmatrix} 1 & 2 & 1 \\ 2 & 1 & 2 \\ -2 & 1 & -2 \end{bmatrix}$

For Exercises 5 and 6, find each determinant for

$A = \begin{bmatrix} 1 & -1 \\ 2 & 0 \end{bmatrix}$ $B = \begin{bmatrix} 1 & 2 & 1 \\ 1 & 2 & -1 \\ 1 & 2 & -1 \end{bmatrix}$ $C = \begin{bmatrix} 1 & 1 & 1 \\ 2 & 1 & 1 \\ 1 & -1 & -1 \end{bmatrix}$

5.(a) det A (b) det A^T 6.(a) (det B)(det C) (b) det BC

(A^T is the transpose of A. If $A = \{a_{ij}\}$, then $A^T = \{a_{ji}\}$. The rows of A are the columns of A^T.)

For Exercises 7 and 8 evaluate for $\mathbf{v}_1 = (12, -5)$, $\mathbf{v}_2 = (-10, 24)$, $\mathbf{v}_3 = (-2, 2, 4)$, and $\mathbf{v}_4 = (1, 2, -1)$.

7.(a) $\mathbf{v}_1 \cdot \mathbf{v}_2$ 8.(a) $\mathbf{v}_3 \times \mathbf{v}_4$

(b) $\mathbf{v}_2 \cdot \mathbf{v}_1$ (b) $\mathbf{v}_4 \times \mathbf{v}_3$

(c) $\mathbf{v}_1 \cdot (\mathbf{v}_1 + \mathbf{v}_2)$ (c) $\mathbf{v}_3 \times \mathbf{v}_3$

9. Find $\dfrac{\mathbf{v}_1 \cdot \mathbf{v}_2}{\|\mathbf{v}_2\|^2}\mathbf{v}_2$.

(a) $\mathbf{v}_1 = (0, 1)$, $\mathbf{v}_2 = (2, 2)$ (b) $\mathbf{v}_1 = (3, 3, 3)$, $\mathbf{v}_2 = (1, 1, 0)$

For Exercises 10 and 11 find the equation of each plane.

10. The plane contains P(-1, 0, -1) with normal $\mathbf{N} = (1, 1, 1)$.

11. The plane contains points $P_1(1, 1, 1)$, $P_2(-1, 1, 0)$, and $P_3(0, -1, 1)$.

For Exercises 12 and 13 find the symmetric and parametric forms of the equation of each line. Find two additional points on each line.

12. The line passing through (1, 2, 3) in the direction of $\mathbf{v} = (\sqrt{3}, -1, 0)$.

13. The line containing the points $P_1(-1, 2, 3)$ and $P_2(1, -2, 3)$.

14. Find the measure of the acute angle between the Normals to the two plane $x + y + z = 1$ and $x - y + z = 1$.

15. The Swing Problem You are swinging on a swing attached with 10-m chains. If you jump off when you are 5 ft off the ground and your velocity is given by $\mathbf{v} = (10, 10)$ measured in ft/sec, how far will you fly?

16. The Work Problem You apply a force of 200 lb by pulling on the tongue of a wagon at an angle of $\theta = 55°$ with the horizontal direction of motion. Find the work done if you move the wagon 12 ft.

Appendix I

Prerequisite Material

Applying Order of Operations

When more than one operation is involved in an expression we apply the order of operations summarized below.

1. Simplify within the innermost grouping symbols (i.e. parentheses).
2. Raise to powers as indicated by any exponents.
3. Multiply and divide from left to right.
4. Add and subtract from left to right.

Example

(a) Calculate $25 + (-11 + 3)^2 \div (-2)$.

(b) Evaluate $x^2(y + z)$ for $x = -17$, $y = 11$, and $z = -12$.

Solution

(a) $25 + (-11 + 3)^2 \div (-2) = 25 + (-8)^2 \div (-2)$

$= 25 + 64 \div (-2)$

$= 25 + (-32) = -7$

(b) $x^2(y + z) = (-17)^2(11 + (-12))$

$= -289$

When using a graphing utility, we use parentheses to dictate the order of operations. We enter

$$25 + (-11 + 3)^2 \div (-2)$$

as

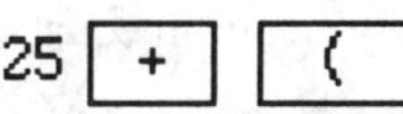 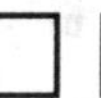 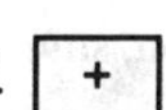

25 [+] [(] [(-)] 11 [+] 3 [)] [^] 2 [÷] [(-)] 2 [ENTER]

Skill Building Exercises Calculate the following.

1. (a) $216 - 16(-120)^2$
   (b) $(216 - 16)(-120)^2$

2. (a) $115 + 85 - 2^2 \div 28$
   (b) $[115 + 85(-2)^2] \div 28$

3. (a) $(115 - 27) - 7$
   (b) $115 - (27 - 7)$

4. (a) $(-24 \div 6) \div 3$
   (b) $-24 \div (6 \div 3)$

5. $16 - 6[12 - 2(-4)^2] - 24 \div 8$

6. $216 - 216(2 + 4^3) \div 6^2$

Evaluate each expression for $x = -12$, $y = 9$, and $z = 27$.

7. (a) $(x - y)^2$
   (b) $x^2 - y^2$
   (c) $x^2 - 2xy + y^2$

8. (a) $(2x - z)^2$
   (b) $(2x)^2 + (-z)^2$
   (c) $4x^2 - 4xz + z^2$

Answers **1.**(a) -230,184 (b) 2,880,000 **2.**(a) $199\frac{6}{7}$ (b) 16.25 **3.**(a) 81 (b) 95 **4.**(a) $-\frac{4}{3}$ (b) -12 **5.** 133 **6.** -180 **7.**(a) 441 (b) 63 (c) 441 **8.**(a) 2601 (b) 1305 (c) 2601

Factoring Binomials

A key to factoring is to recall certain patterns. The following summarizes some common binomial patterns. We check our factoring by multiplication.

1. A common factor:	$ax + ay = a(x + y)$
2. Difference of two squares:	$x^2 - y^2 = (x + y)(x - y)$
3. Difference of two cubes:	$x^3 - y^3 = (x - y)(x^2 + xy + y^2)$
4. Sum of two cubes:	$x^3 + y^3 = (x + y)(x^2 - xy + y^2)$

Example Factor the following.

(a) $2x^2 - 18$ (b) $x^3 + 27$ (c) $(x - 4)x + (x - 4)\cdot 3$

Solution

(a) Factoring the common factor 2 creates the difference of two squares.

$$2x^2 - 18 = 2(x^2 - 9) = 2(x - 3)(x + 3)$$

(b) This is the sum of two cubes. $x^3 + 27 = (x + 3)(x^2 - 3x + 9)$

(c) We use the distributive property to factor the common factor $(x - 4)$.

$$(x - 4)x + (x - 4)\cdot 3 = (x - 4)(x + 3)$$

When using a graphing utility to evaluate an expression, simplifying first may create an equal expression that is calculated more efficiently. For example, to evaluate $(x - 4)x + (x - 4)\cdot 3$ from part (c) for $x = 10$, the equal expression $(x - 4)(x + 3)$ can be evaluated using fewer keystrokes.

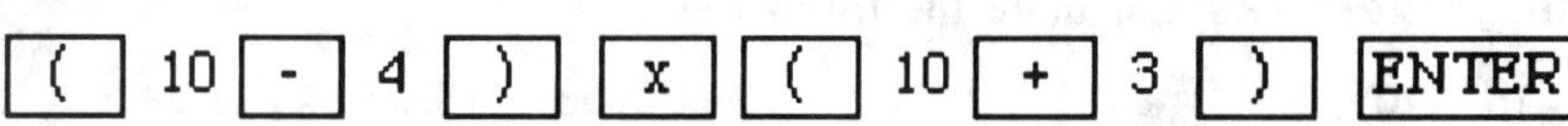

Skill Building Exercises Factor completely.

1.(a) $x^2 - 1$ (b) $1 - x^2$

2.(a) $4x^2 - 25$ (b) $25 - 4x^2$

3.(a) $x^3 - x^2$ (b) $x^2 - x^4$

4.(a) $16x^2 - 64$ (b) $-8x^3 + 64$

5. (a) $8x^3 - 27$ (b) $8x^3 + 27$

6.(a) $125x^3 - 1000$ (b) $1000 + 125x^3$

7.(a) $4x^2 - 24x$ (b) $4(x + 1)^2 - 24(x + 1)$

8.(a) $44x^2 - 33x$ (b) $44(t - 4)^2 - 33(t - 4)$

Answers **1.**(a) $(x + 1)(x - 1)$ (b) $(1 + x)(1 - x)$ **2.**(a) $(2x + 5)(2x - 5)$ (b) $(5 + 2x)(5 - 2x)$ **3.**(a) $x^2(x - 1)$ (b) $x^2(1 + x)(1 - x)$ **4.**(a) $16(x + 2)(x - 2)$ (b) $-8(x - 2)(x^2 + 2x + 4)$ **5.**(a) $(2x - 3)(4x^2 + 6x + 9)$ (b) $(2x + 3)(4x^2 - 6x + 9)$ **6.**(a) $125(x - 2)(x^2 + 2x + 4)$ (b) $125(2 + x)(4 - 2x + x^2)$ **7.**(a) $4x(x - 6)$ (b) $4(x + 1)(x - 5)$ **8.**(a) $11x(4x - 3)$ (b) $11(t - 4)(4t - 19)$

Factoring Trinomials

When factoring a trinomial of the form $ax^2 + bx + c$ into two binomials, the signs of the terms in a trinomial determine the signs of the terms in the binomials. If c is positive, the signs in the two binomials are the same, both positive when b is positive and both negative when b is negative. If c is negative, the signs in the two binomials are different. The two steps involved are as follows.

1. Remove any common factors using the distributive property *2. Determine if there are two binomials that are factors.*

Example Factor completely.

(a) $x^2 + 14x + 48$ (b) $x^2 - 19x + 48$ (c) $2x^2 + 44x - 96$ (d) $10x^2 - x - 2$

Solution

(a) Find two numbers whose product is 48 and sum is 14. Use 6 and 8 to factor the trinomial.

$$x^2 + 14x + 48 = (x + 6)(x + 8)$$

(b) Find two numbers whose product is 48 and sum is -19. Use -3 and -16 to factor the trinomial.

$$x^2 - 19x + 48 = (x - 3)(x - 16)$$

(c) First, remove the common factor, 2. Find two numbers whose product is -48 and sum is 22.

$$2x^2 + 44x - 96 = 2(x^2 + 22x - 48) = 2(x - 2)(x + 24)$$

(d) Rewrite the linear term, -x, as the sum of two terms whose product equals the product of the first and last terms, $(10x^2)(-2) = -20x^2$, and whose sum is the middle term, -x. (Use -5x and 4x.)

$$10x^2 - x - 2$$
$$/ \backslash$$
$$10x^2 \mathbf{- 5x + 4x} - 2$$
$$(10x^2 - 5x) + (4x - 2)$$
$$5x(\mathbf{2x - 1}) + 2(\mathbf{2x - 1})$$
$$(5x + 2)(2x - 1)$$

We can use a graphing utility to compare the expressions $10x^2 - x - 2$ and $(5x + 2)(2x - 1)$ by substituting a value for the variable into each expression. If the values of the expressions differ, we know the factoring is incorrect. Enter $y_1 = 10x^2 - x - 2$ and $y_2 = (5x + 2)(2x - 1)$. Use the value feature. For example, x = 12 yields y = 1426 for both expressions.

Skill Building Exercises Factor completely.

1.(a) $x^2 + 13x + 40$ 2.(a) $x^2 - 18x + 77$ 3.(a) $2x^2 + 7x + 3$ 4.(a) $4x^2 + 11x + 6$

(b) $x^2 + 3x - 40$ (b) $x^2 - 4x - 77$ (b) $2x^2 - 5x + 3$ (b) $4x^2 - 11x + 6$

5. $2x^2 + 5x - 3$ 6. $9x^2 - 12x + 4$ 7. $25x^3 - 30x^2 + 9x$ 8. $x^4 + x^2 - 2$

Answers **1.**(a) $(x + 5)(x + 8)$ (b) $(x - 5)(x + 8)$ **2.**(a) $(x - 7)(x - 11)$ (b) $(x + 7)(x - 11)$ **3.**(a) $(2x + 1)(x + 3)$ (b) $(2x - 3)(x - 1)$ **4.**(a) $(4x + 3)(x + 2)$ (b) $(4x - 3)(x - 2)$ **5.** $(2x - 1)(x + 3)$ **6.** $(3x - 2)^2$ **7.** $x(5x - 3)^2$ **8.** $(x + 1)(x - 1)(x^2 + 2)$

Dividing Polynomials Using Long Division

Long division is a useful tool in factoring polynomials. When the remainder is zero, the divisor is one of the factors of the polynomial and the quotient is the other. We divide as follows.

> *1. Arrange the terms of both polynomials in order of descending power.*
> *2. Divide the first term of the dividend by the first term of the divisor.*
> *3. Multiply the quotient by the divisor.*
> *4. Subtract and bring down the next term.*
> *5. Repeat the process until the degree of the remainder is less than the degree of the divisor.*

Example Divide the polynomials using long division.

(a) $(2x^2 - 7x - 15) \div (x - 3)$ (b) $(4x^3 + 8x^2 - x - 2) \div (x + 2)$

Solution

(a) We first divide $2x^2$ by the first term of the binomial, x. This result, 2x, is the first entry in the quotient.

```
          2x - 1
x - 3 ) 2x^2 - 7x - 15
        2x^2 - 6x
             - x - 15
             - x + 3
                 - 18
```

(b) We first divide $4x^3$ by $4x^2$. The result, x, is the first entry in the quotient.

```
                x + 2
4x^2 - 1 ) 4x^3 + 8x^2 - x - 2
           4x^3        - x
                  8x^2     - 2
                  8x^2     - 2
                             0
```

In part (a) since the remainder is not zero, x - 3 is not a factor of $2x^2 - 7x - 15$. There is a remainder of -18. We write

$$\frac{2x^2 - 7x - 15}{x - 3} = 2x - 1 + \frac{-18}{x - 3} \quad \text{and} \quad 2x^2 - 7x - 15 = (2x - 1)(x - 3) + (-18)$$

In part (b) the remainder is zero and $4x^3 + 8x^2 - x - 2 = (4x^2 - 1)(x + 2)$.

For part (a), evaluating $y_1 = 2x^2 - 7x - 15$ using the value feature yields y = -18 when x = 3. The remainder in dividing $2x^2 - 7x - 15$ by x - 3 is -18. Notice that the remainder is the value of the polynomial for x = 3.

Skill Building Exercises Divide using long division.

1. $(6x^2 - 2x - 4) \div (x - 1)$
2. $(6 + 19x + 10x^2) \div (2x + 3)$
3. $(x^3 + 6x^2 + 12x + 8) \div (x + 2)$
4. $(8x^3 - 36x^2 + 54x - 27) \div (2x - 3)$
5. $(3x^3 + 2x^2 - 27x - 18) \div (x + 3)$
6. $(2x^3 - 3x^2 - 8x + 8) \div (2x - 3)$
7. $(x^4 - 1) \div (x^2 + 1)$
8. $(9x^4 - 3x^3 - 12x^2 - x - 5) \div (3x^2 + 1)$

Answers **1.** $6x + 4$ **2.** $5x + 2$ **3.** $x^2 + 4x + 4$ **4.** $4x^2 - 12x + 9$ **5.** $3x^2 - 7x - 6$ **6.** $x^2 - 4 - \frac{4}{2x - 3}$ **7.** $x^2 - 1$ **8.** $3x^2 - x - 5$

Dividing Polynomials Using Synthetic Division

Synthetic division is division that requires only arithmetic. The divisor must be a binomial of the form x - h. Synthetic division proceeds as follows.

1. Write the coefficients of the dividend in order of descending power on the top line.
2. To the left of the list of coefficients, write the value of h.
3. Drop the leading coefficient down to the bottom row (Figure 1a).
4. Repeatedly multiply by h and add each product to the coefficient in the next column until the constant term is added (Figures 1b and 1c).
5. The bottom row is the list of coefficients of the quotient and remainder.

Example Divide $2x^3 - 3x^2 - 11x + 6$ by $x + 2$.

Solution

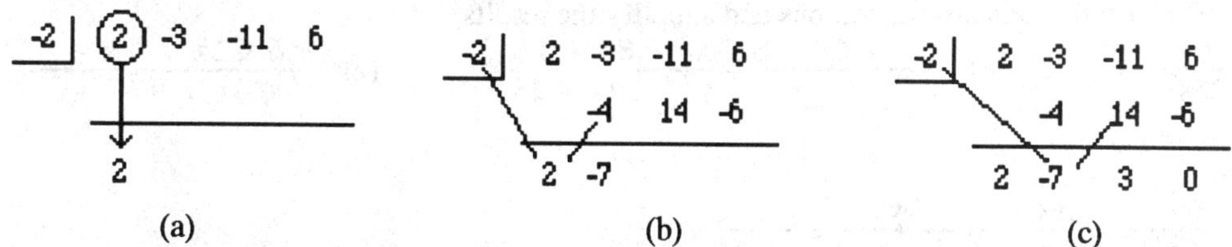

Figure 1 (a) (b) (c)

The last number in the answer row, 0, is the remainder. The first three numbers in the bottom row, 2 -7 3, are the coefficients of the quotient. Since we divided a third-degree polynomial by a first-degree polynomial, the quotient is a second-degree polynomial. Hence,

$$(2x^3 - 3x^2 - 11x + 6) \div (x + 2) = 2x^2 - 7x + 3$$

Evaluating $2x^3 - 3x^2 - 11x + 6$ using the value feature for $x = -2$ yields $y = 0$. Synthetic division can be carried out quickly on a calculator or easily programmed for a computer.

Skill Building Exercises Divide using synthetic division.

1. $(16x^3 + 48x^2 - x - 3) \div (x + 3)$
2. $(10x^3 - x^2 - 5x - 4) \div (x + 1)$
3. $(3x^3 - 2x^2 + 10) \div (x - 2)$
4. $(x^3 - 27) \div (x - 3)$

Factor the following completely, using synthetic division to find the first factor.

5. $x^3 + 3x^2 - 25x - 75$ [Try x + 5.]
6. $4x^3 - 16x^2 - x + 4$ [Try x - 4.]
7. $2x^3 - 9x^2 + x + 12$ [Try x + 1.]
8. $x^3 - 8x^2 + 4x + 48$ [Try x - 6.]

Answers **1.** $16x^2 - 1$ **2.** $10x^2 - 11x + 6 + \frac{-10}{x + 1}$ **3.** $3x^2 + 4x + 8 + \frac{26}{x - 2}$
4. $x^2 + 3x + 9$ **5.** $(x + 5)(x - 5)(x + 3)$ **6.** $(x - 4)(2x + 1)(2x - 1)$ **7.** $(x + 1)(2x - 3)(x - 4)$
8. $(x - 6)(x - 4)(x + 2)$

Performing Operations on Rational Expressions

Rational expressions are quotients of real numbers or algebraic expressions -- fractions. The following summarizes operations involving rational expressions.

Multiplying and Dividing:	*Adding and Subtracting:*
1. Multiply numerators and denominators: $\frac{a}{b}\cdot\frac{d}{c}=\frac{ad}{bc}$	*1. Determine the least common multiple, LCM, of the denominators.*
2. Rewrite a division problem and multiply: $\frac{a}{b}\div\frac{c}{d}=\frac{a}{b}\cdot\frac{d}{c}=\frac{ad}{bc}$	*2. Create fractions that are equal to our original ones but have the LCM as their common denominators.*
	3. Combine the numerators.

Example Perform the indicated operations and simplify the results.

(a) $\frac{5x}{18y^2}\cdot\frac{6y}{25x^2}$ (b) $\frac{x^2+6x+8}{x^2-25}\div\frac{x^2+8x+16}{x^2-10x+25}$ (c) $\frac{6-2x}{x^2-6x+9}-\frac{x+1}{3-x}$

Solution

(a) $\frac{5x}{18y^2}\cdot\frac{6y}{25x^2}=\frac{5\cdot x}{6\cdot 3\cdot y\cdot y}\cdot\frac{6\cdot y}{5\cdot 5\cdot x\cdot x}=\frac{1}{15xy}$

(b)

$$\frac{x^2+6x+8}{x^2-25}\div\frac{x^2+8x+16}{x^2-10x+25}=$$
$$\frac{x^2+6x+8}{x^2-25}\cdot\frac{x^2-10x+25}{x^2+8x+16}=$$
$$\frac{(x+2)(x+4)}{(x+5)(x-5)}\cdot\frac{(x-5)(x-5)}{(x+4)(x+4)}=$$
$$\frac{(x+2)(x-5)}{(x+5)(x+4)}=$$
$$\frac{x^2-3x-10}{x^2+9x+20}$$

(c) The LCM is $(x-3)^2$.

$$\frac{6-2x}{x^2-6x+9}-\frac{x+1}{3-x}=$$
$$\frac{6-2x}{(x-3)^2}-\frac{x+1}{-(x-3)}=$$
$$\frac{6-2x}{(x-3)^2}+\frac{x+1}{x-3}\cdot\frac{x-3}{x-3}=$$
$$\frac{6-2x}{(x-3)^2}+\frac{(x+1)(x-3)}{(x-3)^2}=$$
$$\frac{x^2-4x+3}{(x-3)^2}=$$
$$\frac{(x-1)(x-3)}{(x-3)^2}=\frac{x-1}{x-3}$$

Skill Building Exercises Perform the indicated operations and simplify the result.

1. $\frac{4x-12}{3x+3}\cdot\frac{5x+5}{2x-6}$
2. $\frac{12x-4x^2}{5x^2+10x}\div\frac{x^2-3x}{2x+4}$
3. $\frac{25x^2-1}{x^2+8x+15}\div\frac{25x^2+10x+1}{x^2+2x-15}$
4. $\frac{x^2-25}{x^2+10x+25}\cdot\frac{x^2+x-20}{x^2-9x+20}$
5. $\frac{4}{5x^2}+\frac{1}{15x}$
6. $\frac{7}{12y}-\frac{8}{9y^2}$
7. $\frac{1}{4x-4}-\frac{1}{6x-6}$
8. $\frac{12}{5x-15}-\frac{3}{6-2x}$

Answers 1. $\frac{10}{3}$ 2. $-\frac{8}{5x}$ 3. $\frac{(x-3)(5x-1)}{(x+3)(5x+1)}$ 4. 1 5. $\frac{12+x}{15x^2}$ 6. $\frac{21y-32}{36y^2}$ 7. $\frac{1}{12(x-1)}$ 8. $\frac{39}{10(x-3)}$

Applying Dimensional Analysis

Dimensional analysis is the process of studying units of measure. A conversion factor is a fraction whose numerator and denominator are equal, such as

$$\frac{12\text{ in.}}{1\text{ ft}}, \frac{3\text{ ft}}{1\text{ yd}}, \frac{1\text{ yd}}{3\text{ ft}}, \frac{60\text{ min}}{1\text{ hr}}, \frac{1\text{ m}}{100\text{ cm}}, \text{ and } \frac{144\text{ in.}^2}{1\text{ ft}^2}$$

Changing dimensions is accomplished as follows.

1. *Show the conversion factors. (See the inside front cover for a list of conversions.)*
2. *Multiply the resulting numerators.*
3. *Divide by the denominators.*

Example Convert 88 ft/sec to mi/hr.

Solution

We use the equations 5280 ft = 1 mi and 3600 sec = 1 hr to create the conversion factors

$$\frac{1\text{ mi}}{5280\text{ ft}} \text{ and } \frac{3600\text{ sec}}{1\text{ hr}}$$

$$\frac{88\text{ ft}}{\text{sec}} = \frac{88\text{ ft}}{\text{sec}} \cdot \frac{1\text{ mi}}{5280\text{ ft}} \cdot \frac{3600\text{ sec}}{1\text{ hr}} = \frac{88(3600)}{5280}\ \frac{\text{mi}}{\text{hr}} = 60\text{ mi/hr}$$

Some calculators have a conversion feature that converts dimensions from one unit of measure to another. Using such a calculator to convert 88 feet to miles, we enter

[CONV] 2 [ft] [mi] [ENTER]

Skill Building Exercises

1. Convert (a) 1 square mile to square feet and (b) 1 square foot to square miles.
2. Convert 1 cubic yard to (a) cubic feet and (b) cubic inches.
3. Convert 10 ft to (a) meters and (b) centimeters.
4. Convert (a) 100 km/hr to mi/hr and (b) 100 mi/hr to km/hr.
5. If a rectangle is 20 in. wide and 2 ft long, (a) find its perimeter and (b) find its area.
6. Find the number of seconds it takes a baseball thrown 100 mi/hr to travel 60 ft 6 in.
7. If an aquarium measures 18 in. by 2 ft by 4 ft, find the weight of the water in the filled tank if water weighs about 62.4 lb/ft^3.
8. Find the total time spent traveling if you travel 3/4 mi at 88 ft/sec followed by traveling 1000 ft at 30 mi/hr.

Answers **1.**(a) 1 mi^2 = 27,878,400 ft^2 (b) 1 ft$^2 \approx 3.59 \times 10^{-8}$ mi^2 **2.**(a) 1 yd^3 = 27 ft^3 (b) 1 yd^3 = 46,656 in.^{3} **3.**(a) 10 ft ≈ 3.048 m (b) 10 ft ≈ 304.8 cm **4.**(a) 100 km/hr ≈ 62.14 mi/hr (b) 100 mi/hr ≈ 160.9 km/hr **5.**(a) 88 in. (b) 480 in^2 **6.** about 0.4 sec **7.** about 748.8 lb **8.** about 67.7 sec

Simplifying Exponential Expressions

The following definitions of exponents summarize some basic ideas. Definitions cannot be proved.

For $a > 0$ with natural numbers m and n

$a^{-1} = \frac{1}{a}$ $\quad a^0 = 1$ $\quad a^{1/m} = \sqrt[m]{a}$ $\quad a^{m/n} = \sqrt[n]{a^m} = (\sqrt[n]{a})^m$

Example Simplify.

(a) $(\frac{2}{3})^{-1}$ (b) π^0 (c) $144^{3/2}$ (d) $8^{2/3}$

Solution

(a) $(\frac{2}{3})^{-1} = \frac{1}{2/3} = 1 \cdot \frac{3}{2} = \frac{3}{2}$

(b) $\pi^0 = 1$

(c) $144^{3/2} = (\sqrt{144})^3 = 12^3 = 1728$ also

$144^{3/2} = \sqrt{144^3} = \sqrt{2985984} = 1728$

(d) $8^{2/3} = (\sqrt[3]{8})^2 = (2)^2 = 4$; also

$8^{2/3} = \sqrt[3]{8^2} = \sqrt[3]{64} = 4$

Entering roots other than square roots on a graphing utility requires using the [^] key. We evaluate $144^{3/2}$, the square root of 144 cubed, by entering

The result is 1728. Parentheses are necessary to carry out the division first.

Skill Building Exercises

Simplify each expression in a form with no negative or fractional exponents.

1. (a) -4^{-1} (b) $(-4)^{-1}$
2. (a) -15^0 (b) $(-15)^0$
3. (a) $-64^{1/3}$ (b) $(-64)^{1/3}$
4. (a) $(-\frac{8}{27})^{1/3}$ (b) $(-\frac{8}{27})^{-1/3}$
5. (a) $16^{1/2}$ (b) $-16^{1/2}$
6. (a) $16^{-1/2}$ (b) $-16^{-1/2}$
7. (a) $\sqrt[5]{1024}$ (b) $\sqrt[10]{1024}$
8. (a) $(8 \times 10^9)^{10/3}$ (b) $(8 \times 10^{-9})^{-10/3}$

Answers **1.**(a) -1/4 (b) -1/4 **2.**(a) -1 (b) 1 **3.**(a) -4 (b) - 4 **4.**(a) $-\frac{2}{3}$ (b) $-\frac{3}{2}$ **5.**(a) 4 (b) -4 **6.**(a) 1/4 (b) -1/4 **7.**(a) 4 (b) 2 **8.**(a) $1024 \times 10^{30} = 1.024 \times 10^{33}$ (b) $1/1024 \times 10^{30} \approx 9.8 \times 10^{26}$

Performing Operations Involving Exponents

Expressions with exponents can often be simplified by applying the definitions and following theorems in different orders. The approach you take is your choice.

> *If a, b > 0, and m and n are rational numbers, then*
>
> $a^m \cdot a^n = a^{m+n}$ $\qquad \dfrac{a^m}{a^n} = a^{m-n}$ $\qquad (a^m)^n = a^{mn}$
>
> $(ab)^m = a^m b^m$ $\qquad (\dfrac{a}{b})^m = \dfrac{a^m}{b^m}$

Example Evaluate each expression for x = 5, y = 10 and z ≠ 0.

(a) $\dfrac{x^{-1}}{y^{-1}}$ (b) $(x^{1/3})^2 \cdot x^{1/3}$ (c) $(\dfrac{x^{-2}y^3}{x^2y^{-1}z^0})^{1/2}$

Solution

We simplify each expression first, then substitute x = 5 and y = 10.

(a) $\dfrac{x^{-1}}{y^{-1}} = \dfrac{1/x}{1/y} = \dfrac{1}{x} \div \dfrac{1}{y} = \dfrac{1}{x} \cdot \dfrac{y}{1} = \dfrac{y}{x}$ For x = 5 and y = 10, $\dfrac{y}{x} = \dfrac{10}{5} = 2$.

(b) $(x^{1/3})^2 \cdot x^{1/3} = x^{2/3} \cdot x^{1/3} = x^{(2/3 + 1/3)} = x$ By substitution, x = 5.

(c) $(\dfrac{x^{-2}y^3}{x^2y^{-1}z^0})^{1/2} = (\dfrac{y^4}{x^4})^{1/2} = \dfrac{y^{4(1/2)}}{x^{4(1/2)}} = \dfrac{y^2}{x^2}$ For x = 5 and y = 10, $\dfrac{y^2}{x^2} = \dfrac{100}{25} = 4$.

> We evaluate $(x^{1/3})^2 \cdot x^{1/3}$ from part (b) for x = 5 by entering
>
> 5 [^] [(] 1 [÷] 3 [)] [^] 2 [x] 5 [^] [(] 1 [÷] 3 [)] [ENTER]
>
>   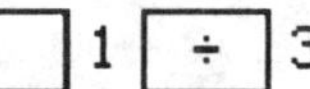 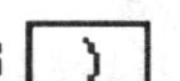 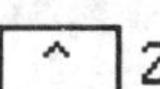  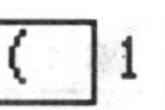 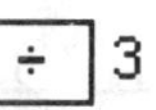  
>
> If we simplify first we see that this expression equals x for all x.

Skill Building Exercises Simplify each expression to a form with no negative or fractional exponents. Then evaluate each expression for x = 2 and y = 1/2. Assume that all expressions represent real numbers.

1. (a) $x^5 \cdot x^7 \cdot x^3$ (b) $y^{-2} \cdot y^{-3} \cdot y^{-4}$
2. (a) $(x^2y^3)^2$ (b) $(x^{-2}y^{-3})^{-2}$
3. (a) $(\dfrac{x^2}{y})^{-1}$ (b) $(\dfrac{x^{-2}}{y^{-1}})^{-1}$
4. (a) $(\dfrac{-y^2}{x^3})^2$ (b) $(\dfrac{-x^{-3}}{y^{-2}})^{-2}$
5. $x^{1/2}(x^{3/2} + x^{3/2})$
6. $(x^{1/2} + x^{-1/2})^2$
7. $(x^{-1/2} - x^{1/2})^2$
8. $\dfrac{x^{-1} - y^{-1}}{(xy)^{-1}}$

Answers 1.(a) $x^{15} = 32{,}768$ (b) $\dfrac{1}{y^9} = 512$ **2.**(a) $x^4y^6 = \dfrac{1}{4}$ (b) $x^4y^6 = \dfrac{1}{4}$ **3.**(a) $\dfrac{y}{x^2} = \dfrac{1}{8}$ (b) $\dfrac{x^2}{y} = 8$ **4.**(a) $\dfrac{y^4}{x^6} = \dfrac{1}{1024}$ (b) $\dfrac{x^6}{y^4} = 1024$ **5.** $2x^2 = 8$ **6.** $x + 2 + \dfrac{1}{x} = \dfrac{9}{2}$ **7.** $\dfrac{1}{x} - 2 + x = \dfrac{1}{2}$ **8.** $y - x = \dfrac{-3}{2}$

Performing Operations on Complex Numbers

The number $\sqrt{-1}$ is called an imaginary number, such that

$$\sqrt{-1} = i \text{ and } i^2 = -1$$

To avoid errors when dealing with complex numbers, we first change any square roots of negative numbers into expressions involving i. If a real number and an imaginary number are added, the result, is neither real nor imaginary. It is called a complex number. A complex number can be written in the form a + bi where a and b are real numbers.

$$\sqrt{-b} = \sqrt{b}\sqrt{-1} = \sqrt{b}\, i,\ b > 0$$

$$(a + bi) + (c + di) = (a + c) + (b + d)i$$

$$(a + bi)(c + di) = (ac - bd) + (ad + bc)i$$

$$\frac{a + bi}{c + di} = \frac{a + bi}{c + di} \cdot \frac{c - di}{c - di} = \frac{ac + bd}{c^2 + d^2} + \frac{bc - ad}{c^2 + d^2} i$$

Example Perform the indicated operations.

(a) $\sqrt{-8} \cdot \sqrt{-2}$ (b) $(3 + 4i) + (2 - 3i)$ (c) $(2 + 3i)(3 - 5i)$ (d) $\frac{2 + i}{4 + 3i}$

Solution

(a) $\sqrt{-8} \cdot \sqrt{-2} = (\sqrt{8}\, i)(\sqrt{2}\, i) = \sqrt{16}\, i^2 = -4$

(b) $(3 + 4i) + (2 - 3i) = (3 + 2) + (4i - 3i) = 5 + i$

(c) $(2 + 3i)(3 - 5i) = 2(3 - 5i) + 3i(3 - 5i) = 6 - 10i + 9i - 15i^2 = 6 - i - 15(-1) = 6 - i + 15 = 21 - i$

(d) $\frac{2 + i}{4 + 3i} = \frac{(2 + i)}{(4 + 3i)} \cdot \frac{(4 - 3i)}{(4 - 3i)} = \frac{8 - 6i + 4i - 3i^2}{16 - 12i + 12i - 9i^2} = \frac{11 - 2i}{25} = \frac{11}{25} - \frac{2}{25} i$

To multiply (2 + 3i) and (3 - 5i) using a graphing utility in COMPLEX mode, we enter

(2,3) x (3,-5)

The result is (21, -1), which represents 21 - i.

Skill Building Exercises Perform the indicated operations.

1. $(\sqrt{-9})(\sqrt{-4})$
2. $\sqrt{-64} + \sqrt[3]{-64}$
3. $(\sqrt{-8})^2$
4. $\frac{1}{\sqrt{-2}}$
5. $(3 + 4i) + (2 + 3i)$
6. $(-2 + 3i) + (-1 - 2i)$
7. (a) $(2 + 3i)(2 - 3i)$ (b) $(2 + 3i) \div (2 - 3i)$
8. (a) $(2 + 4i)(3 - 6i)$ (b) $(2 + 4i) \div (3 - 6i)$

Answers **1.** -6 **2.** -4 + 8i **3.** -8 **4.** $\frac{-\sqrt{2}}{2} i$ **5.** 5 + 7i **6.** -3 + i **7.**(a) 13 (b) $-\frac{5}{13} + \frac{12}{13} i$ **8.**(a) 30 (b) $-\frac{6}{5} + \frac{8}{5} i$

Solving and Checking a Linear Equation

In solving a linear equation we show the necessary steps in writing a short, efficient solution. Each line follows from the preceding one. We solve a linear equation as follows.

> *1. Combine all terms containing the variable on one side of the equation and all constant terms on the other.*
> *2. Divide both sides of the equation by the coefficient of the variable.*
> *3. Check the solution by substituting the answer into each side of the equation.*

Example Solve and check $3(x + 2) = 2x + 8$.

Solution

$3(x + 2) = 2x + 8$ — *Apply the distributive property.*

$3x + 6 = 2x + 8$ — *Subtract 2x from both sides.*

$x + 6 = 8$ — *Subtract 6 from both sides.*

$x = 2$

Check: If $x = \mathbf{2}$, then

$$3(x + 2) = 3(\mathbf{2} + 2) = 12$$

and

$$2x + 8 = 2(\mathbf{2}) + 8 = 12$$

Therefore, $3(x + 2) = 2x + 8$ for $x = 2$.

One way to solve $3(x + 2) = 2x + 8$ using a graphing utility is to let each side of the equation equal y,

$$y_1 = 3(x + 3) \quad \text{and} \quad y_2 = 2x + 8$$

and graph the two equations. Using the intersect feature, we find the point of intersection of the two lines (Figure 2). Notice that the x coordinate is the solution to the original equation and the y coordinate is its check value.

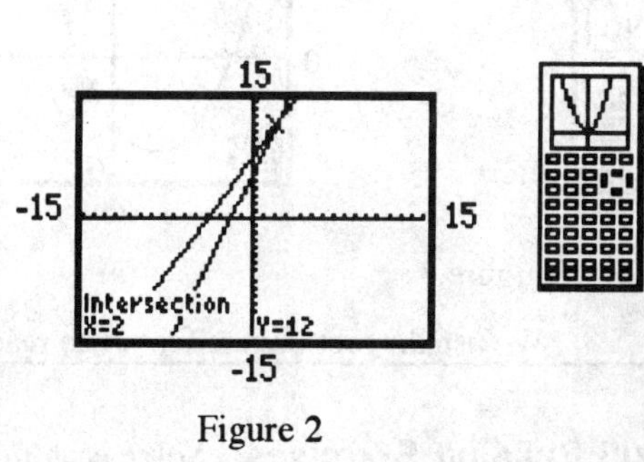

Figure 2

Skill Building Exercises Solve and check each linear equation.

1. $14 - 4(3 - 2x) = x + 2$
2. $5(2x - 1) = 10(2x - 1)$
3. $8x + 2 = 8(x + 2)$
4. $10x - 5 = 5(2x - 1)$
5. $5(4 - x) - 3(x - 4) = 4x - 16 - 15(x - 4)$
6. $4[12 + 2(3x - 5)] = -9x - 5(1 - 4x)$
7. (a) $\frac{x}{2} - \frac{1}{4} = \frac{3}{2}$
   (b) $0.5x - 0.25 = 1.5$
8. (a) $\frac{1}{4} - \frac{1}{2}(x - \frac{1}{2}) = \frac{3x}{4}$
   (b) $0.25 - 0.5(x - 0.5) = 0.75x$

Answers **1.** $x = 0$ **2.** $x = 1/2$ **3.** no solution **4.** all real values of x **5.** $x = 4$ **6.** $x = -1$ **7.**(a) $x = 7/2$ (b) $x = 3.5$ **8.**(a) $x = 2/5$ (b) $x = 0.4$

Solving a Quadratic Equation by Factoring

A quadratic equation is an equation of the form $ax^2 + bx + c = 0, a \neq 0$. A quadratic equation is written in standard form by setting it equal to zero and collecting like terms. If the polynomial in a quadratic equation can be expressed in factored form, we solve it as follows.

> 1. *Write the equation in standard form by setting it equal to zero.*
> 2. *Factor the quadratic expression.*
> 3. *Apply the zero product theorem: If $a \cdot b = 0$, then $a = 0$ or $b = 0$.*
> 4. *Solve the resulting linear equations.*

Example Solve $(x + 1)(x + 3) = 10 - (x - 7)$ by factoring.

Solution

$(x + 1)(x + 3) = 10 - (x - 7)$	*Write in standard form.*
$x^2 + 4x + 3 = 10 - x + 7$	
$x^2 + 5x - 14 = 0$	*Factor the quadratic expression.*
$(x + 7)(x - 2) = 0$	*Apply the zero product theorem.*
$x + 7 = 0$ or $x - 2 = 0$	*Solve for x.*
$x = -7$ or $x = 2$	

One way to solve $(x + 1)(x + 3) = 10 - (x - 7)$ is to write the equation in standard form, $x^2 + 5x - 14 = 0$, and enter $y = x^2 + 5x - 14$. The solutions are the x intercepts of the graph.

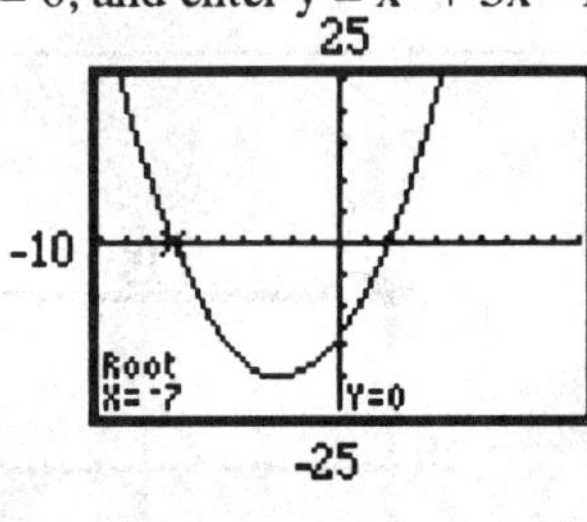

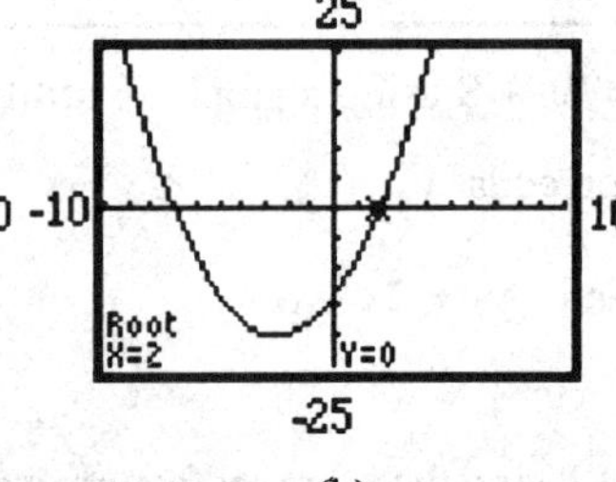

Figure 3 (a) (b)

We use the root or zero feature to read the values $x = -7$ (Figure 3a), $x = 2$ (Figure 3b).

Skill Building Exercises Solve each quadratic equation by factoring.

1.(a) $x^2 - 8x - 9 = 0$

(b) $x^2 + 8x - 9 = 0$

2. (a) $x^2 - 9x + 18 = 0$

(b) $x^2 + 9x + 18 = 0$

3. $x^2 - 10x = 0$

4. $15x = x^2$

5. $x(x - 4) = 5$

6. $(x + 2)(x - 4) = -9$

7. $-9 - x(x + 2) = -8(x + 2)$

8. $2x(x - 2) = 9(x - 2)$

Answers **1.**(a) $x = -1$ or $x = 9$ (b) $x = -9$ or $x = 1$ **2.**(a) $x = 3$ or $x = 6$ (b) $x = -6$ or $x = -3$ **3.** $x = 0$ or $x = 10$ **4.** $x = 0$ or $x = 15$ **5.** $x = -1$ or $x = 5$ **6.** $x = 1$ **7.** $x = -1$ or $x = 7$ **8.** $x = 2$ or $x = 9/2$

Solving a Quadratic Equation by Extracting Roots

A quadratic equation that can be written in the form $x^2 = c$ or $(x - h)^2 = c$ can be solved quickly by taking the square root of both sides of the equation. If a quadratic equation can be expressed in such a form, we solve it as follows.

1. Isolate the constant.
2. Take the square root of both sides. Remember to use the ± symbol.

Example Solve by extracting roots.

(a) $x^2 = 5$

(b) $2(x - 7)^2 - 8 = 0$

Solution

(a)

$$x^2 = 5$$
$$x = \pm\sqrt{5}$$

(b)

$$2(x - 7)^2 - 8 = 0$$
$$2(x - 7)^2 = 8$$
$$(x - 7)^2 = 4$$
$$x - 7 = \pm 2$$
$$x = 7 \pm 2$$
$$x = 7 - 2 = 5 \text{ or } x = 7 + 2 = 9$$

One way to solve $x^2 = 5$ from part (a) using a graphing utility is to let each side of the equation equal y, $y_1 = x^2$ and $y_2 = 5$, and graph the two equations. Using the intersect feature, we find the points of intersection (Figure 4).

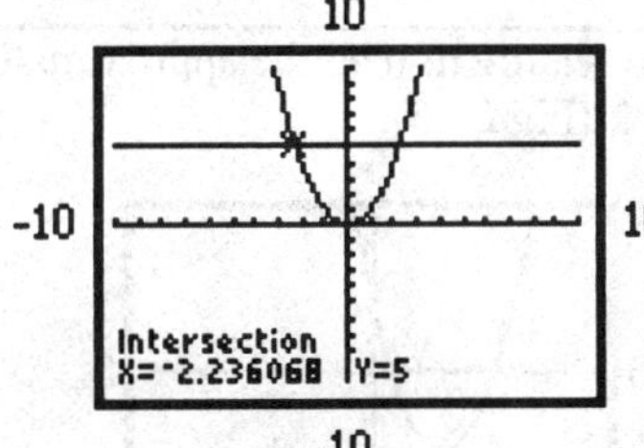

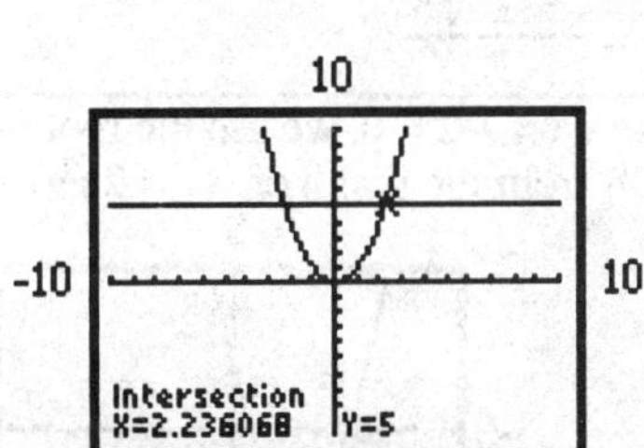

Figure 4

The x coordinate is the solution to the original equation, $x \approx \pm 2.24$.

Skill Building Exercises Solve each quadratic equation by extracting roots.

1. (a) $x^2 = 4$ (b) $2x^2 = 4$
2. (a) $x^2 = 8$ (b) $x^2 = -8$
3. (a) $(x - 2)^2 = 16$ (b) $(x + 2)^2 = 16$
4. (a) $(x + 3)^2 = 0$ (b) $x^2 - 8x + 16 = 0$
5. (a) $(x - 3)^2 - 9 = 0$ (b) $(x^2 - 6x + 9) - 9 = 0$
6. (a) $1 - x^2 = 0$ (b) $1 + x^2 = 0$

Answers **1.**(a) $x = \pm 2$ (b) $x = \pm\sqrt{2}$ **2.**(a) $x = \pm 2\sqrt{2}$ (b) $x = \pm 2\sqrt{2}\,i$ **3.**(a) $x = -2$ or $x = 6$ (b) $x = -6$ or $x = 2$ **4.**(a) $x = -3$ (b) $x = 4$ **5.**(a) $x = 0$ or $x = 6$ (b) $x = 0$ or $x = 6$ **6.**(a) $x = \pm 1$ (b) $x = \pm i$

Solving a Quadratic Equation by Completing the Square

Completing the square means transforming a binomial expression into a perfect square trinomial. For the perfect square trinomials $x^2 + 10x + 25$ and $x^2 - 12x + 36$, notice the relationship between the coefficient of the x terms, 10 and -12 respectively, and the constant terms, 25 and 36. In each case the constant is the square of half the coefficient of the x term, $((1/2)(10))^2 = 25$ and $((1/2)(-12))^2 = 36$.

Example Add a constant to each binomial to make it a perfect square trinomial.

(a) $x^2 - 8x$ (b) $x^2 + 3x$

Solution

(a) The square of half of -8 is $(-4)^2 = 16$. We add 16, $x^2 - 8x + \mathbf{16}$. Notice that $x^2 - 8x + 16 = (x - 4)^2$.

(b) We add $((1/2)(3))^2 = (3/2)^2 = 9/4$, $x^2 + 3x + \mathbf{9/4}$. Notice that $x^2 + 3x + 9/4 = (x + 3/2)^2$.

1. *Write the equation in the form $ax^2 + bx = -c$ to isolate the constant term.*
2. *If $a \neq 1$, divide both sides of the equation by a.*
3. *Complete the square by adding the square of half the coefficient of x, $[(\frac{1}{2})(\frac{b}{a})]^2$, to both sides.*
4. *Take the square root of both sides and solve for x.*

Example Solve the equation $2x^2 + 6x + 2 = 0$ by completing the square.

Solution

$2x^2 + 6x + 2 = 0$ *Isolate the constant term.*

$2x^2 + 6x = -2$ *Divide both sides by 2.*

$x^2 + 3x = -1$ *Add $[(\frac{1}{2})(3)]^2 = \frac{9}{4}$ to both sides.*

$x^2 + 3x + \mathbf{9/4} = -1 + \mathbf{9/4}$ *Factor the quadratic expression.*

$(x + 3/2)^2 = 5/4$ *Take the square root of both sides.*

$x + 3/2 = \pm\sqrt{5/4}$ *Solve for x.*

$$x = \frac{-3 \pm \sqrt{5}}{2}$$

To solve $2x^2 + 6x + 2 = 0$, we use the root or zero feature to read the approximate values $x \approx -2.62$ and $x \approx -0.38$ from the graph of $y_1 = 2x^2 + 6x + 2$ (Figure 5).

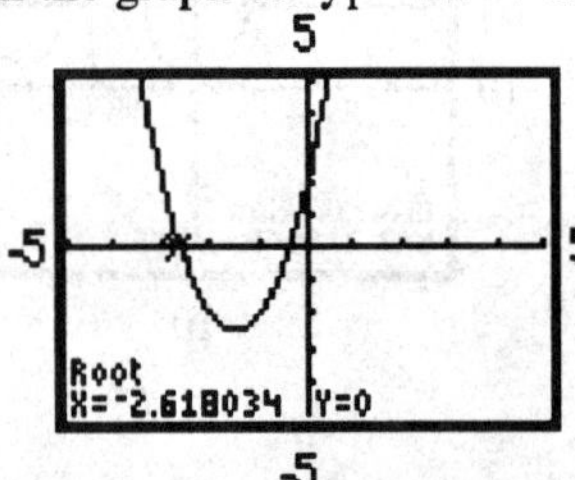

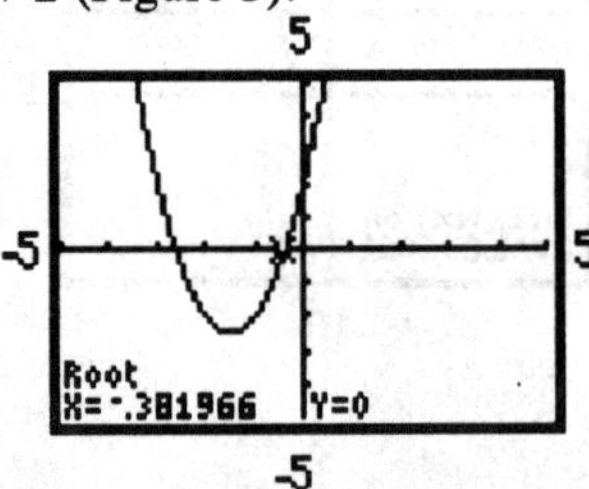

Figure 5

Skill Building Exercises Add a constant to each binomial to make it a perfect square trinomial.

1.(a) $x^2 - 18x$ (b) $x^2 + 18x$ 2.(a) $x^2 - x$ (b) $x^2 + 5x$

Solve each quadratic equation by completing the square.

3.(a) $x^2 - 8x = 0$ (b) $x^2 - 8x = 9$ 4.(a) $x^2 - 10x = 0$ (b) $x^2 - 10x = -16$

5. $x^2 - 3x + 4 = 0$ 6. $x^2 - 3x - 6 = 0$ 7. $4x^2 - 8x = 3$ 8. $2x^2 + 4x = -3$

Answers **1.**(a) $x^2 - 18x + 81$ (b) $x^2 + 18x + 81$ **2.**(a) $x^2 - x + 1/4$ (b) $x^2 + 5x + 25/4$ **3.**(a) $x = 0$ or $x = 8$ (b) $x = -1$ or $x = 9$ **4.**(a) $x = 0$ or $x = 10$ (b) $x = 2$ or $x = 8$ **5.** $x = \frac{3}{2} + \frac{\sqrt{7}}{2}i$ **6.** $\frac{3 \pm \sqrt{33}}{2}$ **7.** $x = 1 \pm \frac{\sqrt{7}}{2}$ **8.** $x = -1 \pm \frac{\sqrt{2}}{2}i$

Solving a Quadratic Equation Using the Quadratic Formula

While the factoring method is a fast way to solve equations that factor easily, the quadratic formula is a method for solving any quadratic equation.

> *1. Write the equation in standard form, $ax^2 + bx + c = 0$.*
> *2. Identify the values for a, b, and c.*
> *3. Substitute into the formula $x = \dfrac{-b \pm \sqrt{b^2 - 4ac}}{2a}$.*
> *4. Solve for x.*

Example Solve $2x^2 - 3x - 4 = 0$ using the quadratic formula.

Solution

$$2x^2 - 3x - 4 = 0$$

$$x = \frac{-b \pm \sqrt{b^2 - 4ac}}{2a}$$

Substitute a = 2, b = -3, and c = -4.

$$x = \frac{3 \pm \sqrt{(-3)^2 - 4(2)(-4)}}{2(2)}$$

$$x = \frac{3 \pm \sqrt{9 + 32}}{4} = \frac{3 \pm \sqrt{41}}{4}$$

$x \approx 2.35$ or $x \approx -0.85$.

Some graphing utilities have a solve feature. To solve the equation $2x^2 - 3x - 4 = 0$, we enter the quadratic expression, the variable, and a guess. Using a guess of x = 5 yields $x \approx 2.35$, while a guess of x = 0 yields $x \approx -0.85$ (Figure 6).

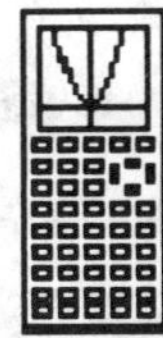

```
solve(2X²-3X-4,X
,5)
       2.350781059
solve(2X²-3X-4,X
,0)
      -.8507810594
```

Figure 6

Skill Building Exercises Solve each quadratic equation using the quadratic formula.

1. $x^2 - 6x = 0$
2. $x^2 - 2x - 6 = 0$
3. $2x^2 + 4x + 1 = 0$
4. $3x^2 + x + 1 = 0$
5. $x^2 - 6x = -1$
6. $3x^2 + 2x - 4 = 0$
7. $x^2 - 2x + 2 = 0$
8. $2x^2 + \frac{7}{2}x - 1 = 0$

Answers **1.** $x = 0$ or $x = 6$ **2.** $x = 1 \pm \sqrt{7} \approx 1 \pm 2.64$; $x \approx -1.64$ or $x \approx 3.64$

3. $x = \dfrac{-2 \pm \sqrt{2}}{2} \approx -1 \pm 0.71$; $x \approx -1.71$ or $x \approx -0.29$

4. $x = \dfrac{-1 \pm \sqrt{-11}}{6}$, $x \approx -0.17 + 0.55i$ or $x \approx -0.17 - 0.55i$ **5.** $x \approx 0.17$ or $x \approx 5.83$

6. $x \approx -1.54$ or $x \approx 0.87$ **7.** $x = 1 \pm i$ **8.** $x = -2$ or $x = 1/4$

Solving a Pair of Linear Equations by Substitution

One way to solve a pair of linear equations with two variables is to use the substitution method to create one equation with one variable. We proceed as follows.

1. Solve one equation for one of the variables.
2. Substitute the result from step 1 into the other equation to obtain one equation with one variable.
3. Solve the equation from step 2.
4. Substitute the result from step 3 into one of the original or an equivalent equations to obtain a value for the other variable.

Example Solve $2x + 3y = 8$
$3x - y = 1$

Solution

$3x - y = 1$ — *Solve for y in terms of x.*
$y = 3x - 1$

$2x + 3y = 8$ — *Substitute y = 3x - 1.*

$2x + 3(\mathbf{3x - 1}) = 8$ — *Solve for x.*

$2x + 9x - 3 = 8$

$11x = 11$

$x = 1$

$y = 3\mathbf{x} - 1$ — *Substitute x = 1.*
$y = 3(\mathbf{1}) - 1 = 2$

The solution is the ordered pair (1, 2). A solution to a pair of linear equations can be expressed as an ordered pair since the solution represents the point of intersection of two lines.

We could use a graphing utility to solve this pair of equations by solving both equations for y and entering $y_1 = (8 - 2x)/3$ and $y_2 = 3x - 1$. Using the intersect feature we find the ordered pair, (1,2) (Figure 7).

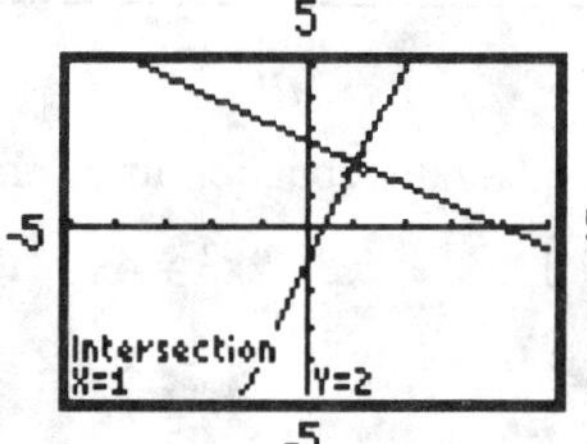

Figure 7

Skill Building Exercises Solve each pair of equations by substitution.

1. $x + y = 10$, $y = x + 2$
2. $x + y = 5$, $y = x + 2$
3. $2x + y = 3$, $3x + 2y = 5$
4. $x - 3y = 4$, $5x + 2y = 3$
5. $y = -2x - 1$, $3x + 2y = 0$
6. $x = 8 - 2y$, $-x + y = 4$
7. $y - 3x = -1$, $6x = 2y$
8. $2x + 4y = 12$, $-x - 2y + 6 = 0$

Answers **1.** (4, 6) **2.** $(\frac{3}{2}, \frac{7}{2})$ **3.** (1, 1) **4.** (1, -1) **5.** (-2, 3) **6.** (0, 4) **7.** no solution **8.** true for all values of x and y such that $x + 2y = 6$

Solving a Pair of Nonlinear Equations by Substitution

The substitution method can be used to solve a pair of equations involving nonlinear equations. The process is similar to solving two linear equations.

1. Solve one equation for one of the variables. *2. Substitute the result from step 1 into the other equation to obtain one equation with one variable.* *3. Solve the equation from step 2.* *4. Substitute the result from step 3 into one of the original or an equivalent equations to obtain a value for the other variable.*

Example Solve the $y = x^2 - 2x$
$y = x + 4$

Solution

$y = x + 4$	*Substitute $x^2 - 2x$ for y.*
$x^2 - 2x = x + 4$	*Write in standard form.*
$x^2 - 3x - 4 = 0$	*Factor the quadratic expression.*
$(x + 1)(x - 4) = 0$	*Apply the zero product theorem.*
$x + 1 = 0$ or $x - 4 = 0$	
$x = -1$ or $x = 4$	

To find the values of y, we substitute $x = -1$ and $x = 4$ into $y = x + 4$. If $x = -1$, then $y = -1 + 4 = 3$. If $x = 4$, then $y = 4 + 4 = 8$. The solution is two ordered pairs, (-1, 3), (4, 8).

To solve this system using a graphing utility, we enter the equations $y_1 = x^2 - 2x$ and $y_2 = x + 4$ and use the intersect feature to read the ordered pairs, (-1, 3) (Figure 8a), (4, 8) (Figure 8b).

10
-5 5
Intersection X=-1 Y=3
-5

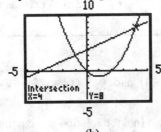

10
-5 5
Intersection X=4 Y=8
-5

Figure 8 (a) (b)

Skill Building Exercises Solve each pair of equations.

1. $y = x^2$
   $y = x$
2. $y = x^2$
   $y = -x$
3. $y = x^2$
   $y = 2x + 3$
4. $y = (x - 1)^2$
   $y = -(x - 1)^2$
5. $y = -x + 10$
   $y = x^2 + 4x + 4$
6. $y = x^2 - 6x + 9$
   $y = 2x - 7$
7. $y = 2x^2 + 5x + 3$
   $y = \frac{1}{2}x + \frac{15}{2}$
8. $y = 2x^2 - 3x - 5$
   $y = -x^2 + 3x + 4$

Answers **1.** (0, 0), (1, 1) **2.** (-1, 1), (0, 0) **3.** (-1, 1), (3, 9) **4.** (1, 0) **5.** (-6, 16), (1, 9) **6.** (4, 1) **7.** (-3, 6), (3/4, 63/8) **8.** (-1, 0), (3, 4)

Solving an Equation Involving Rational Expressions

The easiest way to solve most equations involving fractions is to eliminate the denominators by multiplying both sides of the equation by the least common multiple of all the denominators in the equation. The procedure is outlined as follows.

1. Find the LCM for all the denominators in the equation.
2. Multiply both sides of the equation by the LCM.
3. Solve the resulting equation.
4. If the original equation contains a variable in the denominator, check the solutions from step 3 for division by zero.

Example Solve (a) $\frac{3x}{4} - \frac{2}{3} = \frac{1}{6}$ and (b) $\frac{x - 5}{2} + \frac{1}{x + 1} = \frac{1 - x}{2(x + 1)}$.

Solution

(a) Multiply both sides by 12.

$$\frac{3x}{4} - \frac{2}{3} = \frac{1}{6}$$
$$\mathbf{12}\left[\frac{3x}{4} - \frac{2}{3}\right] = \mathbf{12}\left(\frac{1}{6}\right)$$
$$12\left(\frac{3x}{4}\right) - 12\left(\frac{2}{3}\right) = 12\left(\frac{1}{6}\right)$$
$$9x - 8 = 2$$
$$9x = 10$$
$$x = \frac{10}{9}$$

(b) Multiply both sides by $2(x + 1)$.

$$\frac{x - 5}{2} + \frac{1}{x + 1} = \frac{1 - x}{2(x + 1)}$$
$$\mathbf{2(x + 1)}\left[\frac{x - 5}{2} + \frac{1}{x + 1}\right] = \mathbf{2(x + 1)}\left[\frac{1 - x}{2(x + 1)}\right]$$
$$(x + 1)(x - 5) + 2(1) = 1 - x$$
$$x^2 - 4x - 5 + 2 = 1 - x$$
$$x^2 - 3x - 4 = 0$$
$$(x + 1)(x - 4) = 0$$
$$x + 1 = 0 \text{ or } x - 4 = 0$$
$$x = -1 \text{ or } x = 4$$

We must exclude $x = -1$, hence $x = 4$.

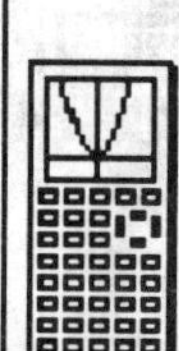

On a graphing utility we enter

$y_1 = (x - 5)/2 + 1/(x + 1)$ and $y_2 = (1 - x)/(2x + 2)$

We use the intersect feature to find the solution at the point of intersection of the two graphs (Figure 9).

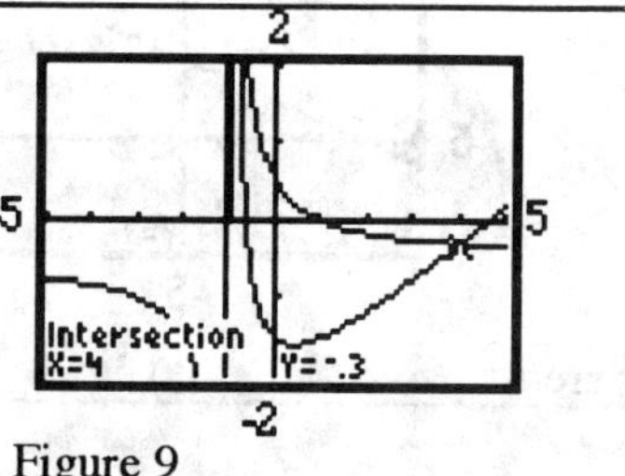

Figure 9

Skill Building Exercises Solve each equation and check the solutions to make sure that none of them cause division by zero in the original equation.

1. $\frac{5x}{4} + \frac{x}{2} = 1$
2. $\frac{x + 1}{3} - \frac{2x - 1}{5} = -\frac{2}{15}$
3. $\frac{3}{x} + \frac{5}{x^2} = 2$
4. $\frac{x}{x + 2} = \frac{2}{-x - 2}$
5. $\frac{4}{x + 4} - \frac{x}{2x + 8} = \frac{1}{3x + 12}$
6. $\frac{3}{12 - 6x} - \frac{1}{x - 2} = \frac{x}{8x - 16}$
7. $\frac{t}{t + 3} - \frac{1}{3 - t} = \frac{2t}{t^2 - 9}$
8. $\frac{3}{6t + 4} + \frac{2}{3t + 2} = \frac{1}{2}$

Answers **1.** $x = 4/7$ **2.** $x = 10$ **3.** $x = -1$ or $x = 5/2$ **4.** no solution ($x = -2$) **5.** $x = 22/3$ **6.** $x = -12$ **7.** $t = 1$ ($t = 3$ causes division by zero) **8.** $t = 5/3$

Solving an Equation Involving Radicals

When solving an equation with radicals we square both sides of the equation. It is not unusual to find a solution for the resulting equation that does not satisfy the original one, an extraneous solution. We proceed as follows.

1. Isolate the radical if possible.
2. Square both sides of the equation using $(\sqrt{u})^2 = u$ *for* $u \geq 0$.
3. If a radical is still present, repeat steps 1 and 2.
4. Solve for the variable.
5. Check for extraneous solutions.

Example Solve (a) $\sqrt{x + 17} = 3 - x$ and (b) $\sqrt{2x + 1} = 1 + \sqrt{x}$.

Solution

(a)

$$\begin{aligned} \sqrt{x + 17} &= 3 - x \\ (\sqrt{x + 17})^2 &= (3 - x)^2 \\ \mathbf{x + 17} &= \mathbf{9 - 6x + x^2} \\ x^2 - 7x - 8 &= 0 \\ (x + 1)(x - 8) &= 0 \\ x + 1 = 0 &\text{ or } x - 8 = 0 \\ x = -1 &\text{ or } x = 8 \end{aligned}$$

Since $x = 8$ is extraneous, the solution is $x = -1$.

(b)

$$\begin{aligned} \sqrt{2x + 1} &= 1 + \sqrt{x} \\ (\sqrt{2x + 1})^2 &= (1 + \sqrt{x})^2 \\ 2x + 1 &= 1 + 2\sqrt{x} + x \\ x &= 2\sqrt{x} \\ x^2 &= (2\sqrt{x})^2 \\ x^2 &= 4x \\ x^2 - 4x &= 0 \\ x(x - 4) &= 0 \\ x = 0 &\text{ or } x = 4 \end{aligned}$$

A graphing utility display of $y_1 = \sqrt{x + 17}$ and $y_2 = 3 - x$ in Figure 10 shows only one point of intersection, confirming our result that there is only one solution, $x = -1$.

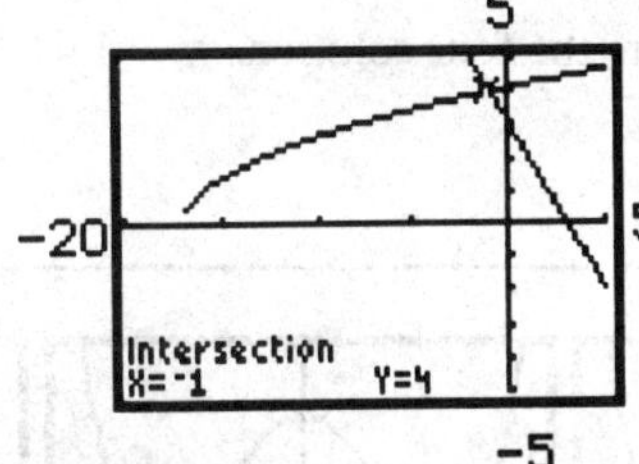

Figure 10

For $y_1 = \sqrt{2x + 1}$ and $y_2 = 1 + \sqrt{x}$ there are two points of intersection, $x = 0$ and $x = 4$ (Figure 11).

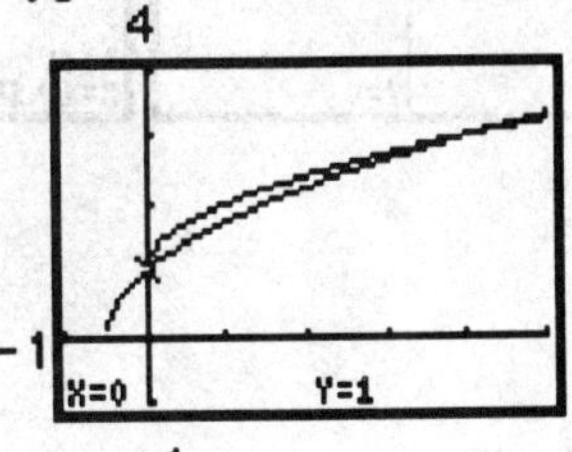

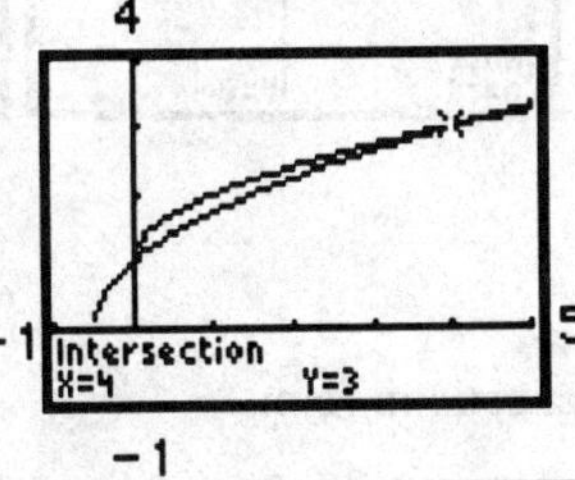

Figure 11

Skill Building Exercises Solve each equation involving radicals and check for extraneous roots.

1. (a) $\sqrt{x} = x$
   (b) $\sqrt{x} = -x$
2. (a) $\sqrt{x + 3} = x - 3$
   (b) $\sqrt{x + 3} = 3 - x$
3. (a) $\sqrt{7 - 6x} = 2x - 1$
   (b) $\sqrt{7 - 6x} = 1 - 2x$
4. $\sqrt{2x - 1} - \sqrt{2 - x} = 0$
5. $\sqrt{x - 1} = 2 - \sqrt{x - 1}$
6. $\dfrac{2 - \sqrt{x}}{3} = \dfrac{1}{2 + \sqrt{x}}$

Answers **1.**(a) $x = 0$ or $x = 1$ (b) $x = 0$ **2.**(a) $x = 6$ $(x = 1)$ (b) $x = 1$ $(x = 6)$ **3.**(a) $x = 1$ $(x = -3/2)$ (b) $x = -3/2$ $(x = 1)$ **4.** $x = 1$ **5.** $x = 2$ **6.** $x = 1$

Solving an Equation That Is Quadratic-in-Form

An equation is **quadratic-in-form** if the variable in one term is the square of the variable in another term. After making a change of variables, we solve the resulting quadratic equation.

1. Write a quadratic equation using the variable u. *2. Solve the equation for u.* *3. Substitute the value(s) for u to find the values for x.*

Example Solve (a) $x^4 - 3x^2 + 2 = 0$ and (b) $x^{2/3} - 7x^{1/3} - 8 = 0$.

Solution

(a) We substitute $u = x^2$ and $u^2 = x^4$.

$$x^4 - 3x^2 + 2 = 0$$
$$u^2 - 3u + 2 = 0$$
$$(u - 1)(u - 2) = 0$$
$$u - 1 = 0 \text{ or } u - 2 = 0$$
$$u = 1 \text{ or } u = 2$$
$$x^2 = 1 \text{ or } x^2 = 2$$
$$x = \pm 1 \text{ or } x = \pm\sqrt{2} \approx \pm 1.41$$

(b) We substitute $u = x^{1/3}$ and $u^2 = x^{2/3}$.

$$x^{2/3} - 7x^{1/3} - 8 = 0$$
$$u^2 - 7u - 8 = 0$$
$$(u + 1)(u - 8) = 0$$
$$u + 1 = 0 \text{ or } u - 8 = 0$$
$$u = -1 \text{ or } u = 8$$
$$x^{1/3} = -1 \text{ or } x^{1/3} = 8$$
$$(x^{1/3})^3 = (-1)^3 \text{ or } (x^{1/3})^3 = 8^3$$
$$x = -1 \text{ or } x = 512$$

To solve the equation in part (a), $x^4 - 3x^2 + 2 = 0$, we graph $y = x^4 - 3x^2 + 2$ and use the root or zero feature four times. The x intercepts are the four solutions $x \approx -1.41$, $x = -1$, $x = 1$, and $x \approx 1.41$ (Figure 12).

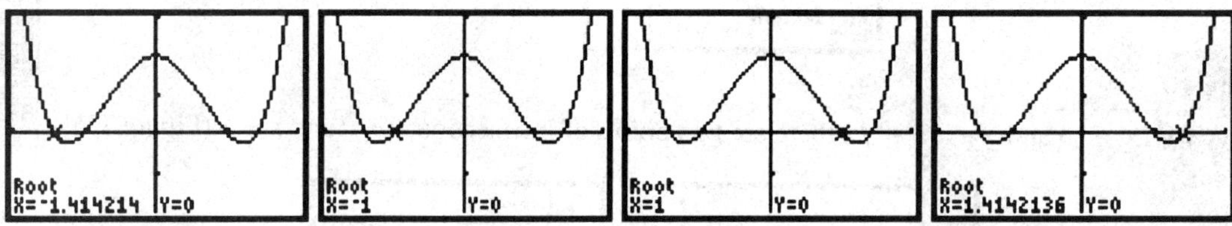

Figure 12

Skill Building Exercises
Solve each equation that is quadratic-in-form.

1. $x^4 - 5x^2 + 4 = 0$
2. $x^4 - 13x^2 + 36 = 0$
3. $x - \sqrt{x} - 2 = 0$
4. $2x - \sqrt{x} - 3 = 0$
5. $x^4 = 14x^2 + 32$
6. $x^6 - 14x^3 = 32$
7. $\frac{1}{x^2} - \frac{2}{x} - 3 = 0$
8. $\frac{1}{(x+1)^2} - \frac{2}{x+1} - 3 = 0$

Answers **1.** $x = \pm 2$ or $x = \pm 1$ **2.** $x = \pm 2$ or $x = \pm 3$ **3.** $x = 4$ **4.** $x = 9/4$ **5.** $x = \pm 4$ **6.** $x = \sqrt[3]{-2}$ or $x = 2\sqrt[3]{2}$ **7.** $x = -1$ or $x = 1/3$ **8.** $x = -2$ or $x = -2/3$

Using Interval Notation

Interval notation is a convenient way of representing inequalities. Parentheses are used to designate open intervals, intervals that do not contain the endpoints. Brackets are used for closed intervals, intervals that contain the endpoints. Half-open (half-closed) intervals, where only one endpoint is contained in the interval, use a combination of a parenthesis and a bracket. In interval notation, we use the ∞ symbol to indicate that the interval continues indefinitely. We always use a parenthesis with the ∞ symbol. The possibilities are summarized in the table.

Interval	Notation	Graph
$-2 < x < 2$	(-2, 2)	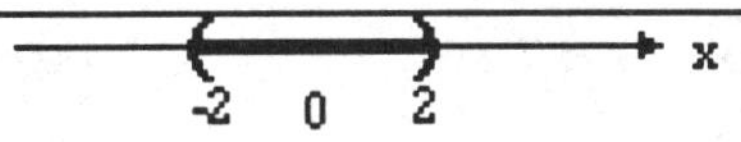 -2 0 2 x
$-2 \leq x \leq 2$	[-2, 2]	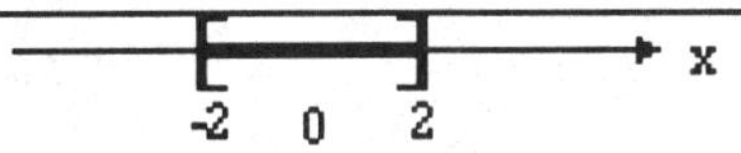 -2 0 2 x
$-2 < x \leq 2$	(-2, 2]	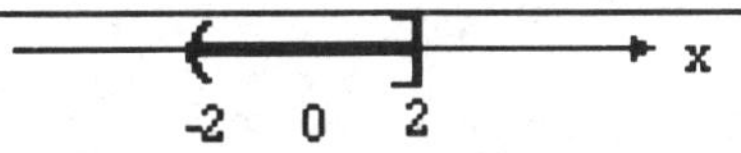 -2 0 2 x
$-2 \leq x < 2$	[-2, 2)	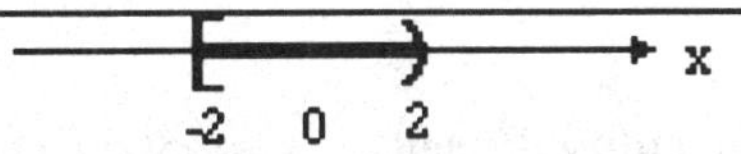 -2 0 2 x
$x > 2$	(2, ∞)	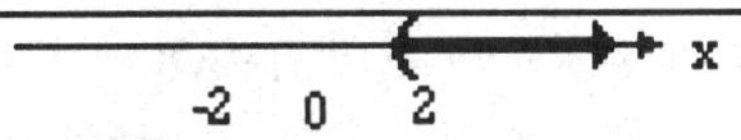 -2 0 2 x
$x \geq 2$	[2, ∞)	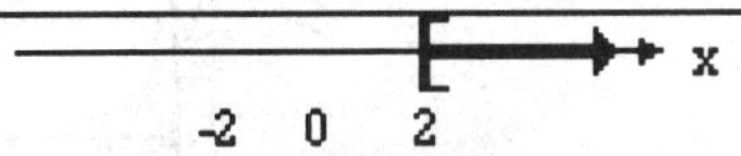 -2 0 2 x
$x < 2$	(-∞, 2)	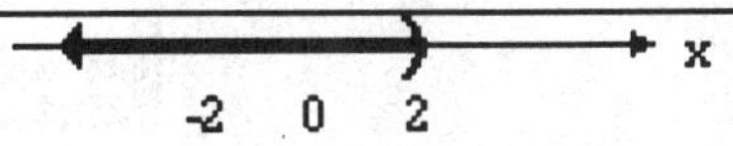 -2 0 2 x
$x \leq 2$	(-∞, 2]	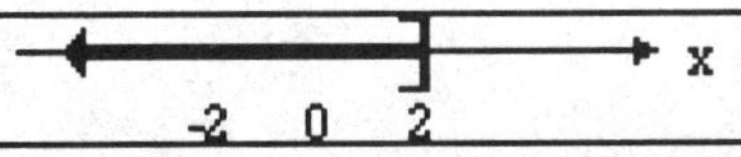 -2 0 2 x

Since intervals are sets, we use the symbol from set notation ∪ (union) to combine inequalities connected with the word or.

Example Express in interval notation.

(a) $x < -1$ or $x \geq 5$ (b) $x \nless 4$ (c) all real numbers

Solution

(a) (-∞, -1) ∪ [5, ∞) (b) [4, ∞) (c) (-∞, ∞)

Skill Building Exercises Express in interval notation.

1. $-1 \leq x \leq 8$
2. $3 < x \leq 9$
3. $x \neq 0$
4. $x \ngeq 4$
5. all positive real numbers
6. all negative real numbers
7. all nonnegative real numbers
8. all nonpositive real numbers

Answers **1.** [-1, 8] **2.** (3, 9] **3.** (-∞, 0) ∪ (0, ∞) **4.** (-∞, 4) **5.** (0, ∞) **6.** (-∞, 0) **7.** [0, ∞) **8.** (-∞, 0]

Solving a Linear Inequality

Solving linear inequalities is similar to solving linear equations except that whenever an inequality is multiplied or divided by a negative number, the order of the inequality is reversed. We proceed as follows.

1. Combine all terms containing the variable on one side of the equation and all constant terms on the other. *2. Divide both sides of the equation by the coefficient of the variable.*

Example Solve the inequality $-5(x + 1) < 7$.

Solution

$-5(x + 1) < 7$	*Apply the distributive property.*
$-5x - 5 < 7$	*Add 5 to both sides.*
$-5x < 12$	*Multiply both sides by $\frac{-1}{5}$. Reverse the inequality.*
$x > -\frac{12}{5}$	

In interval notation, the solution is $(-\frac{12}{5}, \infty)$.

A graphing utility display of $y_1 = -5(x + 1)$ and $y_2 = 7$ shows that $-5(x + 1) < 7$ for $x > -2.4$, since $y_1 < y_2$ whenever the graph of $y_1 = -5(x + 1)$ is <u>below</u> the graph of $y_2 = 7$ (Figure 13).

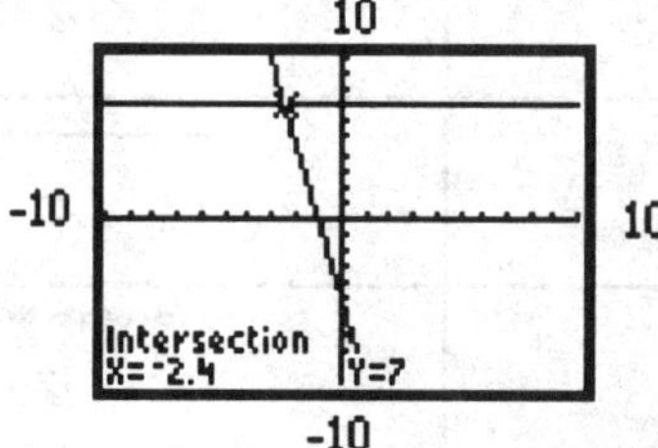

Figure 13

Finding the point of intersection of the two graphs provides the boundary point for the solution to the inequality. This boundary point also yields the solutions to the following inequalities.

$-5(x + 1) \leq 7 \Rightarrow x \leq 2.4$ $\quad -5(x + 1) > 7 \Rightarrow x > 2.4$ $\quad -5(x + 1) \geq 7 \Rightarrow x \geq 2.4$

Skill Building Exercises

Solve each linear inequality and express each solution in interval notation.

1. $3x - 4 \leq 11$
2. $8 - 5x > 15$
3. $7x - 8 \leq 15x + 2$
4. $3 - 2(x - 1) \geq x + 6$
5. $6 - 4(2x - 1) < 10 - 2x$
6. $4x - x(x - 1) \leq x(4 - x)$
7. $(x - 3)(x + 5) > 6x + x(x + 2)$
8. $(2x + 4)(3x - 5) > (6x + 3)(x - 4)$

Answers **1.** $(-\infty, 5]$ **2.** $(-\infty, -7/5)$ **3.** $[-5/4, \infty)$ **4.** $(-\infty, -1/3]$ **5.** $(0, \infty)$ **6.** $(-\infty, 0]$ **7.** $(-\infty, -5/2)$ **8.** $(8/23, \infty)$

Solving a Quadratic Inequality

To solve a quadratic inequality we consider all possible sign combinations of the factors of the quadratic expression. We proceed as follows.

> *1. Replace the inequality symbol with an equal sign and solve the corresponding quadratic equation.*
> *2. Consider all possible sign combinations of the factors of the quadratic expression. If the quadratic expression changes sign, it probably occurs where its value is zero.*

Example Solve the quadratic inequality $x^2 + 2x - 8 > 0$.

Solution First, solve the corresponding quadratic equation.

$$x^2 + 2x - 8 = 0$$
$$(x + 4)(x - 2) = 0$$
$$x + 4 = 0 \text{ or } x - 2 = 0$$
$$x = -4 \text{ or } x = 2$$

Since the quadratic equation equals zero when $x = -4$ or $x = 2$, they are the boundary points in Figure 14. The following table states the sign of each factor for the three intervals.

	$x < -4$	$-4 < x < 2$	$x > 2$
$x + 4$	negative	positive	positive
$x - 2$	negative	negative	positive

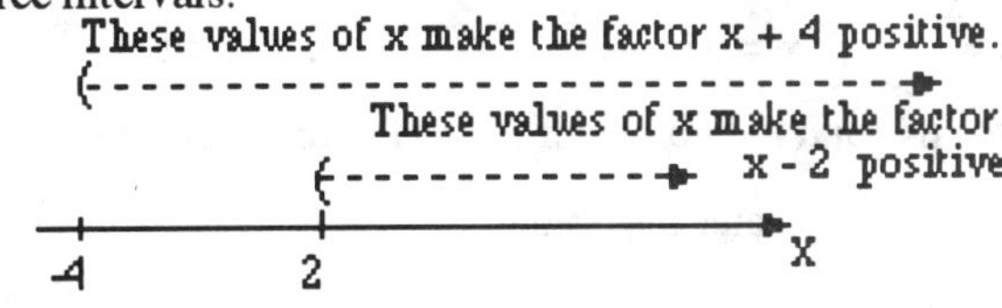

Figure 14

A greater-than inequality requires the product of $x + 4$ and $x - 2$ to be positive. The factors $(x + 4)$ and $(x - 2)$ must be both positive or both negative. The solution is $x < -4$ or $x > 2$. If a quadratic inequality involves less than, such as $x^2 + 2x - 8 < 0$, the algebraic solution requires the product of two factors to be negative. One factor must be positive and the other negative. Notice the solutions to the following inequalities.

$$x^2 + 2x - 8 > 0 \Rightarrow x < -4 \text{ or } x > 2 \qquad x^2 + 2x - 8 < 0 \Rightarrow -4 < x < 2$$
$$x^2 + 2x - 8 \geq 0 \Rightarrow x \leq -4 \text{ or } x \geq 2 \qquad x^2 + 2x - 8 \leq 0 \Rightarrow -4 \leq x \leq 2$$

To use a graphing utility we find the x intercepts of the graph of $y_1 = x^2 + 2x - 8$ (Figure 15).

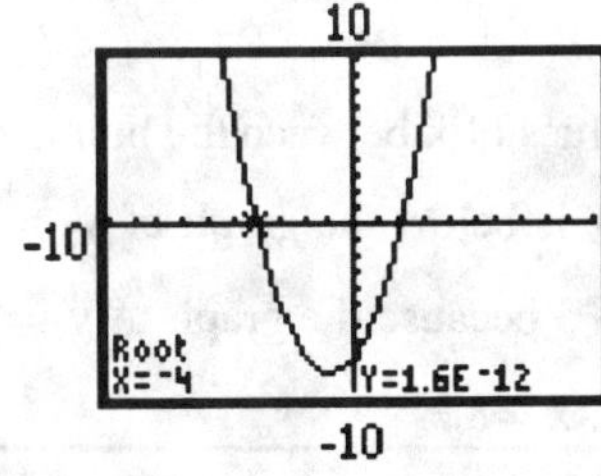

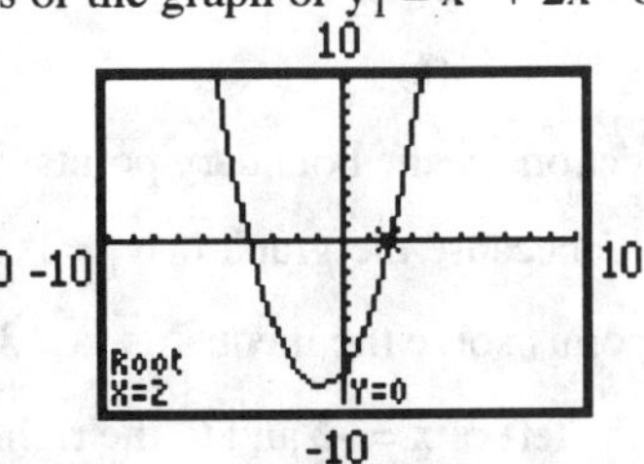

Figure 15
These are the boundary points for the solution to the quadratic inequality. The expression $x^2 + 2x - 8$ is positive whenever the graph is above the x axis.

Skill Building Exercises Solve each quadratic inequality.

1. (a) $x^2 - 25 < 0$ (b) $25 - x^2 < 0$
2. (a) $16 - x^2 > 0$ (b) $16 - x^2 \leq 0$
3. (a) $(x + 1)(x + 2) \leq 0$ (b) $(x + 1)(x + 2) > 0$
4. (a) $x^2 + x - 12 \leq 0$ (b) $x(x + 1) \geq 12$
5. $-x^2 - 7x - 12 \geq 0$
6. $x^2 + 2x + 1 < 0$
7. $4x^2 + 20x + 25 \leq 0$
8. $x^2 + 6x + 9 > 0$

Answers **1.**(a) $-5 < x < 5$ (b) $x < -5$ or $x > 5$ **2.**(a) $-4 < x < 4$ (b) $x \leq -4$ or $x \geq 4$ **3.**(a) $-2 \leq x \leq -1$ (b) $x < -2$ or $x > -1$ **4.**(a) $-4 \leq x \leq 3$ (b) $x \leq -4$ or $x \geq 3$ **5.** $-4 \leq x \leq -3$ **6.** no solution **7.** $x = -5/2$ **8.** $x \neq -3$

Solving an Equation or Inequality Involving Absolute Value

Absolute value is often used for distances on the real number line. We use the following theorems to remove the absolute value symbols.

For k > 0	
1. $|x| = k$ *for* $x = -k$ *or* $x = +k$	*(Both* $x = \pm k$ *are k units from 0.)*
2. $|x| < k$ *for* $-k < x < k$	*(Numbers between* $\pm k$ *are less than k units from 0.)*
3. $|x| > k$ *for* $x < -k$ *or* $x > k$	*(Numbers left of -k or right of +k are more than k units from 0.)*

Example Solve (a) $|x - 3| = 5$, (b) $|x - 3| < 5$, and (c) $|x - 3| > 5$.

Solution

(a) $|x - 3| = 5$

$x - 3 = -5$ or $x - 3 = 5$

$x = -2$ or $x = 8$

(b) $|x - 3| < 5$

$-5 < x - 3 < 5$

$-2 < x < 8$

(c) $|x - 3| > 5$

$x - 3 < -5$ or $x - 3 > 5$

$x < -2$ or $x > 8$

We have found the points on the real number line whose distance form 3 is (a) exactly 5 units, (b) less than 5 units, and (c) greater than 5 units.

On a graphing utility, we enter $y_1 =$ ABS $(x - 3)$ and $y_2 = 5$, and use the intersect feature to find where $|x - 3| = 5$ (Figure 16).

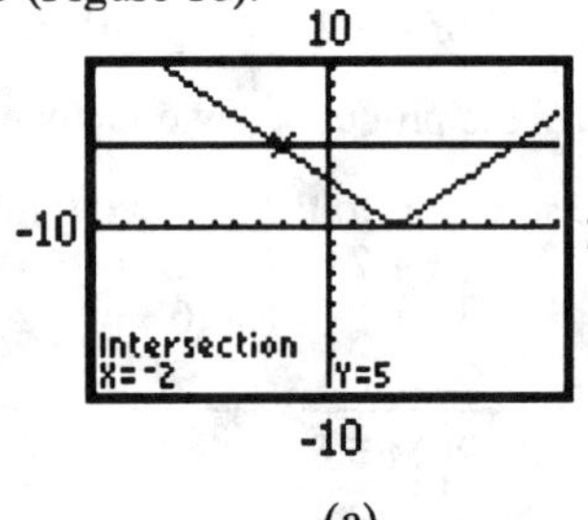

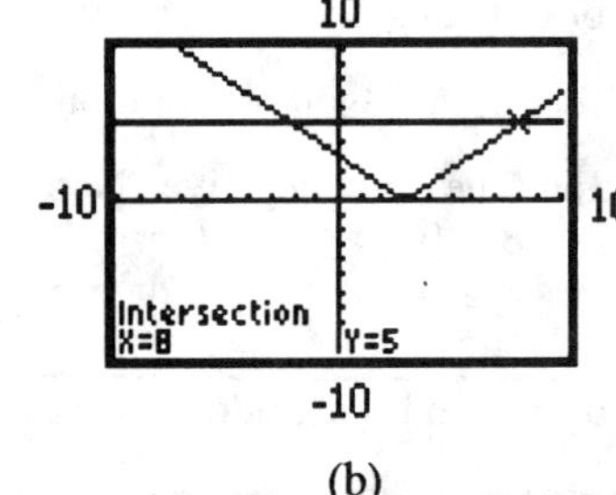

Figure 16 (a) (b)

These points of intersection are the boundary points. Values of x between the boundary points solve the inequality $|x - 3| < 5$ because the graph of $y_1 = |x - 3|$ is below the graph of $y_2 = 5$. Values of x outside the boundary points solve the inequality $|x - 3| > 5$ because the graph of $y_1 = |x - 3|$ is above the graph of $y_2 = 5$ to the left of $x = -2$ and to the right of $x = 8$.

Skill Building Exercises Solve each equation or inequality involving absolute value.

1. (a) $|x| = 5$ (b) $|x| = -5$
2. (a) $|x| > 4$ (b) $|x - 2| > 4$
3. (a) $|x - 4| \leq 10$ (b) $|x - 4| > 10$
4. (a) $|1 - x| \leq 4$ (b) $|x - 1| \leq 4$
5. $|2x - 8| > 6$
6. $|2x - 4| < 12$
7. $|x| > 0$
8. $|x| < 0$

Answers **1.**(a) $x = -5$ or $x = 5$ (b) no solution **2.**(a) $x < -4$ or $x > 4$ (b) $x < -2$ or $x > 6$ **3.**(a) $-6 \leq x \leq 14$ (b) $x < -6$ or $x > 14$ **4.**(a) $-3 \leq x \leq 5$ (b) $-3 \leq x \leq 5$ **5.** $x < 1$ or $x > 7$ **6.** $-4 < x < 8$ **7.** $x \neq 0$ **8.** no solution

Appendix II

Library of Reference Functions

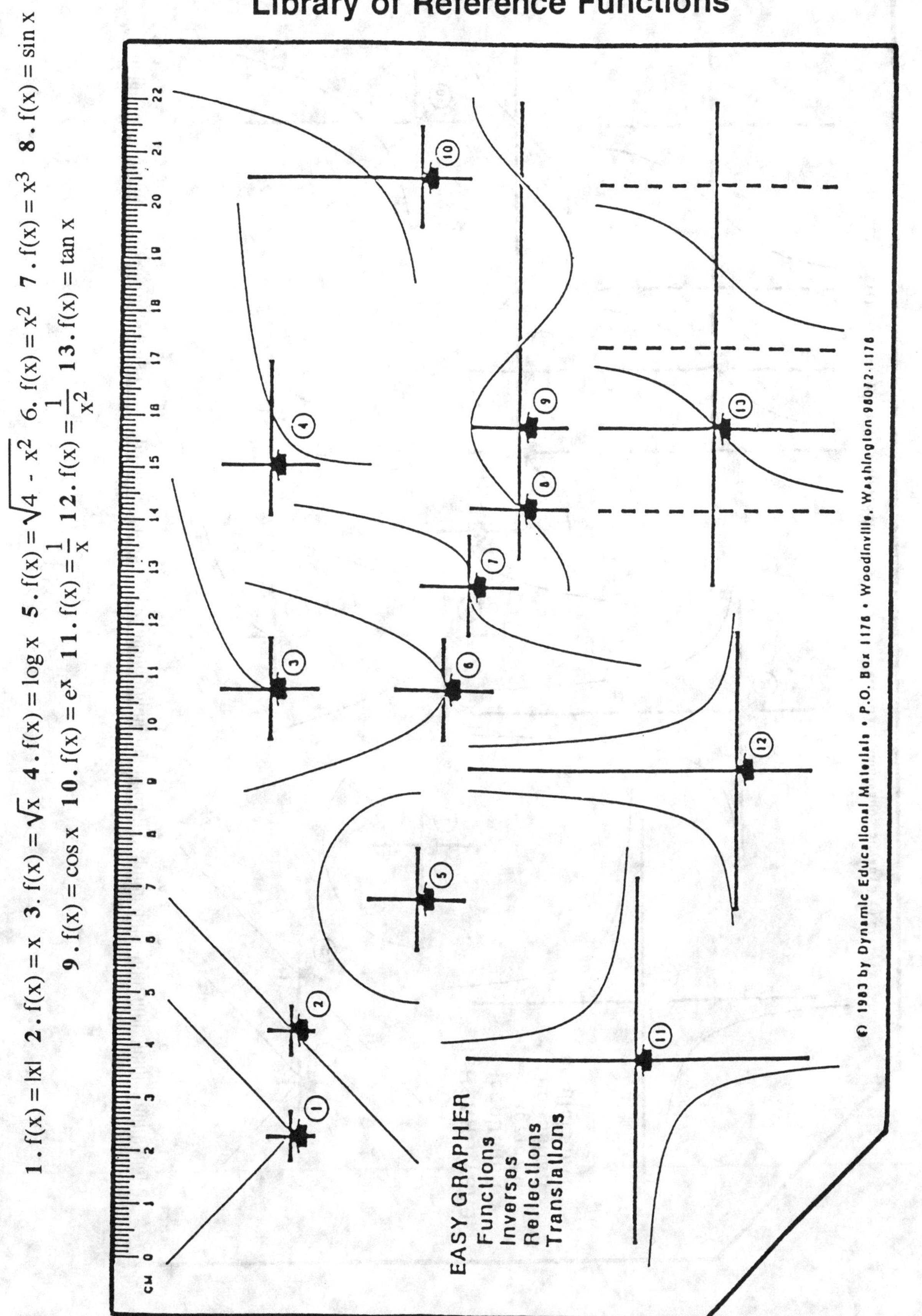

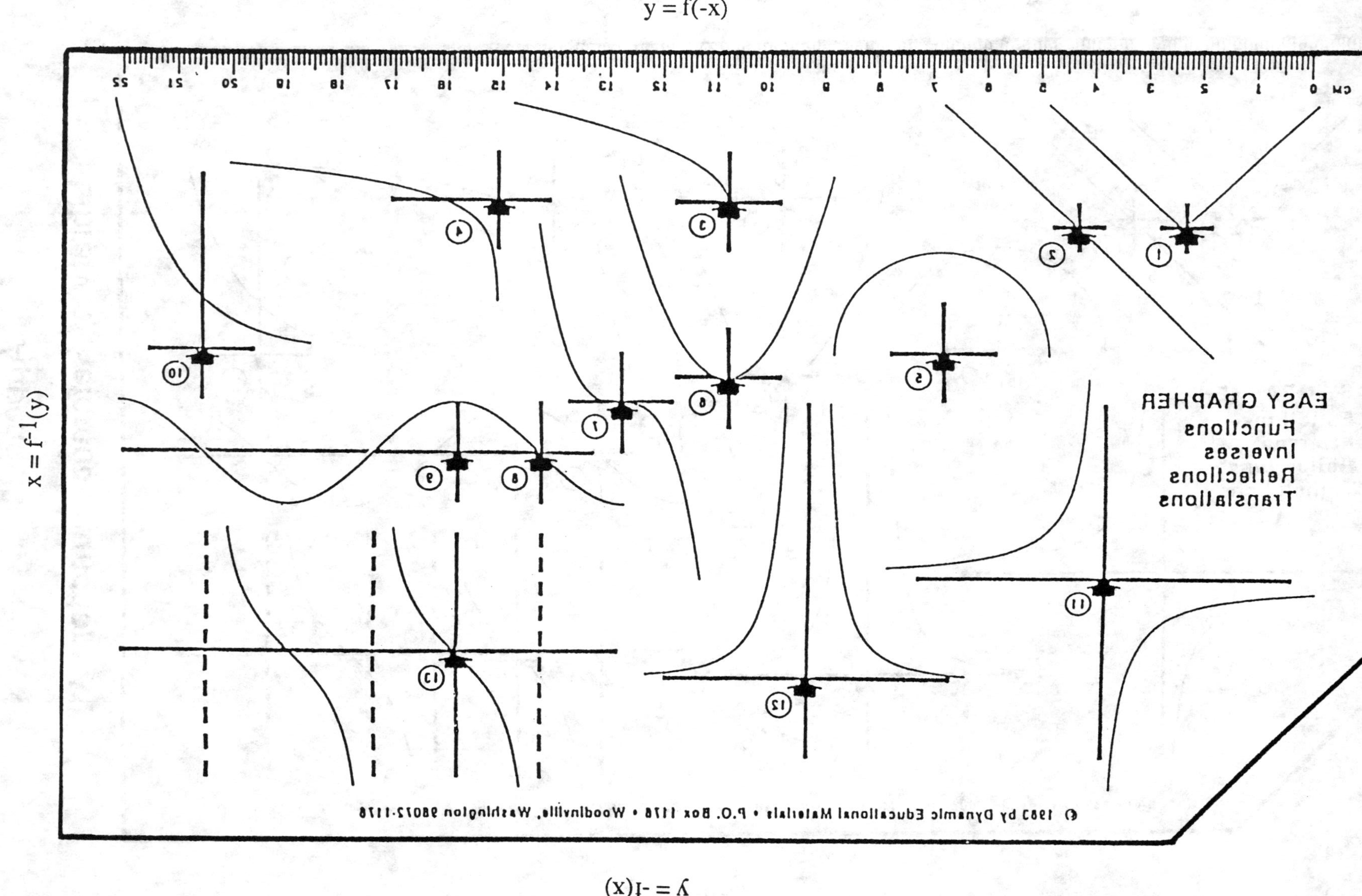
$y = f(-x)$
$x = f^{-1}(y)$
$y = -f(x)$
EASY GRAPHER
Functions
Inverses
Reflections
Translations
© 1983 by Dynamic Educational Materials • P.O. Box 1178 • Woodinville, Washington 98072-1178

Appendix III

Regression Analysis Formulas

Using Formulas to Find the Equation of the Best-Fitting Line

Graphing utilities use the following formulas to find the best-fitting line. As we will see, the formulas are derived by the method of least squares.

Formulas for the Best-Fitting Line and the Correlation Coefficient

If

$$A = n\left(\sum_{i=1}^{n} x_i y_i\right) - \left(\sum_{i=1}^{n} x_i\right)\left(\sum_{i=1}^{n} y_i\right)$$

$$B = n\left(\sum_{i=1}^{n} x_i^2\right) - \left(\sum_{i=1}^{n} x_i\right)\left(\sum_{i=1}^{n} x_i\right)$$

$$C = n\left(\sum_{i=1}^{n} y_i^2\right) - \left(\sum_{i=1}^{n} y_i\right)\left(\sum_{i=1}^{n} y_i\right)$$

then the equation of the **best-fitting line** from a set of data is

$$y = a + bx \quad \text{where } a = \bar{y} - b\bar{x} \text{ and } b = \frac{A}{B}$$

The **correlation coefficient** is $r = \dfrac{A}{\sqrt{BC}}$.

Graphing utilities use the fact that the point $(\bar{x}, \bar{y})$ lies on the best-fitting line to find the y intercept, $a = \bar{y} - b\bar{x}$. In the next example we use these formulas to find the equation of the best-fitting line.

Example 1

(a) Use the regression formulas to show that the equation of the line that best fits the data {(-1,-1), (2,3), (5,2)} is $y = 0.33 + 0.5x$.

(b) Find the correlation coefficient.

Solution

(a) The following table helps organize the data. The sum of each column is listed in the bottom row.

x	y	x^2	xy	y^2
-1	-1	1	1	1
2	3	4	6	9
5	2	25	10	4
6	**4**	30	17	14

Dividing the sum of the x coordinates, 6, and the sum of the y coordinates, 4, by the number of data points, n = 3, yields

$$\bar{x} = \frac{6}{3} = 2 \text{ and } \bar{y} = \frac{4}{3}$$

Substituting the sums from the table allows us to find the values of A and B as follows.

$$A = n\left(\sum_{i=1}^{n} x_i y_i\right) - \left(\sum_{i=1}^{n} x_i\right)\left(\sum_{i=1}^{n} y_i\right) = 3(17) - (6)(4) = 27$$

$$B = n\left(\sum_{i=1}^{n} x_i^2\right) - \left(\sum_{i=1}^{n} x_i\right)\left(\sum_{i=1}^{n} x_i\right) = 3(30) - (6)(6) = 54$$

Therefore, the slope is

$$b = \frac{A}{B} = \frac{27}{54} = \frac{1}{2}$$

and the y intercept is

$$a = \bar{y} - b\bar{x} = \frac{4}{3} - \left(\frac{1}{2}\right)(2) = \frac{1}{3}$$

The equation of the best-fitting line for these data is

$$y = a + bx$$

$$y = \frac{1}{3} + \frac{1}{2}x$$

(b) We find the value of C using the values from the table.

$$C = n\left(\sum_{i=1}^{n} y_i^2\right) - \left(\sum_{i=1}^{n} y_i\right)\left(\sum_{i=1}^{n} y_i\right) = 3(14) - (4)(4) = 42 - 16 = 26$$

Therefore,

$$r = \frac{A}{\sqrt{BC}} = \frac{27}{\sqrt{54 \cdot 26}} \approx 0.7205 \approx 0.72$$

Using a graphing utility to find the best-fitting line for this set of data yields y = 0.33 + 0.5x (Figure 1).

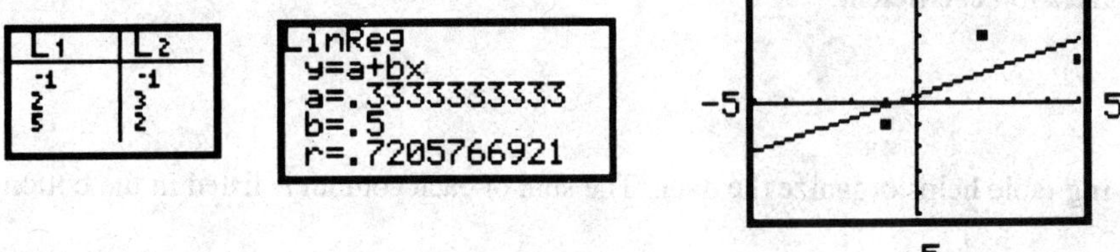

Figure 1

Interpreting the Coefficient of Determination

While graphing utilities sometimes calculate the correlation coefficient r, we are really interested in r^2, the *coefficient of determination*. Now we are ready to study the meaning of the correlation coefficient $r = 0.72$ by interpreting the coefficient of determination $r^2 = 0.52$.

Finding the best-fitting line requires finding the slope and y intercept of the equation that gets closest to all the data points. By *closest* we mean that the sum of the squares of the vertical distances from each data point to the line is minimized. We call this criterion, *least squares*.

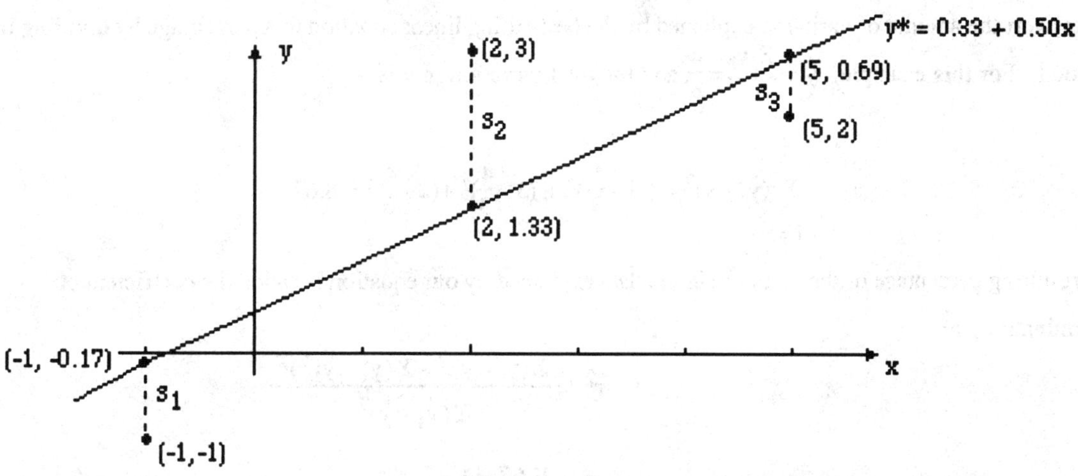

Figure 2

The y values on the line, called y_i^*, are calculated from the equation $y^* = 0.33 + 0.50x$. In Figure 2, the sum of the squares of the indicated distances is 4.17, since

$$S = s_1^2 + s_2^2 + s_3^2$$

$$S = \sum_{i=1}^{n}(y_i - y_i^*)^2 = [-1 - (-0.2)]^2 + (3 - 1.3)^2 + (2 - 2.8)^2 = 4.17$$

This is the variation in y that is <u>not explained</u> by the equation of the best-fitting line. This is a measure of our error. Our linear regression (least squares) method finds the slope and the y intercept that makes the value of S as small as possible.

How would you find a measure of the total variation in the data?

The total amount of variation in y is found by calculating the variation from the mean, $\bar{y} = \dfrac{\Sigma y_i}{n}$.

(If there were no variation in y, each y value would be equal to the mean and $y_i - \bar{y}$ would be zero for each y.)

The total variation in y is given by

$$\sum_{i=1}^{n} (y_i - \bar{y})^2$$

How would you find a measure of the variation in the data that is explained by the equation of the best-fitting line?

If we subtract the amount of variation in y not explained by our equation from the total variation, we find the variation explained by our equation:

$$\sum_{i=1}^{n} (y_i - \bar{y})^2 - \sum_{i=1}^{n} (y_i - y_i^*)^2$$

We convert the amount of variation explained by the best-fitting linear equation to a percentage by dividing by the total. For this example, $\bar{y} = \frac{\Sigma y_i}{n} = \frac{4}{3}$, and the total variation in y is

$$\sum_{i=1}^{n} (y_i - \bar{y})^2 = (-1 - \frac{4}{3})^2 + (3 - \frac{4}{3})^2 + (2 - \frac{4}{3})^2 = 8.67$$

The resulting percentage of the variation in the data explained by our equation is called the coefficient of determination, r^2.

$$r^2 = \frac{\Sigma(y_i - \bar{y})^2 - \Sigma(y_i - y_i^*)^2}{\Sigma(y_i - \bar{y})^2}$$

$$r^2 = \frac{8.67 - 4.17}{8.67} = 0.519$$

Since the coefficient of determination is $r^2 = 0.519$, 52% of the total variation in y is explained by the equation of the best-fitting line. This is near the practical lower limit for r^2. We can interpret $r \approx 0.72$ by squaring r to find the coefficient of determination.

Coefficient of Determination

$$r^2 = \frac{\Sigma(y_i - \bar{y})^2 - \Sigma(y_i - y_i^*)^2}{\Sigma(y_i - \bar{y})^2}$$

is the coefficient of determination where y_i is each given value of y, y_i^* is the corresponding value of y on the best fitting line, and $\bar{y}$ is the mean

$$\bar{y} = \frac{\Sigma y_i}{n}$$

If less than half of the variation in y is explained by the equation of the best-fitting line, would you use the equation to estimate unknown values of y for specified values of x?

It would probably be of little value to use such an equation to estimate missing values of y. If $r^2 < 0.50$, we use $\bar{y}$ to approximate missing y values.

Now let's investigate the derivation of the formulas themselves. We will parallel the steps in the general proof using the three points in Example 1. We use the vertex formula to find the minimum value of two quadratic equations to create two equations with two unknowns, the slope b and y intercept a of the best-fitting linear equation.

We will find a and b in the equation $y^* = a + bx$ that minimizes the sum of the squares of the deviations from the given points (-1, -1), (2, 3), and (5, 2) to this best fitting line (Figure 2).

$$S = s_1^2 + s_2^2 + s_3^2$$

$$\begin{aligned} S &= \Sigma(y_i - y_i^*)^2 \\ &= \Sigma[y_i - (a + bx_i)]^2 \\ &= [-1 - (a + b(-1))]^2 + [3 - (a + b(2))]^2 + [2 - (a + b(5))]^2 \\ &= 14 - 8a - 34b + 3a^2 + 12ab + 30b^2 \end{aligned}$$

We are now going to create two linear equations with two variables, a and b, that we know will minimize S. Treating this as a quadratic equation in a and applying the vertex formula will minimize S as required.

$$S = 3a^2 + (12b - 8)a + (30b^2 - 34b + 14)$$

$$a = -\frac{12b - 8}{2(3)}$$

$$\mathbf{6a + 12b = 8}$$

Treating this as a quadratic equation in b and applying the vertex formula again will also minimize S.

$$S = 30b^2 + (12a - 34)b + (3a^2 - 8a + 14)$$

$$b = -\frac{12a - 34}{2(30)}$$

$$\mathbf{12a + 60b = 34}$$

Solving these two equations simultaneously yields the values of a and b that minimize S.

$$\begin{aligned} 6a + 12b &= 8 \\ 12a + 60b &= 34 \end{aligned} \qquad \begin{aligned} 6a + 12b &= 8 \\ \underline{-6a - 30b} &= \underline{-17} \\ -18b &= -9 \\ b &= 1/2 \end{aligned}$$

Substituting b = 1/2 into either equation yields a = 1/3. Hence, the equation of the best-fitting line is

$$y = \frac{1}{3} + \frac{1}{2}x$$

Do you see why this method is called "least squares"?

We used the vertex formula to find the minimum (least) value of S -- the sum of the squares of the distances between the given points and the corresponding points on the line.

Using Quadratic Regression

We can apply the method of least squares to finding the best-fitting polynomial function of any degree. In the next example we use a graphing utility to find the values of a, b, and c for the quadratic function that minimizes

$$S = \Sigma(y_i - y_i^*)^2$$

where $y^* = ax^2 + bx + c$ is our best -fitting quadratic and the y_is are the y values of the given data. Most important, we calculate the coefficient of determination for our best-fitting quadratic function, a calculation not programmed by most graphing utilities.

Example 2

(a) Find the quadratic equation that best fits the data {(-1,-1), (2,3), (5,2)}.

(b) Find the coefficient of determination.

Solution

(a) The equation of the best-fitting parabola is $y = -0.28x^2 + 1.61x + 0.89$ (Figure 3).

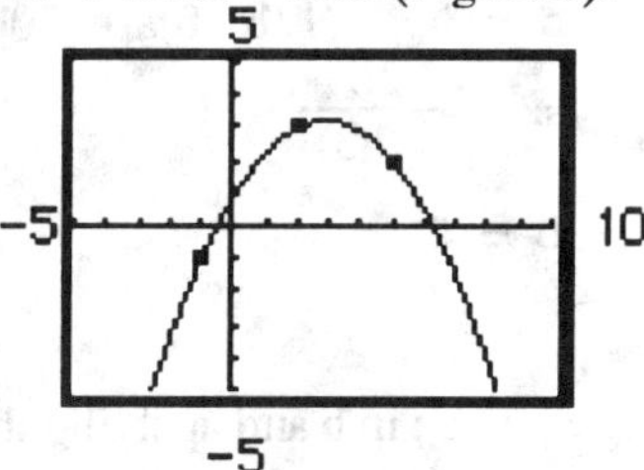

Figure 3

(b) We find the coefficient of determination from its formula, where the y_i^*' values are found from the best-fitting quadratic equation. We know that $\bar{y} = \frac{\Sigma y_i}{n} = \frac{4}{3}$.

x	y	$y^* = -0.28x^2 + 1.61x + 0.89$	$(y_i - y_i^*)^2$	$(y_i - \bar{y})^2$
-1	-1	-1	0	5.43
2	3	2.99	.0001	2.79
5	2	1.94	0.0036	0.45
			0.0037	8.67

$$r^2 = \frac{\Sigma(y_i - \bar{y})^2 - \Sigma(y_i - y_i^*)^2}{\Sigma(y_i - \bar{y})^2}$$

$$\approx \frac{8.67 - 0.0037}{8.67} \approx 0.99957$$

We know that 99.96% of the variation in y is explained by this best-fitting quadratic equation.

Appendix IV

Formulas

Geometric Formulas

A = area, P = perimeter, C = circumference, V = volume, S = surface area

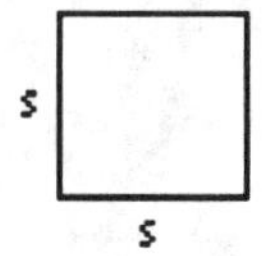	**Square** $A = s^2$ $P = 4s$	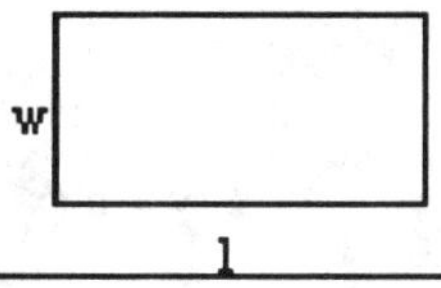	**Rectangle** $A = lw$ $P = 2l + 2w$
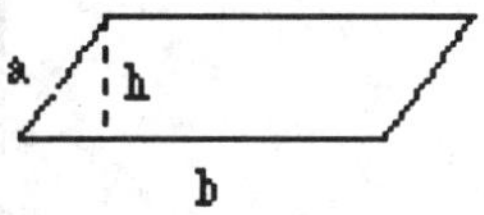	**Parallelogram** $A = bh$ $P = 2a + 2b$	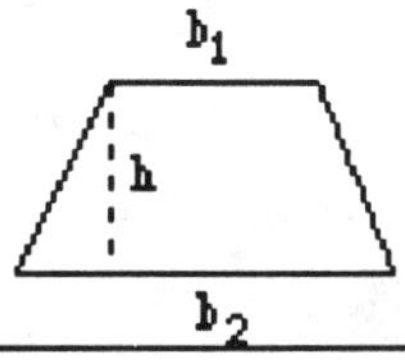	**Trapezoid** $A = \left(\frac{b_1 + b_2}{2}\right)h$
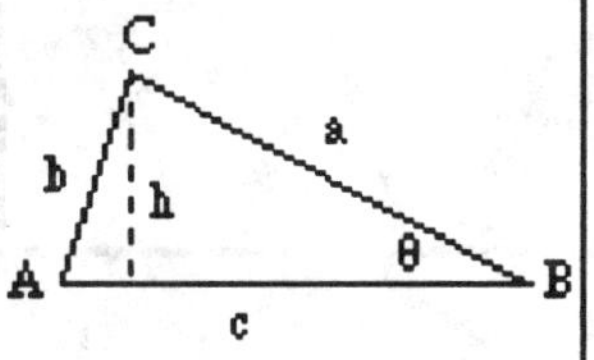	**Triangle** $A = \frac{1}{2}bh$ $A = ac \sin\theta$ $P = a + b + c$	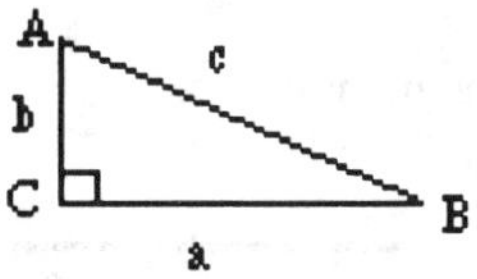	**Right Triangle** $a^2 + b^2 = c^2$
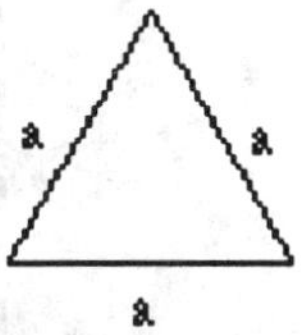	**Equilateral Triangle** $A = \frac{\sqrt{3}}{4}a^2$ $P = 3a$		**Isosceles Triangle** $A = \frac{1}{4}b\sqrt{4a^2 - b^2}$ $P = 2a + b$
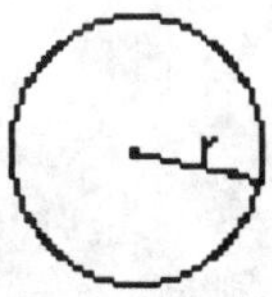	**Circle** $A = \pi r^2$ $C = 2\pi r = \pi d$		**Sphere** $V = \frac{4}{3}\pi r^3$ $S = 4\pi r^2$
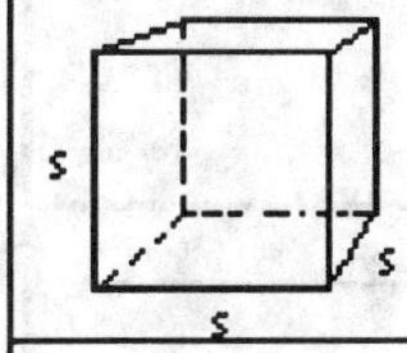	**Cube** $V = s^3$ $S = 6s^2$	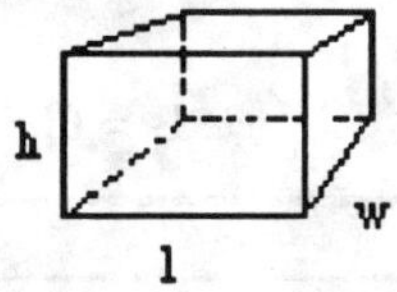	**Rectangular Solid** $V = lwh$ $S = 2lh + 2lw + 2wh$
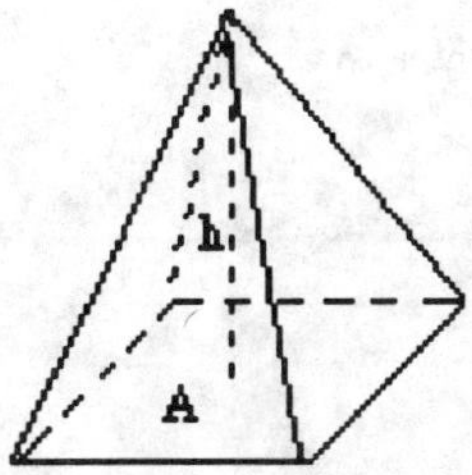	**Pyramid** $V = \frac{1}{3}Ah$	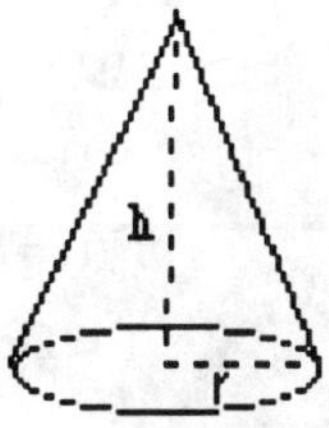	**Cone** $V = \frac{1}{3}\pi r^2h$ $S = \pi rl + \pi r^2$

Algebraic Formulas

midpoint:	$M = (\frac{x_1 + x_2}{2}, \frac{y_1 + y_2}{2})$
distance:	$d(P,Q) = \sqrt{(x_2 - x_1)^2 + (y_2 - y_1)^2}$
slope:	$m = \frac{\Delta y}{\Delta x} = \frac{y_2 - y_1}{x_2 - x_1}, x_1 \neq x_2$
slope-intercept form of a line:	$y = mx + b$ or $y = a + bx$ or $y = a + bx$
point-slope form of a line:	$y = m(x - x_1) + y_1$
quadratic formula:	$x = \frac{-b \pm \sqrt{b^2 - 4ac}}{2a}$
discriminant:	$b^2 - 4ac$
vertex of a parabola:	$x = \frac{-b}{2a},\ y = f(\frac{-b}{2a})$
turning points of a cubic polynomial:	$x = \frac{-b \pm \sqrt{b^2 - 3ac}}{3a}$
point of inflection of a cubic polynomial:	$x = -\frac{b}{3a}$

Finance Formulas

simple interest:	$I = prt$
annually compounded interest:	$A = P(1 + r)^t$
interest compounded k times a year:	$A = P(1 + \frac{r}{k})^{tk}$
continuously compounded interest:	$A = Pe^{rt}$
revenue:	$R = px$
cost:	$C = C_1 + C_2$
profit:	$P = R - C$
annuity:	$S = R[\frac{a_n - 1}{i}], a = 1 + i$
monthly payment:	$M = \frac{iP}{1 - a^{-n}}, a = 1 + i$
monthly balance:	$B = P[\frac{a^n - a^{k-1}}{a^n - 1}], a = 1 + i$

Application Formulas

freefall (in feet and meters):	$s(t) = -16t^2 + v_0 t + s_0$ and $s(t) = -4.9t^2 + v_0 t + s_0$
exponential growth:	$N(t) = N_0 e^{kt}$
exponential decay:	$N(t) = N_0 e^{-kt}$
Newton's law of cooling (heating):	$T = De^{-kt} + T_a$
logistic:	$N = \frac{c}{1 + ae^{-bt}}$

Appendix V

Theory of Equations

Remainder and Factor Theorems

Consider dividing the polynomial $p(x) = 2x^2 - 7x - 15$ by $x - 5$. We write

$$\frac{2x^2 - 7x - 15}{x - 5} = 2x + 3$$

This means that $x - 5$ is a factor of $p(x) = 2x^2 - 7x - 15$, and the resulting quotient, $2x + 3$, is the other factor.

If we divide $p(x) = 2x^2 - 7x - 15$ by $x - 3$, we have a remainder of -18. Hence,

$$\frac{p(x)}{x - 3} = \frac{2x^2 - 7x - 15}{x - 3} = 2x - 1 + \frac{-18}{x - 3}$$

This means that $p(x) = (x - 3)\left[2x - 1 + \frac{-18}{x - 3}\right] = (x - 3)(2x - 1) - 18$. Then $p(3) = 0 - 18 = -18$. Since $p(3) = -18$, $(3, -18)$ is a point on the graph of this polynomial. The remainder theorem summarizes these observations; if a polynomial $p(x)$ is divided by $x - a$, the remainder is $p(a)$.

Remainder Theorem

In the quotient $\frac{p(x)}{x - a}$ there exists a unique polynomial q such that

$$\frac{p(x)}{x - a} = q(x) + \frac{r}{x - a} \quad \text{and} \quad p(x) = (x - a)q(x) + r$$

where r is the remainder and $p(a) = r$.

Example 1 Use synthetic division to find p(2) for $p(x) = 3x^3 - 2x^2 + x - 1$.

Solution

$$\begin{array}{r|l} & 3x^2 + 4x + 9 \\ \hline x - 2 & 3x^3 - 2x^2 + x - 1 \\ & \underline{3x^3 - 6x^2} \\ & \quad 4x^2 + x \\ & \quad \underline{4x^2 - 8x} \\ & \qquad 9x - 1 \\ & \qquad \underline{9x - 18} \\ & \qquad\quad 17 \end{array}$$

Since the remainder is 17, $p(2) = 17$.

An important corollary of the remainder theorem is the factor theorem, which says that if the remainder is zero,

$r = 0$, we have found one of the polynomial's factors.

Factor Theorem

x - a is a factor of a polynomial function p(x) if and only if p(a) = 0.

Examining Possible Rational Roots

To factor a polynomial equation, we need a plan for determining a list of possible factors. Division can be used to check any possibility. Consider the factors of $2x^2 - 7x - 15$, (2x + 3), and (x - 5).

The leading coefficient in each binomial factor, 2 and 1, divide the leading coefficient of the polynomial, 2. The constant terms in each binomial factor, 3 and -5, divide the constant term of the polynomial, -15. Since 2x + 3 is a *factor* of the polynomial, its root is a rational number of the form p/q,

$$2x + 3 = 0$$
$$2x = -3$$
$$x = -3/2$$

If qx - p = 0 then x = p/q is a root (zero).

To obtain a list of possible rational roots p/q of any polynomial, $a_n x^n + a_{n-1} x^{n-1} + \cdots + a_0$, p must divide the constant term a_0 and q must divide the leading coefficient a_n. For example, to factor

$$f(x) = \mathbf{4}x^3 - 24x^2 + 33x + \mathbf{9}$$

we look at divisors of 9 and of 4.

$$p \in \{\pm 1, \pm 3, \pm 9\} \quad \text{and} \quad q \in \{\pm 1, \pm 2, \pm 4\}$$

The possible values of p/q are all the divisors of p divided by each divisor of q.

$$\frac{p}{q} \in \{ \pm 1, \pm 3, \pm 9, \pm \frac{1}{2}, \pm \frac{3}{2}, \pm \frac{9}{2}, \pm \frac{1}{4}, \pm \frac{3}{4}, \pm \frac{9}{4}\}$$

There are no other possible rational roots. The rational root theorem generalizes these results.

Rational Root Theorem

If p/q is a rational root, reduced to lowest terms, of a polynomial function with integer coefficients, $a_0, a_1, \ldots, a_n$,

$$f(x) = a_n x^n + a_{n-1} x^{n-1} + \cdots + a_0$$

then p must divide the constant term a_0 and q must divide the leading coefficient a_n.

From the graph of $f(x) = 4x^3 - 24x^2 + 33x + 9$ we suspect that x = 3 may be a rational root (Figure 1).

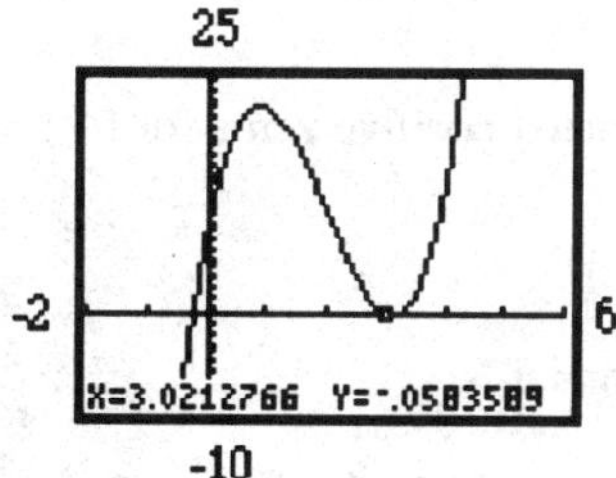

Figure 1

Using division confirms that x = 3 is a root. Since the remainder is 0, the divisor (x - 3) is one of the factors and the quotient $(4x^2 - 12x - 3)$ is the other, $f(x) = (x - 3)(4x^2 - 12x - 3)$.

$$f(x) = 0 \Rightarrow x = 3 \quad \text{or} \quad 4x^2 - 12x - 3 = 0$$

The quadratic formula yields the exact values of the two irrational roots.

$$x = \frac{-b \pm \sqrt{b^2 - 4ac}}{2a} \quad \text{and} \quad a = 4,\ b = -12,\ c = -3 \Rightarrow x = \frac{3 \pm \sqrt{12}}{2}$$

How do you know how many roots (zeros) a polynomial function has?

The Fundamental Theorem of Algebra establishes the existence of solutions, and its corollary establishes their number.

Fundamental Theorem of Algebra

Every polynomial of degree n > 0 with real coefficients has at least one zero in the complex numbers.

Corollary to the Fundamental Theorem of Algebra

Every polynomial of degree n > 0 with real coefficients has exactly n complex zeros.

A zero with multiplicity k is counted k times. A third-degree polynomial has three zeros, a fourth-degree polynomial has four zeros, a fifth-degree polynomial has five zeros, and so on.

Using Descartes' Rule of Signs

Recall that when we learned how to factor quadratic expressions such as $x^2 + 7x + 12$, $x^2 - 7x + 12$, $x^2 + x - 12$, and $x^2 - x - 12$, we studied the four possible sign patterns. The signs of a quadratic expression gave us the possible signs of its factors and hence of the corresponding zeros. Considering sign patterns generalizes to all polynomials.

Descartes' Rule of Signs

The number of positive zeros of a polynomial p(x) is equal to the number of sign changes or is less than the number of sign changes by some even number.

The number of negative zeros of a polynomial is equal to the number of sign changes of p(-x) or is less than the number of sign changes by some even number.

Example 2 Find the number of anticipated positive zeros for $f(x) = x^4 - 5x^3 + 5x^2 + 5x - 6$.

Solution

There are three sign changes in the polynomial

$$f(x) = \underbrace{x^4 - 5x^3}_{1} \underbrace{\;+ 5x^2}_{2} + \underbrace{5x - 6}_{3}$$

By Descartes' rule of signs, there are three positive zeros or one positive zero. If all the zeros of a polynomial are real numbers, the number of sign changes exactly matches the number of positive zeros.

Finding Complex Zeros

For polynomials with real coefficients, complex zeros that are not real numbers occur in *conjugate pairs*. If $a + bi$ is a zero, then its conjugate, $a - bi$, is also a zero. This explains why the number of positive zeros may be less than the number of sign changes by some multiple of 2. Recall that the zeros of $f(x) = ax^2 + bx + c$,

$$x = \frac{-b}{2a} + \frac{\sqrt{b^2 - 4ac}}{2a} \quad \text{and} \quad x = \frac{-b}{2a} - \frac{\sqrt{b^2 - 4ac}}{2a}$$

are complex conjugates whenever $b^2 - 4ac < 0$. For $f(x) = x^2 - 2x + 5$ with two sign changes, there are no real zeros.

$$f(x) = x^2 - 2x + 5 \text{ and } f(x) = 0 \Rightarrow x^2 - 2x + 5 = 0 \Rightarrow x = \frac{2 \pm \sqrt{-16}}{2} = \frac{2 \pm 4\sqrt{-1}}{2} = 1 \pm 2\sqrt{-1} = 1 \pm 2i$$

Both $x_1 = 1 + 2i$ and its conjugate, $x_2 = 1 - 2i$, are zeros of this polynomial.

> **Conjugate Zeros**
> If $x = a + bi$ is a zero of a polynomial with real coefficients, its conjugate $\bar{x} = a - bi$ is also a zero of the polynomial.

Example 3 For $f(x) = 2x^3 + x^2 + 25$, (a) discuss the possible number of x intercepts, (b) list of possible rational roots, (c) discuss the number of positive and negative real zeros, and (d) find the zeros.

Solution

(a) The degree is 3, so there are 3 complex zeros. It may have 1, 2, or 3 x intercepts (real zeros).

(b) $p \in \{\pm 1, \pm 5, \pm 25\}$ and $q \in \{\pm 1, \pm 2\}$ $p/q \in \{\pm 1, \pm 2, \pm 5, \pm 5/2, \pm 25, \pm 25/2\}$.

(c) For $f(x) = 2x^3 + x^2 + 25$ there are no sign changes, therefore no positive real zeros.

For $f(-x) = -2x^3 + x^2 + 25$ there is one sign change, indicating there is one negative real zero.

In all, there must be two complex roots and one negative real zero.

(d) The graphing utility output shows a zero at x = -2.5 (Figure 2).

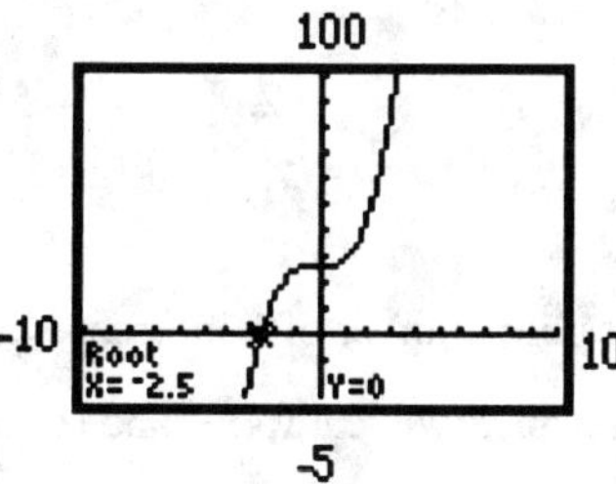

Figure 2

Division yields $f(x) = (x + 5/2)(2x^2 - 4x + 10) = (2x + 5)(x^2 - 2x + 5)$. Therefore, the three zeros of this cubic polynomial are -5/2, 1 + 2i, and 1 - 2i.

Skill Building Exercises

Graph each of the following and label all intercepts.

1. $y = x^3 + 2x^2 - x - 2$
2. $y = x^3 + 5x^2 - 2x - 24$
3. $y = x^4 - 4x^3 + 3x^2 + 4x - 4$
4. $y = -x^4 - 6x^3 - 8x^2 + 6x + 9$

For each of the following, (a) state the possible number of x intercepts, (b) list the set of possible rational roots, (c) discuss the number of positive and negative real roots, and (d) find the exact zeros.

5. $p(x) = x^3 - 3x^2 - 5x$
6. $p(x) = x^4 + 2x^3 - 10x^2$
7. $p(x) = 2x^3 + 7x^2 - 8x - 16$
8. $p(x) = -2x^3 - 7x^2 + 16x - 6$

Find all complex zeros.

9. $y = x^4 + x^3 + 10x^2 + 9x + 9$
10. $y = x^4 - 2x^3 + 6x^2 - 8x + 8$
11. $y = x^4 - 2x^3 - 2x^2 + 6x + 5$
12. $y = x^4 - 2x^3 - 3x^2 + 2x + 2$

Answers

1.

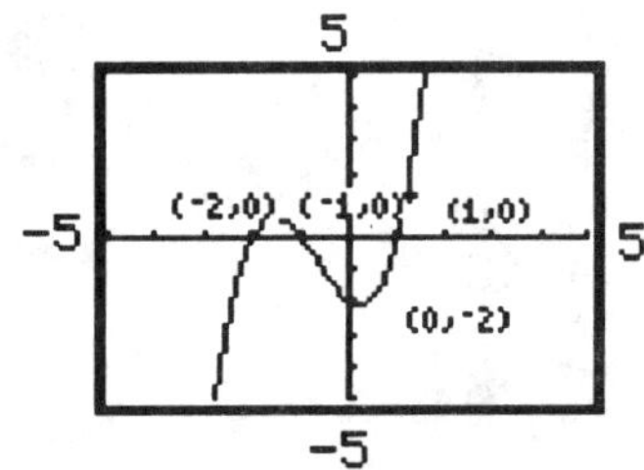

2.

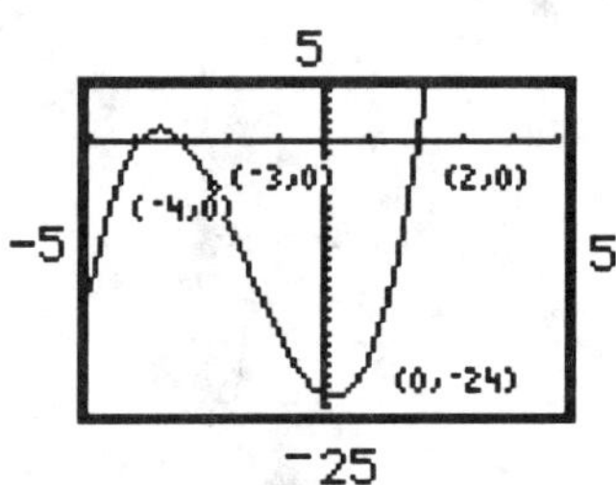

3.

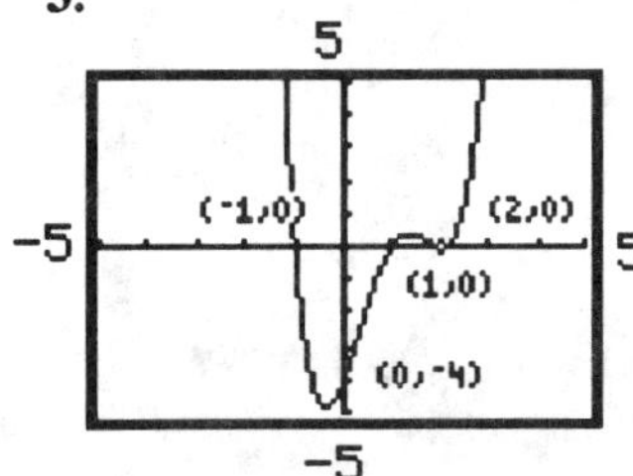

4.

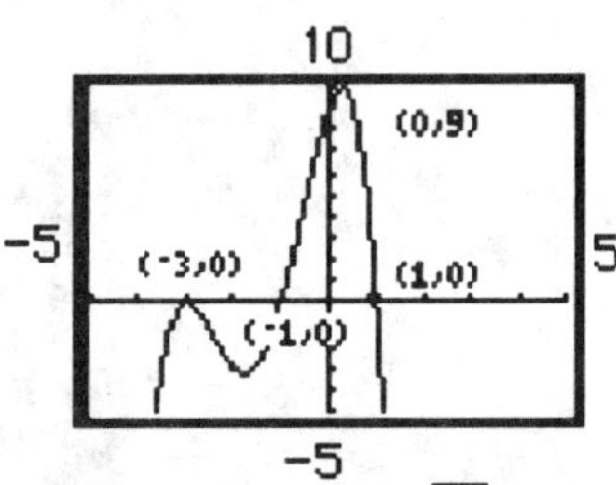

5.(a) 3 (b) $p/q \in \{0\}$ (c) 1 positive real root, 1 negative real root (d) $0, \frac{3 \pm \sqrt{29}}{2}$

6.(a) 4 (b) $p/q \in \{0\}$ (c) 1 positive real root, 1 negative real root (d) $0, -1 \pm \sqrt{11}$

7.(a) 3 (b) $p/q \in \{\pm 1, \pm 2, \pm 4, \pm 8, \pm 16, \pm 1/2\}$

(c) 1 positive real root, 2 or 0 negative real roots (d) $-4, \frac{1 \pm \sqrt{33}}{4}$

8.(a) 3 (b) $p/q \in \{\pm 1, \pm 2, \pm 3, \pm 6, \pm 16, \pm 1/2, \pm 3/2\}$

(c) 2 or 0 positive real roots, 1 negative real root (d) $1/2, -2 \pm \sqrt{10}$

9. $x = \pm 3i$ or $x = \frac{-1 \pm \sqrt{3}i}{2}$

10. $x = 1 \pm i$ or $x = \pm 2i$

11. $x = 2 \pm i$ or $x = -1$

12. $x = -1 \pm \sqrt{2}$ or $x = \pm 1$

Appendix VI

The Binomial Theorem

The binomial theorem is used to expand powers of binomials such as

$$(3x - 1)^3, (x^2 + 2)^{27}, \text{ and } (4x^3 - \frac{3}{5})^5$$

without performing three, twenty-seven, or five multiplications.

Using Factorial Notation

Factorial notation is used in the study of the binomial theorem. Five factorial, written 5!, means

$$1 \cdot 2 \cdot 3 \cdot 4 \cdot 5$$

Therefore, 5! = 120. By noticing that $4! = 1 \cdot 2 \cdot 3 \cdot 4$, we see that $5! = 4! \cdot 5$. In general,

$$n! = 1 \cdot 2 \cdot \ldots \cdot (n - 2) \cdot (n - 1) \cdot n$$

and

$$n! = (n - 1)! \cdot n$$

By definition, 0! = 1.

Example 1 Simplify the following.

(a) $\frac{7!}{5!}$ (b) $\frac{100!}{3! \cdot 97!}$

Solution

(a) $\frac{7!}{5!} = \frac{1 \cdot 2 \cdot 3 \cdot 4 \cdot 5 \cdot 6 \cdot 7}{1 \cdot 2 \cdot 3 \cdot 4 \cdot 5} = 6 \cdot 7 = 42$

(b) $\frac{100!}{3! \cdot 97!} = \frac{97! \cdot 98 \cdot 99 \cdot 100}{6 \cdot 97!} = \frac{98 \cdot 99 \cdot 100}{6} = 161{,}700$

Examining Pascal's Triangle

Before examining the binomial theorem, we investigate the formula for finding the number of combinations of n things taken r at a time. For example, consider the four letters a, b, c, and d. The number of combinations of these four letters taken two at a time is 6. They are ab, ac, ad, bc, bd, and cd. (The combination ab is the same as the combination ba.) We use the formula

$${}_nC_r = \frac{n!}{r!\,(n - r)!}$$

with n = 4 and r = 2.

$${}_4C_5 = \frac{4!}{2!\,(4 - 2)!} = \frac{4!}{2! \cdot 2!} = \frac{1 \cdot 2 \cdot 3 \cdot 4}{1 \cdot 2 \cdot 1 \cdot 2} = 6$$

NOTE: Most calculators have both a factorial key and a combination key.

The combinations ${}_1C_0$, ${}_2C_0$, and ${}_3C_0$ all equal 1. There is only one way to take one, two, or three things none at a time. In all cases you do not take any. The number of combinations for ${}_1C_1$, ${}_2C_1$, and ${}_3C_1$ are 1, 2, and 3 respectively. There is only one way to take one thing one at a time, two ways to take two things one at a time, and three ways to take three things one at a time. If n = r, the number of combinations is always 1. For example, ${}_5C_5 = 1$ because there is only one way to take all five things. Table 1 shows the result of computing the combinations of n things for n = 1, 2, 3, 4, 5, and 6 things for all possible values of r.

Table 1

${}_nC_r$	${}_nC_0$	${}_nC_1$	${}_nC_2$	${}_nC_3$	${}_nC_4$	${}_nC_5$	${}_nC_6$
${}_0C_r$	${}_0C_0 = 1$						
${}_1C_r$	${}_1C_0 = 1$	${}_1C_1 = 1$					
${}_2C_r$	${}_2C_0 = 1$	${}_2C_1 = 2$	${}_2C_2 = 1$				
${}_3C_r$	${}_3C_0 = 1$	${}_3C_1 = 3$	${}_3C_2 = 3$	${}_3C_3 = 1$			
${}_4C_r$	${}_4C_0 = 1$	${}_4C_1 = 4$	${}_4C_2 = 6$	${}_4C_3 = 4$	${}_4C_4 = 1$		
${}_5C_r$	${}_5C_0 = 1$	${}_5C_1 = 5$	${}_5C_2 = 10$	${}_5C_3 = 10$	${}_5C_4 = 5$	${}_5C_5 = 1$	
${}_6C_r$	${}_6C_0 = 1$	${}_6C_1 = 6$	${}_6C_2 = 15$	${}_6C_3 = 20$	${}_6C_4 = 15$	${}_6C_5 = 6$	${}_6C_6 = 1$

Figure 1 shows the results in Table 1 written in a triangular array.

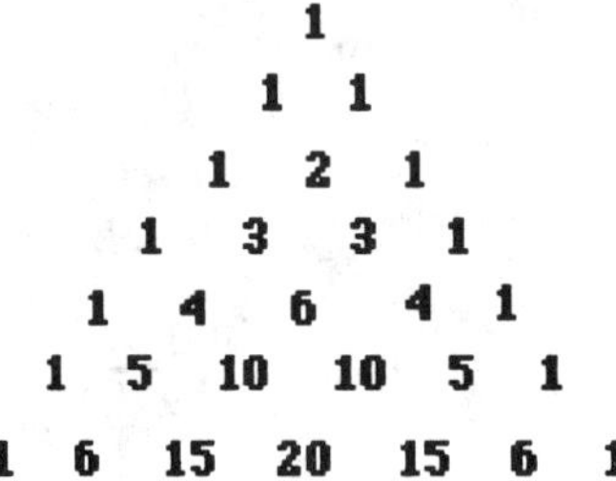

Figure 1

Notice the symmetry of the triangle and that the first and last entry in each row is 1. Each interior entry in the triangle is the sum of the two entries above it. For example, the 6 in the fourth row is the sum, 3 + 3.

What would be the next row in the triangle in Figure 1?

It has eight numbers. The first and last number in the row is 1. The second number is the sum of 1 + 6 = 7, the third is the sum of 6 and 15 = 21, and so on.

Figure 2 shows the binomial (a + b) raised to the first, second, third, fourth, fifth, and sixth powers. See if you can find the triangle pattern from Figure 1 in Figure 2.

$$(a + b)^0 = 1$$
$$(a + b)^1 = 1a + 1b$$
$$(a + b)^2 = 1a^2 + 2ab + 1b^2$$
$$(a + b)^3 = 1a^3 + 3a^2b + 3ab^2 + 1b^3$$
$$(a + b)^4 = 1a^4 + 4a^3b + 6a^2b^2 + 4ab^3 + 1b^4$$
$$(a + b)^5 = 1a^5 + 5a^4b + 10a^3b^2 + 10a^2b^3 + 5ab^4 + 1b^5$$
$$(a + b)^6 = 1a^6 + 6a^5b + 15a^4b^2 + 20a^3b^3 + 15a^2b^4 + 6ab^5 + 1b^6$$

Figure 2

Blaise Pascal (1623 - 1662) identified the relationship between the combination results in Table 1 and the coefficients of the expanded binomials in Figure 2. For this reason, the triangle in Figure 1 is called *Pascal's triangle.*

We can use Pascal's triangle to find the coefficients of the terms in the expansion of $(a + b)^n$, provided that n is small enough to allow us to create enough rows of the triangle. It appears that the number of terms in each expansion is n + 1. For example, $(a + b)^3 = a^3 + 3a^2b + 3ab^2 + b^3$ has four terms. The other factors in each term of the expansion consist of a raised to some power and b raised to some power. Notice in Figure 3 that the sum of the exponents of a and b equals the power of the expansion. The exponents of the first term, a, start at n and decrease to 0. The exponents on the second term b, start at 0 and increase to n.

Sum of exponents in each term is 3.

/\ /\ /\ /\

$$(a + b)^3 = 1a^3b^0 + 3a^2b^1 + 3a^1b^2 + 1a^0b^3$$

exponents of *a* decrease →

exponents of *b* increase →

Figure 3

Example 2 Expand the following using Pascal's triangle to find the coefficient of each term.

(a) $(x - y)^3$ (b) $(x^2 + 2y)^4$

Solution

(a) Since n = 3, there will be four terms in the expansion. From Pascal's triangle, the row with four numbers is 1, 3, 3, and 1. The powers of the first term, x, will be 3, 2, 1, and 0, while the powers of -y will be 0, 1, 2, and 3.

$$\begin{aligned}(x - y)^3 &= 1(x)^3(-y)^0 + 3(x)^2(-y)^1 + 3(x)^1(-y)^2 + 1(x)^0(-y)^3\\ &= x^3 - 3x^2y + 3xy^2 - y^3\end{aligned}$$

(b) Since n = 4, there will be five terms in the expansion. From Pascal's triangle, the row with five numbers is 1, 4, 6, 4, and 1. The powers of the first term, x^2, will be 4, 3, 2, 1, and 0, while the powers of 2y will be 0, 1, 2, 3, and 4.

$$\begin{aligned}(x^2 + 2y)^4 &= 1(x^2)^4(2y)^0 + 4(x^2)^3(2y)^1 + 6(x^2)^2(2y)^2 + 4(x^2)^1(2y)^3 + 1(x^2)^0(2y)^4\\ &= x^8 + 4(x^6)(2y) + 6(x^4)(4y^2) + 4(x^2)(8y^3) + 16y^4\\ &= x^8 + 8x^6y + 24x^4y^2 + 32x^2y^3 + 16y^4\end{aligned}$$

The following theorem states the expansion in terms of combinations.

Binomial Theorem

$$(a + b)^n = \sum_{r=0}^{n} {}_nC_r\, a^{n-r} b^r$$

Finding a Term in a Binomial Expansion

The following corollary to the binomial theorem allows us to find any particular term of an expansion without writing out the entire expansion.

Corollary to the Binomial Theorem

In the expansion of the binomial $(a + b)^n$, the $(r + 1)$st term is given by

$$ {}_nC_r\, a^{n-r} b^r$$

Example 3 Find the 25th term in the expansion of $(2x - 1)^{27}$.

Solution

Using the corollary to find the 25th term, we substitute values of n, r, a, and b into ${}_nC_r\, a^{n-r} b^r$. The value of n is 27 since the binomial is raised to the 27th power. For the 25th term, we let $r + 1 = 25 \Rightarrow r = 24$. The first term of the binomial, 2x, is a and the second term, -1, is b.

${}_nC_r\, a^{n-r} b^r$ and $n = 27$, $r = 24$, $a = 2x$, $b = -1 \Rightarrow$

$$\begin{aligned} {}_{27}C_{24}\,(2x)^{27-24}\,(-1)^{24} &= {}_{27}C_{24}\,(2x)^3\,(-1)^{24} \\ &= \frac{27!}{24!(27-24)!}(8x^3)(1) \\ &= (2925)\,(8x^3)(1) = 23{,}400x^3 \end{aligned}$$

Skill Building Exercises

Expand the following binomials.

1.(a) $(x + y)^5$ (b) $(x - y)^5$

2.(a) $(x + 1)^4$ (b) $(x^2 + 1)^4$

3.(a) $(2x - 3)^4$ (b) $(2x - 3)^5$

4.(a) $(2x - 1)^3$ (b) $(2x^3 - 1)^3$

Find the indicated term.

5.(a) the first term of $(x - y)^{10}$
(b) the last term of $(x - y)^{10}$

6.(a) the second term of $(2x - y)^{10}$
(b) the eighth term of $(2x - y)^{10}$

Answers

1.(a) $x^5 + 5x^4y + 10x^3y^2 + 10x^2y^3 + 5xy^4 + y^5$ (b) $x^5 - 5x^4y + 10x^3y^2 - 10x^2y^3 + 5xy^4 - y^5$
2.(a) $x^4 + 4x^3 + 6x^2 + 4x + 1$ (b) $x^8 + 4x^6 + 6x^4 + 4x^2 + 1$ **3.**(a) $16x^4 - 96x^3 + 216x^2 - 216x + 81$
(b) $32x^5 - 240x^4 + 720x^3 - 1080x^2 + 810x - 243$ **4.**(a) $8x^3 - 12x^2 + 6x - 1$ (b) $8x^9 - 12x^6 + 6x^3 - 1$
5.(a) x^{10} (b) y^{10} **6** (a) $-5120x^9y$ (b) $-960x^3y^7$

Appendix VII

Proof by Math Induction

Mathematical induction (math induction) is the method of proof used whenever we must verify a statement for all natural numbers or for all natural numbers beyond some initial number. The use of the word *induction* is not related to its use in the scientific method, where inductive reasoning involves studying several examples in order to hypothesize the general rule. Math induction involves deductive reasoning. The induction part of the proof stems from the recursiveness of the problems themselves. Recursive definitions generate a statement from a preceding statement. For example, we can define the *arithmetic sequence* (sequence whose successive terms differ by a constant)

$$2, 4, 6, 8, \ldots$$

as

$$a_1 = 2 \text{ and } a_{i+1} = a_i + 2 \text{ for } i = 1, 2, 3, \ldots$$

which says that the successor of ai, namely, a_{i+1}, can be found by adding the common difference, 2, to a_i. For the following *geometric sequence* (sequence whose successive terms have a common ratio)

$$3, 9, 27, 81, \ldots$$

the recursive (inductive) formula is

$$a_1 = 3 \text{ and } a_{i-1} = 3a_i, \text{ for } i = 1, 2, 3, \ldots$$

It is also possible to find an inductive definition for the set of natural numbers. Consider the following two statements describing a set:

1. The number 1 is in the set.
2. If any number k is in the set, its successor, k + 1, must be in the set.

The only set defined completely by both of these statements is the set containing all natural numbers. Since k = 1 is in the set, according to statement 2, its successor, k + 1 = 2, must be there also. Once we have k = 2, its successor, k + 1 = 3, must be in the set also, and so on.

The falling of dominoes is analogous to proving a statement true for all natural numbers. Suppose that we set up a string of dominoes in such a way that if any domino, say the kth, falls, its successor, the (k + 1)st, will be struck, causing it to fall. In addition, if any initial domino is knocked over, by the inductive statement above we know that all the dominoes, from the initial one on, will fall. We generally use the first domino as the initial one.

Proofs by math induction always involve the same two separate steps. If we want to prove the conditional proposition p(n), for all $n \in$ naturals, first we prove that p(n) is true for some initial value, usually n = 1. Then we prove that if p(n) is true for any particular natural number k, p(n) must be true for the next natural number, n = k + 1.

Proof by Math Induction

To prove p(n) for all $n \in$ naturals, we must take the following steps:

1. Prove p(1).
2. Prove $p(k) \Rightarrow p(k + 1),\ 1 \le k \le n$.

Example 1 Prove p(n): $1 + 2 + \cdots + n = \dfrac{n(n + 1)}{2}$.

Solution

1. Prove true for p(1):

$$p(n) = 1 + 2 + \ldots + n = \frac{n(n + 1)}{2} \text{ and } n = 1 \Rightarrow p(1) = \frac{1(1 + 1)}{2} = 1$$

2. Prove $p(k) \Rightarrow p(k + 1)$.

p(k): $$1 + 2 + \cdots + k = \frac{k(k + 1)}{2}$$

p(k + 1): $$1 + 2 + \cdots + (k + 1) = \frac{(k + 1)(k + 1) + 1}{2}$$
$$= \frac{(k + 1)(k + 2)}{2}$$

Proof: $$1 + 2 + \cdots + k = \frac{k(k + 1)}{2}$$
$$1 + 2 + \cdots + k + (k + 1) = \frac{k(k + 1)}{2} + (k + 1)$$
$$= \frac{k(k + 1)}{2} + \frac{2(k + 1)}{2} = \frac{(k + 1)(k + 2)}{2}$$

Therefore, $p(k) \Rightarrow p(k + 1)$ as required. Hence, p(n) is true for all natural numbers n.

Example 2 Prove p(n): $r^0 + r^1 + r^2 + \cdots + r^{n-1} = \dfrac{1 - r^n}{1 - r}$

Solution

1. Prove p(1): $n = 1 \Rightarrow r^0 = 1$ and $\dfrac{1 - r^1}{1 - r} = 1$

2. Prove $p(k) \Rightarrow p(k + 1)$.

p(k): $$r^0 + r^1 + r^2 + \cdots + r^{k-1} = \frac{1 - r^k}{1 - r}$$

p(k + 1): $$r^0 + r^1 + r^2 + \cdots + r^{(k+1)-1} = \frac{1 - r^{(k+1)}}{1 - r}$$
$$r^0 + r^1 + r^2 + \cdots + r^k = \frac{1 - r^{(k+1)}}{1 - r}$$

Proof: $$r^0 + r^1 + r^2 + \cdots + r^{k-1} = \frac{1 - r^k}{1 - r}$$
$$r^0 + r^1 + r^2 + \cdots + r^{k-1} + r^k = \frac{1 - r^k}{1 - r} + r^k$$
$$= \frac{1 - r^k}{1 - r} + \frac{r^k(1 - r)}{1 - r}$$
$$= \frac{1 - r^k}{1 - r} + \frac{r^k - r^{k+1}}{1 - r} = \frac{1 - r^{k+1}}{1 - r}$$

Therefore, $p(k) \Rightarrow p(k + 1)$ as required. Hence, p(n) is true for all natural numbers n.

Skill Building Exercises

Prove p(1), p(2), p(3), and p(4) for each of the following propositions.

1. p(n): $2 + 4 + \cdots + 2n = n(n + 1)$
2. p(n): $1 + 4 + \cdots + n^2 = \frac{n(n + 1)(2n + 1)}{6}$
3. p(n): $1 + 8 + \cdots + n^3 = \frac{n^2(n + 1)^2}{4}$

Use mathematical induction to prove each of the following.

4. p(n): $2 + 4 + 6 + \cdots + 2n = n(n + 1)$
5. p(n): $1 + 4 + 9 + \cdots + n^2 = \frac{n(n + 1)(2n + 1)}{6}$
6. p(n): $1 + \frac{1}{2} + \frac{1}{4} + \cdots + \frac{1}{2^{n-1}} = \frac{2^n - 1}{2^{n-1}}$
7. p(n): $1 + \frac{1}{3} + \frac{1}{9} + \cdots + \frac{1}{3^{n-1}} = \frac{3^n - 1}{2(3^{n-1})}$

Answers

1. p(1) = 2 and p(1) = 1(1 + 1) = 2; p(2) = 2 + 4 = 6 and p(2) = 2(3) = 6; p(3) = 2 + 4 + 6 = 12 and p(3) = 3(4) = 12; p(4) = 2 + 4 + 6 + 8 + 20 and p(4) = 4(5) = 20

2. p(1) = 1 and p(1) = $\frac{1(2)(3)}{6}$ = 1; p(2) = 1 + 4 = 5 and p(2) = $\frac{2(3)(5)}{6}$ = 5;

p(3) = 1 + 4 + 9 = 14 and p(3) = $\frac{3(4)(7)}{6}$ = 14; p(4) = 1 + 4 + 9 + 16 = 30 and p(4) = $\frac{4(5)(9)}{6}$ = 30

3. p(1) = 1 and p(1) = $\frac{1^2(2)^2}{4}$ = 1; p(2) = 1 + 8 = 9 and p(2) = $\frac{2^2(3)^2}{4}$ = 9;

p(3) = 1 + 8 + 27 = 36 and p(3) = $\frac{3^2(4)^2}{4}$ = 36; p(4) = 1 + 8 + 27 + 64 = 100 and p(4) = $\frac{4^2(5)^2}{4}$ = 100

4. Prove true for p(1): $n = 1 \Rightarrow 2n = 2$ and $n(n + 1) = 1(1 + 1) = 2$

Prove p(k) ⇒ p(k + 1).

p(k): $2 + 4 + \cdots + 2k = k(k + 1)$

p(k + 1):
$$\begin{aligned} 2 + 4 + \cdots + 2(k + 1) &= (k + 1)((k+ 1) + 1) \\ &= (k + 1)(k + 2) \end{aligned}$$

Proof:
$$\begin{aligned} 2 + 4 + \cdots + 2k &= k(k + 1) \\ 2 + 4 + \cdots + 2k + 2(k + 1) &= k(k + 1) + 2(k + 1) \\ &= (k + 1)(k + 2) \end{aligned}$$

Therefore, p(k) ⇒ p(k + 1) as required. Hence, p(n) is true for all natural numbers n.

5. Prove true for p(1): $n = 1 \Rightarrow n^2 = 1$ and $\frac{n(n + 1)(2n + 1)}{6} = \frac{1(1 + 1)(2(1) + 1)}{6} = \frac{6}{6} = 1$

Prove p(k) ⇒ p(k + 1).

p(k): $1 + 4 + \cdots + k^2 = \frac{k(k + 1)(2k + 1)}{6}$

p(k + 1):
$$\begin{aligned} 1 + 4 + \cdots + (k + 1)^2 &= \frac{(k + 1)(k + 1 + 1)(2(k + 1) + 1)}{6} \\ &= \frac{(k + 1)(k + 2)(2k + 3)}{6} \end{aligned}$$

Proof:

$$1 + 4 + \cdots + k^2 = \frac{k(k+1)(2k+1)}{6}$$

$$\begin{aligned} 1 + 4 + \cdots + k^2 + (k+1)^2 &= \frac{k(k+1)(2k+1)}{6} + (k+1)^2 \\ &= \frac{k(k+1)(2k+1)}{6} + \frac{6(k+1)^2}{6} \\ &= \frac{(k+1)\,(k(2k+1) + 6(k+1))}{6} \\ &= \frac{(k+1)\,(2k^2 + k + 6k + 6)}{6} \\ &= \frac{(k+1)(k+2)(2k+3)}{6} \end{aligned}$$

Therefore, p(k) ⇒ p(k + 1) as required. Hence, p(n) is true for all natural numbers n.

6. Prove true for p(1): $n = 1 \Rightarrow \frac{1}{2^{n-1}} = \frac{1}{2^0} = 1$ and $\frac{2^n - 1}{2^{n-1}} = \frac{2-1}{2^0} = 1$

Prove p(k) ⇒ p(k + 1).

p(k): $$1 + \frac{1}{2} + \cdots + \frac{1}{2^{k-1}} = \frac{2^k - 1}{2^{k-1}}$$

p(k + 1): $$1 + \frac{1}{2} + \cdots + \frac{1}{2^{k+1-1}} = \frac{2^{k+1} - 1}{2^{k+1-1}}$$

$$1 + \frac{1}{2} + \cdots + \frac{1}{2^k} = \frac{2^{k+1} - 1}{2^k}$$

Proof: $$1 + \frac{1}{2} + \cdots + \frac{1}{2^{k-1}} = \frac{2^k - 1}{2^{k-1}}$$

$$\begin{aligned} 1 + \frac{1}{2} + \cdots + \frac{1}{2^{k-1}} + \frac{1}{2^k} &= \frac{2^k - 1}{2^{k-1}} + \frac{1}{2^k} \\ &= \frac{2}{2} \cdot \frac{2^k - 1}{2^{k-1}} + \frac{1}{2^k} = \frac{2^{k+1} - 2}{2^k} + \frac{1}{2^k} = \frac{2^{k+1} - 1}{2^k} \end{aligned}$$

Therefore, p(k) ⇒ p(k + 1) as required. Hence, p(n) is true for all natural numbers n.

7. Prove true for p(1): $n = 1 \Rightarrow \frac{1}{3^{n-1}} = \frac{1}{3^0} = 1$ and $\frac{3^n - 1}{2(3^{n-1})} = \frac{3-1}{2(3)^0} = \frac{2}{2} = 1$

Prove p(k) ⇒ p(k + 1).

p(k): $$1 + \frac{1}{3} + \cdots + \frac{1}{3^{k-1}} = \frac{3^k - 1}{2(3^{k-1})}$$

p(k + 1): $$1 + \frac{1}{3} + \cdots + \frac{1}{3^{k+1-1}} = \frac{3^{k+1} - 1}{2(3^{k+1-1})}$$

$$1 + \frac{1}{3} + \cdots + \frac{1}{3^k} = \frac{3^{k+1} - 1}{2(3^k)}$$

Proof: $$1 + \frac{1}{3} + \cdots + \frac{1}{3^{k-1}} = \frac{3^k - 1}{2(3^{k-1})}$$

$$\begin{aligned} 1 + \frac{1}{3} + \cdots + \frac{1}{3^{k-1}} + \frac{1}{3^k} &= \frac{3^k - 1}{2(3^{k-1})} + \frac{1}{3^k} \\ &= \frac{3}{3} \cdot \frac{3^k - 1}{2(3^{k-1})} + \frac{2}{2} \cdot \frac{1}{3^k} = \frac{3^{k+1} - 3}{2(3)^k} + \frac{2}{2(3)^k} = \frac{3^{k+1} - 1}{2(3)^k} \end{aligned}$$

Therefore, p(k) ⇒ p(k + 1) as required. Hence, p(n) is true for all natural numbers n.

Answers to Odd-Numbered Exercises

Graphing utility windows use a standard screen ($-10 \le x \le 10$ and $-10 \le y \le 10$) unless otherwise indicated.

Prologue (Page 10)

1.(a)

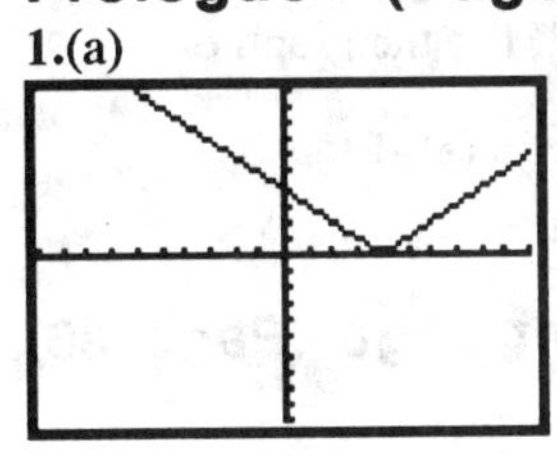

(b)

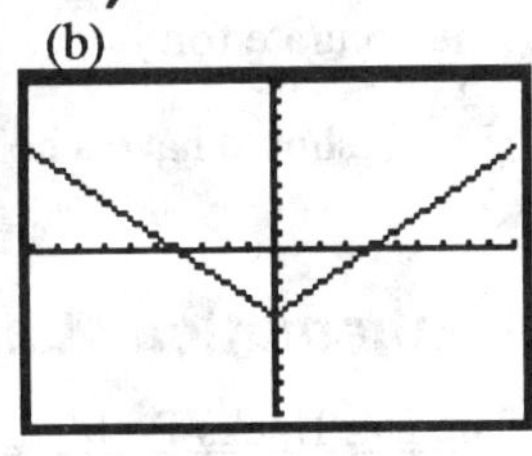

2.(a) $-\sqrt{10} \approx -3.16$ (b) 0 (c) $\sqrt{10} \approx 3.16$
(d) 100 (e) 0

3.(a) (4,24), (-1,9) (b) $(-2.4\tilde{1}, 2.1\tilde{7})$, $(0.4\tilde{1}, 7.8\tilde{3})$

4.(a) (-3,0), (3,0) (b) (0,27) (c) (0,27)

5.(a) (-1,0), (0.5,0), (2,0) (b) (0,1) (c) $(-0.3\tilde{7}, 1.3\tilde{0})$
(d) $(1.3\tilde{7}, -1.3\tilde{0})$ (e) y = -154 (f) y = 1495

6.(a) $(-1.\tilde{4}, 0)$, (0.5, 0), $(1.\tilde{4}, 0)$ (b) (0,-1) (c) (1, 0.5)
(d) $(-0.6\tilde{7}, -1.8\tilde{1})$ (e) $y \approx -0.31$ (f) y = 8159

Section 1.1 Functions and Their Properties (Page 22)

1. no

3. yes, domain $[0,\infty)$, range $[0,\infty)$

5. no

7. yes, domain $(-\infty,\infty)$, range $(-\infty,\infty)$

9. yes domain $(-\infty,\infty)$, range $[-2,\infty)$

11. yes, domain $[-2,\infty)$, range $[0,\infty)$

13. yes, domain $(-\infty,\infty)$, range $[-1,1]$

15. yes, domain $(-\infty,\infty)$, range $(1,\infty)$

17. domain $(-\infty,\infty)$, range $(-\infty,\infty)$, yes

19. domain $[0,\infty)$, range $[0,\infty)$, yes

21. $y = |x|$, $y = x$, $y = x^2$, $y = x^3$

23. $y = x$, $y = \sqrt{x}$, $y = x^3$, $y = 1/x$

25. $y = |x|$, $y = x^2$

27. $y = x$

29.

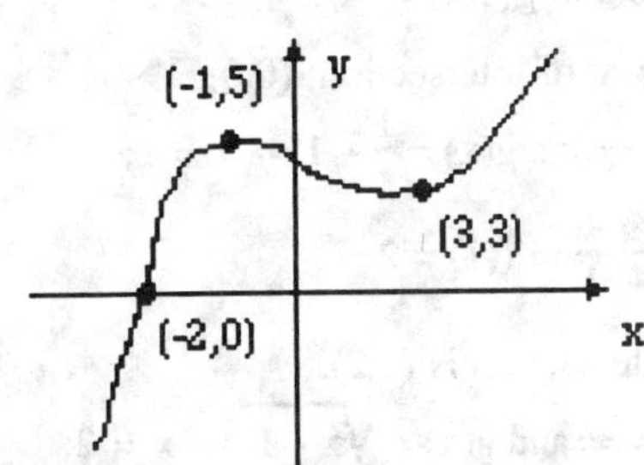

31. decreasing

33. The Hot Tub Problem

(a) A: 4000 D: 13,750

(b) equation: $(-\infty,\infty)$ application: $[0,\infty)$

(c) Enter $y = 650x + 4000$ for $0 \le x \le 15$ and $0 \le y \le 13750$.

(d) B: 7250 C: 9.2

(e) Point A represents the purchase price, $4000.

35. The Rainbow Kite and Balloon Company

(a) A: -400 D: 200

(b) equation $(-\infty,\infty)$, application $[0,200]$

(c) Enter $y = -x^2/5 + 42x - 400$ for $0 \le x \le 200$ and $-400 \le y \le 2000$.

(d) B: 10 C: 105

(e) A: If no kites sell, the company loses $400.

B and D: If 10 or 200 kites sell, there is no profit.

C: The maximum profit of $1805 occurs when 105 kites sell.

37. The Inverse-Square-Law Problem

(a) A: 193.6 D: 28.6

(b) equation $(-\infty,0) \cup (0,\infty)$, application $(0,\infty)$

(c) Enter $y = 121000/x^2$ for $0 \le x \le 65$ and $0 \le y \le 194$.

(d) B: 34.6 C: 54.6

(e) $52.5 \le v \le 58.1$

39. The Celsius-to-Fahrenheit Problem

(a) Enter $y = 9x/5 + 32$ and $y = 2x + 30$ for $0 \le x \le 10$ and $0 \le y \le 50$, A: 32 B: 30 C: (10,50)

(b) The rule of thumb is accurate for temperatures close to 10°C.

41.(a) domain [-1,1], range [0,1] (b) no

43.(a) domain $(0,\infty)$, range $(-\infty,\infty)$ (b) yes

45. yes

47. It is accurate for $y = x - 1$, but the graph of $y = \frac{x^2 - 1}{x + 1}$ should have a hole at (-1,-2).

Section 1.2 Functional Notation and Mathematical Language (Page 36)

1.(a) 2 (b) 28 (c) -3/4

3.(a) 1/3 (b) 28/81 (c) 8

5.(a) $\sqrt{x+1}$ (b) $\sqrt{x} + 1$

7.(a) $x^2 - 3x$ (b) $-x^2 - 3x$

9. $\sqrt{x+4} - 1$ **11.** $1/x + 1$

13.(a) 0 (b) -1000 (c) 20, 30

15.(a) -10 (b) 13 (c) -10

17.(a) 5 (b) 12 (c) 0, -1, 5.8

19.(a) 100 (b) 200 (c) $20 < x \le 30$ (d) none

21. $A = \pi(10)^2$

23. $x = \frac{-(-2) \pm \sqrt{(-2)^2 - 4(1)(-3)}}{2(1)}$

25. $3x + 4 = 2(x + 2)$

27. $f(x) = 5x - 10$ and $f(x) = 0 \Rightarrow x = 2$;

$f(x) = 5x - 10 \Rightarrow f(0) = -10$

29. $f(x) = x^3 - 8$ and $f(x) = 0 \Rightarrow x = 2$;

$f(x) = x^3 - 8 \Rightarrow f(0) = -8$

31.

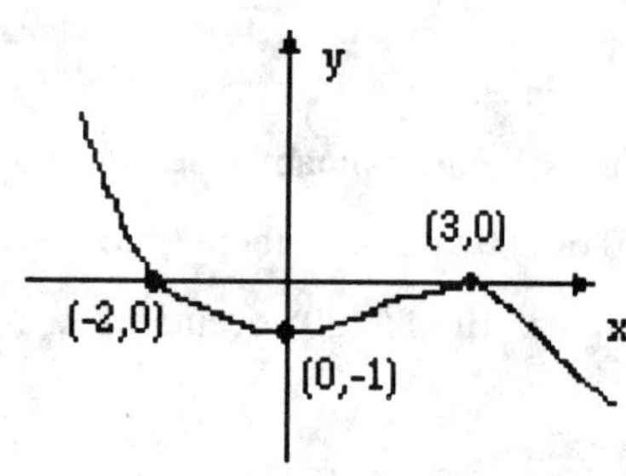

33. B **35.** A

37. Another Billfish Problem

(a) $W(G) = \frac{G^2}{40}$

(b) 22.5 lb and 90 lb

(c) about 42 in.

39. The Big Bucks Problem

(a) $A(r) = 500(1 + r)$ (b) \$536.25

(c) $A(P) = P(1.065)$ (d) about \$11,267.60

41. The Drapery Problem

(a) $F(l) = \frac{5}{3}(l + 20)$ (b) $153.\bar{3}$ in.

(c) $F(w) \approx 4.26w$ (d) about 70 in.

43. The Rainbow Kite and Balloon Company

(a) $R(0) = 0$ (b) $R(10) = 70$

(c) $2R(10) + 10 = 2(70) + 10 = 150$

(d) $p = 0$ or $p = 150$

(e) When the price is zero the revenue is zero. As the price increases the revenue increases to a point, then the price increases to where none are sold and the revenue is zero again.

45. $f(x) = x^2$ and $g(x) = x + 6 \Rightarrow x = -2$ or $x = 3$;

$f(x) = x^2 \Rightarrow f(-2) = 4$ and $f(3) = 9$

The points of intersection are (-2,4) and (3,9).

47. $f(x) = \sqrt{x}$ and $g(x) = -2x \Rightarrow x = 0$ or $x = 1/4$;

$f(x) = \sqrt{x} \Rightarrow f(0) = 0$ and $f(1/4) = 1/2 g(x) = -2x \Rightarrow$

$g(0) = 0$ and $g(1/4) = -1/2$

There is one point of intersection; (0,0).

49. $f(x) = \frac{4}{x + 1}$ and $g(x) = \frac{1}{x} + 1 \Rightarrow x = 1$;

$f(x) = \frac{4}{x + 1} \Rightarrow f(1) = 2$

The point of intersection is (1,2).

51. $f(x) = \frac{1}{\sqrt{x - 1}}$ and $g(x) = \sqrt{x - 1} \Rightarrow x = 2$;

$f(x) = \frac{1}{\sqrt{x - 1}} \Rightarrow f(2) = 1$

The point of intersection is (2,1).

Section 1.3 Translations and Reflections of Graphs of Functions (Page 55)

1.(a) left 2 units (b) up 5 units

3.(a) left 3 units and down 1 unit

(b) right 3 units and up 1 unit

5. about the y axis

7. about the y axis

9.

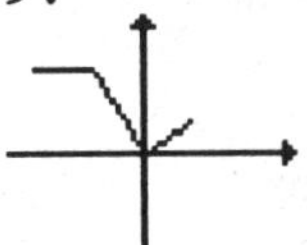
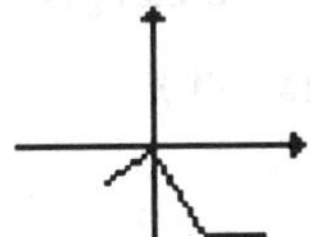
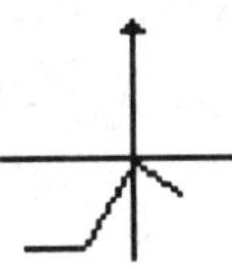

11.

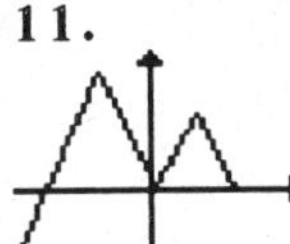
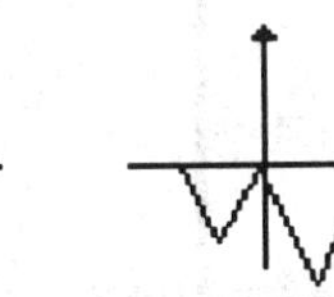
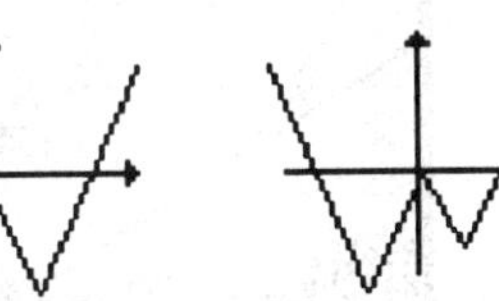

13.(a) $y = \sqrt{x + 1}$ (b) $y = \sqrt{x} + 3$

15. $y = \sqrt{x - 2} - 1$

17.(a) $y = -\sqrt{x} + 1$ (b) $y = \sqrt{-x + 1} - 3$

19.

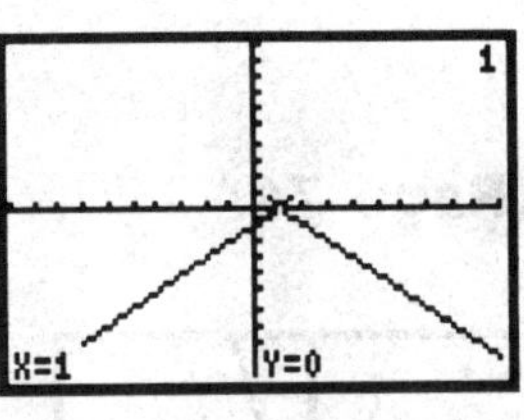

21.

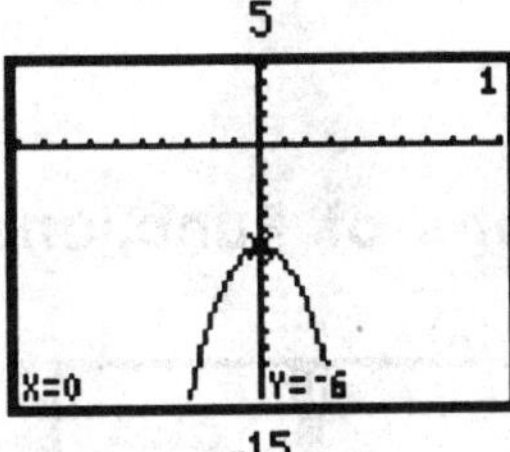

23.

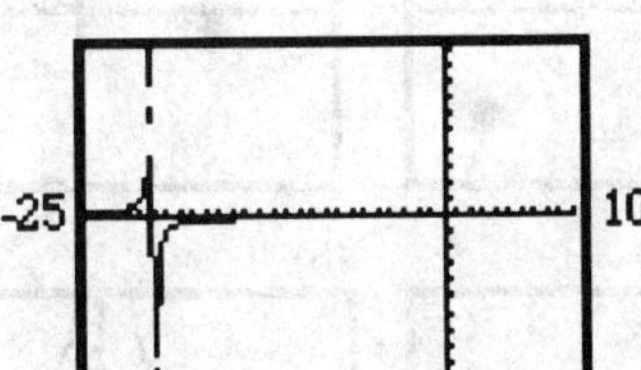

25.

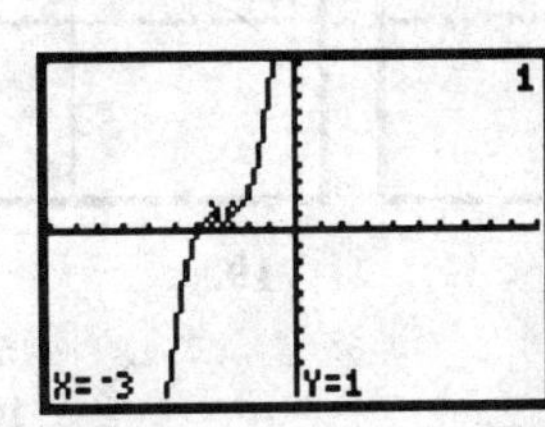

27. $y = \sqrt{x - 30} + 10$

29. $y = |x - 30| + 10$

31. $y = \sqrt{x + 1} - 2$

33. $y = x^3 - 20$

35. $y = -x - 10$ **37.** $y = -\sqrt{x + 20} + 30$

39. $y = -|x + 10| + 60$

41. $y = \dfrac{1}{x - 20} + 30$

43.(a)

x	-10	0	10	20
y = f(x) + 5	0	5	20	30

(b)

x	-10	0	10	20
y = -f(x)	5	0	-15	-25

45.(a)

x	10	0	-10	-20
y = f(-x)	-5	0	15	25

(b)

x	10	0	-10	-20
y = f(-x) + 1	-4	1	16	26

47.

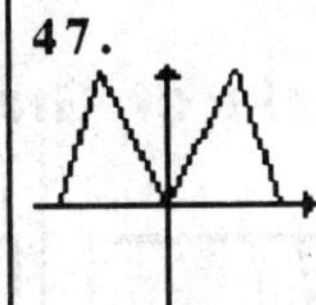
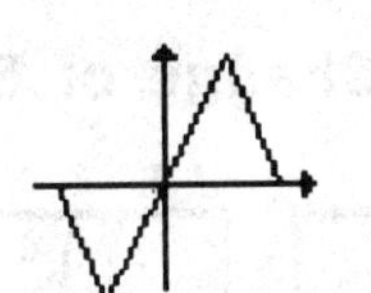
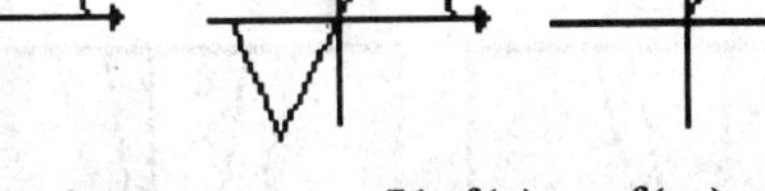

49. f(x) = f(-x)

even function

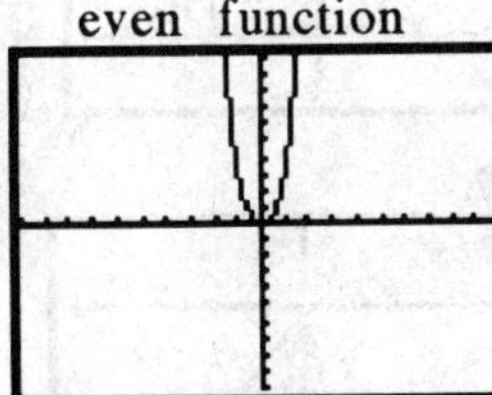

51. f(x) = -f(-x)

odd function

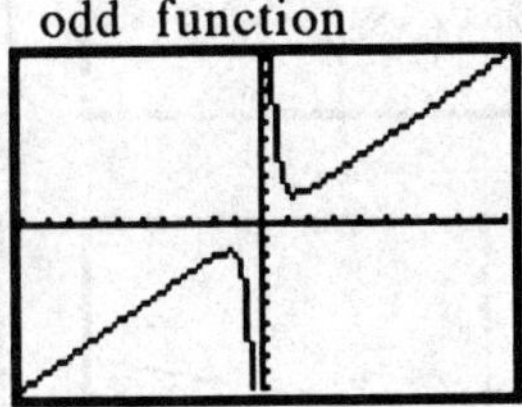

53. <u>A Bell-Shaped Curve Problem</u>

(a) (b) y = f(x + 15) (c) 100

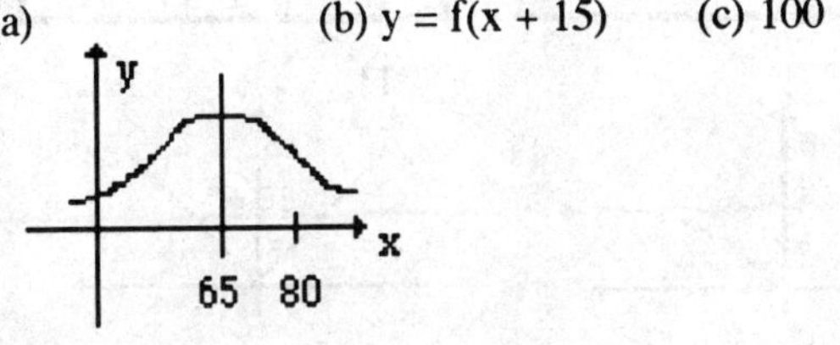

55. The Template Design Problem

(a) $f(x) = |x - 103.5| + 96.5$

(b) [101,106]

57. The Rainbow Kite and Balloon Company

(a) (b) $p = 43.30 - 0.21(x - 120)$

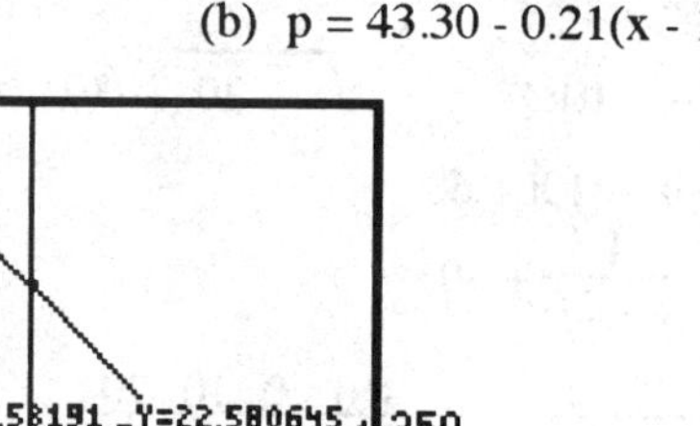

(c)

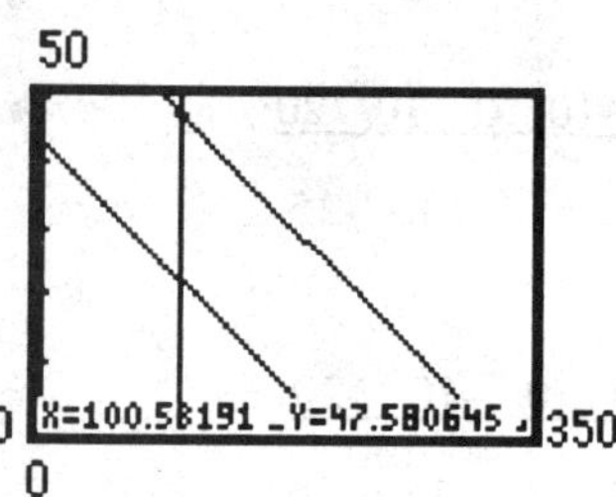

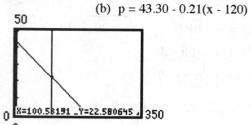

(d) The selling price went up from about $22.58 to $47.58.

59. Wonderworld Amusement Park

(a)

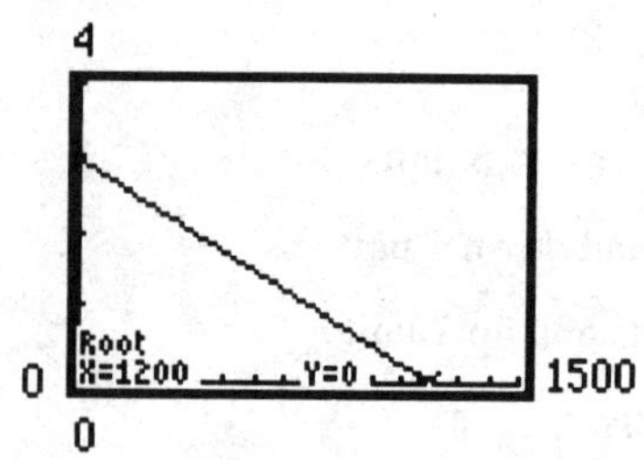

(b) p intercept: 3 At $3, customers won't buy any.

x intercept: 1200 You can give 1200 away.

(c) $p = 3.00 - 0.0025(x - 40)$

(d)

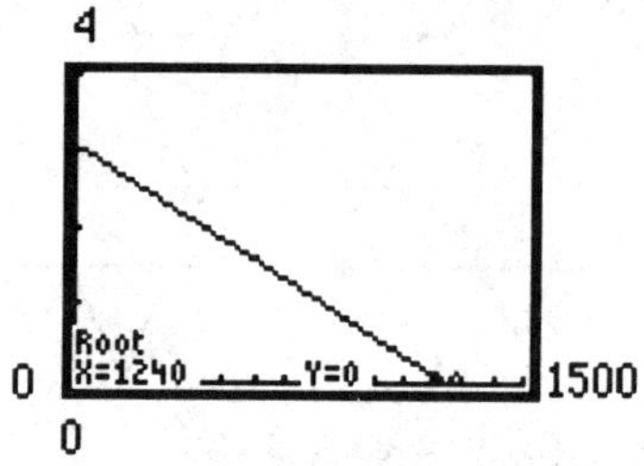

61. (answers vary) For $f(x) = x^2 + 2$, $f(x) = f(-x)$

63. (answers vary) $f(x) = \sqrt{x}$

Section 1.4 Change of Scale of Graphs of Functions (Page 74)

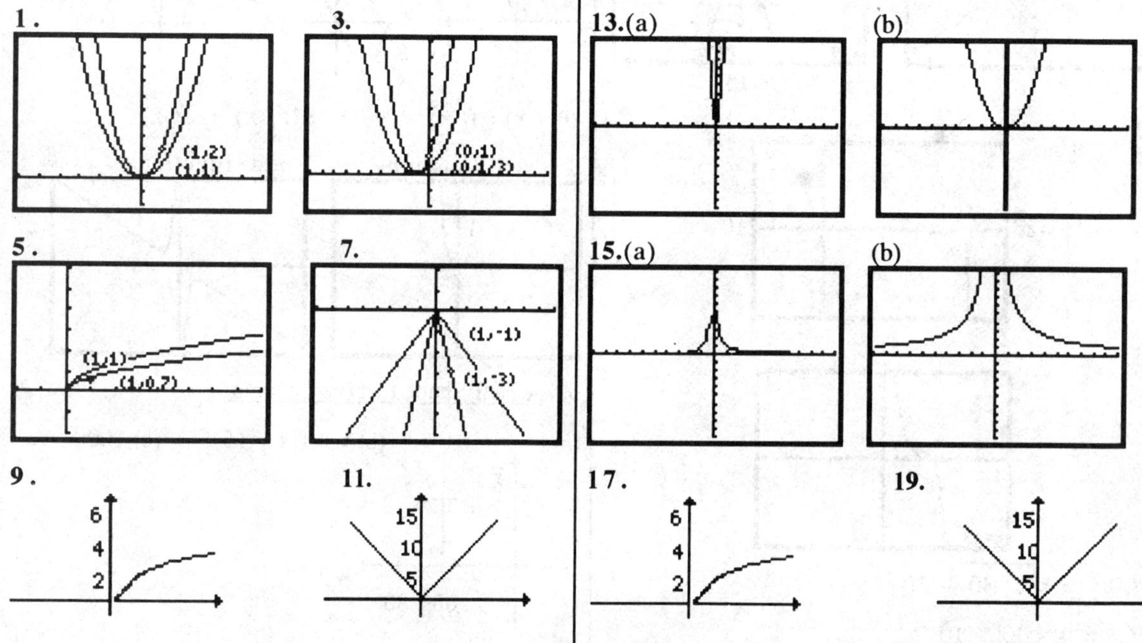

21.(a) (b)

23.(a) (b)

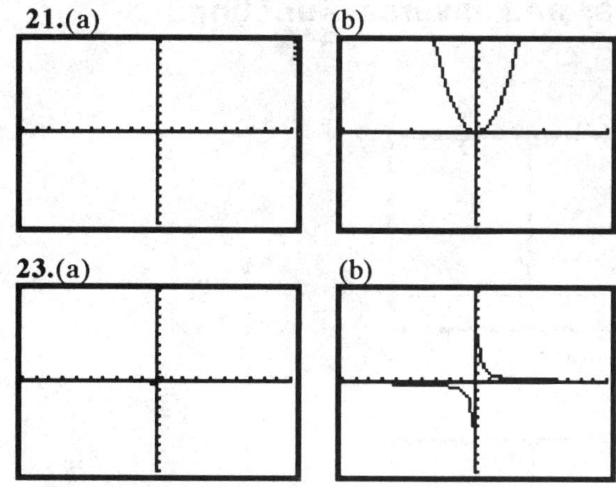

25. $-100 \le x \le 100$, $-1 \le y \le 1$

27. $-120 \le x \le 120$, $-120 \le y \le 120$

29. $-0.25 \le x \le 0.25$, $-200 \le y \le 200$

31. $-20 \le x \le 20$, $-150 \le y \le 150$

33. **35.** **37.** **39.** **41.** **43.**

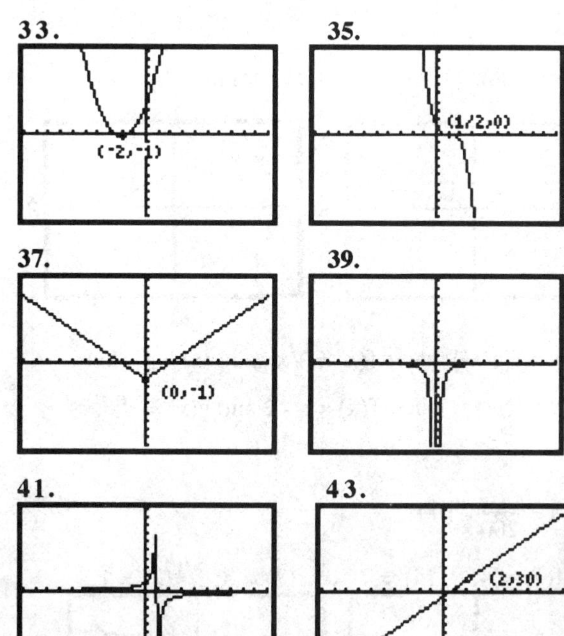

45. B **47.** D

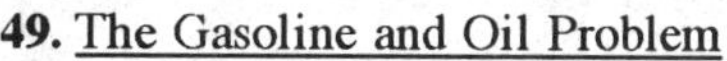

49. The Gasoline and Oil Problem

(a) (b)

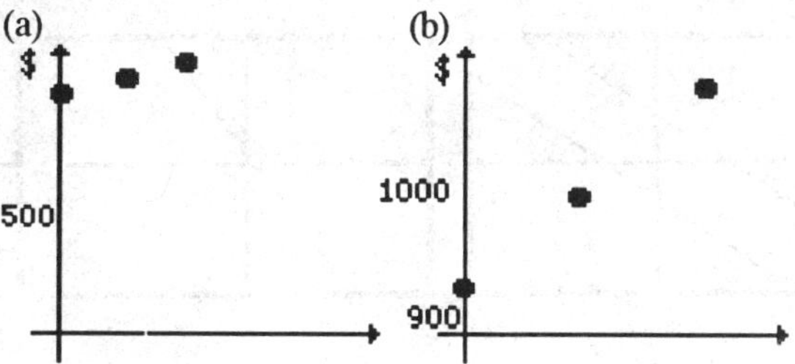

51. The Misleading Car Advertisement Problem

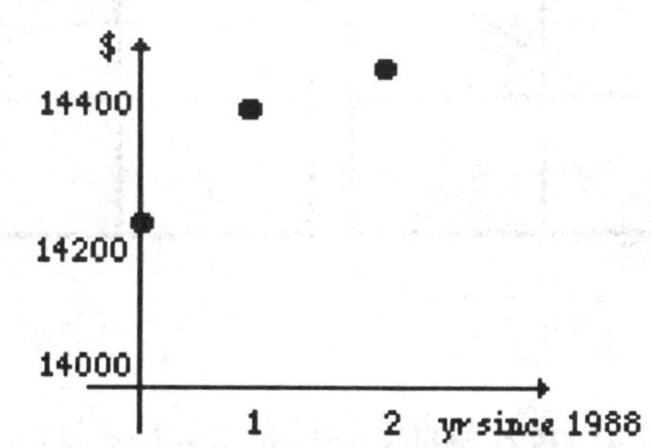

53. The Rainbow Kite and Balloon Company

(a)

d	2	4	6	8
2R(d)	0.64	0.16	0.08	0.04

(b) The rate of increase is doubled for each diameter.

(c)

d	1	2	3	4
R(2d)	0.32	0.08	0.04	0.02

(d) Each rate is reached at smaller diameters (smaller by half).

55. $y = 2|x| - 2$

57. $y = x^2/10$

Section 1.5 Operations, Compositions, and Inverse Functions (Page 92)

1.

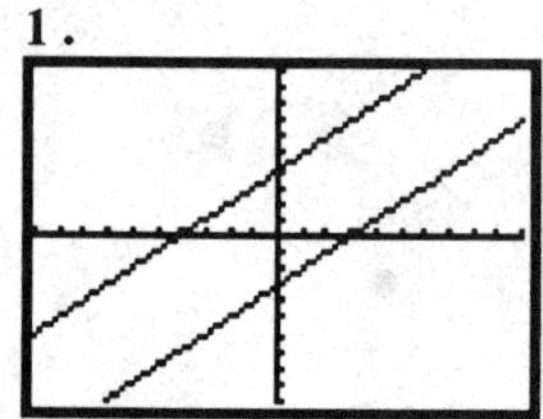 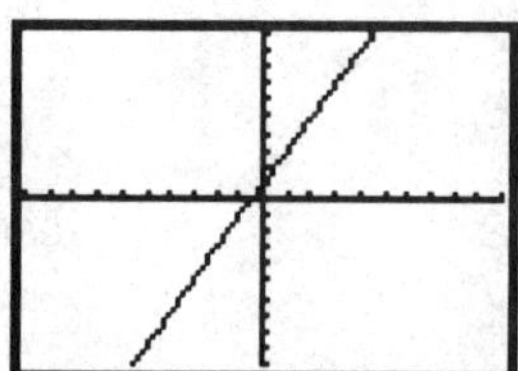 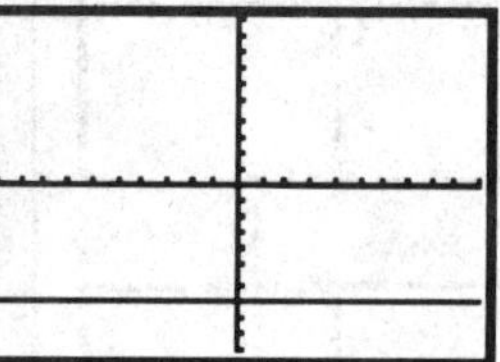

3.

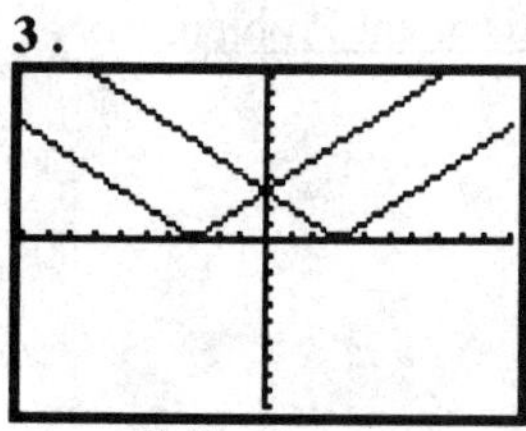 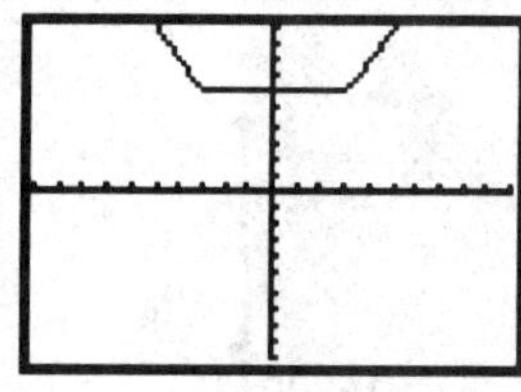 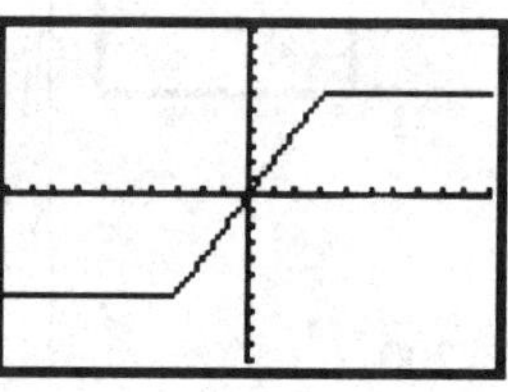

5. Factors are zero at x = -1 and x = 1, as is their product. For x > 1, the factors are both positive as is their product. For all other values of x the factors differ in sign and their product is negative.

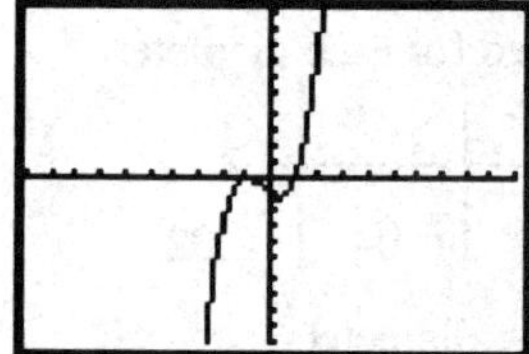 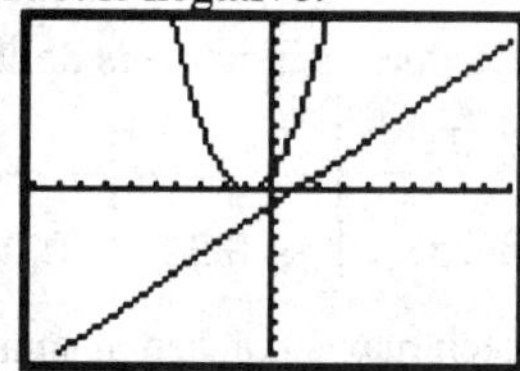

7. Factors are zero at x = -1 and x = 1, as is their product. For x < 1, the factors differ in sign and their product is negative. For 0 < x < and x > 1, the product is positive since both factors are positive.

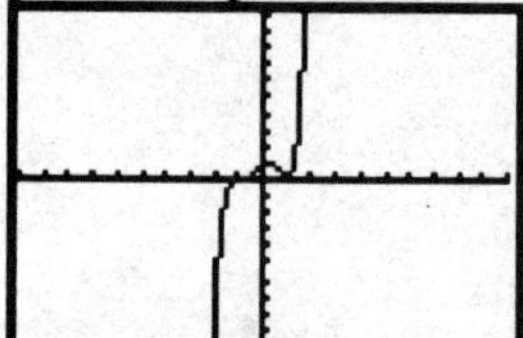 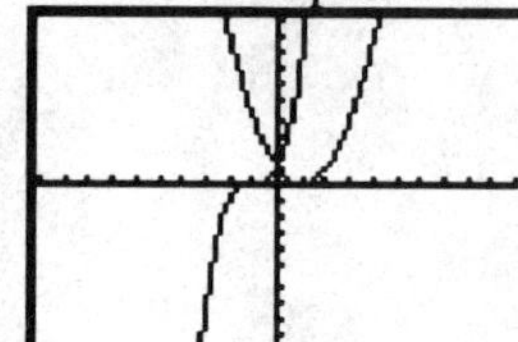

9. $f(g(x)) = x^2 + 4x + 2$; $g(f(x)) = x^2$

11. $f(g(x)) = \sqrt{-x}$; $g(f(x)) = -\sqrt{x}$

13. $f(g(x)) = |1 - x^2|$; $g(f(x)) = 1 - x^2$

15. $f(g(x)) = \sqrt{-x^2 - 2x}$; $g(f(x)) = \sqrt{1 - x^2} + 1$

17. domain: $[1,\infty)$

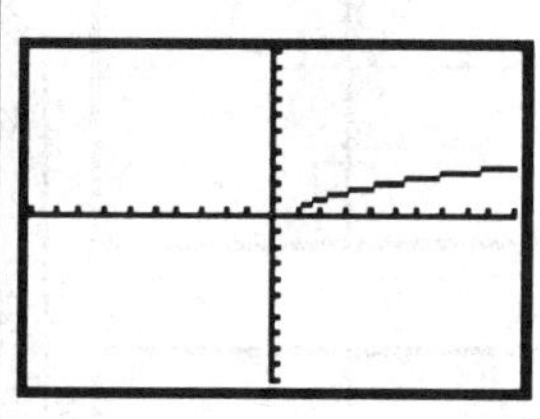

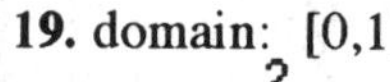

19. domain: $[0,1]$

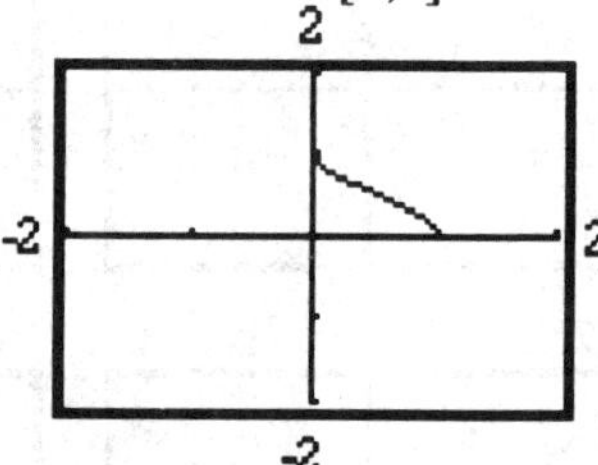

21. $y = f(g(x))$ where $f(x) = \sqrt{x}$ and $g(x) = 1 - 2x$

23. $y = f(g(x))$ where $f(x) = x - 2$ and $g(x) = \sqrt{x}$

25. $F = \frac{9}{5}C + 32$

27. $t = \frac{c}{200} - 100$

29. $f(7) = 4$; $f^{-1}(4) = 7$ **31.** $f(8) = 3$; $f^{-1}(2) = 1$

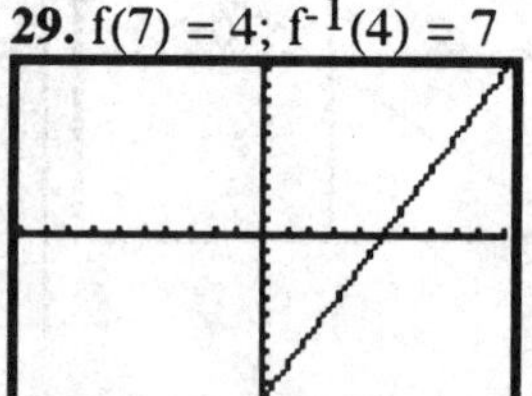 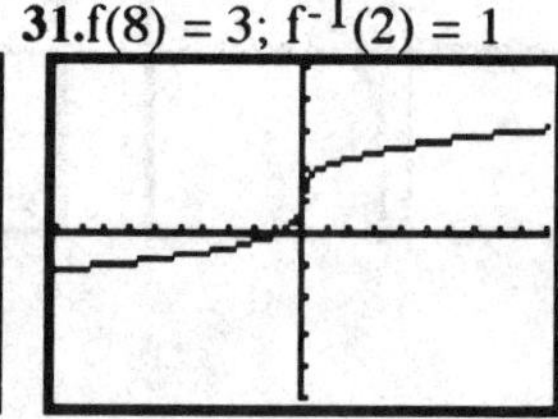

33. (a) 4 (b) 5

35. (a) -5 (b) [10,20]

37.(a) $f^{-1}(y) = \dfrac{4y+3}{2}$

(b) $f^{-1}(f(a)) = \dfrac{4\left(\frac{2a-3}{4}\right)+3}{2} = \dfrac{2a-3+3}{2} = a$

(c) $f(f^{-1}(b)) = \dfrac{2\left(\frac{4b+3}{2}\right)-3}{4} = \dfrac{4b+3-3}{4} = b$

39.(a) $f^{-1}(y) = \sqrt[3]{1-y}$

(b) $f^{-1}(f(a)) = \sqrt[3]{1-(1-a^3)} = \sqrt[3]{1-1+a^3} = a$

(c) $f(f^{-1}(b)) = 1-(\sqrt[3]{1-b})^3 = 1-1+b = b$

41. The Fuel Consumption Problem

gallons	10	6	4
hours	6	4	2.5

43. The Storm Sewer Pipe Problem

(a) 100.5 ft^3/sec (b) 60.29 ft^3/sec

45. The Mountain Climbing Problem

(a) (b) $h = \dfrac{15-T}{5}$

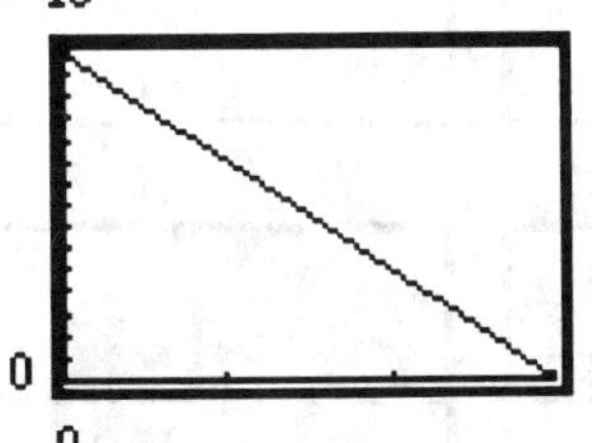

45.(c) 3 km = 3000 m.

(d)

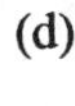

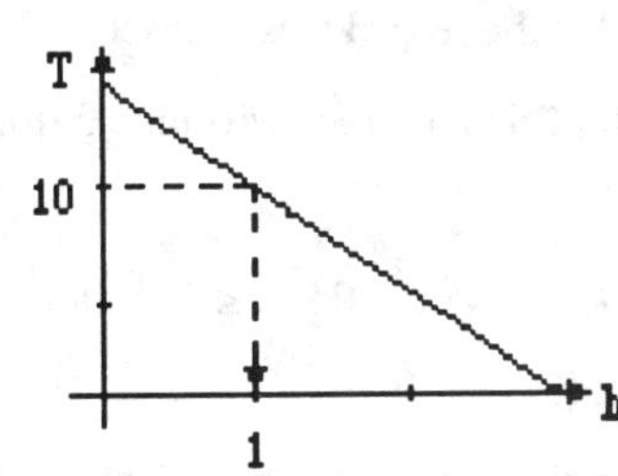

47. The Rainbow Kite and Balloon Company

(a) $r = \dfrac{1}{\sqrt{\pi R}}$ (b) yes (c) $r = \sqrt{\dfrac{0.5}{\pi R_2}}$

49.(a) $(cf(x)^{-1}) = \dfrac{5y+4}{3c}$

(b) $(f(cx))^{-1} = \dfrac{5y+4}{3c}$

51.(a) $f(g(x)) = \sqrt{x-1}$

(b) domain of f $[0,\infty)$; range of f $[0,\infty)$;

domain of g $(-\infty,\infty)$; range of g $(-\infty,\infty)$;

domain of f(g) $[1,\infty)$; range of f(g) $[0,\infty)$

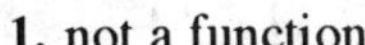

Chapter 1 Review (Page 97)

1. not a function

2. yes, not one-to-one

3. yes, not one-to-one

4. not a function

5.(a) not a function (b) yes, one-to-one

6. yes, not one-to-one

7. $1 - x \geq 0 \Rightarrow x \leq 1$

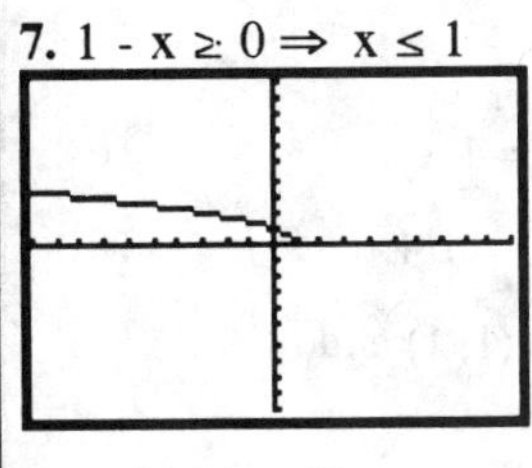

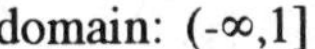

domain: $(-\infty,1]$

8. $x - 1 \neq 0 \Rightarrow x \neq 1$

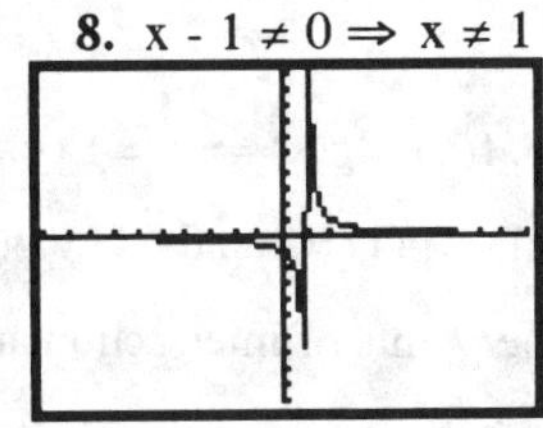

domain: $(-\infty,0) \cup (0,\infty)$

9. The Record-Breaking Mile Run Problem

(a) A: 246, In 1913, the record was 246 sec.

D: 615, In the year 2528 the race will take 0 time to run.

(b) graph $y = -0.4x + 246$ for $0 \le x \le 615$ and $0 \le y \le 250$; B: 240, C: 65

(c) Yes, but by smaller and smaller amounts.

(d) The graph will have an asymptote for some value of T.

10. The Hull Speed Problem

(a) A: 0 D: 40

(b) graph $y = 1.35\sqrt{x}$ for $0 \le x \le 40$ and $0 \le y \le 10$

(c) B: 2.7, C: 25

(d) Building longer and longer boats increases the speed but by smaller and smaller amounts.

11.(a) -8 (b) $-x^2 + 1$ (c) $x^2 + 1$ (d) $-x^2 - 6x - 8$

12. $2x + 3 = 0$

13. $x = \dfrac{0 \pm \sqrt{0^2 - 4(2)(-1)}}{2(2)}$

14. $f(x) = g(x) \Rightarrow x = -2$ or $x = 3$;

$f(-2) = g(-2) = 0$ and $f(3) = g(3) = 5$

The points of intersection are (-2,0) and (3,5).

15. $f(x) = g(x) \Rightarrow x = 6$ or $x = 11$; $f(6) \ne g(6)$ and $f(11) = g(11) = 3$

The point of intersection is (11, 3).

16. $f(x) = g(x) \Rightarrow x = 1$ or $x = 4$;

$f(1) = g(1) = 1$ and $f(4) = g(4) = 1/4$

The points of intersection are (1, 1) and (4, 1/4).

17.(a) $y = x^2 + 2$

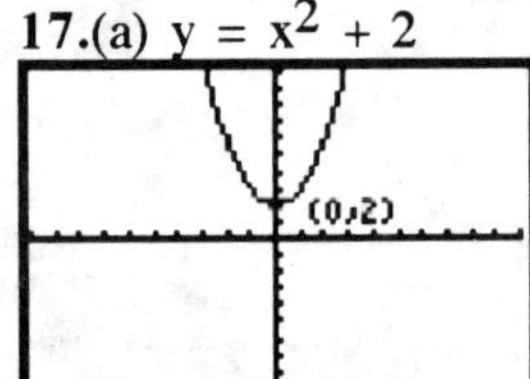

(b) $y = (x + 2)^2$

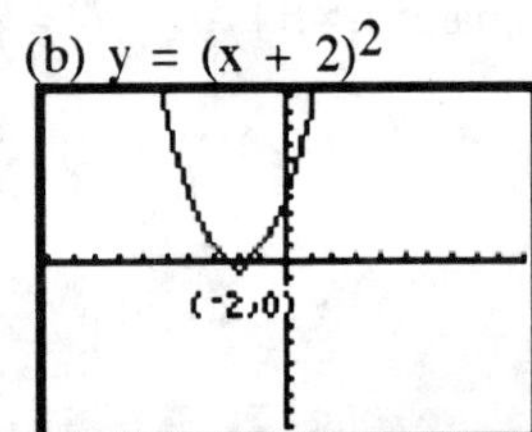

17.(c) $y = (x - 2)^2$

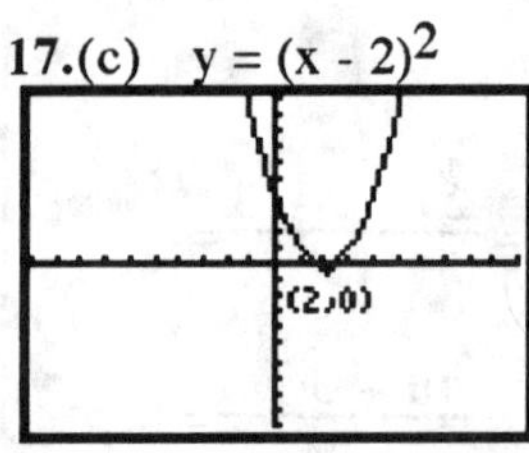

(d) $y = x^2 - 2$

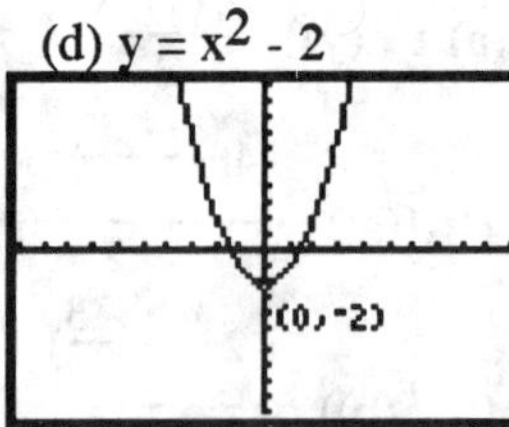

18.(a) $y = -\sqrt{x} - 5$

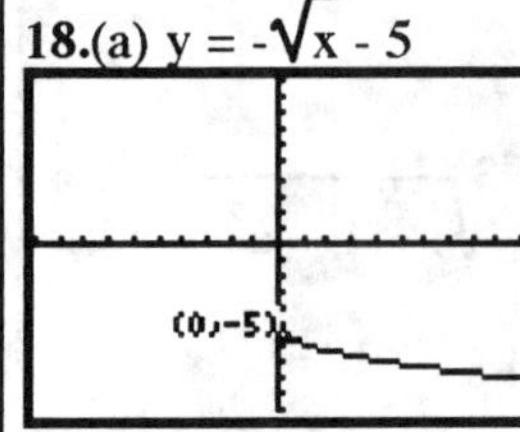

(b) $y = \sqrt{-x} + 5$

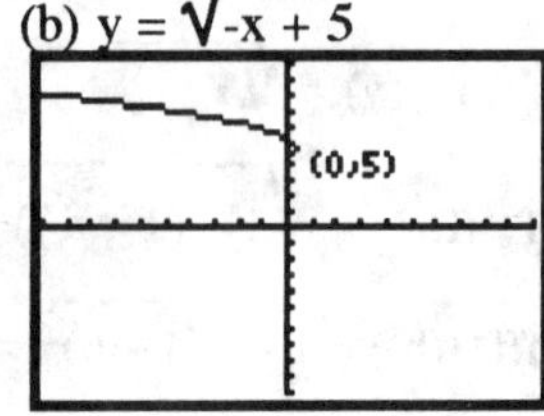

(c) $y = -\sqrt{-x} - 5$

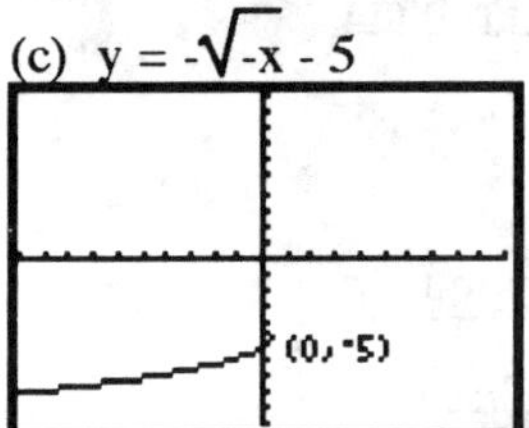

19.

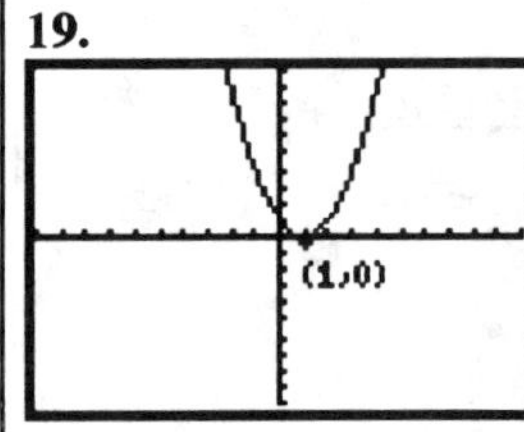

20.

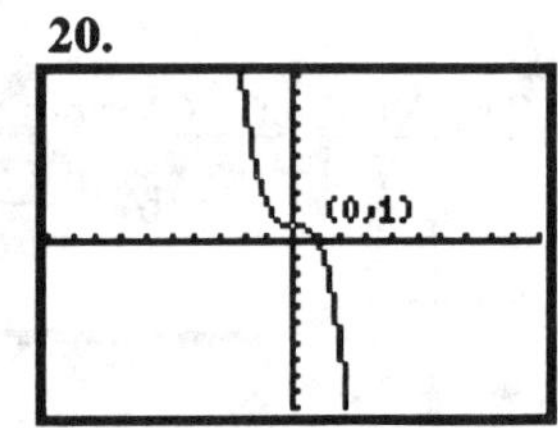

21.

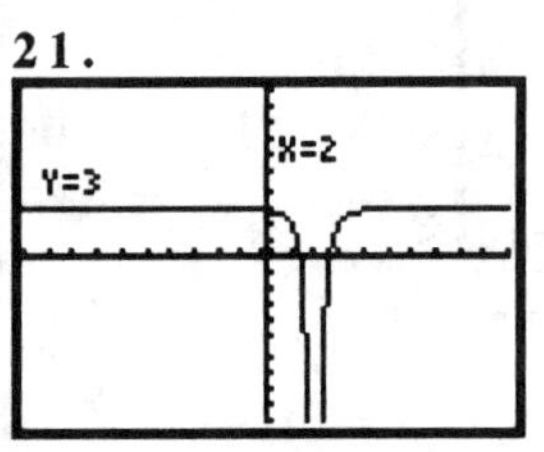

22.

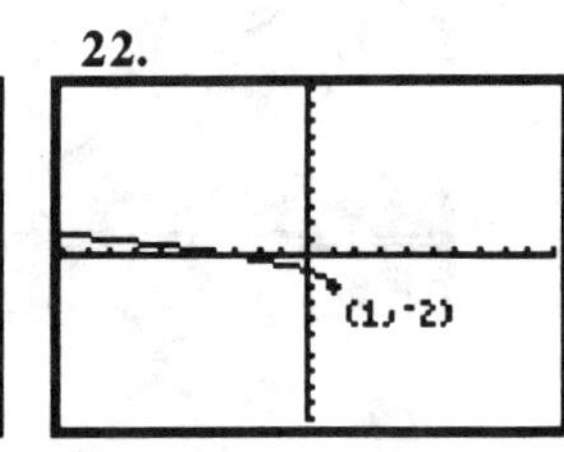

23.(a)

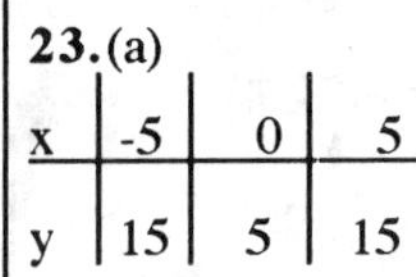

x	-5	0	5
y	15	5	15

(b)

x	0	5	10
y	10	0	10

24.(a)

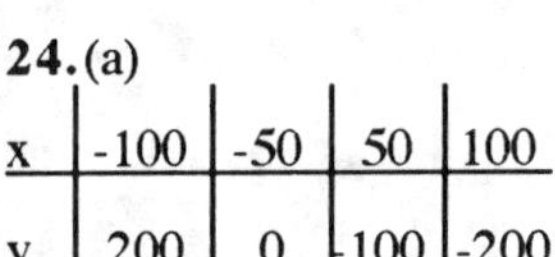

x	-100	-50	50	100
y	200	0	-100	-200

(b)

x	100	50	-50	-100
y	200	0	-100	-200

25. $y = -|x - 10|$

26. $y = -\sqrt{x} + 10$

27. $y = -x^2 + 10$

28. $y = x^2$

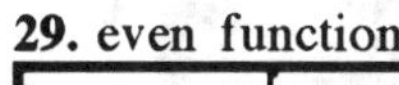

29. even function **30.** neither

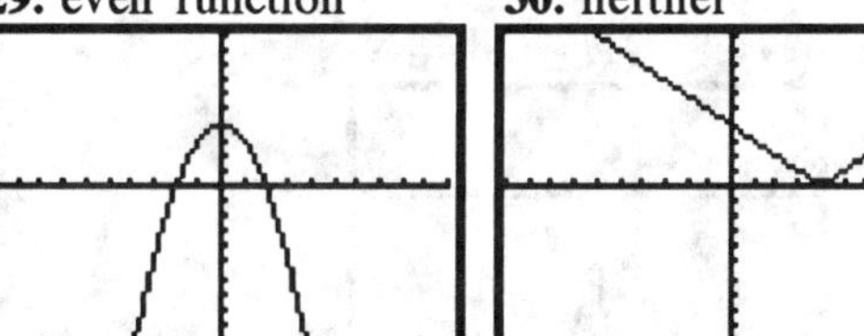

31. odd function

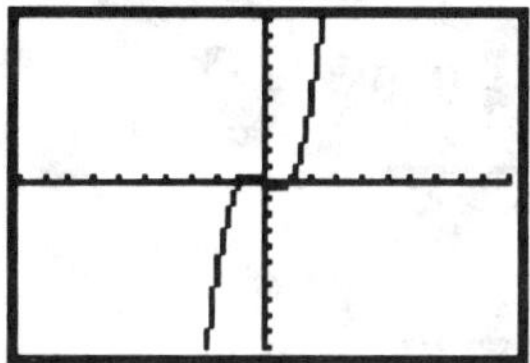

32.

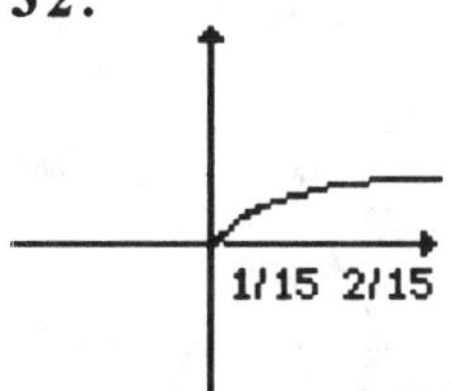

33.

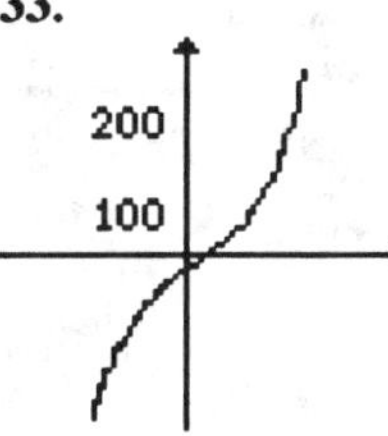

34.

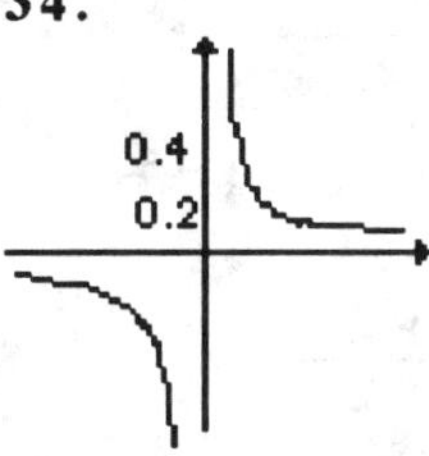

35.

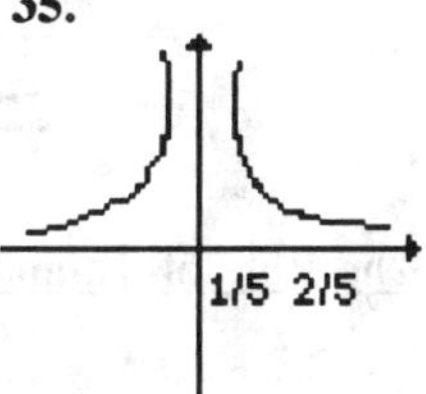

36.

x	-2	0	2	4	8
y	81	27	9	3	1

37.

x	-2	0	2	4	8
y	3	1	1/3	1/9	1/27

38.

x	-1	0	1	2	4
y	9	3	1	1/3	1/9

39.(a)

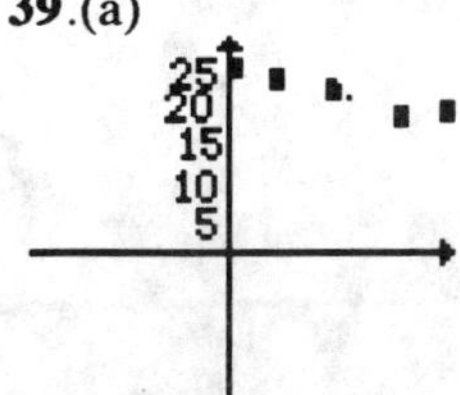

(b)

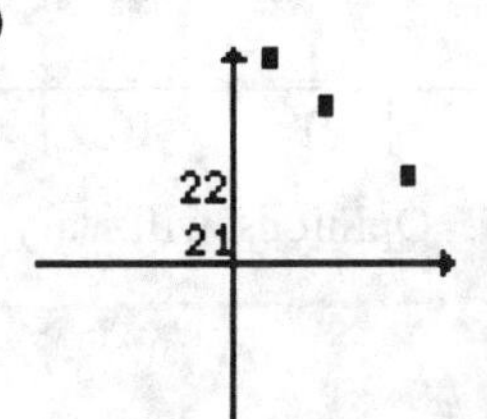

40. **41.**

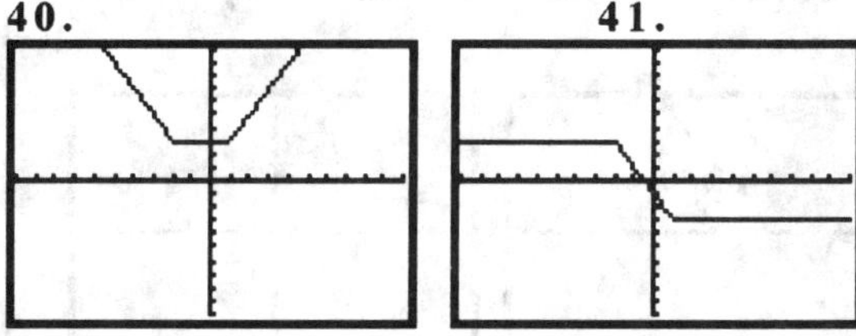

42.

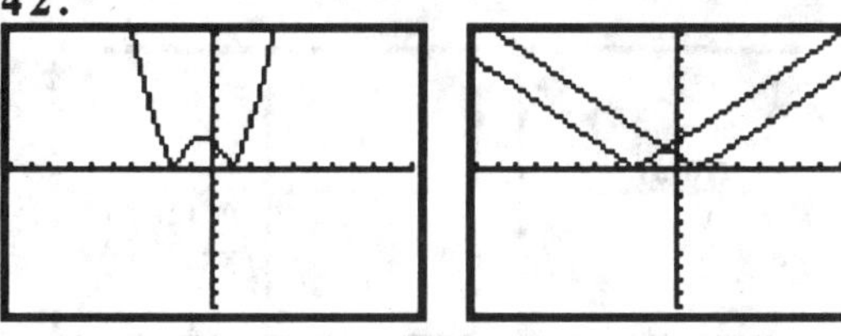

Factors are zero at $x = -2$ and $x = 1$, as is their product. For all other values of x the product is positive since both factors are positive.

43. x **44.** x

45.

x	-10	10
y	0	40

46.

x	-10	0
y	20	10

47. $f(g(b)) = (\sqrt{b})^2 = b$

48. $g(f(a)) = \sqrt{a^2} = |a|$

49.(a) $x = \pm\sqrt{9 - y}$

(b) no because the function $y = 9 - x^2$ is not one-to-one.

50. $f(5) = 4$ $f^{-1}(10) = 2$

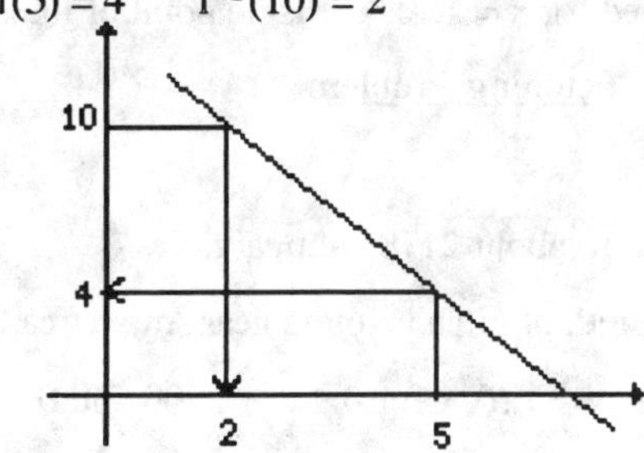

Chapter 1 Test (Page 101)

1. **2.** **3.** **4.**

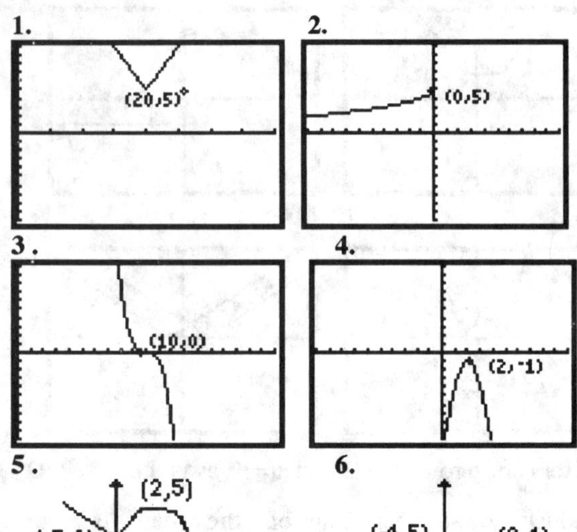

5.

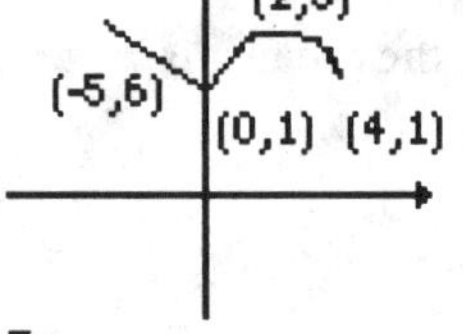

6.

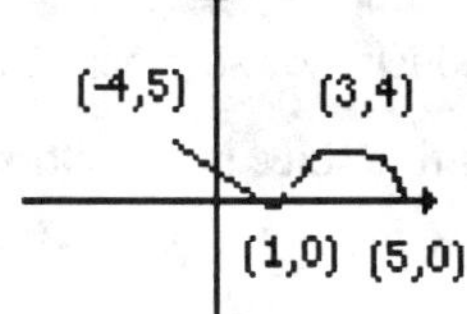

7.

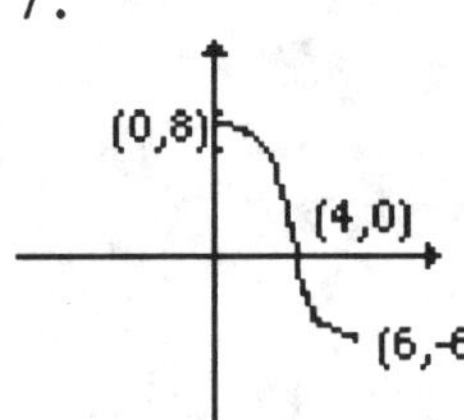

8.

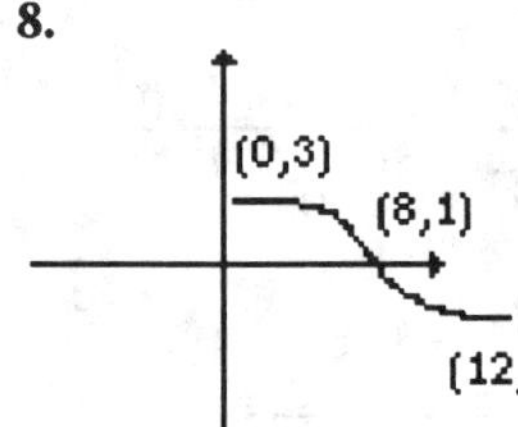

(12,-4)

9. g **10.** [-4,-2] and [0,5]

11. (-2,4) and (0,0) **12.** neither

13. $f(x) = \frac{1}{x - 4} + 2$ and $f(x) = 0 \Rightarrow x = \frac{7}{2}$

$f(x) = \frac{1}{x - 4} + 2 \Rightarrow f(0) = \frac{7}{4}$

The x intercept is $\frac{7}{2}$ and the y intercept is $\frac{7}{4}$.

14. $f(x) = g(x) \Rightarrow x = 1$ but both functions are undefined for $x = 1$. There is no point of intersection.

15. The Fetching Problem

(a) 7.5 ft

(b) 1600 to about 2178 nautical miles

(This length of fetch is found near Antarctica.)

(c) Enter $y = 1.5\sqrt{x}$ for $0 \le x \le 2200$ and $0 \le y \le 70$.

16. 2, -5, 2, -5 **17.** 2, -5, 2, -5

18. $f(x) = f(-x)$ even function

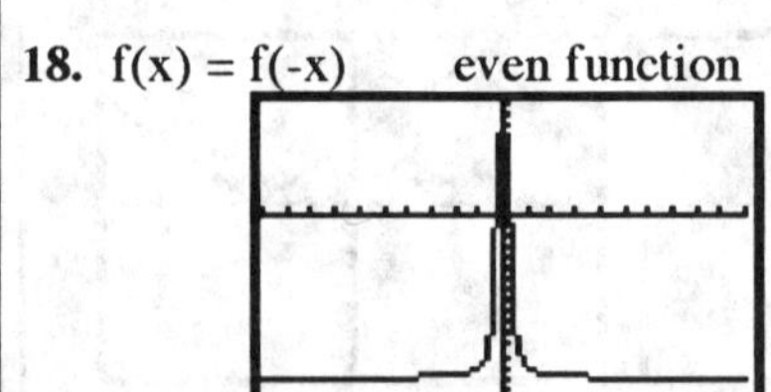

19.(a) (answers vary)

$f(x) = x^2$ and $g(x) = x \Rightarrow y = (f \cdot g)(x) = x^3$

(b)

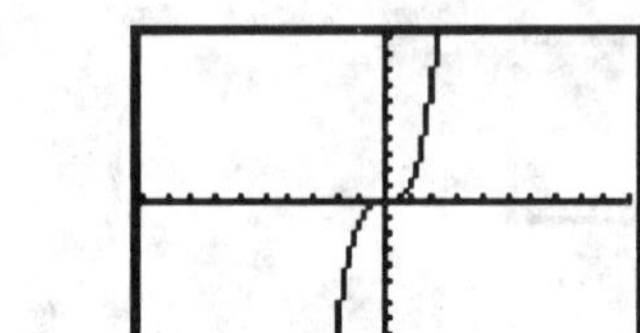

20.

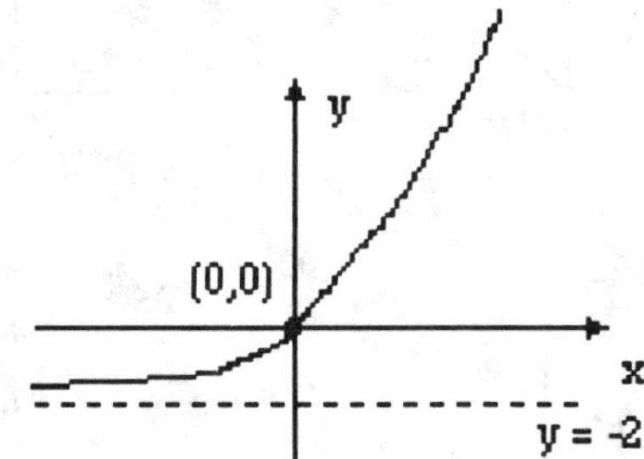

21. The Rule-of-Thumb Problem

(a) $F = 2C + 30$ (b) $C = \frac{F - 30}{2}$

(c) A temperature in degrees Fahrenheit can be converted to degrees Celsius by taking half the difference of the temperature and 30.

22. The Salmon Cookbook Problem

(a) $T = 10F$ (b) $F = T/10$

(c) wanting to know the cooking time of a 2-in. piece of salmon; wanting to know how thick a fish can be cooked in 20 min.

23.(a)

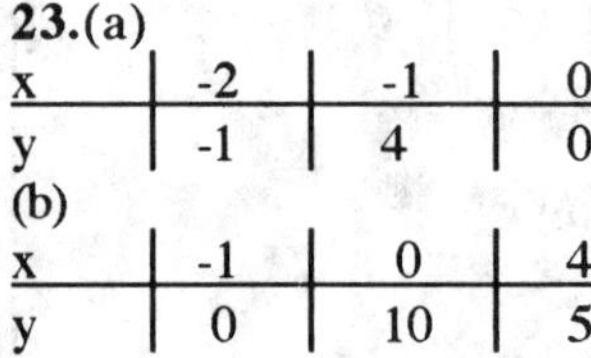

x	-2	-1	0
y	-1	4	0

(b)

x	-1	0	4
y	0	10	5

24. Opinions will vary.

Section 2.1 Linear Functions (Page 117)

1. $y = -\frac{2}{3}x + 4$; $m = -\frac{2}{3}$, $b = 4$

3. $y = \frac{2}{5}x - 3$; $m = \frac{2}{5}$, $b = -3$

5. $y = \frac{3}{4}x + 1$ **7.** $y = 2$

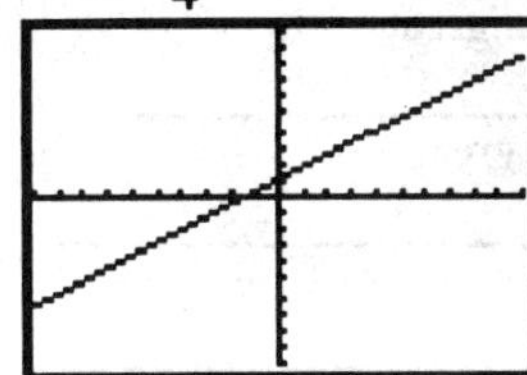

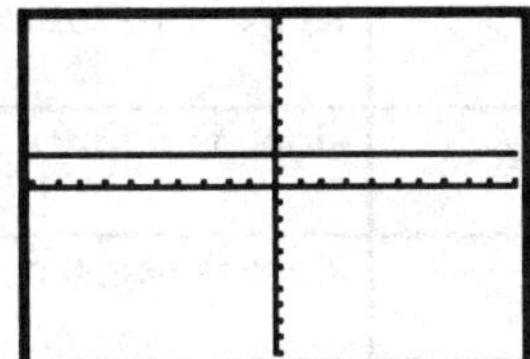

9. $y = 3x + 6$ **11.** $y = 3$

13. $f(x) = -3x + 2$ **15.** $y = \frac{1}{2}x - 2$

17. $y = \frac{2}{3}x - \frac{7}{3}$ **19.** $x = 6$

21. $y = 2x$ **23.** $y = -2x + 1$

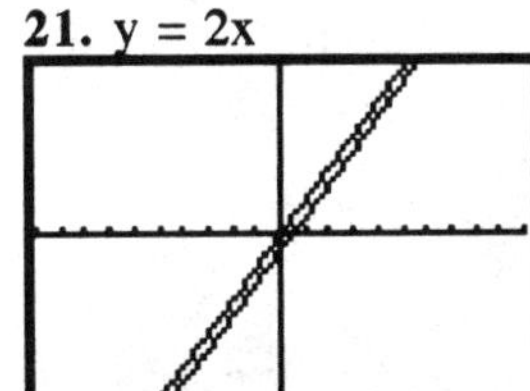

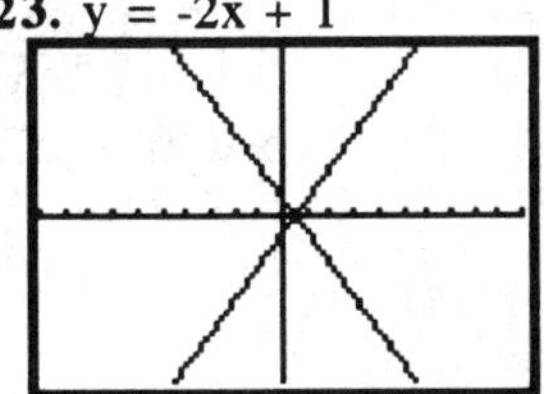

25. $y = -x - 1/2$ **27.** $y = 24x - 79$

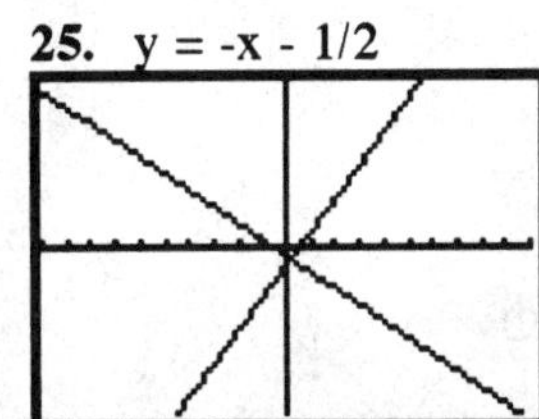

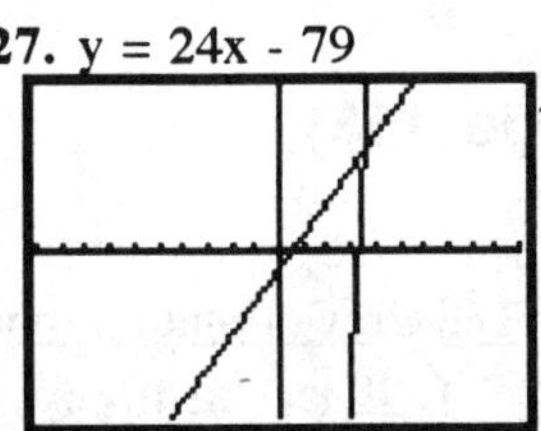

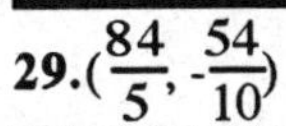

29. $(\frac{84}{5}, -\frac{54}{10})$ **31.** none

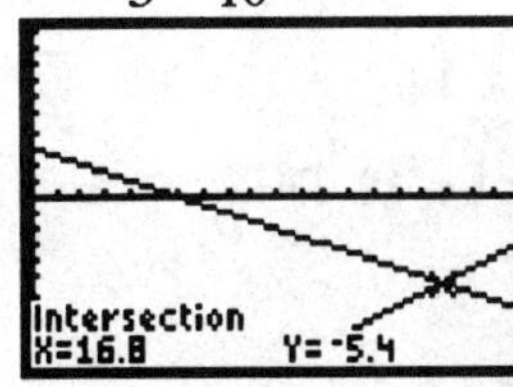

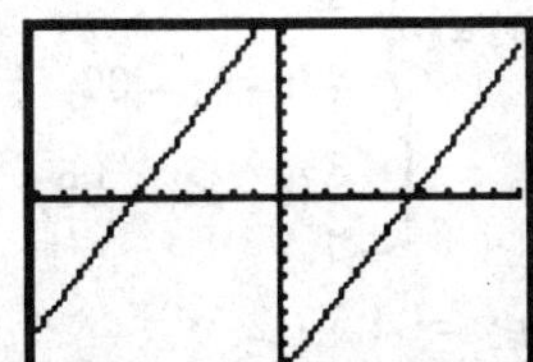

33. (-1/2,0)

35. **37.**

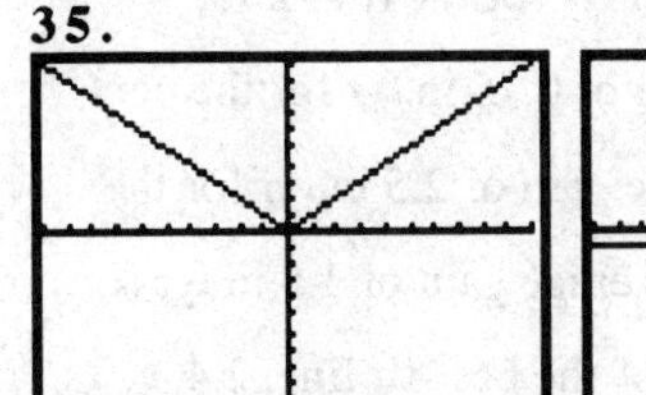

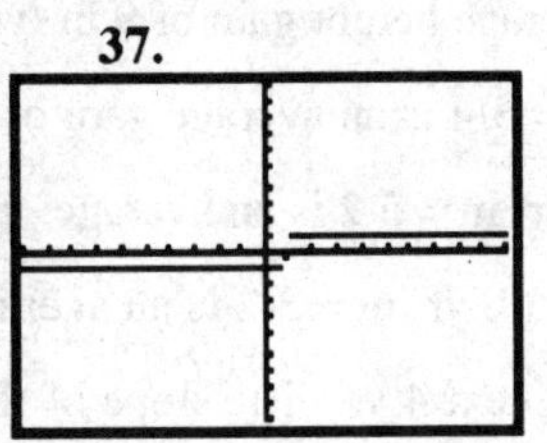

39. The Car Rental Problem

If you're driving less than 39 miles or more than 137 miles the second company is the better choice.

41. The Triathlon Problem

1 km swimming; about 31 km bicycling; about 20 km running

43. The Weight Problem

(a) $6\frac{3}{4}$ lb (b)

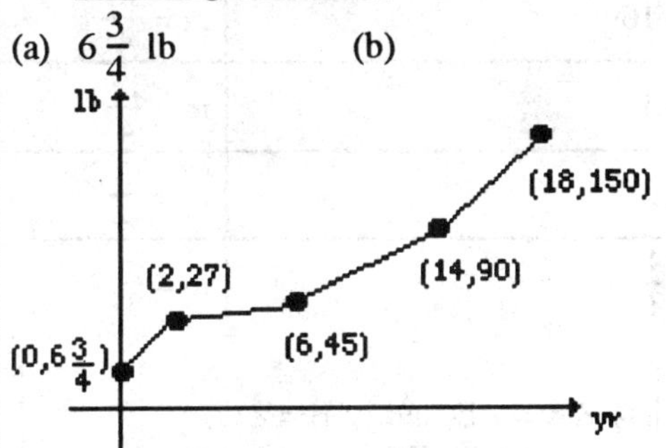

45. The 1993 Tax Law Problem

(a)

$$y = \begin{cases} 0.15x, & 0 \le x \le 22{,}100 \\ 0.28(x - 22{,}100) + 3315, & 22{,}100 \le x \le 53{,}500 \\ 0.31(x - 53{,}500) + 12{,}107, & 53{,}500 \le x \le 115{,}000 \\ 0.36(x - 115{,}000) + 31{,}172, & 115{,}000 \le x \le 250{,}000 \\ 0.396(x - 250{,}000) + 79{,}772, & x \ge 250{,}000 \end{cases}$$

(b) Both brackets 2 and 3 yield a tax of $12,107.

47. An Arachnid Problem

(140, 0.42) When 140 species on the island, the number of species stabilizes because the rates of immigration and extinction are the same, 0.42 species/month.

49. The World Population Problem

(a) $R_b = 138$, $R_d = 21.5N - 66$

(b) $(9.4\bar{9}, 138)$ Zero population growth will occur when world population reaches about 9.5 billion people and the birth and death rates both equal 138 million people/yr.

51. The Rainbow Kite and Balloon Company

(a) $20 (b) yes (c) no (d) (80, 20) Maximum profit will occur when 80 windsocks are produced since MC = MR = 20.

53. 13

55. $f(x_1) = mx_1 + b$, $f(x_2) = mx_2 + b$ and $m = \dfrac{f(x_2) - f(x_1)}{x_2 - x_1} \Rightarrow$

$$m = \frac{mx_2 + b - (mx_1 + b)}{x_2 - x_1} = \frac{mx_2 + b - mx_1 - b}{x_2 - x_1} = \frac{m(x_2 - x_1)}{x_2 - x_1} = m$$

57.

$y = \frac{30}{4}x - 10$	$m = \frac{30}{4}$	slope is 10 times as great
$y = \frac{15}{4}x - 1$	$m = \frac{15}{4}$	slope is 5 times as great
$y = -\frac{3}{4}x - 1$	$m = -\frac{3}{4}$	slope is negated
$x = \frac{4}{3}y + \frac{4}{3}$	$m = \frac{4}{3}$	slope is reciprocal

59.(a) $l_1 \parallel l_2$ $\angle A = \angle D$ and $\angle B = \angle C$

$\Delta AOB \sim \Delta DOC \Rightarrow \dfrac{AO}{DO} = \dfrac{BO}{CO} \Rightarrow$

$AO \cdot CO = DO \cdot BO$

$\dfrac{CO}{DO} = \dfrac{BO}{AO}$

$m_2 = m_1$

(b) $m_1 = m_2 \Rightarrow \dfrac{BO}{AO} = \dfrac{CO}{DO} \Rightarrow$

$BO \cdot DO = AO \cdot CO$

$\dfrac{BO}{CO} = \dfrac{AO}{DO}$

$\Delta AOB \sim \Delta DOC$

$\angle A = \angle D$ and $\angle B = \angle C$

$l_1 \parallel l_2$

Section 2.2 Linear Regression (Page 133)

1.(a) positive

(b) none

3.(a) negative

(b) positive

5. negative

7. negative

9. $y = 1.25 - 0.25x$

11. $y = -0.2\bar{6} + 0.4\bar{6}x$

13.(a) $y \approx -3.03 + 0.886x$

(b) $(-9,-11)$ (c) $y \approx -3.03 + 0.886(-9) = -11.004$

15. $y = 3149 - 592x$; $(5.3,0)$ $(2,1965)$

17. $y = 8.4 + 0.98x$; the year 2000 = 29 then $(29, 36.82)$

19. The Calorie Problem

$y = -1.55 + 0.0008x$; The slope, 0.0008, means an 8-lb change in weight/day for every 10,000 cal. The y intercept, -1.55, means when no calories are taken in, then 1.55 lb/day weight loss.

21. The Growth Study Problem

(a)
$$h(t) = \begin{cases} \frac{18}{2}t + 18, & 0 \le t < 2 \\ \frac{3}{4}(t - 2) + 36, & 2 \le t < 6 \\ \frac{5}{2}(t - 6) + 39, & 6 \le t < 14 \\ \frac{3}{2}(t - 14) + 59, & 14 \le t \le 18 \end{cases}$$

(b) $y = 24.5 + 2.4x$

(c) For the piecewise-defined function, $m = 18/2$ is an average height gain of 9 in./yr for the first 2 yr; $m = 3/4$ is an average gain of 0.75 in./yr for the next 4 yr; $m = 5/2$ is an average gain of 2.5 in./yr for the next 8 yr; $m = 3/2$ is an average gain of 1.5 in./yr for the next 4 yr. The slope of the best-fit line, 2.4, suggests an average growth rate of 2.4 in./yr.

23. The Airfare Problem

about $535

25. The Life Expectancy Problem

about 80 years

27. The Rainbow Kite and Balloon Company

(50.44, 50.92) Market equilibrium occurs at a price of $50.92 when about 50 kites are supplied and bought.

29. Wonderworld Amusement Park

about $7

31. The Rainbow Kite and Balloon Company

(a) 176 ft, 361 ft, 546 ft (b) 4.37 mi

33. The Ring Size Problem

Answers will vary depending on data.

35. The Shirt Size Problem

Answers will vary depending on data.

37.(a)

ht (in.)	65	69	70	72	74	76
wt (lb)	140.5	160.5	164.5	173.5	184.5	195.5

(b) y = -181 + 4.9x

(c) Translating the data down 2.5 units results in a new equation with the same slope but whose y intercept is not translated down exactly 2.5 units.

39. (a)

x	2	4	6	8
y	-10	0	5	10

(b) y = -15 + 3.25x

(c) This equation appears to be the inverse of the original.

(d) The coefficient of determination is identical.

41.(a)

-10	0	5	10
-2	-4	-6	-8

(b) y = -4.63 - 0.297x (c) y = -f(x) (d) The coefficient of determination is identical. (The correlation coefficient is negative.)

Section 2.3 Graphs of Quadratic Functions and Optimization (Page 154)

1. (2, -9) **3.** (-1/6, 49/12)

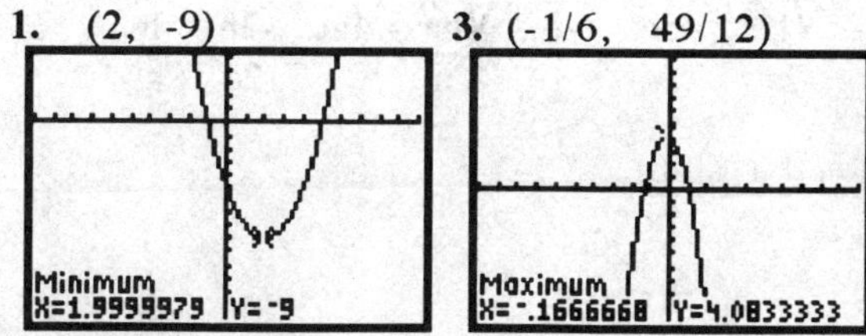

5.(a) 3 (b) 0

7.(a) -1 and -1/2 (b) -1 and -1/2

9.(a) -3 (b) 1 (c) -4

11.(a) 2 (b) -2 (c) 2

13.

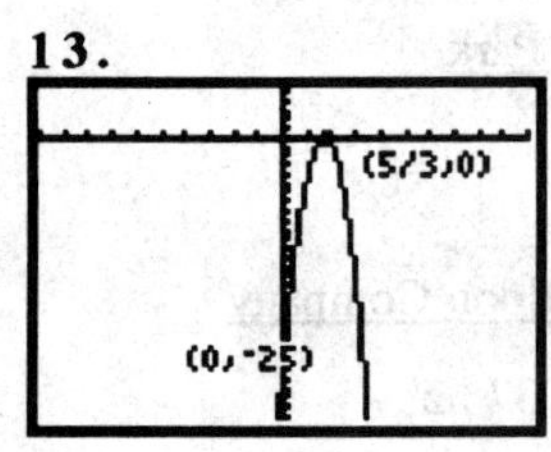

15.

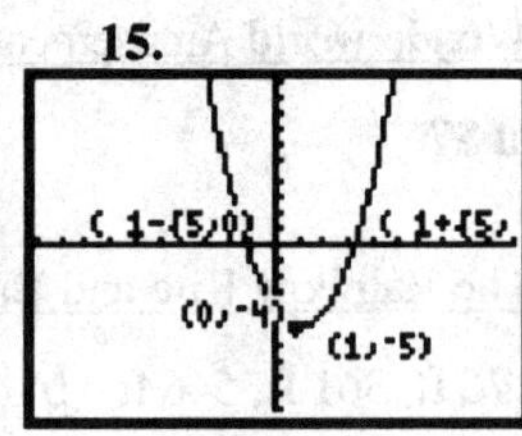

17.(a) $k < 1/3$ (b) $k > 1/3$ (c) $k = 1/3$

19. The Skydiver Problem

Whoops, he's on the ground.

21. The Long Shot Problem

352 yd

23. The Perennial Border Problem

25 ft by 100 ft

25. The Rainbow Kite and Balloon Problem

$x = 36$ in. and $y = 36$ in.

27. The Adjustable Playpen Problem

(a) 36 ft^2 (b) about 45.8 ft^2

29. The Towering Rock Problem

(a) 192 ft (b) The vertex is located at $t = -2$, which is not in the domain of the application.

31. The Rainbow Kite and Balloon Problem

$x = 100/3$ in. and $y = 0$

33. The Discrete Hedge Problem

number of plants, m	0	1	2	3	4
number of plants, m + 8	8	9	10	11	12
number of plants n	28	26	24	22	20
length, l = 1.5n	42	39	36	33	30
width, w = 12 + 1.5m	12	13.5	15	16.5	16
area, A = lw	504	526.5	540	544.5	480

$m = 3$ plants, $8 + m = 11$ plants and $n = 22$ plants

35. The Flowerpot Problem

(a) about 32 pots (b) \$3.14

37. The Taxing Problem

(a) $p(x) = 2x + 50 + t$ (b) $t = 50 - 2.1x$

(c) $TR = 50x - 2.1x^2$ (d) $t = \$25$

39. The Corner Playpen Problem

Let x be one leg of the triangle, y the other leg, and 20 the hypotenuse.

$A = \frac{1}{2}bh$, $b = x$, and $y = h \Rightarrow A = \frac{1}{2}xy$

$x^2 + y^2 = 400$

$A = \frac{1}{2}xy$ and $y = \sqrt{400 - x^2} \Rightarrow A = \frac{x}{2}\sqrt{400 - x^2}$

$x \approx 14.14 \Rightarrow y = \sqrt{400 - x^2} \approx 14.1444$

Since the maximum area is achieved when $x \approx y$, the triangle is approximately isosceles.

41. The Sailboat Problem

about 1/2 hr

43. $r_1 + r_2 =$

$$\frac{-b + \sqrt{b^2 - 4ac}}{2a} + \frac{-b - \sqrt{b^2 - 4ac}}{2a} = \frac{-2b}{2a} = \frac{-b}{a}$$

45. $\frac{r_1 + r_2}{2} = \frac{-b}{2a}$

Section 2.4 Finding Equations of Quadratic Functions (Page 169)

1. $y = 4x^2 + 8x + 4$ **3.** $y = x^2 - 2x + 3$

5. $y = -2x^2 + 8$ **7.** $y = x^2 - 2x - 2$

9. $y = -2x^2 + 3x + 9$

11. $y = 0.04x^2 + 2.7x - 2.375$

13. $y = 2x^2$

15. $y = -1.5x^2 + 11.5x + 1.5$

17. The Rotten Egg Problem

(a) $y = -4.9x^2 + 19.6x$ (b) (2, 19.6)

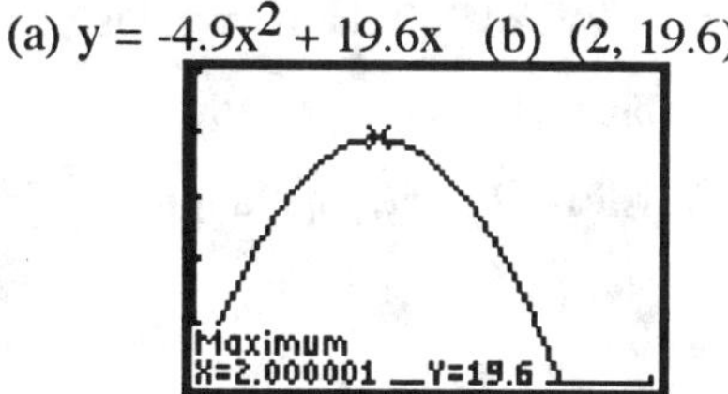

19. A Profit Problem

(a) $y = -8.03x^2 + 127.5x + 2692$ (b) $(7.\tilde{9}, 319\tilde{8})$

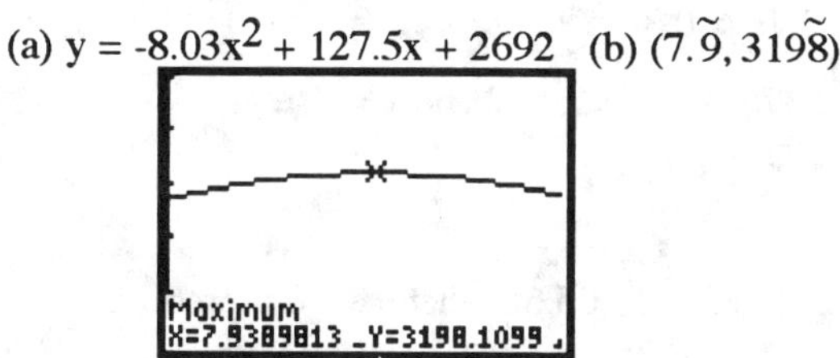

21. The Suspended Cable Problem

The second graph is nearly uniformly loaded horizontally creating a parabola.

23. The Circle Problem

How accurately the radius and area were measured depends on how close the equation is to $A = \pi r^2$.

25. The ZPG and Immigration Problem

Zero population growth will be reached when the population reaches 303.5 million people and the growth rate = death rate ≈ 2.87 million/yr.

27. The Intercom Problem

(a)

n	0	1	2	3	4	5
l	0	0	1	3	6	10

(b) $y = 0.5x^2 - 0.5x$

(c) 15, 45

29. The Rainbow Kite and Balloon Company

Cross Section	Surface Area Function, SA	Constraint	Surface Area Subject to Constraint	Maximum Surface Area
x x x	$SA = 3xy$	$6x + y = 72$	$SA = 3x(72 - 6x)$	x = 6 in. y = 36 in. SA = 648 in.^{2}
x x x x	$SA = 4xy$	$8x + y = 72$	$SA = 4x(72 - 8x)$	x = 4.5 in. y = 36 in. SA = 648 in.^{2}
x x x x x	$SA = 5xy$	$10x + y = 72$	$SA = 5x(72 - 10x)$	x = 3.6 in. y = 36 in. SA = 648 in.^{2}
x	$SA = 2\pi xy$	$4\pi x + y = 72$	$SA = 2\pi x(72 - 4\pi x)$	x = 2.85 in. y = 36 in. SA = 648 in.^{2}

31. The Book Budget Problem 49% refinement

Chapter 2 Review (Page 174)

1. $y = -\frac{1}{2}x - 2$ **2.** $y = -3$

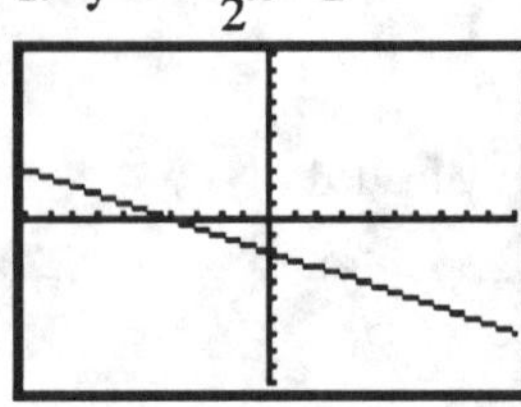

3. $y = 20x + 10$ **4.** $y = \frac{2}{5}x + \frac{24}{5}$

5. $x = -1$ **6.** $y = -\frac{7}{9}x - \frac{10}{9}$

7.

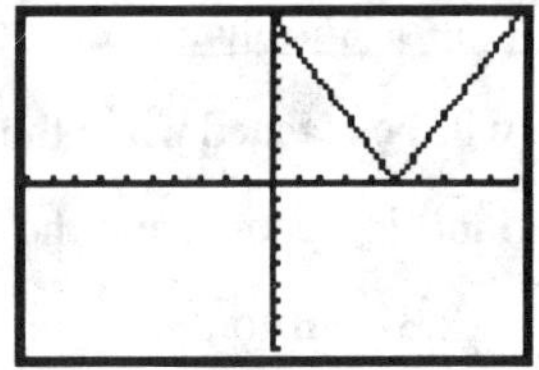

8.

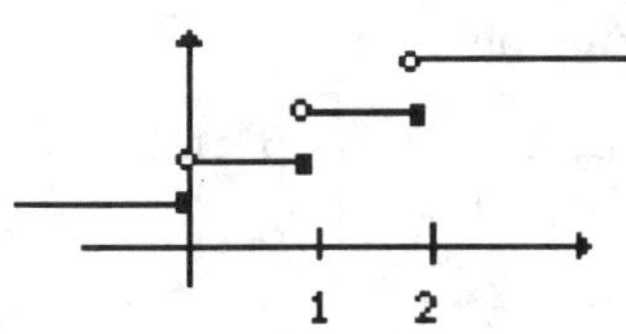

9. The Pyramid Problem

$\frac{\Delta y}{\Delta x} = \frac{177}{700} \approx 0.25, \quad \frac{\Delta y}{\Delta x} = \frac{177}{990} \approx 0.178$

10. The Luxor Hotel Problem

$\frac{\Delta y}{\Delta x} = \frac{350}{247.5} \approx 1.41$

11.(a) $f^{-1}(y) = 2y + 8$

(b) $f(f^{-1}(y)) = \frac{2y+8}{2} - 4 = y + 4 - 4 = y$

(c) $(f^{-1}(f(x)) = 2(\frac{x}{2} - 4) + 8 = x - 8 + 8 = x$

12. The Weather Balloon Problem

(a)

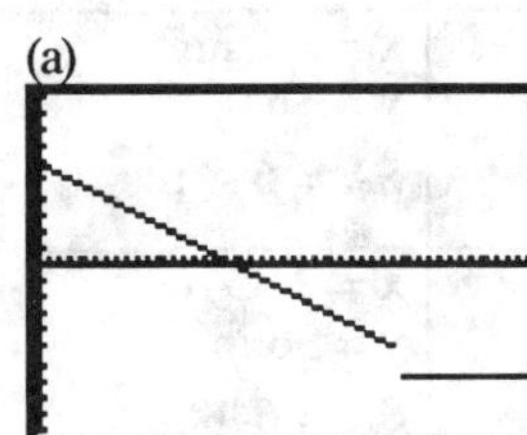

(b) 60°F (c) -67°F

(d) $h = -\frac{1000}{3}(T - 60)$ (e) about 9333 ft

13. The Test Scoring Problem

(a) $y = \frac{15}{16}(x - 45) + 65$ (b) about 88

(c) If a student scored zero on the test, the new score would be 23.

14. The Clamming Problem

(a) $R_g = \frac{-3}{50}N + 1000$, $R_h = 200$ (b) (13,333, 200)

The equilibrium point is 13,333 clams, which are growing and harvested at a rate of 200 clams/yr. This is the maximum sustainable yield since harvesting more than 200 clams/yr will reduce the population.

15.(a) positive (b) positive

16.(a) negative (b) none

17. $y = -0.4x + 0.3$

18. $y = 2.87 + 0.065x$, $\bar{x} = 2$ and $\bar{y} = 3$,

$y = 2.87 + 0.065(2) = 3$

19. $y = 23{,}373.5 + 1{,}005.6x$ where $x = 0$ represents the year 1985, $24,379, half way through 1991

20. The Salt Concentration Problem

$y = 204 + 2.1x$

21. The Skydiver Problem

The equation for the first jumper, $y = 907 - 56.8t$, estimates a jump from 907 ft that took about 16 min. The equation for the second jumper, $y = 1671 - 139t$, estimates a jump from 976 ft when t = 5 min that took 12 min. The second diver caught the first diver 4.3 min after jumping at an altitude of 379 ft.

22.

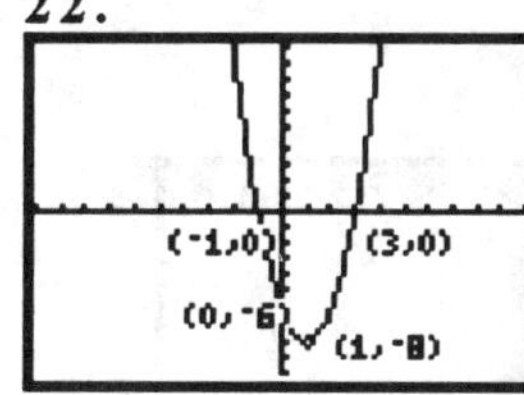

23.

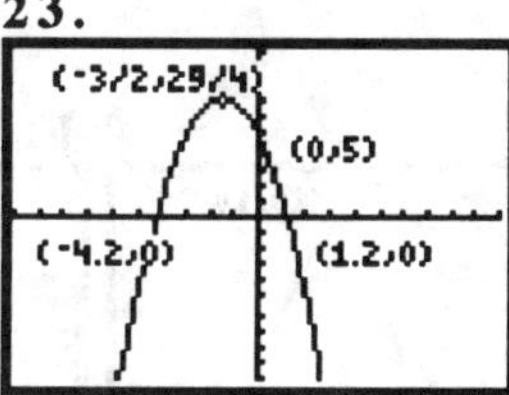

24.(a) $k < 9/8$ (b) $k > 9/8$ (c) $k = 9/8$

25.(a) $k > -9/8$ (b) $k < -9/8$ (c) $k = -9/8$

26. (0.-8) and (4,0)

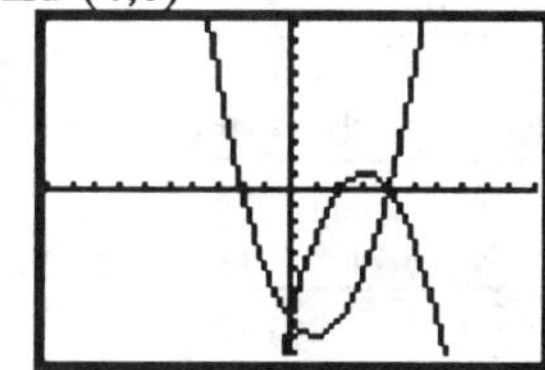

27. $y = a(x - h)^2 + k$ and $y = 0 \Rightarrow$

$$a(x - h)^2 + k = 0$$
$$a(x - h)^2 = -k$$
$$(x - h)^2 = -k/a$$
$$x = h \pm \sqrt{-k/a}$$

28. The Firecracker Problem

about 5.4 sec

29. The Drum Major Problem

about 4 sec

30. $x \approx 15.4$ in., $y \approx 46.4$ in.

31. $x \approx 18.9$ in., $y \approx 51.3$ in.

32. The Ribbon Problem

$x = 1$ ft, $y = 2$ ft

33. The Window Problem

$x \approx 1.4$ m, $y \approx 0.9$ m

34. The Compact Disc Revenue Problem

(a) $R(x) = -0.0099x(x - 1600)$ (b) about 553

35. An Economic Model Problem

(a) \$883 when $x = 38$ (b)

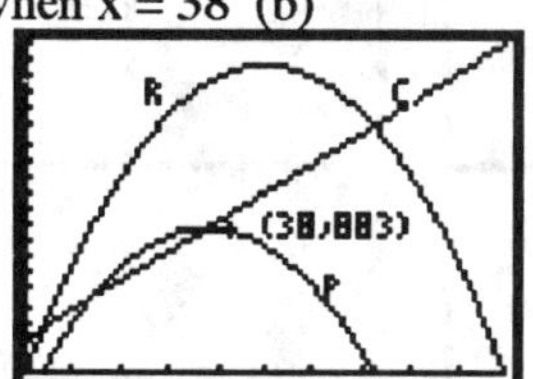

(c) For 38 items, profit is \$883, revenue is \$1767, and cost is \$884. $883 = 1767 - 884$; $P = R - C$

36. The Charter Boat Problem

17 or 18

37.

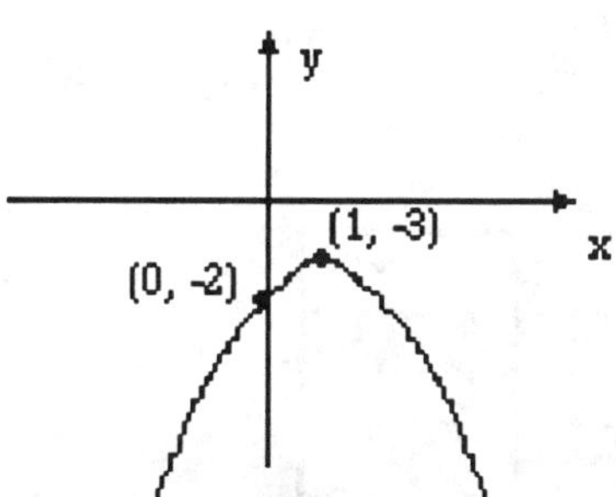

38. The Equilateral Triangle Problem

$A = \frac{\sqrt{3}}{4}1$

39. $y = -0.067x^2 + 0.067x + 2$

40. $y = -x^2 + 10$

41. The Tunnel Problem

(a) 45 mi/hr (b) As the speed limit increases, the space between the cars also increases.

42. The Refined Project Problem

(a) 50% (b) possibly reduced sales

Chapter 2 Test (Page 180)

1.

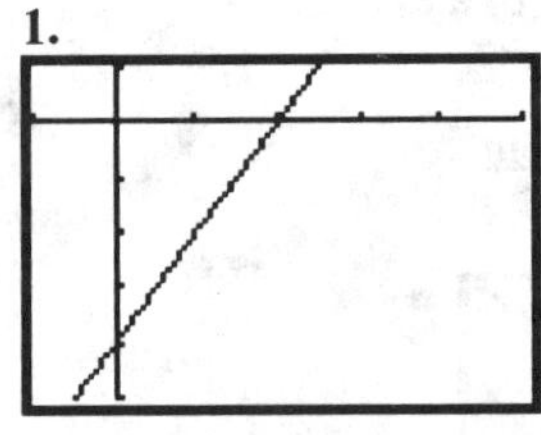

2.

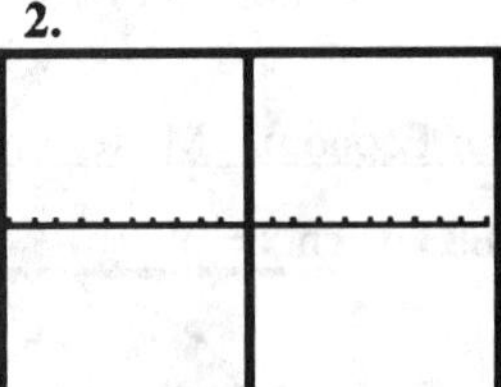

3.

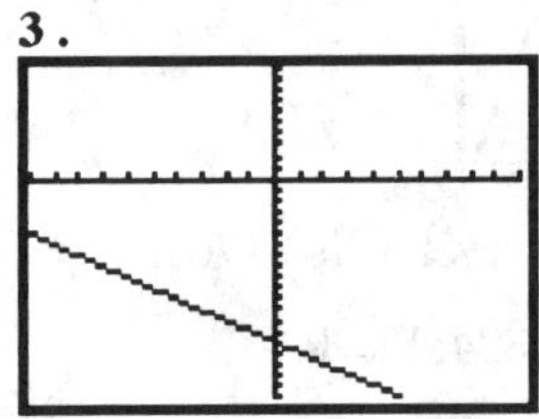

4. $y = \frac{1}{2}x - 2$

5. $y = 3x - 16$

6. $y = -\frac{2}{3}x + \frac{23}{3}$

7.(a) $y = 3.14 + 0.97x$ (b) $x \approx -3.24$, $y = 1.2$

8.

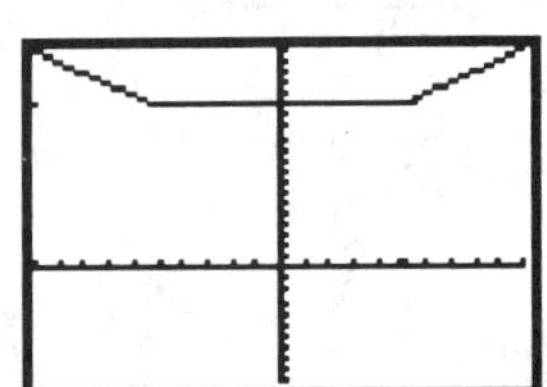

9.(a) $f^{-1}(y) = \frac{4y - 1}{3}$

(b) $f(f^{-1}(y)) = \frac{3\left(\frac{4y-1}{3}\right) + 1}{4} = \frac{4y - 1 + 1}{4} = y$

$f^{-1}(f(x)) = \frac{4\left(\frac{3x+1}{4}\right) - 1}{3} = \frac{3x + 1 - 1}{3} = x$

(c)

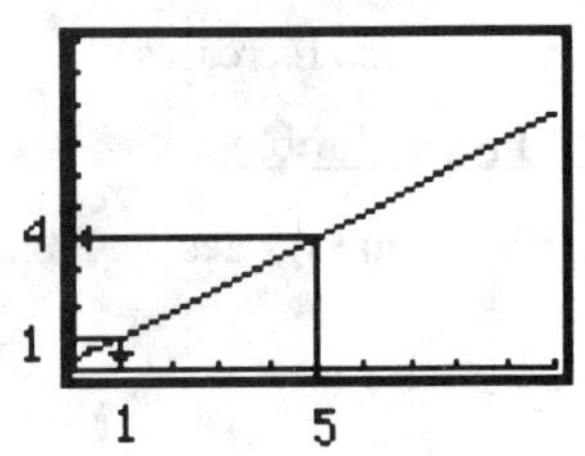

10.

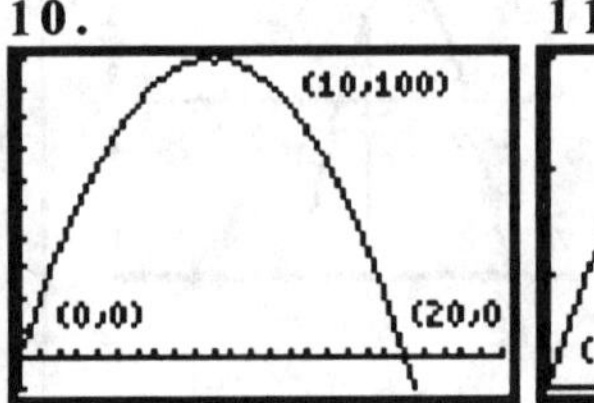

11.

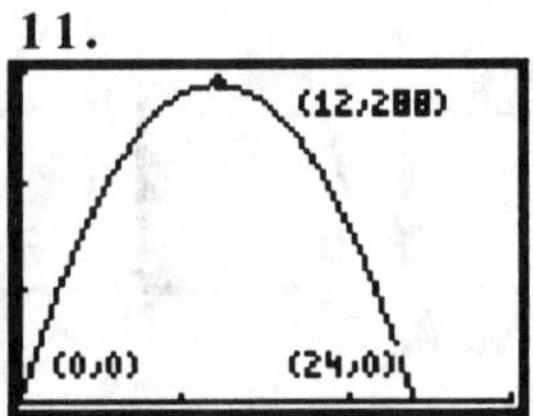

12. $f(x) = a(x + \frac{b}{2a})^2 - \frac{b^2 - 4ac}{4a}$ and $f(x) = 0 \Rightarrow$

$$a(x + \frac{b}{2a})^2 - \frac{b^2 - 4ac}{4a} = 0$$

$$a(x + \frac{b}{2a})^2 = \frac{b^2 - 4ac}{4a}$$

$$(x + \frac{b}{2a})^2 = \frac{b^2 - 4ac}{4a^2}$$

$$x + \frac{b}{2a} = \pm\sqrt{\frac{b^2 - 4ac}{4a^2}}$$

$$x = -\frac{b}{2a} \pm \frac{\sqrt{b^2 - 4ac}}{2a}$$

$$x = \frac{-b \pm \sqrt{b^2 - 4ac}}{2a}$$

13. <u>The Equality Problem</u>

$h_1 = 0.03t + 69$ and $h_2 = 0.05t + 64$, where h is the height in inches and t is time in years since 1972. $h_1 = h_2 \Rightarrow t = 250$, the year 2222. In the year 2222, the average male and female height will both be 76.5 in.

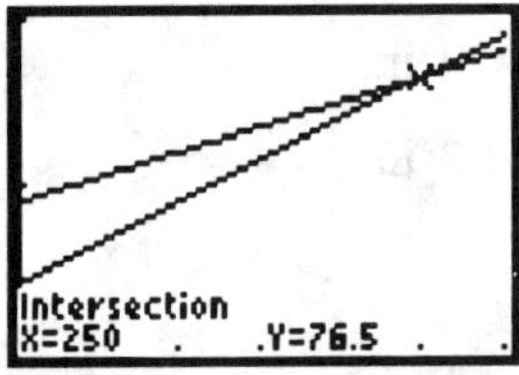

14. <u>The Wishing Well Problem</u>

(a) 16.6 m below the ground (b) 19.6 m

(c) 57 m deep

15. The Deck Problem

$x = 3$ ft, $y = 16$ ft

16. The Windsock Problem

$x \approx 16.4$ in. $y \approx 10.6$ in.

17. The Queuing Problem

(a) about 59 ft/min (b) If they spend less time on the lift, they spend more time in line, since the time on the hill is the same.

18. The U.S. Population Problem

(a) $R = -0.00000016N(N - 500)$

(b) (0,0): If there are no people, there would be no growth. (500,0): If the population ever reaches 500 million, it will quit increasing. (250,0.010025) When the population was 250 million, it was growing at its maximum rate of just over 1%.

19. Discussions will vary.

20. Discussions will vary.

Section 3.1 Graphs of Polynomial Functions (Page 197)

1.(a) up to the right, down to the left

(b) up to the right, down to the left

3.(a) down at both ends

(b) up at both ends

5.(a) -4 (b) 1 and -2

(c) graph appears linear at (1,0) and has a turning point at (-2,0)

7.(a) 16 (b) 2

(c) graph has a turning point at (2,0)

9. B **11.** C

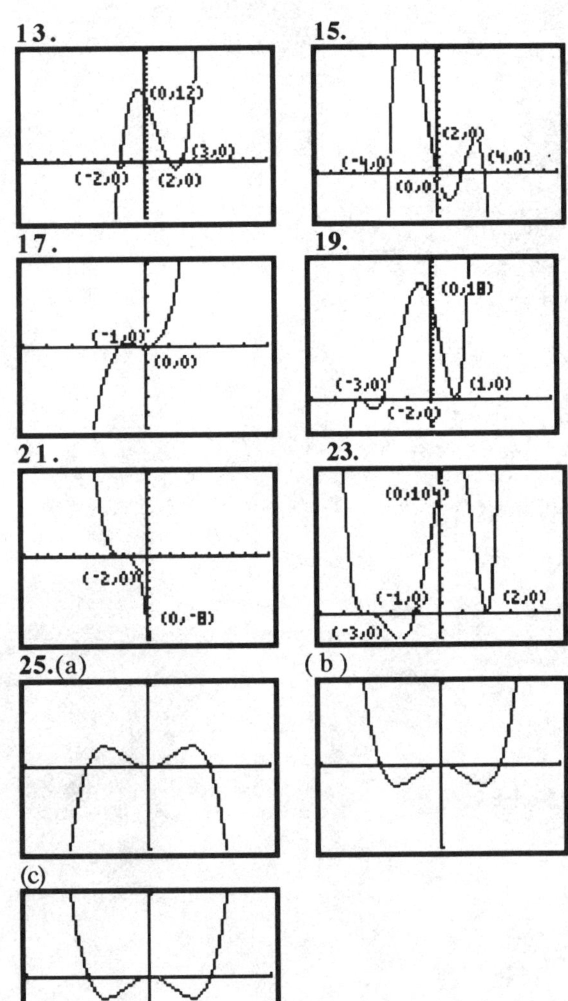

27.(a) (b)

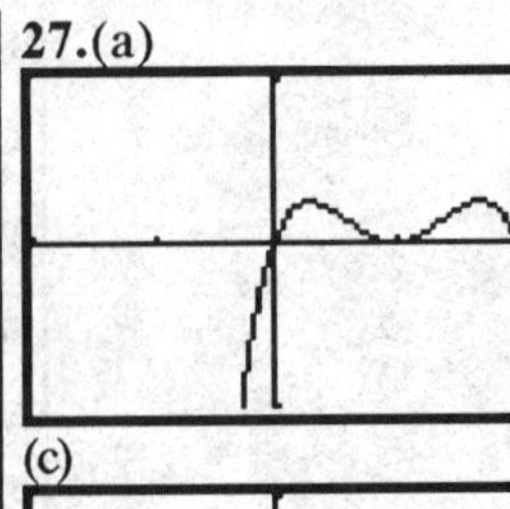

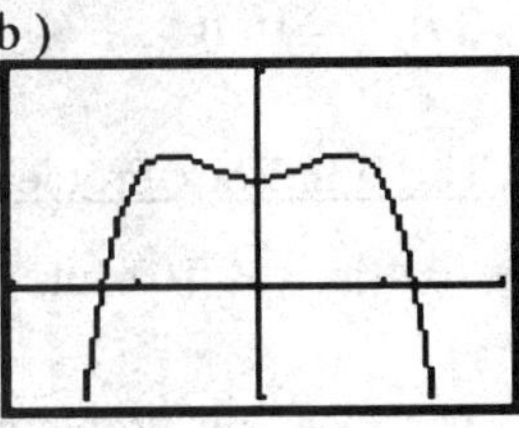

(c)

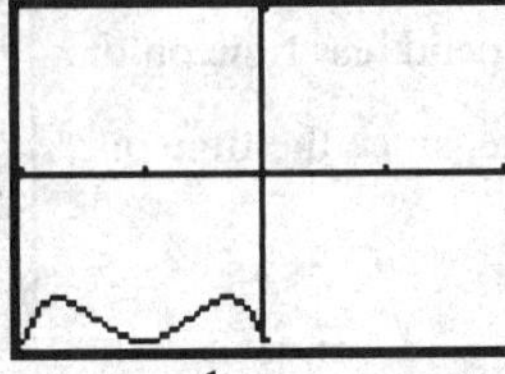

29. $y = \frac{1}{96}(x + 4)(x + 1)^2(x - 4)(x - 6)$

31. $y = -\frac{1}{27}(x + 3)^3(x - 1)^2(x - 3)$

33. $(-\frac{3}{2}, -\frac{45}{8}), (1,0), (2,4)$

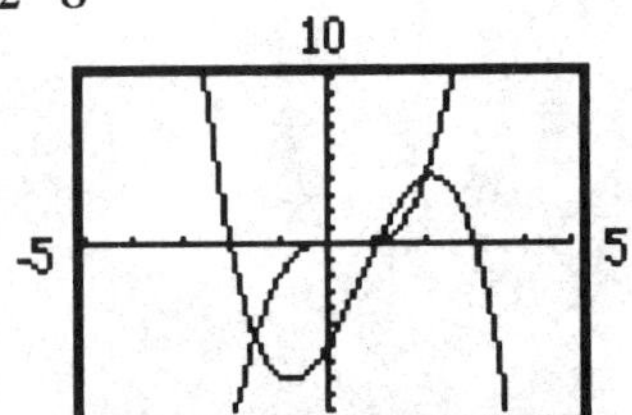

35. $(-10,0), \left(-\sqrt{\frac{125}{2}}, -\frac{5625}{4}\right), \left(\sqrt{\frac{125}{2}}, -\frac{5625}{4}\right), (10,0)$

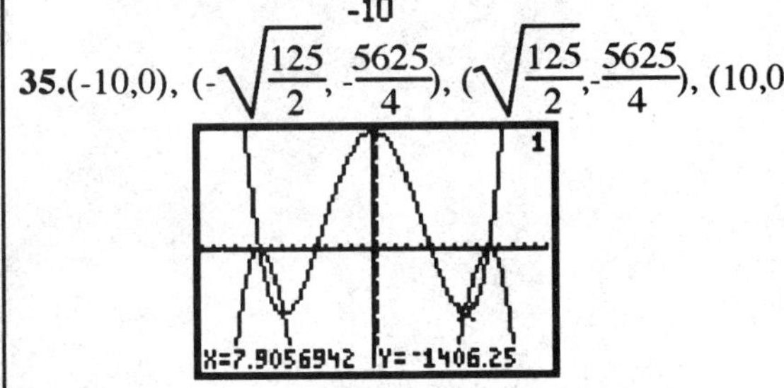

37. turning points: (0,135) and (12,-729)

point of inflection: (6,-297)

39. turning points: (-5, -325) and (7, 539)

point of inflection: (1, 107).

41. The Pizza Box Problem

Cut squares about 3.3 in. on a side.

43. The Rainbow Kite and Balloon Company

Crosssection	Volume Function	Constraint	Volume Subject to Constraint	Maximum Volume x, y, V
Triangle	$V = \frac{\sqrt{3}}{4}x^2y$	$6x + y = 96$	$V = \frac{\sqrt{3}}{4}x^2(96\text{-}6x)$	10.67 in., 32 in.,1576.6 in.^{3}
Square	$V = x^2y$	$8x + y = 96$	$V = x^2(96\text{-}8x)$	8 in., 32 in., 2048 in.^{3}
Circle	$V = \pi x^2y$	$4\pi x + y = 96$	$V = \pi x^2(96\text{-}4\pi x)$	5.09 in., 32 in., 2607.6 in.^{3}

45. The Fisheries Problem

(a)

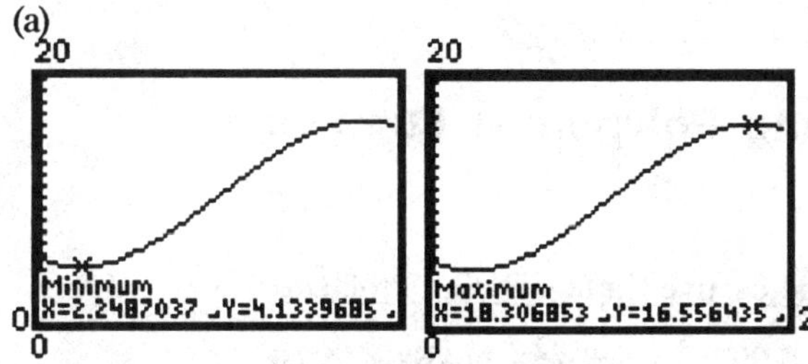

The F intercept, 4.933, represents the 4,933 returning salmon without any help from the Department of Fisheries. The turning points are $(2.2\overline{5}, 4.1\overline{3})$ and $(18.\overline{3}, 16.5\overline{6})$. They represent the fewest number of returning salmon, 4,130, when 22,500 smolt are introduced and the most number of returning salmon, 16,560, when 183,000 smolt are introduced.

(b) The point of inflection is $(10.\overline{3}, 10.\overline{4})$. It represents the number of smolt to introduce to have the returning number of adult salmon increase as fast as possible.

(c) 103,000 smolt.

47. The Rainbow Kite and Balloon Company

(a) turning points: t = 8.07 or t = 18.60
point of inflection: 13.33

(b) Relative maximum at t = 8.07, the best time to sell. Relative minimum at t = 18.06, the best time to buy. Point of inflection t = 13.33 is a transition point of a downward trend, a good time to consider buying.

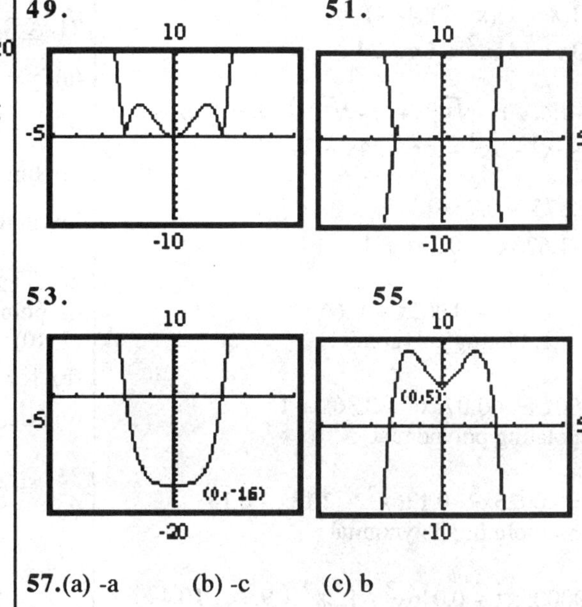

57.(a) -a (b) -c (c) b

Section 3.2 Finite Differences (Page 209)

1.(a) $f(x) = 4x + 3$ (b) $f(4) = 19$

3.(a) $a + b + c$, $4a + 2b + c$, $f(x) = 2x^2 + 3$

(b) $f(4) = 35$

5.(a) $a + b + c + d$, $8a + 4b + 2c + d$, $27a + 9b + 3c + d$, $f(x) = x^3 - 25x$ (b) $f(5) = 0$

7. A Free-Falling Object Problem
$s(t) = -16t^2 + 128t + 144$

9. The Rainbow Kite and Balloon Company
(a) 0, 1, 3, 6 (b) $1 = \frac{1}{2}n^2 - \frac{1}{2}n$

(c) 66, 190

(d) Get a server. Run a line from each building to the server.

11. A Golf Ball Problem

(a) $x(t) = 128t$ (b) $y(t) = -16t^2 + 128t$

(c) In the y directions the average velocity is positive initially, then zero, then negative. The acceleration is a constant. In the x direction the average velocity is a constant and the acceleration is zero.

13. The Big Screen TV Problem

(a) $b = 2199 - 21.50t$ and $b = 0 \Rightarrow b \approx 102$

about 102 mo.

(b) $2199

(c) The first differences are the payments and second differences all equal zero, indicating that the payment does not change.

15. $y = -x^3 - 4x^2 + 4x \Rightarrow y = -x(x + 2)^2$

17. $x = y^3 - 6y^2 + 12y - 8 \Rightarrow x = (y - 2)^3 \Rightarrow$

$y = \sqrt[3]{x} + 2$

19.(a) degree 5 (b) 13, 21

(c) $a_1 + a_2 = a_3$

Section 3.3 Interpolating and Best-Fitting Polynomial Functions (Page 220)

1. $y = 3(x + 1)(x - 2)(x - 4) \Rightarrow$
$y = 3x^3 - 15x^2 + 6x + 24$

3. $y = -3[(x + 1) - \sqrt{3}][(x + 1) + \sqrt{3}]\,(x - 2i)(x + 2i) \Rightarrow$
$y = -3x^4 - 6x^3 - 6x^2 - 24x + 24$

5. $y = 2.75 + 0.75x$

7. $y = -1.625x^2 - 0.25x + 1$

9. $y = -0.07x^2 + 0.002x + 0.66$
not an interpolating polynomial

$y = - 0.002x^3 - 0.076x^2 + 0.36x + 1$
an interpolating polynomial

11. $y = 0.0025x^3 -0.136x^2 + 2.18x - 8.14$
not an interpolating polynomial

$y = -0.00005x^4 + 0.046x^3 - 1.2x^2 + 9.9x - 10$
an interpolating polynomial

13. $y = 0.214x^3 + 0.742x^2 + 0.956x + 0.798$
$x = 0 \Rightarrow y = 0.798$

15. $y = 0.03x^4 - 0.2x^3 - 0.01x^2 + 1.02x$
$x = 6 \Rightarrow y = 1.44$

17. The Interest Problem

(a) $y = 0.255x^3 + 10.505x^2 + 325.04x + 5000$ (b) $6277.26, $6484.56, $6919.70

19. And Another Interest Problem

(a) $y = -0.83x^4 + 5.23x^3 + 1.38x^2 + 330x + 5000$
(b) $6271.59, $6464.32, $6819.5

21. A Radioactive Carbon Dating Problem

(a) $y = -0.097x^3 + 5.39x^2 - 114.8x + 999.4$

(b) $y = 0.00273x^4 - 0.206x^3 + 6.73x^2 - 119.65x + 1000$

(c) third degree: 500.32, 245.68, -146.5

fourth degree: 499.56, 249.72, 91.09

23. The Fasting and Feasting Problem

(a) points vary (0, 2), (3, 8), (4, 4.5), (4.5, 4), and (7, 10)

(b) $h = 0.1c^3 - 1.8c^2 + 10c - 13.6$

(c) $(4.\bar{1}, 4.\bar{4})$ (d) $(2.\bar{3}, 7.\bar{6})$

25. The Smoothing Curve Problem

(a)

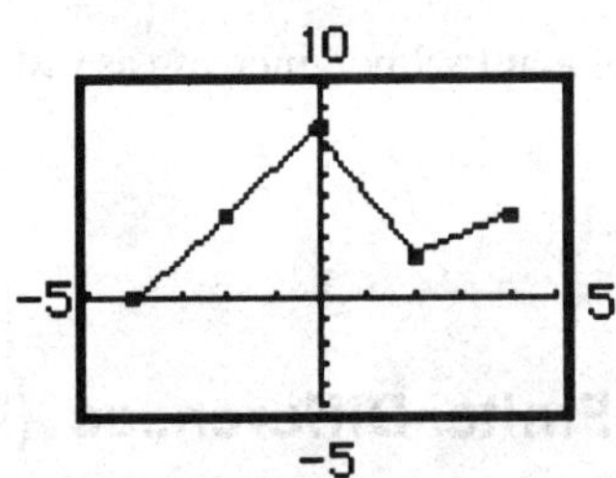

(b) $y = -0.21x^3 - 1.25x^2 + 0.33x + 8$ for $x < 0$ and
$y = 0.375x^3 - 1.25x^2 - 2x + 8$ for $x > 0$

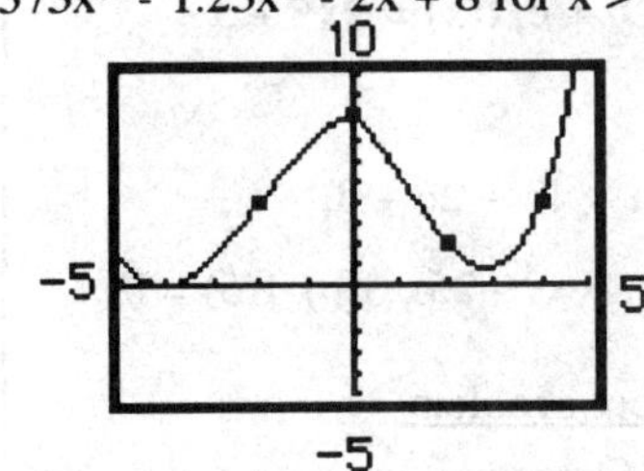

27. The Rainbow Kite and Balloon Company

(a) $p(t) = 0.02t^3 - 0.78t^2 + 9.05t + 12.84$

(b) turning points: $t = 8.07$ or $t = 18.60$

point of inflection: $t = 13.33$

(c) Relative maximum at $t = 8.07$, the best time to sell. Relative minimum at $t = 18.06$, the best time to buy. Point of inflection $t = 13.33$ is a transition point, a good time to consider buying.

29. $y = x^4 - 2x^3 - 3x^2 + 4x + 2$

31. $y = x^4 - 4x^3 + 9x^2 - 16x + 20$

Section 3.4 Graphs of Rational Functions (Page 235)

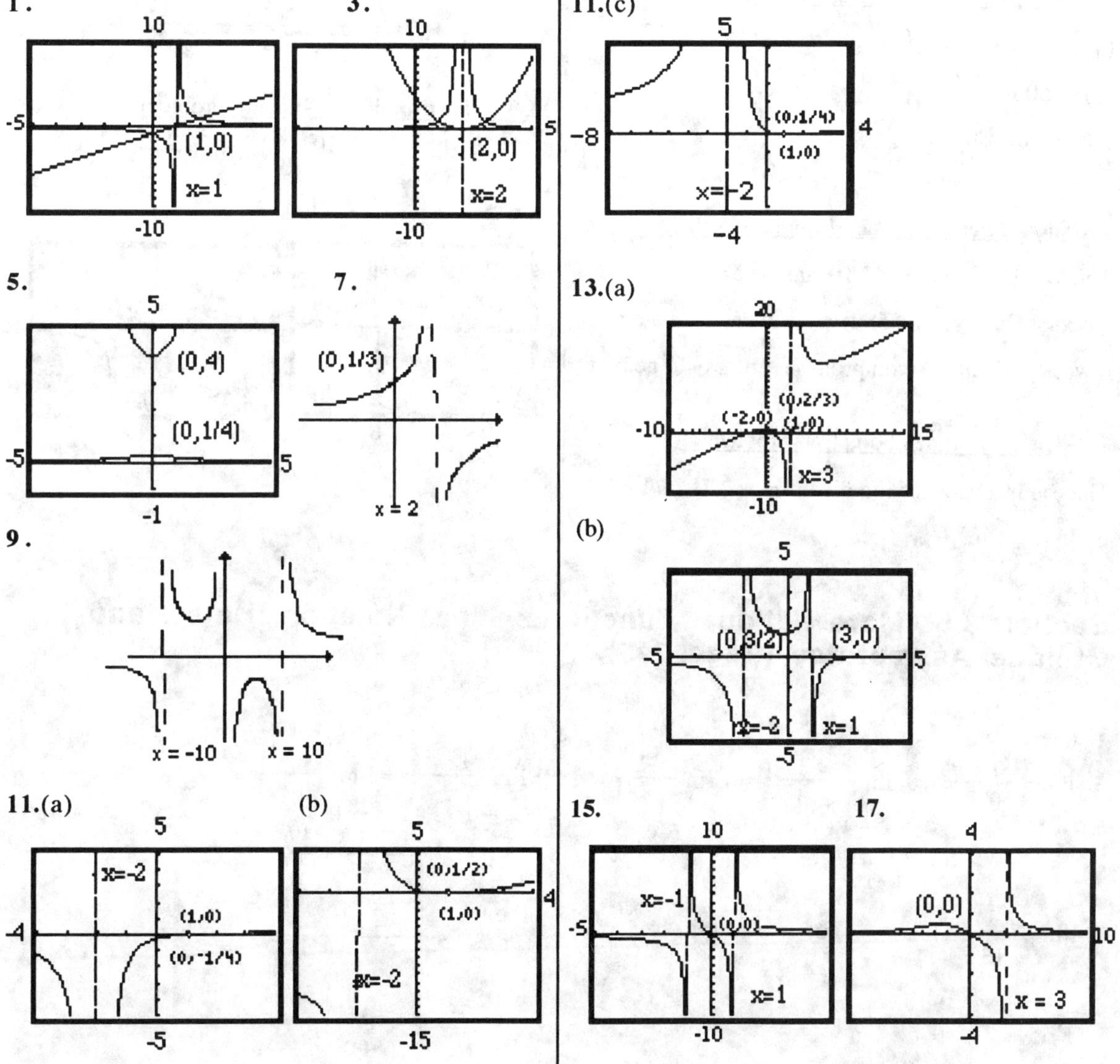

19. **21.** **23.**

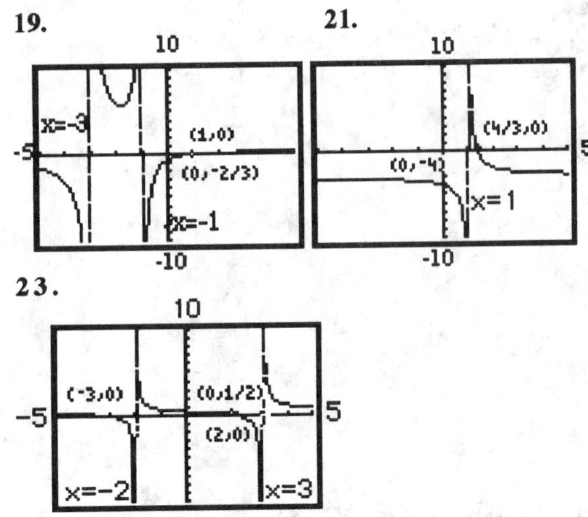

25. The Quick Weight-Loss Problem

(a) $k \approx 6.69 \times 10^{16}$

(b) 150 lb

(c) about 130 lb

27. The Rainbow Kite and Balloon Company

The lowest price is $\$2.13/\text{ft}^2$ and takes about 10.6 weeks. However, $\$4.58/\text{ft}^2$ would get the material in 6 weeks. Perhaps a compromise time would be best.

29. The Redesigned Soda Can Problem

The can is a cube having dimensions 2.71 in. by 2.71 in. by 2.71 in.

31. The Nuclear Fusion Problem

(a) domain: $t \geq 0$, range: $0 \leq v < c$

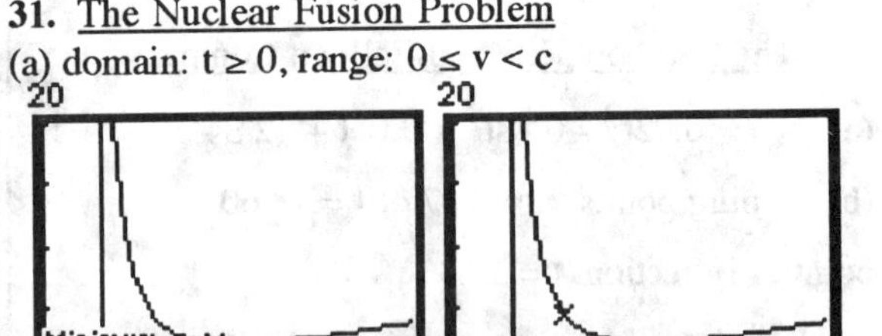

(b) The speed of light is c. The horizontal asymptote $v = c$ is the limit of speed, the speed of light.

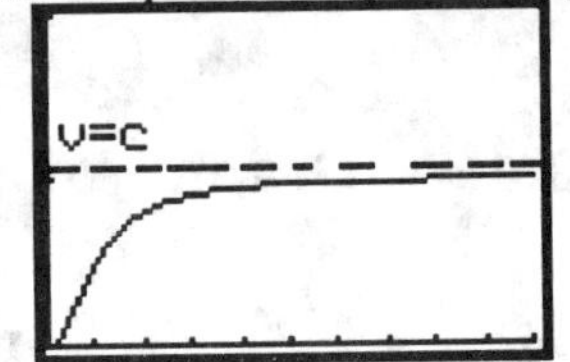

33. The Distribution of Income Problem

(a) 2500 (b) $\$6.25 \times 10^8$ (c) 24,972,222

35. **37.**

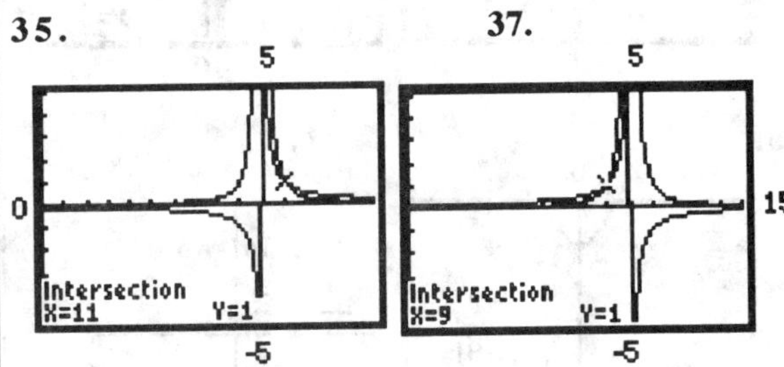

Section 3.5 More Rational Functions: Limit Notation, Holes, and Oblique Asymptotes (Page 245)

1. 0

3. $-\infty$

5. $\lim_{x \to 1^+} \frac{-1}{x+1} = -\infty \quad \lim_{x \to 1^-} \frac{-1}{x+1} = +\infty \quad \lim_{x \to \pm\infty} \frac{-1}{x+1} = 0$

7. $\lim_{x \to 0} \left(\frac{-1}{x^2} + 1\right) = -\infty, \quad \lim_{x \to \infty} \left(\frac{-1}{x^2} + 1\right) = 1$

9. $\lim_{x \to 1} \frac{1 - x^2}{2x - 2} = \lim_{x \to 1} -\frac{x+1}{2} = -1$

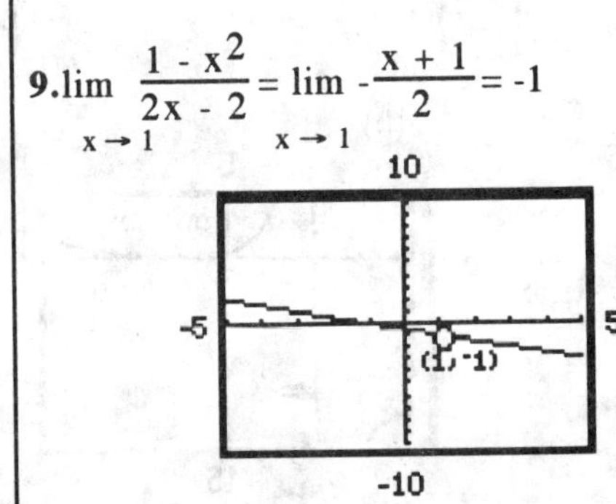

11.(a) $\lim\limits_{x \to 1} \dfrac{x^2 - 2x + 1}{x - 1} = \lim\limits_{x \to 1} (x - 1) = 0$

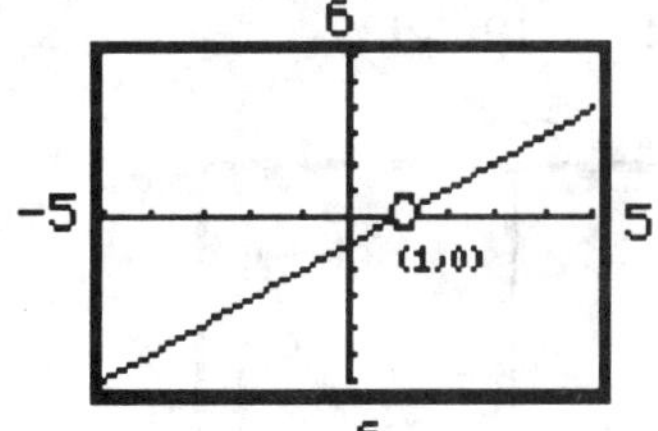

(b) $\lim\limits_{x \to 1} \dfrac{x - 1}{x^2 - 2x + 1} = \lim\limits_{x \to 1} \dfrac{1}{x - 1} = \infty$

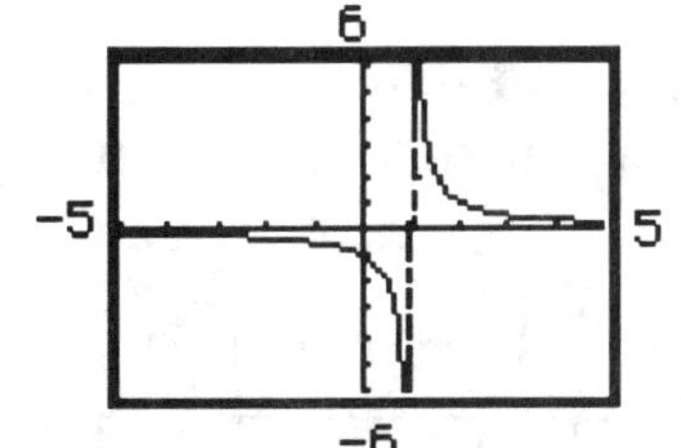

13.

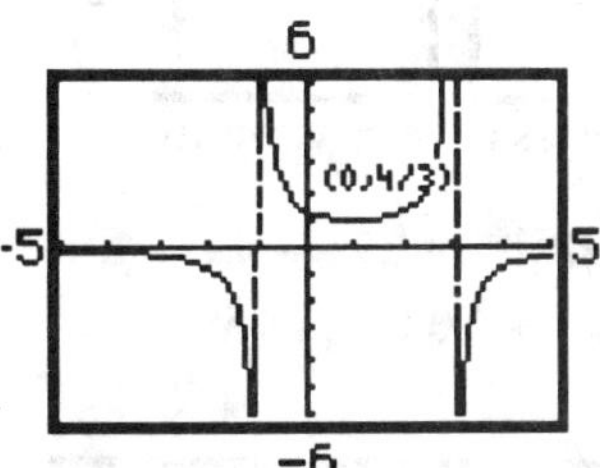

15.

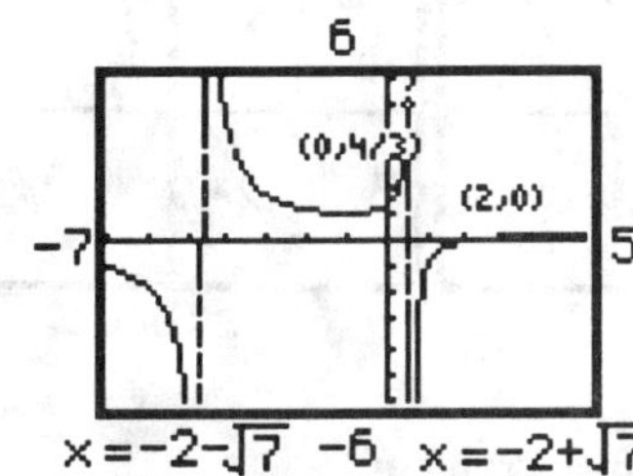

17.

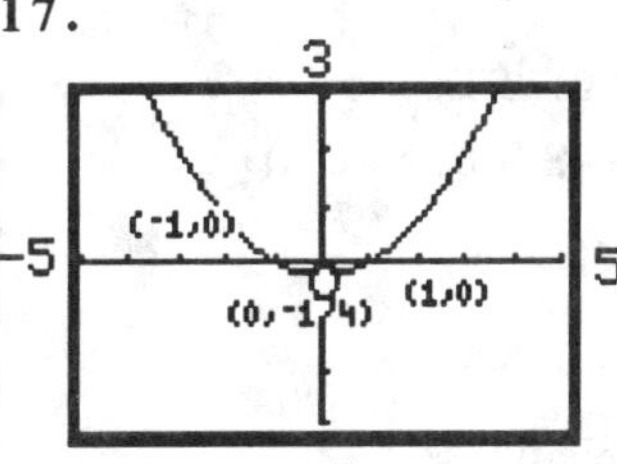

19.

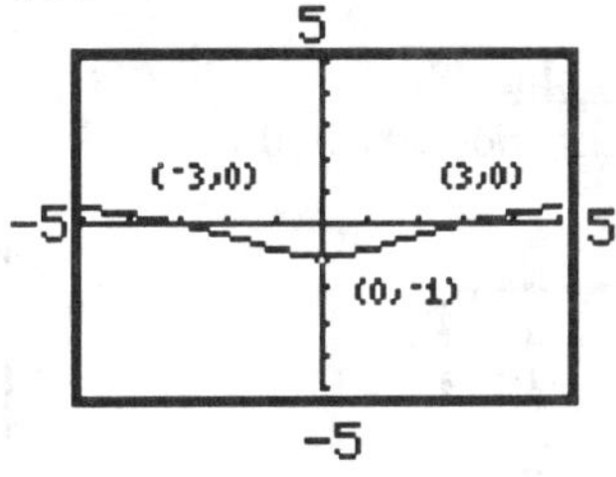

21. $y = x$, $y = x + 1$

23. $f(x) = -x$, $f(x) = -x$

25. The Soda-Can Problem

$S = 2\pi r^2$ represents the surface area approaching the area of the top plus the bottom as the radius increases and the height decreases, approaching zero.

27. The Rainbow Kite and Balloon Company

$c = 0.3t - 3$ represents the costs eventually increasing at 30%, partly due to inflation.

29.

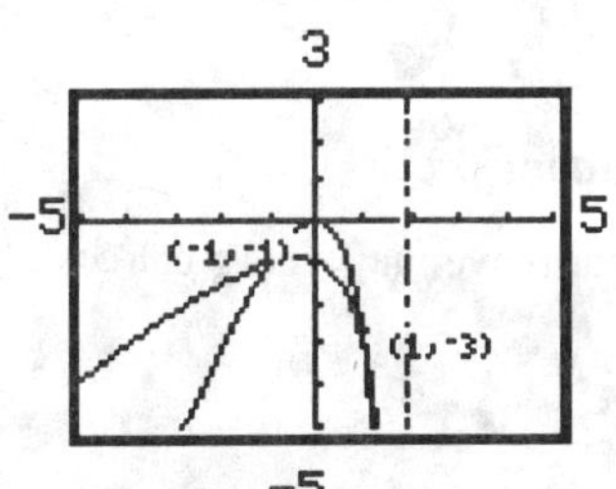

Chapter 3 Review (Page 247)

1.

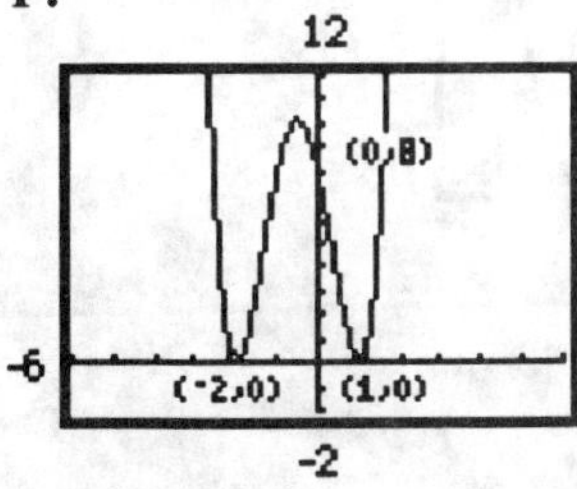

2.

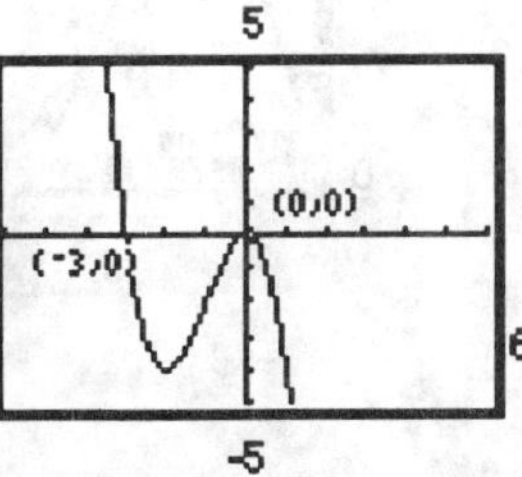

3. $y = (x + 2)(x + 1)(x - 2)(x - 5)$

4. $y = (x + 2)(x + 1)(x - 2)(x - 5)^2$

5. $y = (x + 2)(x + 1)(x - 2)^2(x - 5)$

6. $y = (x + 2)(x + 1)(x - 2)(x - 5)^3$

7. The Largest Triangle Problem
$\sqrt{72}$ ft x $\sqrt{72}$ ft (about 8.48 ft by 8.48 ft)

8. A Big Box Problem
x = a/6

9. Another Big Box Problem
x = a/6

10. The Big Revenue Problem
Work 3.9 hr/day at $5.17/hr and make $20.17 in revenue.

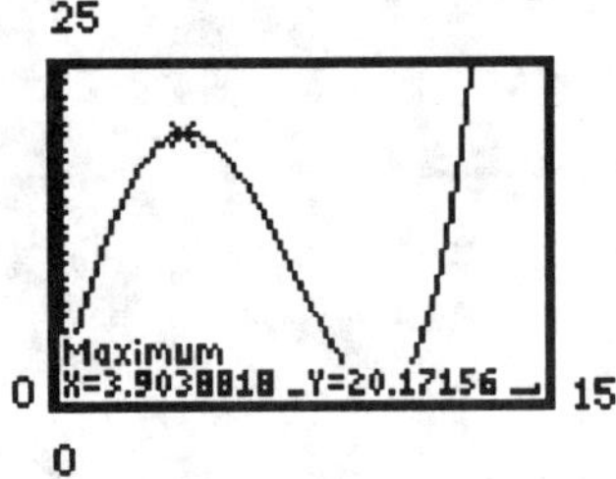

11. turning points: (0, 0) and (2/3, 0.14$\tilde{8}$)
point of inflection: (1/3, 2/27)

12. turning points: (-1,0) and (4,-125)
point of inflection: (3/2,-625)

13. $R(x) = -7.50x^2 + 525.0x$

14. $C(x) = 85x + 1700$

15. $P(x) = -7.50x^2 + 440x - 1700$

16. about 29 kites

17. \$300, \$1500, \$0, and \$-1500

18. all are \$85

19. At maximum profit,

marginal renvenue = marginal cost

about 30 kites

20.(a) $y = 0x^3 + 109.66x^2 + 0x - 89015$,
$y = 23x^4 + 0x^3 + 275x^2 - 144$
(b)

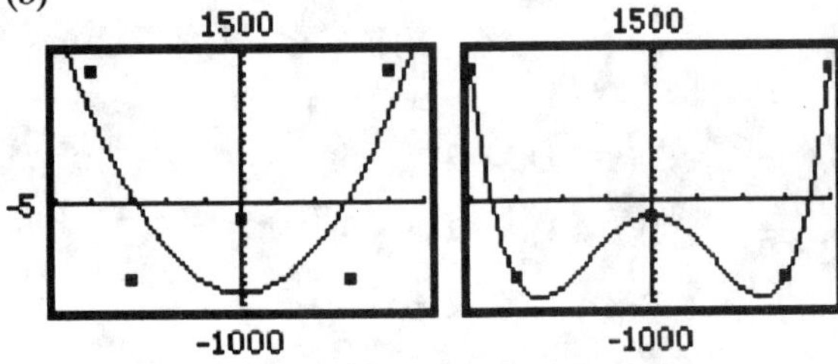

(c) Substitute each x value into the quartic polynomial.

21.(a) $y = 2x^3 - 4x^2 + 10x - 12$,
$y = 0x^4 + 2x^3 - 4x^2 + 10x - 12$
(b)

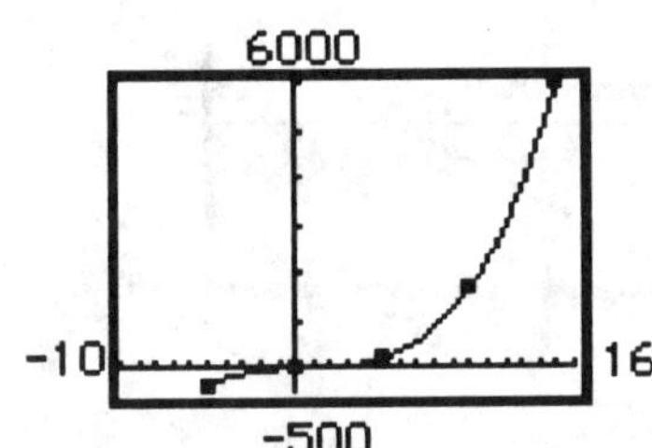

(c) The cubic is an exact fit.

22.(a) $y = -0.21x^3 + 1.79x^2 - 0.95x - 8.71$,
$y = -0.23x^4 + 3.54x^3 - 16.56x^2 + 25.83x - 10$
(b)

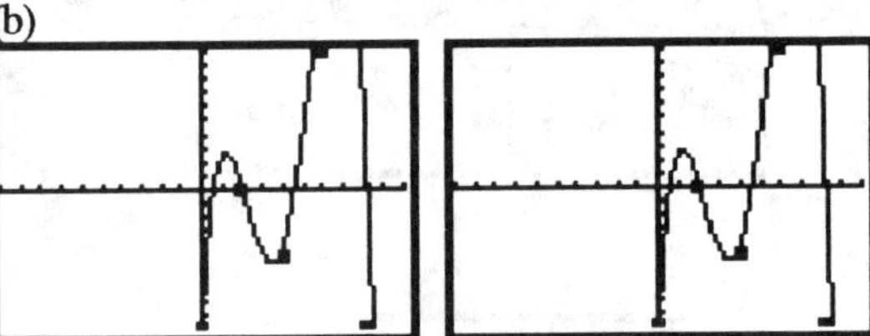

(c) The quartic appears to be an exact fit.

23.(a) $y = x^3 + 1.72x^2 + 1.29x + 3.14$,
$y = 0.02x^4 + 1.02x^3 + 1.3x^2 + 1.15x + 4$
(b)

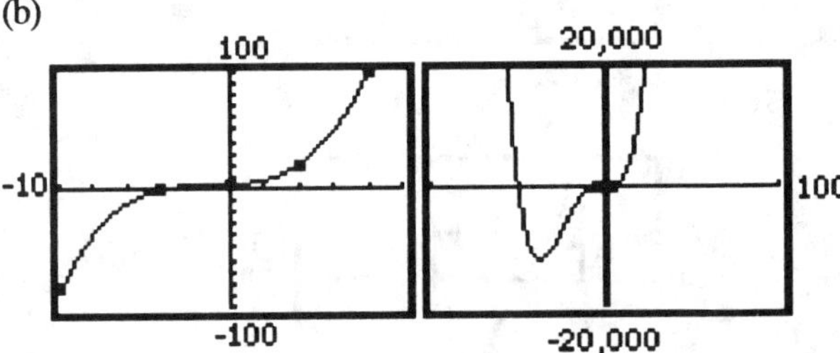

(c) yes

24. The Big Revenue Problem

(a) $p = 0.009x^3 - 0.09x^2 - 1.03x + 10$

(b) $R = px = 0.009x^4 - .09x^3 - 1.03x^2 + 10x$

(c) Charge \$5.15/hr

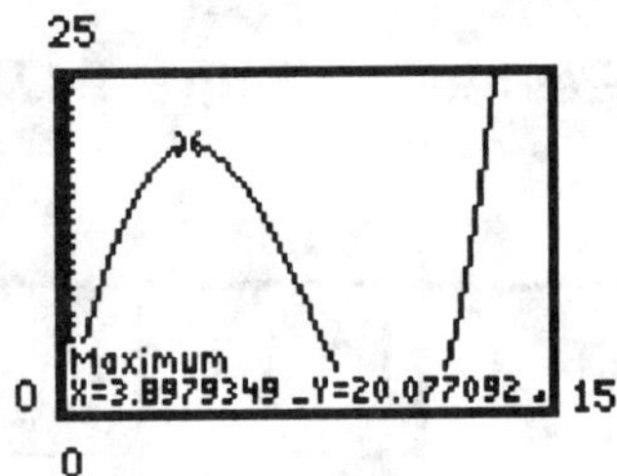

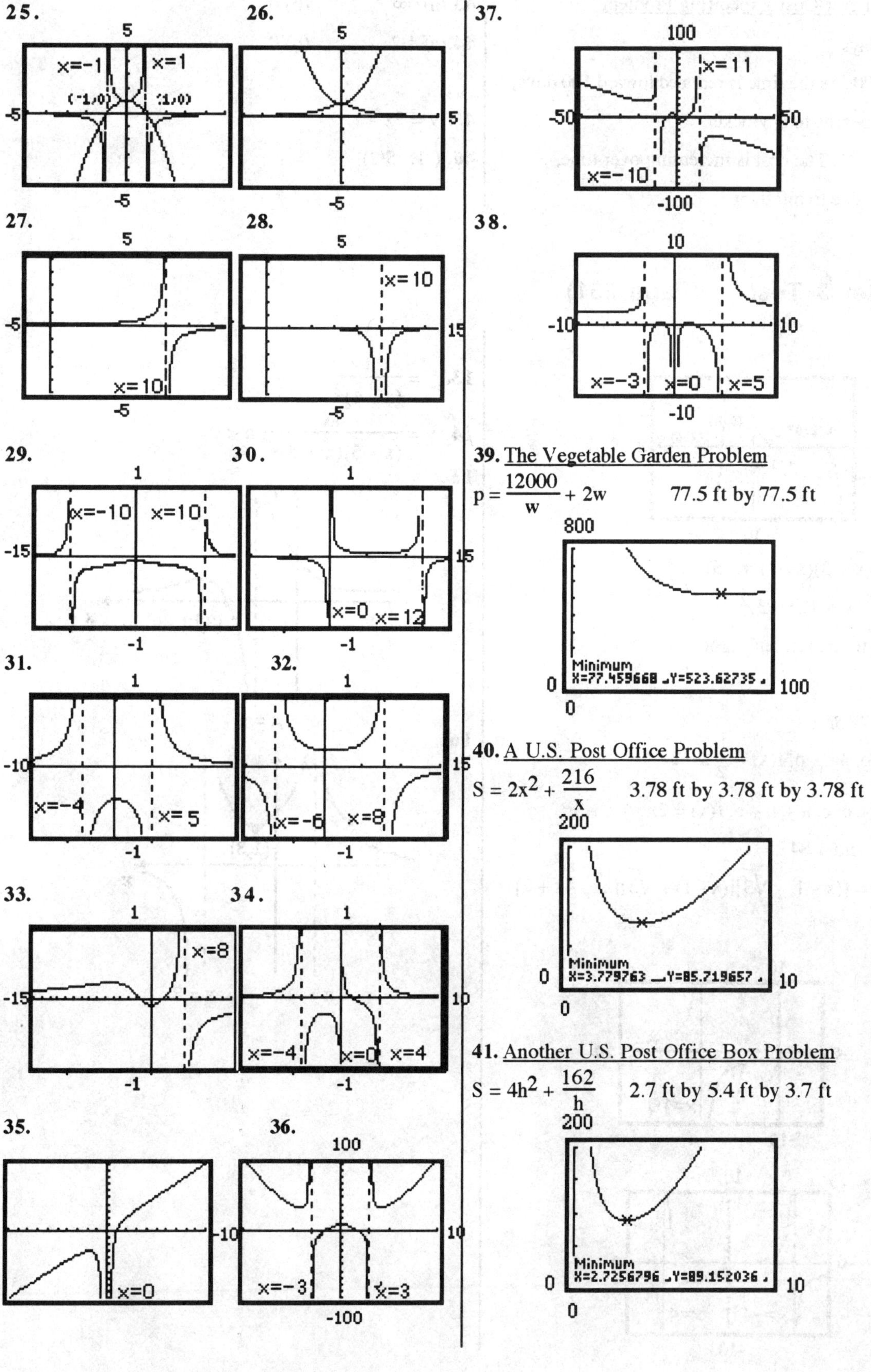
25.
x=-1
x=1
(-1,0)
(1,0)
26.
27.
x=10
28.
x=10
29.
x=-10
x=10
30.
x=0
x=12
31.
x=-4
x=5
32.
x=-6
x=8
33.
x=8
34.
x=-4
x=0
x=4
35.
x=0
36.
x=-3
x=3
37.
x=11
x=-10
38.
x=-3
x=0
x=5
39. The Vegetable Garden Problem
$p = \frac{12000}{w} + 2w$
77.5 ft by 77.5 ft
Minimum
X=77.459668 Y=523.62735
40. A U.S. Post Office Problem
$S = 2x^2 + \frac{216}{x}$
3.78 ft by 3.78 ft by 3.78 ft
Minimum
X=3.779763 Y=85.719657
41. Another U.S. Post Office Box Problem
$S = 4h^2 + \frac{162}{h}$
2.7 ft by 5.4 ft by 3.7 ft
Minimum
X=2.7256796 Y=89.152036

42. The Least-Cost Expediting Problem

(a) 300 days

(b) $t = 100$ As the time is reduced toward 100 days, the cost begins to skyrocket.

(c) $C = 0.04t$ The cost is increasing over time, probably due to inflation.

43.(a) $-\infty$ (b) 0

44.(a) 1/2 (b) 0

45. $y = 2x - 1$

46. $(-1, -5/2)$

Chapter 3 Test (Page 251)

1.

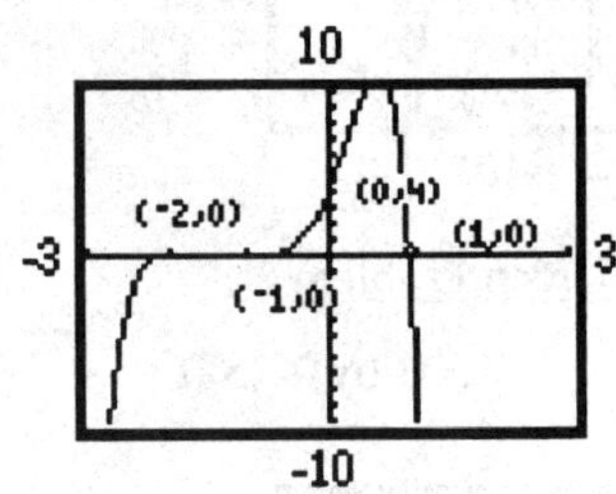

2. $y = 2(x + 3)(x - 1)(x - 5)$

3. $y = x^2(x + 4)(x - 3)^2$

4. down to the left and right

5. 15

6. 4, 3, 2, or 1

7. $2a + b$, $4a + b$, $f(x) = \frac{1}{2}x - 2$

8. $a - b + c$, c, $a + b + c$, $f(x) = 2x^2 - x + 3$

9. -2.22 and 1.84

10. $f(x) = [(x - 1) - \sqrt{3}][(x - 1) + \sqrt{3}][x^2 - 4x + 2]$

11.

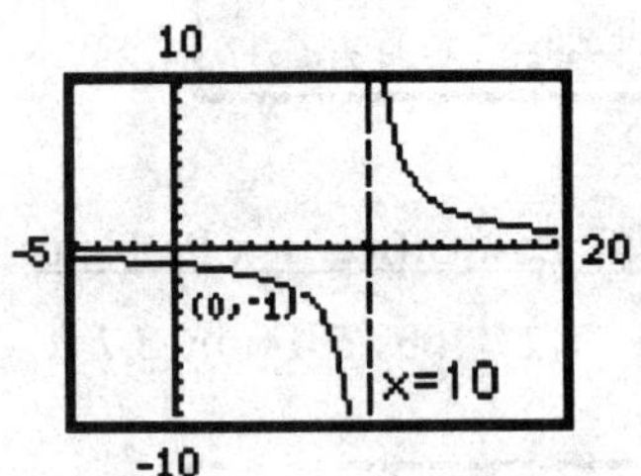

12.

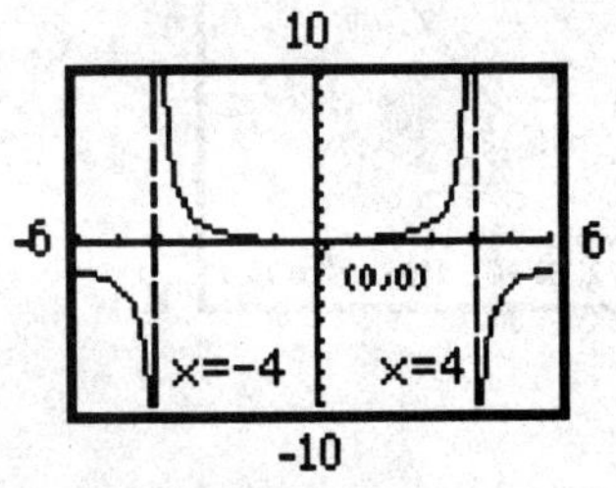

13. $y = \dfrac{1}{(x - 4)^2}$

14. $y = \dfrac{-ax}{(x - 5)(x + 5)^2}$, $a < 0$

15.

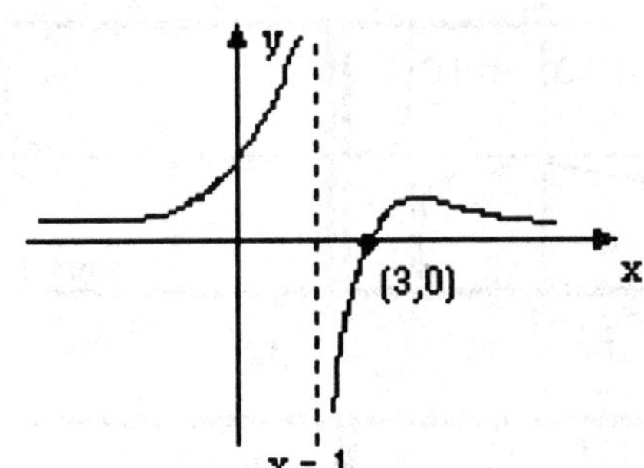

16.

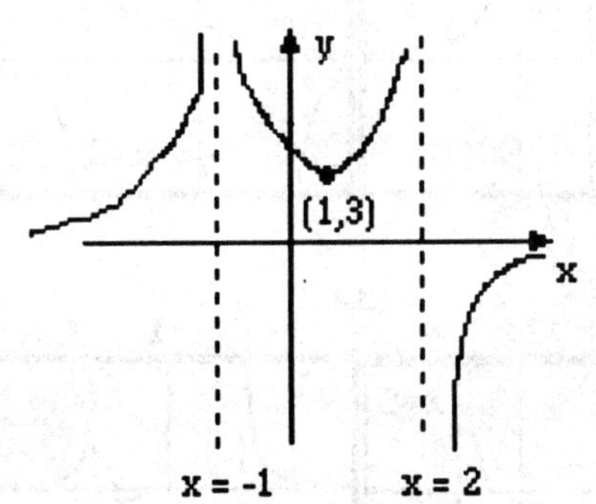

17. The Tent Problem

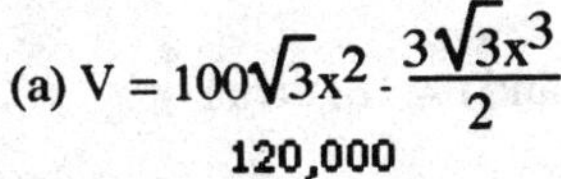

(a) $V = 100\sqrt{3}x^2 - \frac{3\sqrt{3}x^3}{2}$

120,000

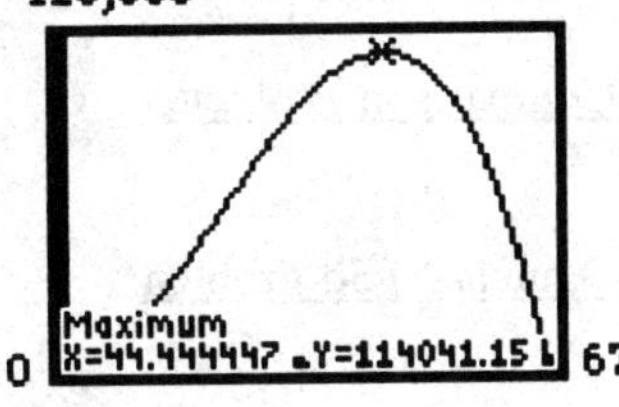

0 … 67

0

x = 44.4 in. y = 133.4 in.

(b) $S = (\frac{\sqrt{3}}{2} - 18)x^2 + 1200x$

25,000

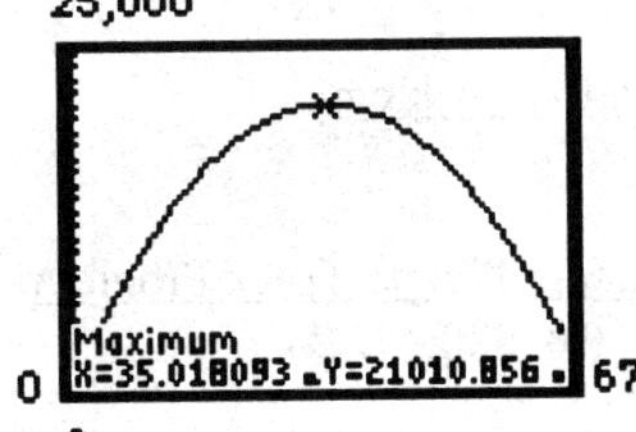

0 … 67

0

35 in. ≈ 2.9 ft by 15.8 ft

18. Another Tent Problem

$SA = \frac{792}{\sqrt{3}x} + \frac{\sqrt{3}}{2}x^2$

1000

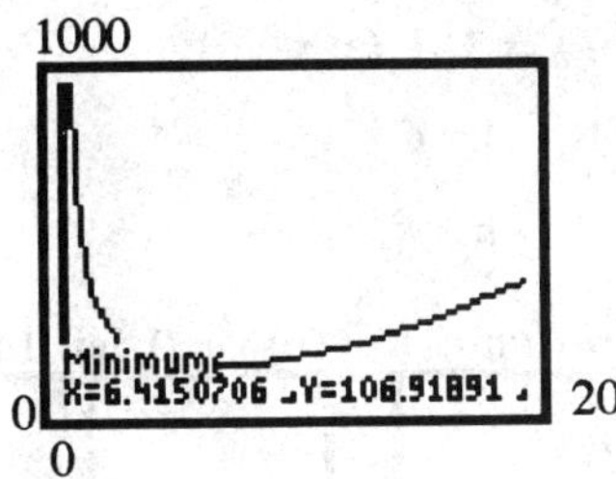

0 … 20

0

x ≈ 6.4 in., y ≈ 3.7 in.

(b) $s = 6x + \frac{264}{\sqrt{3}x^2}$

100

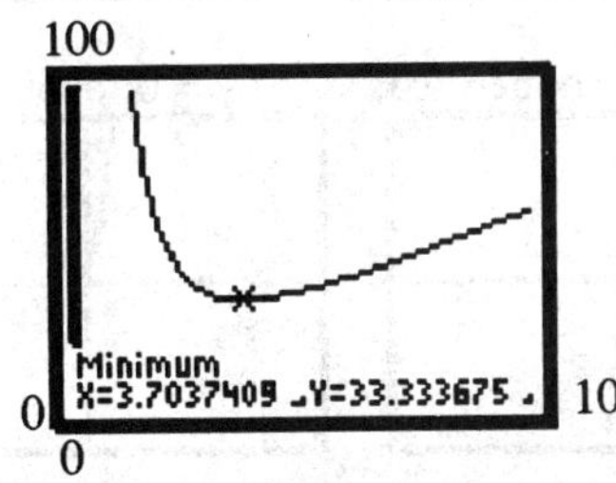

0 … 10

0

x ≈ 3.7 in. y ≈ 11.1 in.

Section 4.1 Exponential Functions (Page 264)

1.(a) $x = 3$ (b) $x = -1/2$ (c) $x = 0$

3.(a) $x = 0$ (b) $x = 2$ (c) $x = -1$

5. f: A, g: B, h: C

7. f: C, g: B, h: A

9.(a) $y = 1$; up 1 unit (b) $y = 0$; left 1 unit

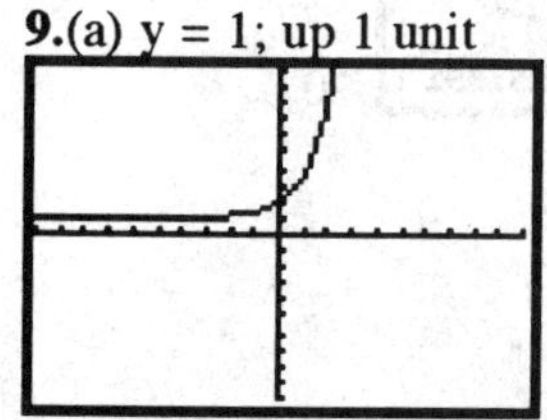

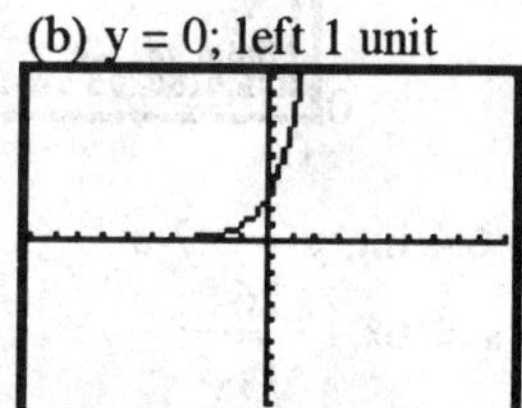

11.(a) $y = 0$; about x axis (b) $y = 0$; about y axis

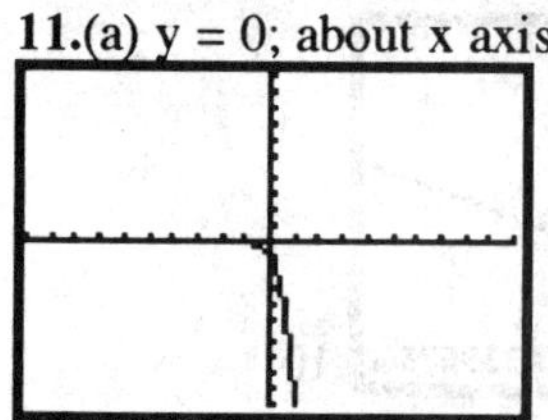

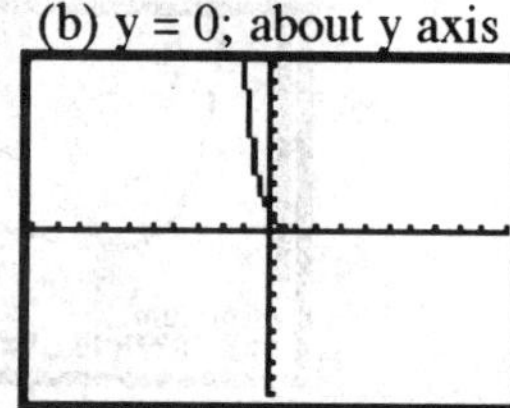

13. $y = 1$ **15.** $y = 1$

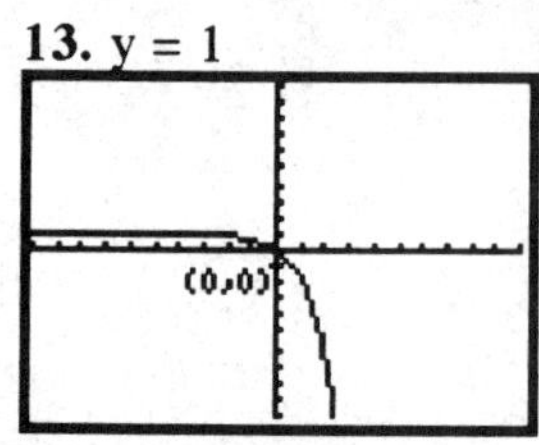

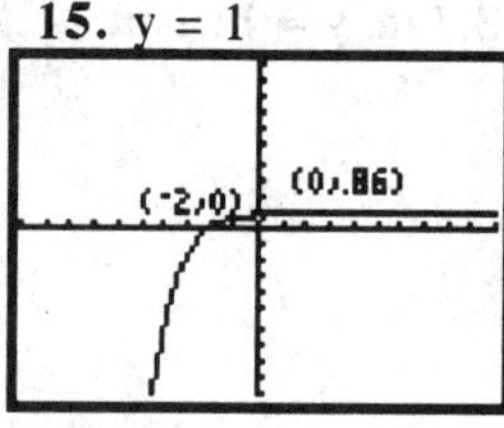

17. $(-1/2, \sqrt{2})$ **19.** $(0,0)$

21.(a) $A \approx \$1711.75$ (b) $A \approx \$1715.51$
(c) $A \approx \$1716.37$ (d) $A \approx \$1716.80$

23. Answers vary: If $r = 8\%$ and $t = 18$, then $A \approx \$42{,}206.96$

25. The Interest Comparison Problem
\$9617.72, \$9869.39

27. An Effective Annual Yield Problem
10.5171%

29. The Rainbow Kite and Balloon Company
about 44%

31. The U.S. Population Problem
275,119,930 (annually compounded)

33. The Immigration Problem
271,049,542

35. The Depreciating Pickup Truck Problem
(a) about 39.5% (b) \$20,442.24

37. The Game of Chess Problem
about 9.22×10^{18}

39. The Origami Problem
(a) over 1 km (b) about 7

41. The Amoeba Population Problem
(a) about 19,221 (b) about 3.6 hr

43. The Baby Boom Problem
(a) $N = 250e^{0.01t}$ (b) about 284.7
(c) about 69 yr

45. $x = 0$ **47.** $x = 0$
49. $e \approx 2.6667$

Section 4.2 Logarithmic Functions (Page 279)

1.(a) $\log 0.001 = -3$ (b) $\ln 1 = 0$

3.(a) $10^{-1} = \frac{1}{10}$ (b) $e^{4.2} \approx 64$

5.(a) 0.3944 (b) 2.3944 (c) 6.3944

7.(a) undefined (b) undefined

9. $x = 10{,}000$ **11.** $x = \pm 1$

13. $x \approx 0.9542$ **15.** $x \approx 0.6931$

17. $x \approx 0.5756$

19.(a) $x = 1$; right 1 unit (b) $x = 0$; down 1 unit

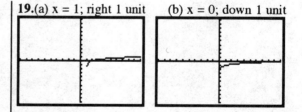

21.(a) $x = 0$; about origin (b) $x = 0$; about x axis

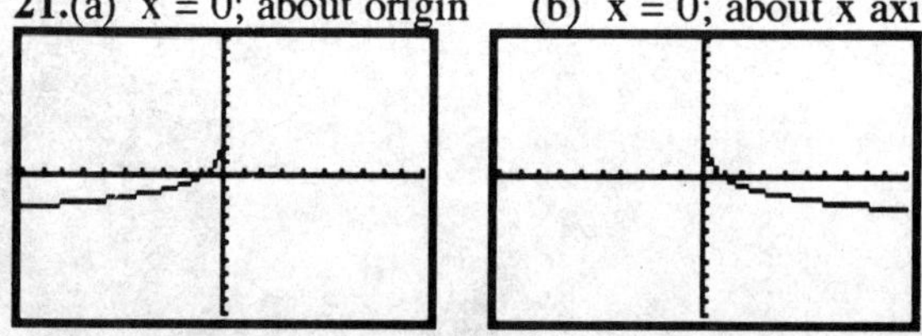

23. $x = -3$

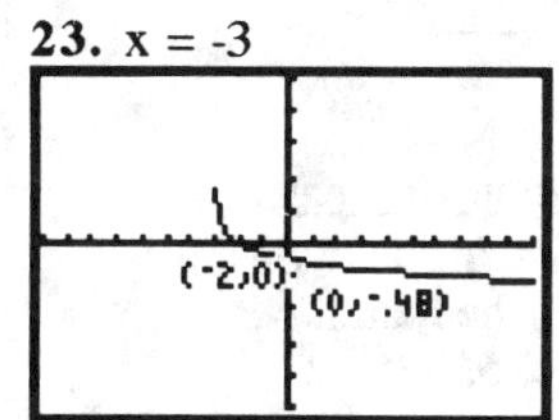

25. $x = 2$

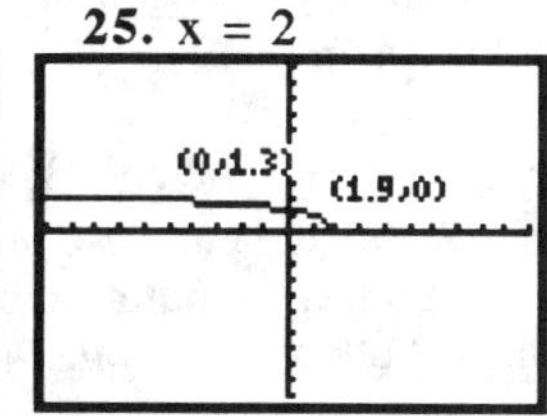

27. $x = 2 + \ln y$

29. $x = e^y + 2$

31. $f^{-1}(y) = \log(y + 3)$

33. $f^{-1}(y) = 10^{y+2} - 1$

35.

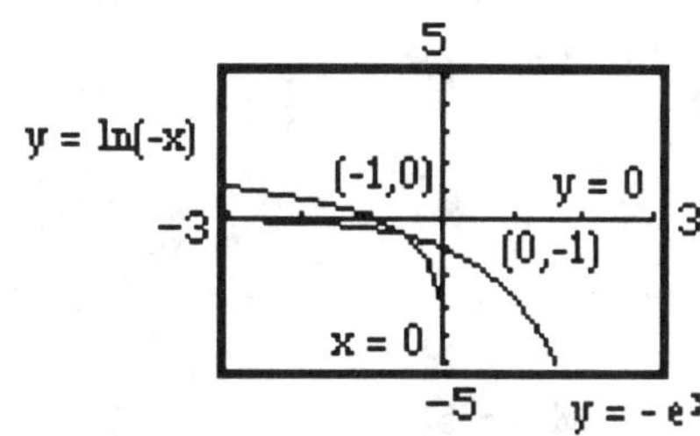

37.

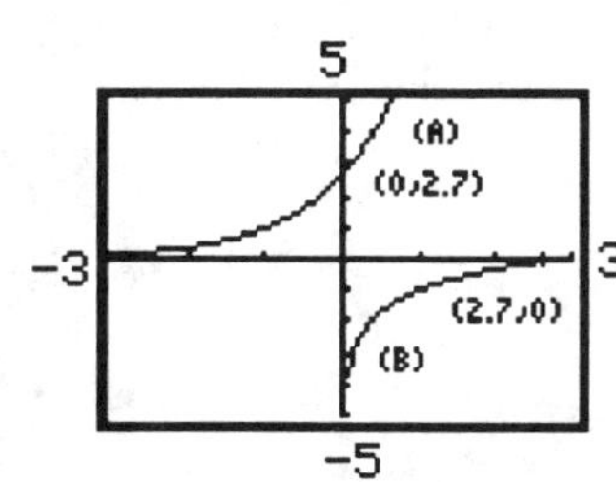

39. Answers vary: If $r = 8\%$ then $t \approx 17.33$

41. <u>The Rainbow Kite and Balloon Company</u>
(a) about 384 kites (b) about 12,288 kites
(c) about 19 yr

43. <u>The Fossil Problem</u>
(a) less than 0.2% (b) less than 0.02%

45. <u>The Rural Population Problem</u>
(a) $N = 350e^{-0.12t}$ (b) about 43 yr

47. <u>The Wrong Half-Life Problem</u>
(a) $k \approx 0.000124$ (b) Multiply by 1.02

49. <u>The Lifesaving-Radiation Problem</u>
about 3.9 yr

51. <u>The Effects-or-the-Error-in-Carbon-Dating Prob.</u>
Mismeasuring the amount of carbon-14 present as 18% rather than 15% causes an error of 14,171.88 - 15,678.68 = -1506.8 yr.

53. $x = 4$ **55.** $x = 26$

57. $x = 1$ **59.** $x = 10^{10^{10}}$ (a googol)

61. $\ln(1 + x) \approx 0.8333$

Section 4.3 Theorems on Logarithms (Page 291)

1.(a) 3.3219 (b) 1.5850 (c) 4.9069

3.(a) 0.6826 (b) 1.3652 (c) 2.0478

5.(a) -2, 1.58 (b) $y = \dfrac{\ln(x + 3)}{\ln 2}$

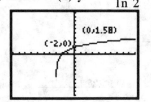

7.(a) 1/8 (b) $y = \dfrac{\ln x}{\ln 2} + 3$

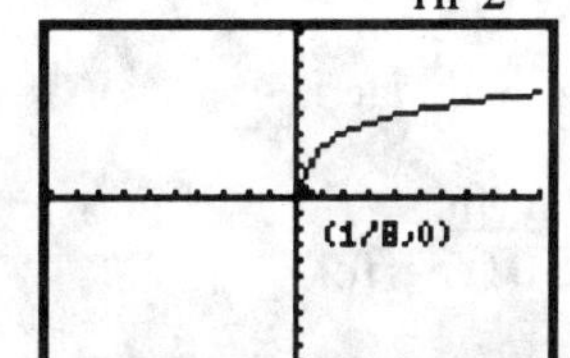

9.(a) $x \approx 2.718$ (b) $x = 1$ (c) $x \approx 0.3679$

11.(a) $x = 5$ (b) $x \approx 3.4657$
(c) $x \approx 3.1546$ (d) $x \approx 1.505$

13. $x = 3$ ($x = -1$)

15. $x = \dfrac{-10 + \sqrt{110}}{10}$ $\left(x = \dfrac{-10 - \sqrt{110}}{10}\right)$

17. $x = -1$ or $x = 3$

19. no solution ($x = 2$ nor $x = 3$)

21. $x = 3/2$ ($x = -2$)

23. $x = 4$ ($x = -1$)

25.(a) $10^x = 10^{-3} \Rightarrow x = -3$
(b) $\log 0.001 = x \Rightarrow x = -3$
(c) $\log 10^x = \log 0.001 \Rightarrow x = -3$
(d) $\ln 10^x = \ln 0.001 \Rightarrow x = -3$

27. 29.

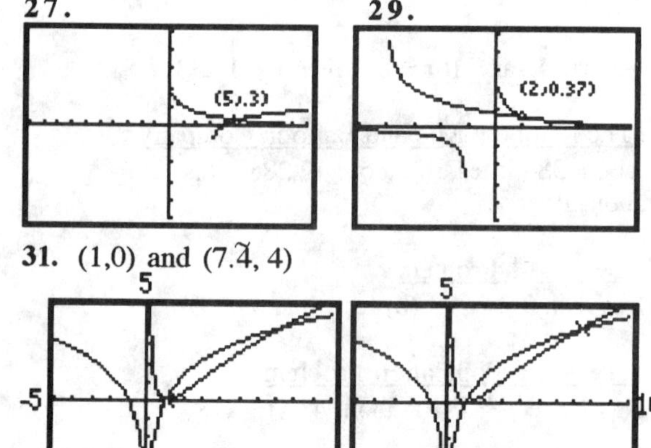

31. (1,0) and (7.4, 4)

33.

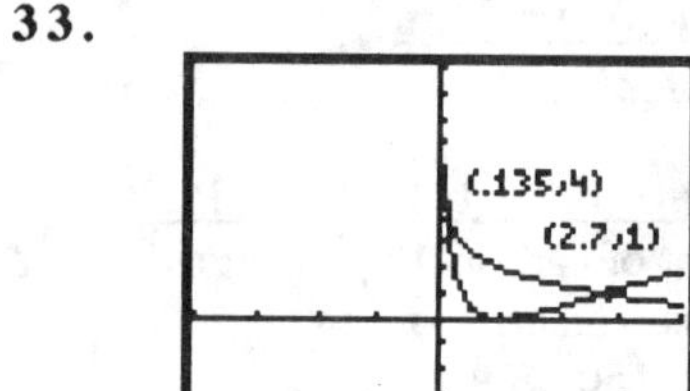

35. The Millionaire Problem
about 66 yr

37. The Rule-of-72 Problem
about 4.19 yr

39. The Growing Population Problem
about 36.9 yr from 1990

41. The Rainbow Kite and Balloon Company
(a) B = 42,000; the purchase price of the two vans is \$42,000. t = 48; the duration of the loan is 48 months (b) about 26 months (c) \$5925.12

43. The E. coli Problem
(a) $N = 2e^{0.035t}$ (b) $N = 2(2^{t/20})$ (c) 243 min and 246 min

45. x = 2

47. x = -1.001 or x = -0.999

49. Let $s = \log_b M$ and $t = \log_b N$, so that $\log_b M + \log_b N = s + t$.
$M = b^s$ and $N = b^t \Rightarrow$

$$M \cdot N = b^s \cdot b^t$$
$$M \cdot N = b^{s+t} \quad \Leftrightarrow$$
$$\log_b MN = s + t$$

Therefore, $\log_b MN = \log_b M + \log_b N$.

Section 4.4 (Page 302) More on Exponential and Logarithmic Functions

1. The Equality Problem
$A = 25{,}000e^{0.055t}$ and $A = 13{,}000(1.11)^x$,
about 13 yr

3. The Matching Population Problem
The populations will be equal 44 yr from 1990; in the year 2034.

5. The Mortgage Problem
(a) \$984.69 and \$768.88
(b) \$177,244.20 and \$276,796.80

7. The Payoff Problem
For P = \$100,000 and r = 6% APR,
10-yr loan totals \$133,221 and 20-yr loan totals \$171,943.45, \$38,719.45 more
(monthly payments \$1110.20 and \$716.43)
You pay more than half the interest if you pay off the loan in half the time. This is fair because initially you were borrowing more money.

9. The Rainbow Kite and Balloon Company
\$50,889.72 and \$188,104.88 Notice how much more money was earned the second 10 yr.

11. An Annuity Problem
about 373.21 months ≈ 31 yr

13. The Populations Projections Problem
(a) about 6.321 billion
(b) $k \approx 0.014$
(c) 8.345 billion
(d) $N = 9.38(1.015)^t$
(e) $N = 5.38e^{0.0149t}$, growth rate: 1.49%

15. The Wind Chill Problem
(a) y = (59.74 - 10) - 15.31 ln x for y = 37,26,..., 4
(b) Using x = 1 yields 51°F, and x = 27.7 mph for y = 0°F.

17. $y \to 0$ **19.** $y \to 0$ **21.** even function **23.** even function

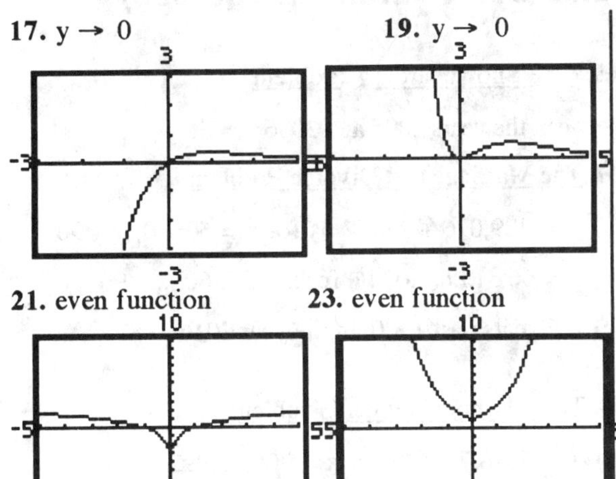

25. They are identical.

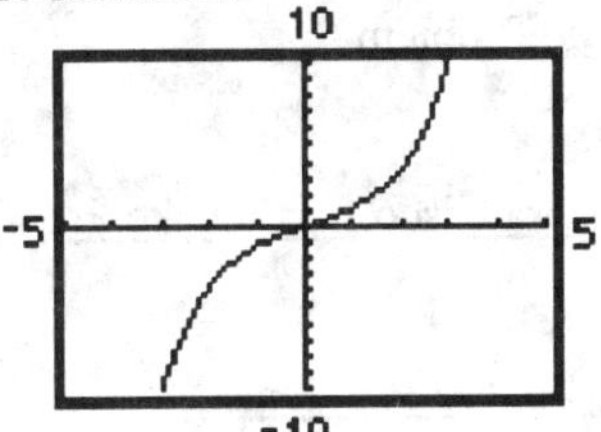

27. $(\sinh x)^2 - (\cosh x)^2 =$

$$\left(\frac{e^x - e^{-x}}{2}\right)^2 - \left(\frac{e^x + e^{-x}}{2}\right)^2 =$$

$$\frac{e^{2x} - 2 + e^{-2x} - e^{2x} - 2 - e^{-2x}}{4} = \frac{-4}{4} = -1$$

Section 4.5 More Applications Including Logistic Equations (Page 316)

1. The Cooling Coffee Problem
$T \approx 176°F$

3. The Hot and Iced Coffee Problem

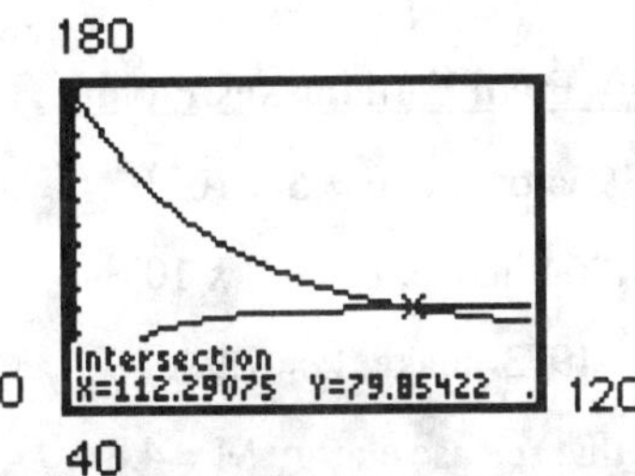

Both coffees will be about 79.8°F in 112 min.

5. The Two-Cars Problem
In 5 yr: $12,624.64 and $6180.38
In 10 yr: $12,600.10 and $3052.75

7. The End of the World Problem

(a) $N(t) = \dfrac{10}{1 + 0.9960e^{-0.0531t}}$

(b) about 6.31 billion and about 10 billion

9. The Epidemic Problem
about 26 days

11. The Rainbow Kite and Balloon Company

(a) $N(t) = \dfrac{6070.81}{1 + 5.334e^{-0.1842t}}$

(b) about 13 months

13. The Barometric Pressure Problem
(a) $P_0 = 760$ (b) $k \approx -0.0824$ (c)

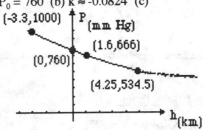

(-3.3,1000) P (mm Hg) (1.6,666) (0,760) (4.25,534.5) h (km)

15. The Heron Problem
(a) 100 fish (b) 97 fish, 75 fish (c) $N \to 0$

17. The Peak Time Problem
(a) $f(t) = 10te^{-t}$, $f(t) = 10te^{-2t}$, $f(t) = 10te^{-4t}$, $f(t) = 10te^{-10t}$ (b) 1, 1/2,1/4, 1/10 (c) $t = 1/b$
(d) A vertical change in scale does not affect the horizontal value at maximum.

19. The Nicotine in the Bloodstream Problem
It takes about 11 min for the concentraion to peak. The half-life is about 1/2 hr.

Section 4.6 Logarithmic Scales and Curve Fitting (Page 333)

1. The Earthquake Problem
10,000 times

3. The 1989 San Francisco Earthquake Problem
1.58 times

5. The Rustling Leaves Problems
10 times

7.
$$y = 100x^{-4}$$
$$\log y = \log(100x^{-4})$$
$$\log y = \log 100 + (-4)\log x$$
$$\log y = 2 - 4\log x$$

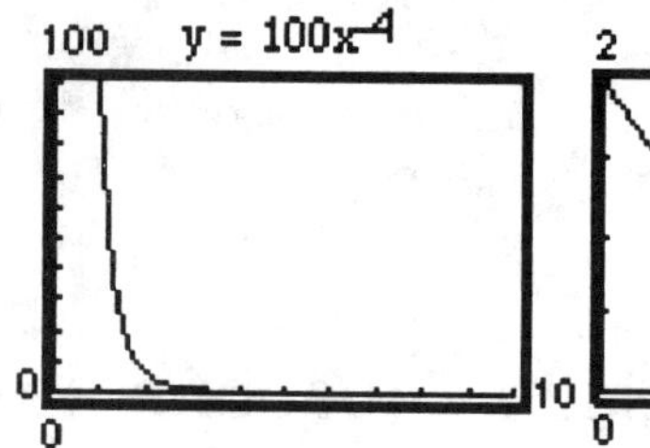

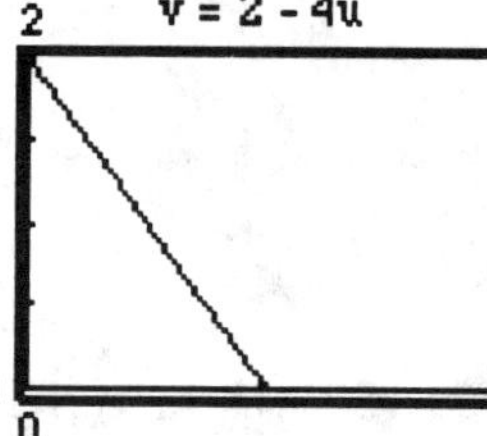

9.
$$y = 2e^{0.5x}$$
$$\log y = \log(2e^{0.5x})$$
$$\log y = \log 2 + 0.5x$$

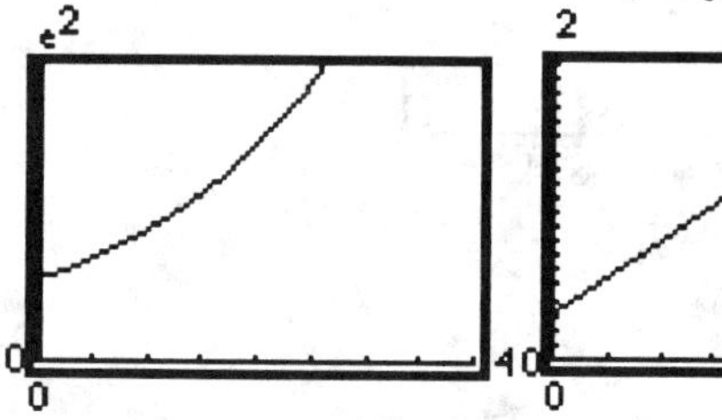

v = 0.3 + 0.5x

2

0

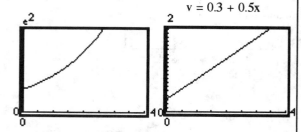

11. The Beanie Baby Problem

(a) $y = 167.5x - 850.4$ for $x = 5, 7, 17$

(b) about \$5347 ($x = 37$)

13. The Taxes by Year Problem

(a) $y = (5.6 \times 10^9)x^{-3.27}$ for $x = 75, 91, 92, 93$.

(b) about \$1915

15. The Snowboarding Problem

between the years 2005 and 2006

17. The Marriage and Divorce Problem

(a) $y = 329{,}076 + 24452.5x$ for $x = 50, 60, \ldots, 90$

(b) $y = 66{,}313.6(1.034^x)$ for $x = 50, 60, \ldots, 90$

(c) 116 years past 1900; in the year 2016

19. The Olympic Winners Problem

(a) $y = 5618.77x^{-0.92}$ for $x = 900, \ldots, 992$

(b) $y = 133{,}700x^{-1.7}$ for $x = 28, \ldots, 92$

21. The Rainbow Kite and Balloon Company

(a) The exponential graph is reflected about the x axis and translated up 100 units. Subtracted 100 from the y values and then negate the y values.

(b) $y = 100 - (15{,}580(0.8932)^x)$ for $x = 50, 55, \ldots$

(c) $x = 69 \Rightarrow y = 93.586$, about 93.6%

(d) $x = 101 \Rightarrow y = 99.8$, about 99.8%

23. The Energy From Earthquakes Problem

(a) Seattle's 1964 quake, $E = 5 \times 10^{20}$

(b) Alaska's 1964 quake, $E = 6.3 \times 10^{22}$

(c) Managua's 1973 quake, $E = 7.9 \times 10^{21}$

(d) a tornado that release energy $M = 4.8$

(e) a one megaton nuclear bomb, $M = 2.8$

25. The Age and Height Problem

Answers will vary depending on the data. The data should be adjusted by adding 10 to each age. The 0-in. height should be massaged by entering the zero as a small positive number such as 1×10^{-100} using scientific notation.

Chapter 4 Review (Page 340)

1. The Stacked Paper Problem

about 14 cuts

2.(a) \$1174.24 (b) \$1178.95 (c) \$1179.29 (d) \$1179.39 (e) \$1179.39

3.(a) log 0.1 = -1 (b) ln 0.37 ≈ -1

4.(a) log 2 ≈ 0.30 (b) ln 2 ≈ 0.69

5.(a) $10^{-2} = 1/100$ (b) $e^{2.30} \approx 10$

6. (a) $10^0 = y$ (b) $10^{-1} = y$

7. x = 0.1 **8.** x = 0

9. x = ±10 **10.** x = 0

11. x = -4 **12.** x ≈ -0.4343

13. x ≈ 1 **14.** x = -1

15. x ≈ 2.3 **16.** x = 1/2

17. y = 1 **18.** x = 1

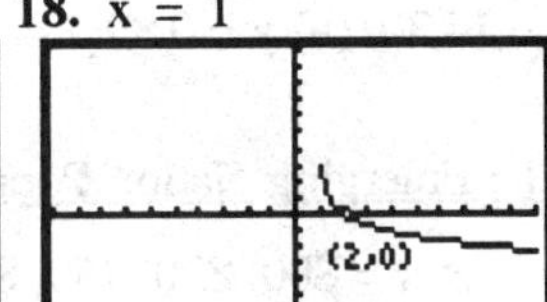

19. x = 10 **20.** x = 0

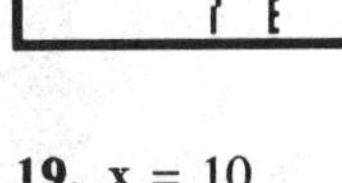

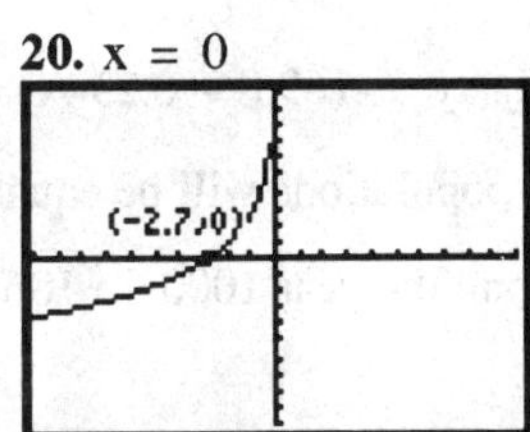

21. x = -ln(y + 3) **22.** $x = e^{y-1}$

23. The Big Bucks Problem

15.4 yr

24. The Populations of France and Germany Prob.

(a) 173.3 yr (b) 15 yr (c) 66.5 yr

25.(a) $x = e^{-2} \approx 0.14$ (b) x = 1 (c) $x = e^2 \approx 7.39$

26.(a) x = 9 (b) x = -1/9 (c) $x = 1 - \sqrt{3}$

27.(a) x ≈ 0.3010 (b) x ≈ 0.6931

28.(a) x ≈ 0.5657 (b) x ≈ 3.3026

29. x = 4 **30.** x = 2

31. no solution **32.** $x = \frac{e}{e-1} \approx 1.58$

33. x = -1 **34.** x = -1

35. x = -1 **36.** x = -1

37. $(1.7\tilde{4}, 1.5\tilde{6})$ **38.** $(8, 1.0\tilde{8})$

39. $(0.5, 0.6\tilde{3})$ **40.** (1,1)

41. The Loan Problem

about 58.4 payments

42. The Annuity Problem

about 406 payments

43. The Cooling Coffee Problem

about 166°F

44. The Turkey Cooking Problem

between 5 hr 27 min and 5 hr 44 min

45. The Epidemic Problem

about 57 days

46. The Cooled-off Corpse Problem

about 1.6 hr ago

47. The Emerging Nation Problem

(a) U.S.: $N = 5e^{0.035t}$ U.K.: $N = \frac{16}{65}t + 16$

(b) The populations were about equal in 1849.

48. The Radioactive Decay Problem

about 16.7 yr.

49.(a) $y = 0.79(1.08^x)$ (b) $y = 0.61(1.09^x)$

(c) If child begins college in 2014, 2-yr college tuition will be \$4996 and 4-yr college tuition will be \$13,384.

50.(a) y = -61 + 0.98x (b) about 37¢

(c) about 34¢

51.(a) 1983 (b) 1989

52. Answers vary depending on data.

Chapter 4 Test (Page 345)

1.(a) $x = \frac{1}{\sqrt{10}}$ (b) $x = \pm\frac{1}{\sqrt{10}}$

2.(a) x = e (b) x = 1/e

3.(a) x = 2.3010 (b) 5.2983

4. (a) x = 0.4343 (b) x = 2.3026

5. y = 1

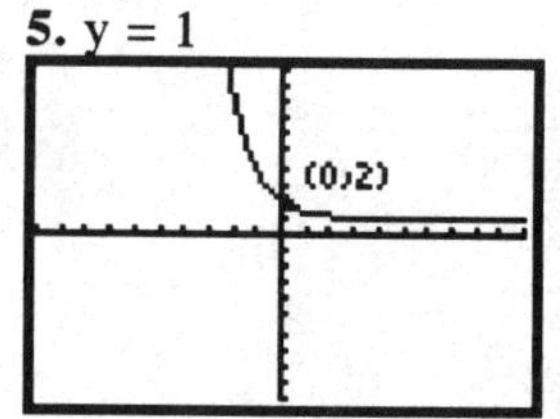

6. x = 1 (0,0)

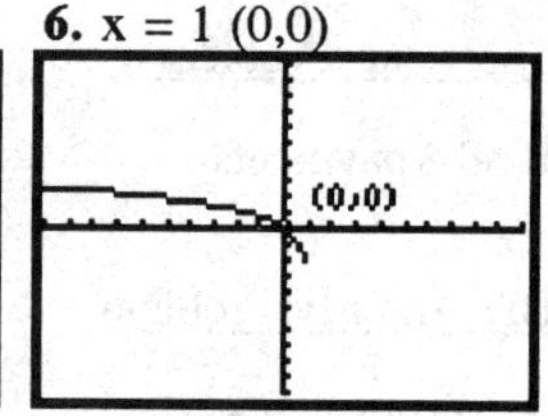

7. (a) x = 0 (b) x = 0

8.(a) x = 2 (b) x = 1/4

9. x = 1/99 **10.** x = 2

11.

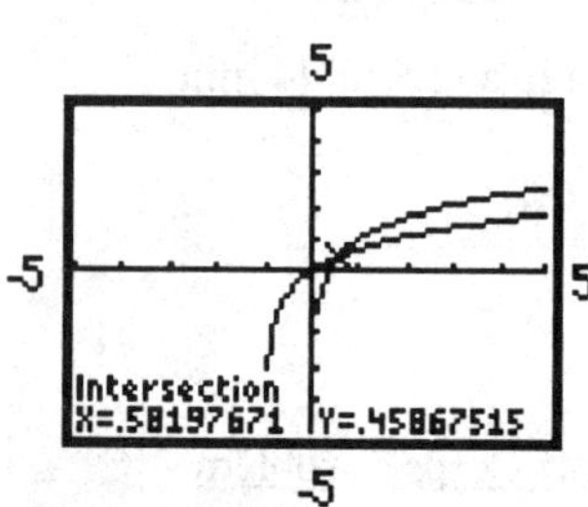

12. x = ln y + 1 **13.** $x = e^{(1-y)/2}$

14. The Doubling and Halving Times Problem
(a) 10 yr (b) 5 yr

15. The Decibel Scale Problem
(a) 0 dB and 110 dB (b) $\frac{A_1}{A_2} = 3{,}162$

16. The Ebola Virus Problem
100 days

17. The Loan Problem
45 months

18. The Annuity Problem
36.5 months

19. The Folding Paper Problem
(a) 2,6,14,30 (b) $y = 0.89(2.45)^x$ (c) about 8 folds

20. The Emerging Nation Population Problem

U.S.: Let x = 800, 830, 850, 890, and 900.

$y = (2.0 \times 10^{-9})(1.028^x)$

U.K.: y = -169.8 + 0.234x

The populations will be equal in about 867 years beyond the year 1000, in 1867.

Section 5.1 Right Triangle Trigonometry (Page 356)

1.(a) $\sqrt{2}/2 \approx 0.707$ (b) $\sqrt{2}/2 \approx 0.707$ (c) 1

3.(a) 0.766 (b) 0.766

(c) -0.766 (d) -0.766

5.(a) 30° (b) 60°

7.(a) 45° (b) 0°

9. $\angle B = 45°$, $b = 21$, $c \approx 29.7$

11. $\angle B = 32.71°$, $b \approx 20.6$, $c \approx 38.1$

13. The Space Needle Problem

about 547 ft

15. The Yellowstone Fire Problem

about 108 ft

17. The Hypotenuse Problem

AD ≈ 15.8

19. The Ninth-Hole Problem

don't try it, about 36°

21. The Airport Problem

about 10.7°

23. The Lighthouse Problem

The boats are about 428 ft apart. They are about 406 ft and 833 ft from the lighthouse.

25. The Rainbow Kite and Balloon Company

about 68.2°

27. The Solar-Heated Room Problem

(a) about 2.4 ft

(b) about 15.3 ft

29. The Rectangular Solid Problem

DG ≈ 14.4, DF ≈ 15.6

31. Pythagoras' Proof of the Pythagorean Theorem

Remove the four triangles from both squares and what remains is $a^2 + b^2 = c^2$.

33.(a) $A \approx 1.3$ (b) $A \approx 5.2$ (c) $1.3 < \pi < 5.2$

35.(a) $A \approx 2.8$ (b) $A \approx 3.3$ (c) $2.8 < \pi < 3.3$

Section 5.2 The Circular Functions (Page 372)

1.(a) 0 (b) $2\pi \approx 6.28$ (c) $4\pi \approx 12.57$

3.(a) -0.35 (b) -1.22 (c) 5.93

5.(a) 90° (b) 270° (c) 450°

7.(a) 85.9° (b) 143.2° (c) 200.5°

9. C

11. A

13.(a) 75° (b) 75°

15.(a) $\pi/6$ (b) $\pi/12$

17.(a) (1.3, 1.6)

(b) $\frac{y}{r} = 0.78$, $\frac{x}{r} = 0.63$, $\frac{y}{x} = 1.23$

(c) sin 50° ≈ 0.77, cos 50° ≈ 0.64, tan 50° ≈ 1.19

19.(a) (-1.1, -0.9) (b) $\frac{y}{r} = -0.63$, $\frac{x}{r} = -0.77$, $\frac{y}{x} = 0.82$

(c) sin 220° ≈ -0.64, cos 220° ≈ -0.77, tan 220° ≈ 0.84

21.(a) **23.**(a)

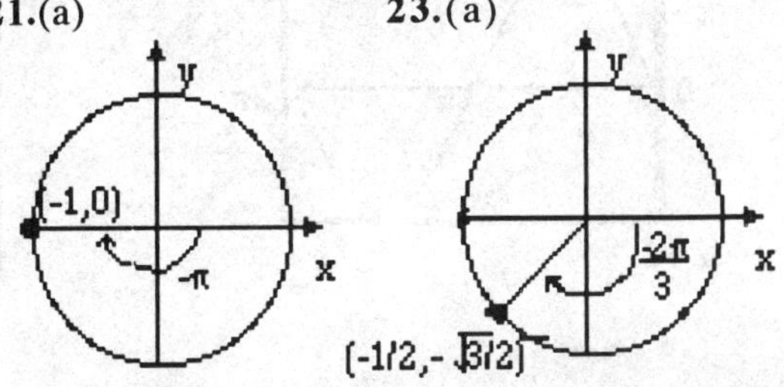

21.(b) $\sin(-\pi) = 0$

$\cos(-\pi) = -1$

$\tan(-\pi) = 0$

23.(b) $\sin(-2\pi/3) = -\sqrt{3}/2$

$\cos(-2\pi/3) = -1/2$

$\tan(-2\pi/3) = \sqrt{3}$

25.(a) 12/13 (b) -12/13

27.(a) 6/10 (b) -6/10

29. The Rainbow Kite and Balloon Company

273 mi, 513 mi, 240 mi

31. The Spinning Earth Problem

(a) $v = 1047.2 \cos \lambda$

(b) domain: $0 \le \lambda \le 90$

range: $0 \le v \le 1047.2$

33.

θ	0	$\frac{\pi}{6}$	$\frac{\pi}{3}$	$\frac{\pi}{2}$	$\frac{2\pi}{3}$	$\frac{5\pi}{6}$	π	$\frac{7\pi}{6}$	$\frac{4\pi}{3}$	$\frac{3\pi}{2}$	$\frac{5\pi}{3}$	$\frac{11\pi}{6}$	2π
sin θ	0	$\frac{1}{2}$	$\frac{\sqrt{3}}{2}$	1	$\frac{\sqrt{3}}{2}$	$\frac{1}{2}$	0	$-\frac{1}{2}$	$-\frac{\sqrt{3}}{2}$	-1	$-\frac{\sqrt{3}}{2}$	$-\frac{1}{2}$	0
cos θ	1	$\frac{\sqrt{3}}{2}$	$\frac{1}{2}$	0	$-\frac{1}{2}$	$-\frac{\sqrt{3}}{2}$	-1	$-\frac{\sqrt{3}}{2}$	$-\frac{1}{2}$	0	$\frac{1}{2}$	$\frac{\sqrt{3}}{2}$	1
tan θ	0	$\frac{1}{\sqrt{3}}$	$\sqrt{3}$	und.	$-\sqrt{3}$	$-\frac{1}{\sqrt{3}}$	0	$\frac{1}{\sqrt{3}}$	$\sqrt{3}$	und.	$-\sqrt{3}$	$-\frac{1}{\sqrt{3}}$	0

35.

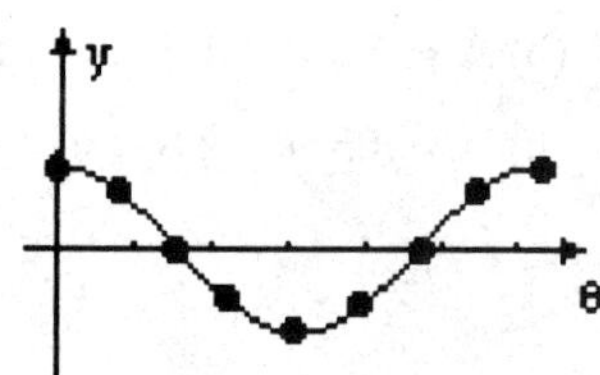

37.(a) 1 rad ≈ 57° 17' 45"

(b) about 0.017 rad, 0.00029 rad, 0.0000048 rad

39.

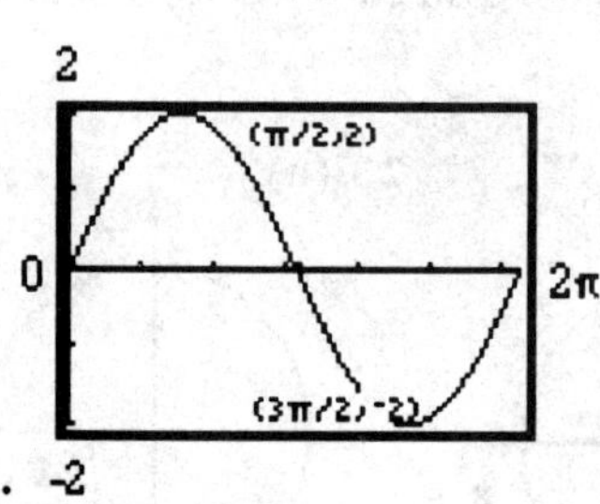

41.

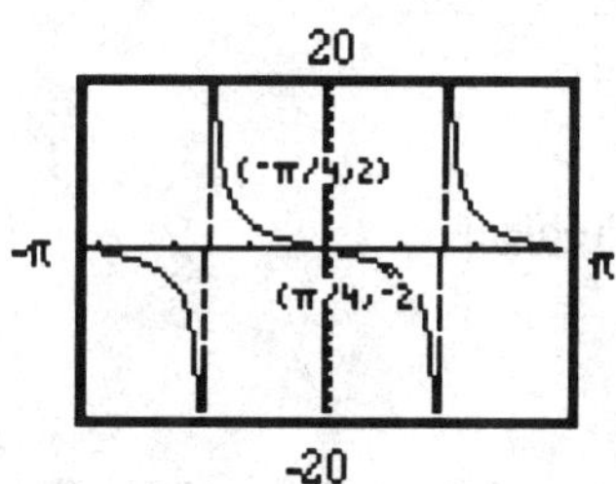

43. $\angle OPP' = \angle OPP' = \pi/3 = 60°$ Hence, Δ OPP' is equilateral and OP = OP'= PP' = 1.

OQ bisects PP' making QP = QP' = 1/2 = y

$x^2 + y^2 = 1^2$ and $y = \frac{1}{2}$ yields $x = \frac{\sqrt{3}}{2}$.

Section 5.3 Graphs of the Circular Functions (Page 386)

1. C **3.** G

5. F **7.** E

9.(a) (b)

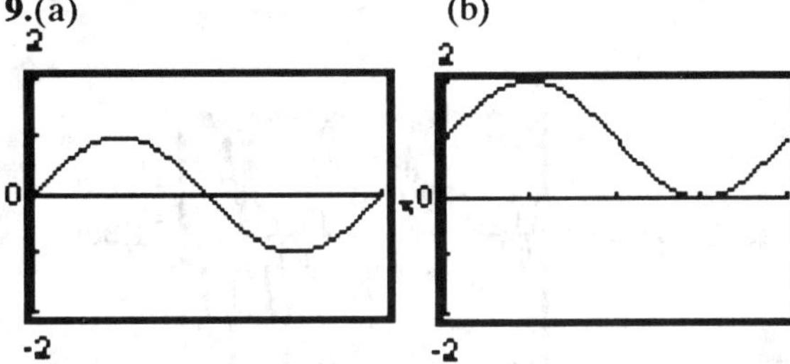

11.(a) (b)

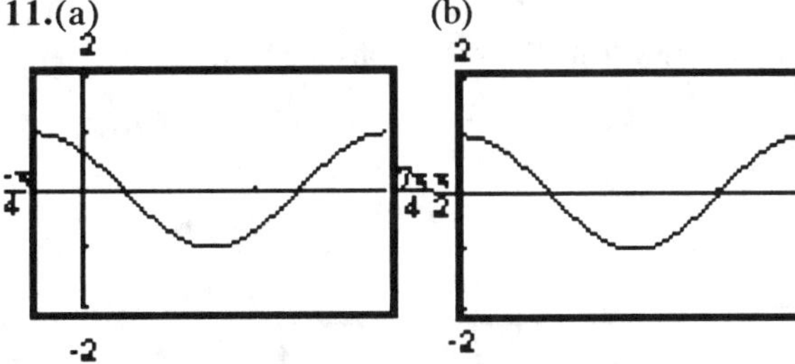

13.

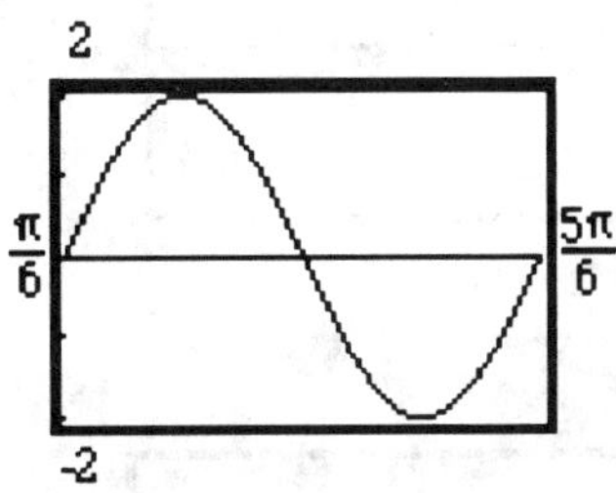

15.

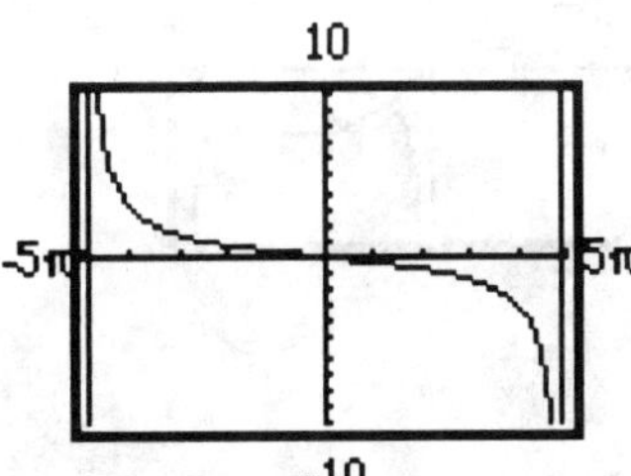

17.

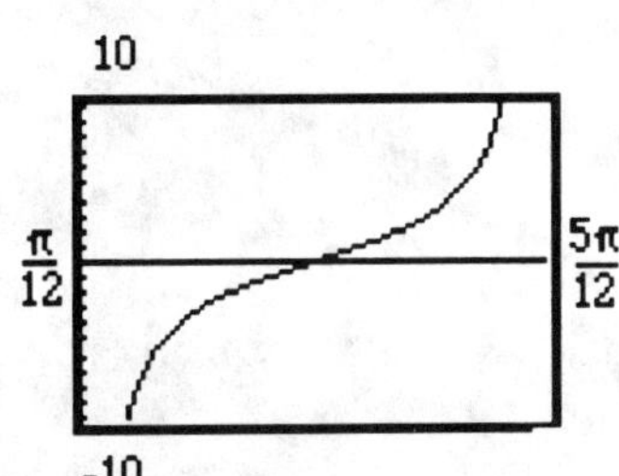

19. 4π **21.** $\pi/2$

23. 2 **25.** 2

27. $x = -\pi/4,\ x = \pi/4,\ x = 3\pi/4$

29. $y = 3\sin(-x)$ **31.** $y = \frac{1}{2}\sin x$

33. $y = -\cos 2x$ **35.** $y = \tan x + 1$

37. neither **39.** odd

41. A Tide Problem

$h(t) = 28 \sin 0.50t + 30$

43. The Prey-Predator Problem

(a)

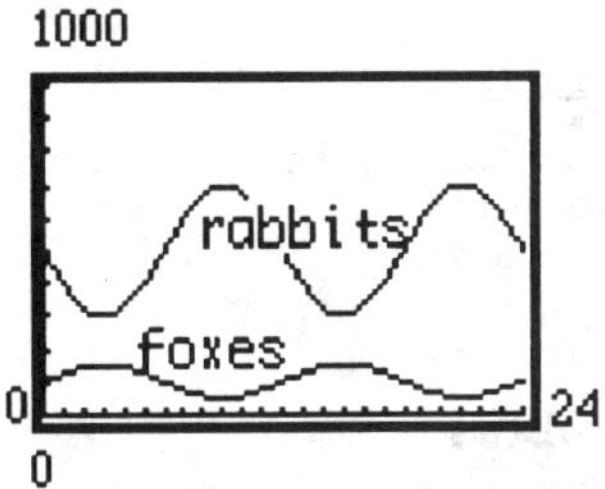

(b) Foxes: $N_f = 100 + 50 \sin \frac{\pi}{6} t$

A = 50 is the possible increase in foxes. The increasing and decreasing population cycle repeats every 12 yr since $B = \pi/6$. D = 100 is the average number of foxes, so that D + A = 150 is the highest population and D - A = 50 is the lowest population of foxes.

Rabbits: $N_r = 500 + 200 \sin\left(\frac{\pi}{6} t - \pi\right)$

A = 200 is the possible increase in rabbits. The population cycle also repeats every 12 yr since $B = \pi/6$. $C = -\pi$ causes a phase shift to the right about 6 yr. D = 500 is the average number of rabbits, so that D + A = 700 is the highest population and D - A = 300 is the lowest population.

(c) As the population of foxes increases, rabbits decrease. Fewer rabbits cause the fox population to decrease, causing the rabbit population to increase. The cycle repeats every 12 yr.

45. The Rainbow Kite and Balloon Company

(a) $A = \frac{1}{2}(2\sin\frac{\theta}{2})(\cos\frac{\theta}{2})$ in square feet (b) $\theta = \pi/2 = 90°$

47. The Biorhythms Problem

(a) $y = \sin\frac{2\pi}{23}t$, $y = \sin\frac{2\pi}{28}t$, $y = \sin\frac{2\pi}{33}t$

(b)

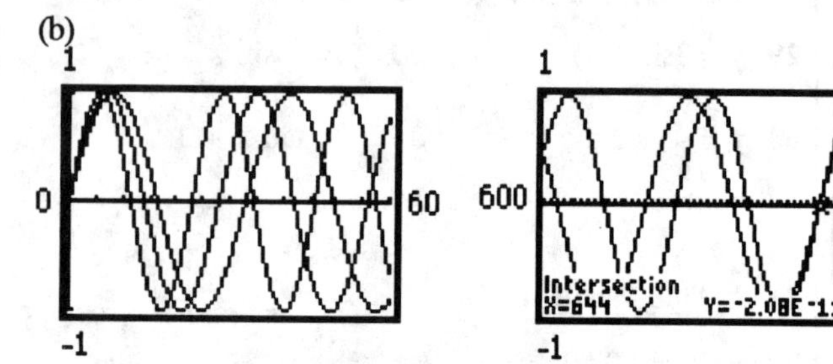

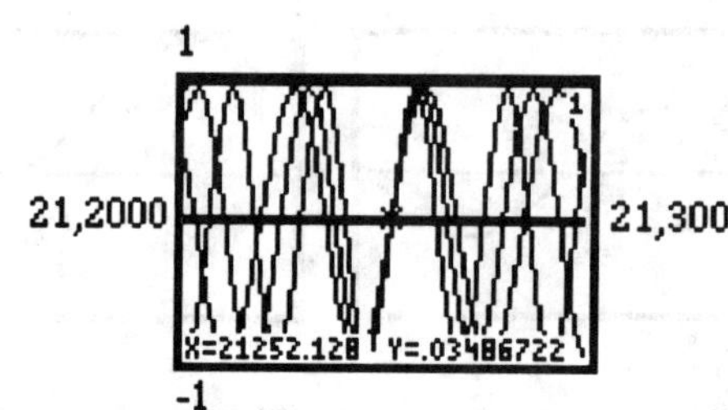

(c) Accept any challenge when all three graphs are high, and sleep in when all three graphs are low.

(d) 644 days (e) 21,252 days

49. The Earth Problem

(a) $H = 12\sin\frac{2\pi}{365.2422}t + 12$

(b)

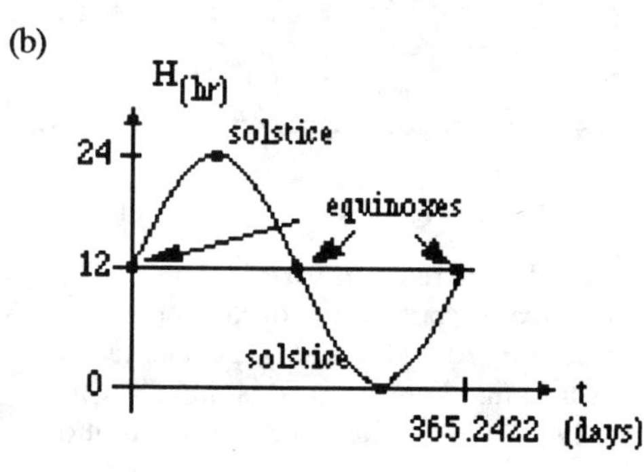

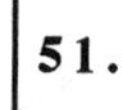

51.

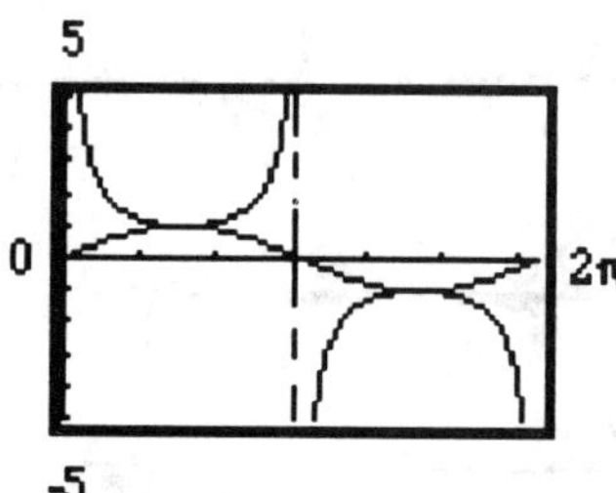

53.

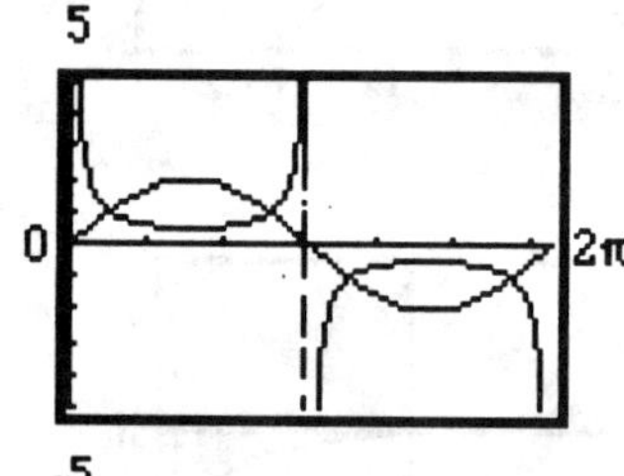

Section 5.4 Reciprocals, Sums, and Products of the Circular Functions (Page 398)

1.(a) 0.42 (b) 2.37

3.(a) -0.53 (b) 0

5. $y = -\sec x$ **7.** $y = \cot(x - \pi/2)$

9.

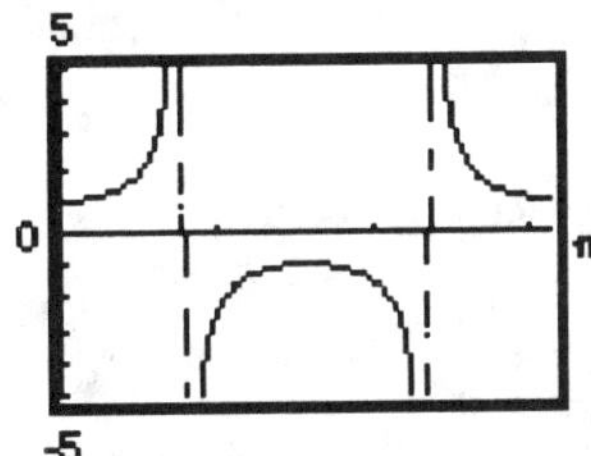

11.

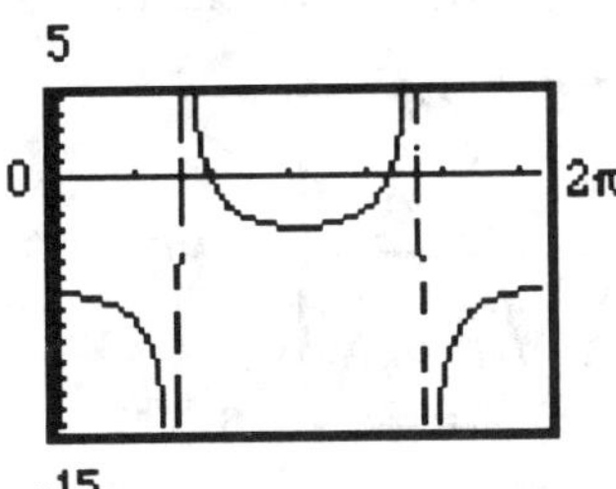

13.

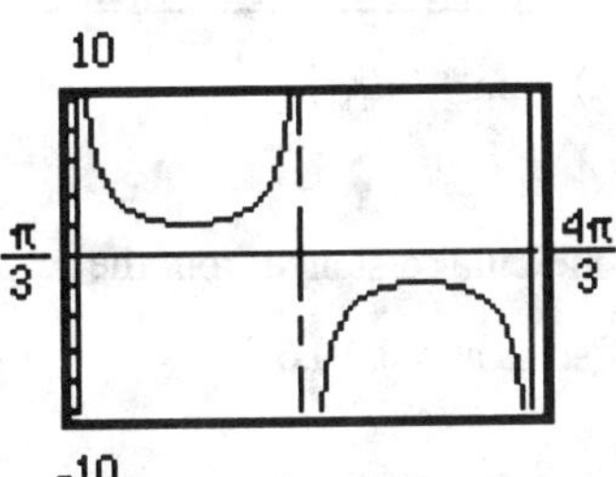

15.

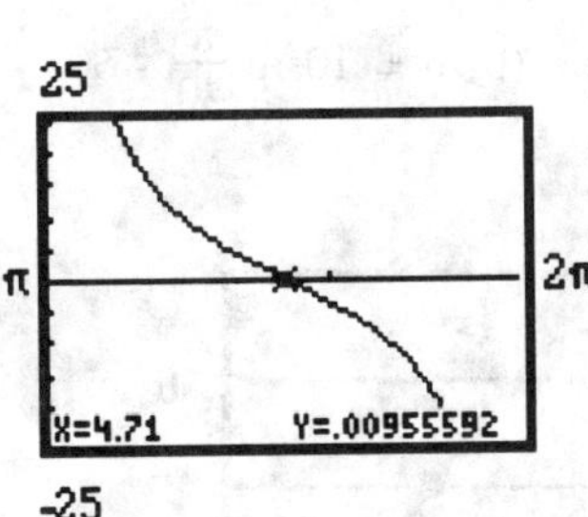

17. $x = 0$, $x = \pi$, $x = 2\pi$, no y intercept

19. $x = -\pi/2$, $x = \pi/2$, $x = 3\pi/2$, 1

21. even **23.** odd

25. The cosine graph along the graph of $y = 1$.

27. The graph of sin 2x along the graph of $y = 1/x$.

29. This graph is called a square wave.

31. This graph is called a sawtooth wave.

33. The varying amplitude is $y = x$. The lines $y = x$ and $y = -x$ form an envelope for the graph.

35. The varying amplitude is $y = x^3$. The curves $y = x^3$ and $y = -x^3$ form an envelope for the graph.

37. The varying amplitude is $y = e^{-x}$. The curves $y = e^x$ and $y = e^{-x}$ form an envelope for the graph.

39. They are the same except for amplitude.

41. They are the same.

43. The Pothole Problem

The cab bounces higher, but stops after 10 sec. This bus never stops bouncing.

45. Another Tide Problem

(a) $h_1 = \sin\frac{\pi}{6}t$, $h_2 = 0.50\sin\frac{\pi}{12}t$

(b) The period is 24 hr and the amplitude is 1.4

47. The Ducks and Geese Problem

(a) $y = 10\sin\frac{\pi}{16}t + 30$ (b) $y = 5\sin\frac{\pi}{10}t + 10$

(c) The population peaks at 55 million and repeats every 160 yr.

49. The AM Radio Problem

The figure shows the amplitude being modulated at a level of $y = \sin 512\pi t$.

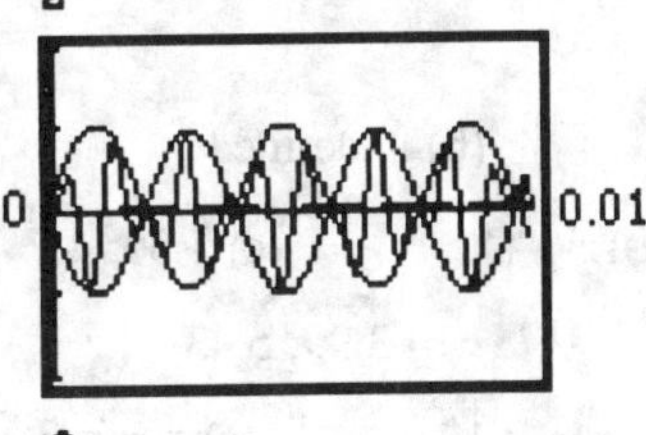

51. period: 2π, amplitude: 3.6, phase shift: 0.59 of a cosine graph

53. period: π, amplitude: 1.4, phase shift: 0.39 of a cosine

Section 5.5 Inverses of the Circular Functions (Page 408)

1. not a function **3.** function

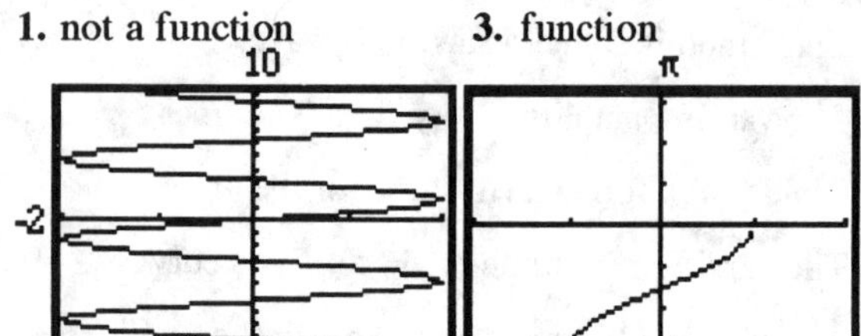

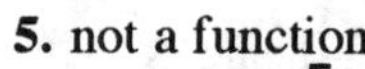

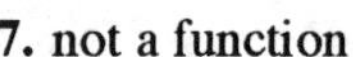

5. not a function **7.** not a function

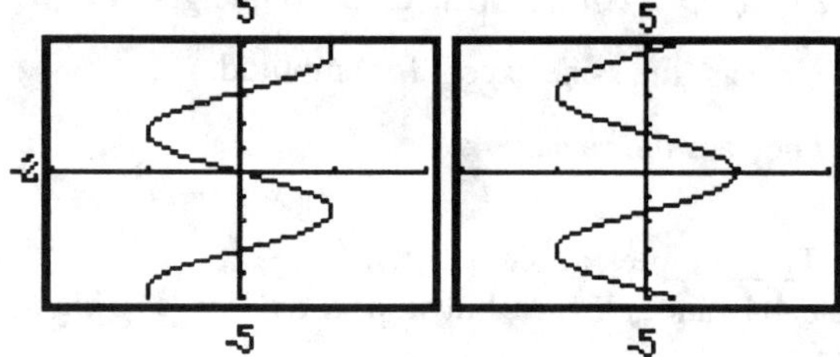

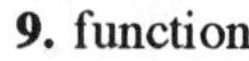

9. function **11.** function

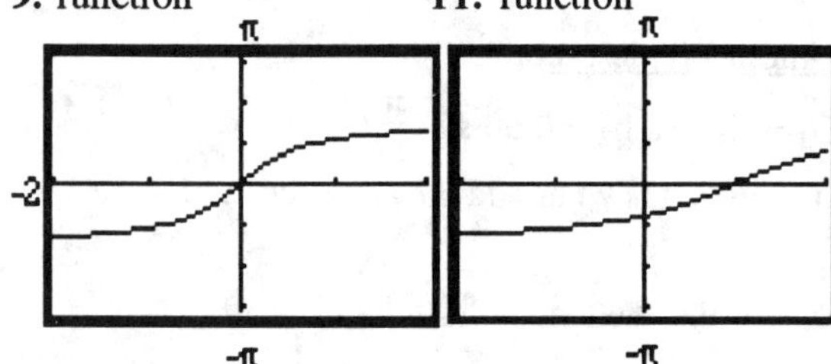

13.(a) 0.78 and 2.37 (b) 5.51 and 3.92

15.(a) 1.34 and 4.52 (b) 4.91 and 1.76

17.(a) 0.80 and 5.48 (b) 2.35 and 3.93

19.(a) 0.72 and 2.42 (b) 0.85 and 5.43

21.(a) 60°

(b) 60° + 360°k, -60° + 360°k

(c) 60°, 300°

23.(a) -63.4° (b) undefined

25.(a) 1.56 (b) $1.56 + k\pi$ (c) 1.56, 4.7

27.(a) 0.85, 2.29 (b) 3.99, 5.43

29.(a) $x = \pi/6$ (b) $x = \pi/12$ (c) $x = \pi/3$

31. $x \approx 0.93$ (b) $x \approx 0.46$ (c) $x \approx 2.54$

33. ± 3/5 **35.** 4/3

37. $\pm\frac{\sqrt{21}}{5}$ **39.** $x = \pi/12 \approx 0.26$

41. $x = -\frac{\pi}{9} + 1 \approx 0.65$ **43.** $x = 0$

45. $x = 0$ or $x = 1$

47. The Ferris Wheel Ride Problem

(a) $y = 25\sin(8\pi t - \pi/2) + 25$

(b) $t = \frac{1}{8\pi}\sin^{-1}\left(\frac{y - 25}{25}\right) + \frac{1}{16}$

(c) about 0.06, 0.19, 0.31, and 0.44 min

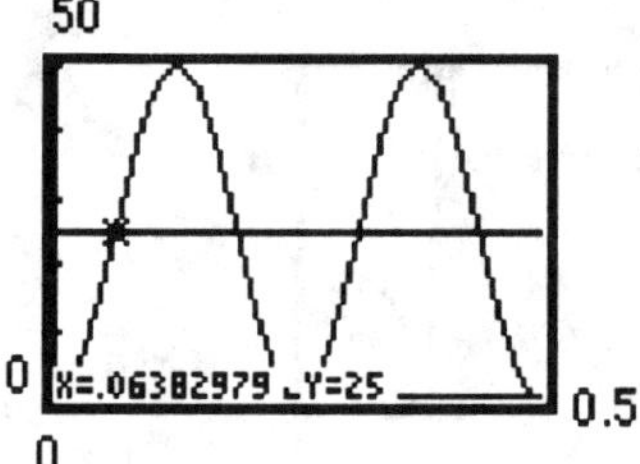

49. The Rainbow Kite and Balloon Company

(a) $\theta = \tan^{-1}(28/x) - \tan^{-1}(4/x)$

(b) $x \approx 10.6$; $\theta \approx 48.7°$

(c) Make the perpendicular distance from the screen to the middle row of seats about 10.6 ft.

51. The Ocean Swells Problem

(a) $h = 10\sin\frac{\pi}{40}t$ (b) $h = 10\sin\frac{\pi}{40}t - 60$

(c)

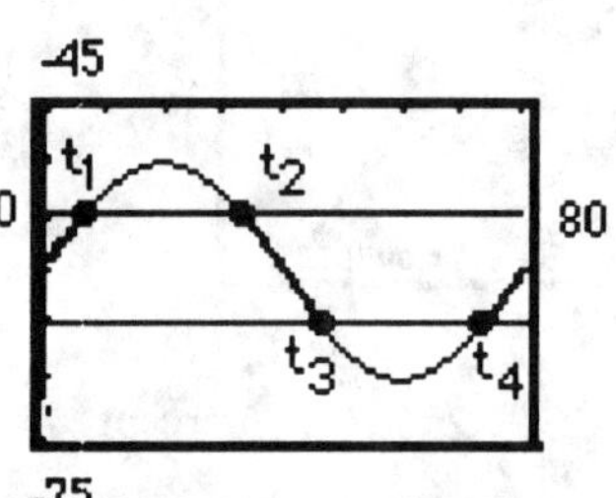

Out of every 80 sec, the bait is within reach of the fish 80 - 2(26.4) = 27.2 sec.

53.

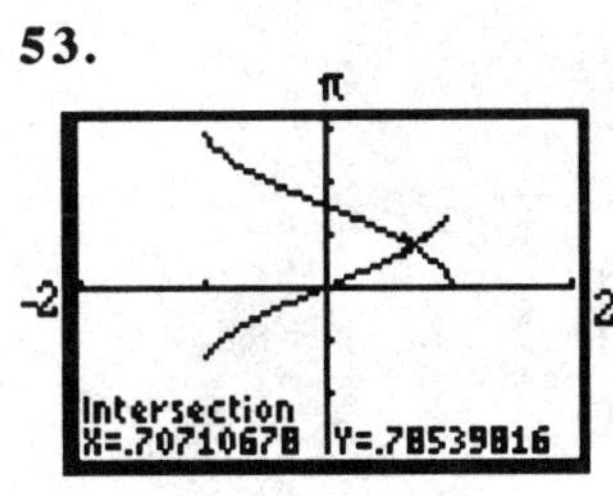

55.

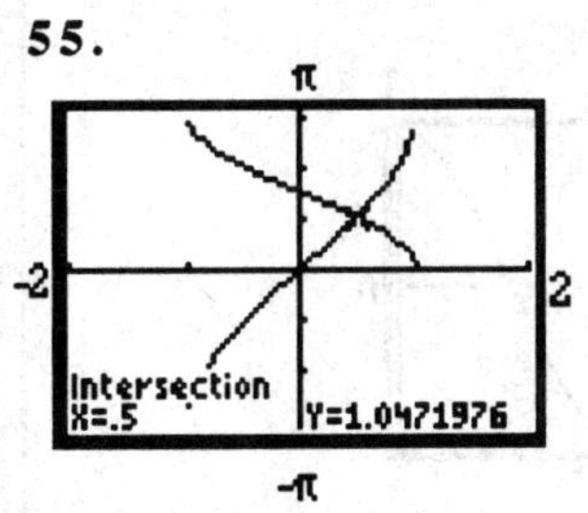

Chapter 5 Review (Page 412)

1.(a) 1/2 (b) $\sqrt{3}/2$ (c) undefined

2.(a) $\sqrt{2}/2$ (b) $\sqrt{2}/2$ (c) -1

3.(a) 60° (b) 60°

4.(a) 0° (b) 0°

5. $a = 1/\sqrt{2} \approx 0.7$, $b = 1/\sqrt{2} \approx 0.7$, $\angle B = 45°$

6. $a = 1/2$, $b = \sqrt{3}/2$, $\angle B = 60°$

7. $a \approx 0.8387$, $b \approx 0.5446$, $\angle B = 33°$

8. $a \approx 0.0174$, $b \approx 0.9998$, $\angle B = 89°$

9. The Short Path on a Box Problem

(a) about 9.8 in. (b) about 7.07 in.

(c) about 9.5 in.

10. The Old-Fashioned Sine Problem

AB = sin θ, OB = cos θ, CD = tan θ

11. The Bald Eagle Problem

(a) about 10,313 ft, almost 2 mi high

(b) $y = \dfrac{1.5}{\tan \alpha/2}$

12. The Luxor Hotel Problem

about 200 ft

13.(a) $2\pi/3 \approx 2.09$ (b) $4\pi/3 \approx 4.19$

(c) $5\pi/3 \approx 5.24$

14.(a) $-\pi/6 \approx -0.52$ (b) $-5\pi/6 \approx -2.62$

(c) $-7\pi/6 \approx -3.66$

15.(a) 60° (b) 240° (c) 300°

16.(a) -225° (b) -315° (c) 585°

17. The CD Problem

(a) about 8.8 rev/sec (b) about 3.4 rev/sec

18. The Speeding Satellite Problem

(a) $\alpha \approx 17.5°$

(c) about 6885 mph

19. D **20.** C

21. E **22.** B

23. F **24.** A

25.(a)

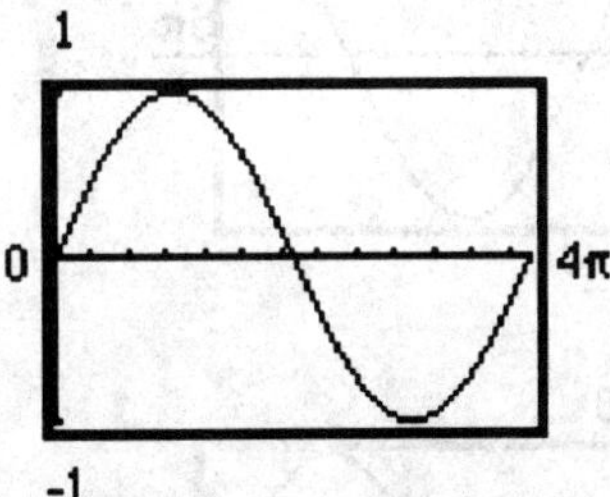

(b)

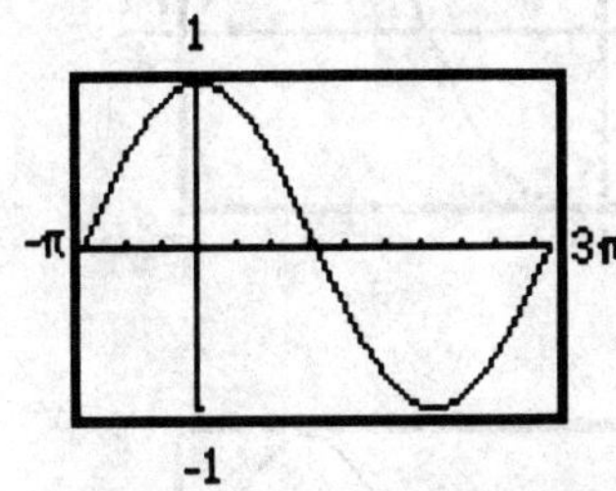

26.(a)

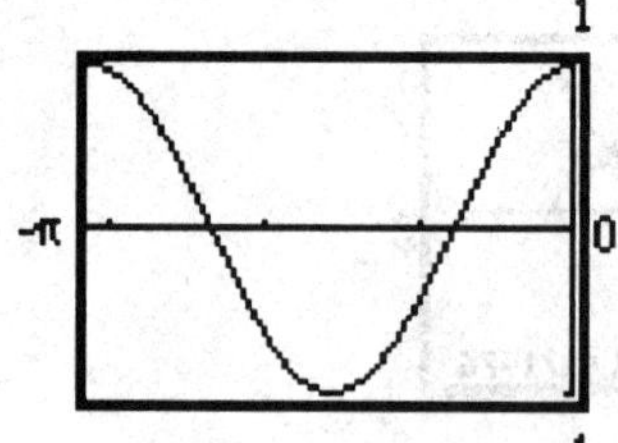

(b)

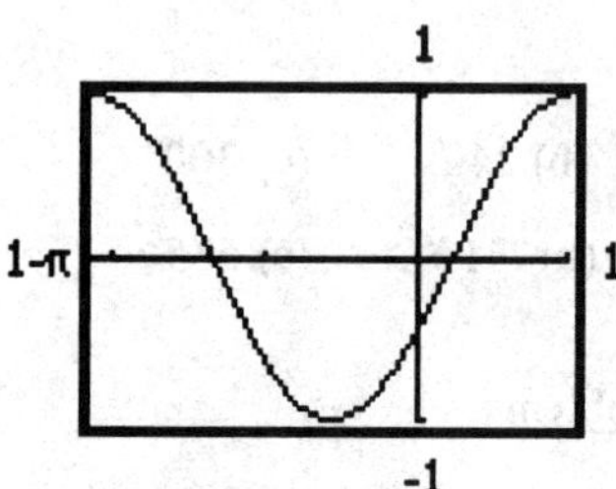

27.(a)

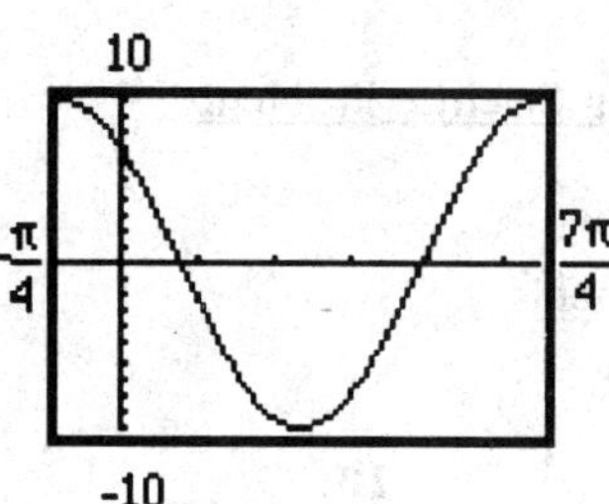

(b)

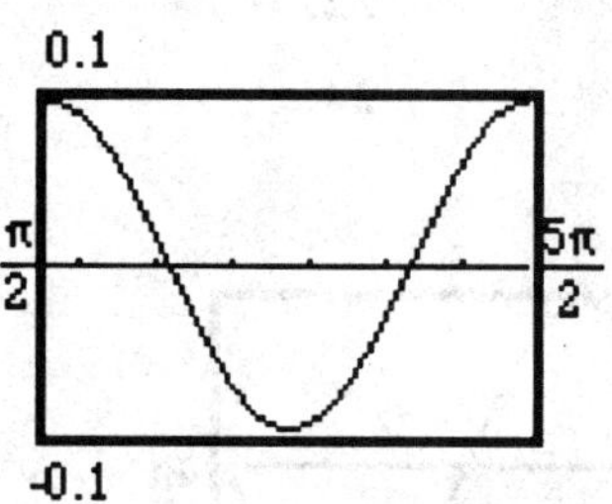

28.(a)

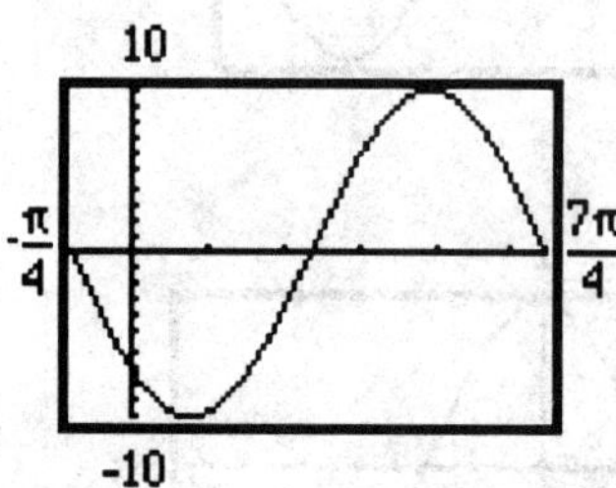

(b)

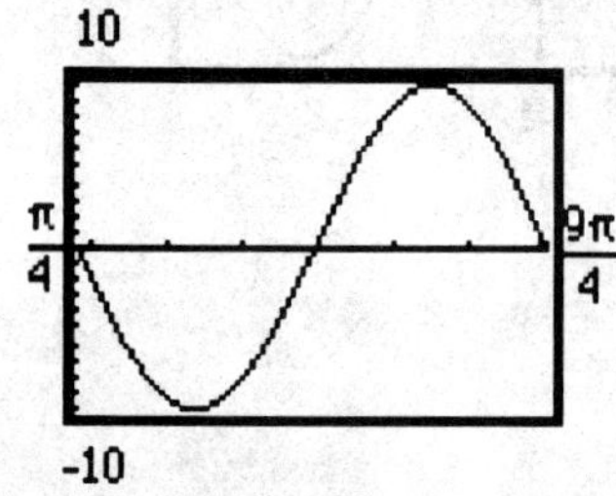

29.(a) $4\pi/3$

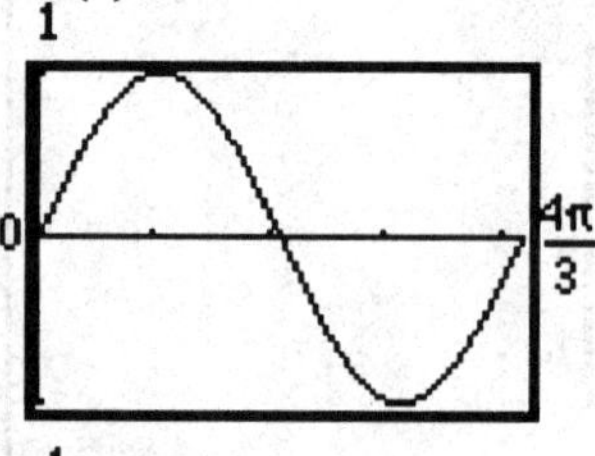

(b) $3\pi/2$

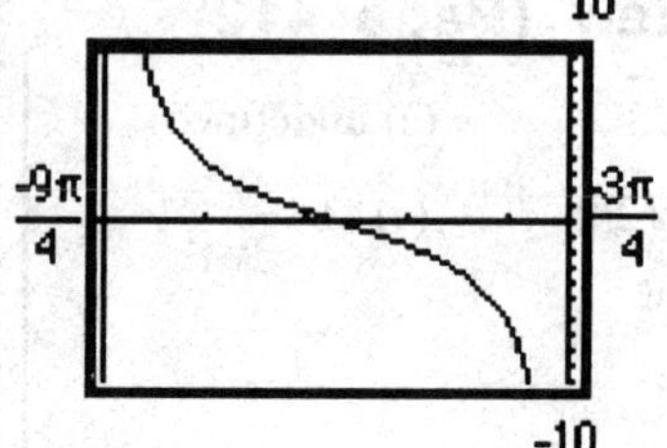

30.(a) $3\pi/2$

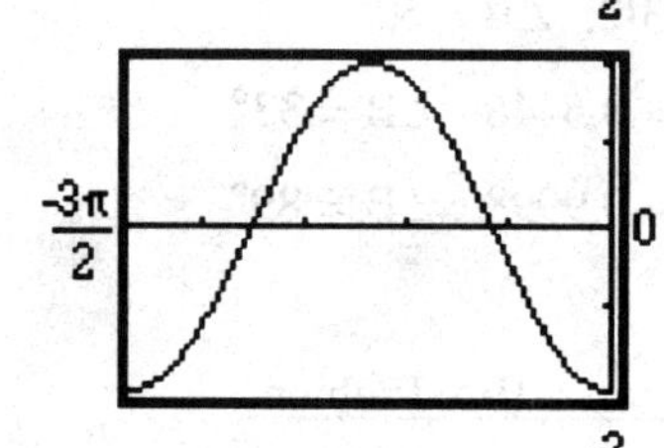

(b) $\pi/10$

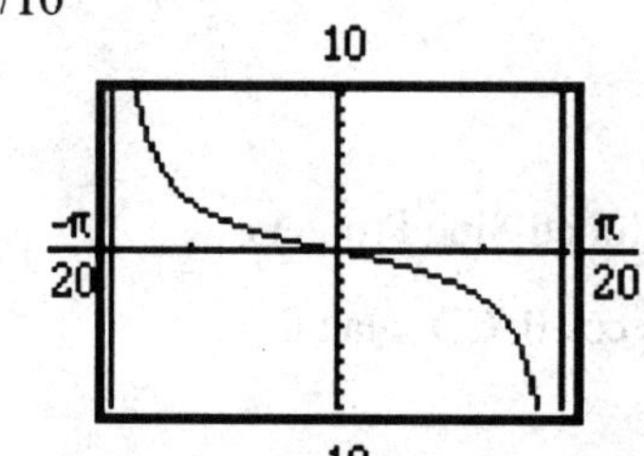

31. $y = 2 \sin 2x$

32. $y = -2 \sin 2x$

33. $y = \frac{1}{2}\sin 2x$

34. $y = \frac{1}{2}\cos 2x$

35. $y = -2 \cos x$

36. $y = -\tan(\frac{x}{4})$

37. <u>The High-Temperature Problem</u>

$y = -15.4 \sin(\frac{\pi}{6}x) + 60.4$ For March, $x = 4$, $T = 47.1°F$, and for November, $x = 10$, $T = 73.7°F$.

38. The Wet Feet Problem

$y = 2.385 \sin(\frac{\pi}{6}x) + 3.525$ For Nov, x = 1 and

y = 4.72 in. For July, x = 9 and y = 1.14 in.

39.(a) -1 (b) -0.7 (c) -1.29

40.

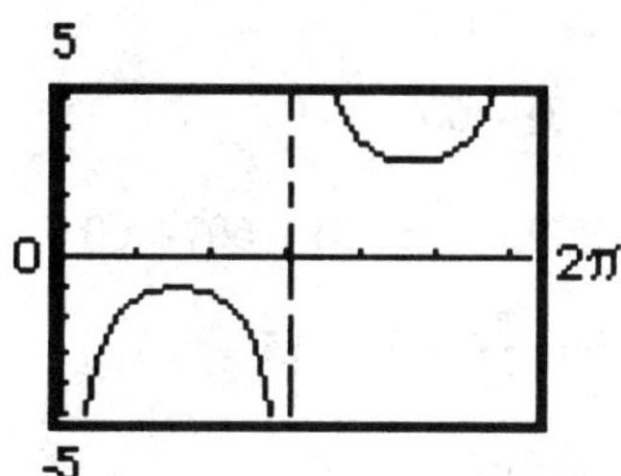

41.

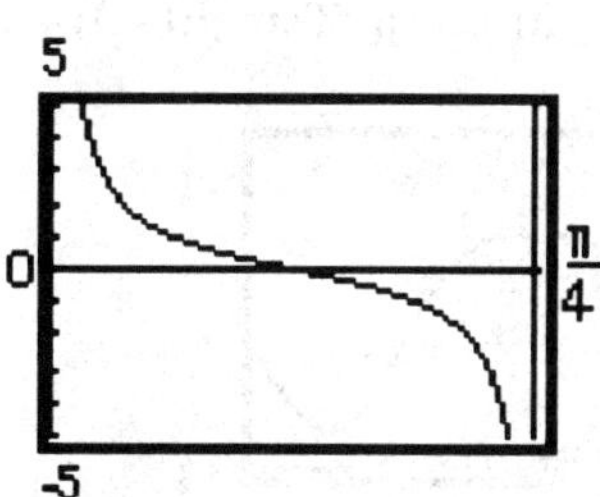

42.

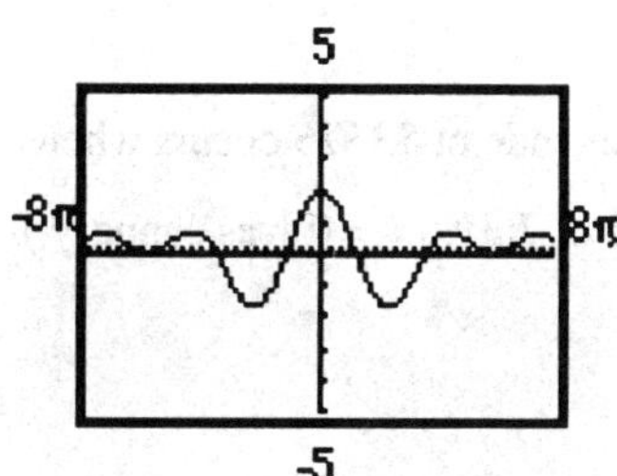

43.

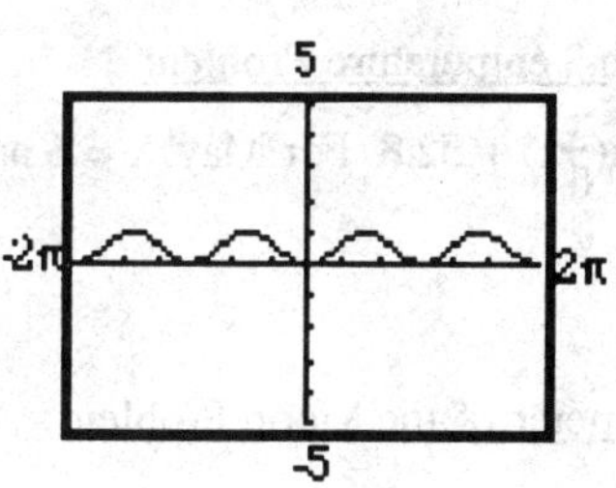

44.(a) $\theta \approx 60°$ and $\theta \approx 120°$

(b) $\theta = 240°$ and $\theta = 300°$

45.(a) $\theta \approx 30°$ and $\theta \approx 330°$

(b) $\theta \approx 150°$ and $\theta \approx 210°$

46. $x = \sin^{-1} y$ requires all values of y whose sine is x. $x = \text{Sin}^{-1} y$ requires only the principal (calculator) value of y whose sine is x.

47.(a) $x \approx 0.8 + 2k\pi$ or $x \approx 5.5 + 2k\pi$, k an integer

(b) $x \approx 0.8$ (c) $x \approx 0.8$ or $x = 2\pi - 0.8 \approx 5.5$

48. $x \approx 1.21$ **49.** $x \approx 2.9$

50. The Classroom Board Problem

about 48.4 in.

Chapter 5 Test (Page 417)

1.(a) < (b) = (c) > (d) <

2.(a) $7\pi/4 \approx 5.5$ (b) -225°

3. (-1,0)

4. A = 30°, a = 1/2, $b = \sqrt{3}/2$

5. A = 26.6°, B = 63.4°, $c \approx 4.47$

6. B **7.** D

8. C **9.** A

10. The Golfers' Dilemma Problem

No. The other golfer is only 73 yd away.

11. The Solar Orbit Problem

(a) 1 rev/yr (b) 66,700 mi/hr

12. The Household Current Problem

$V = 120 \sin 120\pi t$

13. The High-Temperature Problem

$y = -12.7\sin(\frac{\pi}{6}x) + 52.8$ For May, $x = 6$ and

$y = 46.5°F$.

14. The Diameter of the Moon Problem

$\alpha \approx 0.56°$ and $\beta \approx 0.49°$

15. The Golf Ball Problem Revisited

(a) $R = 450\sin 2\theta$

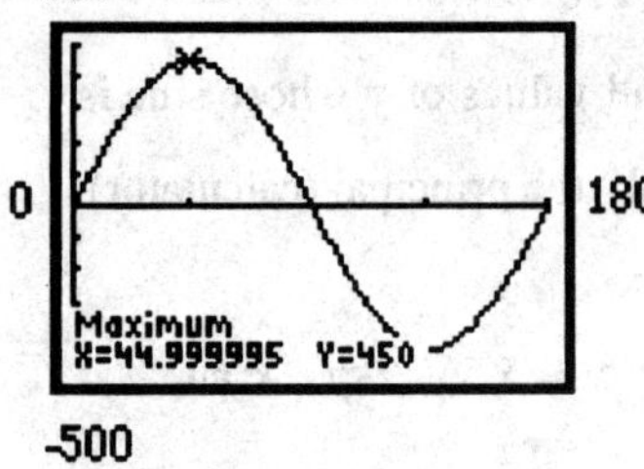

(b) about 45°

(c) 450 ft

16. The Art Exhibit Problem

(a) $\theta = \tan^{-1}(\frac{36}{x}) - \tan^{-1}(\frac{12}{x})$; As $x \to 0$, θ increases to a maximum of 29.8° when $x = 20.8$ ft, then θ decreases to zero.

(b) $\theta = \tan^{-1}(\frac{48}{x})$; As $x \to 0$, $\theta \to 90°$, which is the maximum.

17. The Inexpensive Gutters Problem

(a) $A = 18\sin\theta\cos\theta$, about 45°

(b) $A = 24\sin\theta + 8\sin\theta\cos\theta$, about 74°

18. The Season Problem

(a) $p = -0.25q + 60$

(b) $q = 50\sin\frac{\pi}{6}(t - 3) + 60$

(c) $p(q(t)) = -0.25[50\sin\frac{\pi}{6}(t - 3) + 60] + 60$

$= -12.5\sin\frac{\pi}{6}(t - 3) + 45$

(d) R = (price)(quantity)

$R = [-12.5\sin\frac{\pi}{6}(t - 3) + 45][50\sin\frac{\pi}{6}(t - 3) + 60]$

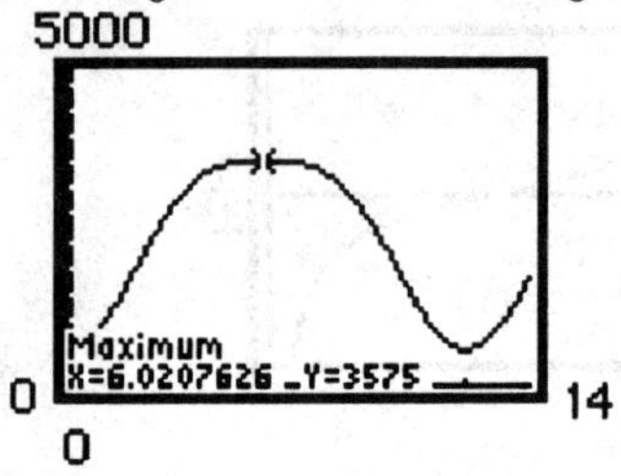

The maximum revenue of $3575 occurs when x = 6, the month of July. (x = 0 was January.)

Section 6.1 Trigonometric Identities (Page 427)

1. If the graph of g is reflected about the x axis and translated up 1 unit it is the graph of f.

3. If the graph of g is translated up 1 unit it is the graph of f.

5. $\sin x \cot x = \sin x \left(\frac{\cos x}{\sin x}\right) = \cos x \therefore \sin x \cot x = \cos x$

7. $\tan x \sin x \cos x + \sin x \cos x \cot x = \frac{\sin x}{\cos x} \cdot \sin x \cos x + \sin x \cos x \cdot \frac{\cos x}{\sin x} = \sin^2 x + \cos^2 x = 1$

$\therefore \tan x \sin x \cos x + \sin x \cos x \cot x = 1$

9. $\sin 2x = 2\sin x \cos x \neq 2\sin x$ We suspect that this is not an identity and find a counterexample.
For $x = \frac{\pi}{2}$, $\sin 2x = 0$ and $2\sin x = 2$ Therefore, $\sin 2x \neq 2\sin x$.

11. $\frac{1}{\sec^2 x} + \frac{1}{\csc^2 x} = \cos^2 x + \sin^2 x = 1 \therefore \frac{1}{\sec^2 x} + \frac{1}{\csc^2 x} = 1$

13. $\frac{\sin x}{1 + \cos x} = \frac{\sin x}{1 + \cos x} \cdot \frac{1 - \cos x}{1 - \cos x} = \frac{\sin x(1 - \cos x)}{\sin^2 x} = \frac{1 - \cos x}{\sin x} \therefore \frac{\sin x}{1 + \cos x} = \frac{1 - \cos x}{\sin x}$

15. $\sin x \tan x + \cos x = \frac{\sin^2 x}{\cos x} + \cos x = \frac{\sin^2 x + \cos^2 x}{\cos x} = \frac{1}{\cos x} = \sec x \therefore \sin x \tan x + \cos x = \sec x$

17. $\frac{\sec x \cot x \sin x}{\cos x \csc x} = \frac{\frac{1}{\cos x}\frac{\cos x}{\sin x}\sin x}{\cos x \frac{1}{\sin x}} = \frac{1}{\frac{\cos x}{\sin x}} = \frac{\sin x}{\cos x} = \tan x \therefore \frac{\sec x \cot x \sin x}{\cos x \csc x} = \tan x$

19. $\sec^2 x \neq \tan x + 1$

21. $\frac{\cos^2 x}{\sin x(1 + \csc x)} = \frac{1 - \sin^2 x}{\sin x + \sin x \csc x} = \frac{(1 + \sin x)(1 - \sin x)}{\sin x + 1} = 1 - \sin x$
$\therefore \frac{\cos^2 x}{\sin x (1 + \csc x)} = 1 - \sin x$

23. $\frac{\tan x}{\sec x - 1} = \frac{\tan x (\sec x + 1)}{(\sec x + 1)(\sec x - 1)} = \frac{\tan x (\sec x + 1)}{\sec^2 x - 1} = \frac{\tan x (\sec x + 1)}{\tan^2 x} =$
$\frac{\sec x + 1}{\tan x} \therefore \frac{\tan x}{\sec x - 1} = \frac{\sec x + 1}{\tan x}$

25. $\frac{1 - \cos x}{\sin x} + \frac{\sin x}{1 - \cos x} = \frac{1 - \cos x}{\sin x}\frac{1 - \cos x}{1 - \cos x} + \frac{\sin x}{1 - \cos x} \cdot \frac{\sin x}{\sin x} = \frac{1 - 2\cos x + \cos^2 x + \sin^2 x}{\sin x(1 - \cos x)}$
$= \frac{1 - 2\cos x + 1}{\sin x(1 - \cos x)} = \frac{2(1 - \cos x)}{\sin x(1 - \cos x)} = \frac{2}{\sin x} \therefore \frac{1 - \cos x}{\sin x} + \frac{\sin x}{1 - \cos x} = \frac{2}{\sin x}$

27. $(\sin x + \cos x)^2 + (\sin x - \cos x)^2 = \sin^2 x + 2\sin x \cos x + \cos^2 x + \sin^2 x - 2\sin x \cos x + \cos^2 x = 2$
$\sin^2 x + 2\cos^2 x = 2(\sin^2 x + \cos^2 x) = 2 \cdot 1 = 2 \therefore (\sin x + \cos x)^2 + (\sin x - \cos x)^2 = 2$

29. $\sin^2 x + 2\sin x \cos x + \cos^2 x \neq 1$

31. From the graphs, we assume that the pair of functions is identical.

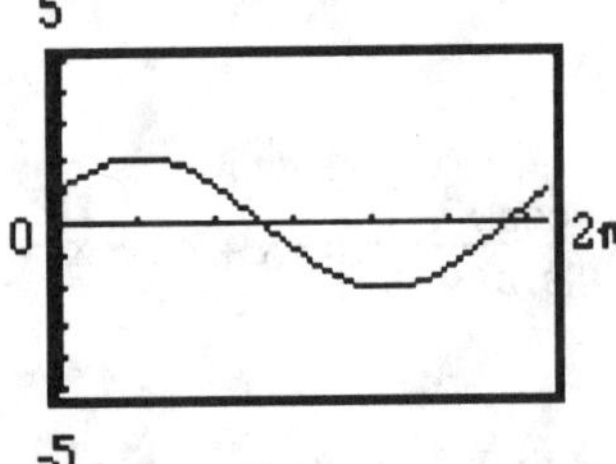

33. identical

35. odd

37. even

39.(a) $y = \sin(-x)$, $y = -\sin x$, $y = -\sin(-x)$

(b) $\sin(-x) = -\sin x$, $\sin x = -\sin(-x)$

41.(a) $y = \tan(-x)$, $y = -\tan x$, $y = -\tan(-x)$

(b) $\tan(-x) = -\tan x$, $\tan x = -\tan(-x)$

43. (a) and (d), (b) and (c)

45. (a) and (c), (b) and (d)

47. identical **49.** identical

Section 6.2 More Trigonometric Identities (Page 436)

1. $$\frac{\cos(\alpha - \beta)}{\cos(\alpha + \beta)} = \frac{\cos\alpha\cos\beta + \sin\alpha\sin\beta}{\cos\alpha\cos\beta - \sin\alpha\sin\beta} = \frac{\dfrac{\cos\alpha\cos\beta}{\cos\alpha\cos\beta} + \dfrac{\sin\alpha\sin\beta}{\cos\alpha\cos\beta}}{\dfrac{\cos\alpha\cos\beta}{\cos\alpha\cos\beta} - \dfrac{\sin\alpha\sin\beta}{\cos\alpha\cos\beta}} = \frac{1 + \tan\alpha\tan\beta}{1 - \tan\alpha\tan\beta}$$

3. $$\cot(x - y) = \frac{1}{\tan(x - y)} = \frac{1 + \tan x \cdot \tan y}{\tan x - \tan y} = \frac{1 + \dfrac{1}{\cot x \cot y}}{\dfrac{1}{\cot x} - \dfrac{1}{\cot y}} = \frac{\cot x \cdot \cot y + 1}{\cot y - \cot x}$$

5. $$\frac{\sin 2x}{\sin x} - \frac{\cos 2x}{\cos x} = \frac{2\sin x \cdot \cos^2 x - 2\cos^2 x \cdot \sin x - \sin x}{\sin x \cdot \cos x} = \sec x$$

7. $$\csc 2x + \cot 2x = \frac{1}{\sin 2\theta} + \frac{\cos 2x}{\sin 2x} = \frac{2\cos^2 x}{\sin 2x} = \frac{2\cos^2 x}{2\sin x \cdot \cos x} = \cot x$$

9. $$\frac{\sin 3x - \sin x}{\cos 3x + \cos x} = \frac{\sin(2x + x) - \sin x}{\cos(2x + x) + \cos x} = \frac{3\sin x \cdot \cos^2 x - \sin^3 x - \sin x}{-3\sin^2 x \cdot \cos x + \cos^3 x + \cos x} =$$
$$\frac{\sin x(3\cos^2 x - \sin^2 x - 1)}{-\cos x(3\sin^2 x - \cos^2 x - 1)} = \frac{\sin x(4 - 4\sin^2 x - 2)}{-\cos x(4\sin^2 x - 2)} = \tan x$$

11. $4\sin x\cos^2 x - \sin 3x = 4\sin x\cos^2 x - \sin(2x + x) = 4\sin x\cos^2 x - [\sin 2x\cos x + \cos 2x\sin x] =$
$4\sin x\cos^2 x - 2\sin x\cos^2 x - \sin x(2\cos^2 x - 1) = 2\sin x\cos^2 x - 2\sin x\cos^2 x + \sin x = \sin x$

13. By the double angle formula, $2\sin 2x\cos 2x = \sin 4x$. Therefore, $\cos 2x = \dfrac{\sin 4x}{2\sin 2x}$.

15.(a) $$\tan\frac{x}{2} = \frac{\sin\frac{x}{2}}{\cos\frac{x}{2}} = \frac{\sqrt{1 - \cos x}}{2} \div \frac{\sqrt{1 + \cos x}}{2} = \sqrt{\frac{1 - \cos x}{1 + \cos x}} \cdot \sqrt{\frac{1 + \cos x}{1 + \cos x}} = \frac{\sin x}{1 + \cos x}$$

(b) $$\frac{\sin x}{1 + \cos x} = \frac{\sin x}{1 + \cos x} \cdot \frac{1/\cos x}{1/\cos x} = \frac{\tan x}{1 + \sec x}$$

17. (a) $\cot \frac{x}{2} = \frac{\cos \frac{x}{2}}{\sin \frac{x}{2}} = \frac{\sqrt{1 + \cos x}}{2} \div \frac{\sqrt{1 - \cos x}}{2} = \sqrt{\frac{1 + \cos x}{1 - \cos x}} \cdot \sqrt{\frac{1 - \cos x}{1 - \cos x}} = \frac{\sin x}{1 - \cos x}$

(b) $\frac{\sin x}{1 - \cos x} = \frac{\sin x}{1 - \cos x} \cdot \frac{1/\sin x}{1/\sin x} = \frac{1}{\csc x - \cot x}$

19. $\frac{2 \sin^2 \frac{x}{2}}{\sin^2 x} = \frac{2\left(\frac{1 - \cos x}{2}\right)}{\sin^2 x} = \frac{1 - \cos x}{1 - \cos^2 x} = \frac{1}{1 + \cos x}$

21. (a) $\frac{\sin 2x}{1 + \cos 2x} = \frac{2 \sin x \cos x}{1 + \cos^2 x - \sin^2 x} = \frac{2 \sin x \cos x}{2 \cos^2 x} = \tan x$

(b) $\frac{\sin 2x}{1 - \cos 2x} = \frac{2 \sin x \cos x}{1 - \cos^2 x + \sin^2 x} = \frac{2 \sin x \cos x}{2 \sin^2 x} = \cot x$

23. $f(t) = \sin (t + \pi/4)$

25. $\sin x + \sin y = 2 \sin \frac{x + y}{2} \cos \frac{x - y}{2}$ $u = \frac{x + y}{2}$ and $v = \frac{x - y}{2} \Rightarrow$

$x = x + v$ and $y = u - v$. Therefore, $\sin (u + v) + \sin (u - v) = 2 \sin u \cos v$.

27. $\tan (x - y) = \frac{\tan x - \tan y}{1 + \tan x \tan y}$ and $\tan \theta = \beta - \alpha$, where $\tan \beta = \frac{6.5}{x}$ and

$\tan \alpha = \frac{4}{x}$ $\tan \theta = \tan (\beta - \alpha) = \frac{6.5/x - 4/x}{1 + 6.5/x \cdot 4/x} \cdot \frac{x^2}{x^2} \Rightarrow \theta = \frac{6.5x - 4x}{x^2 + 26} = \frac{2.5x}{x^2 + 26}$

29. $f(x) = \sin x \cos x$

$f(x) = \frac{1}{2}(2 \sin x \cos x)$

$f(x) = \frac{1}{2} \sin 2x$

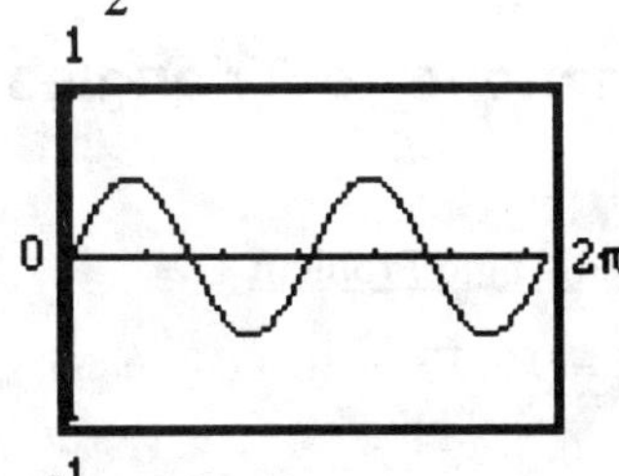

31. $f(x) = \cos 2x$

33. $f(x) = 1/2 - (1/2) \cos x$

35. $f(x) = 1/2 - (1/2) \cos 4x$

37. $(x) = \cos (x - \frac{\pi}{4})$

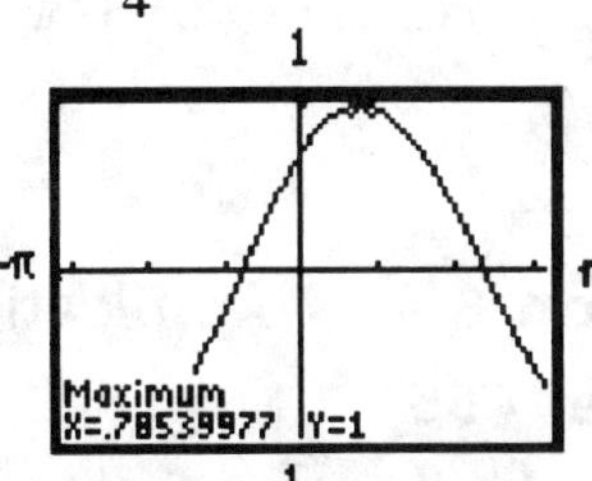

39. $f(x) = \cos (x - \pi/3)$

41. $x = 0$

43. $x = 0$

45. The graph of $y = \tan \frac{x}{2}$ is identical to that of

$y = \frac{\sin x}{1 + \cos x}$.

The other part, $y = -\frac{\sqrt{1 - \cos^2 x}}{1 + \cos x} = \frac{-\sin x}{1 + \cos x}$

Section 6.3 Equations Involving Circular Functions (Page 443)

1. $x \approx 1.53$

3. $x = 0$ or $x = \pi/2 \approx 1.57$

5. $x = 0$ or $x = \pi/4 \approx 0.78$

7. $x = 0$ or $x = \pi/2 \approx 1.57$

9. $x = \pi/3 \approx 1.047$ or $x = -\pi/3 \approx -1.047$ or $x = \pi/4 \approx 0.785$

11. $x = \frac{\pi}{4} \approx 0.785$ or $x = \frac{3\pi}{4} \approx 2.356$ or $x = 0$ or $x = \pi \approx 3.14$

13. $x \approx -0.64$ or $x = \pi/4 \approx 0.785$

15. $x = \pi/3 \approx 1.047$

17. $x = \pi/2 \approx 1.57$ or $x = \pi/6 \approx 0.52$

19. $x = 0$ or $x = \pi \approx 3.14$

21. $x = 0$ or $x = \pi \approx 3.14$

23. $x = \pi/2 \approx 1.57$ or $x = 3\pi/2 \approx 4.71$ or $x = \pi/6 \approx 0.52$ or $x = 5\pi/6$

25. $x = \pi/6 \approx 0.52$ or $x = 5\pi/6$ or $x = 3\pi/2 \approx 4.71$

27. true for all x

29. $x \approx -0.67 + 2k\pi$

31. $x \approx 0.41 + 2k\pi$ or $x \approx 2.73 + 2k\pi$ or $x \approx 0.52 + 2k\pi$ or $x \approx 2.62 + 2k\pi$

33. $(\frac{\pi}{2}, 0)$, $(\frac{5\pi}{6}, 1.5)$, $(\frac{7\pi}{6}, 1.5)$

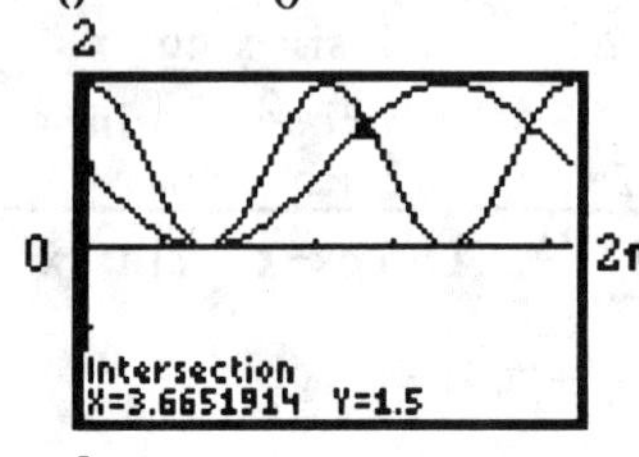

35. $(0, 0)$, $(\pi, 0)$

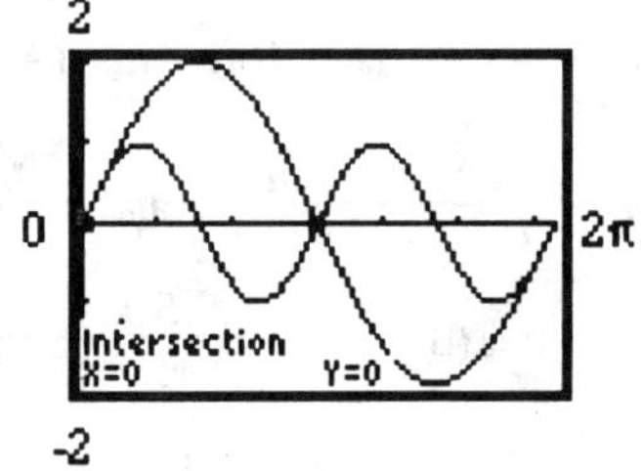

37. $x = 1/4$

39. $y = \pi/2$ or $x = \pi/6$

Section 6.4 Applications of the Circular Functions to Triangles (Page 455)

1. $b \approx 69.9$ **3.** $c \approx 28.1$

5. $c \approx 286.7$ **7.** $c \approx 334.6$ or $c \approx 50.8$

9. $B \approx 64.9°$, $C \approx 75.1°$, $a \approx 19.88$

11. $B \approx 30°$, $C \approx 140°$, $c \approx 370.2$ or $B \approx 150°$, $C \approx 20°$, $c \approx 196.96$

13. no solution

15. $A \approx 27.15°$, $C \approx 102.85°$, $a \approx 65.53$ or $A \approx 52.85°$, $C \approx 77.15°$, $a \approx 114.45$

17. $A = 10$ units^2 **19.** $A \approx 11.6$ units^2

21. The Trisected Angle Problem

CD ≈ 10.4, DE ≈ 9.2, EF ≈ 10.4

23. The Inscribed Octagon Problem

about 282.8 cm^2

25. The Racing-Sloop Problem

21.6 na mi/hr

27. The Rainbow Kite and Balloon Company

90°

Section 6.5 More Applications IncludingSimple Harmonic Motion and Vectors (Page 469)

1.(a) (8,16) (b) (8,16)

3.(a) 5 and 13 (b) $\sqrt{320} \approx 17.89$ and 18

5.(a) $\|\mathbf{v}_1\| = 13$, $\theta = \tan\text{-}1(12/5) \approx 1.18$

(b) $\|\mathbf{v}_2\| = 13$,

$\theta = \tan\text{-}1(-5/-12) \approx 0.39 + 1.18 \approx 3.53$

7.(a) $\|\mathbf{v}_1\| = 1$, $\theta = \pi/4 \approx 0.79$ (b) $\|\mathbf{v}_2\| = 1$, $\theta = 5\pi/4 \approx 3.93$

9.(a) a = 1, b = 0 (b) a = 0, b = 1

11.(a) a ≈ 129.67, b ≈ 201.95 (b) a ≈ -99.88, b ≈ 218.23

13. The Musical Fifth Problem

(a) $y = \sin 523.2\pi t$ (b) $y = \sin 768\pi t$

(c) $y = \sin 523.2\pi t$ $y = \sin 768\pi t$ $y = \sin 523.2\pi t + \sin 768\pi t$

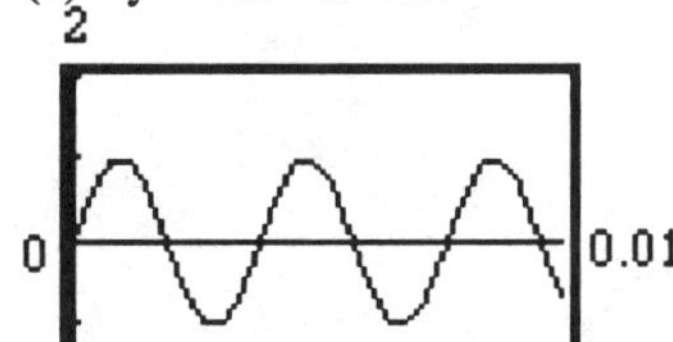

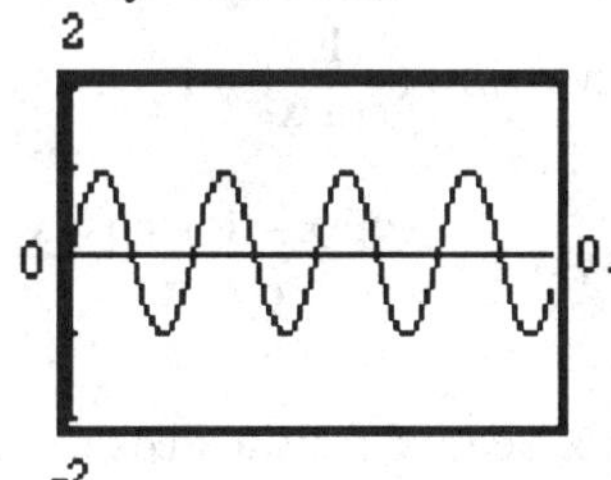

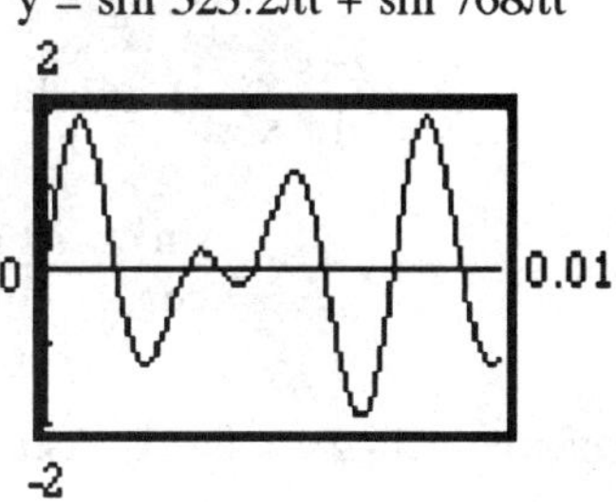

15. The Beating Guitar Problem Revisited

$y = \sin 329\pi t + \sin 320\pi t = 2\sin 324.5\pi t \sin 4.5\pi t$.

period of the beat: $2\pi/4.5\pi = 0.4$

17. The Wagon Pulling Problem

$\theta = \pi/6$: a ≈ 86.6, b = 50

$\theta = \pi/3$: a = 50, b ≈ 86.6

The horizontal component of the force is 50 lb and the vertical component of the force is about 86.6 lb, The horizontal component causes the wagon to roll forward, while the vertical component tries to lift the wagon off the ground.

19. The Rainbow Kite and Balloon Company

162.24 lb at an angle of 15.4°

21. The Santa Claus Problem

275.95°

Chapter 6 Review (Page 472)

1. Translating the graph of 2g down 1 unit yields the graph of f.

2. Reflecting the graph of 2h about the x axis and translating the result down 1 unit yields the graph of f.

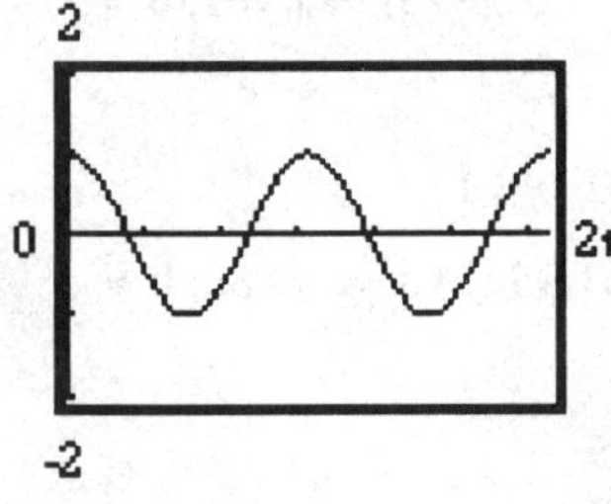

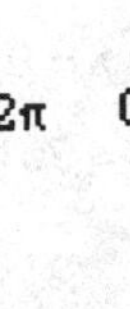

$f(x) = \cos 2x$

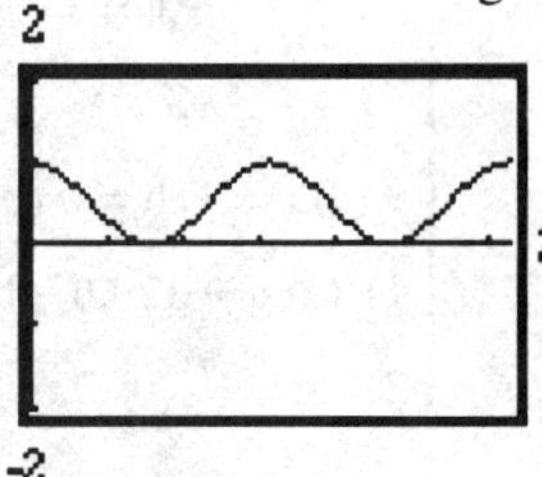

$g(x) = \cos 2x$

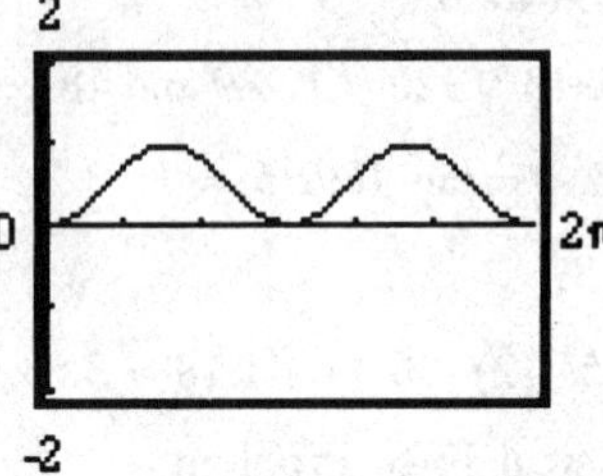

$h(x) = \sin^2 x$

3. $\tan x \cot x = \left(\frac{\sin x}{\cos x}\right)\left(\frac{\cos x}{\sin x}\right) = 1$ and $\sin x \csc x = \sin x \left(\frac{1}{\sin x}\right) = 1.$

4. $\tan x \sin x - \sec x + \cos x = \frac{\sin^2 x}{\cos x} - \frac{1}{\cos x} + \cos x = \frac{\sin^2 x - 1 + \cos^2 x}{\cos x} = 0$

5. $(\sin x - \cos x)^2 + \sin 2x = \sin^2 x - 2\sin x \cos x + \cos^2 x + 2 \sin x \cos x = 1$

6. $\tan x + \cot x = \frac{\sin x}{\cos x} + \frac{\cos x}{\sin x} = \frac{\sin^2 x + \cos^2 x}{\sin x \cos x} = \frac{1}{\sin x \cos x} = \sec x \csc x$

7. They are identical.

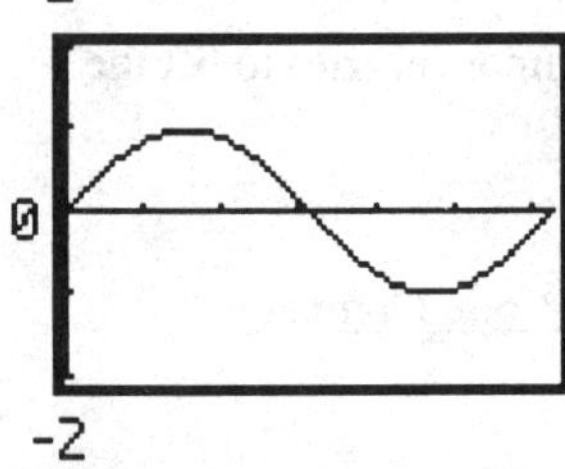

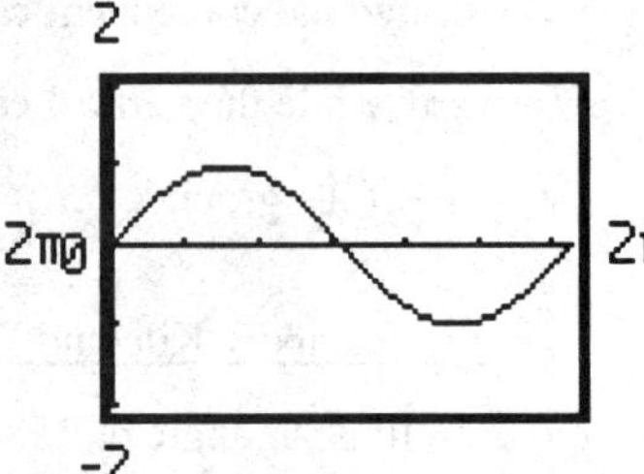

8. They are identical.

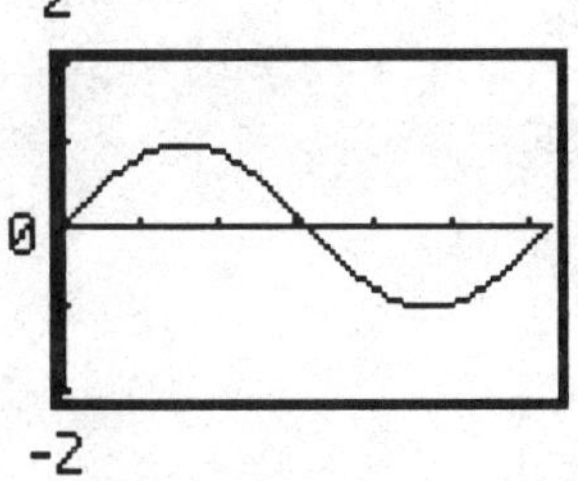

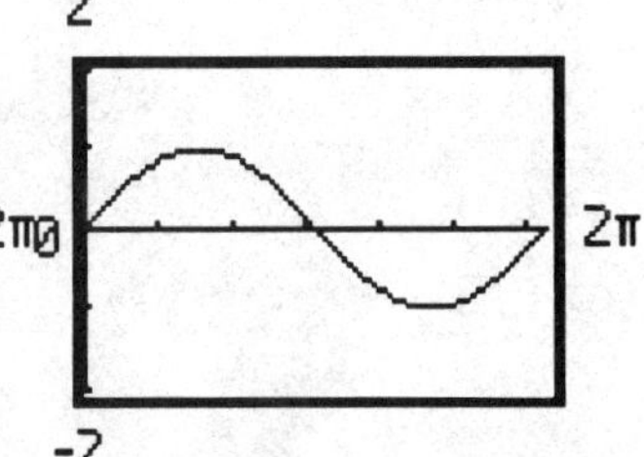

9. $\sin(\alpha + \beta) + \sin(\alpha - \beta) = \sin\alpha \sin\beta + \cos\alpha \cos\beta + \sin\alpha \sin\beta - \cos\alpha \cos\beta = 2 \sin\alpha \sin\beta$

10. $\frac{\cos(\alpha - \beta)}{\sin(\alpha - \beta)} = \frac{\sin\alpha \cos\beta + \cos\alpha \sin\beta}{\sin\alpha \sin\beta - \cos\alpha \cos\beta} \cdot \frac{\frac{1}{(\sin\alpha \sin\beta)}}{\frac{1}{(\sin\alpha \sin\beta)}} = \frac{\cot\alpha \cot\beta - 1}{\cot\beta + \cot\alpha}$

11. $\dfrac{1 - \tan x}{1 + \tan x} = \dfrac{1 - \sin x/\cos x}{1 + \sin x/\cos x} \cdot \dfrac{\cos x}{\cos x} = \dfrac{\cos x - \sin x}{\cos x + \sin x} \cdot \dfrac{\cos x - \sin x}{\cos x - \sin x} = \dfrac{1 - \sin 2x}{\cos 2x}$

12. $\cot x - \tan x = \dfrac{\cos x}{\sin x} - \dfrac{\sin x}{\cos x} = \dfrac{\cos^2 x - \sin^2 x}{\sin x \cos x}$ and $\dfrac{2}{\tan 2x} = \dfrac{2\cos 2x}{\sin 2x} =$
$2\dfrac{\cos^2 x - \sin^2 x}{2\sin x \cos x} = \dfrac{\cos^2 x - \sin^2 x}{\sin x \cos x}$ Therefore, $\cot x - \tan x = \dfrac{2}{\tan 2x}$

13. $\cos(\alpha + \beta)\cos(\alpha - \beta) = [\sin\alpha\cos\beta - \cos\alpha\sin\beta][\sin\alpha\cos\beta + \cos\alpha\sin\beta] =$
$\sin^2\alpha\cos^2\beta - \cos^2\alpha\sin^2\beta = (1 - \cos^2\alpha)(1 - \sin^2\beta) - \cos^2\alpha\sin^2\beta =$
$1 - \cos^2\alpha - \sin^2\beta + \cos^2\alpha\sin^2\beta - \cos^2\alpha\sin^2\beta = \sin^2\alpha - \sin^2\beta.$

14. $-2\sin\left(\dfrac{u + v}{2}\right)\sin\left(\dfrac{u - v}{2}\right) = -2[\sin(u/2)\cos(v/2) + \cos(u/2)\sin(v/2)][\sin(u/2)\cos(v/2) - \cos(u/2)\sin(v/2)] =$
$-2[\sin^2(u/2)\cos^2(v/2) - \cos^2(u/2)\sin^2(v/2)] = -2\left[\dfrac{1 - \cos u}{2}\,\dfrac{1 + \cos v}{2} - \dfrac{1 + \cos u}{2}\,\dfrac{1 - \cos v}{2}\right] =$
$(-1/2)[1 - \cos u + \cos v - \cos u\cos v - 1 - \cos u + \cos v + \cos u\cos v] = \cos u - \cos v$

15. $x = 0$ or $x = \pi$

16. $x \neq \pi/2$

17. $x \approx \pm 1.94$

18. $x = \pi/8$, $5\pi/6$ or $x = 5\pi/8$

19. $x = 0$ or $x = \pi$ or $x = \pi/3 \approx 1.047$ or
$x = -\pi/3 \approx -1.047$

20. $x = 0.33$ or $x = 1.23$

21. $x = k\pi$

22. $x = 1.11 + k\pi$

23. $(0,1)$, $(\pi,1)$ $(\pi/2,0)$ $(3\pi/2)$

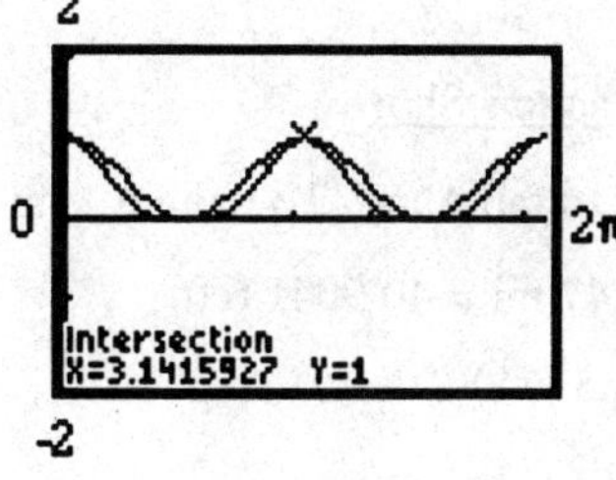

24. $(0, -1)$ $(\pi, -1)$

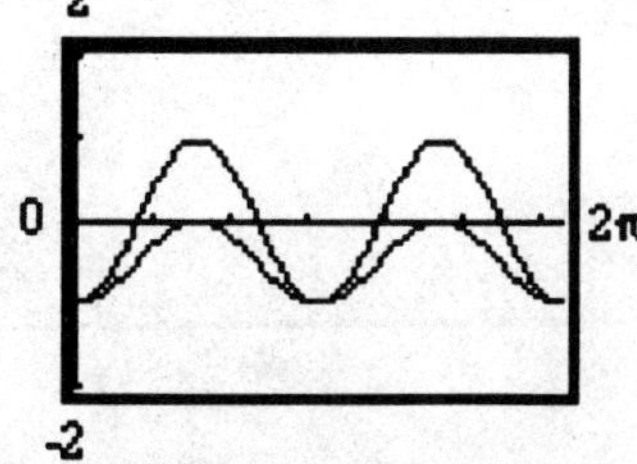

25. $A = C = 30°$, $b \approx 173$

26. $C = 60°$, $a = b = c = 17.6$

27. $A = 38.3°$, $C = 101.7°$, $b \approx 32.8$

28. $C = 90°$, $a \approx 18.0$, $b \approx 28.8$

29. <u>The Titanic II Problem</u>

The captain sailed 5.2 nautical miles off course.

30. (a) (7, 0) (b) (7, 0)

31. (a) (0, 24) (b) (0, 24)

32. (a) 25 (b) 27.8

33. (a) 7 (b) 52.8

34. $a \approx 20.78$, $b = 12$

35. $\|\mathbf{v}_x\| \approx 20.78$, $\|\mathbf{v}_y\| = -12$

36. <u>The Out-of-Tune Duet Problem Revisited</u>
$y = \sin 880\pi t + \sin 900\pi t = 2\sin 10\pi t \sin 890\pi t$
The period of the beat is 1/5.

37. <u>The English Channel Problem</u>

Head S32°E

38. <u>The Wagon Pulling Problem</u>

$\theta = \pi/6$: $\|\mathbf{v}_x\| \approx 86.60$, $\|\mathbf{v}_y\| = 50$

$\theta = \pi/3$: $\|\mathbf{v}_x\| = 50$, $\|\mathbf{v}_y\| \approx 86.60$

39. The Diameter of the Sun Problem

$d \approx 1.4 \times 10^6$ km

40. The Dead Radio Problem

either 119.8 mi or 53.4 mi away at a bearing of 30°

Chapter 6 Test (Page 475)

1. If the graph of $g(x) = \sin^2 2x$ is reflected about the x axis and translated up 1 unit, it will match the graph of $f(x) = \cos^2 2x$.

2. (a) $\tan 2x \cos 2x = \dfrac{\sin 2x}{\cos 2x} \cos 2x = \sin 2x$

(b) $\dfrac{\tan^2 x}{1 - \sec^2 x} + \cos^2 x = \dfrac{\sec^2 x - 1}{1 - \sec^2 x} + \cos^2 x = -1 + \cos^2 x = -\sin^2 x$

(c) For $x = \pi/2$, $\sin 2x = \sin(\pi) = 0$ and $2\sin x$ $2 \sin(\pi/2) = 2$. Hence, $\sin 2x \neq 2\sin x$.

(d) $\dfrac{\sin^2 2x}{1 + \cos 2x} = \dfrac{1 - \cos^2 2x}{1 + \cos 2x} = \dfrac{(1 - \cos 2x)(1 + \cos 2x)}{1 + \cos 2x} = 1 - \cos 2x = 1 - (1 - 2\sin^2 x) = 2\sin^2 x$

3. (a) $x = 0$ or $x = \pi/2$ (b) $x = 0$ or $x = 3\pi/2$

4. $x = 0.28 + 2k\pi$ or $x = \pi - 0.28 + 2k\pi$

particular solution is 0.28.

5.(a) $51\sqrt{2}$ (b) 150

6. $a = \|\mathbf{v}\| \cos\theta = 2 \cos(\pi/3) = 1$

$b = \|\mathbf{v}\| \sin\theta = 2(\sqrt{3}/2) = \sqrt{3}$

7. (a) $A = 62°$, $B = 88°$, $c = 5.66$

(b) $C = 79°$, $a \approx 19.1$, $b \approx 23.8$

8. The D.B. Cooper Caper Problem

From the airport, the parachute is 3.17 mi at a heading of 36.8°.

9. The Out-of Tune-Duet Problem

(a) $y = \sin 659.2\pi t$, $y = \sin 640\pi t$

(b) The period of the beat is about 0.2sec.

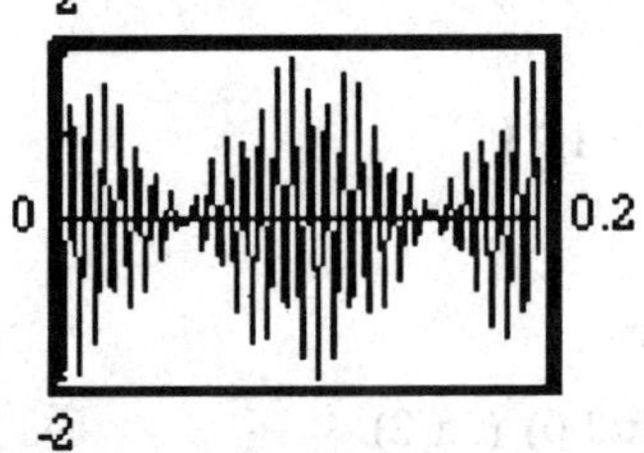

10. The Super Cannon Shot

It takes 18.17 sec to fall 1 mi. To remain 1 mi above the Earth, x = 77.47 mi = 40,9041.6 ft.

Then, $\|\mathbf{v}_x\|$ = 22,516.88 ft/sec.

Section 7.1 Introduction to Polar Coordinates (Page 490)

1. **3.**

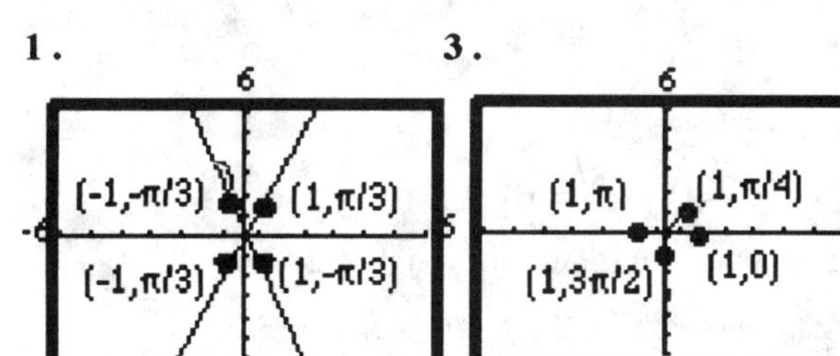

5. $(-\sqrt{29}, -0.38)$ or $(\sqrt{29}, 2.76)$

7. $(-4, 0)$ or $(4, \pi)$

9. $(\frac{\sqrt{2}}{2}, \frac{\sqrt{2}}{2})$

11. $(0, 4)$ **13.** $r = 5$

15. $r = 10 \cos \theta$ **17.** $x^2 + y^2 = 9$

19. $y = (\sqrt{3}/2)x$ **21.** $x^2 + (y - 1)^2 = 1$

23. line **25.** circle

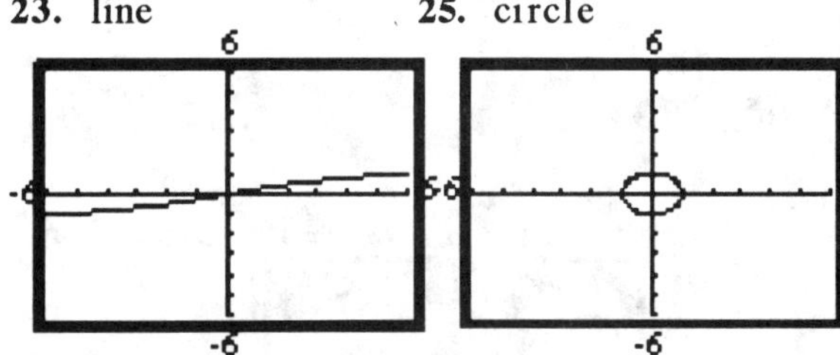

27. circle **29.** circle

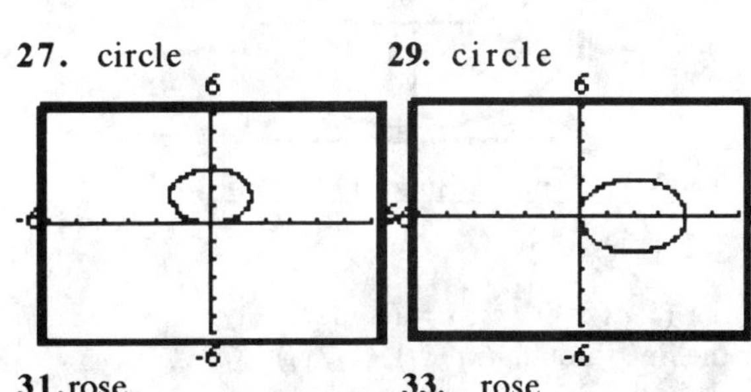

31. rose **33.** rose

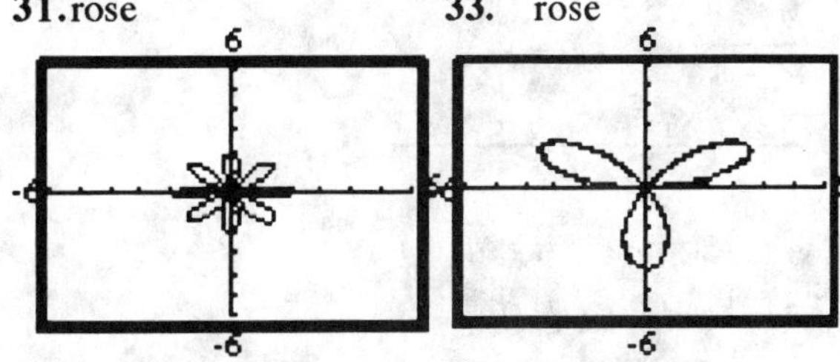

35. cardioid **37.** limacon

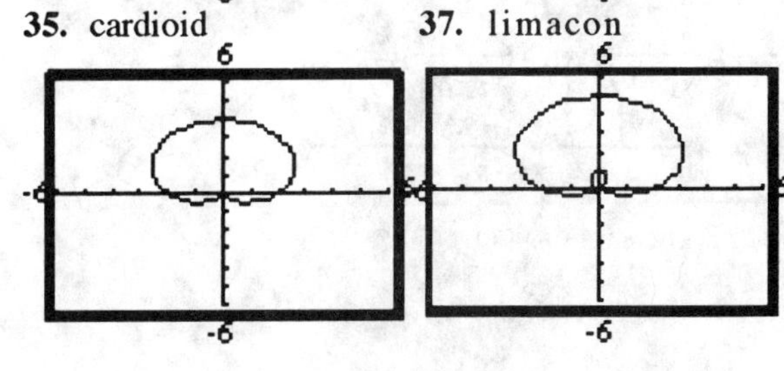

39. limacon **41.**

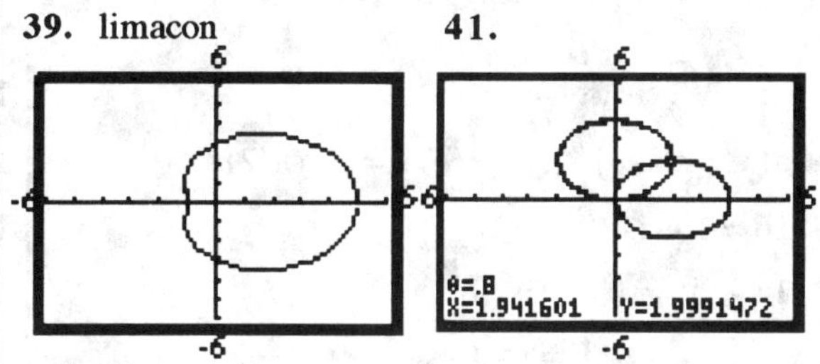

43. $(0, 0)$, $(\frac{\sqrt{3}}{2}, \frac{\pi}{3})$, and $(-\frac{\sqrt{3}}{2}, -\frac{\pi}{3})$.

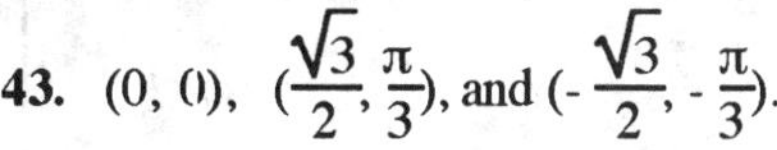

45. $(6,0)$ **47.** $(0,0),(1/2,\pi/6),(1/2,5\pi/6)$

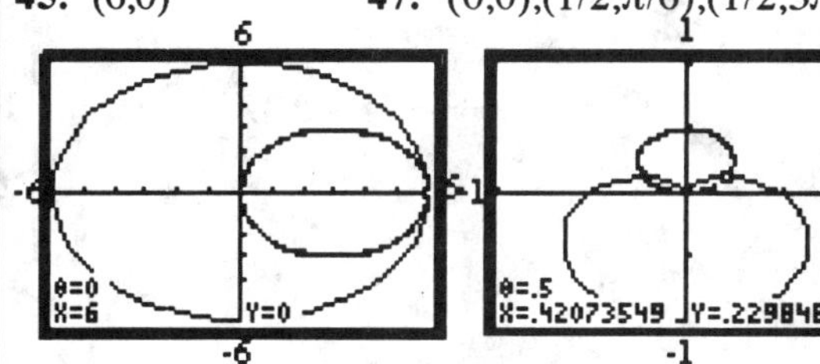

49. $(0,0)$, $(\frac{-1}{2},\frac{\pi}{3})$, $(\frac{-1}{2},\frac{5\pi}{3})$, $(1,\pi)$

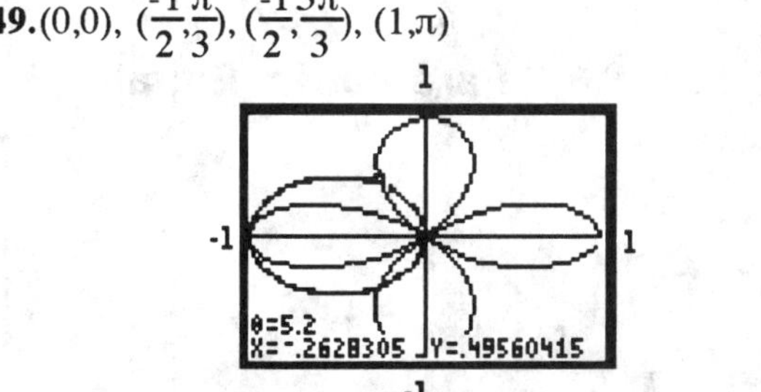

51. The Rainbow Kite and Balloon Company

$(60.5, 90^{\circ})$

53. **55.**

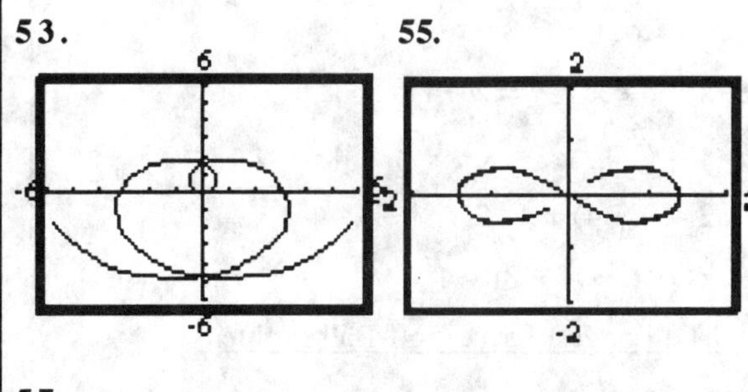

57.

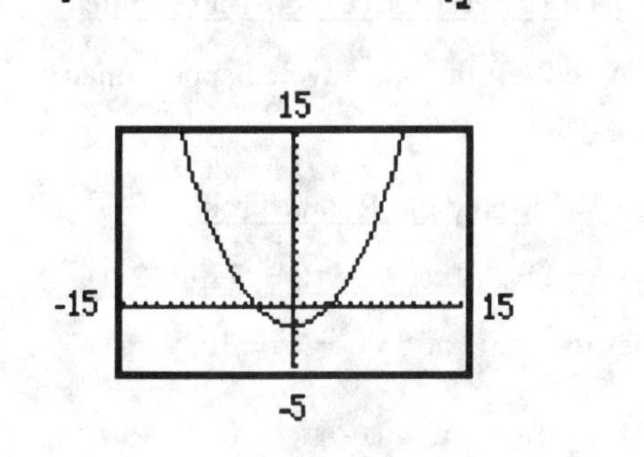

Section 7.2 Polar Form of Complex Numbers and DeMoivre's Theorem (Page 498)

1. $z = \sqrt{2}\,\text{cis}\,(\pi/4)$

3.(a) $z = \text{cis}\,(0)$ (b) $z = \text{cis}\,(\pi/2)$

5. $16 = 16\,\text{cis}\,(0)$

7. $14 - 2i = 10\sqrt{2}\,\text{cis}\,(-8°)$

9. $z = 1$ or $z = \dfrac{-1 \pm \sqrt{3}\,i}{2}$

11. $z \approx \pm 0.95 + 0.31i$, $z = i$, or $z \approx \pm 5.9 - 0.81i$

13. $z \approx 1.18 + 0.43i$, or $z \approx -0.96 + 0.81i$, or

$z \approx -0.22 - 1.24i$

15. $z \approx 0.87 + 3.25i$ or $z \approx -3.25 - 0.87i$ or

$z \approx 2.38 - 2.38i$

17. $\pm 1, \pm i$

19. $(1 + \sqrt{3}i)i = -\sqrt{3} + i$

This product is a rotation through $\pi/2$.

21. $z_1 = r\,\text{cis}\,\theta$, $z_2 = 2\,\text{cis}\,\pi/6$, $z_1/z_2 = (3/2)\text{cis}\,\dfrac{\pi}{3}$

For the quotient z_1/z_2, $r = r_1/r_2$ and $\theta = \theta_1 - \theta_2$.

23. $z = re^{i\theta} \Rightarrow \ln z = \ln re^{i\theta} = \ln r + \ln e^{i\theta}$

25. $z = i^i \Rightarrow \ln z = i \ln i$, where $i = 1 \cdot e^{(\pi/2)i} \Rightarrow$

$\ln z = i \ln e^{(\pi/2i)} \Rightarrow \ln z = -\pi/2 \Rightarrow$

$z = e^{-\pi/2} \approx 0.208 \in$ reals

Section 7.3 Parametric Equations (Page 509)

1. $x^2 + y^2 = 1$

3. $\dfrac{x^2}{4} + \dfrac{y^2}{16} = 1$

5. $(x - 1)^2 + (y - 1)^2 = 1$

7. $y = x^2$, $x > 0$

9. right half for $x \geq 0$

11. up from (1, 1)

13. $x = \cos t$, $y = \sin t$

15. $x = 3 \cos t$, $y = 2 \sin t$

17. $x = 4 - t \Rightarrow t = x - 4$, $y = 2 + t$ and $t = x - 4 \Rightarrow$

$y = 2 + (x - 4) = x - 2$

19. $x = 4t \Rightarrow y = 3t + 4$

21. The Longest Drive Golf Ball Problem

You tied. Both balls traveled approximately 79.5 m.

23. The Merry-Go-Round Problem

$x(t) = 5\cos(12\pi t)$ $y(t) = 5\sin(12\pi t)$

The ride starts at $t = 0 \Rightarrow x = 10, y = 0$.

One revolution at $t = 1/6$ min $\Rightarrow x = 10, y = 0$.

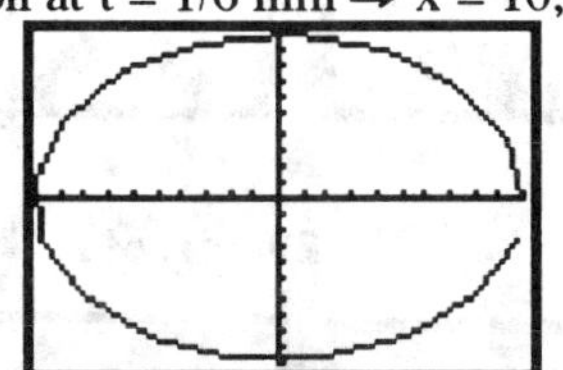

25. The Wheel and Its Hubcap Problem

$x = 2t - \sin t$, $y = 2 - \cos t$

27. For $(1 - t)x_1 + tx_2$, $y = (1 - t)y_1 + ty_2$
For the line through (x_1, y_1), (x_2, y_2),

$$y = \left(\frac{y_2 - y_1}{x_2 - x_1}\right)(x - x_1) + y_1$$

$$= \frac{xy_2 - xy_1 - x_1y_2 + x_1y_1}{x_2 - x_1} + y_1$$

$x = (1 - t)x_1 + tx_2 = x_1 - tx_1 + tx_2$

$t = \dfrac{x_2 - x_1}{x_2 - x_1}$ and $y = (1 - t)y_1 + ty_2$

$$y = \left(1 - \frac{x - x_1}{x_2 - x_1}\right)x_1 + \left(\frac{x - x_1}{x_2 - x_1}\right)y_2$$

$$y = y_1 + \frac{-xy_1 + x_1y_1 + xy_2 - x_1y_2}{x_2 - x_1}$$

which is the same as (1).

Chapter 7 Review (Page 511)

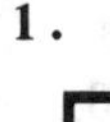

1.

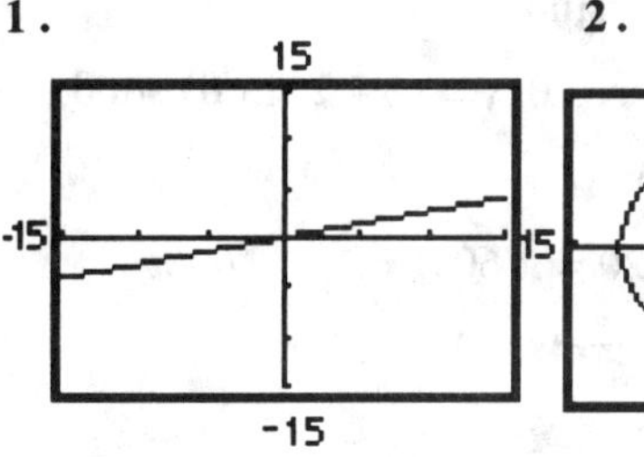

2.

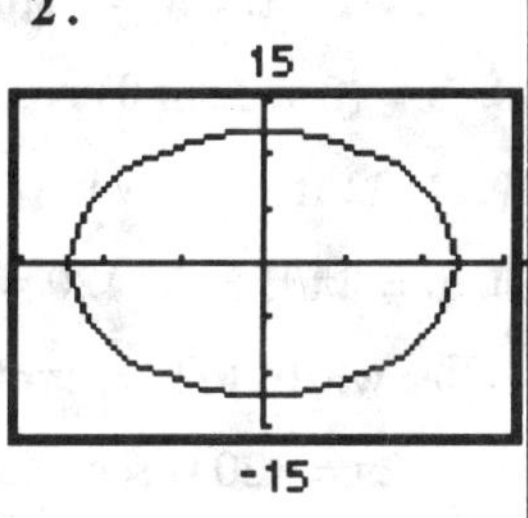

3.

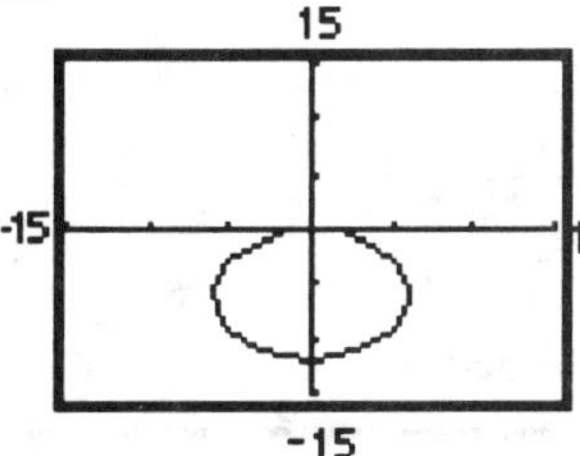

4.

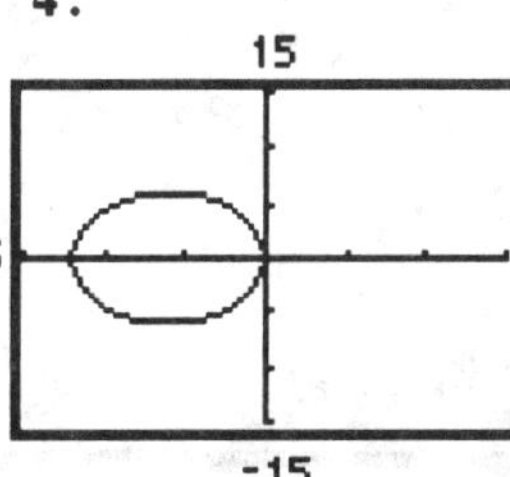
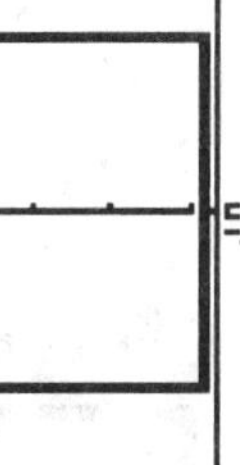

5. $\theta = -\pi/4$ 6. $\theta = -\pi/3$

7. $r = 12$ 8. $r = 2$

9. $x^2 + y^2 = 144$ 10. $x^2 + y^2 = 16$

11. $y = -\sqrt{3}\,x$ 12. $x = 0$

13. $x^2 + (y + 6)^2 = 36$ 14. $(x + 6)^2 + y^2 = 36$

15. $y = \dfrac{x^2 - 1}{2}$ 16. $y^2 = \dfrac{4x + 1}{2}$

17. line 18. spiral

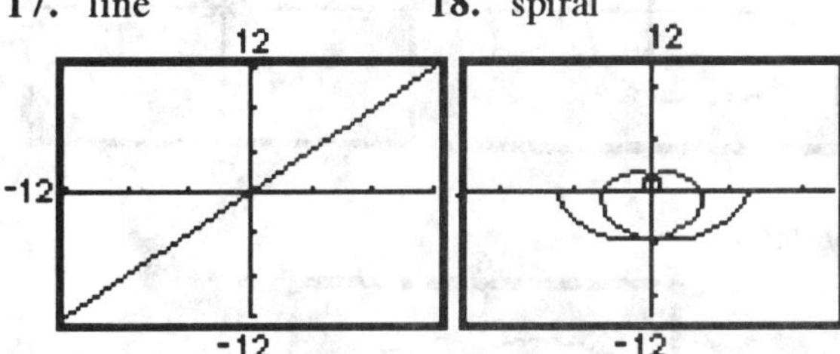
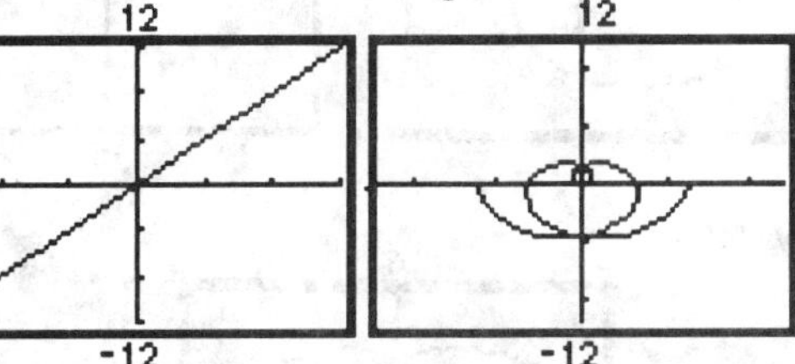

19. circle 20. circle

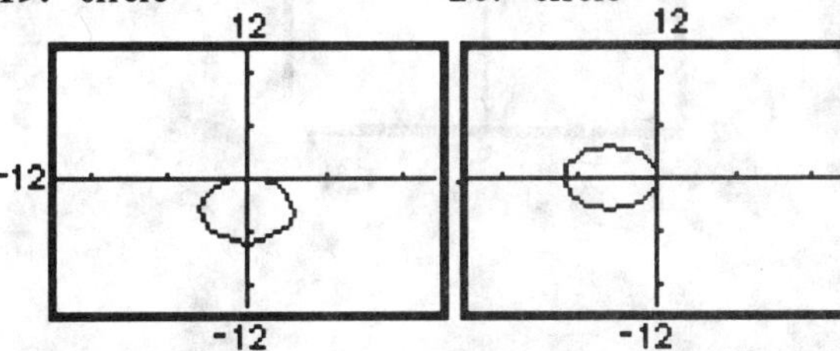

21. cardioid 22. lemniscate

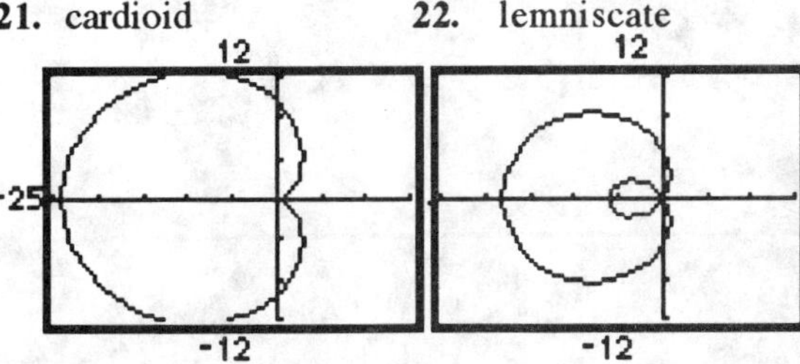

23. lemniscate 24. lemniscate

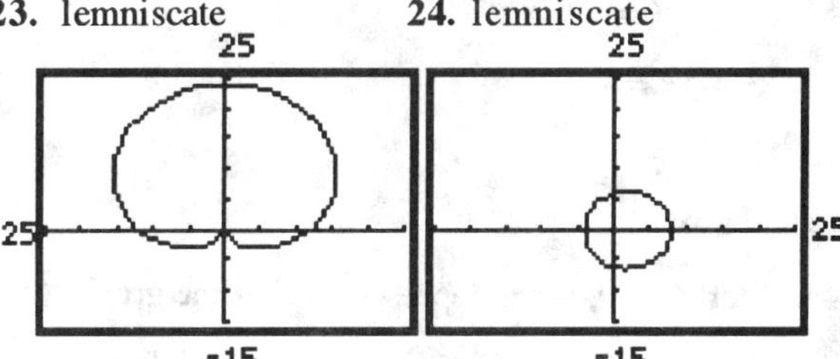

25. rose 26. rose

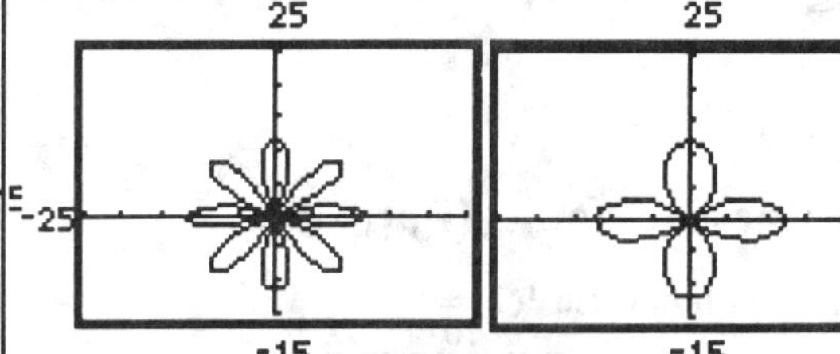

27. limacon 28. limacon

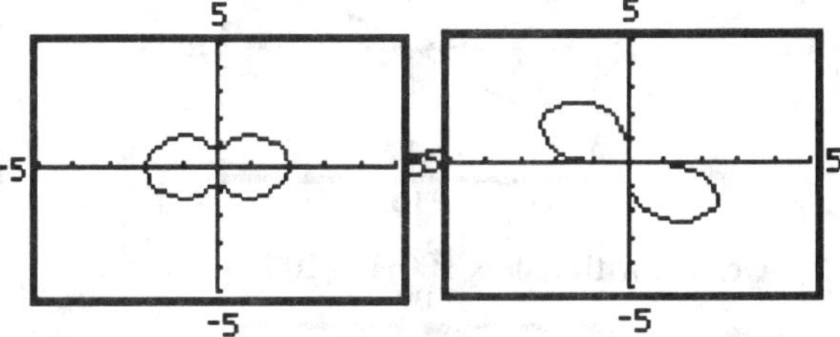

29. (0,0), $(1, 7\pi/4)$ 30 (0,0), $(\sqrt{8}, 1.23)$, $(\sqrt{8}, 4.37)$

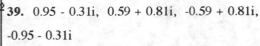

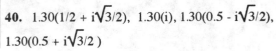

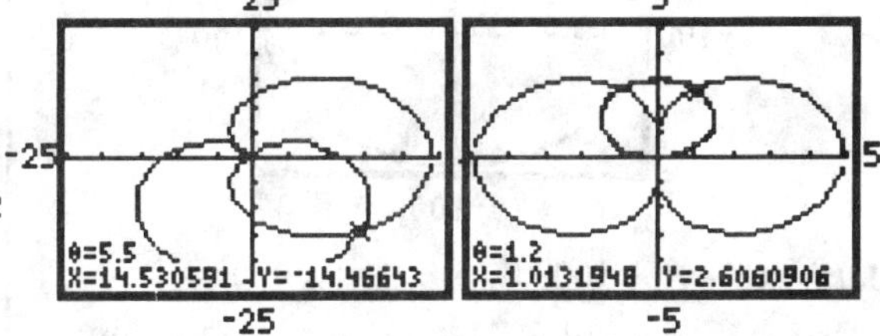

31. $r = 13\text{cis}(-1.17)$

32. $r = 13\text{cis}(1.17)$

33. (a) $r = \text{cis } 0$ (b) $r = \text{cis } \pi$

34. (a) $r = 2\text{cis}(11\pi/6)$ (b) $r = 2\text{cis}(\pi/3)$

35. $r = 4\ \text{cis}(\pi/2)$

36. $r = 4\text{cis}(-\pi/6)$

37. 5, $5(-1/2 + i\sqrt{3}/2)$, $5(-1/2 - i\sqrt{3}/2)$

38. $125\sqrt{2}\,(0.97 + 0.26i)$, $125\sqrt{2}\,(\sqrt{2}/2 - i\sqrt{2}/2)$, $125\sqrt{2}(-.26 - 0.97i)$

39. $0.95 - 0.31i$, $0.59 + 0.81i$, $-0.59 + 0.81i$, $-0.95 - 0.31i$

40. $1.30(1/2 + i\sqrt{3}/2)$, $1.30(i)$, $1.30(0.5 - i\sqrt{3}/2)$, $1.30(0.5 + i\sqrt{3}/2)$

41. $x^2 + y^2 = 1$

42. $x^2 + y^2 = 4$

43. $2x^2 + y^2 = 4$

44. $y = \sqrt{x}$

45. Starts at (0,0) and represents all of the graph

46. Starts at (1,1) and represents the graph for $0 \leq x \leq 1$ and $-1 \leq y \leq 1$.

47. $x = 2 \cos t$, $y = 2 \sin t$

48. $x = (2 + 2 \sin \theta) \cos \theta$, $y = (2 + 2 \sin \theta) \sin \theta$

49. 1632 ft = 541 yd

50. $p = 2\pi/B = 2\pi/120 = \pi/60$, $f = 1/p = 60/\pi$ rev/sec where 1 rev = $2\pi r$ ft and r = 1 ft means $f = \frac{60}{\pi} 2\pi = 120$ ft/sec.

Chapter 7 Test (Page 513)

1. A line with $m = \tan^{-1}(1) = 0.78$

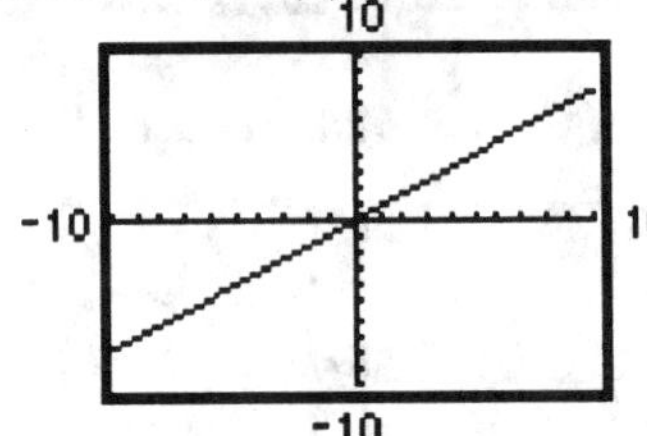

2. A circle with radius 100, r = 100

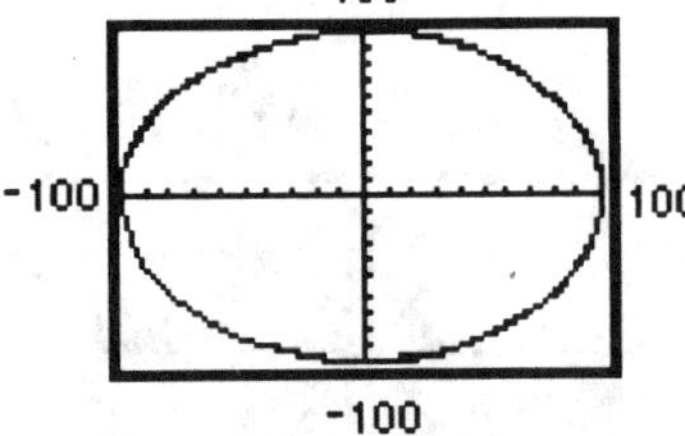

3. (a) $x^2 + y^2 = 10000$ (b) $x^2 + (y + 1)^2 = 1$

4. (a) $r = 2 \cos(-3\theta)$ (b) $r = 2 \cos(3(\frac{\pi}{4} - \theta))$

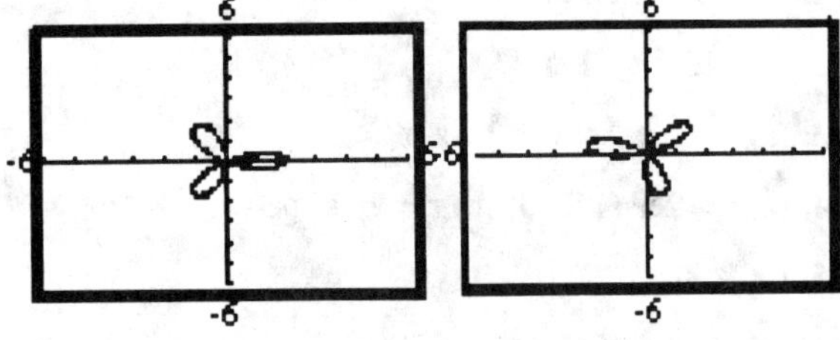

5. (a) circle (b) cardioid

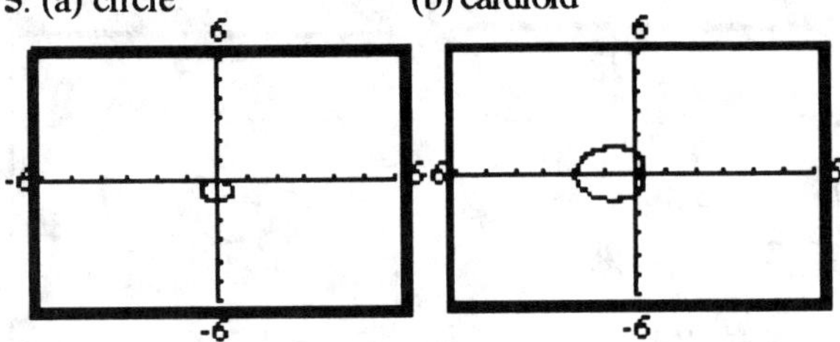

(c) convex limniscate (d) limacon

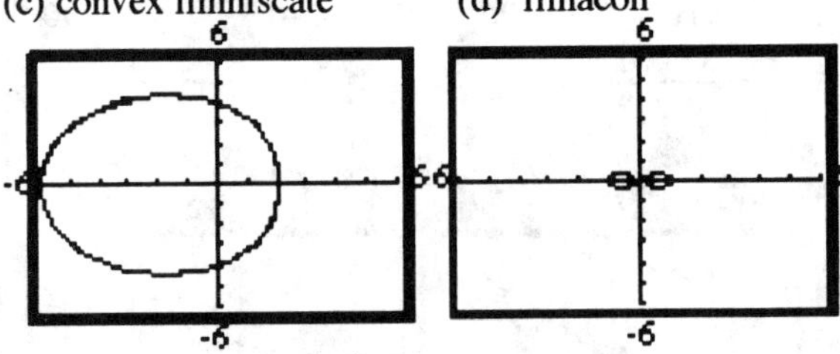

(e) spiral

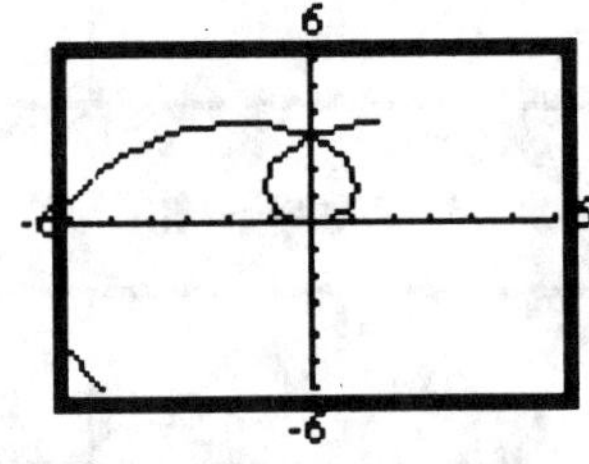

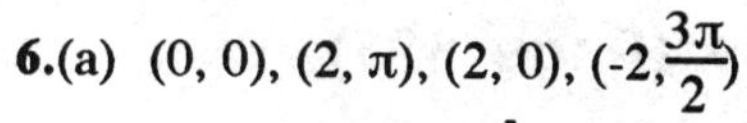

6.(a) (0, 0), (2, π), (2, 0), (-2, $\frac{3\pi}{2}$)

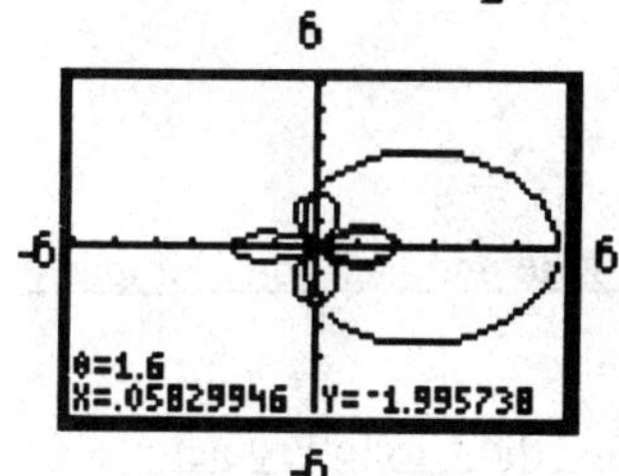

(b) (1, π/4), (1, 5π/4)

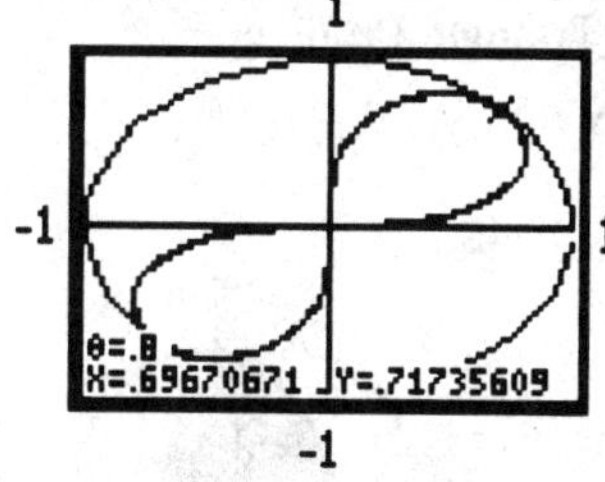

7. (a) $z = \sqrt{2}\text{cis}(\frac{3\pi}{4})$

(b) $z_1 = 2\text{cis}\,(-\frac{\pi}{3})$, $z_2 = 2\text{cis}\,(\frac{-\pi}{6})$, $z_1 z_2 = 4\text{cis}\,(\frac{-\pi}{2})$

8. $x = (2\sin\theta)\cos\theta$, $y = (2\sin\theta)\sin\theta$

$x = r\cos\theta$, $y = r\sin\theta$ for $r = 2\sin\theta$

9. $x = (2\sin\theta)\cos\theta$, $y = (2\sin\theta)\sin\theta$

10. The Golf Ball on the Moon Problem

250 ft

11. The Angular Velocity Problem

(a) $x = \cos 100t$, $y = \sin 100t$

(b) $x = 2\cos(200t)$, $y = 2\sin(200t)$

Section 8.1 The Distance Formula and the Circle (Page 525)

1. 5 **3.** $4\sqrt{2}$

5. yes, $d(A,B) = \sqrt{26}$, $d(B,C) = 3\sqrt{26}$, $d(A,C) = 4\sqrt{26}$

7. no, $d(A,B) = \sqrt{26}$, $d(B,C) = \sqrt{10}$, $d(A,C) = \sqrt{8}$
$a^2 + b^2 = 8 + 10 \neq 26$

9. $x^2 + y^2 = 25$ **11.** $(x + 2)^2 + (y + 1)^2 = 2$

13. C(0,0), r = 1 **15.** C(0,0), r = 0

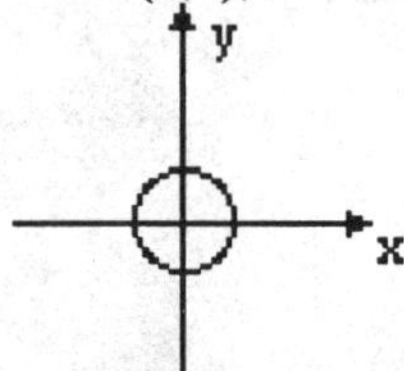

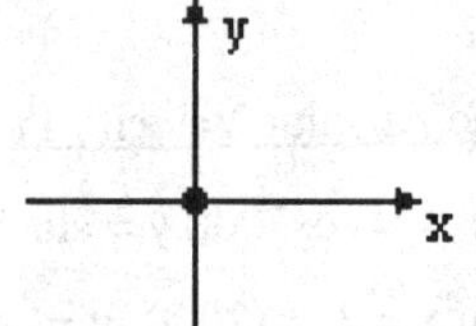

17. C(1,3), r = 1 **19.** C(1,0), r = 1

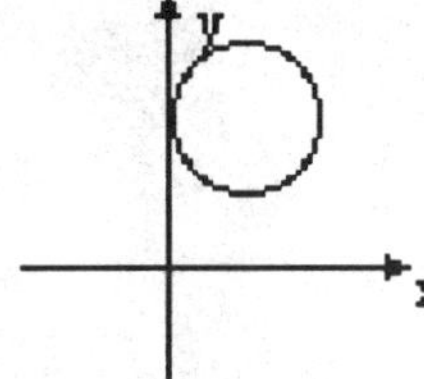

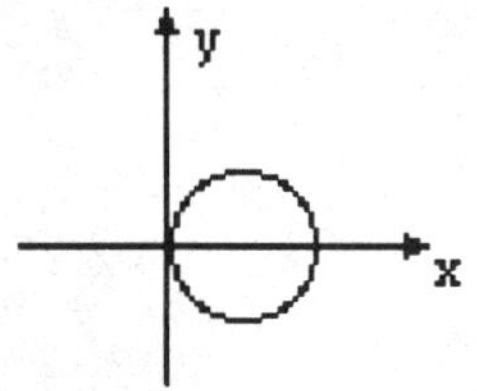

21. $(x + 1)^2 + (y + 2)^2 = 16$, C(-1,-2) r = 4

23. $(x - 2)^2 + (y + 1)^2 = -1$, no circle

25. $(x - 2)^2 + y^2 = 4$, C(2,0) r = 2

27. $(x - \frac{3}{2})^2 + (y + \frac{1}{2})^2 = \frac{5}{2}$, $C(\frac{3}{2}, -\frac{1}{2})$ $r = \sqrt{5/2}$

29. **31.**

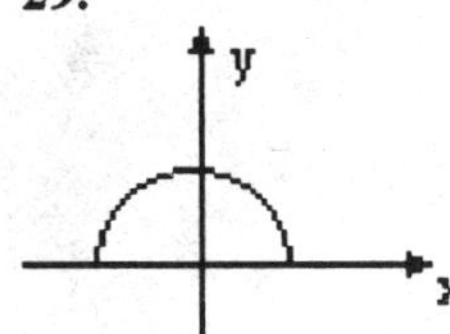

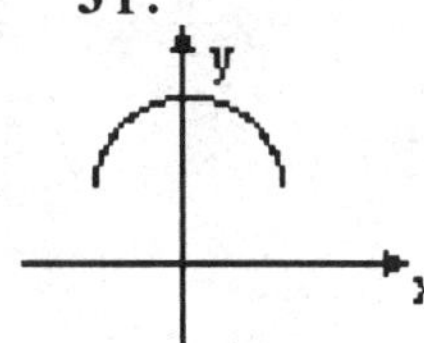

33. $x^2 + y^2 = 25$

35. $(x - \frac{1}{2})^2 + (y - \frac{7}{2})^2 = \frac{5}{2}$

37. <u>The Rainbow Kite and Balloon Company</u>
(100,50)

39. <u>The Tangent-to-the-Circle Problem</u>
(4,2) and (-3,1)

41. <u>The Inscribed Triangle Problem</u>
$y = 4\sqrt{4 - x^2}$ and x = 0, y = 2

43. **45.**

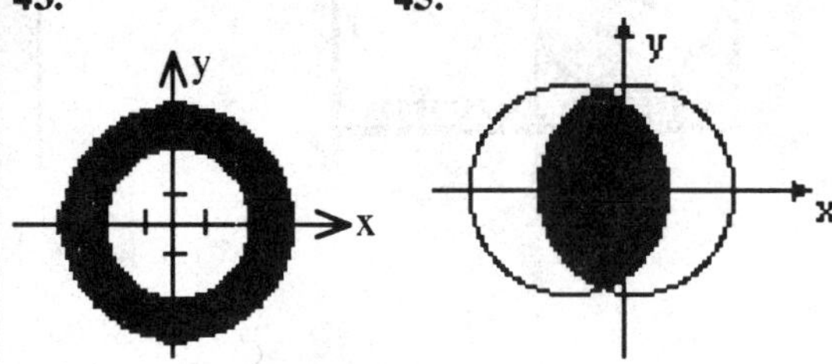

47. $d(A,M) = \sqrt{\frac{(x_1 - x_2)^2}{2} + \frac{(y_1 - y_2)^2}{2}}$ and

$d(M,B) = \sqrt{\frac{(x_1 - x_2)^2}{2} + \frac{(y_1 - y_2)^2}{2}}$

$\therefore$ d(A,M) = d(M,B) and M is the midpoint of segment AB.

49. d(A,C) = 6, d(C,B) = 8 and d(A,B) = 10, The sum of the lengths of any two sides of a triangle is greater than the length of the third side.

51. $x^2 + y^2 = 16$

53. $9x^2 + 25y^2 = 225$

Section 8.2 The Ellipse and the Hyperbola (Page 539)

1. $\frac{x^2}{9} + \frac{y^2}{4} = 1$
C(0,0), 2a = 6, 2b = 4
$F(-\sqrt{5}, 0)$ $F'(\sqrt{5}, 0)$

3. $\frac{(x - 1)^2}{4} + \frac{y^2}{9} = 1$
C(1,0) 2b = 6, 2a = 4
$F(1, -\sqrt{5})$ $F'(1, \sqrt{5})$

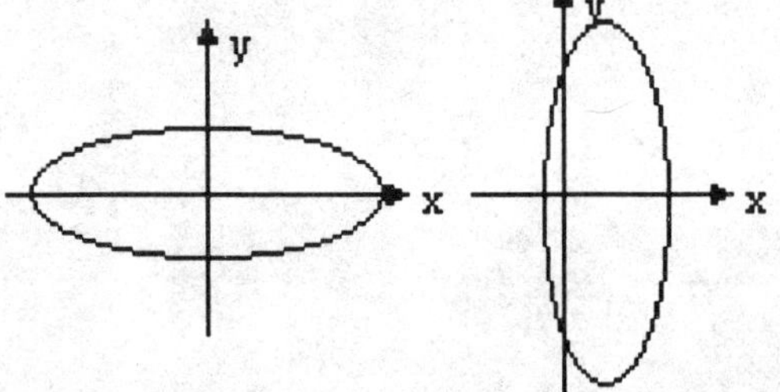

5. $\frac{(x + 2)^2}{9} + \frac{(y - 1)^2}{4} = 1$
C(-2,1), 2a = 6, 2b = 4, $F(-2-\sqrt{5}, 1)$ $F'(-2+\sqrt{5}, 1)$

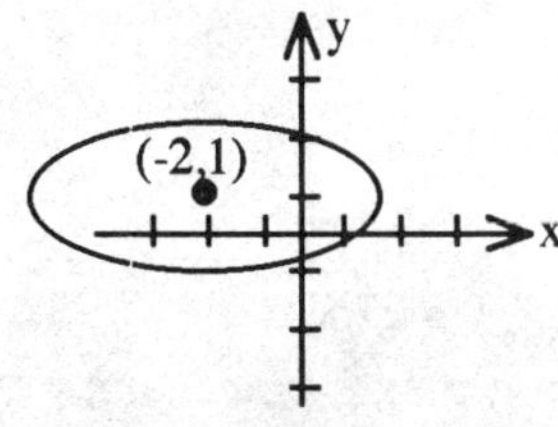

7. $\frac{x^2}{9} + \frac{y^2}{8} = 1$

9. $\frac{x^2}{25} - \frac{y^2}{16} = 1$ C(0, 0), V(±5, 0), F($\pm\sqrt{41}$, 0

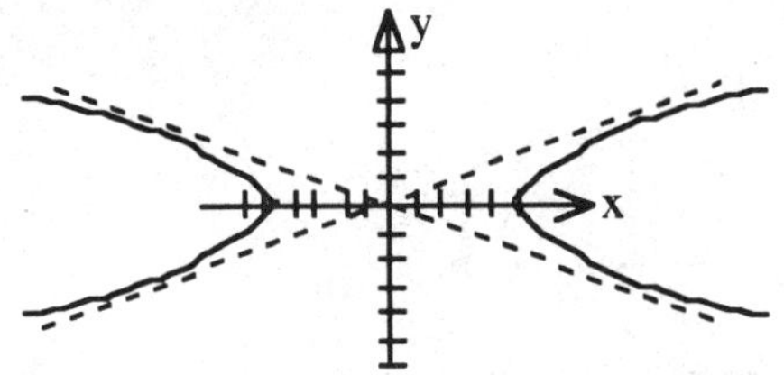

11. $\frac{x^2}{4} - \frac{y^2}{4} = 1$ C(0, 0), V(±2, 0), F($\pm\sqrt{8}$, 0)

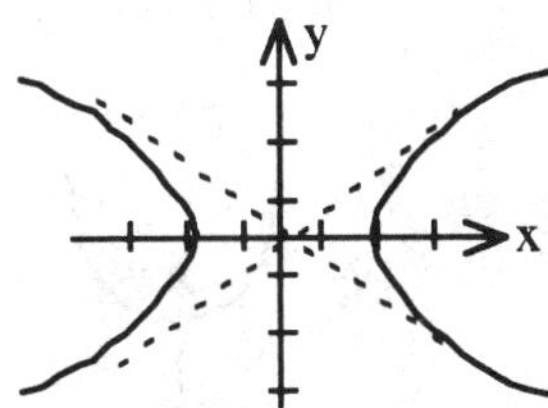

13. $\frac{(x-2)^2}{4} - \frac{(y+3)^2}{16} = 1$

C(2, -3), V(0, -3), V(4, -3), F($2 \pm\sqrt{20}$, -3)

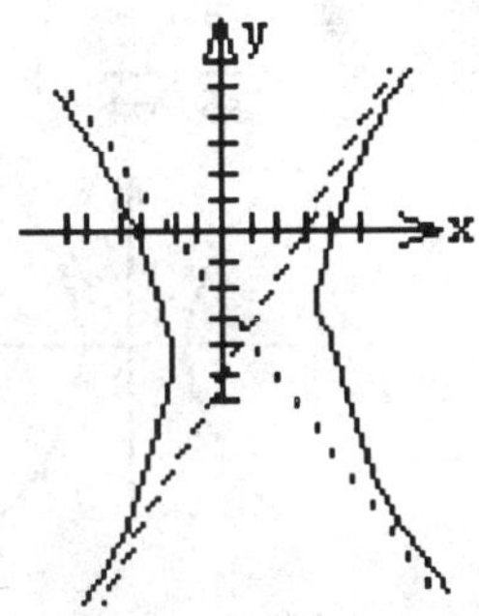

15. $\frac{(y+1)^2}{16/3} - \frac{(x-1)^2}{16} = 1$

C(1, -1), V(1, $-1\pm \sqrt{16/3}$), F(1, $-1\pm\sqrt{64/3}$)

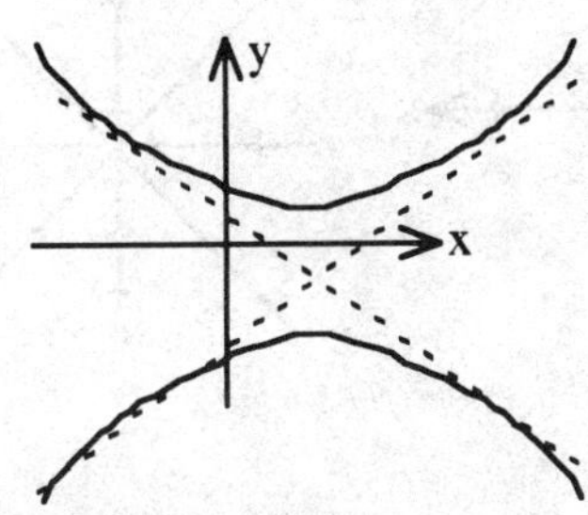

17. $\frac{(x-1)^2}{9} - \frac{(y+2)^2}{16} = 1$

19. $\frac{x^2}{16} + \frac{(y+3)^2}{4} = 1$

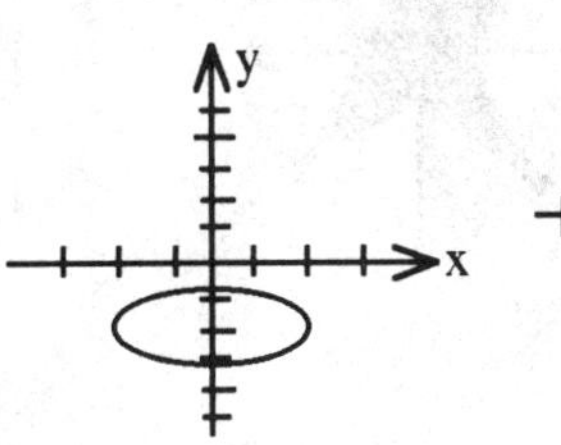

21. $\frac{(x-4)^2}{20} + \frac{(y+2)^2}{27} = 1$

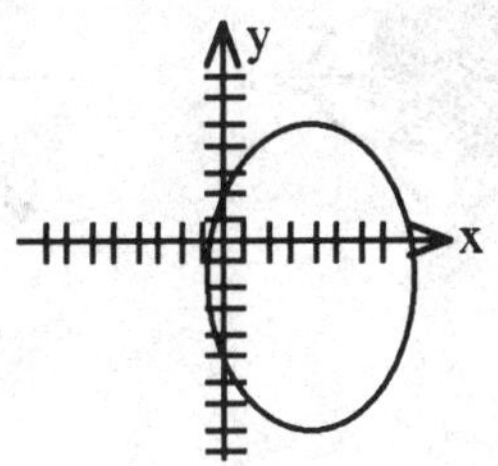

23. $\frac{(x+5/2)^2}{15/2} + \frac{(y+1/2)^2}{3/2} = 1$

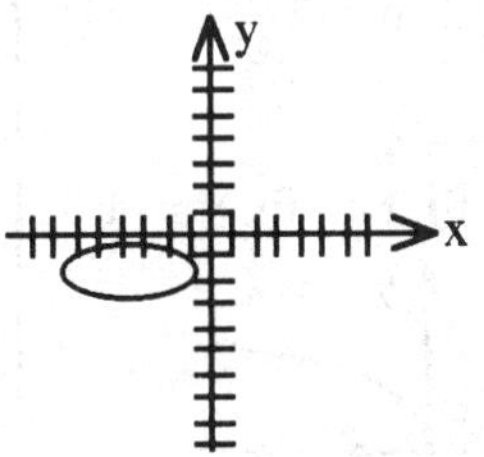

25. $\frac{(y+1)^2}{75/9} - \frac{(x-2)^2}{3} = 1$

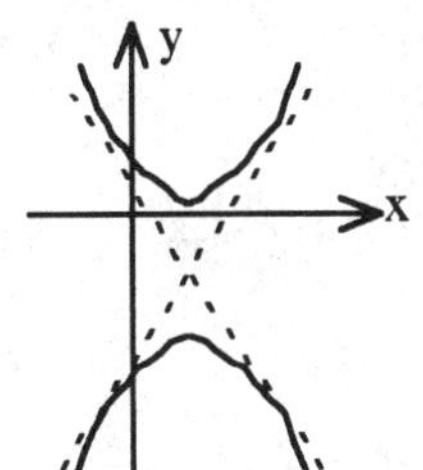

27. The Rainbow Kite and Balloon Company
20 ft long, 6 ft from the center

29. The Earth's-Orbit Problem

$$\frac{x^2}{2.25 \times 10^{16}} + \frac{y^2}{2.249375 \times 10^{16}} = 1$$

31. The Jupiter Problem

$$\frac{x^2}{2.3358 \times 10^{17}} + \frac{y^2}{2.3413 \times 10^{17}} = 1$$

33. The Length of the Orbit Problem
$C \approx 26.206\pi = 82.33$ AU = 7,653,065,007 mi

35. (0.1, 0), (3.9, 0), (0, -1)

37. $OP_1 = OP_2 \Rightarrow s = 2(OP_1) = 2\sqrt{x_1^2 + y_1^2} = 2\sqrt{x_1^2 + y_2^2}$, because of the symmetry, $y_1 = -y_2$.

39. $y = 2x/\sqrt{3} + 2/\sqrt{3} + 2$
$y = -2x/\sqrt{3} - 2/\sqrt{3} + 2$

41.

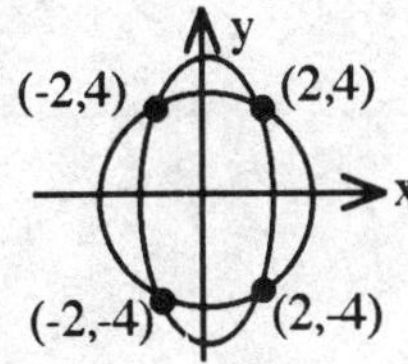

43.

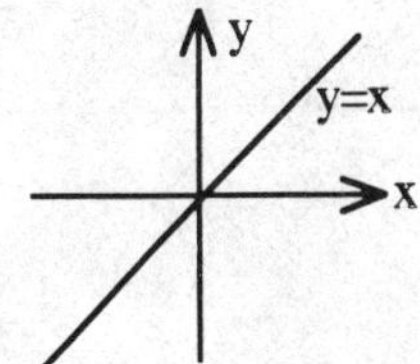

45.

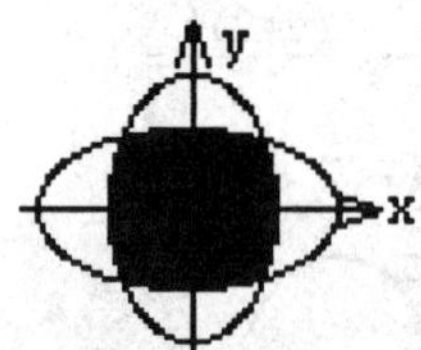

47.

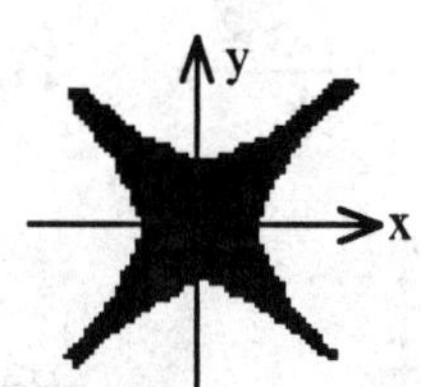

49. no solution

51. One family of branches must be on dry land because two of the hyperbolas may intersect in two points.

Section 8.3 The Parabola and a Review of Conic Sections (Page 550)

1. up, $p = 5$

3. right, $p = 2$

5. $x^2 = 16y$
V(0, 0), F(0, 4), $y = -4$

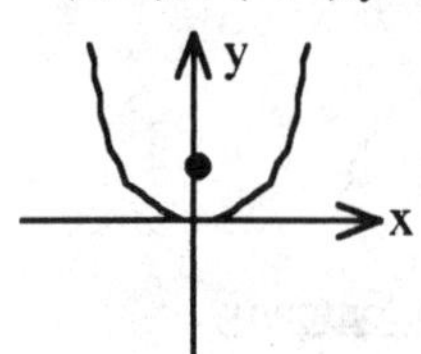

7. $y^2 = 4x$
V(0, 0), F(1, 0), $x = -1$

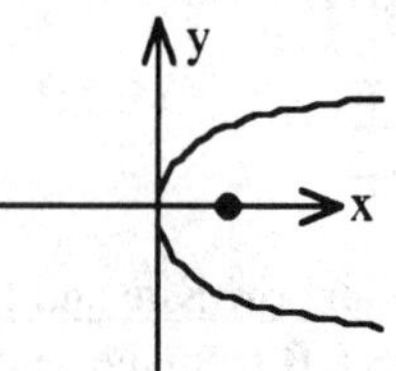

9. $(x - 1)^2 = 8y$
V(1, 0), F(1, 2), $y = -2$

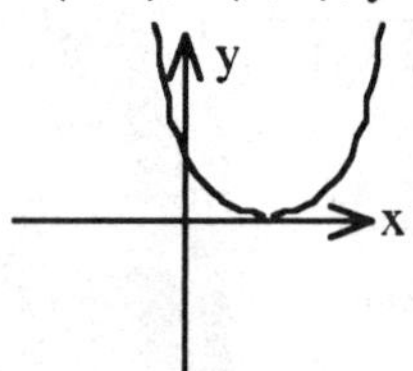

11. $(y - 1)^2 = -8(x + 2)$
V(-2, 1), F(-4, 1), $x = 0$

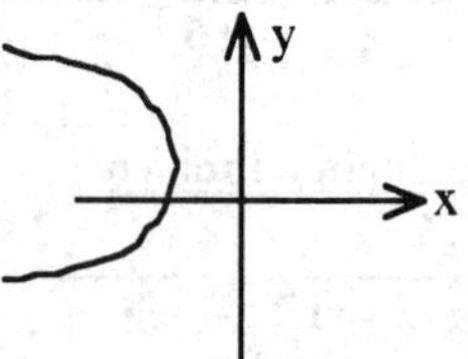

13. $x^2 = -12y$

15. $x^2 = -6(y - \frac{3}{2})^2$

17. $x^2 = -2(y + 6)$

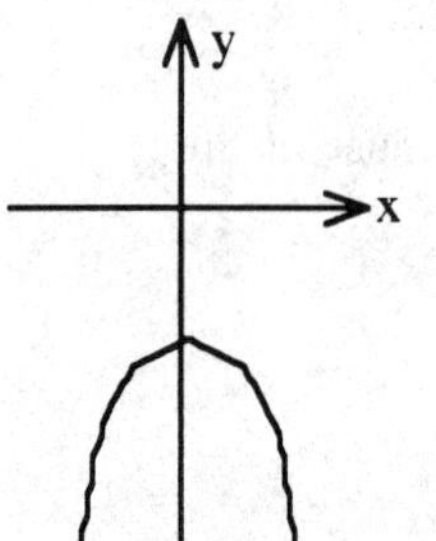

19. $(x - 3)^2 = 10(y + 1)$

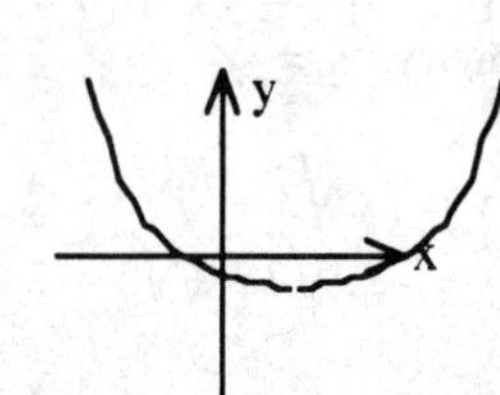

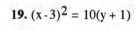

21. hyperbola
$\frac{x^2}{9} - \frac{y^2}{16} = 1$

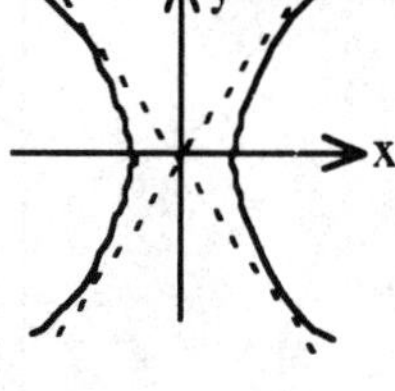

23. parabola
$x^2 = \frac{9}{16}(y - 16)$

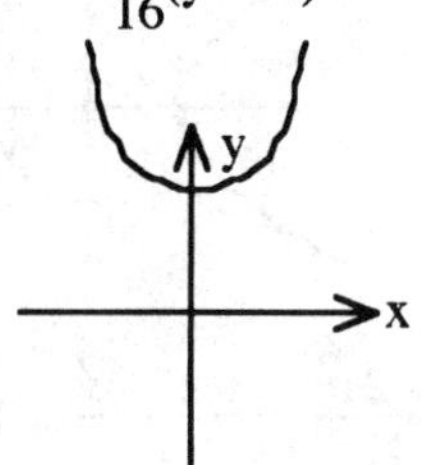

25. hyperbola
$\frac{(x - 2)^2}{144} - \frac{(y - 1)^2}{25} = 1$

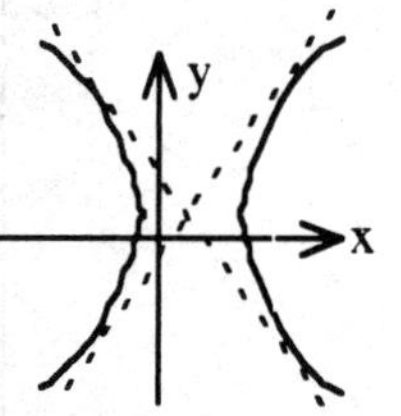

27. hyperbola
$\frac{x^2}{36} - \frac{y^2}{64} = 1$

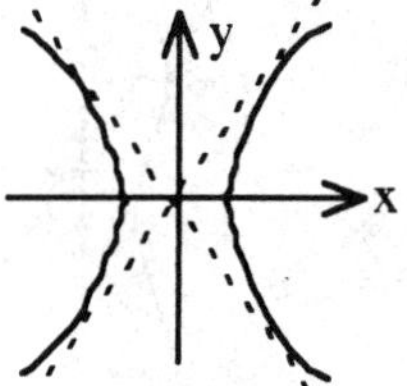

29. ellipse
$\frac{(y + 3)^2}{64} + \frac{(x + 1)^2}{16} = 1$

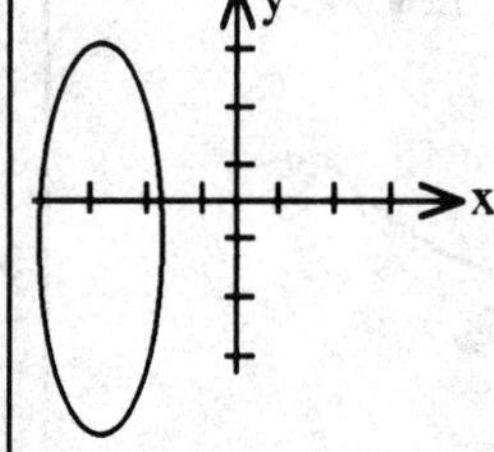

31. degenerate
$y = \pm x$

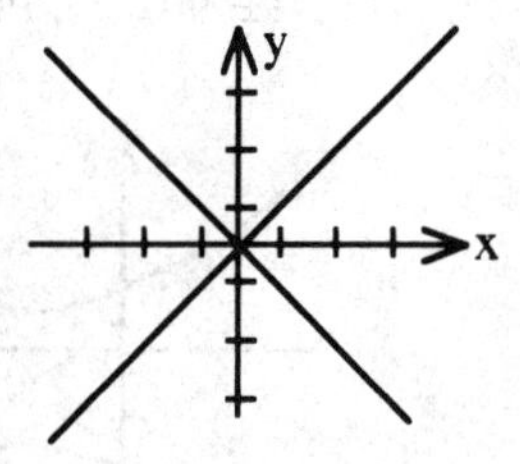

33. point ellipse

$16(x - 1)^2 + 4(y + 2)^2 = 0$

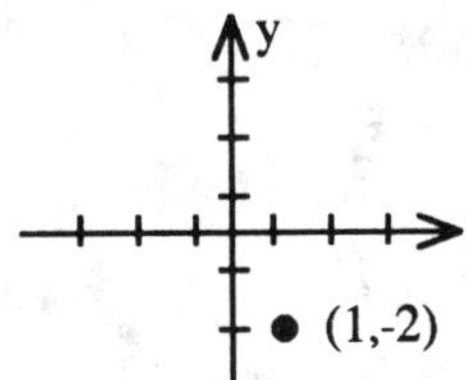

35. hyperbola

$\frac{(x + 2)^2}{4} - \frac{(y - 1)^2}{16/9} = 1$

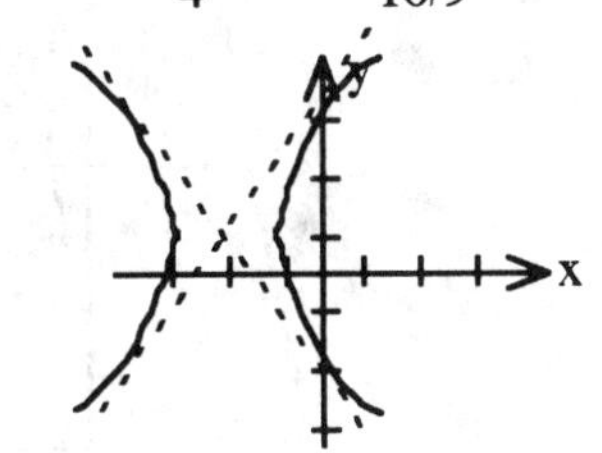

37. The Rainbow Kite and Balloon Company

$(\frac{25}{24}, 0)$

39. The Newtonian Telescope Problem

Mount the camera 1000 in. from the mirror.

41.

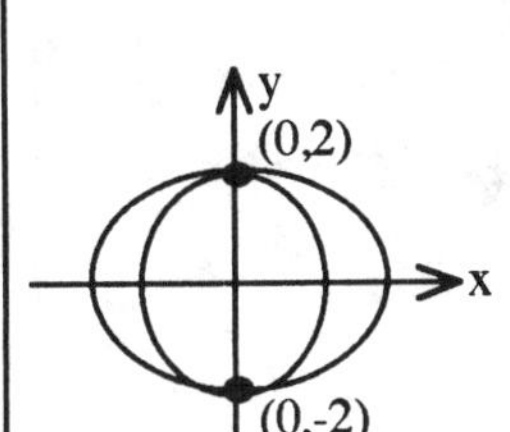

43.

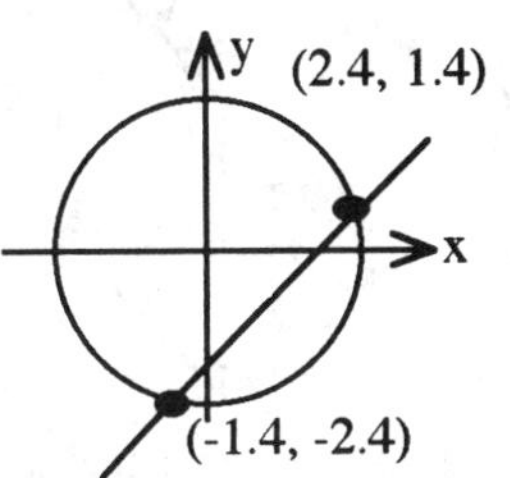

Section 8.4 Rotations and Conic Sections (Page 560)

1. parabola

3. hyperbola

5. 22.5°

7. $-\pi/12$

9.

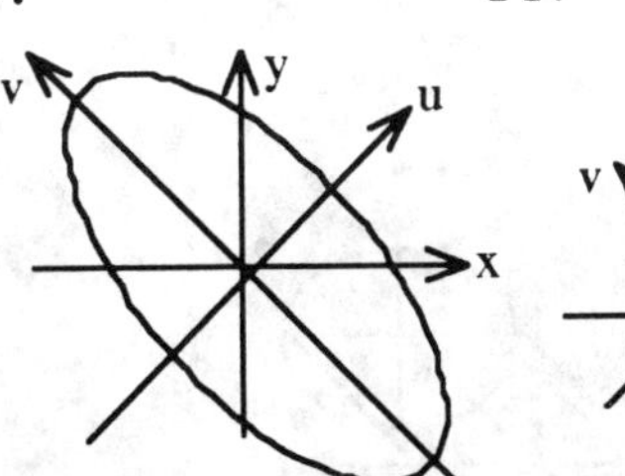

11.

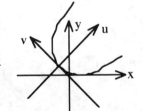

13. $25x^2 + 14xy + 25y^2 = 288$

15. $9(x + y - \sqrt{2})^2 + 4(-x + y - 2\sqrt{2})^2 = 72$

17. $\theta = 45°, \frac{u^2}{16} - \frac{v^2}{36} = 1$

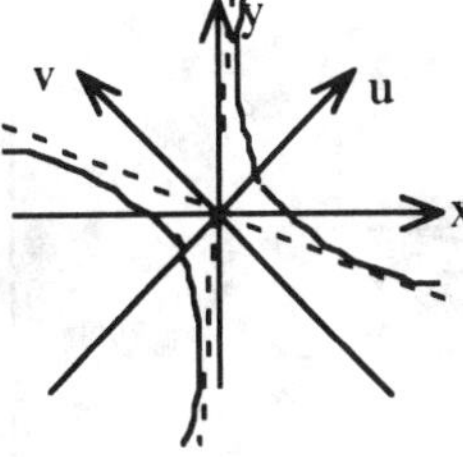

19. $\theta = 11\frac{1}{3}°, u^2 + \frac{v^2}{4} = 1$

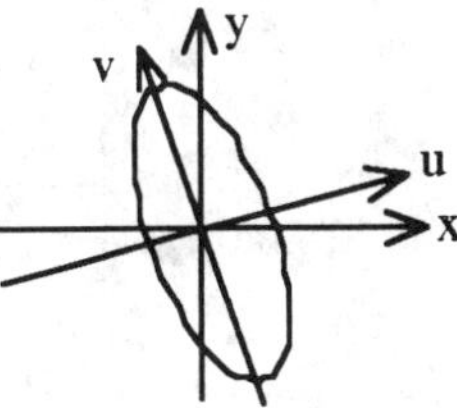

21. $\theta = 45°$

$(u-1)^2 = \frac{(v + 1)^2}{4}$

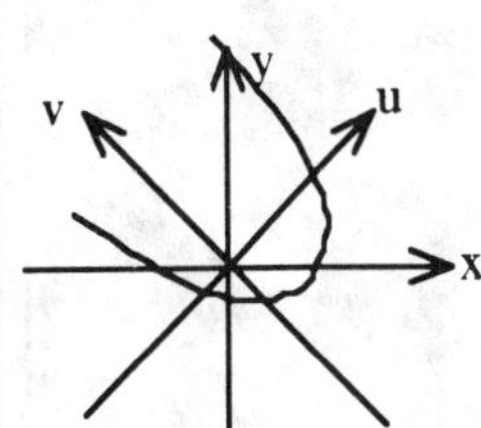

23. $\theta = 45°$

$\frac{(u - 3)^2}{16} + \frac{(v - 2)^2}{4} = 1$

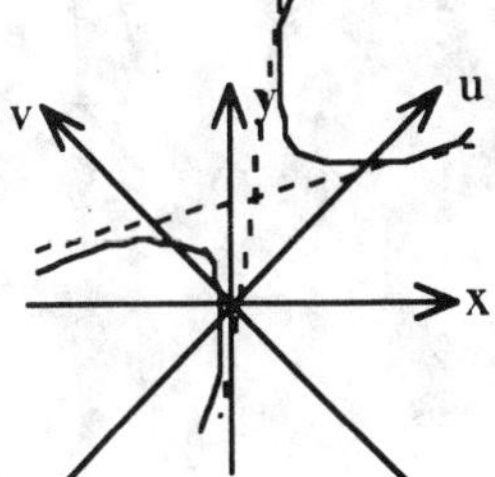

Section 8.5 Graphs in a Three-Dimensional Coordinate System (Page 567)

1. (3, -4, 5) **3.** (1, 4, 0)

5.

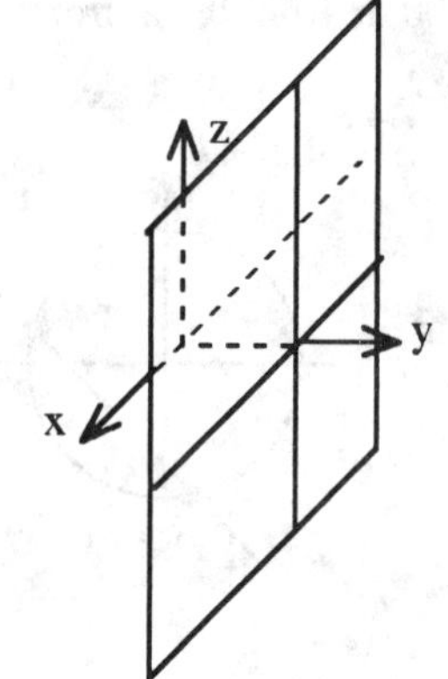

7.

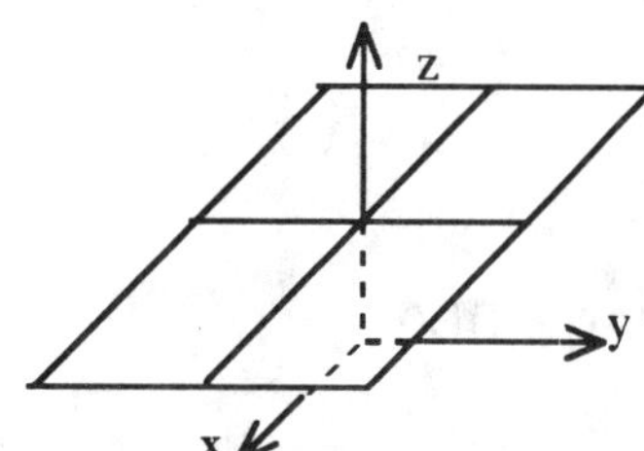

9.

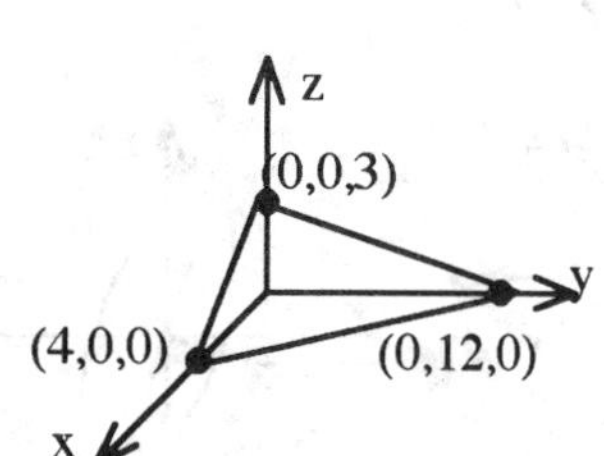

11.

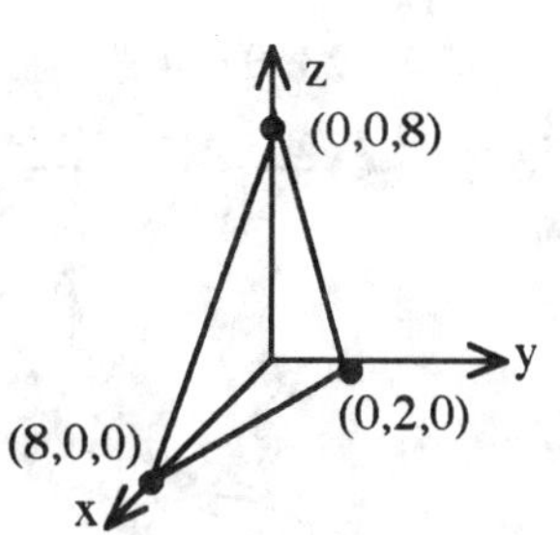

13.(a)

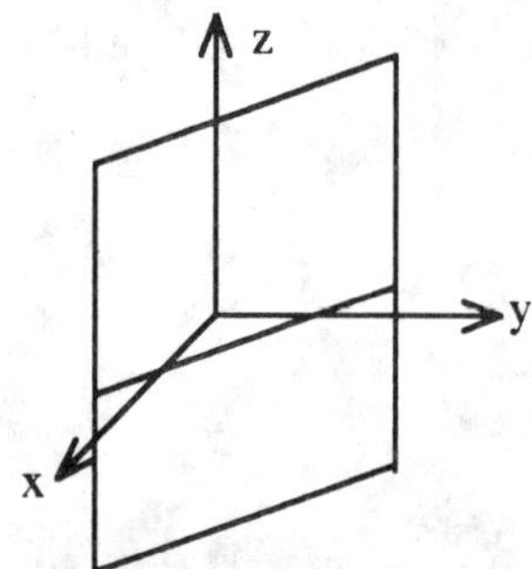

13.(b)

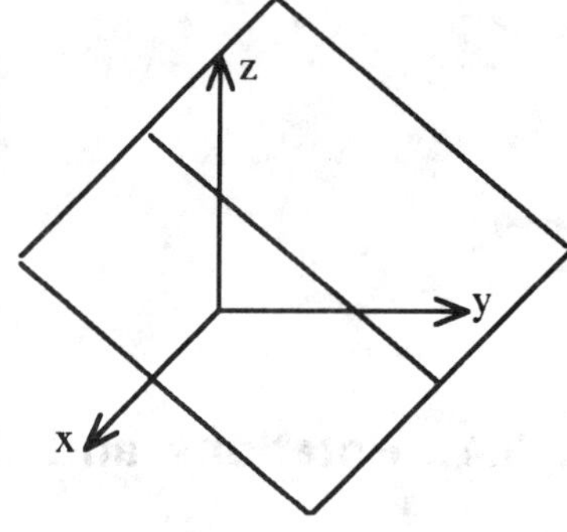

(c)

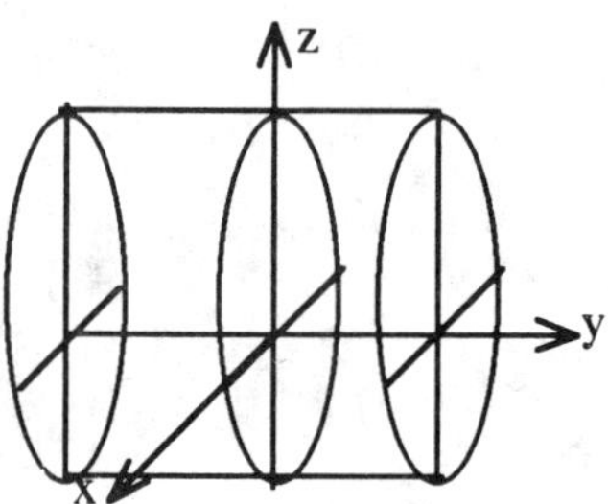

15.(a)

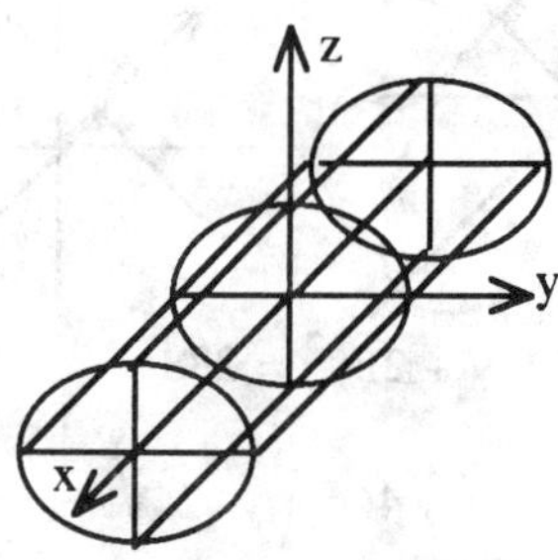

(b)

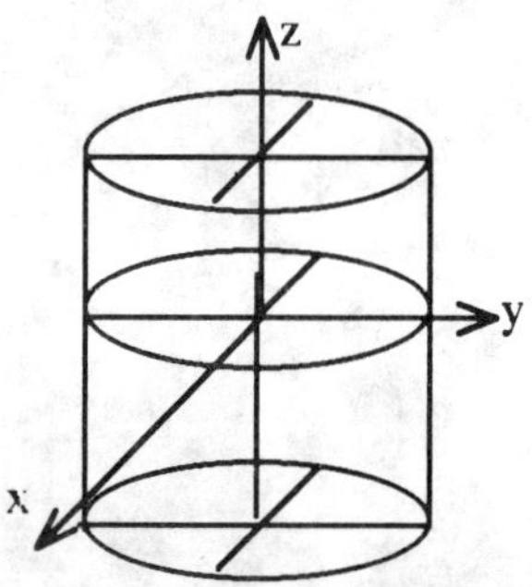

(c)

17.(a)

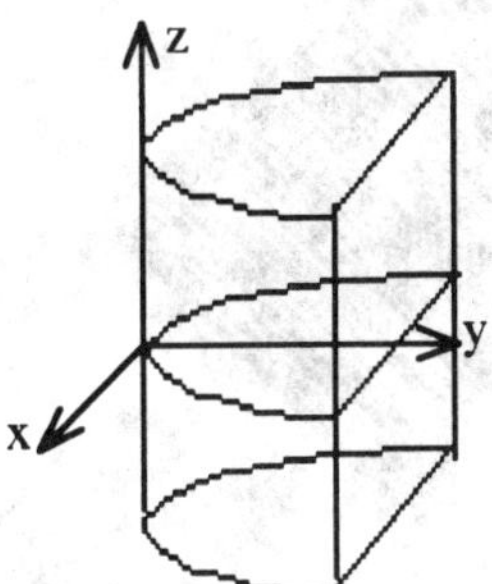

(b)

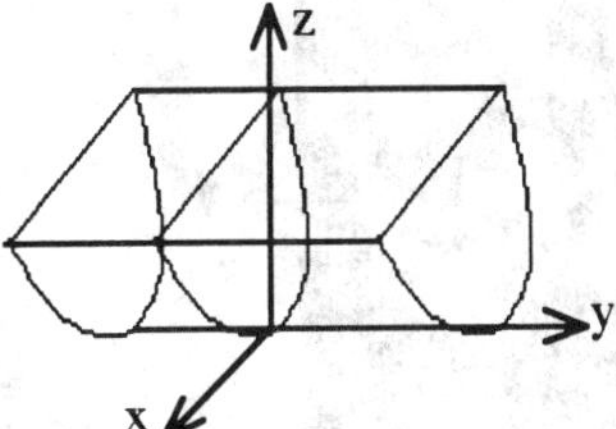

(c)

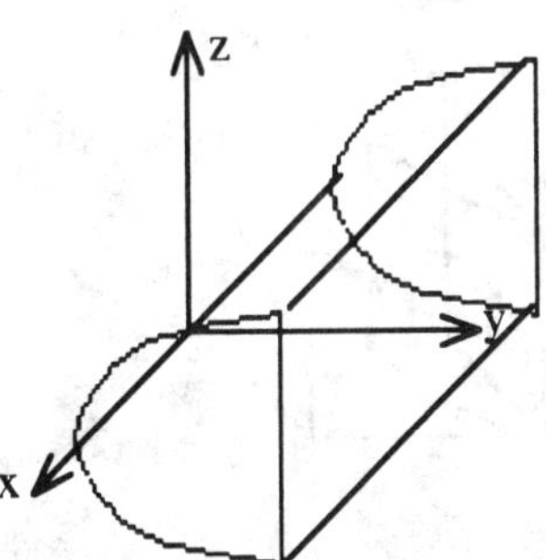

19.(a)

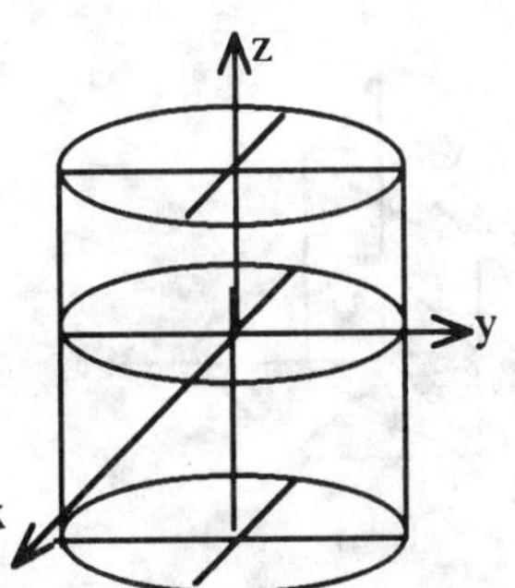

(b)

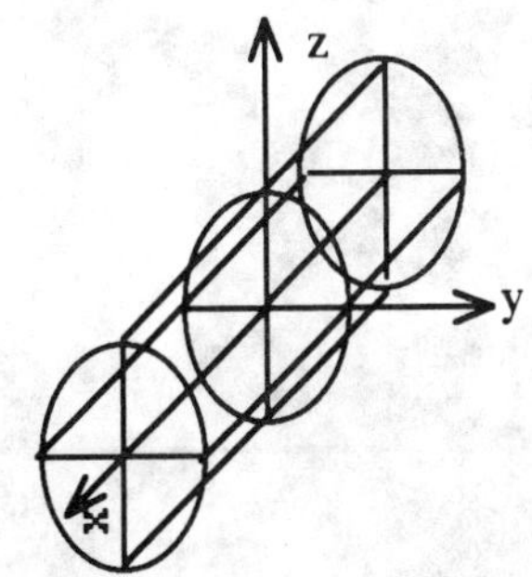

19.(c)

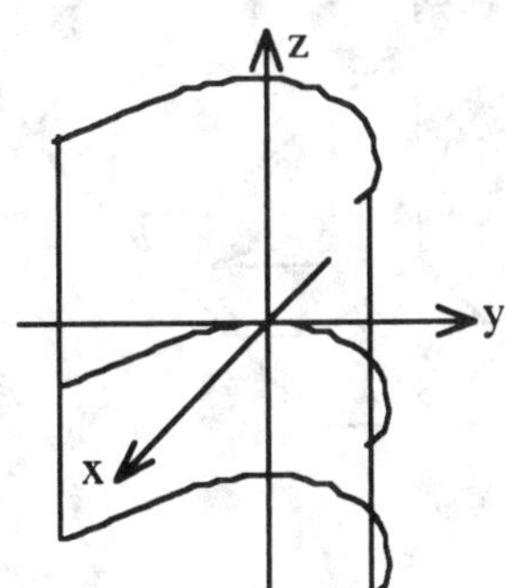

21.

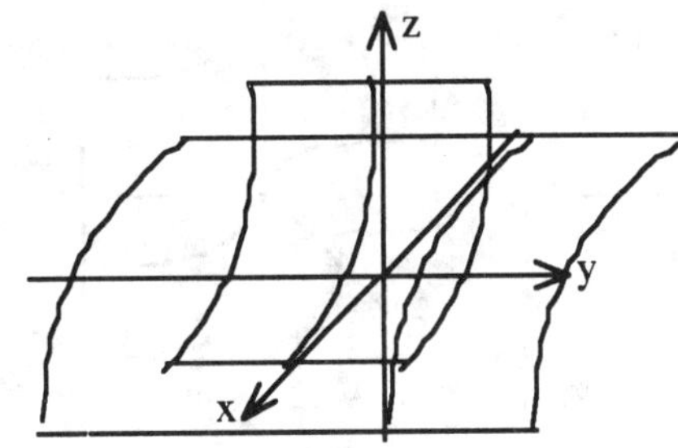

23.

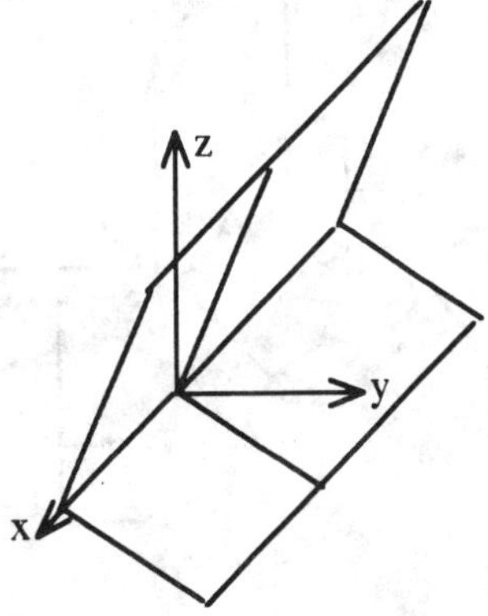

25.

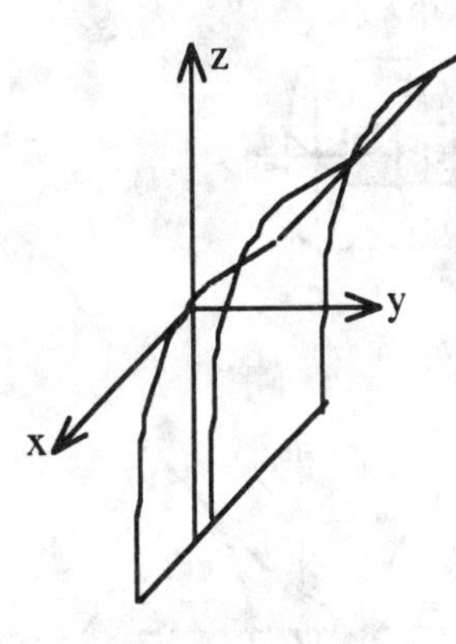

27.

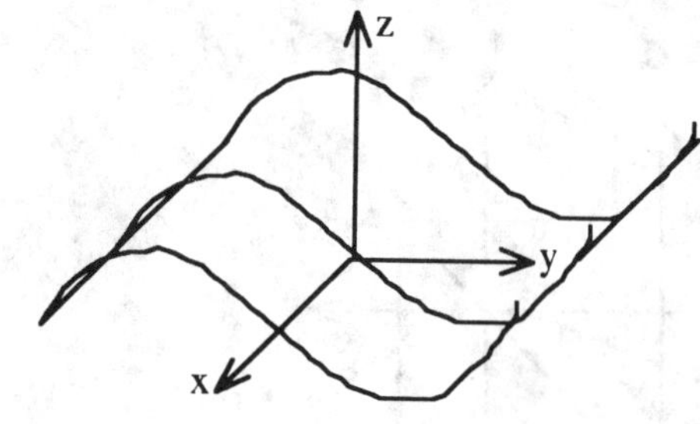

29.

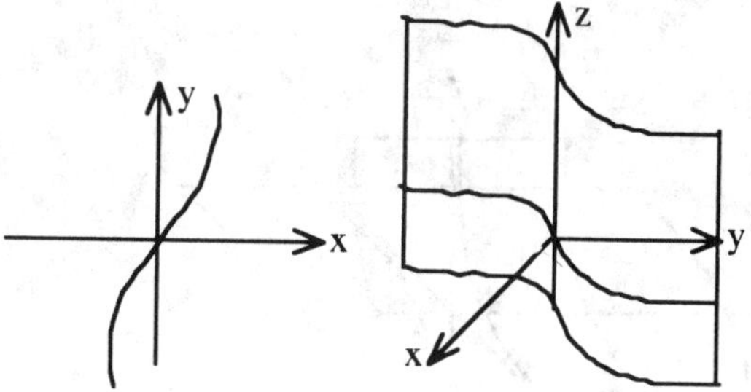

31.

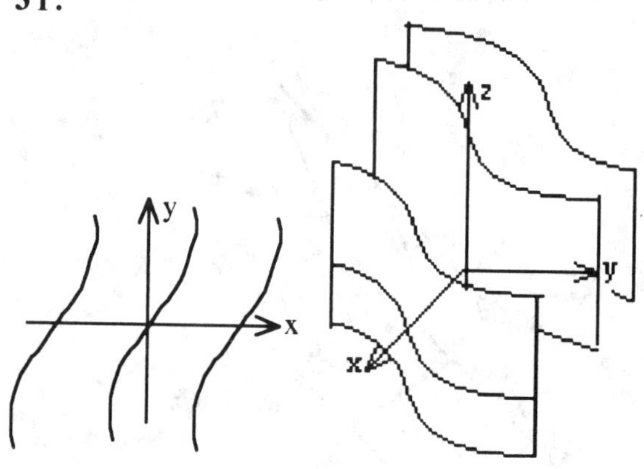

33.

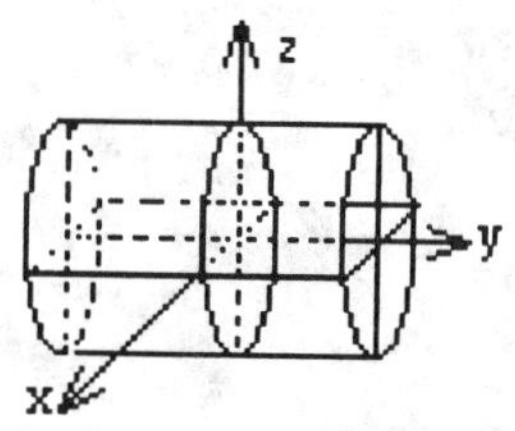

35.

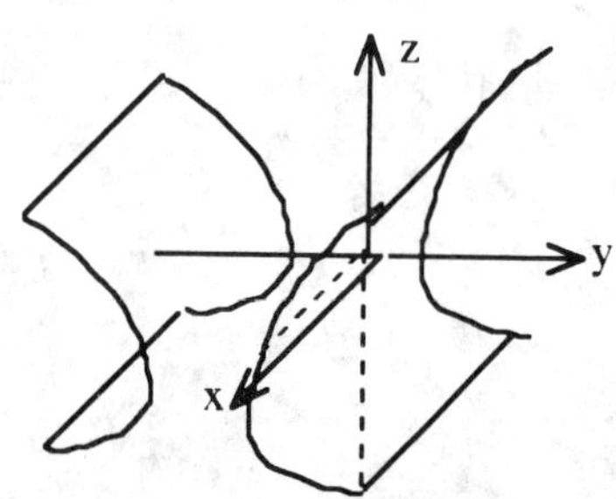

37.

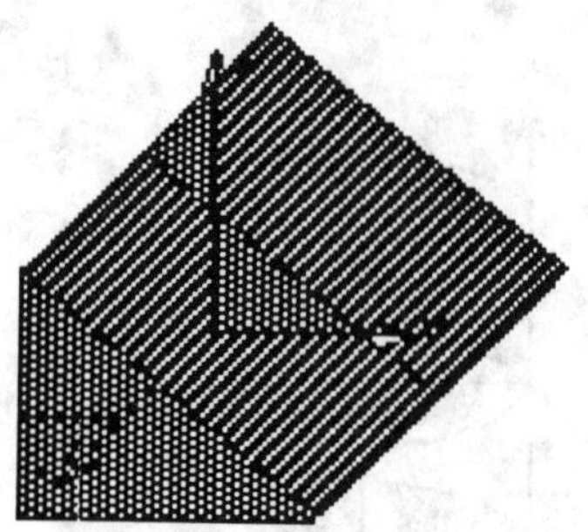

39.

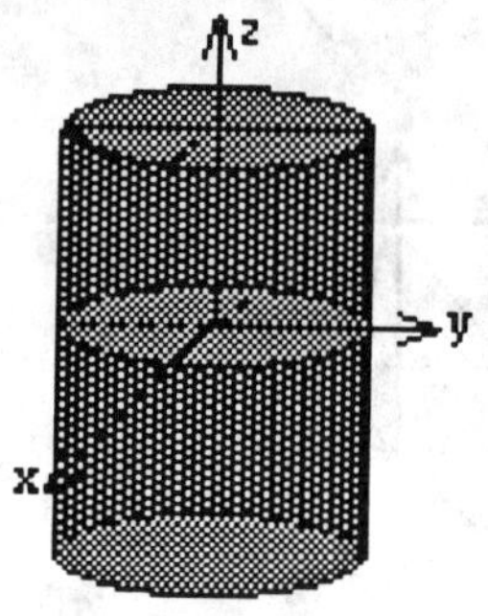

41.

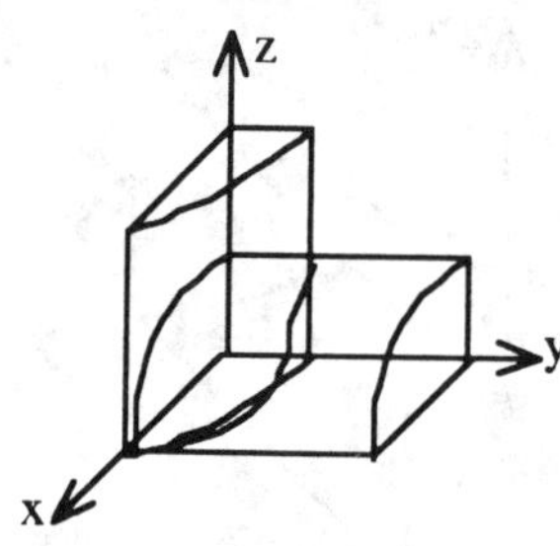

43.

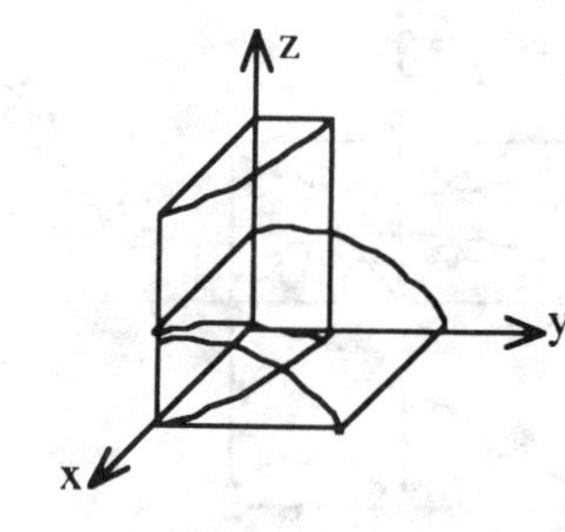

Section 8.6 Quadric Surfaces (Page 576)

1. sphere

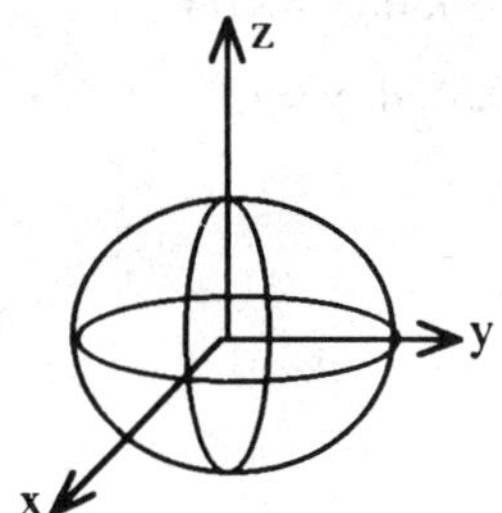

3. ellipsoid

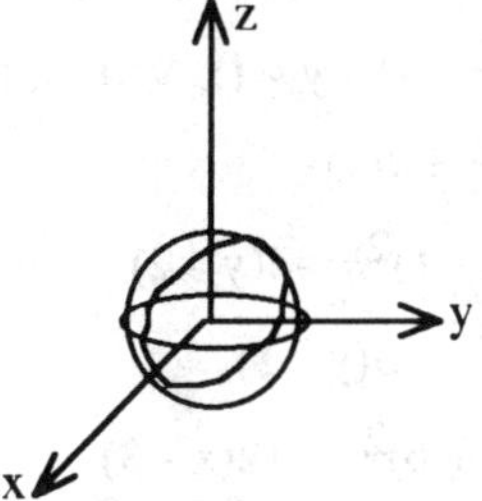
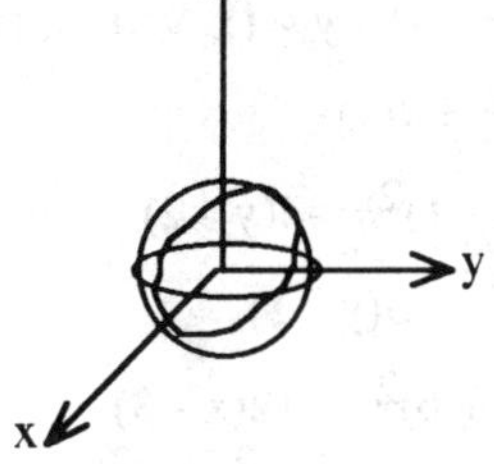

5. hyperboloid

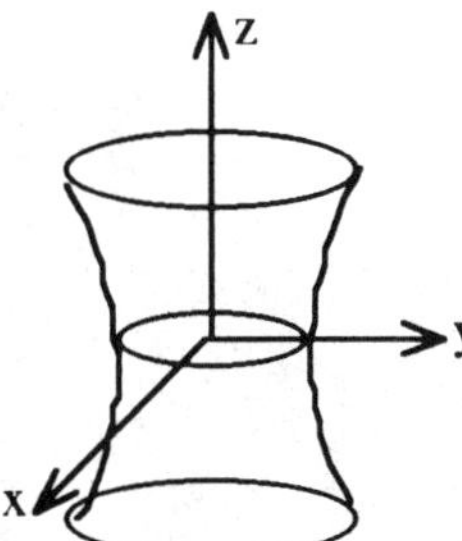

7. hyperboloid

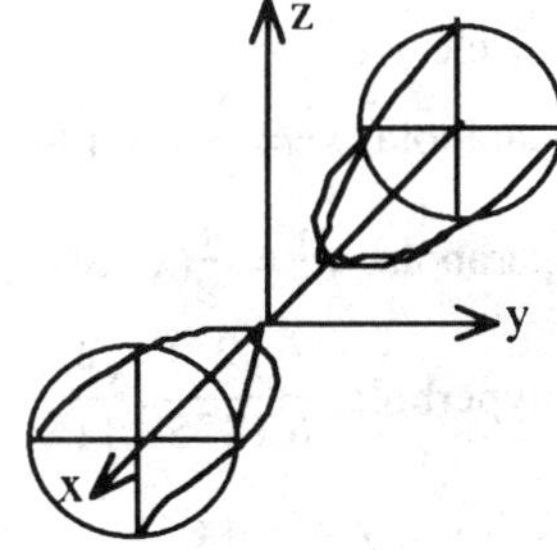

9. paraboloid

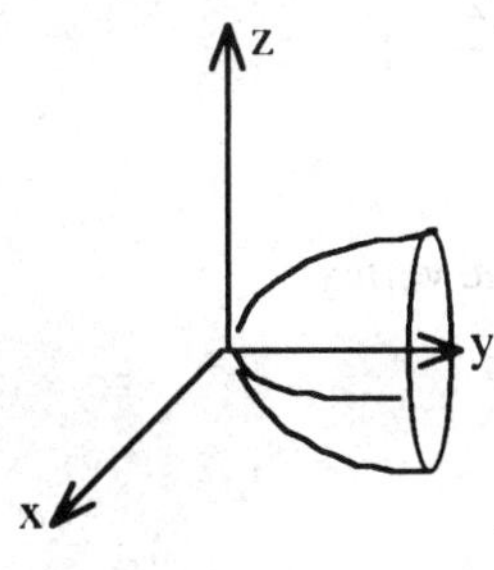

11. cone

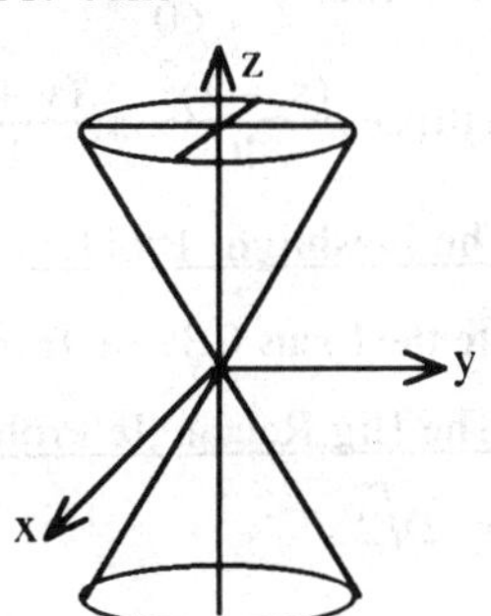

13. sphere

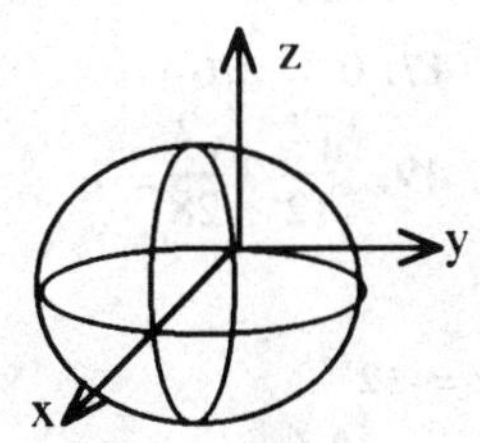

15. ellipsoid

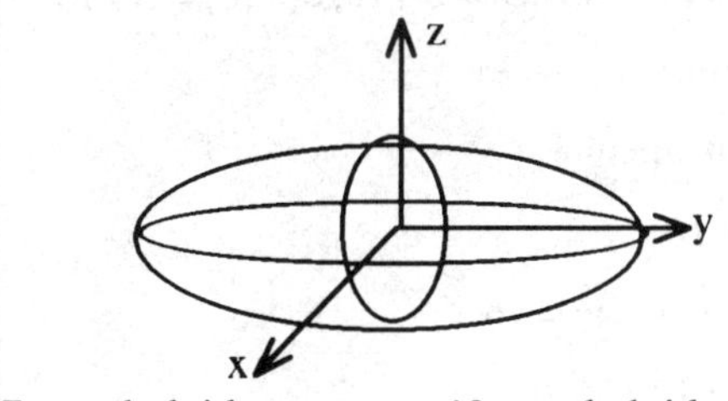

17. paraboloid

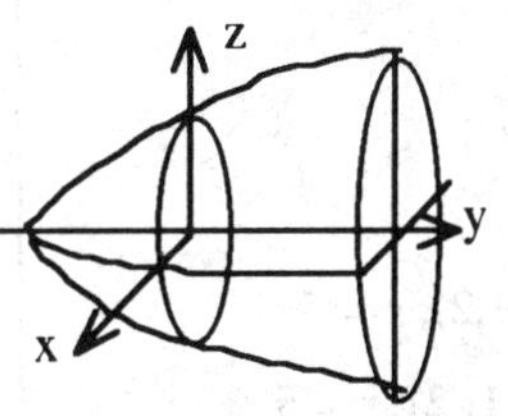

19. paraboloid

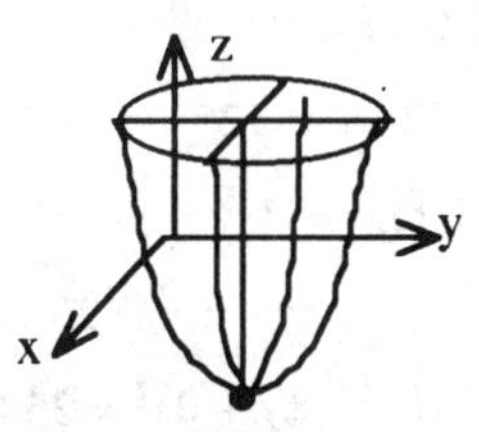

21. paraboloid

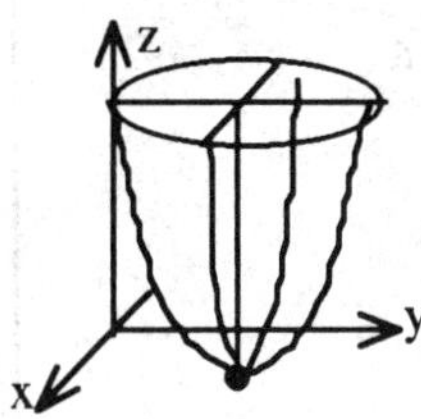

23. cone

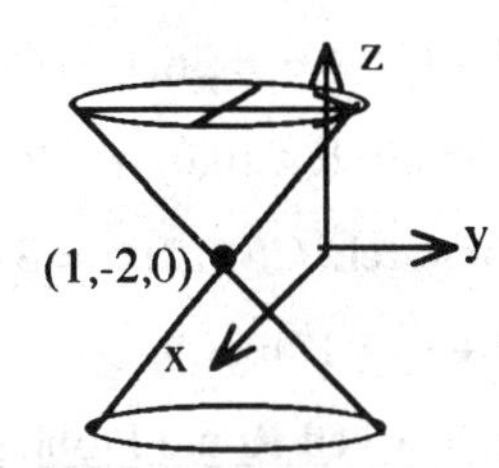

25. sphere

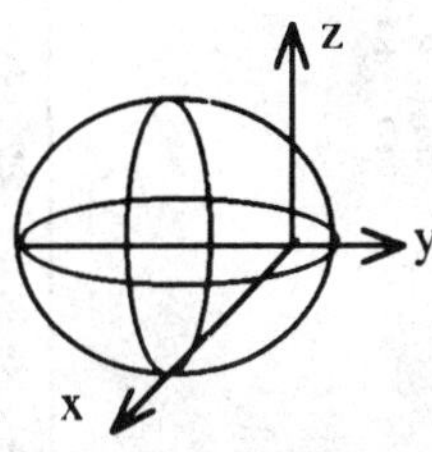

27. paraboloid

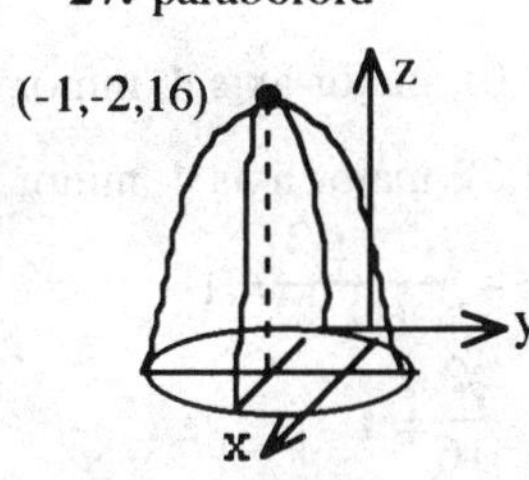

29. paraboloid

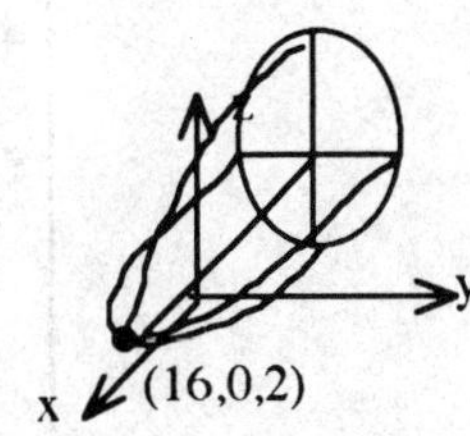

31. hyperboloid

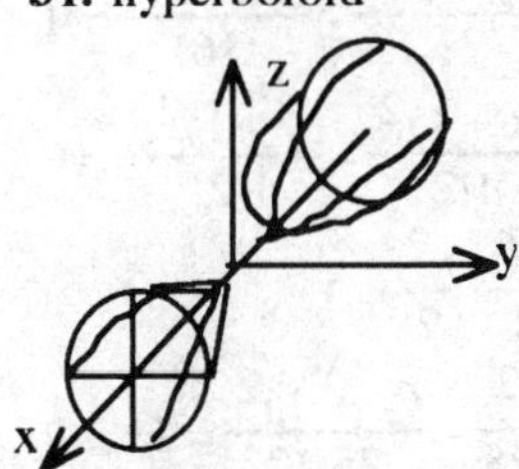

33.

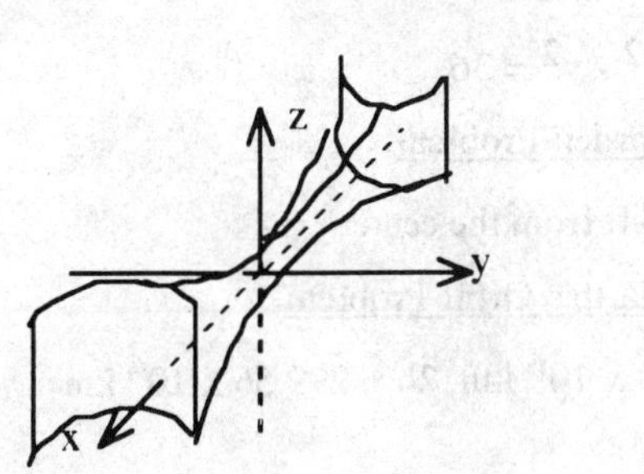

Chapter 8 Review (Page 577)

1. $d(A,B) = 2\sqrt{10}$, $d(B,C) = 3\sqrt{10}$, $d(A,C) = 5\sqrt{10}$

They are collinear.

2. Not a right triangle.

3. $x^2 + y^2 = 16$

4. $(x + 1)^2 + (y - 2)^2 = 16$

5. C(0,0), r = 12

6. C(0,0), r = 0

7. C(0,12), r = 12

8. C(-1,1), r = 1

9. $(x + 1)^2 + (y - 2)^2 = 25$ C(-1,2), r = 5

10. $(x - 1)^2 + (y + 2)^2 = 25$ C(1,-2), r = 5

11. $(x - 6)^2 + y^2 = 36$, C(6,0), r = 6

12. $x^2 + (y - 6)^2 = 36$, C(0,6), r = 6

13. Semicircle C(0,0), r = 2

14. Semicircle C(0,12), r = 3

15. $x^2 + (y - 12)^2 = 9$

16. <u>The Curved Route Problem</u>

C(200,100)

17. C(0,0), major axis 4, minor axis 2

18. C(2,0), major axis 4, minor axis 2

19. $\frac{x^2}{16} + \frac{(y - 4)^2}{12} = 1$

20. $\frac{x^2}{1} - \frac{y^2}{16} = 1$

21. $\frac{(y - 2)^2}{1} - \frac{(x + 2)^2}{16} = 1$

22. $\frac{(x - 1)^2}{1/2} - \frac{(y - 1)^2}{1/2} = 1$

23. $\frac{y^2}{2} - \frac{x^2}{2} = 1$

24. $\frac{x^2}{25} + \frac{(y + 1)^2}{25/12} = 1$

25. $(x - 6)^2 - y^2 = 36$

26. <u>The Garden Problem</u>

about 26.5 ft from the center

27. <u>The Earth's Orbit Problem</u>

$2a = 300.0 \times 10^6$ km, $2b = 299.56 \times 10^6$ km

28. $x^2 = 4(1/4)y$, V(0,0), F(0,1/4), d: y = -1/4

29. $x^2 = -16(y - 1)$, V(0,1), F(0,5), d: y = -3

30. $y^2 = 4(4)x$

31. $(x - 1)^2 = -4(y - 2)$

32. $x^2 = -3(y - 1)$

33. $(y + 6)^2 = -12(x - 3)$

34. hyperbola, $\frac{x^2}{144} - \frac{y^2}{144} = 1$

35. circle

36. parabola, $x^2 = \frac{1}{4}(y + 144)$

37. parabola, $y^2 = -\frac{4}{9}(x - 36)$

38. hyperbola, $\frac{x^2}{6.1875} - \frac{y^2}{11} = 1$

39. 2 lines, $y = \pm 4x$

40. hyperbola, $\frac{(x - 6)^2}{60} - \frac{(y - 1)^2}{5} = 1$

41. ellipse, $\frac{(x + 2)^2}{40} + \frac{(y + 3)^2}{160} = 1$

42. <u>The Flashlight Problem</u>

Locate the focus 0.39 in. from the vertex.

43. <u>The Big Rectangle Problem</u>

$\sqrt{2}$ by $2\sqrt{2}$

44. parabola

45. hyperbola

46. $\theta = \pi/4$

47. $\theta = \pi/6$

48. $\frac{u^2}{1} - \frac{v^2}{3} = 1$

49. $\frac{u^2}{12} + \frac{v^2}{28} = 1$

50. 

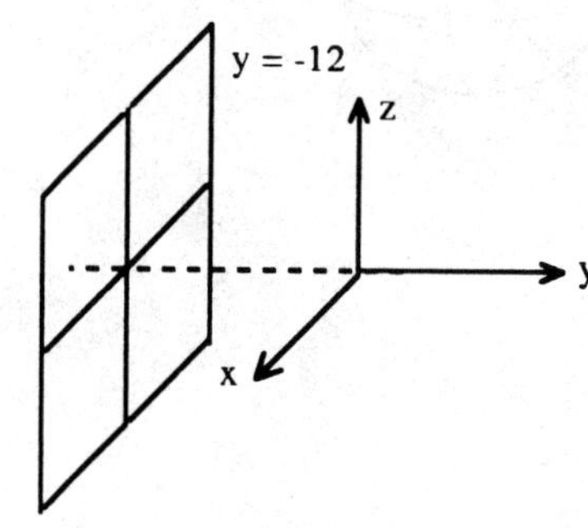

51.

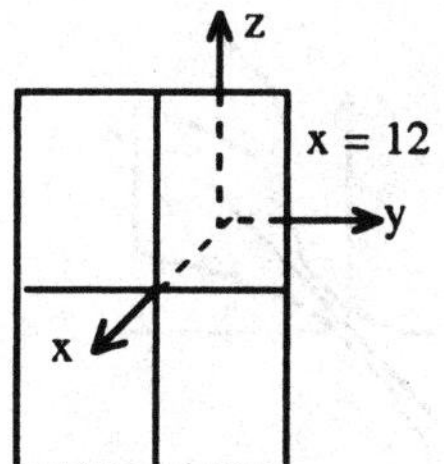

52.

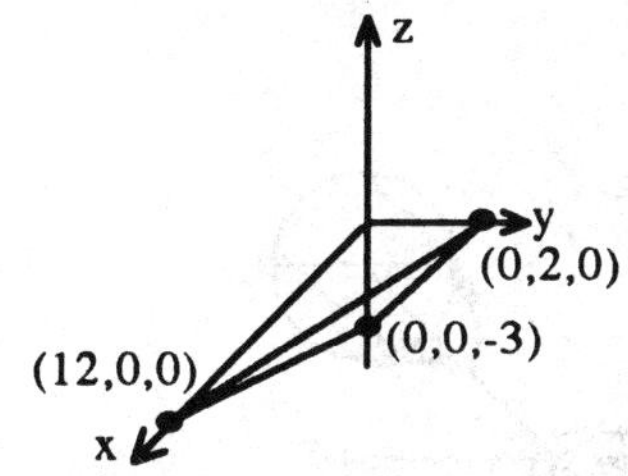

53.

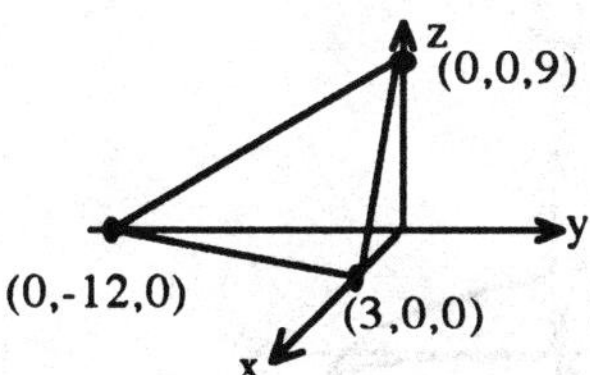

54.(a)

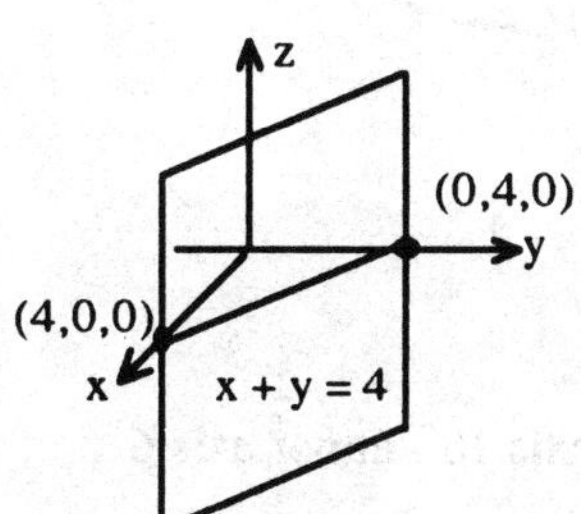

(b)

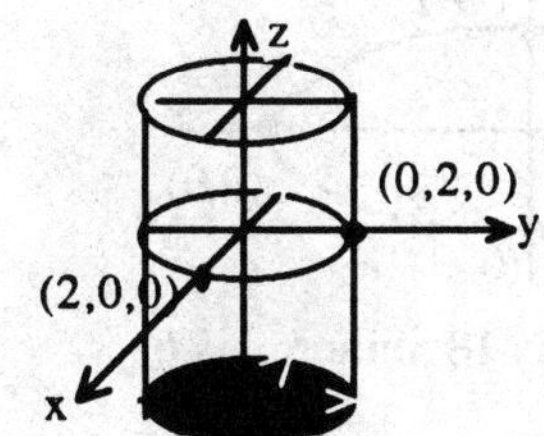

(c)

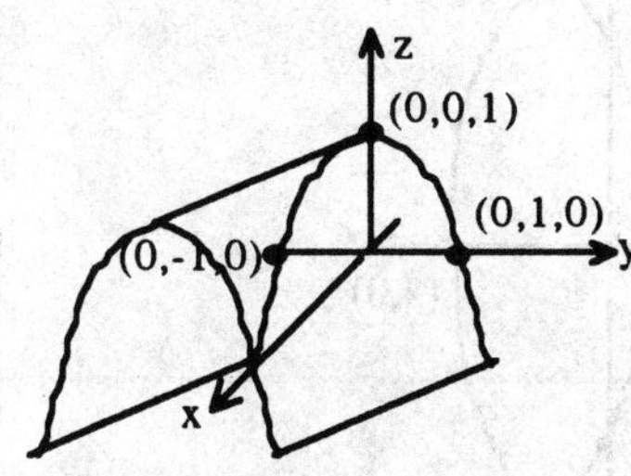

55.(a)

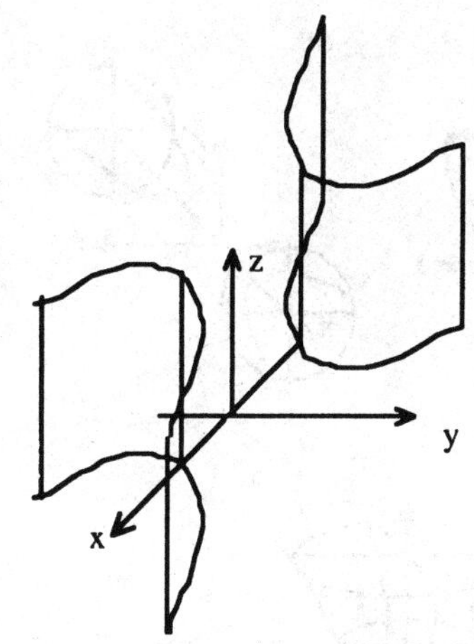

(b)

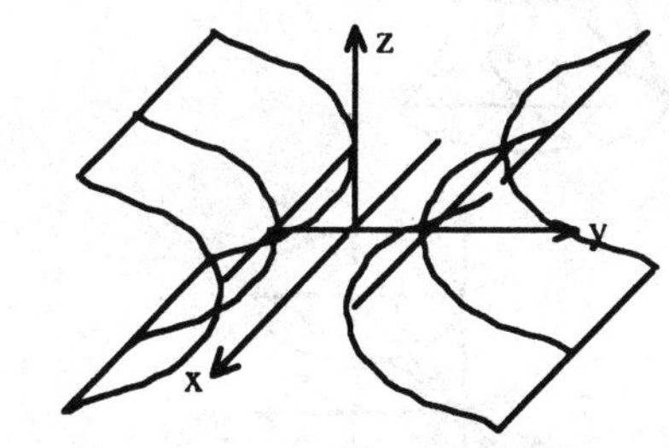

(c)

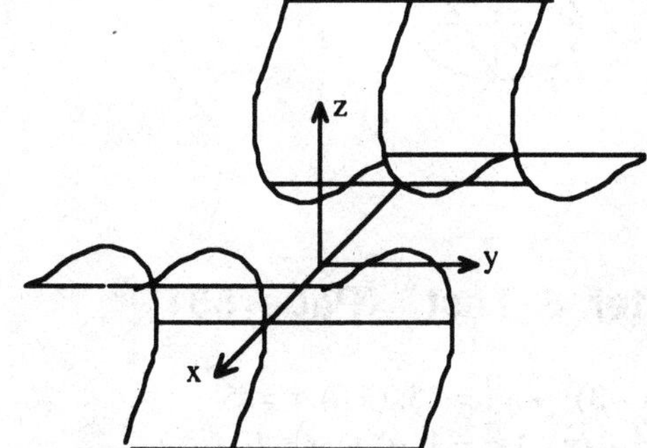

56.

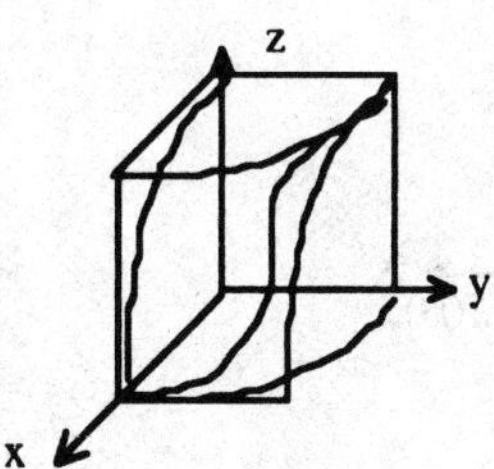

57.

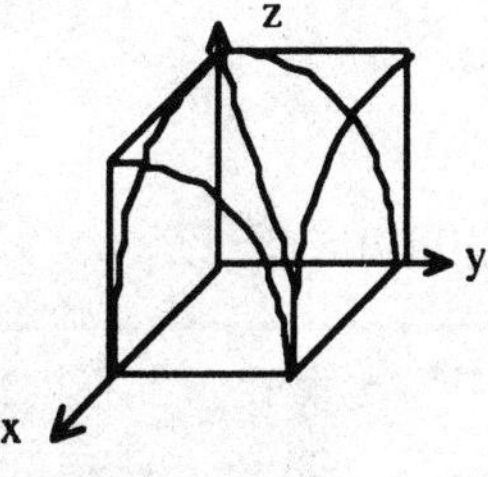

58.

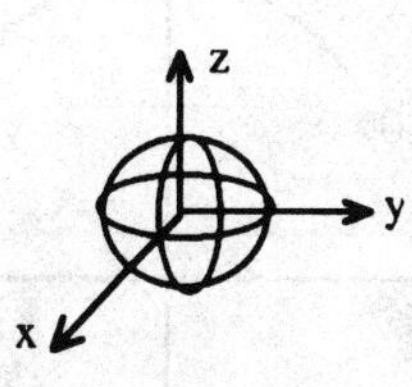

59. **60.**

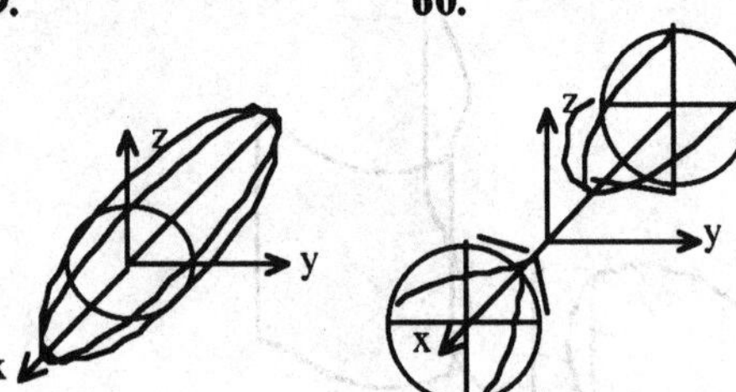

61.

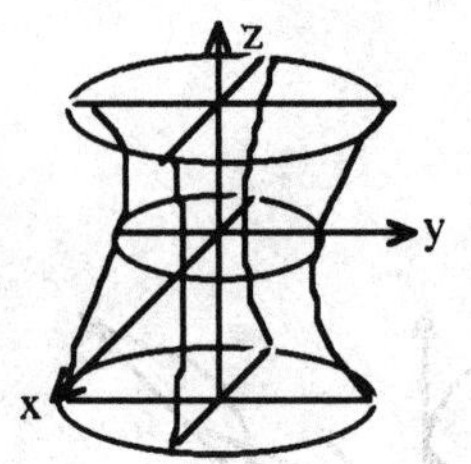

62.

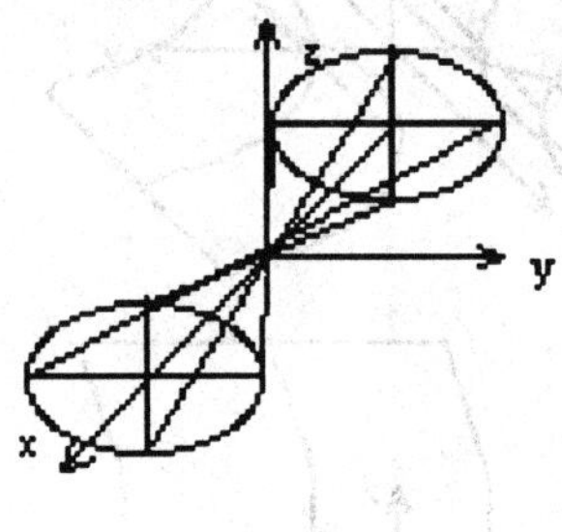

63.

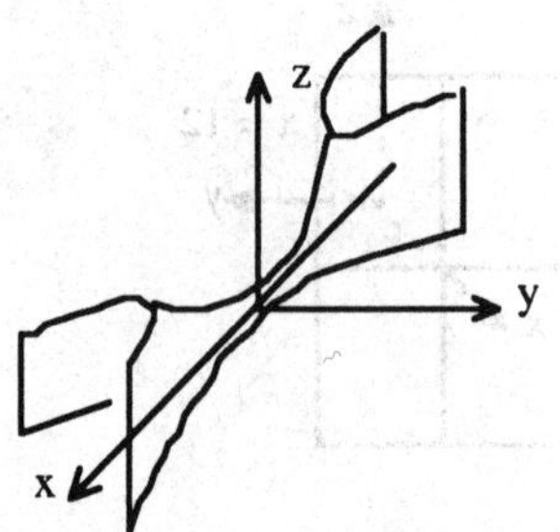

64.

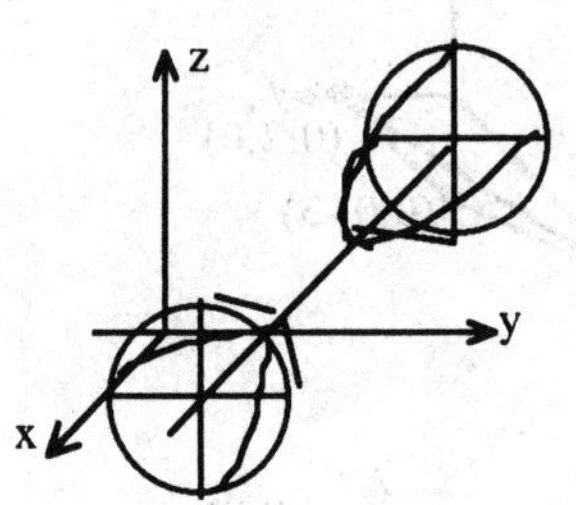

65.

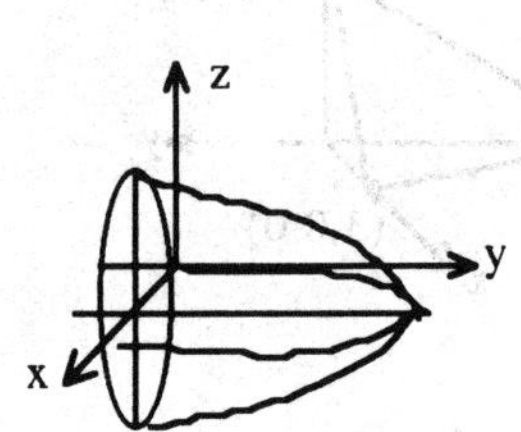

Chapter 8 Test (Page 581)

1.(a) $(x - 3)^2 + y^2 = 25$; (3,0), r = 5
(b) $x^2 + (y - 1)^2 = 1$; (0,1), r = 1

2.(a)

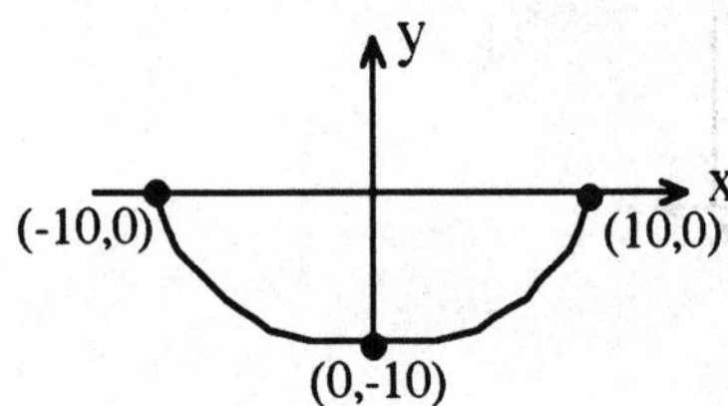

(b)

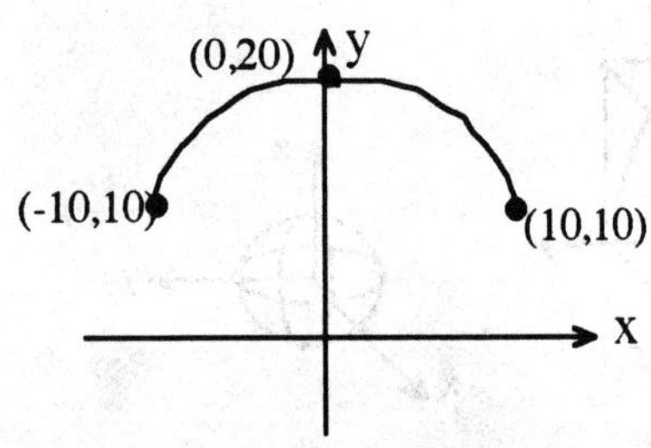

3.(a) (0,0), major axis 18 , minor axis 6

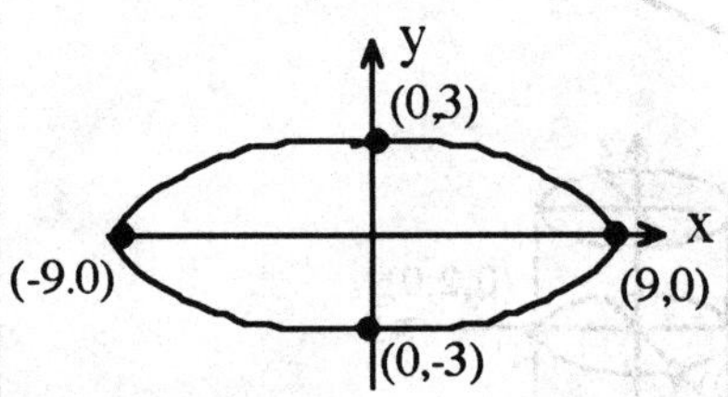

(b) (2,0), major axis 18, minor axis 6

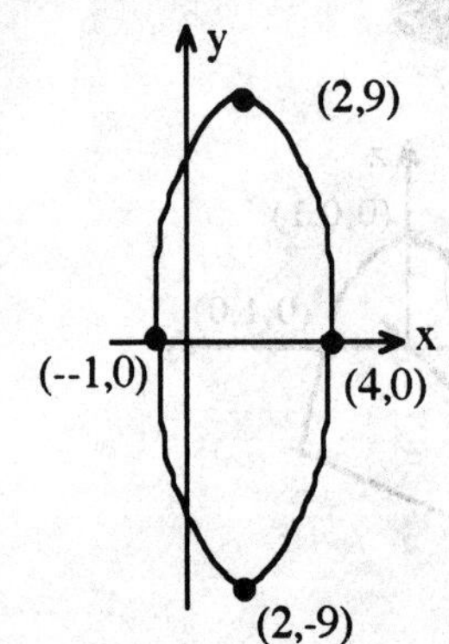

4. $\frac{x^2}{36} + \frac{y^2}{20} = 1$

5.(a) $\frac{x^2}{4} - \frac{y^2}{16} = 1$; center (0,0); vertices (-2,0), (2,0); foci $(\sqrt{12},0)$ $(-\sqrt{12},0)$

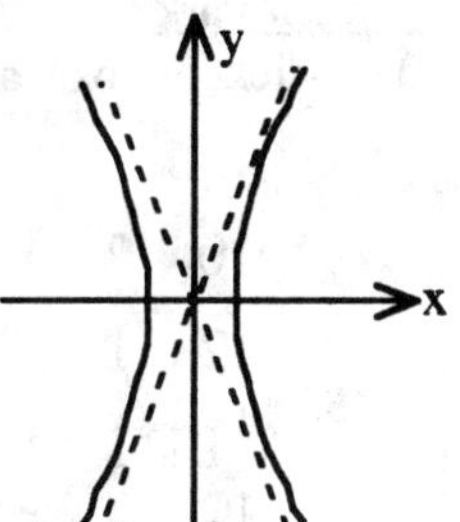

(b) $\frac{(y+2)^2}{1} - \frac{x^2}{36} = 1$

center (0,-2); vertices (0,-1), (0,-3)

foci $(0, -2 + \sqrt{35})$, $(0, -2 - \sqrt{35})$

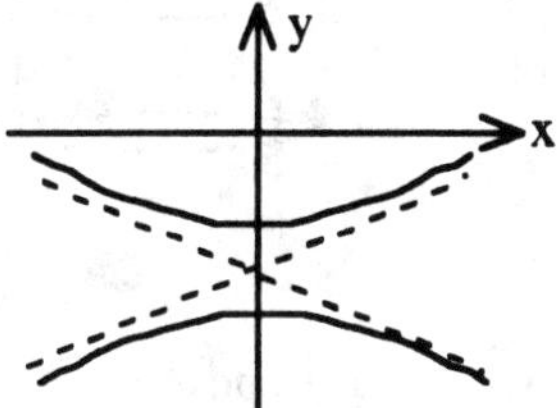

6. $y^2 = 16(x - 1)$

7. (a) $x^2 = -12(y - 2)$

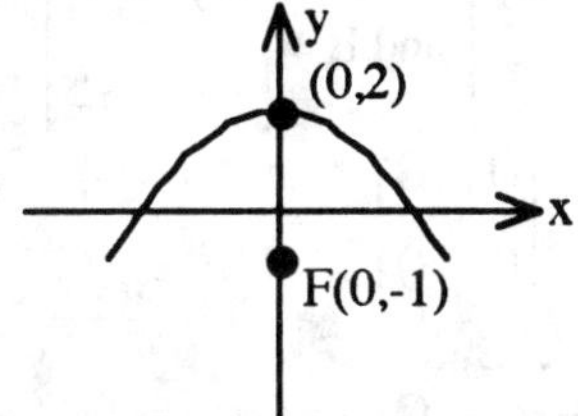

(b) $(y + 1)^2 = -(x - 1)$

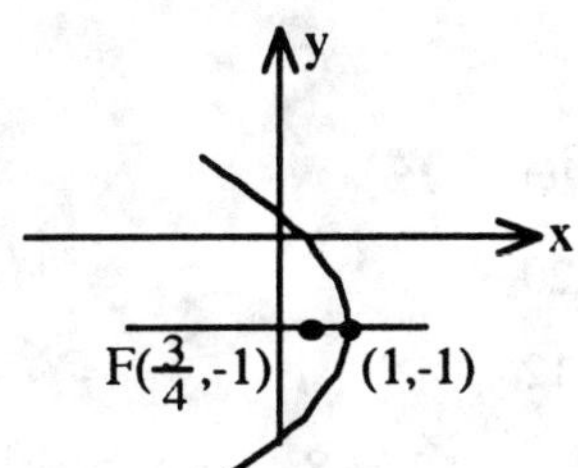

8. (a) hyperbola: $\frac{x^2}{64} - \frac{y^2}{64} = 1$

(b) ellipse: $\frac{x^2}{64} + \frac{y^2}{64} = 1$

(c) parabola: $x^2 = -36(y - 1)$

(d) ellipse: $\frac{(x+2)^2}{8} + \frac{y^2}{48} = 1$

9. 15°

10. $\frac{3}{2}u^2 + \frac{1}{2}v^2 = 1$

11. (a) (b)

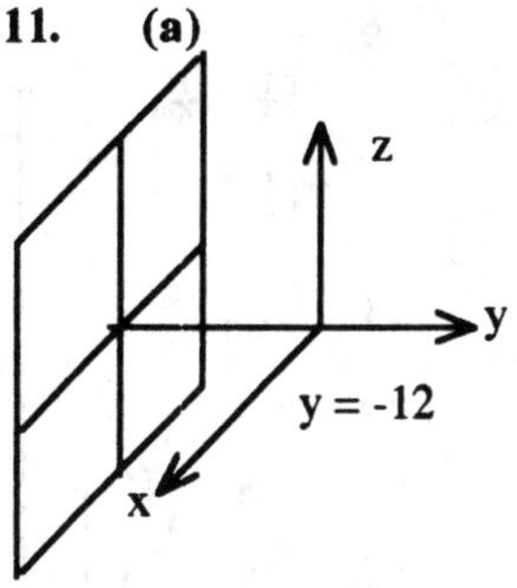

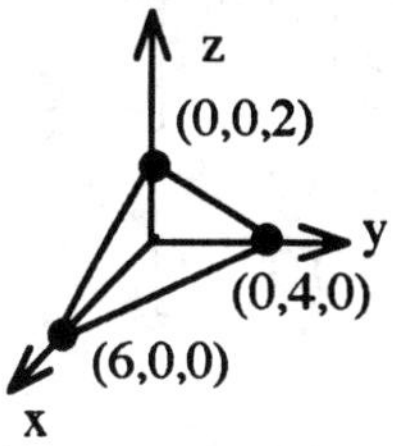

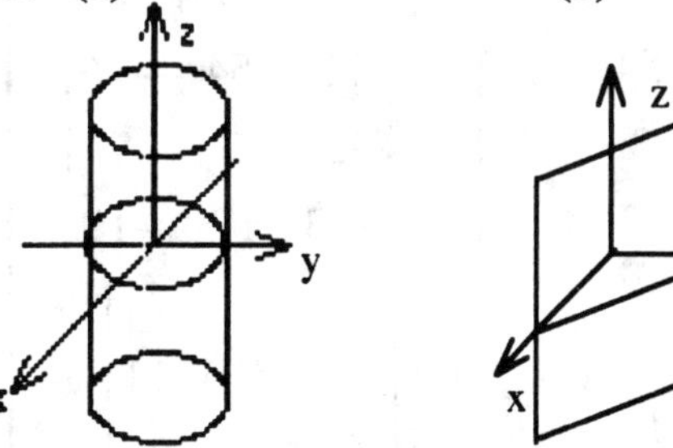

12. (a) (b)

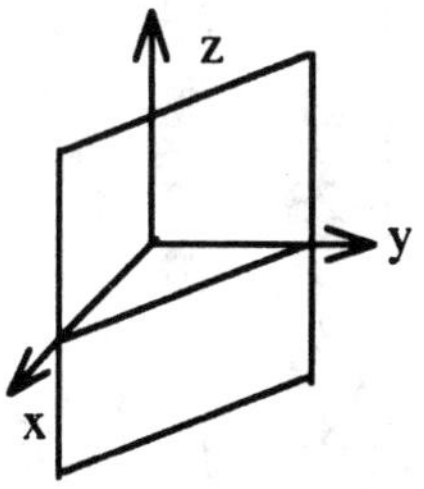

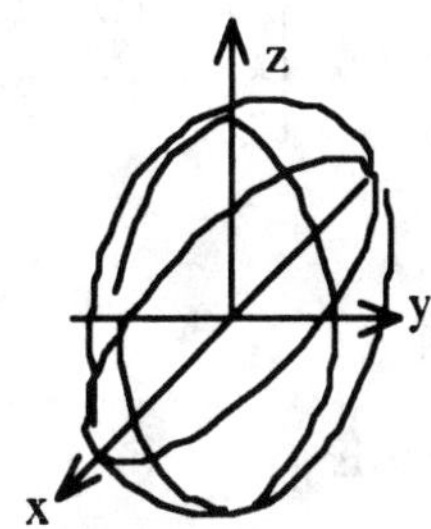

13.(a) xy trace: ellipse, xz trace: circle, yz trace: ellipse

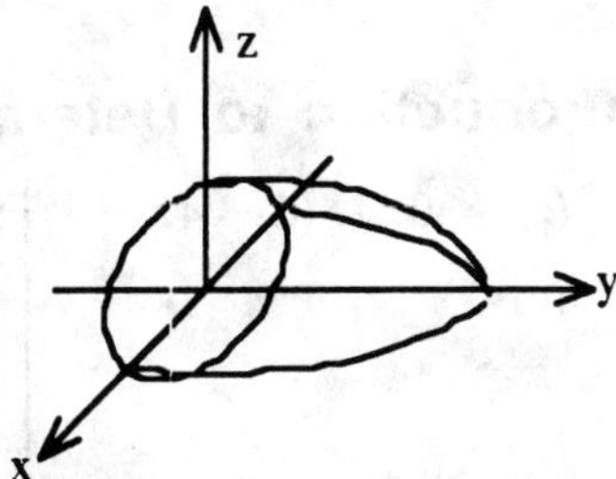

(b) xy trace: parabola, xz trace: ellipse, yz trace: parabola

14. <u>The Big Rectangle Problem</u>

$2\sqrt{2}$ by $\frac{2}{\sqrt{2}}$

15. <u>The Garden Problem</u>

126 ft 8.5 in.

Section 9.1 Introduction to Matrices (Page 592)

1.(a) $\begin{bmatrix} 3 & -4 \\ 5 & 11 \end{bmatrix}$ (b) $\begin{bmatrix} 3 & -4 \\ 5 & 11 \end{bmatrix}$

3 $\begin{bmatrix} 5 \\ 7 \\ -10 \end{bmatrix}$ **5.** $\begin{bmatrix} 4 & 7 & 1 \\ 3 & 2 & -2 \\ 1 & 5 & 7 \end{bmatrix}$ **7.** $\begin{bmatrix} \frac{1}{3} & \frac{1}{6} \\ \frac{1}{2} & \frac{1}{2} \end{bmatrix}$

9. (a) $\begin{bmatrix} -10 & 9 \\ -10 & 7 \end{bmatrix}$ (b) $\begin{bmatrix} -1 & 6 \\ -3 & -2 \end{bmatrix}$

11.(a) $\begin{bmatrix} -1 \end{bmatrix}$ (b) $\begin{bmatrix} 4 & 8 & 12 \\ 2 & 4 & 6 \\ -3 & -6 & -9 \end{bmatrix}$

13.(a) $\begin{bmatrix} 11 & 54 & -16 \\ 13 & 3 & 7 \\ -23 & 17 & -19 \end{bmatrix}$ (b) $\begin{bmatrix} 11 & 54 & -16 \\ 13 & 3 & 7 \\ -23 & 17 & -19 \end{bmatrix}$

15.(a) $\begin{bmatrix} 2 & 3 \\ 3 & 7 \end{bmatrix}$ (b) $\begin{bmatrix} 3 & -1 & 2 \\ 4 & 7 & -11 \\ 10 & 9 & 8 \end{bmatrix}$

17. not conformable

19.(a) $\begin{bmatrix} 5 & 10 & 15 \\ 20 & 25 & 30 \end{bmatrix}$ (b) $\begin{bmatrix} 5 & 10 & 15 \\ 20 & 25 & 30 \end{bmatrix}$

21. same $\begin{bmatrix} -3 & -6 \\ -9 & -12 \end{bmatrix}$

23. reflected about the y axis

25. reflected about the line y = x

27. rotated 30°

29. <u>The Rainbow Kite and Balloon Company</u>
beach \$7575, uptown \$14,875, downtown \$1307.50, totaling \$23,757.50

31. <u>Wonderworld Amusement Park</u>
10 one ticket options, 20 ten-ticket books, and 10 25-ticket books

33. $\begin{bmatrix} 0 & 2 & 2 & 0 \\ 2 & 0 & 0 & 2 \\ 2 & 0 & 0 & 2 \\ 0 & 2 & 2 & 0 \end{bmatrix}$ **35.** $\begin{bmatrix} 0 & 2 & 1 & 1 \\ 2 & 1 & 1 & 1 \\ 1 & 1 & 1 & 0 \\ 1 & 1 & 0 & 0 \end{bmatrix}$

37. **39.**

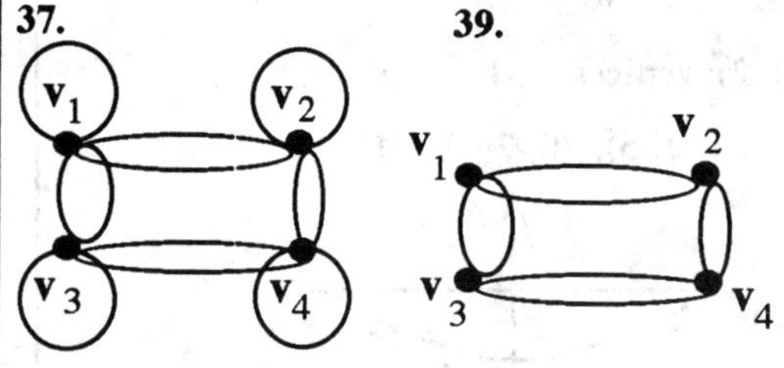

41. $(AB)^T = B^TA^T \neq A^TB^T$ and $(A + B)^T = A^T + B^T$

43. yes

45. For $A = \begin{bmatrix} 1 & -1 \\ 2 & 5 \end{bmatrix}$ and $B = \begin{bmatrix} -1 & 1 \\ -2 & -5 \end{bmatrix}$,

$A + B = B + A = \begin{bmatrix} 0 & 0 \\ 0 & 0 \end{bmatrix}$

47. For matrices, A, B, and C,
A +(B + C) = (A + B) + C.

Section 9.2 Introduction to Determinants (Page 604)

1.(a) - 2 (b) - 4 (c) - 30 (d) - 42

3.(a) -3 (b) 3 (c) 3

5.(a) 14 (b) 14

7.(a) 24 (b) 24 (c) 24

9.(a) 5 (b) 20

11.(a) 4 (b) 12

13. 10 **15.** 8

17.(a) -8 (b) -8 **19.**(a) 128 (b) 26

21.

$B = \begin{bmatrix} 4 & 0 & -2 \\ 1 & 3 & 1 \\ 2 & 1 & 0 \end{bmatrix} \Rightarrow 2B = \begin{bmatrix} 8 & 0 & -4 \\ 2 & 6 & 2 \\ 4 & 2 & 0 \end{bmatrix}$ det B = 6 and det 2B = 48 ⇒ det 2B = 2^3(det B) = 48

23. 3240

25. Add row 1 to row 2 then add the opposite of column 2 to column 1.

27. Interchange rows 3 and 2, then interchange rows 2 and 1. Then interchange columns 1 and 2, then interchange columns 3 and 1.

Section 9.3 Vectors in Two and Three Dimensions (Page 618)

1.(a) 1 (b) 0

3.(a) 14 (b) 14

5.(a) 1 (b) 0 (c) 0

7.(a) **k** (b) **i** (c) **-j**

9.(a) 12 (b) 12

11.(a) 18 (b) 36

13.(a) 6 (b) 6

15.(a) not defined (b) not defined

17. (1, 2, 1) and (-1, -2, -1)

19. (-3, -6, -3) **21.** **O**

23. $\mathbf{v}_1\cdot(\mathbf{v}_2 \times \mathbf{v}_3) = \mathbf{v}_1\cdot(2,4,2) = 0$

$(\mathbf{v}_1 \times \mathbf{v}_2)\cdot\mathbf{v}_3 - (1,2,1)\cdot(-2,-1,4) = 0$

25. k = 4/3 **27.** k = 1

29. (3, 0) **31.** (1, 0, 0)

33. $(\frac{205}{169}, \frac{492}{169})$ **35.** (0, 0, 5)

37. <u>The Swing Problem</u>

6 ft

39. <u>The Work Problem</u>

433 ft/lb

41. $\mathbf{v}_1\cdot(\mathbf{v}_2 + \mathbf{v}_3) =$

$a_1(b_1 + c_1) + a_2(b_2 + c_2) + a_3(b_3 + c_2)$ and

$\mathbf{v}_1\cdot\mathbf{v}_2 + \mathbf{v}_1\cdot\mathbf{v}_3 =$

$(a_1b_1 + a_2b_2 + a_3b_3) + (a_1c_1 + a_2c_2 + a_3c_3) =$

$a_1(b_1 + c_1) + a_2(b_2 + c_2) + a_3(b_3 + c_2)$

Hence, $\mathbf{v}_1\cdot(\mathbf{v}_2 + \mathbf{v}_3) = \mathbf{v}_1\cdot\mathbf{v}_2 + \mathbf{v}_1\cdot\mathbf{v}_3$

43. $\mathbf{v}_1\cdot\mathbf{v}_1 = (a_1,a_2,a_3)\cdot(a_1,a_2,a_3) = a_1{}^2 + a_2{}^2 + a_3{}^2$

and $\|\mathbf{v}_1\|^2 = (\sqrt{a_1{}^2 + a_2{}^2 + a_3{}^2})^2 = a_1{}^2 + a_2{}^2 + a_3{}^2$

Hence, $\mathbf{v}_1\cdot\mathbf{v}_1 = \|\mathbf{v}_1\|^2$

Section 9.4 Equations of Lines and Planes in Three Dimensions (Page 625)

1. x + 2y + 2z = 11 **3.** x + 4y - 2z = 0

5. $\frac{x-1}{2} = \frac{y-2}{4} = \frac{z-3}{-2}$

x = 2t + 1, y = 4t + 2, z = -2t + 3

7. $\frac{x}{3} = \frac{y}{1} = \frac{z}{2}$ x = 3t , y = t , z = 2t

9. x = t + 2, y = 1, z = 3

11. $\frac{x-1}{3} = \frac{y+1}{-2} = \frac{z-1}{-1}$

13. 2x + 4y - z = -11 **15.** 3x - y + 4z = 16

17. 3x + 5y + z = 8 **19.** 41x - 8y + 5z = 12

21. 21° **23.** 3 **25.** 0.6

27.

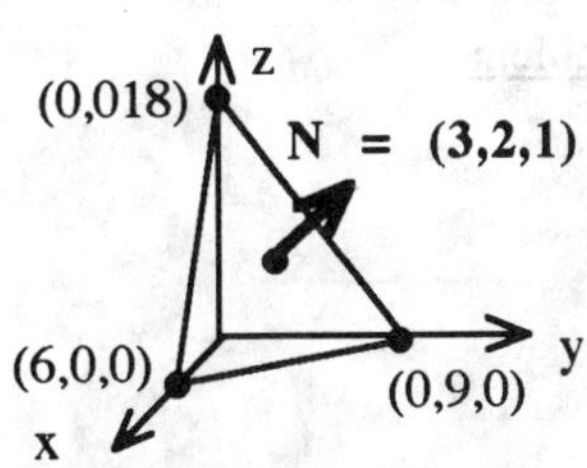

29. $P_1(a,0,0)$, $P_2(0,b,0)$, $P_3(0,0,c)$

$$\mathbf{N} = \mathbf{P_1P_1} \times \mathbf{P_1P_3} = \begin{vmatrix} \mathbf{i} & \mathbf{j} & \mathbf{k} \\ -a & b & 0 \\ -a & 0 & c \end{vmatrix} = (bc,ac,ab)$$

bc(x - a) + ac(y - 0) + ab(z - 0) = 0

$\frac{x}{a} + \frac{y}{b} + \frac{z}{c} = 1$

31. $P = \frac{\mathbf{v}\cdot\mathbf{N}}{\|\mathbf{N}\|}$

$$= \frac{(0,0,\frac{d_1 - d_2}{c})\cdot(a,b,c)}{\sqrt{a^2 + b^2 + c^2}}$$

$$= \frac{d_1 - d_2}{\sqrt{a^2 + b^2 + c^2}}$$

Chapter 9 Review (Page 627)

1.(a) $\begin{bmatrix} 1 & 0 \\ 1 & 0 \end{bmatrix}$ (b) $\begin{bmatrix} 1 & 0 \\ 1 & 10 \end{bmatrix}$

2.(a) $\begin{bmatrix} -3 & 6 & 0 \\ -2 & -2 & 2 \end{bmatrix}$ (b) $\begin{bmatrix} -3 & 6 & 0 \\ -2 & -2 & 2 \end{bmatrix}$

3. They are equal, $\begin{bmatrix} 144 & 0 \\ -144 & 144 \end{bmatrix}$

4. They are equal, $\begin{bmatrix} -10 & -10 \\ 0 & -10 \end{bmatrix}$

5.(a) 1 (b) 0

6.(a) 120 (b) -120

7.(a) 8 (b) 8

8.(a) -32 (b) -32

9.(a) -32 (b) -32

10.(a) 16 (b) 16

11.(a) 1 (b) 0

12.(a) 36 (b) 216

13.(a) 36 (b) 36

14.(a) 101 (b) 101

15.(a) 0 (b) 0 (c) 0

16.(a) 32 (b) 32 (c) 96

17.(a) **k** (b) **O** (c) **O**

18.(a) (-3, 12, 4) (b) (3, -12, -4) (c) **O**

19. k = 1 **20.** k = -8/3

21. (4.2, 5.6) **22.** (1, 1, 1)

23. The Swing Problem
You only flew 3 ft 7 in.

24. The Work Problem
$W = \mathbf{F}\cdot\mathbf{s} = (\frac{100}{\sqrt{2}}, \frac{100}{\sqrt{2}})\cdot(10,0) = 707$ ft-lbs

25. x + y = 3 **26.** x + y = 3

27. x + y + z = 1 **28.** y = 0

29. $\frac{x-1}{1} = \frac{y-1}{\sqrt{3}}$ and z = 1

30. $\frac{x-1}{-\sqrt{3}} = \frac{z-1}{1}$ and y = 2

31. x = y = z **32.** x - 1 = y = z - 1

33. 2x - 2y - z = 1 **34.** x - y = 0

Chapter 9 Test (Page 630)

1. $\begin{bmatrix} 0 & 0 \\ 0 & 0 \end{bmatrix}$

2. $\begin{bmatrix} -2 & 2 & -2 \\ -2 & -2 & 2 \end{bmatrix}$

3.(a) 0 (b) 0

4.(a) 0 (b) 0

5.(a) 2 (b) 2

6.(a) 0 (b) 0

7.(a) -240 (b) -240 (c) -71

8.(a) (-10, 2, -6) (b) (10, -2, 6) (c) (0, 0, 0)

9.(a) (0.5, 0.5) (b) (3, 3, 0)

10. x + y + z = -2

11. 2x - y - 4z = -4

12. $\frac{x-1}{3} = \frac{y-2}{-1}$, z = 3 (-2, 3, 3) and (4, 1, 3)

13. $\frac{x+1}{2} = \frac{y-2}{-4}$, z = 3 (1, -2, 3) and (-3, 6, 3)

14. $\theta = \cos^{-1}(\frac{1}{3}) = 70.5°$

15. The Swing Problem

9.5 ft

16. The Work Problem
1376.4 ft-lb

Index

E

F

G

H

I

L